▼ Formulas from Analytical Geometry: $P_1 \to (x_1, y_1)$, $P_2 \to (x_2, y_2)$

Distance between P_1 and P_2

$$d = \sqrt{(x_2 - x_1)^2 + (y_2 - y_1)^2}$$

Slope of Line Containing P_1 and P_2

$$m = \frac{\Delta y}{\Delta x} = \frac{y_2 - y_1}{x_2 - x_1}$$

Equation of Line Containing P_1 and P_2

Point-Slope Form

$$y - y_1 = m(x - x_1)$$

Equation of Line Containing P_1 and P_2

Slope-Intercept Form (slope m, y-intercept b)

$$y = mx + b, \text{ where } b = y_1 - mx_1$$

Parallel Lines

Slopes Are Equal: $m_1 = m_2$

Perpendicular Lines

Slopes Have a Product of -1: $m_1 m_2 = -1$

Intersecting Lines

Slopes Are Unequal: $m_1 \neq m_2$

Dependent (Coincident) Lines

Slopes and y-Intercepts Are Equal: $m_1 = m_2, b_1 = b_2$

▼ Logarithms and Logarithmic Properties

$$y = \log_b x \Leftrightarrow b^y = x$$

$$\log_b b = 1$$

$$\log_b 1 = 0$$

$$\log_b b^x = x$$

$$b^{\log_b x} = x$$

$$\log_c x = \frac{\log_b x}{\log_b c}$$

$$\log_b MN = \log_b M + \log_b N$$

$$\log_b \frac{M}{N} = \log_b M - \log_b N$$

$$\log_b M^P = P \cdot \log_b M$$

▼ Applications of Exponentials and Logarithms

$A \to$ amount accumulated

$P \to$ initial deposit, $p \to$ periodic payment

$n \to$ compounding periods/year

$r \to$ interest rate per year

$R \to$ interest rate per time period $\left(\dfrac{r}{n}\right)$

$t \to$ time in years

Interest Compounded n Times per Year

$$A = P\left(1 + \frac{r}{n}\right)^{nt}$$

Interest Compounded Continuously

$$A = Pe^{rt}$$

Accumulated Value of an Annuity

$$A = \frac{p}{R}\left[(1 + R)^{nt} - 1\right]$$

Payments Required to Accumulate Amount A

$$p = \frac{AR}{(1 + R)^{nt} - 1}$$

▼ Sequences and Series:

$a_1 \to$ 1st term, $a_n \to$ nth term, $S_n \to$ sum of n terms, $d \to$ common difference, $r \to$ common ratio

Arithmetic Sequences

$$a_1, a_2 = a_1 + d, a_3 = a_1 + 2d, \ldots, a_n = a_1 + (n - 1)d$$

$$S_n = \frac{n}{2}(a_1 + a_n)$$

$$S_n = \frac{n}{2}[2a_1 + (n - 1)d]$$

Geometric Sequences

$$a_1, a_2 = a_1 r, a_3 = a_1 r^2, \ldots, a_n = a_1 r^{n-1}$$

$$S_n = \frac{a_1 - a_1 r^n}{1 - r}$$

$$S_\infty = \frac{a_1}{1 - r}; |r| < 1$$

▼ Binomial Theorem

$$(a + b)^n = \binom{n}{0}a^n b^0 + \binom{n}{1}a^{n-1}b^1 + \binom{n}{2}a^{n-2}b^2 + \cdots + \binom{n}{n-1}a^1 b^{n-1} + \binom{n}{n}a^0 b^n$$

$$n! = n(n - 1)(n - 2) \cdots (3)(2)(1) \qquad \binom{n}{k} = \frac{n!}{k!(n - k)!}; \qquad 0! = 1$$

COLLEGE ALGEBRA

COLLEGE ALGEBRA

SECOND EDITION

JOHN W. COBURN

St. Louis Community College at Florissant Valley

Mc Graw Hill **Higher Education**

Boston   Burr Ridge, IL   Dubuque, IA   New York   San Francisco   St. Louis
Bangkok   Bogotá   Caracas   Kuala Lumpur   Lisbon   London   Madrid   Mexico City
Milan   Montreal   New Delhi   Santiago   Seoul   Singapore   Sydney   Taipei   Toronto

Higher Education

COLLEGE ALGEBRA, SECOND EDITION

1 2 3 4 5 6 7 8 9 0 DOW/DOW 0 9

ISBN 978–0–07–351941–8
MHID 0–07–351941–3

ISBN 978–0–07–334903–9 (Annotated Instructor's Edition)
MHID 0–07–334903–8

Editorial Director: *Stewart K. Mattson*
Sponsoring Editor: *Dawn R. Bercier*
Vice-President New Product Launches: *Michael Lange*
Senior Developmental Editor: *Michelle L. Flomenhoft*
Developmental Editor: *Katie White*
Marketing Manager: *John Osgood*
Senior Project Manager: *Vicki Krug*
Senior Production Supervisor: *Sherry L. Kane*
Senior Media Project Manager: *Sandra M. Schnee*
Designer: *Laurie B. Janssen*
Cover Designer: *Christopher Reese*
(USE) Cover Image: *©Jeff Hunter/Gettyimages*
Senior Photo Research Coordinator: *John C. Leland*
Supplement Producer: *Mary Jane Lampe*
Compositor: *Aptara®, Inc.*
Typeface: *10.5/12 Times Roman*
Printer: *R. R. Donnelley Willard, OH*

Library of Congress Cataloging-in-Publication Data

Coburn, John W.
    College algebra / John Coburn. — 2nd ed.
        p. cm.
    Includes index.
    ISBN   978–0–07–351941–8 — ISBN   0–07–351941–3 (hard copy : acid-free paper)
    ISBN   978–0–07–334903–9 — ISBN   0–07–334903–8 (annotated instructor's ed. : acid-free paper)
    1. Algebra–Textbooks. I. Title.

QA154.3.C594 2010
512.9–dc22   2008028790

www.mhhe.com

Additional Topics Online

(Visit www.mhhe.com/coburn)

Background

John Coburn grew up in the Hawaiian Islands, the seventh of sixteen children. John's mother and father were both teachers. John's mother taught English and his father, as fate would have it, held advanced degrees in physics, chemistry, and mathematics. Whereas John's father was well known, well respected, and a talented mathematician, John had to work very hard to see the connections so necessary for success in mathematics. In many ways, his writing is born of this experience.

Education

In 1979 John received a bachelor's degree in education from the University of Hawaii. After working in the business world for a number of years, John returned to his first love by accepting a teaching position in high school mathematics and in 1987 was recognized as Teacher of the Year. Soon afterward John decided to seek a master's degree, which he received two years later from the University of Oklahoma.

Teaching Experience

John is a full professor at the Florissant Valley campus of St. Louis Community College where he has taught mathematics for the last eighteen years. During his time there he has received numerous nominations as an outstanding teacher by the local chapter of Phi Theta Kappa, and was recognized as Post-Secondary Teacher of the Year in 2004 by Mathematics Educators of Greater St. Louis (MEGSL). John is a member of the following organizations: National Council of Teachers of Mathematics (NCTM), Missouri Council of Teachers of Mathematics (MCTM), Mathematics Educators of Greater Saint Louis (MEGSL), American Mathematical Association of two Year Colleges (AMATYC), Missouri Mathematical Association of two Year Colleges (MoMATYC), Missouri Community College Association (MCCA), and Mathematics Association of America (MAA).

Personal Interests

Some of John's other interests include body surfing, snorkeling, and beach combing whenever he gets the chance. In addition, John's loves include his family, music, athletics, games, and all things beautiful. John hopes that this love of life comes through in the writing, and serves to make the learning experience an interesting and engaging one for all students.

Dedication

To my wife and best friend Helen, whose love, support, and willingness to sacrifice never faltered.

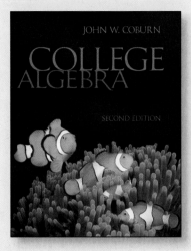

JOHN W. COBURN

COLLEGE ALGEBRA

SECOND EDITION

Coral reefs support an extraordinary biodiversity as they are home to over 4000 species of tropical or reef fish. In addition, coral reefs are immensely beneficial to humans; buffering coastal regions from strong waves and storms, providing millions of people with food and jobs, and prompting advances in modern medicine.

Similar to a reef, a college algebra course is unique because of its diverse population of students. Nearly every major is represented in this course, featuring students with a wide range of backgrounds and skill sets. Just like the variety of the fish in the sea rely on the coral reefs to survive, the assortment of students in college algebra rely on succeeding in this course in order to further pursue their degree, as well as their career goals.

From the Author

This text is the result of a mighty confluence of needs, ideas, desires, and directions. This is easily understandable, as the intended audience is one of the most diverse in all of education. Our students come to us with a wide range of backgrounds, varying degrees of preparation, and interest levels that vary from apathy to excitement. In addition, our classes include those needing only a general education requirement, as well as our country's future engineers and scientists. To say our greatest challenge is meeting the needs of so diverse a population would be an understatement. In reflecting on this diversity, the image of a coral reef came to mind, and I was struck by the strength of the analogy. We have a hugely diverse population, with the reef as a common meeting place, with all the inhabitants depending on the reef for their purpose, nourishment, and direction.

Writing a text for this course has been one of the most daunting and challenging experiences in my life. Long before I began, my teaching experience left a nagging sense that most texts on the market lacked the ability to connect with so diverse an audience. In addition, they appeared to offer too scant a framework to build concepts, too terse a development to make connections, and insufficient support in their exercise sets to develop long-term retention or foster a love of mathematics. In particular, the applications seemed to lack a sense of realism, curious interest, and/or connections to a student's everyday experience.

With all of this in mind and a strong desire to write a better text, I set about the task of creating what I hoped would become a more engaging tool for students, and a more supportive tool for instructors. Drawing on the diversity of my own educational experience, and an early exposure to different cultures, views, and perspectives, I believe has contributed to the text's unique and engaging style, and I hope in the end, to more and better connections with our diverse audience. Having feedback from more than 400 people, including manuscript reviewers, focus group participants, and contributors, was invaluable to helping me hone the connections in the book. As a collateral outgrowth of this experience, I admit there was also a desire to interest and engage ourselves, the instructors—to remind us again and again, why we fell in love with mathematics in the first place.   —John Coburn

College algebra tends to be a challenging course for many students. They don't see the connections that college algebra has to their life or why it is so critical that they take and pass this course for both technical and nontechnical careers alike. Others may enter into this course underprepared or improperly placed and with very little motivation.

Instructors are faced with several challenges as well. They are given the task of improving pass rates and student retention while energizing a classroom full of students comprised of nearly every major. Furthermore, it can be difficult to distinguish between students who are likely to succeed and students who may struggle until after the first test is given.

The goal of the Coburn series is to provide both students and instructors with tools to address these challenges, as well as the diversity of the students taking this course, so that you can experience greater success in college algebra. For instance, the comprehensive exercise sets have a range of difficulty that provides very strong support for weaker students, while advanced students are challenged to reach even further. The rest of this preface further explains the tools that John Coburn and McGraw-Hill have developed and how they can be used to *connect* students to college algebra and *connect* instructors to their students.

The Coburn College Algebra Series provides you with strong tools to achieve better outcomes in your College Algebra course as follows:

▶ *Better Student Preparedness*

▶ *Increased Student Engagement*

▶ *Solid Skill Development*

▶ *Strong Connections*

▶ Better Student Preparedness

No two students have the same strengths and weaknesses in mathematics. Typically students will enter any math course with different preparedness levels. For most students who have trouble retaining or recalling concepts learned in past courses, basic review is simply not enough to sustain them successfully throughout the course. Moreover, instructors whose main focus is to prepare students for the next course do not have adequate time in or out of class to individually help each student with review material.

ALEKS Prep uniquely assesses each student to determine their individual strengths and weaknesses and informs the student of their capabilities using a personalized pie chart. From there, students begin learning through ALEKS via a personalized learning path uniquely designed for each student. ALEKS Prep interacts with students like a private tutor and provides a safe learning environment to remediate their individual knowledge gaps of the course pre-requisite material outside of class.

ALEKS Prep is the only learning tool that empowers students by giving them an opportunity to remediate individual knowledge gaps and improve their chances for success. ALEKS Prep is especially effective when used in conjunction with ALEKS Placement and ALEKS 3.0 course-based software.

▶ Increased Student Engagement

What makes John Coburn's applications unique is that he is constantly thinking mathematically. John's applications are spawned during a trip to Chicago, a phone call with his brother or sister, or even while watching the evening news for the latest headlines. John literally takes notes on things that he sees in everyday life and connects these situations to math. This truly makes for relevant applications that are born from real-life experiences as opposed to applications that can seem fictitious or contrived.

▶ Solid Skill Development

The Coburn series intentionally relates the examples to the exercise sets so there is a strong connection between what students are learning while working through the examples in each section and the homework exercises that they complete. In turn, students who attempt to work the exercises first can surely rely on the examples to offer support as needed. Because of how well the examples and exercises are connected, key concepts are easily understood and students have plenty of help when using the book outside of class.

There are also an abundance of exercise types to choose from to ensure that homework challenges a wide variety of skills. Furthermore, John reconnects students to earlier chapter material with Mid-Chapter Checks; students have praised these exercises for helping them understand what key concepts require additional practice.

▶ Strong Connections

John Coburn's experience in the classroom and his strong connections to how students comprehend the material are evident in his writing style. This is demonstrated by the way he provides a tight weave from topic to topic and fosters an environment that doesn't just focus on procedures but illustrates the big picture, which is something that so often is sacrificed in this course. Moreover, he deploys a clear and supportive writing style, providing the students with a tool they can depend on when the teacher is not available, when they miss a day of class, or simply when working on their own.

Better Student Preparedness...

Experience Student Success!

ALEKS® ALEKS is a unique online math tool that uses adaptive questioning and artificial intelligence to correctly place, prepare, and remediate students . . . all in one product! Institutional case studies have shown that **ALEKS has improved pass rates by over 20% versus traditional online homework and by over 30% compared to using a text alone.**

By offering each student an individualized learning path, ALEKS directs students to work on the math topics that they are ready to learn. Also, to help students keep pace in their course, instructors can correlate ALEKS to their textbook or syllabus in seconds.

To learn more about how ALEKS can be used to boost student performance, please visit **www.aleks.com/highered/math** or contact your McGraw-Hill representative.

Easy Graphing Utility!
Students can answer graphing problems with ease!

ALEKS Pie
Each student is given their own individualized learning path.

Click on the "Dashed" button.

Mastery of College Algebra
(88 of 200 Topics)

► Algebra review (40 of 46)
► Exponential and logarithmic functions (4 of 15)
► Sequences, series, and probability (3 of 22)
► Conic sections (3 of 22)
► Polynomial and rational functions (15 of 37)
► Systems of linear equations and matrices (6 of 25)
► Functions and graphs (17 of 33)

Move your mouse along the ruler's edge, and release your mouse within the blue circle. Notice that you can draw the line past the end of the ruler.

Clear   Undo

My Course: Math 99 PQ MW - Spring 2008 / Intermediate Algebra ▼   Edit this Course | Add a Course | Administrator Center | My Account Settings

| Home | Reports | Gradebook | Homework | Quizzes | Tests | Assessments | Worksheets |

Assignment Status view as list

July 21, 2008
- Quiz on Friday
- Quiz 10
- HW 3

• Create assignment: (Choose one) ▼ Go
• Modify existing assignment by selecting from the below calendar and editing in the action box to the right.

◄ June   **JULY 2008**   August ►

Sun	Mon	Tue	Wed	Thr	Fri	Sat
29	30	1	2	3	4	5
				• Assessment		
6	7	8	9	10	11	
	• Quiz 9					
13	14	15	16	17	18	
			• Homework 2			
20	**21**	22	23	24	25	26
	• Quiz on Friday				• Quiz 10	
27	28	29	30	31	1	2
• HW 3						

Course Calendar
Instructors can schedule assignments and reminders for students.

...Through ALEKS®

New ALEKS Instructor Module

Enhanced Functionality and Streamlined Interface Help to Save Instructor Time

ALEKS® The new ALEKS Instructor Module features enhanced functionality and streamlined interface based on research with ALEKS instructors and homework management instructors. Paired with powerful assignment driven features, textbook integration, and extensive content flexibility, the new ALEKS Instructor Module simplifies administrative tasks and makes ALEKS more powerful than ever.

Gradebook view for all students

Gradebook view for an individual student

New Gradebook!
Instructors can seamlessly track student scores on automatically graded assignments. They can also easily adjust the weighting and grading scale of each assignment.

Track Student Progress Through Detailed Reporting
Instructors can track student progress through automated reports and robust reporting features.

Select topics for each assignment

Automatically Graded Assignments
Instructors can easily assign homework, quizzes, tests, and assessments to all or select students. Deadline extensions can also be created for select students.

Learn more about ALEKS by visiting **www.aleks.com/highered/math** or contact your McGraw-Hill representative.

Increased Student Engagement...

Through Meaningful Applications

Making mathematics meaningful requires that students experience the connection between mathematics and its impact on the world they live in. This text is also the result of a powerful commitment to provide applications of the highest quality, having close ties to the examples, and with carefully monitored levels of difficulty.

   Many of these examples were born of my own diverse life experiences, others came from a curious, lucid, and even visionary folly that allows one to seize upon the every day events of life, and see the significant or meaningful mathematics in the background. My ever-present notebook was used a thousand times to capture that casual observation, or that sudden burst of inspiration that is the genesis for outstanding applications. These were supported at home by a substantial library of reference and research books, an eye toward both history and current events, and of course our modern marvel of a research tool—the Internet. After a (sometimes long) period of thought, reflection, and research, followed by a wording and a rewording of the exercise so that it would resonate with students while filling the need, a significant and meaningful application was born.   —JC

▶ **Chapter Openers** highlight Chapter Connections, an interesting application exercise from the chapter, and provide a list of other real-world connections to give context for students who wonder how math relates to them.

> **❝**I especially like the depth and variety of applications in this textbook. Other College Algebra texts the department considered did not share this strength. In particular, there is a clear effort on the part of the author to include realistic examples showing how such math can be utilized in the real world.**❞**
> —*George Alexander, Madison Area Technical College*

▶ **Examples** throughout the text feature word problems, providing students with a starting point for how to solve these types of problems in their exercise sets.

> **❝**One of this text's strongest features is the wide range of applications exercises. As an instructor, I can choose which exercises fit my teaching style as well as the student interest level.**❞**
> —*Stephen Toner, Victor Valley College*

▶ **Application Exercises** at the end of each section are the hallmark of the Coburn series. Never contrived, always creative, and born out of the author's life and experiences, each application tells a story and appeals to a variety of teaching styles, disciplines, backgrounds, and interests.

> **❝**[The application problems] answered the question, 'When are we ever going to use this?'**❞**
> —*Student class tester at Metropolitan Community College–Longview*

▶ **Math in Action Applets,** located online, enable students to work collaboratively as they manipulate applets that apply mathematical concepts in real-world contexts.

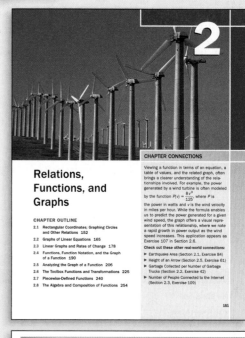

CHAPTER CONNECTIONS

2

Relations, Functions, and Graphs

CHAPTER OUTLINE

2.1 Rectangular Coordinates; Graphing Circles and Other Relations 152
2.2 Graphs of Linear Equations 165
2.3 Linear Graphs and Rates of Change 178
2.4 Functions, Function Notation, and the Graph of a Function 190
2.5 Analyzing the Graph of a Function 206
2.6 The Toolbox Functions and Transformations 225
2.7 Piecewise-Defined Functions 240
2.8 The Algebra and Composition of Functions 254

Viewing a function in terms of an equation, a table of values, and the related graph, often brings a clearer understanding of the relationships involved. For example, the power generated by a wind turbine is often modeled by the function $P(v) = \frac{8v^3}{125}$, where P is the power in watts and v is the wind velocity in miles per hour. While the formula enables us to predict the power generated for a given wind speed, the graph offers a visual representation of this relationship, where we note a rapid growth in power output as the wind speed increases. This application appears as Exercise 107 in Section 2.6.

Check out these other real-world connections:
▶ Earthquake Area (Section 2.1, Exercise 84)
▶ Height of an Arrow (Section 2.5, Exercise 61)
▶ Garbage Collected per Number of Garbage Trucks (Section 2.2, Exercise 42)
▶ Number of People Connected to the Internet (Section 2.3, Exercise 109)

151

EXAMPLE 10 ▶ **Determining the Domain and Range from the Context**

Paul's 1993 Voyager has a 20-gal tank and gets 18 mpg. The number of miles he can drive (his range) depends on how much gas is in the tank. As a function we have $M(g) = 18g$, where $M(g)$ represents the total distance in miles and g represents the gallons of gas in the tank. Find the domain and range.

Solution ▶ Begin evaluating at $x = 0$, since the tank cannot hold less than zero gallons. On a full tank the maximum range of the van is $20 \cdot 18 = 360$ miles or $M(g) \in [0, 360]$. Because of the tank's size, the domain is $g \in [0, 20]$.

☑ **C.** You've just learned how to use function notation and evaluate functions

Now try Exercises 94 through 101 ▶

CA Project 1 – Hooke's Law
Mathematics in Action · Collaborative Learning on the World Wide Web

▸ Click HERE for Instructions

$D = kW$

Through Timely Examples

In mathematics, it would be difficult to overstate the importance of examples that set the stage for learning. Not a few educational experiences have faltered due to an example that was too difficult, a poor fit, out of sequence, or had a distracting result. In this series, a careful and deliberate effort was made to select examples that were timely and clear, with a direct focus on the concept or skill at hand. Everywhere possible, they were further designed to link previous concepts to current ideas, and to lay the groundwork for concepts to come. As a trained educator knows, the best time to answer a question is often before it's ever asked, and a timely sequence of carefully constructed examples can go a long way in this regard, making each new idea simply the next logical, even anticipated step. When successful, the mathematical maturity of a student grows in unnoticed increments, as though it was just supposed to be that way. —JC

▶ **Titles** have been added to Examples in this edition to highlight relevant learning objectives and reinforce the importance of speaking mathematically using vocabulary.

▶ **Annotations** located to the right of the solution sequence help the student recognize which property or procedure is being applied.

▶ **"Now Try"** boxes immediately following Examples guide students to specific matched exercises at the end of the section, helping them identify exactly which homework problems coincide with each discussed concept.

▶ **Graphical Support Boxes,** located after selected examples, visually reinforce algebraic concepts with a corresponding graphing calculator example.

EXAMPLE 3 ▶ Solving a System Using Substitution

Solve using substitution: $\begin{cases} 4x + y = 4 \\ y = x + 2 \end{cases}$

Solution ▶ Since $y = x + 2$, we can replace y with $x + 2$ in the first equation.

$$4x + y = 4 \qquad \text{first equation}$$
$$4x + (x + 2) = 4 \qquad \text{substitute } x + 2 \text{ for } y$$
$$5x + 2 = 4 \qquad \text{simplify}$$
$$x = \frac{2}{5} \qquad \text{result}$$

The x-coordinate is $\frac{2}{5}$. To find the y-coordinate, substitute $\frac{2}{5}$ for x into either of the original equations. Substituting in the second equation gives

$$y = x + 2 \qquad \text{second equation}$$
$$= \frac{2}{5} + 2 \qquad \text{substitute } \frac{2}{5} \text{ for } x$$
$$= \frac{12}{5} \qquad \frac{2}{1} = \frac{10}{5}, \frac{10}{5} + \frac{2}{5} = \frac{12}{5}$$

The solution to the system is $(\frac{2}{5}, \frac{12}{5})$. Verify by substituting $\frac{2}{5}$ for x and $\frac{12}{5}$ for y into both equations.

Now try Exercises 23 through 32 ▶

GRAPHICAL SUPPORT

Graphing the lines from Example 8 as Y1 and Y2 on a graphing calculator, we note the lines do appear to be parallel (they actually *must* be since they have identical slopes). Using the **ZOOM 8:ZInteger** feature of the TI-84 Plus we can quickly verify that Y2 indeed contains the point $(-6, -1)$.

Solid Skill Development...

Through Exercises

I have included a wealth of exercises in support of each section's main ideas. I constructed each set with great care, in an effort to provide strong support for weaker students, while challenging advanced students to reach even further. I also designed the various exercises to support instructors in their teaching endeavors—the quantity and quality of the exercises allow for numerous opportunities to guide students through difficult calculations, and to illustrate important problem-solving techniques.—JC

Mid-Chapter Checks

Mid-Chapter Checks provide students with a good stopping place to assess their knowledge before moving on to the second half of the chapter.

End-of-Section Exercise Sets

▶ **Concepts and Vocabulary** exercises to help students recall and retain important terms.

▶ **Developing Your Skills** exercises to provide practice of relevant concepts just learned with increasing levels of difficulty.

> ❝Some of our instructors would mainly assign the developing your skills and working with formula problems, however, I would focus on the writing, research and decision making [in] extending the concept. The flexibility is one of the things I like about the Coburn text.❞
>
> —Sherry Meier, Illinois State University

▶ **Working with Formulas** exercises to demonstrate contextual applications of well-known formulas.

▶ **Extending the Concept** exercises that require communication of topics, synthesis of related concepts, and the use of higher-order thinking skills.

▶ **Maintaining Your Skills** exercises that address skills from previous sections to help students retain previously learning knowledge.

> ❝He not only has exercises for skill development, but also problems for 'extending the concept' and 'maintaining your skills,' which our current text does not have. I also like the mid-chapter checks provided. All these give Coburn an advantage in my view.❞
>
> —Randy Ross, Morehead State University

MID-CHAPTER CHECK

1. Compute $(x^3 + 8x^2 + 7x - 14) \div (x + 2)$ using long division and write the result in two ways:
   (a) dividend = (quotient)(divisor) + remainder and
   (b) $\dfrac{\text{dividend}}{\text{divisor}} = (\text{quotient}) + \dfrac{\text{remainder}}{\text{divisor}}$

2. Given that $x - 2$ is a factor of $f(x) = 2x^4 - x^3 - 8x^2 + x + 6$, use the rational zeroes theorem to write $f(x)$ in completely factored form.

9. Use the *Guidelines for Graphing* to draw the graph of $q(x) = x^3 + 5x^2 + 2x - 8$.

10. When fighter pilots train for dogfighting, a "hard-deck" is usually established below which no competitive activity can take place. The polynomial graph given shows Maverick's altitude above and below this hard-deck during a 5-sec interval.
    a. What is the minimum

1.3 EXERCISES

▶ **CONCEPTS AND VOCABULARY**

Fill in the blank with the appropriate word or phrase. Carefully reread the section if needed.

1. When multiplying or dividing by a negative quantity, we _____ the inequality to maintain a true statement.

2. To write an absolute value equation or inequality in simplified form, we _____ the absolute value

4. The absolute value inequality $|3x - 6| < 12$ is true when $3x - 6 > $ _____ and $3x - 6 < $ _____.

Describe each solution set (assume $k > 0$). Justify your answer.

5. $|ax + b| < -k$

▶ **DEVELOPING YOUR SKILLS**

Solve each absolute value equation. Write the solution in set notation.

7. $2|m - 1| - 7 = 3$

8. $3|n - 5| - 14 = -2$

9. $-3|x + 5| + 6 = -15$

10. $-2|y + 3| - 4 = -14$

11. $2|4v + 5| - 6.5 = 10.3$

12. $7|2w + 5| + 6.3 = 11.2$

13. $-|7p - 3| + 6 = -5$

14. $-|3q + 4| + 3 = -5$

Solve each absolute value inequality. Write solutions in interval notation.

25. $|x - 2| \le 7$

26. $|y + 1| \le 3$

27. $-3|m| - 2 > 4$

28. $-2|n| + 3 > 7$

29. $\dfrac{|5v + 1|}{4} + 8 < 9$

30. $\dfrac{|3w - 2|}{2} + 6 < 8$

31. $3|p + 4| + 5 < 8$

32. $5|q - 2| - 7 \le 8$

33. $|3b - 11| + 6 \le 9$

34. $|2c + 3| - 5 < 1$

35. $|4 - 3t| + 12 < 7$

36. $|2 - 7u| + 7 \le 4$

37. $\left|\dfrac{4x + 5}{2} - \dfrac{1}{2}\right| \le \dfrac{7}{2}$

38. $\left|\dfrac{2y - 3}{4} - \dfrac{3}{8}\right| < \dfrac{15}{16}$

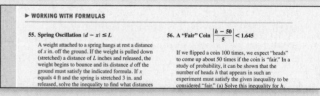

▶ **WORKING WITH FORMULAS**

55. **Spring Oscillation** $|d - x| \le L$

A weight attached to a spring hangs at rest a distance of x in. off the ground. If the weight is pulled down (stretched) a distance of L inches and released, the weight begins to bounce and its distance d off the ground must satisfy the indicated formula. If x equals 4 ft and the spring is stretched 3 in. and released, solve the inequality to find what distances

56. **A "Fair" Coin** $\left|\dfrac{h - 50}{5}\right| < 1.645$

If we flipped a coin 100 times, we expect "heads" to come up about 50 times if the coin is "fair." In a study of probability, it can be shown that the number of heads h that appears in such an experiment must satisfy the given inequality to be considered "fair." (a) Solve this inequality for h.

▶ **EXTENDING THE CONCEPT**

67. Determine the value or values (if any) that will make the equation or inequality true.
   a. $|x| + x = 8$
   b. $|x - 2| \le \dfrac{x}{2}$
   c. $x - |x| = x + |x|$
   d. $|x + 3| \ge 6x$
   e. $|2x + 1| = x - 3$

68. The equation $|5 - 2x| = |3 + 2x|$ has only one solution. Find it and explain why there is only one.

▶ **MAINTAINING YOUR SKILLS**

69. (R.4) Factor the expression completely: $18x^3 + 21x^2 - 60x$.

70. (1.1) Solve $V^2 = \dfrac{2W}{C\rho A}$ for ρ (physics).

72. (1.2) Solve the inequality, then write the solution set in interval notation:
   $-3(2x - 5) > 2(x + 1) - 7$.

> ❝ The strongest feature seems to be the wide variety of exercises included at the end of each section. There are plenty of drill problems along with good applications.❞
>
> —Jason Pallett, Metropolitan Community College–Longview

End-of-Chapter Review Material

Exercises located at the end of the chapter provide students with the tools they need to prepare for a quiz or test. Each chapter features the following:

▶ **Chapter Summary and Concept Review** that presents key concepts with corresponding exercises by section in a format easily used by students.

▶ **Mixed Review** exercises that offer more practice on topics from the entire chapter, arranged in random order requiring students to identify problem types and solution strategies on their own.

▶ **Practice Test** that gives students the opportunity to check their knowledge and prepare for classroom quizzes, tests, and other assessments.

> **❝** We always did reviews and a quiz before the actual test; it helped a lot. **❞**
> —*Melissa Cowan, student class tester Metropolitan Community College–Longview*

▶ **Cumulative Reviews** that are presented at the end of each chapter to help students retain previously learned skills and concepts by revisiting important ideas from earlier chapters (starting with Chapter 2).

> **❝** The cumulative review is very good and is considerably better than some of the books I have reviewed/used. I have found these to be wonderful practice for the final exam. **❞**
> —*Sarah Clifton, Southeastern Louisiana University*

▶ 🖩 **Graphing Calculator** icons appear next to exercises where important concepts can be supported by the use of graphing technology.

> **❝** The summary and concept review was very helpful because it breaks down each section. That is what helps me the most. **❞**
> —*Brittany Pratt, student class tester at Baton Rouge Community College*

Homework Selection Guide

A list of suggested homework exercises has been provided for each section of the text (Annotated Instructor's Edition only). This feature may prove especially useful for departments that encourage consistency among many sections, or those having a large adjunct population. The feature was also designed as a convenience to instructors, enabling them to develop an inventory of exercises that is more in tune with the course as they like to teach it. The Guide provides prescreened and preselected assignments at four different levels: *Core, Standard, Extended,* and *In Depth*.

- **Core:** These assignments go right to the heart of the material, offering a minimal selection of exercises that cover the primary concepts and solution strategies of the section, along with a small selection of the best applications.

- **Standard:** The assignments at this level include the *Core* exercises, while providing for additional practice without excessive drill. A wider assortment of the possible variations on a theme are included, as well as a greater variety of applications.

- **Extended:** Assignments from the *Extended* category expand on the *Standard* exercises to include more applications, as well as some conceptual or theory-based questions. Exercises may include selected items from the *Concepts and Vocabulary, Working with Formulas,* and the *Extending* the Thought categories of the exercise sets.

- **In Depth:** The *In Depth* assignments represent a more comprehensive look at the material from each section, while attempting to keep the assignment manageable for students. These include a selection of the most popular and highest-quality exercises from each category of the exercise set, with an additional emphasis on *Maintaining Your Skills*.

Through a Conversational Writing Style

While examples and applications are arguably the most prominent features of a mathematics text, it's the writing style and readability that binds them together. It may be true that some students don't read the text, and that others open the text only when looking for an example similar to the exercise they're currently working. But when they do and for those students who do (read the text), it's important they have a text that "speaks to them," relating concepts in a form and at a level they understand and can relate to. Ideally this text will draw students in and keep their interest, becoming a positive experience and bringing them back a second and third time, until it becomes habitual. At this point, students might begin to see the true value of their text (as more that just a source of problems—pun intended), and it becomes a resource for learning on equal footing with any other form of supplemental instruction.   —JC

Conversational Writing Style

John Coburn's experience in the classroom and his strong connections to how students comprehend the material are evident in his writing style. He uses a conversational and supportive writing style, providing the students with a tool they can depend on when the teacher is not available, when they miss a day of class, or simply when working on their own. The effort John has put into the writing is representative of his unofficial mantra: "If you want more students to reach the top, you gotta put a few more rungs on the ladder."

Through Student Involvement

How do you design a student-friendly textbook? We decided to get students involved by hosting two separate focus groups. During these sessions we asked students to advise us on how they use their books,

what pedagogical elements are useful, which elements are distracting and not useful, as well as general feedback on page layout. During this process there were times when we thought, "Now why hasn't anyone ever thought of that before?" Clearly these student focus groups were invaluable. Taking direct student feedback and incorporating what is feasible and doesn't detract from instructor use of the text is the best way to design a truly student-friendly text. The next two pages will highlight what we learned from students so you can see for yourself how their feedback played an important role in the development of the Coburn series.

Students said that **Learning Objectives** should clearly define the goals of each section.

Students asked for **Check Points** throughout each section to alert them when a specific learning objective has been covered and to reinforce the use of correct mathematical terms.

Described by students as one of the most useful features in a math text, **Caution Boxes** signal a student to stop and take note in order to avoid mistakes in problem solving.

Students told us that the color red should only be used for things that are really important. Also, anything significant should be included in the body of the text; marginal readings imply optional.

Examples are called out in the margins so they are easy for students to spot.

Examples are "boxed" so students can clearly see where they begin and end

Students told us they liked when the examples were linked to the exercises.

1.1 Linear Equations, Formulas, and Problem Solving

Learning Objectives

In Section 1.1 you will learn how to:

☐ **A.** Solve linear equations using properties of equality

☐ **B.** Recognize equations that are identities or contradictions

☐ **C.** Solve for a specified variable in a formula or literal equation

☐ **D.** Use the problem-solving guide to solve various problem types

In a study of algebra, you will encounter many **families of equations,** or groups of equations that share common characteristics. Of interest to us here is the family of **linear equations in one variable,** a study that lays the foundation for understanding more advanced families. In addition to *solving* linear equations, we'll use the skills we develop to *solve for a specified variable* in a formula, a practice widely used in science, business, industry, and research.

A. Solving Linear Equations Using Properties of Equality

An **equation** is *a statement that two expressions are equal.* From the expressions $3(x - 1) + x$ and $-x + 7$, we can form the equation

$$3(x - 1) + x = -x + 7.$$

which is a **linear equation in one variable.** To solve an equation, we attempt to find a specific input or *x*-value that will make the equation true, meaning the left-hand expression will be equal to the right. Using

Table 1.1

x	$3(x - 1) + x$	$-x + 7$
−2	−11	9
−1	−7	8
0	−3	7
1	1	6
2	5	5

EXAMPLE 2 ► Solving a Linear Equation with Fractional Coefficients

Solve for n: $\frac{1}{4}(n + 8) - 2 = \frac{1}{2}(n - 6)$.

Solution ►

$\frac{1}{4}(n + 8) - 2 = \frac{1}{2}(n - 6)$ original equation

$\frac{1}{4}n + 2 - 2 = \frac{1}{2}n - 3$ distributive property

$\frac{1}{4}n = \frac{1}{2}n - 3$ combine like terms

$4(\frac{1}{4}n) = 4(\frac{1}{2}n - 3)$ multiply both sides by LCD = 4

$n = 2n - 12$ distributive property

$-n = -12$ subtract 2n

$n = 12$ multiply by −1

☑ **A.** You've just learned how to solve linear equations using properties of equality

Verify the solution is $n = 12$ using back-substitution.

Now try Exercises 13 through 30 ►

$$= \frac{3}{6} = \frac{1}{2}$$

The slope of this line is $\frac{1}{2}$.

$$= \frac{-4}{6} = \frac{-2}{3}$$

The slope of this line is $\frac{-2}{3}$.

Now try Exercises 33 through 40 ►

⚠ **CAUTION** ► When using the slope formula, try to avoid these common errors.

1. The order that the *x*- and *y*-coordinates are subtracted must be consistent, since $\frac{y_2 - y_1}{x_2 - x_1} \neq \frac{y_2 - y_1}{x_1 - x_2}$.

2. The vertical change (involving the *y*-values) always occurs in the numerator: $\frac{y_2 - y_1}{x_2 - x_1} \neq \frac{x_2 - x_1}{y_2 - y_1}$.

3. When x_1 or y_1 is negative, use parentheses when substituting into the formula to prevent confusing the negative sign with the subtraction operation.

Actually, the slope value does much more than quantify the slope of a line, it expresses a **rate of change** between the quantities measured along each axis. In applications of slope, the ratio $\frac{\text{change in } y}{\text{change in } x}$ is symbolized as $\frac{\Delta y}{\Delta x}$. The symbol Δ is the Greek letter **delta** and has come to represent a change in some quantity, and the notation $m = \frac{\Delta y}{\Delta x}$ is read, "slope is equal to the *change in y* over the *change in x*." Interpreting slope as a rate of change has many significant applications in college algebra and beyond.

EXAMPLE 8 ► Determining the Domain of an Expression

Determine the domain of the expression $\frac{6}{x - 2}$. State the result in set notation, graphically, and using interval notation.

Solution ► Set the denominator equal to zero and solve: $x - 2 = 0$ yields $x = 2$. This means 2 is outside the domain and *must be excluded*.

• Set notation: $\{x | x \in \mathbb{R}, x \neq 2\}$

• Graph:

• Interval notation: $x \in (-\infty, 2) \cup (2, \infty)$

Now try Exercises 61 through 68 ►

A second area where allowable values are a concern involves the square root operation. Recall that $\sqrt{49} = 7$ since $7 \cdot 7 = 49$. However, $\sqrt{-49}$ cannot be written as the product of two real numbers since $(-7) \cdot (-7) = 49$ and $7 \cdot 7 = 49$. In other words, $\sqrt{X}$ represents a real number only if the radicand is positive or zero. If X represents an algebraic expression, the domain of $\sqrt{X}$ is $\{X | X \geq 0\}$.

EXAMPLE 9 ► Determining the Domain of an Expression

Determine the domain of $\sqrt{x + 3}$. State the domain in set notation, graphically, and in interval notation.

Solution ► The radicand must represent a nonnegative number. Solving $x + 3 \geq 0$ gives $x \geq -3$.

• Set notation: $\{x | x \geq -3\}$

• Graph:

• Interval notation: $x \in [-3, \infty)$

Now try Exercises 69 through 76 ►

Students told us that directions should be in bold so they are easily distinguishable from the problems.

Solve using the zero product property. Be sure each equation is in standard form and factor out any common factors before attempting to solve. Check all answers in the original equation.

7. $22x = x^3 - 9x^2$
8. $x^3 = 13x^2 - 42x$
9. $3x^3 = -7x^2 + 6x$
10. $7x^2 + 15x = 2x^3$
11. $2x^4 - 3x^3 = 9x^2$
12. $-7x^2 = 2x^4 - 9x^3$
13. $2x^4 - 16x = 0$
14. $x^4 + 64x = 0$
15. $x^3 - 4x = 5x^2 - 20$
16. $x^3 - 18 = 9x - 2x^2$
17. $4x - 12 = 3x^2 - x^3$
18. $x - 7 = 7x^2 - x^3$
19. $2x^3 - 12x^2 = 10x - 60$
20. $9x + 81 = 27x^2 + 3x^3$
21. $x^4 - 7x^3 + 4x^2 = 28x$
22. $x^4 + 3x^3 + 9x^2 = -27x$
23. $x^4 - 81 = 0$
24. $x^4 - 1 = 0$
25. $x^4 - 256 = 0$
26. $x^4 - 625 = 0$
27. $x^6 - 2x^4 - x^2 + 2 = 0$
28. $x^6 - 3x^4 - 16x^2 + 48 = 0$
29. $x^5 - x^3 - 8x^2 + 8 = 0$
30. $x^5 - 9x^3 - x^2 + 9 = 0$
31. $x^6 - 1 = 0$
32. $x^6 - 64 = 0$

Solve each equation. Identify any extraneous roots.

33. $\dfrac{2}{x} + \dfrac{1}{x+1} = \dfrac{5}{x^2 + x}$

34. $\dfrac{3}{m+3} - \dfrac{5}{m^2 + 3m} = \dfrac{1}{m}$

35. $\dfrac{21}{a+2} = \dfrac{3}{a-1}$

36. $\dfrac{4}{2y-3} = \dfrac{7}{3y-5}$

39. $x + \dfrac{14}{x-7} = 1 + \dfrac{2x}{x-7}$

40. $\dfrac{10}{x-5} + x = 1 + \dfrac{2x}{x-5}$

41. $\dfrac{6}{n+3} + \dfrac{20}{n^2 + n - 6} = \dfrac{5}{n-2}$

42. $\dfrac{7}{p+2} - \dfrac{1}{p^2 + 5p + 6} = \dfrac{2}{p+3}$

43. $\dfrac{a}{2a+1} - \dfrac{2a^2 + 5}{2a^2 - 5a - 3} = \dfrac{3}{a-3}$

44. $\dfrac{-18}{6n^2 - n - 1} + \dfrac{3n}{2n-1} = \dfrac{4n}{3n+1}$

Solve for the variable indicated.

45. $\dfrac{1}{f} = \dfrac{1}{f_1} + \dfrac{1}{f_2}$; for f
46. $\dfrac{1}{x} - \dfrac{1}{y} = \dfrac{1}{z}$; for z
47. $I = \dfrac{E}{R+r}$; for r
48. $q = \dfrac{pf}{p-f}$; for p
49. $V = \dfrac{1}{3}\pi r^2 h$; for h
50. $s = \dfrac{1}{2}gt^2$; for g
51. $V = \dfrac{4}{3}\pi r^3$; for r^3
52. $V = \dfrac{1}{3}\pi r^2 h$; for r^2

Solve each equation and check your solutions by substitution. Identify any extraneous roots.

53. a. $-3\sqrt{3x-5} = -9$ b. $x = \sqrt{3x+1} + 3$
54. a. $-2\sqrt{4x-1} = -10$ b. $-5 = \sqrt{5x-1} - x$
55. a. $2 = \sqrt[3]{3m-1}$ b. $2\sqrt[3]{7-3x} - 3 = -7$
c. $\dfrac{\sqrt[3]{2m+3}}{-5} + 2 = 3$ d. $\sqrt[3]{2x-9} = \sqrt[3]{3x+7}$
56. a. $-3 = \sqrt[3]{5p+2}$ b. $3\sqrt[3]{3-4x} - 7 = -4$
c. $\dfrac{\sqrt[3]{6x-7}}{4} - 5 = -6$
d. $3\sqrt[3]{x+3} = 2\sqrt[3]{2x+17}$
57. a. $\sqrt{x-9} + \sqrt{x} = 9$
b. $x = 3 + \sqrt{23-x}$
c. $\sqrt{x-2} - \sqrt{2x} = -2$

Because students spend a lot of time in the exercise section of a text, they said that a white background is hard on their eyes...so we used a soft, off-white color for the background

88. **Composite figures—gelatin capsules:** The gelatin capsules manufactured for cold and flu medications are shaped like a cylinder with a hemisphere on each end. The interior volume V of each capsule can be modeled by $V = \frac{4}{3}\pi r^3 + \pi r^2 h$, where h is the height of the cylindrical portion and r is its radius. If the cylindrical portion of the capsule is 8 mm long ($h = 8$ mm), what radius would give the capsule a volume that is numerically equal to 15π times this radius?

89. **Running shoes:** When a popular running shoe is priced at $70, The Shoe House will sell 15 pairs each week. Using a survey, they have determined that for each decrease of $2 in price, 3 additional pairs will be sold each week. What selling price will give a weekly revenue of $2250?

90. **Cell phone charges:** A cell phone service sells 48 subscriptions each month if their monthly fee is $30. Using a survey, they find that for each decrease of $1, 6 additional subscribers will join. What charge(s) will result in a monthly revenue of $2160?

Projectile height: In the absence of resistance, the height of an object that is projected upward can be modeled by the equation $h = -16t^2 + vt + k$, where h represents the height of the object (in feet) t sec after it has been thrown, v represents the initial velocity (in feet per second), and k represents the height of the object when $t = 0$ (before it has

velocity of 700 ft/sec and a height of 240 ft, it runs out of fuel and becomes a projectile.

a. How high is the rocket three seconds later? Four seconds later?
b. How long will it take the rocket to attain a height of 640 ft?
c. How many times is a height of 384 ft attained? When do these occur?
d. How many seconds until the rocket returns to the ground?

93. **Printing newspapers:** The editor of the school newspaper notes the college's new copier can complete the required print run in 20 min, while the back-up copier took 30 min to do the same amount of work. How long would it take if both copiers are used?

94. **Filling a sink:** The cold water faucet can fill a sink in 2 min. The drain can empty a full sink in 3 min. If the faucet were left on and the drain was left open, how long would it take to fill the sink?

95. **Triathalon competition:** As one part of a Mountain-Man triathalon, participants must row a canoe 5 mi down river (with the current), circle a buoy and row 5 mi back up river (against the current) to the starting point. If the current is flowing at a steady rate of 4 mph and Tom Chaney made the round-trip in 3 hr, how fast can he row in still water? (*Hint:* The time rowing down river and the time rowing up river must add up to 3 hr.)

96. **Flight time:** The flight distance from Cincinnati, Ohio, to Chicago, Illinois, is approximately 300 mi. On a recent round-trip between these cities in my private plane, I encountered a steady 25 mph headwind on the way to Chicago, with a 25 mph tailwind on the return trip. If my total flying time

---

Students said having a lot of icons was confusing. The graphing calculator is the only icon used in the exercise sets; no unnecessary icons are used

▶ **WORKING WITH FORMULAS**

79. **Lateral surface area of a cone:** $S = \pi r\sqrt{r^2 + h^2}$

The lateral surface area (surface area excluding the base) S of a cone is given by the formula shown, where r is the radius of the base and h is the height of the cone. (a) Solve the equation for h. (b) Find the surface area of a cone that has a radius of 6 m and a height of 10 m. Answer in simplest form.

80. **Painted area on a canvas:** $A = \dfrac{4x^2 + 60x + 104}{x}$

A rectangular canvas is to contain a small painting with an area of 52 in^2, and requires 2-in. margins on the left and right, with 1-in. margins on the top and bottom for framing. The total area of such a canvas is given by the formula shown, where x is the height of the *painted* area.

a. What is the area A of the canvas if the height of the painting is $x = 10$ in.?
b. If the area of the canvas is $A = 120$ in^2, what are the dimensions of the painted area?

College Algebra, Second Edition

Review ◆ Equations and Inequalities ◆ Relations, Functions, and Graphs ◆ Polynomial and Rational Functions ◆ Exponential and Logarithmic Functions ◆ Systems of Equations and Inequalities ◆ Matrices ◆ Geometry and Conic Sections ◆ Additional Topics in Algebra
ISBN 0-07-351941-3, ISBN 978-0-07351941-8

College Algebra Essentials, Second Edition

Review ◆ Equations and Inequalities ◆ Relations, Functions, and Graphs ◆ Polynomial and Rational Functions ◆ Exponential and Logarithmic Functions ◆ Systems of Equations and Inequalities
ISBN 0-07-351968-5, ISBN 978-0-07351968-5

Algebra and Trigonometry, Second Edition

Review ◆ Equations and Inequalities ◆ Relations, Functions, and Graphs ◆ Polynomial and Rational Functions ◆ Exponential and Logarithmic Functions ◆ Trigonometric Functions ◆ Trigonometric Identities, Inverses and Equations ◆ Applications of Trigonometry ◆ Systems of Equations and Inequalities ◆ Matrices ◆ Geometry and Conic Sections ◆ Additional Topics in Algebra
ISBN 0-07-351952-9, ISBN 978-0-07-351952-4

Precalculus, Second Edition

Equations and Inequalities ◆ Relations, Functions, and Graphs ◆ Polynomial and Rational Functions ◆ Exponential and Logarithmic Functions ◆ Trigonometric Functions ◆ Trigonometric Identities, Inverses and Equations ◆ Applications of Trigonometry ◆ Systems of Equations and Inequalities, and Matrices ◆ Geometry and Conic Sections ◆ Additional Topics in Algebra ◆ Limits
ISBN 0-07-351942-1, ISBN 978-0-07351942-5

Trigonometry, Second Edition—Coming in 2010!

Introduction to Trigonometry ◆ Trigonometric Functions ◆ Trigonometric Identities ◆ Trigonometric Inverses and Equations ◆ Applications of Trigonometry ◆ Conic Sections and Polar Coordinates
ISBN 0-07-351948-0, ISBN 978-0-07351948-7

Making Connections . . .

Through New and Updated Content

New to the Second Edition

▶ An extensive reworking of the narrative and reduction of advanced concepts enhances the clarity of the exposition, improves the student's experience in the text, and decreases the overall length of the text.

▶ A modified interior design based on student and instructor feedback from focus groups features increased font size, improved exercise and example layout, more white space on the page, and the careful use of color to enhance the presentation of pedagogy.

▶ Chapter Openers based on applications bring awareness to students of the relevance of concepts presented in each chapter.

▶ The removal of algebraic proofs from the main body of the text to an appendix provides better focus in the chapter and presents mathematics in a less technical manner.

▶ Checkpoints throughout each section alert students when a specific learning objective has been covered and reinforce the use of correct mathematical terms.

▶ The Homework Selection Guide, appearing in each exercise section in the Annotated Instructor's Edition, provides instructors with suggestions for developing core, standard, extended, and in-depth homework assignments without much prep work.

▶ The Modeling with Technology feature between chapters presents standalone coverage of regression, with pedagogy, exercises, and applications for those instructors who choose to cover this material

Chapter-by-Chapter Changes

CHAPTER R A Review of Basic Concepts and Skills

- Square and cube roots are now covered together.
- Section R.2 features more opportunities for mathematical modeling as well as a summary of exponential properties.
- Examples using radicals have been added to Section R.3 to provide more practice solving, factoring, and simplifying.
- The discussion of factoring in Section R.4 now includes $x^2 - k$, when k is not a perfect square, and higher-degree expressions.
- A Chapter Overview has been added to the end of the chapter, offering students a study tool for the review of prerequisite topics.

CHAPTER 1 Equations and Inequalities

- Chapter 1 now includes coverage of absolute value equations and inequalities in Section 1.3.
- Information on solving quadratics has been consolidated to a single section (1.5) and summary boxes are now used for solving linear equations, solving quadratic equations, and solution methods for quadratic equations.
- Examples and exercises employing the use of a graphing calculator have been added throughout the chapter.

CHAPTER 2 Relations, Functions, and Graphs

- The organization of Chapter 2 has changed from the first edition in an effort to concentrate the introduction of graphs and general functions.
- Coverage of the midpoint formula, the distance formula, and circles has been improved and reorganized (Section 2.1).
- Linear graphs are established early in the chapter (Sections 2.2 and 2.3) before functions are introduced.
- The section on the toolbox (basic parent) functions (2.6) now appears after analyzing graphs (Section 2.5) to improve connections among the material.
- Coverage of rates of change has been consolidated while coverage of the implied domain, distance quotient, end behavior, and even/odd functions has been expanded and improved.
- Additional applications of the floor and ceiling functions and the algebra of functions have been added.
- Regression material in this chapter has been removed and concentrated in the between-chapter Modeling with Technology feature.

CHAPTER 3 Polynomial and Rational Functions

- Chapter 3 has been significantly reorganized to bring focus to and provide a better bridge between general functions and polynomial functions.
- More coverage of completing the square and increased emphasis on graphing using the vertex formula is found in Section 3.1.
- Complex conjugates, zeroes of multiplicity, and number of zeroes have been realigned together in Section 3.2.
- Section 3.3 features an improved description of Descartes' rule of signs, as well as stronger connections between the fundamental theorem of algebra and the linear factorization theorem and its corollaries.
- A better introduction regarding polynomials versus nonpolynomials is found in Section 3.4, in addition to an improved discussion of end behavior.
- Section 3.6 provides better treatment of removable discontinuities and a clearer discussion of pointwise versus asymptotic continuities.
- Section 3.8 presents a clearer, stronger connection between previously covered topics and applications of variation and the toolbox functions.
- Regression material in this chapter has been removed and concentrated in the between-chapter Modeling with Technology feature.

CHAPTER 4 Exponential and Logarithmic Functions

- Chapter 4 now begins with coverage of one-to-one and inverse functions given their applications for exponents and logarithms.
- This section (4.1) includes examples of finding inverses of rational functions, as well as better coverage of restricting the domain to find the inverse.
- Coverage of base e as an alternative to base 10 or b is addressed in one section (4.2) as opposed to two sections as in the first edition.
- Likewise, coverage of properties of logs and log equations is found in the same section (4.4).
- A clear introduction to fundamental logarithmic properties has also been added to Section 4.4.
- Applications have been added and improved throughout the chapter.
- Regression material in this chapter has been removed and concentrated in the between-chapter Modeling with Technology feature.

CHAPTER 5 Systems of Equations and Inequalities

- The coverage of systems and matrices has been split into two chapters for the second edition (Chapter 5 on systems and Chapter 6 on matrices).
- Section 5.1 includes improved coverage of equivalent systems in addition to more examples and exercises having to do with distance and navigation.
- Section 5.2 features improved coverage of dependent and inconsistent systems.
- Section 5.3 now presents nonlinear systems before linear programming and features a better presentation of nonlinear and nonpolynomial systems.
- More business examples and exercises have also been added to Section 5.3.
- New applications of linear programming are found in Section 5.4.

CHAPTER 6 Matrices and Matrix Applications

- Section 6.1 features an added example of Gauss-Jordan Elimination.
- Section 6.2 includes better sequencing of examples and improved coverage of matrix properties.
- Coverage of determinants has been streamlined with more development given to determinants in Section 6.3, while improved coverage of partial fractions and a better introduction of the decomposition template is found in Section 6.4.

CHAPTER 7 Analytical Geometry and the Conic Sections

- New Section 7.1 presents a brief introduction to analytical geometry to provide a better bridge to the conic sections and show why cone/conic connection is important.
- Greater emphasis on the connection between ellipses and circles is featured in section 7.2.
- Exercises requiring the movement from graph to equation have been added throughout the chapter.

CHAPTER 8 Additional Topics in Algebra

- The exposition has been revised throughout Chapter 8 for increased clarity and improved flow of topics.

Through 360° Development

McGraw-Hill's 360° Development Process is an ongoing, never-ending, market-oriented approach to building accurate and innovative print and digital products. It is dedicated to continual large-scale and incremental improvement driven by multiple customer feedback loops and checkpoints. This process is initiated during the early planning stages of our new products, intensifies during the development and production stages, and then begins again on publication, in anticipation of the next edition.

A key principle in the development of any mathematics text is its ability to adapt to teaching specifications in a universal way. The only way to do so is by contacting those universal voices—and learning from their suggestions. We are confident that our book has the most current content the industry has to offer, thus pushing our desire for accuracy to the highest standard possible. In order to accomplish this, we have moved through an arduous road to production. Extensive and open-minded advice is critical in the production of a superior text.

We engaged over 400 instructors and students to provide us guidance in the development of the second edition. By investing in this extensive endeavor, McGraw-Hill delivers to you a product suite that has been created, refined, tested, and validated to be a successful tool in your course.

Board of Advisors

A hand-picked group of trusted teachers active in the College Algebra and Precalculus course areas served as the chief advisors and consultants to the author and editorial team with regards to manuscript development. The Board of Advisors reviewed the manuscript in two drafts; served as a sounding board for pedagogical, media, and design concerns; approved organizational changes; and attended a symposium to confirm the manuscript's readiness for publication.

Bill Forrest, *Baton Rouge Community College*
Marc Grether, *University of North Texas*
Sharon Hamsa, *Metropolitan Community College
    –Longview*
Max Hibbs, *Blinn College*
Terry Hobbs, *Metropolitan Community College–
    Maple Woods*
Klay Kruczek, *Western Oregon University*
Rita Marie O'Brien's , *Navarro College*

Nancy Matthews, *University of Oklahoma*
Rebecca Muller, *Southeastern Louisiana University*
Jason Pallett, *Metropolitan Community College*
Kevin Ratliff, *Blue Ridge Community College*
Stephen Toner, *Victor Valley College*

Accuracy Panel

A selected trio of key instructors served as the chief advisors for the accuracy and clarity of the text and solutions manual. These individuals reviewed the final manuscript, the page proofs in first and revised rounds, as well as the writing and accuracy check of the instructor's solutions manuals. This trio, in addition to several other accuracy professionals, gives you the assurance of accuracy.

J.D. Herdlick, *St. Louis Community College–Meramac*
Richard A. Pescarino, *St. Louis Community College–Florissant Valley*
Nathan G. Wilson, *St. Louis Community College–Meramac*

Student Focus Groups

Two student focus groups were held at Illinois State University and Southeastern Louisiana University to engage students in the development process and provide feedback as to how the design of a textbook impacts homework and study habits in the College Algebra and Precalculus course areas.

Francisco Arceo, *Illinois State University*
Dave Cepko, *Illinois State University*
Andrea Connell, *Illinois State University*
Brian Lau, *Illinois State University*
Daniel Nathan Mielneczek, *Illinois State University*
Mingaile Orakauskaite, *Illinois State University*
Todd Michael Rapnikas, *Illinois State University*
Bethany Rollet, *Illinois State University*
Teddy Schrishuhn, *Illinois State University*
Josh Schultz, *Illinois State University*
Andy Thurman, *Illinois State University*
Candace Banos, *Southeastern Louisiana University*
Nicholas Curtis, *Southeastern Louisiana University*

M. D. "Boots" Feltenberger, *Southeastern Louisiana University*
Regina Foreman, *Southeastern Louisiana University*
Ashley Lae, *Southeastern Louisiana University*
Jessica Smith, *Southeastern Louisiana University*
Ashley Youngblood, *Southeastern Louisiana University*

Special Thanks

Sherry Meier, *Illinois State University*
Rebecca Muller, *Southeastern Louisiana University*
Anne Schmidt, *Illinois State University*

Instructor Focus Groups

Focus groups held at Baton Rouge Community College and ORMATYC provided feedback on the new Connections to Calculus feature in *Precalculus*, and shed light on the coverage of review material in this course. User focus groups at Southeastern Louisiana University and Madison Area Technical College confirmed the organizational changes planned for the second edition, provided feedback on the interior design, and helped us enhance and refine the strengths of the first edition.

Virginia Adelmann, *Southeastern Louisiana University*
George Alexander, Madison Area Technical College
Kenneth R. Anderson, *Chemeketa Community College*
Wayne G.Barber, *Chemeketa Community College*
Thomas Dick, *Oregon State University*
Vickie Flanders, *Baton Rouge Community College*
Bill Forrest, *Baton Rouge Community College*
Susan B. Guidroz, *Southeastern Louisiana University*
Christopher Guillory, *Baton Rouge Community College*
Cynthia Harrison, *Baton Rouge Community College*
Judy Jones, Madison Area Technical College
Lucyna Kabza, *Southeastern Louisiana University*
Ann Kirkpatrick, *Southeastern Louisiana University*
Sunmi Ku, *Bellevue Community College*

Pamela Larson, *Madison Area Technical College*
Jennifer Laveglia, *Bellevue Community College*
DeShea Miller, *Southeastern Louisiana University*
Elizabeth Miller, *Southeastern Louisiana University*
Rebecca Muller, *Southeastern Louisiana University*
Donna W. Newman, *Baton Rouge Community College*
Scott L. Peterson, *Oregon State University*
Ronald Posey, *Baton Rouge Community College*
Ronni Settoon, *Southeastern Louisiana University*
Jeganathan Sriskandarajah, Madison Area Technical College
Martha Stevens, *Bellevue Community College*
Mark J. Stigge, *Baton Rouge Community College*
Nataliya Svyeshnikova, *Southeastern Louisiana University*

Developmental Symposia

McGraw-Hill conducted two symposia directly related to the development of Coburn's second edition. These events were an opportunity for editors from McGraw-Hill to gather information about the needs and challenges of instructors teaching these courses and confirm the direction of the second edition.

Diary Reviews and Class Tests

Users of the first edition, Said Ngobi and Stephen Toner of Victor Valley College, provided chapter-by chapter feedback in diary form based on their experience using the text. Board of Advisors members facilitated class tests of the manuscript for a given topic. Both instructors and students returned questionnaires detailing their thoughts on the effectiveness of the text's features.

Class Tests

Hayley Hentzen, *University of North Texas*
Courtney Hodge, *University of North Texas*
Janice Hollaway, *Navarro College*
Weslon Hull, *Baton Rouge Community College*
Sarah James, *Baton Rouge Community College*
Georlin Johnson, *Baton Rouge Community College*
Michael Jones, *Navarro College*
Robert Koon, *Metropolitan Community College–Longview*
Ben Lenfant, *Baton Rouge Community College*
Colin Luke, *Baton Rouge Community College*
Lester Maloney, *Baton Rouge Community College*
Ana Mariscal, *Navarro College*
Tracy Ann Nguyen, *Baton Rouge Community College*
Alexandra Ortiz, *University of North Texas*
Robert T. R. Paine, *Baton Rouge Community College*
Kade Parent, *Baton Rouge Community College*
Brittany Louise Pratt, *Baton Rouge Community College*

Brittney Pruitt, *Metropolitan Community College–Longview*
Paul Rachal, *Baton Rouge Community College*
Matt Rawls, *Baton Rouge Community College*
Adam Reichert, *Metropolitan Community College–Longview*
Ryan Rodney, *Baton Rouge Community College*
Cody Scallan, *Baton Rouge Community College*
Laura Shafer, *University of North Texas*
Natina Simpson, *Navarro College*
Stephanie Sims, *Metropolitan Community College–Longview*
Cassie Snow, *University of North Texas*
Justin Stewart, *Metropolitan Community College–Longview*
Marjorie Tulana, *Navarro College*
Ashleigh Variest, *Baton Rouge Community College*
James A. Wann, *Navarro College*
Amber Wendleton, *Metropolitan Community College–Longview*
Eric Williams, *Metropolitan Community College–Longview*
Katy Wood, *Metropolitan Community College–Longview*

Developmental Editing

The manuscript has been impacted by numerous developmental editors who edited for clarity and consistency. Efforts resulted in cutting length from the manuscript, while retaining a conversational and casual narrative style. Editorial work also ensured the positive visual impact of art and photo placement.

First Edition Chapter Reviews and Manuscript Reviews

Over 200 instructors participated in postpublication single chapter reviews of the first edition and helped the team build the revision plan for the second edition. Over 100 teachers and academics from across the country reviewed the current edition text, the proposed second edition table of contents, and first-draft second edition manuscript to give feedback on reworked narrative, design changes, pedagogical enhancements, and organizational changes. This feedback was summarized by the book team and used to guide the direction of the second-draft manuscript.

Scott Adamson, *Chandler-Gilbert Community College*
Teresa Adsit, *University of Wisconsin–Green Bay*
Ebrahim Ahmadizadeh, *Northampton Community College*
George M. Alexander, *Madison Area Technical College*
Frances Alvarado, *University of Texas–Pan American*
Deb Anderson, *Antelope Valley College*
Philip Anderson, *South Plains College*
Michael Anderson, *West Virginia State University*
Jeff Anderson, *Winona State University*
Raul Aparicio, *Blinn College*
Judith Barclay, *Cuesta College*
Laurie Battle, *Georgia College and State University*
Annette Benbow, *Tarrant County College–Northwest*
Amy Benvie, *Florida Gulf Coast University*

Scott Berthiaume, *Edison State College*
Wes Black, *Illinois Valley Community College*
Arlene Blasius, *SUNY College of Old Westbury*
Caroline Maher Boulis, *Lee University*
Amin Boumenir, *University of West Georgia*
Terence Brenner, *Hostos Community College*
Gail Brooks, *McLennan Community College*
G. Robert Carlson, *Victor Valley College*
Hope Carr, *East Mississippi Community College*
Denise Chellsen, *Cuesta College*
Kim Christensen, *Metropolitan Community College– Maple Woods*
Lisa Christman, *University of Central Arkansas*
John Church, *Metropolitan Community College–Longview*
Sarah Clifton, *Southeastern Louisiana University*
David Collins, *Southwestern Illinois College*
Sarah V. Cook, *Washburn University*
Rhonda Creech, *Southeast Kentucky Community and Technical College*
Raymond L. Crownover, *Gateway College of Evangelism*
Marc Cullison, *Connors State College*
Steven Cunningham, *San Antonio College*
Callie Daniels, *St. Charles Community College*
John Denney, *Northeast Texas Community College*
Donna Densmore, *Bossier Parish Community College*
Alok Dhital, *University of New Mexico–Gallup*

James Michael *Dubrowsky Wayne Community College*

Brad Dyer, *Hazzard Community & Technical College*

Sally Edwards, *Johnson County Community College*

John Elliott, *St. Louis Community College–Meramec*

Gay Ellis, *Missouri State University*

Barbara Elzey, *Bluegrass Community College*

Dennis Evans, *Concordia University Wisconsin*

Samantha Fay, *University of Central Arkansas*

Victoria Fischer, *California State University–Monterey Bay*

Dorothy French, *Community College of Philadelphia*

Eric Garcia, *South Texas College*

Laurice Garrett, *Edison College*

Ramona Gartman, *Gadsden State Community College–Ayers Campus*

Scott Gaulke, *University of Wisconsin–Eau Claire*

Scott Gordon, *University of West Georgia*

Teri Graville, *Southern Illinois University Edwardsville*

Marc Grether, *University of North Texas*

Shane Griffith, *Lee University*

Gary Grohs, *Elgin Community College*

Peter Haberman, *Portland Community College*

Joseph Harris, *Gulf Coast Community College*

Margret Hathaway, *Kansas City Community College*

Tom Hayes, *Montana State University*

Bill Heider, *Hibbling Community College*

Max Hibbs, *Blinn College*

Terry Hobbs, *Metropolitan Community College–Maple Woods*

Sharon Holmes, *Tarrant County College–Southeast*

Jamie Holtin, *Freed-Hardeman University*

Brian Hons, *San Antonio College*

Kevin Hopkins, *Southwest Baptist University*

Teresa Houston, *East Mississippi Community College*

Keith Hubbard, *Stephen F. Austin State University*

Jeffrey Hughes, *Hinds Community College–Raymond*

Matthew Isom, *Arizona State University*

Dwayne Jennings, *Union University*

Judy Jones, *Madison Area Technical College*

Lucyna Kabza, *Southeastern Louisiana University*

Aida Kadic-Galeb, *University of Tampa*

Cheryl Kane, *University of Nebraska*

Rahim Karimpour, *Southern Illinois University Edwardsville*

Ryan Kasha, *Valencia Community College*

David Kay, *Moorpark College*

Jong Kim, *Long Beach City College*

Lynette King, *Gadsden State Community College*

Carolyn Kistner, *St. Petersburg College*

Barbara Kniepkamp, *Southern Illinois University Edwardsville*

Susan Knights, *Boise State University*

Stephanie Kolitsch, *University of Tennessee at Martin*

Louis Kolitsch, *University of Tennessee at Martin*

William Kirby, *Gadsden State Community College*

Karl Kruczek, *Northeastern State University*

Conrad Krueger, *San Antonio College*

Marcia Lambert, *Pitt Community College*

Rebecca Lanier, *Bluegrass Community College*

Marie Larsen, *Cuesta College*

Pam Larson, *Madison Area Technical College*

Jennifer Lawhon, *Valencia Community College*

John Levko, *University of Scranton*

Mitchel Levy, *Broward Community College*

John Lofberg, *South Dakota School of Mines and Technology*

Mitzi Logan, *Pitt Community College*

Sandra Maldonado, *Florida Gulf Coast University*

Robin C. Manker, *Illinois College*

Manoug Manougian, *University of South Florida*

Nancy Matthews, *University of Oklahoma*

Roger McCoach, *County College of Morris*

James McKinney, *California Polytechnic State University–Pomona*

Jennifer McNeilly, *University of Illinois Urbana Champaign*

Kathleen Miranda, *SUNY College at Old Westbury*

Mary Ann (Molly) Misko, *Gadsden State Community College*

Marianne Morea, *SUNY College of Old Westbury*

Michael Nasab, *Long Beach City College*

Said Ngobi, *Victor Valley College*

Tonie Niblett, *Northeast Alabama Community College*

Gary Nonnemacher, *Bowling Green State University*

Elaine Nye, *Alfred State College*

Rhoda Oden, *Gadsden State Community College*

Jeannette O'Rourke, *Middlesex County College*

Darla Ottman, *Elizabethtown Community & Technical College*

Jason Pallett, *Metropolitan Community College–Longview*

Priti Patel, *Tarrant County College–Southeast*

Judy Pennington-Price, *Midway College*

Susan Pfeifer, *Butler County Community College*

Margaret Poitevint, *North Georgia College & State University*

Tammy Potter, *Gadsden State Community College*

Debra Prescott, *Central Texas College*

Elise Price, *Tarrant County College*

Kevin Ratliff, *Blue Ridge Community College*

Bruce Reid, *Howard Community College*

Jolene Rhodes, *Valencia Community College*

Karen Rollins, *University of West Georgia*

Randy Ross, *Morehead State University*

Michael Sawyer, *Houston Community College*

Richard Schnackenberg, *Florida Gulf Coast University*

Bethany Seto, *Horry-Georgetown Technical College*

Delphy Shaulis, *University of Colorado–Boulder*
Jennifer Simonton, *Southwestern Illinois College*
David Slay, *McNeese State University*
David Snyder, *Texas State University at San Marcos*
Larry L. Southard, *Florida Gulf Coast University*
Lee Ann Spahr, *Durham Technical Community College*
Jeganathan Sriskandarajah, *Madison Area Technical College*
Adam Stinchcombe, *Eastern Arizona College*
Pam Stogsdill, *Bossier Parish Community College*
Eleanor Storey, *Front Range Community College*
Kathy Stover, *College of Southern Idaho*
Mary Teel, *University of North Texas*
Carlie Thompson, *Southeast Kentucky Community & Technical College*
Bob Tilidetzke, *Charleston Southern University*
Stephen Toner, *Victor Valley College*

Thomas Tunnell, *Illinois Valley Community College*
Carol Ulsafer, *University of Montana*
John Van Eps, *California Polytechnic State University–San Luis Obispo*
Andrea Vorwark, *Metropolitan Community College–Maple Woods*
Jim Voss, *Front Range Community College*
Jennifer Walsh, *Daytona State College*
Jiantian Wang, *Kean University*
Sheryl Webb, *Tennessee Technological University*
Bill Weber, *Fort Hays State University*
John Weglarz, *Kirkwood Community College*
Tressa White, *Arkansas State University–Newport*
Cheryl Winter, *Metropolitan Community College–Blue River*
Kenneth Word, *Central Texas College*
Laurie Yourk, *Dickinson State University*

Acknowledgments

I first want to express a deep appreciation for the guidance, comments and suggestions offered by all reviewers of the manuscript. I have once again found their collegial exchange of ideas and experience very refreshing and instructive, and always helping to create a better learning tool for our students.

I would especially like to thank Vicki Krug for her uncanny ability to bring innumerable pieces from all directions into a unified whole; Patricia Steele for her eagle-eyed attention to detail; Katie White and Michelle Flomenhoft for their helpful suggestions, infinite patience, tireless efforts, and steady hand in bringing the manuscript to completion; John Osgood for his ready wit and creative energies, Laurie Janssen and our magnificent design team, and Dawn Bercier, the master of this large ship, whose indefatigable spirit kept the ship on course through trial and tempest, and brought us all safely to port. In truth, my hat is off to all the fine people at McGraw-Hill for their continuing support and belief in this series. A final word of thanks must go to Rick Armstrong, whose depth of knowledge, experience, and mathematical connections seems endless; J. D. Herdlick for his friendship and his ability to fill an instant and sudden need, Anne Marie Mosher for her contributions to various features of the text, Mitch Levy for his consultation on the exercise sets, Stephen Toner for his work on the videos, Rosemary Karr for her meticulous work on the solutions manuals, Jay Miller and Carrie Green for their invaluable ability to catch what everyone else misses; and to Rick Pescarino, Nate Wilson, and all of my colleagues at St. Louis Community College, whose friendship, encouragement and love of mathematics makes going to work each day a joy.

Making Connections . . .

Through Supplements

*All online supplements are available through the book's website: www.mhhe.com/coburn.

Instructor Supplements

- **Computerized Test Bank Online:** Utilizing Brownstone Diploma® algorithm-based testing software enables users to create customized exams quickly.
- **Instructor's Solutions Manual:** Provides comprehensive, worked-out solutions to all exercises in the text.
- **Annotated Instructor's Edition:** Contains all answers to exercises in the text, which are printed in a second color, adjacent to corresponding exercises, for ease of use by the instructor.
- **PowerPoint Slides:** Fully editable slides that follow the textbook.

Student Supplements

- **Student Solutions Manual** provides comprehensive, worked-out solutions to all of the odd-numbered exercises.
- **Videos**
  - Interactive video lectures are provided for each section in the text, which explain to the students how to do key problem types, as well as highlighting common mistakes to avoid.
  - Exercise videos provide step-by-step instruction for the key exercises which students will most wish to see worked out.
  - Graphing calculator videos help students master the most essential calculator skills used in the college algebra course.
  - The videos are closed-captioned for the hearing impaired, subtitled in Spanish, and meet the Americans with Disabilities Act Standards for Accessible Design.

MathZone www.mhhe.com/coburn

McGraw-Hill's MathZone is a complete online homework system for mathematics and statistics. Instructors can assign textbook-specific content from over 40 McGraw-Hill titles as well as customize the level of feedback students receive, including the ability to have students show their work for any given exercise. Assignable content includes an array of videos and other multimedia along with algorithmic exercises, providing study tools for students with many different learning styles.

Within MathZone, a diagnostic assessment tool powered by ALEKS® is available to measure student preparedness and provide detailed reporting and personalized remediation. MathZone also helps ensure consistent assignment delivery across several sections through a course administration function and makes sharing courses with other instructors easy.

For additional study help students have access to NetTutor™, a robust online live tutoring service that incorporates whiteboard technology to communicate mathematics. The tutoring schedules are built around peak homework times to best accommodate student schedules. Instructors can also take advantage of this whiteboard by setting up a Live Classroom for online office hours or a review session with students.

For more information, visit the book's website (**www.mhhe.com/ coburn**) or contact your local McGraw-Hill sales representative (**www.mhhe.com/rep**).

ALEKS® www.aleks.com

ALEKS (**A**ssessment and **LE**arning in **K**nowledge **S**paces) is a dynamic online learning system for mathematics education, available over the Web 24/7. ALEKS assesses students, accurately determines their knowledge, and then guides them to the material that they are most ready to learn. With a variety of reports, Textbook Integration Plus, quizzes, and homework assignment capabilities, ALEKS offers flexibility and ease of use for instructors.

- ALEKS uses artificial intelligence to determine exactly what each student knows and is ready to learn. ALEKS remediates student gaps and provides highly efficient learning and improved learning outcomes
- ALEKS is a comprehensive curriculum that aligns with syllabi or specified textbooks. Used in conjunction with McGraw-Hill texts, students also receive links to text-specific videos, multimedia tutorials, and textbook pages.
- Textbook Integration Plus allows ALEKS to be automatically aligned with syllabi or specified McGraw-Hill textbooks with instructor chosen dates, chapter goals, homework, and quizzes.
- ALEKS with AI-2 gives instructors increased control over the scope and sequence of student learning. Students using ALEKS demonstrate a steadily increasing mastery of the content of the course.
- ALEKS offers a dynamic classroom management system that enables instructors to monitor and direct student progress towards mastery of course objectives.

ALEKS Prep/Remediation:

- Helps instructors meet the challenge of remediating unequally prepared or improperly placed students.
- Assesses students on their pre-requisite knowledge needed for the course they are entering (i.e. Calculus students are tested on Precalculus knowledge).
- Based on the assessment, students are prescribed a unique and efficient learning path specific to address their strengths and weaknesses.
- Students can address pre-requisite knowledge gaps outside of class freeing the instructor to use class time pursuing course outcomes.

Electronic Textbook: CourseSmart is a new way for faculty to find and review eTextbooks. It's also a great option for students who are interested in accessing their course materials digitally and saving money. CourseSmart offers thousands of the most commonly adopted textbooks across hundreds of courses from a wide variety of higher education publishers. It is the only place for faculty to review and compare the full text of a textbook online, providing immediate access without the environmental impact of requesting a print exam copy. At CourseSmart, students can save up to 50% off the cost of a print book, reduce their impact on the environment, and gain access to powerful web tools for learning including full text search, notes and highlighting, and email tools for sharing notes between classmates. **www.CourseSmart.com**

Primis: You can customize this text with McGraw-Hill/Primis Online. A digital database offers you the flexibility to customize your course including material from the largest online collection of textbooks, readings, and cases. Primis leads the way in customized eBooks with hundreds of titles available at prices that save your students over 20% off bookstore prices. Additional information is available at 800-228-0634.

Additional Topics Online

(Visit www.mhhe.com/coburn)

MEDICINE, NURSING, NUTRITION, DIETETICS, HEALTH

METEOROLOGY

MUSIC

PHYSICS, ASTRONOMY, PLANETARY STUDIES

COLLEGE
ALGEBRA

R

Business

A Review of Basic Concepts and Skills

CHAPTER OUTLINE

CHAPTER CONNECTIONS

Jared places a small inheritance of $2475 in a certificate of deposit that earns 6% interest compounded quarterly. The total in the CD after 10 years is given by the expression

$$2475\left(1 + \frac{0.06}{4}\right)^{4\cdot 10}.$$

This chapter reviews the skills required to correctly determine the CD's value, as well as other mathematical skills to be used throughout this course. This expression appears as Exercise 93 in Section R.1.

Check out these other real-world connections:

▶ Pediatric Dosages and Clark's Rule (Section R.1, Exercise 96)

▶ Maximizing Revenue of Video Game Sales (Section R.3, Exercise 143)

▶ Growth of a New Stock Hitting the Market (Section R.5, Exercise 83)

▶ Accident Investigation (Section R.6, Exercise 55)

1

Learning Objectives

In Section R.1 you will review:

☐ **A.** Sets of numbers, graphing real numbers, and set notation

☐ **B.** Inequality symbols and order relations

☐ **C.** The absolute value of a real number

☐ **D.** The Order of Operations

The most fundamental requirement for learning algebra is mastering the words, symbols, and numbers used to express mathematical ideas. "Words are the symbols of knowledge, the keys to accurate learning" (Norman Lewis in *Word Power Made Easy,* Penguin Books).

A. Sets of Numbers, Graphing Real Numbers, and Set Notation

To effectively use mathematics as a problem-solving tool, we must first be familiar with the **sets of numbers** used to quantify (give a numeric value to) the things we investigate. Only then can we make comparisons and develop equations that lead to informed decisions.

Natural Numbers

The most basic numbers are those used to count physical objects: 1, 2, 3, 4, and so on. These are called **natural numbers** and are represented by the capital letter $\mathbb{N}$, often written in the special font shown. We use **set notation** to list or describe a set of numbers. Braces { } are used to group **members** or **elements** of the set, commas separate each member, and three dots (called an *ellipsis*) are used to indicate a pattern that continues indefinitely. The notation $\mathbb{N} = \{1, 2, 3, 4, 5, \ldots\}$ is read, "$\mathbb{N}$ is the set of numbers 1, 2, 3, 4, 5, and so on." To show membership in a set, the symbol $\in$ is used. It is read "is an element of" or "belongs to." The statements $6 \in \mathbb{N}$ (6 is an element of $\mathbb{N}$) and $0 \notin \mathbb{N}$ (0 is not an element of $\mathbb{N}$) are true statements. A set having no elements is called the **empty** or **null set,** and is designated by empty braces { } or the symbol $\varnothing$.

EXAMPLE 1 ▶ **Writing Sets of Numbers Using Set Notation**

List the set of natural numbers that are

   a. negative   **b.** greater than 100

   c. greater than or equal to 5 and less than 12

Solution ▶   **a.** { }; all natural numbers are positive.

       b. $\{101, 102, 103, 104, \ldots\}$

       c. $\{5, 6, 7, 8, 9, 10, 11\}$

Now try Exercises 7 and 8 ▶

Whole Numbers

Combining zero with the natural numbers produces a new set called the **whole numbers** $\mathbb{W} = \{0, 1, 2, 3, 4, \ldots\}$. We say that the natural numbers are a **proper subset** of the whole numbers, denoted $\mathbb{N} \subset \mathbb{W}$, since every natural number is also a whole number. The symbol $\subset$ means "is a proper subset of."

EXAMPLE 2 ▶ **Determining Membership in a Set**

Given $A = \{1, 2, 3, 4, 5, 6\}$, $B = \{2, 4\}$, and $C = \{0, 1, 2, 3, 5, 8\}$, determine whether the following statements are true or false.

   a. $B \subset A$ **b.** $B \subset C$ **c.** $C \subset \mathbb{W}$

   d. $C \subset \mathbb{N}$ **e.** $104 \in \mathbb{W}$ **f.** $0 \in \mathbb{N}$

   g. $2 \notin \mathbb{W}$

Solution ▶   **a.** True: Every element of B is in A.   **b.** False: $4 \notin C$.
             c. True: All elements are whole numbers.   **d.** False: $0 \notin \mathbb{N}$.
             e. True: 104 is a whole number.   **f.** False: $0 \notin \mathbb{N}$.
             g. False: 2 *is* a whole number.

> **Now try Exercises 9 through 14** ▶

Integers

Numbers greater than zero are **positive numbers.** Every positive number has an *opposite* that is a **negative number** (a number less than zero). The set containing zero and the natural numbers with their opposites produces the set of **integers** $\mathbb{Z} = \{\ldots, -3, -2, -1, 0, 1, 2, 3, \ldots\}$. We can illustrate the location of a number (in relation to other numbers) using a **number line** (see Figure R.1).

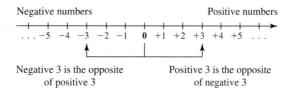

Figure R.1

The number that corresponds to a given point on the number line is called the **coordinate** of that point. When we want to note a specific location on the line, a bold dot "•" is used and we have then **graphed** the number. Since we need only one coordinate to denote a location on the number line, it can be referred to as a **one-dimensional graph.**

> **WORTHY OF NOTE**
>
> The integers are a subset of the rational numbers: $\mathbb{Z} \subset \mathbb{Q}$, since any integer can be written as a fraction using a denominator of one: $-2 = \frac{-2}{1}$ and $0 = \frac{0}{1}$.

Rational Numbers

Fractions and mixed numbers are part of a set called the **rational numbers** $\mathbb{Q}$. A rational number is one that can be written as a fraction with an integer numerator and an integer denominator other than zero. In set notation we write $\mathbb{Q} = \{\frac{p}{q} | p, q \in \mathbb{Z}; q \neq 0\}$. The vertical bar "|" is read "such that" and indicates that a description follows. In words, we say, "$\mathbb{Q}$ is the set of numbers of the form p over q, such that p and q are integers and q is not equal to zero."

EXAMPLE 3 ▶   **Graphing Rational Numbers**

Graph the fractions by converting to decimal form and estimating their location between two integers:

a. $-2\frac{1}{3}$ **b.** $\frac{7}{2}$

Solution ▶   **a.** $-2\frac{1}{3} = -2.3333333\ldots$ or $-2.\overline{3}$ **b.** $\frac{7}{2} = 3.5$

> **Now try Exercises 15 through 18** ▶

Since the division $\frac{7}{2}$ **terminated,** the result is called a **terminating decimal.** The decimal form of $-2\frac{1}{3}$ is called **repeating** and **nonterminating.** Recall that a repeating decimal is written with a horizontal bar over the first block of digit(s) that repeat.

Irrational Numbers

Although any fraction can be written in decimal form, not all decimal numbers can be written as a fraction. One example is the number represented by the Greek letter π (pi), frequently seen in a study of circles. Although we often approximate π using 3.14, its true value has a **nonrepeating** and *nonterminating* decimal form. Other numbers of this type include 2.101001000100001 . . . (there is no block of digits that repeat), and $\sqrt{5} \approx 2.2360679$ . . . (the decimal form never terminates). Numbers with a nonrepeating and nonterminating decimal form belong to the set of irrational numbers $\mathbb{H}$.

EXAMPLE 4 ▶ **Approximating Irrational Numbers**

Use a calculator as needed to approximate the value of each number given (round to 100ths), then graph them on the number line:

   a. $\sqrt{3}$ **b.** π **c.** $\sqrt{19}$ **d.** $-\frac{\sqrt{2}}{2}$

Solution ▶   **a.** $\sqrt{3} \approx 1.73$ **b.** $\pi \approx 3.14$ **c.** $\sqrt{19} \approx 4.36$ **d.** $-\frac{\sqrt{2}}{2} = -0.71$

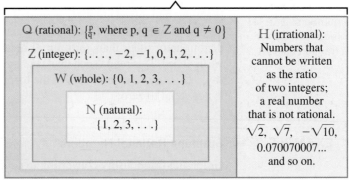

Now try Exercises 19 through 22 ▶

WORTHY OF NOTE

Checking the approximation for $\sqrt{5}$ shown, we obtain $2.2360679^2 = 4.999999653$. While we can find better approximations by using more and more decimal places, we never obtain five *exactly* (although some calculators will say the result is 5 due to limitations in programming).

Real Numbers

The set of rational numbers combined with the set of irrational numbers produces the set of **real numbers** $\mathbb{R}$. Figure R.2 illustrates the relationship between the sets of numbers we've discussed so far. Notice how each subset appears "nested" in a larger set.

$\mathbb{R}$ (real): All rational and irrational numbers

$\mathbb{Q}$ (rational): $\{\frac{p}{q}, \text{ where } p, q \in \mathbb{Z} \text{ and } q \neq 0\}$

$\mathbb{Z}$ (integer): $\{\ldots, -2, -1, 0, 1, 2, \ldots\}$

$\mathbb{W}$ (whole): $\{0, 1, 2, 3, \ldots\}$

$\mathbb{N}$ (natural): $\{1, 2, 3, \ldots\}$

$\mathbb{H}$ (irrational): Numbers that cannot be written as the ratio of two integers; a real number that is not rational. $\sqrt{2}, \sqrt{7}, -\sqrt{10}, 0.070070007\ldots$ and so on.

Figure R.2

EXAMPLE 5 ▶ **Identifying Numbers**

List the numbers in set $A = \{-2, 0, 5, \sqrt{7}, 12, \frac{2}{3}, 4.5, \sqrt{21}, \pi, -0.75\}$ that belong to

   a. $\mathbb{Q}$ **b.** $\mathbb{H}$ **c.** $\mathbb{W}$ **d.** $\mathbb{Z}$

Solution ▶   **a.** $-2, 0, 5, 12, \frac{2}{3}, 4.5, -0.75 \in \mathbb{Q}$ **b.** $\sqrt{7}, \sqrt{21}, \pi \in \mathbb{H}$
         c. $0, 5, 12 \in \mathbb{W}$ **d.** $-2, 0, 5, 12 \in \mathbb{Z}$

Now try Exercises 23 through 26 ▶

EXAMPLE 6 ▶ **Evaluating Statements about Sets of Numbers**

Determine whether the statements are true or false.

   a. $\mathbb{N} \subset \mathbb{Q}$ **b.** $\mathbb{H} \subset \mathbb{Q}$ **c.** $\mathbb{W} \subset \mathbb{Z}$ **d.** $\mathbb{Z} \subset \mathbb{R}$

Solution ▶

☑ **A.** You've just reviewed sets of numbers, graphing real numbers, and set notation

a. True: All natural numbers can be written as a fraction over 1.
b. False: No irrational number can be written in fraction form.
c. True: All whole numbers are integers.
d. True: Every integer is a real number.

Now try Exercises 27 through 38 ▶

B. Inequality Symbols and Order Relations

We compare numbers of different size using **inequality notation,** known as the **greater than** ($>$) and **less than** ($<$) symbols. Note that $-4 < 3$ is the same as saying -4 is to the left of 3 on the number line. In fact, on a number line, any given number is smaller than any number to the right of it (see Figure R.3).

$$-4 < 3$$

$$a < b$$

Figure R.3

Order Property of Real Numbers

Given any two real numbers a and b.
1. $a < b$ if a is to the left of b on the number line.
2. $a > b$ if a is to the right of b on the number line.

Inequality notation is used with numbers and variables to write mathematical statements. A **variable** is a symbol, commonly a letter of the alphabet, used to represent an unknown quantity. Over the years x, y, and n have become most common, although any letter (or symbol) can be used. Often we'll use variables that remind us of the quantities they represent, like L for length, and D for distance.

EXAMPLE 7 ▶ **Writing Mathematical Models Using Inequalities**

Use a variable and an inequality symbol to represent the statement: "To hit a home run out of Jacobi Park, the ball must travel over three hundred twenty-five feet."

Solution ▶ Let D represent distance: $D > 325$ ft.

Now try Exercises 39 through 42 ▶

In Example 7, note the number 325 itself is not a possible value for D. If the ball traveled *exactly* 325 ft, it would hit the fence and stay in play. Numbers that mark the limit or boundary of an inequality are called **endpoints.** If the endpoint(s) are *not* included, the less than ($<$) or greater than ($>$) symbols are used. When the endpoints *are* included, the *less than or equal to symbol* ($\leq$) or the *greater than or equal to symbol* ($\geq$) is used. The decision to *include* or *exclude* an endpoint is often an important one, and many mathematical decisions (and real-life decisions) depend on a clear understanding of the distinction.

☑ **B.** You've just reviewed inequality symbols and order relations

C. The Absolute Value of a Real Number

Any nonzero real number "n" is either a positive number or a negative number. But in some applications, our primary interest is simply the *size* of n, rather than its sign. This is called the **absolute value** of n, denoted $|n|$, and can be thought of as its *distance from* *zero on the number line*, regardless of the direction (see Figure R.4). Since distance is always positive or zero, $|n| \geq 0$.

Figure R.4

| **EXAMPLE 8** ▶ | **Absolute Value Reading and Reasoning** |

In the table shown, the absolute value of a number is given in column 1. Complete the remaining columns.

Solution ▶

Column 1 (In Symbols)	Column 2 (Spoken)	Column 3 (Result)	Column 4 (Reason)
$\|7.5\|$	"the absolute value of seven and five-tenths"	7.5	the distance between 7.5 and 0 is 7.5 units
$\|-2\|$	"the absolute value of negative two"	2	the distance between -2 and 0 is 2 units
$-\|-6\|$	"the opposite of the absolute value of negative six"	-6	the distance between -6 and 0 is 6 units, the opposite of 6 is -6

Now try Exercises 43 through 50 ▶

Example 8 shows the absolute value of a positive number is the number itself, while the absolute value of a negative number is the *opposite of that number* (recall that $-n$ is positive if n itself is negative). For this reason the formal definition of absolute value is stated as follows.

Absolute Value

For any real number n,

$$|n| = \begin{cases} n & \text{if} \quad n \geq 0 \\ -n & \text{if} \quad n < 0 \end{cases}$$

The concept of absolute value can actually be used to find the distance between *any* two numbers on a number line. For instance, we know the distance between 2 and 8 is 6 (by counting). Using absolute values, we write $|8 - 2| = |6| = 6$, or $|2 - 8| = |-6| = 6$. Generally, if a and b are two numbers on the real number line, the distance between them is $|a - b|$ or $|b - a|$.

| **EXAMPLE 9** ▶ | **Using Absolute Value to Find the Distance between Points** |

Find the distance between -5 and 3 on the number line.

Solution ▶ $|-5 - 3| = |-8| = 8$ or $|3 - (-5)| = |8| = 8.$

☑ **C.** You've just reviewed the absolute value of a real number

Now try Exercises 51 through 58 ▶

D. The Order of Operations

The operations of addition, subtraction, multiplication, and division are defined for the set of real numbers, and the concept of absolute value plays an important role. Prior to our study of the order of operations, we will review fundamental concepts related to division and zero, exponential notation, and square roots/cube roots.

Division and Zero

EXAMPLE 10 ▶ **Understanding Division with Zero by Writing the Related Product**

Rewrite each quotient *using the related product.*

    a. $0 \div 8 = p$ **b.** $\frac{16}{0} = q$ **c.** $\frac{0}{12} = n$

Solution ▶   **a.** $0 \div 8 = p$, if $p \cdot 8 = 0$. This shows $p = 0$.

         b. $\frac{16}{0} = q$, if $q \cdot 0 = 16$. There is no such number q.

         c. $\frac{0}{12} = n$, if $n \cdot 12 = 0$. This shows $n = 0$.

Now try Exercises 59 through 62 ▶

WORTHY OF NOTE

When a pizza is delivered to your home, it often has "8 parts to the whole," and in fraction form we have $\frac{8}{8}$. When all 8 pieces are eaten, 0 pieces remain and the fraction form becomes $\frac{0}{8} = 0$. However, the expression $\frac{8}{0}$ is meaningless, since it would indicate a pizza that has "0 parts to the whole (??)." The special case of $\frac{0}{0}$ is said to be indeterminate, as $\frac{0}{0} = n$ is true for all real numbers n (since the check gives $n \cdot 0 = 0$✓).

In Example 10(a), a dividend of 0 and a divisor of 8 means we are going to divide zero into eight groups. The related multiplication shows there will be zero in each group. As in Example 10(b), an expression with a divisor of 0 *cannot be computed or checked*. Although it seems trivial, division by zero has many implications in a study of mathematics, so make an effort to know the facts: The quotient of zero and any nonzero number is zero, but division *by* zero is undefined.

Division and Zero

The quotient of zero and any real number n is zero ($n \neq 0$):

$$0 \div n = 0 \qquad\qquad \frac{0}{n} = 0.$$

The expressions $n \div 0$ and $\dfrac{n}{0}$ are undefined.

Squares, Cubes, and Exponential Form

When a number is repeatedly multiplied by itself as in $(10)(10)(10)(10)$, we write it using **exponential notation** as 10^4. The number used for repeated multiplication (in this case 10) is called the **base,** and the superscript number is called an **exponent.** The exponent tells how many times the base occurs as a factor, and we say 10^4 is written in **exponential form.** Numbers that result from squaring an integer are called **perfect squares,** while numbers that result from cubing an integer are called **perfect cubes.** These are often collected into a table, such as Table R.1, and memorized to help complete many common calculations mentally. Only the square and cube of selected positive integers are shown.

Table R.1

Perfect Squares				Perfect Cubes	
N	N^2	N	N^2	N	N^3
1	1	7	49	1	1
2	4	8	64	2	8
3	9	9	81	3	27
4	16	10	100	4	64
5	25	11	121	5	125
6	36	12	144	6	216

EXAMPLE 11 ▶ **Evaluating Numbers in Exponential Form**

Write each exponential in expanded form, then determine its value.

    a. 4^3 **b.** $(-6)^2$ **c.** -6^2 **d.** $\left(\frac{2}{3}\right)^3$

Solution ▶   **a.** $4^3 = 4 \cdot 4 \cdot 4 = 64$ **b.** $(-6)^2 = (-6) \cdot (-6) = 36$

                   c. $-6^2 = -(6 \cdot 6) = -36$ **d.** $\left(\frac{2}{3}\right)^3 = \frac{2}{3} \cdot \frac{2}{3} \cdot \frac{2}{3} = \frac{8}{27}$

Now try Exercises 63 and 64 ▶

Examples 11(b) and 11(c) illustrate an important distinction. The expression $(-6)^2$ is read, "the square of negative six" and the negative sign is included in both factors. The expression -6^2 is read, "the opposite of six squared," and the square of six is calculated first, then made negative.

Square Roots and Cube Roots

Index $\sqrt[3]{A}$ ——— Radical, Radicand

For the square root operation, either the $\sqrt{}$ or $\sqrt[2]{}$ notation can be used. The $\sqrt{}$ symbol is called a **radical,** the number under the radical is called the **radicand,** and the small case number used is called the **index.** The index tells how many factors are needed to obtain the radicand. For example, $\sqrt{25} = 5$, since $5 \cdot 5 = 5^2 = 25$ (when the $\sqrt{}$ symbol is used, the index is understood to be 2). In general, $\sqrt{a} = b$ only if $b^2 = a$. All numbers greater than zero have one positive and one negative square root. The *positive* or **principal square root** of 49 is 7 ($\sqrt{49} = 7$) since $7^2 = 49$. The *negative* square root of 49 is -7 ($-\sqrt{49} = -7$). The cube root of a number has the form $\sqrt[3]{a} = b$, where $b^3 = a$. This means $\sqrt[3]{27} = 3$ since $3^3 = 27$, and $\sqrt[3]{-8} = -2$ since $(-2)^3 = -8$. The cube root of a real number has one unique real value. In general, we have the following:

Square Roots	Cube Roots
$\sqrt{a} = b$ if $b^2 = a$	$\sqrt[3]{a} = b$ if $b^3 = a$
$(a \geq 0)$	$(a \in \mathbb{R})$
This indicates that	This indicates that
$\sqrt{a} \cdot \sqrt{a} = a$	$\sqrt[3]{a} \cdot \sqrt[3]{a} \cdot \sqrt[3]{a} = a$
or $(\sqrt{a})^2 = a$	or $(\sqrt[3]{a})^3 = a$

WORTHY OF NOTE

It is helpful to note that both 0 and 1 are their own square root, cube root, and *n*th root. That is, $\sqrt{0} = 0$, $\sqrt[3]{0} = 0, \ldots,$ $\sqrt[n]{0} = 0$; and $\sqrt{1} = 1$, $\sqrt[3]{1} = 1, \ldots, \sqrt[n]{1} = 1$.

EXAMPLE 12 ▶ **Evaluating Square Roots and Cube Roots**

Determine the value of each expression.

    a. $\sqrt{49}$ **b.** $\sqrt[3]{125}$ **c.** $\sqrt{\frac{9}{16}}$ **d.** $-\sqrt{16}$ **e.** $\sqrt{-25}$

Solution ▶   **a.** 7 since $7 \cdot 7 = 49$ **b.** 5 since $5 \cdot 5 \cdot 5 = 125$

             c. $\frac{3}{4}$ since $\frac{3}{4} \cdot \frac{3}{4} = \frac{9}{16}$ **d.** -4 since $\sqrt{16} = 4$

             e. not a real number since $5 \cdot 5 = (-5)(-5) = 25$

Now try Exercises 65 through 70 ▶

For square roots, if the radicand is a perfect square or has perfect squares in both the numerator and denominator, the result is a rational number as in Examples 12(a) and 12(c). If the radicand is not a perfect square, the result is an irrational number. Similar statements can be made regarding cube roots [see Example 12(b)].

The Order of Operations

When basic operations are combined into a larger mathematical expression, we use a specified **priority** or **order of operations** to evaluate them.

The Order of Operations

1. Simplify within grouping symbols (parentheses, brackets, braces, etc.). If there are "nested" symbols of grouping, begin with the innermost group. If a fraction bar is used, simplify the numerator and denominator separately.
2. Evaluate all exponents and roots.
3. Compute all multiplications or divisions *in the order they occur from left to right.*
4. Compute all additions or subtractions *in the order they occur from left to right.*

EXAMPLE 13 ▶ **Evaluating Expressions Using the Order of Operations**

Simplify using the order of operations:

a. $5 + 2 \cdot 3$

b. $8 + 36 \div 4(12 - 3^2)$

c. $7500\left(1 + \dfrac{0.075}{12}\right)^{12 \cdot 15}$

d. $\dfrac{-4.5(8) - 3}{\sqrt[3]{125} + 2^3}$

Solution ▶

a. $5 + 2 \cdot 3 = 5 + 6$ multiplication before addition

           $= 11$ result

b. $8 + 36 \div 4(12 - 3^2)$

     $= 8 + 36 \div 4(12 - 9)$ simplify within parentheses

     $= 8 + 36 \div 4(3)$ $12 - 9 = 3$

     $= 8 + 9(3)$ division before multiplication

     $= 8 + 27$ multiply

     $= 35$ result

c. $7500\left(1 + \dfrac{0.075}{12}\right)^{12 \cdot 15}$ original expression

     $= 7500(1.00625)^{12 \cdot 15}$ simplify within the parenthesis (division before addition)

     $= 7500(1.00625)^{180}$ simplify the exponent

     $\approx 7500(3.069451727)$ exponents before multiplication

     $\approx 23{,}020.89$ result (rounded to hundredths)

d. $\dfrac{-4.5(8) - 3}{\sqrt[3]{125} + 2^3}$ original expression

     $= \dfrac{-36 - 3}{5 + 8}$ simplify terms in the numerator and denominator

     $= \dfrac{-39}{13}$ combine terms

     $= -3$ result

☑ **D.** You've just reviewed the order of operations

Now try Exercises 71 through 94 ▶

R.1 EXERCISES

▶ CONCEPTS AND VOCABULARY

Fill in each blank with the appropriate word or phrase. Carefully reread the section, if necessary.

1. The symbol $\subset$ means: is a _____ _____ of and the symbol $\in$ means: is an _____ of.

2. A number corresponding to a point on the number line is called the _____ of that point.

3. Every positive number has two square roots, one _____ and one _____. The two square roots of 49 are ___ and ___; $\sqrt{49}$ represents the _____ square root of 49.

4. The decimal form of $\sqrt{7}$ contains an infinite number of non _____ and non _____ digits. This means that $\sqrt{7}$ is a(n) _____ number.

5. Discuss/Explain why the value of $12 \cdot \frac{1}{3} + \frac{2}{3}$ is $4\frac{2}{3}$ and not 12.

6. Discuss/Explain (a) why $(-5)^2 = 25$, while $-5^2 = -25$; and (b) why $-5^3 = (-5)^3 = -125$.

▶ DEVELOPING YOUR SKILLS

7. List the natural numbers that are
   a. less than 6.
   b. less than 1.

8. List the natural numbers that are
   a. between 0 and 1.
   b. greater than 50.

Identify each of the following statements as either true or false. If false, give an example that shows why.

9. $\mathbb{N} \subset \mathbb{W}$ 10. $\mathbb{W} \not\subset \mathbb{N}$

11. $\{33, 35, 37, 39\} \subset \mathbb{W}$

12. $\{2.2, 2.3, 2.4, 2.5\} \subset \mathbb{W}$

13. $6 \in \{0, 1, 2, 3, \ldots\}$

14. $1297 \notin \{0, 1, 2, 3, \ldots\}$

Convert to decimal form and graph by estimating the number's location between two integers.

15. $\frac{4}{3}$ 16. $-\frac{7}{8}$ 17. $2\frac{5}{9}$ 18. $-1\frac{5}{6}$

Use a calculator to find the principal square root of each number (round to hundredths as needed). Then graph each number by estimating its location between two integers.

19. 7   20. $\frac{75}{4}$ 21. 3   22. $\frac{25\pi}{2}$

For the sets in Exercises 23 through 26:

a. List all numbers that are elements of (i) $\mathbb{N}$, (ii) $\mathbb{W}$, (iii) $\mathbb{Z}$, (iv) $\mathbb{Q}$, (v) $\mathbb{H}$, and (vi) $\mathbb{R}$.

b. Reorder the elements of each set from smallest to largest.

c. Graph the elements of each set on a number line.

23. $\{-1, 8, 0.75, \frac{9}{2}, 5.\overline{6}, 7, \frac{3}{5}, 6\}$

24. $\{-7, 2.\overline{1}, 5.73, -3\frac{5}{6}, 0, -1.12, \frac{7}{8}\}$

25. $\{-5, \sqrt{49}, 2, -3, 6, -1, \sqrt{3}, 0, 4, \pi\}$

26. $\{-8, 5, -2\frac{3}{5}, 1.75, -\sqrt{2}, -0.6, \pi, \frac{7}{2}, \sqrt{64}\}$

State true or false. If false, state why.

27. $\mathbb{R} \subset \mathbb{H}$ 28. $\mathbb{N} \subset \mathbb{R}$

29. $\mathbb{Q} \subset \mathbb{Z}$ 30. $\mathbb{Z} \subset \mathbb{Q}$

31. $\sqrt{25} \in \mathbb{H}$ 32. $\sqrt{19} \in \mathbb{H}$

Match each set with its correct symbol and description/illustration.

33. ___ Irrational numbers   a. $\mathbb{R}$ I. $\{1, 2, 3, 4, \ldots\}$

34. ___ Integers   b. $\mathbb{Q}$ II. $\{\frac{a}{b} \,|\, a, b \in \mathbb{Z}; b \neq 0\}$

35. ___ Real numbers   c. $\mathbb{H}$ III. $\{0, 1, 2, 3, 4, \ldots\}$

36. ___ Rational numbers   d. $\mathbb{W}$ IV. $\{\pi, \sqrt{7}, -\sqrt{13}, \text{etc.}\}$

37. ___ Whole numbers   e. $\mathbb{N}$ V. $\{\ldots -3, -2, -1, 0, 1, 2, 3, \ldots\}$

38. ___ Natural numbers   f. $\mathbb{Z}$ VI. $\mathbb{N}, \mathbb{W}, \mathbb{Z}, \mathbb{Q}, \mathbb{H}$

Use a descriptive variable and an inequality symbol ($<, >, \leq, \geq$) to write a model for each statement.

39. To spend the night at a friend's house, Kylie must be at least 6 years old.

40. Monty can spend at most $2500 on the purchase of a used automobile.

41. If Jerod gets no more than two words incorrect on his spelling test he can play in the soccer game this weekend.

42. Andy must weigh less than 112 lb to be allowed to wrestle in his weight class at the meet.

Evaluate/simplify each expression.

43. $|-2.75|$ 44. $|-7.24|$

45. $-|-4|$ 46. $-|-6|$

47. $\left|\dfrac{1}{2}\right|$ 48. $\left|\dfrac{2}{5}\right|$

49. $\left|-\dfrac{3}{4}\right|$ 50. $\left|-\dfrac{3}{7}\right|$

Use the concept of absolute value to complete Exercises 51 to 58.

51. Write the statement two ways, then simplify. "The distance between -7.5 and 2.5 is . . ."

52. Write the statement two ways, then simplify. "The distance between $13\frac{2}{5}$ and $-2\frac{3}{5}$ is . . ."

53. What two numbers on the number line are five units from negative three?

54. What two numbers on the number line are three units from two?

55. If n is positive, then $-n$ is _____.

56. If n is negative, then $-n$ is _____.

57. If $n < 0$, then $|n|$ = _____.

58. If $n > 0$, then $|n|$ = _____.

Determine which expressions are equal to zero and which are undefined. Justify your responses by writing the related multiplication.

59. $12 \div 0$ 60. $0 \div 12$

61. $\frac{7}{0}$ 62. $\frac{0}{7}$

Without computing the actual answer, state whether the result will be positive or negative. Be careful to note

what power is used and whether the negative sign is included in parentheses.

63. a. $(-7)^2$ b. -7^2
    c. $(-7)^5$ d. -7^5

64. a. $(-7)^3$ b. -7^3
    c. $(-7)^4$ d. -7^4

Evaluate without the aid of a calculator.

65. $-\sqrt{\dfrac{121}{36}}$ 66. $-\sqrt{\dfrac{25}{49}}$

67. $\sqrt[3]{-8}$ 68. $\sqrt[3]{-64}$

69. What perfect square is closest to 78?

70. What perfect cube is closest to -71?

Perform the operation indicated without the aid of a calculator.

71. $-24 - (-31)$ 72. $-45 - (-54)$

73. $7.045 - 9.23$ 74. $0.0762 - 0.9034$

75. $4\frac{5}{6} + (-\frac{1}{2})$ 76. $1\frac{1}{8} + (-\frac{3}{4})$

77. $(-\frac{2}{3})(3\frac{5}{8})$ 78. $(-8)(2\frac{1}{4})$

79. $(12)(-3)(0)$ 80. $(-1)(0)(-5)$

81. $-60 \div 12$ 82. $75 \div (-15)$

83. $\frac{4}{5} \div (-8)$ 84. $-15 \div \frac{1}{2}$

85. $-\frac{2}{3} \div \frac{16}{21}$ 86. $-\frac{3}{4} \div \frac{7}{8}$

Evaluate without a calculator, using the order of operations.

87. $12 - 10 \div 2 \times 5 + (-3)^2$

88. $(5 - 2)^2 - 16 \div 4 \cdot 2 - 1$

89. $\sqrt{\dfrac{9}{16}} - \dfrac{3}{5} \cdot \left(\dfrac{5}{3}\right)^2$ 90. $\left(\dfrac{3}{2}\right)^2 \div \left(\dfrac{9}{4}\right) - \sqrt{\dfrac{25}{64}}$

91. $\dfrac{4(-7) - 6^2}{6 - \sqrt{49}}$ 92. $\dfrac{5(-6) - 3^2}{9 - \sqrt{64}}$

Evaluate using a calculator (round to hundredths).

93. $2475\left(1 + \dfrac{0.06}{4}\right)^{4 \cdot 10}$

94. $5100\left(1 + \dfrac{0.078}{52}\right)^{52 \cdot 20}$

▶ WORKING WITH FORMULAS

95. Pitch diameter: $D = \dfrac{d \cdot n}{n + 2}$

Mesh gears are used to transfer rotary motion and power from one shaft to another. The *pitch diameter D* of a drive gear is given by the formula shown, where *d* is the outer diameter of the gear and *n* is the number of teeth on the gear. Find the pitch diameter of a gear with 12 teeth and an outer diameter of 5 cm.

96. Pediatric dosages and Clark's rule: $D_C = \dfrac{D_A \cdot W}{150}$

The amount of medication prescribed for young children depends on their weight, height, age, body surface area and other factors. **Clark's rule** is a formula that helps estimate the correct child's dose D_C based on the adult dose D_A and the weight W of the child (an average adult weight of 150 lb is assumed). Compute a child's dose if the adult dose is 50 mg and the child weighs 30 lb.

▶ APPLICATIONS

Use positive and negative numbers to model the situation, then compute.

97. Temperature changes: At 6:00 P.M., the temperature was 50°F. A cold front moves through that causes the temperature to *drop* 3°F each hour until midnight. What is the temperature at midnight?

98. Air conditioning: Most air conditioning systems are designed to create a 2° *drop* in the air temperature each hour. How long would it take to reduce the air temperature from 86° to 71°?

99. Record temperatures: The state of California holds the record for the greatest temperature swing between a record high and a record low. The record high was 134°F and the record low was −45°F. How many degrees *difference* are there between the record high and the record low?

100. Cold fronts: In Juneau, Alaska, the temperature was 17°F early one morning. A cold front later moved in and the temperature *dropped* 32°F by lunch time. What was the temperature at lunch time?

▶ EXTENDING THE CONCEPT

101. Here are some historical approximations for π. Which one is closest to the true value?

Archimedes: $3\frac{1}{7}$ Tsu Ch'ung-chih: $\frac{355}{113}$
Aryabhata: $\frac{62,832}{20,000}$ Brahmagupta: $\sqrt{10}$

102. If $A > 0$ and $B < 0$, is the product $A \cdot (-B)$ positive or negative?

103. If $A < 0$ and $B < 0$, is the quotient $-(A \div B)$ positive or negative?

Learning Objectives

In Section R.2 you will review how to:

☐ **A.** Identify terms, coefficients, and expressions

☐ **B.** Create mathematical models

☐ **C.** Evaluate algebraic expressions

☐ **D.** Identify and use properties of real numbers

☐ **E.** Simplify algebraic expressions

To effectively use mathematics as a problem-solving tool, you must develop the ability to translate written or verbal information into a mathematical model. After obtaining a model, many applications require that you work effectively with algebraic terms and expressions. The basic ideas involved are reviewed here.

A. Terms, Coefficients, and Algebraic Expressions

An **algebraic term** is a *collection of factors* that may include numbers, variables, or expressions within parentheses. Here are some examples:

$$(1)\ 3 \qquad (2)\ -6P \qquad (3)\ 5xy \qquad (4)\ -8n^2 \qquad (5)\ n \qquad (6)\ 2(x+3)$$

If a term consists of a single nonvariable number, it is called a **constant** term. In (1), 3 is a constant term. Any term that contains a variable is called a **variable term.** We call the constant factor of a term the **numerical coefficient** or simply the **coefficient.** The coefficients for (1), (2), (3), and (4) are 3, -6, 5, and -8, respectively. In (5), the coefficient of n is 1, since $1 \cdot n = 1n = n$. The term in (6) has two factors as written, 2 and $(x + 3)$. The coefficient is 2.

An **algebraic expression** can be a single term or a sum or difference of terms. To avoid confusion when identifying the coefficient of each term, the expression can be rewritten using algebraic addition if desired: $A - B = A + (-B)$. To identify the coefficient of a rational term, it sometimes helps to **decompose** the term, rewriting it using a unit fraction as in $\frac{n-2}{5} = \frac{1}{5}(n-2)$ and $\frac{x}{2} = \frac{1}{2}x$.

EXAMPLE 1 ► **Identifying Terms and Coefficients**

State the number of terms in each expression as given, then identify the coefficient of each term.

a. $2x - 5y$ **b.** $\dfrac{x+3}{7} - 2x$ **c.** $-(x - 12)$ **d.** $-2x^2 - x + 5$

Solution ►

	a. $2x + (-5y)$	**b.** $\frac{1}{7}(x+3) + (-2x)$	**c.** $-1(x-12)$	**d.** $-2x^2 + (-1x) + 5$
Rewritten:				
Number of terms:	two	two	one	three
Coefficient(s):	2 and -5	$\frac{1}{7}$ and -2	-1	-2, -1, and 5

✓ **A.** You've just reviewed how to identify terms, coefficients, and expressions

Now try Exercises 7 through 14 ►

B. Translating Written or Verbal Information into a Mathematical Model

The key to solving many applied problems is finding an algebraic expression that accurately models the situation. First, we assign a variable to represent an unknown quantity, then build related expressions using words from the English language that suggest a mathematical operation.

As mentioned earlier, variables that remind us of what they represent are often used in the modeling process. Capital letters are also used due to their widespread appearance in other fields.

EXAMPLE 2 ► **Translating English Phrases into Algebraic Expressions**

Assign a variable to the unknown number, then translate each phrase into an algebraic expression.
  a. twice a number, increased by five
  b. six less than three times the width

    c. ten less than triple the payment

    d. two hundred fifty feet more than double the length

Solution ▶
    a. Let n represent the number. Then $2n$ represents twice the number, and $2n + 5$ represents twice a number, increased by five.

    b. Let W represent the width. Then $3W$ represents three times the width, and $3W - 6$ represents six less than three times the width.

    c. Let p represent the payment. Then $3p$ represents a triple payment, and $3p - 10$ represents 10 less than triple the payment.

    d. Let L represent the length in feet. Then $2L$ represents double the length, and $2L + 250$ represents 250 feet more than double the length.

Now try Exercises 15 through 32 ▶

Identifying and translating such phrases *when they occur in context* is an important problem-solving skill. Note how this is done in Example 3.

EXAMPLE 3 ▶ Creating a Mathematical Model

The cost for a rental car is $35 plus 15 cents per mile. Express the cost of renting a car in terms of the number of miles driven.

Solution ▶ Let m represent the number of miles driven. Then $0.15m$ represents the cost for each mile and $C = 35 + 0.15m$ represents the total cost for renting the car.

☑ **B.** You've just reviewed how to create mathematical models

Now try Exercises 33 through 40 ▶

C. Evaluating Algebraic Expressions

We often need to **evaluate** expressions to investigate patterns and note relationships.

> **Evaluating a Mathematical Expression**
>
> 1. Replace each variable with open parentheses ( ).
> 2. Substitute the given values for each variable.
> 3. Simplify using the order of operations.

In this evaluation, it's best to use a **vertical format,** with the original expression written first, the substitutions shown next, followed by the simplified forms and the final result. The numbers substituted or "plugged into" the expression are often called the **input values,** with the resulting values called **outputs.**

EXAMPLE 4 ▶ Evaluating an Algebraic Expression

Evaluate the expression $x^3 - 2x^2 + 5$ for $x = -3$.

Solution ▶ For $x = -3$:
$$
\begin{aligned}
x^3 - 2x^2 + 5 &= (-3)^3 - 2(-3)^2 + 5 &&\text{substitute } -3 \text{ for } x\\
&= -27 - 2(9) + 5 &&\text{simplify: } (-3)^3 = -27, (-3)^2 = 9\\
&= -27 - 18 + 5 &&\text{simplify: } 2(9) = 18\\
&= -40 &&\text{result}
\end{aligned}
$$

Now try Exercises 41 through 60 ▶

If the same expression is evaluated repeatedly, results are often collected and analyzed in a table of values, as shown in Example 5. As a practical matter, the substitutions and simplifications are often done mentally or on scratch paper, with the table showing only the input and output values.

EXAMPLE 5 ▶ **Evaluating an Algebraic Expression**

Evaluate $x^2 - 2x - 3$ to complete the table shown. Which input value(s) of x cause the expression to have an output of 0?

Solution ▶

Input x	Output $x^2 - 2x - 3$
−2	5
−1	0
0	−3
1	−4
2	−3
3	0
4	5

The expression has an output of 0 when $x = -1$ and $x = 3$.

Now try Exercises 61 through 66 ▶

WORTHY OF NOTE

In Example 4, note the importance of the first step in the evaluation process: *replace each variable with open parentheses.* Skipping this step could easily lead to confusion as we try to evaluate the squared term, since $-3^2 = -9$, while $(-3)^2 = 9$. Also see **Exercises 55 and 56.**

☑ **C.** You've just reviewed how to evaluate algebraic expressions

For exercises that combine the skills from Examples 3 through 5, **see Exercises 91 to 98.**

D. Properties of Real Numbers

While the phrase, "an unknown number times five," is accurately modeled by the expression $n5$ for some number n, in algebra we prefer to have numerical coefficients precede variable factors. When we reorder the factors as $5n$, we are using the **commutative property of multiplication.** A reordering of terms involves the **commutative property of addition.**

The Commutative Properties

Given that a and b represent real numbers:

ADDITION: $a + b = b + a$ MULTIPLICATION: $a \cdot b = b \cdot a$

Terms can be combined in any order without changing the sum.

Factors can be multiplied in any order without changing the product.

Each property can be extended to include any number of terms or factors. While the commutative property implies a *reordering* or *movement* of terms (to commute implies back-and-forth movement), the **associative property** implies a *regrouping* or reassociation of terms. For example, the sum $\left(\frac{3}{4} + \frac{3}{5}\right) + \frac{2}{5}$ is easier to compute if we regroup the addends as $\frac{3}{4} + \left(\frac{3}{5} + \frac{2}{5}\right)$. This illustrates the **associative property of addition.** Multiplication is also associative.

The Associative Properties

Given that a, b, and c represent real numbers:

ADDITION: MULTIPLICATION:

$$(a + b) + c = a + (b + c) \qquad\qquad (a \cdot b) \cdot c = a \cdot (b \cdot c)$$

Terms can be regrouped.   Factors can be regrouped.

EXAMPLE 6 ▶ **Simplifying Expressions Using Properties of Real Numbers**

Use the commutative and associative properties to simplify each calculation.

a. $\frac{3}{8} - 19 + \frac{5}{8}$ **b.** $[-2.5 \cdot (-1.2)] \cdot 10$

Solution ▶

a. $\frac{3}{8} - 19 + \frac{5}{8} = -19 + \frac{3}{8} + \frac{5}{8}$ commutative property

$\qquad\qquad\qquad = -19 + \left(\frac{3}{8} + \frac{5}{8}\right)$ associative property

$\qquad\qquad\qquad = -19 + 1$ simplify

$\qquad\qquad\qquad = -18$ result

b. $[-2.5 \cdot (-1.2)] \cdot 10 = -2.5 \cdot [(-1.2) \cdot 10]$ associative property

$\qquad\qquad\qquad = -2.5 \cdot (-12)$ simplify

$\qquad\qquad\qquad = 30$ result

Now try Exercises 67 and 68 ▶

WORTHY OF NOTE

Is subtraction commutative? Consider a situation involving money. If you had $100, you could easily buy an item costing $20: $100 − $20 leaves you with $80. But if you had $20, could you buy an item costing $100? Obviously $100 − $20 is not the same as $20 − $100. Subtraction is *not* commutative. Likewise, 100 ÷ 20 is not the same as 20 ÷ 100, and division is *not* commutative.

For any real number x, $x + 0 = x$ and 0 is called the **additive identity** since the original number was returned or "identified." Similarly, 1 is called the **multiplicative identity** since $1 \cdot x = x$. The identity properties are used extensively in the process of solving equations.

The Additive and Multiplicative Identities

Given that x is a real number,

$$x + 0 = x \qquad\qquad\qquad 1 \cdot x = x$$

Zero is the identity   One is the identity
for addition.   for multiplication.

For any real number x, there is a real number $-x$ such that $x + (-x) = 0$. The number $-x$ is called the **additive inverse** of x, since their sum results in the additive identity. Similarly, the **multiplicative inverse** of any nonzero number x is $\frac{1}{x}$, since $x \cdot \frac{1}{x} = 1$ (the multiplicative identity). This property can also be stated as $\frac{p}{q} \cdot \frac{q}{p} = 1$ ($p, q \neq 0$) for any rational number $\frac{p}{q}$. Note that $\frac{p}{q}$ and $\frac{q}{p}$ are **reciprocals.**

The Additive and Multiplicative Inverses

Given that p, q, and x represent real numbers ($p, q \neq 0$):

$$x + (-x) = 0 \qquad\qquad\qquad \frac{p}{q} \cdot \frac{q}{p} = 1$$

x and $-x$ are $\frac{p}{q}$ and $\frac{q}{p}$ are
additive inverses.   multiplicative inverses.

EXAMPLE 7 ▶   Determining Additive and Multiplicative Inverses

Replace the box to create a true statement:

a. $\square \cdot \dfrac{-3}{5}x = 1 \cdot x$ b. $x + 4.7 + \square = x$

Solution ▶   a. $\square = \dfrac{5}{-3}$, since $\dfrac{5}{-3} \cdot \dfrac{-3}{5} = 1$

b. $\square = -4.7$, since $4.7 + (-4.7) = 0$

Now try Exercises 69 and 70 ▶

The **distributive property of multiplication over addition** is widely used in a study of algebra, because it enables us to rewrite a product as an equivalent sum and vice versa.

The Distributive Property of Multiplication over Addition

Given that a, b, and c represent real numbers:

$a(b + c) = ab + ac$ $\qquad\qquad$ $ab + ac = a(b + c)$

A factor outside a sum can be distributed to each addend in the sum. $\qquad$ A factor common to each addend in a sum can be "undistributed" and written outside a group.

EXAMPLE 8 ▶   Simplifying Expressions Using the Distributive Property

Apply the distributive property as appropriate. Simplify if possible.

a. $7(p + 5.2)$ $\qquad$ b. $-(2.5 - x)$ $\qquad$ c. $7x^3 - x^3$ $\qquad$ d. $\dfrac{5}{2}n + \dfrac{1}{2}n$

Solution ▶   a. $7(p + 5.2) = 7p + 7(5.2)$ $\qquad\qquad$ b. $-(2.5 - x) = -1(2.5 - x)$
$= 7p + 36.4$ $\qquad\qquad\qquad\qquad\qquad = -1(2.5) - (-1)(x)$
$\qquad\qquad\qquad\qquad\qquad\qquad\qquad\qquad\qquad = -2.5 + x$

WORTHY OF NOTE

From Example 8(b) we learn that a negative sign outside a group changes the sign of all terms within the group: $-(2.5 - x) = -2.5 + x$.

c. $7x^3 - x^3 = 7x^3 - 1x^3$ $\qquad\qquad$ d. $\dfrac{5}{2}n + \dfrac{1}{2}n = \left(\dfrac{5}{2} + \dfrac{1}{2}\right)n$
$= (7 - 1)x^3$ $\qquad\qquad\qquad\qquad\qquad = \left(\dfrac{6}{2}\right)n$
$= 6x^3$ $\qquad\qquad\qquad\qquad\qquad\qquad = 3n$

Now try Exercises 71 through 78 ▶

☑ **D.** You've just reviewed how to identify and use properties of real numbers

E. Simplifying Algebraic Expressions

Two terms are **like terms** only if they have the *same variable factors* (the coefficient is not used to identify like terms). For instance, $3x^2$ and $-\frac{1}{7}x^2$ are like terms, while $5x^3$ and $5x^2$ are not. We simplify expressions by **combining like terms** using the distributive property, along with the commutative and associative properties. Many times the distributive property is used to eliminate grouping symbols *and* combine like terms within the same expression.

EXAMPLE 9 ▶ Simplifying an Algebraic Expression

Simplify the expression completely: $7(2p^2 + 1) - (p^2 + 3)$.

Solution ▶ $7(2p^2 + 1) - 1(p^2 + 3)$ original expression; note coefficient of -1

$= 14p^2 + 7 - 1p^2 - 3$ distributive property

$= (14p^2 - 1p^2) + (7 - 3)$ commutative and associative properties (collect like terms)

$= (14 - 1)p^2 + 4$ distributive property

$= 13p^2 + 4$ result

Now try Exercises 79 through 88 ▶

The steps for simplifying an algebraic expression are summarized here:

To Simplify an Expression

1. Eliminate parentheses by applying the distributive property.
2. Use the commutative and associative properties to group like terms.
3. Use the distributive property to combine like terms.

 E. You've just reviewed how to simplify algebraic expressions

As you practice with these ideas, many of the steps will become more automatic. At some point, the distributive property, the commutative and associative properties, as well as the use of algebraic addition will all be performed mentally.

R.2 EXERCISES

▶ **CONCEPTS AND VOCABULARY**

Fill in each blank with the appropriate word or phrase. Carefully reread the section, if necessary.

1. A term consisting of a single number is called a(n) _____ term.

2. A term containing a variable is called a(n) _____ term.

3. The constant factor in a variable term is called the _____.

4. When $3 \cdot 14 \cdot \frac{2}{3}$ is written as $3 \cdot \frac{2}{3} \cdot 14$, the _____ property has been used.

5. Discuss/Explain why the additive inverse of -5 is 5, while the multiplicative inverse of -5 is $-\frac{1}{5}$.

6. Discuss/Explain how we can rewrite the sum $3x + 6y$ as a product, and the product $2(x + 7)$ as a sum.

▶ **DEVELOPING YOUR SKILLS**

Identify the number of terms in each expression and the coefficient of each term.

7. $3x - 5y$

8. $-2a - 3b$

9. $2x + \dfrac{x + 3}{4}$

10. $\dfrac{n - 5}{3} + 7n$

11. $-2x^2 + x - 5$

12. $3n^2 + n - 7$

13. $-(x + 5)$

14. $-(n - 3)$

Translate each phrase into an algebraic expression.

15. seven fewer than a number

16. x decreased by six

17. the sum of a number and four

18. a number increased by nine

19. the difference between a number and five is squared

20. the sum of a number and two is cubed

21. thirteen less than twice a number

22. five less than double a number

23. a number squared plus the number doubled

24. a number cubed less the number tripled

25. five fewer than two-thirds of a number

26. fourteen more than one-half of a number

27. three times the sum of a number and five, decreased by seven

28. five times the difference of a number and two, increased by six

Create a mathematical model using descriptive variables.

29. The length of the rectangle is three meters less than twice the width.

30. The height of the triangle is six centimeters less than three times the base.

31. The speed of the car was fifteen miles per hour more than the speed of the bus.

32. It took Romulus three minutes more time than Remus to finish the race.

33. **Hovering altitude:** The helicopter was hovering 150 ft above the top of the building. Express the altitude of the helicopter in terms of the building's height.

34. **Stacks on a cruise liner:** The smoke stacks of the luxury liner cleared the bridge by 25 ft as it passed beneath it. Express the height of the stacks in terms of the bridge's height.

35. **Dimensions of a city park:** The length of a rectangular city park is 20 m more than twice its width. Express the length of the park in terms of the width.

36. **Dimensions of a parking lot:** In order to meet the city code while using the available space, a

contractor planned to construct a parking lot with a length that was 50 ft less than three times its width. Express the length of the lot in terms of the width.

37. **Cost of milk:** In 2008, a gallon of milk cost two and one-half times what it did in 1990. Express the cost of a gallon of milk in 2008 in terms of the 1990 cost.

38. **Cost of gas:** In 2008, a gallon of gasoline cost one and one-half times what it did in 1990. Express the cost of a gallon of gas in 2008 in terms of the 1990 cost.

39. **Pest control:** In her pest control business, Judy charges $50 per call plus $12.50 per gallon of insecticide for the control of spiders and other insects. Express the total charge in terms of the number of gallons of insecticide used.

40. **Computer repairs:** As his reputation and referral business grew, Keith began to charge $75 per service call plus an hourly rate of $50 for the repair and maintenance of home computers. Express the cost of a service call in terms of the number of hours spent on the call.

Evaluate each algebraic expression given $x = 2$ and $y = -3$.

41. $4x - 2y$

42. $5x - 3y$

43. $-2x^2 + 3y^2$

44. $-5x^2 + 4y^2$

45. $2y^2 + 5y - 3$

46. $3x^2 + 2x - 5$

47. $-2(3y + 1)$

48. $-3(2y + 5)$

49. $3x^2y$

50. $6xy^2$

51. $(-3x)^2 - 4xy - y^2$

52. $(-2x)^2 - 5xy - y^2$

53. $\frac{1}{2}x - \frac{1}{3}y$

54. $\frac{2}{3}x - \frac{1}{2}y$

55. $(3x - 2y)^2$

56. $(2x - 3y)^2$

57. $\dfrac{-12y + 5}{-3x + 1}$

58. $\dfrac{12x + (-3)}{-3y + 1}$

59. $\sqrt{-12y} \cdot 4$

60. $7 \cdot \sqrt{-27y}$

Evaluate each expression for integers from -3 to 3 inclusive. What input(s) give an output of zero?

61. $x^2 - 3x - 4$

62. $x^2 - 2x - 3$

63. $-3(1 - x) - 6$

64. $5(3 - x) - 10$

65. $x^3 - 6x + 4$

66. $x^3 + 5x + 18$

Rewrite each expression using the given property and simplify if possible.

67. Commutative property of addition
- **a.** $-5 + 7$
- **b.** $-2 + n$
- **c.** $-4.2 + a + 13.6$
- **d.** $7 + x - 7$

68. Associative property of multiplication
- **a.** $2 \cdot (3 \cdot 6)$
- **b.** $3 \cdot (4 \cdot b)$
- **c.** $-1.5 \cdot (6 \cdot a)$
- **d.** $-6 \cdot (-\frac{5}{6} \cdot x)$

Replace the box so that a true statement results.

69.
- **a.** $x + (-3.2) + \square = x$
- **b.** $n - \frac{5}{6} + \square = n$

70.
- **a.** $\square \cdot \frac{2}{3}x = 1x$
- **b.** $\square \cdot \dfrac{n}{-3} = 1n$

Simplify by removing all grouping symbols (as needed) and combining like terms.

71. $-5(x - 2.6)$

72. $-12(v - 3.2)$

73. $\frac{2}{3}(-\frac{1}{5}p + 9)$

74. $\frac{5}{6}(-\frac{2}{15}q + 24)$

75. $3a + (-5a)$

76. $13m + (-5m)$

77. $\frac{2}{3}x + \frac{3}{4}x$

78. $\frac{5}{12}y - \frac{3}{8}y$

79. $3(a^2 + 3a) - (5a^2 + 7a)$

80. $2(b^2 + 5b) - (6b^2 + 9b)$

81. $x^2 - (3x - 5x^2)$

82. $n^2 - (5n - 4n^2)$

83. $(3a + 2b - 5c) - (a - b - 7c)$

84. $(x - 4y + 8z) - (8x - 5y - 2z)$

85. $\frac{3}{5}(5n - 4) + \frac{5}{8}(n + 16)$

86. $\frac{2}{3}(2x - 9) + \frac{3}{4}(x + 12)$

87. $(3a^2 - 5a + 7) + 2(2a^2 - 4a - 6)$

88. $2(3m^2 + 2m - 7) - (m^2 - 5m + 4)$

▶ WORKING WITH FORMULAS

89. Electrical resistance: $R = \dfrac{kL}{d^2}$

The electrical resistance in a wire depends on the length and diameter of the wire. This resistance can be modeled by the formula shown, where R is the resistance in ohms, L is the length in feet, and d is the diameter of the wire in inches. Find the resistance if $k = 0.000025$, $d = 0.015$ in., and $L = 90$ ft.

90. Volume and pressure: $P = \dfrac{k}{V}$

If temperature remains constant, the pressure of a gas held in a closed container is related to the volume of gas by the formula shown, where P is the pressure in pounds per square inch, V is the volume of gas in cubic inches, and k is a constant that depends on given conditions. Find the pressure exerted by the gas if $k = 440{,}310$ and $V = 22{,}580$ in^3.

▶ APPLICATIONS

Translate each key phrase into an algebraic expression, then evaluate as indicated.

91. Cruising speed: A turbo-prop airliner has a cruising speed that is one-half the cruising speed of a 767 jet aircraft. (a) Express the speed of the turbo-prop in terms of the speed of the jet, and (b) determine the speed of the airliner if the cruising speed of the jet is 550 mph.

92. Softball toss: Macklyn can throw a softball two-thirds as far as her father. (a) Express the distance that Macklyn can throw a softball in terms of the distance her father can throw. (b) If her father can

throw the ball 210 ft, how far can Macklyn throw the ball?

93. Dimensions of a lawn: The length of a rectangular lawn is 3 ft more than twice its width. (a) Express the length of the lawn in terms of the width. (b) If the width is 52 ft, what is the length?

94. Pitch of a roof: To obtain the proper pitch, the crossbeam for a roof truss must be 2 ft less than three-halves the rafter. (a) Express the length of the crossbeam in terms of the rafter. (b) If the rafter is 18 ft, how long is the crossbeam?

95. Postage costs: In 2004, a first class stamp cost 22¢ more than it did in 1978. Express the cost of a 2004 stamp in terms of the 1978 cost. If a stamp cost 15¢ in 1978, what was the cost in 2004?

96. Minimum wage: In 2004, the federal minimum wage was $2.85 per hour more than it was in 1976. Express the 2004 wage in terms of the 1976 wage. If the hourly wage in 1976 was $2.30, what was it in 2004?

97. Repair costs: The TV repairman charges a flat fee of $43.50 to come to your house and $25 per hour for labor. Express the cost of repairing a TV in terms of the time it takes to repair it. If the repair took 1.5 hr, what was the total cost?

98. Repair costs: At the local car dealership, shop charges are $79.50 to diagnose the problem and $85 per shop hour for labor. Express the cost of a repair in terms of the labor involved. If a repair takes 3.5 hr, how much will it cost?

▶ **EXTENDING THE CONCEPT**

99. If C must be a positive odd integer and D must be a negative even integer, then $C^2 + D^2$ must be a:

 a. positive odd integer.

 b. positive even integer.

 c. negative odd integer.

 d. negative even integer.

 e. Cannot be determined.

100. Historically, several attempts have been made to create metric time using factors of 10, but our current system won out. If 1 day was 10 metric hours, 1 metric hour was 10 metric minutes, and 1 metric minute was 10 metric seconds, what time would it really be if a metric clock read 4:3:5? Assume that each new day starts at midnight.

| **R.3** | **Exponents, Scientific Notation, and a Review of Polynomials** |

Learning Objectives

In Section R.3 you will review how to:

☐ **A.** Apply properties of exponents

☐ **B.** Perform operations in scientific notation

☐ **C.** Identify and classify polynomial expressions

☐ **D.** Add and subtract polynomials

☐ **E.** Compute the product of two polynomials

☐ **F.** Compute special products: binomial conjugates and binomial squares

In this section, we review basic exponential properties and operations on polynomials. Although there are five to eight exponential properties (depending on how you count them), all can be traced back to the basic definition involving repeated multiplication.

A. The Properties of Exponents

As noted in Section R.1, an exponent tells how many times the base occurs as a factor. For $b \cdot b \cdot b = b^3$, we say b^3 is written in *exponential form*. In some cases, we may refer to b^3 as an **exponential term.**

Exponential Notation

For any positive integer n,

$$b^n = \underbrace{b \cdot b \cdot b \cdot \ldots \cdot b}_{n \text{ times}} \quad \text{and} \quad \underbrace{b \cdot b \cdot b \cdot \ldots \cdot b}_{n \text{ times}} = b^n$$

The Product and Power Properties

There are two properties that follow immediately from this definition. When b^3 is multiplied by b^2, we have an uninterrupted string of five factors: $b^3 \cdot b^2 = (b \cdot b \cdot b) \cdot (b \cdot b)$, which can be written as b^5. This is an example of the **product property of exponents.**

WORTHY OF NOTE

In this statement of the product property and the exponential properties that follow, it is assumed that for any expression of the form 0^m, $m > 0$ hence $0^m = 0$.

Product Property Of Exponents

For any base b and positive integers m and n:

$$b^m \cdot b^n = b^{m+n}$$

In words, the property says, *to multiply exponential terms with the* **same base,** *keep the common base and add the exponents.* A special application of the product property uses repeated factors of the *same* exponential term, as in $(x^2)^3$. Using the product property, we have $(x^2)(x^2)(x^2) = x^6$. Notice the same result can be found more quickly by multiplying the inner exponent by the outer exponent: $(x^2)^3 = x^{2\cdot3} = x^6$. We generalize this idea to state the **power property of exponents.** In words the property says, *to raise an exponential term to a power, keep the same base and multiply the exponents.*

Power Property of Exponents

For any base b and positive integers m and n:

$$(b^m)^n = b^{m \cdot n}$$

EXAMPLE 1 ▶ **Multiplying Terms Using Exponential Properties**

Compute each product.

  a. $-4x^3 \cdot \frac{1}{2}x^2$ **b.** $(p^3)^2 \cdot (p^4)^2$

Solution ▶   **a.** $-4x^3 \cdot \frac{1}{2}x^2 = (-4 \cdot \frac{1}{2})(x^3 \cdot x^2)$ commutative and associative properties

$= (-2)(x^{3+2})$ product property; simplify

$= -2x^5$ result

  b. $(p^3)^2 \cdot (p^4)^2 = p^6 \cdot p^8$ power property

$= p^{6+8}$ product property

$= p^{14}$ result

Now try Exercises 7 through 12 ▶

The power property can easily be extended to include more than one factor within the parentheses. This application of the power property is sometimes called the **product to a power property.** We can also raise a quotient of exponential terms to a power. The result is called the **quotient to a power property,** and can be extended to include any number of factors. In words the properties say, to raise a product or quotient of exponential terms to a power, *multiply every exponent inside* the parentheses *by the exponent outside* the parentheses.

Product to a Power Property

For any bases a and b, and positive integers m, n, and p:

$$(a^m b^n)^p = a^{mp} \cdot b^{np}$$

Quotient to a Power Property

For any bases a and $b \neq 0$, and positive integers m, n, and p:

$$\left(\frac{a^m}{b^n}\right)^p = \frac{a^{mp}}{b^{np}}$$

EXAMPLE 2 ▶ **Simplifying Terms Using the Power Properties**

Simplify using the power property (if possible):

a. $(-3a)^2$ b. $-3a^2$ c. $\left(\dfrac{-5a^3}{2b}\right)^2$

Solution ▶ a. $(-3a)^2 = (-3)^2 \cdot (a^1)^2$ b. $-3a^2$ is in simplified form
$$= 9a^2$$

c. $\left(\dfrac{-5a^3}{2b}\right)^2 = \dfrac{(-5)^2(a^3)^2}{2^2 b^2}$
$$= \dfrac{25a^6}{4b^2}$$

WORTHY OF NOTE

Regarding Examples 2(a) and 2(b), note the difference between the expressions $(-3a)^2 = (-3 \cdot a)^2$ and $-3a^2 = -3 \cdot a^2$. In the first, the exponent acts on both the negative 3 *and* the *a*; in the second, the exponent acts on only the *a* and there is no "product to a power."

Now try Exercises 13 through 24 ▶

Applications of exponents sometimes involve linking one exponential term with another using a substitution. The result is then simplified using exponential properties.

EXAMPLE 3 ▶ **Applying the Power Property after a Substitution**

The formula for the volume of a cube is $V = S^3$, where S is the length of one edge. If the length of each edge is $2x^2$:

a. Find a formula for volume in terms of x.

b. Find the volume if $x = 2$.

Solution ▶ a. $V = \quad S^3$
         $\downarrow \quad S = 2x^2$
      $= (2x^2)^3$
      $= 8x^6$

b. For $V = 8x^6$,
$$V = 8(2)^6 \quad \text{substitute 2 for } x$$
$$= 8 \cdot 64 \text{ or } 512 \quad (2)^6 = 64$$
The volume of the cube would be 512 units^3.

Now try Exercises 25 and 26 ▶

The Quotient Property of Exponents

By combining exponential notation and the property $\dfrac{x}{x} = 1$ for $x \neq 0$, we note a pattern that helps to simplify a *quotient* of exponential terms. For $\dfrac{x^5}{x^2} = \dfrac{x \cdot x \cdot x \cdot x \cdot x}{x \cdot x}$ or x^3, the exponent of the final result appears to be the *difference between the exponent in the numerator and the exponent in the denominator*. This seems reasonable since the subtraction would indicate a removal of the factors that reduce to 1. Regardless of how many factors are used, we can generalize the idea and state the **quotient property of exponents.** In words the property says, to divide two exponential terms with the same base, *keep the common base and subtract the exponent of the denominator* from *the exponent of the numerator*.

Quotient Property of Exponents

For any base $b \neq 0$ and positive integers m and n: $\dfrac{b^m}{b^n} = b^{m-n}$

Zero and Negative Numbers as Exponents

If the exponent of the denominator is *greater* than the exponent in the numerator, the quotient property yields a negative exponent: $\dfrac{x^2}{x^5} = x^{2-5} = x^{-3}$. To help understand what a negative exponent *means*, let's look at the expanded form of the expression: $\dfrac{x^2}{x^5} = \dfrac{x \cdot x^1}{x \cdot x \cdot x \cdot x \cdot x} = \dfrac{1}{x^3}$. A negative exponent can literally be interpreted as "write the factors as a reciprocal." A good way to remember this is

$$2^{-3} \overset{\text{three factors of 2}}{\underset{\text{written as a reciprocal}}{\qquad}} \qquad \dfrac{2^{-3}}{1} = \dfrac{1}{2^3} = \dfrac{1}{8}$$

Since the result would be similar regardless of the base used, we can generalize this idea and state the **property of negative exponents.**

WORTHY OF NOTE

The use of zero as an exponent should not strike you as strange or odd; it's simply a way of saying that *no factors of the base remain,* since all terms have been reduced to 1. For $\dfrac{2^3}{2^3}$, we have $\dfrac{8}{8} = 1$, or $\dfrac{2 \cdot 2 \cdot 2}{2 \cdot 2 \cdot 2}^{\frac{1}{\ }\frac{1}{\ }\frac{1}{\ }} = 1$, or $2^{3-3} = 2^0 = 1$.

Property of Negative Exponents

For any base $b \neq 0$ and integer n:

$$\dfrac{b^{-n}}{1} = \dfrac{1}{b^n} \qquad \dfrac{1}{b^{-n}} = \dfrac{b^n}{1} \qquad \left(\dfrac{a}{b}\right)^{-n} = \left(\dfrac{b}{a}\right)^{n}; a \neq 0$$

Finally, when we consider that $\dfrac{x^3}{x^3} = 1$ by division, and $\dfrac{x^3}{x^3} = x^{3-3} = x^0$ using the quotient property, we conclude that $x^0 = 1$ as long as $x \neq 0$. We can also generalize this observation and state the meaning of zero as an exponent. In words the property says, *any nonzero quantity raised to an exponent of zero is equal to 1.*

Zero Exponent Property

For any base $b \neq 0$: $b^0 = 1$

EXAMPLE 4 ▶ Simplifying Expressions Using Exponential Properties

Simplify using exponential properties. Answer using positive exponents only.

a. $\left(\dfrac{2a^3}{b^2}\right)^{-2}$

b. $(3hk^{-2})^3(6h^{-2}k^{-3})^{-2}$

c. $(3x)^0 + 3x^0 + 3^{-2}$

d. $\dfrac{(-2m^2n^3)^5}{(4mn^2)^3}$

Solution ▶ a. $\left(\dfrac{2a^3}{b^2}\right)^{-2} = \left(\dfrac{b^2}{2a^3}\right)^2$ property of negative exponents

$= \dfrac{(b^2)^2}{2^2(a^3)^2}$ power property

$= \dfrac{b^4}{4a^6}$ result

b. $(3hk^{-2})^3(6h^{-2}k^{-3})^{-2} = (3^3h^3k^{-6})(6^{-2}h^4k^6)$ power property

$= 3^3 \cdot 6^{-2} \cdot h^{3+4} \cdot k^{-6+6}$ product property

$= \dfrac{27h^7k^0}{36}$ simplify $\left(6^{-2} = \dfrac{1}{6^2} = \dfrac{1}{36}\right)$

$= \dfrac{3h^7}{4}$ result ($k^0 = 1$)

WORTHY OF NOTE

Notice in Example 4(c), we have $(3x)^0 = (3 \cdot x)^0 = 1$, while $3x^0 = 3 \cdot x^0 = 3(1)$. This is another example of operations and grouping symbols working together: $(3x)^0 = 1$ because any *quantity* to the zero power is 1. However, for $3x^0$ there are no grouping symbols, so the exponent 0 acts only on the x and not the 3.

c. $(3x)^0 + 3x^0 + 3^{-2} = 1 + 3(1) + \dfrac{1}{3^2}$ zero exponent property; property of negative exponents

$$= 4 + \dfrac{1}{9}$$ simplify

$$= 4\dfrac{1}{9} = \dfrac{37}{9}$$ result

d. $\dfrac{(-2m^2n^3)^5}{(4mn^2)^3} = \dfrac{(-2)^5(m^2)^5(n^3)^5}{4^3 m^3 (n^2)^3}$ power property

$$= \dfrac{-32m^{10}n^{15}}{64m^3 n^6}$$ simplify

$$= \dfrac{-1m^7 n^9}{2}$$ quotient property

$$= -\dfrac{m^7 n^9}{2}$$ result

Now try Exercises 27 through 66 ▶

Summary of Exponential Properties

For real numbers a and b, and integers m, n, and p (excluding 0 raised to a nonpositive power)

Product property: $\quad b^m \cdot b^n = b^{m+n}$

Power property: $\quad (b^m)^n = b^{m \cdot n}$

Product to a power: $\quad (a^m b^n)^p = a^{mp} \cdot b^{np}$

Quotient to a power: $\quad \left(\dfrac{a^m}{b^n}\right)^p = \dfrac{a^{mp}}{b^{np}} \; (b \neq 0)$

Quotient property: $\quad \dfrac{b^m}{b^n} = b^{m-n} \; (b \neq 0)$

Zero exponents: $\quad b^0 = 1 \; (b \neq 0)$

Negative exponents: $\quad \dfrac{b^{-n}}{1} = \dfrac{1}{b^n}, \; \dfrac{1}{b^{-n}} = b^n, \; \left(\dfrac{a}{b}\right)^{-n} = \left(\dfrac{b}{a}\right)^n \; (a, b \neq 0)$

☑ **A.** You've just reviewed how to apply properties of exponents

B. Exponents and Scientific Notation

In many technical and scientific applications, we encounter numbers that are either extremely large or very, very small. For example, the mass of the moon is over 73 quintillion kilograms (73 followed by 18 zeroes), while the constant for universal gravitation contains 10 zeroes before the first nonzero digit. When computing with numbers of this size, scientific notation has a distinct advantage over the common decimal notation (base-10 place values).

Scientific Notation

A non-zero number written in scientific notation has the form

$$N \times 10^k$$

where $1 \leq |N| < 10$ and k is an integer.

WORTHY OF NOTE

Recall that multiplying by 10's (or multiplying by 10^k, $k > 0$) shifts the decimal to the right k places, making the number larger. Dividing by 10's (or multiplying by 10^{-k}, $k > 0$) shifts the decimal to the left k places, making the number smaller.

To convert a number from decimal notation into scientific notation, we begin by placing the decimal point to the immediate right of the first nonzero digit (creating a number less than 10 but greater than or equal to 1) and multiplying by 10^k. Then we determine the power of 10 (the value of k) needed to ensure that the two forms are equivalent. When writing large or small numbers in scientific notation, we sometimes round the value of N to two or three decimal places.

EXAMPLE 5 ▶ **Converting from Decimal Notation to Scientific Notation**

The mass of the moon is about 73,000,000,000,000,000,000 kg. Write this number in scientific notation.

Solution ▶ Place decimal to the right of first nonzero digit (7) and multiply by 10^k.

$$73,000,000,000,000,000,000 = 7.3 \times 10^k$$

To return the decimal to its original position would require 19 shifts to the *right*, so k must be *positive* 19.

$$73,000,000,000,000,000,000 = 7.3 \times 10^{19}$$

The mass of the moon is 7.3×10^{19} kg.

Now try Exercises 67 and 68 ▶

Converting a number from scientific notation to decimal notation is simply an application of multiplication or division with powers of 10.

EXAMPLE 6 ▶ **Converting from Scientific Notation to Decimal Notation**

The constant of gravitation is 6.67×10^{-11}. Write this number in common decimal form.

Solution ▶ Since the exponent is *negative* 11, shift the decimal 11 *places to the left*, using placeholder zeroes as needed to return the decimal to its original position:

$$6.67 \times 10^{-11} = 0.000\ 000\ 000\ 066\ 7$$

☑ **B.** You've just reviewed how to perform operations in scientific notation

Now try Exercises 69 through 72 ▶

C. Identifying and Classifying Polynomial Expressions

A **monomial** is a term using *only whole number exponents* on variables, with no variables in the denominator. One important characteristic of a monomial is its **degree.** For a monomial in one variable, the degree is the same as the exponent *on the variable.* The degree of a monomial in two or more variables is the sum of exponents occurring on variable factors. A **polynomial** is a monomial or any sum or difference of monomial terms. For instance, $\frac{1}{2}x^2 - 5x + 6$ is a polynomial, while $3n^{-2} + 2n - 7$ is not (the exponent -2 is not a whole number). Identifying polynomials is an important skill because they represent a very different kind of real-world model than nonpolynomials. In addition, there are different **families of polynomials,** with each family having different characteristics. We classify polynomials according to their *degree* and *number of terms.* The **degree of a polynomial** in one variable is the largest exponent occurring on the variable. The degree of a polynomial in more than one variable is the largest sum of exponents in any one term. A polynomial with two terms is called a **binomial** (*bi* means two) and a polynomial with three terms is called a **trinomial** (*tri* means three). There are special names for polynomials with four or more terms, but for these, we simply use the general name *polynomial* (*poly* means many).

EXAMPLE 7 ▶ **Classifying and Describing Polynomials**

For each expression:
  a. Classify as a monomial, binomial, trinomial, or polynomial.
  b. State the degree of the polynomial.
  c. Name the coefficient of each term.

Solution ▶

Expression	Classification	Degree	Coefficients
$5x^2y - 2xy$	binomial	three	$5, -2$
$x^2 - 0.81$	binomial	two	$1, -0.81$
$z^3 - 3z^2 + 9z - 27$	polynomial (four terms)	three	$1, -3, 9, -27$
$\frac{-3}{4}x + 5$	binomial	one	$\frac{-3}{4}, 5$
$2x^2 + x - 3$	trinomial	two	$2, 1, -3$

Now try Exercises 73 through 78 ▶

A polynomial expression is in **standard form** when the terms of the polynomial are written in *descending order of degree,* beginning with the highest-degree term. The coefficient of the highest-degree term is called the **leading coefficient.**

EXAMPLE 8 ▶ **Writing Polynomials in Standard Form**

Write each polynomial in standard form, then identify the leading coefficient.

Solution ▶

Polynomial	Standard Form	Leading Coefficient
$9 - x^2$	$-x^2 + 9$	-1
$5z + 7z^2 + 3z^3 - 27$	$3z^3 + 7z^2 + 5z - 27$	3
$2 + (\frac{-3}{4})x$	$\frac{-3}{4}x + 2$	$\frac{-3}{4}$
$-3 + 2x^2 + x$	$2x^2 + x - 3$	2

☑ **C.** You've just reviewed how to identify and classify polynomial expressions

Now try Exercises 79 through 84 ▶

D. Adding and Subtracting Polynomials

Adding polynomials simply involves using the distributive, commutative, and associative properties to combine like terms (at this point, the properties are usually applied mentally). As with real numbers, the subtraction of polynomials involves adding the opposite of the second polynomial using algebraic addition. This can be viewed as distributing -1 to the second polynomial and combining like terms.

EXAMPLE 9 ▶ **Adding and Subtracting Polynomials**

Perform the indicated operations:

$(0.7n^3 + 4n^2 + 8) + (0.5n^3 - n^2 - 6n) - (3n^2 + 7n - 10)$.

Solution ▶ $0.7n^3 + 4n^2 + 8 + 0.5n^3 - n^2 - 6n - 3n^2 - 7n + 10$ eliminate parentheses (distributive property)

$= 0.7n^3 + 0.5n^3 + 4n^2 - 1n^2 - 3n^2 - 6n - 7n + 8 + 10$ use real number properties to collect like terms

$= 1.2n^3 - 13n + 18$ combine like terms

Now try Exercises 85 through 90 ▶

Sometimes it's easier to add or subtract polynomials using a vertical format and aligning like terms. Note the use of a placeholder zero in Example 10.

EXAMPLE 10 ▶ **Subtracting Polynomials Using a Vertical Format**

Compute the difference of $x^3 - 5x + 9$ and $x^3 + 3x^2 + 2x - 8$ using a vertical format.

Solution ▶
$$
\begin{array}{r}
x^3 + \mathbf{0}\,x^2 - 5x + 9 \\
-(x^3 + 3x^2 + 2x - 8)
\end{array}
\longrightarrow
\begin{array}{r}
x^3 + \mathbf{0}x^2 - 5x + 9 \\
-x^3 - 3x^2 - 2x + 8 \\
\hline
-3x^2 - 7x + 17
\end{array}
$$

☑ **D.** You've just reviewed how to add and subtract polynomials

The difference is $-3x^2 - 7x + 17$.

Now try Exercises 91 and 92 ▶

E. The Product of Two Polynomials

Monomial Times Monomial

The simplest case of polynomial multiplication is the product of monomials shown in Example 1(a). These were computed using exponential properties and the properties of real numbers.

Monomial Times Polynomial

To compute the product of a monomial and a polynomial, we use the distributive property.

EXAMPLE 11 ▶ **Multiplying a Monomial by a Polynomial**

Find the product: $-2a^2(a^2 - 2a + 1)$.

Solution ▶
$$
\begin{aligned}
-2a^2(a^2 - 2a + 1) &= -2a^2(a^2) - (-2a^2)(2a^1) + (-2a^2)(1) \quad \text{distribute} \\
&= -2a^4 + 4a^3 - 2a^2 \quad \text{simplify}
\end{aligned}
$$

Now try Exercises 93 and 94 ▶

Binomial Times Polynomial

For products involving binomials, we still use a version of the distributive property—this time to distribute one polynomial to each term of the other polynomial factor. Note the distribution can be performed either from the left or from the right.

EXAMPLE 12 ▶ **Multiplying a Binomial by a Polynomial**

Multiply as indicated:

   a. $(2z + 1)(z - 2)$ **b.** $(2v - 3)(4v^2 + 6v + 9)$

Solution ▶
a. $\begin{aligned}(2z + 1)(z - 2) &= 2z(z - 2) + 1(z - 2) \quad \text{distribute to every term in the first binomial} \\ &= 2z^2 - 4z + 1z - 2 \quad \text{eliminate parentheses (distribute again)} \\ &= 2z^2 - 3z - 2 \quad \text{simplify}\end{aligned}$

b. $\begin{aligned}(2v - 3)(4v^2 + 6v + 9) &= 2v(4v^2 + 6v + 9) - 3(4v^2 + 6v + 9) \quad \text{distribute} \\ &= 8v^3 + 12v^2 + 18v - 12v^2 - 18v - 27 \quad \text{simplify} \\ &= 8v^3 - 27 \quad \text{combine like terms}\end{aligned}$

Now try Exercises 95 through 100 ▶

The F-O-I-L Method

By observing the product of two binomials in Example 12(a), we note a pattern that can make the process more efficient. We illustrate here using the product $(2x - 1)(3x + 2)$.

The F-O-I-L Method for Multiplying Binomials

The product of two binomials can quickly be computed by multiplying:

$$6x^2 + 4x - 3x - 2$$
$$\underline{\text{First}}\ \ \underline{\text{Outer}}\ \ \underline{\text{Inner}}\ \ \underline{\text{Last}}$$

and combining like terms

$$6x^2 + x - 2$$

The first term of the result will always be the product of the first terms from each binomial, and the last term of the result is the product of their last terms. We also note that here, the middle term is found by adding the *outermost product* with the *innermost product*. As you practice with the F-O-I-L process, much of the work can be done mentally and you can often compute the entire product without writing anything down except the answer.

EXAMPLE 13 ▶ **Multiplying Binomials Using F-O-I-L**

Compute each product mentally:

  a. $(5n - 1)(n + 2)$

  b. $(2b + 3)(5b - 6)$

☑ E. You've just reviewed how to compute the product of two polynomials

Solution ▶

  a. $(5n - 1)(n + 2)$: $5n^2 + 9n - 2$

                product of   sum of   product of
            first two terms   outer and inner last two terms

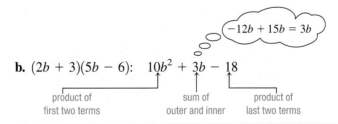

  b. $(2b + 3)(5b - 6)$: $10b^2 + 3b - 18$

                product of   sum of   product of
            first two terms   outer and inner last two terms

Now try Exercises 101 through 116 ▶

WORTHY OF NOTE

Consider the product $(x + 3)(x + 2)$ in the context of *area*. If we view $x + 3$ as the length of a rectangle (an unknown length plus 3 units), and $x + 2$ as its width (the same unknown length plus 2 units), a diagram of the total area would look like the following, with the result $x^2 + 5x + 6$ clearly visible.

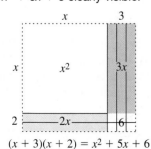

$(x + 3)(x + 2) = x^2 + 5x + 6$

F. Special Polynomial Products

Certain polynomial products are considered "special" for two reasons: (1) the product follows a predictable pattern, and (2) the result can be used to simplify expressions, graph functions, solve equations, and/or develop other skills.

Binomial Conjugates

Expressions like $x + 7$ and $x - 7$ are called **binomial conjugates.** For any given binomial, its conjugate is found by using the same two terms with the opposite sign between them. Example 14 shows that when we multiply a binomial and its conjugate, the "outers" and "inners" sum to zero and the result is a **difference of two squares.**

EXAMPLE 14 ▶ **Multiplying Binomial Conjugates**

Compute each product mentally:

a. $(x + 7)(x - 7)$ **b.** $(2x - 5y)(2x + 5y)$ **c.** $\left(x + \dfrac{2}{5}\right)\left(x - \dfrac{2}{5}\right)$

$\boxed{-7x + 7x = 0x}$

Solution ▶ **a.** $(x + 7)(x - 7) = x^2 - 49$ difference of squares $(x)^2 - (7)^2$

$\boxed{10xy + (-10xy) = 0xy}$

b. $(2x - 5y)(2x + 5y) = 4x^2 - 25y^2$ difference of squares: $(2x)^2 - (5y)^2$

$\boxed{-\dfrac{2}{5}x + \dfrac{2}{5}x = 0}$

c. $\left(x + \dfrac{2}{5}\right)\left(x - \dfrac{2}{5}\right) = x^2 - \dfrac{4}{25}$ difference of squares: $x^2 - \left(\dfrac{2}{5}\right)^2$

Now try Exercises 117 through 124 ▶

The Product of a Binomial and Its Conjugate

Given any expression that can be written in the form $A + B$, the conjugate of the expression is $A - B$ and their product is a difference of two squares:

$$(A + B)(A - B) = A^2 - B^2$$

Binomial Squares

Expressions like $(x + 7)^2$ are called **binomial squares** and are useful for solving many equations and sketching a number of basic graphs. Note $(x + 7)^2 = (x + 7)(x + 7) = x^2 + 14x + 49$ using the F-O-I-L process. The expression $x^2 + 14x + 49$ is called a **perfect square trinomial** because it is the result of expanding a binomial square. If we write a binomial square in the more general form $(A + B)^2 = (A + B)(A + B)$ and compute the product, we notice a pattern that helps us write the expanded form more quickly.

$$\begin{aligned} (A + B)^2 &= (A + B)(A + B) & \text{repeated multiplication} \\ &= A^2 + AB + AB + B^2 & \text{F-O-I-L} \\ &= A^2 + 2AB + B^2 & \text{simplify (perfect square trinomial)} \end{aligned}$$

The first and last terms of the trinomial are squares of the terms A and B. Also, the middle term of the trinomial is *twice the product of these two terms*: $AB + AB = 2AB$. The F-O-I-L process shows us why. Since the outer and inner products are identical, we always end up with two. A similar result holds for $(A - B)^2$ and the process can be summarized for both cases using the $\pm$ symbol.

The Square of a Binomial

Given any expression that can be written in the form $(A \pm B)^2$,

1. $(A + B)^2 = A^2 + 2AB + B^2$

2. $(A - B)^2 = A^2 - 2AB + B^2$

LOOKING AHEAD

Although a binomial square can always be found using repeated factors and F-O-I-L, learning to expand them using the pattern is a valuable skill. Binomial squares occur often in a study of algebra and it helps to find the expanded form quickly.

⚠ **CAUTION** ▶ Note the square of a binomial always results in a trinomial (three terms). Specifically $(A + B)^2 \neq A^2 + B^2$.

EXAMPLE 15 ▶   Find each binomial square without using F-O-I-L:

     a. $(a + 9)^2$ **b.** $(3x - 5)^2$ **c.** $(3 + \sqrt{x})^2$

Solution ▶   **a.** $(a + 9)^2 = a^2 + 2(a \cdot 9) + 9^2$ $(A + B)^2 = A^2 + 2AB + B^2$

                 $= a^2 + 18a + 81$ simplify

     b. $(3x - 5)^2 = (3x)^2 - 2(3x \cdot 5) + 5^2$ $(A - B)^2 = A^2 - 2AB + B^2$

                 $= 9x^2 - 30x + 25$ simplify

     c. $(3 + \sqrt{x})^2 = 9 + 2(3 \cdot \sqrt{x}) + x$ $(A + B)^2 = A^2 + 2AB + B^2$

                 $= 9 + 6\sqrt{x} + x$ simplify

Now try Exercises 125 through 136 ▶

☑ **F.** You've just reviewed how to compute special products: binomial conjugates and binomial squares

With practice, you will be able to go directly from the binomial square to the resulting trinomial.

R.3 EXERCISES

▶ **CONCEPTS AND VOCABULARY**

Fill in each blank with the appropriate word or phrase. Carefully reread the section, if necessary.

1. The equation $(3x^2)^3 = 27x^6$ is an example of the _____ property of exponents.

2. The equation $(2x^3)^{-2} = \dfrac{1}{4x^6}$ is an example of the property of _____ exponents.

3. The sum of the "outers" and "inners" for $(2x + 5)^2$ is _____, while the sum of outers and inners for $(2x + 5)(2x - 5)$ is _____.

4. The expression $2x^2 - 3x - 10$ can be classified as a _____ of degree _____, with a leading coefficient of _____.

5. Discuss/Explain why one of the following expressions can be simplified further, while the other cannot: (a) $-7n^4 + 3n^2$; (b) $-7n^4 \cdot 3n^2$.

6. Discuss/Explain why the degree of $2x^2y^3$ is greater than the degree of $2x^2 + y^3$. Include additional examples for contrast and comparison.

▶ **DEVELOPING YOUR SKILLS**

Determine each product using the product and/or power properties.

7. $\dfrac{2}{3}n^2 \cdot 21n^5$ 8. $24g^5 \cdot \dfrac{3}{8}g^9$

9. $(-6p^2q)(2p^3q^3)$ 10. $(-1.5vy^2)(-8v^4y)$

11. $(a^2)^4 \cdot (a^3)^2 \cdot b^2 \cdot b^5$ 12. $d^2 \cdot d^4 \cdot (c^5)^2 \cdot (c^3)^2$

Simplify each expression using the product to a power property.

13. $(6pq^2)^3$ 14. $(-3p^2q)^2$

15. $(3.2hk^2)^3$ 16. $(-2.5h^5k)^2$

17. $\left(\dfrac{p}{2q}\right)^2$ 18. $\left(\dfrac{b}{3a}\right)^3$

19. $(-0.7c^4)^2(10c^3d^2)^2$ **20.** $(-2.5a^3)^2(3a^2b^2)^3$

21. $\left(\frac{3}{4}x^3y\right)^2$ **22.** $\left(\frac{4}{5}x^3\right)^2$

23. $\left(-\frac{3}{8}x\right)^2\left(16xy^2\right)$ **24.** $\left(\frac{2}{3}m^2n\right)^2 \cdot \left(\frac{1}{2}mn^2\right)$

25. Volume of a cube: The formula for the volume of a cube is $V = S^3$, where S is the length of one edge. If the length of each edge is $3x^2$,

  a. Find a formula for volume in terms of the variable x.

  b. Find the volume of the cube if $x = 2$.

26. Area of a circle: The formula for the area of a circle is $A = \pi r^2$, where r is the length of the radius. If the radius is given as $5x^3$,

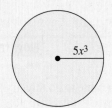

  a. Find a formula for area in terms of the variable x.

  b. Find the area of the circle if $x = 2$.

Simplify using the quotient property or the property of negative exponents. Write answers using positive exponents only.

27. $\dfrac{-6w^5}{-2w^2}$ **28.** $\dfrac{8z^7}{16z^5}$

29. $\dfrac{-12a^3b^5}{4a^2b^4}$ **30.** $\dfrac{5m^3n^5}{10mn^2}$

31. $\left(\frac{2}{3}\right)^{-3}$ **32.** $\left(\frac{5}{6}\right)^{-1}$

33. $\dfrac{2}{h^{-3}}$ **34.** $\dfrac{3}{m^{-2}}$

35. $(-2)^{-3}$ **36.** $(-4)^{-2}$

37. $\left(\frac{-1}{2}\right)^{-3}$ **38.** $\left(\frac{-2}{3}\right)^{-2}$

Simplify each expression using the quotient to a power property.

39. $\left(\dfrac{2p^4}{q^3}\right)^2$ **40.** $\left(\dfrac{-5v^4}{7w^3}\right)^2$

41. $\left(\dfrac{0.2x^2}{0.3y^3}\right)^3$ **42.** $\left(\dfrac{-0.5a^3}{0.4b^2}\right)^2$

43. $\left(\dfrac{5m^2n^3}{2r^4}\right)^2$ **44.** $\left(\dfrac{4p^3}{3x^2y}\right)^3$

45. $\left(\dfrac{5p^2q^3r^4}{-2pq^2r^4}\right)^2$ **46.** $\left(\dfrac{9p^3q^2r^3}{12p^5qr^2}\right)^3$

Use properties of exponents to simplify the following. Write the answer using positive exponents only.

47. $\dfrac{9p^6q^4}{-12p^4q^6}$ **48.** $\dfrac{5m^5n^2}{10m^5n}$

49. $\dfrac{20h^{-2}}{12h^5}$ **50.** $\dfrac{5k^3}{20k^{-2}}$

51. $\dfrac{(a^2)^3}{a^4 \cdot a^5}$ **52.** $\dfrac{(5^3)^4}{5^9}$

53. $\left(\dfrac{a^{-3} \cdot b}{c^{-2}}\right)^{-4}$ **54.** $\dfrac{(p^{-4}q^8)^2}{p^5q^{-2}}$

55. $\dfrac{-6(2x^{-3})^2}{10x^{-2}}$ **56.** $\dfrac{18n^{-3}}{-8(3n^{-2})^3}$

57. $\dfrac{14a^{-3}bc^0}{-7(3a^2b^{-2}c)^3}$ **58.** $\dfrac{-3(2x^3y^{-4}z)^2}{18x^{-2}yz^0}$

59. $4^0 + 5^0$ **60.** $(-3)^0 + (-7)^0$

61. $2^{-1} + 5^{-1}$ **62.** $4^{-1} + 8^{-1}$

63. $3^0 + 3^{-1} + 3^{-2}$ **64.** $2^{-2} + 2^{-1} + 2^0$

65. $-5x^0 + (-5x)^0$ **66.** $-2n^0 + (-2n)^0$

Convert the following numbers to scientific notation.

67. In mid-2007, the U.S. Census Bureau estimated the world population at nearly 6,600,000,000 people.

68. The mass of a proton is generally given as 0.000 000 000 000 000 000 000 000 001 670 kg.

Convert the following numbers to decimal notation.

69. As of 2006, the smallest microprocessors in common use measured 6.5×10^{-9} m across.

70. In 2007, the estimated net worth of Bill Gates, the founder of Microsoft, was 5.6×10^{10} dollars.

Compute using scientific notation. Show all work.

71. The average distance between the Earth and the planet Jupiter is 465,000,000 mi. How many hours would it take a satellite to reach the planet if it traveled an average speed of 17,500 mi per hour? How many days? Round to the nearest whole.

72. In fiscal terms, a nation's debt-per-capita is the ratio of its total debt to its total population. In the year 2007, the total U.S. debt was estimated at $9,010,000,000,000, while the population was estimated at 303,000,000. What was the U.S. debt-per-capita ratio for 2007? Round to the nearest whole dollar.

Identify each expression as a polynomial or nonpolynomial (if a nonpolynomial, state why); classify each as a monomial, binomial, trinomial, or none of these; and state the degree of the polynomial.

73. $-35w^3 + 2w^2 + (-12w) + 14$

74. $-2x^3 + \frac{2}{3}x^2 - 12x + 1.2$

75. $5n^{-2} + 4n + \sqrt{17}$ **76.** $\frac{4}{r^3} + 2.7r^2 + r + 1$

77. $p^3 - \frac{2}{5}$ **78.** $q^3 + 2q^{-2} - 5q$

Write the polynomial in standard form and name the leading coefficient.

79. $7w + 8.2 - w^3 - 3w^2$

80. $-2k^2 - 12 - k$

81. $c^3 + 6 + 2c^2 - 3c$

82. $-3v^3 + 14 + 2v^2 + (-12v)$

83. $12 - \frac{2}{3}x^2$

84. $8 + 2n^2 + 7n$

Find the indicated sum or difference.

85. $(3p^3 - 4p^2 + 2p - 7) + (p^2 - 2p - 5)$

86. $(5q^2 - 3q + 4) + (-3q^2 + 3q - 4)$

87. $(5.75b^2 + 2.6b - 1.9) + (2.1b^2 - 3.2b)$

88. $(0.4n^2 + 5n - 0.5) + (0.3n^2 - 2n + 0.75)$

89. $(\frac{3}{4}x^2 - 5x + 2) - (\frac{1}{2}x^2 + 3x - 4)$

90. $(\frac{5}{9}n^2 + 4n - \frac{1}{2}) - (\frac{2}{3}n^2 - 2n + \frac{3}{4})$

91. Subtract $q^5 + 2q^4 + q^2 + 2q$ from $q^6 + 2q^5 + q^4 + 2q^3$ using a vertical format.

92. Find $x^4 + 2x^3 + x^2 + 2x$ decreased by $x^4 - 3x^3 + 4x^2 - 3x$ using a vertical format.

Compute each product.

93. $-3x(x^2 - x - 6)$

94. $-2v^2(v^2 + 2v - 15)$

95. $(3r - 5)(r - 2)$

96. $(s - 3)(5s + 4)$

97. $(x - 3)(x^2 + 3x + 9)$

98. $(z + 5)(z^2 - 5z + 25)$

99. $(b^2 - 3b - 28)(b + 2)$

100. $(2h^2 - 3h + 8)(h - 1)$

101. $(7v - 4)(3v - 5)$ **102.** $(6w - 1)(2w + 5)$

103. $(3 - m)(3 + m)$ **104.** $(5 + n)(5 - n)$

105. $(p - 2.5)(p + 3.6)$ **106.** $(q - 4.9)(q + 1.2)$

107. $(x + \frac{1}{2})(x + \frac{1}{4})$ **108.** $(z + \frac{1}{3})(z + \frac{5}{6})$

109. $(m + \frac{3}{4})(m - \frac{3}{4})$ **110.** $(n - \frac{2}{5})(n + \frac{2}{5})$

111. $(3x - 2y)(2x + 5y)$ **112.** $(6a + b)(a + 3b)$

113. $(4c + d)(3c + 5d)$ **114.** $(5x + 3y)(2x - 3y)$

115. $(2x^2 + 5)(x^2 - 3)$ **116.** $(3y^2 - 2)(2y^2 + 1)$

For each binomial, determine its conjugate and then find the product of the binomial with its conjugate.

117. $4m - 3$ **118.** $6n + 5$

119. $7x - 10$ **120.** $c + 3$

121. $6 + 5k$ **122.** $11 - 3r$

123. $x + \sqrt{6}$ **124.** $p - \sqrt{2}$

Find each binomial square.

125. $(x + 4)^2$ **126.** $(a - 3)^2$

127. $(4g + 3)^2$ **128.** $(5x - 3)^2$

129. $(4p - 3q)^2$ **130.** $(5c + 6d)^2$

131. $(4 - \sqrt{x})^2$ **132.** $(\sqrt{x} + 7)^2$

Compute each product.

133. $(x - 3)(y + 2)$

134. $(a + 3)(b - 5)$

135. $(k - 5)(k + 6)(k + 2)$

136. $(a + 6)(a - 1)(a + 5)$

▶ WORKING WITH FORMULAS

137. Medication in the bloodstream:
$M = 0.5t^4 + 3t^3 - 97t^2 + 348t$

If 400 mg of a pain medication are taken orally, the number of milligrams in the bloodstream is modeled by the formula shown, where M is the number of milligrams and t is the time in hours, $0 \le t < 5$. Construct a table of values for $t = 1$ through 5, then answer the following.

a. How many milligrams are in the bloodstream after 2 hr? After 3 hr?

b. Based on parts a and b, would you expect the number of milligrams in the bloodstream after 4 hr to be less or more than in part b? Why?

c. Approximately how many hours until the medication wears off (the number of milligrams in the bloodstream is 0)?

138. Amount of a mortgage payment:

$$M = \frac{A\left(\dfrac{r}{12}\right)\left(1 + \dfrac{r}{12}\right)^n}{\left(1 + \dfrac{r}{12}\right)^n - 1}$$

The monthly mortgage payment required to pay off (or amortize) a loan is given by the formula shown,

where M is the monthly payment, A is the original amount of the loan, r is the annual interest rate, and n is the term of the loan in months. Find the monthly payment (to the nearest cent) required to purchase a \$198,000 home, if the interest rate is 6.5% and the home is financed over 30 yr.

▶ **APPLICATIONS**

139. Attraction between particles: In electrical theory, the force of attraction between two particles P and Q with opposite charges is modeled by $F = \dfrac{kPQ}{d^2}$, where d is the distance between them and k is a constant that depends on certain conditions. This is known as Coulomb's law. Rewrite the formula using a negative exponent.

140. Intensity of light: The intensity of illumination from a light source depends on the distance from the source according to $I = \dfrac{k}{d^2}$, where I is the intensity measured in footcandles, d is the distance from the source in feet, and k is a constant that depends on the conditions. Rewrite the formula using a negative exponent.

141. Rewriting an expression: In advanced mathematics, negative exponents are widely used because they are easier to work with than rational expressions. Rewrite the expression $\dfrac{5}{x^3} + \dfrac{3}{x^2} + \dfrac{2}{x^1} + 4$ using negative exponents.

142. Swimming pool hours: A swimming pool opens at 8 A.M. and closes at 6 P.M. In summertime, the

number of people in the pool at any time can be approximated by the formula $S(t) = -t^2 + 10t$, where S is the number of swimmers and t is the number of hours the pool has been open (8 A.M.: $t = 0$, 9 A.M.: $t = 1$, 10 A.M.: $t = 2$, etc.).

  a. How many swimmers are in the pool at 6 P.M.? Why?

  b. Between what times would you expect the largest number of swimmers?

  c. Approximately how many swimmers are in the pool at 3 P.M.?

  d. Create a table of values for $t = 1, 2, 3, 4, \ldots$ and check your answer to part b.

 143. Maximizing revenue: A sporting goods store finds that if they price their video games at \$20, they make 200 sales per day. For each decrease of \$1, 20 additional video games are sold. This means the store's revenue can be modeled by the formula $R = (20 - 1x)(200 + 20x)$, where x is the number of \$1 decreases. Multiply out the binomials and use a table of values to determine what price will give the most revenue.

144. Maximizing revenue: Due to past experience, a jeweler knows that if they price jade rings at \$60, they will sell 120 each day. For each decrease of \$2, five additional sales will be made. This means the jeweler's revenue can be modeled by the formula $R = (60 - 2x)(120 + 5x)$, where x is the number of \$2 decreases. Multiply out the binomials and use a table of values to determine what price will give the most revenue.

▶ **EXTENDING THE CONCEPT**

145. If $(3x^2 + kx + 1) - (kx^2 + 5x - 7) + (2x^2 - 4x - k) = -x^2 - 3x + 2$, what is the value of k?

146. If $\left(2x + \dfrac{1}{2x}\right)^2 = 5$, then the expression $4x^2 + \dfrac{1}{4x^2}$ is equal to what number?

Learning Objectives

In Section R.4 you will review:

☐ **A.** Factoring out the greatest common factor

☐ **B.** Common binomial factors and factoring by grouping

☐ **C.** Factoring quadratic polynomials

☐ **D.** Factoring special forms and quadratic forms

It is often said that knowing which tool to use is just as important as knowing how to use the tool. In this section, we review the tools needed to factor an expression, an important part of solving polynomial equations. This section will also help us decide which factoring tool is appropriate when many different factorable expressions are presented.

A. The Greatest Common Factor

To **factor** an expression means to *rewrite the expression as an equivalent product*. The distributive property is an example of factoring in action. To factor $2x^2 + 6x$, we might first rewrite each term using the common factor $2x$: $2x^2 + 6x = 2x \cdot x + 2x \cdot 3$, then apply the distributive property to obtain $2x(x + 3)$. We commonly say that we have *factored out $2x$*. The **greatest common factor** (or GCF) is the largest factor common to *all* terms in the polynomial.

EXAMPLE 1 ▶ **Factoring Polynomials**

Factor each polynomial:

    a. $12x^2 + 18xy - 30y$ **b.** $x^5 + x^2$

Solution ▶   **a.** 6 is common to all three terms:

$$12x^2 + 18xy - 30y \qquad \text{mentally: } 6 \cdot 2x^2 + 6 \cdot 3xy - 6 \cdot 5y$$
$$= 6(2x^2 + 3xy - 5y)$$

    b. x^2 is common to both terms:

$$x^5 + x^2 \qquad \text{mentally: } x^2 \cdot x^3 + x^2 \cdot 1$$
$$= x^2(x^3 + 1)$$

☑ **A.** You've just reviewed how to factor out the greatest common factor

Now try Exercises 7 and 8 ▶

B. Common Binomial Factors and Factoring by Grouping

If the terms of a polynomial have a **common *binomial* factor,** it can also be factored out using the distributive property.

EXAMPLE 2 ▶ **Factoring Out a Common Binomial Factor**

Factor:

    a. $(x + 3)x^2 + (x + 3)5$ **b.** $x^2(x - 2) - 3(x - 2)$

Solution ▶   **a.** $(x + 3)x^2 + (x + 3)5$ **b.** $x^2(x - 2) - 3(x - 2)$
$$\qquad = (x + 3)(x^2 + 5) \qquad\qquad = (x - 2)(x^2 - 3)$$

Now try Exercises 9 and 10 ▶

One application of removing a binomial factor involves **factoring by grouping.** At first glance, the expression $x^3 + 2x^2 + 3x + 6$ appears unfactorable. But by grouping the terms (applying the associative property), we can remove a monomial factor from each subgroup, which then reveals a common binomial factor.

EXAMPLE 3 ▶ Factoring by Grouping

Factor $3t^3 + 15t^2 - 6t - 30$.

Solution ▶ Notice that all four terms have a common factor of 3. Begin by factoring it out.

$$3t^3 + 15t^2 - 6t - 30 \qquad \text{original polynomial}$$
$$= 3(t^3 + 5t^2 - 2t - 10) \qquad \text{factor out 3}$$
$$= 3(\underline{t^3 + 5t^2} - \underline{2t - 10}) \qquad \text{group remaining terms}$$
$$= 3[t^2(t + 5) - 2(t + 5)] \qquad \text{factor common } \textit{monomial}$$
$$= 3(t + 5)(t^2 - 2) \qquad \text{factor common } \textit{binomial}$$

Now try Exercises 11 and 12 ▶

☑ **B.** You've just reviewed how to factor by grouping

When asked to factor an expression, first look for common factors. The resulting expression will be easier to work with and help ensure the final answer is written in **completely factored form.** If a four-term polynomial cannot be factored as written, try rearranging the terms to find a combination that enables factoring by grouping.

C. Factoring Quadratic Polynomials

A quadratic polynomial is one that can be written in the form $ax^2 + bx + c$, where a, $b, c \in \mathbb{R}$ and $a \neq 0$. One common form of factoring involves quadratic trinomials such as $x^2 + 7x + 10$ and $2x^2 - 13x + 15$. While we know $(x + 5)(x + 2) = x^2 + 7x + 10$ and $(2x - 3)(x - 5) = 2x^2 - 13x + 15$ using F-O-I-L, how can we factor these trinomials without seeing the original problem in advance? First, it helps to place the trinomials in two families—those with a leading coefficient of 1 and those with a leading coefficient other than 1.

WORTHY OF NOTE

Similarly, a cubic polynomial is one of the form $ax^3 + bx^2 + cx + d$. It's helpful to note that a cubic polynomial can be factored by grouping only when $ad = bc$, where a, b, c, and d are the coefficients shown. This is easily seen in Example 3, where $(3)(-30) = (15)(-6)$ gives $-90 = -90✓$.

$ax^2 + bx + c$, where $a = 1$

When $a = 1$, the only factor pair for x^2 (other than $1 \cdot x^2$) is $x \cdot x$ and the first term in each binomial will be x: $(x \quad)(x \quad)$. The following observation helps guide us to the complete factorization. Consider the product $(x + b)(x + a)$:

$$(x + b)(x + a) = x^2 + ax + bx + ab \qquad \text{F-O-I-L}$$
$$= x^2 + (a + b)x + ab \qquad \text{distributive property}$$

Note the last term is the product ab (the *lasts*), while the coefficient of the middle term is $a + b$ (the sum of the *outers* and *inners*). Since the last term of $x^2 - 8x + 7$ is 7 and the coefficient of the middle term is -8, we are seeking two numbers with a product of positive 7 and a sum of negative 8. The numbers are -7 and -1, so the factored form is $(x - 7)(x - 1)$. It is also helpful to note that if the constant term is positive, the binomials will have *like* signs, since only *the product of like signs is positive.* If the constant term is negative, the binomials will have *unlike* signs, since only *the product of unlike signs is negative.* This means we can use the sign of the linear term (the term with degree 1) to guide our choice of factors.

Factoring Trinomials with a Leading Coefficient of 1

If the constant term is positive, the binomials will have *like* signs:

$$(x + \quad)(x + \quad) \text{ or } (x - \quad)(x - \quad),$$

to match the sign of the linear (middle) term.

If the constant term is negative, the binomials will have *unlike* signs:

$$(x + \quad)(x - \quad),$$

with the larger factor placed in the binomial whose sign *matches* the linear (middle) term.

EXAMPLE 4 ▶ **Factoring Trinomials**

Factor these expressions:

a. $-x^2 + 11x - 24$ **b.** $x^2 - 10 - 3x$

Solution ▶ **a.** First rewrite the trinomial in standard form as $-1(x^2 - 11x + 24)$. For $x^2 - 11x + 24$, the constant term is positive so the binomials will have like signs. Since the linear term is negative,

$$-1(x^2 - 11x + 24) = -1(x - \quad)(x - \quad) \quad \text{like signs, both negative}$$
$$= -1(x - 8)(x - 3) \quad (-8)(-3) = 24; -8 + (-3) = -11$$

b. First rewrite the trinomial in standard form as $x^2 - 3x - 10$. The constant term is negative so the binomials will have unlike signs. Since the linear term is negative,

$$x^2 - 3x - 10 = (x + \quad)(x - \quad) \quad \begin{array}{l} \text{unlike signs, one positive and one negative} \\ 5 > 2, 5 \text{ is placed in the second binomial;} \end{array}$$
$$= (x + 2)(x - 5) \quad (2)(-5) = -10; 2 + (-5) = -3$$

Now try Exercises 13 and 14 ▶

Sometimes we encounter **prime polynomials,** or polynomials that cannot be factored. For $x^2 + 9x + 15$, the factor pairs of 15 are $1 \cdot 15$ and $3 \cdot 5$, with neither pair having a sum of $+9$. We conclude that $x^2 + 9x + 15$ is prime.

$ax^2 + bx + c$, where $a \neq 1$

If the leading coefficient is not one, the possible combinations of outers and inners are more numerous. Furthermore, their sum will change depending on the position of the possible factors. Note that $(2x + 3)(x + 9) = 2x^2 + 21x + 27$ and $(2x + 9)(x + 3) = 2x^2 + 15x + 27$ result in a different middle term, even though identical numbers were used.

To factor $2x^2 - 13x + 15$, note the constant term is positive so the binomials *must have like signs*. The negative linear term indicates these signs will be negative. We then list possible factors for the first and last terms of each binomial, then sum the outer and inner products.

Possible First and Last Terms for $2x^2$ and 15	Sum of Outers and Inners
1. $(2x - 1)(x - 15)$	$-30x - 1x = -31x$
2. $(2x - 15)(x - 1)$	$-2x - 15x = -17x$
3. $(2x - 3)(x - 5)$	$-10x - 3x = -13x$ ←
4. $(2x - 5)(x - 3)$	$-6x - 5x = -11x$

As you can see, only possibility 3 yields a linear term of $-13x$, and the correct factorization is then $(2x - 3)(x - 5)$. With practice, this **trial-and-error** process can be completed very quickly.

If the constant term is negative, the number of possibilities can be reduced by finding a factor pair with a sum *or* difference equal to the *absolute value* of the linear coefficient, as we can then arrange the sign of each binomial to obtain the needed result (see Example 5).

EXAMPLE 5 ▶ Factoring a Trinomial Using Trial and Error

Factor $6z^2 - 11z - 35$.

Solution ▶ Note the constant term is negative (binomials will have unlike signs), $|-11| = 11$, and the factors of 35 are $1 \cdot 35$ and $5 \cdot 7$. Two possible first terms are: $(6z\quad)(z\quad)$ and $(3z\quad)(2z\quad)$, and we begin with 5 and 7 as factors of 35.

$(6z\quad)(z\quad)$	Outers/Inners		$(3z\quad)(2z\quad)$	Outers/Inners	
	Sum	Diff		Sum	Diff
1. $(6z\quad5)(z\quad7)$	$47z$	$37z$	3. $(3z\quad5)(2z\quad7)$	$31z$	$11z$ ←
2. $(6z\quad7)(z\quad5)$	$37z$	$23z$	4. $(3z\quad7)(2z\quad5)$	$29z$	$1z$

☑ **C.** You've just reviewed how to factor quadratic polynomials

Since possibility 3 yields the linear term of $11z$, we need not consider other factors of 35 and write the factored form as $6z^2 - 11z - 35 = (3z\quad5)(2z\quad7)$. The signs can then be arranged to obtain a middle term of $-11z$: $(3z + 5)(2z - 7)$, $-21z + 10z = -11z$ ✓.

Now try Exercises 15 and 16 ▶

WORTHY OF NOTE

In an attempt to factor a *sum* of two perfect squares, say $v^2 + 49$, let's list all possible binomial factors. These are (1) $(v + 7)(v + 7)$, (2) $(v - 7)(v - 7)$, and (3) $(v + 7)(v - 7)$. Note that (1) and (2) are the binomial squares $(v + 7)^2$ and $(v - 7)^2$, with each product resulting in a "middle" term, whereas (3) is a binomial times its conjugate, resulting in a *difference* of squares: $v^2 - 49$. With all possibilities exhausted, we conclude that *the sum of two squares is prime!*

D. Factoring Special Forms and Quadratic Forms

Next we consider methods to factor each of the special products we encountered in Section R.3.

The Difference of Two Squares

Multiplying and factoring are inverse processes. Since $(x - 7)(x + 7) = x^2 - 49$, we know that $x^2 - 49 = (x - 7)(x + 7)$. In words, *the difference of two squares will factor into a binomial and its conjugate*. To find the terms of the factored form, rewrite each term in the original expression as a square: $(\quad)^2$.

Factoring the Difference of Two Perfect Squares

Given any expression that can be written in the form $A^2 - B^2$,
$$A^2 - B^2 = (A + B)(A - B)$$

Note that the *sum* of two perfect squares $A^2 + B^2$ *cannot be factored* using real numbers (the expression is prime). As a reminder, always check for a common factor first and be sure to write all results in completely factored form. See Example 6(c).

EXAMPLE 6 ▶ Factoring the Difference of Two Perfect Squares

Factor each expression completely.

a. $4w^2 - 81$ **b.** $v^2 + 49$ **c.** $-3n^2 + 48$ **d.** $z^4 - \frac{1}{81}$ **e.** $x^2 - 7$

Solution ▶ **a.** $4w^2 - 81 = (2w)^2 - 9^2$ write as a difference of squares
$\qquad\qquad = (2w + 9)(2w - 9)$ $A^2 - B^2 = (A + B)(A - B)$

b. $v^2 + 49$ is prime.

c. $-3n^2 + 48 = -3(n^2 - 16)$ factor out -3
$\qquad\qquad = -3[n^2 - (4)^2]$ write as a difference of squares
$\qquad\qquad = -3(n + 4)(n - 4)$ $A^2 - B^2 = (A + B)(A - B)$

d. $z^4 - \frac{1}{81} = (z^2)^2 - \left(\frac{1}{9}\right)^2$ write as a difference of squares

$\quad\quad\quad\quad = \left(z^2 + \frac{1}{9}\right)\left(z^2 - \frac{1}{9}\right)$ $A^2 - B^2 = (A + B)(A - B)$

$\quad\quad\quad\quad = \left[z^2 - \left(\frac{1}{3}\right)^2\right]\left(z^2 + \frac{1}{9}\right)$ write as a difference of squares

$\quad\quad\quad\quad = \left(z + \frac{1}{3}\right)\left(z - \frac{1}{3}\right)\left(z^2 + \frac{1}{9}\right)$ result

e. $x^2 - 7 = (x)^2 - (\sqrt{7})^2$ write as a difference of squares

$\quad\quad\quad\quad = (x + \sqrt{7})(x - \sqrt{7})$ $A^2 - B^2 = (A + B)(A - B)$

Now try Exercises 17 and 18 ▶

Perfect Square Trinomials

Since $(x + 7)^2 = x^2 + 14x + 49$, we know that $x^2 + 14x + 49 = (x + 7)^2$. In words, *a perfect square trinomial will factor into a binomial square.* To use this idea effectively, we must learn to *identify* perfect square trinomials. Note that the first and last terms of $x^2 + 14x + 49$ are *the squares* of x and 7, and the middle term is *twice the product of these two terms*: $2(7x) = 14x$. These are the characteristics of a perfect square trinomial.

Factoring Perfect Square Trinomials

Given any expression that can be written in the form $A^2 \pm 2AB + B^2$,
  1. $A^2 + 2AB + B^2 = (A + B)^2$
  2. $A^2 - 2AB + B^2 = (A - B)^2$

EXAMPLE 7 ▶   **Factoring a Perfect Square Trinomial**

Factor $12m^3 - 12m^2 + 3m$.

Solution ▶ $12m^3 - 12m^2 + 3m$ check for common factors: GCF = $3m$

$\quad\quad = 3m(4m^2 - 4m + 1)$ factor out $3m$

For the remaining trinomial $4m^2 - 4m + 1 \ldots$

  1. Are the first and last terms perfect squares?

$$4m^2 = (2m)^2 \text{ and } 1 = (1)^2 \ \checkmark \quad \text{Yes.}$$

  2. Is the linear term twice the product of $2m$ and 1?

$$2 \cdot 2m \cdot 1 = 4m \ \checkmark \quad \text{Yes.}$$

Factor as a binomial square: $4m^2 - 4m + 1 = (2m - 1)^2$

This shows $12m^3 - 12m^2 + 3m = 3m(2m - 1)^2$.

Now try Exercises 19 and 20 ▶

⚠ **CAUTION ▶** As shown in Example 7, be sure to include the GCF in your final answer. It is a common error to "leave the GCF behind."

In actual practice, the tests for a perfect square trinomial are performed mentally, with only the factored form being written down.

Sum or Difference of Two Perfect Cubes

Recall that the *difference* of two perfect squares is factorable, but the *sum* of two perfect squares is prime. In contrast, *both the sum and difference of two perfect* cubes *are factorable.* For either $A^3 + B^3$ or $A^3 - B^3$ we have the following:

1. Each will factor into the product of a binomial and a trinomial:

$$(\underset{\text{binomial}}{\quad})(\underset{\text{trinomial}}{\qquad})$$

2. The terms of the binomial are the quantities being cubed:

$$(A \quad B)(\qquad)$$

3. The terms of the trinomial are the square of A, the product AB, and the square of B, respectively:

$$(A \quad B)(A^2 \quad AB \quad B^2)$$

4. The binomial takes the same sign as the original expression

$$(A \pm B)(A^2 \quad AB \quad B^2)$$

5. The middle term of the trinomial takes the opposite sign of the original exercise (the last term is always positive):

$$(A \pm B)(A^2 \mp AB + B^2)$$

Factoring the Sum or Difference of Two Perfect Cubes: $A^3 \pm B^3$

$$A^3 + B^3 = (A + B)(A^2 - AB + B^2)$$
$$A^3 - B^3 = (A - B)(A^2 + AB + B^2)$$

EXAMPLE 8 ▶ **Factoring the Sum and Difference of Two Perfect Cubes**

Factor completely:

a. $x^3 + 125$ **b.** $-5m^3n + 40n^4$

Solution ▶ **a.** $\quad x^3 + 125 = x^3 + 5^3$ write terms as perfect cubes

Use $A^3 + B^3 = (A + B)(A^2 - AB + B^2)$ factoring template

$\quad x^3 + 5^3 = (x + 5)(x^2 - 5x + 25)$ $A \to x$ and $B \to 5$

b. $\quad -5m^3n + 40n^4 = -5n(m^3 - 8n^3)$ check for common factors (GCF = $-5n$)

$\quad = -5n[m^3 - (2n)^3]$ write terms as perfect cubes

Use $A^3 - B^3 = (A - B)(A^2 + AB + B^2)$ factoring template

$m^3 - (2n)^3 = (m - 2n)[m^2 + m(2n) + (2n)^2]$ $A \to m$ and $B \to 2n$

$\quad = (m - 2n)(m^2 + 2mn + 4n^2)$ simplify

$\Rightarrow -5m^3n + 40n^4 = -5n(m - 2n)(m^2 + 2mn + 4n^2).$ factored form

The results for parts (a) and (b) can be checked using multiplication.

Now try Exercises 21 and 22 ▶

Using *u*-Substitution to Factor Quadratic Forms

For any quadratic expression $ax^2 + bx + c$ in standard form, the degree of the leading term is twice the degree of the middle term. Generally, a trinomial is in **quadratic form** if it can be written as $a(\_\_)^2 + b(\_\_) + c$, where the parentheses "hold" the same factors. The equation $x^4 - 13x^2 + 36 = 0$ is in quadratic form since $(x^2)^2 - 13(x^2) + 36 = 0$. In many cases, we can factor these expressions using a **placeholder substitution** that transforms these expressions into a more recognizable form. In a study of algebra, the letter "u" often plays this role. If we let u represent x^2, the expression $(x^2)^2 - 13(x^2) + 36$ becomes $u^2 - 13u + 36$, which can be factored into $(u - 9)(u - 4)$. After "unsubstituting" (replace u with x^2), we have $(x^2 - 9)(x^2 - 4) = (x + 3)(x - 3)(x + 2)(x - 2)$.

EXAMPLE 9 ▶ **Factoring a Quadratic Form**

Write in completely factored form: $(x^2 - 2x)^2 - 2(x^2 - 2x) - 3$.

Solution ▶ Expanding the binomials would produce a fourth-degree polynomial that would be very difficult to factor. Instead we note the expression is in *quadratic form*. Letting u represent $x^2 - 2x$ (the variable part of the "middle" term), $(x^2 - 2x)^2 - 2(x^2 - 2x) - 3$ becomes $u^2 - 2u - 3$.

$$u^2 - 2u - 3 = (u - 3)(u + 1) \quad \text{\small factor}$$

To finish up, write the expression in terms of x, substituting $x^2 - 2x$ for u.

$$= (x^2 - 2x - 3)(x^2 - 2x + 1) \quad \text{\small substitute } x^2 - 2x \text{ for } u$$

The resulting trinomials can be further factored.

$$= (x - 3)(x + 1)(x - 1)^2 \quad \text{\small } x^2 - 2x + 1 = (x - 1)^2$$

Now try Exercises 23 and 24 ▶

☑ **D. You've just reviewed how to factor special forms and quadratic forms**

It is well known that information is retained longer and used more effectively when it's placed in an organized form. The "factoring flowchart" provided in Figure R.5 offers a streamlined and systematic approach to factoring and the concepts involved. However, with some practice the process tends to "flow" more naturally than following a chart, with many of the decisions becoming automatic.

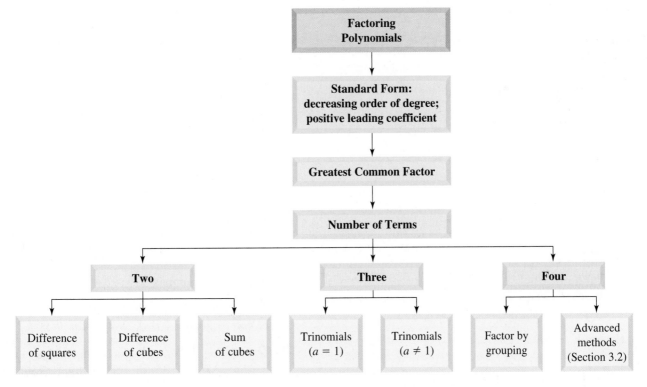

• Can any result be factored further?   • Polynomials that cannot be factored are said to be *prime*.

Figure R.5

R.4 EXERCISES

▶ CONCEPTS AND VOCABULARY

Fill in each blank with the appropriate word or phrase. Carefully reread the section, if necessary.

1. To factor an expression means to rewrite the expression as an equivalent _____.

2. If a polynomial will not factor, it is said to be a(n) _____ polynomial.

3. The difference of two perfect squares always factors into the product of a(n) _____ and its _____.

4. The expression $x^2 + 6x + 9$ is said to be a(n) _____ _____ trinomial, since its factored form is a perfect (binomial) square.

5. Discuss/Explain why $4x^2 - 36 = (2x - 6)(2x + 6)$ is not written in completely factored form, then rewrite it so it is factored completely.

6. Discuss/Explain why $a^3 + b^3$ is factorable, but $a^2 + b^2$ is not. Demonstrate by writing $x^3 + 64$ in factored form, and by exhausting all possibilities for $x^2 + 64$ to show it is prime.

▶ DEVELOPING YOUR SKILLS

Factor each expression using the method indicated.

Greatest Common Factor

7. **a.** $-17x^2 + 51$ **b.** $21b^3 - 14b^2 + 56b$
   c. $-3a^4 + 9a^2 - 6a^3$

8. **a.** $-13n^2 - 52$ **b.** $9p^2 + 27p^3 - 18p^4$
   c. $-6g^5 + 12g^4 - 9g^3$

Common Binomial Factor

9. **a.** $2a(a + 2) + 3(a + 2)$
   b. $(b^2 + 3)3b + (b^2 + 3)2$
   c. $4m(n + 7) - 11(n + 7)$

10. **a.** $5x(x - 3) - 2(x - 3)$
    b. $(v - 5)2v + (v - 5)3$
    c. $3p(q^2 + 5) + 7(q^2 + 5)$

Grouping

11. **a.** $9q^3 + 6q^2 + 15q + 10$
    b. $h^5 - 12h^4 - 3h + 36$
    c. $k^5 - 7k^3 - 5k^2 + 35$

12. **a.** $6h^3 - 9h^2 - 2h + 3$
    b. $4k^3 + 6k^2 - 2k - 3$
    c. $3x^2 - xy - 6x + 2y$

Trinomial Factoring where $|a| = 1$

13. **a.** $-p^2 + 5p + 14$ **b.** $q^2 - 4q - 45$
    c. $n^2 + 20 - 9n$

14. **a.** $-m^2 + 13m - 42$ **b.** $x^2 + 12 + 13x$
    c. $v^2 + 10v + 15$

Trinomial Factoring where $a \neq 1$

15. **a.** $3p^2 - 13p - 10$ **b.** $4q^2 + 7q - 15$
    c. $10u^2 - 19u - 15$

16. **a.** $6v^2 + v - 35$ **b.** $20x^2 + 53x + 18$
    c. $15z^2 - 22z - 48$

Difference of Perfect Squares

17. **a.** $4s^2 - 25$ **b.** $9x^2 - 49$
    c. $50x^2 - 72$ **d.** $121h^2 - 144$
    e. $b^2 - 5$

18. **a.** $9v^2 - \frac{1}{25}$ **b.** $25w^2 - \frac{1}{49}$
    c. $v^4 - 1$ **d.** $16z^4 - 81$
    e. $x^2 - 17$

Perfect Square Trinomials

19. **a.** $a^2 - 6a + 9$ **b.** $b^2 + 10b + 25$
    c. $4m^2 - 20m + 25$ **d.** $9n^2 - 42n + 49$

20. **a.** $x^2 + 12x + 36$ **b.** $z^2 - 18z + 81$
    c. $25p^2 - 60p + 36$ **d.** $16q^2 + 40q + 25$

Sum/Difference of Perfect Cubes

21. **a.** $8p^3 - 27$ **b.** $m^3 + \frac{1}{8}$
    c. $g^3 - 0.027$ **d.** $-2t^4 + 54t$

22. a. $27q^3 - 125$ **b.** $n^3 + \frac{8}{27}$

     c. $b^3 - 0.125$ **d.** $3r^4 - 24r$

u-Substitution

23. a. $x^4 - 10x^2 + 9$ **b.** $x^4 + 13x^2 + 36$

     c. $x^6 - 7x^3 - 8$

24. a. $x^6 - 26x^3 - 27$

     b. $3(n + 5)^2 + (2n + 10) - 21$

     c. $2(z + 3)^2 + (3z + 9) - 54$

25. Completely factor each of the following (recall that "1" is its own perfect square and perfect cube).

     a. $n^2 - 1$ **b.** $n^3 - 1$

     c. $n^3 + 1$ **d.** $28x^3 - 7x$

26. Carefully factor each of the following trinomials, if possible. Note differences and similarities.

     a. $x^2 - x + 6$ **b.** $x^2 + x - 6$

     c. $x^2 + x + 6$ **d.** $x^2 - x - 6$

     e. $x^2 - 5x + 6$ **f.** $x^2 + 5x - 6$

Factor each expression completely, if possible. Rewrite the expression in standard form (factor out "−1" if needed) and factor out the GCF if one exists. If you believe the expression will not factor, write "prime."

27. $a^2 + 7a + 10$ **28.** $b^2 + 9b + 20$

29. $2x^2 - 24x + 40$ **30.** $10z^2 - 140z + 450$

31. $64 - 9m^2$ **32.** $25 - 16n^2$

33. $-9r + r^2 + 18$ **34.** $28 + s^2 - 11s$

35. $2h^2 + 7h + 6$ **36.** $3k^2 + 10k + 8$

37. $9k^2 - 24k + 16$ **38.** $4p^2 - 20p + 25$

39. $-6x^3 + 39x^2 - 63x$ **40.** $-28z^3 + 16z^2 + 80z$

41. $12m^2 - 40m + 4m^3$ **42.** $-30n - 4n^2 + 2n^3$

43. $a^2 - 7a - 60$ **44.** $b^2 - 9b - 36$

45. $8x^3 - 125$ **46.** $27r^3 + 64$

47. $m^2 + 9m - 24$ **48.** $n^2 - 14n - 36$

49. $x^3 - 5x^2 - 9x + 45$ **50.** $x^3 + 3x^2 - 4x - 12$

51. Match each expression with the description that fits *best*.

     ____ **a.** prime polynomial

     ____ **b.** standard trinomial $a = 1$

     ____ **c.** perfect square trinomial

     ____ **d.** difference of cubes

     ____ **e.** binomial square

     ____ **f.** sum of cubes

     ____ **g.** binomial conjugates

     ____ **h.** difference of squares

     ____ **i.** standard trinomial $a \neq 1$

     A. $x^3 + 27$ **B.** $(x + 3)^2$

     C. $x^2 - 10x + 25$ **D.** $x^2 - 144$

     E. $x^2 - 3x - 10$ **F.** $8s^3 - 125t^3$

     G. $2x^2 - x - 3$ **H.** $x^2 + 9$

     I. $(x - 7)$ and $(x + 7)$

52. Match each polynomial to its factored form. Two of them are prime.

     ____ **a.** $4x^2 - 9$

     ____ **b.** $4x^2 - 28x + 49$

     ____ **c.** $x^3 - 125$

     ____ **d.** $8x^3 + 27$

     ____ **e.** $x^2 - 3x - 10$

     ____ **f.** $x^2 + 3x + 10$

     ____ **g.** $2x^2 - x - 3$

     ____ **h.** $2x^2 + x - 3$

     ____ **i.** $x^2 + 25$

     A. $(x - 5)(x^2 + 5x + 25)$

     B. $(2x - 3)(x + 1)$

     C. $(2x + 3)(2x - 3)$

     D. $(2x - 7)^2$

     E. prime trinomial

     F. prime binomial

     G. $(2x + 3)(x - 1)$

     H. $(2x + 3)(4x^2 - 6x + 9)$

     I. $(x - 5)(x + 2)$

▶ WORKING WITH FORMULAS

53. Surface area of a cylinder: $2\pi r^2 + 2\pi rh$

The surface area of a cylinder is given by the formula shown, where h is the height of the cylinder and r is the radius. Factor out the GCF and use the result to find the surface area of a cylinder where $r = 35$ cm and $h = 65$ cm. Answer in exact form and in approximate form rounded to the nearest whole number.

 54. Volume of a cylindrical shell: $\pi R^2 h - \pi r^2 h$

The volume of a cylindrical shell (a larger cylinder with a smaller cylinder removed) can be found using the formula shown, where R is the radius of the larger cylinder and r is the radius of the smaller. Factor the expression completely and use the result to find the volume of a shell where $R = 9$ cm, $r = 3$ cm, and $h = 10$ cm (use $\pi \approx 3.14$).

▶ APPLICATIONS

In many cases, factoring an expression can make it easier to evaluate as in the following applications.

55. Conical shells: The volume of a conical shell (like the shell of an ice cream cone) is given by the formula $V = \dfrac{1}{3}\pi R^2 h - \dfrac{1}{3}\pi r^2 h$, where R is the outer radius and r is the inner radius of the cone. Write the formula in completely factored form, then find the volume of a shell when $R = 5.1$ cm, $r = 4.9$ cm, and $h = 9$ cm. Answer in exact form and in approximate form rounded to the nearest tenth.

56. Spherical shells: The volume of a spherical shell (like the outer shell of a cherry cordial) is given by the formula $V = \dfrac{4}{3}\pi R^3 - \dfrac{4}{3}\pi r^3$, where R is the outer radius and r is the inner radius of the shell. Write the right-hand side in completely factored form, then find the volume of a shell where $R = 1.8$ cm and $r = 1.5$ cm.

57. Volume of a box: The volume of a rectangular box x inches in height is given by the relationship $V = x^3 + 8x^2 + 15x$. Factor the right-hand side to determine: (a) The number of inches that the width exceeds the height, (b) the number of inches the length exceeds the height, and (c) the volume given the height is 2 ft.

58. Shipping textbooks: A publisher ships paperback books stacked x copies high in a box. The total number of books shipped per box is given by the relationship $B = x^3 - 13x^2 + 42x$. Factor the right-hand side to determine (a) how many more

or fewer books fit the width of the box (than the height), (b) how many more or fewer books fit the length of the box (than the height), and (c) the number of books shipped per box if they are stacked 10 high in the box.

59. Space-Time relationships: Due to the work of Albert Einstein and other physicists who labored on space-time relationships, it is known that the faster an object moves the shorter it appears to become. This phenomenon is modeled by the **Lorentz transformation** $L = L_0\sqrt{1 - \left(\dfrac{v}{c}\right)^2}$, where L_0 is the length of the object at rest, L is the relative length when the object is moving at velocity v, and c is the speed of light. Factor the radicand and use the result to determine the relative length of a 12-in. ruler if it is shot past a stationary observer at 0.75 times the speed of light ($v = 0.75c$).

60. Tubular fluid flow: As a fluid flows through a tube, it is flowing faster at the center of the tube than at the sides, where the tube exerts a backward drag. **Poiseuille's law** gives the velocity of the flow at any point of the cross section: $v = \dfrac{G}{4\eta}(R^2 - r^2)$, where R is the inner radius of the tube, r is the distance from the center of the tube to a point in the flow, G represents what is called the pressure gradient, and η is a constant that depends on the viscosity of the fluid. Factor the right-hand side and find v given $R = 0.5$ cm, $r = 0.3$ cm, $G = 15$, and $\eta = 0.25$.

▶ EXTENDING THE CONCEPT

61. Factor out a constant that leaves integer coefficients for each term:

    a. $\frac{1}{2}x^4 + \frac{1}{8}x^3 - \frac{3}{4}x^2 + 4$

    b. $\frac{2}{3}b^5 - \frac{1}{6}b^3 + \frac{4}{9}b^2 - 1$

62. If $x = 2$ is substituted into $2x^3 + hx + 8$, the result is zero. What is the value of h?

63. Factor the expression: $192x^3 - 164x^2 - 270x$.

64. As an alternative to evaluating polynomials by direct substitution, **nested factoring** can be used. The method has the advantage of using only products and sums—no powers. For $P = x^3 + 3x^2 + 1x + 5$, we begin by grouping all variable terms and factoring x: $P = [x^3 + 3x^2 + 1x] + 5 = x[x^2 + 3x + 1] + 5$. Then we group the inner terms with x and factor again: $P = x[x^2 + 3x + 1] + 5 = x[x(x + 3) + 1] + 5$. The expression can now be evaluated using any input and the order of operations. If $x = 2$, we quickly find that $P = 27$. Use this method to evaluate $H = x^3 + 2x^2 + 5x - 9$ for $x = -3$.

Factor each expression completely.

65. $x^4 - 81$

66. $16n^4 - 1$

67. $p^6 - 1$

68. $m^6 - 64$

69. $q^4 - 28q^2 + 75$

70. $a^4 - 18a^2 + 32$

R.5 | Rational Expressions

Learning Objectives

In Section R.5 you will learn how to:

☐ **A.** Write a rational expression in simplest form

☐ **B.** Multiply and divide rational expressions

☐ **C.** Add and subtract rational expressions

☐ **D.** Simplify compound fractions

☐ **E.** Rewrite formulas and algebraic models

A rational number is one that can be written as the quotient of two integers. Similarly, a *rational expression* is one that can be written as the quotient of two polynomials. We can apply the skills developed in a study of fractions (how to reduce, add, subtract, multiply, and divide) to **rational expressions,** sometimes called **algebraic fractions.**

A. Writing a Rational Expression in Simplest Form

A rational expression is in **simplest form** when the numerator and denominator have no common factors (other than 1). After factoring the numerator and denominator, we apply the **fundamental property of rational expressions.**

Fundamental Property of Rational Expressions

If P, Q, and R are polynomials, with $Q, R \neq 0$,

$$(1) \quad \frac{P \cdot R}{Q \cdot R} = \frac{P}{Q} \quad \text{and} \quad (2) \quad \frac{P}{Q} = \frac{P \cdot R}{Q \cdot R}$$

In words, the property says (1) a rational expression can be simplified by canceling common factors in the numerator and denominator, and (2) an equivalent expression can be formed by multiplying numerator and denominator by the same nonzero polynomial.

EXAMPLE 1 ▶ **Simplifying a Rational Expression**

Write the expression in simplest form: $\dfrac{x^2 - 1}{x^2 - 3x + 2}$.

Solution ▶

$$\dfrac{x^2 - 1}{x^2 - 3x + 2} = \dfrac{(x - 1)(x + 1)}{(x - 1)(x - 2)} \quad \text{factor numerator and denominator}$$

$$= \dfrac{\cancel{(x - 1)}(x + 1)}{\cancel{(x - 1)}(x - 2)} \quad \text{cancel common factors}$$

$$= \dfrac{x + 1}{x - 2} \quad \text{simplest form}$$

Now try Exercises 7 through 10 ▶

WORTHY OF NOTE

If we view a and b as two points on the number line, we note that they are the same distance apart, regardless of the order they are subtracted. This tells us the numerator and denominator will have the same absolute value but be opposite in sign, giving a value of -1 (check using a few test values).

When simplifying rational expressions, we sometimes encounter expressions of the form $\dfrac{a - b}{b - a}$. If we factor -1 from the numerator, we see that $\dfrac{a - b}{b - a} = \dfrac{-1\cancel{(b - a)}}{\cancel{b - a}} = -1$.

⚠ **CAUTION** ▶ When reducing rational numbers or expressions, only common *factors* can be reduced. It is incorrect to reduce (or divide out) individual terms: $\dfrac{-6 + 4\sqrt{3}}{2} \neq -3 + 4\sqrt{3}$, and $\dfrac{x + 1}{x + 2} \neq \dfrac{1}{2}$ (except for $x = 0$)

EXAMPLE 2 ▶ **Simplifying a Rational Expression**

Write the expression in simplest form: $\dfrac{(6 - 2x)}{x^2 - 9}$.

Solution ▶

$$\dfrac{(6 - 2x)}{x^2 - 9} = \dfrac{2(3 - x)}{(x - 3)(x + 3)} \quad \text{factor numerator and denominator}$$

$$= \dfrac{(2)(-1)}{x + 3} \quad \text{reduce: } \dfrac{(3 - x)}{(x - 3)} = -1$$

$$= \dfrac{-2}{x + 3} \quad \text{simplest form}$$

 A. You've just reviewed how to write a rational expression in simplest form

Now try Exercises 11 through 16 ▶

B. Multiplication and Division of Rational Expressions

Operations on rational expressions use the factoring skills reviewed earlier, along with much of what we know about rational numbers.

Multiplying Rational Expressions

Given that P, Q, R, and S are polynomials with $Q, S \neq 0$,

$$\dfrac{P}{Q} \cdot \dfrac{R}{S} = \dfrac{PR}{QS}$$

1. Factor all numerators and denominators completely.
2. Reduce common factors.
3. Multiply numerator × numerator and denominator × denominator.

EXAMPLE 3 ▶ **Multiplying Rational Expressions**

Compute the product: $\dfrac{2a + 2}{3a - 3a^2} \cdot \dfrac{3a^2 - a - 2}{9a^2 - 4}$.

Solution ▶ $\dfrac{2a + 2}{3a - 3a^2} \cdot \dfrac{3a^2 - a - 2}{9a^2 - 4} = \dfrac{2(a + 1)}{3a(1 - a)} \cdot \dfrac{(3a + 2)(a - 1)}{(3a - 2)(3a + 2)}$ factor

$= \dfrac{2(a + 1)}{3a(1 - a)} \cdot \dfrac{\cancel{(3a + 2)}(a - 1)}{(3a - 2)\cancel{(3a + 2)}}$ reduce: $\dfrac{a - 1}{1 - a} = -1$

$= \dfrac{-2(a + 1)}{3a(3a - 2)}$ simplest form

Now try Exercises 17 through 20 ▶

To divide fractions, we multiply the first expression by the *reciprocal of the second*. The quotient of two rational expressions is computed in the same way.

Dividing Rational Expressions

Given that P, Q, R, and S are polynomials with $Q, R, S \neq 0$,

$$\frac{P}{Q} \div \frac{R}{S} = \frac{P}{Q} \cdot \frac{S}{R} = \frac{PS}{QR}$$

Invert the divisor and multiply.

EXAMPLE 4 ▶ **Dividing Rational Expressions**

Compute the quotient $\dfrac{4m^3 - 12m^2 + 9m}{m^2 - 49} \div \dfrac{10m^2 - 15m}{m^2 + 4m - 21}$.

Solution ▶ $\dfrac{4m^3 - 12m^2 + 9m}{m^2 - 49} \div \dfrac{10m^2 - 15m}{m^2 + 4m - 21}$

$= \dfrac{4m^3 - 12m^2 + 9m}{m^2 - 49} \cdot \dfrac{m^2 + 4m - 21}{10m^2 - 15m}$ invert and multiply

$= \dfrac{m(4m^2 - 12m + 9)}{(m + 7)(m - 7)} \cdot \dfrac{(m + 7)(m - 3)}{5m(2m - 3)}$ factor

$= \dfrac{\cancel{m}(2m - 3)\cancel{(2m - 3)}}{\cancel{(m + 7)}(m - 7)} \cdot \dfrac{\cancel{(m + 7)}(m - 3)}{5\cancel{m}\cancel{(2m - 3)}}$ factor and reduce

$= \dfrac{(2m - 3)(m - 3)}{5(m - 7)}$ lowest terms

Note that we sometimes refer to simplest form as *lowest terms*.

Now try Exercises 21 through 42 ▶

⚠ CAUTION ▶ For products like $\dfrac{(w+7)(w-7)}{(w-7)(w-2)} \cdot \dfrac{(w-2)}{(w+7)}$, it is a common mistake to think that all factors "cancel," leaving an answer of zero. Actually, all factors *reduce to 1*, and the result is a value of 1 for all inputs where the product is defined.

☑ **B.** You've just reviewed how to multiply and divide rational expressions

$$\dfrac{\overset{1}{(w+7)}\overset{1}{(w-7)}}{\underset{1}{(w-7)}\underset{1}{(w-2)}} \cdot \dfrac{\overset{1}{(w-2)}}{\underset{1}{(w+7)}} = 1$$

C. Addition and Subtraction of Rational Expressions

Recall that the addition and subtraction of *fractions* requires finding the lowest common denominator (LCD) and building equivalent fractions. The sum or difference of the numerators is then placed over this denominator. The procedure for the addition and subtraction of *rational expressions* is very much the same.

Addition and Subtraction of Rational Expressions

1. Find the LCD of all rational expressions.
2. Build equivalent expressions using the LCD.
3. Add or subtract numerators as indicated.
4. Write the result in lowest terms.

EXAMPLE 5 ▶ Adding and Subtracting Rational Expressions

Compute as indicated:

a. $\dfrac{7}{10x} + \dfrac{3}{25x^2}$ **b.** $\dfrac{10x}{x^2-9} - \dfrac{5}{x-3}$

Solution ▶ a. The LCD for $10x$ and $25x^2$ is $50x^2$. find the LCD

$$\dfrac{7}{10x} + \dfrac{3}{25x^2} = \dfrac{7}{10x}\cdot\dfrac{(5x)}{(5x)} + \dfrac{3}{25x^2}\cdot\dfrac{(2)}{(2)}$$ write equivalent expressions

$$= \dfrac{35x}{50x^2} + \dfrac{6}{50x^2}$$ simplify

$$= \dfrac{35x+6}{50x^2}$$ add the numerators and write the result over the LCD

The result is in simplest form.

b. The LCD for x^2-9 and $x-3$ is $(x-3)(x+3)$. find the LCD

$$\dfrac{10x}{x^2-9} - \dfrac{5}{x-3} = \dfrac{10x}{(x-3)(x+3)} - \dfrac{5}{x-3}\cdot\dfrac{(x+3)}{(x+3)}$$ write equivalent expressions

$$= \dfrac{10x-5(x+3)}{(x-3)(x+3)}$$ subtract numerators, write the result over the LCD

$$= \dfrac{10x-5x-15}{(x-3)(x+3)}$$ distribute

$$= \dfrac{5x-15}{(x-3)(x+3)}$$ combine like terms

$$= \dfrac{5\overset{1}{(x-3)}}{\underset{1}{(x-3)}(x+3)} = \dfrac{5}{x+3}$$ factor and reduce

Now try Exercises 43 through 48 ▶

EXAMPLE 6 ▶ **Adding and Subtracting Rational Expressions**

Perform the operations indicated:

a. $\dfrac{5}{n+2} - \dfrac{n-3}{n^2-4}$

b. $\dfrac{b^2}{4a^2} - \dfrac{c}{a}$

Solution ▶ a. The LCD for $n+2$ and n^2-4 is $(n+2)(n-2)$.

$$\dfrac{5}{n+2} - \dfrac{n-3}{n^2-4} = \dfrac{5}{(n+2)} \cdot \dfrac{(n-2)}{(n-2)} - \dfrac{n-3}{(n+2)(n-2)}$$ write equivalent expressions

$$= \dfrac{5(n-2) - (n-3)}{(n+2)(n-2)}$$ subtract numerators, write the result over the LCD

$$= \dfrac{5n - 10 - n + 3}{(n+2)(n-2)}$$ distribute

$$= \dfrac{4n - 7}{(n+2)(n-2)}$$ result

b. The LCD for a and $4a^2$ is $4a^2$: $\dfrac{b^2}{4a^2} - \dfrac{c}{a} = \dfrac{b^2}{4a^2} - \dfrac{c}{a} \cdot \dfrac{(4a)}{(4a)}$ write equivalent expressions

$$= \dfrac{b^2}{4a^2} - \dfrac{4ac}{4a^2}$$ simplify

$$= \dfrac{b^2 - 4ac}{4a^2}$$ subtract numerators, write the result over the LCD

 C. You've just reviewed how to add and subtract rational expressions

Now try Exercises 49 through 64 ▶

⚠ **CAUTION** ▶ When the second term in a subtraction has a binomial numerator as in Example 6(a), be sure the subtraction *is applied to both terms*. It is a common error to write $\dfrac{5(n-2)}{(n+2)(n-2)} - \dfrac{n-3}{(n+2)(n-2)} = \dfrac{5n - 10 \boxed{-n-3}}{(n+2)(n-2)}$✗ in which the subtraction is applied to the first term only. This is incorrect!

D. Simplifying Compound Fractions

Rational expressions whose numerator or denominator contain a fraction are called **compound fractions.** The expression $\dfrac{\dfrac{2}{3m} - \dfrac{3}{2}}{\dfrac{3}{4m} - \dfrac{1}{3m^2}}$ is a compound fraction with a numerator of $\dfrac{2}{3m} - \dfrac{3}{2}$ and a denominator of $\dfrac{3}{4m} - \dfrac{1}{3m^2}$. The two methods commonly used to simplify compound fractions are summarized in the following boxes.

Simplifying Compound Fractions (Method I)

1. Add/subtract fractions in the numerator, writing them as a single expression.
2. Add/subtract fractions in the denominator, also writing them as a single expression.
3. Multiply the numerator by the reciprocal of the denominator and simplify if possible.

> **Simplifying Compound Fractions (Method II)**
>
> 1. Find the LCD of all fractions in the numerator and denominator.
> 2. Multiply the numerator and denominator by this LCD and simplify.
> 3. Simplify further if possible.

Method II is illustrated in Example 7.

EXAMPLE 7 ▶ **Simplifying a Compound Fraction**

Simplify the compound fraction:

$$\frac{\dfrac{2}{3m} - \dfrac{3}{2}}{\dfrac{3}{4m} - \dfrac{1}{3m^2}}$$

Solution ▶ The LCD for all fractions is $12m^2$.

$$\frac{\dfrac{2}{3m} - \dfrac{3}{2}}{\dfrac{3}{4m} - \dfrac{1}{3m^2}} = \frac{\left(\dfrac{2}{3m} - \dfrac{3}{2}\right)\left(\dfrac{12m^2}{1}\right)}{\left(\dfrac{3}{4m} - \dfrac{1}{3m^2}\right)\left(\dfrac{12m^2}{1}\right)}$$

multiply numerator and denominator by $12m^2 = \dfrac{12m^2}{1}$

$$= \frac{\left(\dfrac{2}{3m}\right)\left(\dfrac{12m^2}{1}\right) - \left(\dfrac{3}{2}\right)\left(\dfrac{12m^2}{1}\right)}{\left(\dfrac{3}{4m}\right)\left(\dfrac{12m^2}{1}\right) - \left(\dfrac{1}{3m^2}\right)\left(\dfrac{12m^2}{1}\right)}$$

distribute

$$= \frac{8m - 18m^2}{9m - 4}$$

simplify

$$= \frac{2m(4 \overset{-1}{\cancel{-9m}})}{\cancel{9m - 4}} = -2m$$

factor and write in lowest terms

☑ **D.** You've just reviewed how to simplify compound fractions

Now try Exercises 65 through 74 ▶

E. Rewriting Formulas and Algebraic Models

In many fields of study, formulas and algebraic models involve rational expressions and we often need to write them in an alternative form.

EXAMPLE 8 ▶ **Rewriting a Formula**

In an electrical circuit with two resistors in parallel, the total resistance R is related to resistors R_1 and R_2 by the formula $\dfrac{1}{R} = \dfrac{1}{R_1} + \dfrac{1}{R_2}$. Rewrite the right-hand side as a single term.

Solution ▶

$$\frac{1}{R} = \frac{1}{R_1} + \frac{1}{R_2}$$

LCD for the right-hand side is R_1R_2

$$= \frac{R_2}{R_1R_2} + \frac{R_1}{R_1R_2}$$

build equivalent expressions using LCD

$$= \frac{R_2 + R_1}{R_1R_2}$$

write as a single expression

Now try Exercises 75 and 76 ▶

EXAMPLE 9 ▶ **Simplifying an Algebraic Model**

When studying rational expressions and rates of change, we encounter the

expression $\dfrac{\dfrac{1}{x + h} - \dfrac{1}{x}}{h}$. Simplify the compound fraction.

Solution ▶ Using Method I gives:

$$\frac{\dfrac{1}{x + h} - \dfrac{1}{x}}{h} = \frac{\dfrac{x}{x(x + h)} - \dfrac{x + h}{x(x + h)}}{h} \qquad \text{LCD for the numerator is } x(x + h)$$

$$= \frac{\dfrac{x - (x + h)}{x(x + h)}}{h} \qquad \text{write numerator as a single expression}$$

$$= \frac{\dfrac{-h}{x(x + h)}}{h} \qquad \text{simplify}$$

$$= \frac{-h}{x(x + h)} \cdot \frac{1}{h} \qquad \text{invert and multiply}$$

$$= \frac{-1}{x(x + h)} \qquad \text{result}$$

☑ **E.** You've just reviewed how to rewrite formulas and algebraic models

Now try Exercises 77 through 80 ▶

R.5 EXERCISES

▶ **CONCEPTS AND VOCABULARY**

Fill in each blank with the appropriate word or phrase. Carefully reread the section, if necessary.

1. In simplest form, $(a - b)/(a - b)$ is equal to ____, while $(a - b)/(b - a)$ is equal to ____.

2. A rational expression is in _____ ____ when the numerator and denominator have no common factors, other than ____.

3. As with numeric fractions, algebraic fractions require a _____ _____ for addition and subtraction.

4. Since $x^2 + 9$ is prime, the expression $(x^2 + 9)/(x + 3)$ is already written in _____ ____.

State T or F and discuss/explain your response.

5. $\dfrac{x}{x + 3} - \dfrac{x + 1}{x + 3} = \dfrac{1}{x + 3}$

6. $\dfrac{(x + 3)(x - 2)}{(x - 2)(x + 3)} = 0$

▶ **DEVELOPING YOUR SKILLS**

Reduce to lowest terms.

7. **a.** $\dfrac{a - 7}{-3a + 21}$ **b.** $\dfrac{2x + 6}{4x^2 - 8x}$

8. **a.** $\dfrac{x - 4}{-7x + 28}$ **b.** $\dfrac{3x - 18}{6x^2 - 12x}$

9. **a.** $\dfrac{x^2 - 5x - 14}{x^2 + 6x - 7}$ **b.** $\dfrac{a^2 + 3a - 28}{a^2 - 49}$

10. **a.** $\dfrac{r^2 + 3r - 10}{r^2 + r - 6}$ **b.** $\dfrac{m^2 + 3m - 4}{m^2 - 4m}$

11. a. $\dfrac{x-7}{7-x}$ **b.** $\dfrac{5-x}{x-5}$

12. a. $\dfrac{v^2-3v-28}{49-v^2}$ **b.** $\dfrac{u^2-10u+25}{25-u^2}$

13. a. $\dfrac{-12a^3b^5}{4a^2b^{-4}}$ **b.** $\dfrac{7x+21}{63}$

c. $\dfrac{y^2-9}{3-y}$ **d.** $\dfrac{m^3n-m^3}{m^4-m^4n}$

14. a. $\dfrac{5m^{-3}n^5}{-10mn^2}$ **b.** $\dfrac{-5v+20}{25}$

c. $\dfrac{n^2-4}{2-n}$ **d.** $\dfrac{w^4-w^4v}{w^3v-w^3}$

15. a. $\dfrac{2n^3+n^2-3n}{n^3-n^2}$ **b.** $\dfrac{6x^2+x-15}{4x^2-9}$

c. $\dfrac{x^3+8}{x^2-2x+4}$ **d.** $\dfrac{mn^2+n^2-4m-4}{mn+n+2m+2}$

16. a. $\dfrac{x^3+4x^2-5x}{x^3-x}$ **b.** $\dfrac{5p^2-14p-3}{5p^2+11p+2}$

c. $\dfrac{12y^2-13y+3}{27y^3-1}$ **d.** $\dfrac{ax^2-5x^2-3a+15}{ax-5x+5a-25}$

Compute as indicated. Write final results in lowest terms.

17. $\dfrac{a^2-4a+4}{a^2-9}\cdot\dfrac{a^2-2a-3}{a^2-4}$

18. $\dfrac{b^2+5b-24}{b^2-6b+9}\cdot\dfrac{b}{b^2-64}$

19. $\dfrac{x^2-7x-18}{x^2-6x-27}\cdot\dfrac{2x^2+7x+3}{2x^2+5x+2}$

20. $\dfrac{6v^2+23v+21}{4v^2-4v-15}\cdot\dfrac{4v^2-25}{3v+7}$

21. $\dfrac{p^3-64}{p^3-p}\div\dfrac{p^2+4p+16}{p^2-5p+4}$

22. $\dfrac{a^2+3a-28}{a^2+5a-14}\div\dfrac{a^3-4a^2}{a^3-8}$

23. $\dfrac{3x-9}{4x+12}\div\dfrac{3-x}{5x+15}$

24. $\dfrac{5b-10}{7b-28}\div\dfrac{2-b}{5b-20}$

25. $\dfrac{a^2+a}{a^2-3a}\cdot\dfrac{3a-9}{2a+2}$

26. $\dfrac{p^2-36}{2p}\cdot\dfrac{4p^2}{2p^2+12p}$

27. $\dfrac{8}{a^2-25}\cdot(a^2-2a-35)$

28. $(m^2-16)\cdot\dfrac{m^2-5m}{m^2-m-20}$

29. $\dfrac{xy-3x+2y-6}{x^2-3x-10}\div\dfrac{xy-3x}{xy-5y}$

30. $\dfrac{2a-ab+7b-14}{b^2-14b+49}\div\dfrac{ab-2a}{ab-7a}$

31. $\dfrac{m^2+2m-8}{m^2-2m}\div\dfrac{m^2-16}{m^2}$

32. $\dfrac{18-6x}{x^2-25}\div\dfrac{2x^2-18}{x^3-2x^2-25x+50}$

33. $\dfrac{y+3}{3y^2+9y}\cdot\dfrac{y^2+7y+12}{y^2-16}\div\dfrac{y^2+4y}{y^2-4y}$

34. $\dfrac{x^2+4x-5}{x^2-5x-14}\div\dfrac{x^2-1}{x^2-4}\cdot\dfrac{x+1}{x+5}$

35. $\dfrac{x^2-0.49}{x^2+0.5x-0.14}\div\dfrac{x^2-0.10x+0.21}{x^2-0.09}$

36. $\dfrac{x^2-0.25}{x^2+0.1x-0.2}\div\dfrac{x^2-0.8x+0.15}{x^2-0.16}$

37. $\dfrac{n^2-\frac{4}{9}}{n^2-\frac{13}{15}n+\frac{2}{15}}\div\dfrac{n^2+\frac{4}{3}n+\frac{4}{9}}{n^2-\frac{1}{25}}$

38. $\dfrac{q^2-\frac{9}{25}}{q^2-\frac{1}{10}q-\frac{3}{10}}\div\dfrac{q^2+\frac{17}{20}q+\frac{3}{20}}{q^2-\frac{1}{16}}$

39. $\dfrac{3a^3-24a^2-12a+96}{a^2-11a+24}\div\dfrac{6a^2-24}{3a^3-81}$

40. $\dfrac{p^3+p^2-49p-49}{p^2+6p-7}\div\dfrac{p^2+p+1}{p^3-1}$

41. $\dfrac{4n^2-1}{12n^2-5n-3}\cdot\dfrac{6n^2+5n+1}{2n^2+n}\cdot\dfrac{12n^2-17n+6}{6n^2-7n+2}$

42. $\left(\dfrac{4x^2-25}{x^2-11x+30}\div\dfrac{2x^2-x-15}{x^2-9x+18}\right)\dfrac{4x^2+25x-21}{12x^2-5x-3}$

Compute as indicated. Write answers in lowest terms [recall that $a-b=-1(b-a)$].

43. $\dfrac{3}{8x^2}+\dfrac{5}{2x}$ **44.** $\dfrac{15}{16y}-\dfrac{7}{2y^2}$

45. $\dfrac{7}{4x^2y^3}-\dfrac{1}{8xy^4}$ **46.** $\dfrac{3}{6a^3b}+\dfrac{5}{9ab^3}$

47. $\dfrac{4p}{p^2-36}-\dfrac{2}{p-6}$ **48.** $\dfrac{3q}{q^2-49}-\dfrac{3}{2q-14}$

49. $\dfrac{m}{m^2 - 16} + \dfrac{4}{4 - m}$ **50.** $\dfrac{2}{4 - p^2} + \dfrac{p}{p - 2}$

51. $\dfrac{2}{m - 7} - 5$ **52.** $\dfrac{4}{x - 1} - 9$

53. $\dfrac{y + 1}{y^2 + y - 30} - \dfrac{2}{y + 6}$

54. $\dfrac{4n}{n^2 - 5n} - \dfrac{3}{4n - 20}$

55. $\dfrac{1}{a + 4} + \dfrac{a}{a^2 - a - 20}$

56. $\dfrac{2x - 1}{x^2 + 3x - 4} - \dfrac{x - 5}{x^2 + 3x - 4}$

57. $\dfrac{3y - 4}{y^2 + 2y + 1} - \dfrac{2y - 5}{y^2 + 2y + 1}$

58. $\dfrac{-2}{3a + 12} - \dfrac{7}{a^2 + 4a}$

59. $\dfrac{2}{m^2 - 9} + \dfrac{m - 5}{m^2 + 6m + 9}$

60. $\dfrac{m + 2}{m^2 - 25} - \dfrac{m + 6}{m^2 - 10m + 25}$

61. $\dfrac{y + 2}{5y^2 + 11y + 2} + \dfrac{5}{y^2 + y - 6}$

62. $\dfrac{m - 4}{3m^2 - 11m + 6} + \dfrac{m}{2m^2 - m - 15}$

Write each term as a rational expression. Then compute the sum or difference indicated.

63. a. $p^{-2} - 5p^{-1}$ **b.** $x^{-2} + 2x^{-3}$

64. a. $3a^{-1} + (2a)^{-1}$ **b.** $2y^{-1} - (3y)^{-1}$

Simplify each compound rational expression. Use either method.

65. $\dfrac{\dfrac{5}{a} - \dfrac{1}{4}}{\dfrac{25}{a^2} - \dfrac{1}{16}}$ **66.** $\dfrac{\dfrac{8}{x^3} - \dfrac{1}{27}}{\dfrac{2}{x} - \dfrac{1}{3}}$

67. $\dfrac{p + \dfrac{1}{p - 2}}{1 + \dfrac{1}{p - 2}}$ **68.** $\dfrac{1 + \dfrac{3}{y - 6}}{y + \dfrac{9}{y - 6}}$

69. $\dfrac{\dfrac{2}{3 - x} + \dfrac{3}{x - 3}}{\dfrac{4}{x} + \dfrac{5}{x - 3}}$ **70.** $\dfrac{\dfrac{1}{y - 5} - \dfrac{2}{5 - y}}{\dfrac{3}{y - 5} - \dfrac{2}{y}}$

71. $\dfrac{\dfrac{2}{y^2 - y - 20}}{\dfrac{3}{y + 4} - \dfrac{4}{y - 5}}$ **72.** $\dfrac{\dfrac{2}{x^2 - 3x - 10}}{\dfrac{6}{x + 2} - \dfrac{4}{x - 5}}$

Rewrite each expression as a compound fraction. Then simplify using either method.

73. a. $\dfrac{1 + 3m^{-1}}{1 - 3m^{-1}}$ **b.** $\dfrac{1 + 2x^{-2}}{1 - 2x^{-2}}$

74. a. $\dfrac{4 - 9a^{-2}}{3a^{-2}}$ **b.** $\dfrac{3 + 2n^{-1}}{5n^{-2}}$

Rewrite each expression as a single term.

75. $\dfrac{1}{f_1} + \dfrac{1}{f_2}$ **76.** $\dfrac{1}{w} + \dfrac{1}{x} - \dfrac{1}{y}$

77. $\dfrac{\dfrac{a}{x + h} - \dfrac{a}{x}}{h}$ **78.** $\dfrac{\dfrac{a}{h - x} - \dfrac{a}{-x}}{h}$

79. $\dfrac{\dfrac{1}{2(x + h)^2} - \dfrac{1}{2x^2}}{h}$ **80.** $\dfrac{\dfrac{a}{(x + h)^2} - \dfrac{a}{x^2}}{h}$

▶ **WORKING WITH FORMULAS**

81. Cost to seize illegal drugs: $C = \dfrac{450P}{100 - P}$

The cost C, in millions of dollars, for a government to find and seize $P\%$ ($0 \le P < 100$) of a certain illegal drug is modeled by the rational equation shown. Complete the table (round to the nearest dollar) and answer the following questions.

a. What is the cost of seizing 40% of the drugs? Estimate the cost at 85%.

b. Why does cost increase dramatically the closer you get to 100%?

c. Will 100% of the drugs ever be seized?

P	$\dfrac{450P}{100 - P}$
40	
60	
80	
90	
93	
95	
98	
100	

82. Chemicals in the bloodstream: $C = \dfrac{200H^2}{H^3 + 40}$

Rational equations are often used to model chemical concentrations in the bloodstream. The percent concentration C of a certain drug H hours after injection into muscle tissue can be modeled by the equation shown ($H \geq 0$). Complete the table (round to the nearest tenth of a percent) and answer the following questions.

   a. What is the percent concentration of the drug 3 hr after injection?

b. Why is the concentration virtually equal at $H = 4$ and $H = 5$?

c. Why does the concentration begin to decrease?

d. How long will it take for the concentration to become less than 10%?

H	$\dfrac{200H^2}{H^3 + 40}$
0	
1	
2	
3	
4	
5	
6	
7	

► **APPLICATIONS**

83. Stock prices: When a hot new stock hits the market, its price will often rise dramatically and then taper off over time. The equation

$$P = \frac{50(7d^2 + 10)}{d^3 + 50}$$

models the price of stock XYZ d days after it has "hit the market." Create a table of values showing the price of the stock for the first 10 days and comment on what you notice. Find the opening price of the stock—does the stock ever return to its original price?

84. Population growth: The Department of Wildlife introduces 60 elk into a new game reserve. It is projected that the size of the herd will grow according to the equation $N = \dfrac{10(6 + 3t)}{1 + 0.05t}$, where N is the number of elk and t is the time in years. Approximate the population of elk after 14 yr.

85. Typing speed: The number of words per minute that a beginner can type is approximated by the equation $N = \dfrac{60t - 120}{t}$, where N is the number of words per minute after t weeks, $2 < t < 12$. Use a table to determine how many weeks it takes for a student to be typing an average of forty-five words per minute.

86. Memory retention: A group of students is asked to memorize 50 Russian words that are unfamiliar to them. The number N of these words that the average student remembers D days later is modeled by the equation $N = \dfrac{5D + 35}{D}$ ($D \geq 1$). How many words are remembered after (a) 1 day? (b) 5 days? (c) 12 days? (d) 35 days? (e) 100 days? According to this model, is there a certain number of words that the average student never forgets? How many?

► **EXTENDING THE CONCEPT**

87. One of these expressions is *not* equal to the others. Identify which and explain why.

   a. $\dfrac{20n}{10n}$

   b. $20 \cdot n \div 10 \cdot n$

   c. $20n \cdot \dfrac{1}{10n}$

   d. $\dfrac{20}{10} \cdot \dfrac{n}{n}$

88. The average of A and B is x. The average of C, D, and E is y. The average of A, B, C, D, and E is

   a. $\dfrac{3x + 2y}{5}$

   b. $\dfrac{2x + 3y}{5}$

   c. $\dfrac{2(x + y)}{5}$

   d. $\dfrac{3(x + y)}{5}$

89. Given the rational numbers $\dfrac{2}{5}$ and $\dfrac{3}{4}$, what is the reciprocal of the sum of their reciprocals? Given that $\dfrac{a}{b}$ and $\dfrac{c}{d}$ are *any* two numbers—what is the reciprocal of the sum of their reciprocals?

R.6 | Radicals and Rational Exponents

Learning Objectives

In Section R.6 you will learn how to:

☐ **A.** Simplify radical expressions of the form $\sqrt[n]{a^n}$

☐ **B.** Rewrite and simplify radical expressions using rational exponents

☐ **C.** Use properties of radicals to simplify radical expressions

☐ **D.** Add and subtract radical expressions

☐ **E.** Multiply and divide radical expressions; write a radical expression in simplest form

☐ **F.** Evaluate formulas involving radicals

Square roots and cube roots come from a much larger family called **radical expressions.** Expressions containing radicals can be found in virtually every field of mathematical study, and are an invaluable tool for modeling many real-world phenomena.

A. Simplifying Radical Expressions of the Form $\sqrt[n]{a^n}$

In Section R.1 we noted $\sqrt{a} = b$ only if $b^2 = a$. The expression $\sqrt{-16}$ does not represent a real number because there is no number b such that $b^2 = -16$, showing $\sqrt{a}$ is a real number only if $a \geq 0$. Of particular interest to us now is an inverse operation for a^2. In other words, what operation can be applied to a^2 to return a? Consider the following.

EXAMPLE 1 ▶ **Evaluating a Radical Expression**

Evaluate $\sqrt{a^2}$ for the values given:

  a. $a = 3$ **b.** $a = 5$ **c.** $a = -6$

Solution ▶   **a.** $\sqrt{3^2} = \sqrt{9}$ **b.** $\sqrt{5^2} = \sqrt{25}$ **c.** $\sqrt{(-6)^2} = \sqrt{36}$
                 $= 3$ $= 5$ $= 6$

Now try Exercises 7 and 8 ▶

The pattern seemed to indicate that $\sqrt{a^2} = a$ and that our search for an inverse operation was complete—until Example 1(c), where we found that $\sqrt{(-6)^2} \neq -6$. Using the absolute value concept, we can repair this apparent discrepancy and state a general rule for simplifying these expressions: $\sqrt{a^2} = |a|$. For expressions like $\sqrt{49x^2}$ and $\sqrt{y^6}$, the radicands can be rewritten as perfect squares and simplified in the same manner: $\sqrt{49x^2} = \sqrt{(7x)^2} = 7|x|$ and $\sqrt{y^6} = \sqrt{(y^3)^2} = |y^3|$.

The Square Root of a^2: $\sqrt{a^2}$

For any real number a, $\sqrt{a^2} = |a|$.

EXAMPLE 2 ▶ **Simplifying Square Root Expressions**

Simplify each expression.
    a. $\sqrt{169x^2}$ **b.** $\sqrt{x^2 - 10x + 25}$

Solution ▶   **a.** $\sqrt{169x^2} = |13x|$
                       $= 13|x|$ since *x* could be negative

          b. $\sqrt{x^2 - 10x + 25} = \sqrt{(x - 5)^2}$
                               $= |x - 5|$ since *x* − 5 could be negative

Now try Exercises 9 and 10 ▶

⚠ **CAUTION** ▶ In Section R.3, we noted that $(A + B)^2 \neq A^2 + B^2$, indicating that you cannot square the individual terms in a sum (the square of a binomial results in a perfect square trinomial). In a similar way, $\sqrt{A^2 + B^2} \neq A + B$, and you cannot take the square root of individual terms. There is a big difference between the expressions $\sqrt{A^2 + B^2}$ and $\sqrt{(A + B)^2} = |A + B|$. Try evaluating each when $A = 3$ and $B = 4$.

To investigate expressions like $\sqrt[3]{x^3}$, note the radicand in both $\sqrt[3]{8}$ and $\sqrt[3]{-64}$ can be written as a perfect cube. From our earlier definition of cube roots we know $\sqrt[3]{8} = \sqrt[3]{(2)^3} = 2$, $\sqrt[3]{-64} = \sqrt[3]{(-4)^3} = -4$, and that every real number has only one real cube root. For this reason, absolute value notation is not used or needed when taking cube roots.

The Cube Root of a^3: $\sqrt[3]{a^3}$

For any real number a, $\sqrt[3]{a^3} = a$.

EXAMPLE 3 ▶ **Simplifying Cube Root Expressions**

Simplify each expression.

   a. $\sqrt[3]{-27x^3}$ **b.** $\sqrt[3]{-64n^6}$

Solution ▶   **a.** $\sqrt[3]{-27x^3} = \sqrt[3]{(-3x)^3}$ **b.** $\sqrt[3]{-64n^6} = \sqrt[3]{(-4n^2)^3}$
                     $= -3x$ $= -4n^2$

Now try Exercises 11 and 12 ▶

WORTHY OF NOTE

Just as $\sqrt[2]{-16}$ is not a real number, $\sqrt[4]{-16}$ or $\sqrt[6]{-16}$ do not represent real numbers. An even number of repeated factors is always positive!

We can extend these ideas to fourth roots, fifth roots, and so on. For example, the fifth root of a is b only if $b^5 = a$. In symbols, $\sqrt[5]{a} = b$ implies $b^5 = a$. Since an odd number of negative factors is always negative: $(-2)^5 = -32$, and an even number of negative factors is always positive: $(-2)^4 = 16$, we must take the index into account when evaluating expressions like $\sqrt[n]{a^n}$. If n is even and the radicand is unknown, absolute value notation must be used.

The nth Root of a^n: $\sqrt[n]{a^n}$

For any real number a,
   1. $\sqrt[n]{a^n} = |a|$ when n is even.   **2.** $\sqrt[n]{a^n} = a$ when n is odd.

EXAMPLE 4 ▶ **Simplifying Radical Expressions**

Simplify each expression.

   a. $\sqrt[4]{81}$ **b.** $\sqrt[4]{-81}$ **c.** $\sqrt[5]{32}$ **d.** $\sqrt[5]{-32}$
   e. $\sqrt[4]{16m^4}$ **f.** $\sqrt[5]{32p^5}$ **g.** $\sqrt[6]{(m+5)^6}$ **h.** $\sqrt[7]{(x-2)^7}$

Solution ▶   **a.** $\sqrt[4]{81} = 3$ **b.** $\sqrt[4]{-81}$ is not a real number
             c. $\sqrt[5]{32} = 2$ **d.** $\sqrt[5]{-32} = -2$
             e. $\sqrt[4]{16m^4} = \sqrt[4]{(2m)^4}$ **f.** $\sqrt[5]{32p^5} = \sqrt[5]{(2p)^5}$
                        $= |2m|$ or $2|m|$ $= 2p$
             g. $\sqrt[6]{(m+5)^6} = |m+5|$ **h.** $\sqrt[7]{(x-2)^7} = x-2$

☑ **A.** You've just reviewed how to simplify radical expressions of the form $\sqrt[n]{a^n}$

Now try Exercises 13 and 14 ▶

B. Radical Expressions and Rational Exponents

As an alternative to radical notation, a rational (fractional) exponent can be used, along with the power property of exponents. For $\sqrt[3]{a^3} = a$, notice that an exponent of one-third can replace the cube root notation and produce the same result: $\sqrt[3]{a^3} = (a^3)^{\frac{1}{3}} = a^{\frac{3}{3}} = a$. In the same way, an exponent of one-half can replace the square root notation: $\sqrt{a^2} = (a^2)^{\frac{1}{2}} = a^{\frac{2}{2}} = |a|$. In general, we have the following:

Rational Exponents

If a is a real number and n is an integer greater than 1,

$$\text{then } \sqrt[n]{a} = \sqrt[n]{a^1} = a^{\frac{1}{n}}$$

provided $\sqrt[n]{a}$ represents a real number.

EXAMPLE 5 ▶ **Simplifying Radical Expressions Using Rational Exponents**

Simplify by rewriting each radicand as a perfect nth power and converting to rational exponent notation.

a. $\sqrt[3]{-125}$ **b.** $-\sqrt[4]{16x^{20}}$ **c.** $\sqrt[4]{-81}$ **d.** $\sqrt[3]{\dfrac{8w^3}{27}}$

Solution ▶

a. $\sqrt[3]{-125} = \sqrt[3]{(-5)^3}$
$= \left[(-5)^3\right]^{\frac{1}{3}}$
$= (-5)^{\frac{3}{3}}$
$= -5$

b. $-\sqrt[4]{16x^{20}} = -\sqrt[4]{(2x^5)^4}$
$= -\left[(2x^5)^4\right]^{\frac{1}{4}}$
$= -(2x^5)^{\frac{4}{4}}$
$= -2|x|^5$

c. $\sqrt[4]{-81} = (-81)^{\frac{1}{4}}$
is not a real number

d. $\sqrt[3]{\dfrac{8w^3}{27}} = \sqrt[3]{\left(\dfrac{2w}{3}\right)^3}$
$= \left[\left(\dfrac{2w}{3}\right)^3\right]^{\frac{1}{3}}$
$= \left(\dfrac{2w}{3}\right)^{\frac{3}{3}}$
$= \dfrac{2w}{3}$

Now try Exercises 15 and 16 ▶

WORTHY OF NOTE

Any rational number can be decomposed into the product of a unit fraction and an integer: $\dfrac{m}{n} = \dfrac{1}{n} \cdot m$.

When a rational exponent is used, as in $\sqrt[n]{a} = \sqrt[n]{a^1} = a^{\frac{1}{n}}$, the denominator of the exponent represents the index number, while the numerator of the exponent represents the original power on a. *This is true even when the exponent on a is something other than one!* In other words, the radical expression $\sqrt[4]{16^3}$ can be rewritten as $\left(16^3\right)^{\frac{1}{4}} = \left(16^{\frac{1}{4}}\right)^3$ or $16^{\frac{3}{4}}$. This is further illustrated in Figure R.6 where we see the rational exponent has the form, "power over root." To evaluate this expression without the aid of a calculator, we use the commutative property to rewrite $\left(16^{\frac{3}{1}}\right)^{\frac{1}{4}}$ as $\left(16^{\frac{1}{4}}\right)^{\frac{3}{1}}$ and begin with the fourth root of 16: $\left(16^{\frac{1}{4}}\right)^{\frac{3}{1}} = 2^3 = 8$.

In general, if m and n have no common factors (other than 1) the expression $a^{\frac{m}{n}}$ can be interpreted in the following two ways.

Figure R.6

Rational Exponents

If $\dfrac{m}{n}$ is a rational number expressed in lowest terms with $n \geq 2$, then

$$(1)\ a^{\frac{m}{n}} = \left(\sqrt[n]{a}\right)^m \qquad\qquad \text{or} \qquad\qquad (2)\ a^{\frac{m}{n}} = \sqrt[n]{a^m}$$

(compute $\sqrt[n]{a}$, then take the mth power)   (compute a^m, then take the nth root)

provided $\sqrt[n]{a}$ represents a real number.

EXAMPLE 6 ▶ Simplifying Expressions with Rational Exponents

Find the value of each expression without a calculator, by rewriting the exponent as the product of a unit fraction and an integer.

   a. $27^{\frac{2}{3}}$ **b.** $(-8)^{\frac{4}{3}}$ **c.** $\left(\dfrac{4x^6}{9}\right)^{\frac{5}{2}}$

Solution ▶

a. $27^{\frac{2}{3}} = 27^{\frac{1}{3}\cdot 2}$ **b.** $(-8)^{\frac{4}{3}} = (-8)^{\frac{1}{3}\cdot 4}$
$\quad = \left(27^{\frac{1}{3}}\right)^2$ $= \left[(-8)^{\frac{1}{3}}\right]^4$
$\quad = 3^2$ or 9 $= [-2]^4 = 16$

c. $\left(\dfrac{4x^6}{9}\right)^{\frac{5}{2}} = \left(\dfrac{4x^6}{9}\right)^{\frac{1}{2}\cdot 5}$

$\qquad = \left[\left(\dfrac{4x^6}{9}\right)^{\frac{1}{2}}\right]^5$

$\qquad = \left[\dfrac{2|x|^3}{3}\right]^5 = \dfrac{32|x|^{15}}{243}$

WORTHY OF NOTE

While the expression $(-8)^{\frac{1}{3}} = \sqrt[3]{-8}$ represents the real number -2, the expression $(-8)^{\frac{2}{6}} = (\sqrt[6]{-8})^2$ is not a real number, even though $\dfrac{1}{3} = \dfrac{2}{6}$. Note that the second exponent is not in lowest terms.

Now try Exercises 17 and 18 ▶

Expressions with rational exponents are generally easier to evaluate if we compute the root first, then apply the exponent. Computing the root first also helps us determine whether or not an expression represents a real number.

EXAMPLE 7 ▶ Simplifying Expressions with Rational Exponents

Simplify each expression, if possible.

  a. $-49^{\frac{3}{2}}$ **b.** $(-49)^{\frac{3}{2}}$ **c.** $(-8)^{\frac{2}{3}}$ **d.** $-8^{-\frac{2}{3}}$

Solution ▶

a. $-49^{\frac{3}{2}} = -\left(49^{\frac{1}{2}}\right)^3$ **b.** $(-49)^{\frac{3}{2}} = \left[(-49)^{\frac{1}{2}}\right]^3$,
$\quad = -(\sqrt{49})^3$ $= (\sqrt{-49})^3$
$\quad = -(7)^3$ or -343 not a real number

c. $(-8)^{\frac{2}{3}} = \left[(-8)^{\frac{1}{3}}\right]^2$ **d.** $-8^{-\frac{2}{3}} = -\left(8^{\frac{1}{3}}\right)^{-2}$
$\quad = (\sqrt[3]{-8})^2$ $= -(\sqrt[3]{8})^{-2}$
$\quad = (-2)^2$ or 4 $= -2^{-2}$ or $-\dfrac{1}{4}$

☑ **B.** You've just reviewed how to rewrite and simplify radical expressions using rational exponents

Now try Exercises 19 through 22 ▶

C. Using Properties of Radicals to Simplify Radical Expressions

The properties used to simplify radical expressions are closely connected to the properties of exponents. For instance, the product to a power property holds even when n is a rational number. This means $(xy)^{\frac{1}{2}} = x^{\frac{1}{2}}y^{\frac{1}{2}}$ and $(4 \cdot 25)^{\frac{1}{2}} = 4^{\frac{1}{2}} \cdot 25^{\frac{1}{2}}$. When the second statement is expressed in radical form, we have $\sqrt{4 \cdot 25} = \sqrt{4} \cdot \sqrt{25}$, with both forms having a value of 10. This suggests the **product property of radicals,** which can be extended to include cube roots, fourth roots, and so on.

Product Property of Radicals

If $\sqrt[n]{A}$ and $\sqrt[n]{B}$ represent real-valued expressions, then
$$\sqrt[n]{AB} = \sqrt[n]{A} \cdot \sqrt[n]{B} \quad \text{and} \quad \sqrt[n]{A} \cdot \sqrt[n]{B} = \sqrt[n]{AB}.$$

⚠ CAUTION	Note that this property applies only to a *product* of two terms, not to a sum or difference. In other words, while $\sqrt{9x^2} = |3x|$, $\sqrt{9 + x^2} \neq |3 + x|$!

One application of the product property is to simplify radical expressions. In general, the expression $\sqrt[n]{a}$ is in simplified form if a has no factors (other than 1) that are perfect nth roots.

EXAMPLE 8 ▶ **Simplifying Radical Expressions**

Write each expression in simplest form using the product property.

    a. $\sqrt{18}$ **b.** $5\sqrt[3]{125x^4}$ **c.** $\dfrac{-4 + \sqrt{20}}{2}$

Solution ▶   **a.** $\sqrt{18} = \sqrt{9 \cdot 2}$ **b.** $5\sqrt[3]{125x^4} = 5 \cdot \sqrt[3]{125 \cdot x^4}$

$$= \sqrt{9}\sqrt{2}$$

$$= 3\sqrt{2}$$

These steps can be done mentally.
$$\begin{cases} = 5 \cdot \sqrt[3]{125} \cdot \sqrt[3]{x^3} \cdot \sqrt[3]{x^1} \\ = 5 \cdot 5 \cdot x \cdot \sqrt[3]{x} \\ = 25x\sqrt[3]{x} \end{cases}$$

WORTHY OF NOTE

For expressions like those in Example 8(c), students must resist the "temptation" to reduce individual terms as in
$$\dfrac{-4 + \sqrt{20}}{2} \not= -2 + \sqrt{20}.$$
Remember, only *factors* can be reduced.

  c. $\dfrac{-4 + \sqrt{20}}{2} = \dfrac{-4 + \sqrt{4 \cdot 5}}{2}$ look for perfect square factors of 20

$$= \dfrac{-4 + 2\sqrt{5}}{2}$$ product property

$$= \dfrac{2(-2 + \sqrt{5})}{2}$$ factor and reduce

$$= -2 + \sqrt{5}$$ result

Now try Exercises 23 and 24 ▶

When radicals are *combined* using the product property, the result may contain a perfect nth root, which should be simplified. Note that the *index numbers must be the same* in order to use this property.

EXAMPLE 9 ▶ **Simplifying Radical Expressions**

Combine factors using the product property and simplify: $1.2\sqrt[3]{16n^4} \sqrt[3]{4n^5}$.

Solution ▶ $1.2\sqrt[3]{16n^4} \sqrt[3]{4n^5} = 1.2\sqrt[3]{64 \cdot n^9}$ product property

Since the index is 3, we look for perfect cube factors in the radicand.

$$= 1.2\sqrt[3]{64} \sqrt[3]{n^9}$$ product property

$$= 1.2\sqrt[3]{64} \sqrt[3]{(n^3)^3}$$ rewrite n^9 as a perfect cube

$$= 1.2(4)n^3$$ simplify

$$= 4.8n^3$$ result

Now try Exercises 25 and 26 ▶

WORTHY OF NOTE

Rational exponents also could have been used to simplify the expression from Example 9, since
$1.2\sqrt[3]{64} \sqrt[3]{n^9} = 1.2(4)n^{\frac{9}{3}} = 4.8n^3$. Also see Example 11.

The **quotient property of radicals** can also be established using exponential properties. The fact that $\dfrac{\sqrt{100}}{\sqrt{25}} = \sqrt{\dfrac{100}{25}} = 2$ suggests the following:

Quotient Property of Radicals

If $\sqrt[n]{A}$ and $\sqrt[n]{B}$ represent real-valued expressions with $B \neq 0$, then

$$\sqrt[n]{\frac{A}{B}} = \frac{\sqrt[n]{A}}{\sqrt[n]{B}} \quad \text{and} \quad \frac{\sqrt[n]{A}}{\sqrt[n]{B}} = \sqrt[n]{\frac{A}{B}}.$$

Many times the product and quotient properties must work together to simplify a radical expression, as shown in Example 10.

EXAMPLE 10 ▶ **Simplifying Radical Expressions**

Simplify each expression:

a. $\dfrac{\sqrt{18a^5}}{\sqrt{2a}}$ b. $\sqrt[3]{\dfrac{81}{125x^3}}$

Solution ▶ a. $\dfrac{\sqrt{18a^5}}{\sqrt{2a}} = \sqrt{\dfrac{18a^5}{2a}}$ b. $\sqrt[3]{\dfrac{81}{125x^3}} = \dfrac{\sqrt[3]{81}}{\sqrt[3]{125x^3}}$

$\qquad\qquad\qquad = \sqrt{9a^4}$ $\qquad\qquad\qquad\qquad = \dfrac{\sqrt[3]{27 \cdot 3}}{5x}$

$\qquad\qquad\qquad = 3a^2$ $\qquad\qquad\qquad\qquad\qquad = \dfrac{3\sqrt[3]{3}}{5x}$

Now try Exercises 27 and 28 ▶

Radical expressions can also be simplified using rational exponents.

EXAMPLE 11 ▶ **Using Rational Exponents to Simplify Radical Expressions**

Simplify using rational exponents:

a. $\sqrt{36p^4q^5}$ b. $v\sqrt[3]{v^4}$ c. $\sqrt[3]{\sqrt{x}}$ d. $\sqrt[3]{m}\sqrt{m}$

Solution ▶ a. $\sqrt{36p^4q^5} = (36p^4q^5)^{\frac{1}{2}}$ $\qquad$ b. $v\sqrt[3]{v^4} = v^1 \cdot v^{\frac{4}{3}}$

$\qquad\qquad\qquad = 36^{\frac{1}{2}}p^{\frac{4}{2}}q^{\frac{5}{2}}$ $\qquad\qquad\qquad\qquad = v^{\frac{3}{3}} \cdot v^{\frac{4}{3}}$

$\qquad\qquad\qquad = 6p^2q^{\left(\frac{4}{2}+\frac{1}{2}\right)}$ $\qquad\qquad\qquad\qquad = v^{\frac{7}{3}}$

$\qquad\qquad\qquad = 6p^2q^2q^{\frac{1}{2}}$ $\qquad\qquad\qquad\qquad = v^{\frac{6}{3}}v^{\frac{1}{3}}$

$\qquad\qquad\qquad = 6p^2q^2\sqrt{q}$ $\qquad\qquad\qquad\qquad = v^2\sqrt[3]{v}$

☑ **C.** You've just reviewed how to use properties of radicals to simplify radical expressions

c. $\sqrt[3]{\sqrt{x}} = \sqrt[3]{x^{\frac{1}{2}}}$ $\qquad\qquad$ d. $\sqrt[3]{m}\sqrt{m} = m^{\frac{1}{3}}m^{\frac{1}{2}}$

$\qquad\qquad\quad = \left(x^{\frac{1}{2}}\right)^{\frac{1}{3}}$ $\qquad\qquad\qquad\qquad = m^{\frac{1}{3}+\frac{1}{2}}$

$\qquad\qquad\quad = x^{\frac{1}{2}\cdot\frac{1}{3}}$ $\qquad\qquad\qquad\qquad = m^{\frac{5}{6}}$

$\qquad\qquad\quad = x^{\frac{1}{6}} \text{ or } \sqrt[6]{x}$ $\qquad\qquad\qquad = \sqrt[6]{m^5}$

Now try Exercises 29 and 30 ▶

D. Addition and Subtraction of Radical Expressions

Since $3x$ and $5x$ are like terms, we know $3x + 5x = 8x$. If $x = \sqrt[3]{7}$, the sum becomes $3\sqrt[3]{7} + 5\sqrt[3]{7} = 8\sqrt[3]{7}$, illustrating how *like* radical expressions can be combined. Like radicals are those that have *the same index and radicand*. In some cases, we can identify like radicals only after radical terms have been simplified.

EXAMPLE 12 ▶ **Adding and Subtracting Radical Expressions**

Simplify and combine (if possible).

a. $\sqrt{45} + 2\sqrt{20}$ b. $\sqrt[3]{16x^5} - x\sqrt[3]{54x^2}$

Solution ▶ a. $\sqrt{45} + 2\sqrt{20} = 3\sqrt{5} + 2(2\sqrt{5})$ simplify radicals: $\sqrt{45} = \sqrt{9 \cdot 5}$; $\sqrt{20} = \sqrt{4 \cdot 5}$
$= 3\sqrt{5} + 4\sqrt{5}$ like radicals
$= 7\sqrt{5}$ result

b. $\sqrt[3]{16x^5} - x\sqrt[3]{54x^2} = \sqrt[3]{8 \cdot 2 \cdot x^3 \cdot x^2} - x\sqrt[3]{27 \cdot 2 \cdot x^2}$
$= 2x\sqrt[3]{2x^2} - 3x\sqrt[3]{2x^2}$ simplify radicals
$= -x\sqrt[3]{2x^2}$ result

☑ **D.** You've just reviewed how to add and subtract radical expressions

Now try Exercises 31 through 34 ▶

E. Multiplication and Division of Radical Expressions; Radical Expressions in Simplest Form

Multiplying radical expressions is simply an extension of our earlier work. The multiplication can take various forms, from the distributive property to any of the special products reviewed in Section R.3. For instance, $(A \pm B)^2 = A^2 \pm 2AB + B^2$, even if A or B is a radical term.

EXAMPLE 13 ▶ **Multiplying Radical Expressions**

Compute each product and simplify.

a. $5\sqrt{3}(\sqrt{6} - 4\sqrt{3})$ b. $(2\sqrt{2} + 6\sqrt{3})(3\sqrt{10} + \sqrt{15})$
c. $(x + \sqrt{7})(x - \sqrt{7})$ d. $(3 - \sqrt{2})^2$

Solution ▶ a. $5\sqrt{3}(\sqrt{6} - 4\sqrt{3}) = 5\sqrt{18} - 20(\sqrt{3})^2$ distribute; $(\sqrt{3})^2 = 3$
$= 5(3)\sqrt{2} - (20)(3)$ simplify: $\sqrt{18} = 3\sqrt{2}$
$= 15\sqrt{2} - 60$ result

b. $(2\sqrt{2} + 6\sqrt{3})(3\sqrt{10} + \sqrt{15}) = 6\sqrt{20} + 2\sqrt{30} + 18\sqrt{30} + 6\sqrt{45}$ F-O-I-L
$= 12\sqrt{5} + 20\sqrt{30} + 18\sqrt{5}$ extract roots *and* simplify
$= 30\sqrt{5} + 20\sqrt{30}$ result

c. $(x + \sqrt{7})(x - \sqrt{7}) = x^2 - (\sqrt{7})^2$ $(A + B)(A - B) = A^2 - B^2$
$= x^2 - 7$ result

d. $(3 - \sqrt{2})^2 = (3)^2 - 2(3)(\sqrt{2}) + (\sqrt{2})^2$ $(A + B)^2 = A^2 + 2AB + B^2$
$= 9 - 6\sqrt{2} + 2$ simplify each term
$= 11 - 6\sqrt{2}$ result

LOOKING AHEAD

Notice that the answer for Example 13(c) contains no radical terms, since the outer and inner products sum to zero. This result will be used to simplify certain radical expressions in this section and later in Chapter 1.

Now try Exercises 35 through 38 ▶

One application of products and powers of radical expressions is to evaluate certain quadratic expressions, as illustrated in Example 14.

EXAMPLE 14 ▶ **Evaluating a Quadratic Expression**

Show that when $x^2 - 4x + 1$ is evaluated at $x = 2 + \sqrt{3}$, the result is zero.

Solution ▶ $x^2 - 4x + 1$ original expression
$(2 + \sqrt{3})^2 - 4(2 + \sqrt{3}) + 1$ substitute $2 + \sqrt{3}$ for x
$4 + 4\sqrt{3} + 3 - 8 - 4\sqrt{3} + 1$ multiply
$(4 + 3 - 8 + 1) + (4\sqrt{3} - 4\sqrt{3})$ commutative and associative properties
$0 \checkmark$

Now try Exercises 39 through 42 ▶

When we applied the quotient property in Example 10, we obtained a denominator free of radicals. Sometimes the denominator is not automatically free of radicals, and the need to write radical expressions in *simplest form* comes into play. This process is called **rationalizing the denominator.**

Radical Expressions in Simplest Form
A radical expression is in simplest form if: **1.** The radicand has no perfect nth root factors. **2.** The radicand contains no fractions. **3.** No radicals occur in a denominator.

As with other types of simplification, the desired form can be achieved in various ways. If the denominator is a single radical term, we multiply the numerator and denominator by the factors required to eliminate the radical in the denominator [see Examples 15(a) and 15(b)]. If the radicand is a rational expression, it is generally easier to build an equivalent fraction *within the radical* having perfect nth root factors in the denominator [see Example 15(c)].

EXAMPLE 15 ▶ **Simplifying Radical Expressions**

Simplify by rationalizing the denominator. Assume $a, x \neq 0$.

$$\textbf{a. } \frac{2}{5\sqrt{3}} \qquad \textbf{b. } \frac{-7}{\sqrt[3]{x}} \qquad \textbf{c. } \sqrt[3]{\frac{3}{4a^4}}$$

Solution ▶

a. $\dfrac{2}{5\sqrt{3}} = \dfrac{2}{5\sqrt{3}} \cdot \dfrac{\sqrt{3}}{\sqrt{3}}$ multiply numerator and denominator by $\sqrt{3}$

$\qquad = \dfrac{2\sqrt{3}}{5(\sqrt{3})^2} = \dfrac{2\sqrt{3}}{15}$ simplify—denominator is now rational

b. $\dfrac{-7}{\sqrt[3]{x}} = \dfrac{-7(\sqrt[3]{x})(\sqrt[3]{x})}{\sqrt[3]{x}(\sqrt[3]{x})(\sqrt[3]{x})}$ multiply using two additional factors of $\sqrt[3]{x}$

$\qquad = \dfrac{-7\sqrt[3]{x^2}}{\sqrt[3]{x^3}}$ product property

$\qquad = \dfrac{-7\sqrt[3]{x^2}}{x}$ $\sqrt[3]{x^3} = x$

c. $\sqrt[3]{\dfrac{3}{4a^4}} = \sqrt[3]{\dfrac{3}{4a^4} \cdot \dfrac{2a^2}{2a^2}}$ $4 \cdot 2 = 8$ is the smallest perfect cube with 4 as a factor; $a^4 \cdot a^2 = a^6$ is the smallest perfect cube with a^4 as a factor

$\qquad = \sqrt[3]{\dfrac{6a^2}{8a^6}}$ the denominator is now a perfect cube—simplify

$\qquad = \dfrac{\sqrt[3]{6a^2}}{2a^2}$ result

Now try Exercises 43 and 44 ▶

In some applications, the denominator may be a sum or difference containing a radical term. In this case, the methods from Example 15 are ineffective, and instead we multiply by a conjugate since $(A + B)(A - B) = A^2 - B^2$. If either A or B is a square root, the result will be a denominator free of radicals.

EXAMPLE 16 ▶ Simplifying Radical Expressions Using a Conjugate

Simplify the expression by rationalizing the denominator. Write the answer in exact form and approximate form rounded to three decimal places.

$$\frac{2 + \sqrt{3}}{\sqrt{6} - \sqrt{2}}$$

Solution ▶

$$\frac{2 + \sqrt{3}}{\sqrt{6} - \sqrt{2}} = \frac{2 + \sqrt{3}}{\sqrt{6} - \sqrt{2}} \cdot \frac{\sqrt{6} + \sqrt{2}}{\sqrt{6} + \sqrt{2}} \qquad \text{multiply by the conjugate of the denominator}$$

$$= \frac{2\sqrt{6} + 2\sqrt{2} + \sqrt{18} + \sqrt{6}}{(\sqrt{6})^2 - (\sqrt{2})^2} \qquad \begin{array}{l}\text{FOIL}\\[4pt]\text{difference of squares}\end{array}$$

$$= \frac{3\sqrt{6} + 2\sqrt{2} + 3\sqrt{2}}{6 - 2} \qquad \text{simplify}$$

$$= \frac{3\sqrt{6} + 5\sqrt{2}}{4} \qquad \text{exact form}$$

$$\approx 3.605 \qquad \text{approximate form}$$

 E. You've just reviewed how to multiply and divide radical expressions and write a radical expression in simplest form

Now try Exercises 45 through 48 ▶

F. Formulas and Radicals

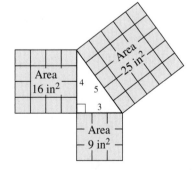

A right triangle is one that has a 90° angle. The longest side (opposite the right angle) is called the **hypotenuse,** while the other two sides are simply called "legs." The **Pythagorean theorem** is a formula that says if you add the square of each leg, the result will be equal to the square of the hypotenuse. Furthermore, we note the converse of this theorem is also true.

Pythagorean Theorem

1. For any right triangle with legs a and b and hypotenuse c,
$$a^2 + b^2 = c^2$$
2. For any triangle with sides a, b, and c, if $a^2 + b^2 = c^2$, then the triangle is a right triangle.

A geometric interpretation of the theorem is given in the figure, which shows $3^2 + 4^2 = 5^2$.

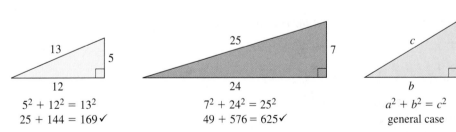

$$5^2 + 12^2 = 13^2 \qquad\qquad 7^2 + 24^2 = 25^2 \qquad\qquad a^2 + b^2 = c^2$$
$$25 + 144 = 169 \checkmark \qquad 49 + 576 = 625 \checkmark \qquad \text{general case}$$

EXAMPLE 17 ▶ Applying the Pythagorean Theorem

An extension ladder is placed 9 ft from the base of a building in an effort to reach a third-story window that is 27 ft high. What is the minimum length of the ladder required? Answer in exact form using radicals, and approximate form by rounding to one decimal place.

Solution ▶ We can assume the building makes a 90° angle with the ground, and use the Pythagorean theorem to find the required length. Let c represent this length.

$$c^2 = a^2 + b^2 \qquad \text{Pythagorean theorem}$$
$$c^2 = (9)^2 + (27)^2 \qquad \text{substitute 9 for } a \text{ and 27 for } b$$
$$c^2 = 81 + 729 \qquad 9^2 = 81, \, 27^2 = 729$$
$$c^2 = 810 \qquad \text{add}$$
$$c = \sqrt{810} \qquad \text{definition of square root; } c > 0$$
$$c = 9\sqrt{10} \qquad \text{exact form: } \sqrt{810} = \sqrt{81 \cdot 10} = 9\sqrt{10}$$
$$c \approx 28.5 \text{ ft} \qquad \text{approximate form}$$

The ladder must be at least 28.5 ft tall.

Now try Exercises 51 and 52 ▶

☑ **F.** You've just reviewed how to evaluate formulas involving radicals

R.6 EXERCISES

▶ CONCEPTS AND VOCABULARY

Fill in each blank with the appropriate word or phrase. Carefully reread the section, if necessary.

1. $\sqrt[n]{a^n} = |a|$ if $n > 0$ is a(n) _____ integer.

2. The conjugate of $2 - \sqrt{3}$ is _____.

3. By decomposing the rational exponent, we can rewrite $16^{\frac{3}{4}}$ as $(16^{\frac{?}{?}})^?$.

4. $(x^{\frac{3}{2}})^{\frac{2}{3}} = x^{\frac{3}{2} \cdot \frac{2}{3}} = x^1$ is an example of the _____ property of exponents.

5. Discuss/Explain what it means when we say an expression like $\sqrt{A}$ has been written in simplest form.

6. Discuss/Explain why it would be easier to simplify the expression given using rational exponents rather than radicals:

$$\frac{x^{\frac{1}{2}}}{x^{\frac{1}{3}}}$$

▶ DEVELOPING YOUR SKILLS

Evaluate the expression $\sqrt{x^2}$ for the values given.

7. **a.** $x = 9$ **b.** $x = -10$

8. **a.** $x = 7$ **b.** $x = -8$

Simplify each expression, assuming that variables can represent any real number.

9. **a.** $\sqrt{49p^2}$ **b.** $\sqrt{(x-3)^2}$
   c. $\sqrt{81m^4}$ **d.** $\sqrt{x^2 - 6x + 9}$

10. **a.** $\sqrt{25n^2}$ **b.** $\sqrt{(y+2)^2}$
    c. $\sqrt{v^{10}}$ **d.** $\sqrt{4a^2 + 12a + 9}$

11. **a.** $\sqrt[3]{64}$ **b.** $\sqrt[3]{-125x^3}$
    c. $\sqrt[3]{216z^{12}}$ **d.** $\sqrt[3]{\dfrac{v^3}{-8}}$

12. **a.** $\sqrt[3]{-8}$ **b.** $\sqrt[3]{-125p^3}$
    c. $\sqrt[3]{27q^9}$ **d.** $\sqrt[3]{\dfrac{w^3}{-64}}$

13. **a.** $\sqrt[6]{64}$ **b.** $\sqrt[6]{-64}$
    c. $\sqrt[5]{243x^{10}}$ **d.** $\sqrt[5]{-243x^5}$
    e. $\sqrt[5]{(k-3)^5}$ **f.** $\sqrt[6]{(h+2)^6}$

14. a. $\sqrt[4]{216}$ **b.** $\sqrt[4]{-216}$

    c. $\sqrt[5]{1024z^{15}}$ **d.** $\sqrt[5]{-1024z^{20}}$

    e. $\sqrt[5]{(q-9)^5}$ **f.** $\sqrt[6]{(p+4)^6}$

15. a. $\sqrt[3]{-125}$ **b.** $-\sqrt[4]{81n^{12}}$

    c. $\sqrt{-36}$ **d.** $\sqrt{\dfrac{49v^{10}}{36}}$

16. a. $\sqrt[3]{-216}$ **b.** $-\sqrt[4]{16m^{24}}$

    c. $\sqrt{-121}$ **d.** $\sqrt{\dfrac{25x^6}{4}}$

17. a. $8^{\frac{2}{3}}$ **b.** $\left(\dfrac{16}{25}\right)^{\frac{3}{2}}$

    c. $\left(\dfrac{4}{25}\right)^{-\frac{3}{2}}$ **d.** $\left(\dfrac{-27p^6}{8q^3}\right)^{\frac{2}{3}}$

18. a. $9^{\frac{3}{2}}$ **b.** $\left(\dfrac{4}{9}\right)^{\frac{3}{2}}$

    c. $\left(\dfrac{16}{81}\right)^{-\frac{3}{4}}$ **d.** $\left(\dfrac{-125v^9}{27w^6}\right)^{\frac{2}{3}}$

19. a. $-144^{\frac{3}{2}}$ **b.** $\left(-\dfrac{4}{25}\right)^{\frac{3}{2}}$

    c. $(-27)^{-\frac{2}{3}}$ **d.** $-\left(\dfrac{27x^3}{64}\right)^{-\frac{4}{3}}$

20. a. $-100^{\frac{3}{2}}$ **b.** $\left(-\dfrac{49}{36}\right)^{\frac{3}{2}}$

    c. $(-125)^{-\frac{2}{3}}$ **d.** $-\left(\dfrac{x^9}{8}\right)^{-\frac{4}{3}}$

Use properties of exponents to simplify. Answer in exponential form without negative exponents.

21. a. $\left(2n^2p^{-\frac{2}{5}}\right)^5$ **b.** $\left(\dfrac{8y^{\frac{3}{4}}}{64y^{\frac{3}{2}}}\right)^{\frac{1}{3}}$

22. a. $\left(\dfrac{24x^{\frac{3}{8}}}{4x^{\frac{1}{2}}}\right)^2$ **b.** $\left(2x^{-\frac{1}{4}}y^{\frac{3}{4}}\right)^4$

Simplify each expression. Assume all variables represent non-negative real numbers.

23. a. $\sqrt{18m^2}$ **b.** $-2\sqrt[3]{-125p^3q^7}$

    c. $\dfrac{3}{8}\sqrt[3]{64m^3n^5}$ **d.** $\sqrt{32p^3q^6}$

    e. $\dfrac{-6+\sqrt{28}}{2}$ **f.** $\dfrac{27-\sqrt{72}}{6}$

24. a. $\sqrt{8x^6}$ **b.** $3\sqrt[3]{128a^4b^2}$

    c. $\dfrac{2}{9}\sqrt[3]{27a^2b^6}$ **d.** $\sqrt{54m^6n^8}$

    e. $\dfrac{12-\sqrt{48}}{8}$ **f.** $\dfrac{-20+\sqrt{32}}{4}$

25. a. $2.5\sqrt{18a}\sqrt{2a^3}$ **b.** $-\dfrac{2}{3}\sqrt{3b}\sqrt{12b^2}$

    c. $\sqrt{\dfrac{x^3y}{3}}\sqrt{\dfrac{4x^5y}{12y}}$ **d.** $\sqrt[3]{9v^2u}\sqrt[3]{3u^5v^2}$

26. a. $5.1\sqrt{2p}\sqrt{32p^5}$ **b.** $-\dfrac{4}{5}\sqrt{5q}\sqrt{20q^3}$

    c. $\sqrt{\dfrac{ab^2}{3}}\sqrt{\dfrac{25ab^4}{27}}$ **d.** $\sqrt[3]{5cd^2}\,\sqrt[3]{25cd}$

27. a. $\dfrac{\sqrt{8m^5}}{\sqrt{2m}}$ **b.** $\dfrac{\sqrt[3]{108n^4}}{\sqrt[3]{4n}}$

    c. $\sqrt{\dfrac{45}{16x^2}}$ **d.** $12\sqrt[3]{\dfrac{81}{8z^9}}$

28. a. $\dfrac{\sqrt{27y^7}}{\sqrt{3y}}$ **b.** $\dfrac{\sqrt[3]{72b^5}}{\sqrt[3]{3b^2}}$

    c. $\sqrt{\dfrac{20}{4x^4}}$ **d.** $-9\sqrt[3]{\dfrac{125}{27x^6}}$

29. a. $\sqrt[5]{32x^{10}y^{15}}$ **b.** $x\sqrt[4]{x^5}$

    c. $\sqrt[4]{\sqrt[3]{b}}$ **d.** $\dfrac{\sqrt[3]{6}}{\sqrt{6}}$

    e. $\sqrt{b}\sqrt[4]{b}$

30. a. $\sqrt[4]{81a^{12}b^{16}}$ **b.** $a\sqrt[5]{a^6}$

    c. $\sqrt{\sqrt[4]{a}}$ **d.** $\dfrac{\sqrt[3]{3}}{\sqrt[4]{3}}$

    e. $\sqrt[3]{c}\sqrt[4]{c}$

Simplify and add (if possible).

31. a. $12\sqrt{72}-9\sqrt{98}$

    b. $8\sqrt{48}-3\sqrt{108}$

    c. $7\sqrt{18m}-\sqrt{50m}$

    d. $2\sqrt{28p}-3\sqrt{63p}$

32. a. $-3\sqrt{80}+2\sqrt{125}$

    b. $5\sqrt{12}+2\sqrt{27}$

    c. $3\sqrt{12x}-5\sqrt{75x}$

    d. $3\sqrt{40q}+9\sqrt{10q}$

33. **a.** $3x\sqrt[3]{54x} - 5\sqrt[3]{16x^4}$

   b. $\sqrt{4} + \sqrt{3x} - \sqrt{12x} + \sqrt{45}$

   c. $\sqrt{72x^3} + \sqrt{50} - \sqrt{7x} + \sqrt{27}$

34. **a.** $5\sqrt[3]{54m^3} - 2m\sqrt[3]{16m^3}$

   b. $\sqrt{10b} + \sqrt{200b} - \sqrt{20} + \sqrt{40}$

   c. $\sqrt{75r^3} + \sqrt{32} - \sqrt{27r} + \sqrt{38}$

Compute each product and simplify the result.

35. **a.** $(7\sqrt{2})^2$ **b.** $\sqrt{3}(\sqrt{5} + \sqrt{7})$

   c. $(n + \sqrt{5})(n - \sqrt{5})$ **d.** $(6 - \sqrt{3})^2$

36. **a.** $(0.3\sqrt{5})^2$ **b.** $\sqrt{5}(\sqrt{6} - \sqrt{2})$

   c. $(4 + \sqrt{3})(4 - \sqrt{3})$ **d.** $(2 + \sqrt{5})^2$

37. **a.** $(3 + 2\sqrt{7})(3 - 2\sqrt{7})$

   b. $(\sqrt{5} - \sqrt{14})(\sqrt{2} + \sqrt{13})$

   c. $(2\sqrt{2} + 6\sqrt{6})(3\sqrt{10} + \sqrt{7})$

38. **a.** $(5 + 4\sqrt{10})(1 - 2\sqrt{10})$

   b. $(\sqrt{3} + \sqrt{2})(\sqrt{10} + \sqrt{11})$

   c. $(3\sqrt{5} + 4\sqrt{2})(\sqrt{15} + \sqrt{6})$

Use a substitution to verify the solutions to the quadratic equation given.

39. $x^2 - 4x + 1 = 0$

   a. $x = 2 + \sqrt{3}$ **b.** $x = 2 - \sqrt{3}$

40. $x^2 - 10x + 18 = 0$

   a. $x = 5 - \sqrt{7}$ **b.** $x = 5 + \sqrt{7}$

41. $x^2 + 2x - 9 = 0$

   a. $x = -1 + \sqrt{10}$ **b.** $x = -1 - \sqrt{10}$

42. $x^2 - 14x + 29 = 0$

   a. $x = 7 - 2\sqrt{5}$ **b.** $x = 7 + 2\sqrt{5}$

Rationalize each expression by building perfect nth root factors for each denominator. Assume all variables represent positive quantities.

43. **a.** $\dfrac{3}{\sqrt{12}}$ **b.** $\sqrt{\dfrac{20}{27x^3}}$

   c. $\sqrt{\dfrac{27}{50b}}$ **d.** $\sqrt[3]{\dfrac{1}{4p}}$ **e.** $\dfrac{5}{\sqrt[3]{a}}$

44. **a.** $\dfrac{-4}{\sqrt{20}}$ **b.** $\sqrt{\dfrac{125}{12n^3}}$

   c. $\sqrt{\dfrac{5}{12x}}$ **d.** $\sqrt[3]{\dfrac{3}{2m^2}}$ **e.** $\dfrac{-8}{3\sqrt[3]{5}}$

Simplify the following expressions by rationalizing the denominators. Where possible, state results in exact form and approximate form, rounded to hundredths.

45. **a.** $\dfrac{8}{3 + \sqrt{11}}$ **b.** $\dfrac{6}{\sqrt{x} - \sqrt{2}}$

46. **a.** $\dfrac{7}{\sqrt{7} + 3}$ **b.** $\dfrac{12}{\sqrt{x} + \sqrt{3}}$

47. **a.** $\dfrac{\sqrt{10} - 3}{\sqrt{3} + \sqrt{2}}$ **b.** $\dfrac{7 + \sqrt{6}}{3 - 3\sqrt{2}}$

48. **a.** $\dfrac{1 + \sqrt{2}}{\sqrt{6} + \sqrt{14}}$ **b.** $\dfrac{1 + \sqrt{6}}{5 + 2\sqrt{3}}$

▶ WORKING WITH FORMULAS

49. **Fish length to weight relationship:** $L = 1.13(W)^{\frac{1}{3}}$

   The length to weight relationship of a female Pacific halibut can be approximated by the formula shown, where W is the weight in pounds and L is the length in feet. A fisherman lands a halibut that weighs 400 lb. Approximate the length of the fish (round to two decimal places).

50. **Timing a falling object:** $t = \dfrac{\sqrt{s}}{4}$

   The time it takes an object to fall a certain distance is given by the formula shown, where t is the time in seconds and s is the distance the object has fallen. Approximate the time it takes an object to hit the ground, if it is dropped from the top of a building that is 80 ft in height (round to hundredths).

▶ APPLICATIONS

51. Length of a cable: A radio tower is secured by cables that are anchored in the ground 8 m from its base. If the cables are attached to the tower 24 m above the ground, what is the length of each cable? Answer in (a) exact form using radicals, and (b) approximate form by rounding to one decimal place.

52. Height of a kite: Benjamin Franklin is flying his kite in a storm once again. John Adams has walked to a position directly under the kite and is 75 m from Ben. If the kite is 50 m above John Adams' head, how much string S has Ben let out? Answer in (a) exact form using radicals, and (b) approximate form by rounding to one decimal place.

The time T (in days) required for a planet to make one revolution around the sun is modeled by the function $T = 0.407R^{\frac{3}{2}}$, where R is the maximum radius of the planet's orbit (in millions of miles). This is known as *Kepler's third law of planetary motion*. Use the equation given to approximate the number of days required for one complete orbit of each planet, given its maximum orbital radius.

53. a. Earth: 93 million mi

   b. Mars: 142 million mi

   c. Mercury: 36 million mi

54. a. Venus: 67 million mi

   b. Jupiter: 480 million mi

   c. Saturn: 890 million mi

55. Accident investigation: After an accident, police officers will try to determine the approximate velocity V that a car was traveling using the formula $V = 2\sqrt{6L}$, where L is the length of the skid marks in feet and V is the velocity in miles

per hour. (a) If the skid marks were 54 ft long, how fast was the car traveling? (b) Approximate the speed of the car if the skid marks were 90 ft long.

56. Wind-powered energy: If a wind-powered generator is delivering P units of power, the velocity V of the wind (in miles per hour) can be determined using $V = \sqrt[3]{\dfrac{P}{k}}$, where k is a constant that depends on the size and efficiency of the generator. Rationalize the radical expression and use the new version to find the velocity of the wind if $k = 0.004$ and the generator is putting out 13.5 units of power.

57. Surface area: The lateral surface area (surface area excluding the base) S of a cone is given by the formula $S = \pi r \sqrt{r^2 + h^2}$, where r is the radius of the base and h is the height of the cone. Find the surface area of a cone that has a radius of 6 m and a height of 10 m. Answer in simplest form.

58. Surface area: The lateral surface area S of a frustum (a truncated cone) is given by the formula $S = \pi(a + b)\sqrt{h^2 + (b - a)^2}$, where a is the radius of the upper base, b is the radius of the lower base, and h is the height. Find the surface area of a frustum where $a = 6$ m, $b = 8$ m, and $h = 10$ m. Answer in simplest form.

The expression $x^2 - 7$ is not factorable using *integer values*. But the expression *can be written* in the form $x^2 - (\sqrt{7})^2$, enabling us to factor it as a "binomial" and its conjugate: $(x + \sqrt{7})(x - \sqrt{7})$. Use this idea to factor the following expressions.

59. a. $x^2 - 5$ **b.** $n^2 - 19$

60. a. $4v^2 - 11$ **b.** $9w^2 - 11$

▶ EXTENDING THE CONCEPT

61. The following terms $\sqrt{3x} + \sqrt{9x} + \sqrt{27x} + \ldots$ form a pattern that continues until the sixth term is found. **(a)** Compute the sum of all six terms; **(b)** develop a system (investigate the pattern further) that will enable you to find the sum of 12 such terms *without actually writing out the terms*.

62. Find a quick way to simplify the expression without the aid of a calculator.

63. If $\left(x^{\frac{1}{2}} + x^{-\frac{1}{2}}\right)^2 = \dfrac{9}{2}$, find the value of $x^{\frac{1}{2}} + x^{-\frac{1}{2}}$.

64. Rewrite by rationalizing the *numerator*:
$$\dfrac{\sqrt{x + h} - \sqrt{x}}{h}$$

OVERVIEW OF CHAPTER R

Important Definitions, Properties, Formulas, and Relationships

R.1 Notation and Relations

concept	notation	description	example
• Set notation:	{*members*}	braces enclose the members of a set	set of even whole numbers $A = \{0, 2, 4, 6, 8, \ldots\}$
• Is an element of	$\in$	indicates membership in a set	$14 \in A$
• Empty set	Ø or { }	a set having no elements	odd numbers in A
• Is a proper subset of	$\subset$	indicates the elements of one set are entirely contained in another	$S = \{0, 6, 12, 18, 24, \ldots\}$ $S \subset A$
• Defining a set	$\{x \mid x \ldots\}$	the set of all x, *such that $x \ldots$*	$S = \{x \mid x = 6n \text{ for } n \in \mathbb{W}\}$

R.1 Sets of Numbers

- Natural: $\mathbb{N} = \{1, 2, 3, 4, \ldots\}$
- Integers: $\mathbb{Z} = \{\ldots, -3, -2, -1, 0, 1, 2, 3, \ldots\}$
- Irrational: $\mathbb{H} = \{$numbers with a nonterminating, nonrepeating decimal form$\}$

- Whole: $\mathbb{W} = \{0, 1, 2, 3, \ldots\}$
- Rational: $\mathbb{Q} = \left\{\dfrac{p}{q}, \text{where } p, q \in \mathbb{Z}; q \neq 0\right\}$
- Real: $\mathbb{R} = \{$all rational and irrational numbers$\}$

R.1 Absolute Value of a Number

$$|n| = \begin{cases} n & \text{if } n \geq 0 \\ -n & \text{if } n < 0 \end{cases}$$

R.1 Distance between a and b on a number line

$$|a - b| \text{ or } |b - a|$$

R.2 Properties of Real Numbers: For real numbers a, b, and c,

Commutative Property
- Addition: $a + b = b + a$
- Multiplication: $a \cdot b = b \cdot a$

Identities
- Additive: $0 + a = a$
- Multiplicative: $1 \cdot a = a$

Associative Property
- Addition: $(a + b) + c = a + (b + c)$
- Multiplication: $(a \cdot b) \cdot c = a \cdot (b \cdot c)$

Inverses
- Additive: $a + (-a) = 0$
- Multiplicative: $\dfrac{p}{q} \cdot \dfrac{q}{p} = 1; p, q \neq 0$

R.3 Properties of Exponents: For real numbers a and b, and integers m, n, and p
(excluding 0 raised to a nonpositive power),

- Product property: $b^m \cdot b^n = b^{m+n}$
- Product to a power: $(a^m b^n)^p = a^{mp} \cdot b^{np}$
- Quotient property: $\dfrac{b^m}{b^n} = b^{m-n} \ (b \neq 0)$
- Negative exponents: $b^{-n} = \dfrac{1}{b^n}; \left(\dfrac{a}{b}\right)^{-n} = \left(\dfrac{b}{a}\right)^{n}$
  $(a, b \neq 0)$

- Power property: $(b^m)^n = b^{mn}$
- Quotient to a power: $\left(\dfrac{a^m}{b^n}\right)^p = \dfrac{a^{mp}}{b^{np}} \ (b \neq 0)$
- Zero exponents: $b^0 = 1 \ (b \neq 0)$
- Scientific notation: $N \times 10^k; 1 \leq |N| < 10, k \in \mathbb{Z}$

R.3 Special Products

- $(A + B)(A - B) = A^2 - B^2$
- $(A - B)(A^2 + AB + B^2) = A^3 - B^3$

- $(A + B)^2 = A^2 + 2AB + B^2;$
  $(A - B)^2 = A^2 - 2AB + B^2$
- $(A + B)(A^2 - AB + B^2) = A^3 + B^3$

R.4 Special Factorizations

- $A^2 - B^2 = (A + B)(A - B)$
- $A^3 - B^3 = (A - B)(A^2 + AB + B^2)$

- $A^2 \pm 2AB + B^2 = (A \pm B)^2$
- $A^3 + B^3 = (A + B)(A^2 - AB + B^2)$

R.5 Rational Expressions: For polynomials P, Q, R, and S with no denominator of zero,

- Lowest terms: $\dfrac{P \cdot R}{Q \cdot R} = \dfrac{P}{Q}$
- Multiplication: $\dfrac{P}{Q} \cdot \dfrac{R}{S} = \dfrac{P \cdot R}{Q \cdot S} = \dfrac{PR}{QS}$
- Addition: $\dfrac{P}{R} + \dfrac{Q}{R} = \dfrac{P + Q}{R}$

- Equivalence: $\dfrac{P}{Q} = \dfrac{P \cdot R}{Q \cdot R}$
- Division: $\dfrac{P}{Q} \div \dfrac{R}{S} = \dfrac{P}{Q} \cdot \dfrac{S}{R} = \dfrac{PS}{QR}$
- Subtraction: $\dfrac{P}{R} - \dfrac{Q}{R} = \dfrac{P - Q}{R}$

- Addition/subtraction with unlike denominators:
  1. Find the LCD of all rational expressions.
  2. Build equivalent expressions using LCD.
  3. Add/subtract numerators as indicated.
  4. Write the result in lowest terms.

R.6 Properties of Radicals

- $\sqrt{a}$ is a real number only for $a \geq 0$
- $\sqrt[n]{a} = b$, only if $b^n = a$
- For any real number a, $\sqrt[n]{a^n} = |a|$ when n is even

- $\sqrt{a} = b$, only if $b^2 = a$
- If n is even, $\sqrt[n]{a}$ represents a real number only if $a \geq 0$
- For any real number a, $\sqrt[n]{a^n} = a$ when n is odd

- If a is a real number and n is an integer greater than 1, then $\sqrt[n]{a} = a^{\frac{1}{n}}$ provided $\sqrt[n]{a}$ represents a real number

- If $\sqrt[n]{A}$ and $\sqrt[n]{B}$ represent real numbers, $\sqrt[n]{AB} = \sqrt[n]{A} \cdot \sqrt[n]{B}$

- If $\frac{m}{n}$ is a rational number written in lowest terms with $n \geq 2$, then $a^{\frac{m}{n}} = (\sqrt[n]{a})^m$ and $a^{\frac{m}{n}} = \sqrt[n]{a^m}$ provided $\sqrt[n]{a}$ represents a real number.

- If $\sqrt[n]{A}$ and $\sqrt[n]{B}$ represent real numbers and $B \neq 0$, $\sqrt[n]{\dfrac{A}{B}} = \dfrac{\sqrt[n]{A}}{\sqrt[n]{B}}$

- A radical expression is in simplest form when:

1. the radicand has no factors that are perfect nth roots,
2. the radicand contains no fractions, and
3. no radicals occur in a denominator.

R.6 Pythagorean Theorem

- For any right triangle with legs a and b and hypotenuse c: $a^2 + b^2 = c^2$.

- For any triangle with sides a, b, and c, if $a^2 + b^2 = c^2$, then the triangle is a right triangle.

PRACTICE TEST

1. State true or false. If false, state why.

   a. $\mathbb{H} \subset \mathbb{R}$ **b.** $\mathbb{N} \subset \mathbb{Q}$

   c. $\sqrt{2} \in \mathbb{Q}$ **d.** $\dfrac{1}{2} \notin \mathbb{W}$

2. State the value of each expression.

   a. $\sqrt{121}$ **b.** $\sqrt[3]{-125}$

   c. $\sqrt{-36}$ **d.** $\sqrt{400}$

3. Evaluate each expression:

   a. $\dfrac{7}{8} - \left(-\dfrac{1}{4}\right)$ **b.** $-\dfrac{1}{3} - \dfrac{5}{6}$

   c. $-0.7 + 1.2$ **d.** $1.3 + (-5.9)$

4. Evaluate each expression:

   a. $(-4)\left(-2\dfrac{1}{3}\right)$ **b.** $(-0.6)(-1.5)$

   c. $\dfrac{-2.8}{-0.7}$ **d.** $4.2 \div (-0.6)$

 5. Evaluate using a calculator: $2000\left(1 + \dfrac{0.08}{12}\right)^{12 \cdot 10}$

6. State the value of each expression, if possible.

   a. $0 \div 6$ **b.** $6 \div 0$

7. State the number of terms in each expression and identify the coefficient of each.

   a. $-2v^2 + 6v + 5$ **b.** $\dfrac{c + 2}{3} + c$

8. Evaluate each expression given $x = -0.5$ and $y = -2$. Round to hundredths as needed.

   a. $2x - 3y^2$ **b.** $\sqrt{2} - x(4 - x^2) + \dfrac{y}{x}$

9. Translate each phrase into an algebraic expression.

   a. Nine less than twice a number is subtracted from the number cubed.

   b. Three times the square of half a number is subtracted from twice the number.

10. Create a mathematical model using descriptive variables.

   a. The radius of the planet Jupiter is approximately 119 mi less than 11 times the radius of the Earth. Express the radius of Jupiter in terms of the Earth's radius.

   b. Last year, Video Venue Inc. earned $1.2 million more than four times what it earned this year. Express last year's earnings of Video Venue Inc. in terms of this year's earnings.

11. Simplify by combining like terms.

   a. $8v^2 + 4v - 7 + v^2 - v$

   b. $-4(3b - 2) + 5b$

   c. $4x - (x - 2x^2) + x(3 - x)$

12. Factor each expression completely.

   a. $9x^2 - 16$ **b.** $4v^3 - 12v^2 + 9v$

   c. $x^3 + 5x^2 - 9x - 45$

13. Simplify using the properties of exponents.

   a. $\dfrac{5}{b^{-3}}$ **b.** $(-2a^3)^2(a^2b^4)^3$

   c. $\left(\dfrac{m^2}{2n}\right)^3$ **d.** $\left(\dfrac{5p^2q^3r^4}{-2pq^2r^4}\right)^2$

14. Simplify using the properties of exponents.

   a. $\dfrac{-12a^3b^5}{3a^2b^4}$

   b. $(3.2 \times 10^{-17}) \times (2.0 \times 10^{15})$

   c. $\left(\dfrac{a^{-3} \cdot b}{c^{-2}}\right)^{-4}$

   d. $-7x^0 + (-7x)^0$

15. Compute each product.

   a. $(3x^2 + 5y)(3x^2 - 5y)$

   b. $(2a + 3b)^2$

16. Add or subtract as indicated.

   a. $(-5a^3 + 4a^2 - 3) + (7a^4 + 4a^2 - 3a - 15)$

   b. $(2x^2 + 4x - 9) - (7x^4 - 2x^2 - x - 9)$

Simplify or compute as indicated.

17. a. $\dfrac{x - 5}{5 - x}$

   b. $\dfrac{4 - n^2}{n^2 - 4n + 4}$

   c. $\dfrac{x^3 - 27}{x^2 + 3x + 9}$

   d. $\dfrac{3x^2 - 13x - 10}{9x^2 - 4}$

   e. $\dfrac{x^2 - 25}{3x^2 - 11x - 4} \div \dfrac{x^2 + x - 20}{x^2 - 8x + 16}$

   f. $\dfrac{m + 3}{m^2 + m - 12} - \dfrac{2}{5(m + 4)}$

18. a. $\sqrt{(x + 11)^2}$

   b. $\sqrt[3]{\dfrac{-8}{27v^3}}$

   c. $\left(\dfrac{25}{16}\right)^{\frac{-3}{2}}$

   d. $\dfrac{-4 + \sqrt{32}}{8}$

   e. $7\sqrt{40} - \sqrt{90}$

   f. $(x + \sqrt{5})(x - \sqrt{5})$

   g. $\sqrt{\dfrac{2}{5x}}$

   h. $\dfrac{8}{\sqrt{6} - \sqrt{2}}$

19. **Maximizing revenue:** Due to past experience, the manager of a video store knows that if a popular video game is priced at $30, the store will sell 40 each day. For each decrease of $0.50, one additional sale will be made. The formula for the store's revenue is then $R = (30 - 0.5x)(40 + x)$, where x represents the number of times the price is decreased. Multiply the binomials and use a table of values to determine (a) the number of 50¢ decreases that will give the most revenue and (b) the maximum amount of revenue.

20. **Diagonal of a rectangular prism:** Use the Pythagorean theorem to determine the length of the diagonal of the rectangular prism shown in the figure. (*Hint:* First find the diagonal of the base.)

42 cm

24 cm

32 cm

Equations and Inequalities

CHAPTER OUTLINE

CHAPTER CONNECTIONS

The more you understand equations, the better you can apply them in context. Impressed by a friend's 3-hr time in a 10-mi kayaking event (5-mi up river, 5-mi down river), you wish to determine the speed of the kayak in still water knowing only that the river current runs at 4 mph. The techniques illustrated in this chapter will assist you in answering this question. This application appears as Exercise 95 in Section 1.6.

Check out these other real-world connections:

Learning Objectives

In Section 1.1 you will learn how to:

☐ **A.** Solve linear equations using properties of equality

☐ **B.** Recognize equations that are identities or contradictions

☐ **C.** Solve for a specified variable in a formula or literal equation

☐ **D.** Use the problem-solving guide to solve various problem types

In a study of algebra, you will encounter many **families of equations,** or groups of equations that share common characteristics. Of interest to us here is the family of **linear equations in one variable,** a study that lays the foundation for understanding more advanced families. In addition to *solving* linear equations, we'll use the skills we develop to *solve for a specified variable* in a formula, a practice widely used in science, business, industry, and research.

A. Solving Linear Equations Using Properties of Equality

An **equation** is *a statement that two expressions are equal.* From the expressions $3(x - 1) + x$ and $-x + 7$, we can form the equation

$$3(x - 1) + x = -x + 7,$$

which is a **linear equation in one variable.** To solve an equation, we attempt to find a specific input or x-value that will make the equation true, meaning the left-hand expression will be equal to the right. Using Table 1.1, we find that $3(x - 1) + x = -x + 7$ is a true equation when x is replaced by 2, and is a false equation otherwise. Replacement values that make the equation true are called **solutions** or **roots** of the equation.

Table 1.1

x	$3(x - 1) + x$	$-x + 7$
-2	-11	9
-1	-7	8
0	-3	7
1	1	6
2	**5**	**5**
3	9	4
4	13	3

⚠ **CAUTION** ▶ From Section R.6, an algebraic *expression* is a sum or difference of algebraic terms. Algebraic expressions can be simplified, evaluated or written in an equivalent form, but cannot be *"solved,"* since we're not seeking a specific value of the unknown.

Solving equations using a table is too time consuming to be practical. Instead we attempt to write a sequence of **equivalent equations,** each one simpler than the one before, until we reach a point where the solution is obvious. Equivalent equations are those that have the same solution set, and are obtained by using the distributive property to simplify the expressions on each side of the equation, and the additive and multiplicative properties of equality to obtain an equation of the form $x = $ constant.

The Additive Property of Equality	The Multiplicative Property of Equality
If A, B, and C represent algebraic expressions and $A = B$,	If A, B, and C represent algebraic expressions and $A = B$,
then $A + C = B + C$	then $AC = BC$ and $\dfrac{A}{C} = \dfrac{B}{C}, (C \neq 0)$

In words, the additive property says that like quantities, numbers or terms can be added to both sides of an equation. A similar statement can be made for the multiplicative property. These properties are combined into a general guide for solving linear equations, which you've likely encountered in your previous studies. Note that not all steps in the guide are required to solve every equation.

> **Guide to Solving Linear Equations in One Variable**
> - Eliminate parentheses using the distributive property, then combine any like terms.
> - Use the additive property of equality to write the equation with all variable terms on one side, and all constants on the other. Simplify each side.
> - Use the multiplicative property of equality to obtain an equation of the form $x = $ constant.
> - For applications, answer in a complete sentence and include any units of measure indicated.

For our first example, we'll use the equation $3(x - 1) + x = -x + 7$ from our initial discussion.

EXAMPLE 1 ▶ **Solving a Linear Equation Using Properties of Equality**

Solve for x: $3(x - 1) + x = -x + 7$.

Solution ▶

$3(x - 1) + x = -x + 7$	original equation
$3x - 3 + x = -x + 7$	distributive property
$4x - 3 = -x + 7$	combine like terms
$5x - 3 = 7$	add x to both sides (additive property of equality)
$5x = 10$	add 3 to both sides (additive property of equality)
$x = 2$	multiply both sides by $\frac{1}{5}$ or divide both sides by 5 (multiplicative property of equality)

As we noted in Table 1.1, the solution is $x = 2$.

Now try Exercises 7 through 12 ▶

To check a solution by substitution means we substitute the solution back into the original equation (this is sometimes called **back-substitution**), and verify the left-hand side is equal to the right. For Example 1 we have:

$3(x - 1) + x = -x + 7$	original equation
$3(2 - 1) + 2 = -2 + 7$	substitute 2 for x
$3(1) + 2 = 5$	simplify
$5 = 5$ ✓	solution checks

If any coefficients in an equation are fractional, multiply both sides by the least common denominator (LCD) to *clear the fractions*. Since any decimal number can be written in fraction form, the same idea can be applied to decimal coefficients.

EXAMPLE 2 ▶ **Solving a Linear Equation with Fractional Coefficients**

Solve for n: $\frac{1}{4}(n + 8) - 2 = \frac{1}{2}(n - 6)$.

Solution ▶

$\frac{1}{4}(n + 8) - 2 = \frac{1}{2}(n - 6)$	original equation
$\frac{1}{4}n + 2 - 2 = \frac{1}{2}n - 3$	distributive property
$\frac{1}{4}n = \frac{1}{2}n - 3$	combine like terms
$4(\frac{1}{4}n) = 4(\frac{1}{2}n - 3)$	multiply both sides by LCD = 4
$n = 2n - 12$	distributive property
$-n = -12$	subtract $2n$
$n = 12$	multiply by -1

☑ **A.** You've just learned how to solve linear equations using properties of equality

Verify the solution is $n = 12$ using back-substitution.

Now try Exercises 13 through 30 ▶

B. Identities and Contradictions

Example 1 illustrates what is called a **conditional equation,** since the equation is true for $x = 2$, but false for all other values of x. The equation in Example 2 is also conditional. An **identity** is an equation that is *always true,* no matter what value is substituted for the variable. For instance, $2(x + 3) = 2x + 6$ is an identity with a solution set of all real numbers, written as $\{x|x \in \mathbb{R}\}$, or $x \in (-\infty, \infty)$ in interval notation. **Contradictions** are equations that are *never true,* no matter what real number is substituted for the variable. The equations $x - 3 = x + 1$ and $-3 = 1$ are contradictions. To state the solution set for a contradiction, we use the symbol "$\varnothing$" (the null set) or "$\{\ \}$" (the empty set). Recognizing these special equations will prevent some surprise and indecision in later chapters.

EXAMPLE 3 ▶ **Solving an Equation That Is a Contradiction**

Solve for x: $2(x - 4) + 10x = 8 + 4(3x + 1)$, and state the solution set.

Solution ▶
$$2(x - 4) + 10x = 8 + 4(3x + 1) \quad \text{original equation}$$
$$2x - 8 + 10x = 8 + 12x + 4 \quad \text{distributive property}$$
$$12x - 8 = 12x + 12 \quad \text{combine like terms}$$
$$-8 = 12 \quad \text{subtract } 12x$$

Since -8 is never equal to 12, the original equation is a contradiction. The solution is the empty set $\{\ \}$.

Now try Exercises 31 through 36 ▶

☑ **B.** You've just learned how to recognize equations that are identities or contradictions

In Example 3, our attempt to solve for x ended with all variables being eliminated, leaving an equation that is *always false*—a contradiction (-8 is never equal to 12). There is nothing wrong with the solution process, the result is simply telling us the original equation has *no solution.* In other equations, the variables may once again be eliminated, but leave a result that is *always true*—an identity.

C. Solving for a Specified Variable in Literal Equations

A **formula** is an equation that models a known relationship between two or more quantities. A **literal equation** is simply one that has two or more variables. Formulas are a type of literal equation, but not every literal equation is a formula. For example, the formula $A = P + PRT$ models the growth of money in an account earning simple interest, where A represents the total amount accumulated, P is the initial deposit, R is the annual interest rate, and T is the number of years the money is left on deposit. To *describe* $A = P + PRT$, we might say the formula has been "solved for A" or that "A is written in terms of P, R, and T." In some cases, before using a formula it may be convenient to solve for one of the other variables, say P. In this case, P is called the **object variable.**

EXAMPLE 4 ▶ **Solving for Specified Variable**

Given $A = P + PRT$, write P in terms of A, R, and T (solve for P).

Solution ▶ Since the object variable occurs in more than one term, we first apply the distributive property.

$$A = P + PRT \qquad \text{focus on } \boldsymbol{P} \text{—the object variable}$$

$$A = P(1 + RT) \qquad \text{factor out } \boldsymbol{P}$$

$$\frac{A}{1 + RT} = \frac{P(1 + RT)}{(1 + RT)} \qquad \text{solve for } \boldsymbol{P} \text{ [divide by } (1 + RT)\text{]}$$

$$\frac{A}{1 + RT} = P \qquad \text{result}$$

Now try Exercises 37 through 48 ▶

We solve literal equations for a specified variable using the same methods we used for other equations and formulas. Remember that it's good practice to *focus on the object variable* to help guide you through the solution process, as again shown in Example 5.

EXAMPLE 5 ▶ **Solving for a Specified Variable**

Given $2x + 3y = 15$, write y in terms of x (solve for y).

Solution ▶

$$2x + 3y = 15 \qquad \text{focus on the object variable}$$

$$3y = -2x + 15 \qquad \text{subtract } 2x \text{ (isolate term with } y\text{)}$$

$$\tfrac{1}{3}(3y) = \tfrac{1}{3}(-2x + 15) \qquad \text{multiply by } \tfrac{1}{3} \text{ (solve for } y\text{)}$$

$$y = \tfrac{-2}{3}x + 5 \qquad \text{distribute and simplify}$$

Now try Exercises 49 through 54 ▶

WORTHY OF NOTE

In Example 5, notice that in the second step we wrote the subtraction of $2x$ as $-2x + 15$ instead of $15 - 2x$. For reasons that become clear later in this chapter, we generally write variable terms before constant terms.

Literal Equations and General Solutions

Solving literal equations for a specified variable can help us develop the general solution for an entire family of equations. This is demonstrated here for the family of linear equations written in the form $ax + b = c$. A side-by-side comparison with a specific linear equation demonstrates that identical ideas are used.

Specific Equation   **Literal Equation**

$$2x + 3 = 15 \qquad \text{focus on object variable} \qquad ax + b = c$$

$$2x = 15 - 3 \qquad \text{subtract constant} \qquad ax = c - b$$

$$x = \frac{15 - 3}{2} \qquad \text{divide by coefficient} \qquad x = \frac{c - b}{a}$$

Of course the solution on the left would be written as $x = 6$ and checked in the original equation. On the right we now have a general formula for all equations of the form $ax + b = c$.

EXAMPLE 6 ▶ **Solving Equations of the Form $ax + b = c$ Using the General Formula**

Solve $6x - 1 = -25$ using the formula just developed, and check your solution in the original equation.

Solution ▶ For this equation, $a = 6$, $b = -1$, and $c = -25$, this gives

$$x = \frac{c - b}{a}$$

$$= \frac{-25 - (-1)}{6}$$

$$= \frac{-24}{6}$$

$$= -4$$

Check: $6x - 1 = -25$

$6(-4) - 1 = -25$

$-24 - 1 = -25$

$-25 = -25$ ✓

Now try Exercises 55 through 60 ▶

☑ **C.** You've just learned how to solve for a specified variable in a formula or literal equation

D. Using the Problem-Solving Guide

Becoming a good problem solver is an evolutionary process. Over time and with continued effort, your problem-solving skills grow, as will your ability to solve a wider range of applications. Most good problem solvers develop the following characteristics:

- A positive attitude
- A mastery of basic facts
- Strong mental arithmetic skills
- Good mental-visual skills
- Good estimation skills
- A willingness to persevere

These characteristics form a solid basis for applying what we call the **Problem-Solving Guide,** which simply organizes the basic elements of good problem solving. Using this guide will help save you from two common stumbling blocks—indecision and not knowing where to start.

Problem-Solving Guide

- **Gather and organize information.**
  Read the problem several times, forming a mental picture as you read. *Highlight key phrases.* List given information, including any related formulas. *Clearly identify what you are asked to find.*

- **Make the problem visual.**
  Draw and label a diagram or create a table of values, as appropriate. This will help you see how different parts of the problem fit together.

- **Develop an equation model.**
  Assign a variable to represent what you are asked to find and build any related expressions referred to in the exercise. Write an equation model from the information given in the exercise. *Carefully reread the exercise to double-check your equation model.*

- **Use the model and given information to solve the problem.**
  Substitute given values, then simplify and solve. State the answer in sentence form, and check that the answer is reasonable. Include any units of measure indicated.

General Modeling Exercises

In Section R.2, we learned to translate word phrases into symbols. This skill is used to build equations from information given in paragraph form. Sometimes the variable *occurs more than once* in the equation, because two different items in the same exercise are related. If the relationship involves a comparison of size, we often use line segments or bar graphs to model the relative sizes.

EXAMPLE 7 ▶ **Solving an Application Using the Problem-Solving Guide**

The largest state in the United States is Alaska (AK), which covers an area that is 230 square miles (mi^2) more than 500 times that of the smallest state, Rhode Island (RI). If they have a combined area of 616,460 mi^2, how many square miles does each cover?

Solution ▶ Combined area is 616,460 mi^2, AK covers   gather and organize information
230 more than 500 times the area of RI.   highlight any key phrases

                                                                     make the problem visual

Let R represent the area of Rhode Island.   assign a variable
Then $500R + 230$ represents Alaska's area.   build related expressions

Rhode Island's area + Alaska's area = Total
$$R + (500R + 230) = 616{,}460$$ write the equation model
$$501R = 616{,}230$$ combine like terms, subtract 230
$$R = 1230$$ divide by 501

Rhode Island covers an area of 1230 mi^2, while Alaska covers an area of $500(1230) + 230 = 615{,}230$ mi^2.

Now try Exercises 63 through 68 ▶

Consecutive Integer Exercises

Exercises involving **consecutive integers** offer excellent practice in assigning variables to unknown quantities, building related expressions, and the problem-solving process in general. We sometimes work with consecutive **odd** integers or consecutive **even** integers as well.

EXAMPLE 8 ▶ **Solving a Problem Involving Consecutive Odd Integers**

The sum of three consecutive *odd* integers is 69. What are the integers?

Solution ▶ The sum of three consecutive odd integers . . .   gather/organize information
                                                                      highlight any key phrases

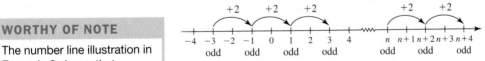
                                                                      make the problem visual

WORTHY OF NOTE

The number line illustration in Example 8 shows that consecutive odd integers are *two units* apart and the related expressions were built accordingly: $n, n + 2, n + 4$, and so on. In particular, *we cannot use $n, n + 1, n + 3, \ldots$* because n and $n + 1$ are *not two units* apart. If we know the exercise involves *even* integers instead, the same model is used, since even integers are also two units apart. For *consecutive* integers, the labels are $n, n + 1, n + 2$, and so on.

Let n represent the smallest consecutive odd integer,   assign a variable
then $n + 2$ represents the second odd integer and   build related expressions
$(n + 2) + 2 = n + 4$ represents the third.

In words: first + second + third odd integer = 69   write the equation model
$$n + (n + 2) + (n + 4) = 69$$ equation model
$$3n + 6 = 69$$ combine like terms
$$3n = 63$$ subtract 6
$$n = 21$$ divide by 3

The odd integers are $n = 21$, $n + 2 = 23$, and $n + 4 = 25$.

$21 + 23 + 25 = 69$ ✓

Now try Exercises 69 through 72 ▶

Uniform Motion (Distance, Rate, Time) Exercises

Uniform motion problems have many variations, and it's important to draw a good diagram when you get started. Recall that if speed is constant, the distance traveled is equal to the rate of speed multiplied by the time in motion: $D = RT$.

EXAMPLE 9 ▶ Solving a Problem Involving Uniform Motion

I live 260 mi from a popular mountain retreat. On my way there to do some mountain biking, my car had engine trouble—forcing me to bike the rest of the way. If I drove 2 hr longer than I biked and averaged 60 miles per hour driving and 10 miles per hour biking, how many hours did I spend pedaling to the resort?

Solution ▶ The sum of the two distances must be 260 mi. The **rates** are given, and the driving time is *2 hr more than biking* time.

gather/organize information
highlight any key phrases
make the problem visual

$$D_1 + D_2 = \text{Total distance}$$
260 miles

Let t represent the biking time,
then $T = t + 2$ represents time spent driving.

$$D_1 + D_2 = 260$$
$$RT + rt = 260$$
$$60(t + 2) + 10t = 260$$
$$70t + 120 = 260$$
$$70t = 140$$
$$t = 2$$

assign a variable
build related expressions

write the equation model

$RT = D_1, rt = D_2$

substitute $t + 2$ for T, 60 for R, 10 for r

distribute and combine like terms

subtract 120

divide by 70

I rode my bike for $t = 2$ hr, after driving $t + 2 = 4$ hr.

Now try Exercises 73 through 76 ▶

Exercises Involving Mixtures

Mixture problems offer another opportunity to refine our problem-solving skills while using many elements from the problem-solving guide. They also lend themselves to a very useful mental-visual image and have many practical applications.

EXAMPLE 10 ▶ Solving an Application Involving Mixtures

As a nasal decongestant, doctors sometimes prescribe saline solutions with a concentration between 6% and 20%. In "the old days," pharmacists had to create different mixtures, but only needed to stock these concentrations, since any percentage in between could be obtained using a mixture. An order comes in for a 15% solution. How many milliliters (mL) of the 20% solution must be mixed with 10 mL of the 6% solution to obtain the desired 15% solution?

Solution ▶ Only 6% and 20% concentrations are available;   gather/organize information
              mix a 20% solution with 10 mL of a 6% solution   highlight any key phrases

make the problem visual

Let x represent the amount of 20% solution, then   assign a variable
$10 + x$ represents the total amount of 15% solution.   build related expressions

1st quantity times its concentration		2nd quantity times its concentration		1st+2nd quantity times desired concentration	
$10(0.06)$	$+$	$x(0.2)$	$=$	$(10 + x)(0.15)$	write equation model
0.6	$+$	$0.2x$	$=$	$1.5 + 0.15x$	distribute/simplify
		$0.2x$	$=$	$0.9 + 0.15x$	subtract 0.6
		$0.05x$	$=$	0.9	subtract 0.15x
		x	$=$	18	divide by 0.05

☑ **D.** You've just learned how to use the problem-solving guide to solve various problem types

To obtain a 15% solution, 18 mL of the 20% solution must be mixed with 10 mL of the 6% solution.

Now try Exercises 77 through 84 ▶

TECHNOLOGY HIGHLIGHT

Using a Graphing Calculator as an Investigative Tool

The mixture concept can be applied in a wide variety of ways, including mixing zinc and copper to get bronze, different kinds of nuts for the holidays, diversifying investments, or mixing two acid solutions in order to get a desired concentration. Whether the value of each part in the mix is monetary or a percent of concentration, the general mixture equation has this form:

Quantity 1 · Value I + Quantity 2 · Value II = Total quantity · Desired value

Graphing calculators are a great tool for exploring this relationship, because the TABLE feature enables us to test the result of various mixtures in an instant. Suppose 10 oz of an 80% glycerin solution are to be mixed with an unknown amount of a 40% solution. How much of the 40% solution is used if a 56% solution is needed? To begin, we might consider that using equal amounts of the 40% and 80% solutions would result in a 60% concentration (halfway between 40% and 80%). To illustrate, let C represent the final concentration of the mix.

$$10(0.8) + 10(0.4) = (10 + 10)C \quad \text{equal amounts}$$
$$8 + 4 = 20C \quad \text{simplify}$$
$$12 = 20C \quad \text{add}$$
$$0.6 = C \quad \text{divide by 20}$$

Figure 1.1	Figure 1.2	Figure 1.3

Figure 1.1
```
Plot1 Plot2 Plot3
\Y1◼.8(10)+.4X
\Y2◼.56(10+X)
\Y3=
\Y4=
\Y5=
\Y6=
\Y7=
```

Figure 1.2
```
TABLE SETUP
 TblStart=10
 ΔTbl=1
Indpnt: Auto Ask
Depend: Auto Ask
```

Figure 1.3
X	Y1	Y2
10	12	11.2
11	12.4	11.76
12	12.8	12.32
13	13.2	12.88
14	13.6	13.44
15	14	14
16	14.4	14.56

X=10

Since this is too high a concentration (a 56% = 0.56 solution is desired), we know more of the weaker solution should be used. To explore the relationship further, assume x oz of the 40% solution are used and enter the resulting equation on the **Y=** screen as $Y_1 = .8(10) + .4X$. Enter the result of the mix as $Y_2 = .56(10 + X)$ (see Figure 1.1). Next, set up a TABLE using **2nd** **WINDOW** (**TBLSET**) with **TblStart** = 10, Δ**Tbl** = 1, and the calculator set in Indpnt: **AUTO** mode (see Figure 1.2). Finally, access the TABLE results using **2nd** **GRAPH** (**TABLE**). The resulting screen is shown in Figure 1.3, where we note that 15 oz of the 40% solution should be used (the equation is true when X is 15: $Y_1 = Y_2$).

Exercise 1:   Use this idea to solve Exercises 81 and 82 from the Exercises.

1.1 EXERCISES

▶ **CONCEPTS AND VOCABULARY**

Fill in each blank with the appropriate word or phrase. Carefully reread the section, if necessary.

1. A(n) _____ is an equation that is always true, regardless of the _____ value.

2. A(n) _____ is an equation that is always false, regardless of the _____ value.

3. A(n) _____ equation is an equation having _____ or more unknowns.

4. For the equation $S = 2\pi r^2 + 2\pi rh$, we can say that S is written in terms of ____ and ____.

5. Discuss/Explain the three tests used to identify a linear equation. Give examples and counterexamples in your discussion.

6. Discuss/Explain each of the four basic parts of the *problem-solving guide*. Include a solved example in your discussion.

▶ **DEVELOPING YOUR SKILLS**

Solve each equation. Check your answer by substitution.

7. $4x + 3(x - 2) = 18 - x$

8. $15 - 2x = -4(x + 1) + 9$

9. $21 - (2v + 17) = -7 - 3v$

10. $-12 - 5w = -9 - (6w + 7)$

11. $8 - (3b + 5) = -5 + 2(b + 1)$

12. $2a + 4(a - 1) = 3 - (2a + 1)$

Solve each equation.

13. $\frac{1}{5}(b + 10) - 7 = \frac{1}{3}(b - 9)$

14. $\frac{1}{6}(n - 12) = \frac{1}{4}(n + 8) - 2$

15. $\frac{2}{3}(m + 6) = \frac{-1}{2}$

16. $\frac{4}{5}(n - 10) = \frac{-8}{9}$

17. $\frac{1}{2}x + 5 = \frac{1}{3}x + 7$ 18. $-4 + \frac{2}{3}y = \frac{1}{2}y - 5$

19. $\frac{x+3}{5} + \frac{x}{3} = 7$ 20. $\frac{z-4}{6} - 2 = \frac{z}{2}$

21. $15 = -6 - \frac{3p}{8}$ 22. $-15 - \frac{2q}{9} = -21$

23. $0.2(24 - 7.5a) - 6.1 = 4.1$

24. $0.4(17 - 4.25b) - 3.15 = 4.16$

25. $6.2v - (2.1v - 5) = 1.1 - 3.7v$

26. $7.9 - 2.6w = 1.5w - (9.1 + 2.1w)$

27. $\frac{n}{2} + \frac{n}{5} = \frac{2}{3}$

28. $\frac{m}{3} - \frac{2}{5} = \frac{m}{4}$

29. $3p - \frac{p}{4} - 5 = \frac{p}{6} - 2p + 6$

30. $\frac{q}{6} + 1 - 3q = 2 - 4q + \frac{q}{8}$

Identify the following equations as an identity, a contradiction, or a conditional equation, then state the solution.

31. $-3(4z + 5) = -15z - 20 + 3z$

32. $5x - 9 - 2 = -5(2 - x) - 1$

33. $8 - 8(3n + 5) = -5 + 6(1 + n)$

34. $2a + 4(a - 1) = 1 + 3(2a + 1)$

35. $-4(4x + 5) = -6 - 2(8x + 7)$

36. $-(5x - 3) + 2x = 11 - 4(x + 2)$

Solve for the specified variable in each formula or literal equation.

37. $P = C + CM$ for C (retail)

38. $S = P - PD$ for P (retail)

39. $C = 2\pi r$ for r (geometry)

40. $V = LWH$ for W (geometry)

41. $\frac{P_1 V_1}{T_1} = \frac{P_2 V_2}{T_2}$ for T_2 (science)

42. $\frac{C}{P_2} = \frac{P_1}{d^2}$ for P_2 (communication)

43. $V = \frac{4}{3}\pi r^2 h$ for h (geometry)

44. $V = \frac{1}{3}\pi r^2 h$ for h (geometry)

45. $S_n = n\left(\frac{a_1 + a_n}{2}\right)$ for n (sequences)

46. $A = \frac{h(b_1 + b_2)}{2}$ for h (geometry)

47. $S = B + \frac{1}{2}PS$ for P (geometry)

48. $s = \frac{1}{2}gt^2 + vt$ for g (physics)

49. $Ax + By = C$ for y

50. $2x + 3y = 6$ for y

51. $\frac{5}{6}x + \frac{3}{8}y = 2$ for y

52. $\frac{2}{3}x - \frac{7}{9}y = 12$ for y

53. $y - 3 = \frac{-4}{5}(x + 10)$ for y

54. $y + 4 = \frac{-2}{15}(x + 10)$ for y

The following equations are given in $ax + b = c$ form. Solve by identifying the value of $a, b,$ and c, then using the formula $x = \frac{c - b}{a}$.

55. $3x + 2 = -19$

56. $7x + 5 = 47$

57. $-6x + 1 = 33$

58. $-4x + 9 = 43$

59. $7x - 13 = -27$

60. $3x - 4 = -25$

▶ WORKING WITH FORMULAS

61. Surface area of a cylinder: $SA = 2\pi r^2 + 2\pi rh$

The surface area of a cylinder is given by the formula shown, where h is the height of the cylinder and r is the radius of the base. Find the height of a cylinder that has a radius of 8 cm and a surface area of 1256 cm^2. Use $\pi \approx 3.14$.

62. Using the equation-solving process for Exercise 61 as a model, solve the formula $SA = 2\pi r^2 + 2\pi rh$ for h.

▶ **APPLICATIONS**

Solve by building an equation model and using the
problem-solving guidelines as needed.

General Modeling Exercises

63. Two spelunkers (cave explorers) were exploring
different branches of an underground cavern. The
first was able to descend 198 ft farther than twice
the second. If the first spelunker descended a
1218 ft, how far was the second spelunker able
to descend?

64. The area near the joining of the Tigris and
Euphrates Rivers (in modern Iraq) has often been
called the *Cradle of Civilization,* since the area has
evidence of many ancient cultures. The length of
the Euphrates River exceeds that of the Tigris by
620 mi. If they have a combined length of 2880 mi,
how long is each river?

65. U.S. postal regulations
require that a package
can have a maximum
combined length and
girth (distance around)
of 108 in. A shipping
carton is constructed so
that it has a width of
14 in., a height of 12
in., and can be cut or folded to various lengths.
What is the maximum length that can be used?
Source: www.USPS.com

Girth

H   *L*

W

66. Hi-Tech Home Improvements buys a fleet of
identical trucks that cost $32,750 each. The company
is allowed to depreciate the value of their trucks for
tax purposes by $5250 per year. If company policies
dictate that older trucks must be sold once their value
declines to $6500, approximately how many years
will they keep these trucks?

67. The longest suspension bridge in the world is the
Akashi Kaikyo (Japan) with a length of 6532 feet.
Japan is also home to the Shimotsui Straight
bridge. The Akashi Kaikyo bridge is 364 ft more
than twice the length of the Shimotsui bridge. How
long is the Shimotsui bridge?
Source: www.guinnessworldrecords.com

68. The Mars rover *Spirit* landed on January 3, 2004.
Just over 1 yr later, on January 14, 2005, the
Huygens probe landed on Titan (one of Saturn's
moons). At their closest approach, the distance
from the Earth to Saturn is 29 million mi more than
21 times the distance from the Earth to Mars. If the
distance to Saturn is 743 million mi, what is the
distance to Mars?

Consecutive Integer Exercises

69. Find two consecutive even integers such that the
sum of twice the smaller integer plus the larger
integer is one hundred forty-six.

70. When the smaller of two consecutive integers is
added to three times the larger, the result is fifty-
one. Find the smaller integer.

71. Seven times the first of two consecutive odd
integers is equal to five times the second. Find each
integer.

72. Find three consecutive even integers where the sum
of triple the first and twice the second is eight more
than four times the third.

Uniform Motion Exercises

73. At 9:00 A.M., Linda leaves work on a business trip,
gets on the interstate, and sets her cruise control at
60 mph. At 9:30 A.M., Bruce notices she's left her
briefcase and cell phone, and immediately starts
after her driving 75 mph. At what time will Bruce
catch up with Linda?

74. A plane flying at 300 mph has a 3-hr head start on
a "chase plane," which has a speed of 800 mph.
How far from the airport will the chase plane
overtake the first plane?

75. Jeff had a job interview in a nearby city 72 mi away.
On the first leg of the trip he drove an average of
30 mph through a long construction zone, but was
able to drive 60 mph after passing through this zone.
If driving time for the trip was $1\frac{1}{2}$ hr, how long was
he driving in the construction zone?

76. At a high-school cross-country meet, Jared jogged
8 mph for the first part of the race, then increased
his speed to 12 mph for the second part. If the race
was 21 mi long and Jared finished in 2 hr, how far
did he jog at the faster pace?

Mixture Exercises

Give the total amount of the mix that results and the percent concentration or worth of the mix.

77. Two quarts of 100% orange juice are mixed with 2 quarts of water (0% juice).

78. Ten pints of a 40% acid are combined with 10 pints of an 80% acid.

79. Eight pounds of premium coffee beans worth $2.50 per pound are mixed with 8 lb of standard beans worth $1.10 per pound.

80. A rancher mixes 50 lb of a custom feed blend costing $1.80 per pound, with 50 lb of cheap cottonseed worth $0.60 per pound.

Solve each application of the mixture concept.

81. To help sell more of a lower grade meat, a butcher mixes some premium ground beef worth $3.10/lb, with 8 lb of lower grade ground beef worth $2.05/lb. If the result was an intermediate grade of ground beef worth $2.68/lb, how much premium ground beef was used?

82. Knowing that the camping/hiking season has arrived, a nutrition outlet is mixing GORP (Good Old Raisins and Peanuts) for the anticipated customers. How many pounds of peanuts worth $1.29/lb, should be mixed with 20 lb of deluxe raisins worth $1.89/lb, to obtain a mix that will sell for $1.49/lb?

83. How many pounds of walnuts at 84¢/lb should be mixed with 20 lb of pecans at $1.20/lb to give a mixture worth $1.04/lb?

84. How many pounds of cheese worth 81¢/lb must be mixed with 10 lb cheese worth $1.29/lb to make a mixture worth $1.11/lb?

▶ **EXTENDING THE THOUGHT**

85. Look up and read the following article. Then turn in a one page summary. "Don't Give Up!," William H. Kraus, *Mathematics Teacher*, Volume 86, Number 2, February 1993: pages 110–112.

86. A chemist has four solutions of a very rare and expensive chemical that are 15% acid (cost $120 per ounce), 20% acid (cost $180 per ounce), 35% acid (cost $280 per ounce) and 45% acid (cost $359 per ounce). She requires 200 oz of a 29% acid solution. Find the combination of any two of these concentrations that will minimize the total cost of the mix.

87. P, Q, R, S, T, and U represent numbers. The arrows in the figure show the sum of the two or three numbers added in the indicated direction

(Example: $Q + T = 23$). Find $P + Q + R + S + T + U$.

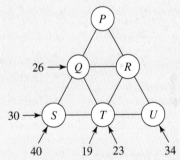

88. Given a sphere circumscribed by a cylinder, verify the volume of the sphere is $\frac{2}{3}$ that of the cylinder.

▶ **MAINTAINING YOUR SKILLS**

89. (R.1) Simplify the expression using the order of operations.

$$-2 - 6^2 \div 4 + 8$$

90. (R.3) Name the coefficient of each term in the expression:

$$-3v^3 + v^2 - \frac{v}{3} + 7$$

91. (R.4) Factor each expression:

    a. $4x^2 - 9$ **b.** $x^3 - 27$

92. (R.2) Identify the property illustrated:

$$\frac{6}{7} \cdot 5 \cdot 21 = \frac{6}{7} \cdot 21 \cdot 5$$

Learning Objectives

In Section 1.2 you will learn how to:

☐ **A.** Solve inequalities and state solution sets

☐ **B.** Solve linear inequalities

☐ **C.** Solve compound inequalities

☐ **D.** Solve applications of inequalities

There are many real-world situations where the mathematical model leads to a statement of *inequality* rather than equality. Here are a few examples:

Clarice wants to buy a house costing $85,000 or less.

To earn a "B," Shantë must score more than 90% on the final exam.

To escape the Earth's gravity, a rocket must travel 25,000 mph or more.

While conditional linear equations in one variable have a single solution, linear inequalities often have an *infinite number of solutions*—which means we must develop additional methods for writing a solution set.

A. Inequalities and Solution Sets

WORTHY OF NOTE

Some texts will use an open dot "○" to mark the location of an endpoint that is not included, and a closed dot "●" for an included endpoint.

The set of numbers that satisfy an inequality is called the **solution set.** Instead of using a simple inequality to write solution sets, we will often use (1) a form of **set notation,** (2) a **number line** graph, or (3) **interval notation.** Interval notation is a symbolic way of indicating a selected interval of the real number line. When a number acts as the **boundary point** for an interval (also called an **endpoint**), we use a left bracket "[" or a right bracket "]" to indicate **inclusion** of the endpoint. If the boundary point is **not included,** we use a left parenthesis "(" or right parenthesis ")."

EXAMPLE 1 ▶ **Using Inequalities in Context**

Model the given phrase using the correct inequality symbol. Then state the result in set notation, graphically, and in interval notation: "If the ball had traveled at least one more foot in the air, it would have been a home run."

Solution ▶ Let d represent additional distance: $d \geq 1$.

WORTHY OF NOTE

Since infinity is really a *concept* and not a number, it is *never included* (using a bracket) as an endpoint for an interval.

- Set notation: $\{d \mid d \geq 1\}$
- Graph
- Interval notation: $d \in [1, \infty)$

Now try Exercises 7 through 18 ▶

The "$\in$" symbol says the number d is *an element of the set or interval* given. The "∞" symbol represents positive infinity and indicates the interval continues forever to the right. Note that the endpoints of an interval must occur in the same order as on the number line *(smaller value on the left; larger value on the right).*

A short summary of other possibilities is given here. Many variations are possible.

A. You've just learned how to solve inequalities and state solution sets

Conditions ($a < b$)	Set Notation	Number Line	Interval Notation
x is greater than k	$\{x \mid x > k\}$		$x \in (k, \infty)$
x is less than or equal to k	$\{x \mid x \leq k\}$		$x \in (-\infty, k]$
x is less than b and greater than a	$\{x \mid a < x < b\}$		$x \in (a, b)$
x is less than b and greater than or equal to a	$\{x \mid a \leq x < b\}$		$x \in [a, b)$
x is less than a or x is greater than b	$\{x \mid x < a$ or $x > b\}$		$x \in (-\infty, a) \cup (b, \infty)$

B. Solving Linear Inequalities

A linear *inequality* resembles a linear *equality* in many respects:

	Linear Inequality	**Related Linear Equation**
(1)	$x < 3$	$x = 3$
(2)	$\dfrac{3}{8}p - 2 \geq -12$	$\dfrac{3}{8}p - 2 = -12$

A linear inequality in one variable is one that can be written in the form $ax + b < c$, where a, b, and $c \in \mathbb{R}$ and $a \neq 0$. This definition and the following properties also apply when other inequality symbols are used. Solutions to simple inequalities are easy to spot. For instance, $x = -2$ is a solution to $x < 3$ since $-2 < 3$. For more involved inequalities we use the **additive property of inequality** and the **multiplicative property of inequality.** Similar to solving equations, we solve inequalities by isolating the variable on one side to obtain a solution form such as *variable < number*.

The Additive Property of Inequality

If A, B, and C represent algebraic expressions and $A < B$,

$$\text{then} \quad A + C < B + C$$

Like quantities (numbers or terms) can be added to both sides of an inequality.

While there is little difference between the additive property of *equality* and the additive property of *inequality,* there is an *important difference* between the multiplicative property of *equality* and the multiplicative property of *inequality.* To illustrate, we begin with $-2 < 5$. Multiplying both sides by positive three yields $-6 < 15$, a true inequality. But notice what happens when we **multiply both sides by negative three:**

$$-2 < 5 \qquad \text{original inequality}$$
$$-2(-3) < 5(-3) \qquad \text{multiply by negative three}$$
$$6 < -15 \qquad \text{false}$$

This is a *false* inequality, because 6 is *to the right* of -15 on the number line. Multiplying (or dividing) an inequality by a negative quantity *reverses the order relationship between two quantities* (we say it changes the *sense* of the inequality). We must compensate for this by reversing the inequality symbol.

$$6 > -15 \qquad \text{change direction of symbol to maintain a true statement}$$

For this reason, the multiplicative property of inequality is stated in two parts.

The Multiplicative Property of Inequality

If A, B, and C represent algebraic expressions and $A < B$,	If A, B, and C represent algebraic expressions and $A < B$,
then $\quad AC < BC$	then $\quad AC > BC$
if C is a *positive quantity* (inequality symbol remains the same).	if C is a *negative quantity* (inequality symbol must be reversed).

EXAMPLE 2 ▶ **Solving an Inequality**

Solve the inequality, then graph the solution set and write it in interval notation: $\frac{-2}{3}x + \frac{1}{2} \leq \frac{5}{6}$.

Solution ▶

$$\frac{-2}{3}x + \frac{1}{2} \leq \frac{5}{6} \qquad \text{original inequality}$$

$$6\left(\frac{-2}{3}x + \frac{1}{2}\right) \leq (6)\frac{5}{6} \qquad \text{clear fractions (multiply by LCD)}$$

$$-4x + 3 \leq 5 \qquad \text{simplify}$$

$$-4x \leq 2 \qquad \text{subtract 3}$$

$$x \geq -\frac{1}{2} \qquad \text{divide by } -4, \textit{reverse inequality sign}$$

WORTHY OF NOTE

As an alternative to multiplying or dividing by a negative value, the additive property of inequality can be used to ensure the variable term will be positive. From Example 2, the inequality $-4x \leq 2$ can be written as $-2 \leq 4x$ by adding $4x$ to both sides and subtracting 2 from both sides. This gives the solution $-\frac{1}{2} \leq x$, which is equivalent to $x \geq -\frac{1}{2}$.

• Graph:

$-\frac{1}{2}$

$$\begin{array}{c} \quad\quad\quad\quad -\frac{1}{2} \\ \longleftarrow\!\!+\!\!-\!\!+\!\!-\!\![\!\!-\!\!+\!\!-\!\!+\!\!-\!\!+\!\!-\!\!+\!\!\longrightarrow \\ -3 \;-2\; -1 \quad 0 \quad 1 \quad 2 \quad 3 \quad 4 \end{array}$$

• Interval notation: $x \in \left[-\frac{1}{2}, \infty\right)$

Now try Exercises 19 through 28 ▶

To check a linear inequality, you often have an infinite number of choices—any number from the solution set/interval. If a test value from the solution interval results in a true inequality, all numbers in the interval are solutions. For Example 2, using $x = 0$ results in the true statement $\frac{1}{2} \leq \frac{5}{6}$ ✓.

Some inequalities have all real numbers as the solution set: $\{x | x \in \mathbb{R}\}$, while other inequalities have no solutions, with the answer given as the empty set: $\{\quad\}$.

EXAMPLE 3 ▶ **Solving Inequalities**

Solve the inequality and write the solution in set notation:
a. $7 - (3x + 5) \geq 2(x - 4) - 5x$ **b.** $3(x + 4) - 5 < 2(x - 3) + x$

Solution ▶

a. $7 - (3x + 5) \geq 2(x - 4) - 5x$ original inequality
$\quad 7 - 3x - 5 \geq 2x - 8 - 5x$ distributive property
$\quad\quad -3x + 2 \geq -3x - 8$ combine like terms
$\quad\quad\quad\quad 2 \geq -8$ add $3x$

Since the resulting statement is always true, the original inequality is true for all real numbers. The solution is $\{x | x \in \mathbb{R}\}$.

b. $3(x + 4) - 5 < 2(x - 3) + x$ original inequality
$\quad 3x + 12 - 5 < 2x - 6 + x$ distribute
$\quad\quad 3x + 7 < 3x - 6$ combine like terms
$\quad\quad\quad\quad 7 < -6$ subtract $3x$

☑ **B.** You've just learned how to solve linear inequalities

Since the resulting statement is always false, the original inequality is false for all real numbers. The solution is $\{\quad\}$.

Now try Exercises 29 through 34 ▶

C. Solving Compound Inequalities

In some applications of inequalities, we must consider more than one solution interval. These are called **compound inequalities,** and require us to take a close look at the operations of **union** "∪" and **intersection** "∩". The intersection of two sets A and B, written $A \cap B$, is the set of all elements *common to both sets*. The union of two sets A and B, written $A \cup B$, is the set of all elements *that are in either set*. When stating the union of two sets, repetitions are unnecessary.

EXAMPLE 4 ▶ **Finding the Union and Intersection of Two Sets**

For set $A = \{-2, -1, 0, 1, 2, 3\}$ and set $B = \{1, 2, 3, 4, 5\}$, determine $A \cap B$ and $A \cup B$.

Solution ▶ $A \cap B$ is the set of all elements in *both A and B*:
$A \cap B = \{1, 2, 3\}$.
$A \cup B$ is the set of all elements in *either A or B*:
$A \cup B = \{-2, -1, 0, 1, 2, 3, 4, 5\}$.

WORTHY OF NOTE

For the long term, it may help to rephrase the distinction as follows. The intersection is a *selection* of elements that are common to two sets, while the union is a *collection* of the elements from two sets (with no repetitions).

Now try Exercises 35 through 40 ▶

Notice the intersection of two sets is described using the word "and," while the union of two sets is described using the word "or." When compound inequalities are formed using these words, the solution is modeled after the ideas from Example 4. If "and" is used, the solutions must satisfy *both* inequalities. If "or" is used, the solutions can satisfy *either* inequality.

EXAMPLE 5 ▶ **Solving a Compound Inequality**

Solve the compound inequality, then write the solution in interval notation:
$-3x - 1 < -4$ **or** $4x + 3 < -6$.

Solution ▶ Begin with the statement as given:

WORTHY OF NOTE

The graphs from Example 5 clearly show the solution consists of two disjoint (disconnected) intervals. This is reflected in the "or" statement: $x < -\frac{9}{4}$ or $x > 1$, and in the interval notation. Also, note the solution $x < -\frac{9}{4}$ or $x > 1$ is not equivalent to $-\frac{9}{4} > x > 1$, as there is no single number that is both greater than 1 and less than $-\frac{9}{4}$ at the same time.

$-3x - 1 < -4$	or	$4x + 3 < -6$	original statement
$-3x < -3$	or	$4x < -9$	isolate variable term
$x > 1$	or	$x < -\dfrac{9}{4}$	solve for x, reverse first inequality symbol

The solution $x > 1$ **or** $x < -\frac{9}{4}$ is better understood by graphing each interval separately, *then selecting both intervals (the union).*

Interval notation: $x \in \left(-\infty, -\dfrac{9}{4}\right) \cup (1, \infty)$.

Now try Exercises 41 and 42 ▶

EXAMPLE 6 ▶ **Solving a Compound Inequality**

Solve the compound inequality, then write the solution in interval notation:
$3x + 5 > -13$ **and** $3x + 5 < -1$.

Solution ▶ Begin with the statement as given:

$3x + 5 > -13$	and	$3x + 5 < -1$	original statement
$3x > -18$	and	$3x < -6$	subtract five
$x > -6$	and	$x < -2$	divide by 3

WORTHY OF NOTE

The inequality $a < b$ (a is less than b) can equivalently be written as $b > a$ (b is greater than a). In Example 6, the solution is read, "$x > -6$ and $x < -2$," but if we rewrite the first inequality as $-6 < x$ (with the "arrowhead" still pointing at -6), we have $-6 < x$ and $x < -2$ and can clearly see that x must be in the single interval between -6 and -2.

The solution $x > -6$ **and** $x < -2$ can best be understood by graphing each interval separately, then *noting where they intersect.*

Interval notation: $x \in (-6, -2)$.

Now try Exercises 43 through 54 ▶

The solution from Example 6 consists of the single interval $(-6, -2)$, indicating the original inequality could actually be *joined* and written as $-6 < x < -2$, called a **joint** or **compound inequality** (see Worthy of Note). We solve joint inequalities in much the same way as linear inequalities, but must remember they *have three parts (left, middle, and right)*. This means operations must be applied to *all three parts* in each step of the solution process, to obtain a solution form such as *smaller number $< x <$ larger number*. The same ideas apply when other inequality symbols are used.

EXAMPLE 7 ▶   Solving a Compound Inequality

Solve the compound inequality, then graph the solution set and write it in interval notation: $1 > \dfrac{2x + 5}{-3} \geq -6$.

Solution ▶

$1 > \dfrac{2x + 5}{-3} \geq -6$ original inequality

$-3 < 2x + 5 \leq 18$ multiply all parts by -3; reverse the inequality symbols

$-8 < 2x \leq 13$ subtract 5 from all parts

$-4 < x \leq \dfrac{13}{2}$ divide all parts by 2

☑ C. You've just learned how to solve compound inequalities

• Graph:

• Interval notation: $x \in \left(-4, \frac{13}{2}\right]$

Now try Exercises 55 through 60 ▶

D. Applications of Inequalities

Domain and Allowable Values

One application of inequalities involves the concept of allowable values. Consider the expression $\frac{24}{x}$. As Table 1.2 suggests, we can evaluate this expression using any real number *other than zero*, since the expression $\frac{24}{0}$ is undefined. Using set notation the allowable values are written $\{x | x \in \mathbb{R}, x \neq 0\}$. To graph the solution we must be careful to exclude zero, as shown in Figure 1.4.

Table 1.2

x	$\frac{24}{x}$
6	4
-12	-2
$\frac{1}{2}$	48
0	error

The graph gives us a snapshot of the solution using interval notation, which is written as a union of two **disjoint (disconnected) intervals** so as to exclude zero: $x \in (-\infty, 0) \cup (0, \infty)$. The set of allowable values is referred to as the **domain** of the expression. Allowable values are said to be "*in the domain*" of the expression; values that are not allowed are said to be "*outside the domain.*" When the denominator of a fraction contains a variable expression, values that cause a denominator of zero are outside the domain.

Figure 1.4

EXAMPLE 8 ▶ **Determining the Domain of an Expression**

Determine the domain of the expression $\dfrac{6}{x-2}$. State the result in set notation, graphically, and using interval notation.

Solution ▶ Set the denominator equal to zero and solve: $x - 2 = 0$ yields $x = 2$. This means 2 is outside the domain and *must be excluded.*

- Set notation: $\{x \mid x \in \mathbb{R}, x \neq 2\}$

- Graph:

- Interval notation: $x \in (-\infty, 2) \cup (2, \infty)$

Now try Exercises 61 through 68 ▶

A second area where allowable values are a concern involves the square root operation. Recall that $\sqrt{49} = 7$ since $7 \cdot 7 = 49$. However, $\sqrt{-49}$ cannot be written as the product of two real numbers since $(-7) \cdot (-7) = 49$ and $7 \cdot 7 = 49$. In other words, $\sqrt{X}$ represents a real number only if the radicand is positive or zero. If X represents an algebraic expression, the domain of $\sqrt{X}$ is $\{X \mid X \geq 0\}$.

EXAMPLE 9 ▶ **Determining the Domain of an Expression**

Determine the domain of $\sqrt{x + 3}$. State the domain in set notation, graphically, and in interval notation.

Solution ▶ The radicand must represent a nonnegative number. Solving $x + 3 \geq 0$ gives $x \geq -3$.

- Set notation: $\{x \mid x \geq -3\}$

- Graph:

- Interval notation: $x \in [-3, \infty)$

Now try Exercises 69 through 76 ▶

Inequalities are widely used to help gather information, and to make comparisons that will lead to informed decisions. Here, the problem-solving guide is once again a valuable tool.

EXAMPLE 10 ▶ **Using an Inequality to Compute Desired Test Scores**

Justin earned scores of 78, 72, and 86 on the first three out of four exams. What score must he earn on the fourth exam to have an average of at least 80?

Solution ▶ **Gather and organize information;** highlight any key phrases.
First the scores: 78, 72, 86. An average of *at least* 80 means $A \geq 80$.
Make the problem visual.

Test 1	Test 2	Test 3	Test 4	Computed Average	Minimum
78	72	86	x	$\dfrac{78 + 72 + 86 + x}{4}$	80

Assign a variable; build related expressions.

Let x represent Justin's score on the fourth exam, then $\dfrac{78 + 72 + 86 + x}{4}$ represents his average score.

$$\dfrac{78 + 72 + 86 + x}{4} \geq 80 \quad \text{average must be greater than or equal to 80}$$

Write the equation model and solve.

$$78 + 72 + 86 + x \geq 320 \quad \text{multiply by 4}$$
$$236 + x \geq 320 \quad \text{simplify}$$
$$x \geq 84 \quad \text{solve for } x \text{ (subtract 236)}$$

Justin must score at least an 84 on the last exam to earn an 80 average.

Now try Exercises 79 through 86 ▶

As your problem-solving skills improve, the process outlined in the problem-solving guide naturally becomes less formal, as we work more directly toward the equation model. See Example 11.

EXAMPLE 11 ▶   **Using an Inequality to Make a Financial Decision**

As Margaret starts her new job, her employer offers two salary options. Plan 1 is base pay of $1475/mo plus 3% of sales. Plan 2 is base pay of $500/mo plus 15% of sales. What level of monthly sales is needed for her to earn more under Plan 2?

Solution ▶   Let x represent her monthly sales in dollars. The equation model for Plan 1 would be $0.03x + 1475$; for Plan 2 we have $0.15x + 500$. To find the sales volume needed for her to earn more under Plan 2, we solve the inequality

$$0.15x + 500 > 0.03x + 1475 \quad \text{Plan 2} > \text{Plan 1}$$
$$0.12x + 500 > 1475 \quad \text{subtract } 0.03x$$
$$0.12x > 975 \quad \text{subtract 500}$$
$$x > 8125 \quad \text{divide by 0.12}$$

☑ **D.** You've just learned how to solve applications of inequalities

If Margaret can generate more than $8125 in monthly sales, she will earn more under Plan 2.

Now try Exercises 87 and 88 ▶

1.2 EXERCISES

▶ CONCEPTS AND VOCABULARY

Fill in each blank with the appropriate word or phrase. Carefully reread the section, if necessary.

1. For inequalities, the three ways of writing a solution set are _____ notation, a number line graph, and _____ notation.

2. The mathematical sentence $3x + 5 < 7$ is a(n) _____ inequality, while $-2 < 3x + 5 < 7$ is a(n) _____ inequality.

3. The _____ of sets A and B is written $A \cap B$. The _____ of sets A and B is written $A \cup B$.

4. The intersection of set A with set B is the set of elements in A _____ B. The union of set A with set B is the set of elements in A _____ B.

5. Discuss/Explain how the concept of domain and allowable values relates to rational and radical expressions. Include a few examples.

6. Discuss/Explain why the inequality symbol must be reversed when multiplying or dividing by a negative quantity. Include a few examples.

▶ DEVELOPING YOUR SKILLS

Use an inequality to write a mathematical model for each statement.

7. To qualify for a secretarial position, a person must type at least 45 words per minute.

8. The balance in a checking account must remain above $1000 or a fee is charged.

9. To bake properly, a turkey must be kept between the temperatures of 250° and 450°.

10. To fly effectively, the airliner must cruise at or between altitudes of 30,000 and 35,000 ft.

Graph each inequality on a number line.

11. $y < 3$ **12.** $x > -2$

13. $m \leq 5$ **14.** $n \geq -4$

15. $x \neq 1$ **16.** $x \neq -3$

17. $5 > x > 2$ **18.** $-3 < y \leq 4$

Write the solution set illustrated on each graph in set notation and interval notation.

19.

20.

21.

22.

Solve the inequality and write the solution in set notation. Then graph the solution and write it in interval notation.

23. $5a - 11 \geq 2a - 5$

24. $-8n + 5 > -2n - 12$

25. $2(n + 3) - 4 \leq 5n - 1$

26. $-5(x + 2) - 3 < 3x + 11$

27. $\dfrac{3x}{8} + \dfrac{x}{4} < -4$ **28.** $\dfrac{2y}{5} + \dfrac{y}{10} < -2$

Solve each inequality and write the solution in set notation.

29. $7 - 2(x + 3) \geq 4x - 6(x - 3)$

30. $-3 - 6(x - 5) \leq 2(7 - 3x) + 1$

31. $4(3x - 5) + 18 < 2(5x + 1) + 2x$

32. $8 - (6 + 5m) > -9m - (3 - 4m)$

33. $-6(p - 1) + 2p \leq -2(2p - 3)$

34. $9(w - 1) - 3w \geq -2(5 - 3w) + 1$

Determine the intersection and union of sets A, B, C, and D as indicated, given $A = \{-3, -2, -1, 0, 1, 2, 3\}$, $B = \{2, 4, 6, 8\}$, $C = \{-4, -2, 0, 2, 4\}$, and $D = \{4, 5, 6, 7\}$.

35. $A \cap B$ and $A \cup B$ **36.** $A \cap C$ and $A \cup C$

37. $A \cap D$ and $A \cup D$ **38.** $B \cap C$ and $B \cup C$

39. $B \cap D$ and $B \cup D$ **40.** $C \cap D$ and $C \cup D$

Express the compound inequalities graphically and in interval notation.

41. $x < -2$ or $x > 1$ **42.** $x < -5$ or $x > 5$

43. $x < 5$ and $x \geq -2$ **44.** $x \geq -4$ and $x < 3$

45. $x \geq 3$ and $x \leq 1$ **46.** $x \geq -5$ and $x \leq -7$

Solve the compound inequalities and graph the solution set.

47. $4(x - 1) \leq 20$ or $x + 6 > 9$

48. $-3(x + 2) > 15$ or $x - 3 \leq -1$

49. $-2x - 7 \leq 3$ and $2x \leq 0$

50. $-3x + 5 \leq 17$ and $5x \leq 0$

51. $\dfrac{3}{5}x + \dfrac{1}{2} > \dfrac{3}{10}$ and $-4x > 1$

52. $\dfrac{2}{3}x - \dfrac{5}{6} \leq 0$ and $-3x < -2$

53. $\dfrac{3x}{8} + \dfrac{x}{4} < -3$ or $x + 1 > -5$

54. $\dfrac{2x}{5} + \dfrac{x}{10} < -2$ or $x - 3 > 2$

55. $-3 \leq 2x + 5 < 7$ **56.** $2 < 3x - 4 \leq 19$

57. $-0.5 \leq 0.3 - x \leq 1.7$

58. $-8.2 < 1.4 - x < -0.9$

59. $-7 < -\frac{3}{4}x - 1 \le 11$

60. $-21 \le -\frac{2}{3}x + 9 < 7$

65. $\dfrac{a + 5}{6a - 3}$

66. $\dfrac{m + 5}{8m + 4}$

67. $\dfrac{15}{3x - 12}$

68. $\dfrac{7}{2x + 6}$

Determine the domain of each expression. Write your answer in interval notation.

61. $\dfrac{12}{m}$

62. $\dfrac{-6}{n}$

63. $\dfrac{5}{y + 7}$

64. $\dfrac{4}{x - 3}$

Determine the domain for each expression. Write your answer in interval notation.

69. $\sqrt{x - 2}$

70. $\sqrt{y + 7}$

71. $\sqrt{3n - 12}$

72. $\sqrt{2m + 5}$

73. $\sqrt{b - \frac{4}{3}}$

74. $\sqrt{a + \frac{3}{4}}$

75. $\sqrt{8 - 4y}$

76. $\sqrt{12 - 2x}$

▶ WORKING WITH FORMULAS

77. Body mass index: $B = \dfrac{704W}{H^2}$

The U.S. government publishes a body mass index formula to help people consider the risk of heart disease. An index "B" of 27 or more means that a person is at risk. Here W represents weight in pounds and H represents height in inches. (a) Solve the formula for W. (b) If your height is 5′8″ what range of weights will help ensure you remain safe from the risk of heart disease?

Source: www.surgeongeneral.gov/topics.

78. Lift capacity: $75S + 125B \le 750$

The capacity in pounds of the lift used by a roofing company to place roofing shingles and buckets of roofing nails on rooftops is modeled by the formula shown, where S represents packs of shingles and B represents buckets of nails. Use the formula to find (a) the largest number of shingle packs that can be lifted, (b) the largest number of nail buckets that can be lifted, and (c) the largest number of shingle packs that can be lifted along with three nail buckets.

▶ APPLICATIONS

Write an inequality to model the given information and solve.

79. Exam scores: Jacques is going to college on an academic scholarship that requires him to maintain at least a 75% average in all of his classes. So far he has scored 82%, 76%, 65%, and 71% on four exams. What scores are possible on his last exam that will enable him to keep his scholarship?

80. Timed trials: In the first three trials of the 100-m butterfly, Johann had times of 50.2, 49.8, and 50.9 sec. How fast must he swim the final timed trial to have an average time of 50 sec?

81. Checking account balance: If the average daily balance in a certain checking account drops below $1000, the bank charges the customer a $7.50 service fee. The table gives the daily balance for one customer. What must the daily balance be for Friday to avoid a service charge?

Weekday	Balance
Monday	$1125
Tuesday	$850
Wednesday	$625
Thursday	$400

82. Average weight: In the National Football League, many consider an offensive line to be "small" if the average weight of the five down linemen is less than 325 lb. Using the table, what must the weight of the right tackle be so that the line will not be considered too small?

Lineman	Weight
Left tackle	318 lb
Left guard	322 lb
Center	326 lb
Right guard	315 lb
Right tackle	?

83. Area of a rectangle: Given the rectangle shown, what is the range of values for the width, in order to keep the area less than 150 m^2?

20 m

w

84. Area of a triangle: Using the triangle shown, find the height that will guarantee an area equal to or greater than 48 in^2.

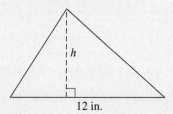

h

12 in.

85. Heating and cooling subsidies: As long as the outside temperature is over 45°F and less than 85°F ($45 < F < 85$), the city does not issue heating or cooling subsidies for low-income families. What is the corresponding range of Celsius temperatures C? Recall that $F = \frac{9}{5}C + 32$.

86. U.S. and European shoe sizes: To convert a European male shoe size "E" to an American male shoe size "A," the formula $A = 0.76E - 23$ can be used. Lillian has five sons in the U.S. military, with shoe sizes ranging from size 9 to size 14 ($9 \leq A \leq 14$). What is the corresponding range of European sizes? Round to the nearest half-size.

87. Power tool rentals: Sunshine Equipment Co. rents its power tools for a $20 fee, plus $4.50/hr. Kealoha's Rentals offers the same tools for an $11 fee plus $6.00/hr. How many hours h must a tool be rented to make the cost at Sunshine a better deal?

88. Moving van rentals: Davis Truck Rentals will rent a moving van for $15.75/day plus $0.35 per mile. Bertz Van Rentals will rent the same van for $25/day plus $0.30 per mile. How many miles m must the van be driven to make the cost at Bertz a better deal?

▶ **EXTENDING THE CONCEPT**

89. Use your local library, the Internet, or another resource to find the highest and lowest point on each of the seven continents. Express the range of altitudes for each continent as a joint inequality. Which continent has the greatest range?

90. The sum of two consecutive even integers is greater than or equal to 12 and less than or equal to 22. List all possible values for the two integers.

Place the correct inequality symbol in the blank to make the statement true.

91. If $m > 0$ and $n < 0$, then mn _____ 0.

92. If $m > n$ and $p > 0$, then mp _____ np.

93. If $m < n$ and $p > 0$, then mp _____ np.

94. If $m \leq n$ and $p < 0$, then mp _____ np.

95. If $m > n$, then $-m$ _____ $-n$.

96. If $m < n$, then $\frac{1}{m}$ _____ $\frac{1}{n}$.

97. If $m > 0$ and $n < 0$, then m^2 _____ n.

98. If $m < 0$, then m^3 _____ 0.

▶ **MAINTAINING YOUR SKILLS**

99. (R.2) Translate into an algebraic expression: eight subtracted from twice a number.

100. (1.1) Solve: $-4(x - 7) - 3 = 2x + 1$

101. (R.3) Simplify the algebraic expression: $2(\frac{5}{9}x - 1) - (\frac{1}{6}x + 3)$.

102. (1.1) Solve: $\frac{4}{5}m + \frac{2}{3} = \frac{1}{2}$

Learning Objectives

In Section 1.3 you will learn how to:

☐ **A.** Solve absolute value equations

☐ **B.** Solve "less than" absolute value inequalities

☐ **C.** Solve "greater than" absolute value inequalities

☐ **D.** Solve applications involving absolute value

While the equations $x + 1 = 5$ and $|x + 1| = 5$ are similar in many respects, note the first has only the solution $x = 4$, while either $x = 4$ or $x = -6$ will satisfy the second. The fact there are two solutions shouldn't surprise us, as it's a natural result of how absolute value is defined.

A. Solving Absolute Value Equations

The absolute value of a number x can be thought of as its distance from zero on the number line, regardless of direction. This means $|x| = 4$ will have *two solutions*, since there are two numbers that are four units from zero: $x = -4$ and $x = 4$ (see Figure 1.5).

Figure 1.5

WORTHY OF NOTE

Note if $k < 0$, the equation $|X| = k$ has no solutions since the absolute value of any quantity is always positive or zero. On a related note, we can verify that if $k = 0$, the equation $|X| = 0$ has only the solution $X = 0$.

This basic idea can be extended to include situations where the quantity within absolute value bars *is an algebraic expression,* and suggests the following property.

Property of Absolute Value Equations

If X represents an algebraic expression and k is a positive real number,

then $|X| = k$

implies $X = -k$ or $X = k$

As the statement of this property suggests, it can only be applied *after* the absolute value expression has been isolated on one side.

EXAMPLE 1 ▶ **Solving an Absolute Value Equation**

Solve: $-5|x - 7| + 2 = -13$.

Solution ▶ Begin by isolating the absolute value expression.

$$-5|x - 7| + 2 = -13 \quad \text{original equation}$$
$$-5|x - 7| = -15 \quad \text{subtract 2}$$
$$|x - 7| = 3 \quad \text{divide by } -5 \text{ (simplified form)}$$

Now consider $x - 7$ as the variable expression "X" in the property of absolute value equations, giving

$$x - 7 = -3 \quad \text{or} \quad x - 7 = 3 \quad \text{apply the property of absolute value equations}$$
$$x = 4 \quad \text{or} \quad x = 10 \quad \text{add 7}$$

Substituting into the original equation verifies the solution set is {4, 10}.

Now try Exercises 7 through 18 ▶

⚠ **CAUTION** ▶ For equations like those in Example 1, be careful not to treat the absolute value bars as simple grouping symbols. The equation $-5(x - 7) + 2 = -13$ has only the solution $x = 10$, and "misses" the second solution since it yields $x - 7 = 3$ in simplified form. The equation $-5|x - 7| + 2 = -13$ simplifies to $|x - 7| = 3$ and there are actually *two* solutions.

Absolute value equations come in many different forms. Always begin by isolating the absolute value expression, then apply the property of absolute value equations to solve.

EXAMPLE 2 ▶ **Solving an Absolute Value Equation**

Solve: $\left| 5 - \frac{2}{3}x \right| - 9 = 8$

Solution ▶

$$\left| 5 - \frac{2}{3}x \right| - 9 = 8 \qquad \text{original equation}$$

$$\left| 5 - \frac{2}{3}x \right| = 17 \qquad \text{add 9}$$

$$5 - \frac{2}{3}x = -17 \quad \text{or} \quad 5 - \frac{2}{3}x = 17 \qquad \text{apply the property of absolute value equations}$$

$$-\frac{2}{3}x = -22 \quad \text{or} \quad -\frac{2}{3}x = 12 \qquad \text{subtract 5}$$

$$x = 33 \quad \text{or} \quad x = -18 \qquad \text{multiply by } -\frac{3}{2}$$

Check ▶ For $x = 33$: $\left| 5 - \frac{2}{3}(33) \right| - 9 = 8$ For $x = -18$: $\left| 5 - \frac{2}{3}(-18) \right| - 9 = 8$

$$|5 - 2(11)| - 9 = 8 \qquad\qquad\qquad |5 - 2(-6)| - 9 = 8$$

$$|5 - 22| - 9 = 8 \qquad\qquad\qquad\qquad |5 + 12| - 9 = 8$$

$$|-17| - 9 = 8 \qquad\qquad\qquad\qquad\qquad |17| - 9 = 8$$

$$17 - 9 = 8 \qquad\qquad\qquad\qquad\qquad\quad 17 - 9 = 8$$

$$8 = 8 ✓ \qquad\qquad\qquad\qquad\qquad\qquad 8 = 8 ✓$$

Both solutions check. The solution set is $\{-18, 33\}$.

Now try Exercises 19 through 22 ▶

> **WORTHY OF NOTE**
>
> As illustrated in both Examples 1 and 2, the property we use to solve absolute value equations can only be applied *after* the absolute value term has been isolated. As you will see, the same is true for the properties used to solve absolute value inequalities.

For some equations, it's helpful to apply the **multiplicative property of absolute value:**

Multiplicative Property of Absolute Value

If A and B represent algebraic expressions,

$$\text{then } |AB| = |A||B|.$$

Note that if $A = -1$ the property says $|-B| = |-1| \, |B| = |B|$. More generally the property is applied where A is any constant.

EXAMPLE 3 ▶ **Solving Equations Using the Multiplicative Property of Absolute Value**

Solve: $|-2x| + 5 = 13$.

Solution ▶

$$|-2x| + 5 = 13 \qquad \text{original equation}$$

$$|-2x| = 8 \qquad \text{subtract 5}$$

$$|-2||x| = 8 \qquad \text{apply multiplicative property of absolute value}$$

$$2|x| = 8 \qquad \text{simplify}$$

$$|x| = 4 \qquad \text{divide by 2}$$

$$x = -4 \quad \text{or} \quad x = 4 \qquad \text{apply property of absolute value equations}$$

☑ **A. You've just learned how to solve absolute value equations**

Both solutions check. The solution set is $\{-4, 4\}$.

Now try Exercises 23 and 24 ▶

B. Solving "Less Than" Absolute Value Inequalities

Absolute value *inequalities* can be solved using the basic concept underlying the property of absolute value equalities. Whereas the equation $|x| = 4$ asks for all numbers x whose distance from zero is *equal* to 4, the inequality $|x| < 4$ asks for all numbers x whose distance from zero is *less than* 4.

Distance from zero is less than 4

Figure 1.6

As Figure 1.6 illustrates, the solutions are $x > -4$ and $x < 4$, which can be written as the joint inequality $-4 < x < 4$. This idea can likewise be extended to include the absolute value of an algebraic expression X as follows.

Property I: Absolute Value Inequalities

If X represents an algebraic expression and k is a positive real number,

then $|X| < k$

implies $-k < X < k$

EXAMPLE 4 ▶ Solving "Less Than" Absolute Value Inequalities

Solve the inequalities:

a. $\dfrac{|3x + 2|}{4} \le 1$ **b.** $|2x - 7| < -5$

Solution ▶ **a.**

$$\dfrac{|3x + 2|}{4} \le 1 \quad \text{original inequality}$$
$$|3x + 2| \le 4 \quad \text{multiply by 4}$$
$$-4 \le 3x + 2 \le 4 \quad \text{apply Property I}$$
$$-6 \le 3x \le 2 \quad \text{subtract 2 from all three parts}$$
$$-2 \le x \le \dfrac{2}{3} \quad \text{divide all three parts by 3}$$

The solution interval is $\left[-2, \frac{2}{3}\right]$.

b. $|2x - 7| < -5$ original inequality

Since the absolute value of any quantity is always positive or zero, the solution for this inequality is the empty set: { }.

WORTHY OF NOTE

As with the inequalities from Section 1.2, solutions to absolute value inequalities can be checked using a test value. For Example 4(a), substituting $x = 0$ from the solution interval yields:

$$\dfrac{1}{2} \le 1 ✓$$

☑ **B.** You've just learned how to solve less than absolute value inequalities

Now try Exercises 25 through 38 ▶

C. Solving "Greater Than" Absolute Value Inequalities

For "greater than" inequalities, consider $|x| > 4$. Now we're asked to find all numbers x whose distance from zero is *greater than* 4. As Figure 1.7 shows, solutions are found in the interval to the left of -4, or to the right of 4. The fact the intervals are disjoint (disconnected) is reflected in this graph, in the inequalities $x < -4$ **or** $x > 4$, as well as the interval notation $x \in (-\infty, -4) \cup (4, \infty)$.

Figure 1.7

As before, we can extend this idea to include algebraic expressions, as follows:

Property II: Absolute Value Inequalities

If X represents an algebraic expression and k is a positive real number,

$$\text{then } |X| > k$$
$$\text{implies } X < -k \quad \text{or} \quad X > k$$

EXAMPLE 5 ▶   Solving "Greater Than" Absolute Value Inequalities

Solve the inequalities:

a. $-\dfrac{1}{3}\left|3 + \dfrac{x}{2}\right| < -2$ **b.** $|5x + 2| \geq -\dfrac{3}{2}$

Solution ▶   **a.** Note the exercise is given as a *less than* inequality, but as we multiply both sides by -3, we must *reverse the inequality symbol*.

$$-\frac{1}{3}\left|3 + \frac{x}{2}\right| < -2 \qquad \text{original inequality}$$

$$\left|3 + \frac{x}{2}\right| > 6 \qquad \text{multiply by } -3,\ \textit{reverse the symbol}$$

$$3 + \frac{x}{2} < -6 \quad \text{or} \quad 3 + \frac{x}{2} > 6 \qquad \text{apply Property II}$$

$$\frac{x}{2} < -9 \quad \text{or} \qquad \frac{x}{2} > 3 \qquad \text{subtract 3}$$

$$x < -18 \quad \text{or} \qquad x > 6 \qquad \text{multiply by 2}$$

Property II yields the disjoint intervals $x \in (-\infty, -18) \cup (6, \infty)$ as the solution.

—30 −24 −18 −12 −6   0   6   12   18   24   30

b. $|5x + 2| \geq -\dfrac{3}{2}$ original inequality

☑ **C.** You've just learned how to solve greater than absolute value inequalities

Since the absolute value of any quantity is always positive or zero, the solution for this inequality is all real numbers: $x \in \mathbb{R}$.

Now try Exercises 39 through 54 ▶

⚠ **CAUTION ▶**   Be sure you note the difference between the individual solutions of an absolute value equation, and the solution intervals that often result from solving absolute value inequalities. The solution $\{-2, 5\}$ indicates that both $x = -2$ and $x = 5$ are solutions, while the solution $[-2, 5)$ indicates that all numbers between -2 and 5, including -2, are solutions.

D. Applications Involving Absolute Value

Applications of absolute value often involve finding a range of values for which a given statement is true. Many times, the equation or inequality used must be modeled after a given description or from given information, as in Example 6.

EXAMPLE 6 ▶ **Solving Applications Involving Absolute Value Inequalities**

For new cars, the number of miles per gallon (mpg) a car will get is heavily dependent on whether it is used mainly for short trips and city driving, or primarily on the highway for longer trips. For a certain car, the number of miles per gallon that a driver can expect varies by no more than 6.5 mpg above or below its field tested average of 28.4 mpg. What range of mileage values can a driver expect for this car?

Solution ▶ Field tested average: 28.4 mpg   gather information
mileage varies by no more than 6.5 mpg   highlight key phrases

```
        −6.5        +6.5
   ←─────────|─────────→
─┼──┼──┼──┼──┼──┼──┼──┼──┼──┼──→
              28.4
```
make the problem visual

Let m represent the miles per gallon a driver can expect.   assign a variable
Then the difference between m and 28.4 can be no more
than 6.5, or $|m - 28.4| \le 6.5$.   write an equation model

$$|m - 28.4| \le 6.5$$ equation model
$$-6.5 \le m - 28.4 \le 6.5$$ apply Property I
$$21.9 \le m \le 34.9$$ add 28.4 to all three parts

☑ **D. You've just learned how to solve applications involving absolute value**

The mileage that a driver can expect ranges from a low of 21.9 mpg to a high of 34.9 mpg.

Now try Exercises 57 through 64 ▶

TECHNOLOGY HIGHLIGHT

Absolute Value Equations and Inequalities

Graphing calculators can explore and solve inequalities in many different ways. Here we'll use a table of values and a *relational test*. To begin we'll consider the equation $2|x - 3| + 1 = 5$ by entering the left-hand side as Y_1 on the [Y=] screen. The calculator does not use absolute value bars the way they're written, and the equation is actually entered as $Y_1 = 2$ **abs** $(X - 3) + 1$ (see Figure 1.8). The **"abs("** notation is accessed by pressing [MATH], [▶] (NUM) [1] (option 1 gives only the left parenthesis, you must supply the right). Preset the TABLE as in the previous Highlight (page 81). By scrolling through the table (use the up [▲] and down [▼] arrows), we find $Y_1 = 5$ when $x = 1$ or $x = 5$ (see Figure 1.9).

Figure 1.8

Although we could also solve the *inequality* $2|x - 3| + 1 \le 5$ using the table (the solution interval is $x \in [1, 5]$), a relational test can help. Relational tests have the calculator return a "1" if a given statement is true, and a "0" otherwise. Enter $Y_2 = Y_1 \le 5$, by accessing Y_1 using [VARS] [▶] (Y-VARS) 1:Function [ENTER], and the "≤" symbol using [2nd] [MATH] (TEST) [the "less than or equal to" symbol is option 6]. Returning to the table shows $Y_1 \le 5$ is true for $1 \le x \le 5$ (see Figure 1.9).

Figure 1.9

Use a table and a relational test to help solve the following inequalities. Verify the result algebraically.

Exercise 1: $3|x + 1| - 2 \ge 7$ **Exercise 2:** $-2|x + 2| + 5 \ge -1$ **Exercise 3:** $-1 \le 4|x - 3| - 1$

1.3 EXERCISES

▶ CONCEPTS AND VOCABULARY

Fill in the blank with the appropriate word or phrase. Carefully reread the section if needed.

1. When multiplying or dividing by a negative quantity, we _____ the inequality to maintain a true statement.

2. To write an absolute value equation or inequality in simplified form, we _____ the absolute value expression on one side.

3. The absolute value equation $|2x + 3| = 7$ is true when $2x + 3 =$ _____ or when $2x + 3 =$ _____.

4. The absolute value inequality $|3x - 6| < 12$ is true when $3x - 6 >$ _____ and $3x - 6 <$ _____.

Describe each solution set (assume $k > 0$). Justify your answer.

5. $|ax + b| < -k$

6. $|ax + b| > -k$

▶ DEVELOPING YOUR SKILLS

Solve each absolute value equation. Write the solution in set notation.

7. $2|m - 1| - 7 = 3$

8. $3|n - 5| - 14 = -2$

9. $-3|x + 5| + 6 = -15$

10. $-2|y + 3| - 4 = -14$

11. $2|4v + 5| - 6.5 = 10.3$

12. $7|2w + 5| + 6.3 = 11.2$

13. $-|7p - 3| + 6 = -5$

14. $-|3q + 4| + 3 = -5$

15. $-2|b| - 3 = -4$

16. $-3|c| - 5 = -6$

17. $-2|3x| - 17 = -5$

18. $-5|2y| - 14 = 6$

19. $-3\left|\dfrac{w}{2} + 4\right| - 1 = -4$

20. $-2\left|3 - \dfrac{v}{3}\right| + 1 = -5$

21. $8.7|p - 7.5| - 26.6 = 8.2$

22. $5.3|q + 9.2| + 6.7 = 43.8$

23. $8.7|-2.5x| - 26.6 = 8.2$

24. $5.3|1.25n| + 6.7 = 43.8$

Solve each absolute value inequality. Write solutions in interval notation.

25. $|x - 2| \leq 7$

26. $|y + 1| \leq 3$

27. $-3|m| - 2 > 4$

28. $-2|n| + 3 > 7$

29. $\dfrac{|5v + 1|}{4} + 8 < 9$

30. $\dfrac{|3w - 2|}{2} + 6 < 8$

31. $3|p + 4| + 5 < 8$

32. $5|q - 2| - 7 \leq 8$

33. $|3b - 11| + 6 \leq 9$

34. $|2c + 3| - 5 < 1$

35. $|4 - 3z| + 12 < 7$

36. $|2 - 7u| + 7 \leq 4$

37. $\left|\dfrac{4x + 5}{3} - \dfrac{1}{2}\right| \leq \dfrac{7}{6}$

38. $\left|\dfrac{2y - 3}{4} - \dfrac{3}{8}\right| < \dfrac{15}{16}$

39. $|n + 3| > 7$

40. $|m - 1| > 5$

41. $-2|w| - 5 \leq -11$

42. $-5|v| - 3 \leq -23$

43. $\dfrac{|q|}{2} - \dfrac{5}{6} \geq \dfrac{1}{3}$

44. $\dfrac{|p|}{5} + \dfrac{3}{2} \geq \dfrac{9}{4}$

45. $3|5 - 7d| + 9 \geq 15$

46. $5|2c + 7| + 1 \geq 11$

47. $|4z - 9| + 6 \geq 4$

48. $|5u - 3| + 8 > 6$

49. $4|5 - 2h| - 9 > 11$

50. $3|7 + 2k| - 11 > 10$

51. $-3.9|4q - 5| + 8.7 \leq -22.5$

52. $0.9|2p + 7| - 16.11 \geq 10.89$

53. $2 < \left|-3m + \dfrac{4}{5}\right| - \dfrac{1}{5}$

54. $4 \leq \left|\dfrac{5}{4} - 2n\right| - \dfrac{3}{4}$

▶ WORKING WITH FORMULAS

55. Spring Oscillation $|d - x| \leq L$

A weight attached to a spring hangs at rest a distance of x in. off the ground. If the weight is pulled down (stretched) a distance of L inches and released, the weight begins to bounce and its distance d off the ground must satisfy the indicated formula. If x equals 4 ft and the spring is stretched 3 in. and released, solve the inequality to find what distances from the ground the weight will oscillate between.

56. A "Fair" Coin $\left|\dfrac{h - 50}{5}\right| < 1.645$

If we flipped a coin 100 times, we expect "heads" to come up about 50 times if the coin is "fair." In a study of probability, it can be shown that the number of heads h that appears in such an experiment must satisfy the given inequality to be considered "fair." (a) Solve this inequality for h. (b) If you flipped a coin 100 times and obtained 40 heads, is the coin "fair"?

▶ APPLICATIONS

Solve each application of absolute value.

57. Altitude of jet stream: To take advantage of the jet stream, an airplane must fly at a height h (in feet) that satisfies the inequality $|h - 35{,}050| \leq 2550$. Solve the inequality and determine if an altitude of 34,000 ft will place the plane in the jet stream.

58. Quality control tests: In order to satisfy quality control, the marble columns a company produces must earn a stress test score S that satisfies the inequality $|S - 17{,}750| \leq 275$. Solve the inequality and determine if a score of 17,500 is in the passing range.

59. Submarine depth: The sonar operator on a submarine detects an old World War II submarine net and must decide to detour over or under the net. The computer gives him a depth model $|d - 394| - 20 > 164$, where d is the depth in feet that represents safe passage. At what depth should the submarine travel to go under or over the net? Answer using simple inequalities.

60. Optimal fishing depth: When deep-sea fishing, the optimal depths d (in feet) for catching a certain type of fish satisfy the inequality $28|d - 350| - 1400 < 0$. Find the range of depths that offer the best fishing. Answer using simple inequalities.

For Exercises 61 through 64, (a) develop a model that uses an absolute value inequality, and (b) solve.

61. Stock value: My stock in MMM Corporation fluctuated a great deal in 2009, but never by more than \$3.35 from its current value. If the stock is worth \$37.58 today, what was its range in 2009?

62. Traffic studies: On a given day, the volume of traffic at a busy intersection averages 726 cars per hour (cph). During rush hour the volume is much higher, during "off hours" much lighter. Find the range of this volume if it never varies by more than 235 cph from the average.

63. Physical training for recruits: For all recruits in the 3rd Armored Battalion, the average number of sit-ups is 125. For an individual recruit, the amount varies by no more than 23 sit-ups from the battalion average. Find the range of sit-ups for this battalion.

64. Computer consultant salaries: The national average salary for a computer consultant is \$53,336. For a large computer firm, the salaries offered to their employees varies by no more than \$11,994 from this national average. Find the range of salaries offered by this company.

65. According to the official rules for golf, baseball, pool, and bowling, (a) golf balls must be within 0.03 mm of $d = 42.7$ mm, (b) baseballs must be within 1.01 mm of $d = 73.78$ mm, (c) billiard balls must be within 0.127 mm of $d = 57.150$ mm, and (d) bowling balls must be within 12.05 mm of $d = 2171.05$ mm. Write each statement using an absolute value inequality, then (e) determine which sport gives the least tolerance t $\left(t = \dfrac{\text{width of interval}}{\text{average value}}\right)$ for the diameter of the ball.

66. The machines that fill boxes of breakfast cereal are programmed to fill each box within a certain tolerance. If the box is overfilled, the company loses money. If it is underfilled, it is considered unsuitable for sale. Suppose that boxes marked "14 ounces" of cereal must be filled to within 0.1 oz. Write this relationship as an absolute value inequality, then solve the inequality and explain what your answer means. Let W represent weight.

▶ **EXTENDING THE CONCEPT**

67. Determine the value or values (if any) that will make the equation or inequality true.

    a. $|x| + x = 8$ **b.** $|x - 2| \leq \dfrac{x}{2}$

    c. $x - |x| = x + |x|$ **d.** $|x + 3| \geq 6x$

    e. $|2x + 1| = x - 3$

68. The equation $|5 - 2x| = |3 + 2x|$ has only one solution. Find it and explain why there is only one.

▶ **MAINTAINING YOUR SKILLS**

69. (R.4) Factor the expression completely: $18x^3 + 21x^2 - 60x$.

70. (1.1) Solve $V^2 = \dfrac{2W}{C\rho A}$ for ρ (physics).

71. (R.6) Simplify $\dfrac{-1}{3 + \sqrt{3}}$ by rationalizing the denominator. State the result in exact form and approximate form (to hundredths):

72. (1.2) Solve the inequality, then write the solution set in interval notation:

$$-3(2x - 5) > 2(x + 1) - 7.$$

MID-CHAPTER CHECK

1. Solve each equation. If the equation is an identity or contradiction, so state and name the solution set.

    a. $\dfrac{r}{3} + 5 = 2$

    b. $5(2x - 1) + 4 = 9x - 7$

    c. $m - 2(m + 3) = 1 - (m + 7)$

    d. $\dfrac{1}{5}y + 3 = \dfrac{3}{2}y - 2$

    e. $\dfrac{1}{2}(5j - 2) = \dfrac{3}{2}(j - 4) + j$

    f. $0.6(x - 3) + 0.3 = 1.8$

Solve for the variable specified.

2. $H = -16t^2 + v_0 t$; for v_0

3. $S = 2\pi x^2 + \pi x^2 y$; for x

4. Solve each inequality and graph the solution set.

    a. $-5x + 16 \leq 11$ or $3x + 2 \leq -4$

    b. $\dfrac{1}{2} < \dfrac{1}{12}x - \dfrac{5}{6} \leq \dfrac{3}{4}$

5. Determine the domain of each expression. Write your answer in interval notation.

    a. $\dfrac{3x + 1}{2x - 5}$ **b.** $\sqrt{17 - 6x}$

6. Solve the following absolute value equations. Write the solution in set notation.

    a. $\dfrac{2}{3}|d - 5| + 1 = 7$ **b.** $5 - |s + 3| = \dfrac{11}{2}$

7. Solve the following absolute value inequalities. Write solutions in interval notation.

    a. $3|q + 4| - 2 < 10$

    b. $\left|\dfrac{x}{3} + 2\right| + 5 \leq 5$

8. Solve the following absolute value inequalities. Write solutions in interval notation.

    a. $3.1|d - 2| + 1.1 \geq 7.3$

    b. $\dfrac{|1 - y|}{3} + 2 > \dfrac{11}{2}$

    c. $-5|k - 2| + 3 < 4$

9. Motocross: An enduro motocross motorcyclist averages 30 mph through the first part of a 115-mi course, and 50 mph though the second part. If the rider took 2 hr and 50 min to complete the course, how long was she on the first part?

10. Kiteboarding: With the correct sized kite, a person can kiteboard when the wind is blowing at a speed w (in mph) that satisfies the inequality $|w - 17| \leq 9$. Solve the inequality and determine if a person can kiteboard with a windspeed of 9 mph.

REINFORCING BASIC CONCEPTS

Using Distance to Understand Absolute Value Equations and Inequalities

In Section R.1 we noted that for any two numbers a and b on the number line, *the distance between a and b* is written $|a - b|$ or $|b - a|$. In exactly the same way, the equation $|x - 3| = 4$ can be read, "the distance between 3 and an unknown number is equal to 4." The advantage of reading it in this way (instead of *the absolute value of x minus 3 is 4*), is that a much clearer *visualization* is formed, giving a constant reminder there are two solutions. In diagram form we have Figure 1.10.

Figure 1.10

From this we note the solution is $x = -1$ or $x = 7$.

In the case of an inequality such as $|x + 2| \leq 3$, we rewrite the inequality as $|x - (-2)| \leq 3$ and read it, "the distance between -2 and an unknown number is less than or equal to 3." With some practice, visualizing this relationship mentally enables a quick statement of the solution: $x \in [-5, 1]$. In diagram form we have Figure 1.11.

Figure 1.11

Equations and inequalities where the coefficient of x is not 1 still lend themselves to this form of conceptual understanding. For $|2x - 1| \geq 3$ we read, "the distance between 1 and twice an unknown number is greater than or equal to 3." On the number line (Figure 1.12), the number 3 units to the right of 1 is 4, and the number 3 units to the left of 1 is -2.

Figure 1.12

For $2x \leq -2$, $x \leq -1$, and for $2x \geq 4$, $x \geq 2$, and the solution is $x \in (-\infty, -1] \cup [2, \infty)$.

Attempt to solve the following equations and inequalities by visualizing a number line. Check all results algebraically.

Exercise 1: $|x - 2| = 5$

Exercise 2: $|x + 1| \leq 4$

Exercise 3: $|2x - 3| \geq 5$

Learning Objectives

In Section 1.4 you will learn how to:

☐ **A.** Identify and simplify imaginary and complex numbers

☐ **B.** Add and subtract complex numbers

☐ **C.** Multiply complex numbers and find powers of i

☐ **D.** Divide complex numbers

For centuries, even the most prominent mathematicians refused to work with equations like $x^2 + 1 = 0$. Using the principal of square roots gave the "solutions" $x = \sqrt{-1}$ and $x = -\sqrt{-1}$, which they found baffling and mysterious, since there is no real number whose square is -1. In this section, we'll see how this "mystery" was finally resolved.

A. Identifying and Simplifying Imaginary and Complex Numbers

The equation $x^2 = -1$ has no real solutions, since the square of any real number is positive. But if we apply the principle of square roots we get $x = \sqrt{-1}$ and $x = -\sqrt{-1}$, which seem to check when substituted into the original equation:

$$x^2 + 1 = 0 \qquad \text{original equation}$$

$$(1) \qquad (\sqrt{-1})^2 + 1 = 0 \qquad \text{substitute } \sqrt{-1} \text{ for } x$$

$$-1 + 1 = 0 \checkmark \qquad \text{answer "checks"}$$

$$(2) \qquad (-\sqrt{-1})^2 + 1 = 0 \qquad \text{substitute } -\sqrt{-1} \text{ for } x$$

$$-1 + 1 = 0 \checkmark \qquad \text{answer "checks"}$$

This observation likely played a part in prompting Renaissance mathematicians to study such numbers in greater depth, as they reasoned that while these were not *real number* solutions, they must be *solutions of a new and different kind*. Their study eventually resulted in the introduction of the set of **imaginary numbers** and the **imaginary unit *i*,** as follows.

Imaginary Numbers and the Imaginary Unit

- Imaginary numbers are those of the form $\sqrt{-k}$, where k is a positive real number.
- The imaginary unit i represents the number whose square is -1:

$$i^2 = -1 \text{ and } i = \sqrt{-1}$$

WORTHY OF NOTE

It was René Descartes (in 1637) who first used the term *imaginary* to describe these numbers; Leonhard Euler (in 1777) who introduced the letter i to represent $\sqrt{-1}$; and Carl F. Gauss (in 1831) who first used the phrase *complex number* to describe solutions that had both a real number part and an imaginary part. For more on complex numbers and their story, see www.mhhe.com/coburn

As a convenience to understanding and working with imaginary numbers, we rewrite them in terms of i, allowing that the product property of radicals $(\sqrt{AB} = \sqrt{A}\sqrt{B})$ still applies if *only one* of the radicands is negative. For $\sqrt{-3}$, we have $\sqrt{-1 \cdot 3} = \sqrt{-1}\sqrt{3} = i\sqrt{3}$. In general, we simply state the following property.

Rewriting Imaginary Numbers

- For any positive real number k, $\sqrt{-k} = i\sqrt{k}$.

For $\sqrt{-20}$ we have:

$$\sqrt{-20} = i\sqrt{20}$$
$$= i\sqrt{4 \cdot 5}$$
$$= 2i\sqrt{5},$$

and we say the expression has been *simplified and written in terms of i*. Note that we've written the result with the unit "i" *in front of the radical* to prevent it being interpreted as being *under the radical*. In symbols, $2i\sqrt{5} = 2\sqrt{5}i \neq 2\sqrt{5i}$.

The solutions to $x^2 = -1$ also serve to illustrate that for $k > 0$, there are two solutions to $x^2 = -k$, namely, $i\sqrt{k}$ and $-i\sqrt{k}$. In other words, every negative number has two square roots, one positive and one negative. The first of these, $i\sqrt{k}$, is called the **principal square root** of $-k$.

EXAMPLE 1 ▶ **Simplifying Imaginary Numbers**

Rewrite the imaginary numbers in terms of i and simplify if possible.

a. $\sqrt{-7}$ **b.** $\sqrt{-81}$ **c.** $\sqrt{-24}$ **d.** $-3\sqrt{-16}$

Solution ▶   **a.** $\sqrt{-7} = i\sqrt{7}$ **b.** $\sqrt{-81} = i\sqrt{81}$
$$= 9i$$

c. $\sqrt{-24} = i\sqrt{24}$ **d.** $-3\sqrt{-16} = -3i\sqrt{16}$
$$= i\sqrt{4\cdot 6} \qquad\qquad\qquad = -3i(4)$$
$$= 2i\sqrt{6} \qquad\qquad\qquad\quad = -12i$$

Now try Exercises 7 through 12 ▶

EXAMPLE 2 ▶ **Writing an Expression in Terms of i**

The numbers $x = \dfrac{-6 + \sqrt{-16}}{2}$ and $x = \dfrac{-6 - \sqrt{-16}}{2}$ are not real, but are known

to be solutions of $x^2 + 6x + 13 = 0$. Simplify $\dfrac{-6 + \sqrt{-16}}{2}$.

Solution ▶ Using the i notation, we have

$$\frac{-6 + \sqrt{-16}}{2} = \frac{-6 + i\sqrt{16}}{2} \qquad \text{write in } i \text{ notation}$$
$$= \frac{-6 + 4i}{2} \qquad \text{simplify}$$
$$= \frac{2(-3 + 2i)}{2} \qquad \text{factor numerator}$$
$$= -3 + 2i \qquad \text{reduce}$$

Now try Exercises 13 through 16 ▶

WORTHY OF NOTE

The expression $\dfrac{-6 + 4i}{2}$ from the solution of Example 2 can also be simplified by rewriting it as two separate terms, then simplifying each term:
$$\frac{-6 + 4i}{2} = \frac{-6}{2} + \frac{4i}{2}$$
$$= -3 + 2i.$$

The result in Example 2 contains both a **real number part** (-3) and an **imaginary part** $(2i)$. Numbers of this type are called **complex numbers.**

Complex Numbers

Complex numbers are numbers that can be written in the form $a + bi$, where a and b are real numbers and $i = \sqrt{-1}$.

The expression $a + bi$ is called the **standard form** of a complex number. From this definition we note that all real numbers are also complex numbers, since $a + 0i$ is complex with $b = 0$. In addition, all imaginary numbers are complex numbers, since $0 + bi$ is a complex number with $a = 0$.

EXAMPLE 3 ▶ **Writing Complex Numbers in Standard Form**

Write each complex number in the form $a + bi$, and identify the values of a and b.

a. $2 + \sqrt{-49}$ **b.** $\sqrt{-12}$ **c.** 7 **d.** $\dfrac{4 + 3\sqrt{-25}}{20}$

Solution ▶   **a.** $2 + \sqrt{-49} = 2 + i\sqrt{49}$ **b.** $\sqrt{-12} = 0 + i\sqrt{12}$
$$= 2 + 7i \qquad\qquad\qquad\qquad = 0 + 2i\sqrt{3}$$
$$a = 2, b = 7 \qquad\qquad\qquad\quad a = 0, b = 2\sqrt{3}$$

c. $7 = 7 + 0i$
$a = 7, b = 0$

d. $\dfrac{4 + 3\sqrt{-25}}{20} = \dfrac{4 + 3i\sqrt{25}}{20}$

$= \dfrac{4 + 15i}{20}$

$= \dfrac{1}{5} + \dfrac{3}{4}i$

$a = \dfrac{1}{5}, b = \dfrac{3}{4}$

> **Now try Exercises 17 through 24 ▶**

☑ **A.** You've just learned how to identify and simplify imaginary and complex numbers

Complex numbers complete the development of our "numerical landscape." Sets of numbers and their relationships are represented in Figure 1.13, which shows how some sets of numbers are nested within larger sets and highlights the fact that complex numbers consist of a real number part (any number within the orange rectangle), and an imaginary number part (any number within the yellow rectangle).

C (complex): Numbers of the form $a + bi$, where $a, b \in \mathbb{R}$ and $i = \sqrt{-1}$.

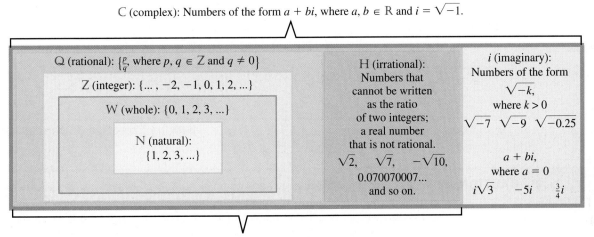

R (real): All rational and irrational numbers: $a + bi$, where $a \in \mathbb{R}$ and $b = 0$.

Figure 1.13

B. Adding and Subtracting Complex Numbers

The sum and difference of two polynomials is computed by identifying and combining like terms. The sum or difference of two complex numbers is computed in a similar way, by adding the real number parts from each, and the imaginary parts from each. Notice in Example 4 that the commutative, associative, and distributive properties also apply to complex numbers.

EXAMPLE 4 ▶   **Adding and Subtracting Complex Numbers**

Perform the indicated operation and write the result in $a + bi$ form.

a. $(2 + 3i) + (-5 + 2i)$ **b.** $(-5 - 4i) - (-2 - \sqrt{2}i)$

Solution ▶

a. $(2 + 3i) + (-5 + 2i)$ original sum

$= 2 + 3i + (-5) + 2i$ distribute

$= 2 + (-5) + 3i + 2i$ commute terms

$= [2 + (-5)] + (3i + 2i)$ group like terms

$= -3 + 5i$ result

b. $(-5 - 4i) - (-2 - \sqrt{2}i)$ original difference

$= -5 - 4i + 2 + \sqrt{2}i$ distribute

$= -5 + 2 + (-4i) + \sqrt{2}i$ commute terms

$= (-5 + 2) + [(-4i) + \sqrt{2}i]$ group like terms

$= -3 + (-4 + \sqrt{2})i$ result

> **Now try Exercises 25 through 30 ▶**

☑ **B. You've just learned how to add and subtract complex numbers**

C. Multiplying Complex Numbers; Powers of i

The product of two complex numbers is computed using the distributive property and the F-O-I-L process in the same way we apply these to binomials. If any result gives a factor of i^2, remember that $i^2 = -1$.

EXAMPLE 5 ▶ **Multiplying Complex Numbers**

Find the indicated product and write the answer in $a + bi$ form.

 a. $\sqrt{-4}\sqrt{-9}$ b. $\sqrt{-6}\,(2 + \sqrt{-3})$

 c. $(6 - 5i)(4 + i)$ d. $(2 + 3i)(2 - 3i)$

Solution ▶ a. $\sqrt{-4}\sqrt{-9} = i\sqrt{4} \cdot i\sqrt{9}$ rewrite in terms of i b. $\sqrt{-6}\,(2 + \sqrt{-3}) = i\sqrt{6}(2 + i\sqrt{3})$ rewrite in terms of i

 $\qquad\qquad\quad = 2i \cdot 3i$ simplify $\qquad\qquad\qquad\qquad = 2i\sqrt{6} + i^2\sqrt{18}$ distribute

 $\qquad\qquad\quad = 6i^2$ multiply $\qquad\qquad\qquad\qquad = 2i\sqrt{6} + (-1)\sqrt{9}\sqrt{2}$ $i^2 = -1$

 $\qquad\qquad\quad = -6 + 0i$ result ($i^2 = -1$) $\qquad\qquad\qquad\qquad = 2i\sqrt{6} - 3\sqrt{2}$ simplify

 $\qquad\qquad\qquad\qquad\qquad\qquad\qquad\qquad\qquad\qquad\qquad\qquad\; = -3\sqrt{2} + 2i\sqrt{6}$ standard form

 c. $(6 - 5i)(4 + i)$ d. $(2 + 3i)(2 - 3i)$

 $\quad = (6)(4) + 6i + (-5i)(4) + (-5i)(i)$ F-O-I-L $\quad = (2)^2 - (3i)^2$ $(A + B)(A - B) = A^2 - B^2$

 $\quad = 24 + 6i + (-20i) + (-5)i^2$ $i \cdot i = i^2$ $\quad = 4 - 9i^2$ $(3i)^2 = 9i^2$

 $\quad = 24 + 6i + (-20i) + (-5)(-1)$ $i^2 = -1$ $\quad = 4 - 9(-1)$ $i^2 = -1$

 $\quad = 29 - 14i$ result $\quad = 13 + 0i$ result

Now try Exercises 31 through 48 ▶

⚠ **CAUTION** ▶ When computing with imaginary and complex numbers, always write the square root of a negative number in terms of i before you begin, as shown in Examples 5(a) and 5(b). Otherwise we get conflicting results, since $\sqrt{-4}\,\sqrt{-9} = \sqrt{36} = 6$ if we multiply the radicands first, which is an incorrect result because the original factors were imaginary. **See Exercise 80.**

Recall that expressions $2x + 5$ and $2x - 5$ are called binomial conjugates. In the same way, $a + bi$ and $a - bi$ are called **complex conjugates.** Note from Example 5(d) that the *product* of the complex number $a + bi$ with its complex conjugate $a - bi$ is *a real number.* This relationship is useful when rationalizing expressions with a complex number in the denominator, and we generalize the result as follows:

WORTHY OF NOTE

Notice that the product of a complex number and its conjugate also gives us a method for *factoring the sum of two squares* using complex numbers! For the expression $x^2 + 4$, the factored form would be $(x + 2i)(x - 2i)$. For more on this idea, **see Exercise 79.**

Product of Complex Conjugates

For a complex number $a + bi$ and its conjugate $a - bi$,
their product $(a + bi)(a - bi)$ is the real number $a^2 + b^2$;

$$(a + bi)(a - bi) = a^2 + b^2$$

Showing that $(a + bi)(a - bi) = a^2 + b^2$ is left as an exercise (see Exercise 79), but from here on, when asked to compute the product of complex conjugates, simply refer to the formula as illustrated here: $(-3 + 5i)(-3 - 5i) = (-3)^2 + 5^2$ or 34.

These operations on complex numbers enable us to verify complex solutions by substitution, in the same way we verify solutions for real numbers. In Example 2 we stated that $x = -3 + 2i$ was one solution to $x^2 + 6x + 13 = 0$. This is verified here.

EXAMPLE 6 ▶ **Checking a Complex Root by Substitution**

Verify that $x = -3 + 2i$ is a solution to $x^2 + 6x + 13 = 0$.

Solution ▶

$$x^2 + 6x + 13 = 0 \quad \text{original equation}$$
$$(-3 + 2i)^2 + 6(-3 + 2i) + 13 = 0 \quad \text{substitute } -3 + 2i \text{ for } x$$
$$(-3)^2 + 2(-3)(2i) + (2i)^2 - 18 + 12i + 13 = 0 \quad \text{square and distribute}$$
$$9 - 12i + 4i^2 + 12i - 5 = 0 \quad \text{simplify}$$
$$9 + (-4) - 5 = 0 \quad \text{combine terms } (12i - 12i = 0; \, i^2 = -1)$$
$$0 = 0 ✔$$

Now try Exercises 49 through 56 ▶

EXAMPLE 7 ▶ **Checking a Complex Root by Substitution**

Show that $x = 2 - i\sqrt{3}$ is a solution of $x^2 - 4x = -7$.

Solution ▶

$$x^2 - 4x = -7 \quad \text{original equation}$$
$$(2 - i\sqrt{3})^2 - 4(2 - i\sqrt{3}) = -7 \quad \text{substitute } 2 - i\sqrt{3} \text{ for } x$$
$$4 - 4i\sqrt{3} + (i\sqrt{3})^2 - 8 + 4i\sqrt{3} = -7 \quad \text{square and distribute}$$
$$4 - 4i\sqrt{3} - 3 - 8 + 4i\sqrt{3} = -7 \quad (i\sqrt{3})^2 = -3$$
$$-7 = -7 ✔ \quad \text{solution checks}$$

Now try Exercises 57 through 60 ▶

The imaginary unit i has another interesting and useful property. Since $i = \sqrt{-1}$ and $i^2 = -1$, we know that $i^3 = i^2 \cdot i = (-1)i = -i$ and $i^4 = (i^2)^2 = 1$. We can now simplify any *higher power of i* by rewriting the expression in terms of i^4.

$$i^5 = i^4 \cdot i = i$$
$$i^6 = i^4 \cdot i^2 = -1$$
$$i^7 = i^4 \cdot i^3 = -i$$
$$i^8 = (i^4)^2 = 1$$

Notice the powers of i "cycle through" the four values i, -1, $-i$ and 1. In more advanced classes, powers of complex numbers play an important role, and next we learn to reduce higher powers using the power property of exponents and $i^4 = 1$. Essentially, we divide the exponent on i by 4, then use the remainder to compute the value of the expression. For i^{35}, $35 \div 4 = 8$ remainder 3, showing $i^{35} = (i^4)^8 \cdot i^3 = -i$.

EXAMPLE 8 ▶ **Simplifying Higher Powers of i**

Simplify:

    a. i^{22} **b.** i^{28} **c.** i^{57} **d.** i^{75}

Solution ▶   **a.** $i^{22} = (i^4)^5 \cdot (i^2)$ **b.** $i^{28} = (i^4)^7$

                          $= (1)^5(-1)$ $= (1)^7$

                          $= -1$ $= 1$

☑ **C.** You've just learned how to multiply complex numbers and find powers of i

c. $i^{57} = (i^4)^{14} \cdot i$
$= (1)^{14}i$
$= i$

d. $i^{75} = (i^4)^{18} \cdot (i^3)$
$= (1)^{18}(-i)$
$= -i$

Now try Exercises 61 and 62 ▶

D. Division of Complex Numbers

Since $i = \sqrt{-1}$, expressions like $\dfrac{3-i}{2+i}$ actually have a radical in the denominator. To divide complex numbers, we simply apply our earlier method of rationalizing denominators (Section R.6), but this time using a *complex* conjugate.

EXAMPLE 9 ▶   Dividing Complex Numbers

Divide and write each result in $a + bi$ form.

a. $\dfrac{2}{5-i}$ **b.** $\dfrac{3-i}{2+i}$ **c.** $\dfrac{6+\sqrt{-36}}{3+\sqrt{-9}}$

Solution ▶   **a.** $\dfrac{2}{5-i} = \dfrac{2}{5-i} \cdot \dfrac{5+i}{5+i}$

$= \dfrac{2(5+i)}{5^2+1^2}$

$= \dfrac{10+2i}{26}$

$= \dfrac{10}{26} + \dfrac{2}{26}i$

$= \dfrac{5}{13} + \dfrac{1}{13}i$

b. $\dfrac{3-i}{2+i} = \dfrac{3-i}{2+i} \cdot \dfrac{2-i}{2-i}$

$= \dfrac{6-3i-2i+i^2}{2^2+1^2}$

$= \dfrac{6-5i+(-1)}{5}$

$= \dfrac{5-5i}{5} = \dfrac{5}{5} - \dfrac{5i}{5}$

$= 1 - i$

c. $\dfrac{6+\sqrt{-36}}{3+\sqrt{-9}} = \dfrac{6+i\sqrt{36}}{3+i\sqrt{9}}$ convert to i notation

$= \dfrac{6+6i}{3+3i}$ simplify

The expression can be further simplified by reducing common factors.

$= \dfrac{6(1+i)}{3(1+i)} = 2$ factor and reduce

Now try Exercises 63 through 68 ▶

Operations on complex numbers can be checked using inverse operations, just as we do for real numbers. To check the answer $1 - i$ from Example 9(b), we multiply it by the divisor:

$$(1-i)(2+i) = 2 + i - 2i - i^2$$
$$= 2 - i - (-1)$$
$$= 2 - i + 1$$
$$= 3 - i\checkmark$$

☑ **D.** You've just learned how to divide complex numbers

1.4 EXERCISES

▶ CONCEPTS AND VOCABULARY

Fill in each blank with the appropriate word or phrase. Carefully reread the section, if necessary.

1. Given the complex number $3 + 2i$, its complex conjugate is _____.

2. The product $(3 + 2i)(3 - 2i)$ gives the real number _____.

3. If the expression $\dfrac{4 + 6i\sqrt{2}}{2}$ is written in the standard form $a + bi$, then $a =$ _____ and $b =$ _____.

4. For $i = \sqrt{-1}$, $i^2 =$ _____, $i^4 =$ _____, $i^6 =$ _____, and $i^8 =$ _____, $i^3 =$ _____, $i^5 =$ _____, $i^7 =$ _____, and $i^9 =$ _____.

5. Discuss/Explain which is correct:
   a. $\sqrt{-4} \cdot \sqrt{-9} = \sqrt{(-4)(-9)} = \sqrt{36} = 6$
   b. $\sqrt{-4} \cdot \sqrt{-9} = 2i \cdot 3i = 6i^2 = -6$

6. Compare/Contrast the product $(1 + \sqrt{2})(1 - \sqrt{3})$ with the product $(1 + i\sqrt{2})(1 - i\sqrt{3})$. What is the same? What is different?

▶ DEVELOPING YOUR SKILLS

Simplify each radical (if possible). If imaginary, rewrite in terms of i and simplify.

7. a. $\sqrt{-16}$ b. $\sqrt{-49}$
   c. $\sqrt{27}$ d. $\sqrt{72}$

8. a. $\sqrt{-81}$ b. $\sqrt{-169}$
   c. $\sqrt{64}$ d. $\sqrt{98}$

9. a. $-\sqrt{-18}$ b. $-\sqrt{-50}$
   c. $3\sqrt{-25}$ d. $2\sqrt{-9}$

10. a. $-\sqrt{-32}$ b. $-\sqrt{-75}$
    c. $3\sqrt{-144}$ d. $2\sqrt{-81}$

11. a. $\sqrt{-19}$ b. $\sqrt{-31}$
    c. $\sqrt{\dfrac{-12}{25}}$ d. $\sqrt{\dfrac{-9}{32}}$

12. a. $\sqrt{-17}$ b. $\sqrt{-53}$
    c. $\sqrt{\dfrac{-45}{36}}$ d. $\sqrt{\dfrac{-49}{75}}$

Write each complex number in the standard form $a + bi$ and clearly identify the values of a and b.

13. a. $\dfrac{2 + \sqrt{-4}}{2}$ b. $\dfrac{6 + \sqrt{-27}}{3}$

14. a. $\dfrac{16 - \sqrt{-8}}{2}$ b. $\dfrac{4 + 3\sqrt{-20}}{2}$

15. a. $\dfrac{8 + \sqrt{-16}}{2}$ b. $\dfrac{10 - \sqrt{-50}}{5}$

16. a. $\dfrac{6 - \sqrt{-72}}{4}$ b. $\dfrac{12 + \sqrt{-200}}{8}$

17. a. 5 b. $3i$

18. a. -2 b. $-4i$

19. a. $2\sqrt{-81}$ b. $\dfrac{\sqrt{-32}}{8}$

20. a. $-3\sqrt{-36}$ b. $\dfrac{\sqrt{-75}}{15}$

21. a. $4 + \sqrt{-50}$ b. $-5 + \sqrt{-27}$

22. a. $-2 + \sqrt{-48}$ b. $7 + \sqrt{-75}$

23. a. $\dfrac{14 + \sqrt{-98}}{8}$ b. $\dfrac{5 + \sqrt{-250}}{10}$

24. a. $\dfrac{21 + \sqrt{-63}}{12}$ b. $\dfrac{8 + \sqrt{-27}}{6}$

Perform the addition or subtraction. Write the result in $a + bi$ form.

25. a. $(12 - \sqrt{-4}) + (7 + \sqrt{-9})$
    b. $(3 + \sqrt{-25}) + (-1 - \sqrt{-81})$
    c. $(11 + \sqrt{-108}) - (2 - \sqrt{-48})$

26. a. $(-7 - \sqrt{-72}) + (8 + \sqrt{-50})$
    b. $(\sqrt{3} + \sqrt{-2}) - (\sqrt{12} + \sqrt{-8})$
    c. $(\sqrt{20} - \sqrt{-3}) + (\sqrt{5} - \sqrt{-12})$

27. a. $(2 + 3i) + (-5 - i)$
    b. $(5 - 2i) + (3 + 2i)$
    c. $(6 - 5i) - (4 + 3i)$

28. a. $(-2 + 5i) + (3 - i)$
    b. $(7 - 4i) - (2 - 3i)$
    c. $(2.5 - 3.1i) + (4.3 + 2.4i)$

29. a. $(3.7 + 6.1i) - (1 + 5.9i)$

   b. $\left(8 + \dfrac{3}{4}i\right) - \left(-7 + \dfrac{2}{3}i\right)$

   c. $\left(-6 - \dfrac{5}{8}i\right) + \left(4 + \dfrac{1}{2}i\right)$

30. a. $(9.4 - 8.7i) - (6.5 + 4.1i)$

   b. $\left(3 + \dfrac{3}{5}i\right) - \left(-11 + \dfrac{7}{15}i\right)$

   c. $\left(-4 - \dfrac{5}{6}i\right) + \left(13 + \dfrac{3}{8}i\right)$

Multiply and write your answer in $a + bi$ form.

31. a. $5i \cdot (-3i)$ **b.** $(4i)(-4i)$

32. a. $3(2 - 3i)$ **b.** $-7(3 + 5i)$

33. a. $-7i(5 - 3i)$ **b.** $6i(-3 + 7i)$

34. a. $(-4 - 2i)(3 + 2i)$ **b.** $(2 - 3i)(-5 + i)$

35. a. $(-3 + 2i)(2 + 3i)$ **b.** $(3 + 2i)(1 + i)$

36. a. $(5 + 2i)(-7 + 3i)$ **b.** $(4 - i)(7 + 2i)$

For each complex number, name the complex conjugate. Then find the product.

37. a. $4 + 5i$ **b.** $3 - i\sqrt{2}$

38. a. $2 - i$ **b.** $-1 + i\sqrt{5}$

39. a. $7i$ **b.** $\frac{1}{2} - \frac{2}{3}i$

40. a. $-5i$ **b.** $\frac{3}{4} + \frac{1}{5}i$

Compute the special products and write your answer in $a + bi$ form.

41. a. $(4 - 5i)(4 + 5i)$

   b. $(7 - 5i)(7 + 5i)$

42. a. $(-2 - 7i)(-2 + 7i)$

   b. $(2 + i)(2 - i)$

43. a. $(3 - i\sqrt{2})(3 + i\sqrt{2})$

   b. $(\frac{1}{6} + \frac{2}{3}i)(\frac{1}{6} - \frac{2}{3}i)$

44. a. $(5 + i\sqrt{3})(5 - i\sqrt{3})$

   b. $(\frac{1}{2} + \frac{3}{4}i)(\frac{1}{2} - \frac{3}{4}i)$

45. a. $(2 + 3i)^2$ **b.** $(3 - 4i)^2$

46. a. $(2 - i)^2$ **b.** $(3 - i)^2$

47. a. $(-2 + 5i)^2$ **b.** $(3 + i\sqrt{2})^2$

48. a. $(-2 - 5i)^2$ **b.** $(2 - i\sqrt{3})^2$

Use substitution to determine if the value shown is a solution to the given equation.

49. $x^2 + 36 = 0;\ x = -6$

50. $x^2 + 16 = 0;\ x = -4$

51. $x^2 + 49 = 0;\ x = -7i$

52. $x^2 + 25 = 0;\ x = -5i$

53. $(x - 3)^2 = -9;\ x = 3 - 3i$

54. $(x + 1)^2 = -4;\ x = -1 + 2i$

55. $x^2 - 2x + 5 = 0;\ x = 1 - 2i$

56. $x^2 + 6x + 13 = 0;\ x = -3 + 2i$

57. $x^2 - 4x + 9 = 0;\ x = 2 + i\sqrt{5}$

58. $x^2 - 2x + 4 = 0;\ x = 1 - \sqrt{3}\,i$

59. Show that $x = 1 + 4i$ is a solution to $x^2 - 2x + 17 = 0$. Then show its complex conjugate $1 - 4i$ is also a solution.

60. Show that $x = 2 - 3\sqrt{2}\,i$ is a solution to $x^2 - 4x + 22 = 0$. Then show its complex conjugate $2 + 3\sqrt{2}\,i$ is also a solution.

Simplify using powers of i.

61. a. i^{48} **b.** i^{26} **c.** i^{39} **d.** i^{53}

62. a. i^{36} **b.** i^{50} **c.** i^{19} **d.** i^{65}

Divide and write your answer in $a + bi$ form. Check your answer using multiplication.

63. a. $\dfrac{-2}{\sqrt{-49}}$ **b.** $\dfrac{4}{\sqrt{-25}}$

64. a. $\dfrac{2}{1 - \sqrt{-4}}$ **b.** $\dfrac{3}{2 + \sqrt{-9}}$

65. a. $\dfrac{7}{3 + 2i}$ **b.** $\dfrac{-5}{2 - 3i}$

66. a. $\dfrac{6}{1 + 3i}$ **b.** $\dfrac{7}{7 - 2i}$

67. a. $\dfrac{3 + 4i}{4i}$ **b.** $\dfrac{2 - 3i}{3i}$

68. a. $\dfrac{-4 + 8i}{2 - 4i}$ **b.** $\dfrac{3 - 2i}{-6 + 4i}$

▶ WORKING WITH FORMULAS

69. Absolute value of a complex number:
$$|a + bi| = \sqrt{a^2 + b^2}$$

The absolute value of any complex number $a + bi$ (sometimes called the *modulus* of the number) is computed by taking the square root of the sum of the squares of a and b. Find the absolute value of the given complex numbers.

   a. $|2 + 3i|$　　　　　　**b.** $|4 - 3i|$

   c. $|3 + \sqrt{2}\,i|$

70. Binomial cubes:
$$(A + B)^3 = A^3 + 3A^2B + 3AB^2 + B^3$$

The cube of any binomial can be found using the formula shown, where A and B are the terms of the binomial. Use the formula to compute $(1 - 2i)^3$ (note $A = 1$ and $B = -2i$).

▶ APPLICATIONS

71. Dawn of imaginary numbers: In a day when imaginary numbers were imperfectly understood, Girolamo Cardano (1501–1576) once posed the problem, "Find two numbers that have a sum of 10 and whose product is 40." In other words, $A + B = 10$ and $AB = 40$. Although the solution is routine today, at the time the problem posed an enormous challenge. Verify that $A = 5 + \sqrt{15}i$ and $B = 5 - \sqrt{15}i$ satisfy these conditions.

72. Verifying calculations using i: Suppose Cardano had said, "Find two numbers that have a sum of 4 and a product of 7" (see Exercise 71). Verify that $A = 2 + \sqrt{3}i$ and $B = 2 - \sqrt{3}i$ satisfy these conditions.

Although it may seem odd, imaginary numbers have several applications in the real world. Many of these involve a study of electrical circuits, in particular *alternating current* or AC circuits. Briefly, the components of an AC circuit are current I (in amperes), voltage V (in volts), and the impedance Z (in ohms). The impedance of an electrical circuit is a measure of the total opposition to the flow of current through the circuit and is calculated as $Z = R + iX_L - iX_C$ where R represents a pure resistance, X_C represents the capacitance, and X_L represents the inductance. Each of these is also measured in ohms (symbolized by Ω).

73. Find the impedance Z if $R = 7\ \Omega$, $X_L = 6\ \Omega$, and $X_C = 11\ \Omega$.

74. Find the impedance Z if $R = 9.2\ \Omega$, $X_L = 5.6\ \Omega$, and $X_C = 8.3\ \Omega$.

The voltage V (in volts) across any element in an AC circuit is calculated as a product of the current I and the impedance Z: $V = IZ$.

75. Find the voltage in a circuit with a current $I = 3 - 2i$ amperes and an impedance of $Z = 5 + 5i\ \Omega$.

76. Find the voltage in a circuit with a current $I = 2 - 3i$ amperes and an impedance of $Z = 4 + 2i\ \Omega$.

In an AC circuit, the total impedance (in ohms) is given by $Z = \dfrac{Z_1 Z_2}{Z_1 + Z_2}$, where Z represents the total impedance of a circuit that has Z_1 and Z_2 wired in parallel.

77. Find the total impedance Z if $Z_1 = 1 + 2i$ and $Z_2 = 3 - 2i$.

78. Find the total impedance Z if $Z_1 = 3 - i$ and $Z_2 = 2 + i$.

▶ EXTENDING THE CONCEPT

79. Up to this point, we've said that expressions like $x^2 - 9$ and $p^2 - 7$ are factorable:

$$x^2 - 9 = (x + 3)(x - 3) \quad \text{and}$$
$$p^2 - 7 = (p + \sqrt{7})(p - \sqrt{7}),$$

while $x^2 + 9$ and $p^2 + 7$ are prime. More correctly, we should state that $x^2 + 9$ and $p^2 + 7$ are nonfactorable *using real numbers,* since they actually *can* be factored if complex numbers are used. From $(a + bi)(a - bi) = a^2 + b^2$ we note $a^2 + b^2 = (a + bi)(a - bi)$, showing

$$x^2 + 9 = (x + 3i)(x - 3i) \quad \text{and}$$
$$p^2 + 7 = (p + i\sqrt{7})(p - i\sqrt{7}).$$

Use this idea to factor the following.

a. $x^2 + 36$ b. $m^2 + 3$

c. $n^2 + 12$ d. $4x^2 + 49$

80. In this section, we noted that the product property of radicals $\sqrt{AB} = \sqrt{A}\sqrt{B}$, can still be applied when at most one of the factors is negative. So what happens if *both* are negative? First consider the expression $\sqrt{-4 \cdot -25}$. What happens if you first multiply in the radicand, then compute the square root? Next consider the product $\sqrt{-4} \cdot \sqrt{-25}$. Rewrite each factor using the *i* notation, then compute the product. Do you get the same result as before? What can you say about $\sqrt{-4 \cdot -25}$ and $\sqrt{-4} \cdot \sqrt{-25}$?

81. Simplify the expression
$i^{17}(3 - 4i) - 3i^3(1 + 2i)^2$.

82. While it is a simple concept for real numbers, the square root of a complex number is much more involved due to the interplay between its real and imaginary parts. For $z = a + bi$ the square root of z can be found using the formula:

$$\sqrt{z} = \frac{\sqrt{2}}{2}\left(\sqrt{|z| + a} \pm i\sqrt{|z| - a}\right),$$ where the sign

is chosen to match the sign of b (see Exercise 69). Use the formula to find the square root of each complex number, then check by squaring.

a. $z = -7 + 24i$ b. $z = 5 - 12i$

c. $z = 4 + 3i$

▶ **MAINTAINING YOUR SKILLS**

83. (R.7) State the perimeter and area formulas for: (a) squares, (b) rectangles, (c) triangles, and (d) circles.

84. (R.1) Write the symbols in words and state True/False.

a. $6 \notin \mathbb{Q}$ b. $\mathbb{Q} \subset \mathbb{R}$

c. $103 \in \{3, 4, 5, \ldots\}$ d. $\mathbb{R} \not\subset \mathbb{C}$

85. (1.1) John can run 10 m/sec, while Rick can only run 9 m/sec. If Rick gets a 2-sec head start, who will hit the 200-m finish line first?

86. (R.4) Factor the following expressions completely.

a. $x^4 - 16$ b. $n^3 - 27$

c. $x^3 - x^2 - x + 1$ d. $4n^2m - 12nm^2 + 9m^3$

1.5 | Solving Quadratic Equations

Learning Objectives

In Section 1.5 you will learn how to:

☐ **A.** Solve quadratic equations using the zero product property

☐ **B.** Solve quadratic equations using the square root property of equality

☐ **C.** Solve quadratic equations by completing the square

☐ **D.** Solve quadratic equations using the quadratic formula

☐ **E.** Use the discriminant to identify solutions

☐ **F.** Solve applications of quadratic equations

In Section 1.1 we solved the equation $ax + b = c$ for x to establish a general solution for all linear equations of this form. In this section, we'll establish a general solution for the quadratic equation $ax^2 + bx + c = 0$, $(a \neq 0)$ using a process known as *completing the square*. Other applications of completing the square include the graphing of parabolas, circles, and other relations from the family of *conic sections*.

A. Quadratic Equations and the Zero Product Property

A **quadratic equation** is one that can be written in the form $ax^2 + bx + c = 0$, where a, b, and c are real numbers and $a \neq 0$. As shown, the equation is written in **standard form,** meaning the terms are in decreasing order of degree and the equation is set equal to zero.

Quadratic Equations

A quadratic equation can be written in the form

$$ax^2 + bx + c = 0,$$

with $a, b, c \in \mathbb{R}$, and $a \neq 0$.

Notice that a is the leading coefficient, b is the coefficient of the linear (first degree) term, and c is a constant. All quadratic equations have degree two, but can have one, two, or three terms. The equation $n^2 - 81 = 0$ is a quadratic equation with two terms, where $a = 1$, $b = 0$, and $c = -81$.

EXAMPLE 1 ▶ **Determining Whether an Equation Is Quadratic**

State whether the given equation is quadratic. If yes, identify coefficients a, b, and c.

a. $2x^2 - 18 = 0$ **b.** $z - 12 - 3z^2 = 0$ **c.** $\dfrac{-3}{4}x + 5 = 0$

d. $z^3 - 2z^2 + 7z = 8$ **e.** $0.8x^2 = 0$

Solution ▶

WORTHY OF NOTE

The word *quadratic* comes from the Latin word *quadratum*, meaning square. The word historically refers to the "four sidedness" of a square, but mathematically to the *area* of a square. Hence its application to polynomials of the form $ax^2 + bx + c$— the variable of the leading term is *squared*.

	Standard Form	Quadratic	Coefficients
a.	$2x^2 - 18 = 0$	yes, deg 2	$a = 2$ $b = 0$ $c = -18$
b.	$-3z^2 + z - 12 = 0$	yes, deg 2	$a = -3$ $b = 1$ $c = -12$
c.	$\dfrac{-3}{4}x + 5 = 0$	no, deg 1	(linear equation)
d.	$z^3 - 2z^2 + 7z - 8 = 0$	no, deg 3	(cubic equation)
e.	$0.8x^2 = 0$	yes, deg 2	$a = 0.8$ $b = 0$ $c = 0$

Now try Exercises 7 through 18 ▶

With quadratic and other polynomial equations, we generally cannot isolate the variable on one side using only properties of equality, because the variable is raised to different powers. Instead we attempt to solve the equation by factoring and applying the **zero product property.**

Zero Product Property

If A and B represent real numbers or real valued expressions

and $A \cdot B = 0$,

then $A = 0$ or $B = 0$.

In words, the property says, *If the product of any two (or more) factors is equal to zero, then at least one of the factors must be equal to zero.* We can use this property to solve higher degree equations after rewriting them in terms of equations with lesser degree. As with linear equations, values that make the original equation true are called *solutions* or *roots* of the equation.

EXAMPLE 2 ▶ **Solving Equations Using the Zero Product Property**

Solve by writing the equations in factored form and applying the zero product property.

a. $3x^2 = 5x$ **b.** $-5x + 2x^2 = 3$ **c.** $4x^2 = 12x - 9$

Solution ▶ **a.**

$\qquad 3x^2 = 5x \qquad$ given equation

$\qquad 3x^2 - 5x = 0 \qquad$ standard form

$\qquad x(3x - 5) = 0 \qquad$ factor

$\qquad x = 0 \quad \text{or} \quad 3x - 5 = 0 \quad$ set factors equal to zero (zero product property)

$\qquad x = 0 \quad \text{or} \quad x = \dfrac{5}{3} \quad$ result

b.

$\qquad -5x + 2x^2 = 3 \quad$ given equation

$\qquad 2x^2 - 5x - 3 = 0 \quad$ standard form

$\qquad (2x + 1)(x - 3) = 0 \quad$ factor

$\qquad 2x + 1 = 0 \quad \text{or} \quad x - 3 = 0 \quad$ set factors equal to zero (zero product property)

$\qquad x = -\dfrac{1}{2} \quad \text{or} \quad x = 3 \quad$ result

c.

$$4x^2 = 12x - 9 \qquad \text{given equation}$$

$$4x^2 - 12x + 9 = 0 \qquad \text{standard form}$$

$$(2x - 3)(2x - 3) = 0 \qquad \text{factor}$$

$$2x - 3 = 0 \quad \text{or} \quad 2x - 3 = 0 \qquad \text{set factors equal to zero (zero product property)}$$

$$x = \frac{3}{2} \quad \text{or} \qquad x = \frac{3}{2} \qquad \text{result}$$

This equation has only the solution $x = \dfrac{3}{2}$, which we call a *repeated root*.

Now try Exercises 19 through 42 ▶

⚠ **CAUTION** ▶ Consider the equation $x^2 - 2x - 3 = 12$. While the left-hand side is factorable, the result is $(x - 3)(x + 1) = 12$ and finding a solution becomes a "guessing game" because the equation is not set equal to zero. If you *misapply* the zero factor property and say that $x - 3 = 12$ or $x + 1 = 12$, the "solutions" are $x = 15$ or $x = 11$, which are both incorrect! After subtracting 12 from both sides $x^2 - 2x - 3 = 12$ becomes $x^2 - 2x - 15 = 0$, giving $(x - 5)(x + 3) = 0$ with solutions $x = 5$ or $x = -3$.

 A. You've just learned how to solve quadratic equations using the zero product property

B. Solving Quadratic Equations Using the Square Root Property of Equality

The equation $x^2 = 9$ can be solved by factoring. In standard form we have $x^2 - 9 = 0$ (note $b = 0$), then $(x - 3)(x + 3) = 0$. The solutions are $x = -3$ or $x = 3$, which are simply the *positive and negative square roots of 9*. This result suggests an alternative method for solving equations of the form $X^2 = k$, known as the **square root property of equality.**

Square Root Property of Equality

If X represents an algebraic expression

$$\text{and } X^2 = k,$$

$$\text{then } X = \sqrt{k} \text{ or } X = -\sqrt{k};$$

$$\text{also written as } X = \pm\sqrt{k}$$

EXAMPLE 3 ▶ **Solving an Equation Using the Square Root Property of Equality**

Use the square root property of equality to solve each equation.

a. $-4x^2 + 3 = -6$ **b.** $x^2 + 12 = 0$ **c.** $(x - 5)^2 = 24$

Solution ▶ **a.** $-4x^2 + 3 = -6 \qquad \text{original equation}$

$$x^2 = \frac{9}{4} \qquad \text{subtract 3, divide by } -4$$

$$x = \sqrt{\frac{9}{4}} \quad \text{or} \quad x = -\sqrt{\frac{9}{4}} \qquad \text{square root property of equality}$$

$$x = \frac{3}{2} \quad \text{or} \quad x = -\frac{3}{2} \qquad \text{simplify radicals}$$

This equation has two rational solutions.

b. $x^2 + 12 = 0$ original equation

$\qquad x^2 = -12$ subtract 12

$x = \sqrt{-12} \quad \text{or} \quad x = -\sqrt{12}$ square root property of equality

$x = 2i\sqrt{3} \quad \text{or} \quad x = -2i\sqrt{3}$ simplify radicals

This equation has two complex solutions.

c. $(x - 5)^2 = 24$ original equation

$x - 5 = \sqrt{24} \quad \text{or} \quad x - 5 = -\sqrt{24}$ square root property of equality

$x = 5 + 2\sqrt{6} \qquad x = 5 - 2\sqrt{6}$ solve for x and simplify radicals

This equation has two irrational solutions.

 B. You've just learned how to solve quadratic equations using the square root property of equality

Now try Exercises 43 through 58 ▶

⚠ **CAUTION ▶** For equations of the form $(x + d)^2 = k$ [see Example 3(c)], you should resist the temptation to expand the binomial square in an attempt to simplify the equation and solve by factoring—many times the result is nonfactorable. *Any* equation of the form $(x + d)^2 = k$ can quickly be solved using the square root property of equality.

WORTHY OF NOTE

In Section R.6 we noted that for any real number a, $\sqrt{a^2} = |a|$. From Example 3(a), solving the equation by taking the square root of both sides produces $\sqrt{x^2} = \sqrt{\frac{9}{4}}$. This is equivalent to $|x| = \sqrt{\frac{9}{4}}$, again showing this equation must have two solutions, $x = -\sqrt{\frac{9}{4}}$ and $x = \sqrt{\frac{9}{4}}$.

Answers written using radicals are called **exact** or **closed form** solutions. Actually checking the exact solutions is a nice application of fundamental skills. Let's check $x = 5 + 2\sqrt{6}$ from Example 3(c).

check: $(x - 5)^2 = 24$ original equation

$(5 + 2\sqrt{6} - 5)^2 = 24$ substitute $5 + 2\sqrt{6}$ for x

$(2\sqrt{6})^2 = 24$ simplify

$4(6) = 24$ $(2\sqrt{6})^2 = 4(6)$

$24 = 24 ✓$ result checks ($x = 5 - 2\sqrt{6}$ also checks)

C. Solving Quadratic Equations by Completing the Square

Again consider $(x - 5)^2 = 24$ from Example 3(c). If we had first expanded the binomial square, we would have obtained $x^2 - 10x + 25 = 24$, then $x^2 - 10x + 1 = 0$ in standard form. Note that this equation *cannot be solved by factoring*. Reversing this process leads us to a strategy for solving nonfactorable quadratic equations, by creating a *perfect square trinomial* from the quadratic and linear terms. This process is known as **completing the square.** To transform $x^2 - 10x + 1 = 0$ back into $x^2 - 10x + 25 = 24$ [which we would then rewrite as $(x - 5)^2 = 24$ and solve], we subtract 1 from both sides, then add 25:

$$x^2 - 10x + 1 = 0$$

$$x^2 - 10x = -1 \qquad \text{subtract 1}$$

$$x^2 - 10x + 25 = -1 + 25 \qquad \text{add 25}$$

$$(x - 5)^2 = 24 \qquad \text{factor, simplify}$$

In general, after subtracting the constant term, the number that "completes the square" is found by squaring $\frac{1}{2}$ the coefficient of the linear term: $\left[\frac{1}{2}(10)\right]^2 = 25$. **See Exercises 59 through 64 for additional practice.**

EXAMPLE 4 ▶   Solving a Quadratic Equation by Completing the Square

Solve by completing the square: $x^2 + 13 = 6x$.

Solution ▶

$$x^2 + 13 = 6x \qquad \text{original equation}$$
$$x^2 - 6x + 13 = 0 \qquad \text{standard form}$$
$$x^2 - 6x + \underline{\quad} = -13 + \underline{\quad} \qquad \text{subtract 13 to make room for new constant}$$
$$\left[\left(\tfrac{1}{2}\right)(-6)\right]^2 = 9 \qquad \text{compute } \left[\left(\tfrac{1}{2}\right)(\textit{linear coefficient})\right]^2$$
$$x^2 - 6x + 9 = -13 + 9 \qquad \text{add 9 to both sides (completing the square)}$$
$$(x - 3)^2 = -4 \qquad \text{factor and simplify}$$
$$x - 3 = \sqrt{-4} \quad \text{or} \quad x - 3 = -\sqrt{-4} \qquad \text{square root property of equality}$$
$$x = 3 + 2i \quad \text{or} \quad x = 3 - 2i \qquad \text{simplify radicals and solve for } x$$

Now try Exercises 65 through 74 ▶

The process of completing the square can be applied to any quadratic equation with a leading coefficient of 1. If the leading coefficient is not 1, we simply divide through by a before beginning, which brings us to this summary of the process.

WORTHY OF NOTE

It's helpful to note that the number you're squaring in step three, $\left[\dfrac{1}{2} \cdot \dfrac{b}{a}\right] = \dfrac{b}{2a}$, turns out to be the constant term in the factored form. From Example 4, the number we squared was $\left(\tfrac{1}{2}\right)(-6) = -3$, and the binomial square was $(x - 3)^2$.

Completing the Square to Solve a Quadratic Equation

To solve $ax^2 + bx + c = 0$ by completing the square:
1. Subtract the constant c from both sides.
2. Divide both sides by the leading coefficient a.
3. Compute $\left[\dfrac{1}{2} \cdot \dfrac{b}{a}\right]^2$ and add the result to both sides.
4. Factor left-hand side as a binomial square; simplify right-hand side.
5. Solve using the square root property of equality.

EXAMPLE 5 ▶ **Solving a Quadratic Equation by Completing the Square**

Solve by completing the square: $-3x^2 + 1 = 4x$.

Solution ▶

$$-3x^2 + 1 = 4x \qquad \text{original equation}$$
$$-3x^2 - 4x + 1 = 0 \qquad \text{standard form (nonfactorable)}$$
$$-3x^2 - 4x = -1 \qquad \text{subtract 1}$$
$$x^2 + \frac{4}{3}x + \quad = \frac{1}{3} \qquad \text{divide by } -3$$
$$x^2 + \frac{4}{3}x + \frac{4}{9} = \frac{1}{3} + \frac{4}{9} \qquad \left[\frac{1}{2}\frac{b}{a}\right]^2 = \left[\left(\frac{1}{2}\right)\left(\frac{4}{3}\right)\right]^2 = \frac{4}{9}; \text{ add } \frac{4}{9}$$
$$\left(x + \frac{2}{3}\right)^2 = \frac{7}{9} \qquad \text{factor and simplify } \left(\frac{1}{3} = \frac{3}{9}\right)$$
$$x + \frac{2}{3} = \sqrt{\frac{7}{9}} \quad \text{or} \quad x + \frac{2}{3} = -\sqrt{\frac{7}{9}} \qquad \text{square root property of equality}$$
$$x = -\frac{2}{3} + \frac{\sqrt{7}}{3} \quad \text{or} \quad x = -\frac{2}{3} - \frac{\sqrt{7}}{3} \qquad \text{solve for } x \text{ and simplify (exact form)}$$
$$x \approx 0.22 \quad \text{or} \quad x \approx -1.55 \qquad \text{approximate form (to hundredths)}$$

☑ **C.** You've just learned how to solve quadratic equations by completing the square

Now try Exercises 75 through 82 ▶

⚠ **CAUTION** ▶ For many of the skills/processes needed in a study of algebra, it's actually easier to work with the fractional form of a number, rather than the decimal form. For example, computing $\left(\tfrac{2}{3}\right)^2$ is easier than computing $(0.\overline{6})^2$, and finding $\sqrt{\tfrac{9}{16}}$ is much easier than finding $\sqrt{0.5625}$.

D. Solving Quadratic Equations Using the Quadratic Formula

In Section 1.1 we found a general solution for the linear equation $ax + b = c$ by comparing it to $2x + 3 = 15$. Here we'll use a similar idea to find a general solution for quadratic equations. In a side-by-side format, we'll solve the equations $2x^2 + 5x + 3 = 0$ and $ax^2 + bx + c = 0$ by completing the square. Note the similarities.

$2x^2 + 5x + 3 = 0$	given equations	$ax^2 + bx + c = 0$
$2x^2 + 5x + = -3$	subtract constant term	$ax^2 + bx + = -c$
$x^2 + \dfrac{5}{2}x + \underline{} = -\dfrac{3}{2}$	divide by lead coefficient	$x^2 + \dfrac{b}{a}x + \underline{} = -\dfrac{c}{a}$
$\left[\dfrac{1}{2}\left(\dfrac{5}{2}\right)\right]^2 = \dfrac{25}{16}$	$\left[\dfrac{1}{2}(\text{linear coefficient})\right]^2$	$\left[\dfrac{1}{2}\left(\dfrac{b}{a}\right)\right]^2 = \dfrac{b^2}{4a^2}$
$x^2 + \dfrac{5}{2}x + \dfrac{25}{16} = \dfrac{25}{16} - \dfrac{3}{2}$	add to both sides	$x^2 + \dfrac{b}{a}x + \dfrac{b^2}{4a^2} = \dfrac{b^2}{4a^2} - \dfrac{c}{a}$
$\left(x + \dfrac{5}{4}\right)^2 = \dfrac{25}{16} - \dfrac{3}{2}$	left side factors as a binomial square	$\left(x + \dfrac{b}{2a}\right)^2 = \dfrac{b^2}{4a^2} - \dfrac{c}{a}$
$\left(x + \dfrac{5}{4}\right)^2 = \dfrac{25}{16} - \dfrac{24}{16}$	determine LCDs	$\left(x + \dfrac{b}{2a}\right)^2 = \dfrac{b^2}{4a^2} - \dfrac{4ac}{4a^2}$
$\left(x + \dfrac{5}{4}\right)^2 = \dfrac{1}{16}$	simplify right side	$\left(x + \dfrac{b}{2a}\right)^2 = \dfrac{b^2 - 4ac}{4a^2}$
$x + \dfrac{5}{4} = \pm\sqrt{\dfrac{1}{16}}$	square root property of equality	$x + \dfrac{b}{2a} = \pm\sqrt{\dfrac{b^2 - 4ac}{4a^2}}$
$x + \dfrac{5}{4} = \pm\dfrac{1}{4}$	simplify radicals	$x + \dfrac{b}{2a} = \pm\dfrac{\sqrt{b^2 - 4ac}}{2a}$
$x = -\dfrac{5}{4} \pm \dfrac{1}{4}$	solve for x	$x = -\dfrac{b}{2a} \pm \dfrac{\sqrt{b^2 - 4ac}}{2a}$
$x = \dfrac{-5 \pm 1}{4}$	combine terms	$x = \dfrac{-b \pm \sqrt{b^2 - 4ac}}{2a}$
$x = \dfrac{-5 + 1}{4}$ or $x = \dfrac{-5 - 1}{4}$	solutions	$x = \dfrac{-b + \sqrt{b^2 - 4ac}}{2a}$ or $x = \dfrac{-b - \sqrt{b^2 - 4ac}}{2a}$

On the left, our final solutions are $x = -1$ or $x = -\dfrac{3}{2}$. The general solution is called the **quadratic formula,** which can be used to solve *any equation belonging to the quadratic family.*

Quadratic Formula

If $ax^2 + bx + c = 0$, with a, b, and $c \in \mathbb{R}$ and $a \neq 0$, then

$$x = \frac{-b + \sqrt{b^2 - 4ac}}{2a} \quad \text{or} \quad x = \frac{-b - \sqrt{b^2 - 4ac}}{2a};$$

also written $x = \dfrac{-b \pm \sqrt{b^2 - 4ac}}{2a}$.

⚠ **CAUTION** ▶ It's very important to note the values of a, b, and c come from an equation *written in standard form.* For $3x^2 - 5x = -7$, $a = 3$ and $b = -5$, but $c \neq -7$! In standard form we have $3x^2 - 5x + 7 = 0$, and note the value for use in the formula is actually $c = 7$.

EXAMPLE 6 ▶ Solving Quadratic Equations Using the Quadratic Formula

Solve $4x^2 + 1 = 8x$ using the quadratic formula. State the solution(s) in both exact and approximate form. Check one of the exact solutions in the original equation.

Solution ▶ Begin by writing the equation in standard form and identifying the values of a, b, and c.

$$4x^2 + 1 = 8x \qquad \text{original equation}$$
$$4x^2 - 8x + 1 = 0 \qquad \text{standard form}$$
$$a = 4, b = -8, c = 1$$
$$x = \frac{-(-8) \pm \sqrt{(-8)^2 - 4(4)(1)}}{2(4)} \qquad \text{substitute 4 for } a, -8 \text{ for } b, \text{ and 1 for } c$$
$$x = \frac{8 \pm \sqrt{64 - 16}}{8} = \frac{8 \pm \sqrt{48}}{8} \qquad \text{simplify}$$
$$x = \frac{8 \pm 4\sqrt{3}}{8} = \frac{8}{8} \pm \frac{4\sqrt{3}}{8} \qquad \text{rationalize the radical (see following Caution)}$$
$$x = 1 + \frac{\sqrt{3}}{2} \quad \text{or} \quad x = 1 - \frac{\sqrt{3}}{2} \qquad \text{exact solutions}$$
$$x \approx 1.87 \quad \text{or} \quad x \approx 0.13 \qquad \text{approximate solutions}$$

Check ▶
$$4x^2 + 1 = 8x \qquad \text{original equation}$$
$$4\left(1 + \frac{\sqrt{3}}{2}\right)^2 + 1 = 8\left(1 + \frac{\sqrt{3}}{2}\right) \qquad \text{substitute } 1 + \frac{\sqrt{3}}{2} \text{ for } x$$
$$4\left[1 + 2\left(\frac{\sqrt{3}}{2}\right) + \frac{3}{4}\right] + 1 = 8 + 4\sqrt{3} \qquad \text{square binomial; distribute}$$
$$4 + 4\sqrt{3} + 3 + 1 = 8 + 4\sqrt{3} \qquad \text{distribute}$$
$$8 + 4\sqrt{3} = 8 + 4\sqrt{3}✓ \qquad \text{result checks}$$

☑ **D.** You've just learned how to solve quadratic equations using the quadratic formula

Now try Exercises 83 through 112 ▶

⚠ **CAUTION** ▶ For $\dfrac{8 \pm 4\sqrt{3}}{8}$, be careful not to incorrectly "cancel the eights" as in $\dfrac{\overset{1}{\cancel{8}} \pm 4\sqrt{3}}{\underset{1}{\cancel{8}}} \ne 1 \pm 4\sqrt{3}$.

No! Use a calculator to verify that the results are not equivalent. Both terms in the numerator are divided by 8 and we must either rewrite the expression as separate terms (as above) or factor the numerator to see if the expression simplifies further:
$$\frac{8 \pm 4\sqrt{3}}{8} = \frac{\overset{1}{\cancel{4}}(2 \pm \sqrt{3})}{\underset{2}{\cancel{8}}} = \frac{2 \pm \sqrt{3}}{2}, \text{ which is equivalent to } 1 \pm \frac{\sqrt{3}}{2}.$$

E. The Discriminant of the Quadratic Formula

Recall that $\sqrt{X}$ represents a real number only for $X \geq 0$. Since the quadratic formula contains the radical $\sqrt{b^2 - 4ac}$, the expression $b^2 - 4ac$, called the **discriminant,** will determine the nature (real or complex) and the number of solutions to a given quadratic equation.

The Discriminant of the Quadratic Formula

For $ax^2 + bx + c = 0$, $a \neq 0$,
1. If $b^2 - 4ac = 0$, the equation has one real root.
2. If $b^2 - 4ac > 0$, the equation has two real roots.
3. If $b^2 - 4ac < 0$, the equation has two complex roots.

Further analysis of the discriminant reveals even more concerning the nature of quadratic solutions. If a, b, and c are rational and the discriminant is a perfect square, there will be two *rational* roots, which means the original equation can be solved by factoring. If the discriminant is not a perfect square, there will be two *irrational* roots that are conjugates. If the discriminant is zero there is one rational root, and the original equation is a perfect square trinomial.

EXAMPLE 7 ▶ **Using the Discriminant to Analyze Solutions**

Use the discriminant to determine if the equation given has any real root(s). If so, state whether the roots are rational or irrational, and whether the quadratic expression is factorable.

    a. $2x^2 + 5x + 2 = 0$ **b.** $x^2 - 4x + 7 = 0$ **c.** $4x^2 - 20x + 25 = 0$

Solution ▶   **a.** $a = 2, b = 5, c = 2$ **b.** $a = 1, b = -4, c = 7$ **c.** $a = 4, b = -20, c = 25$

$$b^2 - 4ac = (5)^2 - 4(2)(2) \qquad b^2 - 4ac = (-4)^2 - 4(1)(7) \qquad b^2 - 4ac = (-20)^2 - 4(4)(25)$$
$$= 9 \qquad\qquad\qquad\qquad = -12 \qquad\qquad\qquad\qquad = 0$$

      Since $9 > 0$,   Since $-12 < 0$,   Since $b^2 - 4ac = 0$,
  → two rational roots,   → two complex roots,   → one rational root,
        factorable   nonfactorable   factorable

Now try Exercises 113 through 124 ▶

In Example 7(b), $b^2 - 4ac = -12$ and the quadratic formula shows $x = \dfrac{4 \pm \sqrt{-12}}{2}$. After simplifying, we find the solutions are the complex conjugates $x = 2 + i\sqrt{3}$ or $x = 2 - i\sqrt{3}$. In general, when $b^2 - 4ac < 0$, the solutions *will be complex conjugates*.

Complex Solutions

The complex solutions of a quadratic equation with real coefficients occur in conjugate pairs.

EXAMPLE 8 ▶ **Solving Quadratic Equations Using the Quadratic Formula**

Solve: $2x^2 - 6x + 5 = 0$.

Solution ▶   With $a = 2$, $b = -6$, and $c = 5$, the discriminant becomes $(-6)^2 - 4(2)(5) = -4$, showing there will be two complex roots. The quadratic formula then yields

$$x = \frac{-b \pm \sqrt{b^2 - 4ac}}{2a} \qquad \text{quadratic formula}$$

$$x = \frac{-(-6) \pm \sqrt{-4}}{2(2)} \qquad b^2 - 4ac = -4, \text{ substitute 2 for } a, \text{ and } -6 \text{ for } b$$

$$x = \frac{6 \pm 2i}{4} \qquad \text{simplify, write in } i \text{ form}$$

☑ **E. You've just learned how to use the discriminant to identify solutions**

$$x = \frac{3}{2} \pm \frac{1}{2}i \qquad \text{solutions are complex conjugates}$$

Now try Exercises 125 through 130 ▶

WORTHY OF NOTE

While it's possible to solve by completing the square if $\dfrac{b}{a}$ is a fraction or an odd number (see Example 5), the process is usually most efficient when $\dfrac{b}{a}$ is an even number. This is one observation you could use when selecting a solution method.

Summary of Solution Methods for $ax^2 + bx + c = 0$

1. If $b = 0$, isolate x and use the square root property of equality.
2. If $c = 0$, factor out the GCF and solve using the zero product property.
3. If no coefficient is zero, you can attempt to solve by
   a. factoring the trinomial
   b. completing the square
   c. using the quadratic formula

F. Applications of the Quadratic Formula

A projectile is any object that is thrown, shot, or *projected* upward with no sustaining source of propulsion. The height of the projectile at time t is modeled by the equation $h = -16t^2 + vt + k$, where h is the height of the object in feet, t is the elapsed time in seconds, and v is the initial velocity in feet per second. The constant k represents the initial height of the object above ground level, as when a person releases an object 5 ft above the ground in a throwing motion. If the person were on a cliff 60 ft high, k would be 65 ft.

EXAMPLE 9 ▶ **Solving an Application of Quadratic Equations**

A person standing on a cliff 60 ft high, throws a ball upward with an initial velocity of 102 ft/sec (assume the ball is released 5 ft above where the person is standing). Find (a) the height of the object after 3 sec and (b) how many seconds until the ball hits the ground at the base of the cliff.

Solution ▶ Using the given information, we have $h = -16t^2 + 102t + 65$. To find the height after 3 sec, substitute $t = 3$.

  a. $h = -16t^2 + 102t + 65$ original equation
    $ = -16(3)^2 + 102(3) + 65$ substitute 3 for t
    $ = 227$ result

  After 3 sec, the ball is 227 ft above the ground.

  b. When the ball hits the ground at the base of the cliff, it has a height of zero. Substitute $h = 0$ and solve using the quadratic formula.

$$0 = -16t^2 + 102t + 65 \qquad \text{\small $a = -16$, $b = 102$, $c = 65$}$$

$$t = \frac{-b \pm \sqrt{b^2 - 4ac}}{2a} \qquad \text{\small quadratic formula}$$

$$t = \frac{-(102) \pm \sqrt{(102)^2 - 4(-16)(65)}}{2(-16)} \qquad \text{\small substitute -16 for a, 102 for b, 65 for c}$$

$$t = \frac{-102 \pm \sqrt{14{,}564}}{-32} \qquad \text{\small simplify}$$

Since we're trying to find the time in seconds, we go directly to the approximate form of the answer.

$$t \approx -0.58 \quad \text{or} \quad t \approx 6.96 \qquad \text{\small approximate solutions}$$

The ball will strike the base of the cliff about 7 sec later. Since t represents time, the solution $t \approx -0.58$ does not apply.

Now try Exercises 133 through 140 ▶

EXAMPLE 10 ▶ Solving Applications Using the Quadratic Formula

For the years 1995 to 2002, the amount A of annual international telephone traffic (in billions of minutes) can be modeled by $A = 0.3x^2 + 8.9x + 61.8$, where $x = 0$ represents the year 1995 [*Source:* Data from the *2005 Statistical Abstract of the United States,* Table 1372, page 870]. If this trend continues, in what year will the annual number of minutes reach or surpass 275 billion minutes?

 Solution ▶ We are essentially asked to solve $A = 0.3x^2 + 8.9x + 61.8$, when $A = 275$.

$$275 = 0.3x^2 + 8.9x + 61.8 \qquad \text{given equation}$$
$$0 = 0.3x^2 + 8.9x - 213.2 \qquad \text{subtract 275}$$

For $a = 0.3$, $b = 8.9$, and $c = -213.2$, the quadratic formula gives

$$x = \frac{-b \pm \sqrt{b^2 - 4ac}}{2a} \qquad \text{quadratic formula}$$

$$x = \frac{-8.9 \pm \sqrt{(8.9)^2 - 4(0.3)(-213.2)}}{2(0.3)} \qquad \text{substitute known values}$$

$$x = \frac{-8.9 \pm \sqrt{335.05}}{0.6} \qquad \text{simplify}$$

$$x \approx 15.7 \quad \text{or} \quad x \approx -45.3 \qquad \text{result}$$

☑ **F. You've just learned how to solve applications of quadratic equations**

We disregard the negative solution (since x represents time), and find the annual number of international telephone minutes will reach or surpass 275 billion 15.7 years after 1995, or in the year 2010.

Now try Exercises 141 and 142 ▶

TECHNOLOGY HIGHLIGHT

The Discriminant

Quadratic equations play an important role in a study of College Algebra, forming a bridge between our previous and current studies, and the more advanced equations to come. As seen in this section, the discriminant of the quadratic formula ($b^2 - 4ac$) reveals the type and number of solutions, and whether the original equation can be solved by factoring (the discriminant is a perfect square). It will often be helpful to have this information in advance of trying to solve or graph the equation. Since this will be done for each new equation, the discriminant is a prime candidate for a short program. To begin a new program press [PRGM] [▶] [▶](NEW) [ENTER]. The calculator will prompt you to name the program using the green [ALPHA] letters (eight letters max), then allow you to start entering program lines. In [PRGM] mode, pressing [PRGM] once again will bring up menus that contain all needed commands. For very basic programs, these commands will be in the **I/O** (Input/Output) submenu, with the most common options being **2:Prompt, 3:Disp,** and **8:CLRHOME.** As you can see, we have named our program *DISCRMNT.*

PROGRAM:DISCRMNT

:CLRHOME	Clears the home screen, places cursor in upper left corner
:DISP "DISCRIMINANT"	Displays the word *DISCRIMINANT* as user information
:DISP "B²−4AC"	Displays $B^2 - 4AC$ as user information
:DISP ""	Displays a blank line (for formatting)
:Prompt A, B, C	Prompts the user to enter the values of A, B, and C
:B²−4AC → D	Computes $B^2 - 4AC$ using given values and stores result in memory location D

—continued

:CLRHOME	Clears the home screen, places cursor in upper left corner
:DISP "DISCRIMINANT IS:"	Displays the words *DISCRIMINANT IS* as user information
:DISP D	Displays the computed value of *D*

Exercise 1: Run the program for $x^2 - 3x - 10 = 0$ and $x^2 + 5x - 14 = 0$ to verify that both can be solved by factoring. What do you notice?

Exercise 2: Run the program for $25x^2 - 90x + 81 = 0$ and $4x^2 + 20x + 25 = 0$, then check to see if each is a perfect square trinomial. What do you notice?

Exercise 3: Run the program for $y = x^2 + 2x + 10$ and $y = x^2 - 2x + 5$. Do these equations have real number solutions? Why or why not?

Exercise 4: Once the discriminant D is known, the quadratic formula becomes $x = \dfrac{-b \pm \sqrt{D}}{2a}$ and

solutions can quickly be found. Solve the equations in Exercises 1–3 above.

1.5 EXERCISES

▶ CONCEPTS AND VOCABULARY

Fill in each blank with the appropriate word or phrase. Carefully reread the section, if necessary.

1. A polynomial equation is in standard form when written in _____ order of degree and set equal to _____.

2. The solution $x = 2 + \sqrt{3}$ is called an _____ form of the solution. Using a calculator, we find the _____ form is $x \approx 3.732$.

3. To solve a quadratic equation by completing the square, the coefficient of the _____ term must be a _____.

4. The quantity $b^2 - 4ac$ is called the _____ of the quadratic equation. If $b^2 - 4ac > 0$, there are _____ real roots.

5. According to the summary on page 122, what method should be used to solve $4x^2 - 5x = 0$? What are the solutions?

6. Discuss/Explain why this version of the quadratic formula is incorrect:
$$x = -b \pm \frac{\sqrt{b^2 - 4ac}}{2a}$$

▶ DEVELOPING YOUR SKILLS

Determine whether each equation is quadratic. If so, identify the coefficients a, b, and c. If not, discuss why.

7. $2x - 15 - x^2 = 0$ 8. $21 + x^2 - 4x = 0$

9. $\dfrac{2}{3}x - 7 = 0$ 10. $12 - 4x = 9$

11. $\dfrac{1}{4}x^2 = 6x$ 12. $0.5x = 0.25x^2$

13. $2x^2 + 7 = 0$ 14. $5 = -4x^2$

15. $-3x^2 + 9x - 5 + 2x^3 = 0$

16. $z^2 - 6z + 9 - z^3 = 0$

17. $(x - 1)^2 + (x - 1) + 4 = 9$

18. $(x + 5)^2 - (x + 5) + 4 = 17$

Solve using the zero factor property. Be sure each equation is in standard form and factor out any common factors before attempting to solve. Check all answers in the original equation.

19. $x^2 - 15 = 2x$ 20. $z^2 - 10z = -21$

21. $m^2 = 8m - 16$ 22. $-10n = n^2 + 25$

23. $5p^2 - 10p = 0$ **24.** $6q^2 - 18q = 0$

25. $-14h^2 = 7h$ **26.** $9w = -6w^2$

27. $a^2 - 17 = -8$ **28.** $b^2 + 8 = 12$

29. $g^2 + 18g + 70 = -11$

30. $h^2 + 14h - 2 = -51$

31. $m^3 + 5m^2 - 9m - 45 = 0$

32. $n^3 - 3n^2 - 4n + 12 = 0$

33. $(c - 12)c - 15 = 30$

34. $(d - 10)d + 10 = -6$

35. $9 + (r - 5)r = 33$

36. $7 + (s - 4)s = 28$

37. $(t + 4)(t + 7) = 54$

38. $(g + 17)(g - 2) = 20$

39. $2x^2 - 4x - 30 = 0$

40. $-3z^2 + 12z + 36 = 0$

41. $2w^2 - 5w = 3$

42. $-3v^2 = -v - 2$

Solve the following equations using the square root property of equality. Write answers in exact form and approximate form rounded to hundredths. If there are no real solutions, so state.

43. $m^2 = 16$ **44.** $p^2 = 49$

45. $y^2 - 28 = 0$ **46.** $m^2 - 20 = 0$

47. $p^2 + 36 = 0$ **48.** $n^2 + 5 = 0$

49. $x^2 = \frac{21}{16}$ **50.** $y^2 = \frac{13}{9}$

51. $(n - 3)^2 = 36$ **52.** $(p + 5)^2 = 49$

53. $(w + 5)^2 = 3$ **54.** $(m - 4)^2 = 5$

55. $(x - 3)^2 + 7 = 2$ **56.** $(m + 11)^2 + 5 = 3$

57. $(m - 2)^2 = \frac{18}{49}$ **58.** $(x - 5)^2 = \frac{12}{25}$

Fill in the blank so the result is a perfect square trinomial, then factor into a binomial square.

59. $x^2 + 6x +$ _____   **60.** $y^2 + 10y +$ _____

61. $n^2 + 3n +$ _____   **62.** $x^2 - 5x +$ _____

63. $p^2 + \frac{2}{3}p +$ _____   **64.** $x^2 - \frac{3}{2}x +$ _____

Solve by completing the square. Write your answers in both exact form and approximate form rounded to the hundredths place. If there are no real solutions, so state.

65. $x^2 + 6x = -5$ **66.** $m^2 + 8m = -12$

67. $p^2 - 6p + 3 = 0$ **68.** $n^2 = 4n + 10$

69. $p^2 + 6p = -4$ **70.** $x^2 - 8x - 1 = 0$

71. $m^2 + 3m = 1$ **72.** $n^2 + 5n - 2 = 0$

73. $n^2 = 5n + 5$ **74.** $w^2 - 7w + 3 = 0$

75. $2x^2 = -7x + 4$ **76.** $3w^2 - 8w + 4 = 0$

77. $2n^2 - 3n - 9 = 0$ **78.** $2p^2 - 5p = 1$

79. $4p^2 - 3p - 2 = 0$ **80.** $3x^2 + 5x - 6 = 0$

81. $m^2 = 7m - 4$ **82.** $a^2 - 15 = 4a$

Solve each equation using the most efficient method: factoring, square root property of equality, or the quadratic formula. Write your answer in both exact and approximate form (rounded to hundredths). Check one of the exact solutions in the original equation.

83. $x^2 - 3x = 18$ **84.** $w^2 + 6w - 1 = 0$

85. $4m^2 - 25 = 0$ **86.** $4a^2 - 4a = 1$

87. $4n^2 - 8n - 1 = 0$ **88.** $2x^2 - 4x + 5 = 0$

89. $6w^2 - w = 2$ **90.** $3a^2 - 5a + 6 = 0$

91. $4m^2 = 12m - 15$ **92.** $3p^2 + p = 0$

93. $4n^2 - 9 = 0$ **94.** $4x^2 - x = 3$

95. $5w^2 = 6w + 8$ **96.** $3m^2 - 7m - 6 = 0$

97. $3a^2 - a + 2 = 0$ **98.** $3n^2 - 2n - 3 = 0$

99. $5p^2 = 6p + 3$ **100.** $2x^2 + x + 3 = 0$

101. $5w^2 - w = 1$ **102.** $3m^2 - 2 = 5m$

103. $2a^2 + 5 = 3a$ **104.** $n^2 + 4n - 8 = 0$

105. $2p^2 - 4p + 11 = 0$ **106.** $8x^2 - 5x - 1 = 0$

107. $w^2 + \frac{2}{3}w = \frac{1}{9}$ **108.** $\frac{5}{4}m^2 - \frac{8}{3}m + \frac{1}{6} = 0$

109. $0.2a^2 + 1.2a + 0.9 = 0$

110. $-5.4n^2 + 8.1n + 9 = 0$

111. $\frac{2}{7}p^2 - 3 = \frac{8}{21}p$

112. $\frac{5}{9}x^2 - \frac{16}{15}x = \frac{3}{2}$

Use the discriminant to determine whether the given equation has irrational, rational, repeated, or complex roots. Also state whether the original equation is factorable using integers, but do not solve for x.

113. $-3x^2 + 2x + 1 = 0$ **114.** $2x^2 - 5x - 3 = 0$

115. $-4x + x^2 + 13 = 0$ **116.** $-10x + x^2 + 41 = 0$

117. $15x^2 - x - 6 = 0$ **118.** $10x^2 - 11x - 35 = 0$

119. $-4x^2 + 6x - 5 = 0$ **120.** $-5x^2 - 3 = 2x$

121. $2x^2 + 8 = -9x$ **122.** $x^2 + 4 = -7x$

123. $4x^2 + 12x = -9$ **124.** $9x^2 + 4 = 12x$

Solve the quadratic equations given. Simplify each result.

125. $-6x + 2x^2 + 5 = 0$ **126.** $17 + 2x^2 = 10x$

127. $5x^2 + 5 = -5x$ **128.** $x^2 = -2x - 19$

129. $-2x^2 = -5x + 11$ **130.** $4x - 3 = 5x^2$

▶ WORKING WITH FORMULAS

131. Height of a projectile: $h = -16t^2 + vt$

If an object is projected vertically upward from ground level with no continuing source of propulsion, the height of the object (in feet) is modeled by the equation shown, where v is the initial velocity, and t is the time in seconds. Use the quadratic formula to solve for t in terms of v and h. (*Hint:* Set the equation equal to zero and identify the coefficients as before.)

132. Surface area of a cylinder: $A = 2\pi r^2 + 2\pi rh$

The surface area of a cylinder is given by the formula shown, where h is the height and r is the radius of the base. The equation can be considered a quadratic in the variable r. Use the quadratic formula to solve for r in terms of h and A. (*Hint:* Rewrite the equation in standard form and identify the coefficients as before.)

▶ APPLICATIONS

133. Height of a projectile: The height of an object thrown upward from the roof of a building 408 ft tall, with an initial velocity of 96 ft/sec, is given by the equation $h = -16t^2 + 96t + 408$, where h represents the height of the object after t seconds. How long will it take the object to hit the ground? Answer in exact form and decimal form rounded to the nearest hundredth.

134. Height of a projectile: The height of an object thrown upward from the floor of a canyon 106 ft deep, with an initial velocity of 120 ft/sec, is given by the equation $h = -16t^2 + 120t - 106$, where h represents the height of the object after t seconds. How long will it take the object to rise to the height of the canyon wall? Answer in exact form and decimal form rounded to hundredths.

135. Cost, revenue, and profit: The revenue for a manufacturer of microwave ovens is given by the equation $R = x(40 - \frac{1}{3}x)$, where revenue is in thousands of dollars and x thousand ovens are manufactured and sold. What is the minimum number of microwave ovens that must be sold to bring in a revenue of $900,000?

136. Cost, revenue, and profit: The revenue for a manufacturer of computer printers is given by the equation $R = x(30 - 0.4x)$, where revenue is in thousands of dollars and x thousand printers are manufactured and sold. What is the minimum number of printers that must be sold to bring in a revenue of $440,000?

137. Cost, revenue, and profit: The cost of raw materials to produce plastic toys is given by the cost equation $C = 2x + 35$, where x is the number of toys in hundreds. The total income (revenue) from the sale of these toys is given by $R = -x^2 + 122x - 1965$. (a) Determine the profit equation (profit = revenue − cost). During the Christmas season, the owners of the company decide to manufacture and donate as many toys as they can, without taking a loss (i.e., they break even: profit or $P = 0$). (b) How many toys will they produce for charity?

138. Cost, revenue, and profit: The cost to produce bottled spring water is given by the cost equation $C = 16x + 63$, where x is the number of bottles in thousands. The total revenue from the sale of these bottles is given by the equation $R = -x^2 + 326x - 18,463$. (a) Determine the profit equation (profit = revenue − cost). (b) After a bad flood contaminates the drinking water of a nearby community, the owners decide to bottle and donate as many bottles of water as they can, without taking a loss (i.e., they break even: profit or $P = 0$). How many bottles will they produce for the flood victims?

139. Height of an arrow: If an object is projected vertically upward from ground level with no continuing source of propulsion, its height (in feet) is modeled by the equation $h = -16t^2 + vt$, where v is the initial velocity and t is the time in seconds. Use the quadratic formula to solve for t, given an arrow is shot into the air with $v = 144$ ft/sec and $h = 260$ ft. See Exercise 131.

140. Surface area of a cylinder: The surface area of a cylinder is given by $A = 2\pi r^2 + 2\pi rh$, where h is the height and r is the radius of the base. The equation can be considered a quadratic in the variable r. Use the quadratic formula to solve for r, given $A = 4710$ cm^2 and $h = 35$ cm. See Exercise 132.

141. Cell phone subscribers: For the years 1995 to 2002, the number N of cellular phone subscribers (in millions) can be modeled by the equation $N = 17.4x^2 + 36.1x + 83.3$, where $x = 0$ represents the year 1995 [*Source:* Data from the *2005 Statistical Abstract of the United States,* Table 1372, page 870]. If this trend continued, in what year did the number of subscribers reach or surpass 3750 million?

142. U.S. international trade balance: For the years 1995 to 2003, the international trade balance B (in millions of dollars) can be approximated by the equation $B = -3.1x^2 + 4.5x - 19.9$, where $x = 0$ represents the year 1995 [*Source:* Data from the *2005 Statistical Abstract of the United States*, Table 1278, page 799]. If this trend continues, in what year will the trade balance reach a deficit of $750 million dollars or more?

143. Tennis court dimensions: A regulation tennis court for a doubles match is laid out so that its length is 6 ft more than two times its width. The area of the doubles court is 2808 ft^2. What is the length and width of the doubles court?

Exercises 143 and 144

144. Tennis court dimensions: A regulation tennis court for a singles match is laid out so that its length is 3 ft less than three times its width. The area of the singles court is 2106 ft^2. What is the length and width of the singles court?

Singles

Doubles

▶ **EXTENDING THE CONCEPT**

145. Using the discriminant: Each of the following equations can easily be solved by factoring, since $a = 1$. Using the discriminant, we can create factorable equations with identical values for b and c, but where $a \ne 1$. For instance, $x^2 - 3x - 10 = 0$ and $4x^2 - 3x - 10 = 0$ can both be solved by factoring. Find similar equations ($a \ne 1$) for the quadratics given here. (*Hint:* The discriminant $b^2 - 4ac$ must be a perfect square.)

    a. $x^2 + 6x - 16 = 0$

    b. $x^2 + 5x - 14 = 0$

    c. $x^2 - x - 6 = 0$

146. Using the discriminant: For what values of c will the equation $9x^2 - 12x + c = 0$ have

    a. no real roots   **b.** one rational root

    c. two real roots   **d.** two integer roots

Complex polynomials: Many techniques applied to solve polynomial equations with real coefficients can be applied to solve polynomial equations with *complex coefficients*. Here we apply the idea to carefully chosen quadratic equations, as a more general application must wait until a future course, when the square root of a complex number is fully developed. Solve each equation using the quadratic formula, noting that $\dfrac{1}{i} = -i$.

147. $z^2 - 3iz = -10$

148. $z^2 - 9iz = -22$

149. $4iz^2 + 5z + 6i = 0$

150. $2iz^2 - 9z + 26i = 0$

151. $0.5z^2 + (7 + i)z + (6 + 7i) = 0$

152. $0.5z^2 + (4 - 3i)z + (-9 - 12i) = 0$

► **MAINTAINING YOUR SKILLS**

153. (R.7) State the formula for the perimeter and area of each figure illustrated.

a.

b.

c.

d.

154. (1.3) Factor and solve the following equations:
   a. $x^2 - 5x - 36 = 0$ **b.** $4x^2 - 25 = 0$
   c. $x^3 + 6x^2 - 4x - 24 = 0$

155. (1.1) A total of 900 tickets were sold for a recent concert and $25,000 was collected. If good seats were $30 and cheap seats were $20, how many of each type were sold?

156. (1.1) Solve for C: $P = C + Ct$.

1.6 Solving Other Types of Equations

Learning Objectives

In Section 1.6 you will learn how to:

☐ **A.** Solve polynomial equations of higher degree

☐ **B.** Solve rational equations

☐ **C.** Solve radical equations and equations with rational exponents

☐ **D.** Solve equations in quadratic form

☐ **E.** Solve applications of various equation types

The ability to solve linear and quadratic equations is the foundation on which a large percentage of our future studies are built. Both are closely linked to the solution of other equation types, as well as to the graphs of these equations. In this section, we get our first glimpse of these connections, as we learn to solve certain polynomial, rational, radical, and other equations.

A. Polynomial Equations of Higher Degree

In standard form, linear and quadratic equations have a known number of terms, so we commonly represent their coefficients using the early letters of the alphabet, as in $ax^2 + bx + c = 0$. However, these equations belong to the larger family of **polynomial equations.** To write a general polynomial, where the number of terms is unknown, we often represent the coefficients using subscripts on a single variable, such as a_1, a_2, a_3, and so on. A *polynomial equation of degree n* has the form

$$a_n x^n + a_{n-1} x^{n-1} + \cdots + a_1 x^1 + a_0 = 0$$

where $a_n, a_{n-1}, \ldots, a_1, a_0$ are real numbers and $a_n \neq 0$. Factorable polynomials of degree 3 and higher can also be solved using the zero product property and fundamental algebra skills. As with linear equations, values that make an equation true are called *solutions* or *roots* to the equation.

EXAMPLE 1 ► **Solving Polynomials by Factoring**

Solve by factoring: $2x^3 - 20x = 3x^2$.

Solution ►

$$2x^3 - 20x = 3x^2 \quad \text{given equation}$$
$$2x^3 - 3x^2 - 20x = 0 \quad \text{standard form}$$
$$x(2x^2 - 3x - 20) = 0 \quad \text{common factor is } x$$
$$x(2x + 5)(x - 4) = 0 \quad \text{factored form}$$
$$x = 0 \quad \text{or} \quad 2x + 5 = 0 \quad \text{or} \quad x - 4 = 0 \quad \text{zero product property}$$
$$x = 0 \quad \text{or} \quad x = \frac{-5}{2} \quad \text{or} \quad x = 4 \quad \text{result}$$

Substituting these values into the original equation verifies they are solutions.

Now try Exercises 7 through 14 ►

EXAMPLE 2 ▶ **Solving Higher Degree Equations**

Solve each equation by factoring:

    a. $x^3 - 7x + 21 = 3x^2$ **b.** $x^4 - 16 = 0$

Solution ▶   **a.**

$$x^3 - 7x + 21 = 3x^2 \qquad \text{given equation}$$

$$x^3 - 3x^2 - 7x + 21 = 0 \qquad \text{standard form; factor by grouping}$$

$$x^2(x - 3) - 7(x - 3) = 0 \qquad \text{remove common factors from each group}$$

$$(x - 3)(x^2 - 7) = 0 \qquad \text{factored form}$$

$$x - 3 = 0 \quad \text{or} \quad x^2 - 7 = 0 \qquad \text{zero product property}$$

$$x = 3 \quad \text{or} \quad x^2 = 7 \qquad \text{isolate variables}$$

$$x = \pm\sqrt{7} \qquad \text{square root property of equality}$$

The solutions are $x = 3$, $x = \sqrt{7}$, and $x = -\sqrt{7}$.

  b.

$$x^4 - 16 = 0 \qquad \text{given equation}$$

$$(x^2 + 4)(x^2 - 4) = 0 \qquad \text{factor as a difference of squares}$$

$$(x^2 + 4)(x + 2)(x - 2) = 0 \qquad \text{factor } x^2 - 4$$

$$x^2 + 4 = 0 \quad \text{or} \quad x + 2 = 0 \quad \text{or} \quad x - 2 = 0 \qquad \text{zero product property}$$

$$x^2 = -4 \quad \text{or} \quad x = -2 \quad \text{or} \quad x = 2 \qquad \text{isolate variables}$$

$$x = \pm\sqrt{-4} \qquad \text{square root property of equality}$$

Since $\pm\sqrt{-4} = \pm 2i$, the solutions are $x = 2i$, $x = -2i$, $x = 2$, and $x = -2$.

Now try Exercises 15 through 32 ▶

In Examples 1 and 2, we were able to solve higher degree polynomials by "breaking them down" into linear and quadratic forms. This basic idea can be applied to other kinds of equations as well, by rewriting them as equivalent linear and/or quadratic equations. For future use, it will be helpful to note that for a third-degree equation in the standard form $ax^3 + bx^2 + cx + d = 0$, a solution using factoring by grouping is always possible when $ad = bc$.

☑ **A.** You've just learned how to solve polynomial equations of higher degree

B. Rational Equations

In Section 1.1 we solved linear equations using basic properties of equality. If any equation contained fractional terms, we "cleared the fractions" using the least common denominator (LCD). We can also use this idea to solve **rational equations,** or equations that contain rational *expressions*.

Solving Rational Equations

1. Identify and exclude any values that cause a zero denominator.
2. Multiply both sides by the LCD and simplify (this will eliminate all denominators).
3. Solve the resulting equation.
4. Check all solutions in the original equation.

EXAMPLE 3 ▶ **Solving a Rational Equation**

Solve for m: $\dfrac{2}{m} - \dfrac{1}{m - 1} = \dfrac{4}{m^2 - m}$.

Solution ▶ Since $m^2 - m = m(m - 1)$, the LCD is $m(m - 1)$, where $m \neq 0$ and $m \neq 1$.

$$m(m - 1)\left(\frac{2}{m} - \frac{1}{m - 1}\right) = m(m - 1)\left[\frac{4}{m(m - 1)}\right] \quad \text{multiply by LCD}$$

$$2(m - 1) - m = 4 \quad \text{simplify—denominators are eliminated}$$

$$2m - 2 - m = 4 \quad \text{distribute}$$

$$m = 6 \quad \text{solve for } m$$

Checking by substitution we have:

$$\frac{2}{m} - \frac{1}{m - 1} = \frac{4}{m^2 - m} \quad \text{original equation}$$

$$\frac{2}{(6)} - \frac{1}{(6) - 1} = \frac{4}{(6)^2 - (6)} \quad \text{substitute 6 for } m$$

$$\frac{1}{3} - \frac{1}{5} = \frac{4}{30} \quad \text{simplify}$$

$$\frac{5}{15} - \frac{3}{15} = \frac{2}{15} \quad \text{common denominator}$$

$$\frac{2}{15} = \frac{2}{15} \checkmark \quad \text{result}$$

Now try Exercises 33 through 38 ▶

Multiplying both sides of an equation by a variable sometimes introduces a solution that satisfies the *resulting equation,* but not the original equation—the one we're trying to solve. Such "solutions" are called **extraneous roots** and illustrate the need to check all apparent solutions in the original equation. In the case of rational equations, we are particularly aware that any value that causes a zero denominator is outside the domain and cannot be a solution.

EXAMPLE 4 ▶ **Solving a Rational Equation**

Solve: $x + \dfrac{12}{x - 3} = 1 + \dfrac{4x}{x - 3}$.

Solution ▶ The LCD is $x - 3$, where $x \neq 3$.

$$(x - 3)\left(x + \frac{12}{x - 3}\right) = (x - 3)\left(1 + \frac{4x}{x - 3}\right) \quad \text{multiply both sides by LCD}$$

$$x^2 - 3x + 12 = x - 3 + 4x \quad \text{simplify—denominators are eliminated}$$

$$x^2 - 8x + 15 = 0 \quad \text{set equation equal to zero}$$

$$(x - 3)(x - 5) = 0 \quad \text{factor}$$

$$x = 3 \quad \text{or} \quad x = 5 \quad \text{zero factor property}$$

Checking shows $x = 3$ is an extraneous root, and $x = 5$ is the only valid solution.

Now try Exercises 39 through 44 ▶

In many fields of study, formulas involving rational expressions are used as equation models. Frequently, we need to solve these equations for one variable in terms of others, a skill closely related to our work in Section 1.1.

EXAMPLE 5 ▶ **Solving for a Specified Variable in a Formula**

Solve for the indicated variable: $S = \dfrac{a}{1 - r}$ for r.

Solution ▶

$$S = \frac{a}{1 - r} \qquad \text{LCD is } 1 - r$$

$$(1 - r)S = (1 - r)\left(\frac{a}{1 - r}\right) \qquad \text{multiply both sides by } (1 - r)$$

$$S - Sr = a \qquad \text{simplify—denominator is eliminated}$$

$$-Sr = a - S \qquad \text{isolate term with } r$$

$$r = \frac{a - S}{-S} \qquad \text{solve for } r \text{ (divide both sides by } -S)$$

$$r = \frac{S - a}{S}; S \neq 0 \qquad \text{multiply numerator/denominator by } -1$$

> **WORTHY OF NOTE**
>
> Generally, we should try to write rational answers with the fewest number of negative signs possible. Multiplying the numerator and denominator in Example 5 by -1 gave $r = \frac{S - a}{S}$, a more acceptable answer.

Now try Exercises 45 through 52 ▶

☑ **B.** You've just learned how to solve rational equations

C. Radical Equations and Equations with Rational Exponents

A **radical equation** is any equation that contains terms with a variable in the radicand. To solve a radical equation, we attempt to isolate a radical term on one side, then apply the appropriate nth power to free up the radicand and solve for the unknown. This is an application of the **power property of equality.**

> **The Power Property of Equality**
>
> If $\sqrt[n]{u}$ and v are real-valued expressions and $\sqrt[n]{u} = v$,
> $$\text{then } (\sqrt[n]{u})^n = v^n$$
> $$u = v^n$$
> for n an integer, $n \geq 2$.

Raising both sides of an equation to an *even* power can also introduce a false solution (extraneous root). Note that by inspection, the equation $x - 2 = \sqrt{x}$ has only the solution $x = 4$. But the equation $(x - 2)^2 = x$ (obtained by squaring both sides) has both $x = 4$ *and* $x = 1$ as solutions, yet $x = 1$ does not satisfy the original equation. This means we should *check all solutions of an equation where an even power is applied.*

EXAMPLE 6 ▶ **Solving Radical Equations**

Solve each radical equation:

    a. $\sqrt{3x - 2} + 12 = x + 10$ **b.** $2\sqrt[3]{x - 5} + 4 = 0$

Solution ▶   **a.** $\sqrt{3x - 2} + 12 = x + 10$ original equation

$$\sqrt{3x - 2} = x - 2 \qquad \text{isolate radical term (subtract 12)}$$

$$(\sqrt{3x - 2})^2 = (x - 2)^2 \qquad \text{apply power property, power is even}$$

$$3x - 2 = x^2 - 4x + 4 \qquad \text{simplify; square binomial}$$

$$0 = x^2 - 7x + 6 \qquad \text{set equal to zero}$$

$$0 = (x - 6)(x - 1) \qquad \text{factor}$$

$$x - 6 = 0 \quad \text{or} \quad x - 1 = 0 \qquad \text{apply zero product property}$$

$$x = 6 \quad \text{or} \quad x = 1 \qquad \text{result, check for extraneous roots}$$

Check ▶ $x = 6$: $\sqrt{3(6) - 2} + 12 = (6) + 10$
$$\sqrt{16} + 12 = 16$$
$$16 = 16 ✓$$

Check ▶ $x = 1$: $\sqrt{3(1) - 2} + 12 = (1) + 10$
$$\sqrt{1} + 12 = 11$$
$$13 = 11 ✗$$

The only solution is $x = 6$; $x = 1$ is extraneous.

b. $2\sqrt[3]{x - 5} + 4 = 0$ original equation
$$\sqrt[3]{x - 5} = -2$$ isolate radical term (subtract 4, divide by 2)
$$(\sqrt[3]{x - 5})^3 = (-2)^3$$ apply power property, power is odd
$$x - 5 = -8$$ simplify: $(\sqrt[3]{x - 5})^3 = x - 5$
$$x = -3$$ solve

Substituting -3 for x in the original equation verifies it is a solution.

Now try Exercises 53 through 56 ▶

Sometimes squaring both sides of an equation still results in an equation with a radical term, but often there is *one fewer* than before. In this case, we simply repeat the process, as indicated by the flowchart in Figure 1.14.

Figure 1.14

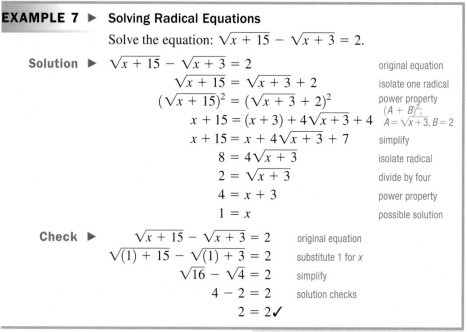

EXAMPLE 7 ▶ **Solving Radical Equations**
Solve the equation: $\sqrt{x + 15} - \sqrt{x + 3} = 2$.

Solution ▶ $\sqrt{x + 15} - \sqrt{x + 3} = 2$ original equation
$$\sqrt{x + 15} = \sqrt{x + 3} + 2$$ isolate one radical
$$(\sqrt{x + 15})^2 = (\sqrt{x + 3} + 2)^2$$ power property $(A + B)^2$; $A = \sqrt{x+3}, B = 2$
$$x + 15 = (x + 3) + 4\sqrt{x + 3} + 4$$
$$x + 15 = x + 4\sqrt{x + 3} + 7$$ simplify
$$8 = 4\sqrt{x + 3}$$ isolate radical
$$2 = \sqrt{x + 3}$$ divide by four
$$4 = x + 3$$ power property
$$1 = x$$ possible solution

Check ▶ $\sqrt{x + 15} - \sqrt{x + 3} = 2$ original equation
$$\sqrt{(1) + 15} - \sqrt{(1) + 3} = 2$$ substitute 1 for x
$$\sqrt{16} - \sqrt{4} = 2$$ simplify
$$4 - 2 = 2$$ solution checks
$$2 = 2 ✓$$

Now try Exercises 57 and 58 ▶

Since rational exponents are so closely related to radicals, the solution process for each is very similar. The goal is still to "undo" the radical (rational exponent) and solve for the unknown.

Power Property of Equality

For real-valued expression u and v, with positive integers m, n, and $\frac{m}{n}$ in lowest terms:

If m is odd	**If m is even**
and $u^{\frac{m}{n}} = v$,	and $u^{\frac{m}{n}} = v \, (v > 0)$,
then $\left(u^{\frac{m}{n}}\right)^{\frac{n}{m}} = v^{\frac{n}{m}}$	then $\left(u^{\frac{m}{n}}\right)^{\frac{n}{m}} = \pm v^{\frac{n}{m}}$
$u = v^{\frac{n}{m}}$	$u = \pm v^{\frac{n}{m}}$

EXAMPLE 8 ▶ **Solving Equations with Rational Exponents**

Solve each equation:

   a. $3(x + 1)^{\frac{3}{4}} - 9 = 15$ **b.** $(x - 3)^{\frac{2}{3}} = 4$

Solution ▶   **a.** $3(x + 1)^{\frac{3}{4}} - 9 = 15$ original equation; $\frac{m}{n} = \frac{3}{4}$

$(x + 1)^{\frac{3}{4}} = 8$ isolate variable term (add 9, divide by 3)

$\left[(x + 1)^{\frac{3}{4}}\right]^{\frac{4}{3}} = 8^{\frac{4}{3}}$ apply power property, note m is odd

$x + 1 = 16$ simplify $\left[8^{\frac{4}{3}} = \left(8^{\frac{1}{3}}\right)^4 = 16\right]$

$x = 15$ result

Check ▶ $3(15 + 1)^{\frac{3}{4}} - 9 = 15$ substitute 15 for x in the original equation

$3\left(16^{\frac{1}{4}}\right)^3 - 9 = 15$ simplify, rewrite exponent

$3(2)^3 - 9 = 15$ $\sqrt[4]{16} = 2$

$3(8) - 9 = 15$ $2^3 = 8$

$15 = 15 \checkmark$ solution checks

b. $(x - 3)^{\frac{2}{3}} = 4$ original equation; $\frac{m}{n} = \frac{2}{3}$

$\left[(x - 3)^{\frac{2}{3}}\right]^{\frac{3}{2}} = \pm 4^{\frac{3}{2}}$ apply power property, note m is even

$x - 3 = \pm 8$ simplify $\left[4^{\frac{3}{2}} = \left(4^{\frac{1}{2}}\right)^3 = \pm 8\right]$

$x = 3 \pm 8$ result

 C. You've just learned how to solve radical equations and equations with rational exponents

The solutions are $3 + 8 = 11$ and $3 - 8 = -5$.
Verify by checking both in the original equation.

Now try Exercises 59 through 64 ▶

⚠ **CAUTION** ▶ As you continue solving equations with radicals and rational exponents, be careful not to arbitrarily place the "$\pm$" sign in front of terms *given* in radical form. The expression $\sqrt{18}$ indicates the positive square root of 18, where $\sqrt{18} = 3\sqrt{2}$. The equation $x^2 = 18$ becomes $x = \pm\sqrt{18}$ after applying the power property, with solutions $x = \pm 3\sqrt{2}$ $(x = -3\sqrt{2}, x = 3\sqrt{2})$, since the square of either number produces 18.

D. Equations in Quadratic Form

In Section R.4 we used a technique called *u-substitution* to factor expressions in quadratic form. The following equations are in quadratic form since the degree of the leading term is twice the degree of the middle term: $x^{\frac{2}{3}} - 3x^{\frac{1}{3}} - 10 = 0$, $(x^2 + x)^2 - 8(x^2 + x) + 12 = 0$ and $x - 3\sqrt{x + 4} + 4 = 0$ [*Note:* The last equation can be rewritten as $(x + 4) - 3(x + 4)^{\frac{1}{2}} = 0$]. A *u*-substitution will help to solve these equations by factoring. The first equation appears in Example 9, the other two are in Exercises 70 and 74, respectively.

EXAMPLE 9 ▶ Solving Equations in Quadratic Form

Solve using a u-substitution:

a. $x^{\frac{2}{3}} - 3x^{\frac{1}{3}} - 10 = 0$ **b.** $x^4 - 36 = 5x^2$

Solution ▶ **a.** This equation is in quadratic form since it can be rewritten as:
$\left(x^{\frac{1}{3}}\right)^2 - 3\left(x^{\frac{1}{3}}\right)^1 - 10 = 0$, where the degree of leading term is twice that of second term. If we let $u = x^{\frac{1}{3}}$, then $u^2 = x^{\frac{2}{3}}$ and the equation becomes $u^2 - 3u^1 - 10 = 0$ which is factorable.

$\begin{array}{llll} (u - 5)(u + 2) = 0 & & & \text{factor} \\ u = 5 & \text{or} & u = -2 & \text{solution in terms of } u \\ x^{\frac{1}{3}} = 5 & \text{or} & x^{\frac{1}{3}} = -2 & \text{resubstitute } x^{\frac{1}{3}} \text{ for } u \\ \left(x^{\frac{1}{3}}\right)^3 = 5^3 & \text{or} & \left(x^{\frac{1}{3}}\right)^3 = (-2)^3 & \text{cube both sides: } \frac{1}{3}(3) = 1 \\ x = 125 & \text{or} & x = -8 & \text{solve for } x \end{array}$

Both solutions check.

b. In the standard form $x^4 - 5x^2 - 36 = 0$, we note the equation is also in quadratic form, since it can be written as $(x^2)^2 - 5(x^2)^1 - 36 = 0$. If we let $u = x^2$, then $u^2 = x^4$ and the equation becomes $u^2 - 5u - 36 = 0$, which is factorable.

$\begin{array}{llll} (u - 9)(u + 4) = 0 & & & \text{factor} \\ u = 9 & \text{or} & u = -4 & \text{solution in terms of } u \\ x^2 = 9 & \text{or} & x^2 = -4 & \text{resubstitute } x^2 \text{ for } u \\ x = \pm\sqrt{9} & \text{or} & x = \pm\sqrt{-4} & \text{square root property} \\ x = \pm 3 & \text{or} & x = \pm 2i & \text{simplify} \end{array}$

☑ **D.** You've just learned how to solve equations in quadratic form

The solutions are $x = -3$, $x = 3$, $x = -2i$, and $x = 2i$. Verify that all solutions check.

Now try Exercises 65 through 78 ▶

E. Applications

Applications of the skills from this section come in many forms. **Number puzzles** and **consecutive integer** exercises help develop the ability to translate written information into algebraic forms **(see Exercises 81 through 84).** Applications involving **geometry** or a stated relationship between two quantities often depend on these skills, and in many scientific fields, equation models involving radicals and rational exponents are commonplace **(see Exercises 99 and 100).**

EXAMPLE 10 ▶ Solving a Geometry Application

A legal size sheet of typing paper has a length equal to 3 in. less than twice its width. If the area of the paper is 119 in^2, find the length and width.

Solution ▶ Let W represent the width of the paper.
Then $2W$ represents twice the width, and $2W - 3$ represents three less than twice the width: $L = 2W - 3$:

$\begin{array}{ll} (\text{length})(\text{width}) = \text{area} & \text{verbal model} \\ (2W - 3)(W) = 119 & \text{substitute } 2W - 3 \text{ for length} \end{array}$

Since the equation is not set equal to zero, multiply and write the equation in standard form.

$$2W^2 - 3W = 119 \quad \text{distribute}$$
$$2W^2 - 3W - 119 = 0 \quad \text{subtract 119}$$
$$(2W - 17)(W + 7) = 0 \quad \text{factor}$$
$$W = \tfrac{17}{2} \text{ or } W = -7 \quad \text{solve}$$

We ignore $W = -7$, since the width cannot be negative. The width of the paper is $\frac{17}{2} = 8\frac{1}{2}$ in. and the length is $L = 2\left(\frac{17}{2}\right) - 3$ or 14 in.

Now try Exercises 85 and 86 ▶

EXAMPLE 11 ▶   Solving a Geometry Application

A hemispherical wash basin has a radius of 6 in. The volume of water in the basin can be modeled by $V = 6\pi h^2 - \frac{\pi}{3}h^3$, where h is the height of the water (see diagram). At what height h is the volume of water numerically equal to 15π times the height h?

Solution ▶   We are essentially asked to solve $V = 6\pi h^2 - \frac{\pi}{3}h^3$ when $V = 15\pi h$.

The equation becomes

$$15\pi h = 6\pi h^2 - \frac{\pi}{3}h^3 \quad \text{original equation, substitute } 15\pi h \text{ for } V$$

$$\frac{\pi}{3}h^3 - 6\pi h^2 + 15\pi h = 0 \quad \text{standard form}$$

$$h^3 - 18h^2 + 45h = 0 \quad \text{multiply by } \tfrac{3}{\pi}$$
$$h(h^2 - 18h + 45) = 0 \quad \text{factor out } h$$
$$h(h - 3)(h - 15) = 0 \quad \text{factored form}$$
$$h = 0 \quad \text{or} \quad h = 3 \quad \text{or} \quad h = 15 \quad \text{result}$$

The "solution" $h = 0$ can be discounted since there would be no water in the basin, and $h = 15$ is too large for this context (the radius is only 6 in.). The only solution that fits this context is $h = 3$.

Check ▶ $$15\pi h = 6\pi h^2 - \frac{\pi}{3}h^3 \quad \text{resulting equation}$$

$$15\pi(3) = 6\pi(3)^2 - \frac{\pi}{3}(3)^3 \quad \text{substitute 3 for } h$$

$$45\pi = 6\pi(9) - \frac{\pi}{3}(27) \quad \text{apply exponents}$$

$$45\pi = 54\pi - 9\pi \quad \text{simplify}$$
$$45\pi = 45\pi \checkmark \quad \text{result checks}$$

Now try Exercises 87 and 88 ▶

In this section, we noted that extraneous roots can occur when (1) both sides of an equation are multiplied by a variable term (as when solving rational equations) and (2) when both sides of an equation are raised to an even power (as when solving certain radical equations or equations with rational exponents). Example 11 illustrates a third way that extraneous roots can occur, as when a solution checks out fine algebraically, but does not fit the context or physical constraints of the situation.

Revenue Models

In a free-market economy, we know that if the price of an item is decreased, more people will buy it. This is why stores have sales and bargain days. But if the item is sold too cheaply, revenue starts to decline because less money is coming in—even though more sales are made. This phenomenon is analyzed in Example 12, where we use the revenue formula *revenue = price · number of sales* or $R = P \cdot S$.

EXAMPLE 12 ▶ **Solving a Revenue Application**

When a popular printer is priced at $300, Compu-Store will sell 15 printers per week. Using a survey, they find that for each decrease of $8, two additional sales will be made. What price will result in weekly revenue of $6500?

Solution ▶ Let x represent the number of times the price is decreased by $8. Then $300 - 8x$ represents the new price. Since sales increase by 2 each time the price is decreased, $15 + 2x$ represents the total sales.

$$R = P \cdot S \quad \text{revenue model}$$
$$6500 = (300 - 8x)(15 + 2x) \quad R = 6500, P = 300 - 8x, S = 15 + 2x$$
$$6500 = 4500 + 600x - 120x - 16x^2 \quad \text{multiply binomials}$$
$$0 = -16x^2 + 480x - 2000 \quad \text{simplify and write in standard form}$$
$$0 = x^2 - 30x + 125 \quad \text{divide by } -16$$
$$0 = (x - 5)(x - 25) \quad \text{factor}$$
$$x = 5 \quad \text{or} \quad x = 25 \quad \text{result}$$

Surprisingly, the store's weekly revenue will be $6500 after 5 decreases of $8 each ($40 total), or 25 price decreases of $8 each ($200 total). The related selling prices are $300 - 5(8) = \$260$ and $300 - 25(8) = \$100$. To maximize profit, the manager of Compu-Store decides to go with the $260 selling price.

Now try Exercises 89 and 90 ▶

Applications of rational equations can also take many forms. Work and uniform motion exercises help us develop important skills that can be used with more complex equation models. A work example follows here. For more on uniform motion, see **Exercises 95 and 96.**

EXAMPLE 13 ▶ **Solving a Work Application**

Lyf can clean a client's house in 5 hr, while it takes his partner Angie 4 hr to clean the same house. Both of them want to go to the Cubs' game today, which starts in $2\frac{1}{2}$ hr. If they work together, will they see the first pitch?

Solution ▶ After 1 hr, Lyf has cleaned $\frac{1}{5}$ and Angie has cleaned $\frac{1}{4}$ of the house, so together $\frac{1}{5} + \frac{1}{4} = \frac{9}{20}$ or 45% of the house has been cleaned. After 2 hr, $2\left(\frac{1}{5}\right) + 2\left(\frac{1}{4}\right)$ or $\frac{2}{5} + \frac{1}{2} = \frac{9}{10}$ or 90% of the house is clean. We can use these two illustrations to form an equation model where H represents hours worked:

$$H\left(\frac{1}{5}\right) + H\left(\frac{1}{4}\right) = 1 \text{ clean house } (1 = 100\%).$$

$$H\left(\frac{1}{5}\right) + H\left(\frac{1}{4}\right) = 1 \qquad \text{equation model}$$

$$20H\left(\frac{1}{5}\right) + 20H\left(\frac{1}{4}\right) = 1(20) \qquad \text{multiply by LCD of 20}$$

$$4H + 5H = 20 \qquad \text{simplify, denominators are eliminated}$$

$$9H = 20 \qquad \text{combine like terms}$$

$$H = \frac{20}{9} \qquad \text{solve for } H$$

It will take Lyf and Angie $2\frac{2}{9}$ hr (about 2 hr and 13 min) to clean the house. Yes! They will make the first pitch, since Wrigley Field is only 10 min away.

> **Now try Exercises 93 and 94 ▶**

EXAMPLE 14 ▶   **Solving an Application Involving a Rational Equation**

In Verano City, the cost C to remove industrial waste from drinking water is given by the equation $C = \dfrac{80P}{100 - P}$, where P is the percent of total pollutants removed and C is the cost in thousands of dollars. If the City Council budgets $1,520,000 for the removal of these pollutants, what percentage of the waste will be removed?

Solution ▶

$$C = \frac{80P}{100 - P} \qquad \text{equation model}$$

$$1520 = \frac{80P}{100 - P} \qquad \text{substitute 1520 for } C$$

$$1520(100 - P) = 80P \qquad \text{multiply by LCD of } (100 - P)$$

$$152{,}000 = 1600P \qquad \text{distribute and simplify}$$

$$95 = P \qquad \text{result}$$

☑ E. You've just learned how to solve applications of various equation types

On a budget of $1,520,000, 95% of the pollutants will be removed.

> **Now try Exercises 97 and 98 ▶**

1.6 EXERCISES

▶ CONCEPTS AND VOCABULARY

Fill in each blank with the appropriate word or phrase. Carefully reread the section, if necessary.

1. For rational equations, values that cause a zero denominator must be _____.

2. The equation or formula for revenue models is revenue = _____.

3. "False solutions" to a rational or radical equation are also called _____ roots.

4. Factorable polynomial equations can be solved using the _____ _____ property.

5. Discuss/Explain the power property of equality as it relates to rational exponents and properties of reciprocals. Use the equation $(x - 2)^{\frac{2}{3}} = 9$ for your discussion.

6. One factored form of an equation is shown. Discuss/Explain why $x = -8$ and $x = 1$ are not solutions to the equation, and what must be done to find the actual solutions: $2(x + 8)(x - 1) = -16$.

▶ DEVELOPING YOUR SKILLS

Solve using the zero product property. Be sure each equation is in standard form and factor out any common factors before attempting to solve. Check all answers in the original equation.

7. $22x = x^3 - 9x^2$ **8.** $x^3 = 13x^2 - 42x$

9. $3x^3 = -7x^2 + 6x$ **10.** $7x^2 + 15x = 2x^3$

11. $2x^4 - 3x^3 = 9x^2$ **12.** $-7x^2 = 2x^4 - 9x^3$

13. $2x^4 - 16x = 0$ **14.** $x^4 + 64x = 0$

15. $x^3 - 4x = 5x^2 - 20$ **16.** $x^3 - 18 = 9x - 2x^2$

17. $4x - 12 = 3x^2 - x^3$ **18.** $x - 7 = 7x^2 - x^3$

19. $2x^3 - 12x^2 = 10x - 60$

20. $9x + 81 = 27x^2 + 3x^3$

21. $x^4 - 7x^3 + 4x^2 = 28x$

22. $x^4 + 3x^3 + 9x^2 = -27x$

23. $x^4 - 81 = 0$

24. $x^4 - 1 = 0$

25. $x^4 - 256 = 0$

26. $x^4 - 625 = 0$

27. $x^6 - 2x^4 - x^2 + 2 = 0$

28. $x^6 - 3x^4 - 16x^2 + 48 = 0$

29. $x^5 - x^3 - 8x^2 + 8 = 0$

30. $x^5 - 9x^3 - x^2 + 9 = 0$

31. $x^6 - 1 = 0$

32. $x^6 - 64 = 0$

Solve each equation. Identify any extraneous roots.

33. $\dfrac{2}{x} + \dfrac{1}{x + 1} = \dfrac{5}{x^2 + x}$

34. $\dfrac{3}{m + 3} - \dfrac{5}{m^2 + 3m} = \dfrac{1}{m}$

35. $\dfrac{21}{a + 2} = \dfrac{3}{a - 1}$

36. $\dfrac{4}{2y - 3} = \dfrac{7}{3y - 5}$

37. $\dfrac{1}{3y} - \dfrac{1}{4y} = \dfrac{1}{y^2}$

38. $\dfrac{3}{5x} - \dfrac{1}{2x} = \dfrac{1}{x^2}$

39. $x + \dfrac{14}{x - 7} = 1 + \dfrac{2x}{x - 7}$

40. $\dfrac{10}{x - 5} + x = 1 + \dfrac{2x}{x - 5}$

41. $\dfrac{6}{n + 3} + \dfrac{20}{n^2 + n - 6} = \dfrac{5}{n - 2}$

42. $\dfrac{7}{p + 2} - \dfrac{1}{p^2 + 5p + 6} = -\dfrac{2}{p + 3}$

43. $\dfrac{a}{2a + 1} - \dfrac{2a^2 + 5}{2a^2 - 5a - 3} = \dfrac{3}{a - 3}$

44. $\dfrac{-18}{6n^2 - n - 1} + \dfrac{3n}{2n - 1} = \dfrac{4n}{3n + 1}$

Solve for the variable indicated.

45. $\dfrac{1}{f} = \dfrac{1}{f_1} + \dfrac{1}{f_2}$; for f **46.** $\dfrac{1}{x} - \dfrac{1}{y} = \dfrac{1}{z}$; for z

47. $I = \dfrac{E}{R + r}$; for r **48.** $q = \dfrac{pf}{p - f}$; for p

49. $V = \dfrac{1}{3}\pi r^2 h$; for h **50.** $s = \dfrac{1}{2}gt^2$; for g

51. $V = \dfrac{4}{3}\pi r^3$; for r^3 **52.** $V = \dfrac{1}{3}\pi r^2 h$; for r^2

Solve each equation and check your solutions by *substitution*. Identify any extraneous roots.

53. a. $-3\sqrt{3x - 5} = -9$ **b.** $x = \sqrt{3x + 1} + 3$

54. a. $-2\sqrt{4x - 1} = -10$ **b.** $-5 = \sqrt{5x - 1} - x$

55. a. $2 = \sqrt[3]{3m - 1}$ **b.** $2\sqrt[3]{7 - 3x} - 3 = -7$

   c. $\dfrac{\sqrt[3]{2m + 3}}{-5} + 2 = 3$ **d.** $\sqrt[3]{2x - 9} = \sqrt[3]{3x + 7}$

56. a. $-3 = \sqrt[3]{5p + 2}$ **b.** $3\sqrt[3]{3 - 4x} - 7 = -4$

   c. $\dfrac{\sqrt[3]{6x - 7}}{4} - 5 = -6$

   d. $3\sqrt[3]{x + 3} = 2\sqrt[3]{2x + 17}$

57. a. $\sqrt{x - 9} + \sqrt{x} = 9$

   b. $x = 3 + \sqrt{23 - x}$

   c. $\sqrt{x - 2} - \sqrt{2x} = -2$

   d. $\sqrt{12x + 9} - \sqrt{24x} = -3$

58. a. $\sqrt{x + 7} - \sqrt{x} = 1$

   b. $\sqrt{2x + 31} + x = 2$

   c. $\sqrt{3x} = \sqrt{x - 3} + 3$

   d. $\sqrt{3x + 4} - \sqrt{7x} = -2$

Write the equation in simplified form, then solve. Check all answers by substitution.

59. $x^{\frac{3}{5}} + 17 = 9$ **60.** $-2x^{\frac{3}{4}} + 47 = -7$

61. $0.\overline{3}x^{\frac{5}{2}} - 39 = 42$ **62.** $0.\overline{5}x^{\frac{5}{3}} + 92 = -43$

63. $2(x + 5)^{\frac{2}{3}} - 11 = 7$

64. $-3(x - 2)^{\frac{4}{5}} + 29 = -19$

Solve each equation using a u-substitution. Check all answers.

65. $x^{\frac{2}{3}} - 2x^{\frac{1}{3}} - 15 = 0$ **66.** $x^3 - 9x^{\frac{3}{2}} + 8 = 0$

67. $x^4 - 24x^2 - 25 = 0$ **68.** $x^4 - 37x^2 + 36 = 0$

69. $(x^2 - 3)^2 + (x^2 - 3) - 2 = 0$

70. $(x^2 + x)^2 - 8(x^2 + x) + 12 = 0$

71. $x^{-2} - 3x^{-1} - 4 = 0$

72. $x^{-2} - 2x^{-1} - 35 = 0$

73. $x^{-4} - 13x^{-2} + 36 = 0$

Use a u-substitution to solve each radical equation.

74. $x - 3\sqrt{x + 4} + 4 = 0$

75. $x + 4 = 7\sqrt{x + 4}$

76. $2(x + 1) = 5\sqrt{x + 1} - 2$

77. $2\sqrt{x + 10} + 8 = 3(x + 10)$

78. $4\sqrt{x - 3} = 3(x - 3) - 4$

► WORKING WITH FORMULAS

 79. Lateral surface area of a cone: $S = \pi r \sqrt{r^2 + h^2}$

The lateral surface area (surface area excluding the base) S of a cone is given by the formula shown, where r is the radius of the base and h is the height of the cone. (a) Solve the equation for h. (b) Find the surface area of a cone that has a radius of 6 m and a height of 10 m. Answer in simplest form.

80. Painted area on a canvas: $A = \dfrac{4x^2 + 60x + 104}{x}$

A rectangular canvas is to contain a small painting with an area of 52 in², and requires 2-in. margins on the left and right, with 1-in. margins on the top and bottom for framing. The total area of such a canvas is given by the formula shown, where x is the height of the *painted* area.

a. What is the area A of the canvas if the height of the painting is $x = 10$ in.?

b. If the area of the canvas is $A = 120$ in², what are the dimensions of the painted area?

► APPLICATIONS

Find all real numbers that satisfy the following descriptions.

81. When the cube of a number is added to twice its square, the result is equal to 18 more than 9 times the number.

82. Four times a number decreased by 20 is equal to the cube of the number decreased by 5 times its square.

83. Find three consecutive even integers such that 4 times the largest plus the fourth power of the smallest is equal to the square of the remaining even integer increased by 24.

84. Find three consecutive integers such that the sum of twice the largest and the fourth power of the smallest is equal to the square of the remaining integer increased by 75.

85. Envelope sizes: Large mailing envelopes often come in standard sizes, with 5- by 7-in. and 9- by

12-in. envelopes being the most common. The next larger size envelope has an area of 143 in², with a length that is 2 in. longer than the width. What are the dimensions of the larger envelope?

86. Paper sizes: Letter size paper is 8.5 in. by 11 in. Legal size paper is $8\frac{1}{2}$ in. by 14 in. The next larger (common) size of paper has an area of 187 in², with a length that is 6 in. longer than the width. What are the dimensions of the Ledger size paper?

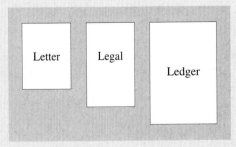

87. **Composite figures—grain silos:** Grain silos can be described as a hemisphere sitting atop a cylinder. The interior volume V of the silo can be modeled by $V = \frac{2}{3}\pi r^3 + \pi r^2 h$, where h is the height of a cylinder with radius r. For a cylinder 6 m tall, what radius would give the silo a volume that is numerically equal to 24π times this radius?

88. **Composite figures—gelatin capsules:** The gelatin capsules manufactured for cold and flu medications are shaped like a cylinder with a hemisphere on each end. The interior volume V of each capsule can be modeled by $V = \frac{4}{3}\pi r^3 + \pi r^2 h$, where h is the height of the cylindrical portion and r is its radius. If the cylindrical portion of the capsule is 8 mm long ($h = 8$ mm), what radius would give the capsule a volume that is numerically equal to 15π times this radius?

89. **Running shoes:** When a popular running shoe is priced at $70, The Shoe House will sell 15 pairs each week. Using a survey, they have determined that for each decrease of $2 in price, 3 additional pairs will be sold each week. What selling price will give a weekly revenue of $2250?

90. **Cell phone charges:** A cell phone service sells 48 subscriptions each month if their monthly fee is $30. Using a survey, they find that for each decrease of $1, 6 additional subscribers will join. What charge(s) will result in a monthly revenue of $2160?

Projectile height: In the absence of resistance, the height of an object that is projected upward can be modeled by the equation $h = -16t^2 + vt + k$, where h represents the height of the object (in feet) t sec after it has been thrown, v represents the initial velocity (in feet per second), and k represents the height of the object when $t = 0$ (before it has

been thrown). Use this information to complete the following problems.

91. From the base of a canyon that is 480 feet deep (*below* ground level → −480), a slingshot is used to shoot a pebble upward toward the canyon's rim. If the initial velocity is 176 ft per second:
    a. How far is the pebble below the rim after 4 sec?
    b. How long until the pebble returns to the bottom of the canyon?
    c. What happens at $t = 5$ and $t = 6$ sec? Discuss and explain.

92. A model rocket blasts off. A short time later, at a velocity of 160 ft/sec and a height of 240 ft, it runs out of fuel and becomes a projectile.
    a. How high is the rocket three seconds later? Four seconds later?
    b. How long will it take the rocket to attain a height of 640 ft?
    c. How many times is a height of 384 ft attained? When do these occur?
    d. How many seconds until the rocket returns to the ground?

93. **Printing newspapers:** The editor of the school newspaper notes the college's new copier can complete the required print run in 20 min, while the back-up copier took 30 min to do the same amount of work. How long would it take if both copiers are used?

94. **Filling a sink:** The cold water faucet can fill a sink in 2 min. The drain can empty a full sink in 3 min. If the faucet were left on and the drain was left open, how long would it take to fill the sink?

95. **Triathalon competition:** As one part of a Mountain-Man triathalon, participants must row a canoe 5 mi down river (with the current), circle a buoy and row 5 mi back up river (against the current) to the starting point. If the current is flowing at a steady rate of 4 mph and Tom Chaney made the round-trip in 3 hr, how fast can he row in still water? (*Hint:* The time rowing down river and the time rowing up river must add up to 3 hr.)

96. **Flight time:** The flight distance from Cincinnati, Ohio, to Chicago, Illinois, is approximately 300 mi. On a recent round-trip between these cities in my private plane, I encountered a steady 25 mph headwind on the way to Chicago, with a 25 mph tailwind on the return trip. If my total flying time

came to exactly 5 hr, what was my flying time to Chicago? What was my flying time back to Cincinnati? (*Hint:* The flight time between the two cities must add up to 5 hr.)

97. Pollution removal: For a steel mill, the cost C (in millions of dollars) to remove toxins from the resulting sludge is given by $C = \dfrac{92P}{100 - P}$, where P is the percent of the toxins removed. What percent can be removed if the mill spends $100,000,000 on the cleanup? Round to tenths of a percent.

98. Wildlife populations: The Department of Wildlife introduces 60 elk into a new game reserve. It is projected that the size of the herd will grow according to the equation $N = \dfrac{10(6 + 3t)}{1 + 0.05t}$, where N is the number of elk and t is the time in years. If recent counts find 225 elk, approximately how many years have passed? (See Section R.5, Exercise 82.)

99. Planetary motion: The time T (in days) for a planet to make one revolution around the sun is modeled by $T = 0.407R^{\frac{3}{2}}$, where R is the maximum radius of the planet's orbit in millions of miles (*Kepler's third law of planetary motion*). Use the equation to approximate the maximum radius of each orbit, given the number of days it takes for one revolution. (See Section R.6, Exercises 53 and 54.)

  a. Mercury: 88 days

  b. Venus: 225 days

  c. Earth: 365 days

  d. Mars: 687 days

  e. Jupiter: 4,333 days

  f. Saturn: 10,759 days

100. Wind-powered energy: If a wind-powered generator is delivering P units of power, the velocity V of the wind (in miles per hour) can be determined using $V = \sqrt[3]{\dfrac{P}{k}}$, where k is a constant that depends on the size and efficiency of the generator. Given $k = 0.004$, approximately how many units of power are being delivered if the wind is blowing at 27 miles per hour? (See Section R.6, Exercise 56.)

▶ **EXTENDING THE CONCEPT**

101. To solve the equation $3 - \dfrac{8}{x + 3} = \dfrac{1}{x}$, a student multiplied by the LCD $x(x + 3)$, simplified, and got this result: $3 - 8x = (x + 3)$. Identify and fix the mistake, then find the correct solution(s).

102. The expression $x^2 - 7$ is not factorable using *integer values*. But the expression *can be written* in the form $x^2 - (\sqrt{7})^2$, enabling us to factor it as a binomial and its conjugate: $(x + \sqrt{7})(x - \sqrt{7})$. Use this idea to solve the following equations:

  a. $x^2 - 5 = 0$ **b.** $n^2 - 19 = 0$

  c. $4v^2 - 11 = 0$ **d.** $9w^2 - 11 = 0$

Determine the values of x for which each expression represents a real number.

103. $\dfrac{\sqrt{x - 1}}{x^2 - 4}$ **104.** $\dfrac{x^2 - 4}{\sqrt{x - 1}}$

105. As an extension of working with absolute values, try the following exercises.

  Recall that for $|X| = k$, $X = -k$ or $X = k$.

  a. $|x^2 - 2x - 25| = 10$

  b. $|x^2 - 5x - 10| = 4$

  c. $|x^2 - 4| = x + 2$

  d. $|x^2 - 9| = -x + 3$

  e. $|x^2 - 7x| = -x + 7$

  f. $|x^2 - 5x - 2| = x + 5$

▶ **MAINTAINING YOUR SKILLS**

106. (1.1) Two jets take off on parallel runways going in opposite directions. The first travels at a rate of 250 mph and the second at 325 mph. How long until they are 980 miles apart?

107. (R.6) Find the missing side.

12 cm

10 cm

108. (R.3) Simplify using properties of exponents:

$$2^{-1} + (2x)^0 + 2x^0$$

109. (1.2) Graph the relation given:

$$2x - 3 < 7 \text{ and } x + 2 > 1$$

SUMMARY AND CONCEPT REVIEW

SECTION 1.1   Linear Equations, Formulas, and Problem Solving

KEY CONCEPTS

- An equation is a statement that two expressions are equal.
- Replacement values that make an equation true are called solutions or roots.
- Equivalent equations are those that have the same solution set.
- To solve an equation we use the distributive property and the properties of equality to write a sequence of simpler, equivalent equations until the solution is obvious. A guide for solving linear equations appears on page 75.
- If an equation contains fractions, multiply both sides by the LCD of all denominators, then solve.
- Solutions to an equation can be checked using back-substitution, by replacing the variable with the proposed solution and verifying the left-hand expression is equal to the right.
- An equation can be:
  1. an identity, one that is always true, with a solution set of all real numbers.
  2. a contradiction, one that is never true, with the empty set as the solution set.
  3. conditional, or one that is true/false depending on the value(s) input.
- To solve formulas for a specified variable, focus on the object variable and apply properties of equality to write this variable in terms of all others.
- The basic elements of good problem solving include:
  1. Gathering and organizing information
  2. Making the problem visual
  3. Developing an equation model
  4. Using the model to solve the application

For a complete review, see the problem-solving guide on page 78.

EXERCISES

1. Use substitution to determine if the indicated value is a solution to the equation given.

  a. $6x - (2 - x) = 4(x - 5), x = -6$ **b.** $\frac{3}{4}b + 2 = \frac{5}{2}b + 16, b = -8$ **c.** $4d - 2 = -\frac{1}{2} + 3d, d = \frac{3}{2}$

Solve each equation.

2. $-2b + 7 = -5$ **3.** $3(2n - 6) + 1 = 7$ **4.** $4m - 5 = 11m + 2$

5. $\frac{1}{2}x + \frac{2}{3} = \frac{3}{4}$ **6.** $6p - (3p + 5) - 9 = 3(p - 3)$ **7.** $-\frac{g}{6} = 3 - \frac{1}{2} - \frac{5g}{12}$

Solve for the specified variable in each formula or literal equation.

8. $V = \pi r^2 h$ for h

9. $P = 2L + 2W$ for L

10. $ax + b = c$ for x

11. $2x - 3y = 6$ for y

Use the problem-solving guidelines (page 78) to solve the following applications.

12. At a large family reunion, two kegs of lemonade are available. One is 2% sugar (too sour) and the second is 7% sugar (too sweet). How many gallons of the 2% keg, must be mixed with 12 gallons of the 7% keg to get a 5% mix?

13. A rectangular window with a width of 3 ft and a height of 4 ft is topped by a semi-circular window. Find the total area of the window.

14. Two cyclists start from the same location and ride in opposite directions, one riding at 15 mph and the other at 18 mph. If their radio phones have a range of 22 mi, how many minutes will they be able to communicate?

SECTION 1.2   Linear Inequalities in One Variable

KEY CONCEPTS

- Inequalities are solved using properties similar to those for solving equalities (see page 87). The one exception is the multiplicative property of inequality, since the truth of the resulting statement depends on whether a positive or negative quantity is used.
- Solutions to an inequality can be graphed on a number line, stated using a simple inequality, or expressed using set or interval notation.
- For two sets A and B: A intersect B ($A \cap B$) is the set of elements in both A **and** B (i.e., *elements common to both sets*). A union B ($A \cup B$) is the set of elements in either A **or** B (i.e., *all elements from either set*).
- Compound inequalities are formed using the conjunctions "and"/"or." These can be either a joint inequality as in $-3 < x \leq 5$, or a disjoint inequality, as in $x < -2$ or $x > 7$.

EXERCISES

Use inequality symbols to write a mathematical model for each statement.

15. You must be 35 yr old or older to run for president of the United States.

16. A child must be under 2 yr of age to be admitted free.

17. The speed limit on many interstate highways is 65 mph.

18. Our caloric intake should not be less than 1200 calories per day.

Solve the inequality and write the solution using interval notation.

19. $7x > 35$

20. $-\dfrac{3}{5}m < 6$

21. $2(3m - 2) \leq 8$

22. $-1 < \dfrac{1}{3}x + 2 \leq 5$

23. $-4 < 2b + 8$ and $3b - 5 > -32$

24. $-5(x + 3) > -7$ or $x - 5.2 > -2.9$

25. Find the allowable values for each of the following. Write your answer in interval notation.

   a. $\dfrac{7}{n - 3}$

   b. $\dfrac{5}{2x - 3}$

   c. $\sqrt{x + 5}$

   d. $\sqrt{-3n + 18}$

26. Latoya has earned grades of 72%, 95%, 83%, and 79% on her first four exams. What grade must she make on her fifth and last exam so that her average is 85% or more?

SECTION 1.3   Absolute Value Equations and Inequalities

KEY CONCEPTS

- To solve absolute value equations and inequalities, begin by writing the equation in simplified form, with the absolute value isolated on one side.
- If X represents an algebraic expression and k is a nonnegative constant:
  - Absolute value equations: $|X| = k$ is equivalent to $X = -k$ or $X = k$
  - "Less than" inequalities: $|X| < k$ is equivalent to $-k < X < k$
  - "Greater than" inequalities: $|X| > k$ is equivalent to $X < -k$ or $X > k$
- These properties also apply when the symbols "≤" or "≥"are used.
- If the absolute value quantity has been isolated on the left, the solution to a less-than inequality will be a single interval, while the solution to a greater-than inequality will consist of two disjoint intervals.
- The multiplicative property states that for algebraic expressions A and B, $|AB| = |A||B|$.

EXERCISES

Solve each equation or inequality. Write solutions to inequalities in interval notation.

27. $7 = |x - 3|$

28. $-2|x + 2| = -10$

29. $|-2x + 3| = 13$

30. $\dfrac{|2x + 5|}{3} + 8 = 9$

31. $-3|x + 2| - 2 < -14$

32. $\left|\dfrac{x}{2} - 9\right| \leq 7$

33. $|3x + 5| = -4$

34. $3|x + 1| < -9$

35. $2|x + 1| > -4$

36. $5|m - 2| - 12 \leq 8$

37. $\dfrac{|3x - 2|}{2} + 6 \geq 10$

38. Monthly rainfall received in Omaha, Nebraska, rarely varies by more than 1.7 in. from an average of 2.5 in. per month. (a) Use this information to write an absolute value inequality model, then (b) solve the inequality to find the highest and lowest amounts of monthly rainfall for this city.

SECTION 1.4   Complex Numbers

KEY CONCEPTS

- The italicized i represents the number whose square is -1. This means $i^2 = -1$ and $i = \sqrt{-1}$.
- Larger powers of i can be simplified using $i^4 = 1$.
- For $k > 0$, $\sqrt{-k} = i\sqrt{k}$ and we say the expression has been *written in terms of i*.
- The standard form of a *complex number* is $a + bi$, where a is the *real number part* and bi is the *imaginary number part*.
- To add or subtract complex numbers, combine the like terms.
- For any complex number $a + bi$, its *complex conjugate* is $a - bi$.
- The *product* of a complex number and its conjugate is a real number.

- The commutative, associative, and distributive properties also apply to complex numbers and are used to perform basic operations.
- To multiply complex numbers, use the F-O-I-L method and simplify.
- To find a *quotient* of complex numbers, multiply the numerator and denominator by the conjugate of the denominator.

EXERCISES

Simplify each expression and write the result in standard form.

39. $\sqrt{-72}$

40. $6\sqrt{-48}$

41. $\dfrac{-10 + \sqrt{-50}}{5}$

42. $\sqrt{3}\sqrt{-6}$

43. i^{57}

Perform the operation indicated and write the result in standard form.

44. $(5 + 2i)^2$

45. $\dfrac{5i}{1 - 2i}$

46. $(-3 + 5i) - (2 - 2i)$

47. $(2 + 3i)(2 - 3i)$

48. $4i(-3 + 5i)$

Use substitution to show the given complex number and its conjugate are solutions to the equation shown.

49. $x^2 - 9 = -34; x = 5i$

50. $x^2 - 4x + 9 = 0; x = 2 + i\sqrt{5}$

SECTION 1.5   Solving Quadratic Equations

KEY CONCEPTS

- The standard form of a quadratic equation is $ax^2 + bx + c = 0$, where a, b, and c are real numbers and $a \neq 0$. In words, we say the equation is written in decreasing order of degree and set equal to zero.
- The coefficient of the squared term a is called the *leading coefficient*, b is called the *linear coefficient*, and c is called the *constant term*.
- The square root property of equality states that if $X^2 = k$, where $k \geq 0$, then $X = \sqrt{k}$ or $X = -\sqrt{k}$.
- Factorable quadratics can be solved using the zero product property, which states that if the product of two factors is zero, then one, the other, or both must be equal to zero. Symbolically, if $A \cdot B = 0$, then $A = 0$ or $B = 0$.
- Quadratic equations can also be solved by *completing the square,* or using the *quadratic formula.*
- If the discriminant $b^2 - 4ac = 0$, the equation has one real (repeated) root. If $b^2 - 4ac > 0$, the equation has two real roots; and if $b^2 - 4ac < 0$, the equation has two complex roots.

EXERCISES

51. Determine whether the given equation is quadratic. If so, write the equation in standard form and identify the values of a, b, and c.

   a. $-3 = 2x^2$ **b.** $7 = -2x + 11$ **c.** $99 = x^2 - 8x$ **d.** $20 = 4 - x^2$

52. Solve by factoring.

   a. $x^2 - 3x - 10 = 0$ **b.** $2x^2 - 50 = 0$ **c.** $3x^2 - 15 = 4x$ **d.** $x^3 - 3x^2 = 4x - 12$

53. Solve using the square root property of equality.

   a. $x^2 - 9 = 0$ **b.** $2(x - 2)^2 + 1 = 11$ **c.** $3x^2 + 15 = 0$ **d.** $-2x^2 + 4 = -46$

54. Solve by completing the square. Give real number solutions in exact and approximate form.

   a. $x^2 + 2x = 15$ **b.** $x^2 + 6x = 16$ **c.** $-4x + 2x^2 = 3$ **d.** $3x^2 - 7x = -2$

55. Solve using the quadratic formula. Give solutions in both exact and approximate form.

   a. $x^2 - 4x = -9$ **b.** $4x^2 + 7 = 12x$ **c.** $2x^2 - 6x + 5 = 0$

Solve the following quadratic applications. For 56 and 57, recall the height of a projectile is modeled by $h = -16t^2 + v_0t + k$.

56. A projectile is fired upward from ground level with an initial velocity of 96 ft/sec. (a) To the nearest tenth of a second, how long until the object first reaches a height of 100 ft? (b) How long until the object is again at 100 ft? (c) How many seconds until it returns to the ground?

57. A person throws a rock upward from the top of an 80-ft cliff with an initial velocity of 64 ft/sec. (a) To the nearest tenth of a second, how long until the object is 120 ft high? (b) How long until the object is again at 120 ft? (c) How many seconds until the object hits the ground at the base of the cliff?

58. The manager of a large, 14-screen movie theater finds that if he charges $2.50 per person for the matinee, the average daily attendance is 4000 people. With every increase of 25 cents the attendance drops an average of 200 people. (a) What admission price will bring in a revenue of $11,250? (b) How many people will purchase tickets at this price?

59. After a storm, the Johnson's basement flooded and the water needed to be pumped out. A cleanup crew is sent out with two powerful pumps to do the job. Working alone (if one of the pumps were needed at another job), the larger pump would be able to clear the basement in 3 hr less time than the smaller pump alone. Working together, the two pumps can clear the basement in 2 hr. How long would it take the smaller pump alone?

SECTION 1.6   Solving Other Types of Equations

KEY CONCEPTS

- Certain equations of higher degree can be solved using factoring skills and the zero product property.
- To solve rational equations, clear denominators using the LCD, noting values that must be excluded.
- Multiplying an equation by a variable quantity sometimes introduces extraneous solutions. Check all results in the original equation.
- To solve radical equations, isolate the radical on one side, then apply the appropriate "nth power" to free up the radicand. Repeat the process if needed. See flowchart on page 132.
- For equations with a rational exponent $\frac{m}{n}$, isolate the variable term and raise both sides to the $\frac{n}{m}$ power. If m is even, there will be two real solutions.
- Any equation that can be written in the form $u^2 + bu + c = 0$, where u represents an algebraic expression, is said to be in quadratic form and can be solved using u-substitution and standard approaches.

EXERCISES

Solve by factoring.

60. $x^3 - 7x^2 = 3x - 21$

61. $3x^3 + 5x^2 = 2x$

62. $x^4 - 8x = 0$

63. $x^4 - \dfrac{1}{16} = 0$

Solve each equation.

64. $\dfrac{3}{5x} + \dfrac{7}{10} = \dfrac{1}{4x}$

65. $\dfrac{3h}{h+3} - \dfrac{7}{h^2+3h} = \dfrac{1}{h}$

66. $\dfrac{2n}{n+2} - \dfrac{3}{n-4} = \dfrac{n^2+20}{n^2-2n-8}$

67. $\dfrac{\sqrt{x^2+7}}{2} + 3 = 5$

68. $3\sqrt{x+4} = x + 4$

69. $\sqrt{3x+4} = 2 - \sqrt{x+2}$

70. $3\left(x - \dfrac{1}{4}\right)^{-\frac{3}{2}} = \dfrac{8}{9}$

71. $-2(5x+2)^{\frac{2}{3}} + 17 = -1$

72. $(x^2 - 3x)^2 - 14(x^2 - 3x) + 40 = 0$

73. $x^4 - 7x^2 = 18$

74. The science of *allometry* studies the growth of one aspect of an organism relative to the entire organism or to a set standard. Allometry tells us that the amount of food F (in kilocalories per day) an herbivore must eat to survive is related to its weight W (in grams) and can be approximated by the equation $F \approx 1.5W^{\frac{3}{4}}$.

   a. How many kilocalories per day are required by a 160-kg gorilla (160 kg = 160,000 g)?

   b. If an herbivore requires 40,500 kilocalories per day, how much does it weigh?

75. The area of a common stenographer's tablet, commonly called a *steno book,* is 54 in². The length of the tablet is 3 in. more than the width. Model the situation with a quadratic equation and find the dimensions of the tablet.

76. A batter has just flied out to the catcher, who catches the ball while standing on home plate. If the batter made contact with the ball at a height of 4 ft and the ball left the bat with an initial velocity of 128 ft/sec, how long will it take the ball to reach a height of 116 ft? How high is the ball 5 sec after contact? If the catcher catches the ball at a height of 4 ft, how long was it airborne?

77. Using a survey, a firewood distributor finds that if they charge $50 per load, they will sell 40 loads each winter month. For each decrease of $2, five additional loads will be sold. What selling price(s) will result in new monthly revenue of $2520?

MIXED REVIEW

1. Find the allowable values for each expression. Write your response in interval notation.

   a. $\dfrac{10}{\sqrt{x-8}}$ b. $\dfrac{-5}{3x+4}$

2. Perform the operations indicated.

   a. $\sqrt{-18} + \sqrt{-50}$ b. $(1-2i)^2$

   c. $\dfrac{3i}{1+i}$ d. $(2+i\sqrt{3})(2-i\sqrt{3})$

3. Solve each equation or inequality.

   a. $-2x^3 + 4x^2 = 50x - 100$

   b. $-3x^4 - 375x = 0$ c. $-2|3x+1| = -12$

   d. $-3\left|\dfrac{x}{3} - 5\right| \le -12$ e. $v^{\frac{4}{3}} = 81$

   f. $-2(x+1)^{\frac{1}{4}} = -6$

Solve for the variable indicated.

4. $V = \dfrac{1}{3}\pi r^2 h + \dfrac{2}{3}\pi r^3$; for h 5. $3x + 4y = -12$; for y

Solve as indicated, using the method of your choice.

6. a. $-20 \le 4x + 8 < 56$

   b. $-2x + 7 \le 12$ and $3 - 4x > -5$

7. a. $5x - (2x-3) + 3x = -4(5+x) + 3$

   b. $\dfrac{n}{5} - 2 = 2 - \dfrac{5}{3} - \dfrac{4}{15}n$

8. $5x(x-10)(x+1) = 0$

9. $x^2 - 18x + 77 = 0$ 10. $3x^2 - 10 = 5 - x + x^2$

11. $4x^2 - 5 = 19$ 12. $3(x+5)^2 - 3 = 30$

13. $25x^2 + 16 = 40x$ 14. $3x^2 - 7x + 3 = 0$

15. $2x^4 - 50 = 0$

16. a. $\dfrac{2}{x} - \dfrac{x}{5x+12} = 0$ b. $\dfrac{1}{n-1} - \dfrac{2}{n^2-1} = -\dfrac{1}{2}$

   c. $\dfrac{2x}{x+3} - \dfrac{36}{x^2-9} = \dfrac{x}{x-3}$

17. a. $\sqrt{2v-3} + 3 = v$

   b. $\sqrt[3]{x^2-9} + \sqrt[3]{x-11} = 0$

   c. $\sqrt{x+7} - \sqrt{2x} = 1$

18. The local Lion's Club rents out two banquet halls for large meetings and other events. The records show that when they charge $250 per day for use of the halls, there are an average of 156 bookings per year. For every increase of $20 per day, there will be three less bookings. (a) What price per day will bring in $61,950 for the year? (b) How many bookings will there be at the price from part (a)?

19. The Jefferson College basketball team has two guards who are 6'3" tall and two forwards who are 6'7" tall. How tall must their center be to ensure the "starting five" will have an average height of at least 6'6"?

20. The volume of an inflatable hot-air balloon can be approximated using the formulas for a hemisphere and a cone: $V = \frac{2}{3}\pi r^3 + \frac{1}{3}\pi r^2 h$. Assume the conical portion has height $h = 24$ ft. During inflation, what is the radius of the balloon at the moment the volume of air is numerically equal to 126π times this radius?

PRACTICE TEST

1. Solve each equation.

   a. $-\dfrac{2}{3}x - 5 = 7 - (x+3)$

   b. $-5.7 + 3.1x = 14.5 - 4(x + 1.5)$

   c. $P = C + kC$; for C

   d. $2|2x + 5| - 17 = -11$

2. How much water that is 102°F must be mixed with 25 gal of water at 91°F, so that the resulting temperature of the water will be 97°F?

3. Solve each equation or inequality.

   a. $-\dfrac{2}{5}x + 7 < 19$

   b. $-1 < 3 - x \le 8$

   c. $\dfrac{1}{2}x + 3 < 9$ or $\dfrac{2}{3}x - 1 \ge 3$

   d. $\dfrac{1}{2}|x - 3| + \dfrac{5}{4} = \dfrac{7}{4}$

   e. $-\dfrac{2}{3}|x + 1| - 5 < -7$

4. To make the bowling team, Jacques needs a three-game average of 160. If he bowled 141 and 162 for the first two games, what score S must be obtained in the third game so that his average is at least 160?

Solve each equation.

5. $z^2 - 7z - 30 = 0$ 6. $x^2 + 25 = 0$

7. $(x - 1)^2 + 3 = 0$ 8. $x^4 + 16 = 17x^2$

9. $3x^2 - 20x = -12$

10. $4x^3 + 8x^2 - 9x - 18 = 0$

11. $\dfrac{2}{x - 3} + \dfrac{2x}{x + 2} = \dfrac{x^2 + 16}{x^2 - x - 6}$

12. $\dfrac{4}{x - 3} + 2 = \dfrac{5x}{x^2 - 9}$

13. $\sqrt{x + 1} = \sqrt{2x - 7}$

14. $(x + 3)^{\frac{-2}{3}} = \dfrac{1}{4}$

15. The Spanish Club at Rock Hill Community College has decided to sell tins of gourmet popcorn as a fundraiser. The suggested selling price is $3.00 per tin, but Maria, who also belongs to the Math Club, decides to take a survey to see if they can increase "the fruits of their labor." The survey shows it's likely that 120 tins will be sold on campus at the $3.00 price, and for each price increase of $0.10, 2 fewer tins will be sold. (a) What price per tin will bring in a

revenue of $405? (b) How many tins will be sold at the price from part (a)?

16. Due to the seasonal nature of the business, the revenue of Wet Willey's Water World can be modeled by the equation $r = -3t^2 + 42t - 135$, where t is the time in months ($t = 1$ corresponds to January) and r is the dollar revenue in thousands. (a) What month does Wet Willey's open? (b) What month does Wet Willey's close? (c) Does Wet Willey's bring in more revenue in July or August? How much more?

Simplify each expression.

17. $\dfrac{-8 + \sqrt{-20}}{6}$ 18. i^{39}

19. Given $x = \dfrac{1}{2} + \dfrac{\sqrt{3}}{2}i$ and $y = \dfrac{1}{2} - \dfrac{\sqrt{3}}{2}i$ find

    a. $x + y$ b. $x - y$ c. xy

20. Compute the quotient: $\dfrac{3i}{1 - i}$.

21. Find the product: $(3i + 5)(5 - 3i)$.

22. Show $x = 2 - 3i$ is a solution of $x^2 - 4x + 13 = 0$.

23. Solve by completing the square.
    a. $2x^2 - 20x + 49 = 0$
    b. $2x^2 - 5x = -4$

24. Solve using the quadratic formula.
    a. $3x^2 + 2 = 6x$ b. $x^2 = 2x - 10$

25. Allometric studies tell us that the necessary food intake F (in grams per day) of nonpasserine birds (birds other than song birds and other small birds) can be modeled by the equation $F \approx 0.3W^{\frac{3}{4}}$, where W is the bird's weight in grams. (a) If my Green-winged macaw weighs 1296 g, what is her anticipated daily food intake? (b) If my blue-headed pionus consumes 19.2 g per day, what is his estimated weight?

CALCULATOR EXPLORATION AND DISCOVERY

Evaluating Expressions and Looking for Patterns

These "explorations" are designed to explore the full potential of a graphing calculator, as well as to use this potential to investigate patterns and discover connections that might otherwise be overlooked. In this *Exploration and Discovery,* we point out the various ways an expression can be evaluated on a graphing calculator. Some ways seem easier, faster, and/or better than others, but each has

advantages and disadvantages depending on the task at hand, and it will help to be aware of them all for future use.

One way to evaluate an expression is to use the TABLE feature of a graphing calculator, with the expression entered as Y_1 on the ⟨Y=⟩ screen. If you want the calculator to generate inputs, use the ⟨2nd⟩ ⟨WINDOW⟩ (TBLSET) screen to indicate a starting value

(**TblStart=**) and an increment value (**ΔTbl=**), and set the calculator in **Indpnt:** AUTO ASK mode (to input specific values, the calculator should be in **Indpnt: AUTO** ASK mode). After pressing 2nd GRAPH (**TABLE**), the calculator shows the corresponding input and output values. For help with the basic TABLE feature of the TI-84 Plus, you can visit Section R.7 at www.mhhe.com/coburn.

Expressions can also be evaluated on the home screen for a single value or a series of values. Enter the expression $-\frac{3}{4}x + 5$ on the Y= screen (see Figure 1.15) and use 2nd MODE (**QUIT**) to get back to the home screen. To evaluate this expression, access Y_1 using VARS ▶ (**Y-VARS**), and use the first option **1:Function** ENTER. This brings us to a submenu where any of the equations Y_1 through Y_0 (actually Y_{10}) can be accessed. Since the default setting is the one we need **1:Y1**, simply press ENTER and Y_1 appears on the home screen. To evaluate a single input, simply enclose it in parentheses. To evaluate more than one input, enter the numbers as a set of values with the set enclosed in parentheses. In Figure 1.16, Y_1 has been evaluated for $x = -4$, then simultaneously for $x = -4, -2, 0,$ and 2.

A third way to evaluate expressions is using a list, with the desired inputs entered in List 1 (L1), and List 2 (L2) defined in terms of L1. For example, L2 = $-\frac{3}{4}$L1 + 5 will return the same values for inputs of $-4, -2, 0,$ and 2 seen previously on the home screen (remember to clear the lists first). Lists are accessed by pressing STAT **1:Edit.** Enter the numbers $-4, -2, 0$ and 2 in L1, then use the right arrow ▶ to move to L2. It is important to note that you *next press the up arrow key* ▲ so that the cursor overlies L2. The bottom of the screen now reads **L2=** (see Figure 1.17) and the calculator is waiting for us to define L2. After entering L2 = $-\frac{3}{4}$L1 + 5 and pressing ENTER we obtain the same outputs as before (see Figure 1.18).

The advantage of using the "list" method is that we can *further explore or experiment with the output values* in a search for patterns.

Exercise 1: Evaluate the expression 0.2L1 + 3 on the list screen, using consecutive integer inputs from -6 to 6 inclusive. What do you notice about the outputs?

Exercise 2: Evaluate the expression $\sqrt{2}$L1 $- \sqrt{9.1}$ on the list screen, using consecutive integer inputs from -6 to 6 inclusive. We suspect there is a pattern to the output values, but this time the pattern is very difficult to see. Compute the difference between a few successive outputs from L2 [for Example L2(1) $-$ L2(2)]. What do you notice?

Figure 1.15

Figure 1.16

Figure 1.17

L1	▨	L3	2
-4		------	------
-2			
0			
2			

L2 = (-3/4)L1+5

Figure 1.18

L1	L2	L3	3
-4	8	------	
-2	6.5		
0	5		
2	3.5		
	------	------	

STRENGTHENING CORE SKILLS

An Alternative Method for Checking Solutions to Quadratic Equations

To solve $x^2 - 2x - 15 = 0$ by factoring, students will often begin by looking for two numbers whose product is -15 (the constant term) and whose sum is -2 (the linear coefficient). The two numbers are -5 and 3 since $(-5)(3) = -15$ and $-5 + 3 = -2$. In factored form, we have $(x - 5)(x + 3) = 0$ with solutions $x_1 = 5$ and $x_2 = -3$. When these solutions are compared *to the original coefficients,* we can still see the sum/product relationship, but note that while $(5)(-3) = -15$ still gives the constant term, $5 + (-3) = 2$ gives the linear coefficient *with opposite sign.* Although more difficult to accomplish,

this method can be applied to *any* factorable quadratic equation $ax^2 + bx + c = 0$ if we divide through by a, giving $x^2 + \frac{b}{a}x + \frac{c}{a} = 0$. For $2x^2 - x - 3 = 0$, we divide both sides by 2 and obtain $x^2 - \frac{1}{2}x - \frac{3}{2} = 0$, then look for two numbers whose product is $-\frac{3}{2}$ and whose sum is $-\frac{1}{2}$. The numbers are $-\frac{3}{2}$ and 1 since $\left(-\dfrac{3}{2}\right)(1) = -\dfrac{3}{2}$ and $-\dfrac{3}{2} + 1 = -\dfrac{1}{2}$, showing the

solutions are $x_1 = \dfrac{3}{2}$ and $x_2 = -1$. We again note the

product of the solutions is the constant $-\dfrac{3}{2} = \dfrac{c}{a}$, and the

sum of the solutions is the linear coefficient *with opposite*

sign: $\dfrac{1}{2} = -\dfrac{b}{a}$. No one actually promotes this method for

solving trinomials where $a \neq 1$, but it does illustrate an important and useful concept:

If x_1 and x_2 are the two roots of $x^2 + \dfrac{b}{a}x + \dfrac{c}{a} = 0$,

then $x_1 x_2 = \dfrac{c}{a}$ and $x_1 + x_2 = -\dfrac{b}{a}$.

Justification for this can be found by taking the product and sum of the general solutions $x_1 = \dfrac{-b}{2a} + \dfrac{\sqrt{b^2 - 4ac}}{2a}$

and $x_2 = \dfrac{-b}{2a} - \dfrac{\sqrt{b^2 - 4ac}}{2a}$. Although the computation

looks impressive, the product can be computed as a binomial times its conjugate, and the radical parts add to zero for the sum, each yielding the results as already stated.

This observation provides a useful technique for checking solutions to a quadratic equation, *even those having irrational or complex roots!* Check the solutions shown in these exercises.

Exercise 1: $2x^2 - 5x - 7 = 0$

$$x_1 = \dfrac{7}{2}$$

$$x_2 = -1$$

Exercise 2: $2x^2 - 4x - 7 = 0$

$$x_1 = \dfrac{2 + 3\sqrt{2}}{2}$$

$$x_2 = \dfrac{2 - 3\sqrt{2}}{2}$$

Exercise 3: $x^2 - 10x + 37 = 0$

$$x_1 = 5 + 2\sqrt{3}\,i$$

$$x_2 = 5 - 2\sqrt{3}\,i$$

Exercise 4: Verify this sum/product check by computing the sum and product of the general solutions.

Relations, Functions, and Graphs

CHAPTER OUTLINE

CHAPTER CONNECTIONS

Viewing a function in terms of an equation, a table of values, and the related graph, often brings a clearer understanding of the relationships involved. For example, the power generated by a wind turbine is often modeled by the function $P(v) = \dfrac{8v^3}{125}$, where P is the power in watts and v is the wind velocity in miles per hour. While the formula enables us to predict the power generated for a given wind speed, the graph offers a visual representation of this relationship, where we note a rapid growth in power output as the wind speed increases. This application appears as Exercise 107 in Section 2.6.

Check out these other real-world connections:

Learning Objectives

In Section 2.1 you will learn how to:

☐ **A.** Express a relation in mapping notation and ordered pair form

☐ **B.** Graph a relation

☐ **C.** Develop the equation of a circle using the distance and midpoint formulas

☐ **D.** Graph circles

WORTHY OF NOTE

From a purely practical standpoint, we note that while it is possible for two different people to share the same birthday, it is quite impossible for the same person to have two different birthdays. Later, this observation will help us mark the difference between a relation and a function.

In everyday life, we encounter a large variety of relationships. For instance, the time it takes us to get to work is related to our average speed; the monthly cost of heating a home is related to the average outdoor temperature; and in many cases, the amount of our charitable giving is related to changes in the cost of living. In each case we say that a relation exists between the two quantities.

A. Relations, Mapping Notation, and Ordered Pairs

In the most general sense, a **relation** is simply a correspondence between two sets. Relations can be represented in many different ways and may even be very "unmathematical," like the one shown in Figure 2.1 between a set of people and the set of their corresponding birthdays. If P represents the set of people and B represents the set of birthdays, we say that elements of P correspond to elements of B, or the birthday relation maps elements of P to elements of B. Using what is called **mapping notation,** we might simply write $P \rightarrow B$.

Figure 2.1

The bar graph in Figure 2.2 is also an example of a relation. In the graph, each year is related to average annual consumer spending on Internet media (music downloads, Internet radio, Web-based news articles, etc.). As an alternative to mapping or a bar graph, the relation could also be represented using **ordered pairs.** For example, the ordered pair (3, 98) would indicate that in 2003, spending per person on Internet media averaged $98 in the United States. Over a long period of time, we could collect many ordered pairs of the

Figure 2.2

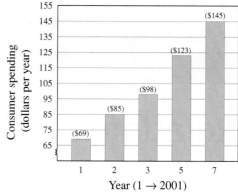

Source: 2006 Statistical Abstract of the United States

form (t, s), where consumer spending s *depends* on the time t. For this reason we often call the second coordinate of an ordered pair (in this case s) the **dependent variable,** with the first coordinate designated as the **independent variable.** In this form, the set of all first coordinates is called the **domain** of the relation. The set of all second coordinates is called the **range.**

EXAMPLE 1 ▶ **Expressing a Relation as a Mapping and in Ordered Pair Form**

Represent the relation from Figure 2.2 in mapping notation and ordered pair form, then state its domain and range.

Solution ▶ Let t represent the year and s represent consumer spending. The mapping $t \rightarrow s$ gives the diagram shown. In ordered pair form we have (1, 69), (2, 85), (3, 98), (5, 123), and (7, 145). The domain is {1, 2, 3, 5, 7}, the range is {69, 85, 98, 123, 145}.

t		s
1	→	69
2	→	85
3	→	98
5	→	123
7	→	145

☑ **A.** You've just learned how to express a relation in mapping notation and ordered pair form

Now try Exercises 7 through 12 ▶

For more on this relation, **see Exercise 81.**

Table 2.1 $y = x - 1$

x	y
−4	−5
−2	−3
0	−1
2	1
4	3

Table 2.2 $x = |y|$

x	y
2	−2
1	−1
0	0
1	1
2	2

B. The Graph of a Relation

Relations can also be stated in **equation form.** The equation $y = x - 1$ expresses a relation where each y-value is one less than the corresponding x-value (see Table 2.1). The equation $x = |y|$ expresses a relation where each x-value corresponds to the absolute value of y (see Table 2.2). In each case, the relation is the set of all ordered pairs (x, y) that create a true statement when substituted, and a few ordered pair solutions are shown in the tables for each equation.

Relations can be expressed graphically using a **rectangular coordinate system.** It consists of a horizontal number line (the x-axis) and a vertical number line (the y-axis) intersecting at their zero marks. The point of intersection is called the *origin*. The x- and y-axes create a flat, two-dimensional surface called the **xy-plane** and divide the plane into four regions called **quadrants.** These are labeled using a capital "Q" (for quadrant) and the Roman numerals I through IV, beginning in the upper right and moving counterclockwise (Figure 2.3). The **grid lines** shown denote the integer values on each axis and further divide the plane into a **coordinate grid,** where every point in the plane corresponds to an ordered pair. Since a point at the origin has not moved along either axis, it has coordinates $(0, 0)$. To plot a point (x, y) means we place a dot at its location in the xy-plane. A few of the ordered pairs from $y = x - 1$ are plotted in Figure 2.4, where a noticeable pattern emerges—the points seem to lie along a straight line.

If a relation is defined by a set of ordered pairs, the graph of the relation is simply the plotted points. The graph of a relation *in equation form,* such as $y = x - 1$, is the set of *all* ordered pairs (x, y) that make the equation true. We generally use only a few select points to determine the shape of a graph, then draw a straight line or smooth curve through these points, as indicated by any patterns formed.

Figure 2.3

Figure 2.4

EXAMPLE 2 ▶ Graphing Relations

Graph the relations $y = x - 1$ and $x = |y|$ using the ordered pairs given earlier.

Solution ▶ For $y = x - 1$, we plot the points then connect them with a straight line (Figure 2.5). For $x = |y|$, the plotted points form a V-shaped graph made up of two half lines (Figure 2.6).

Figure 2.5

Figure 2.6

Now try Exercises 13 through 16 ▶

While we used only a few points to graph the relations in Example 2, they are actually made up of an *infinite number of ordered pairs* that satisfy each equation, including those that might be rational or irrational. All of these points together make these graphs **continuous,** which for our purposes means you can draw the entire graph without lifting your pencil from the paper.

Actually, a majority of graphs cannot be drawn using only a straight line or directed line segments. In these cases, we rely on a "sufficient number" of points to outline the basic shape of the graph, then connect the points with a smooth curve. As your experience with graphing increases, this "sufficient number of points" tends to get smaller as you learn to anticipate what the graph of a given relation should look like.

EXAMPLE 3 ▶ Graphing Relations

Graph the following relations by completing the tables given.

a. $y = x^2 - 2x$ **b.** $y = \sqrt{9 - x^2}$ **c.** $x = y^2$

Solution ▶ For each relation, we use each x-input in turn to determine the related y-output(s), if they exist. Results can be entered in a table and the ordered pairs used to draw the graph.

a. $y = x^2 - 2x$

x	y	(x, y) Ordered Pairs
-4	24	$(-4, 24)$
-3	15	$(-3, 15)$
-2	8	$(-2, 8)$
-1	3	$(-1, 3)$
0	0	$(0, 0)$
1	-1	$(1, -1)$
2	0	$(2, 0)$
3	3	$(3, 3)$
4	8	$(4, 8)$

Figure 2.7

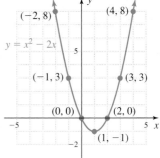

The result is a fairly common graph (Figure 2.7), called a **vertical parabola.** Although $(-4, 24)$ and $(-3, 15)$ cannot be plotted here, the arrowheads indicate an infinite extension of the graph, which will include these points.

b. $y = \sqrt{9 - x^2}$

x	y	(x, y) Ordered Pairs
-4	not real	—
-3	0	$(-3, 0)$
-2	$\sqrt{5}$	$(-2, \sqrt{5})$
-1	$2\sqrt{2}$	$(-1, 2\sqrt{2})$
0	3	$(0, 3)$
1	$2\sqrt{2}$	$(1, 2\sqrt{2})$
2	$\sqrt{5}$	$(2, \sqrt{5})$
3	0	$(3, 0)$
4	not real	—

Figure 2.8

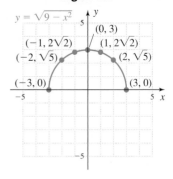

The result is the graph of a **semicircle** (Figure 2.8). The points with irrational coordinates were graphed by estimating their location. Note that when $x < -3$ or $x > 3$, the relation $y = \sqrt{9 - x^2}$ does not represent a real number and no points can be graphed. Also note that no arrowheads are used since the graph terminates at $(-3, 0)$ and $(3, 0)$.

c. Similar to $x = |y|$, the relation $x = y^2$ is defined only for $x \geq 0$ since y^2 is always nonnegative ($-1 = y^2$ has no real solutions). In addition, we reason that each positive x-value will correspond to two y-values. For example, given $x = 4$, $(4, -2)$ and $(4, 2)$ are both solutions.

$x = y^2$

Figure 2.9

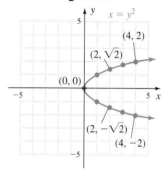

x	y	(x, y) Ordered Pairs
-2	not real	—
-1	not real	—
0	0	$(0, 0)$
1	$-1, 1$	$(1, -1)$ and $(1, 1)$
2	$-\sqrt{2}, \sqrt{2}$	$(2, -\sqrt{2})$ and $(2, \sqrt{2})$
3	$-\sqrt{3}, \sqrt{3}$	$(3, -\sqrt{3})$ and $(3, \sqrt{3})$
4	$-2, 2$	$(4, -2)$ and $(4, 2)$

☑ **B.** You've just learned how to graph a relation

This is the graph of a **horizontal parabola** (Figure 2.9).

Now try Exercises 17 through 24 ▶

C. The Equation of a Circle

Using the midpoint and distance formulas, we can develop the equation of another very important relation, that of a circle. As the name suggests, the **midpoint of a line segment** is located halfway between the endpoints. On a standard number line, the midpoint of the line segment with endpoints 1 and 5 is 3, but more important, note that 3 is the **average distance** (from zero) of 1 unit and 5 units: $\dfrac{1 + 5}{2} = \dfrac{6}{2} = 3$. This observation can be extended to find the midpoint between any two points (x_1, y_1) and (x_2, y_2). We simply find the average distance between the x-coordinates and the average distance between the y-coordinates.

The Midpoint Formula

Given any line segment with endpoints $P_1 = (x_1, y_1)$ and $P_2 = (x_2, y_2)$, the midpoint M is given by

$$M: \left(\frac{x_1 + x_2}{2}, \frac{y_1 + y_2}{2} \right)$$

The midpoint formula can be used in many different ways. Here we'll use it to find the coordinates of the center of a circle.

EXAMPLE 4 ▶ **Using the Midpoint Formula**

The diameter of a circle has endpoints at $P_1 = (-3, -2)$ and $P_2 = (5, 4)$. Use the midpoint formula to find the coordinates of the center, then plot this point.

Solution ▶ Midpoint: $\left(\dfrac{x_1 + x_2}{2}, \dfrac{y_1 + y_2}{2}\right)$

$$M: \left(\frac{-3 + 5}{2}, \frac{-2 + 4}{2}\right)$$

$$M: \left(\frac{2}{2}, \frac{2}{2}\right) = (1, 1)$$

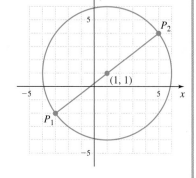

The center is at $(1, 1)$, which we graph directly on the diameter as shown.

Now try Exercises 25 through 34 ▶

Figure 2.10

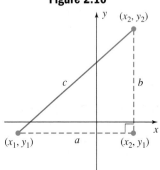

The Distance Formula

In addition to a line segment's midpoint, we are often interested in the *length* of the segment. For any two points (x_1, y_1) and (x_2, y_2) not lying on a horizontal or vertical line, a right triangle can be formed as in Figure 2.10. Regardless of the triangle's orientation, the length of side a (the horizontal segment or base of the triangle) will have length $|x_2 - x_1|$ units, with side b (the vertical segment or height) having length $|y_2 - y_1|$ units. From the Pythagorean theorem (Section R.6), we see that $c^2 = a^2 + b^2$ corresponds to $c^2 = (|x_2 - x_1|)^2 + (|y_2 - y_1|)^2$. By taking the square root of both sides we obtain the length of the hypotenuse, *which is identical to the distance between these two points*: $c = \sqrt{(x_2 - x_1)^2 + (y_2 - y_1)^2}$. The result is called the **distance formula**, although it's most often written using d for **d**istance, rather than c. Note the absolute value bars are dropped from the formula, since the square of any quantity is always nonnegative. This also means that *either* point can be used as the initial point in the computation.

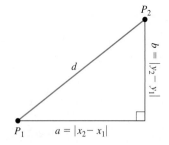

The Distance Formula

Given any two points $P_1 = (x_1, y_1)$ and $P_2 = (x_2, y_2)$, the straight line distance between them is

$$d = \sqrt{(x_2 - x_1)^2 + (y_2 - y_1)^2}$$

EXAMPLE 5 ▶ **Using the Distance Formula**

Use the distance formula to find the diameter of the circle from Example 4.

Solution ▶ For $(x_1, y_1) = (-3, -2)$ and $(x_2, y_2) = (5, 4)$, the distance formula gives

$$d = \sqrt{(x_2 - x_1)^2 + (y_2 - y_1)^2}$$
$$= \sqrt{[5 - (-3)]^2 + [4 - (-2)]^2}$$
$$= \sqrt{8^2 + 6^2}$$
$$= \sqrt{100} = 10$$

The diameter of the circle is 10 units long.

Now try Exercises 35 through 38 ▶

EXAMPLE 6 ▶ **Determining if Three Points Form a Right Triangle**

Use the distance formula to determine if the following points are the vertices of a right triangle: $(-8, 1)$, $(-2, 9)$, and $(10, 0)$

Solution ▶ We begin by finding the distance between each pair of points, then attempt to apply the Pythagorean theorem.

For $(x_1, y_1) = (-8, 1)$, $(x_2, y_2) = (-2, 9)$:

$$d = \sqrt{(x_2 - x_1)^2 + (y_2 - y_1)^2}$$
$$= \sqrt{[-2 - (-8)]^2 + (9 - 1)^2}$$
$$= \sqrt{6^2 + 8^2}$$
$$= \sqrt{100} = 10$$

For $(x_2, y_2) = (-2, 9)$, $(x_3, y_3) = (10, 0)$:

$$d = \sqrt{(x_3 - x_2)^2 + (y_3 - y_2)^2}$$
$$= \sqrt{[10 - (-2)]^2 + (0 - 9)^2}$$
$$= \sqrt{12^2 + (-9)^2}$$
$$= \sqrt{225} = 15$$

For $(x_1, y_1) = (-8, 1)$, $(x_3, y_3) = (10, 0)$:

$$d = \sqrt{(x_3 - x_1)^2 + (y_3 - y_1)^2}$$
$$= \sqrt{[10 - (-8)]^2 + (0 - 1)^2}$$
$$= \sqrt{18^2 + (-1)^2}$$
$$= \sqrt{325} = 5\sqrt{13}$$

Using the unsimplified form, we clearly see that $a^2 + b^2 = c^2$ corresponds to $(\sqrt{100})^2 + (\sqrt{225})^2 = (\sqrt{325})^2$, a true statement. Yes, the triangle is a right triangle.

Now try Exercises 39 through 44 ▶

A circle can be defined as the set of all points in a plane that are a *fixed distance* called the **radius,** from a *fixed point* called the **center.** Since the definition involves *distance,* we can construct the general equation of a circle using the distance formula. Assume the center has coordinates (h, k), and let (x, y) represent any point on the graph. Since the distance between these points is equal to the radius r, the distance formula yields: $\sqrt{(x - h)^2 + (y - k)^2} = r$. Squaring both sides gives the equation of a circle in **standard form:** $(x - h)^2 + (y - k)^2 = r^2$.

The Equation of a Circle

A circle of radius r with center at (h, k) has the equation $(x - h)^2 + (y - k)^2 = r^2$

If $h = 0$ and $k = 0$, the circle is centered at $(0, 0)$ and the graph is a **central circle** with equation $x^2 + y^2 = r^2$. At other values for h or k, the center is at (h, k) with no change in the radius. Note that an open dot is used for the center, as it's actually a point of reference and not a part of the actual graph.

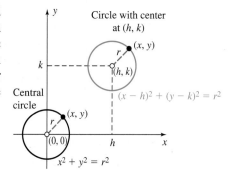

EXAMPLE 7 ▶ **Finding the Equation of a Circle**

Find the equation of a circle with center $(0, -1)$ and radius 4.

Solution ▶ Since the center is at $(0, -1)$ we have $h = 0$, $k = -1$, and $r = 4$. Using the standard form $(x - h)^2 + (y - k)^2 = r^2$ we obtain

$$(x - 0)^2 + [y - (-1)]^2 = 4^2 \quad \text{substitute 0 for } h, -1 \text{ for } k, \text{ and 4 for } r$$
$$x^2 + (y + 1)^2 = 16 \quad \text{simplify}$$

The graph of $x^2 + (y + 1)^2 = 16$ is shown in the figure.

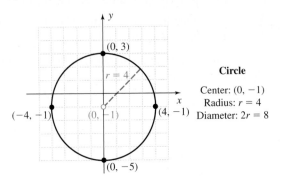

Circle

Center: $(0, -1)$
Radius: $r = 4$
Diameter: $2r = 8$

☑ **C.** You've just learned how to develop the equation of a circle using the distance and midpoint formulas

Now try Exercises 45 through 62 ▶

D. The Graph of a Circle

The graph of a circle can be obtained by first identifying the coordinates of the center and the length of the radius from the equation in standard form. After plotting the center point, we count a distance of r units left and right of center in the horizontal direction, and up and down from center in the vertical direction, obtaining four points on the circle. Neatly graph a circle containing these four points.

EXAMPLE 8 ▶ **Graphing a Circle**

Graph the circle represented by $(x - 2)^2 + (y + 3)^2 = 12$. Clearly label the center and radius.

Solution ▶ Comparing the given equation with the standard form, we find the center is at $(2, -3)$ and the radius is $r = 2\sqrt{3} \approx 3.5$.

$$(x - h)^2 + (y - k)^2 = r^2 \qquad \text{standard form}$$

$$(x - 2)^2 + (y + 3)^2 = 12 \qquad \text{given equation}$$

$$-h = -2 \qquad -k = 3 \qquad r^2 = 12$$

$$h = 2 \qquad k = -3 \qquad r = \sqrt{12} = 2\sqrt{3} \qquad \text{radius must be positive}$$

$$\approx 3.5$$

Plot the center $(2, -3)$ and count approximately 3.5 units in the horizontal and vertical directions. Complete the circle by freehand drawing or using a compass. The graph shown is obtained.

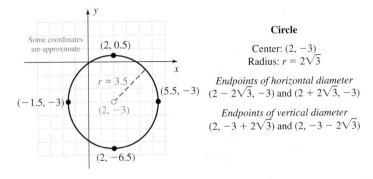

Circle

Center: $(2, -3)$
Radius: $r = 2\sqrt{3}$

Endpoints of horizontal diameter
$(2 - 2\sqrt{3}, -3)$ and $(2 + 2\sqrt{3}, -3)$

Endpoints of vertical diameter
$(2, -3 + 2\sqrt{3})$ and $(2, -3 - 2\sqrt{3})$

Now try Exercises 63 through 68 ▶

WORTHY OF NOTE

After writing the equation in standard form, it is possible to end up with a constant that is zero or negative. In the first case, the graph is a single point. In the second case, no graph is possible since roots of the equation will be complex numbers. These are called *degenerate cases*. **See Exercise 91.**

In Example 8, note the equation is composed of binomial squares in both x and y. By expanding the binomials and collecting like terms, we can write the equation of the circle in the general form:

$$(x - 2)^2 + (y + 3)^2 = 12 \quad \text{standard form}$$
$$x^2 - 4x + 4 + y^2 + 6y + 9 = 12 \quad \text{expand binomials}$$
$$x^2 + y^2 - 4x + 6y + 1 = 0 \quad \text{combine like terms—general form}$$

For future reference, observe the general form contains a *sum* of second-degree terms in x and y, and that *both terms have the same coefficient* (in this case, "1").

Since this form of the equation was derived by squaring binomials, it seems reasonable to assume we can go back to the standard form by creating binomial squares in x and y. This is accomplished by *completing the square*.

EXAMPLE 9 ▶ **Finding the Center and Radius of a Circle**

Find the center and radius of the circle with equation $x^2 + y^2 + 2x - 4y - 4 = 0$. Then sketch its graph and label the center and radius.

Solution ▶ To find the center and radius, we complete the square in both x and y.

$$x^2 + y^2 + 2x - 4y - 4 = 0 \quad \text{given equation}$$
$$(x^2 + 2x + \underline{}) + (y^2 - 4y + \underline{}) = 4 \quad \text{group } x\text{-terms and } y\text{-terms; add 4}$$
$$(x^2 + 2x + 1) + (y^2 - 4y + 4) = 4 + 1 + 4 \quad \text{complete each binomial square}$$
$$\underset{\text{adds 1 to left side}}{} \quad \underset{\text{adds 4 to left side}}{} \quad \underset{\text{add } 1 + 4 \text{ to right side}}{}$$
$$(x + 1)^2 + (y - 2)^2 = 9 \quad \text{factor and simplify}$$

The center is at $(-1, 2)$ and the radius is $r = \sqrt{9} = 3$.

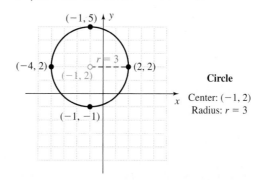

Circle

Center: $(-1, 2)$
Radius: $r = 3$

Now try Exercises 69 through 80 ▶

EXAMPLE 10 ▶ **Applying the Equation of a Circle**

To aid in a study of nocturnal animals, some naturalists install a motion detector near a popular watering hole. The device has a range of 10 m in any direction. Assume the water hole has coordinates $(0, 0)$ and the device is placed at $(2, -1)$.

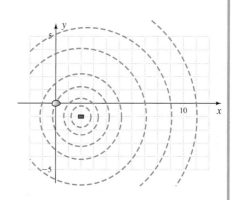

a. Write the equation of the circle that models the maximum effective range of the device.

b. Use the distance formula to determine if the device will detect a badger that is approaching the water and is now at coordinates $(11, -5)$.

Solution ▶   **a.** Since the device is at $(2, -1)$ and the radius (or reach) of detection is 10 m, any movement in the interior of the circle defined by $(x - 2)^2 + (y + 1)^2 = 10^2$ will be detected.

b. Using the points $(2, -1)$ and $(11, -5)$ in the distance formula yields:

$$d = \sqrt{(x_2 - x_1)^2 + (y_2 - y_1)^2} \qquad \text{distance formula}$$

$$= \sqrt{(11 - 2)^2 + [-5 - (-1)]^2} \qquad \text{substitute given values}$$

$$= \sqrt{9^2 + (-4)^2} \qquad \text{simplify}$$

$$= \sqrt{81 + 16} \qquad \text{compute squares}$$

$$= \sqrt{97} \approx 9.85 \qquad \text{result}$$

 D. You've just learned how to graph circles

Since $9.85 < 10$, the badger is within range of the device and will be detected.

Now try Exercises 83 through 88 ▶

TECHNOLOGY HIGHLIGHT

The Graph of a Circle

When using a graphing calculator to study circles, it is important to keep two things in mind. First, we must modify the equation of the circle before it can be graphed using this technology. Second, most standard viewing windows have the x- and y-values preset at $[-10, 10]$ even though the calculator screen is not square. This tends to compress the y-values and give a skewed image of the graph. Consider the *relation* $x^2 + y^2 = 25$, which we know is the equation of a circle centered at $(0, 0)$ with radius $r = 5$. To enable the calculator to graph this relation, we must define it in two pieces by solving for y:

$$x^2 + y^2 = 25 \qquad \text{original equation}$$

$$y^2 = 25 - x^2 \qquad \text{isolate } y^2$$

$$y = \pm\sqrt{25 - x^2} \qquad \text{solve for } y$$

Figure 2.11

Note that we can separate this result into two parts, enabling the calculator to draw the circle: $Y_1 = \sqrt{25 - x^2}$ gives the "upper half" of the circle, and $Y_2 = -\sqrt{25 - x^2}$ gives the "lower half." Enter these on the [Y=] screen (note that $Y_2 = -Y_1$ can be used instead of reentering the entire expression: [VARS] [▶] [ENTER]). But if we graph Y_1 and Y_2 on the standard screen, the result appears more oval than circular (Figure 2.11). One way to fix this is to use the [ZOOM] **5:ZSquare** option, which places the tick marks equally spaced on both axes, instead of trying to force both to display points from -10 to 10 (see Figure 2.12). Although it is a much improved graph, the circle does not appear "closed" as the calculator lacks sufficient pixels to show the proper curvature. A second alternative is to manually set a "friendly" window. Using Xmin $= -9.4$, Xmax $= 9.4$, Ymin $= -6.2$, and Ymax $= 6.2$ will generate a better graph, which we can use to study the relation more closely. Note that we can jump between the upper and lower halves of the circle using the up [▲] or down [▼] arrows.

Figure 2.12

Exercise 1: Graph the circle defined by $x^2 + y^2 = 36$ using a friendly window, then use the [TRACE] feature to find the value of y when $x = 3.6$. Now find the value of y when $x = 4.8$. Explain why the values seem "interchangeable."

Exercise 2: Graph the circle defined by $(x - 3)^2 + y^2 = 16$ using a friendly window, then use the [TRACE] feature to find the value of the y-intercepts. Show you get the same intercept by computation.

2.1 EXERCISES

▶ CONCEPTS AND VOCABULARY

Fill in each blank with the appropriate word or phrase. Carefully reread the section if needed.

1. If a relation is defined by a set of ordered pairs, the domain is the set of all _____ components, the range is the set of all _____ components.

2. For the equation $y = x + 5$ and the ordered pair (x, y), x is referred to as the input or _____ variable, while y is called the _____ or dependent variable.

3. A circle is defined as the set of all points that are an equal distance, called the _____, from a given point, called the _____ .

4. For $x^2 + y^2 = 25$, the center of the circle is at _____ and the length of the radius is _____ units. The graph is called a _____ circle.

5. Discuss/Explain how to find the center and radius of the circle defined by the equation $x^2 + y^2 - 6x = 7$. How would this circle differ from the one defined by $x^2 + y^2 - 6y = 7$?

6. In Example 3b we graphed the semicircle defined by $y = \sqrt{9 - x^2}$. Discuss how you would obtain the equation of the full circle from this equation, and how the two equations are related.

▶ DEVELOPING YOUR SKILLS

Represent each relation in mapping notation, then state the domain and range.

7.

8.
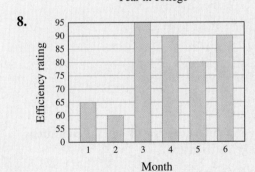

State the domain and range of each relation.

9. $\{(1, 2), (3, 4), (5, 6), (7, 8), (9, 10)\}$

10. $\{(-2, 4), (-3, -5), (-1, 3), (4, -5), (2, -3)\}$

11. $\{(4, 0), (-1, 5), (2, 4), (4, 2), (-3, 3)\}$

12. $\{(-1, 1), (0, 4), (2, -5), (-3, 4), (2, 3)\}$

Complete each table using the given equation. For Exercises 15 and 16, each input may correspond to two outputs (be sure to find both if they exist). Use these points to graph the relation.

13. $y = -\dfrac{2}{3}x + 1$

x	y
-6	
-3	
0	
3	
6	
8	

14. $y = -\dfrac{5}{4}x + 3$

x	y
-8	
-4	
0	
4	
8	
10	

15. $x + 2 = |y|$

x	y
-2	
0	
1	
3	
6	
7	

16. $|y + 1| = x$

x	y
0	
1	
3	
5	
6	
7	

17. $y = x^2 - 1$

x	y
-3	
-2	
0	
2	
3	
4	

18. $y = -x^2 + 3$

x	y
-2	
-1	
0	
1	
2	
3	

19. $y = \sqrt{25 - x^2}$

x	y
-4	
-3	
0	
2	
3	
4	

20. $y = \sqrt{169 - x^2}$

x	y
-12	
-5	
0	
3	
5	
12	

21. $x - 1 = y^2$

x	y
10	
5	
4	
2	
1.25	
1	

22. $y^2 + 2 = x$

x	y
2	
3	
4	
5	
6	
11	

23. $y = \sqrt[3]{x} + 1$

x	y
-9	
-2	
-1	
0	
4	
7	

24. $y = (x - 1)^3$

x	y
-2	
-1	
0	
1	
2	
3	

Find the midpoint of each segment with the given endpoints.

25. $(1, 8), (5, -6)$ **26.** $(5, 6), (6, -8)$

27. $(-4.5, 9.2), (3.1, -9.8)$ **28.** $(5.2, 7.1), (6.3, -7.1)$

29. $\left(\dfrac{1}{5}, -\dfrac{2}{3}\right), \left(-\dfrac{1}{10}, \dfrac{3}{4}\right)$ **30.** $\left(-\dfrac{3}{4}, -\dfrac{1}{3}\right), \left(\dfrac{3}{8}, \dfrac{5}{6}\right)$

Find the midpoint of each segment.

31.   **32.**

Find the center of each circle with the diameter shown.

33.   **34.**

35. Use the distance formula to find the length of the line segment in Exercise 31.

36. Use the distance formula to find the length of the line segment in Exercise 32.

37. Use the distance formula to find the length of the diameter for the circle in Exercise 33.

38. Use the distance formula to find the length of the diameter for the circle in Exercise 34.

In Exercises 39 to 44, three points that form the vertices of a triangle are given. Use the distance formula to determine if any of the triangles are right triangles.

39. $(5, 2), (0, -3), (4, -4)$

40. $(7, 0), (-1, 0), (7, 4)$

41. $(-4, 3), (-7, -1), (3, -2)$

42. $(-3, 7), (2, 2), (5, 5)$

43. $(-3, 2), (-1, 5), (-6, 4)$

44. $(0, 0), (-5, 2), (2, -5)$

Find the equation of a circle satisfying the conditions given, then sketch its graph.

45. center $(0, 0)$, radius 3

46. center $(0, 0)$, radius 6

47. center $(5, 0)$, radius $\sqrt{3}$

48. center $(0, 4)$, radius $\sqrt{5}$

49. center $(4, -3)$, radius 2

50. center $(3, -8)$, radius 9

51. center $(-7, -4)$, radius $\sqrt{7}$

52. center $(-2, -5)$, radius $\sqrt{6}$

53. center $(1, -2)$, diameter 6

54. center $(-2, 3)$, diameter 10

55. center $(4, 5)$, diameter $4\sqrt{3}$

56. center $(5, 1)$, diameter $4\sqrt{5}$

57. center at $(7, 1)$, graph contains the point $(1, -7)$

58. center at $(-8, 3)$, graph contains the point $(-3, 15)$

59. center at $(3, 4)$, graph contains the point $(7, 9)$

60. center at $(-5, 2)$, graph contains the point $(-1, 3)$

61. diameter has endpoints $(5, 1)$ and $(5, 7)$

62. diameter has endpoints $(2, 3)$ and $(8, 3)$

Identify the center and radius of each circle, then graph. Also state the domain and range of the relation.

63. $(x - 2)^2 + (y - 3)^2 = 4$

64. $(x - 5)^2 + (y - 1)^2 = 9$

65. $(x + 1)^2 + (y - 2)^2 = 12$

66. $(x - 7)^2 + (y + 4)^2 = 20$

67. $(x + 4)^2 + y^2 = 81$

68. $x^2 + (y - 3)^2 = 49$

Write each equation in standard form to find the center and radius of the circle. Then sketch the graph.

69. $x^2 + y^2 - 10x - 12y + 4 = 0$

70. $x^2 + y^2 + 6x - 8y - 6 = 0$

71. $x^2 + y^2 - 10x + 4y + 4 = 0$

72. $x^2 + y^2 + 6x + 4y + 12 = 0$

73. $x^2 + y^2 + 6y - 5 = 0$

74. $x^2 + y^2 - 8x + 12 = 0$

75. $x^2 + y^2 + 4x + 10y + 18 = 0$

76. $x^2 + y^2 - 8x - 14y - 47 = 0$

77. $x^2 + y^2 + 14x + 12 = 0$

78. $x^2 + y^2 - 22y - 5 = 0$

79. $2x^2 + 2y^2 - 12x + 20y + 4 = 0$

80. $3x^2 + 3y^2 - 24x + 18y + 3 = 0$

▶ WORKING WITH FORMULAS

81. Spending on Internet media: $s = 12.5t + 59$

The data from Example 1 is closely modeled by the formula shown, where t represents the year ($t = 0$ corresponds to the year 2000) and s represents the average amount spent per person, per year in the United States. (a) List five ordered pairs for this relation using $t = 1, 2, 3, 5, 7$. Does the model give a good approximation of the actual data? (b) According to the model, what will be the average amount spent on Internet media in the year 2008? (c) According to the model, in what year will annual spending surpass \$196? (d) Use the table to graph this relation.

82. Area of an inscribed square: $A = 2r^2$

The area of a square inscribed in a circle is found by using the formula given where r is the radius of the circle. Find the area of the inscribed square shown.

▶ APPLICATIONS

83. Radar detection: A luxury liner is located at map coordinates $(5, 12)$ and has a radar system with a range of 25 nautical miles in any direction. (a) Write the equation of the circle that models the range of the ship's radar, and (b) Use the distance formula to determine if the radar can pick up the liner's sister ship located at coordinates $(15, 36)$.

84. Earthquake range: The epicenter (point of origin) of a large earthquake was located at map coordinates $(3, 7)$, with the quake being felt up to 12 mi away. (a) Write the equation of the circle that models the range of the earthquake's effect. (b) Use the distance formula to determine if a person living at coordinates $(13, 1)$ would have felt the quake.

85. **Inscribed circle:** Find the equation for both the red and blue circles, then find the area of the region shaded in blue.

$(-2, 0)$

86. **Inscribed triangle:** The area of an equilateral triangle inscribed in a circle is given by the formula $A = \dfrac{3\sqrt{3}}{4}r^2$, where r is the radius of the circle. Find the area of the equilateral triangle shown.

$(3, 4)$

87. **Radio broadcast range:** Two radio stations may not use the same frequency if their broadcast areas *overlap*. Suppose station KXRQ has a broadcast area bounded by $x^2 + y^2 + 8x - 6y = 0$ and WLRT has a broadcast area bounded by $x^2 + y^2 - 10x + 4y = 0$. Graph the circle representing each broadcast area on the same grid to determine if both stations may broadcast on the same frequency.

88. **Radio broadcast range:** The emergency radio broadcast system is designed to alert the population by relaying an emergency signal to all points of the country. A signal is sent from a station whose broadcast area is bounded by $x^2 + y^2 = 2500$ (x and y in miles) and the signal is picked up and relayed by a transmitter with range $(x - 20)^2 + (y - 30)^2 = 900$. Graph the circle representing each broadcast area on the same grid to determine the greatest distance from the original station that this signal can be received. Be sure to scale the axes appropriately.

▶ **EXTENDING THE THOUGHT**

89. Although we use the word "domain" extensively in mathematics, it is also commonly seen in literature and heard in everyday conversation. Using a college-level dictionary, look up and write out the various meanings of the word, noting how closely the definitions given are related to its mathematical use.

90. Consider the following statement, then determine whether it is true or false and discuss why. *A graph will exhibit some form of symmetry if, given a point that is h units from the x-axis, k units from the y-axis, and d units from the origin, there is a second point*

on the graph that is a like distance from the origin and each axis.

91. When completing the square to find the center and radius of a circle, we sometimes encounter a value for r^2 that is negative or zero. These are called **degenerate cases.** If $r^2 < 0$, no circle is possible, while if $r^2 = 0$, the "graph" of the circle is simply the point (h, k). Find the center and radius of the following circles (if possible).
    a. $x^2 + y^2 - 12x + 4y + 40 = 0$
    b. $x^2 + y^2 - 2x - 8y - 8 = 0$
    c. $x^2 + y^2 - 6x - 10y + 35 = 0$

▶ **MAINTAINING YOUR SKILLS**

92. (1.3) Solve the absolute value inequality and write the solution in interval notation.
$$\frac{|w - 2|}{3} + \frac{1}{4} \geq \frac{5}{6}$$

93. (R.1) Give an example of each of the following:
    a. a whole number that is not a natural number
    b. a natural number that is not a whole number
    c. a rational number that is not an integer

d. an integer that is not a rational number
e. a rational number that is not a real number
f. a real number that is not a rational number.

94. (1.5) Solve $x^2 + 13 = 6x$ using the quadratic equation. Simplify the result.

95. (1.6) Solve $1 - \sqrt{n + 3} = -n$ and check solutions by substitution. If a solution is extraneous, so state.

Learning Objectives

In Section 2.2 you will learn how to:

☐ **A.** Graph linear equations using the intercept method

☐ **B.** Find the slope of a line

☐ **C.** Graph horizontal and vertical lines

☐ **D.** Identify parallel and perpendicular lines

☐ **E.** Apply linear equations in context

In preparation for sketching graphs of other relations, we'll first consider the characteristics of linear graphs. While linear graphs are fairly simple models, they have many substantive and meaningful applications. For instance, most of us are aware that music and video downloads have been increasing in popularity since they were first introduced. A close look at Example 1 of Section 2.1 reveals that spending on music downloads and Internet radio increased from $69 per person per year in 2001 to $145 in 2007 (Figure 2.13). From an investor's or a producer's point of view, there is a very high interest in the questions, How fast are sales increasing? Can this relationship be modeled mathematically to help predict sales in future years? Answers to these and other questions are precisely what our study in this section is all about.

Figure 2.13

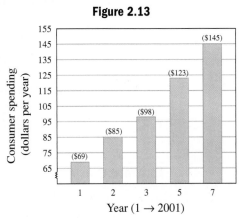

Source: 2006 SAUS

A. The Graph of a Linear Equation

A linear equation can be identified using these three tests: (1) the exponent on any variable is one, (2) no variable occurs in a denominator, and (3) no two variables are multiplied together. The equation $3y = 9$ is a linear equation in one variable, while $2x + 3y = 12$ and $y = -\frac{2}{3}x + 4$ are linear equations in two variables. In general, we have the following definition:

Linear Equations

A linear equation is one that can be written in the form

$$ax + by = c$$

where a and b are not simultaneously zero.

The most basic method for graphing a line is to simply plot a few points, then draw a straight line through the points.

EXAMPLE 1 ▶ **Graphing a Linear Equation in Two Variables**

Graph the equation $3x + 2y = 4$ by plotting points.

Solution ▶ Selecting $x = -2$, $x = 0$, $x = 1$, and $x = 4$ as inputs, we compute the related outputs and enter the ordered pairs in a table. The result is

WORTHY OF NOTE

If you cannot draw a straight line through the plotted points, a computational error has been made. All points satisfying a linear equation *lie on a straight line.*

x input	y output	(x, y) ordered pairs
-2	5	$(-2, 5)$
0	2	$(0, 2)$
1	0.5	$(1, \frac{1}{2})$
4	-4	$(4, -4)$

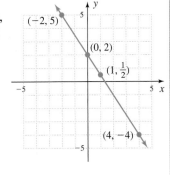

Now try Exercises 7 through 12 ▶

Note the line in Example 1 crosses the y-axis at $(0, 2)$, and this point is called the **y-intercept** of the line. In general, y-intercepts have the form $(0, y)$. Although difficult to see graphically, substituting 0 for y and solving for x shows the line crosses the x-axis at $(\frac{4}{3}, 0)$ and this point is called the **x-intercept.** In general, x-intercepts have the form $(x, 0)$. The x- and y-intercepts are usually easier to calculate than other points (since $y = 0$ or $x = 0$, respectively) and we often graph linear equations using only these two points. This is called the **intercept method** for graphing linear equations.

The Intercept Method

1. Substitute 0 for x and solve for y. This will give the y-intercept $(0, y)$.
2. Substitute 0 for y and solve for x. This will give the x-intercept $(x, 0)$.
3. Plot the intercepts and use them to graph a straight line.

EXAMPLE 2 ▶ **Graphing Lines Using the Intercept Method**

Graph $3x + 2y = 9$ using the intercept method.

Solution ▶ Substitute 0 for x (y-intercept)   Substitute 0 for y (x-intercept)

$$3(0) + 2y = 9 \qquad\qquad 3x + 2(0) = 9$$
$$2y = 9 \qquad\qquad\qquad 3x = 9$$
$$y = \frac{9}{2} \qquad\qquad\qquad x = 3$$
$$\qquad\qquad\qquad\qquad (3, 0)$$
$$\left(0, \frac{9}{2}\right)$$

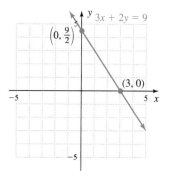

☑ **A.** You've just learned how to graph linear equations using the intercept method

Now try Exercises 13 through 32 ▶

B. The Slope of a Line

After the x- and y-intercepts, we next consider the **slope of a line.** We see applications of the concept in many diverse occupations, including the *grade* of a highway (trucking), the *pitch* of a roof (carpentry), the *climb* of an airplane (flying), the *drainage* of a field (landscaping), and the *slope* of a mountain (parks and recreation). While the general concept is an intuitive one, we seek to quantify the concept (assign it a numeric value) for purposes of comparison and decision making. In each of the preceding examples, slope is a measure of "steepness," as defined by the ratio $\frac{\text{vertical change}}{\text{horizontal change}}$. Using a line segment through arbitrary points $P_1 = (x_1, y_1)$ and $P_2 = (x_2, y_2)$, we can create the right triangle shown in Figure 2.14. The figure illustrates that the **vertical change** or the

Figure 2.14

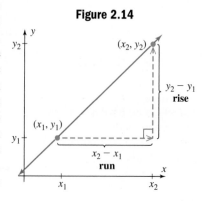

WORTHY OF NOTE

While the original reason that "*m*" was chosen for slope is uncertain, some have speculated that it was because in French, the verb for "to climb" is *monter*. Others say it could be due to the "modulus of slope," the word *modulus* meaning a numeric measure of a given property, in this case the inclination of a line.

change in *y* (also called the **rise**) is simply the difference in *y*-coordinates: $y_2 - y_1$. The **horizontal change** or **change in *x*** (also called the **run**) is the difference in *x*-coordinates: $x_2 - x_1$. In algebra, we typically use the letter "*m*" to represent slope, giving $m = \frac{y_2 - y_1}{x_2 - x_1}$ as the $\frac{\text{change in } y}{\text{change in } x}$. The result is called the **slope formula.**

The Slope Formula

Given two points $P_1 = (x_1, y_1)$ and $P_2 = (x_2, y_2)$, the slope of any nonvertical line through P_1 and P_2 is

$$m = \frac{y_2 - y_1}{x_2 - x_1}$$

where $x_2 \neq x_1$.

EXAMPLE 3 ▶ **Using the Slope Formula**

Find the slope of the line through the given points.

   a. $(2, 1)$ and $(8, 4)$ **b.** $(-2, 6)$ and $(4, 2)$

Solution ▶   **a.** For $P_1 = (2, 1)$ and $P_2 = (8, 4)$,   **b.** For $P_1 = (-2, 6)$ and $P_2 = (4, 2)$,

$$m = \frac{y_2 - y_1}{x_2 - x_1} \qquad\qquad m = \frac{y_2 - y_1}{x_2 - x_1}$$

$$= \frac{4 - 1}{8 - 2} \qquad\qquad = \frac{2 - 6}{4 - (-2)}$$

$$= \frac{3}{6} = \frac{1}{2} \qquad\qquad = \frac{-4}{6} = \frac{-2}{3}$$

The slope of this line is $\frac{1}{2}$.   The slope of this line is $\frac{-2}{3}$.

Now try Exercises 33 through 40 ▶

⚠ **CAUTION** ▶ When using the slope formula, try to avoid these common errors.

1. The order that the *x*- and *y*-coordinates are subtracted must be consistent, since $\frac{y_2 - y_1}{x_2 - x_1} \neq \frac{y_2 - y_1}{x_1 - x_2}$.

2. The vertical change (involving the *y*-values) always occurs in the numerator: $\frac{y_2 - y_1}{x_2 - x_1} \neq \frac{x_2 - x_1}{y_2 - y_1}$.

3. When x_1 or y_1 is negative, use parentheses when substituting into the formula to prevent confusing the negative sign with the subtraction operation.

   Actually, the slope value does much more than quantify the slope of a line, it expresses a **rate of change** between the quantities measured along each axis. In applications of slope, the ratio $\frac{\text{change in } y}{\text{change in } x}$ is symbolized as $\frac{\Delta y}{\Delta x}$. The symbol Δ is the Greek letter **delta** and has come to represent a change in some quantity, and the notation $m = \frac{\Delta y}{\Delta x}$ is read, "slope is equal to the *change in y* over the *change in x*." Interpreting slope as a rate of change has many significant applications in college algebra and beyond.

EXAMPLE 4 ▶ **Interpreting the Slope Formula as a Rate of Change**

Jimmy works on the assembly line for an auto parts remanufacturing company. By 9:00 A.M. his group has assembled 29 carburetors. By 12:00 noon, they have completed 87 carburetors. Assuming the relationship is linear, find the slope of the line and discuss its meaning in this context.

Solution ▶ First write the information as ordered pairs using c to represent the carburetors assembled and t to represent time. This gives $(t_1, c_1) = (9, 29)$ and $(t_2, c_2) = (12, 87)$. The slope formula then gives:

$$\frac{\Delta c}{\Delta t} = \frac{c_2 - c_1}{t_2 - t_1} = \frac{87 - 29}{12 - 9}$$

$$= \frac{58}{3} \text{ or } 19.\overline{3}$$

Here the slope ratio measures $\frac{\text{carburetors assembled}}{\text{hours}}$, and we see that Jimmy's group can assemble 58 carburetors every 3 hr, or about $19\frac{1}{3}$ carburetors per hour.

Now try Exercises 41 through 44 ▶

> **WORTHY OF NOTE**
>
> Actually, the assignment of (t_1, c_1) to (9, 29) and (t_2, c_2) to (12, 87) was arbitrary. The slope ratio will be the same *as long as the order of subtraction is the same.* In other words, if we reverse this assignment and use $(t_1, c_1) = (12, 87)$ and $(t_2, c_2) = (9, 29)$, we have $m = \frac{29 - 87}{9 - 12} = \frac{-58}{-3} = \frac{58}{3}$.

Positive and Negative Slope

If you've ever traveled by air, you've likely heard the announcement, "Ladies and gentlemen, please return to your seats and fasten your seat belts as we begin our descent." For a time, the descent of the airplane follows a linear path, but now the *slope of the line is negative* since the altitude of the plane is decreasing. Positive and negative slopes, as well as the rate of change they represent, are important characteristics of linear graphs. In Example 3a, the slope was a positive number ($m > 0$) and the line will slope upward from left to right since the y-values are increasing. If $m < 0$, the slope of the line is negative and the line slopes downward as you move left to right since y-values are decreasing.

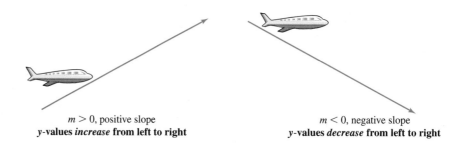

$m > 0$, positive slope
y-values *increase* from left to right

$m < 0$, negative slope
y-values *decrease* from left to right

EXAMPLE 5 ▶ **Applying Slope to Changes in Altitude**

At a horizontal distance of 10 mi after take-off, an airline pilot receives instructions to decrease altitude from their current level of 20,000 ft. A short time later, they are 17.5 mi from the airport at an altitude of 10,000 ft. Find the slope ratio for the descent of the plane and discuss its meaning in this context. Recall that 1 mi = 5280 ft.

Solution ▶ Let a represent the altitude of the plane and d its horizontal distance from the airport. Converting all measures to feet, we have $(d_1, a_1) = (52{,}800, 20{,}000)$ and $(d_2, a_2) = (92{,}400, 10{,}000)$, giving

$$\frac{\Delta a}{\Delta d} = \frac{a_2 - a_1}{d_2 - d_1} = \frac{10{,}000 - 20{,}000}{92{,}400 - 52{,}800}$$

$$= \frac{-10{,}000}{39{,}600} = \frac{-25}{99}$$

✓ **B. You've just learned how to find the slope of a line**

Since this slope ratio measures $\frac{\Delta \text{altitude}}{\Delta \text{distance}}$, we note the plane decreased 25 ft in altitude for every 99 ft it traveled horizontally.

Now try Exercises 45 through 48 ▶

C. Horizontal Lines and Vertical Lines

Horizontal and vertical lines have a number of important applications, from finding the boundaries of a given graph, to performing certain tests on nonlinear graphs. To better understand them, consider that in *one dimension,* the graph of $x = 2$ is a single point (Figure 2.15), indicating a location on the number line 2 units from zero in the positive direction. In *two dimensions,* the equation $x = 2$ represents **all points** with an x-coordinate of 2. A few of these are graphed in Figure 2.16, but since there are an infinite number, we end up with a solid *vertical line* whose equation is $x = 2$ (Figure 2.17).

Figure 2.15

Figure 2.16

Figure 2.17

The same idea can be applied to horizontal lines. In *two dimensions,* the equation $y = 4$ represents *all points* with a y-coordinate of positive 4, and there are an infinite number of these as well. The result is a solid horizontal line whose equation is $y = 4$. **See Exercises 49–54.**

Vertical Lines	Horizontal Lines
The equation of a vertical line is	The equation of a horizontal line is
$x = h$	$y = k$
where $(h, 0)$ is the x-intercept.	where $(0, k)$ is the y-intercept.

So far, the slope formula has only been applied to lines that were nonhorizontal or nonvertical. So what *is* the slope of a horizontal line? On an intuitive level, we expect that a perfectly level highway would have an incline or slope of zero. In general, for any two points on a horizontal line, $y_2 = y_1$ and $y_2 - y_1 = 0$, giving a slope of $m = \frac{0}{x_2 - x_1} = 0$. For any two points on a vertical line, $x_2 = x_1$ and $x_2 - x_1 = 0$, making the slope ratio undefined: $m = \frac{y_2 - y_1}{0}$.

The Slope of a Vertical Line	The Slope of a Horizontal Line
The slope of any vertical line is undefined.	The slope of any horizontal line is zero.

EXAMPLE 6 ▶ **Calculating Slopes**

The federal minimum wage remained constant from 1997 through 2006. However, the buying power (in 1996 dollars) of these wage earners fell each year due to inflation (see Table 2.3). This decrease in buying power is approximated by the red line shown.

a. Using the data or graph, find the slope of the line segment representing the minimum wage.

b. Select two points on the line representing buying power to approximate the slope of the line segment, and explain what it means in this context.

Table 2.3

Time t (years)	Minimum wage w	Buying power p
1997	5.15	5.03
1998	5.15	4.96
1999	5.15	4.85
2000	5.15	4.69
2001	5.15	4.56
2002	5.15	4.49
2003	5.15	4.39
2004	5.15	4.28
2005	5.15	4.14
2006	5.15	4.04

Solution ▶

a. Since the minimum wage did not increase or decrease from 1997 to 2006, the line segment has slope $m = 0$.

b. The points (1997, 5.03) and (2006, 4.04) from the table appear to be on or close to the line drawn. For buying power p and time t, the slope formula yields:

$$\frac{\Delta p}{\Delta t} = \frac{p_2 - p_1}{t_2 - t_1}$$

$$= \frac{4.04 - 5.03}{2006 - 1997}$$

$$= \frac{-0.99}{9} = \frac{-0.11}{1}$$

The buying power of a minimum wage worker decreased by 11¢ per year during this time period.

> **WORTHY OF NOTE**
>
> In the context of lines, try to avoid saying that a horizontal line has "no slope," since it's unclear whether a slope of zero or an undefined slope is intended.

☑ **C.** You've just learned how to graph horizontal and vertical lines

Now try Exercises 55 and 56 ▶

D. Parallel and Perpendicular Lines

Two lines in the same plane that never intersect are called **parallel lines.** When we place these lines on the coordinate grid, we find that "never intersect" is equivalent to saying "the lines have equal slopes but different y-intercepts." In Figure 2.18, notice the rise and run of each line is identical, and that by counting $\frac{\Delta y}{\Delta x}$ both lines have slope $m = \frac{3}{4}$.

Figure 2.18

Coordinate plane

Parallel Lines

Given L_1 and L_2 are distinct, nonvertical lines with slopes of m_1 and m_2, respectively.
1. If $m_1 = m_2$, then L_1 is parallel to L_2.
2. If L_1 is parallel to L_2, then $m_1 = m_2$.

In symbols we write $L_1 \| L_2$.

Any two vertical lines (undefined slope) are parallel.

EXAMPLE 7A ▶ **Determining Whether Two Lines Are Parallel**

Teladango Park has been mapped out on a rectangular coordinate system, with a ranger station at (0, 0). BJ and Kapi are at coordinates $(-24, -18)$ and have set a direct course for the pond at (11, 10). Dave and Becky are at $(-27, 1)$ and are heading straight to the lookout tower at $(-2, 21)$. Are they hiking on parallel or nonparallel courses?

Solution ▶ To respond, we compute the slope of each trek across the park.

For BJ and Kapi:

$$m = \frac{y_2 - y_1}{x_2 - x_1}$$

$$= \frac{10 - (-18)}{11 - (-24)}$$

$$= \frac{28}{35} = \frac{4}{5}$$

For Dave and Becky:

$$m = \frac{y_2 - y_1}{x_2 - x_1}$$

$$= \frac{21 - 1}{-2 - (-27)}$$

$$= \frac{20}{25} = \frac{4}{5}$$

Since the slopes are equal, the couples are hiking on parallel courses.

Two lines in the same plane that intersect at right angles are called **perpendicular lines.** Using the coordinate grid, we note that *intersect at right angles* suggests that *their slopes are negative reciprocals*. From Figure 2.19, the ratio $\frac{\text{rise}}{\text{run}}$ for L_1 is $\frac{4}{3}$, the ratio $\frac{\text{rise}}{\text{run}}$ for L_2 is $\frac{-3}{4}$. Alternatively, we can say their **slopes have a product of -1,** since $m_1 \cdot m_2 = -1$ implies $m_1 = -\frac{1}{m_2}$.

Figure 2.19

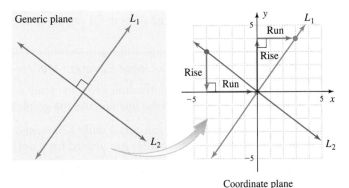

Generic plane

Coordinate plane

WORTHY OF NOTE

Since $m_1 \cdot m_2 = -1$ implies $m_1 = -\frac{1}{m_2}$, we can easily find the slope of a line perpendicular to a second line whose slope is given—just find the reciprocal and make it negative. For $m_1 = -\frac{3}{7}$, $m_2 = \frac{7}{3}$, and for $m_1 = -5$, $m_2 = \frac{1}{5}$.

Perpendicular Lines

Given L_1 and L_2 are distinct, nonvertical lines with slopes of m_1 and m_2, respectively.
1. If $m_1 \cdot m_2 = -1$, then L_1 is perpendicular to L_2.
2. If L_1 is perpendicular to L_2, then $m_1 \cdot m_2 = -1$.

In symbols we write $L_1 \perp L_2$.

Any vertical line (undefined slope) is perpendicular to any horizontal line (slope $m = 0$).

EXAMPLE 7B ▶ Determining Whether Two Lines Are Perpendicular

The three points $P_1 = (5, 1)$, $P_2 = (3, -2)$, and $P_3 = (-3, 2)$ form the vertices of a triangle. Use these points to draw the triangle, then use the slope formula to determine if they form a *right* triangle.

Solution ▶ For a right triangle to be formed, two of the lines through these points must be perpendicular (forming a right angle). From Figure 2.20, it *appears* a right triangle is formed, but we must *verify* that two of the sides are perpendicular. Using the slope formula, we have:

Figure 2.20

For P_1 and P_2

$$m_1 = \frac{-2 - 1}{3 - 5}$$

$$= \frac{-3}{-2} = \frac{3}{2}$$

For P_1 and P_3

$$m_2 = \frac{2 - 1}{-3 - 5}$$

$$= \frac{1}{-8}$$

For P_2 and P_3

$$m_3 = \frac{2 - (-2)}{-3 - 3}$$

$$= \frac{4}{-6} = \frac{2}{-3}$$

Since $m_1 \cdot m_3 = -1$, the triangle has a right angle and must be a right triangle.

☑ **D.** You've just learned how to identify parallel and perpendicular lines

Now try Exercises 57 through 68 ▶

E. Applications of Linear Equations

The graph of a linear equation can be used to help solve many applied problems. If the numbers you're working with are either very small or very large, **scale the axes** appropriately. This can be done by letting each tic mark represent a smaller or larger unit so the data points given will fit on the grid. Also, many applications use only nonnegative values and although points with negative coordinates may be used to graph a line, only ordered pairs in QI can be meaningfully interpreted.

EXAMPLE 8 ▶ Applying a Linear Equation Model—Commission Sales

Use the information given to create a linear equation model in two variables, then graph the line and use the graph to answer the question:

A salesperson gets a daily $20 meal allowance plus $7.50 for every item she sells. How many sales are needed for a daily income of $125?

Solution ▶ Let x represent sales and y represent income. This gives

verbal model: Daily income (y) equals $7.5 per sale ($x$) + $20 for meals

equation model: $y = 7.5x + 20$

Using $x = 0$ and $x = 10$, we find $(0, 20)$ and $(10, 95)$ are points on this graph. From the graph, we estimate that 14 sales are needed to generate a daily income of $125.00. Substituting $x = 14$ into the equation verifies that $(14, 125)$ is indeed on the graph:

E. You've just learned how to apply linear equations in context

$$y = 7.5x + 20$$
$$= 7.5(14) + 20$$
$$= 105 + 20$$
$$= 125 \checkmark$$

Now try Exercises 71 through 74 ▶

TECHNOLOGY HIGHLIGHT

Linear Equations, Window Size, and Friendly Windows

To graph linear equations on the TI-84 Plus, we (1) solve the equation for the variable y, (2) enter the equation on the $\boxed{\text{Y=}}$ screen, and (3) $\boxed{\text{GRAPH}}$ the equation and adjust the $\boxed{\text{WINDOW}}$ if necessary.

1. Solve the equation for y.

   For the equation $2x - 3y = -3$, we have

   $2x - 3y = -3$ given equation

   $-3y = -2x - 3$ subtract $2x$ from each side

   $y = \dfrac{2}{3}x + 1$ divide both sides by -3

2. Enter the equation on the $\boxed{\text{Y=}}$ screen.

   On the $\boxed{\text{Y=}}$ screen, enter $\frac{2}{3}x + 1$. Note that for some calculators parentheses are needed to group $(2 \div 3)x$, to prevent the calculator from interpreting this term as $2 \div (3x)$.

3. $\boxed{\text{GRAPH}}$ the equation, adjust the $\boxed{\text{WINDOW}}$.

   Since much of our work is centered at $(0, 0)$ on the coordinate grid, the calculator's default settings have a domain of $x \in [-10, 10]$ and a range of $y \in [-10, 10]$, as shown in Figure 2.21. This is referred to as the $\boxed{\text{WINDOW}}$ size. To graph the line in this window, it is easiest to use the $\boxed{\text{ZOOM}}$ key and select **6:ZStandard,** which resets the window to these default settings. The graph is shown in Figure 2.22. The Xscl and Yscl

Figure 2.21

Figure 2.22

entries give the scale used on each axis, indicating that each "tic mark" represents 1 unit. Graphing calculators have many features that enable us to find ordered pairs on a line. One is the ($\boxed{\text{2nd}}$ $\boxed{\text{GRAPH}}$) **(TABLE)** feature we have seen previously. We can also use the calculator's $\boxed{\text{TRACE}}$ feature. As the name implies, this feature enables us to trace along the line by moving a blinking cursor using the left $\boxed{\blacktriangleleft}$ and right $\boxed{\blacktriangleright}$ arrow keys. The calculator simultaneously displays the coordinates of the current location of the cursor. After pressing the $\boxed{\text{TRACE}}$ button, the cursor appears automatically—usually at the y-intercept. Moving the cursor left and right, note the coordinates changing at the bottom of the screen. The point $(3.4042553, 3.2695035)$ is on the line and satisfies the equation of the line. The calculator is displaying decimal values because the screen is exactly 95 pixels wide, 47 pixels to the left of the y-axis, and 47 pixels to the right. This means that each time you press the left or right arrow, the x-value changes by 1/47—which is *not* a nice round number. To $\boxed{\text{TRACE}}$ through "friendlier" values, we can use the $\boxed{\text{ZOOM}}$ **4:ZDecimal** feature, which sets Xmin $= -4.7$ and Xmax $= 4.7$, or **8:Zinteger,** which sets Xmin $= -47$ and Xmax $= 47$. Press $\boxed{\text{ZOOM}}$ **4:ZDecimal** and the calculator will automatically regraph the line. Now when you $\boxed{\text{TRACE}}$ the line, "friendly" decimal values are displayed.

Exercise 1:   Use the $\boxed{\text{ZOOM}}$ **4:ZDecimal** and **TRACE** features to identify the x- and y-intercepts for $Y_1 = \frac{2}{3}x + 1$.

Exercise 2:   Use the $\boxed{\text{ZOOM}}$ **8:Zinteger** and **TRACE** features to graph the line $79x - 55y = 869$, then identify the x- and y-intercepts.

2.2 EXERCISES

▶ **CONCEPTS AND VOCABULARY**

Fill in each blank with the appropriate word or phrase. Carefully reread the section if needed.

1. To find the x-intercept of a line, substitute _____ for y and solve for x. To find the y-intercept, substitute _____ for x and solve for y.

2. The slope formula is $m =$ _____ = _____, and indicates a rate of change between the x- and y-variables.

3. If $m < 0$, the slope of the line is _____ and the line slopes _____ from left to right.

4. The slope of a horizontal line is _____, the slope of a vertical line is _____, and the slopes of two parallel lines are _____.

5. Discuss/Explain If $m_1 = 2.1$ and $m_2 = 2.01$, will the lines intersect? If $m_1 = \frac{2}{3}$ and $m_2 = -\frac{2}{3}$, are the lines perpendicular?

6. Discuss/Explain the relationship between the slope formula, the Pythagorean theorem, and the distance formula. Include several illustrations.

▶ **DEVELOPING YOUR SKILLS**

Create a table of values for each equation and sketch the graph.

7. $2x + 3y = 6$

x	y

8. $-3x + 5y = 10$

x	y

9. $y = \frac{3}{2}x + 4$

x	y

10. $y = \frac{5}{3}x - 3$

x	y

11. If you completed Exercise 9, verify that $(-3, -0.5)$ and $(\frac{1}{2}, \frac{19}{4})$ also satisfy the equation given. Do these points appear to be on the graph you sketched?

12. If you completed Exercise 10, verify that $(-1.5, -5.5)$ and $(\frac{11}{2}, \frac{37}{6})$ also satisfy the equation given. Do these points appear to be on the graph you sketched?

Graph the following equations using the intercept method. Plot a third point as a check.

13. $3x + y = 6$

14. $-2x + y = 12$

15. $5y - x = 5$

16. $-4y + x = 8$

17. $-5x + 2y = 6$

18. $3y + 4x = 9$

19. $2x - 5y = 4$

20. $-6x + 4y = 8$

21. $2x + 3y = -12$

22. $-3x - 2y = 6$

23. $y = -\frac{1}{2}x$

24. $y = \frac{2}{3}x$

25. $y - 25 = 50x$

26. $y + 30 = 60x$

27. $y = -\frac{2}{5}x - 2$

28. $y = \frac{3}{4}x + 2$

29. $2y - 3x = 0$

30. $y + 3x = 0$

31. $3y + 4x = 12$

32. $-2x + 5y = 8$

Compute the slope of the line through the given points, then graph the line and use $m = \frac{\Delta y}{\Delta x}$ to find two additional points on the line. Answers may vary.

33. $(3, 5), (4, 6)$

34. $(-2, 3), (5, 8)$

35. $(10, 3), (4, -5)$

36. $(-3, -1), (0, 7)$

37. $(1, -8), (-3, 7)$

38. $(-5, 5), (0, -5)$

39. $(-3, 6), (4, 2)$

40. $(-2, -4), (-3, -1)$

41. The graph shown models the relationship between the cost of a new home and the size of the home in square feet. (a) Determine the slope of the line and

interpret what the slope ratio means in this context and (b) estimate the cost of a 3000 ft² home.

Exercise 41

Exercise 42

42. The graph shown models the relationship between the volume of garbage that is dumped in a landfill and the number of commercial garbage trucks that enter the site. (a) Determine the slope of the line and interpret what the slope ratio means in this context and (b) estimate the number of trucks entering the site daily if 1000 m³ of garbage is dumped per day.

43. The graph shown models the relationship between the distance of an aircraft carrier from its home port and the number of hours since departure. (a) Determine the slope of the line and interpret what the slope ratio means in this context and (b) estimate the distance from port after 8.25 hours.

Exercise 43

Exercise 44

44. The graph shown models the relationship between the number of circuit boards that have been assembled at a factory and the number of hours since starting time. (a) Determine the slope of the line and interpret what the slope ratio means in this context and (b) estimate how many hours the factory has been running if 225 circuit boards have been assembled.

45. **Height and weight:** While there are many exceptions, numerous studies have shown a close relationship between an average height and average weight. Suppose a person 70 in. tall weighs 165 lb, while a person 64 in. tall weighs 142 lb. Assuming the relationship is linear, (a) find the slope of the line and discuss its meaning in this context and (b) determine how many pounds are added for each inch of height.

46. **Rate of climb:** Shortly after takeoff, a plane increases altitude at a constant (linear) rate. In 5 min the altitude is 10,000 feet. Fifteen minutes after takeoff, the plane has reached its cruising altitude of 32,000 ft. (a) Find the slope of the line and discuss its meaning in this context and (b) determine how long it takes the plane to climb from 12,200 feet to 25,400 feet.

47. **Sewer line slope:** Fascinated at how quickly the plumber was working, Ryan watched with great interest as the new sewer line was laid from the house to the main line, a distance of 48 ft. At the edge of the house, the sewer line was six in. under ground. If the plumber tied in to the main line at a depth of 18 in., what is the slope of the (sewer) line? What does this slope indicate?

48. **Slope (pitch) of a roof:** A contractor goes to a lumber yard to purchase some trusses (the triangular frames) for the roof of a house. Many sizes are available, so the contractor takes some measurements to ensure the roof will have the desired slope. In one case, the height of the truss (base to ridge) was 4 ft, with a width of 24 ft (eave to eave). Find the slope of the roof if these trusses are used. What does this slope indicate?

Graph each line using two or three ordered pairs that satisfy the equation.

49. $x = -3$ 50. $y = 4$

51. $x = 2$ 52. $y = -2$

Write the equation for each line L_1 and L_2 shown. Specifically state their point of intersection.

53.   54.

55. The table given shows the total number of justices j sitting on the Supreme Court of the United States for selected time periods t (in decades), along with the number of nonmale, nonwhite justices n for the same years. (a) Use the data to graph the linear relationship between t and j, then determine the slope of the line and discuss its meaning in this context. (b) Use the data to graph the linear relationship between t and n, then determine the slope of the line and discuss its meaning.

Exercise 55

Time t (1960 → 0)	Justices j	Nonwhite, nonmale n
0	9	0
10	9	1
20	9	2
30	9	3
40	9	4
50	9	5 (est)

56. The table shown gives the boiling temperature t of water as related to the altitude h. Use the data to graph the linear relationship between h and t, then determine the slope of the line and discuss its meaning in this context.

Exercise 56

Altitude h (ft)	Boiling Temperature t (°F)
0	212.0
1000	210.2
2000	208.4
3000	206.6
4000	204.8
5000	203.0
6000	201.2

Two points on L_1 and two points on L_2 are given. Use the slope formula to determine if lines L_1 and L_2 are parallel, perpendicular, or neither.

57. L_1: $(-2, 0)$ and $(0, 6)$
   L_2: $(1, 8)$ and $(0, 5)$

58. L_1: $(1, 10)$ and $(-1, 7)$
   L_2: $(0, 3)$ and $(1, 5)$

59. L_1: $(-3, -4)$ and $(0, 1)$
   L_2: $(0, 0)$ and $(-4, 4)$

60. L_1: $(6, 2)$ and $(8, -2)$
   L_2: $(5, 1)$ and $(3, 0)$

61. L_1: $(6, 3)$ and $(8, 7)$
   L_2: $(7, 2)$ and $(6, 0)$

62. L_1: $(-5, -1)$ and $(4, 4)$
   L_2: $(4, -7)$ and $(8, 10)$

In Exercises 63 to 68, three points that form the vertices of a triangle are given. Use the points to draw the triangle, then use the slope formula to determine if any of the triangles are right triangles. Also see Exercises 39–44 in Section 2.1.

63. $(5, 2)$, $(0, -3)$, $(4, -4)$

64. $(7, 0)$, $(-1, 0)$, $(7, 4)$

65. $(-4, 3)$, $(-7, -1)$, $(3, -2)$

66. $(-3, 7)$, $(2, 2)$, $(5, 5)$

67. $(-3, 2)$, $(-1, 5)$, $(-6, 4)$

68. $(0, 0)$, $(-5, 2)$, $(2, -5)$

▶ **WORKING WITH FORMULAS**

69. Human life expectancy: $L = 0.11T + 74.2$

The average number of years that human beings live has been steadily increasing over the years due to better living conditions and improved medical care. This relationship is modeled by the formula shown, where L is the average life expectancy and T is number of years since 1980. (a) What was the life expectancy in the year 2000? (b) In what year will average life expectancy reach 77.5 yr?

70. Interest earnings: $I = \left(\dfrac{7}{100}\right)(5000)T$

If $5000 dollars is invested in an account paying 7% simple interest, the amount of interest earned is given by the formula shown, where I is the interest and T is the time in years. (a) How much interest is earned in 5 yr? (b) How much is earned in 10 yr? (c) Use the two points (5 yr, interest) and (10 yr, interest) to calculate the slope of this line. What do you notice?

▶ **APPLICATIONS**

For exercises 71 to 74, use the information given to build a linear equation model, then use the equation to respond.

71. Business depreciation: A business purchases a copier for $8500 and anticipates it will depreciate in value $1250 per year.

   a. What is the copier's value after 4 yr of use?

   b. How many years will it take for this copier's value to decrease to $2250?

72. Baseball card value: After purchasing an autographed baseball card for $85, its value increases by $1.50 per year.

   a. What is the card's value 7 yr after purchase?

   b. How many years will it take for this card's value to reach $100?

73. Water level: During a long drought, the water level in a local lake decreased at a rate of 3 in. per month. The water level before the drought was 300 in.

a. What was the water level after 9 months of drought?

b. How many months will it take for the water level to decrease to 20 ft?

74. Gas mileage: When empty, a large dump-truck gets about 15 mi per gallon. It is estimated that for each 3 tons of cargo it hauls, gas mileage decreases by $\frac{3}{4}$ mi per gallon.

a. If 10 tons of cargo is being carried, what is the truck's mileage?

b. If the truck's mileage is down to 10 mi per gallon, how much weight is it carrying?

75. Parallel/nonparallel roads: Aberville is 38 mi north and 12 mi west of Boschertown, with a straight road "farm and machinery road" (FM 1960) connecting the two cities. In the next county, Crownsburg is 30 mi north and 9.5 mi west of Dower, and these cities are likewise connected by a straight road (FM 830). If the two roads continued indefinitely in both directions, would they intersect at some point?

76. Perpendicular/nonperpendicular course headings: Two shrimp trawlers depart Charleston Harbor at the same time. One heads for the shrimping grounds located 12 mi north and 3 mi east of the harbor. The other heads for a point 2 mi south and 8 mi east of the harbor. Assuming the harbor is at (0, 0), are the routes of the trawlers perpendicular? If so, how far apart are the boats when they reach their destinations (to the nearest one-tenth mi)?

77. Cost of college: For the years 1980 to 2000, the cost of tuition and fees per semester (in constant dollars) at a public 4-yr college can be approximated by the equation $y = 144x + 621$, where y represents the cost in dollars and $x = 0$

represents the year 1980. Use the equation to find: (a) the cost of tuition and fees in 2002 and (b) the year this cost will exceed $5250.

Source: 2001 New York Times Almanac, p. 356

78. Female physicians: In 1960 only about 7% of physicians were female. Soon after, this percentage began to grow dramatically. For the years 1980 to 2002, the percentage of physicians that were female can be approximated by the equation $y = 0.72x + 11$, where y represents the percentage (as a whole number) and $x = 0$ represents the year 1980. Use the equation to find: (a) the percentage of physicians that were female in 1992 and (b) the projected year this percentage will exceed 30%.

Source: Data from the 2004 Statistical Abstract of the United States, Table 149

79. Decrease in smokers: For the years 1980 to 2002, the percentage of the U.S. adult population who were smokers can be approximated by the equation $y = -\frac{7}{15}x + 32$, where y represents the percentage of smokers (as a whole number) and $x = 0$ represents 1980. Use the equation to find: (a) the percentage of adults who smoked in the year 2000 and (b) the year the percentage of smokers is projected to fall below 20%.

Source: Statistical Abstract of the United States, various years

80. Temperature and cricket chirps: Biologists have found a strong relationship between temperature and the number of times a cricket chirps. This is modeled by the equation $T = \frac{N}{4} + 40$, where N is the number of times the cricket chirps per minute and T is the temperature in Fahrenheit. Use the equation to find: (a) the outdoor temperature if the cricket is chirping 48 times per minute and (b) the number of times a cricket chirps if the temperature is 70°.

▶ **EXTENDING THE CONCEPT**

81. If the lines $4y + 2x = -5$ and $3y + ax = -2$ are perpendicular, what is the value of a?

82. Let m_1, m_2, m_3, and m_4 be the slopes of lines L_1, L_2, L_3, and L_4, respectively. Which of the following statements is true?

a. $m_4 < m_1 < m_3 < m_2$

b. $m_3 < m_2 < m_4 < m_1$

c. $m_3 < m_4 < m_2 < m_1$

d. $m_1 < m_3 < m_4 < m_2$

e. $m_1 < m_4 < m_3 < m_2$

83. An *arithmetic sequence* is a sequence of numbers where each successive term is found by adding a

fixed constant, called the common difference d, to the preceding term. For instance 3, 7, 11, 15, . . . is an arithmetic sequence with $d = 4$. The formula for the "nth term" t_n of an arithmetic sequence is a linear equation of the form $t_n = t_1 + (n - 1)d$, where d is the common difference and t_1 is the first term of the sequence. Use the equation to find the term specified for each sequence.

a. 2, 9, 16, 23, 30, . . . ; 21st term

b. 7, 4, 1, −2, −5, . . . ; 31st term

c. 5.10, 5.25, 5.40, 5.55, . . . ; 27th term

d. $\frac{3}{2}$, $\frac{9}{4}$, 3, $\frac{15}{4}$, $\frac{9}{2}$, . . . ; 17th term

▶ MAINTAINING YOUR SKILLS

84. (1.1) Simplify the equation, then solve. Check your answer by substitution:
$$3x^2 - 3 + 4x + 6 = 4x^2 - 3(x + 5)$$

85. (R.7) Identify the following formulas:

$$P = 2L + 2W \qquad V = LWH$$
$$V = \pi r^2 h \qquad C = 2\pi r$$

86. (1.1) How many gallons of a 35% brine solution must be mixed with 12 gal of a 55% brine solution in order to get a 45% solution?

87. (1.1) Two boats leave the harbor at Lahaina, Maui, going in opposite directions. One travels at 15 mph and the other at 20 mph. How long until they are 70 mi apart?

2.3 | Linear Graphs and Rates of Change

Learning Objectives

In Section 2.3 you will learn how to:

☐ **A.** Write a linear equation in slope-intercept form

☐ **B.** Use slope-intercept form to graph linear equations

☐ **C.** Write a linear equation in point-slope form

☐ **D.** Apply the slope-intercept form and point-slope form in context

The concept of slope is an important part of mathematics, because it gives us a way to measure and compare change. The value of an automobile changes with time, the circumference of a circle increases as the radius increases, and the tension in a spring grows the more it is stretched. The real world is filled with examples of how one change affects another, and slope helps us understand how these changes are related.

A. Linear Equations and Slope-Intercept Form

In Section 1.1, formulas and literal equations were written in an alternate form by solving for an object variable. The new form made using the formula more efficient. Solving for y in equations of the form $ax + by = c$ offers similar advantages to linear graphs and their applications.

EXAMPLE 1 ▶ **Solving for y in $ax + by = c$**

Solve $2y - 6x = 4$ for y, then evaluate at $x = 4$, $x = 0$, and $x = -\frac{1}{3}$.

Solution ▶
$$2y - 6x = 4 \qquad \text{given equation}$$
$$2y = 6x + 4 \qquad \text{add } 6x$$
$$y = 3x + 2 \qquad \text{divide by 2}$$

Since the coefficients are integers, evaluate the function mentally. Inputs are multiplied by 3, then increased by 2, yielding the ordered pairs (4, 14), (0, 2), and $\left(-\frac{1}{3}, 1\right)$.

Now try Exercises 7 through 12 ▶

This form of the equation (where y has been written in terms of x) enables us to quickly identify what operations are performed on x in order to obtain y. For $y = 3x + 2$, *multiply inputs by 3, then add 2.*

EXAMPLE 2 ▶ **Solving for y in $ax + by = c$**

Solve the linear equation $3y - 2x = 6$ for y, then identify the new coefficient of x and the constant term.

Solution ▶ $3y - 2x = 6$ given equation

$3y = 2x + 6$ add 2x

$y = \dfrac{2}{3}x + 2$ divide by 3

The new coefficient of x is $\frac{2}{3}$ and the constant term is 2.

Now try Exercises 13 through 18 ▶

WORTHY OF NOTE

In Example 2, the final form can be written $y = \frac{2}{3}x + 2$ as shown (inputs are multiplied by two-thirds, then increased by 2), or written as $y = \dfrac{2x}{3} + 2$ (inputs are multiplied by two, the result divided by 3 and this amount increased by 2). The two forms are equivalent.

When the coefficient of x is rational, it's helpful to select inputs that are multiples of the denominator if the context or application requires us to evaluate the equation. This enables us to perform most operations mentally. For $y = \frac{2}{3}x + 2$, possible inputs might be $x = -9, -6, 0, 3, 6$, and so on. **See Exercises 19 through 24.**

In Section 2.2, linear equations were graphed using the intercept method. When a linear equation is written with y in terms of x, we notice a powerful connection between the graph and its equation, and one that highlights the primary characteristics of a linear graph.

EXAMPLE 3 ▶ **Noting Relationships between an Equation and Its Graph**

Find the intercepts of $4x + 5y = -20$ and use them to graph the line. Then,

a. Use the intercepts to calculate the slope of the line, then

b. Write the equation with y in terms of x and compare the calculated slope and y-intercept to the equation in this form. Comment on what you notice.

Solution ▶ Substituting 0 for x in $4x + 5y = -20$, we find the y-intercept is $(0, -4)$. Substituting 0 for y gives an x-intercept of $(-5, 0)$. The graph is displayed here.

a. By calculation or counting $\dfrac{\Delta y}{\Delta x}$, the slope is $m = -\frac{4}{5}$.

b. Solving for y:

$4x + 5y = -20$ given equation

$5y = -4x - 20$ subtract 4x

$y = -\dfrac{4}{5}x - 4$ divide by 5

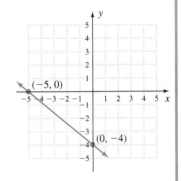

☑ **A.** You've just learned how to write a linear equation in slope-intercept form

The slope value seems to be the coefficient of x, while the y-intercept is the constant term.

Now try Exercises 25 through 30 ▶

B. Slope-Intercept Form and the Graph of a Line

After solving a linear equation for y, an input of $x = 0$ causes the "x-term" to become zero, so the y-intercept is automatically the constant term. As Example 3 illustrates, we can also identify the slope of the line—it is the coefficient of x. In general, a linear equation of the form $y = mx + b$ is said to be in **slope-intercept form,** since the slope of the line is m and the y-intercept is $(0, b)$.

Slope-Intercept Form

For a nonvertical line whose equation is $y = \boldsymbol{mx + b,}$
the slope of the line is $\boldsymbol{m}$ and the y-intercept is $(0, \boldsymbol{b}).$

EXAMPLE 4 ▶   **Finding the Slope-Intercept Form**

Write each equation in slope-intercept form and identify the slope and y-intercept of each line.

    a. $3x - 2y = 9$ **b.** $y + x = 5$ **c.** $2y = x$

Solution ▶   **a.** $3x - 2y = 9$ **b.** $y + x = 5$ **c.** $2y = x$

$$-2y = -3x + 9 \qquad y = -x + 5 \qquad y = \frac{x}{2}$$

$$y = \frac{3}{2}x - \frac{9}{2} \qquad y = -1x + 5 \qquad y = \frac{1}{2}x$$

$$m = \frac{3}{2}, b = -\frac{9}{2} \qquad m = -1, b = 5 \qquad m = \frac{1}{2}, b = 0$$

$$y\text{-intercept}\left(0, -\frac{9}{2}\right) \qquad y\text{-intercept }(0, 5) \qquad y\text{-intercept }(0, 0)$$

Now try Exercises 31 through 38 ▶

If the slope and y-intercept of a linear equation are known or can be found, we can construct its equation by substituting these values directly into the slope-intercept form $y = mx + b$.

EXAMPLE 5 ▶   **Finding the Equation of a Line from Its Graph**

Find the slope-intercept form of the line shown.

Solution ▶   Using $(-3, -2)$ and $(-1, 2)$ in the slope formula,

or by simply counting $\dfrac{\Delta y}{\Delta x}$, the slope is $m = \frac{4}{2}$ or $\frac{2}{1}$.

By inspection we see the y-intercept is $(0, 4)$.
Substituting $\frac{2}{1}$ for m and 4 for b in the slope-intercept form we obtain the equation
$y = 2x + 4$.

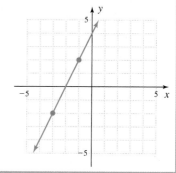

Now try Exercises 39 through 44 ▶

Actually, if the slope is known and we have *any* point (x, y) on the line, we can still construct the equation since the given point *must satisfy the equation of the line*. In this case, we're treating $y = mx + b$ as a simple formula, solving for b after substituting known values for m, x, and y.

EXAMPLE 6 ▶   **Using $y = mx + b$ as a Formula**

Find the equation of a line that has slope $m = \frac{4}{5}$ and contains $(-5, 2)$.

Solution ▶   Using $y = mx + b$ as a "formula," we have $m = \frac{4}{5}$, $x = -5$, and $y = 2$.

$$y = mx + b \qquad \text{slope-intercept form}$$
$$2 = \tfrac{4}{5}(-5) + b \qquad \text{substitute } \tfrac{4}{5} \text{ for } m, -5 \text{ for } x, \text{ and 2 for } y$$
$$2 = -4 + b \qquad \text{simplify}$$
$$6 = b \qquad \text{solve for } b$$

The equation of the line is $y = \frac{4}{5}x + 6$.

Now try Exercises 45 through 50 ▶

Writing a linear equation in slope-intercept form enables us to draw its graph with a minimum of effort, since we can easily locate the y-intercept and a second point using $m = \dfrac{\Delta y}{\Delta x}$. For instance, $\dfrac{\Delta y}{\Delta x} = \dfrac{-2}{3}$ means count down 2 and right 3 from a known point.

EXAMPLE 7 ▶ **Graphing a Line Using Slope-Intercept Form**

Write $3y - 5x = 9$ in slope-intercept form, then graph the line using the y-intercept and slope.

Solution ▶

$$3y - 5x = 9 \qquad \text{given equation}$$
$$3y = 5x + 9 \qquad \text{isolate } y \text{ term}$$
$$y = \tfrac{5}{3}x + 3 \qquad \text{divide by 3}$$

The slope is $m = \tfrac{5}{3}$ and the y-intercept is $(0, 3)$.

Plot the y-intercept, then use $\dfrac{\Delta y}{\Delta x} = \dfrac{5}{3}$ (up 5 and

right 3—shown in blue) to find another point on the line (shown in red). Finish by drawing a line through these points.

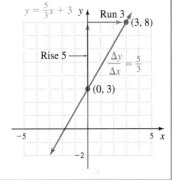

Now try Exercises 51 through 62 ▶

WORTHY OF NOTE

Noting the fraction $\tfrac{5}{3}$ is equal to $\tfrac{-5}{-3}$, we could also begin at $(0, 3)$ and count $\dfrac{\Delta y}{\Delta x} = \dfrac{-5}{-3}$ (down 5 and left 3) to find an additional point on the line: $(-3, -2)$. Also, for any negative slope $\dfrac{\Delta y}{\Delta x} = -\dfrac{a}{b}$, note $-\dfrac{a}{b} = \dfrac{-a}{b} = \dfrac{a}{-b}$.

For a discussion of what graphing method might be most efficient for a given linear equation, **see Exercises 103 and 115.**

Parallel and Perpendicular Lines

From Section 2.2 we know parallel lines have equal slopes: $m_1 = m_2$, and perpendicular lines have slopes with a product of -1: $m_1 \cdot m_2 = -1$ or $m_1 = -\dfrac{1}{m_2}$. In some applications, we need to find the equation of a second line parallel or perpendicular to a given line, through a given point. Using the slope-intercept form makes this a simple four-step process.

Finding the Equation of a Line Parallel or Perpendicular to a Given Line

1. Identify the slope m_1 of the given line.
2. Find the slope m_2 of the new line using the parallel or perpendicular relationship.
3. Use m_2 with the point (x, y) in the "formula" $y = mx + b$ and solve for b.
4. The desired equation will be $y = m_2x + b$.

EXAMPLE 8 ▶ **Finding the Equation of a Parallel Line**

Find the equation of a line that goes through $(-6, -1)$ and is parallel to $2x + 3y = 6$.

Solution ▶ Begin by writing the equation in slope-intercept form to identify the slope.

$$2x + 3y = 6 \qquad \text{given line}$$
$$3y = -2x + 6 \qquad \text{isolate } y \text{ term}$$
$$y = \tfrac{-2}{3}x + 2 \qquad \text{result}$$

The original line has slope $m_1 = \frac{-2}{3}$ and this will also be the slope of any line parallel to it. Using $m_2 = \frac{-2}{3}$ with $(x, y) \rightarrow (-6, -1)$ we have

$$y = mx + b \qquad \text{slope-intercept form}$$

$$-1 = \frac{-2}{3}(-6) + b \qquad \text{substitute } \tfrac{-2}{3} \text{ for } m, -6 \text{ for } x, \text{ and } -1 \text{ for } y$$

$$-1 = 4 + b \qquad \text{simplify}$$

$$-5 = b \qquad \text{solve for } b$$

The equation of the new line is $y = \frac{-2}{3}x - 5$.

> **Now try Exercises 63 through 76** ▶

GRAPHICAL SUPPORT

Graphing the lines from Example 8 as Y1 and Y2 on a graphing calculator, we note the lines do appear to be parallel (they actually *must* be since they have identical slopes). Using the ZOOM 8:ZInteger feature of the TI-84 Plus we can quickly verify that Y2 indeed contains the point $(-6, -1)$.

 For any nonlinear graph, a straight line drawn through two points on the graph is called a **secant line.** The slope of the secant line, and lines parallel and perpendicular to this line, play fundamental roles in the further development of the rate-of-change concept.

EXAMPLE 9 ▶ **Finding Equations for Parallel and Perpendicular Lines**

A secant line is drawn using the points $(-4, 0)$ and $(2, -2)$ on the graph of the function shown. Find the equation of a line that is:

a. parallel to the secant line through $(-1, -4)$

b. perpendicular to the secant line through $(-1, -4)$.

Solution ▶ Either by using the slope formula or counting $\frac{\Delta y}{\Delta x}$, we find the secant line has slope

$$m = \frac{-2}{6} = \frac{-1}{3}.$$

a. For the parallel line through $(-1, -4)$, $m_2 = \frac{-1}{3}$.

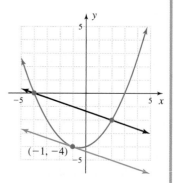

$$y = mx + b \qquad \text{slope-intercept form}$$

$$-4 = \frac{-1}{3}(-1) + b \qquad \text{substitute } \tfrac{-1}{3} \text{ for } m, \\ -1 \text{ for } x, \text{ and } -4 \text{ for } y$$

$$-\frac{12}{3} = \frac{1}{3} + b \qquad \text{simplify}$$

$$-\frac{13}{3} = b \qquad \text{result}$$

The equation of the parallel line (in blue) is $y = \frac{-1}{3}x - \frac{13}{3}$.

WORTHY OF NOTE

The word "secant" comes from the Latin word *secare*, meaning "to cut." Hence a secant line is one that cuts through a graph, as opposed to a tangent line, which touches the graph at only one point.

b. For the line perpendicular through $(-1, -4)$, $m_2 = 3$.

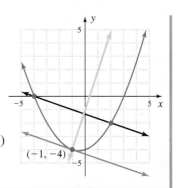

$$y = mx + b \qquad \text{slope-intercept form}$$
$$-4 = 3(-1) + b \qquad \text{substitute 3 for } m, -1 \text{ for } x, \text{ and } -4 \text{ for } y$$
$$-4 = -3 + b \qquad \text{simplify}$$
$$-1 = b \qquad \text{result}$$

The equation of the perpendicular line (in yellow) is $y = 3x - 1$.

☑ **B. You've just learned how to use the slope-intercept form to graph linear equations**

Now try Exercises 77 through 82 ▶

C. Linear Equations in Point-Slope Form

As an alternative to using $y = mx + b$, we can find the equation of the line using the slope formula $\dfrac{y_2 - y_1}{x_2 - x_1} = m$, and the fact that *the slope of a line is constant*. For a given slope m, we can let (x_1, y_1) represent a *given* point on the line and (x, y) represent *any other point* on the line, and the formula becomes $\dfrac{y - y_1}{x - x_1} = m$. Isolating the "y" terms on one side gives a new form for the equation of a line, called the **point-slope form:**

$$\frac{y - y_1}{x - x_1} = m \qquad \text{slope formula}$$

$$\frac{(x - x_1)}{1}\left(\frac{y - y_1}{x - x_1}\right) = m(x - x_1) \quad \text{multiply both sides by } (x - x_1)$$

$$y - y_1 = m(x - x_1) \quad \text{simplify} \rightarrow \text{point-slope form}$$

The Point-Slope Form of a Linear Equation

For a nonvertical line whose equation is $y - y_1 = m(x - x_1)$, the slope of the line is m and (x_1, y_1) is a point on the line.

While using $y = mx + b$ as in Example 6 may appear to be easier, both the y-intercept form and point-slope form have their own advantages and it will help to be familiar with both.

EXAMPLE 10 ▶ **Using $y - y_1 = m(x - x_1)$ as a Formula**

Find the equation of a line in point-slope form, if $m = \frac{2}{3}$ and $(-3, -3)$ is on the line. Then graph the line.

Solution ▶

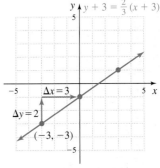

$$y - y_1 = m(x - x_1) \qquad \text{point-slope form}$$

$$y - (-3) = \frac{2}{3}[x - (-3)] \quad \substack{\text{substitute } \frac{2}{3} \text{ for } m; (-3, -3) \\ \text{for } (x_1, y_1)}$$

$$y + 3 = \frac{2}{3}(x + 3) \qquad \text{simplify, point-slope form}$$

☑ **C. You've just learned how to write a linear equation in point-slope form**

To graph the line, plot $(-3, -3)$ and use $\dfrac{\Delta y}{\Delta x} = \dfrac{2}{3}$ to find additional points on the line.

Now try Exercises 83 through 94 ▶

D. Applications of Linear Equations

As a mathematical tool, linear equations rank among the most common, powerful, and versatile. In all cases, it's important to remember that slope represents a *rate of change*. The notation $m = \dfrac{\Delta y}{\Delta x}$ literally means the quantity measured along the *y*-axis, is changing with respect to changes in the quantity measured along the *x*-axis.

EXAMPLE 11 ▶   **Relating Temperature to Altitude**

In meteorological studies, atmospheric temperature depends on the altitude according to the formula $T = -3.5h + 58.6$, where T represents the approximate Fahrenheit temperature at height h (in thousands of feet).

a. Interpret the meaning of the slope in this context.

b. Determine the temperature at an altitude of 12,000 ft.

c. If the temperature is $-10°$F what is the approximate altitude?

Solution ▶   **a.** Notice that h is the input variable and T is the output. This shows $\dfrac{\Delta T}{\Delta h} = \dfrac{-3.5}{1}$, meaning the temperature drops 3.5°F for every 1000-ft increase in altitude.

b. Since height is in thousands, use $h = 12$.

$$T = -3.5h + 58.6 \qquad \text{original function}$$
$$= -3.5(12) + 58.6 \qquad \text{substitute 12 for } h$$
$$= 16.6 \qquad \text{result}$$

At a height of 12,000 ft, the temperature is about 17°F.

c. Replacing T with -10 and solving gives

$$-10 = -3.5h + 58.6 \qquad \text{substitute } -10 \text{ for } T$$
$$-68.6 = -3.5h \qquad \text{simplify}$$
$$19.6 = h \qquad \text{result}$$

The temperature is $-10°$F at a height of $19.6 \times 1000 = 19{,}600$ ft.

Now try Exercises 105 and 106 ▶

In some applications, the relationship is known to be linear but only a few points on the line are given. In this case, we can use two of the known data points to calculate the slope, then the point-slope form to find an equation model. One such application is *linear depreciation*, as when a government allows businesses to depreciate vehicles and equipment over time (the less a piece of equipment is worth, the less you pay in taxes).

EXAMPLE 12A ▶   **Using Point-Slope Form to Find an Equation Model**

Five years after purchase, the auditor of a newspaper company estimates the value of their printing press is $60,000. Eight years after its purchase, the value of the press had depreciated to $42,000. Find a linear equation that models this depreciation and discuss the slope and *y*-intercept in context.

Solution ▶ Since the value of the press depends on time, the ordered pairs have the form (time, value) or (t, v) where *time* is the input, and *value* is the output. This means the ordered pairs are (5, 60,000) and (8, 42,000).

$$m = \frac{v_2 - v_1}{t_2 - t_1} \qquad \text{slope formula}$$

$$= \frac{42,000 - 60,000}{8 - 5} \qquad (t_1, v_1) = (5, 60,000); \ (t_2, v_2) = (8, 42,000)$$

$$= \frac{-18,000}{3} = \frac{-6000}{1} \qquad \text{simplify and reduce}$$

WORTHY OF NOTE

Actually, it doesn't matter which of the two points are used in Example 12A. Once the point (5, 60,000) is plotted, a constant slope of $m = -6000$ will "drive" the line through (8, 42,000). If we first graph (8, 42,000), the same slope would "drive" the line through (5, 60,000). Convince yourself by reworking the problem using the other point.

The slope of the line is $\dfrac{\Delta \text{value}}{\Delta \text{time}} = \dfrac{-6000}{1}$, indicating the printing press loses $6000 in value with each passing year.

$$v - v_1 = m(t - t_1) \qquad \text{point-slope form}$$

$$v - 60,000 = -6000(t - 5) \qquad \text{substitute } -6000 \text{ for } m; \ (5, 60,000) \text{ for } (t_1, v_1)$$

$$v - 60,000 = -6000t + 30,000 \quad \text{simplify}$$

$$v = -6000t + 90,000 \quad \text{solve for } v$$

The depreciation equation is $v = -6000t + 90,000$. The v-intercept (0, 90,000) indicates the original value (cost) of the equipment was $90,000.

Once the depreciation equation is found, it represents the (time, value) relationship for all future (and intermediate) ages of the press. In other words, we can now predict the value of the press for any given year. However, note that some equation models are valid for only a set period of time, and each model should be used with care.

EXAMPLE 12B ▶ **Using an Equation Model to Gather Information**

From Example 12A,
   a. How much will the press be worth after 11 yr?
   b. How many years until the value of the equipment is less than $9,000?
   c. Is this equation model valid for $t = 18$ yr (why or why not)?

Solution ▶ **a.** Find the value v when $t = 11$:

$$v = -6000t + 90,000 \qquad \text{equation model}$$

$$v = -6000(11) + 90,000 \quad \text{substitute 11 for } t$$

$$= 24,000 \qquad\qquad \text{result (11, 24,000)}$$

After 11 yr, the printing press will only be worth $24,000.
   b. ". . . value is less than $9000" means $v < 9000$:

$$v < 9000 \qquad \text{value at time } t$$

$$-6000t + 90,000 < 9000 \qquad \text{substitute } -6000t + 90,000 \text{ for } v$$

$$-6000t < -81,000 \quad \text{subtract 90,000}$$

$$t > 13.5 \qquad\qquad \text{divide by } -6000, \textbf{ reverse inequality symbol}$$

After 13.5 yr, the printing press will be worth less than $9000.

☑ **D. You've just learned how to apply the slope-intercept form and point-slope form in context**

   c. Since substituting 18 for t gives a negative quantity, the equation model is not valid for $t = 18$. In the current context, the model is only valid while $v \geq 0$ and we note the domain of the function is $t \in [0, 15]$.

Now try Exercises 107 through 112 ▶

2.3 EXERCISES

▶ **CONCEPTS AND VOCABULARY**

Fill in each blank with the appropriate word or phrase. Carefully reread the section if needed.

1. For the equation $y = -\frac{7}{4}x + 3$, the slope

   is_____ and the y-intercept is _____.

2. The notation $\frac{\Delta \text{cost}}{\Delta \text{time}}$ indicates the _____ is

   changing in response to changes in _____.

3. Line 1 has a slope of -0.4. The slope of any line perpendicular to line 1 is _____.

4. The equation $y - y_1 = m(x - x_1)$ is called the _____ form of a line.

5. Discuss/Explain how to graph a line using only the slope and a point on the line (no equations).

6. Given $m = -\frac{3}{5}$ and $(-5, 6)$ is on the line. Compare and contrast finding the equation of the line using $y = mx + b$ versus $y - y_1 = m(x - x_1)$.

▶ **DEVELOPING YOUR SKILLS**

Solve each equation for y and evaluate the result using $x = -5, x = -2, x = 0, x = 1,$ and $x = 3$.

7. $4x + 5y = 10$
8. $3y - 2x = 9$
9. $-0.4x + 0.2y = 1.4$
10. $-0.2x + 0.7y = -2.1$
11. $\frac{1}{3}x + \frac{1}{5}y = -1$
12. $\frac{1}{7}y - \frac{1}{3}x = 2$

For each equation, solve for y and identify the new coefficient of x and new constant term.

13. $6x - 3y = 9$
14. $9y - 4x = 18$
15. $-0.5x - 0.3y = 2.1$
16. $-0.7x + 0.6y = -2.4$
17. $\frac{5}{6}x + \frac{1}{7}y = -\frac{4}{7}$
18. $\frac{7}{12}y - \frac{4}{15}x = \frac{7}{6}$

Evaluate each equation by selecting three inputs that will result in integer values. Then graph each line.

19. $y = -\frac{4}{3}x + 5$
20. $y = \frac{5}{4}x + 1$
21. $y = -\frac{3}{2}x - 2$
22. $y = \frac{2}{5}x - 3$
23. $y = -\frac{1}{6}x + 4$
24. $y = -\frac{1}{3}x + 3$

Find the x- and y-intercepts for each line, then (a) use these two points to calculate the slope of the line, (b) write the equation with y in terms of x (solve for y) and compare the calculated slope and y-intercept to the equation from part (b). Comment on what you notice.

25. $3x + 4y = 12$
26. $3y - 2x = -6$
27. $2x - 5y = 10$
28. $2x + 3y = 9$
29. $4x - 5y = -15$
30. $5y + 6x = -25$

Write each equation in slope-intercept form (solve for y), then identify the slope and y-intercept.

31. $2x + 3y = 6$
32. $4y - 3x = 12$
33. $5x + 4y = 20$
34. $y + 2x = 4$
35. $x = 3y$
36. $2x = -5y$
37. $3x + 4y - 12 = 0$
38. $5y - 3x + 20 = 0$

For Exercises 39 to 50, use the slope-intercept form to state the equation of each line.

39.

40.

41.

42. $m = -2$; y-intercept $(0, -3)$
43. $m = 3$; y-intercept $(0, 2)$
44. $m = -\frac{3}{2}$; y-intercept $(0, -4)$

45.   **46.**

47.

48. $m = -4$; $(-3, 2)$ is on the line

49. $m = 2$; $(5, -3)$ is on the line

50. $m = -\frac{3}{2}$; $(-4, 7)$ is on the line

Write each equation in slope-intercept form, then use the slope and intercept to graph the line.

51. $3x + 5y = 20$ **52.** $2y - x = 4$

53. $2x - 3y = 15$ **54.** $-3x + 2y = 4$

Graph each linear equation using the y-intercept and slope determined from each equation.

55. $y = \frac{2}{3}x + 3$ **56.** $y = \frac{5}{2}x - 1$

57. $y = \frac{-1}{3}x + 2$ **58.** $y = \frac{-4}{5}x + 2$

59. $y = 2x - 5$ **60.** $y = -3x + 4$

61. $y = \frac{1}{2}x - 3$ **62.** $y = \frac{-3}{2}x + 2$

Find the equation of the line using the information given. Write answers in slope-intercept form.

63. parallel to $2x - 5y = 10$, through the point $(-5, 2)$

64. parallel to $6x + 9y = 27$, through the point $(-3, -5)$

65. perpendicular to $5y - 3x = 9$, through the point $(6, -3)$

66. perpendicular to $x - 4y = 7$, through the point $(-5, 3)$

67. parallel to $12x + 5y = 65$, through the point $(-2, -1)$

68. parallel to $15y - 8x = 50$, through the point $(3, -4)$

69. parallel to $y = -3$, through the point $(2, 5)$

70. perpendicular to $y = -3$ through the point $(2, 5)$

Write the lines in slope-intercept form and state whether they are parallel, perpendicular, or neither.

71. $4y - 5x = 8$ **72.** $3y - 2x = 6$
    $5y + 4x = -15$ $-2x + 3y = -3$

73. $2x - 5y = 20$ **74.** $5y = 11x + 135$
    $4x - 3y = 18$ $11y + 5x = -77$

75. $-4x + 6y = 12$ **76.** $3x + 4y = 12$
    $2x + 3y = 6$ $6x + 8y = 2$

A *secant line* is one that intersects a graph at two or more points. For each graph given, find the equation of the line (a) parallel and (b) perpendicular to the secant line, through the point indicated.

77.   **78.**

79.   **80.**

81.   **82.**

Find the equation of the line in point-slope form, then graph the line.

83. $m = 2$; $P_1 = (2, -5)$

84. $m = -1$; $P_1 = (2, -3)$

85. $P_1 = (3, -4)$, $P_2 = (11, -1)$

86. $P_1 = (-1, 6)$, $P_2 = (5, 1)$

87. $m = 0.5$; $P_1 = (1.8, -3.1)$

88. $m = 1.5$; $P_1 = (-0.75, -0.125)$

Find the equation of the line in point-slope form, and state the meaning of the slope in context—what information is the slope giving us?

89.

90.

91.

92.

93.

94.

Using the concept of slope, match each description with the graph that best illustrates it. Assume time is scaled on the horizontal axes, and height, speed, or distance from the origin (as the case may be) is scaled on the vertical axis.

95. While driving today, I got stopped by a state trooper. After she warned me to slow down, I continued on my way.

96. After hitting the ball, I began trotting around the bases shouting, "Ooh, ooh, ooh!" When I saw it wasn't a home run, I began sprinting.

97. At first I ran at a steady pace, then I got tired and walked the rest of the way.

98. While on my daily walk, I had to run for a while when I was chased by a stray dog.

99. I climbed up a tree, then I jumped out.

100. I steadily swam laps at the pool yesterday.

101. I walked toward the candy machine, stared at it for a while then changed my mind and walked back.

102. For practice, the girls' track team did a series of 25-m sprints, with a brief rest in between.

▶ **WORKING WITH FORMULAS**

103. General linear equation: $ax + by = c$

The general equation of a line is shown here, where a, b, and c are real numbers, with a and b not simultaneously zero. Solve the equation for y and note the slope (coefficient of x) and y-intercept (constant term). Use these to find the slope and y-intercept of the following lines, without solving for y or computing points.

   a. $3x + 4y = 8$ **b.** $2x + 5y = -15$

   c. $5x - 6y = -12$ **d.** $3y - 5x = 9$

104. Intercept/Intercept form of a linear equation: $\dfrac{x}{h} + \dfrac{y}{k} = 1$

The x- and y-intercepts of a line can also be found by writing the equation in the form shown (with the equation set equal to 1). The x-intercept will be $(h, 0)$ and the y-intercept will be $(0, k)$. Find the x- and y-intercepts of the following lines using this method: (a) $2x + 5y = 10$, (b) $3x - 4y = -12$, and (c) $5x + 4y = 8$. How is the slope of each line related to the values of h and k?

▶ **APPLICATIONS**

105. Speed of sound: The speed of sound as it travels through the air depends on the temperature of the air according to the function $V = \frac{3}{5}C + 331$, where V represents the velocity of the sound waves in meters per second (m/s), at a temperature of $C°$ Celsius.

   a. Interpret the meaning of the slope and y-intercept in this context.

   b. Determine the speed of sound at a temperature of 20°C.

   c. If the speed of sound is measured at 361 m/s, what is the temperature of the air?

106. Acceleration: A driver going down a straight highway is traveling 60 ft/sec (about 41 mph) on cruise control, when he begins accelerating at a rate of 5.2 ft/sec^2. The final velocity of the car is given by $V = \frac{26}{5}t + 60$, where V is the velocity at time t. (a) Interpret the meaning of the slope and y-intercept in this context. (b) Determine the velocity of the car after 9.4 seconds. (c) If the car is traveling at 100 ft/sec, for how long did it accelerate?

107. Investing in coins: The purchase of a "collector's item" is often made in hopes the item will increase in value. In 1998, Mark purchased a 1909-S VDB Lincoln Cent (in fair condition) for $150. By the year 2004, its value had grown to $190. (a) Use the relation (time since purchase, value) with $t = 0$ corresponding to 1998 to find a linear equation modeling the value of the coin. (b) Discuss what the slope and y-intercept indicate in this context. (c) How much will the penny be worth in 2009? (d) How many years after purchase will the penny's value exceed $250? (e) If the penny is now worth $170, how many years has Mark owned the penny?

108. Depreciation: Once a piece of equipment is put into service, its value begins to depreciate. A business purchases some computer equipment for $18,500. At the end of a 2-yr period, the value of the equipment has decreased to $11,500. (a) Use the relation (time since purchase, value) to find a linear equation modeling the value of the equipment. (b) Discuss what the slope and y-intercept indicate in this context. (c) What is the equipment's value after 4 yr? (d) How many years after purchase will the value decrease to $6000? (e) Generally, companies will sell used equipment while it still has value and use the funds to purchase new equipment. According to the function, how many years will it take this equipment to depreciate in value to $1000?

109. Internet connections: The number of households that are hooked up to the Internet (homes that are online) has been increasing steadily in recent years. In 1995, approximately 9 million homes were online. By 2001 this figure had climbed to about 51 million. (a) Use the relation (year, homes online) with $t = 0$ corresponding to 1995 to find an

equation model for the number of homes online. (b) Discuss what the slope indicates in this context. (c) According to this model, in what year did the first homes begin to come online? (d) If the rate of change stays constant, how many households will be on the Internet in 2006? (e) How many years after 1995 will there be over 100 million households connected? (f) If there are 115 million households connected, what year is it?

Source: 2004 Statistical Abstract of the United States, Table 965

110. Prescription drugs: Retail sales of prescription drugs have been increasing steadily in recent years. In 1995, retail sales hit $72 billion. By the year 2000, sales had grown to about $146 billion. (a) Use the relation (year, retail sales of prescription drugs) with $t = 0$ corresponding to 1995 to find a linear equation modeling the growth of retail sales. (b) Discuss what the slope indicates in this context. (c) According to this model, in what year will sales reach $250 billion? (d) According to the model, what was the value of retail prescription drug sales in 2005? (e) How many years after 1995 will retail sales exceed $279 billion? (f) If yearly sales totaled $294 billion, what year is it?

Source: 2004 Statistical Abstract of the United States, Table 122

111. Prison population: In 1990, the number of persons sentenced and serving time in state and federal institutions was approximately 740,000. By the year 2000, this figure had grown to nearly 1,320,000. (a) Find a linear equation with $t = 0$ corresponding to 1990 that models this data, (b) discuss the slope ratio in context, and (c) use the equation to estimate the prison population in 2007 if this trend continues.

Source: Bureau of Justice Statistics at www.ojp.usdoj.gov/bjs

112. Eating out: In 1990, Americans bought an average of 143 meals per year at restaurants. This phenomenon continued to grow in popularity and in the year 2000, the average reached 170 meals per year. (a) Find a linear equation with $t = 0$ corresponding to 1990 that models this growth, (b) discuss the slope ratio in context, and (c) use the equation to estimate the average number of times an American will eat at a restaurant in 2006 if the trend continues.

Source: The NPD Group, Inc., National Eating Trends, 2002

▶ EXTENDING THE CONCEPT

113. Locate and read the following article. Then turn in a one-page summary. "Linear Function Saves Carpenter's Time," Richard Crouse, *Mathematics Teacher,* Volume 83, Number 5, May 1990: pp. 400–401.

114. The general form of a linear equation is $ax + by = c$, where a and b are not simultaneously zero. (a) Find the x- and y-intercepts using the general form (substitute 0 for x, then 0 for y). Based on what you see, when does the intercept method work most efficiently? (b) Find the slope and y-intercept using the general form (solve for y). Based on what you see, when does the intercept method work most efficiently?.

115. Match the correct graph to the conditions stated for m and b. There are more choices than graphs.

   a. $m < 0, b < 0$ **b.** $m > 0, b < 0$

   c. $m < 0, b > 0$ **d.** $m > 0, b > 0$

   e. $m = 0, b > 0$ **f.** $m < 0, b = 0$

   g. $m > 0, b = 0$ **h.** $m = 0, b < 0$

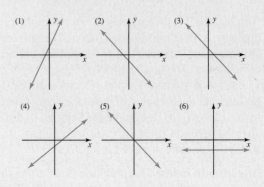

▶ **MAINTAINING YOUR SKILLS**

116. (2.2) Determine the domain:

   a. $y = \sqrt{2x - 5}$

   b. $y = \dfrac{5}{2x^2 + 3x - 2}$

117. (1.5) Solve using the quadratic formula. Answer in exact and approximate form: $3x^2 - 10x = 9$.

118. (1.1) Three equations follow. One is an identity, another is a contradiction, and a third has a solution. State which is which.

$$2(x - 5) + 13 - 1 = 9 - 7 + 2x$$

$$2(x - 4) + 13 - 1 = 9 + 7 - 2x$$

$$2(x - 5) + 13 - 1 = 9 + 7 + 2x$$

119. (R.7) Compute the area of the circular sidewalk shown here. Use your calculator's value of π and round the answer (only) to hundredths.

10 yd

8 yd

2.4 | Functions, Function Notation, and the Graph of a Function

Learning Objectives

In Section 2.4 you will learn how to:

☐ **A.** Distinguish the graph of a function from that of a relation

☐ **B.** Determine the domain and range of a function

☐ **C.** Use function notation and evaluate functions

☐ **D.** Apply the rate-of-change concept to nonlinear functions

In this section we introduce one of the most central ideas in mathematics—the concept of a function. Functions can model the cause-and-effect relationship that is so important to using mathematics as a decision-making tool. In addition, the study will help to unify and expand on many ideas that are already familiar.

A. Functions and Relations

There is a special type of relation that merits further attention. A **function** is a relation where each element of the domain corresponds to exactly one element of the range. In other words, for each first coordinate or input value, there is only one possible second coordinate or output.

> **Functions**
>
> A *function* is a relation that pairs each element from the *domain* with exactly one element from the *range*.

If the relation is defined by a mapping, we need only check that each element of the domain is mapped to exactly one element of the range. This is indeed the case for the mapping $P \rightarrow B$ from Figure 2.1 (page 152), where we saw that each person corresponded to only one birthday, and that it was impossible for one person to be born on two different days. For the relation $x = |y|$ shown in Figure 2.6 (page 153), each element of the domain except zero is paired with *more than one* element of the range. The relation $x = |y|$ is *not* a function.

EXAMPLE 1 ▶ **Determining Whether a Relation Is a Function**

Three different relations are given in mapping notation below. Determine whether each relation is a function.

a.

Person	Room
Marie	270
Pesky	268
Bo	274
Johnny	276
Rick	272
Annie	282
Reece	

b.

Pet	Weight (lbs)
Fido	450
Bossy	550
Silver	2
	40
Frisky	8
Polly	3

c.

War	Year
Civil War	1963
World War I	1950
World War II	1939
Korean War	1917
Vietnam War	1861

Solution ▶ Relation (a) is a function, since each person corresponds to exactly one room. This relation pairs math professors with their respective office numbers. Notice that while two people can be in one office, it is impossible for one person to physically be in two different offices. Relation (b) is not a function, since we cannot tell whether Polly the Parrot weighs 2 lb or 3 lb (one element of the domain is mapped to two elements of the range). Relation (c) is a function, where each major war is paired with the year it began.

Now try Exercises 7 through 10 ▶

If the relation is defined by a set of ordered pairs or a set of individual and distinct plotted points, we need only check that no two points have the same first coordinate with a different second coordinate.

EXAMPLE 2 ▶ **Identifying Functions**

Two relations named f and g are given; f is stated as a set of ordered pairs, while g is given as a set of plotted points. Determine whether each is a function.

$$f: (-3, 0), (1, 4), (2, -5), (4, 2), (-3, -2), (3, 6), (0, -1), (4, -5), \text{ and } (6, 1)$$

Solution ▶ The relation f is not a function, since -3 is paired with two different outputs: $(-3, 0)$ and $(-3, -2)$.

The relation g shown in the figure *is* a function. Each input corresponds to exactly one output, otherwise one point would be directly above the other and have the same first coordinate.

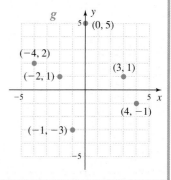

Now try Exercises 11 through 18 ▶

The graphs of $y = x - 1$ and $x = |y|$ from Section 2.1 offer additional insight into the definition of a function. Figure 2.23 shows the line $y = x - 1$ with emphasis on the plotted points $(4, 3)$ and $(-3, -4)$. The vertical movement shown from the x-axis to a point on the graph illustrates *the pairing of a given x-value with one related y-value*. Note the vertical line shows *only one related y-value* ($x = 4$ is paired with only $y = 3$). Figure 2.24 gives the graph of $x = |y|$, highlighting the points $(4, 4)$ and $(4, -4)$. The vertical movement shown here branches in two directions, associating one x-value with more than one y-value. This shows the relation $y = x - 1$ is also a function, while the relation $x = |y|$ is not.

Figure 2.23

Figure 2.24

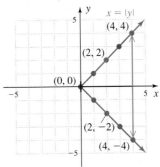

This "vertical connection" of a location on the x-axis to a point on the graph can be generalized into a **vertical line test for functions.**

Vertical Line Test

A given graph is the graph of a function, if and only if every vertical line intersects the graph in at most one point.

Applying the test to the graph in Figure 2.23 helps to illustrate that the graph of any nonvertical line is a function.

EXAMPLE 3 ▶ **Using the Vertical Line Test**

Use the vertical line test to determine if any of the relations shown (from Section 2.1) are functions.

Solution ▶ Visualize a vertical line on each coordinate grid (shown in solid blue), then mentally shift the line to the left and right as shown in Figures 2.25, 2.26, and 2.27 (dashed lines). In Figures 2.25 and 2.26, every vertical line intersects the graph only once, indicating both $y = x^2 - 2x$ and $y = \sqrt{9 - x^2}$ are functions. In Figure 2.27, a vertical line intersects the graph twice for any $x > 0$. The relation $x = y^2$ is not a function.

Figure 2.25

Figure 2.26

Figure 2.27

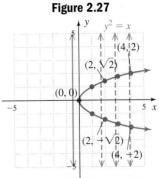

Now try Exercises 19 through 30 ▶

EXAMPLE 4 ▶ **Using the Vertical Line Test**

Use a table of values to graph the relations defined by

a. $y = |x|$ **b.** $y = \sqrt{x}$,

then use the vertical line test to determine whether each relation is a function.

Solution ▶ **a.** For $y = |x|$, using input values from $x = -4$ to $x = 4$ produces the following table and graph (Figure 2.28). Note the result is a V-shaped graph that "opens upward." The point $(0, 0)$ of this absolute value graph is called the **vertex**. Since any vertical line will intersect the graph in at most one point, this is the graph of a function.

WORTHY OF NOTE

For relations and functions, a good way to view the distinction is to consider a mail carrier. It is possible for the carrier to put more than one letter into the same mailbox (more than one x going to the same y), but quite impossible for the carrier to place the same letter in two different boxes (one x going to two y's).

$y = |x|$

Figure 2.28

x	$y = |x|$
-4	4
-3	3
-2	2
-1	1
0	0
1	1
2	2
3	3
4	4

b. For $y = \sqrt{x}$, values less than zero do not produce a real number, so our graph actually begins at $(0, 0)$ (see Figure 2.29). Completing the table for nonnegative values produces the graph shown, which appears to rise to the right and remains in the first quadrant. Since any vertical line will intersect this graph in at most one place, $y = \sqrt{x}$ is also a function.

$y = \sqrt{x}$

Figure 2.29

x	$y = \sqrt{x}$
0	0
1	1
2	$\sqrt{2} \approx 1.4$
3	$\sqrt{3} \approx 1.7$
4	2

☑ **A.** You've just learned how to distinguish the graph of a function from that of a relation

Now try Exercises 31 through 34 ▶

B. The Domain and Range of a Function

Vertical Boundary Lines and the Domain

In addition to its use as a graphical test for functions, a vertical line can help determine the domain of a function from its graph. For the graph of $y = \sqrt{x}$ (Figure 2.29), a vertical line will not intersect the graph until $x = 0$, and then will intersect the graph for all values $x \geq 0$ (showing the function is defined for these values). These **vertical boundary lines** indicate the domain is $x \in [0, \infty)$. For the graph of $y = |x|$ (Figure 2.28), a vertical line will intersect the graph (or its infinite extension) for *all values* of x, and the domain is $x \in (-\infty, \infty)$. Using vertical lines in this way also affirms the domain of $y = x - 1$ (Section 2.1, Figure 2.5) is $x \in (-\infty, \infty)$ while the domain of the relation $x = |y|$ (Section 2.1, Figure 2.6) is $x \in [0, \infty)$.

Range and Horizontal Boundary Lines

The range of a relation can be found using a **horizontal "boundary line,"** since it will associate a value on the y-axis with a point on the graph (if it exists). Simply visualize a horizontal line and move the line up or down until you determine the graph will always intersect the line, or will no longer intersect the line. This will give you the boundaries of the range. Mentally applying this idea to the graph of $y = \sqrt{x}$ (Figure 2.29) shows the range is $y \in [0, \infty)$. Although shaped very differently, a horizontal boundary line shows the range of $y = |x|$ (Figure 2.28) is also $y \in [0, \infty)$.

EXAMPLE 5 ▶ **Determining the Domain and Range of a Function**

Use a table of values to graph the functions defined by

a. $y = x^2$ **b.** $y = \sqrt[3]{x}$

Then use boundary lines to determine the domain and range of each.

Solution ▶ **a.** For $y = x^2$, it seems convenient to use inputs from $x = -3$ to $x = 3$, producing the following table and graph. Note the result is a basic parabola that "opens upward" (both ends point in the positive y direction), with a vertex at $(0, 0)$. Figure 2.30 shows a vertical line will intersect the graph or its extension anywhere it is placed. The domain is $x \in (-\infty, \infty)$. Figure 2.31 shows a horizontal line will intersect the graph only for values of y that are greater than or equal to 0. The range is $y \in [0, \infty)$.

Squaring Function

x	$y = x^2$
-3	9
-2	4
-1	1
0	0
1	1
2	4
3	9

Figure 2.30

Figure 2.31

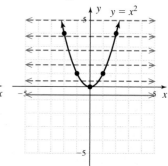

b. For $y = \sqrt[3]{x}$, we select points that are perfect cubes where possible, then a few others to round out the graph. The resulting table and graph are shown, and we notice there is a "pivot point" at $(0, 0)$ called a **point of inflection,** and the ends of the graph point in opposite directions. Figure 2.32 shows a vertical line will intersect the graph or its extension anywhere it is placed. Figure 2.33 shows a horizontal line will likewise always intersect the graph. The domain is $x \in (-\infty, \infty)$, and the range is $y \in (-\infty, \infty)$.

Cube Root Function

x	$y = \sqrt[3]{x}$
-8	-2
-4	≈ -1.6
-1	-1
0	0
1	1
4	≈ 1.6
8	2

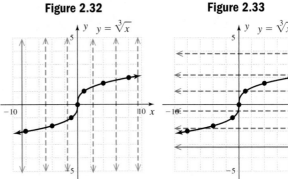

Figure 2.32   Figure 2.33

Now try Exercises 35 through 46 ▶

Implied Domains

When stated in equation form, the domain of a function is implicitly given by the expression used to define it, since the expression will dictate the allowable values (Section 1.2). The **implied domain** is the set of all real numbers for which the function represents a real number. If the function involves a rational expression, the domain will exclude any input that causes a denominator of zero. If the function involves a square root expression, the domain will exclude inputs that create a negative radicand.

EXAMPLE 6 ▶ Determining Implied Domains

State the domain of each function using interval notation.

     a. $y = \dfrac{3}{x + 2}$ **b.** $y = \sqrt{2x + 3}$

     c. $y = \dfrac{x - 5}{x^2 - 9}$ **d.** $y = x^2 - 5x + 7$

Solution ▶   **a.** By inspection, we note an x-value of -2 gives a zero denominator and must be excluded. The domain is $x \in (-\infty, -2) \cup (-2, \infty)$.

     b. Since the radicand must be nonnegative, we solve the inequality $2x + 3 \geq 0$, giving $x \geq \frac{-3}{2}$. The domain is $x \in [\frac{-3}{2}, \infty)$.

     c. To prevent division by zero, inputs of -3 and 3 must be excluded (set $x^2 - 9 = 0$ and solve by factoring). The domain is $x \in (-\infty, -3) \cup (-3, 3) \cup (3, \infty)$. Note that $x = 5$ *is in the domain* since $\frac{0}{16} = 0$ is defined.

     d. Since squaring a number and multiplying a number by a constant are defined for all reals, the domain is $x \in (-\infty, \infty)$.

Now try Exercises 47 through 64 ▶

EXAMPLE 7 ▶ Determining Implied Domains

Determine the domain of each function:

     a. $y = \sqrt{\dfrac{7}{x + 3}}$ **b.** $y = \dfrac{2x}{\sqrt{4x + 5}}$

Solution ▶ **a.** For $y = \sqrt{\dfrac{7}{x + 3}}$, we must have $\dfrac{7}{x + 3} \geq 0$ (for the radicand) **and** $x + 3 \neq 0$ (for the denominator). Since the numerator is *always* positive, we need $x + 3 > 0$, which gives $x > -3$. The domain is $x \in (-3, \infty)$.

☑ **B.** You've just learned how to determine the domain and range of a function

b. For $y = \dfrac{2x}{\sqrt{4x + 5}}$, we must have $4x + 5 \geq 0$ **and** $\sqrt{4x + 5} \neq 0$. This indicates $4x + 5 > 0$ or $x > -\frac{5}{4}$. The domain is $x \in \left(-\frac{5}{4}, \infty\right)$.

Now try Exercises 65 through 68 ▶

C. Function Notation

In our study of functions, you've likely noticed that the relationship between input and output values is an important one. To highlight this fact, think of a function as a simple machine, which can *process inputs* using a stated sequence of operations, then deliver a single output (see Figure 2.34). The inputs are x-values, a program we'll name f performs the operations on x, and y is the resulting output. Once again we see that "the value of y depends on the value of x," or simply "y is a function of x." Notationally, we write "y is a function of x" as $y = f(x)$ using **function notation.** You are already familiar with letting a variable represent a number. Here we do something quite different, as the letter f is used to represent *a sequence of operations to be performed on x.* Consider the function $y = \frac{x}{2} + 1$, which we'll now write as $f(x) = \frac{x}{2} + 1$ [since $y = f(x)$]. In words the function says, "divide inputs by 2, then add 1." To evaluate the function at $x = 4$ (Figure 2.35) we have:

Figure 2.34

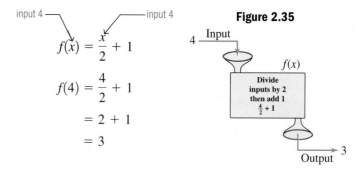

$$f(x) = \frac{x}{2} + 1$$

$$f(4) = \frac{4}{2} + 1$$

$$= 2 + 1$$

$$= 3$$

Figure 2.35

Instead of saying, ". . . when $x = 4$, the value of the function is 3," we simply say "f of 4 is 3," or write $f(4) = 3$. Note that the ordered pair $(4, 3)$ is equivalent to $(4, f(4))$.

⚠ **CAUTION** ▶ Although $f(x)$ is the favored notation for a "function of x," other letters can also be used. For example, $g(x)$ and $h(x)$ also denote functions of x, where g and h represent a different sequence of operations on the x-inputs. It is also important to remember that these represent *function values* and not the product of two variables: $f(x) \neq f \cdot (x)$.

EXAMPLE 8 ▶ **Evaluating a Function**

Given $f(x) = -2x^2 + 4x$, find

a. $f(-2)$ **b.** $f\left(\dfrac{3}{2}\right)$ **c.** $f(2a)$ **d.** $f(a + 1)$

Solution ▶ **a.** $f(x) = -2x^2 + 4x$
$f(-2) = -2(-2)^2 + 4(-2)$
$= -8 + (-8) = -16$

b. $f(x) = -2x^2 + 4x$
$f\left(\dfrac{3}{2}\right) = -2\left(\dfrac{3}{2}\right)^2 + 4\left(\dfrac{3}{2}\right)$
$= -\dfrac{9}{2} + 6 = \dfrac{3}{2}$

c. $f(x) = -2x^2 + 4x$
$f(2a) = -2(2a)^2 + 4(2a)$
$= -2(4a^2) + 8a$
$= -8a^2 + 8a$

d. $f(x) = -2x^2 + 4x$
$f(a + 1) = -2(a + 1)^2 + 4(a + 1)$
$= -2(a^2 + 2a + 1) + 4a + 4$
$= -2a^2 - 4a - 2 + 4a + 4$
$= -2a^2 + 2$

Now try Exercises 69 through 84 ▶

Graphs are an important part of studying functions, and learning to read and interpret them correctly is a high priority. A graph highlights and emphasizes the all-important input/output relationship that defines a function. In this study, we hope to firmly establish that the following statements are synonymous:

1. $f(-2) = 5$
2. $(-2, f(-2)) = (-2, 5)$
3. $(-2, 5)$ is on the graph of f, and
4. When $x = -2, f(x) = 5$

EXAMPLE 9A ▶ **Reading a Graph**

For the functions $f(x)$ and $g(x)$ whose graphs are shown in Figures 2.36 and 2.37
    a. State the domain of the function.
    b. Evaluate the function at $x = 2$.
    c. Determine the value(s) of x for which $y = 3$.
    d. State the range of the function.

Figure 2.36

Figure 2.37

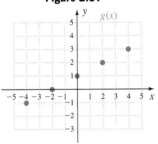

Solution ▶ For $f(x)$,
    a. The graph is a continuous line segment with endpoints at $(-4, -3)$ and $(5, 3)$, so we state the domain in interval notation. Using a vertical boundary line we note the smallest input is -4 and the largest is 5. The domain is $x \in [-4, 5]$.
    b. The graph shows an input of $x = 2$ corresponds to $y = 1$: $f(2) = 1$ since $(2, 1)$ is a point on the graph.
    c. For $f(x) = 3$ (or $y = 3$) the input value must be $x = 5$ since $(5, 3)$ is the point on the graph.
    d. Using a horizontal boundary line, the smallest output value is -3 and the largest is 3. The range is $y \in [-3, 3]$.

For $g(x)$,

a. Since the graph is pointwise defined, we state the domain as the set of first coordinates: $D = \{-4, -2, 0, 2, 4\}$.

b. An input of $x = 2$ corresponds to $y = 2$: $g(2) = 2$ since $(2, 2)$ is on the graph.

c. For $g(x) = 3$ (or $y = 3$) the input value must be $x = 4$, since $(4, 3)$ is a point on the graph.

d. The range is the set of all second coordinates: $R = \{-1, 0, 1, 2, 3\}$.

EXAMPLE 9B ▶ **Reading a Graph**

Use the graph of $f(x)$ given to answer the following questions:

a. What is the value of $f(-2)$?

b. What value(s) of x satisfy $f(x) = 1$?

Solution ▶ **a.** The notation $f(-2)$ says to find the value of the function f when $x = -2$. Expressed graphically, we go to $x = -2$, locate the corresponding point on the graph of f (blue arrows), and find that $f(-2) = 4$.

b. For $f(x) = 1$, we're looking for x-inputs that result in an output of $y = 1$ [since $y = f(x)$]. From the graph, we note there are two points with a y-coordinate of 1, namely, $(-3, 1)$ and $(0, 1)$. This shows $f(-3) = 1, f(0) = 1$, and the required x-values are $x = -3$ and $x = 0$.

Now try Exercises 85 through 90 ▶

In many applications involving functions, the domain and range can be determined by the context or situation given.

EXAMPLE 10 ▶ **Determining the Domain and Range from the Context**

Paul's 1993 Voyager has a 20-gal tank and gets 18 mpg. The number of miles he can drive (his range) depends on how much gas is in the tank. As a function we have $M(g) = 18g$, where $M(g)$ represents the total distance in miles and g represents the gallons of gas in the tank. Find the domain and range.

Solution ▶ Begin evaluating at $x = 0$, since the tank cannot hold less than zero gallons. On a full tank the maximum range of the van is $20 \cdot 18 = 360$ miles or $M(g) \in [0, 360]$. Because of the tank's size, the domain is $g \in [0, 20]$.

☑ **C. You've just learned how to use function notation and evaluate functions**

Now try Exercises 94 through 101 ▶

D. Average Rates of Change

As noted in Section 2.3, one of the defining characteristics of a linear function is that the rate of change $m = \dfrac{\Delta y}{\Delta x}$ is constant. For nonlinear functions the rate of change is not constant, but we can use a related concept called the **average rate of change** to study these functions.

Average Rate of Change

For a function that is smooth and continuous on the interval containing x_1 and x_2, the average rate of change between x_1 and x_2 is given by

$$\frac{\Delta y}{\Delta x} = \frac{y_2 - y_1}{x_2 - x_1}$$

which is the slope of the secant line through (x_1, y_1) and (x_2, y_2)

EXAMPLE 11 ▶ **Calculating Average Rates of Change**

The graph shown displays the number of units shipped of vinyl records, cassette tapes, and CDs for the period 1980 to 2005.

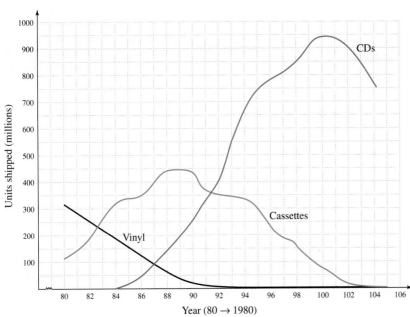

Units shipped in millions

Year	Vinyl	Cassette	CDs
1980	323	110	0
1982	244	182	0
1984	205	332	6
1986	125	345	53
1988	72	450	150
1990	12	442	287
1992	2	366	408
1994	2	345	662
1996	3	225	779
1998	3	159	847
2000	2	76	942
2004	1	5	767
2005	1	3	705

Source: Swivel.com

a. Find the average rate of change in CDs shipped and in cassettes shipped from 1994 to 1998. What do you notice?

b. Does it appear that the rate of increase in CDs shipped was greater from 1986 to 1992, or from 1992 to 1996? Compute the average rate of change for each period and comment on what you find.

Solution ▶ Using 1980 as year zero (1980 → 0), we have the following:

a.   **CDs**   **Cassettes**

1994: $(14, 662)$, 1998: $(18, 847)$ 1994: $(14, 345)$, 1998: $(18, 159)$

$$\frac{\Delta y}{\Delta x} = \frac{847 - 662}{18 - 14} \qquad\qquad \frac{\Delta y}{\Delta x} = \frac{159 - 345}{18 - 14}$$

$$= \frac{185}{4} \qquad\qquad\qquad\qquad = -\frac{186}{4}$$

$$= 46.25 \qquad\qquad\qquad\qquad = -46.5$$

The decrease in the number of cassettes shipped was roughly equal to the increase in the number of CDs shipped (about 46,000,000 per year).

b. From the graph, the secant line for 1992 to 1996 appears to have a greater slope.

1986–1992 CDs	**1992–1996 CDs**

1986: (6, 53), 1992: (12, 408)   1992: (12, 408), 1996: (16, 779)

$$\frac{\Delta y}{\Delta x} = \frac{408 - 53}{12 - 6}$$ $$\frac{\Delta y}{\Delta x} = \frac{779 - 408}{16 - 12}$$

$$= \frac{355}{6}$$ $$= \frac{371}{4}$$

$$= 59.1\overline{6}$$ $$= 92.75$$

 D. You've just learned how to apply the rate-of-change concept to nonlinear functions

For 1986 to 1992: $m \approx 59.2$; for 1992 to 1996: $m = 92.75$, a growth rate much higher than the earlier period.

Now try Exercises 102 and 103 ▶

2.4 EXERCISES

▶ CONCEPTS AND VOCABULARY

Fill in each blank with the appropriate word or phrase. Carefully reread the section if needed.

1. If a relation is given in ordered pair form, we state the domain by listing all of the _____ coordinates in a set.

2. A relation is a function if each element of the _____ is paired with _____ _____ element of the range.

3. The set of output values for a function is called the _____ of the function.

4. Write using function notation: The function f evaluated at 3 is negative 5: _____

5. Discuss/Explain why the relation $y = x^2$ is a function, while the relation $x = y^2$ is not. Justify your response using graphs, ordered pairs, and so on.

6. Discuss/Explain the process of finding the domain and range of a function given its graph, using vertical and horizontal boundary lines. Include a few illustrative examples.

▶ DEVELOPING YOUR SKILLS

Determine whether the mappings shown represent functions or nonfunctions. If a nonfunction, explain how the definition of a function is violated.

7.

8.
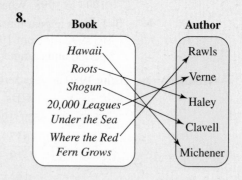

9.

Basketball star	Reported height

MJ
The Mailman
The Doctor
The Iceman
The Shaq

7'1"
6'6"
6'7"
6'9"
7'2"

10.

Country	Language

Canada
Japan
Brazil
Tahiti
Ecuador

Japanese
Spanish
French
Portuguese
English

Ex:✓

FIRST COMPONENT CANNOT REPEAT

NON-FUNCTION CANADA HAS 2 LANGUAGES

Determine whether the relations indicated represent functions or nonfunctions. If the relation is a nonfunction, explain how the definition of a function is violated.

11. $(-3, 0), (1, 4), (2, -5), (4, 2), (-5, 6), (3, 6), (0, -1),$ $(4, -5),$ and $(6, 1)$

12. $(-7, -5), (-5, 3), (4, 0), (-3, -5), (1, -6),$ $(0, 9), (2, -8), (3, -2),$ and $(-5, 7)$

13. $(9, -10), (-7, 6), (6, -10), (4, -1), (2, -2),$ $(1, 8), (0, -2), (-2, -7),$ and $(-6, 4)$

14. $(1, -81), (-2, 64), (-3, 49), (5, -36), (-8, 25),$ $(13, -16), (-21, 9), (34, -4),$ and $(-55, 1)$

Ex:✓   *FUNCTION*

15.

(−3, 4)   •(2, 4)
(−1, 1)   (4, 2)•
(−5, 0)•

16.

(−3, 5)•   (3, 4)•
          •(1, 3)
(−4, −2)• •(0, −2)
              (5, −3)•

17.

(3, 4)•
(−2, 3)•
•(1, 2)
•(−5, 1)

(−2, −4)•
       (4, −5)•

18.

•(−3, 4)
       (3, 3)•
       •(1, 1)
•(−5, −2) (3, −2)•
(−1, −4)•

Determine whether or not the relations given represent a function. If not, explain how the definition of a function is violated.

19.

20.

VERT LINE TEST 21-30   *VLT*

21.

Ex:✓

22.

FUNCTION VLT

23.

Ex:✓

24.

NON-FUNCTION VLT

25.

26.

27.

28.

29.

30.

Graph each relation using a table, then use the vertical line test to determine if the relation is a function.

31. $y = x$

Ex: **32.** $y = \sqrt[3]{x}$ *FUNCTION (VLT) GRAPH*

33. $y = (x + 2)^2$ **34.** $x = |y - 2|$

Determine whether or not the relations indicated represent a function, then determine the domain and range of each.

35.

36.

37.

38.

39.

40.

41.

42.

43.

44.

45.

46.

Determine the domain of the following functions.

47. $f(x) = \dfrac{3}{x - 5}$

48. $g(x) = \dfrac{-2}{3 + x}$

49. $h(a) = \sqrt{3a + 5}$

50. $p(a) = \sqrt{5a - 2}$

51. $v(x) = \dfrac{x + 2}{x^2 - 25}$

52. $w(x) = \dfrac{x - 4}{x^2 - 49}$

53. $u = \dfrac{v - 5}{v^2 - 18}$

54. $p = \dfrac{q + 7}{q^2 - 12}$

55. $y = \dfrac{17}{25}x + 123$

56. $y = \dfrac{11}{19}x - 89$

57. $m = n^2 - 3n - 10$

58. $s = t^2 - 3t - 10$

59. $y = 2|x| + 1$

60. $y = |x - 2| + 3$

61. $y_1 = \dfrac{x}{x^2 - 3x - 10}$

62. $y_2 = \dfrac{x - 4}{x^2 + 2x - 15}$

63. $y = \dfrac{\sqrt{x - 2}}{2x - 5}$

64. $y = \dfrac{\sqrt{x + 1}}{3x + 2}$

65. $f(x) = \sqrt{\dfrac{5}{x - 2}}$

66. $g(x) = \sqrt{\dfrac{-4}{3 - x}}$

67. $h(x) = \dfrac{-2}{\sqrt{4 + x}}$

68. $p(x) = \dfrac{-7}{\sqrt{5 - x}}$

Determine the value of $f(-6)$, $f(\frac{3}{2})$, $f(2c)$, and $f(c + 1)$, then simplify as much as possible.

69. $f(x) = \dfrac{1}{2}x + 3$

70. $f(x) = \dfrac{2}{3}x - 5$

71. $f(x) = 3x^2 - 4x$

72. $f(x) = 2x^2 + 3x$

Determine the value of $h(3)$, $h(-\frac{2}{3})$, $h(3a)$, and $h(a - 2)$, then simplify as much as possible.

73. $h(x) = \dfrac{3}{x}$

74. $h(x) = \dfrac{2}{x^2}$

75. $h(x) = \dfrac{5|x|}{x}$

76. $h(x) = \dfrac{4|x|}{x}$

Determine the value of $g(4)$, $g(\frac{3}{2})$, $g(2c)$, and $g(c + 3)$, then simplify as much as possible.

77. $g(r) = 2\pi r$

78. $g(r) = 2\pi rh$

79. $g(r) = \pi r^2$

80. $g(r) = \pi r^2 h$

Determine the value of $p(5)$, $p(\frac{3}{2})$, $p(3a)$, and $p(a - 1)$, then simplify as much as possible.

81. $p(x) = \sqrt{2x + 3}$

82. $p(x) = \sqrt{4x - 1}$

83. $p(x) = \dfrac{3x^2 - 5}{x^2}$

84. $p(x) = \dfrac{2x^2 + 3}{x^2}$

Use the graph of each function given to (a) state the domain, (b) state the range, (c) evaluate $f(2)$, and (d) find the value(s) x for which $f(x) = k$ (k a constant). Assume all results are integer-valued.

85. $k = 4$

86. $k = 3$

87. $k = 1$ **88.** $k = -3$ **89.** $k = 2$ **90.** $k = -1$

▶ WORKING WITH FORMULAS

91. Ideal weight for males: $W(H) = \frac{9}{2}H - 151$

The ideal weight for an adult male can be modeled by the function shown, where W is his weight in pounds and H is his height in inches. (a) Find the ideal weight for a male who is 75 in. tall. (b) If I am 72 in. tall and weigh 210 lb, how much weight should I lose?

92. Celsius to Fahrenheit conversions: $C = \frac{5}{9}(F - 32)$

The relationship between Fahrenheit degrees and degrees Celsius is modeled by the function shown. (a) What is the Celsius temperature if °F = 41? (b) Use the formula to solve for F in terms of C, then substitute the result from part (a). What do you notice?

93. Pick's theorem: $A = \frac{1}{2}B + I - 1$

Picks theorem is an interesting yet little known formula for computing the area of a polygon drawn in the Cartesian coordinate system. The formula can be applied as long as the vertices of the polygon are lattice points (both x and y are integers). If B represents the number of lattice points lying directly on the boundary of the polygon (including the vertices), and I represents the number of points in the interior, the area of the polygon is given by the formula shown. Use some graph paper to carefully draw a triangle with vertices at $(-3, 1)$, $(3, 9)$, and $(7, 6)$, then use Pick's theorem to compute the triangle's area.

▶ APPLICATIONS

94. Gas mileage: John's old '87 LeBaron has a 15-gal gas tank and gets 23 mpg. The number of miles he can drive is a function of how much gas is in the tank. (a) Write this relationship in equation form and (b) determine the domain and range of the function in this context.

95. Gas mileage: Jackie has a gas-powered model boat with a 5-oz gas tank. The boat will run for 2.5 min on each ounce. The number of minutes she can operate the boat is a function of how much gas is in the tank. (a) Write this relationship in equation form and (b) determine the domain and range of the function in this context.

96. Volume of a cube: The volume of a cube depends on the length of the sides. In other words, volume is a function of the sides: $V(s) = s^3$. (a) In practical terms, what is the domain of this function? (b) Evaluate $V(6.25)$ and (c) evaluate the function for $s = 2x^2$.

97. Volume of a cylinder: For a fixed radius of 10 cm, the volume of a cylinder depends on its height. In other words, volume is a function of height:

$V(h) = 100\pi h$. (a) In practical terms, what is the domain of this function? (b) Evaluate $V(7.5)$ and (c) evaluate the function for $h = \dfrac{8}{\pi}$.

98. Rental charges: Temporary Transportation Inc. rents cars (local rentals only) for a flat fee of $19.50 and an hourly charge of $12.50. This means that cost is a function of the hours the car is rented plus the flat fee. (a) Write this relationship in equation form; (b) find the cost if the car is rented for 3.5 hr; (c) determine how long the car was rented if the bill came to $119.75; and (d) determine the domain and range of the function in this context, if your budget limits you to paying a maximum of $150 for the rental.

99. Cost of a service call: Paul's Plumbing charges a flat fee of $50 per service call plus an hourly rate of $42.50. This means that cost is a function of the hours the job takes to complete plus the flat fee. (a) Write this relationship in equation form; (b) find the cost of a service call that takes $2\frac{1}{2}$ hr; (c) find the number of hours the job took if the charge came to $262.50; and (d) determine the domain and range of the function in this context, if your insurance company has agreed to pay for all charges over $500 for the service call.

100. Predicting tides: The graph shown approximates the height of the tides at Fair Haven, New Brunswick, for a 12-hr period. (a) Is this the graph of a function? Why? (b) Approximately what time did high tide occur? (c) How high is the tide at 6 P.M.? (d) What time(s) will the tide be 2.5 m?

101. Predicting tides: The graph shown approximates the height of the tides at Apia, Western Samoa, for a 12-hr period. (a) Is this the graph of a function? Why? (b) Approximately what time did low tide occur? (c) How high is the tide at 2 A.M.? (d) What time(s) will the tide be 0.7 m?

102. Weight of a fetus: The growth rate of a fetus in the mother's womb (by weight in grams) is modeled by the graph shown here, beginning with the 25th week of

gestation. (a) Calculate the average rate of change (slope of the secant line) between the 25th week and the 29th week. Is the slope of the secant line positive or negative? Discuss what the slope means in this context. (b) Is the fetus gaining weight faster between the 25th and 29th week, or between the 32nd and 36th week? Compare the slopes of both secant lines and discuss.

103. Fertility rates: Over the years, fertility rates for women in the United States (average number of children per woman) have varied a great deal, though in the twenty-first century they've begun to level out. The graph shown models this fertility rate for most of the twentieth century. (a) Calculate the average rate of change from the years 1920 to 1940. Is the slope of the secant line positive or negative? Discuss what the slope means in this context. (b) Calculate the average rate of change from the year 1940 to 1950. Is the slope of the secant line positive or negative? Discuss what the slope means in this context. (c) Was the fertility rate increasing faster from 1940 to 1950, or from 1980 to 1990? Compare the slope of both secant lines and comment.

Source: Statistical History of the United States from Colonial Times to Present

▶ **EXTENDING THE CONCEPT**

104. A father challenges his son to a 400-m race, depicted in the graph shown here.

Father: ——   Son: ——

a. Who won and what was the approximate winning time?

b. Approximately how many meters behind was the second place finisher?

c. Estimate the number of seconds the father was in the lead in this race.

d. How many times during the race were the father and son tied?

105. Sketch the graph of $f(x) = x$, then discuss how you could use this graph to obtain the graph of $F(x) = |x|$ without computing additional points. What would the graph of $g(x) = \dfrac{|x|}{x}$ look like?

106. Sketch the graph of $f(x) = x^2 - 4$, then discuss how you could use this graph to obtain the graph of $F(x) = |x^2 - 4|$ without computing additional points. Determine what the graph of $g(x) = \dfrac{|x^2 - 4|}{x^2 - 4}$ would look like.

107. If the equation of a function is given, the domain is implicitly defined by input values that generate real-valued outputs. But unless the graph is given or can be easily sketched, we must attempt to find the range analytically *by solving for x in terms of y*. We should note that sometimes this is an easy task, while at other times it is virtually impossible and we must rely on other methods. For the following functions, determine the implicit domain and find the range by solving for x in terms of y. **a.** $y = \frac{x-3}{x+2}$ **b.** $y = x^2 - 3$

▶ **MAINTAINING YOUR SKILLS**

108. (2.2) Which line has a steeper slope, the line through $(-5, 3)$ and $(2, 6)$, or the line through $(0, -4)$ and $(9, 4)$?

109. (R.6) Compute the sum and product indicated:
   a. $\sqrt{24} + 6\sqrt{54} - \sqrt{6}$
   b. $(2 + \sqrt{3})(2 - \sqrt{3})$

110. (1.5) Solve the equation using the quadratic formula, then check the result(s) using substitution:
$$x^2 - 4x + 1 = 0$$

111. (R.4) Factor the following polynomials completely:
   a. $x^3 - 3x^2 - 25x + 75$
   b. $2x^2 - 13x - 24$
   c. $8x^3 - 125$

MID-CHAPTER CHECK

1. Sketch the graph of the line $4x - 3y = 12$. Plot and label at least three points.

Exercises 5 and 6

2. Find the slope of the line passing through the given points: $(-3, 8)$ and $(4, -10)$.

3. In 2002, Data.com lost $2 million. In 2003, they lost $0.5 million. Will the slope of the line through these points be positive or negative? Why? Calculate the slope. Were you correct? Write the slope as a unit rate and explain what it means in this context.

Exercises 7 and 8

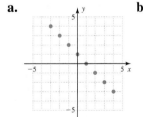

4. Sketch the line passing through $(1, 4)$ with slope $m = \frac{-2}{3}$ (plot and label at least two points). Then find the equation of the line *perpendicular to this line* through $(1, 4)$.

5. Write the equation for line L_1 shown. Is this the graph of a function? Discuss why or why not.

6. Write the equation for line L_2 shown. Is this the graph of a function? Discuss why or why not.

7. For the graph of function $h(x)$ shown, (a) determine the value of $h(2)$; (b) state the domain; (c) determine the value of x for which $h(x) = -3$; and (d) state the range.

8. Judging from the appearance of the graph alone, compare the average rate of change from $x = 1$ to $x = 2$ to the rate of change from $x = 4$ to $x = 5$. Which rate of change is larger? How is that demonstrated graphically?

Exercise 9

9. Find a linear function that models the graph of $F(p)$ given. Explain the slope of the line in this context, then use your model to predict the fox population when the pheasant population is 20,000.

10. State the domain and range for each function below.
   a.   **b.**

   c.

REINFORCING BASIC CONCEPTS

The Various Forms of a Linear Equation

In a study of mathematics, getting a glimpse of the "big picture" can be an enormous help. Learning mathematics is like building a skyscraper: The final height of the skyscraper ultimately depends on the strength of the foundation and quality of the frame supporting each new floor as it is built. Our work with linear functions and their graphs, while having a number of useful applications, is actually the foundation on which *much of your future work will be built*. The study of quadratic and polynomial functions and their applications all have their roots in linear equations. For this reason, it's important that you gain a certain fluency with linear functions—even to a point where things come to you effortlessly and automatically. This level of performance requires a strong desire and a sustained effort. We begin by reviewing the basic facts a student MUST know to reach this level. MUST is an acronym for memorize, understand, synthesize, and teach others. Don't be satisfied until you've done all four. Given points (x_1, y_1) and (x_2, y_2):

Forms and Formulas

slope formula	point-slope form	slope-intercept form	standard form
$m = \dfrac{y_2 - y_1}{x_2 - x_1}$	$y - y_1 = m(x - x_1)$	$y = mx + b$	$Ax + By = C$
given any two points on the line	given slope m and any point (x_1, y_1)	given slope m and y-intercept $(0, b)$	also used in linear systems (Chapter 6)

Characteristics of Lines

y-intercept	x-intercept	increasing	decreasing
$(0, y)$	$(x, 0)$	$m > 0$	$m < 0$
let $x = 0$, solve for y	let $y = 0$, solve for x	line slants upward from left to right	line slants downward from left to right

Practice for Speed and Accuracy

For the two points given, (a) compute the slope of the line and state whether the line is increasing or decreasing; (b) find the equation of the line using point-slope form; (c) write the equation in slope-intercept form; (d) write the equation in standard form; and (e) find the x- and y-intercepts and graph the line.

1. $P_1(0, 5); P_2(6, 7)$ **2.** $P_1(3, 2); P_2(0, 9)$ **3.** $P_1(3, 2); P_2(9, 5)$

4. $P_1(-5, -4); P_2(3, 2)$ **5.** $P_1(-2, 5); P_2(6, -1)$ **6.** $P_1(2, -7); P_2(-8, -2)$

2.5 | Analyzing the Graph of a Function

Learning Objectives

In Section 2.5 you will learn how to:

☐ **A.** Determine whether a function is even, odd, or neither

☐ **B.** Determine intervals where a function is positive or negative

☐ **C.** Determine where a function is increasing or decreasing

☐ **D.** Identify the maximum and minimum values of a function

☐ **E.** Develop a formula to calculate rates of change for any function

In this section, we'll consolidate and refine many of the ideas we've encountered related to functions. When functions and graphs are applied as real-world models, we create a numeric and visual representation that enables an informed response to questions involving *maximum* efficiency, *positive* returns, *increasing* costs, and other relationships that can have a great impact on our lives.

A. Graphs and Symmetry

While the domain and range of a function will remain dominant themes in our study, for the moment we turn our attention to other characteristics of a function's graph. We begin with the concept of symmetry.

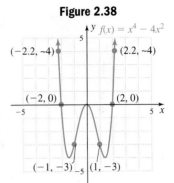

Symmetry with Respect to the *y*-Axis

Figure 2.38

Consider the graph of $f(x) = x^4 - 4x^2$ shown in Figure 2.38, where the portion of the graph to the left of the *y*-axis appears to be a mirror image of the portion to the right. A function is **symmetric to the *y*-axis** if, given any point (x, y) on the graph, the point $(-x, y)$ is also on the graph. We note that $(-1, -3)$ is on the graph, as is $(1, -3)$, and that $(-2, 0)$ is an *x*-intercept of the graph, as is $(2, 0)$. Functions that are symmetric to the *y*-axis are also known as **even functions** and in general we have:

Even Functions: *y*-Axis Symmetry

A function *f* is an *even function* if and only if, for each point (x, y) on the graph of *f*, the point $(-x, y)$ is also on the graph. *In function notation*

$$f(-x) = f(x)$$

Symmetry can be a great help in graphing new functions, enabling us to plot fewer points, and to complete the graph using properties of symmetry.

EXAMPLE 1 ▶ **Graphing an Even Function Using Symmetry**

a. The function $g(x)$ in Figure 2.39 is known to be even. Draw the complete graph (only the left half is shown).

Figure 2.39

b. Show that $h(x) = x^{\frac{2}{3}}$ is an even function using the arbitrary value $x = k$ [show $h(-k) = h(k)$], then sketch the complete graph using $h(0)$, $h(1)$, $h(8)$, and *y*-axis symmetry.

Solution ▶ **a.** To complete the graph of *g* (see Figure 2.39) use the points $(-4, 1)$, $(-2, -3)$, $(-1, 2)$, and *y*-axis symmetry to find additional points. The corresponding ordered pairs are $(4, 1)$, $(2, -3)$, and $(1, 2)$, which we use to help draw a "mirror image" of the partial graph given.

b. To prove that $h(x) = x^{\frac{2}{3}}$ is an even function, we must show $h(-k) = h(k)$ for any constant *k*. After writing $x^{\frac{2}{3}}$ as $\left[x^2\right]^{\frac{1}{3}}$, we have:

Figure 2.40

$$h(-k) \overset{?}{=} h(k) \qquad \text{first step of proof}$$

$$\left[(-k)^2\right]^{\frac{1}{3}} \overset{?}{=} \left[(k)^2\right]^{\frac{1}{3}} \qquad \text{evaluate } h(-k) \text{ and } h(k)$$

$$\sqrt[3]{(-k)^2} \overset{?}{=} \sqrt[3]{(k)^2} \qquad \text{radical form}$$

$$\sqrt[3]{k^2} = \sqrt[3]{k^2} \checkmark \qquad \text{result: } (-k)^2 = k^2$$

Using $h(0) = 0$, $h(1) = 1$, and $h(8) = 4$ with *y*-axis symmetry produces the graph shown in Figure 2.40.

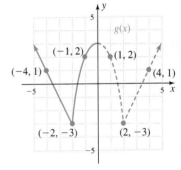

WORTHY OF NOTE

The proof can also be demonstrated by writing $x^{\frac{2}{3}}$ as $\left(x^{\frac{1}{3}}\right)^2$, and you are asked to complete this proof in Exercise 82.

Now try Exercises 7 through 12 ▶

Symmetry with Respect to the Origin

Another common form of symmetry is known as **symmetry to the origin.** As the name implies, the graph is somehow "centered" at (0, 0). This form of symmetry is easy to see for closed figures with their center at (0, 0), like certain polygons, circles, and ellipses (these will exhibit both *y*-axis symmetry *and* symmetry to the origin). Note the relation graphed in Figure 2.41 contains the points (−3, 3) and (3, −3), along with (−1, −4) and (1, 4). But the function *f*(*x*) in Figure 2.42 also contains these points and is, in the same sense, symmetric to the origin (the paired points are on opposite sides of the *x*- and *y*-axes, and a like distance from the origin).

Figure 2.41

Figure 2.42

Functions symmetric to the origin are known as **odd functions** and in general we have:

Odd Functions: Symmetry about the Origin

A function *f* is an *odd function* if and only if, for each point (*x*, *y*) on the graph of *f*, the point (−*x*, −*y*) is also on the graph. *In function notation*

$$f(-x) = -f(x)$$

EXAMPLE 2 ▶ **Graphing an Odd Function Using Symmetry**

  a. In Figure 2.43, the function *g*(*x*) given is known to be *odd*. Draw the complete graph (only the left half is shown).
  b. Show that $h(x) = x^3 - 4x$ is an odd function using the arbitrary value $x = k$ [show $h(-x) = -h(x)$], then sketch the graph using $h(-2)$, $h(-1)$, $h(0)$, and odd symmetry.

Solution ▶ **a.** To complete the graph of *g*, use the points (−6, 3), (−4, 0), and (−2, 2) and odd symmetry to find additional points. The corresponding ordered pairs are (6, −3), (4, 0), and (2, −2), which we use to help draw a "mirror image" of the partial graph given (see Figure 2.43).

Figure 2.43

Figure 2.44

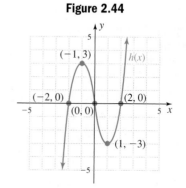

WORTHY OF NOTE

While the graph of an even function may or may not include the point (0, 0), the graph of an odd function will *always* contain this point.

b. To prove that $h(x) = x^3 - 4x$ is an odd function, we must show that $h(-k) = -h(k)$.

$$h(-k) \overset{?}{=} -h(k)$$
$$(-k)^3 - 4(-k) \overset{?}{=} -[k^3 - 4k]$$
$$-k^3 + 4k = -k^3 + 4k \checkmark$$

Using $h(-2) = 0$, $h(-1) = 3$, and $h(0) = 0$ with symmetry about the origin produces the graph shown in Figure 2.44.

> ☑ **A.** You've just learned how to determine whether a function is even, odd, or neither

> **Now try Exercises 13 through 24** ▶

B. Intervals Where a Function Is Positive or Negative

Consider the graph of $f(x) = x^2 - 4$ shown in Figure 2.45, which has x-intercepts at $(-2, 0)$ and $(2, 0)$. Since x-intercepts have the form $(x, 0)$ they are also called the **zeroes** of the function (the x-input causes an output of 0). Just as zero on the number line separates negative numbers from positive numbers, the zeroes of a function that crosses the x-axis separate x-intervals where a function is negative from x-intervals where the function is positive. Noting that outputs (y-values) are positive in Quadrants I and II, $f(x) > 0$ in intervals where its graph is *above the x-axis*. Conversely, $f(x) < 0$ in x-intervals where its graph is *below the x-axis*. To illustrate, compare the graph of f in Figure 2.45, with that of g in Figure 2.46.

Figure 2.45

Figure 2.46

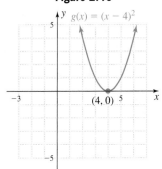

WORTHY OF NOTE

These observations form the basis for studying polynomials of higher degree, where we extend the idea to factors of the form $(x - r)^n$ in a study of **roots of multiplicity** (also see the *Calculator Exploration and Discovery* feature in this chapter).

The graph of f is a parabola, with x-intercepts of $(-2, 0)$ and $(2, 0)$. Using our previous observations, we note $f(x) \geq 0$ for $x \in (-\infty, -2] \cup [2, \infty)$ and $f(x) < 0$ for $x \in (-2, 2)$. The graph of g is also a parabola, but is entirely above or on the x-axis, showing $g(x) \geq 0$ for $x \in \mathbb{R}$. The difference is that zeroes coming from factors of the form $(x - r)$ (with degree 1) allow the graph to cross the x-axis. The zeroes of f came from $(x + 2)(x - 2) = 0$. Zeroes that come from factors of the form $(x - r)^2$ (with degree 2) cause the graph to "bounce" off the x-axis since all outputs must be nonnegative. The zero of g came from $(x - 4)^2 = 0$.

EXAMPLE 3 ▶ **Solving an Inequality Using a Graph**

Use the graph of $g(x) = x^3 - 2x^2 - 4x + 8$ given to solve the inequalities
a. $g(x) \geq 0$
b. $g(x) < 0$

Solution ▶ From the graph, the zeroes of g (x-intercepts) occur at $(-2, 0)$ and $(2, 0)$. a) For $g(x) \geq 0$, the graph must be on or above the x-axis, meaning the solution is $x \in [-2, \infty)$. b) For $g(x) < 0$, the graph must be below the x-axis, and the solution is $x \in (-\infty, -2)$. As we might have anticipated from the graph, factoring by grouping gives $g(x) = (x + 2)(x - 2)^2$, with the graph crossing the x-axis at -2, and bouncing off the x-axis (intersects without crossing) at $x = 2$.

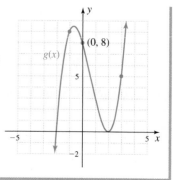

Now try Exercises 25 through 28 ▶

Even if the function is not a polynomial, the zeroes can still be used to find x-intervals where the function is positive or negative.

EXAMPLE 4 ▶   **Solving an Inequality Using a Graph**

For the graph of $r(x) = \sqrt{x + 1} - 2$ shown, solve

  a. $r(x) \leq 0$

  b. $r(x) > 0$

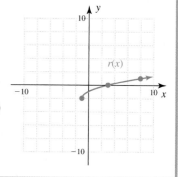

Solution ▶   **a.** The only zero of r is at $(3, 0)$. The graph is on or below the x-axis for $x \in [-1, 3]$, so $r(x) \leq 0$ in this interval.

☑ **B.** You've just learned how to determine intervals where a function is positive or negative

  b. The graph is above the x-axis for $x \in (3, \infty)$, and $r(x) > 0$ in this interval.

Now try Exercises 29 through 32 ▶

C. Intervals Where a Function Is Increasing or Decreasing

In our study of linear graphs, we said a graph was increasing if it "rose" when viewed from left to right. More generally, we say the graph of a function is increasing *on a given interval* if larger and larger x-values produce larger and larger y-values. This suggests the following tests for intervals where a function is increasing or decreasing.

Increasing and Decreasing Functions

Given an interval I that is a subset of the domain, with x_1 and x_2 in I and $x_2 > x_1$,

  1. A function is increasing on I if $f(x_2) > f(x_1)$ for all x_1 and x_2 in I (larger inputs produce larger outputs).

  2. A function is decreasing on I if $f(x_2) < f(x_1)$ for all x_1 and x_2 in I (larger inputs produce smaller outputs).

  3. A function is constant on I if $f(x_2) = f(x_1)$ for all x_1 and x_2 in I (larger inputs produce identical outputs).

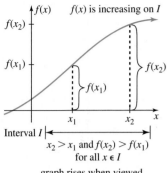

$f(x)$ is increasing on I

$x_2 > x_1$ and $f(x_2) > f(x_1)$
for all $x \in I$

graph rises when viewed
from left to right

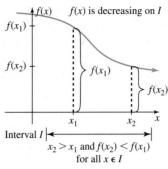

$f(x)$ is decreasing on I

$x_2 > x_1$ and $f(x_2) < f(x_1)$
for all $x \in I$

graph falls when viewed
from left to right

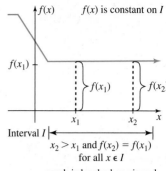

$f(x)$ is constant on I

$x_2 > x_1$ and $f(x_2) = f(x_1)$
for all $x \in I$

graph is level when viewed
from left to right

WORTHY OF NOTE

Questions about the behavior
of a function are asked with
respect to the *y* outputs:
where is the *function* positive,
where is the *function* increasing, etc. Due to the input/
output, cause/effect nature of
functions, the response is
given in terms of *x*, that is,
what is *causing* outputs to be
negative, or to be decreasing.

Consider the graph of $f(x) = -x^2 + 4x + 5$
in Figure 2.47. Since the graph opens downward
with the vertex at $(2, 9)$, the function must increase
until it reaches this maximum value at $x = 2$, and
decrease thereafter. Notationally we'll write this as
$f(x){\uparrow}$ for $x \in (-\infty, 2)$ and $f(x){\downarrow}$ for $x \in (2, \infty)$.
Using the interval $(-3, 2)$ shown, we see that any
larger input value from the interval will indeed
produce a larger output value, and $f(x){\uparrow}$ on the
interval. For instance,

$$1 > -2 \qquad\qquad x_2 > x_1$$

$$\text{and} \qquad\qquad\qquad \text{and}$$

$$f(1) > f(-2) \qquad f(x_2) > f(x_1)$$
$$8 > -7$$

Figure 2.47

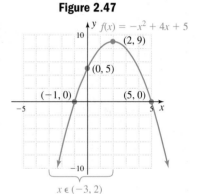

EXAMPLE 5 ▶ **Finding Intervals Where a Function Is Increasing
or Decreasing**

Use the graph of $v(x)$ given to name the interval(s)
where v is increasing, decreasing, or constant.

Solution ▶ From left to right, the graph of v increases until
leveling off at $(-2, 2)$, then it remains constant
until reaching $(1, 2)$. The graph then increases
once again until reaching a peak at $(3, 5)$ and
decreases thereafter. The result is $v(x){\uparrow}$ for
$x \in (-\infty, -2) \cup (1, 3)$, $v(x){\downarrow}$ for $x \in (3, \infty)$, and
$v(x)$ is constant for $x \in (-2, 1)$.

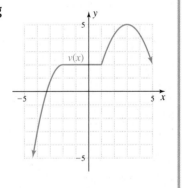

Now try Exercises 33 through 36 ▶

Notice the graph of f in Figure 2.47 and the graph of v in Example 5 have something in common. It appears that both the far left and far right branches of each graph
point downward (in the negative *y*-direction). We say that the **end behavior** of both
graphs is identical, which is the term used to describe what happens to a graph as $|x|$
becomes very large. For $x > 0$, we say a graph is, "up on the right" or "down on the
right," depending on the direction the "end" is pointing. For $x < 0$, we say the graph
is "up on the left" or "down on the left," as the case may be.

EXAMPLE 6 ▶ **Describing the End Behavior of a Graph**

The graph of $f(x) = x^3 - 3x$ is shown. Use the graph to name intervals where f is increasing or decreasing, and comment on the end-behavior of the graph.

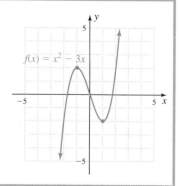

Solution ▶ From the graph we observe that: $f(x)\uparrow$ for $x \in (-\infty, -1) \cup (1, \infty)$, and $f(x)\downarrow$ for $x \in (-1, 1)$. The end behavior of the graph is down on the left, up on the right (down/up).

 C. You've just learned how to determine where a function is increasing or decreasing

Now try Exercises 37 through 40 ▶

D. More on Maximum and Minimum Values

The y-coordinate of the vertex of a parabola where $a < 0$, and the y-coordinate of "peaks" from other graphs are called **maximum values**. A **global maximum** (also called an *absolute* maximum) names the largest range value over the entire domain. A local **maximum** (also called a *relative* maximum) gives the largest range value in a specified interval; and an **endpoint maximum** can occur at an endpoint of the domain. The same can be said for the corresponding minimum values.

We will soon develop the ability to locate maximum and minimum values for quadratic and other functions. In future courses, methods are developed to help locate maximum and minimum values for almost *any* function. For now, our work will rely chiefly on a function's graph.

EXAMPLE 7 ▶ **Analyzing Characteristics of a Graph**

Analyze the graph of function f shown in Figure 2.48. Include specific mention of

   a. domain and range,
   b. intervals where f is increasing or decreasing,
   c. maximum (max) and minimum (min) values,
   d. intervals where $f(x) \geq 0$ and $f(x) < 0$,
   e. whether the function is even, odd, or neither.

Figure 2.48

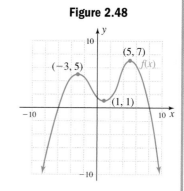

Solution ▶ **a.** Using vertical and horizontal boundary lines show the domain is $x \in \mathbb{R}$, with range: $y \in (-\infty, 7]$.

   b. $f(x)\uparrow$ for $x \in (-\infty, -3) \cup (1, 5)$ shown in blue in Figure 2.49, and $f(x)\downarrow$ for $x \in (-3, 1) \cup (5, \infty)$ as shown in red.

   c. From Part (b) we find that $y = 5$ at $(-3, 5)$ and $y = 7$ at $(5, 7)$ are local maximums, with a local minimum of $y = 1$ at $(1, 1)$. The point $(5, 7)$ is also a global maximum (there is no global minimum).

   d. $f(x) \geq 0$ for $x \in [-6, 8]$; $f(x) < 0$ for $x \in (-\infty, -6) \cup (8, \infty)$

   e. The function is neither even nor odd.

Figure 2.49

 D. You've just learned how to identify the maximum and minimum values of a function

Now try Exercises 41 through 48 ▶

The ideas presented here can be applied to functions of all kinds, including rational functions, piecewise-defined functions, step functions, and so on. There is a wide variety of applications in **Exercises 51 through 58.**

E. Rates of Change and the Difference Quotient

We complete our study of graphs by revisiting the concept of average rates of change. In many business, scientific, and economic applications, it is this attribute of a function that draws the most attention. In Section 2.4 we computed average rates of change by selecting two points from a graph, and computing the slope of the secant line: $m = \dfrac{\Delta y}{\Delta x} = \dfrac{y_2 - y_1}{x_2 - x_1}$. With a simple change of notation, we can *use the function's equation* rather than relying on a graph. Note that y_2 corresponds to the function evaluated at x_2: $y_2 = f(x_2)$. Likewise, $y_1 = f(x_1)$. Substituting these into the slope formula yields $\dfrac{\Delta y}{\Delta x} = \dfrac{f(x_2) - f(x_1)}{x_2 - x_1}$, giving the average rate of change between x_1 and x_2 *for any function f* (assuming the function is smooth and continuous between x_1 and x_2).

Average Rate of Change

For a function f and $[x_1, x_2]$ a subset of the domain, the average rate of change between x_1 and x_2 is

$$\frac{\Delta y}{\Delta x} = \frac{f(x_2) - f(x_1)}{x_2 - x_1}, x_1 \neq x_2$$

Average Rates of Change Applied to Projectile Velocity

A projectile is any object that is thrown, shot, or cast upward, with no continuing source of propulsion. The object's height (in feet) after t sec is modeled by the function $h(t) = -16t^2 + vt + k$, where v is the initial velocity of the projectile, and k is the height of the object at contact. For instance, if a soccer ball is kicked upward from ground level ($k = 0$) with an initial speed of 64 ft/sec, the height of the ball t sec later is $h(t) = -16t^2 + 64t$. From Section 2.5, we recognize the graph will be a parabola and evaluating the function for $t = 0$ to 4 produces Table 2.4 and the graph shown in Figure 2.50. Experience tells us the ball is traveling at a faster rate immediately after being kicked, as compared to when it nears its maximum height where it momentarily stops, then begins its descent. In other words, the rate of change $\dfrac{\Delta \text{height}}{\Delta \text{time}}$ has a larger value at any time prior to reaching its maximum height. To quantify this we'll compute the average rate of change between $t = 0.5$ and $t = 1$, and compare it to the average rate of change between $t = 1$ and $t = 1.5$.

WORTHY OF NOTE

Keep in mind the graph of h represents the relationship between the soccer ball's height in feet and the elapsed time t. It does not model the actual path of the ball.

Table 2.4

Time in seconds	Height in feet
0	0
1	48
2	64
3	48
4	0

Figure 2.50

EXAMPLE 8 ▶ **Calculating Average Rates of Change**

For the projectile function $h(t) = -16t^2 + 64t$, find

a. the average rate of change for $t \in [0.5, 1]$

b. the average rate of change for $t \in [1, 1.5]$.

Then graph the secant lines representing these average rates of change and comment.

Solution ▶ Using the given intervals in the formula $\dfrac{\Delta h}{\Delta t} = \dfrac{h(t_2) - h(t_1)}{t_2 - t_1}$ yields

a. $\dfrac{\Delta h}{\Delta t} = \dfrac{h(1) - h(0.5)}{1 - (0.5)}$

$= \dfrac{48 - 28}{0.5}$

$= 40$

b. $\dfrac{\Delta h}{\Delta t} = \dfrac{h(1.5) - h(1)}{1.5 - 1}$

$= \dfrac{60 - 48}{0.5}$

$= 24$

For $t \in [0.5, 1]$, the average rate of change is $\frac{40}{1}$, meaning the height of the ball is increasing at an average rate of 40 ft/sec. For $t \in [1, 1.5]$, the average rate of change has slowed to $\frac{24}{1}$, and the soccer ball's height is increasing at only 24 ft/sec. The secant lines representing these rates of change are shown in the figure, where we note the line from the first interval (in **red**), has a steeper slope than the line from the second interval (in **blue**).

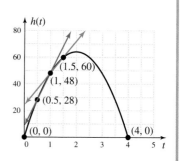

Now try Exercises 59 through 64 ▶

The approach in Example 8 works very well, but requires us to recalculate $\dfrac{\Delta y}{\Delta x}$ for each new interval. Using a slightly different approach, we can develop a general *formula* for the average rate of change. This is done by selecting a point $x_1 = x$ from the domain, then a point $x_2 = x + h$ that is very close to x. Here, $h \neq 0$ is assumed to be a small, arbitrary constant, meaning the interval $[x, x + h]$ is very small as well. Substituting $x + h$ for x_2 and x for x_1 in the rate of change formula gives $\dfrac{\Delta y}{\Delta x} = \dfrac{f(x + h) - f(x)}{(x + h) - x} = \dfrac{f(x + h) - f(x)}{h}$. The result is called the **difference quotient** and represents the average rate of change between x and $x + h$, or equivalently, the slope of the secant line for this interval.

The Difference Quotient

For a function $f(x)$ and constant $h \neq 0$, if f is smooth and continuous on the interval containing x and $x + h$,

$$\frac{f(x + h) - f(x)}{h}$$

is the difference quotient for f.

Note the formula has three parts: (1) the function f evaluated at $x + h \rightarrow f(x + h)$, (2) the function f itself, and (3) the constant h. For convenience, the expression $f(x + h)$ can be evaluated and simplified prior to its use in the difference quotient.

$$\overset{(1)}{f(x + h)} - \overset{(2)}{f(x)}$$
$$\overline{\underset{(3)}{h}}$$

EXAMPLE 9 ▶ **Computing a Difference Quotient and Average Rates of Change**

For $f(x) = x^2 - 4x$,

a. Compute the difference quotient.

b. Find the average rate of change in the intervals [1.9, 2.0] and [3.6, 3.7].

c. Sketch the graph of f along with the secant lines and comment on what you notice.

Solution ▶ **a.** For $f(x) = x^2 - 4x, f(x + h) = (x + h)^2 - 4(x + h)$
$$= x^2 + 2xh + h^2 - 4x - 4h$$

Using this result in the difference quotient yields,

$$\frac{f(x + h) - f(x)}{h} = \frac{(x^2 + 2xh + h^2 - 4x - 4h) - (x^2 - 4x)}{h} \qquad \text{substitute into the difference quotient}$$

$$= \frac{x^2 + 2xh + h^2 - 4x - 4h - x^2 + 4x}{h} \qquad \text{eliminate parentheses}$$

$$= \frac{2xh + h^2 - 4h}{h} \qquad \text{combine like terms}$$

$$= \frac{h(2x + h - 4)}{h} \qquad \text{factor out } h$$

$$= 2x - 4 + h \qquad \text{result}$$

b. For the interval [1.9, 2.0], $x = 1.9$ and $h = 0.1$. The slope of the secant line is
$\dfrac{\Delta y}{\Delta x} = 2(1.9) - 4 + 0.1 = -0.1$. For the interval [3.6, 3.7], $x = 3.6$ and $h = 0.1$. The slope of this secant line is
$\dfrac{\Delta y}{\Delta x} = 2(3.6) - 4 + 0.1 = 3.3$.

c. After sketching the graph of f and the secant lines from each interval (see the figure), we note the slope of the first line (in **red**) is negative and very near zero, while the slope of the second (in **blue**) is positive and very steep.

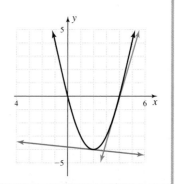

Now try Exercises 65 through 76 ▶

You might be familiar with Galileo Galilei and his studies of gravity. According to popular history, he demonstrated that unequal weights will fall equal distances in equal time periods, by dropping cannonballs from the upper floors of the Leaning Tower of Pisa. Neglecting air resistance, this distance an object falls is modeled by the function $d(t) = 16t^2$, where $d(t)$ represents the distance fallen after t sec. Due to the effects of gravity, the velocity of the object increases as it falls. In other words, the velocity or the average rate of change $\dfrac{\Delta \textbf{distance}}{\Delta \textbf{time}}$ is a nonconstant (increasing) rate of change. We can analyze this rate of change using the difference quotient.

EXAMPLE 10 ▶ **Applying the Difference Quotient in Context**

A construction worker drops a heavy wrench from atop the girder of new skyscraper. Use the function $d(t) = 16t^2$ to

a. Compute the distance the wrench has fallen after 2 sec and after 7 sec.

b. Find a formula for the velocity of the wrench (average rate of change in distance per unit time).

c. Use the formula to find the rate of change in the intervals [2, 2.01] and [7, 7.01].

d. Graph the function and the secant lines representing the average rate of change. Comment on what you notice.

Solution ▶ **a.** Substituting $t = 2$ and $t = 7$ in the given function yields

$$d(2) = 16(2)^2 \qquad d(7) = 16(7)^2 \quad \text{evaluate } d(t) = 16t^2$$
$$= 16(4) \qquad\qquad = 16(49) \quad \text{square input}$$
$$= 64 \qquad\qquad = 784 \quad \text{multiply}$$

After 2 sec, the wrench has fallen 64 ft; after 7 sec, the wrench has fallen 784 ft.

b. For $d(t) = 16t^2$, $d(t + h) = 16(t + h)^2$, which we compute separately.

$$d(t + h) = 16(t + h)^2 \qquad \text{substitute } t + h \text{ for } t$$
$$= 16(t^2 + 2th + h^2) \qquad \text{square binomial}$$
$$= 16t^2 + 32th + 16h^2 \qquad \text{distribute 16}$$

Using this result in the difference quotient yields

$$\frac{d(t + h) - d(t)}{h} = \frac{(16t^2 + 32th + 16h^2) - 16t^2}{h} \qquad \text{substitute into the difference quotient}$$

$$= \frac{16t^2 + 32th + 16h^2 - 16t^2}{h} \qquad \text{eliminate parentheses}$$

$$= \frac{32th + 16h^2}{h} \qquad \text{combine like terms}$$

$$= \frac{h(32t + 16h)}{h} \qquad \text{factor out } h \text{ and simplify}$$

$$= 32t + 16h \qquad \text{result}$$

For any number of seconds t and h a small increment of time thereafter, the velocity of the wrench is modeled by $\dfrac{\Delta\text{distance}}{\Delta\text{time}} = \dfrac{32t + 16h}{1}$.

c. For the interval $[t, t + h] = [2, 2.01]$, $t = 2$ and $h = 0.01$:

$$\frac{\Delta\text{distance}}{\Delta\text{time}} = \frac{32(2) + 16(0.01)}{1} \qquad \text{substitute 2 for } t \text{ and 0.01 for } h$$
$$= 64 + 0.16 = 64.16$$

Two seconds after being dropped, the velocity of the wrench is approximately 64.16 ft/sec. For the interval $[t, t + h] = [7, 7.01]$, $t = 7$ and $h = 0.01$:

$$\frac{\Delta\text{distance}}{\Delta\text{time}} = \frac{32(7) + 16(0.01)}{1} \qquad \text{substitute 7 for } t \text{ and 0.01 for } h$$
$$= 224 + 0.16 = 224.16$$

Seven seconds after being dropped, the velocity of the wrench is approximately 224.16 ft/sec (about 153 mph).

d.

 E. You've learned how to develop a formula to calculate rates of change for any function

The velocity increases with time, as indicated by the steepness of each secant line.

Now try Exercises 77 and 78 ▶

TECHNOLOGY HIGHLIGHT

Locating Zeroes, Maximums, and Minimums

Graphically, the **zeroes** of a function appear as x-intercepts with coordinates $(x, 0)$. An estimate for these zeroes can easily be found using a graphing calculator. To illustrate, enter the function $y = x^2 - 8x + 9$ on the [**Y =**] screen and graph it using the standard window ([**ZOOM**] 6). We access the option for finding zeroes by pressing [**2nd**] [**TRACE**] (**CALC**), which displays the screen shown in Figure 2.51. Pressing the number "2" selects **2:zero** and returns you to the graph, where you're asked to enter a "Left Bound." The calculator is asking you to narrow the area it has to search. Select any number conveniently to the left of the x-intercept you're interested in. For this graph, we entered a left bound of "0" (press [**ENTER**]). The calculator marks this choice with a "▶" marker (pointing to the right), then asks you to enter a "Right Bound." Select any value to the right of the x-intercept, but be sure the value you enter *bounds only one intercept* (see Figure 2.52). For this graph, a choice of 10 would include both x-intercepts, while a choice of 3 would bound only the intercept on the left. After entering 3, the calculator asks for a "Guess." This option is used when there is more than one zero in the interval, and most of the time we'll bypass this option by pressing [**ENTER**] again. The calculator then finds the zero in the selected interval (if it exists), with the coordinates displayed at the bottom of the screen (Figure 2.53).

The maximum and minimum values of a function are located in the same way. Enter $y = x^3 - 3x - 2$ on the [**Y =**] screen and graph the function. As seen in Figure 2.54, it appears a local maximum occurs near $x = -1$. To check, we access the **CALC 4:maximum** option, which returns you to the graph and asks you for a *Left Bound*, a *Right Bound*, and a *Guess* as before. After entering a left bound of "−3" and a right bound of "0," and

Figure 2.51

Figure 2.52

Figure 2.53

Figure 2.54

Figure 2.55

bypassing the Guess option (note the "▶" and "◀" markers), the calculator locates the maximum you selected, and again displays the coordinates. Due to the algorithm used by the calculator to find these values, a decimal number is sometimes displayed, even if the actual value is an integer (see Figure 2.55).

Use a calculator to find all zeroes and to locate the local maximum and minimum values. Round to the nearest hundredth as needed.

Exercise 1: $y = 2x^2 + 4x - 5$

Exercise 2: $y = w^3 - 3w + 1$

Exercise 3: $y = x^2 - 8x + 9$

Exercise 4: $y = x^3 - 2x^2 - 4x + 8$

Exercise 5: $y = x^4 - 5x^2 - 2x$

Exercise 6: $y = x\sqrt{x + 4}$

2.5 EXERCISES

▶ **CONCEPTS AND VOCABULARY**

Fill in each blank with the appropriate word or phrase. Carefully reread the section if needed.

1. The graph of a polynomial will cross through the x-axis at zeroes of _____ factors of degree 1, and _____ off the x-axis at the zeroes from linear factors of degree 2.

2. If $f(-x) = f(x)$ for all x in the domain, we say that f is an _____ function and symmetric to the _____ axis. If $f(-x) = -f(x)$, the function is _____ and symmetric to the _____.

3. If $f(x_2) > f(x_1)$ for $x_1 < x_2$ for all x in a given interval, the function is _____ in the interval.

4. If $f(c) \geq f(x)$ for all x in a specified interval, we say that $f(c)$ is a local _____ for this interval.

5. Discuss/Explain the following statement and give an example of the conclusion it makes. "If a function f is decreasing to the left of $(c, f(c))$ and increasing to the right of $(c, f(c))$, then $f(c)$ is either a local or a global minimum."

6. Without referring to notes or textbook, list as many features/attributes as you can that are related to analyzing the graph of a function. Include details on how to locate or determine each attribute.

▶ **DEVELOPING YOUR SKILLS**

The following functions are known to be even. Complete each graph using symmetry.

7.

8.

Determine whether the following functions are even:
$f(-k) = f(k)$.

9. $f(x) = -7|x| + 3x^2 + 5$ **10.** $p(x) = 2x^4 - 6x + 1$

11. $g(x) = \frac{1}{3}x^4 - 5x^2 + 1$ **12.** $q(x) = \frac{1}{x^2} - |x|$

The following functions are known to be odd. Complete each graph using symmetry.

13.

14.

Determine whether the following functions are odd:
$f(-k) = -f(k)$.

15. $f(x) = 4\sqrt[3]{x} - x$ **16.** $g(x) = \frac{1}{2}x^3 - 6x$

17. $p(x) = 3x^3 - 5x^2 + 1$ **18.** $q(x) = \frac{1}{x} - x$

Determine whether the following functions are even, odd, or neither.

19. $w(x) = x^3 - x^2$ **20.** $q(x) = \frac{3}{4}x^2 + 3|x|$

21. $p(x) = 2\sqrt[3]{x} - \frac{1}{4}x^3$ **22.** $g(x) = x^3 + 7x$

23. $v(x) = x^3 + 3|x|$ **24.** $f(x) = x^4 + 7x^2 - 30$

Use the graphs given to solve the inequalities indicated. Write all answers in interval notation.

25. $f(x) = x^3 - 3x^2 - x + 3; f(x) \geq 0$

26. $f(x) = x^3 - 2x^2 - 4x + 8; f(x) > 0$

27. $f(x) = x^4 - 2x^2 + 1; f(x) > 0$

28. $f(x) = x^3 + 2x^2 - 4x - 8; f(x) \geq 0$

29. $p(x) = \sqrt[3]{x - 1} - 1; p(x) \geq 0$

30. $q(x) = \sqrt{x + 1} - 2; q(x) > 0$

31. $f(x) = (x - 1)^3 - 1; f(x) \leq 0$

32. $g(x) = -(x + 1)^3 - 1; g(x) < 0$

Name the interval(s) where the following functions are increasing, decreasing, or constant. Write answers using interval notation. Assume all endpoints have integer values.

33. $y = V(x)$

34. $y = H(x)$

35. $y = f(x)$

36. $y = g(x)$

For Exercises 37 through 40, determine (a) interval(s) where the function is increasing, decreasing or constant, and (b) comment on the end behavior.

37. $p(x) = 0.5(x + 2)^3$

38. $q(x) = -\sqrt[3]{x + 1}$

39. $y = f(x)$

40. $y = g(x)$

For Exercises 41 through 48, determine the following (answer in interval notation as appropriate): (a) domain and range of the function; (b) zeroes of the function; (c) interval(s) where the function is greater than or equal to zero, or less than or equal to zero; (d) interval(s) where the function is increasing, decreasing, or constant; and (e) location of any local max or min value(s).

41. $y = H(x)$

42. $y = f(x)$

43. $y = g(x)$

44. $y = h(x)$

45. $y = Y_1$

46. $y = Y_2$

47. $p(x) = (x + 3)^3 + 1$

48. $q(x) = |x - 5| + 3$

▶ **WORKING WITH FORMULAS**

49. Conic sections—hyperbola: $y = \frac{1}{3}\sqrt{4x^2 - 36}$

While the conic sections are not covered in detail until later in the course, we've already developed a number of tools that will help us understand these relations and their graphs. The equation here gives the "upper branches" of a hyperbola, as shown in the figure. Find the following by analyzing the equation: (a) the domain and range; (b) the zeroes of the relation; (c) interval(s) where y is increasing or decreasing; and (d) whether the relation is even, odd, or neither.

50. Trigonometric graphs: $y = \sin(x)$ and $y = \cos(x)$

The trigonometric functions are also studied at some future time, but we can apply the same tools to analyze the graphs of these functions as well. The graphs of $y = \sin x$ and $y = \cos x$ are given, graphed over the interval $x \in [-180, 360]$ degrees. Use them to find (a) the range of the functions; (b) the zeroes of the functions; (c) interval(s) where

y is increasing/decreasing; (d) location of minimum/maximum values; and (e) whether each relation is even, odd, or neither.

▶ APPLICATIONS

51. Catapults and projectiles: Catapults have a long and interesting history that dates back to ancient times, when they were used to launch javelins, rocks, and other projectiles. The diagram given illustrates the path of the projectile after release, which follows a parabolic arc. Use the graph to determine the following:

a. State the domain and range of the projectile.

b. What is the maximum height of the projectile?

c. How far from the catapult did the projectile reach its maximum height?

d. Did the projectile clear the castle wall, which was 40 ft high and 210 ft away?

e. On what interval was the height of the projectile increasing?

f. On what interval was the height of the projectile decreasing?

52. Profit and loss: The profit of DeBartolo Construction Inc. is illustrated by the graph shown. Use the graph to estimate the point(s) or the interval(s) for which the profit P was:

a. increasing

b. decreasing

c. constant

d. a maximum

e. a minimum

f. positive

g. negative

h. zero

53. Functions and rational exponents: The graph of $f(x) = x^{\frac{2}{3}} - 1$ is shown. Use the graph to find:

a. domain and range of the function

b. zeroes of the function

c. interval(s) where $f(x) \geq 0$ or $f(x) < 0$

d. interval(s) where $f(x)$ is increasing, decreasing, or constant

e. location of any max or min value(s)

Exercise 53   **Exercise 54**

54. Analyzing a graph: Given $h(x) = |x^2 - 4| - 5$, whose graph is shown, use the graph to find:

a. domain and range of the function

b. zeroes of the function

c. interval(s) where $h(x) \geq 0$ or $h(x) \leq 0$

d. interval(s) where $f(x)$ is increasing, decreasing, or constant

e. location of any max or min value(s)

55. Analyzing interest rates: The graph shown approximates the average annual interest rates on 30-yr fixed mortgages, rounded to the nearest $\frac{1}{4}\%$. Use the graph to estimate the following (write all answers in interval notation).

   a. domain and range

   b. interval(s) where $I(t)$ is increasing, decreasing, or constant

   c. location of the maximum and minimum values

   d. the one-year period with the greatest rate of increase and the one-year period with the greatest rate of decrease

Source: 1998 Wall Street Journal Almanac, p. 446; 2004 Statistical Abstract of the United States, Table 1178

Year (1972 → 72)

56. Analyzing the deficit: The following graph approximates the federal deficit of the United States. Use the graph to estimate the following (write answers in interval notation).

   a. the domain and range

   b. interval(s) where $D(t)$ is increasing, decreasing, or constant

   c. the location of the maximum and minimum values

   d. the one-year period with the greatest rate of increase, and the one-year period with the greatest rate of decrease

Source: 2005 Statistical Abstract of the United States, Table 461

Year (1975 → 75)

57. Constructing a graph: Draw the function f that has the following characteristics, then state the zeroes and the location of all maximum and minimum values. [*Hint:* Write them as $(c, f(c))$.]

   a. Domain: $x \in (-10, \infty)$

   b. Range: $y \in (-6, \infty)$

   c. $f(0) = 0; f(4) = 0$

   d. $f(x)\uparrow$ for $x \in (-10, -6) \cup (-2, 2) \cup (4, \infty)$

   e. $f(x)\downarrow$ for $x \in (-6, -2) \cup (2, 4)$

   f. $f(x) \geq 0$ for $x \in [-8, -4] \cup [0, \infty)$

   g. $f(x) < 0$ for $x \in (-\infty, -8) \cup (-4, 0)$

58. Constructing a graph: Draw the function g that has the following characteristics, then state the zeroes and the location of all maximum and minimum values. [*Hint:* Write them as $(c, g(c))$.]

   a. Domain: $x \in (-\infty, 8)$

   b. Range: $y \in [-6, \infty)$

   c. $g(0) = 4.5; g(6) = 0$

   d. $g(x)\uparrow$ for $x \in (-6, 3) \cup (6, 8)$

   e. $g(x)\downarrow$ for $x \in (-\infty, -6) \cup (3, 6)$

   f. $g(x) \geq 0$ for $x \in (-\infty, -9] \cup [-3, 8)$

   g. $g(x) < 0$ for $x \in (-9, -3)$

For Exercises 59 to 64, use the formula for the average rate of change $\dfrac{f(x_2) - f(x_1)}{x_2 - x_1}$.

59. **Average rate of change:** For $f(x) = x^3$, (a) calculate the average rate of change for the interval $x = -2$ and $x = -1$ and (b) calculate the average rate of change for the interval $x = 1$ and $x = 2$. (c) What do you notice about the answers from parts (a) and (b)? (d) Sketch the graph of this function along with the lines representing these average rates of change and comment on what you notice.

60. **Average rate of change:** Knowing the general shape of the graph for $f(x) = \sqrt[3]{x}$, (a) is the average rate of change greater between $x = 0$ and $x = 1$ or between $x = 7$ and $x = 8$? Why? (b) Calculate the rate of change for these intervals and verify your response. (c) Approximately how many times greater is the rate of change?

61. **Height of an arrow:** If an arrow is shot vertically from a bow with an initial speed of 192 ft/sec, the height of the arrow can be modeled by the function $h(t) = -16t^2 + 192t$, where $h(t)$ represents the height of the arrow after t sec (assume the arrow was shot from ground level).

 a. What is the arrow's height at $t = 1$ sec?

 b. What is the arrow's height at $t = 2$ sec?

 c. What is the average rate of change from $t = 1$ to $t = 2$?

 d. What is the rate of change from $t = 10$ to $t = 11$? Why is it the same as (c) except for the sign?

62. **Height of a water rocket:** Although they have been around for decades, water rockets continue to be a popular toy. A plastic rocket is filled with water and then pressurized using a handheld pump. The rocket is then released and off it goes! If the rocket has an initial velocity of 96 ft/sec, the height of the rocket can be modeled by the function $h(t) = -16t^2 + 96t$, where $h(t)$ represents the

height of the rocket after t sec (assume the rocket was shot from ground level).

 a. Find the rocket's height at $t = 1$ and $t = 2$ sec.

 b. Find the rocket's height at $t = 3$ sec.

 c. Would you expect the average rate of change to be greater between $t = 1$ and $t = 2$, or between $t = 2$ and $t = 3$? Why?

 d. Calculate each rate of change and discuss your answer.

63. **Velocity of a falling object:** The impact velocity of an object dropped from a height is modeled by $v = \sqrt{2gs}$, where v is the velocity in feet per second (ignoring air resistance), g is the acceleration due to gravity (32 ft/sec^2 near the Earth's surface), and s is the height from which the object is dropped.

 a. Find the velocity at $s = 5$ ft and $s = 10$ ft.

 b. Find the velocity at $s = 15$ ft and $s = 20$ ft.

 c. Would you expect the average rate of change to be greater between $s = 5$ and $s = 10$, or between $s = 15$ and $s = 20$?

 d. Calculate each rate of change and discuss your answer.

64. **Temperature drop:** One day in November, the town of Coldwater was hit by a sudden winter storm that caused temperatures to plummet. During the storm, the temperature T (in degrees Fahrenheit) could be modeled by the function $T(h) = 0.8h^2 - 16h + 60$, where h is the number of hours since the storm began. Graph the function and use this information to answer the following questions.

 a. What was the temperature as the storm began?

 b. How many hours until the temperature dropped below zero degrees?

 c. How many hours did the temperature remain below zero?

 d. What was the coldest temperature recorded during this storm?

Compute and simplify the difference quotient $\dfrac{f(x + h) - f(x)}{h}$ **for each function given.**

65. $f(x) = 2x - 3$ 66. $g(x) = 4x + 1$

67. $h(x) = x^2 + 3$ 68. $p(x) = x^2 - 2$

69. $q(x) = x^2 + 2x - 3$ 70. $r(x) = x^2 - 5x + 2$

71. $f(x) = \dfrac{2}{x}$ 72. $g(x) = \dfrac{-3}{x}$

Use the difference quotient to find: **(a)** a rate of change formula for the functions given and **(b)/(c)** calculate the rate of change in the intervals shown. Then **(d)** sketch the graph of each function along with the secant lines and comment on what you notice.

73. $g(x) = x^2 + 2x$
$[-3.0, -2.9], [0.50, 0.51]$

74. $h(x) = x^2 - 6x$
$[1.9, 2.0], [5.0, 5.01]$

75. $g(x) = x^3 + 1$
$[-2.1, -2], [0.40, 0.41]$

76. $r(x) = \sqrt{x}$ (*Hint:* Rationalize the numerator.)
$[1, 1.1], [4, 4.1]$

77. The distance that a person can see depends on how high they're standing above level ground. On a clear day, the distance is approximated by the function $d(h) = 1.5\sqrt{h}$, where $d(h)$ represents the viewing distance (in miles) at height h (in feet). Find the average rate of change in the intervals (a) $[9, 9.01]$ and (b) $[225, 225.01]$. Then (c) graph the function along with the lines representing the average rates of change and comment on what you notice.

78. A special magnifying lens is crafted and installed in an overhead projector. When the projector is x ft from the screen, the size $P(x)$ of the projected image is x^2. Find the average rate of change for $P(x) = x^2$ in the intervals (a) $[1, 1.01]$ and (b) $[4, 4.01]$. Then (c) graph the function along with the lines representing the average rates of change and comment on what you notice.

▶ **EXTENDING THE THOUGHT**

79. Does the function shown have a maximum value? Does it have a minimum value? Discuss/explain/justify why or why not.

c. By approximately how many seconds?

d. Who was leading at $t = 40$ seconds?

e. During the race, how many seconds was the daughter in the lead?

f. During the race, how many seconds was the mother in the lead?

80. The graph drawn here depicts a 400-m race between a mother and her daughter. Analyze the graph to answer questions (a) through (f).

  a. Who wins the race, the mother or daughter?

  b. By approximately how many meters?

81. Draw a general function $f(x)$ that has a local *maximum* at $(a, f(a))$ and a local *minimum* at $(b, f(b))$ but with $f(a) < f(b)$.

82. Verify that $h(x) = x^{\frac{2}{3}}$ is an even function, by first rewriting h as $h(x) = (x^{\frac{1}{3}})^2$.

▶ **MAINTAINING YOUR SKILLS**

83. (1.5) Solve the given quadratic equation three different ways: (a) factoring, (b) completing the square, and (c) using the quadratic formula:
$x^2 - 8x - 20 = 0$

84. (R.5) Find the (a) sum and (b) product of the rational expressions $\dfrac{3}{x+2}$ and $\dfrac{3}{2-x}$.

85. (2.3) Write the equation of the line shown, in the form $y = mx + b$.

Exercise 85

86. (R.7) Find the surface area and volume of the cylinder shown.

Learning Objectives

In Section 2.6 you will learn how to:

☐ **A.** Identify basic characteristics of the toolbox functions

☐ **B.** Perform vertical/horizontal shifts of a basic graph

☐ **C.** Perform vertical/horizontal reflections of a basic graph

☐ **D.** Perform vertical stretches and compressions of a basic graph

☐ **E.** Perform transformations on a general function $f(x)$

Many applications of mathematics require that we select a function known to fit the context, or build a function model from the information supplied. So far we've looked extensively at linear functions, and have introduced the absolute value, squaring, square root, cubing, and cube root functions. These are the six **toolbox functions,** so called because they give us a variety of "tools" to model the real world. In the same way a study of arithmetic depends heavily on the multiplication table, a study of algebra and mathematical modeling depends (in large part) on a solid working knowledge of these functions.

A. The Toolbox Functions

While we can accurately graph a line using only two points, most toolbox functions require more points to show all of the graph's important features. However, our work is greatly simplified in that each function belongs to a **function family,** in which all graphs from a given family share the characteristics of one basic graph, called the **parent function.** This means the number of points required for graphing will quickly decrease as we start anticipating what the graph of a given function should look like. The parent functions and their identifying characteristics are summarized here.

The Toolbox Functions

Identity function

x	$f(x) = x$
-3	-3
-2	-2
-1	-1
0	0
1	1
2	2
3	3

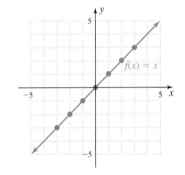

Domain: $x \in (-\infty, \infty)$, Range: $y \in (-\infty, \infty)$
Symmetry: odd
Increasing: $x \in (-\infty, \infty)$
End behavior: down on the left/up on the right

Absolute value function

x	$f(x) = |x|$
-3	3
-2	2
-1	1
0	0
1	1
2	2
3	3

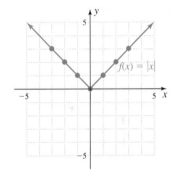

Domain: $x \in (-\infty, \infty)$, Range: $y \in [0, \infty)$
Symmetry: even
Decreasing: $x \in (-\infty, 0)$; Increasing: $x \in (0, \infty)$
End behavior: up on the left/up on the right
Vertex at $(0, 0)$

Squaring function

x	$f(x) = x^2$
-3	9
-2	4
-1	1
0	0
1	1
2	4
3	9

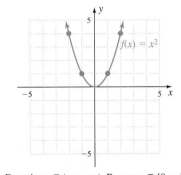

Domain: $x \in (-\infty, \infty)$, Range: $y \in [0, \infty)$
Symmetry: even
Decreasing: $x \in (-\infty, 0)$; Increasing: $x \in (0, \infty)$
End behavior: up on the left/up on the right
Vertex at $(0, 0)$

Square root function

x	$f(x) = \sqrt{x}$
-2	$-$
-1	$-$
0	0
1	1
2	≈ 1.41
3	≈ 1.73
4	2

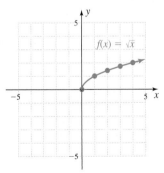

Domain: $x \in [0, \infty)$, Range: $y \in [0, \infty)$
Symmetry: neither even nor odd
Increasing: $x \in (0, \infty)$
End behavior: up on the right
Initial point at $(0, 0)$

Cubing function

x	$f(x) = x^3$
-3	-27
-2	-8
-1	-1
0	0
1	1
2	8
3	27

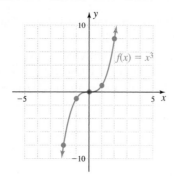

Domain: $x \in (-\infty, \infty)$, Range: $y \in (-\infty, \infty)$
Symmetry: odd
Increasing: $x \in (-\infty, \infty)$
End behavior: down on the left/up on the right
Point of inflection at $(0, 0)$

Cube root function

x	$f(x) = \sqrt[3]{x}$
-27	-3
-8	-2
-1	-1
0	0
1	1
8	2
27	3

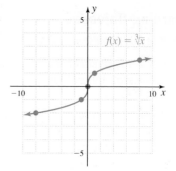

Domain: $x \in (-\infty, \infty)$, Range: $y \in (-\infty, \infty)$
Symmetry: odd
Increasing: $x \in (-\infty, \infty)$
End behavior: down on the left/up on the right
Point of inflection at $(0, 0)$

In applications of the toolbox functions, the parent graph may be altered and/or shifted from its original position, yet the graph will still retain its basic shape and features. The result is called a **transformation** of the parent graph. Analyzing the new graph (as in Section 2.5) will often provide the answers needed.

EXAMPLE 1 ▶ **Identifying the Characteristics of a Transformed Graph**

The graph of $f(x) = x^2 - 2x - 3$ is given.
Use the graph to identify each of the features
or characteristics indicated.

   a. function family

   b. domain and range

   c. vertex

   d. max or min value(s)

   e. end behavior

   f. x- and y-intercept(s)

Solution ▶   **a.** The graph is a parabola, from the squaring function family.

   b. domain: $x \in (-\infty, \infty)$; range: $y \in [-4, \infty)$

   c. vertex: $(1, -4)$

   d. minimum value $y = -4$ at $(1, -4)$

   e. end-behavior: up/up

   f. y-intercept: $(0, -3)$; x-intercepts: $(-1, 0)$ and $(3, 0)$

Now try Exercises 7 through 34 ▶

☑ **A.** You've just learned how to identify basic characteristics of the toolbox functions

Note that we can algebraically verify the x-intercepts by substituting 0 for $f(x)$ and solving the equation by factoring. This gives $0 = (x + 1)(x - 3)$, with solutions $x = -1$ and $x = 3$. It's also worth noting that while the parabola is no longer symmetric to the y-axis, it *is* symmetric to the vertical line $x = 1$. This line is called the **axis of symmetry** for the parabola, and will always be a vertical line that goes through the vertex.

B. Vertical and Horizontal Shifts

As we study specific transformations of a graph, try to develop a *global view* as the transformations can be applied to any function. When these are applied to the toolbox functions, we rely on characteristic features of the parent function to assist in completing the transformed graph.

Vertical Translations

We'll first investigate vertical translations or vertical shifts of the toolbox functions, using the absolute value function to illustrate.

EXAMPLE 2 ▶ **Graphing Vertical Translations**

Construct a table of values for $f(x) = |x|$, $g(x) = |x| + 1$, and $h(x) = |x| - 3$ and graph the functions on the same coordinate grid. Then discuss what you observe.

Solution ▶ A table of values for all three functions is given, with the corresponding graphs shown in the figure.

x	$f(x) = |x|$	$g(x) = |x| + 1$	$h(x) = |x| - 3$
-3	3	4	0
-2	2	3	-1
-1	1	2	-2
0	0	1	-3
1	1	2	-2
2	2	3	-1
3	3	4	0

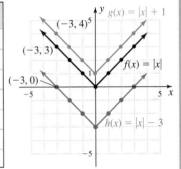

Note that outputs of $g(x)$ are one more than the outputs for $f(x)$, and that each point on the graph of f has been shifted *upward 1 unit* to form the graph of g. Similarly, each point on the graph of f has been shifted *downward 3 units* to form the graph of h. Since $h(x) = f(x) - 3$.

Now try Exercises 35 through 42 ▶

We describe the transformations in Example 2 as a **vertical shift** or **vertical translation** of a basic graph. The graph of g is the graph of f *shifted up 1 unit,* and the graph of h is the graph of f *shifted down 3 units*. In general, we have the following:

Vertical Translations of a Basic Graph

Given $k > 0$ and any function whose graph is determined by $y = f(x)$,
1. The graph of $y = f(x) + k$ is the graph of $f(x)$ shifted upward k units.
2. The graph of $y = f(x) - k$ is the graph of $f(x)$ shifted downward k units.

Horizontal Translations

The graph of a parent function can also be shifted left or right. This happens when we *alter the inputs to the basic function,* as opposed to adding or subtracting something to the basic function itself. For $Y_1 = x^2 + 2$ note that we first square inputs, then add 2, which results in a vertical shift. For $Y_2 = (x + 2)^2$, we add 2 to x *prior to squaring* and since the input values are affected, we might anticipate the graph will shift along the x-axis—horizontally.

EXAMPLE 3 ▶ **Graphing Horizontal Translations**

Construct a table of values for $f(x) = x^2$ and $g(x) = (x + 2)^2$, then graph the functions on the same grid and discuss what you observe.

Solution ▶ Both f and g belong to the quadratic family and their graphs are parabolas. A table of values is shown along with the corresponding graphs.

x	$f(x) = x^2$	$g(x) = (x + 2)^2$
-3	9	1
-2	4	0
-1	1	1
0	0	4
1	1	9
2	4	16
3	9	25

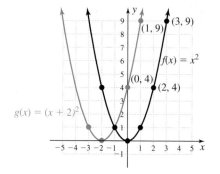

It is apparent the graphs of g and f are identical, but the graph of g has been shifted horizontally 2 units left.

Now try Exercises 43 through 46 ▶

We describe the transformation in Example 3 as a **horizontal shift** or **horizontal translation** of a basic graph. The graph of g is the graph of f, *shifted 2 units to the left.* Once again it seems reasonable that since *input* values were altered, the shift must be horizontal rather than vertical. From this example, we also learn the direction of the shift is **opposite the sign:** $y = (x + 2)^2$ is 2 units *to the left* of $y = x^2$. Although it may seem counterintuitive, the shift *opposite the sign* can be "seen" by locating the new x-intercept, which in this case is also the vertex. Substituting 0 for y gives $0 = (x + 2)^2$ with $x = -2$, as shown in the graph. In general, we have

Horizontal Translations of a Basic Graph

Given $h > 0$ and any function whose graph is determined by $y = f(x)$,
1. The graph of $y = f(x + h)$ is the graph of $f(x)$ shifted *to the left h* units.
2. The graph of $y = f(x - h)$ is the graph of $f(x)$ shifted *to the right h* units.

EXAMPLE 4 ▶ **Graphing Horizontal Translations**

Sketch the graphs of $g(x) = |x - 2|$ and $h(x) = \sqrt{x + 3}$ using a horizontal shift of the parent function and a few characteristic points (not a table of values).

Solution ▶ The graph of $g(x) = |x - 2|$ (Figure 2.56) is the absolute value function shifted 2 units to the right (shift the vertex and two other points from $y = |x|$). The graph of $h(x) = \sqrt{x + 3}$ (Figure 2.57) is a square root function, shifted 3 units to the left (shift the initial point and one or two points from $y = \sqrt{x}$).

Figure 2.56

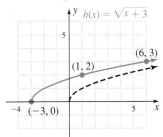

Figure 2.57

✓ **B.** You've just learned how to perform vertical/horizontal shifts of a basic graph

Now try Exercises 47 through 50 ▶

C. Vertical and Horizontal Reflections

The next transformation we investigate is called a **vertical reflection,** in which we compare the function $Y_1 = f(x)$ with the negative of the function: $Y_2 = -f(x)$.

Vertical Reflections

EXAMPLE 5 ▶   Graphing Vertical Reflections

Construct a table of values for $Y_1 = x^2$ and $Y_2 = -x^2$, then graph the functions on the same grid and discuss what you observe.

Solution ▶   A table of values is given for both functions, along with the corresponding graphs.

x	$Y_1 = x^2$	$Y_2 = -x^2$
-2	4	-4
-1	1	-1
0	0	0
1	1	-1
2	4	-4

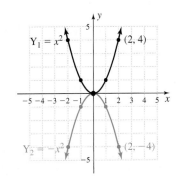

As you might have anticipated, the outputs for f and g differ only in sign. Each output is a **reflection** of the other, being an equal distance from the x-axis but on opposite sides.

Now try Exercises 51 and 52 ▶

The vertical reflection in Example 5 is called a **reflection across the x-axis.** In general,

Vertical Reflections of a Basic Graph

For any function $y = f(x)$, the graph of $y = -f(x)$
is the graph of $f(x)$ reflected across the x-axis.

Horizontal Reflections

It's also possible for a graph to be reflected horizontally *across the y-axis*. Just as we noted that $f(x)$ versus $-f(x)$ resulted in a vertical reflection, $f(x)$ versus $f(-x)$ results in a horizontal reflection.

EXAMPLE 6 ▶ **Graphing a Horizontal Reflection**

Construct a table of values for $f(x) = \sqrt{x}$ and $g(x) = \sqrt{-x}$, then graph the functions on the same coordinate grid and discuss what you observe.

Solution ▶ A table of values is given here, along with the corresponding graphs.

x	$f(x) = \sqrt{x}$	$g(x) = \sqrt{-x}$
-4	not real	2
-2	not real	$\sqrt{2} \approx 1.41$
-1	not real	1
0	0	0
1	1	not real
2	$\sqrt{2} \approx 1.41$	not real
4	2	not real

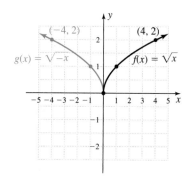

The graph of g is the same as the graph of f, but it has been reflected across the y-axis. A study of the domain shows why—f represents a real number only for nonnegative inputs, so its graph occurs to the right of the y-axis, while g represents a real number for nonpositive inputs, so its graph occurs to the left.

Now try Exercises 53 and 54 ▶

The transformation in Example 6 is called a **horizontal reflection** of a basic graph. In general,

> **Horizontal Reflections of a Basic Graph**
>
> For any function $y = f(x)$, the graph of $y = f(-x)$ is the graph of $f(x)$ reflected across the y-axis.

☑ **C.** You've just learned how to perform vertical/horizontal reflections of a basic graph

D. Vertically Stretching/Compressing a Basic Graph

As the words "stretching" and "compressing" imply, the graph of a basic function can also become elongated or flattened after certain transformations are applied. However, even these transformations preserve the key characteristics of the graph.

EXAMPLE 7 ▶ **Stretching and Compressing a Basic Graph**

Construct a table of values for $f(x) = x^2$, $g(x) = 3x^2$, and $h(x) = \frac{1}{3}x^2$, then graph the functions on the same grid and discuss what you observe.

Solution ▶ A table of values is given for all three functions, along with the corresponding graphs.

x	$f(x) = x^2$	$g(x) = 3x^2$	$h(x) = \frac{1}{3}x^2$
-3	9	27	3
-2	4	12	$\frac{4}{3}$
-1	1	3	$\frac{1}{3}$
0	0	0	0
1	1	3	$\frac{1}{3}$
2	4	12	$\frac{4}{3}$
3	9	27	3

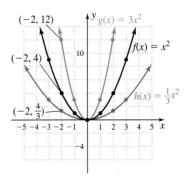

The outputs of g are triple those of f, making these outputs farther from the x-axis and *stretching* g upward (making the graph more narrow). The outputs of h are one-third those of f, and the graph of h is *compressed* downward, with its outputs closer to the x-axis (making the graph wider).

Now try Exercises 55 through 62 ▶

The transformations in Example 7 are called **vertical stretches** or **compressions** of a basic graph. In general,

---

Stretches and Compressions of a Basic Graph

For any function $y = f(x)$, the graph of $y = af(x)$ is
 1. the graph of $f(x)$ stretched vertically if $|a| > 1$,
 2. the graph of $f(x)$ compressed vertically if $0 < |a| < 1$.

☑ **D. You've just learned how to perform vertical stretches and compressions of a basic graph**

E. Transformations of a General Function

If more than one transformation is applied to a basic graph, it's helpful to use the following sequence for graphing the new function.

General Transformations of a Basic Graph

Given a function $y = f(x)$, the graph of $y = af(x \pm h) \pm k$ can be obtained by applying the following sequence of transformations:
 1. horizontal shifts   2. reflections
 3. stretches or compressions   4. vertical shifts

We generally use a few characteristic points to track the transformations involved, then draw the transformed graph through the new location of these points.

EXAMPLE 8 ▶ **Graphing Functions Using Transformations**

Use transformations of a parent function to sketch the graphs of
 a. $g(x) = -(x + 2)^2 + 3$ **b.** $h(x) = 2\sqrt[3]{x - 2} - 1$

Solution ▶ **a.** The graph of g is a parabola, shifted left 2 units, reflected across the x-axis, and shifted up 3 units. This sequence of transformations in shown in Figures 2.58 through 2.60.

Figure 2.58

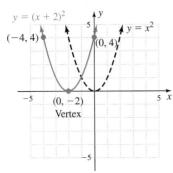

Shifted left 2 units

Figure 2.59

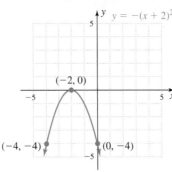

Reflected across the x-axis

Figure 2.60

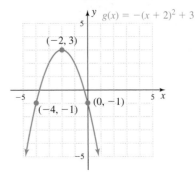

Shifted up 3

b. The graph of h is a cube root function, shifted right 2, stretched by a factor of 2, then shifted down 1. This sequence is shown in Figures 2.61 through 2.63.

Figure 2.61

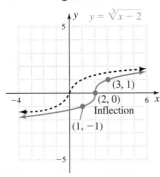

Shifted right 2

Figure 2.62

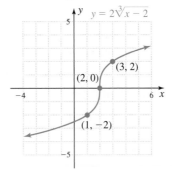

Stretched by a factor of 2

Figure 2.63

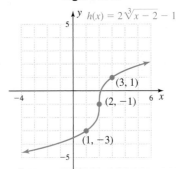

Shifted down 1

Now try Exercises 63 through 92 ▶

It's important to note that the transformations can actually be applied to *any function,* even those that are new and unfamiliar. Consider the following pattern:

Parent Function	Transformation of Parent Function
quadratic: $y = x^2$	$y = -2(x - 3)^2 + 1$
absolute value: $y = |x|$	$y = -2|x - 3| + 1$
cube root: $y = \sqrt[3]{x}$	$y = -2\sqrt[3]{x - 3} + 1$
general: $y = f(x)$	$y = -2f(x - 3) + 1$

In each case, the transformation involves a horizontal shift right 3, a vertical reflection, a vertical stretch, and a vertical shift up 1. Since the shifts are the same regardless of the initial function, we can generalize the results to any function $f(x)$.

General Function

$$y = f(x)$$

Transformed Function

$$y = af(x \pm h) \pm k$$

vertical reflections
vertical stretches and compressions

horizontal shift
h units, opposite direction of sign

vertical shift
k units, same direction as sign

Also bear in mind that the graph will be reflected across the *y*-axis (horizontally) if *x* is replaced with $-x$. Use this illustration to complete Exercise 9. Remember—if the graph of a function is shifted, the *individual points* on the graph are likewise shifted.

EXAMPLE 9 ▶ **Graphing Transformations of a General Function**

Given the graph of $f(x)$ shown in Figure 2.64, graph $g(x) = -f(x + 1) - 2$.

Solution ▶ For *g*, the graph of *f* is (1) shifted horizontally 1 unit left, (2) reflected across the *x*-axis, and (3) shifted vertically 2 units down. The final result is shown in Figure 2.65.

Figure 2.64

Figure 2.65

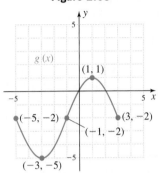

Now try Exercises 93 through 96 ▶

Using the general equation $y = af(x \pm h) \pm k$, we can identify the vertex, initial point, or inflection point of any toolbox function and sketch its graph. Given the *graph* of a toolbox function, we can likewise identify these points and reconstruct its equation. We first identify the function family and the location (h, k) of the characteristic point. By selecting one other point (x, y) on the graph, we then use the general equation as a formula (substituting *h*, *k*, and the *x*- and *y*-values of the second point) to solve for *a* and complete the equation.

EXAMPLE 10 ▶ **Writing the Equation of a Function Given Its Graph**

Find the equation of the toolbox function $f(x)$ shown in Figure 2.66.

Figure 2.66

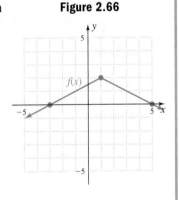

Solution ▶ The function *f* belongs to the absolute value family. The vertex (h, k) is at $(1, 2)$. For an additional point, choose the *x*-intercept $(-3, 0)$ and work as follows:

$y = a|x - h| + k$ general equation

$0 = a|(-3) - 1| + 2$ substitute 1 for *h* and 2 for *k*, substitute -3 for *x* and 0 for *y*

$0 = 4a + 2$ simplify

$-2 = 4a$ subtract 2

$-\dfrac{1}{2} = a$ solve for *a*

The equation for *f* is $y = -\frac{1}{2}|x - 1| + 2$.

☑ **E.** You've just learned how to perform transformations on a general function *f(x)*

Now try Exercises 97 through 102 ▶

TECHNOLOGY HIGHLIGHT

Function Families

Graphing calculators are able to display a number of graphs simultaneously, making them a wonderful tool for studying families of functions. Let's begin by entering the function $y = |x|$ [actually $y = abs(x)$ MATH ▶] as Y_1 on the Y= screen. Next, we enter different variations of the function, but always in terms of its variable name "Y_1." This enables us to simply change the basic function, and observe how the changes affect the graph. Recall that to access the function name Y_1 press VARS ▶ (to access the Y-VARS menu) ENTER (to access the function variables menu) and ENTER (to select Y_1). Enter the functions $Y_2 = Y_1 + 3$ and $Y_3 = Y_1 - 6$ (see Figure 2.67). Graph all three functions in the ZOOM 6:ZStandard window. The calculator draws each graph in the order they were entered and you can always identify the functions by pressing the TRACE key and then the up arrow ▲ or down arrow ▼ keys. In the upper left corner of the window shown in Figure 2.68, the calculator identifies which function the cursor is currently on. Most importantly, note that all functions in this family maintain the same "V" shape.

Next, change Y_1 to $Y_1 = abs(x - 3)$, leaving Y_2 and Y_3 as is. What do you notice when these are graphed again?

Figure 2.67

Figure 2.68

Exercise 1: Change Y_1 to $Y_1 = \sqrt{x}$ and graph, then enter $Y_1 = \sqrt{x - 3}$ and graph once again. What do you observe? What comparisons can be made with the translations of $Y_1 = abs(x)$?

Exercise 2: Change Y_1 to $Y_1 = x^2$ and graph, then enter $Y_1 = (x - 3)^2$ and graph once again. What do you observe? What comparisons can be made with the translations of $Y_1 = abs(x)$ and $Y_1 = \sqrt{x}$?

2.6 EXERCISES

▶ CONCEPTS AND VOCABULARY

Fill in each blank with the appropriate word or phrase. Carefully reread the section if needed.

1. After a vertical _____, points on the graph are farther from the x-axis. After a vertical _____, points on the graph are closer to the x-axis.

2. Transformations that change only the location of a graph and not its shape or form, include _____ and _____.

3. The vertex of $h(x) = 3(x + 5)^2 - 9$ is at _____ and the graph opens _____.

4. The inflection point of $f(x) = -2(x - 4)^3 + 11$ is at _____ and the end behavior is _____, _____.

5. Given the graph of a general function $f(x)$, discuss/explain how the graph of $F(x) = -2f(x + 1) - 3$ can be obtained. If $(0, 5)$, $(6, 7)$, and $(-9, -4)$ are on the graph of f, where do they end up on the graph of F?

6. Discuss/Explain why the shift of $f(x) = x^2 + 3$ is a *vertical shift* of 3 units in the *positive* direction, while the shift of $g(x) = (x + 3)^2$ is a *horizontal shift* 3 units in the *negative* direction. Include several examples linked to a table of values.

▶ DEVELOPING YOUR SKILLS

By carefully inspecting each graph given, (a) indentify the function family; (b) describe or identify the end behavior, vertex, axis of symmetry, and x- and y-intercepts; and (c) determine the domain and range. Assume required features have integer values.

For each graph given, (a) indentify the function family; (b) describe or identify the end behavior, initial point, and x- and y-intercepts; and (c) determine the domain and range. Assume required features have integer values.

7. $f(x) = x^2 + 4x$ **8.** $g(x) = -x^2 + 2x$

13. $p(x) = 2\sqrt{x + 4} - 2$ **14.** $q(x) = -2\sqrt{x + 4} + 2$

9. $p(x) = x^2 - 2x - 3$ **10.** $q(x) = -x^2 + 2x + 8$

15. $r(x) = -3\sqrt{4 - x} + 3$ **16.** $f(x) = 2\sqrt{x + 1} - 4$

11. $f(x) = x^2 - 4x - 5$ **12.** $g(x) = x^2 + 6x + 5$

17. $g(x) = 2\sqrt{4 - x}$ **18.** $h(x) = -2\sqrt{x + 1} + 4$

For each graph given, (a) indentify the function family; (b) describe or identify the end behavior, vertex, axis of symmetry, and x- and y-intercepts; and (c) determine the domain and range. Assume required features have integer values.

19. $p(x) = 2|x + 1| - 4$ **20.** $q(x) = -3|x - 2| + 3$

21. $r(x) = -2|x + 1| + 6$ **22.** $f(x) = 3|x - 2| - 6$

23. $g(x) = -3|x| + 6$ **24.** $h(x) = 2|x + 1|$

For each graph given, (a) indentify the function family; (b) describe or identify the end behavior, inflection point, and x- and y-intercepts; and (c) determine the domain and range. Assume required features have integer values. Be sure to note the scaling of each axis.

25. $f(x) = -(x - 1)^3$ **26.** $g(x) = (x + 1)^3$

27. $h(x) = x^3 + 1$ **28.** $p(x) = -\sqrt[3]{x} + 1$

29. $q(x) = \sqrt[3]{x - 1} - 1$ **30.** $r(x) = -\sqrt[3]{x + 1} - 1$

For Exercises 31–34, identify and state the characteristic features of each graph, including (as applicable) the function family, domain, range, intercepts, vertex, point of inflection, and end behavior.

31.   **32.**

33.   **34.**

Use a table of values to graph the functions given on the same grid. Comment on what you observe.

35. $f(x) = \sqrt{x}$, $g(x) = \sqrt{x} + 2$, $h(x) = \sqrt{x} - 3$

36. $f(x) = \sqrt[3]{x}$, $g(x) = \sqrt[3]{x} - 3$, $h(x) = \sqrt[3]{x} + 1$

37. $p(x) = |x|$, $q(x) = |x| - 5$, $r(x) = |x| + 2$

38. $p(x) = x^2$, $q(x) = x^2 - 4$, $r(x) = x^2 + 1$

Sketch each graph using transformations of a parent function (without a table of values).

39. $f(x) = x^3 - 2$ **40.** $g(x) = \sqrt{x} - 4$

41. $h(x) = x^2 + 3$ **42.** $Y_1 = |x| - 3$

Use a table of values to graph the functions given on the same grid. Comment on what you observe.

43. $p(x) = x^2$, $q(x) = (x + 3)^2$

44. $f(x) = \sqrt{x}$, $g(x) = \sqrt{x + 4}$

45. $Y_1 = |x|$, $Y_2 = |x - 1|$

46. $h(x) = x^3$, $H(x) = (x - 2)^3$

Sketch each graph using transformations of a parent function (without a table of values).

47. $p(x) = (x - 3)^2$ **48.** $Y_1 = \sqrt{x - 1}$

49. $h(x) = |x + 3|$ **50.** $f(x) = \sqrt[3]{x + 2}$

51. $g(x) = -|x|$ **52.** $Y_2 = -\sqrt{x}$

53. $f(x) = \sqrt[3]{-x}$ **54.** $g(x) = (-x)^3$

Use a table of values to graph the functions given on the same grid. Comment on what you observe.

55. $p(x) = x^2$, $q(x) = 2x^2$, $r(x) = \frac{1}{2}x^2$

56. $f(x) = \sqrt{-x}$, $g(x) = 4\sqrt{-x}$, $h(x) = \frac{1}{4}\sqrt{-x}$

57. $Y_1 = |x|$, $Y_2 = 3|x|$, $Y_3 = \frac{1}{3}|x|$

58. $u(x) = x^3$, $v(x) = 2x^3$, $w(x) = \frac{1}{5}x^3$

Sketch each graph using transformations of a parent function (without a table of values).

59. $f(x) = 4\sqrt[3]{x}$ **60.** $g(x) = -2|x|$

61. $p(x) = \frac{1}{3}x^3$ **62.** $q(x) = \frac{3}{4}\sqrt{x}$

Use the characteristics of each function family to match a given function to its corresponding graph. The graphs are not scaled—make your selection based on a careful comparison.

63. $f(x) = \frac{1}{2}x^3$ **64.** $f(x) = \frac{-2}{3}x + 2$

65. $f(x) = -(x - 3)^2 + 2$ **66.** $f(x) = -\sqrt[3]{x - 1} - 1$

67. $f(x) = |x + 4| + 1$ **68.** $f(x) = -\sqrt{x + 6}$

69. $f(x) = -\sqrt{x + 6} - 1$ **70.** $f(x) = x + 1$

71. $f(x) = (x - 4)^2 - 3$ **72.** $f(x) = |x - 2| - 5$

73. $f(x) = \sqrt{x + 3} - 1$ **74.** $f(x) = -(x + 3)^2 + 5$

a.

b.

c.

d.

e.

f.

g.

h.

i.

j.

k.

l.

Graph each function using shifts of a parent function and a few characteristic points. *Clearly state and indicate the transformations used* and identify the location of all vertices, initial points, and/or inflection points.

75. $f(x) = \sqrt{x + 2} - 1$ **76.** $g(x) = \sqrt{x - 3} + 2$

77. $h(x) = -(x + 3)^2 - 2$ **78.** $H(x) = -(x - 2)^2 + 5$

79. $p(x) = (x + 3)^3 - 1$ **80.** $q(x) = (x - 2)^3 + 1$

81. $Y_1 = \sqrt[3]{x + 1} - 2$ **82.** $Y_2 = \sqrt[3]{x - 3} + 1$

83. $f(x) = -|x + 3| - 2$ **84.** $g(x) = -|x - 4| - 2$

85. $h(x) = -2(x + 1)^2 - 3$ **86.** $H(x) = \frac{1}{2}|x + 2| - 3$

87. $p(x) = -\frac{1}{3}(x + 2)^3 - 1$ **88.** $q(x) = 5\sqrt[3]{x + 1} + 2$

89. $Y_1 = -2\sqrt{-x - 1} + 3$ **90.** $Y_2 = 3\sqrt{-x + 2} - 1$

91. $h(x) = \frac{1}{5}(x - 3)^2 + 1$ **92.** $H(x) = -2|x - 3| + 4$

Apply the transformations indicated for the graph of the general functions given.

93.

94.

a. $f(x - 2)$

b. $-f(x) - 3$

c. $\frac{1}{2}f(x + 1)$

d. $f(-x) + 1$

a. $g(x) - 2$

b. $-g(x) + 3$

c. $2g(x + 1)$

d. $\frac{1}{2}g(x - 1) + 2$

95.

96.

a. $h(x) + 3$

b. $-h(x - 2)$

c. $h(x - 2) - 1$

d. $\frac{1}{4}h(x) + 5$

a. $H(x - 3)$

b. $-H(x) + 1$

c. $2H(x - 3)$

d. $\frac{1}{3}H(x - 2) + 1$

Use the graph given and the points indicated to determine the equation of the function shown using the general form $y = af(x \pm h) \pm k$.

97.

98.

99.

100.

101.

102.

▶ **WORKING WITH FORMULAS**

---

103. Volume of a sphere: $V(r) = \frac{4}{3}\pi r^3$

The volume of a sphere is given by the function shown, where $V(r)$ is the volume in cubic units and r is the radius. Note this function belongs to the *cubic family* of functions. Approximate the value of $\frac{4}{3}\pi$ to one decimal place, then graph the function on the interval $[0, 3]$. From your *graph*, estimate the volume of a sphere with radius 2.5 in. Then compute the actual volume. Are the results close?

104. Fluid motion: $V(h) = -4\sqrt{h} + 20$

Suppose the velocity of a fluid flowing from an open tank (no top) through an opening in its side is given by the function shown, where $V(h)$ is the velocity of the fluid (in feet per second) at water height h (in feet). Note this function belongs to the *square root family* of functions. An open tank is 25 ft deep and filled to the brim with fluid. Use a table of values to graph the function on the interval $[0, 25]$. From your graph, estimate the velocity of the fluid when the water level is 7 ft, then find the actual velocity. Are the answers close? If the fluid velocity is 5 ft/sec, how high is the water in the tank?

▶ **APPLICATIONS**

---

105. Gravity, distance, time: After being released, the time it takes an object to fall x ft is given by the function $T(x) = \frac{1}{4}\sqrt{x}$, where $T(x)$ is in seconds. Describe the transformation applied to obtain the graph of T from the graph of $y = \sqrt{x}$, then sketch the graph of T for $x \in [0, 100]$. How long would it take an object to hit the ground if it were dropped from a height of 81 ft?

106. Stopping distance: In certain weather conditions, accident investigators will use the function $v(x) = 4.9\sqrt{x}$ to estimate the speed of a car (in miles per hour) that has been involved in an accident, based on the length of the skid marks x (in feet). Describe the transformation applied to obtain the graph of v from the graph of $y = \sqrt{x}$, then sketch the graph of v for $x \in [0, 400]$. If the skid marks were 225 ft long, how fast was the car traveling? Is this point on your graph?

107. Wind power: The power P generated by a certain wind turbine is given by the function $P(v) = \frac{8}{125}v^3$ where $P(v)$ is the power in watts at wind velocity v (in miles per hour). (a) Describe the transformation applied to obtain the graph of P from the graph of $y = v^3$, then sketch the graph of P for $v \in [0, 25]$ (scale the axes appropriately). (b) How much power is being generated when the wind is blowing at 15 mph? (c) Calculate the rate of change $\frac{\Delta P}{\Delta v}$ in the intervals $[8, 10]$ and $[28, 30]$. What do you notice?

108. Wind power: If the power P (in watts) being generated by a wind turbine is known, the velocity of the wind can be determined using the function

$v(P) = \left(\frac{5}{2}\right)\sqrt[3]{P}$. Describe the transformation applied to obtain the graph of v from the graph of $y = \sqrt[3]{P}$, then sketch the graph of v for $P \in [0, 512]$ (scale the axes appropriately). How fast is the wind blowing if 343W of power is being generated?

109. Acceleration due to gravity: The *distance* a ball rolls down an inclined plane is given by the function $d(t) = 2t^2$, where $d(t)$ represents the distance in feet after t sec. (a) Describe the transformation applied to obtain the graph of d from the graph of $y = t^2$, then sketch the graph of d for $t \in [0, 3]$. (b) How far has the ball rolled after 2.5 sec? (c) Calculate the rate of change $\frac{\Delta d}{\Delta t}$ in the intervals $[1, 1.5]$ and $[3, 3.5]$. What do you notice?

110. Acceleration due to gravity: The *velocity* of a steel ball bearing as it rolls down an inclined plane is given by the function $v(t) = 4t$, where $v(t)$ represents the velocity in feet per second after t sec. Describe the transformation applied to obtain the graph of v from the graph of $y = t$, then sketch the graph of v for $t \in [0, 3]$. What is the velocity of the ball bearing after 2.5 sec?

▶ **EXTENDING THE CONCEPT**

111. Carefully graph the functions $f(x) = |x|$ and $g(x) = 2\sqrt{x}$ on the same coordinate grid. From the graph, in what interval is the graph of $g(x)$ *above* the graph of $f(x)$? Pick a number (call it h) from this interval and substitute it in both functions. Is $g(h) > f(h)$? In what interval is the graph of $g(x)$ below the graph of $f(x)$? Pick a number from this interval (call it k) and substitute it in both functions. Is $g(k) < f(k)$?

112. Sketch the graph of $f(x) = -2|x - 3| + 8$ using transformations of the parent function, then determine the area of the region in quadrant I that is beneath the graph and bounded by the vertical lines $x = 0$ and $x = 6$.

113. Sketch the graph of $f(x) = x^2 - 4$, then sketch the graph of $F(x) = |x^2 - 4|$ using your intuition and the meaning of absolute value (not a table of values). What happens to the graph?

▶ **MAINTAINING YOUR SKILLS**

114. (2.1) Find the distance between the points $(-13, 9)$ and $(7, -12)$, and the slope of the line containing these points.

115. (R.7) Find the perimeter and area of the figure shown (note the units).

32 in.

32 in.

38 in.

2 ft

116. (1.1) Solve for x: $\frac{2}{3}x + \frac{1}{4} = \frac{1}{2}x - \frac{7}{12}$.

117. (2.5) Without graphing, state intervals where $f(x)\uparrow$ and $f(x)\downarrow$ for $f(x) = (x - 4)^2 + 3$.

Learning Objectives

In Section 2.7 you will learn how to:

☐ **A.** State the equation and domain of a piecewise-defined function

☐ **B.** Graph functions that are piecewise-defined

☐ **C.** Solve applications involving piecewise-defined functions

Most of the functions we've studied thus far have been smooth and continuous. Although "smooth" and "continuous" are defined more formally in advanced courses, for our purposes *smooth* simply means the graph has no sharp turns or jagged edges, and *continuous* means you can draw the entire graph without lifting your pencil. In this section, we study a special class of functions, called **piecewise-defined functions,** whose graphs may be various combinations of smooth/not smooth and continuous/not continuous. The absolute value function is one example (see Exercise 31). Such functions have a tremendous number of applications in the real world.

A. The Domain of a Piecewise-Defined Function

For the years 1990 to 2000, the American bald eagle remained on the nation's endangered species list, although the number of breeding pairs was growing slowly. After 2000, the population of eagles grew at a much faster rate, and they were removed from the list soon afterward. From Table 2.5 and plotted points modeling this growth (see Figure 2.69), we observe that a linear model would fit the period from 1992 to 2000 very well, but a line with greater slope would be needed for the years 2000 to 2006 and (perhaps) beyond.

Table 2.5

Year	Bald Eagle Breeding Pairs	Year	Bald Eagle Breeding Pairs
2	3700	10	6500
4	4400	12	7600
6	5100	14	8700
8	5700	16	9800

Source: www.fws.gov/midwest/eagle/population
1990 corresponds to year 0.

Figure 2.69

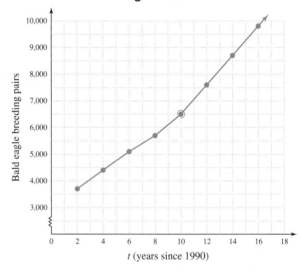

t (years since 1990)

WORTHY OF NOTE

For the years 1992 to 2000, we can estimate the growth in breeding pairs $\frac{\Delta \text{pairs}}{\Delta \text{time}}$ using the points (2, 3700) and (10, 6500) in the slope formula. The result is $\frac{350}{1}$, or 350 pairs per year. For 2000 to 2006, using (10, 6500) and (16, 9800) shows the rate of growth is significantly larger: $\frac{\Delta \text{pairs}}{\Delta \text{years}} = \frac{550}{1}$ or 550 pairs per year.

The combination of these two lines would be a single function that modeled the population of breeding pairs from 1990 to 2006, but it would be *defined in two pieces.* This is an example of a **piecewise-defined function.**

The notation for these functions is a large "left brace" indicating the equations it groups are part of a single function. Using selected data points and techniques from Section 2.3, we find equations that could represent each piece are $p(t) = 350t + 3000$

for $0 \le t \le 10$ and $p(t) = 550t + 1000$ for $t > 10$, where $p(t)$ is the number of breeding pairs in year t. The complete function is then written:

$$\underset{\text{function name}}{p(t)} = \begin{cases} \underset{\text{function pieces}}{350t + 3000} & \underset{\text{domain of each piece}}{2 \le t \le 10} \\ 550t + 1000 & t > 10 \end{cases}$$

EXAMPLE 1 ▶ **Writing the Equation and Domain of a Piecewise-Defined Function**

The linear piece of the function shown has an equation of $y = -2x + 10$. The equation of the quadratic piece is $y = -x^2 + 9x - 14$. Write the related piecewise-defined function, and state the domain of each piece by inspecting the graph.

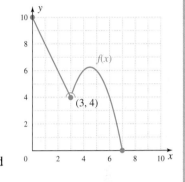

Solution ▶ From the graph we note the linear portion is defined between 0 and 3, with these endpoints included as indicated by the closed dots. The domain here is $0 \le x \le 3$. The quadratic portion begins at $x = 3$ *but does not include 3*, as indicated by the half-circle notation. The equation is

$$\underset{\text{function name}}{f(x)} = \begin{cases} \underset{\text{function pieces}}{-2x + 10} & \underset{\text{domain}}{0 \le x \le 3} \\ -x^2 + 9x - 14 & 3 < x \le 7 \end{cases}$$

☑ **A. You've just learned how to state the equation and domain of a piecewise-defined function**

Now try Exercises 7 and 8 ▶

Piecewise-defined functions can be composed of more than two pieces, and can involve functions of many kinds.

B. Graphing Piecewise-Defined Functions

As with other functions, piecewise-defined functions can be graphed by simply plotting points. Careful attention must be paid to the domain of each piece, both to evaluate the function correctly and to consider the inclusion/exclusion of endpoints. In addition, try to keep the transformations of a basic function in mind, as this will often help graph the function more efficiently.

EXAMPLE 2 ▶ **Graphing a Piecewise-Defined Function**

Graph the function by plotting points, then state its domain and range:

$$h(x) = \begin{cases} -x - 2 & -5 \le x < -1 \\ 2\sqrt{x + 1} - 1 & x \ge -1 \end{cases}$$

Solution ▶ The first piece of h is a line with negative slope, while the second is a transformed square root function. Using the endpoints of each domain specified and a few additional points, we obtain the following:

For $h(x) = -x - 2$, $-5 \le x < -1$,   For $h(x) = 2\sqrt{x + 1} - 1$, $x \ge -1$,

x	$h(x)$
-5	3
-3	1
-1	-1

x	$h(x)$
-1	-1
0	1
3	3

After plotting the points from the first piece, we connect them with a line segment noting the left endpoint is included, while the right endpoint is not (indicated using a semicircle around the point). Then we plot the points from the second piece and draw a square root graph, noting the left endpoint here *is* included, and the graph rises to the right. From the graph we note the complete domain of h is $x \in [-5, \infty)$, and the range is $y \in [-1, \infty)$.

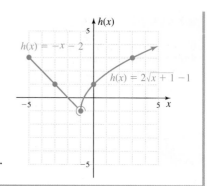

Now try Exercises 9 through 14 ▶

 As an alternative to plotting points, we can graph each piece of the function using transformations of a basic graph, then erase those parts that are outside of the corresponding domain. Repeat this procedure for each piece of the function. One interesting and highly instructive aspect of these functions is the opportunity to investigate restrictions on their domain and the ranges that result.

Piecewise and Continuous Functions

EXAMPLE 3 ▶ **Graphing a Piecewise-Defined Function**

Graph the function and state its domain and range:

$$f(x) = \begin{cases} -(x-3)^2 + 12 & 0 < x \le 6 \\ 3 & x > 6 \end{cases}$$

Solution ▶ The first piece of f is a basic parabola, shifted three units right, reflected across the x-axis (opening downward), and shifted 12 units up. The vertex is at $(3, 12)$ and the axis of symmetry is $x = 3$, producing the following graphs.

1. Graph first piece of f (Figure 2.70).

2. Erase portion outside domain of $0 < x \le 6$ (Figure 2.71).

Figure 2.70

Figure 2.71

The second function is simply a horizontal line through $(0, 3)$.

3. Graph second piece of f (Figure 2.72).

4. Erase portion outside domain of $x > 6$ (Figure 2.73).

Figure 2.72

Figure 2.73

The domain of f is $x \in (0, \infty)$, and the corresponding range is $y \in [3, 12]$.

> **Now try Exercises 15 through 18** ▶

Piecewise and Discontinuous Functions

Notice that although the function in Example 3 was piecewise-defined, the graph was actually continuous—we could draw the entire graph without lifting our pencil. Piecewise graphs also come in the *discontinuous* variety, which makes the domain and range issues all the more important.

EXAMPLE 4 ▶   Graphing a Discontinuous Piecewise-Defined Function

Graph $g(x)$ and state the domain and range:

$$g(x) = \begin{cases} -\frac{1}{2}x + 6 & 0 \le x \le 4 \\ -|x - 6| + 10 & 4 < x \le 9 \end{cases}$$

Solution ▶   The first piece of g is a line, with y-intercept $(0, 6)$ and slope $\frac{\Delta y}{\Delta x} = -\frac{1}{2}$.

1. Graph first piece of g (Figure 2.74).

2. Erase portion outside domain of $0 \le x \le 4$ (Figure 2.75).

Figure 2.74

Figure 2.75

The second is an absolute value function, shifted right 6 units, reflected across the x-axis, then shifted up 10 units.

3. Graph second piece of g (Figure 2.76).

Figure 2.76

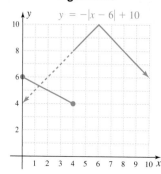

$y = -|x - 6| + 10$

4. Erase portion outside domain of $4 < x \le 9$ (Figure 2.77).

Figure 2.77

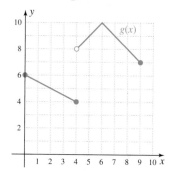

$g(x)$

Note that the left endpoint of the absolute value portion is not included (this piece is not defined at $x = 4$), signified by the open dot. The result is a discontinuous graph, as there is no way to draw the graph other than by jumping the pencil from where one piece ends to where the next begins. Using a vertical boundary line, we note the domain of g includes all values between 0 and 9 inclusive: $x \in [0, 9]$. Using a horizontal boundary line shows the smallest y-value is 4 and the largest is 10, but no range values exist between 6 and 7. The range is $y \in [4, 6] \cup [7, 10]$.

Now try Exercises 19 through 22 ▶

EXAMPLE 5 ▶ Graphing a Discontinuous Function

The given piecewise-defined function is not continuous. Graph $h(x)$ to see why, then comment on what could be done to make it continuous.

$$h(x) = \begin{cases} \dfrac{x^2 - 4}{x - 2} & x \ne 2 \\ 1 & x = 2 \end{cases}$$

Solution ▶ The first piece of h is unfamiliar to us, so we elect to graph it by plotting points, noting $x = 2$ is outside the domain. This produces the table shown in Figure 2.78. After connecting the points, the graph of h turns out to be a straight line, but with no corresponding y-value for $x = 2$. This leaves a "hole" in the graph at (2, 4), as designated by the open dot.

Figure 2.78

x	h(x)
−4	−2
−2	0
0	2
2	—
4	6

Figure 2.79

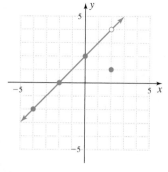

The second piece is point-wise defined, and its graph is simply the point (2, 1) shown in Figure 2.79. It's interesting to note that while the domain of h is all real numbers (h *is* defined at all points), the range is $y \in (-\infty, 4) \cup (4, \infty)$ as the function never takes on the value $y = 4$. In order for h to be continuous, we would need to redefine the second piece as $y = 4$ when $x = 2$.

Now try Exercises 23 through 26 ▶

To develop these concepts more fully, it will help to practice finding the equation of a piecewise-defined function *given its graph,* a process similar to that of Example 10 in Section 2.6.

EXAMPLE 6 ▶ **Determining the Equation of a Piecewise-Defined Function**

Determine the equation of the piecewise-defined function shown, including the domain for each piece.

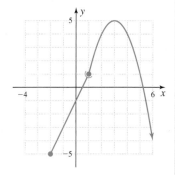

Solution ▶ By counting $\frac{\Delta y}{\Delta x}$ from $(-2, -5)$ to $(1, 1)$, we find the linear portion has slope $m = 2$, and the y-intercept must be $(0, -1)$. The equation of the line is $y = 2x - 1$. The second piece appears to be a parabola with vertex (h, k) at $(3, 5)$. Using this vertex with the point $(1, 1)$ in the general form $y = a(x - h)^2 + k$ gives

$y = a(x - h)^2 + k$	general form
$1 = a(1 - 3)^2 + 5$	substitute 1 for x, 1 for y, 3 for h, 5 for k
$-4 = a(-2)^2$	simplify; subtract 5
$-4 = 4a$	$(-2)^2 = 4$
$-1 = a$	divide by 4

The equation of the parabola is $y = -(x - 3)^2 + 5$. Considering the domains shown in the figure, the equation of this piecewise-defined function must be

$$p(x) = \begin{cases} 2x - 1 & -2 \le x \le 1 \\ -(x - 3)^2 + 5 & x > 1 \end{cases}$$

☑ **B.** You've just learned how to graph functions that are piecewise-defined

Now try Exercises 27 through 30 ▶

C. Applications of Piecewise-Defined Functions

The number of applications for piecewise-defined functions is practically limitless. It is actually fairly rare for a single function to accurately model a situation over a long period of time. Laws change, spending habits change, and technology can bring abrupt alterations in many areas of our lives. To accurately model these changes often requires a piecewise-defined function.

EXAMPLE 7 ▶ **Modeling with a Piecewise-Defined Function**

For the first half of the twentieth century, per capita spending on police protection can be modeled by $S(t) = 0.54t + 12$, where $S(t)$ represents per capita spending on police protection in year t (1900 corresponds to year 0). After 1950, perhaps due to the growth of American cities, this spending greatly increased: $S(t) = 3.65t - 144$. Write these as a piecewise-defined function $S(t)$, state the domain for each piece, then graph the function. According to this model, how much was spent (per capita) on police protection in 2000? How much will be spent in 2010?

Source: Data taken from the *Statistical Abstract of the United States* for various years.

Solution ▶

function name   function pieces   effective domain

$$S(t) = \begin{cases} 0.54t + 12 & 0 \le t \le 50 \\ 3.65t - 144 & t > 50 \end{cases}$$

Since both pieces are linear, we can graph each part using two points. For the first function, $S(0) = 12$ and $S(50) = 39$. For the second function $S(50) \approx 39$ and $S(80) = 148$. The graph for each piece is shown in the figure. Evaluating S at $t = 100$:

$$S(t) = 3.65t - 144$$
$$S(100) = 3.65(100) - 144$$
$$= 365 - 144$$
$$= 221$$

About $221 per capita was spent on police protection in the year 2000. For 2010, the model indicates that $257.50 per capita will be spent: $S(110) = 257.5$.

Now try Exercises 33 through 44 ▶

Step Functions

The last group of piecewise-defined functions we'll explore are the **step functions,** so called because the pieces of the function form a series of horizontal steps. These functions find frequent application in the way consumers are charged for services, and have a number of applications in number theory. Perhaps the most common is called the **greatest integer function,** though recently its alternative name, **floor function,** has gained popularity (see Figure 2.80). This is in large part due to an improvement in notation and as a better contrast to **ceiling functions.** The floor function of a real number x, denoted $f(x) = \lfloor x \rfloor$ or $[\![x]\!]$ (we will use the first), is the largest integer less than or equal to x. For instance, $\lfloor 5.9 \rfloor = 5$, $\lfloor 7 \rfloor = 7$, and $\lfloor -3.4 \rfloor = -4$.

In contrast, the ceiling function $C(x) = \lceil x \rceil$ is the smallest integer greater than or equal to x, meaning $\lceil 5.9 \rceil = 6$, $\lceil 7 \rceil = 7$, and $\lceil -3.4 \rceil = -3$ (see Figure 2.81). In simple terms, for any noninteger value on the number line, the floor function returns the integer to the left, while the ceiling function returns the integer to the right. A graph of each function is shown.

Figure 2.80

Figure 2.81

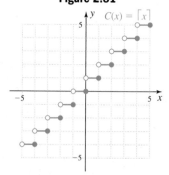

One common application of floor functions is the price of theater admission, where children 12 and under receive a discounted price. Right up until the day they're 13, they qualify for the lower price: $\lfloor 12\frac{364}{365} \rfloor = 12$. Applications of ceiling functions would include how phone companies charge for the minutes used (charging the 12-min rate for a phone call that only lasted 11.3 min: $\lceil 11.3 \rceil = 12$), and postage rates, as in Example 8.

EXAMPLE 8 ▶ **Modeling Using a Step Function**

As of May 2007, the first-class postage rate for large envelopes sent through the U.S. mail was 80¢ for the first ounce, then an additional 17¢ per ounce thereafter, up to 13 ounces. Graph the function and state its domain and range. Use the graph to state the cost of mailing a report weighing (a) 7.5 oz, (b) 8 oz, and (c) 8.1 oz in a large envelope.

Solution ▶ The 80¢ charge applies to letters weighing between 0 oz and 1 oz. Zero is not included since we have to mail *something*, but 1 is included since a large envelope and its contents weighing exactly one ounce still costs 80¢. The graph will be a horizontal line segment.

The function is defined for all weights between 0 and 13 oz, excluding zero and including 13: $x \in (0, 13]$. The range consists of single outputs corresponding to the step intervals: $R \in \{80, 97, 114, \ldots, 267, 284\}$.

a. The cost of mailing a 7.5-oz report is 199¢.

b. The cost of mailing an 8.0-oz report is still 199¢.

c. The cost of mailing an 8.1-oz report is $199 + 17 = 216$¢, since this brings you up to the next step.

☑ **C. You've just learned how to solve applications involving piecewise-defined functions**

Now try Exercises 45 through 48 ▶

TECHNOLOGY HIGHLIGHT

Piecewise-Defined Functions

Most graphing calculators are able to graph piecewise-defined functions. Consider the function f shown here:

$$f(x) = \begin{cases} x + 2 & x < 2 \\ (x - 4)^2 + 3 & x \geq 2 \end{cases}$$

Both "pieces" are well known—the first is a line with slope $m = 1$ and y-intercept $(0, 2)$. The second is a parabola that opens upward, shifted 4 units to the right and 3 units up. If we attempt to graph $f(x)$ using $Y_1 = x + 2$ and $Y_2 = (x - 4)^2 + 3$ as they stand, the resulting graph may be difficult to analyze because the pieces conflict and intersect (Figure 2.82). To graph the functions we must indicate the domain for each piece, separated by a slash and enclosed in parentheses. For instance, for the first piece we enter $Y_1 = x + 2/(x < 2)$, and for the second, $Y_2 = (x - 4)^2 + 3/(x \geq 2)$ (Figure 2.83). The slash looks like (is) the division symbol, but in this context, the calculator interprets it as a means of separating the function from the domain. The inequality symbols are accessed using the $\boxed{2nd}$ $\boxed{MATH}$ (TEST) keys. The graph is shown on Figure 2.84, where we see the function is linear for $x \in (-\infty, 2)$ and quadratic for $x \in [2, \infty)$. How does the calculator remind us the function is defined only for $x = 2$ on the second piece? Using the $\boxed{2nd}$ $\boxed{GRAPH}$ (TABLE) feature reveals the calculator will give an **ERR:** (ERROR) message for inputs outside of its domain (Figure 2.85).

Figure 2.82

Figure 2.83

We can also use the calculator to investigate endpoints of the domain. For instance, we know that $Y_1 = x + 2$ is not defined for $x = 2$, but what about numbers very close to 2? Go to $\boxed{2nd}$ $\boxed{WINDOW}$ (TBLSET) and place the calculator in the Indpnt: Auto $\boxed{ASK}$ mode. With both Y_1 and Y_2 enabled, use the $\boxed{2nd}$ $\boxed{GRAPH}$ (TABLE) feature to evaluate the functions at numbers very near 2. Use $x = 1.9, 1.99, 1.999$, and so on.

Figure 2.84

Figure 2.85

X	Y₁	Y₂
0	2	ERR:
.5	2.5	ERR:
1	3	ERR:
1.5	3.5	ERR:
2	ERR:	7
2.5	ERR:	5.25
3	ERR:	4

X=3

Exercise 1: What appears to be happening to the output values for Y_1? What about Y_2?

Exercise 2: What do you notice about the output values when 1.99999 is entered? Use the right arrow key $\boxed{\blacktriangleright}$ to move the cursor into columns Y_1 and Y_2. Comment on what you think the calculator is doing. Will Y_1 ever really have an output equal to 4?

2.7 EXERCISES

▶ CONCEPTS AND VOCABULARY

Fill in each blank with the appropriate word or phrase. Carefully reread the section if needed.

1. A function whose entire graph can be drawn without lifting your pencil is called a _____ function.

2. The input values for which each part of a piecewise function is defined is the _____ of the function.

3. A graph is called _____ if it has no sharp turns or jagged edges.

4. When graphing $2x + 3$ over a domain of $x > 0$, we leave an _____ dot at $(0, 3)$.

5. Discuss/Explain how to determine if a piecewise-defined function is continuous, without having to graph the function. Illustrate with an example.

6. Discuss/Explain how it is possible for the domain of a function to be defined for all real numbers, but have a range that is defined on more than one interval. Construct an illustrative example.

▶ **DEVELOPING YOUR SKILLS**

For Exercises 7 and 8, (a) use the correct notation to write them as a single piecewise-defined function, state the domain for each piece by inspecting the graph, and (b) state the range of the function.

7. $Y_1 = x^2 - 6x + 10$; $Y_2 = \frac{3}{2}x - \frac{5}{2}$

8. $Y_1 = -1.5|x - 5| + 10$; $Y_2 = -\sqrt{x - 7} + 5$

Evaluate each piecewise-defined function as indicated (if possible).

9. $h(x) = \begin{cases} -2 & x < -2 \\ |x| & -2 \le x < 3 \\ 5 & x \ge 3 \end{cases}$

$h(-5), h(-2), h(-\frac{1}{2}), h(0), h(2.999)$, and $h(3)$

10. $H(x) = \begin{cases} 2x + 3 & x < 0 \\ x^2 + 1 & 0 \le x < 2 \\ 5 & x > 2 \end{cases}$

$H(-3), H(-\frac{3}{2}), H(-0.001), H(1), H(2)$, and $H(3)$

11. $p(x) = \begin{cases} 5 & x < -3 \\ x^2 - 4 & -3 \le x \le 3 \\ 2x + 1 & x > 3 \end{cases}$

$p(-5), p(-3), p(-2), p(0), p(3)$, and $p(5)$

12. $q(x) = \begin{cases} -x - 3 & x < -1 \\ 2 & -1 \le x < 2 \\ -\frac{1}{2}x^2 + 3x - 2 & x \ge 2 \end{cases}$

$q(-3), q(-1), q(0), q(1.999), q(2)$, and $q(4)$

Graph each piecewise-defined function by plotting points, then state its domain and range.

13. $p(x) = \begin{cases} x + 2 & -6 \le x \le 2 \\ 2|x - 4| & x > 2 \end{cases}$

14. $q(x) = \begin{cases} \sqrt{x + 4} & -4 \le x \le 0 \\ |x - 2| & 0 < x \le 7 \end{cases}$

Graph each piecewise-defined function and state its domain and range. Use transformations of the toolbox functions where possible.

15. $g(x) = \begin{cases} -(x - 1)^2 + 5 & -2 \le x \le 4 \\ 2x - 12 & x > 4 \end{cases}$

16. $h(x) = \begin{cases} \frac{1}{2}x + 1 & x \le 0 \\ (x - 2)^2 - 3 & 0 < x \le 5 \end{cases}$

17. $p(x) = \begin{cases} \frac{1}{2}x + 1 & x \ne 4 \\ 2 & x = 4 \end{cases}$

18. $q(x) = \begin{cases} \frac{1}{2}(x - 1)^3 - 1 & x \ne 3 \\ -2 & x = 3 \end{cases}$

19. $H(x) = \begin{cases} -x + 3 & x < 1 \\ -|x - 5| + 6 & 1 \le x < 9 \end{cases}$

20. $w(x) = \begin{cases} \sqrt[3]{x + 1} & x < 1 \\ (x - 3)^2 - 2 & 1 \le x \le 6 \end{cases}$

21. $f(x) = \begin{cases} -x - 3 & x < -3 \\ 9 - x^2 & -3 \le x < 2 \\ 4 & x \ge 2 \end{cases}$

22. $h(x) = \begin{cases} -\frac{1}{2}x - 1 & x < -3 \\ -|x| + 5 & -3 \le x \le 5 \\ 3\sqrt{x - 5} & x > 5 \end{cases}$

Each of the following functions has a pointwise discontinuity. Graph the first piece of each function, then find the value of c so that a continuous function results.

23. $f(x) = \begin{cases} \dfrac{x^2 - 9}{x + 3} & x \neq -3 \\ c & x = -3 \end{cases}$

24. $f(x) = \begin{cases} \dfrac{x^2 - 3x - 10}{x - 5} & x \neq 5 \\ c & x = 5 \end{cases}$

25. $f(x) = \begin{cases} \dfrac{x^3 - 1}{x - 1} & x \neq 1 \\ c & x = 1 \end{cases}$

26. $f(x) = \begin{cases} \dfrac{4x - x^3}{x + 2} & x \neq -2 \\ c & x = -2 \end{cases}$

Determine the equation of each piecewise-defined function shown, including the domain for each piece. Assume all pieces are toolbox functions.

27.

28.

29.

30.

▶ WORKING WITH FORMULAS

31. **Definition of absolute value:** $|x| = \begin{cases} -x & x < 0 \\ x & x \geq 0 \end{cases}$

The absolute value function can be stated as a piecewise-defined function, a technique that is sometimes useful in graphing variations of the function or solving absolute value equations and inequalities. How does this definition ensure that the absolute value of a number is always positive? Use this definition to help sketch the graph of $f(x) = \frac{|x|}{x}$. Discuss what you notice.

32. **Sand dune function:**

$f(x) = \begin{cases} -|x - 2| + 1 & 1 \leq x < 3 \\ -|x - 4| + 1 & 3 \leq x < 5 \\ -|x - 2k| + 1 & 2k - 1 \leq x < 2k + 1, \text{ for } k \in N \end{cases}$

There are a number of interesting graphs that can be created using piecewise-defined functions, and these functions have been the basis for more than one piece of modern art. (a) Use the descriptive name and the pieces given to graph the function f. Is the function accurately named? (b) Use any combination of the toolbox functions to explore your own creativity by creating a piecewise-defined function with some interesting or appealing characteristics.

▶ APPLICATIONS

For Exercises 33 and 34, a. write the information given as a piecewise-defined function, and state the domain for each piece by inspecting the graph. b. Give the range of each.

33. Due to heavy advertising, initial sales of the Lynx Digital Camera grew very rapidly, but started to decline once the advertising blitz was over. During the advertising campaign, sales were modeled by the

function $S(t) = -t^2 + 6t$, where $S(t)$ represents hundreds of sales in month t. However, as Lynx Inc. had hoped, the new product secured a foothold in the market and sales leveled out at a steady 500 sales per month.

34. From the turn of the twentieth century, the number of newspapers (per thousand population) grew rapidly until the 1930s, when the growth slowed down and then declined. The years 1940 to 1946 saw a "spike" in growth, but the years 1947 to 1954 saw an almost equal decline. Since 1954 the number has continued to decline, but at a slower rate. The number of papers

N per thousand population for each period, respectively, can be approximated by
$N_1(t) = -0.13t^2 + 8.1t + 208,$
$N_2(t) = -5.75|t - 46| + 374,$ and
$N_3(t) = -2.45t + 460.$

Source: Data from the Statistical Abstract of the United States, various years; data from The First Measured Century, The AEI Press, Caplow, Hicks, and Wattenberg, 2001.

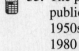 **35.** The percentage of American households that own publicly traded stocks began rising in the early 1950s, peaked in 1970, then began to decline until 1980 when there was a dramatic increase due to easy access over the Internet, an improved economy, and other factors. This phenomenon is modeled by the function $P(t)$, where $P(t)$ represents the percentage of households owning stock in year t, with 1950 corresponding to year 0.

$$P(t) = \begin{cases} -0.03t^2 + 1.28t + 1.68 & 0 \le t \le 30 \\ 1.89t - 43.5 & t > 30 \end{cases}$$

a. According to this model, what percentage of American households held stock in the years 1955, 1965, 1975, 1985, and 1995? If this pattern continues, what percentage held stock in 2005?

b. Why is there a discrepancy in the outputs of each piece of the function for the year 1980 ($t = 30$)? According to how the function is defined, which output should be used?

Source: 2004 Statistical Abstract of the United States, Table 1204; various other years.

 36. America's dependency on foreign oil has always been a "hot" political topic, with the amount of imported oil fluctuating over the years due to political climate, public awareness, the economy, and other factors. The amount of crude oil imported can be approximated by the function given, where $A(t)$ represents the number of barrels imported in year t (in billions), with 1980 corresponding to year 0.

$$A(t) = \begin{cases} 0.047t^2 - 0.38t + 1.9 & 0 \le t < 8 \\ -0.075t^2 + 1.495t - 5.265 & 8 \le t \le 11 \\ 0.133t + 0.685 & t > 11 \end{cases}$$

a. Use $A(t)$ to estimate the number of barrels imported in the years 1983, 1989, 1995, and 2005.

b. What was the minimum number of barrels imported between 1980 and 1988?

Source: 2004 Statistical Abstract of the United States, Table 897; various other years.

37. Energy rationing: In certain areas of the United States, power blackouts have forced some counties to ration electricity. Suppose the cost is $0.09 per kilowatt (kW) for the first 1000 kW a household uses. After 1000 kW, the cost increases to 0.18 per kW: Write these charges for electricity in the form of a piecewise-defined function $C(h)$, where $C(h)$ is the cost for h kilowatt hours. State the domain for each piece. Then sketch the graph and determine the cost for 1200 kW.

38. Water rationing: Many southwestern states have a limited water supply, and some state governments try to control consumption by manipulating the cost of water usage. Suppose for the first 5000 gal a household uses per month, the charge is $0.05 per gallon. Once 5000 gal is used the charge doubles to $0.10 per gallon. Write these charges for water usage in the form of a piecewise-defined function $C(w)$, where $C(w)$ is the cost for w gallons of water and state the domain for each piece. Then sketch the graph and determine the cost to a household that used 9500 gal of water during a very hot summer month.

39. Pricing for natural gas: A local gas company charges $0.75 per therm for natural gas, up to 25 therms. Once the 25 therms has been exceeded, the charge doubles to $1.50 per therm due to limited supply and great demand. Write these charges for natural gas consumption in the form of a piecewise-defined function $C(t)$, where $C(t)$ is the charge for t therms and state the domain for each piece. Then sketch the graph and determine the cost to a household that used 45 therms during a very cold winter month.

40. Multiple births:
The number of multiple births has steadily increased in the United States during the twentieth century and beyond. Between 1985 and 1995 the number of twin births could be modeled by the function $T(x) = -0.21x^2 + 6.1x + 52$, where x is the

number of years since 1980 and T is in thousands. After 1995, the incidence of twins becomes more linear, with $T(x) = 4.53x + 28.3$ serving as a better model. Write the piecewise-defined function modeling the incidence of twins for these years, including the domain of each piece. Then sketch the graph and use the function to estimate the incidence of twins in 1990, 2000, and 2005. If this trend continues, how many sets of twins will be born in 2010?

Source: National Vital Statistics Report, Vol. 50, No. 5, February 12, 2002

41. **U.S. military expenditures:** Except for the year 1991 when military spending was cut drastically, the amount spent by the U.S. government on national defense and veterans' benefits rose steadily from 1980 to 1992. These expenditures can be modeled by the function $S(t) = -1.35t^2 + 31.9t + 152$, where $S(t)$ is in billions of dollars and 1980 corresponds to $t = 0$.

Source: 1992 Statistical Abstract of the United States, Table 525

From 1992 to 1996 this spending declined, then began to rise in the following years. From 1992 to 2002, military-related spending can be modeled by $S(t) = 2.5t^2 - 80.6t + 950$.

Source: 2004 Statistical Abstract of the United States, Table 492

Write $S(t)$ as a single piecewise-defined function, stating the domain for each piece. Then sketch the graph and use the function to find the projected amount the United States will spend on its military in 2005, 2008, and 2010 if this trend continues.

42. **Amusement arcades:** At a local amusement center, the owner has the SkeeBall machines programmed to reward very high scores. For scores of 200 or less, the function $T(x) = \frac{x}{10}$ models the number of tickets awarded (rounded to the nearest whole). For scores over 200, the number of tickets is modeled by $T(x) = 0.001x^2 - 0.3x + 40$. Write these equation models of the number of tickets awarded in the form of a piecewise-defined function and state the domain for each piece. Then sketch the graph and find the number of tickets awarded to a person who scores 390 points.

43. **Phone service charges:** When it comes to phone service, a large number of calling plans are available. Under one plan, the first 30 min of any phone call costs only 3.3¢ per minute. The charge increases to 7¢ per minute thereafter. Write this information in the form of a piecewise-defined function and state the domain for each piece. Then sketch the graph and find the cost of a 46-min phone call.

44. **Overtime wages:** Tara works on an assembly line, putting together computer monitors. She is paid $9.50 per hour for regular time (0, 40 hr], $14.25 for overtime (40, 48 hr], and when demand for computers is high, $19.00 for double-overtime (48, 84 hr]. Write this information in the form of a simplified piecewise-defined function, and state the domain for each piece. Then sketch the graph and find the gross amount of Tara's check for the week she put in 54 hr.

45. **Admission prices:** At Wet Willy's Water World, infants under 2 are free, then admission is charged according to age. Children 2 and older but less than 13 pay $2, teenagers 13 and older but less than 20 pay $5, adults 20 and older but less than 65 pay $7, and senior citizens 65 and older get in at the teenage rate. Write this information in the form of a piecewise-defined function and state the domain for each piece. Then sketch the graph and find the cost of admission for a family of nine which includes: one grandparent (70), two adults (44/45), 3 teenagers, 2 children, and one infant.

46. **Demographics:** One common use of the floor function $y = \lfloor x \rfloor$ is the reporting of ages. As of 2007, the record for longest living human is 122 yr, 164 days for the life of Jeanne Calment, formerly of France. While she actually lived $x = 122\frac{164}{365}$ years, ages are normally reported using the floor function, or the greatest integer number of years less than or equal to the actual age: $\lfloor 122\frac{164}{365} \rfloor = 122$ years. (a) Write a function $A(t)$ that gives a person's age, where $A(t)$ is the reported age at time t. (b) State the domain of the function (be sure to consider Madame Calment's record). Report the age of a person who has been living for (c) 36 years; (d) 36 years, 364 days; (e) 37 years; and (f) 37 years, 1 day.

47. **Postage rates:** The postal charge function from Example 8 is simply a transformation of the basic ceiling function $y = \lceil x \rceil$. Using the ideas from Section 2.6, (a) write the postal charges as a step function $C(w)$, where $C(w)$ is the cost of mailing a large envelope weighing w ounces, and (b) state the domain of the function. Then use the function to find the cost of mailing reports weighing: (c) 0.7 oz, (d) 5.1 oz, (e) 5.9 oz; (f) 6 oz, and (g) 6.1 oz.

48. **Cell phone charges:** A national cell phone company advertises that calls of 1 min or less do not count toward monthly usage. Calls lasting longer than 1 min are calculated normally using a ceiling function, meaning a call of 1 min, 1 sec will be counted as a 2-min call. Using the ideas

from Section 2.6, (a) write the cell phone charges as a piecewise-defined function $C(m)$, where $C(m)$ is the cost of a call lasting m minutes, and include the domain of the function. Then (b) graph the function, and (c) use the graph or function to determine if a cell phone subscriber has exceeded the 30 free minutes granted by her calling plan for calls lasting 2 min 3 sec, 13 min 46 sec, 1 min 5 sec, 3 min 59 sec, 8 min 2 sec. (d) What was the actual usage in minutes and seconds?

49. Combined absolute value graphs: Carefully graph the function $h(x) = |x - 2| - |x + 3|$ using a

table of values over the interval $x \in [-5, 5]$. Is the function continuous? Write this function in piecewise-defined form and state the domain for each piece.

50. Combined absolute value graphs: Carefully graph the function $H(x) = |x - 2| + |x + 3|$ using a table of values over the interval $x \in [-5, 5]$. Is the function continuous? Write this function in piecewise-defined form and state the domain for each piece.

▶ **EXTENDING THE CONCEPT**

51. You've heard it said, "*any number divided by itself is one.*" Consider the functions $Y_1 = \frac{x + 2}{x + 2}$, and $Y_2 = \frac{|x + 2|}{x + 2}$. Are these functions continuous?

52. Find a linear function $h(x)$ that will make the function shown a *continuous* function. Be sure to include its domain.

$$f(x) = \begin{cases} x^2 & x < 1 \\ h(x) & \\ 2x + 3 & x > 3 \end{cases}$$

▶ **MAINTAINING YOUR SKILLS**

53. (1.3) Solve: $\dfrac{3}{x - 2} + 1 = \dfrac{30}{x^2 - 4}$.

54. (R.5) Compute the following and write the result in lowest terms:

$$\frac{x^3 + 3x^2 - 4x - 12}{x - 3} \cdot \frac{2x - 6}{x^2 + 5x + 6} \div (3x - 6)$$

55. (R.7) For the figure shown, (a) find the length of the missing side, (b) state the area of the

triangular base, and (c) compute the volume of the prism.

56. (2.4) Find the equation of the line perpendicular to $3x + 4y = 8$, and through the point $(0, -2)$. Write the result in slope-intercept form.

Learning Objectives

In Section 2.8 you will learn how to:

☐ **A.** Compute a sum or difference of functions and determine the domain of the result

☐ **B.** Compute a product or quotient of functions and determine the domain

☐ **C.** Compose two functions and determine the domain; decompose a function

☐ **D.** Interpret operations on functions graphically

☐ **E.** Apply the algebra and composition of functions in context

In Section 2.5, we created new functions *graphically* by applying transformations to basic functions. In this section, we'll use two (or more) functions to create new functions *algebraically*. Previous courses often contain material on the sum, difference, product, and quotient of polynomials. Here we'll combine these functions with the basic operations, noting the result is also a function that can be evaluated, graphed, and analyzed. We call these basic operations on functions the **algebra of functions.**

A. Sums and Differences of Functions

This section introduces the notation used for basic operations on functions. Here we'll note the result is also a function whose domain depends on the original functions. In general, if f and g are functions *with overlapping domains*, $f(x) + g(x) = (f + g)(x)$ and $f(x) - g(x) = (f - g)(x)$.

Sums and Differences of Functions

For functions f and g with domains P and Q respectively, the sum and difference of f and g are defined by:

$$\begin{array}{cc} & \textbf{Domain of result} \\ (f + g)(x) = f(x) + g(x) & P \cap Q \\ (f - g)(x) = f(x) - g(x) & P \cap Q \end{array}$$

EXAMPLE 1A ▶ Evaluating a Difference of Functions

Given $f(x) = x^2 - 5x$ and $g(x) = 2x - 9$,

a. Determine the domain of $h(x) = (f - g)(x)$. **b.** Find $h(3)$ using the definition.

Solution ▶

a. Since the domain of both f and g is $\mathbb{R}$, their intersection is $\mathbb{R}$, so the domain of h is also $\mathbb{R}$.

b.
$$\begin{aligned} h(x) &= (f - g)(x) && \text{given difference} \\ &= f(x) - g(x) && \text{by definition} \\ h(3) &= f(3) - g(3) && \text{substitute 3 for } x \\ &= \left[(3)^2 - 5(3)\right] - \left[2(3) - 9\right] && \text{evaluate} \\ &= [9 - 15] - [6 - 9] && \text{multiply} \\ &= -6 - [-3] && \text{subtract} \\ &= -3 && \text{result} \end{aligned}$$

If the function h is to be graphed or evaluated numerous times, it helps to compute a *new function rule* for h, rather than repeatedly apply the definition.

EXAMPLE 1B ▶ For the functions f, g, and h, as defined in Example 1A,

a. Find a new function rule for h. **b.** Use the result to find $h(3)$.

Solution ▶

a.
$$\begin{aligned} h(x) &= (f - g)(x) && \text{given difference} \\ &= f(x) - g(x) && \text{by definition} \\ &= (x^2 - 5x) - (2x - 9) && \text{replace } f(x) \text{ with } (x^2 - 5x) \text{ and } g(x) \text{ with } (2x - 9) \\ &= x^2 - 7x + 9 && \text{distribute and combine like terms} \end{aligned}$$

b. $h(3) = (3)^2 - 7(3) + 9$ substitute 3 for x

       $= 9 - 21 + 9$ multiply

       $= -3$ result

Notice the result from Part (b) is identical to that in Example 1A.

Now try Exercises 7 through 10 ▶

⚠ **CAUTION ▶** From Example 1A, note the importance of using grouping symbols with the algebra of functions. Without them, we could easily confuse the signs of g when computing the difference. Also, note that any operation applied to the functions f and g simply results in an *expression* representing a new function rule for h, and is not an *equation* that needs to be factored or solved.

EXAMPLE 2 ▶ Evaluating a Sum of Functions

For $f(x) = x^2$ and $g(x) = \sqrt{x - 2}$,

     a. Determine the domain of $h(x) = (f + g)(x)$.

     b. Find a new function rule for h.

     c. Evaluate $h(3)$.

     d. Evaluate $h(-1)$.

Solution ▶

a. The domain of f is $\mathbb{R}$, while the domain of g is $x \in [2, \infty)$. Since their intersection is $[2, \infty)$, this is the domain of the new function h.

b. $h(x) = (f + g)(x)$ given sum

       $= f(x) + g(x)$ by definition

       $= x^2 + \sqrt{x - 2}$ substitute x^2 for $f(x)$ and $\sqrt{x-2}$ for $g(x)$ (no other simplifications possible)

c. $h(3) = (3)^2 + \sqrt{3 - 2}$ substitute 3 for x

       $= 10$ result

d. $x = -1$ is outside the domain of h.

Now try Exercises 11 through 14 ▶

WORTHY OF NOTE

If we *did* try to evaluate $h(-1)$, the result would be $1 + \sqrt{-3}$, which is not a real number. While it's true we could write $1 + \sqrt{-3}$ as $1 + i\sqrt{3}$ and consider it an "answer," our study here focuses on real numbers and the graphs of functions in a coordinate system where x and y are both real.

☑ **A.** You've just learned how to compute a sum or difference of functions and determine the domain of the result

This "intersection of domains" is illustrated in Figure 2.86 using ideas from Section 1.2.

Figure 2.86

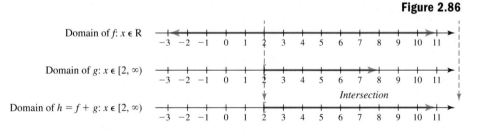

B. Products and Quotients of Functions

The product and quotient of two functions is defined in a manner similar to that for sums and differences. For example, if f and g are functions *with overlapping domains,*

$(f \cdot g)(x) = f(x) \cdot g(x)$ and $\left(\dfrac{f}{g}\right)(x) = \dfrac{f(x)}{g(x)}$. As you might expect, for quotients we must

stipulate $g(x) \neq 0$.

Products and Quotients of Functions

For functions f and g with domains P and Q, respectively, the product and quotient of f and g are defined by:

Domain of result

$$(f \cdot g)(x) = f(x) \cdot g(x) \qquad P \cap Q$$

$$\left(\frac{f}{g}\right)(x) = \frac{f(x)}{g(x)} \qquad P \cap Q, \text{ for all } g(x) \neq 0$$

EXAMPLE 3 ▶ **Computing a Product of Functions**

Given $f(x) = \sqrt{1 + x}$ and $g(x) = \sqrt{3 - x}$,

a. Determine the domian of $h(x) = (f \cdot g)(x)$.

b. Find a new function rule for h.

c. Use the result from part (b) to evaluate $h(2)$ and $h(4)$.

Solution ▶ a. The domain of f is $x \in [-1, \infty)$ and the domain of g is $x \in (-\infty, 3]$. The intersection of these domains gives $x \in [-1, 3]$, which is the domain for h.

b. $h(x) = (f \cdot g)(x)$ given product

$= f(x) \cdot g(x)$ by definition

$= \sqrt{1 + x} \cdot \sqrt{3 - x}$ substitute $\sqrt{1 + x}$ for f and $\sqrt{3 + x}$ for g

$= \sqrt{3 + 2x - x^2}$ combine using properties of radicals

c. $h(2) = \sqrt{3 + 2(2) - (2)^2}$ substitute 2 for x

$= \sqrt{3} \approx 1.732$ result

$h(4) = \sqrt{3 + 2(4) - (4)^2}$ substitute 4 for x

$= \sqrt{-5}$ not a real number

The second result of Part (c) is not surprising, since $x = 4$ is not in the domain of h [meaning $h(4)$ is not defined for this function].

Now try Exercises 15 through 18 ▶

In future sections, we use polynomial division as a tool for factoring, an aid to graphing, and to determine whether two expressions are equivalent. Understanding the notation and domain issues related to division will strengthen our ability in these areas.

EXAMPLE 4 ▶ **Computing a Quotient of Functions**

Given $f(x) = x^3 - 3x^2 + 2x - 6$ and $g(x) = x - 3$,

a. Determine the domain of $h(x) = \left(\frac{f}{g}\right)(x)$.

b. Find a new function rule for h.

c. Use the result from part (b) to evaluate $h(3)$ and $h(0)$.

Solution ▶ a. While the domain of both f and g is $\mathbb{R}$ and their intersection is also $\mathbb{R}$, we know from the definition (and past experience) *that $g(x)$ cannot be zero*. The domain of h is $x \in (-\infty, 3) \cup (3, \infty)$.

b. $h(x) = \left(\frac{f}{g}\right)(x)$ given quotient

$= \frac{f(x)}{g(x)}$ by definition

$= \frac{x^3 - 3x^2 + 2x - 6}{x - 3}$ replace f with $x^3 - 3x^2 + 2x - 6$ and g with $x - 3$

c. Recall that $x = 3$ is not in the domain of h. For $h(0)$ we have:

$$h(0) = \frac{(0)^3 - 3(0)^2 + 2(0) - 6}{(0) - 3} \quad \text{replace } x \text{ with } 0$$

$$= \frac{-6}{-3} = 2 \qquad h(0) = 2$$

Now try Exercises 19 through 34 ▶

☑ B. You've just learned how to compute a product or quotient of functions and determine the domain

From our work with rational expressions in Section R.5, the expression that defines h can be simplified: $\dfrac{x^3 - 3x^2 + 2x - 6}{x - 3} = \dfrac{x^2(x - 3) + 2(x - 3)}{x - 3} = \dfrac{(x^2 + 2)(x \cancel{-3})}{\cancel{x - 3}} =$ $x^2 + 2$. But from the original expression, h is not defined if $g(x) = 3$, *even if the result for h is a polynomial*. In this case, we write the simplified form as $h(x) = x^2 + 2, x \neq 3$.

For additional practice with the algebra of functions, **see Exercises 35 through 46.**

C. Composition of Functions

The composition of functions is best understood by studying the "input/output" nature of a function. Consider $g(x) = x^2 - 3$. For $g(x)$ we might say, "inputs are squared, then decreased by three." In diagram form we have:

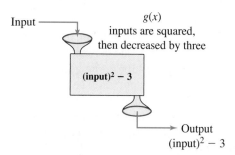

In many respects, a function box can be regarded as a very simple machine, running a simple program. It doesn't matter what the input is, this machine is going to *square the input then subtract three.*

EXAMPLE 5 ▶ **Evaluating a Function**

For $g(x) = x^2 - 3$, find

a. $g(-5)$

b. $g(5t)$

c. $g(t - 4)$

Solution ▶ **a.** $\qquad g(x) = x^2 - 3$ $\qquad$ original function

input -5

$\qquad g(-5) = (-5)^2 - 3$ $\qquad$ square input, then subtract 3

$\qquad\qquad\quad = 25 - 3$ $\qquad$ simplify

$\qquad\qquad\quad = 22$ $\qquad$ result

b. $\qquad g(x) = x^2 - 3$ $\qquad$ original function

input $5t$

$\qquad g(5t) = (5t)^2 - 3$ $\qquad$ square input, then subtract 3

$\qquad\qquad\quad = 25t^2 - 3$ $\qquad$ result

c.

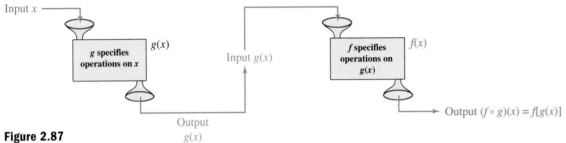

$$g(x) = x^2 - 3 \qquad \text{original function}$$

input $t - 4$

$$g(t - 4) = (t - 4)^2 - 3 \qquad \text{square input, then subtract 3}$$
$$= t^2 - 8t + 16 - 3 \qquad \text{expand binomial}$$
$$= t^2 - 8t + 13 \qquad \text{result}$$

Now try Exercises 47 and 48 ▶

When the input value is itself a function (rather than a single number or variable), this process is called the **composition of functions.** The evaluation method is exactly the same, we are simply using a function input. Using a general function $g(x)$ and a function diagram as before, we illustrate the process in Figure 2.87.

Input x

g specifies operations on x

$g(x)$

Input $g(x)$

f specifies operations on $g(x)$

$f(x)$

Output $g(x)$

Output $(f \circ g)(x) = f[g(x)]$

Figure 2.87

The notation used for the composition of f with g is an open dot "∘" placed between them, and is read, "f composed with g." The notation $(f \circ g)(x)$ indicates that $g(x)$ is an input for f: $(f \circ g)(x) = f[g(x)]$. If the order is reversed, as in $(g \circ f)(x)$, $f(x)$ becomes the input for g: $(g \circ f)(x) = g[f(x)]$. Figure 2.87 also helps us determine the domain of a composite function, in that the first function g can operate only if x is a valid input for g, and the second function f can operate only if $g(x)$ is a valid input for f. In other words, $(f \circ g)(x)$ is defined for *all x in the domain of g, such that $g(x)$ is in the domain of f.*

⚠ **CAUTION ▶** Try not to confuse the new "open dot" notation for the *composition* of functions, with the multiplication dot used to indicate the *product* of two functions: $(f \cdot g)(x) = (fg)(x)$ or the product of f with g; $(f \circ g)(x) = f[g(x)]$ or f composed with g.

The Composition of Functions

Given two functions f and g, the composition of f with g is defined by

$$(f \circ g)(x) = f[g(x)]$$

The domain of the composition is all x in the domain of g for which $g(x)$ is in the domain of f.

In Figure 2.88, these ideas are displayed using mapping notation, as we consider the simple case where $g(x) = x$ and $f(x) = \sqrt{x}$.

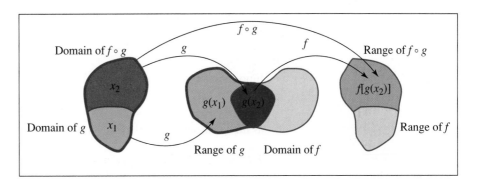

Figure 2.88

The domain of g (all real numbers) is shown within the red border, with g taking the negative inputs represented by x_1 (light red), to a like-colored portion of the range—the negative outputs $g(x_1)$. The nonnegative inputs represented by x_2 (dark red) are also mapped to a like-colored portion of the range—the nonnegative outputs $g(x_2)$. While the range of g is also all real numbers, function f can only use the nonnegative inputs represented by $g(x_2)$. This restricts the domain of $(f \circ g)(x)$ to only the inputs from g, where $g(x)$ is in the domain of f.

EXAMPLE 6 ▶ **Finding a Composition of Functions**

Given $f(x) = \sqrt{x - 4}$ and $g(x) = 3x + 2$, find

a. $(f \circ g)(x)$

b. $(g \circ f)(x)$

Also determine the domain for each.

Solution ▶ **a.** $f(x) = \sqrt{x - 4}$ says "decrease inputs by 4, and take the square root of the result."

$$(f \circ g)(x) = f[g(x)] \qquad \text{\small $g(x)$ is an input for f}$$
$$= \sqrt{g(x) - 4} \qquad \text{\small decrease input by 4, and take the square root of the result}$$
$$= \sqrt{(3x + 2) - 4} \quad \text{\small substitute $3x + 2$ for $g(x)$}$$
$$= \sqrt{3x - 2} \qquad \text{\small result}$$

While g is defined for all real numbers, f is defined only for nonnegative numbers. Since $f[g(x)] = \sqrt{3x - 2}$, we need $3x - 2 \geq 0$, $x \geq \frac{2}{3}$. In interval notation, the domain of $(f \circ g)(x)$ is $x \in [\frac{2}{3}, \infty)$.

b. The function g says "inputs are multiplied by 3, then increased by 2."

$$(g \circ f)(x) = g[f(x)] \qquad \text{\small $f(x)$ is an input for g}$$
$$= 3f(x) + 2 \qquad \text{\small multiply input by 3, then increase by 2}$$
$$= 3\sqrt{x - 4} + 2 \quad \text{\small substitute $\sqrt{x-4}$ for $f(x)$}$$

For $g[f(x)]$, g can accept any real number input, but f can supply only those where $x \geq 4$. The domain of $(g \circ f)(x)$ is $x \in [4, \infty)$.

WORTHY OF NOTE

Example 6 shows that $(f \circ g)(x)$ is generally not equal to $(g \circ f)(x)$. On those occasions when they *are* equal, the functions have a unique relationship that we'll study in Section 4.1.

Now try Exercises 49 through 58 ▶

EXAMPLE 7 ▶ **Finding a Composition of Functions**

For $f(x) = \dfrac{3x}{x-1}$ and $g(x) = \dfrac{2}{x}$, analyze the domain of

a. $(f \circ g)(x)$.

b. $(g \circ f)(x)$.

c. Find the actual compositions and comment.

Solution ▶ **a.** $(f \circ g)(x)$: For g to be defined, $x \neq 0$ is our first restriction. Once $g(x)$ is used as the input, we have $f[g(x)] = \dfrac{3g(x)}{g(x)-1}$, and additionally note that $g(x)$ cannot equal 1. This means $\dfrac{2}{x} \neq 1$, so $x \neq 2$. The domain of $f \circ g$ is $\{x \mid x \neq 0, x \neq 2\}$.

b. $(g \circ f)(x)$: For f to be defined, $x \neq 1$ is our first restriction. Once $f(x)$ is used as the input, we have $g[f(x)] = \dfrac{2}{f(x)}$, and additionally note that $f(x)$ cannot be 0. This means $\dfrac{3x}{x-1} \neq 0$, so $x \neq 0$. The domain of $(g \circ f)(x)$ is $\{x \mid x \neq 0, x \neq 1\}$.

c. For $(f \circ g)(x)$:

$$f[g(x)] = \dfrac{3g(x)}{g(x)-1} \qquad \text{composition of } f \text{ with } g$$

$$= \dfrac{\left(\dfrac{3}{1}\right)\left(\dfrac{2}{x}\right)}{\left(\dfrac{2}{x}\right)-1} \qquad \text{substitute } \dfrac{2}{x} \text{ for } g(x)$$

$$= \dfrac{\dfrac{6}{x}}{\dfrac{2-x}{x}} = \dfrac{6}{x} \cdot \dfrac{x}{2-x} \qquad \text{simplify denominator; invert and multiply}$$

$$= \dfrac{6}{2-x} \qquad \text{result}$$

WORTHY OF NOTE

As Example 7 illustrates, the domain of $h(x) = (f \circ g)(x)$ cannot simply be taken from the new function rule for h. It *must* be determined from the functions composed to obtain h.

Notice the function rule for $(f \circ g)(x)$ has an implied domain of $x \neq 2$, but does not show that g (the inner function) is undefined when $x = 0$ (see Part a). The domain of $(f \circ g)(x)$ is actually $x \neq 2$ **and** $x \neq 0$.

For $(g \circ f)(x)$ we have:

$$g[f(x)] = \dfrac{2}{f(x)} \qquad \text{composition of } g \text{ with } f$$

$$\dfrac{2}{f(x)} = \dfrac{2}{\dfrac{3x}{x-1}} \qquad \text{substitute } \dfrac{3x}{x-1} \text{ for } f(x)$$

$$= \dfrac{2}{1} \cdot \dfrac{x-1}{3x} \qquad \text{invert and multiply}$$

$$= \dfrac{2(x-1)}{3x} \qquad \text{result}$$

Similarly, the function rule for $(g \circ f)(x)$ has an implied domain of $x \neq 0$, but does not show that f (the inner function) is undefined when $x = 1$ (see Part a). The domain of $(g \circ f)(x)$ is actually $x \neq 0$ **and** $x \neq 1$.

Now try Exercises 59 through 64 ▶

To further explore concepts related to the domain of a composition, **see Exercises 92 through 94.**

Decomposing a Composite Function

Based on Figure 2.89, would you say that the circle is inside the square or the square is inside the circle? The decomposition of a composite function is related to a similar question, as we ask ourselves what function (of the composition) is on the "inside"—the input quantity—and what function is on the "outside." For instance, consider $h(x) = \sqrt{x - 4}$, where we see that $x - 4$ is "inside" the radical. Letting $g(x) = x - 4$ and $f(x) = \sqrt{x}$, we have $h(x) = (f \circ g)(x)$ or $f[g(x)]$.

Figure 2.89

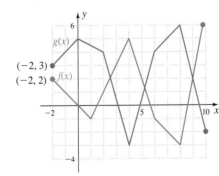

EXAMPLE 8 ▶   **Decomposing a Composite Function**

Given $h(x) = (\sqrt[3]{x} + 1)^2 - 3$, identify two functions f and g so that $(f \circ g)(x) = h(x)$, then check by composing the functions to obtain $h(x)$.

Solution ▶   Noting that $\sqrt[3]{x} + 1$ is inside the squaring function, we assign $g(x)$ as this inner function: $g(x) = \sqrt[3]{x} + 1$. The outer function is the squaring function decreased by 3, so $f(x) = x^2 - 3$.

☑ **C. You've just learned how to compose two functions and determine the domain, and decompose a function**

Check: $(f \circ g)(x) = f[g(x)]$ $g(x)$ is an input for f

$\qquad\qquad = [g(x)]^2 - 3$ f squares inputs, then decreases the result by 3

$\qquad\qquad = [\sqrt[3]{x} + 1]^2 - 3$ substitute $\sqrt[3]{x} + 1$ for $g(x)$

$\qquad\qquad = h(x)$ ✓

Now try Exercises 65 through 68 ▶

D. A Graphical View of Operations on Functions

The algebra and composition of functions also has an instructive *graphical interpretation,* in which values for $f(k)$ and $g(k)$ are read from a graph (k is a given constant), with operations like $(f + g)(k) = f(k) + g(k)$ then computed and lodged. Once the value of $g(k)$ is known, $(f \circ g)(k) = f[g(k)]$ is likewise interpreted and computed (also **see Exercise 95**).

EXAMPLE 9 ▶   **Interpreting Operations on Functions Graphically**

Use the graph given to find the value of each expression:

a. $(f + g)(-2)$

b. $(f \circ g)(7)$

c. $(g - f)(6)$

d. $\left(\dfrac{g}{f}\right)(8)$

e. $(f \cdot g)(4)$

Solution ▶ Since the needed input values for this example are $x = -2, 4, 6, 7$, and 8, we begin by reading the value of $f(x)$ and $g(x)$ at each point. From the graph, we note that $f(-2) = 2$ and $g(-2) = 3$. The other values are likewise found and appear in the table. For $(f + g)(-2)$ we have:

x	$f(x)$	$g(x)$
-2	2	3
4	5	-3
6	-1	4
7	-2	5
8	-3	6

a. $(f + g)(-2) = f(-2) + g(-2)$ definition

$\qquad\qquad\quad = 2 + 3$ substitute 2 for $f(-2)$ and 3 for $g(-2)$

$\qquad\qquad\quad = 5$ result

b. $(f \circ g)(7) = f[g(7)]$ definition

$\qquad\qquad\quad = f(5)$ substitute 5 for $g(7)$

$\qquad\qquad\quad = 2$ result read from graph: $f(5) = 2$

With some practice, the computations can be done mentally and we have

c. $(g - f)(6) = g(6) - f(6)$

$\qquad\qquad\quad = 4 - (-1) = 5$

☑ **D.** You've just learned how to interpret operations on functions graphically

d. $\left(\dfrac{g}{f}\right)(8) = \dfrac{g(8)}{f(8)}$

$\qquad\qquad\quad = \dfrac{6}{-3} = -2$

e. $(f \cdot g)(4) = f(4) \cdot g(4)$

$\qquad\qquad\quad = 5(-3) = -15$

Now try Exercises 69 through 78 ▶

E. Applications of the Algebra and Composition of Functions

The algebra of functions plays an important role in the business world. For example, the cost to manufacture an item, the revenue a company brings in, and the profit a company earns are all functions of the number of items made and sold. Further, we know a company "breaks even" (making $0 profit) when the difference between their revenue R and their cost C, is zero.

EXAMPLE 10 ▶ **Applying Operations on Functions in Context**

The fixed costs to publish *Relativity Made Simple* (by N.O. Way) is $2500, and the variable cost is $4.50 per book. Marketing studies indicate the best selling price for the book is $9.50 per copy.

a. Find the cost, revenue, and profit functions for this book.

b. Determine how many copies must be sold for the company to break even.

Solution ▶ **a.** Let x represent the number of books published and sold. The cost of publishing is $4.50 per copy, plus fixed costs (labor, storage, etc.) of $2500. The cost function is $C(x) = 4.50x + 2500$. If the company charges $9.50 per book, the revenue function will be $R(x) = 9.50x$. Since profit equals revenue minus costs,

$P(x) = R(x) - C(x)$

$\qquad = 9.50x - (4.50x + 2500)$ substitute $9.50x$ for R and $4.50x + 2500$ for C

$\qquad = 9.50x - 4.50x - 2500$ distribute

$\qquad = 5x - 2500$ result

The profit function is $P(x) = 5x - 2500$.

b. When a company "breaks even," the profit is zero: $P(x) = 0$.

$$P(x) = 5x - 2500 \qquad \text{profit function}$$
$$0 = 5x - 2500 \qquad \text{substitute 0 for } P(x)$$
$$2500 = 5x \qquad \text{add 2500}$$
$$500 = x \qquad \text{divide by 5}$$

In order for the company to break even, 500 copies must be sold.

Now try Exercises 81 through 84 ▶

Suppose that due to a collision, an oil tanker is spewing oil into the open ocean. The oil is spreading outward in a shape that is roughly circular, with the radius of the circle modeled by the function $r(t) = 2\sqrt{t}$, where t is the time in minutes and r is measured in feet. How could we determine the *area* of the oil slick in terms of t? As you can see, the radius depends on the time and the area depends on the radius. In diagram form we have:

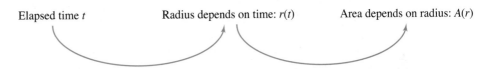

Elapsed time t Radius depends on time: $r(t)$ Area depends on radius: $A(r)$

It is possible to create a direct relationship between the elapsed time and the area of the circular spill using a composition of functions.

EXAMPLE 11 ▶   **Applying a Composition in Context**

Given $r(t) = 2\sqrt{t}$ and $A(r) = \pi r^2$,

    a. Write A directly as a function of t by computing $(A \circ r)(t)$.

    b. Find the area of the oil spill after 30 min.

Solution ▶   **a.** The function A squares inputs, then multiplies by π.

$$(A \circ r)(t) = A[r(t)] \qquad r(t) \text{ is the input for } A$$
$$= [r(t)]^2 \cdot \pi \qquad \text{square input, multiply by } \pi$$
$$= [2\sqrt{t}]^2 \cdot \pi \qquad \text{substitute } 2\sqrt{t} \text{ for } r(t)$$
$$= 4\pi t \qquad \text{result}$$

Since the result contains no variable r, we can now compute the area of the spill directly, given the elapsed time t (in minutes): $A(t) = 4\pi t$.

    b. To find the area after 30 min, use $t = 30$.

$$A(t) = 4\pi t \qquad \text{composite function}$$
$$A(30) = 4\pi(30) \qquad \text{substitute 30 for } t$$
$$= 120\pi \qquad \text{simplify}$$
$$\approx 377 \qquad \text{result (rounded to the nearest unit)}$$

☑ **E.** You've just learned how to apply the algebra and composition of functions in context

After 30 min, the area of the spill is approximately 377 ft^2.

Now try Exercises 85 through 90 ▶

TECHNOLOGY HIGHLIGHT

Composite Functions

The graphing calculator is truly an amazing tool when it comes to studying composite functions. Using this powerful tool, composite functions can be graphed, evaluated, and investigated with ease. To begin, enter the functions $y = x^2$ and $y = x - 5$ as Y_1 and Y_2 on the $\boxed{Y=}$ screen. Enter the composition $(Y_1 \circ Y_2)(x)$ as $Y_3 = Y_1(Y_2(X))$, as shown in Figure 2.90 [in our standard notation we have $f(x) = x^2$, $g(x) = x - 5$, and $h(x) = (f + g)(x) = f[g(x)]$. On the TI-84 *Plus*, we access the function variables Y_1, Y_2, Y_3, and so on by pressing $\boxed{\text{VARS}}$

Figure 2.90

$\boxed{\blacktriangleright}$ $\boxed{\text{ENTER}}$ and selecting the function desired. Pressing $\boxed{\text{ZOOM}}$ **6:ZStandard** will graph all three functions in the standard window. Let's look at the relationship between Y_1 and Y_3. Deactivate Y_2 and regraph Y_1 and Y_3. What do you notice about the graphs? Y_3 is the same as the graph of Y_1, but shifted 5 units to the right! Does this have any connection to $Y_2 = x - 5$? Try changing Y_2 to $Y_2 = x + 4$, then regraph Y_1 and Y_3. Use what you notice to complete the following exercises and continue the exploration.

Exercise 1: Change Y_1 to $Y_1 = \sqrt{x}$, then experiment by changing Y_2 to $x + 3$, then to $x - 6$. Did you notice anything similar? What would happen if we changed Y_2 to $Y_2 = x + 7$?

Exercise 2: Change Y_1 to $Y_1 = x^3$, then experiment by changing Y_2 to $x + 5$, then to $x - 1$. Did the same "shift" occur? What would happen if we changed Y_1 to $Y_1 = |x|$?

2.8 EXERCISES

▶ CONCEPTS AND VOCABULARY

Fill in each blank with the appropriate word or phrase. Carefully reread the section, if necessary.

1. Given function f with domain A and function g with domain B, the sum $f(x) + g(x)$ can also be written _____. The domain of the result is _____.

2. For the product $h(x) = f(x) \cdot g(x)$, $h(5)$ can be found by evaluating f and g then multiplying the result, or multiplying $f \cdot g$ and evaluating the result. Notationally these are written _____ and _____.

3. When combining functions f and g using basic operations, the domain of the result is the _____ of the domains of f and g. For division, we further stipulate that _____ cannot equal zero.

4. When evaluating functions, if the input value is a function itself, the process is called the _____ of functions. The notation $(f \circ g)(x)$ indicates that _____ is the input value for _____, which we can also write as _____.

5. For $f(x) = 2x^3 - 50x$ and $g(x) = x - 5$, discuss/explain why the domain of $h(x) = \left(\dfrac{f}{g}\right)(x)$ must exclude $x = 5$, even though the resulting quotient is the polynomial $2x^2 + 10x$.

6. For $f(x) = \sqrt{2x + 7}$ and $g(x) = \dfrac{2}{x - 1}$, discuss/explain how the domain of $h(x) = (f \circ g)(x)$ is determined. In particular, why is $h(1)$ not defined even though $f(1) = 3$?

▶ DEVELOPING YOUR SKILLS

7. Given $f(x) = 2x^2 - x - 3$ and $g(x) = x^2 + 5x$,
(a) determine the domain for $h(x) = f(x) - g(x)$ and
(b) find $h(-2)$ using the definition.

8. Given $f(x) = 2x^2 - 18$ and $g(x) = -3x - 7$,
(a) determine the domain for $h(x) = f(x) + g(x)$
and (b) find $h(5)$ using the definition.

9. For the functions f, g, and h, as defined in Exercise 7,
(a) find a new function rule for h, and (b) use the
result to find $h(-2)$. (c) How does the result
compare to that of Exercise 7?

10. For the functions f, g, and h as defined in
Exercise 8, (a) find a new function rule for h, and
(b) use the result to find $h(5)$. (c) How does the
result compare to that in Exercise 8?

11. For $f(x) = \sqrt{x - 3}$ and $g(x) = 2x^3 - 54$,
(a) determine the domain of $h(x) = (f + g)(x)$,
(b) find a new function rule for h, and
(c) evaluate $h(4)$ and $h(2)$, if possible.

12. For $f(x) = 4x^2 - 2x + 3$ and $g(x) = \sqrt{2x - 5}$,
(a) determine the domain of $h(x) = (f - g)(x)$,
(b) find a new function rule for h, and (c) evaluate
$h(7)$ and $h(2)$, if possible.

13. For $p(x) = \sqrt{x + 5}$ and $q(x) = \sqrt{3 - x}$,
(a) determine the domain of $r(x) = (p + q)(x)$,
(b) find a new function rule for r, and
(c) evaluate $r(2)$ and $r(4)$, if possible.

14. For $p(x) = \sqrt{6 - x}$ and $q(x) = \sqrt{x + 2}$,
(a) determine the domain of $r(x) = (p - q)(x)$,
(b) find a new function rule for r, and
(c) evaluate $r(-3)$ and $r(2)$, if possible.

15. For $f(x) = \sqrt{x + 4}$ and $g(x) = 2x + 3$,
(a) determine the domain of $h(x) = (f \cdot g)(x)$,
(b) find a new function rule for h, and
(c) evaluate $h(-4)$ and $h(21)$, if possible.

16. For $f(x) = -3x + 5$ and $g(x) = \sqrt{x - 7}$,
(a) determine the domain of $h(x) = (f \cdot g)(x)$,
(b) find a new function rule for h, and
(c) evaluate $h(8)$ and $h(11)$, if possible.

17. For $p(x) = \sqrt{x + 1}$ and $q(x) = \sqrt{7 - x}$,
(a) determine the domain of $r(x) = (p \cdot q)(x)$,
(b) find a new function rule for r, and
(c) evaluate $r(15)$ and $r(3)$, if possible.

18. For $p(x) = \sqrt{4 - x}$ and $q(x) = \sqrt{x + 4}$,
(a) determine the domain of $r(x) = (p \cdot q)(x)$,
(b) find a new function rule for r, and
(c) evaluate $r(-5)$ and $r(-3)$, if possible.

For the functions f and g given, (a) determine the
domain of $h(x) = \left(\dfrac{f}{g}\right)(x)$ and (b) find a new function rule
for h in simplified form (if possible), noting the domain
restrictions along side.

19. $f(x) = x^2 - 16$ and $g(x) = x + 4$

20. $f(x) = x^2 - 49$ and $g(x) = x - 7$

21. $f(x) = x^3 + 4x^2 - 2x - 8$ and $g(x) = x + 4$

22. $f(x) = x^3 - 5x^2 + 2x - 10$ and $g(x) = x - 5$

23. $f(x) = x^3 - 7x^2 + 6x$ and $g(x) = x - 1$

24. $f(x) = x^3 - 1$ and $g(x) = x - 1$

25. $f(x) = x + 1$ and $g(x) = x - 5$

26. $f(x) = x + 3$ and $g(x) = x - 7$

For the functions p and q given, (a) determine the
domain of $r(x) = \left(\dfrac{p}{q}\right)(x)$, (b) find a new function rule
for r, and (c) use it to evaluate $r(6)$ and $r(-6)$, if possible.

27. $p(x) = 2x - 3$ and $q(x) = \sqrt{-2 - x}$

28. $p(x) = 1 - x$ and $q(x) = \sqrt{3 - x}$

29. $p(x) = x - 5$ and $q(x) = \sqrt{x - 5}$

30. $p(x) = x + 2$ and $q(x) = \sqrt{x + 3}$

31. $p(x) = x^2 - 36$ and $q(x) = \sqrt{2x + 13}$

32. $p(x) = x^2 - 6x$ and $q(x) = \sqrt{7 + 3x}$

For the functions f and g given, (a) find a new function
rule for $h(x) = \left(\dfrac{f}{g}\right)(x)$ in simplified form. (b) If $h(x)$ were
the original function, what would be its domain?
(c) Since we know $h(x) = \left(\dfrac{f}{g}\right)(x) = \dfrac{f(x)}{g(x)}$, what additional
values are excluded from the domain of h?

33. $f(x) = \dfrac{6x}{x - 3}$ and $g(x) = \dfrac{3x}{x + 2}$

34. $f(x) = \dfrac{4x}{x + 1}$ and $g(x) = \dfrac{2x}{x - 2}$

For each pair of functions f and g given, determine the
sum, difference, product, and quotient of f and g, then
determine the domain in each case.

35. $f(x) = 2x + 3$ and $g(x) = x - 2$

36. $f(x) = x - 5$ and $g(x) = 2x - 3$

37. $f(x) = x^2 + 7$ and $g(x) = 3x - 2$

38. $f(x) = x^2 - 3x$ and $g(x) = x + 4$

39. $f(x) = x^2 + 2x - 3$ and $g(x) = x - 1$

40. $f(x) = x^2 - 2x - 15$ and $g(x) = x + 3$

41. $f(x) = 3x + 1$ and $g(x) = \sqrt{x - 3}$

42. $f(x) = x + 2$ and $g(x) = \sqrt{x + 6}$

43. $f(x) = 2x^2$ and $g(x) = \sqrt{x + 1}$

44. $f(x) = x^2 + 2$ and $g(x) = \sqrt{x - 5}$

45. $f(x) = \dfrac{2}{x - 3}$ and $g(x) = \dfrac{5}{x + 2}$

46. $f(x) = \dfrac{4}{x - 3}$ and $g(x) = \dfrac{1}{x + 5}$

47. Given $f(x) = x^2 - 5x - 14$, find $f(-2), f(7)$, $f(2a)$, and $f(a - 2)$.

48. Given $g(x) = x^3 - 9x$, find $g(-3), g(2), g(3t)$, and $g(t + 1)$.

For each pair of functions below, find (a) $h(x) = (f \circ g)(x)$ and (b) $H(x) = (g \circ f)(x)$, and (c) determine the domain of each result.

49. $f(x) = \sqrt{x + 3}$ and $g(x) = 2x - 5$

50. $f(x) = x + 3$ and $g(x) = \sqrt{9 - x^2}$

51. $f(x) = \sqrt{x - 3}$ and $g(x) = 3x + 4$

52. $f(x) = \sqrt{x + 5}$ and $g(x) = 4x - 1$

53. $f(x) = x^2 - 3x$ and $g(x) = x + 2$

54. $f(x) = 2x^2 - 1$ and $g(x) = 3x + 2$

55. $f(x) = x^2 + x - 4$ and $g(x) = x + 3$

56. $f(x) = x^2 - 4x + 2$ and $g(x) = x - 2$

57. $f(x) = |x| - 5$ and $g(x) = -3x + 1$

58. $f(x) = |x - 2|$ and $g(x) = 3x - 5$

For the functions $f(x)$ and $g(x)$ given, analyze the domain of (a) $(f \circ g)(x)$ and (b) $(g \circ f)(x)$, then (c) find the actual compositions and comment.

59. $f(x) = \dfrac{2x}{x + 3}$ and $g(x) = \dfrac{5}{x}$

60. $f(x) = \dfrac{-3}{x}$ and $g(x) = \dfrac{x}{x - 2}$

61. $f(x) = \dfrac{4}{x}$ and $g(x) = \dfrac{1}{x - 5}$

62. $f(x) = \dfrac{3}{x}$ and $g(x) = \dfrac{1}{x - 2}$

63. For $f(x) = x^2 - 8, g(x) = x + 2$, and $h(x) = (f \circ g)(x)$, find $h(5)$ in two ways:
  a. $(f \circ g)(5)$ **b.** $f[g(5)]$

64. For $p(x) = x^2 - 8, q(x) = x + 2$, and $H(x) = (p \circ q)(x)$, find $H(-2)$ in two ways:
  a. $(p \circ q)(-2)$ **b.** $p[q(-2)]$

65. For $h(x) = (\sqrt{x - 2} + 1)^3 - 5$, find two functions f and g such that $(f \circ g)(x) = h(x)$.

66. For $H(x) = \sqrt[3]{x^2 - 5} + 2$, find two functions p and q such that $(p \circ q)(x) = h(x)$.

67. Given $f(x) = 2x - 1, g(x) = x^2 - 1$, and $h(x) = x + 4$, find $p(x) = f[g([h(x)])]$ and $q(x) = g[f([h(x)])]$.

68. Given $f(x) = 2x + 3$ and $g(x) = \dfrac{x - 3}{2}$, find
  (a) $(f \circ f)(x)$, (b) $(g \circ g)(x)$, (c) $(f \circ g)(x)$, and (d) $(g \circ f)(x)$.

69. Reading a graph: The graph given shows the number of cars $C(t)$ and trucks $T(t)$ sold by Ullery Used Autos for the years 2000 to 2010. Use the graph to estimate the number of

Exercise 69

  a. cars sold in 2005: $C(5)$

  b. trucks sold in 2008: $T(8)$

  c. vehicles sold in 2009: $C(9) + T(9)$

  d. In function notation, how would you determine how many more cars than trucks were sold in 2009? What was the actual number?

70. Reading a graph: The graph given shows a government's investment in its military $M(t)$ over time, versus its investment in public works $P(t)$, in millions of dollars. Use the graph to estimate the amount of investment in

Exercise 70

  a. the military in 2002: $M(2)$

  b. public works in 2005: $P(5)$

  c. public works and the military in 2009: $M(9) + P(9)$

  d. In function notation, how would you determine how much more will be invested in public works than the military in 2010? What is the actual number?

71. Reading a graph: The graph given shows the revenue $R(t)$ and operating costs $C(t)$ of Space Travel Resources (STR), for the years 2000 to 2010. Use the graph to find the

Exercise 71

   a. revenue in 2002: $R(2)$

   b. costs in 2008: $C(8)$

   c. years STR broke even: $R(t) = C(t)$

   d. years costs exceeded revenue: $C(t) > R(t)$

   e. years STR made a profit: $R(t) > C(t)$

   f. For the year 2005, use function notation to write the profit equation for STR. What was their profit?

72. Reading a graph: The graph given shows a large corporation's investment in research and development $R(t)$ over time, and the amount paid to investors as dividends $D(t)$, in billions of dollars. Use the graph to find the

Exercise 72

   a. dividend payments in 2002: $D(2)$

   b. investment in 2006: $R(6)$

   c. years where $R(t) = D(t)$

   d. years where $R(t) > D(t)$

   e. years where $R(t) < D(t)$

   f. Use function notation to write an equation for the total expenditures of the corporation in year t. What was the total for 2010?

73. Reading a graph: Use the given graph to find the result of the operations indicated.

Note $f(-4) = 5$, $g(-4) = -1$, and so on.

Exercise 73

   a. $(f + g)(-4)$

   b. $(f \cdot g)(1)$ **c.** $(f - g)(4)$

   d. $(f + g)(0)$ **e.** $\left(\dfrac{f}{g}\right)(2)$

   f. $(f \cdot g)(-2)$ **g.** $(g \cdot f)(2)$

   h. $(f - g)(-1)$ **i.** $(f + g)(8)$

   j. $\left(\dfrac{f}{g}\right)(7)$ **k.** $(g \circ f)(4)$

   l. $(f \circ g)(4)$

74. Reading a graph: Use the given graph to find the result of the operations indicated.

Exercise 74

Note $p(-1) = 3$, $q(5) = 6$, and so on.

   a. $(p + q)(-4)$

   b. $(p \cdot q)(1)$

   c. $(p - q)(4)$ **d.** $(p + q)(0)$

   e. $\left(\dfrac{p}{q}\right)(5)$ **f.** $(p \cdot q)(-2)$

   g. $(q \cdot p)(2)$ **h.** $(p - q)(-1)$

   i. $(p + q)(7)$ **j.** $\left(\dfrac{p}{q}\right)(6)$

   k. $(q \circ p)(4)$ **l.** $(p \circ q)(-1)$

Some advanced applications require that we use the algebra of functions to find a function rule for the vertical distance between two graphs. For $f(x) = 3$ and $g(x) = -2$ (two horizontal lines), we "see" this vertical distance is 5 units, or in function form: $d(x) = f(x) - g(x) = 3 - (-2) = 5$ units. However, $d(x) = f(x) - g(x)$ also serves as a *general formula* for the vertical distance between two curves (even those that are not horizontal lines), so long as $f(x) > g(x)$ in a chosen interval. Find a function rule in simplified form, for the vertical distance $h(x)$ between the graphs of f and g shown, for the interval indicated.

75. $x \in [0, 6]$ **76.** $x \in [1, 7]$

77. $x \in [0, 4]$ **78.** $x \in [0, 5]$

▶ WORKING WITH FORMULAS

79. Surface area of a cylinder: $A = 2\pi rh + 2\pi r^2$

If the height of a cylinder is fixed at 20 cm, the formula becomes $A = 40\pi r + 2\pi r^2$. Write this formula in factored form and find two functions $f(r)$ and $g(r)$ such that $A(r) = (f \cdot g)(r)$. Then find $A(5)$ by direct calculation and also by computing the product of $f(5)$ and $g(5)$, then comment on the results.

80. Compound annual growth: $A(r) = P(1 + r)^t$

The amount of money A in a savings account t yr after an initial investment of P dollars depends on the interest rate r. If $1000 is invested for 5 yr, find $f(r)$ and $g(r)$ such that $A(r) = (f \circ g)(r)$.

▶ APPLICATIONS

81. Boat manufacturing: Giaro Boats manufactures a popular recreational vessel, the *Revolution*. To plan for expanded production and increased labor costs, the company carefully tracks current costs and income. The fixed cost to produce this boat is $108,000 and the variable costs are $28,000 per boat. If the *Revolution* sells for $40,000, (a) find the profit function and (b) determine how many boats must be sold for the company to break even.

82. Non-profit publications: Adobe Hope, a nonprofit agency, publishes the weekly newsletter *Community Options*. In doing so, they provide useful information to the surrounding area while giving high school dropouts valuable work experience. The fixed cost for publishing the newsletter is $900 per week, with a variable cost of $0.25 per newsletter. If the newsletter is sold for $1.50 per copy, (a) find the profit function for the newsletter, (b) determine how many newsletters must be sold to break even, and (c) determine how much money will be returned to the community if 1000 newsletters are sold (to preserve their status as a nonprofit organization).

83. Cost, revenue, and profit: Suppose the total cost of manufacturing a certain computer component can be modeled by the function $C(n) = 0.1n^2$, where n is the number of components made and $C(n)$ is in dollars. If each component is sold at a price of $11.45, the revenue is modeled by $R(n) = 11.45n$. Use this information to complete the following.

a. Find the function that represents the total profit made from sales of the components.

b. How much profit is earned if 12 components are made and sold?

c. How much profit is earned if 60 components are made and sold?

d. Explain why the company is making a "negative profit" after the 114th component is made and sold.

84. Cost, revenue, and profit: For a certain manufacturer, revenue has been increasing but so has the cost of materials and the cost of employee benefits. Suppose revenue can be modeled by $R(t) = 10\sqrt{t}$, the cost of materials by $M(t) = 2t + 1$, and the cost of benefits by $C(t) = 0.1t^2 + 2$, where t represents the number of months since operations began and outputs are in thousands of dollars. Use this information to complete the following.

a. Find the function that represents the total manufacturing costs.

b. Find the function that represents how much more the operating costs are than the cost of materials.

c. What was the cost of operations in the 10th month after operations began?

d. How much less were the operating costs than the cost of materials in the 10th month?

e. Find the function that represents the profit earned by this company.

f. Find the amount of profit earned in the 5th month and 10th month. Discuss each result.

85. International shoe sizes: Peering inside her athletic shoes, Morgan notes the following shoe sizes: *US 8.5, UK 6, EUR 40*. The function that relates the U.S. sizes to the European (EUR) sizes is $g(x) = 2x + 23$ where x represents the U.S. size and $g(x)$ represents the EUR size. The function that relates European sizes to sizes in the United Kingdom (UK) is $f(x) = 0.5x - 14$ where x represents the EUR size and $f(x)$ represents the UK size. Find the function $h(x)$ that relates the U.S. measurement directly to the UK measurement by finding $h(x) = (f \circ g)(x)$. Find The UK size for a shoe that has a U.S. size of 13.

86. Currency conversion: On a trip to Europe, Megan had to convert American dollars to euros using the

function $E(x) = 1.12x$, where x represents the number of dollars and $E(x)$ is the equivalent number of euros. Later, she converts her euros to Japanese yen using the function $Y(x) = 1061x$, where x represents the number of euros and $Y(x)$ represents the equivalent number of yen.
(a) Convert 100 U.S. dollars to euros. (b) Convert the answer from part (a) into Japanese yen.
(c) Express yen as a function of dollars by finding $M(x) = (Y \circ E)(x)$, then use $M(x)$ to convert 100 dollars directly to yen. Do parts (b) and (c) agree?

Source: 2005 World Almanac, p. 231

87. **Currency conversion:** While traveling in the Far East, Timi must convert U.S. dollars to Thai baht using the function $T(x) = 41.6x$, where x represents the number of dollars and $T(x)$ is the equivalent number of baht. Later she needs to convert her baht to Malaysian ringgit using the function $R(x) = 10.9x$. (a) Convert 100 dollars to baht.
(b) Convert the result from part (a) to ringgit.
(c) Express ringgit as a function of dollars using $M(x) = (R \circ T)(x)$, then use $M(x)$ to convert 100 dollars to ringgit directly. Do parts (b) and (c) agree?

Source: 2005 World Almanac, p. 231

88. **Spread of a fire:** Due to a lightning strike, a forest fire begins to burn and is spreading outward in a shape that is roughly circular. The radius of the circle is modeled by the function $r(t) = 2t$, where t is the time in minutes and r is measured in meters.
(a) Write a function for the area burned by the fire directly as a function of t by computing $(A \circ r)(t)$.
(b) Find the area of the circular burn after 60 min.

89. **Radius of a ripple:** As Mark drops firecrackers into a lake one 4th of July, each "pop" caused a circular ripple that expanded with time. The radius of the circle is a function of time t. Suppose the function is $r(t) = 3t$, where t is in seconds and r is

in feet. (a) Find the radius of the circle after 2 sec.
(b) Find the area of the circle after 2 sec.
(c) Express the area as a function of time by finding $A(t) = (A \circ r)(t)$ and use $A(t)$ to find the area of the circle after 2 sec. Do the answers agree?

90. **Expanding supernova:** The surface area of a star goes through an expansion phase prior to going *supernova*. As the star begins expanding, the radius becomes a function of time. Suppose this function is $r(t) = 1.05t$, where t is in days and $r(t)$ is in gigameters (Gm). (a) Find the radius of the star two days after the expansion phase begins. (b) Find the surface area after two days. (c) Express the surface area as a function of time by finding $h(t) = (S \circ r)(t)$, then use $h(t)$ to compute the surface area after two days directly. Do the answers agree?

▶ **EXTENDING THE CONCEPT**

91. In a certain country, the function $C(x) = 0.0345x^4 - 0.8996x^3 + 7.5383x^2 - 21.7215x + 40$ approximates the number of Conservatives in the senate for the years 1995 to 2007, where $x = 0$ corresponds to 1995. The function $L(x) = -0.0345x^4 + 0.8996x^3 - 7.5383x^2 + 21.7215x + 10$ gives the number of Liberals for these years. Use this information to answer the following. (a) During what years did the

Conservatives control the senate? (b) What was the greatest difference between the number of seats held by each faction in any one year? In what year did this occur? (c) What was the minimum number of seats held by the Conservatives? In what year? (d) Assuming no independent or third-party candidates are elected, what information does the function $T(x) = C(x) + L(x)$ give us? What information does $t(x) = |C(x) - L(x)|$ give us?

92. Given $f(x) = x^3 + 2$ and $g(x) = \sqrt[3]{x - 2}$, graph each function on the same axes by plotting the points that correspond to integer inputs for $x \in [-3, 3]$. Do you notice anything? Next, find $h(x) = (f \circ g)(x)$ and $H(x) = (g \circ f)(x)$. What happened? Look closely at the functions f and g to see how they are related. Can you come up with two additional functions where the same thing occurs?

93. Given $f(x) = \sqrt{1 - x}$ and $g(x) = \sqrt{x - 2}$, what can you say about the domain of $(f + g)(x)$? Enter the functions as Y_1 and Y_2 on a graphing calculator, then enter $Y_3 = Y_1 + Y_2$. See if you can determine why the calculator gives an error message for Y_3, regardless of the input.

94. Given $f(x) = \dfrac{1}{x^2 - 4}$, $g(x) = \sqrt{x + 1}$, and

$h(x) = (f \circ g)(x)$, (a) find the new function rule for h and (b) determine the implied domain of h. Does this *implied* domain include $x = 2$, $x = -2$, and $x = -3$ as valid inputs? (c) Determine the actual domain for $h(x) = (f \circ g)(x)$ and discuss the result.

95. Instead of calculating the result of an operation on two functions at a *specific point* as in Exercises 69–74, we can actually *graph the function* that

results from the operation. This skill, called the **addition of ordinates,** is widely applied in a study of tides and other areas. For $f(x) = (x - 3)^2 + 2$ and $g(x) = 4|x - 3| - 5$, complete a table of values like the one shown for $x \in [-2, 8]$. For the last column, remember that $(f - g)(x) = f(x) - g(x)$, and use this relation to complete the column. Finally, use the ordered pairs $(x, (f - g)(x))$ to graph the new function. Is the new function smooth? Is the new function continuous?

Exercise 95

x	$f(x)$	$g(x)$	$(f - g)(x)$
-2			
-1			
0			
1			
2			
3			
4			
5			
6			
7			
8			

▶ MAINTAINING YOUR SKILLS

96. (1.4) Find the sum and product of the complex numbers $2 + 3i$ and $2 - 3i$.

97. (2.4) Draw a sketch of the functions (a) $f(x) = \sqrt{x}$, (b) $g(x) = \sqrt[3]{x}$, and (c) $h(x) = |x|$ *from memory.*

98. (1.5) Use the quadratic formula to solve $2x^2 - 3x + 4 = 0$.

99. (2.3) Find the equation of the line perpendicular to $-2x + 3y = 9$, that also goes through the origin.

SUMMARY AND CONCEPT REVIEW

SECTION 2.1   Rectangular Coordinates; Graphing Circles and Other Relations

KEY CONCEPTS

- A relation is a collection of ordered pairs (x, y) and can be given in set or equation form.
- As a set of ordered pairs, the domain of the relation is the set of all first coordinates, and the range is the set of all corresponding second coordinates.
- A relation can be expressed in mapping notation $x \rightarrow y$, indicating an element from the domain is mapped to (corresponds to or is associated with) an element from the range.
- The graph of a relation in equation form is the set of all ordered pairs (x, y) that satisfy the equation. We plot a sufficient number of points and connect them with a straight line or smooth curve, depending on the pattern formed.
- The midpoint of a line segment with endpoints (x_1, y_1) and (x_2, y_2) is $\left(\dfrac{x_1 + x_2}{2}, \dfrac{y_1 + y_2}{2} \right)$.
- The distance between the points (x_1, y_1) and (x_2, y_2) is $d = \sqrt{(x_2 - x_1)^2 + (y_2 - y_1)^2}$.
- The equation of a circle centered at (h, k) with radius r is $(x - h)^2 + (y - k)^2 = r^2$.

EXERCISES

1. Represent the relation in mapping notation, then state the domain and range.

$\{(-7, 3), (-4, -2), (5, 1), (-7, 0), (3, -2), (0, 8)\}$

2. Graph the relation $y = \sqrt{25 - x^2}$ by completing the table, then state the domain and range of the relation.

x	y
-5	
-4	
-2	
0	
2	
4	
5	

Mr. Northeast and Mr. Southwest live in Coordinate County and are good friends. Mr. Northeast lives at *19 East 25 North* or (19, 25), while Mr. Southwest lives at *14 West and 31 South* or $(-14, -31)$. If the streets in Coordinate County are laid out in one mile squares,

3. Use the distance formula to find how far apart they live.

4. If they agree to meet halfway between their homes, what are the coordinates of their meeting place?

5. Sketch the graph of $x^2 + y^2 = 16$.

6. Sketch the graph of $x^2 + y^2 + 6x + 4y + 9 = 0$. Clearly state the center and radius.

7. Find the equation of the circle whose diameter has the endpoints $(-3, 0)$ and $(0, 4)$.

SECTION 2.2   Graphs of Linear Equations

KEY CONCEPTS

- A linear equation can be written in the form $ax + by = c$, where a and b are not simultaneously equal to 0.

- The slope of the line through (x_1, y_1) and (x_2, y_2) is $m = \dfrac{y_2 - y_1}{x_2 - x_1}$, where $x_1 \neq x_2$.

- Other designations for slope are $m = \dfrac{\text{rise}}{\text{run}} = \dfrac{\text{change in } y}{\text{change in } x} = \dfrac{\Delta y}{\Delta x} = \dfrac{\text{vertical change}}{\text{horizontal change}}$.

- Lines with positive slope ($m > 0$) rise from left to right; lines with negative slope ($m < 0$) fall from left to right.

- The equation of a horizontal line is $y = k$; the slope is $m = 0$.

- The equation of a vertical line is $x = h$; the slope is undefined.

- Lines can be graphed using the intercept method. First determine $(x, 0)$ (substitute 0 for y and solve for x), then $(0, y)$ (substitute 0 for x and solve for y). Then draw a straight line through these points.

- Parallel lines have equal slopes ($m_1 = m_2$); perpendicular lines have slopes that are negative reciprocals $(m_1 = -\dfrac{1}{m_2}$ or $m_1 \cdot m_2 = -1)$.

EXERCISES

8. Plot the points and determine the slope, then use the ratio $\dfrac{\Delta y}{\Delta x} = \dfrac{\text{rise}}{\text{run}}$ to find an additional point on the line:

   a. $(-4, 3)$ and $(5, -2)$ and **b.** $(3, 4)$ and $(-6, 1)$.

9. Use the slope formula to determine if lines L_1 and L_2 are parallel, perpendicular, or neither:

   a. L_1: $(-2, 0)$ and $(0, 6)$; L_2: $(1, 8)$ and $(0, 5)$
   b. L_1: $(1, 10)$ and $(-1, 7)$: L_2: $(-2, -1)$ and $(1, -3)$

10. Graph each equation by plotting points: (a) $y = 3x - 2$ and (b) $y = -\frac{3}{2}x + 1$.

11. Find the intercepts for each line and sketch the graph: (a) $2x + 3y = 6$ and (b) $y = \frac{4}{3}x - 2$.

12. Identify each line as either horizontal, vertical, or neither, and graph each line.

   a. $x = 5$ **b.** $y = -4$ **c.** $2y + x = 5$

13. Determine if the triangle with the vertices given is a right triangle: $(-5, -4)$, $(7, 2)$, $(0, 16)$.

14. Find the slope and y-intercept of the line shown and discuss the slope ratio in this context.

Hawk population (100s)

Rodent population (1000s)

SECTION 2.3   Linear Graphs and Rates of Change

KEY CONCEPTS

- The equation of a nonvertical line in slope-intercept form is $y = mx + b$ or $f(x) = mx + b$. The slope of the line is m and the y-intercept is $(0, b)$.

- To graph a line given its equation in slope-intercept form, plot the y-intercept, then use the slope ratio $m = \dfrac{\Delta y}{\Delta x}$ to find a second point, and draw a line through these points.

- If the slope m and a point (x_1, y_1) on the line are known, the equation of the line can be written in point-slope form: $y - y_1 = m(x - x_1)$.

- A secant line is the straight line drawn through two points on a nonlinear graph.

- The notation $m = \dfrac{\Delta y}{\Delta x}$ literally means the quantity measured along the y-axis is changing with respect to changes in the quantity measured along the x-axis.

- The average rate of change on the interval containing x_1 and x_2 is the slope of the secant line through (x_1, y_1) and (x_2, y_2), or $\dfrac{\Delta y}{\Delta x} = \dfrac{y_2 - y_1}{x_2 - x_1}$.

EXERCISES

15. Write each equation in slope-intercept form, then identify the slope and y-intercept.

   a. $4x + 3y - 12 = 0$ **b.** $5x - 3y = 15$

16. Graph each equation using the slope and y-intercept.

   a. $f(x) = -\frac{2}{3}x + 1$ **b.** $h(x) = \frac{5}{2}x - 3$

17. Graph the line with the given slope through the given point.

   a. $m = \frac{2}{3}; (1, 4)$ **b.** $m = -\frac{1}{2}; (-2, 3)$

18. What are the equations of the horizontal line and the vertical line passing through $(-2, 5)$? Which line is the point $(7, 5)$ on?

19. Find the equation of the line passing through $(1, 2)$ and $(-3, 5)$. Write your final answer in slope-intercept form.

20. Find the equation for the line that is parallel to $4x - 3y = 12$ and passes through the point $(3, 4)$. Write your final answer in slope-intercept form.

21. Determine the slope and y-intercept of the line shown. Then write the equation of the line in slope-intercept form and interpret the slope ratio $m = \dfrac{\Delta W}{\Delta R}$ in the context of this exercise.

Wolf population (100s)

Rabbit population (100s)

22. For the graph given, (a) find the equation of the line in point-slope form, (b) use the equation to predict the x- and y-intercepts, (c) write the equation in slope-intercept form, and (d) find y when x = 20, and the value of x for which y = 15.

SECTION 2.4   Functions, Function Notation, and the Graph of a Function

KEY CONCEPTS

- A function is a relation, rule, or equation that pairs each element from the domain with exactly one element of the range.
- The vertical line test says that if every vertical line crosses the graph of a relation in at most one point, the relation is a function.
- On a graph, vertical boundary lines can be used to identify the domain, or the set of "allowable inputs" for a function.
- On a graph, horizontal boundary lines can be used to identify the range, or the set of y-values (outputs) generated by the function.
- When a function is stated as an equation, the implied domain is the set of x-values that yield real number outputs.
- x-values that cause a denominator of zero or that cause the radicand of a square root expression to be negative must be excluded from the domain.
- *The phrase "y is a function of x," is written as y = f(x).* This notation enables us to evaluate functions while tracking corresponding x- and y-values.

EXERCISES

23. State the implied domain of each function:

    a. $f(x) = \sqrt{4x + 5}$

    b. $g(x) = \dfrac{x - 4}{x^2 - x - 6}$

24. Determine $h(-2)$, $h(-\frac{2}{3})$, and $h(3a)$ for $h(x) = 2x^2 - 3x$.

25. Determine if the mapping given represents a function. If not, explain how the definition of a function is violated.

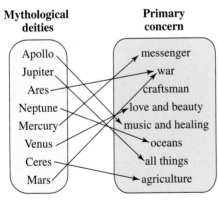

26. For the graph of each function shown, (a) state the domain and range, (b) find the value of $f(2)$, and (c) determine the value(s) of x for which $f(x) = 1$.

I.

II.

III.

SECTION 2.5   Analyzing the Graph of a Function

KEY CONCEPTS

- A function f is even (symmetric to the y-axis), if and only if when a point (x, y) is on the graph, then $(-x, y)$ is also on the graph. In function notation: $f(-x) = f(x)$.
- A function f is odd (symmetric to the origin), if and only if when a point (x, y) is on the graph, then $(-x, -y)$ is also on the graph. In function notation: $f(-x) = -f(x)$.

Intuitive descriptions of the characteristics of a graph are given here. The formal definitions can be found within Section 2.5.

- A function is *increasing* in an interval if the graph rises from left to right (larger inputs produce larger outputs).
- A function is *decreasing* in an interval if the graph falls from left to right (larger inputs produce smaller outputs).
- A function is *positive* in an interval if the graph is above the x-axis in that interval.
- A function is *negative* in an interval if the graph is below the x-axis in that interval.
- A function is *constant* in an interval if the graph is parallel to the x-axis in that interval.
- A maximum value can be a *local* maximum, or *global* maximum. An *endpoint* maximum can occur at the endpoints of the domain. Similar statements can be made for minimum values.
- For any function f, the average rate of change in the interval $[x_1, x_2]$ is $\dfrac{f(x_2) - f(x_1)}{x_2 - x_1}$.
- The difference quotient for a function $f(x)$ is $\dfrac{f(x + h) - f(x)}{h}$.

EXERCISES

State the domain and range for each function $f(x)$ given. Then state the intervals where f is increasing or decreasing and intervals where f is positive or negative. Assume all endpoints have integer values.

27.

28.

29.

30. Determine which of the following are even $[f(-k) = f(k)]$, odd $[f(-k) = -f(k)]$, or neither.

a. $f(x) = 2x^5 - \sqrt[3]{x}$ **b.** $g(x) = x^4 - \dfrac{\sqrt[3]{x}}{x}$

c. $p(x) = |3x| - x^3$ **d.** $q(x) = \dfrac{x^2 - |x|}{x}$

31a. Given $f(x) = \sqrt{x + 4}$, find the average rate of change in the interval $[-3, 5]$. What does the result confirm about the graph of this toolbox function?

31b. Use the difference quotient to find a rate of change formula for the function given, then calculate the rate of change for the interval indicated: $j(x) = x^2 - x$; [2.00, 2.01].

32. Draw the function f that has all of the following characteristics, then name the zeroes of the function and the location of all maximum and minimum values. [*Hint:* Write them in the form $(c, f(c))$.]

a. Domain: $x \in [-6, 10)$ **b.** Range: $y \in (-8, 6)$
c. $f(0) = 0$ **d.** $f(x)\downarrow$ for $x \in (-6, -3) \cup (3, 7.5)$
e. $f(x)\uparrow$ for $x \in (-3, 3) \cup (7.5, 10)$ **f.** $f(x) < 0$ for $x \in (-6, 0) \cup (6, 9)$
g. $f(x) > 0$ for $x \in (0, 6) \cup (9, 10)$

SECTION 2.6 The Toolbox Functions and Transformations

KEY CONCEPTS

- The *toolbox functions* and graphs commonly used in mathematics are
  - the identity function $f(x) = x$
  - square root function: $f(x) = \sqrt{x}$
  - cubing function: $f(x) = x^3$
  - squaring function: $f(x) = x^2$, parabola
  - absolute value function: $f(x) = |x|$
  - cube root function: $f(x) = \sqrt[3]{x}$

- For a basic or parent function $y = f(x)$, the general equation of the transformed function is $y = af(x \pm h) \pm k$. For any function $y = f(x)$ and $h, k > 0$,
  - the graph of $y = f(x) + k$ is the graph of $y = f(x)$ shifted upward k units
  - the graph of $y = f(x + h)$ is the graph of $y = f(x)$ shifted left h units
  - the graph of $y = -f(x)$ is the graph of $y = f(x)$ reflected across the x-axis
  - $y = af(x)$ results in a vertical stretch when $a > 1$

  - the graph of $y = f(x) - k$ is the graph of $y = f(x)$ shifted downward k units
  - the graph of $y = f(x - h)$ is the graph of $y = f(x)$ shifted right h units
  - the graph of $y = f(-x)$ is the graph of $y = f(x)$ reflected across the y-axis
  - $y = af(x)$ results in a vertical compression when $0 < a < 1$

- Transformations are applied in the following order: (1) horizontal shifts, (2) reflections, (3) stretches or compressions, and (4) vertical shifts.

EXERCISES

Identify the function family for each graph given, then (a) describe the end behavior; (b) name the x- and y-intercepts; (c) identify the vertex, initial point, or point of inflection (as applicable); and (d) state the domain and range.

33.

34.

35.

36.

37.
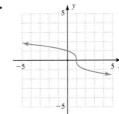

Identify each function as belonging to the linear, quadratic, square root, cubic, cube root, or absolute value family. Then sketch the graph using shifts of a parent function and a few characteristic points.

38. $f(x) = -(x + 2)^2 - 5$

39. $f(x) = 2|x + 3|$

40. $f(x) = x^3 - 1$

41. $f(x) = \sqrt{x - 5} + 2$

42. $f(x) = \sqrt[3]{x} + 2$

43. Apply the transformations indicated for the graph of $f(x)$ given.

  a. $f(x - 2)$
  b. $-f(x) + 4$
  c. $\frac{1}{2}f(x)$

SECTION 2.7   Piecewise-Defined Functions

KEY CONCEPTS

- Each piece of a piecewise-defined function has a domain over which that piece is defined.
- To evaluate a piecewise-defined function, identify the domain interval containg the input value, then use the piece of the function corresponding to this interval.
- To graph a piecewise-defined function you can plot points, or graph each piece in its entirety, then erase portions of the graph outside the domain indicated for each piece.
- If the graph of a function can be drawn without lifting your pencil from the paper, the function is continuous.
- A pointwise discontinuity is said to be removable because we can redefine the function to "fill the hole."
- Step functions are discontinuous and formed by a series of horizontal steps.
- The floor function $\lfloor x \rfloor$ gives the first integer less than or equal to x.
- The ceiling function $\lceil x \rceil$ is the first integer greater than or equal to x.

EXERCISES

44. For the graph and functions given, (a) use the correct notation to write the relation as a single piecewise-defined function, stating the effective domain for each piece by inspecting the graph; and (b) state the range of the function: $Y_1 = 5$, $Y_2 = -x + 1$, $Y_3 = 3\sqrt{x - 3} - 1$.

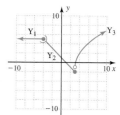

45. Use a table of values as needed to graph $h(x)$, then state its domain and range. If the function has a pointwise discontinuity, state how the second piece could be redefined so that a continuous function results.

$$h(x) = \begin{cases} \dfrac{x^2 - 2x - 15}{x + 3} & x \neq -3 \\ -6 & x = -3 \end{cases}$$

46. Evaluate the piecewise-defined function $p(x)$: $p(-4)$, $p(-2)$, $p(2.5)$, $p(2.99)$, $p(3)$, and $p(3.5)$

$$p(x) = \begin{cases} -4 & x < -2 \\ -|x| - 2 & -2 \leq x < 3 \\ 3\sqrt{x} - 9 & x \geq 3 \end{cases}$$

47. Sketch the graph of the function and state its domain and range. Use transformations of the toolbox functions where possible.

$$q(x) = \begin{cases} 2\sqrt{-x - 3} - 4 & x \leq -3 \\ -2|x| + 2 & -3 < x < 3 \\ 2\sqrt{x - 3} - 4 & x \geq 3 \end{cases}$$

48. Many home improvement outlets now rent flatbed trucks in support of customers that purchase large items. The cost is $20 per hour for the first 2 hr, $30 for the next 2 hr, then $40 for each hour afterward. Write this information as a piecewise-defined function, then sketch its graph. What is the total cost to rent this truck for 5 hr?

SECTION 2.8   The Algebra and Composition of Functions

KEY CONCEPTS

- The notation used to represent the basic operations on two functions is
  - $(f + g)(x) = f(x) + g(x)$ • $(f - g)(x) = f(x) - g(x)$
  - $(f \cdot g)(x) = f(x) \cdot g(x)$ • $\left(\dfrac{f}{g}\right)(x) = \dfrac{f(x)}{g(x)}$; $g(x) \neq 0$
- The result of these operations is a new function $h(x)$. The domain of h is the intersection of domains for f and g, excluding values that make $g(x) = 0$ for $h(x) = \left(\dfrac{f}{g}\right)(x)$.

- The composition of two functions is written $(f \circ g)(x) = f[g(x)]$ (g is an input for f).
- The domain of $f \circ g$ is all x in the domain of g, such that $g(x)$ is in the domain of f.
- To evaluate $(f \circ g)(2)$, we find $(f \circ g)(x)$ then substitute $x = 2$. Alternatively, we can find $g(2) = k$, then find $f(k)$.
- A composite function $h(x) = (f \circ g)(x)$ can be "decomposed" into individual functions by identifying functions f and g such that $(f \circ g)(x) = h(x)$. The decomposition is not unique.

EXERCISES

For $f(x) = x^2 + 4x$ and $g(x) = 3x - 2$, find the following:

49. $(f + g)(a)$ **50.** $(f \cdot g)(3)$ **51.** the domain of $\left(\dfrac{f}{g}\right)(x)$

Given $p(x) = 4x - 3$, $q(x) = x^2 + 2x$, and $r(x) = \dfrac{x + 3}{4}$ find:

52. $(p \circ q)(x)$ **53.** $(q \circ p)(3)$ **54.** $(p \circ r)(x)$ and $(r \circ p)(x)$

For each function here, find functions $f(x)$ and $g(x)$ such that $h(x) = f[g(x)]$:

55. $h(x) = \sqrt{3x - 2} + 1$ **56.** $h(x) = x^{\frac{2}{3}} - 3x^{\frac{1}{3}} - 10$

57. A stone is thrown into a pond causing a circular ripple to move outward from the point of entry. The radius of the circle is modeled by $r(t) = 2t + 3$, where t is the time in seconds. Find a function that will give the area of the circle directly as a function of time. In other words, find $A(t)$.

58. Use the graph given to find the value of each expression:

     a. $(f + g)(-2)$
     b. $(g \circ f)(5)$
     c. $(g - f)(7)$
     d. $\left(\dfrac{g}{f}\right)(10)$
     e. $(f \cdot g)(3)$

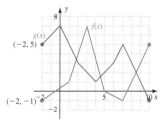

MIXED REVIEW

1. Write the given equation in slope-intercept form:
$4x + 3y = 12$

2. Find the equation of the line perpendicular to $x - 2y = 8$ that passes through $(1, 3)$.

3. Find the implied domain of:

     a. $f(x)\dfrac{x + 1}{x^2 - 5x + 4}$ **b.** $g(x) = \dfrac{1}{\sqrt{2x - 3}}$

4. Given $p(x) = -x^2 + 3x - 1$, find

     a. $p\left(\dfrac{-1}{3}\right)$ **b.** $p(3a)$ **c.** $p(a - 1)$

5. State the equation of the line shown, in slope-intercept form.

6. For the function g whose graph is given, find (a) domain, (b) $g(2)$, and (c) k if $g(k) = -3$.

7. The following three points form a right triangle: $(-3, 7)$, $(2, 2)$ and $(5, 5)$. Use the distance formula to help determine which point is at the vertex of the right angle. Then find the equation of the smallest circle, centered at that point, that encloses the triangle.

8. Discuss the end behavior of $F(x)$ and name the vertex, axis of symmetry, and all intercepts.

9. Graph by plotting the y-intercept, then counting $m = \dfrac{\Delta y}{\Delta x}$ to find additional points:

$$y = \frac{3}{5}x - 2$$

10. Solve the inequality using the graph provided:

$$f(x) = 4x - \frac{4}{3}x^2;\ f(x) < 0.$$

11. a. Graph the function $p(x) = -2x^2 + 8x$. By observing the graph, is the average rate of change positive or negative in the interval $[-2, -1]$? Why? Do you expect the rate of change in $[1, 2]$ to be greater or less than the rate of change in $[-2, -1]$? Calculate the average rate of change in each interval and comment.

b. If $1000 is deposited in an account paying 7% interest compounded continuously, the function model is $A(t) = 1000e^{0.07t}$. Use the average rate of change formula to determine if the amount of interest added to the account exceeds $200 per year $\left(\dfrac{\Delta A}{\Delta t} > 200\right)$ in the 10th, 15th, or 20th year. Use the intervals [10, 10.01], [15, 15.01], and [20, 20.01].

Given $f(x) = \dfrac{3}{x^2 - 1}$ and $g(x) = 3x - 2$, find

12. $\dfrac{g}{f}\left(\dfrac{1}{2}\right)$

13. $(f \circ g)(x)$ and its domain

14. Sketch the function h as defined.

$$h(x) = \begin{cases} 5 & 0 \le x < 8 \\ x - 3 & 8 \le x \le 15 \\ -2x + 40 & x > 15 \end{cases}$$

15. Given $f(x) = x^2 + 1$ and $g(x) = 3x - 2$, calculate the difference quotient for each function and use the results to estimate the value of x for which their rates of change are equal.

16. Identify the function family for the function $g(x) = -2|x + 3| + 4$. Then sketch the graph using transformations of a parent function and a few characteristic points.

17. For the graph shown, determine
  a. the domain and range of g,
  b. intervals where g is increasing, decreasing, or constant,
  c. intervals where g is positive or negative,
  d. any maximum or minimum values for g.

18. Draw a function f that has the following characteristics, then write the zeroes of the function and the location of all maximum and minimum values.
  a. domain: $x \in [0, 30]$
  b. range: $y \in [-10, 12]$
  c. $f(2) = f(10) = 0$
  d. $f(x)\downarrow$ from $x \in (0, 5) \cup (15, 20)$
  e. $f(x)\uparrow$ for $x \in (5, 15)$
  f. $f(x) < 0$ for $x \in (2, 10)$
  g. $f(x) > 0$ for $x \in (0, 2) \cup (10, 30)$
  h. $f(x) = 5$ for $x \in [20, 30]$

19. Find the equation of the function $f(x)$ whose graph is given.

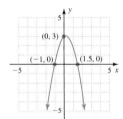

20. Since 1975, the number of deaths in the United States due to heart disease has been declining at a rate that is close to linear. Find an equation model if there were 431 thousand deaths in 1975 and 257 thousand deaths in 2000 (let x represent years since 1975 and $f(x)$ deaths in thousands). How many deaths due to heart disease does the model predict for 2008?

Source: 2004 Statistical Abstract of the United States, Table 102

PRACTICE TEST

1. Two relations here are functions and two are not. Identify the nonfunctions (justify your response).
   a. $x = y^2 + 2y$ **b.** $y = \sqrt{5 - 2x}$
   c. $|y| + 1 = x$ **d.** $y = x^2 + 2x$

2. Determine if the lines are parallel, perpendicular, or neither:
   $L_1: 2x + 5y = -15$ and $L_2: y = \frac{2}{5}x + 7$.

3. Graph the line using the slope and y-intercept:
   $x + 4y = 8$

4. Find the center and radius of the circle defined by $x^2 + y^2 - 4x + 6y - 3 = 0$, then sketch its graph.

5. Find the equation of the line parallel to $6x + 5y = 3$, containing the point $(2, -2)$. Answer in slope-intercept form.

6. My partner and I are at coordinates $(-20, 15)$ on a map. If our destination is at coordinates $(35, -12)$, (a) what are the coordinates of the rest station located halfway to our destination? (b) How far away is our destination? Assume that each unit is 1 mi.

7. Write the equations for lines L_1 and L_2 shown.

8. State the domain and range for the relations shown on graphs 8(a) and 8(b).

Exercise 8(a)   **Exercise 8(b)**

9. For the linear function shown,
   a. Determine the value of $W(24)$ from the graph.
   b. What input h will give an output of $W(h) = 375$?
   c. Find a linear function for the graph.

   d. What does the slope indicate in this context?
   e. State the domain and range of h.

10. Each function graphed here is from a toolbox function family. For each graph, (a) identify the function family, (b) state the domain and range, (c) identify x- and y-intercepts, (d) discuss the end behavior, and (e) solve the inequality $f(x) > 0$, and (f) solve $f(x) < 0$.

I.   II.

III.   IV.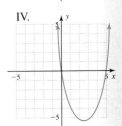

11. Given $f(x) = \dfrac{2 - x^2}{x^2}$, evaluate and simplify:
    (a) $f(\frac{2}{3})$ (b) $f(a + 3)$ (c) $f(1 + 2i)$

12. Given $f(x) = x^2 + 2$ and $g(x) = \sqrt{3x - 1}$, determine $(f \circ g)(x)$ and its domain.

13. Monthly sales volume for a successful new company is modeled by $S(t) = 2t^2 - 3t$, where $S(t)$ represents sales volume in thousands in month t ($t = 0$ corresponds to January 1). (a) Would you expect the average rate of change from May to June to be greater than that from June to July? Why? (b) Calculate the rates of change in these intervals to verify your answer. (c) Calculate the difference quotient for $S(t)$ and use it to estimate the sales volume rate of change after 10, 18, and 24 months.

Sketch each graph using a transformation.

14. $f(x) = |x - 2| + 3$

15. $g(x) = -(x + 3)^2 - 2$

16. A snowball increases in size as it rolls downhill. The snowball is roughly spherical with a radius that can be modeled by the function $r(t) = \sqrt{t}$, where t

is time in seconds and r is measured in inches. The volume of the snowball is given by the function $V(r) = \frac{4}{3}\pi r^3$. Use a composition to (a) write V directly as a function of t and (b) find the volume of the snowball after 9 sec.

17. Determine the following from the graph shown.
    a. the domain and range
    b. estimate the value of $f(-1)$
    c. interval(s) where $f(x)$ is negative or positive
    d. interval(s) where $f(x)$ is increasing, decreasing, or constant
    e. an equation for $f(x)$

18. Given $h(x) = \begin{cases} 4 & x < -2 \\ 2x & -2 \le x \le 2 \\ x^2 & x > 2 \end{cases}$
    a. Find $h(-3)$, $h(-2)$, and $h(\frac{5}{2})$.
    b. Sketch the graph of h. Label important points.

For the function $h(x)$ whose partial graph is given,
19. complete the graph if h is known to be even.
20. complete the graph if h is known to be odd.

CALCULATOR EXPLORATION AND DISCOVERY

Using a Simple Program to Explore Transformations

In Section 2.6 we studied transformations of the toolbox functions. On page 231, an organized sequence for applying these transformations was given. Since the transformations are identical regardless of the function used, a simple program is an efficient way to explore these transformations further. As a good programming practice, clear all functions on the Y = screen, and preset the graphing window to ZOOM 6:ZStandard. Begin by pressing the PRGM key, then the ▶ key twice to name the new program. At the prompt, enter TRANSFRM. The specific functions we use for programming are all accessed in sub menus of the PRGM key. Recall that the relational operators ($=$, $<$, $>$, $\le$, $\ge$) are accessed using 2nd   MATH (TEST).

PROGRAM: TRANSFRM

```
:ClrHome
:FnOff 1,2,3,4,5,6,7,8,9
:Disp "FUNCTION FAMILY"
:Disp "1:SQUARING"
:Disp "2:SQUARE ROOT"
:Disp "3:ABSOLUTE VALUE"
:Disp "4:CUBING"
:Disp "5:CUBE ROOT"
:Input T
```

```
:If T=1:"X^2"→Y1
:If T=2:"√(X)"→Y1
:If T=3:"abs(X)"→Y1
:If T=4:"X^3"→Y1
:If T=5:"X^(1/3)"→Y1
:DispGraph:Pause
:ClrHome
:Disp "HORIZONTAL SHIFT"
:Disp "ENTER 0 IF NONE"
:Prompt H
:"Y1(X + H)"→Y2
:DispGraph
:FnOff 1
:DispGraph:Pause
:ClrHome
:Disp "STRETCH FACTOR A"
:Disp "(A>0)"
:Disp "ENTER 1 IF A=1"
:Input A
:"A*Y2"→Y3
:DispGraph
:FnOff 2
:DispGraph:Pause
```

:ClrHome
:Disp "REFLECTIONS?"
:Disp "0:NONE"
:Disp "1:ACROSS X-AXIS"
:Disp "2:ACROSS Y-AXIS"
:Disp "3:ACROSS BOTH"
:Input B
:If B=0:"Y3"→Y4
:If B=1:"-Y3"→Y4
:If B=2:"Y3(-X)"→Y4
:If B=3:"-Y3(-X)"→Y4
:DispGraph
:FnOff 3
:DispGraph:Pause
:ClrHome
:Disp "VERTICAL SHIFT"
:Disp "ENTER 0 IF NONE"
:Prompt V
:"Y4 + V"→Y5
:DispGraph

:FnOff 4
:DispGraph:Pause
:Stop

Enter the TRANSFRM program into your calculator. Note that as you are writing or editing a program:

1. The "FnOff" command is located at [VARS] Y–VARS 4:On/Off.
2. The "ClrHome" command is located at [PRGM] CTL 8.
3. The "Pause" command is located at [PRGM] I/O 8.

All other needed commands are visible as Options 1 through 7 on the CTL and I/O menus.

Exercise 1: Use the TRANSFRM program to apply the following transformations to $y = x^2$: (1) shift left 4 units, (2) stretch by a factor of 5, (3) reflect across the x-axis, (4) shift up 6 units. What is the equation of the final graph? Where is the vertex located?

Exercise 2: Use TRANSFRM to graph the function $y = -4\sqrt[3]{x} - 2 + 3$. Where is the point of inflection? Estimate the y-intercept from the graph, then compare the estimate to the computed value.

STRENGTHENING CORE SKILLS

Transformations via Composition

Historically, many of the transformations studied in this chapter played a fundamental role in the development of modern algebra. To make the connection, we note that many transformations can be viewed as a composition of functions. For instance, for $f(x) = x^2 + 2$ (a parabola shifted two units up) and $g(x) = (x - 3)$, the composition $h(x) = f[g(x)]$ yields $(x - 3)^2 + 2$, a parabola shifted 2 units up *and* 3 units right. Enter $f(x)$ as Y_1 and $h(x)$ as Y_2 on your graphing calculator, then graph and inspect the results. As you see, we do obtain the same parabola shifted 3 units to the right (see figure). But now, notice what happens when we compose using $g(x) = x + 2$. After simplification, the result is $h(x) = x^2 - 9$ or *a quadratic function whose zeroes can easily be solved by taking square roots,* since the linear term is eliminated. The zeroes of h (the shifted quadratic) are $x = -3$ and $x = 3$, which means the zeroes of f (the original function) can be found by shifting two units *right*, returning them to their original position. The zeroes are $x = -3 + 2 = -1$ and $x = 3 + 2 = 5$ (verify by factoring). Transformations of this type are especially insightful when the zeroes of a

quadratic equation are irrational, since it enables us to find the radical portion by taking square roots, and the rational portion by addition. The key is to shift the quadratic function $y = ax^2 + bx + c$ using $x - \dfrac{b}{2a}$. Let's find the zeroes of $f(x) = x^2 + 6x - 11$ in this way. We find that $\dfrac{b}{2a} = 3$, giving $g(x) = x - 3$. This gives $h(x) = f[g(x)] = (x - 3)^2 + 6(x - 3) - 11$, which simplifies to $h(x) = x^2 - 20$. The zeroes of h are $x = -2\sqrt{5}$ and $x = 2\sqrt{5}$, so the solutions to the original equation must be $x = -2\sqrt{5} - 3$ and $x = 2\sqrt{5} - 3$. For Exercises 1–3, use this method to: (a) find such functions $h(x)$, and (b) use the zeroes of h to find the zeroes of f. Verify each solution using a calculator.

Exercise 1: $f(x) = x^2 - 8x - 12$
Exercise 2: $f(x) = x^2 + 4x + 5$
Exercise 3: $f(x) = 2x^2 - 10x + 11$

CUMULATIVE REVIEW CHAPTERS 1–2

1. Perform the division by factoring the numerator: $(x^3 - 5x^2 + 2x - 10) \div (x - 5)$.

2. Find the solution set for: $2 - x < 5$ **and** $3x + 2 < 8$.

3. The area of a circle is 69 cm². Find the circumference of the same circle.

4. The surface area of a cylinder is $A = 2\pi r^2 + 2\pi rh$. Write r in terms of A and h (solve for r).

5. Solve for x: $-2(3 - x) + 5x = 4(x + 1) - 7$.

6. Evaluate without using a calculator: $\left(\dfrac{27}{8}\right)^{-\frac{2}{3}}$.

7. Find the slope of each line:
  a. through the points: $(-4, 7)$ and $(2, 5)$.
  b. a line with equation $3x - 5y = 20$.

8. Graph using transformations of a parent function.
  a. $f(x) = \sqrt{x - 2} + 3$.
  b. $f(x) = -|x + 2| - 3$.

9. Graph the line passing through $(-3, 2)$ with a slope of $m = \frac{1}{2}$, then state its equation.

10. Show that $x = 1 + 5i$ is a solution to $x^2 - 2x + 26 = 0$.

11. Given $f(x) = 3x^2 - 6x$ and $g(x) = x - 2$ find: $(f \cdot g)(x)$, $(f \div g)(x)$, and $(g \circ f)(-2)$.

12. Graph by plotting the y-intercept, then counting $m = \dfrac{\Delta y}{\Delta x}$ to find additional points: $y = \frac{1}{3}x - 2$

13. Graph the piecewise defined function
$$f(x) = \begin{cases} x^2 - 4 & x < 2 \\ x - 1 & 2 \le x \le 8 \end{cases}$$ and determine the following:
  a. the domain and range
  b. the value of $f(-3), f(-1), f(1), f(2)$, and $f(3)$
  c. the zeroes of the function
  d. interval(s) where $f(x)$ is negative/positive
  e. location of any max/min values
  f. interval(s) where $f(x)$ is increasing/decreasing

14. Given $f(x) = x^2$ and $g(x) = x^3$, use the formula for average rate of change to determine which of these functions is increasing faster in the intervals:
  a. $[0.5, 0.6]$ **b.** $[1.5, 1.6]$.

15. Add the rational expressions:
  a. $\dfrac{-2}{x^2 - 3x - 10} + \dfrac{1}{x + 2}$
  b. $\dfrac{b^2}{4a^2} - \dfrac{c}{a}$

16. Simplify the radical expressions:
  a. $\dfrac{-10 + \sqrt{72}}{4}$ **b.** $\dfrac{1}{\sqrt{2}}$

17. Determine which of the following statements are false, and state why.
  a. $\mathbb{N} \subset \mathbb{Z} \subset \mathbb{W} \subset \mathbb{Q} \subset \mathbb{R}$
  b. $\mathbb{W} \subset \mathbb{N} \subset \mathbb{Z} \subset \mathbb{Q} \subset \mathbb{R}$
  c. $\mathbb{N} \subset \mathbb{W} \subset \mathbb{Z} \subset \mathbb{Q} \subset \mathbb{R}$
  d. $\mathbb{N} \subset \mathbb{R} \subset \mathbb{Z} \subset \mathbb{Q} \subset \mathbb{W}$

18. Determine if the following relation is a function. If not, how is the definition of a function violated?

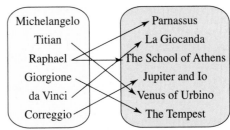

19. Solve by completing the square. Answer in both exact and approximate form: $2x^2 + 49 = -20x$

20. Solve using the quadratic formula. If solutions are complex, write them in $a + bi$ form. $2x^2 + 20x = -51$

21. The *National Geographic Atlas of the World* is a very large, rectangular book with an almost inexhaustible panoply of information about the world we live in. The length of the front cover is 16 cm more than its width, and the area of the cover is 1457 cm². Use this information to write an equation model, then use the quadratic formula to determine the length and width of the Atlas.

22. Compute as indicated:
  a. $(2 + 5i)^2$ **b.** $\dfrac{1 - 2i}{1 + 2i}$

23. Solve by factoring:
  a. $6x^2 - 7x = 20$
  b. $x^3 + 5x^2 - 15 = 3x$

24. A theorem from elementary geometry states, *"A line tangent to a circle is perpendicular to the radius at the point of tangency."* Find the equation of the tangent line for the circle and radius shown.

25. A triangle has its vertices at $(-4, 5), (4, -1)$, and $(0, 8)$. Find the perimeter of the triangle and determine whether or not it is a *right* triangle.

Modeling With Technology I   Linear and Quadratic Equation Models

Learning Objectives

In this section you will learn how to:

☐ **A.** Draw a scatter-plot and identify positive and negative associations

☐ **B.** Use a scatter-plot to identify linear and nonlinear associations

☐ **C.** Use linear regression to find the line of best fit

☐ **D.** Use quadratic regression to find the parabola of best fit

Collecting and analyzing data is a tremendously important mathematical endeavor, having applications throughout business, industry, science, and government. The link between classroom mathematics and real-world mathematics is called a **regression,** in which we attempt to find an equation that will act as a model for the raw data. In this section, we focus on linear and quadratic equation models.

A. Scatter-Plots and Positive/Negative Association

In this section, we continue our study of ordered pairs and functions, but this time using data collected from various sources or from observed real-world relationships. You can hardly pick up a newspaper or magazine without noticing it contains a large volume of data presented in graphs, charts, and tables. In addition, there are many simple experiments or activities that enable you to collect your own data. We begin analyzing the collected data using a **scatter-plot,** which is simply a graph of all of the ordered pairs in a data set. Often, real data (sometimes called **raw data**) is not very "well behaved" and the points may be somewhat scattered—the reason for the name.

Positive and Negative Associations

Earlier we noted that lines with positive slope rise from left to right, while lines with negative slope fall from left to right. We can extend this idea to the data from a scatter-plot. The data points in Example 1A seem to *rise* as you move from left to right, with larger input values generally resulting in larger outputs. In this case, we say there is a **positive association** between the variables. If the data seems to decrease or fall as you move left to right, we say there is a **negative association.**

EXAMPLE 1A ▶ Drawing a Scatter-Plot and Observing Associations

The ratio of the federal debt to the total population is known as the *per capita debt.* The per capita debt of the United States is shown in the table for the odd-numbered years from 1997 to 2007. Draw a scatter-plot of the data and state whether the association is positive or negative.

Source: Data from the Bureau of Public Debt at www.publicdebt.treas.gov

Year	Per Capita Debt ($1000s)
1997	20.0
1999	20.7
2001	20.5
2003	23.3
2005	27.6
2007	30.4

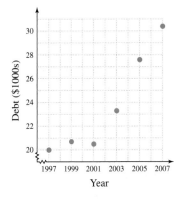

Solution ▶ Since the amount of debt depends on the year, *year* is the input *x* and *per capita debt* is the output *y.* Scale the *x*-axis from 1997 to 2007 and the *y*-axis from 20 to 30 to comfortably fit the data (the "squiggly lines," near the 20 and 1997 in the graph are used to show that some initial values have been skipped). The graph indicates a positive association between the variables, meaning the debt is generally *increasing* as time goes on.

EXAMPLE 1B ▶ **Drawing a Scatter-Plot and Observing Associations**

A cup of coffee is placed on a table and allowed to cool. The temperature of the coffee is measured every 10 min and the data are shown in the table. Draw the scatter-plot and state whether the association is positive or negative.

Elapsed Time (minutes)	Temperature (°F)
0	110
10	89
20	76
30	72
40	71

Solution ▶

☑ **A.** You've just learned how to draw a scatter-plot and identify positive and negative associations

Since temperature depends on cooling time, *time* is the input x and *temperature* is the output y. Scale the x-axis from 0 to 40 and the y-axis from 70 to 110 to comfortably fit the data. As you see in the figure, there is a negative association between the variables, meaning the temperature *decreases* over time.

Now try Exercises 1 and 2 ▶

B. Scatter-Plots and Linear/Nonlinear Associations

The data in Example 1A had a positive association, while the association in Example 1B was negative. But the data from these examples differ in another important way. In Example 1A, the data seem to cluster about an imaginary line. This indicates a linear equation model might be a good approximation for the data, and we say there is a **linear association** between the variables. The data in Example 1B could not accurately be modeled using a straight line, and we say the variables *time* and *cooling temperature* exhibit a **nonlinear association.**

EXAMPLE 2 ▶ **Drawing a Scatter-Plot and Observing Associations**

A college professor tracked her annual salary for 2002 to 2009 and the data are shown in the table. Draw the scatter-plot and determine if there is a linear or nonlinear association between the variables. Also state whether the association is positive, negative, or cannot be determined.

Year	Salary ($1000s)
2002	30.5
2003	31
2004	32
2005	33.2
2006	35.5
2007	39.5
2008	45.5
2009	52

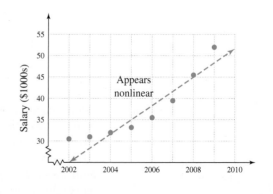

Solution ▶

Since salary earned depends on a given year, *year* is the input *x* and *salary* is the output *y*. Scale the *x*-axis from 1996 to 2005, and the *y*-axis from 30 to 55 to comfortably fit the data. A line doesn't seem to model the data very well, and the association appears to be nonlinear. The data rises from left to right, indicating a positive association between the variables. This makes good sense, since we expect our salaries to increase over time.

☑ **B. You've just learned how to use a scatter-plot to identify linear and nonlinear associations**

Now try Exercises 3 and 4 ▶

C. Linear Regression and the Line of Best Fit

There is actually a sophisticated method for calculating the equation of a line that best fits a data set, called the **regression line.** The method minimizes the vertical distance between all data points and the line itself, making it the unique **line of best fit.** Most graphing calculators have the ability to perform this calculation quickly. The process involves these steps: (1) clearing old data, (2) entering new data; (3) displaying the data; (4) calculating the regression line; and (5) displaying and using the regression line. We'll illustrate by finding the regression line for the data shown in Table MWT 1.1, which gives the men's 400-m freestyle gold medal times (in seconds) for the 1980 through the 2004 Olympics, with 1980→0.

Table MWT 1.1

Year (x) (1980→0)	Time (y) (sec)
0	231
4	231
8	227
12	225
16	228
20	221
24	223

Step 1: Clear Old Data

To prepare for the new data, we first clear out any old data. Press the `STAT` key and select option **4:ClrList.** This places the **ClrList** command on the home screen. We tell the calculator which lists to clear by pressing `2nd` 1 to indicate List1 (L1), then enter a comma using the `,` key, and continue entering other lists we want to clear: `2nd` 2 `,` `2nd` 3 `ENTER` will clear List1 (L1), List2 (L2), and List3 (L3).

Step 2: Enter New Data

Press the `STAT` key and select option **1:Edit.** Move the cursor to the first position of List1, and simply enter the data from the first column of Table MWT 1.1 in order: 0 `ENTER` 4 `ENTER` 8 `ENTER`, and so on. Then use the right arrow `▶` to navigate to List2, and enter the data from the second column: 231 `ENTER` 231 `ENTER` 227 `ENTER`, and so on. When finished, you should obtain the screen shown in Figure MWT 1.1.

Figure MWT 1.1

Step 3: Display the Data

With the data held in these lists, we can now display the related ordered pairs on the coordinate grid. First press the `Y=` key and `CLEAR` any existing equations. Then press `2nd` `Y=` to access the "**STATPLOTS**" screen. With the cursor on **1:Plot1,** press `ENTER` and be sure the options shown in Figure MWT 1.2 are highlighted. If you need to make any changes, navigate the cursor to the desired option and press `ENTER`. Note the data in L1 ranges from 0 to 24, while the data in L2 ranges from 221 to 231. This means an appropriate viewing window might be [0, 30] for the *x*-values, and [200, 250] for the *y*-values. Press the `WINDOW` key and set up the window accordingly. After you're finished, pressing the `GRAPH` key should produce the graph shown in Figure MWT 1.3.

Figure MWT 1.2

Figure MWT 1.3

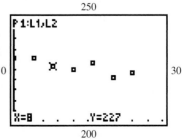

Step 4: Calculate the Regression Equation

To have the calculator compute the regression equation, press the [STAT] and [▶] keys to move the cursor over to the **CALC** options (see Figure MWT 1.4). Since it appears the data is best modeled by a linear equation, we choose option **4:LinReg(ax + b).** Pressing the number 4 places this option on the home screen, and pressing [ENTER] computes the values of a and b (the calculator automatically uses the values in L1 and L2 unless instructed otherwise). Rounded to hundredths, the linear regression model is $y = -0.38x + 231.18$ (Figure MWT 1.5).

Step 5: Display and Use the Results

Although graphing calculators have the ability to paste the regression equation directly into Y_1 on the [Y=] screen, for now we'll enter $Y_1 = -0.38x + 231.18$ by hand. Afterward, pressing the [GRAPH] key will plot the data points (if Plot1 is still active) and graph the line. Your display screen should now look like the one in Figure MWT 1.6. The regression line is the best estimator for the set of data as a whole, but there will still be some difference between the values it generates and the values from the set of raw data (the output in Figure MWT 1.6 shows the estimated time for the 1996 Olympics is 225.1 sec, while the actual time was 228 sec).

Figure MWT 1.4

Figure MWT 1.5

LinReg
 y=ax+b
 a=-.3839285714
 b=231.1785714

Figure MWT 1.6

EXAMPLE 3 ▶ **Using Regression to Model Employee Performance**

Riverside Electronics reviews employee performance semiannually, and awards increases in their hourly rate of pay based on the review. The table shows Thomas' hourly wage for the last 4 yr (eight reviews). Find the regression equation for the data and use it to project his hourly wage for the year 2011, after his fourteenth review.

Year (x)	Wage (y)
(2004) 1	$9.58
2	$9.75
(2005) 3	$10.54
4	$11.41
(2006) 5	$11.60
6	$11.91
(2007) 7	$12.11
8	$13.02

Solution ▶ Following the prescribed sequence produces the equation $y = 0.48x + 9.09$. For $x = 14$ we obtain $y = 0.48(14) + 9.09$ or a wage of $15.81. According to this model, Thomas will be earning $15.81 per hour in 2011.

☑ **C. You've just learned how to use a linear regression to find the line of best fit**

Now try Exercises 9 through 14 ▶

D. Quadratic Regression and the Parabola of Best Fit

Once the data have been entered, graphing calculators have the ability to find many different regression equations. The choice of regression depends on the context of the data, patterns formed by the scatter-plot, and/or some foreknowledge of how the data are related. Earlier we focused on linear regression equations. We now turn our attention to quadratic regression equations.

EXAMPLE 4A ▶ **Drawing a Scatter-Plot to Sketch a Best-Fit Curve**

Since 1990, the number of *new* books published each year has been growing at a rate that can be approximated by a quadratic function. The table shows the number of books published in the United States for selected years. Draw a scatter-plot and sketch an estimated parabola of best fit by hand.

Source: 1998, 2000, 2002, and 2004 Statistical Abstract of the United States.

Year (1990→0)	Books Published (1000s)
0	46.7
2	49.2
3	49.8
4	51.7
5	62.0
6	68.2
7	65.8
9	102.0
10	122.1

Solution ▶ Begin by drawing the scatter-plot, being sure to scale the axes appropriately. The data appear to form a quadratic pattern, and we sketch a parabola that seems to best fit the data (see graph).

The regression abilities of a graphing calculator can be used to find a **parabola of best fit** and the steps are identical to those for linear regression.

EXAMPLE 4B ▶ **Calculating a Nonlinear Regression Model from a Data Set**

Use the data from Example 4A to calculate a quadratic regression equation, then display the data and graph. How well does the equation match the data?

Solution ▶ Begin by entering the data in L1 and L2 as shown in Figure MWT 1.7. Press 2nd Y= to be sure that Plot 1 is still active and is using L1 and L2 with the desired point type. Set the window size to comfortably fit the data (see Figure MWT 1.9—window size is indicated along the perimeter). Finally, press STAT and the right arrow ► to overlay the **CALC** option. The quadratic regression option is number **5:QuadReg.** Pressing 5 places this option directly on the home screen. Lists L1 and L2 are the default lists, so pressing ENTER will have the calculator compute the regression equation for the data in L1 and L2. After "chewing on the data" for a short while, the calculator returns the regression equation in the form shown in Figure MWT 1.8. To maintain a higher degree of accuracy, we can actually paste the entire regression equation in Y_1. Recall the last

Figure MWT 1.7

L1	L2	L3	2
3	49.8		
4	51.7		
5	62		
6	68.2		
7	65.8		
9	102		
10	**122.1**		

L2(9) =122.1

Figure MWT 1.8

```
QuadReg
 y=ax²+bx+c
 a=1.04386823
 b=-3.505801827
 c=49.41433895
```

WORTHY OF NOTE

The TI-84 Plus can round all coefficients to any desired number of decimal places. For three decimal places, press MODE and change the **Float** setting to "3." Also, be aware that there are additional methods for pasting the equation in Y1.

operation using [2nd] [ENTER] , and **QuadReg** should (re)appear. Then enter the function Y_1 after the QuadReg option by pressing [VARS] [▶] (**Y-Vars**) and [ENTER] (**1:Function**) and [ENTER] (Y_1). After pressing [ENTER] once again, the full equation is automatically pasted in Y_1. To compare this equation model with the data, simply press [GRAPH] and both the graph and plotted data will appear. The graph and data seem to match very well (Figure MWT 1.9).

Figure MWT 1.9

EXAMPLE 4C ▶ Using a Regression Model to Predict Trends

Use the equation from Example 4B to answer the following questions: According to the function model, how many new books were published in 1991? If this trend continues, how many new books will be published in 2011?

Solution ▶ Since the year 1990 corresponds to 0 in this data set, we use an input value of 1 for 1991, and an input of 21 for 2011. Accessing the table ([2nd] [GRAPH]) feature and inputting 1 and 21 gives the screen shown. Approximately 47,000 new books were published in 1991, and about 436,000 will be published in the year 2011.

☑ **D. You just learned how to use quadratic regression to find the parabola of best fit**

Now try Exercises 15 through 18 ▶

MODELING WITH TECHNOLOGY EXERCISES

▶ DEVELOPING YOUR SKILLS

1. For mail with a high priority, "Express Mail" offers next day delivery by 12:00 noon to most destinations, 365 days of the year. The service was first offered by the U.S. Postal Service in the early 1980s and has been growing in use ever since. The cost of the service (in cents) for selected years is shown in the table. Draw a scatter-plot of the data, then decide if the association is positive, negative, or cannot be determined.

Source: 2004 Statistical Abstract of the United States

x	y
1981	935
1985	1075
1988	1200
1991	1395
1995	1500
1999	1575
2002	1785

2. After the Surgeon General's first warning in 1964, cigarette consumption began a steady decline as advertising was banned from television and radio, and public awareness of the dangers of cigarette smoking grew. The percentage of the U.S. adult population who considered themselves smokers is shown in the table for selected years. Draw a scatter-plot of the data, then decide if the association is positive, negative, or cannot be determined.

Source: 1998 Wall Street Journal Almanac and 2004 Statistical Abstract of the United States, Table 188

x	y
1965	42.4
1974	37.1
1979	33.5
1985	29.9
1990	25.3
1995	24.6
2000	23.1
2002	22.4

3. Since the 1970s women have made tremendous gains in the political arena, with more and more female candidates running and winning seats in the U.S. Senate and U.S. Congress. The number of women candidates for the U.S. Congress is shown in the table for selected years. Draw a scatter-plot of the data and then decide (a) if the association is linear or nonlinear and (b) if the association is positive or negative.

x	y
1972	32
1978	46
1984	65
1992	106
1998	121
2004	141

Source: Center for American Women and Politics at www.cawp.rutgers.edu/Facts3.html

4. The number of shares traded on the New York Stock Exchange experienced dramatic change in the 1990s as more and more individual investors gained access to the stock market via the Internet and online brokerage houses. The volume is shown in the table for 2002, and the odd numbered years from 1991 to 2001 (in billions of shares). Draw a scatter-plot of the data then decide (a) if the association is linear or nonlinear; and (b) if the association is positive, negative, or cannot be determined.

x	y
1991	46
1993	67
1995	88
1997	134
1999	206
2001	311
2002	369

Source: 2000 and 2004 *Statistical Abstract of the United States,* Table 1202

The data sets in Exercises 5 and 6 are known to be linear.

5. The total value of the goods and services produced by a nation is called its gross domestic product or GDP. The *GDP per capita* is the ratio of the GDP for a given year to the population that year, and is one of many indicators of economic health. The GDP per capita (in $1000s) for the United States is shown in the table for selected years. (a) Draw a scatter-plot using a scale that appropriately fits the data; (b) sketch an estimated line of best fit and decide if the association is positive or negative, then (c) approximate the slope of the line.

x 1970 → 0	y
0	5.1
5	7.6
10	12.3
15	17.7
20	23.3
25	27.7
30	35.0
33	37.8

Source: 2004 *Statistical Abstract of the United States,* Tables 2 and 641

6. Real estate brokers carefully track sales of new homes looking for trends in location, price, size, and other factors. The table relates the average selling price within a price range (homes in the $120,000 to $140,000 range are represented by the $130,000 figure), to the number of new homes sold by Homestead Realty in 2004. (a) Draw a scatter-plot using a scale that appropriately fits the data; (b) sketch an estimated line of best fit and decide if the association is positive or negative, then (c) approximate the slope of the line.

Price	Sales
130's	126
150's	95
170's	103
190's	75
210's	44
230's	59
250's	21

7. In most areas of the country, law enforcement has become a major concern. The number of law enforcement officers employed by the federal government and having the authority to carry firearms and make arrests is shown in the table for selected years. (a) Draw a scatter-plot using a scale that appropriately fits the data. (b) Sketch an estimated line of best fit and decide if the association is positive or negative. (c) Choose two points on or near the estimated line of best fit, and use them to find an equation model and predict the number of federal law enforcement officers in 1995 and the projected number for 2011. Answers may vary.

x (1990→0)	y (1000s)
3	68.8
6	74.5
8	83.1
10	88.5
14	93.4

Source: U.S. Bureau of Justice, Statistics at www.ojp.usdoj.gov/bjs/fedle.htm

8. Due to atmospheric pressure, the temperature at which water will boil varies predictably with the altitude. Using special equipment designed to duplicate atmospheric pressure, a lab experiment is set up to study this relationship for altitudes up to 8000 ft. The set of data collected is shown to the right, with the boiling temperature y in degrees Fahrenheit, depending on the altitude x in feet. (a) Draw a scatter-plot using a scale that appropriately fits the data. (b) Sketch an estimated line of best fit and decide if the association is

x	y
−1000	213.8
0	212.0
1000	210.2
2000	208.4
3000	206.5
4000	204.7
5000	202.9
6000	201.0
7000	199.2
8000	197.4

positive or negative. (c) Choose two points on or near the estimated line of best fit, and use them to find an equation model and predict the boiling point of water on the summit of Mt. Hood in

Washington State (11,239 ft height), and along the shore of the Dead Sea (approximately 1312 ft below sea level). Answers may vary.

▶ APPLICATIONS

Use the regression capabilities of a graphing calculator to complete Exercises 9 through 18.

9. **Height versus wingspan:** Leonardo da Vinci's famous diagram is an illustration of how the human body comes in predictable proportions. One such comparison is a person's wingspan to their height. Careful measurements were taken on eight students and the set of data is shown here. Using the data, (a) draw the scatter-plot; (b) determine whether the association is linear or nonlinear; (c) determine whether the association is positive or negative; and (d) find the regression equation and use it to predict the height of a student with a wingspan of 65 in.

Height (x)	Wingspan (y)
61	60.5
61.5	62.5
54.5	54.5
73	71.5
67.5	66
51	50.75
57.5	54
52	51.5

10. **Patent applications:** Every year the United States Patent and Trademark Office (USPTO) receives thousands of applications from scientists and inventors. The table given shows the number of appplications received for the odd years from 1993 to 2003 (1990 → 0). Use the data to (a) draw the scatter-plot; (b) determine whether the association is linear or nonlinear; (c) determine whether the association is positive or negative; and (d) find the

regression equation and use it to predict the number of applications that will be received in 2011.

Source: United States Patent and Trademark Office at www.uspto.gov/web

Year (1990→0)	Applications (1000s)
3	188.0
5	236.7
7	237.0
9	278.3
11	344.7
13	355.4

11. **Patents issued:** An increase in the number of patent applications (see Exercise 10), typically brings an increase in the number of patents issued, though many applications are denied due to improper filing, lack of scientific support, and other reasons. The table given shows the number of patents issued for the odd years from 1993 to 2003 (1999 → 0). Use the data to (a) draw the scatter-plot; (b) determine whether the association is linear or nonlinear; (c) determine whether the association is positive or negative; and (d) find the regression equation and use it to predict the number of applications that will be approved in 2011. Which is increasing faster, the number of patent applications or the number of patents issued? How can you tell for sure?

Source: United States Patent and Trademark Office at www.uspto.gov/web

Year (1990→0)	Patents (1000s)
3	107.3
5	114.2
7	122.9
9	159.2
11	187.8
13	189.6

12. High jump records: In the sport of track and field, the high jumper is an unusual athlete. They seem to defy gravity as they launch their bodies over the high bar. The winning height at the summer Olympics (to the nearest unit) has steadily increased over time, as shown in the table for selected years. Using the data, (a) draw the scatter-plot, (b) determine whether the association is linear or nonlinear, (c) determine whether the association is positive or negative, and (d) find the regression equation using $t = 0$ corresponding to 1900 and predict the winning height for the 2004 and 2008 Olympics. How close did the model come to the actual heights?

Source: athens2004.com

Year (x)	Height (in.) (y)
0	75
12	76
24	78
36	80
56	84
68	88
80	93
88	94
92	92
96	94
100	93

13. Females/males in the workforce: Over the last 4 decades, the percentage of the female population in the workforce has been increasing at a fairly steady rate. At the same time, the percentage of the male population in the workforce has been declining. The set of data is shown in the tables. Using the data, (a) draw scatter-plots for both data sets, (b) determine whether the associations are linear or nonlinear, (c) determine whether the associations are positive or negative, and (d) determine if the percentage of females in the workforce is increasing faster than the percentage of males is decreasing. Discuss/Explain how you can tell for sure.

Source: 1998 *Wall Street Journal Almanac*, p. 316

Exercise 13 (women)

Year (x) (1950→0)	Percent
5	36
10	38
15	39
20	43
25	46
30	52
35	55
40	58
45	59
50	60

Exercise 13 (men)

Year (x) (1950→0)	Percent
5	85
10	83
15	81
20	80
25	78
30	77
35	76
40	76
45	75
50	73

14. Height versus male shoe size: While it seems reasonable that taller people should have larger feet, there is actually a wide variation in the relationship between height and shoe size. The data in the table show the height (in inches) compared to the shoe size worn for a random sample of 12 male chemistry students. Using the data, (a) draw the scatter-plot, (b) determine whether the association is linear or nonlinear, (c) determine whether the association is positive or negative, and (d) find the regression equation and use it to predict the shoe size of a man 80 in. tall and another that is 60 in. tall.

Height	Shoe Size
66	8
69	10
72	9
75	14
74	12
73	10.5
71	10
69.5	11.5
66.5	8.5
73	11
75	14
65.5	9

15. Plastic money: The total amount of business transacted using credit cards has been changing rapidly over the last 15 to 20 years. The total volume (in billions of dollars) is shown in the table for selected years. (a) Use a graphing calculator to draw a scatter-plot of the data and decide on an appropriate form of regression. (b) Calculate a regression equation with $x = 0$ corresponding to 1990 and display the scatter-plot and graph on the same screen. (c) According to the equation model, how many billions of dollars were transacted in 2003? How much will be transacted in the year 2011?

Source: Statistical Abstract of the United States, various years

x (1990→0)	y
1	481
2	539
4	731
7	1080
8	1157
9	1291
10	1458
12	1638

16. Homeschool education:
Since the early 1980s the number of parents electing to homeschool their children has been steadily increasing. Estimates for the number of children homeschooled (in 1000s) are given in the table for selected years.

x (1985→0)	y
0	183
3	225
5	301
7	470
8	588
9	735
10	800
11	920
12	1100

(a) Use a graphing calculator to draw a scatter-plot of the data and decide on an appropriate form of regression. (b) Calculate a regression equation with $x = 0$ corresponding to 1985 and display the scatter-plot and graph on the same screen. (c) According to the equation model, how many children were homeschooled in 1991? If growth continues at the same rate, how many children will be homeschooled in 2010?

Source: National Home Education Research Institute

▶ **EXTENDING THE CONCEPT**

 17. The height of a projectile: $h(t) = -\frac{1}{2}gt^2 + vt$

The height of a projectile thrown upward from ground level depends primarily on two things—the object's initial velocity and the acceleration due to gravity. This is modeled by the formula shown, where $h(t)$ represents the height of the object at time t, v represents the initial velocity, and g represents the acceleration due to gravity. Suppose an astronaut on one of the inner planets threw a surface rock upward and used hand-held radar to collect the data shown. Given that on Mercury $g = 12$ ft/sec^2, Venus $g = 29$ ft/sec^2, and Earth $g = 32$ ft/sec^2, (a) use your calculator to find an appropriate regression model for the data, (b) use the model to determine the initial velocity of the object, and (c) name the planet on which the astronaut is standing.

Time	Height
1	75.5
2	122
3	139.5
4	128
5	87.5
6	18

 18. In his book *Gulliver's Travels,* Jonathan Swift describes how the Lilliputians were able to measure Gulliver for new clothes, even though he was a giant compared to them. According to the text, "Then they measured my right thumb, and desired no more . . . for by mathematical computation, once around the thumb is twice around the wrist, and so on to the neck and waist." Is it true that once around the neck is twice around the waist? Find at least 10 willing subjects and take measurements of their necks and waists in millimeters. Arrange the data in ordered pair form (circumference of neck, circumference of waist). Draw the scatter-plot for this data. Does the association appear to be linear? Find the equation of the best fit line for this set of data. What is the slope of this line? Is the slope near $m = 2$?

3

Polynomial and Rational Functions

CHAPTER OUTLINE

CHAPTER CONNECTIONS

In a study of demographics, the population density of a city and its surrounding area is measured using a unit called *people per square mile*. The population density is much greater near the city's center, and tends to decrease as you move out into suburban and rural areas. The density can be modeled using the formula $D(x) = \dfrac{ax}{x^2 + b}$, where $D(x)$ represents the density at a distance of x mi from the center of a city, and a and b are constants related to a particular city and its sprawl. Using this equation, city planners can determine how far from the city's center the population drops below a certain level, and answer other important questions to help plan for future growth. This application appears as Exercise 71 in Section 3.5.

Check out these other real-world connections:

Learning Objectives

In Section 3.1 you will learn how to:

☐ **A.** Graph quadratic functions by completing the square

☐ **B.** Graph quadratic functions using the vertex formula

☐ **C.** Find the equation of a quadratic function from its graph

☐ **D.** Solve applications involving extreme values

As our knowledge of functions grows, our ability to apply mathematics in new ways likewise grows. In this section, we'll build on the foundation laid in Chapter 2, as we introduce additional function families and the tools needed to apply them effectively. We begin with the family of quadratic functions.

A. Graphing Quadratic Functions by Completing the Square

The squaring function $f(x) = x^2$ is actually a member of the family of **quadratic functions,** defined as follows.

> **Quadratic Functions**
>
> A quadratic function is one of the form
> $$f(x) = ax^2 + bx + c,$$
> where a, b, and c are real numbers and $a \neq 0$.

As shown in Figure 3.1, the function is written in **standard form.** For $f(x) = x^2$, $a = 1$ with b and c equal to to 0. The function $f(x) = 2x^2 + x - 3$ is also quadratic, with $a = 2$, $b = 1$ and $c = -3$. Our earlier work suggests the graph of *any* quadratic function will be a parabola. Figure 3.1 provides a summary of the characteristic features of this graph. As pictured, the parabola opens upward with the vertex at (h, k), so k is a global minimum. Since the vertex is below the x-axis, the graph has two x-intercepts. The axis of symmetry goes through the vertex, and has equation $x = h$. The y-intercept is $(0, c)$, since $f(0) = c$.

Figure 3.1 $f(x) = ax^2 + bx + c$

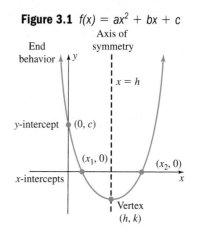

In Section 2.6, we graphed transformations of $f(x) = x^2$, using $y = a(x \pm h)^2 \pm k$. Here, we'll show that by completing the square, we can graph *any* quadratic function as a transformation of this basic graph.

When completing the square on a quadratic *equation* (Section 1.5), we applied the standard properties of equality to both sides of the equation. When completing the square on a *quadratic function,* the process is altered slightly in that we operate on only one side.

EXAMPLE 1 ▶ **Graphing a Quadratic Function by Completing the Square**

Given $g(x) = x^2 - 6x + 5$, complete the square to rewrite g as a transformation of $f(x) = x^2$, then graph the function.

Solution ▶ To begin we note the leading coefficient is $a = 1$.

$$g(x) = x^2 - 6x + 5 \qquad \text{given function}$$
$$= 1(x^2 - 6x + \underline{\quad}) + 5 \qquad \text{group variable terms, note } a = 1$$
$$= 1(\underbrace{x^2 - 6x + 9}_{\text{adds } 1 \cdot 9 = 9}) - 9 + 5 \qquad \text{subtract 9} \quad \left[\left(\tfrac{1}{2}\right)(-6)\right]^2 = 9$$
$$= (x - 3)^2 - 4 \qquad \text{factor and simplify}$$

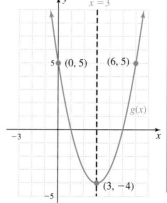

The graph of g is the graph of f shifted 3 units right, and 4 units down. The graph opens upward ($a > 0$) with the vertex at $(3, -4)$, and axis of symmetry $x = 3$. From the original equation we find $g(0) = 5$, giving a y-intercept of $(0, 5)$. The point $(6, 5)$ was obtained using the axis of symmetry. The graph is shown in the figure.

Now try Exercises 7 through 10 ▶

Note that by **adding** 9 and simultaneously **subtracting** 9 (essentially adding "0"), we changed only the *form* of the function, not its value. In other words, the resulting expression is equivalent to the original. If the leading coefficient is not 1, we factor it out from the variable terms, but take it into account when we add the constant needed to maintain an equivalent expression.

EXAMPLE 2 ▶   **Graphing a Quadratic Function by Completing the Square**

Given $p(x) = -2x^2 - 8x - 3$, complete the square to rewrite p as a transformation of $f(x) = x^2$, then graph the function.

Solution ▶

$$p(x) = -2x^2 - 8x - 3 \qquad \text{given function}$$
$$= (-2x^2 - 8x + \underline{\quad}) - 3 \qquad \text{group variable terms}$$
$$= -2(x^2 + 4x + \underline{\quad}) - 3 \qquad \text{factor out } a = -2 \text{ (notice sign change)}$$
$$= -2(x^2 + 4x + 4) - (-8) - 3 \qquad \left[\left(\tfrac{1}{2}\right)(4)\right]^2 = 4$$
$$\underbrace{}_{\text{adds } -2 \cdot 4 = -8} \quad \text{subtract } -8$$
$$= -2(x + 2)^2 + 8 - 3 \qquad \text{factor trinomial, simplify}$$
$$= -2(x + 2)^2 + 5 \qquad \text{result}$$

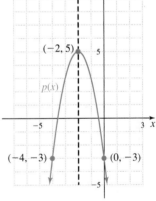

The graph of p is a parabola, shifted 2 units left, stretched by a factor of 2, reflected across the x-axis (opens downward), and shifted up 5 units. The vertex is $(-2, 5)$, and the axis of symmetry is $x = -2$. From the original function, the y-intercept is $(0, -3)$. The point $(-4, -3)$ was obtained using the axis of symmetry. The graph is shown in the figure.

Now try Exercises 11 through 14 ▶

By adding 4 to the variable terms within parentheses, we actually added $-2 \cdot 4 = -8$ to the value of the function. To adjust for this we subtracted -8. The basic ideas are summarized here.

WORTHY OF NOTE

In cases like $f(x) = 3x^2 - 10x + 5$, where the linear coefficient has no integer factors of a, we factor out 3 and *simultaneously divide the linear coefficient by 3*. This yields

$$h(x) = 3\left(x^2 - \frac{10}{3}x + \underline{\quad}\right) + 5,$$

and the process continues as before: $\left[\left(\tfrac{1}{2}\right)\left(\tfrac{10}{3}\right)\right]^2 = \left(\tfrac{5}{3}\right)^2 = \tfrac{25}{9}$, and so on. For more on this idea, **see Exercises 15 through 20.**

☑ **A.** You've just learned how to graph quadratic functions by completing the square

Graphing $f(x) = ax^2 + bx + c$ by Completing the Square

1. Group the variable terms apart from the constant c.
2. Factor out the leading coefficient a.
3. Compute $\left[\tfrac{1}{2}\left(\tfrac{b}{a}\right)\right]^2$ and add the result to the grouped terms,

   then subtract $a \cdot \left[\tfrac{1}{2}\left(\tfrac{b}{a}\right)\right]^2$ to maintain an equivalent expression.
4. Factor the grouped terms as a binomial square and simplify.
5. Graph using transformations of $f(x) = x^2$.

B. Graphing Quadratic Functions Using the Vertex Formula

When the process of completing the square is applied to $f(x) = ax^2 + bx + c$, we obtain a very useful result. Notice the close similarities to Example 2.

$$f(x) = ax^2 + bx + c \qquad \text{quadratic function}$$

$$= (ax^2 + bx + \underline{}) + c \qquad \text{group variable terms apart from the constant } c$$

$$= a\left(x^2 + \frac{b}{a}x + \underline{}\right) + c \qquad \text{factor out } a$$

$$= a\left(x^2 + \frac{b}{a}x + \frac{b^2}{4a^2}\right) - a\left(\frac{b^2}{4a^2}\right) + c \qquad \left[\left(\frac{1}{2}\right)\left(\frac{b}{a}\right)\right]^2 = \frac{b^2}{4a^2}$$

$$= a\left(x + \frac{b}{2a}\right)^2 - \frac{b^2}{4a} + c \qquad \text{factor the trinomial, simplify}$$

$$= a\left(x + \frac{b}{2a}\right)^2 + \frac{4ac - b^2}{4a} \qquad \text{result}$$

By comparing this result with previous transformations, we note the x-coordinate of the vertex is $h = \dfrac{-b}{2a}$ (since the graph shifts horizontally "opposite the sign"). While we could use the expression $\dfrac{4ac - b^2}{4a}$ to find k, we find it easier to substitute $\dfrac{-b}{2a}$ back into the function: $k = f\left(\dfrac{-b}{2a}\right)$. The result is called the **vertex formula.**

Vertex Formula

For the quadratic function $f(x) = ax^2 + bx + c$, the coordinates of the vertex are
$$(h, k) = \left(\frac{-b}{2a}, f\left(\frac{-b}{2a}\right)\right)$$

Since all characteristic features of the graph (end-behavior, vertex, axis of symmetry, x-intercepts, and y-intercept) can now be determined using the original equation, we'll rely on these features to sketch quadratic graphs, rather than having to complete the square.

EXAMPLE 3 ▶ **Graphing a Quadratic Function Using the Vertex Formula**

Graph $f(x) = 2x^2 + 8x + 3$ using the vertex formula and other features of a quadratic graph.

Solution ▶ The graph will open upward since $a > 0$.
The y-intercept is $(0, 3)$.
The vertex formula gives

$$h = \frac{-b}{2a} \qquad \text{x-coordinate of vertex}$$

$$= \frac{-8}{2(2)} \qquad \text{substitute 2 for } a \text{ and 8 for } b$$

$$= -2 \qquad \text{simplify}$$

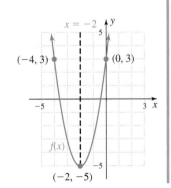

Computing $f(-2)$ to find the y-coordinate of the vertex yields

$$f(-2) = 2(-2)^2 + 8(-2) + 3 \quad \text{substitute } -2 \text{ for } x$$
$$= 2(4) - 16 + 3 \quad \text{multiply}$$
$$= 8 - 13 \quad \text{simplify}$$
$$= -5 \quad \text{result}$$

☑ **B.** You've just learned how to graph quadratic functions using the vertex formula

The vertex is $(-2, -5)$. The graph is shown in the figure, with the point $(-4, 3)$ obtained using symmetry.

Now try Exercises 21 through 32 ▶

C. Finding the Equation of a Quadratic Function from Its Graph

While most of our emphasis so far has centered on graphing quadratic functions, it would be hard to overstate the importance of the reverse process—determining the equation of the function from its graph (as in Section 2.6). This reverse process, which began with our study of lines, will be a continuing theme each time we consider a new function.

EXAMPLE 4 ▶   Finding the Equation of a Quadratic Function

The graph shown is a transformation of $f(x) = x^2$. What function defines this graph?

Solution ▶   Compared to the graph of $f(x) = x^2$, the vertex has been shifted left 1 and up 2, so the function will have the form $F(x) = a(x + 1)^2 + 2$. Since the graph opens downward, we know a will be negative. As before, we select one additional point on the graph and substitute to find the value of a. Using $(x, y) \rightarrow (1, 0)$ we obtain

WORTHY OF NOTE

It helps to remember that any point (x, y) on the parabola can be used. To verify this, try the calculation again using $(-3, 0)$.

$$F(x) = a(x + 1)^2 + 2 \quad \text{transformation}$$
$$0 = a(1 + 1)^2 + 2 \quad \substack{\text{substitute 1 for } x \text{ and 0} \\ \text{for } F(x): (x, y) \rightarrow (1, 0)}$$
$$0 = 4a + 2 \quad \text{simplify}$$
$$-2 = 4a \quad \text{subtract 2}$$
$$-\frac{1}{2} = a \quad \text{solve for } a$$

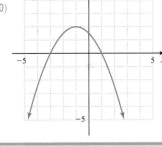

☑ **C.** You've just learned how to find the equation of a quadratic function from its graph

The equation of this function is

$$F(x) = -\frac{1}{2}(x + 1)^2 + 2.$$

Now try Exercises 33 through 38 ▶

D. Quadratic Functions and Extreme Values

If $a > 0$, the parabola opens upward, and the y-coordinate of the vertex is a global minimum, the smallest value attained by the function anywhere in its domain. Conversely, if $a < 0$ the parabola opens downward and the vertex yields a global maximum. These greatest and least points are known as **extreme values** and have a number of significant applications.

EXAMPLE 5 ▶   Applying a Quadratic Model to Manufacturing

An airplane manufacturer can produce up to 15 planes per month. The profit made from the sale of these planes is modeled by $P(x) = -0.2x^2 + 4x - 3$, where $P(x)$ is the profit in hundred-thousands of dollars per month, and x is the number of planes sold. Based on this model,

   a. Find the y-intercept and explain what it means in this context.
   b. How many planes should be made and sold to maximize profit?
   c. What is the maximum profit?

Solution ▶ **a.** $P(0) = -3$, which means the manufacturer loses $300,000 each month if the company produces no planes.

b. Since $a < 0$, we know the graph opens downward and has a maximum value. To find the required number of sales needed to "maximize profit," we use the vertex formula with $a = -0.2$ and $b = 4$:

$$x = \frac{-b}{2a} \qquad \text{vertex formula}$$

$$= \frac{-4}{2(-0.2)} \qquad \text{substitute } -0.2 \text{ for } a \text{ and } 4 \text{ for } b$$

$$= 10 \qquad \text{result}$$

The result shows 10 planes should be sold each month for maximum profit.

c. Evaluating $P(10)$ we find that a maximum profit of 17 "hundred thousand dollars" will be earned ($1,700,000).

Now try Exercises 41 through 45 ▶

Note that if the leading coefficient is positive and the vertex is below the x-axis ($k < 0$), the graph will have two x-intercepts (see Figure 3.2). If $a > 0$ and the vertex is above the x-axis ($k > 0$), the graph will not cross the x-axis (Figure 3.3). Similar statements can be made for the case where a is negative.

Figure 3.2

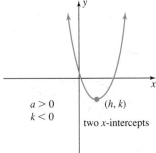

$a > 0$
$k < 0$
(h, k)
two x-intercepts

Figure 3.3

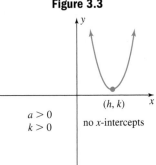

$a > 0$
$k > 0$
(h, k)
no x-intercepts

In some applications of quadratic functions, our interest includes the x-intercepts of the graph. Drawing on our previous work, we note that the following statements are equivalent, meaning if any one statement is true, then all four statements are true.

- $(r, 0)$ is an **x-intercept** of the graph of $f(x)$.
- $x = r$ is a **solution** or **root** of the equation $f(x) = 0$.
- $(x - r)$ is a **factor** of $f(x)$.
- r is a **zero** of $f(x)$.

When the quadratic function is in standard form, our primary tool for finding the zeroes is the quadratic formula. If the function is expressed as a transformation, we will often solve for x using inverse operations.

EXAMPLE 6 ▶ **Modeling the Height of a Projectile**

In the 1976 Pro Bowl, NFL punter Ray Guy of the Oakland Raiders kicked the ball so high it hit the scoreboard hanging from the roof of the New Orleans SuperDome. If we let $h(t)$ represent the height of the football (in feet) after t sec, the function $h(t) = -22t^2 + 132t + 1$ models the relationship (time, height of ball).

a. What does the y-intercept of this function represent?

b. After how many seconds did the football reach its maximum height?

c. What was the maximum height of this kick?

d. To the nearest hundredth of a second, how long until the ball returns to the ground (what was the hang time)?

Solution ▶   a. $h(0) = 1$, meaning the ball was 1 ft off the ground when Ray Guy kicked it.

b. Since $a < 0$, we know the graph opens downward and has a maximum value. To find the time needed to reach the maximum height, we use the vertex formula with $a = -22$ and $b = 132$:

$$t = \frac{-b}{2a} \qquad \text{vertex formula}$$

$$= \frac{-132}{2(-22)} \qquad \text{substitute } -22 \text{ for } a \text{ and } 132 \text{ for } b$$

$$= 3 \qquad \text{result}$$

The ball reached its maximum height after 3 sec.

c. To find the maximum height, we substitute 3 for t [evaluate $h(3)$]:

$$h(t) = -22t^2 + 132t + 1 \qquad \text{given function}$$

$$h(3) = -22(3)^2 + 132(3) + 1 \qquad \text{substitute 3 for } t$$

$$= 199 \qquad \text{result}$$

The ball reached a maximum height of 199 ft.

 d. When the ball returns to the ground it has a height of 0 ft. Substituting 0 for $h(t)$ gives $0 = -22t^2 + 132t + 1$, which we solve using the quadratic formula.

$$t = \frac{-b \pm \sqrt{b^2 - 4ac}}{2a} \qquad \text{quadratic formula}$$

$$= \frac{-132 \pm \sqrt{132^2 - 4(-22)(1)}}{2(-22)} \qquad \text{substitute } -22 \text{ for } a, 132 \text{ for } b, \text{ and 1 for } c$$

$$= \frac{-132 \pm \sqrt{17512}}{-44} \qquad \text{simplify}$$

$$t \approx -0.01 \quad \text{or} \quad t \approx 6.01$$

☑ **D.** You've just learned how to solve applications involving extreme values

The punt had a hang time of just over 6 sec.

> **Now try Exercises 46 through 49** ▶

TECHNOLOGY HIGHLIGHT

Estimating Irrational Zeroes

Once a function is entered into a graphing calculator, an estimate for irrational zeroes can easily be found. Enter the function $y = x^2 - 8x + 9$ on the Y= screen and graph using the standard window (ZOOM 6). Pressing 2nd TRACE (CALC) displays the screen in Figure 3.4. Pressing the number "2" selects the **2:zero** option and returns you to the graph, where you are asked to enter a "Left Bound." The calculator is asking you to narrow down the area it has to search for the x-intercept. Select any number that is conveniently to the left of the x-intercept you're interested in. For this graph, we entered a left bound of "0" (press ENTER), which the calculator indicates with a "▶" marker. It then asks you to enter a "Right Bound."

Figure 3.4

```
CALCULATE
1:value
2:zero
3:minimum
4:maximum
5:intersect
6:dy/dx
7:∫f(x)dx
```

—continued

Figure 3.5

−10

Figure 3.6

−10

Select any value to the right of this *x*-intercept, but be sure the value *bounds only one intercept* (see Figure 3.5). For this graph, a choice of 10 would include both *x*-intercepts, while a choice of 3 would bound only the *x*-intercept on the left. After entering 3, the calculator asks for a "guess." This option is used only when there are many different zeroes close by or if you entered a large interval. Most of the time we'll simply bypass this option by pressing [ENTER]. The cursor will be located at the zero you chose, with the coordinates displayed at the bottom of the screen (see Figure 3.6). The *x*-value is an approximation of the irrational zero. Find the zeroes of these functions using the [2nd] [TRACE] (CALC) **2:Zero** feature.

Exercise 1: $y = x^2 - 8x + 9$

Exercise 2: $y = 3a^2 - 5a - 6$

Exercise 3: $y = 2x^2 + 4x - 5$

Exercise 4: $y = 9w^2 + 6w - 1$

3.1 EXERCISES

▶ CONCEPTS AND VOCABULARY

Fill in each blank with the appropriate word or phrase. Carefully reread the section if needed.

1. Fill in the blank to complete the square, given $f(x) = -2x^2 - 10x - 7$:
   $f(x) = -2(x^2 + 5x + \frac{25}{4}) - 7 + \underline{\hspace{1cm}}$.

2. The maximum and minimum values are called _____ values and can be found using the _____ formula.

3. To find the zeroes of $f(x) = ax^2 + bx + c$, we substitute _____ for _____ and solve.

4. If the leading coefficient is positive and the vertex (h, k) is in Quadrant IV, the graph will have _____ *x*-intercepts.

5. Compare/Contrast how to complete the square on an *equation,* versus how to complete the square on a function. Use the equation $2x^2 + 6x - 3 = 0$ and the function $f(x) = 2x^2 + 6x - 3 = 0$ to illustrate.

6. Discuss/Explain why the graph of a quadratic function has no *x*-intercepts if *a* and *k* [vertex (h, k)] have like signs. Under what conditions will the function have a single real root?

▶ DEVELOPING YOUR SKILLS

Graph each function using end behavior, intercepts, and completing the square to write the function in shifted form. Clearly state the transformations used to obtain the graph, and label the vertex and all intercepts (if they exist). Use the quadratic formula to find the *x*-intercepts.

7. $f(x) = x^2 + 4x - 5$ 8. $g(x) = x^2 - 6x - 7$

9. $h(x) = -x^2 + 2x + 3$ 10. $H(x) = -x^2 + 8x - 7$

11. $Y_1 = 3x^2 + 6x - 5$

12. $Y_2 = 4x^2 - 24x + 15$

13. $f(x) = -2x^2 + 8x + 7$

14. $g(x) = -3x^2 + 12x - 7$

15. $p(x) = 2x^2 - 7x + 3$

16. $q(x) = 4x^2 - 9x + 2$

17. $f(x) = -3x^2 - 7x + 6$

18. $g(x) = -2x^2 + 9x - 7$

19. $p(x) = x^2 - 5x + 2$

20. $q(x) = x^2 + 7x + 4$

Graph each function using the vertex formula and other features of a quadratic graph. Label all important features.

21. $f(x) = x^2 + 2x - 6$ **22.** $g(x) = x^2 + 8x + 11$

23. $h(x) = -x^2 + 4x + 2$

24. $H(x) = -x^2 + 10x - 19$

25. $Y_1 = 0.5x^2 + 3x + 7$ **26.** $Y_2 = 0.2x^2 - 2x + 8$

27. $Y_1 = -2x^2 + 10x - 7$ **28.** $Y_2 = -2x^2 + 8x - 3$

29. $f(x) = 4x^2 - 12x + 3$ **30.** $g(x) = 3x^2 + 12x + 5$

31. $p(x) = \frac{1}{2}x^2 + 3x - 5$ **32.** $q(x) = \frac{1}{3}x^2 - 2x - 4$

▶ WORKING WITH FORMULAS

39. Vertex/intercept formula: $x = h \pm \sqrt{-\dfrac{k}{a}}$

As an alternative to using the quadratic formula *prior* to completing the square, the x-intercepts can more easily be found using the vertex/intercept formula *after* completing the square, when the coordinates of the vertex are known. (a) Beginning with the shifted form $y = a(x - h)^2 + k$, substitute 0 for y and solve for x to derive the formula, and (b) use the formula to find zeroes, real or complex, of the following functions.

  i. $y = (x + 3)^2 - 5$ **ii.** $y = -(x - 4)^2 + 3$

  iii. $y = 2(x + 4)^2 - 7$ **iv.** $y = -3(x - 2)^2 + 6$

State the equation of the function whose graph is shown.

33.

34.

35.

36.

37.

38.

  v. $s(t) = 0.2(t + 0.7)^2 - 0.8$

  vi. $r(t) = -0.5(t - 0.6)^2 + 2$

40. Surface area of a rectangular box with square ends: $S = 2h^2 + 4Lh$

The surface area of a rectangular box with square ends is given by the formula shown, where h is the height and width of the square ends, and L is the length of the box. (a) If L is 3 ft and the box must have a surface area of 32 ft^2, find the dimensions of the square ends. (b) Solve for L, then find the length if the height is 1.5 ft and surface area is 22.5 ft^2.

▶ APPLICATIONS

41. Maximum profit: An automobile manufacturer can produce up to 300 cars per day. The profit made from the sale of these vehicles can be modeled by the function $P(x) = -10x^2 + 3500x - 66,000$, where $P(x)$ is the profit in dollars and x is the number of automobiles made and sold. Based on this model:

  a. Find the y-intercept and explain what it means in this context.

  b. Find the x-intercepts and explain what they mean in this context.

c. How many cars should be made and sold to maximize profit?

d. What is the maximum profit?

42. Maximum profit: The profit for a manufacturer of collectible grandfather clocks is given by the function shown here, where $P(x)$ is the profit in dollars and x is the number of clocks made and sold. Answer the following questions based on this model: $P(x) = -1.6x^2 + 240x - 375$.

a. Find the y-intercept and explain what it means in this context.

b. Find the x-intercepts and explain what they mean in this context.

c. How many clocks should be made and sold to maximize profit?

d. What is the maximum profit?

43. Depth of a dive: As it leaves its support harness, a minisub takes a deep dive toward an underwater exploration site. The dive path is modeled by the function $d(x) = x^2 - 12x$, where $d(x)$ represents the depth of the minisub in hundreds of feet at a distance of x mi from the surface ship.

a. How far from the mother ship did the minisub reach its deepest point?

b. How far underwater was the submarine at its deepest point?

c. At $x = 4$ mi, how deep was the minisub explorer?

d. How far from its entry point did the minisub resurface?

44. Optimal pricing strategy: The director of the Ferguson Valley drama club must decide what to charge for a ticket to the club's performance of *The Music Man.* If the price is set too low, the club will lose money; and if the price is too high, people won't come. From past experience she estimates that the profit P from sales (in hundreds) can be approximated by $P(x) = -x^2 + 46x - 88$, where x is the cost of a ticket and $0 \le x \le 50$.

a. Find the lowest cost of a ticket that would allow the club to break even.

b. What is the highest cost that the club can charge to break even?

c. If the theater were to close down before any tickets are sold, how much money would the club lose?

d. How much should the club charge to maximize their profits? What is the maximum profit?

45. Maximum profit: A kitchen appliance manufacturer can produce up to 200 appliances per

day. The profit made from the sale of these machines can be modeled by the function $P(x) = -0.5x^2 + 175x - 3300$, where $P(x)$ is the profit in dollars, and x is the number of appliances made and sold. Based on this model,

a. Find the y-intercept and explain what it means in this context.

b. Find the x-intercepts and explain what they mean in this context.

c. Determine the domain of the function and explain its significance.

d. How many should be sold to maximize profit? What is the maximum profit?

The projectile function: $h(t) = -16t^2 + vt + k$ applies to any object projected upward with an initial velocity v, from a height k but not to objects under propulsion (such as a rocket). Consider this situation and answer the questions that follow.

46. Model rocketry: A member of the local rocketry club launches her latest rocket from a large field. At the moment its fuel is exhausted, the rocket has a velocity of 240 ft/sec and an altitude of 544 ft (t is in seconds).

a. Write the function that models the height of the rocket.

b. How high is the rocket at $t = 0$? If it took off from the ground, why is it this high at $t = 0$?

c. How high is the rocket 5 sec after the fuel is exhausted?

d. How high is the rocket 10 sec after the fuel is exhausted?

e. How could the rocket be at the same height at $t = 5$ and at $t = 10$?

f. What is the maximum height attained by the rocket?

g. How many seconds was the rocket airborne *after* its fuel was exhausted?

47. Height of a projectile: A projectile is thrown upward with an initial velocity of 176 ft/sec. After t sec, its height $h(t)$ above the ground is given by the function $h(t) = -16t^2 + 176t$.

a. Find the projectile's height above the ground after 2 sec.

b. Sketch the graph modeling the projectile's height.

c. What is the projectile's maximum height? What is the value of t at this height?

d. How many seconds after it is thrown will the projectile strike the ground?

48. Height of a projectile: In the movie *The Court Jester* (1956; Danny Kaye, Basil Rathbone, Angela Lansbury, and Glynis Johns), a catapult is used to toss the nefarious adviser to the king into a river. Suppose the path flown by the king's adviser is modeled by the function $h(d) = -0.02d^2 + 1.64d + 14.4$, where $h(d)$ is the height of the adviser in feet at a distance of d ft from the base of the catapult.

   a. How high was the release point of this catapult?

   b. How far from the catapult did the adviser reach a maximum altitude?

   c. What was this maximum altitude attained by the adviser?

   d. How far from the catapult did the adviser splash into the river?

49. Blanket toss competition: The Fraternities at Steele Head University are participating in a blanket toss competition, an activity borrowed from the whaling villages of the Inuit Eskimos. If the person being tossed is traveling at 32 ft/sec as he is projected into the air, and the Frat members are holding the canvas blanket at a height of 5 ft,

   a. Write the function that models the height at time t of the person being tossed.

   b. How high is the person when (i) $t = 0.5$, (ii) $t = 1.5$?

   c. From part (b) what do you know about *when* the maximum height is reached?

   d. To the nearest tenth of a second, when is the maximum height reached?

   e. To the nearest one-half foot, what was the maximum height?

   f. To the nearest tenth of a second, how long was this person airborne?

32 ft/s

5 ft

50. Cost of production: The cost of producing a plastic toy is given by the function $C(x) = 2x + 35$, where x is the number of hundreds of toys. The revenue from toy sales is given by $R(x) = -x^2 + 122x - 365$. Since profit = revenue − cost, the profit function must be $P(x) = -x^2 + 120x - 400$ (verify). How many toys sold will produce the maximum profit? What is the maximum profit?

51. Cost of production: The cost to produce bottled spring water is given by $C(x) = 16x - 63$, where x is the number of thousands of bottles. The total income (revenue) from the sale of these bottles is given by the function $R(x) = -x^2 + 326x - 7463$. Since profit = revenue − cost, the profit function must be $P(x) = -x^2 + 310x - 7400$ (verify). How many bottles sold will produce the maximum profit? What is the maximum profit?

52. Fencing a backyard: Tina and Imai have just purchased a purebred German Shepherd, and need to fence in their backyard so the dog can run. What is the maximum rectangular area they can enclose with 200 ft of fencing, if (a) they use fencing material along all four sides? What are the dimensions of the rectangle? (b) What is the maximum area if they use the house as one of the sides? What are the dimensions of *this* rectangle?

53. Building sheep pens: It's time to drench the sheep again, so Chance and Chelsea-Lou are fencing off a large rectangular area to build some temporary holding pens. To prep the males, females, and kids, they are separated into three smaller and equal-size pens partitioned within the large rectangle. If 384 ft of fencing is available and the maximum area is desired, what will be (a) the dimensions of the larger, outer rectangle? (b) the dimensions of the smaller holding pens?

► **EXTENDING THE CONCEPT**

54. Use the general solutions from the quadratic formula to show that the average value of the x-intercepts is $\dfrac{-b}{2a}$. Explain/Discuss why the result is valid even if the roots are complex.

$$x_1 = \frac{-b + \sqrt{b^2 - 4ac}}{2a} \qquad x_2 = \frac{-b - \sqrt{b^2 - 4ac}}{2a}$$

55. Write the equation of a quadratic function whose x-intercepts are given by $x = 2 \pm 3i$.

56. Write the equation for the parabola given.

57. Referring to Exercise 39, discuss the nature (real or complex, rational or irrational) and number of zeroes (0, 1, or 2) given by the vertex/intercept formula if (a) a and k have like signs, (b) a and k have unlike signs, (c) k is zero, (d) the ratio $-\dfrac{k}{a}$ is positive and a perfect square, and (e) the ratio $-\dfrac{k}{a}$ is positive and not a perfect square.

► **MAINTAINING YOUR SKILLS**

58. (2.3) Identify the slope and y-intercept for $-4x + 3y = 9$. Do not graph.

59. (R.5) Multiply: $\dfrac{x^2 - 4x + 4}{x^2 + 3x - 10} \cdot \dfrac{x^2 - 25}{x^2 - 10x + 25}$

60. (2.8) Given $f(x) = \sqrt[3]{x + 3}$ and $g(x) = x^3 - 3$, find $(f \circ g)(x)$ and $(g \circ f)(x)$.

61. (2.7) Given $f(x) = 3x^2 + 7x - 6$, solve $f(x) \leq 0$ using the x-intercepts and concavity of f.

3.2 | Synthetic Division; the Remainder and Factor Theorems

Learning Objectives

In Section 3.2 you will learn how to:

☐ **A.** Divide polynomials using long division and synthetic division

☐ **B.** Use the remainder theorem to evaluate polynomials

☐ **C.** Use the factor theorem to factor and build polynomials

☐ **D.** Solve applications using the remainder theorem

To find the zero of a linear function, we can use properties of equality to isolate x. To find the zeroes of a quadratic function, we can factor or use the quadratic formula. To find the zeroes of higher degree polynomials, we must first develop additional tools, including synthetic division and the remainder and factor theorems. These will help us to write a higher degree polynomial in terms of linear and quadratic polynomials, whose zeroes can easily be found.

A. Long Division and Synthetic Division

To help understand **synthetic division** and its use as a mathematical tool, we first review the process of **long division.**

Long Division

Polynomial long division closely resembles the division of whole numbers, with the main difference being that *we group each partial product* in parentheses to prevent errors in subtraction.

EXAMPLE 1 ▶ **Dividing Polynomials Using Long Division**

Divide $x^3 - 4x^2 + x + 6$ by $x - 1$.

Solution ▶ The divisor is $(x - 1)$ and the dividend is $(x^3 - 4x^2 + x + 6)$. To find the first multiplier, we compute *the ratio of leading terms* from each expression. Here the ratio $\dfrac{x^3 \text{ from dividend}}{x \text{ from divisor}}$ shows our first multiplier will be "x^2," with $x^2(x - 1) = x^3 - x^2$.

$$
\begin{array}{r}
x^2 \\
x - 1 \overline{)x^3 - 4x^2 + x + 6} \\
\underline{-(x^3 - x^2)} \quad \text{subtraction}
\end{array}
$$

$\longrightarrow$

$$
\begin{array}{r}
x^2 \\
x - 1 \overline{)x^3 - 4x^2 + x + 6} \\
\underline{-x^3 + x^2} \quad \text{algebraic addition} \\
-3x^2 + x
\end{array}
$$

At each stage, after writing the subtraction as algebraic addition (distributing the negative) we compute the sum in each column and "bring down" the next term. Each following multiplier is found as before, using the ratio $\dfrac{ax^k \text{ next leading term}}{x \text{ from divisor}}$.

$$
\begin{array}{r}
x^2 - 3x - 2 \\
x - 1 \overline{)x^3 - 4x^2 + x + 6} \\
\underline{-(x^3 - x^2)} \\
-3x^2 + x \\
\underline{-(-3x^2 + 3x)} \\
-2x + 6 \\
\underline{-(-2x + 2)} \\
4
\end{array}
$$

next multiplier: $\frac{-3x^2}{x} = -3x$ (ratio of leading terms)

next multiplier: $\frac{-2x}{x} = -2$

subtract $-3x(x - 1) = -3x^2 + 3x$

algebraic addition, bring down next term

subtract $-2(x - 1) = -2x + 2$

algebraic addition, remainder is 4

The result shows $\dfrac{x^3 - 4x^2 + x + 6}{x - 1} = x^2 - 3x - 2 + \dfrac{4}{x - 1}$, or after multiplying both sides by $x - 1$, $x^3 - 4x^2 + x + 6 = (x - 1)(x^2 - 3x - 2) + 4$.

Now try Exercises 7 through 12 ▶

The process illustrated is called the **division algorithm,** and like the division of whole numbers, the final result can be checked by multiplication.

$$
\begin{array}{ll}
\overset{\text{dividend}}{x^3 - 4x^2 + x + 6} = \overset{\text{divisor}}{(x - 1)}\overset{\text{quotient}}{(x^2 - 3x - 2)} + \overset{\text{remainder}}{4} \\
 = (x^3 - 3x^2 - 2x - x^2 + 3x + 2) + 4 & \text{divisor} \cdot \text{quotient} \\
 = (x^3 - 4x^2 + x + 2) + 4 & \text{combine like terms} \\
 = x^3 - 4x^2 + x + 6 \checkmark & \text{add remainder}
\end{array}
$$

check:

In general, the division algorithm for polynomials says

Division of Polynomials

Given polynomials $p(x)$ and $d(x) \neq 0$, there exists unique polynomials $q(x)$ and $r(x)$ such that

$$p(x) = d(x)q(x) + r(x),$$

where $r(x) = 0$ or the degree of $r(x)$ is less than the degree of $d(x)$. Here, $d(x)$ is called the *divisor,* $q(x)$ is the *quotient,* and $r(x)$ is the *remainder.*

In other words, "a polynomial of greater degree can be divided by a polynomial of equal or lesser degree to obtain a quotient and a remainder." As with whole numbers, if the remainder is zero, the divisor is a factor of the dividend.

Synthetic Division

As the word "synthetic" implies, synthetic division *simulates* the long division process, but condenses it and makes it more efficient when the divisor is linear. The process works by capitalizing on the repetition found in the division algorithm. First, the polynomials involved are written in decreasing order of degree, so the variable part of each term is unnecessary as we can let the *position of each coefficient* indicate the degree of the term. For the dividend from Example 1, $1 \quad -4 \quad 1 \quad 6$ would represent the polynomial $1x^3 - 4x^2 + 1x + 6$. Also, each stage of the algorithm involves a product of the divisor with the next multiplier, followed by a subtraction. These can likewise be computed using the coefficients only, as the degree of each term is still determined by its position. Here is the division from Example 1 in the synthetic division format. Note that we must use the *zero of the divisor* (as in $x = \frac{3}{2}$ for a divisor of $2x - 3$, or in this case, "1" from $x - 1 = 0$) and the coefficients of the dividend in the following format:

As this template indicates, the quotient and remainder will be read from the last row.

The arrow indicates we begin by "dropping the leading coefficient into place." We then multiply this coefficient by the "divisor," and place the result in the next column and add. Note that using the zero of the divisor enables us to *add in each column directly*, rather than subtracting then changing to algebraic addition as in long division.

In a sense, we "multiply in the diagonal direction," and "add in the vertical direction." Repeat the process until the division is complete.

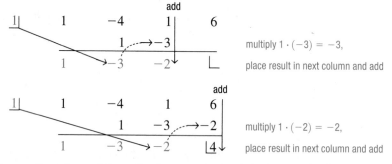

The quotient is read from the last row by noting the remainder is 4, leaving the coefficients $1 \quad -3 \quad -2$, which translates back into the polynomial $x^2 - 3x - 2$. The final result is identical to that in Example 1, but the new process is more efficient, since all stages are actually computed on a single template as shown here:

EXAMPLE 2 ▶ **Dividing Polynomials Using Synthetic Division**

Compute the quotient of $(x^3 + 3x^2 - 4x - 12)$ and $(x + 2)$, then check your answer.

Solution ▶ Using -2 as our "divisor" (from $x + 2 = 0$), we set up the synthetic division template and begin.

use -2 as a "divisor" $\underline{-2|}$

	1	3	-4	-12	drop lead coefficient into place;
$\downarrow$		-2	-2	12	multiply by divisor, place result
	1	1	-6	0	in next column and add

The result shows $\dfrac{x^3 + 3x^2 - 4x - 12}{x + 2} = x^2 + x - 6$, with no remainder.

Check ▶
$$x^3 + 3x^2 - 4x - 12 = (x + 2)(x^2 + x - 6)$$
$$= (x^3 + x^2 - 6x + 2x^2 + 2x - 12)$$
$$= x^3 + 3x^2 - 4x - 12 \checkmark$$

Now try Exercises 13 through 20 ▶

Since the division process is so dependent on the place value (degree) of each term, polynomials such as $2x^3 + 3x + 7$, which has no term of degree 2, must be written using a zero *placeholder:* $2x^3 + \mathbf{0}x^2 + 3x + 7$. This ensures that like place values "line up" as we carry out the division.

EXAMPLE 3 ▶ **Dividing Polynomials Using a Zero Placeholder**

Compute the quotient $\dfrac{2x^3 + 3x + 7}{x - 3}$ and check your answer.

Solution ▶ use 3 as a "divisor" $\underline{3|}$

	2	0	3	7	note place holder $\mathbf{0}x^2$ for "x^2" term
$\downarrow$		6	18	63	
	2	6	21	70	

The result shows $\dfrac{2x^3 + 3x + 7}{x - 3} = 2x^2 + 6x + 21 + \dfrac{70}{x - 3}$. Multiplying by $x - 3$ gives

$$2x^3 + 3x + 7 = (2x^2 + 6x + 21)(x - 3) + 70$$

Check ▶
$$2x^3 + 3x + 7 = (x - 3)(2x^2 + 6x + 21) + 70$$
$$= (2x^3 + 6x^2 + 21x - 6x^2 - 18x - 63) + 70$$
$$= 2x^3 + 3x + 7 \checkmark$$

WORTHY OF NOTE

Many corporations now pay their employees monthly to save on payroll costs. If your monthly salary was $2037/mo, but you received a check for only $237, would you complain? Just as placeholder zeroes ensure the correct value of each digit, they also ensure the correct valuation of each term in the division process.

Now try Exercises 21 through 30 ▶

As noted earlier, for synthetic division the divisor must be a linear polynomial and the zero of this divisor is used. This means for the quotient $\dfrac{2x^3 - 3x^2 - 8x + 12}{2x - 3}$, $\dfrac{3}{2}$ would be used for synthetic division (**see Exercises 43 and 44**). If the divisor is non-linear, long division must be used.

EXAMPLE 4 ▶ Division with a Nonlinear Divisor

Compute the quotient: $\dfrac{2x^4 + x^3 - 7x^2 + 3}{x^2 - 2}$.

Solution ▶ Write the dividend as $2x^4 + x^3 - 7x^2 + \mathbf{0}x + 3$, and the divisor as $x^2 + \mathbf{0}x - 2$.

The quotient of leading terms gives $\dfrac{2x^4 \text{ from dividend}}{x^2 \text{ from divisor}} = 2x^2$ as our first multiplier.

$$
\begin{array}{r}
2x^2 + x - 3 \\
x^2 + 0x - 2 \overline{\smash{\big)}\ 2x^4 + x^3 - 7x^2 + 0x + 3} \\
\end{array}
$$

divisor → $x^2 + 0x - 2$

Multiply $2x^2(x^2 + 0x - 2)$ $-(2x^4 + 0x^3 - 4x^2)$ subtract (algebraic addition)

$x^3 - 3x^2 + 0x$ bring down next term

Multiply $x(x^2 + 0x - 2)$ $-(x^3 + 0x^2 - 2x)$ subtract (algebraic addition)

$-3x^2 + 2x + 3$ bring down next term

Multiply $-3(x^2 + 0x - 2)$ $-(-\mathbf{3}x^2 + \mathbf{0}x + \mathbf{6})$ subtract

$2x - 3$ remainder is $2x - 3$

☑ **A.** You've just learned how to divide polynomials using long division and synthetic division

Since the degree of $2x - 3$ (degree 1) is less than the degree of the divisor (degree 2), the process is complete.

$$\frac{2x^4 + x^3 - 7x^2 + 3}{x^2 - 2} = (2x^2 + x - 3) + \frac{(2x - 3)}{x^2 - 2}$$

Now try Exercises 31 through 34 ▶

B. The Remainder Theorem

In Example 2, we saw that $(x^3 + 3x^2 - 4x - 12) \div (x + 2) = x^2 + x - 6$, with no remainder. Similar to whole number division, this means $x + 2$ must be a factor of $x^3 + 3x^2 - 4x - 12$, a fact made clear as we checked our answer: $x^3 + 3x^2 - 4x - 12 = (x + 2)(x^2 + x - 6)$. To help us find the factors of higher degree polynomials, we combine synthetic division with a relationship known as the **remainder theorem.** Consider the functions $p(x) = x^3 + 5x^2 + 2x - 8$, $d(x) = x + 3$, and their quotient $\dfrac{p(x)}{d(x)} = \dfrac{x^3 + 5x^2 + 2x - 8}{x + 3}$. Using -3 as the divisor in synthetic division gives

use -3 as a "divisor" $\underline{-3}$
$$
\begin{array}{r|rrrr}
 & 1 & 5 & 2 & -8 \\
\downarrow & & -3 & -6 & 12 \\
\hline
 & 1 & 2 & -4 & \underline{4}
\end{array}
$$

This shows $x + 3$ is *not* a factor of $P(x)$, since it didn't "divide evenly." However, from the result $p(x) = (x + 3)(x^2 + 2x - 4) + 4$, we make a remarkable observation—if we evaluate $p(-3)$, *the quotient portion becomes zero,* showing $p(-3) = 4$—*which is the remainder.*

$$p(-3) = (-3 + 3)[(-3)^2 + 2(-3) - 4] + 4$$
$$= (\mathbf{0})(-1) + 4$$
$$= 4$$

This can also be seen by evaluating $p(-3)$ in its original form:

$$p(x) = x^3 + 5x^2 + 2x - 8$$
$$p(-3) = (-3)^3 + 5(-3)^2 + 2(-3) - 8$$
$$= -27 + 45 + (-6) - 8$$
$$= 4$$

The result is no coincidence, and illustrates the conclusion of the remainder theorem.

The Remainder Theorem

If a polynomial $p(x)$ is divided by $(x - c)$ using synthetic division, the remainder is equal to $p(c)$.

This gives us a powerful tool for evaluating polynomials. Where a direct evaluation involves powers of numbers and a long series of calculations, synthetic division reduces the process to simple products and sums.

EXAMPLE 5 ▶ **Using the Remainder Theorem to Evaluate Polynomials**

Use the remainder theorem to find $p(-5)$ for $p(x) = x^4 + 3x^3 - 8x^2 + 5x - 6$. Verify the result using a substitution.

Solution ▶

use -5 as a "divisor" $\underline{-5|}$

$$
\begin{array}{c|ccccc}
-5 & 1 & 3 & -8 & 5 & -6 \\
   &   & -5 & 10 & -10 & 25 \\
\hline
   & 1 & -2 & 2 & -5 & \underline{|19} \\
\end{array}
$$

The result shows $p(-5) = 19$, which we verify directly:

$$p(-5) = (-5)^4 + 3(-5)^3 - 8(-5)^2 + 5(-5) - 6$$
$$= 625 - 375 - 200 - 25 - 6$$
$$= 19 \checkmark$$

☑ **B. You've just learned how to use the Remainder Theorem to evaluate polynomials**

Now try Exercises 35 through 44 ▶

WORTHY OF NOTE

Since $p(-5) = 19$, we know $(-5, 19)$ must be a point of the graph of $p(x)$. The ability to quickly evaluate polynomial functions using the remainder theorem will be used extensively in the sections that follow.

C. The Factor Theorem

As a consequence of the remainder theorem, when $p(x)$ is divided by $x - c$ and the remainder is 0, $p(c) = 0$, and c is a zero of the polynomial. The relationship between $x - c$, c, and $p(c) = 0$ are summarized into the **factor theorem.**

The Factor Theorem

For a polynomial $p(x)$,

1. If $p(c) = 0$, then $x - c$ is a factor of $p(x)$.
2. If $x - c$ is a factor of $p(x)$, then $p(c) = 0$.

The remainder and factor theorems often work together to help us find factors of higher degree polynomials.

EXAMPLE 6 ▶ **Using the Factor Theorem to Find Factors of a Polynomial**

Use the factor theorem to determine if

a. $x - 2$ 　　　　**b.** $x + 1$

are factors of $p(x) = x^4 + x^3 - 10x^2 - 4x + 24$.

Solution ▶ **a.** If $x - 2$ is a factor, then $p(2)$ must be 0. Using the remainder theorem we have

$$
\begin{array}{c|ccccc}
2| & 1 & 1 & -10 & -4 & 24 \\
  \downarrow & & 2 & 6 & -8 & -24 \\
\hline
  & 1 & 3 & -4 & -12 & \underline{|0} \\
\end{array}
$$

Since the remainder is zero, we know $p(2) = 0$ (remainder theorem) and $(x - 2)$ is a factor (factor theorem).

b. Similarly, if $x + 1$ is a factor, then $p(-1)$ must be 0.

$$
\begin{array}{r|rrrrr}
-1 & 1 & 1 & -10 & -4 & 24 \\
   &   & -1 & 0 & 10 & -6 \\
\hline
   & 1 & 0 & -10 & 6 & \underline{18}
\end{array}
$$

Since the remainder is not zero, $(x + 1)$ is not a factor of p.

Now try Exercises 45 through 56 ▶

EXAMPLE 7 ▶ **Building a Polynomial Using the Factor Theorem**

A polynomial $p(x)$ has the zeroes 3, $\sqrt{2}$, and $-\sqrt{2}$. Use the factor theorem to find the polynomial.

Solution ▶ Using the factor theorem, the factors of $p(x)$ must be $(x - 3)$, $(x - \sqrt{2})$, and $(x + \sqrt{2})$. Computing the product will yield the polynomial.

$$
\begin{aligned}
p(x) &= (x - 3)(x - \sqrt{2})(x + \sqrt{2}) \\
     &= (x - 3)(x^2 - 2) \\
     &= x^3 - 3x^2 - 2x + 6
\end{aligned}
$$

Now try Exercises 57 through 64 ▶

As the following *Graphical Support* feature shows, the result obtained in Example 7 is not unique, since any polynomial of the form $a(x^3 - 3x^2 - 2x + 6)$ will also have the same three roots for $a \in \mathbb{R}$.

GRAPHICAL SUPPORT

A graphing calculator helps to illustrate there are actually many different polynomials that have the three roots required by Example 7. Figure 3.7 shows the graph of $Y_1 = p(x)$, as well as graph of $Y_2 = 2p(x)$. The only difference is $2p(x)$ has been vertically stretched. Likewise, the graph of $-1p(x)$ would be a vertical reflection, *but still with the same zeroes.*

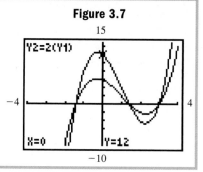

Figure 3.7

EXAMPLE 8 ▶ **Finding Zeroes Using the Factor Theorem**

Given that 2 is a zero of $p(x) = x^4 + x^3 - 10x^2 - 4x + 24$, use the factor theorem to help find all other zeroes.

Solution ▶ Using synthetic division gives:

$$
\begin{array}{r|rrrrr}
\text{use 2 as a "divisor"} \quad 2 & 1 & 1 & -10 & -4 & 24 \\
   &   & 2 & 6 & -8 & -24 \\
\hline
   & 1 & 3 & -4 & -12 & \underline{0}
\end{array}
$$

Since the remainder is zero, $(x - 2)$ is a factor and p can be written:

$$
x^4 + x^3 - 10x^2 - 4x + 24 = (x - 2)(x^3 + 3x^2 - 4x - 12)
$$

WORTHY OF NOTE

In Section R.4 we noted a third degree polynomial $ax^3 + bx^2 + cx + d$ is factorable if $ad = bc$. In Example 8, $1(-12) = 3(-4)$ and the polynomial is factorable.

Note the quotient polynomial can be factored by grouping to find the remaining factors of p.

$$x^4 + x^3 - 10x^2 - 4x + 24 = (x - 2)(x^3 + 3x^2 - 4x - 12) \quad \text{group terms (in color)}$$
$$= (x - 2)[x^2(x + 3) - 4(x + 3)] \quad \text{remove common factors from each group}$$
$$= (x - 2)[(x + 3)(x^2 - 4)] \quad \text{factor common binomial}$$
$$= (x - 2)(x + 3)(x + 2)(x - 2) \quad \text{factor difference of squares}$$
$$= (x + 3)(x + 2)(x - 2)^2 \quad \text{completely factored form}$$

☑ **C. You've just learned how to use the factor theorem to factor and build polynomials**

The final result shows $(x - 2)$ is actually a repeated factor, and the remaining zeroes of p are -3 and -2.

 Now try Exercises 65 through 78 ▶

D. Applications

While the factor and remainder theorems are valuable tools for factoring higher degree polynomials, each has applications that extend beyond this use.

EXAMPLE 9 ▶ **Using the Remainder Theorem to Solve a Discharge Rate Application**

 The *discharge rate* of a river is a measure of the river's water flow as it empties into a lake, sea, or ocean. The rate depends on many factors, but is primarily influenced by the precipitation in the surrounding area and is often seasonal. Suppose the discharge rate of the Shimote River was modeled by $D(m) = -m^4 + 22m^3 - 147m^2 + 317m + 150$, where $D(m)$ represents the discharge rate in thousands of cubic meters of water per second in month $m (m = 1 \rightarrow \text{Jan})$.

  a. What was the discharge rate in June (summer heat)?

  b. Is the discharge rate higher in February (winter runoff) or October (fall rains)?

Solution ▶ **a.** To find the discharge rate in June, we evaluate D at $m = 6$.
    Using the remainder theorem gives

$$
\begin{array}{r|rrrrr}
6 & -1 & 22 & -147 & 317 & 150 \\
  & \downarrow & -6 & 96 & -306 & 66 \\
\hline
  & -1 & 16 & -51 & 11 & \underline{216}
\end{array}
$$

In June, the discharge rate is 216,000 m³/sec.

  b. For the discharge rates in February ($m = 2$) and October ($m = 10$), we have

$$
\begin{array}{r|rrrrr}
2 & -1 & 22 & -147 & 317 & 150 \\
  & \downarrow & -2 & 40 & -214 & 206 \\
\hline
  & -1 & 20 & -107 & 103 & \underline{356}
\end{array}
\qquad
\begin{array}{r|rrrrr}
10 & -1 & 22 & -147 & 317 & 150 \\
   & \downarrow & -10 & 120 & -270 & 470 \\
\hline
   & -1 & 12 & -27 & 47 & \underline{620}
\end{array}
$$

☑ **D. You've just learned how to solve applications using the remainder theorem**

The discharge rate during the fall rains in October is much higher.

Now try Exercises 81 through 84 ▶

3.2 EXERCISES

▶ CONCEPTS AND VOCABULARY

Fill in each blank with the appropriate word or phrase. Carefully reread the section if needed.

1. For _____ division, we use the _____ of the divisor to begin.

2. If the _____ is zero after division, then the _____ is a factor of the dividend.

3. If polynomial $P(x)$ is divided by a linear divisor of the form $x - c$, the remainder is identical to _____. This is a statement of the _____ theorem.

4. If $P(c) = 0$, then _____ must be a factor of $P(x)$. Conversely, if _____ is a factor of $P(x)$, then $P(c) = 0$. These are statements from the _____ theorem.

5. Discuss/Explain how to write the quotient and remainder using the last line from a synthetic division.

6. Discuss/Explain why (a, b) is a point on the graph of P, given b was the remainder after P was divided by a using synthetic division.

Divide using long division. Write the result as dividend = (divisor)(quotient) + remainder.

7. $\dfrac{x^3 - 5x^2 - 4x + 23}{x - 2}$ 8. $\dfrac{x^3 + 5x^2 - 17x - 26}{x + 7}$

9. $(2x^3 + 5x^2 + 4x + 17) \div (x + 3)$

10. $(3x^3 + 14x^2 - 2x - 37) \div (x + 4)$

11. $(x^3 - 8x^2 + 11x + 20) \div (x - 5)$

12. $(x^3 - 5x^2 - 22x - 16) \div (x + 2)$

Divide using synthetic division. Write answers in two ways: (a) $\dfrac{\text{dividend}}{\text{divisor}}$ = quotient + $\dfrac{\text{remainder}}{\text{divisor}}$, and (b) dividend = (divisor)(quotient) + remainder. For Exercises 13–18, check answers using multiplication.

13. $\dfrac{2x^2 - 5x - 3}{x - 3}$ 14. $\dfrac{3x^2 + 13x - 10}{x + 5}$

15. $(x^3 - 3x^2 - 14x - 8) \div (x + 2)$

16. $(x^3 - 6x^2 - 25x - 17) \div (x + 1)$

17. $\dfrac{x^3 - 5x^2 - 4x + 23}{x - 2}$ 18. $\dfrac{x^3 + 12x^2 + 34x - 7}{x + 7}$

19. $(2x^3 - 5x^2 - 11x - 17) \div (x - 4)$

20. $(3x^3 - x^2 - 7x + 27) \div (x - 1)$

Divide using synthetic division. Note that some terms of a polynomial may be "missing." Write answers as dividend = (divisor)(quotient) + remainder.

21. $(x^3 + 5x^2 + 7) \div (x + 1)$

22. $(x^3 - 3x^2 - 37) \div (x - 5)$

23. $(x^3 - 13x - 12) \div (x - 4)$

24. $(x^3 - 7x + 6) \div (x + 3)$

25. $\dfrac{3x^3 - 8x + 12}{x - 1}$ 26. $\dfrac{2x^3 + 7x - 81}{x - 3}$

27. $(n^3 + 27) \div (n + 3)$ 28. $(m^3 - 8) \div (m - 2)$

29. $(x^4 + 3x^3 - 16x - 8) \div (x - 2)$

30. $(x^4 + 3x^2 + 29x - 21) \div (x + 3)$

Compute each indicated quotient. Write answers in the form $\dfrac{\text{dividend}}{\text{divisor}}$ = quotient + $\dfrac{\text{remainder}}{\text{divisor}}$.

31. $\dfrac{2x^3 + 7x^2 - x + 26}{x^2 + 3}$ 32. $\dfrac{x^4 + 3x^3 + 2x^2 - x - 5}{x^2 - 2}$

33. $\dfrac{x^4 - 5x^2 - 4x + 7}{x^2 - 1}$ 34. $\dfrac{x^4 + 2x^3 - 8x - 16}{x^2 + 5}$

▶ DEVELOPING YOUR SKILLS

Use the remainder theorem to evaluate $P(x)$ as given.

35. $P(x) = x^3 - 6x^2 + 5x + 12$
   a. $P(-2)$ b. $P(5)$

36. $P(x) = x^3 + 4x^2 - 8x - 15$
   a. $P(-2)$ b. $P(3)$

37. $P(x) = 2x^3 - x^2 - 19x + 4$
   a. $P(-3)$ b. $P(2)$

38. $P(x) = 3x^3 - 8x^2 - 14x + 9$

    a. $P(-2)$ **b.** $P(4)$

39. $P(x) = x^4 - 4x^2 + x + 1$

    a. $P(-2)$ **b.** $P(2)$

40. $P(x) = x^4 + 3x^3 - 2x - 4$

    a. $P(-2)$ **b.** $P(2)$

41. $P(x) = 2x^3 - 7x + 33$

    a. $P(-2)$ **b.** $P(-3)$

42. $P(x) = -2x^3 + 9x^2 - 11$

    a. $P(-2)$ **b.** $P(-1)$

43. $P(x) = 2x^3 + 3x^2 - 9x - 10$

    a. $P(\frac{3}{2})$ **b.** $P(-\frac{5}{2})$

44. $P(x) = 3x^3 + 11x^2 + 2x - 16$

    a. $P(\frac{1}{3})$ **b.** $P(-\frac{8}{3})$

Use the factor theorem to determine if the factors given are factors of $f(x)$.

45. $f(x) = x^3 - 3x^2 - 13x + 15$

    a. $(x + 3)$ **b.** $(x - 5)$

46. $f(x) = x^3 + 2x^2 - 11x - 12$

    a. $(x + 4)$ **b.** $(x - 3)$

47. $f(x) = x^3 - 6x^2 + 3x + 10$

    a. $(x + 2)$ **b.** $(x - 5)$

48. $f(x) = x^3 + 2x^2 - 5x - 6$

    a. $(x - 2)$ **b.** $(x + 4)$

49. $f(x) = -x^3 + 7x - 6$

    a. $(x + 3)$ **b.** $(x - 2)$

50. $f(x) = -x^3 + 13x - 12$

    a. $(x + 4)$ **b.** $(x - 3)$

Use the factor theorem to show the given value is a zero of $P(x)$.

51. $P(x) = x^3 + 2x^2 - 5x - 6$

    $x = -3$

52. $P(x) = x^3 + 3x^2 - 16x + 12$

    $x = -6$

53. $P(x) = x^3 - 7x + 6$

    $x = 2$

54. $P(x) = x^3 - 13x + 12$

    $x = -4$

55. $P(x) = 9x^3 + 18x^2 - 4x - 8$

    $x = \dfrac{2}{3}$

56. $P(x) = 5x^3 + 13x^2 - 9x - 9$

    $x = -\dfrac{3}{5}$

A polynomial P with integer coefficients has the zeroes and degree indicated. Use the factor theorem to write the function in factored form and standard form.

57. $-2, 3, -5$; degree 3   **58.** $1, -4, 2$; degree 3

59. $-2, \sqrt{3}, -\sqrt{3}$;   **60.** $\sqrt{5}, -\sqrt{5}, 4$;
    degree 3   degree 3

61. $-5, 2\sqrt{3}, -2\sqrt{3}$;   **62.** $4, 3\sqrt{2}, -3\sqrt{2}$;
    degree 3   degree 3

63. $1, -2, \sqrt{10}, -\sqrt{10}$;   **64.** $\sqrt{7}, -\sqrt{7}, 3, -1$;
    degree 4   degree 4

In Exercises 65 through 70, a known zero of the polynomial is given. Use the factor theorem to write the polynomial in completely factored form.

65. $P(x) = x^3 - 5x^2 - 2x + 24$; $x = -2$

66. $Q(x) = x^3 - 7x^2 + 7x + 15$; $x = 3$

67. $p(x) = x^4 + 2x^3 - 12x^2 - 18x + 27$; $x = -3$

68. $q(x) = x^4 + 4x^3 - 6x^2 - 4x + 5$; $x = 1$

69. $f(x) = 2x^3 + 11x^2 - x - 30$; $x = \frac{3}{2}$

70. $g(x) = 3x^3 + 2x^2 - 75x - 50$; $x = -\frac{2}{3}$

If $p(x)$ is a polynomial with rational coefficients and a leading coefficient of $a = 1$, the rational zeroes of p (if they exist) *must be factors of the constant term*. Use this property of polynomials with the factor and remainder theorems to factor each polynomial completely.

71. $p(x) = x^3 - 3x^2 - 9x + 27$

72. $p(x) = x^3 - 4x^2 - 16x + 64$

73. $p(x) = x^3 - 6x^2 + 12x - 8$

74. $p(x) = x^3 - 15x^2 + 75x - 125$

75. $p(x) = (x^2 - 6x + 9)(x^2 - 9)$

76. $p(x) = (x^2 - 1)(x^2 - 2x + 1)$

77. $p(x) = (x^3 + 4x^2 - 9x - 36)(x^2 + x - 12)$

78. $p(x) = (x^3 - 3x^2 + 3x - 1)(x^2 - 3x + 2)$

▶ WORKING WITH FORMULAS

Volume of an open box: $V(x) = 4x^3 - 84x^2 + 432x$

An open box is constructed by cutting square corners from a 24 in. by 18 in. sheet of cardboard and folding up the sides. Its volume is given by the formula shown, where x represents the size of the square cut.

79. Given a volume of 640 in³, use synthetic division and the

remainder theorem to determine if the squares were 2-, 3-, 4-, or 5-in. squares and state the dimensions of the box. (*Hint:* Write as a function $v(x)$ and use synthetic division.)

80. Given the volume is 357.5 in³, use synthetic division and the remainder theorem to determine if the squares were 5.5-, 6.5-, or 7.5-in. squares and state the dimensions of the box. (*Hint:* Write as a function $v(x)$ and use synthetic division.)

▶ APPLICATIONS

81. Tourist population: During the 12 weeks of summer, the population of tourists at a popular beach resort is modeled by the polynomial $P(w) = -0.1w^4 + 2w^3 - 14w^2 + 52w + 5$, where $P(w)$ is the tourist population (in 1000s) during week w. Use the remainder theorem to help answer the following questions.

  a. Were there more tourists at the resort in week 5 ($w = 5$) or week 10? How many more tourists?

  b. Were more tourists at the resort one week after opening ($w = 1$) or one week before closing ($w = 11$). How many more tourists?

  c. The tourist population peaked (reached its highest) between weeks 7 and 10. Use the remainder theorem to determine the peak week.

82. Debt load: Due to a fluctuation in tax revenues, a county government is projecting a deficit for the next 12 months, followed by a quick recovery and the repayment of all debt near the end of this period. The projected debt can be modeled by the polynomial $D(m) = 0.1m^4 - 2m^3 + 15m^2 - 64m - 3$, where $D(m)$ represents the amount of debt (in millions of dollars) in month m. Use the remainder theorem to help answer the following questions.

  a. Was the debt higher in month 5 ($m = 5$) or month 10 of this period? How much higher?

  b. Was the debt higher in the first month of this period (one month into the deficit) or after the eleventh month (one month before the expected recovery)? How much higher?

  c. The total debt reached its maximum between months 7 and 10. Use the remainder theorem to determine which month.

83. Volume of water: The volume of water in a rectangular, in-ground, swimming pool is given by $V(x) = x^3 + 11x^2 + 24x$, where $v(x)$ is the volume in cubic feet when the water is x ft high. (a) Use the remainder theorem to find the volume when $x = 3$ ft. (b) If the volume is 100 ft³ of water, what is the height x? (c) If the maximum capacity of the pool is 1000 ft³, what is the maximum depth (to the nearest integer)?

84. Amusement park attendance: Attendance at an amusement park depends on the weather. After opening in spring, attendance rises quickly, slows during the summer, soars in the fall, then quickly falls with the approach of winter when the park closes. The model for attendance is given by $A(m) = -\frac{1}{4}m^4 + 6m^3 - 52m^2 + 196m - 260$, where $A(m)$ represents the number of people attending in month m (in thousands). (a) Did more people go to the park in April ($m = 4$) or June ($m = 6$)? (b) In what month did maximum attendance occur? (c) When did the park close?

In these applications, synthetic division is applied in the usual way, treating k as an unknown constant.

85. Find a value of k that will make $x = -2$ a zero of $f(x) = x^3 - 3x^2 - 5x + k$.

86. Find a value of k that will make $x - 3$ a factor of $g(x) = x^3 + 2x^2 - 7x + k$.

87. For what value(s) of k will $x - 2$ be a factor of $p(x) = x^3 - 3x^2 + kx + 10$?

88. For what value(s) of k will $x + 5$ be a factor of $q(x) = x^3 + 6x^2 + kx + 50$?

▶ **EXTENDING THE CONCEPT**

89. To investigate whether the remainder and factor theorems can be applied when the coefficients or zeroes of a polynomial are complex, try using the factor theorem to find a polynomial with degree 3, whose zeroes are $x = 2i$, $x = -2i$, and $x = 3$. Then see if the result can be verified using the remainder theorem and these zeroes. What does the result suggest?

90. Since we use a base-10 number system, numbers like 1196 can be written in polynomial form as $p(x) = 1x^3 + 1x^2 + 9x + 6$, where $x = 10$. Divide $p(x)$ by $x + 3$ using synthetic division and write your answer as $\frac{x^3 + x^2 + 9x + 6}{x + 3} =$ quotient $+$ $\frac{remainder}{divisor}$. For $x = 10$, what is the value of quotient $+ \frac{remainder}{divisor}$? What is the result of dividing 1196 by $10 + 3 = 13$? What can you conclude?

91. The sum of the first n perfect cubes is given by the formula $S = \frac{1}{4}(n^4 + 2n^3 + n^2)$. Use the remainder theorem on S to find the sum of (a) the first three

perfect cubes (divide by $n - 3$) and (b) the first five perfect cubes (divide by $n - 5$). Check results by adding the perfect cubes manually. To avoid working with fractions you can initially ignore the $\frac{1}{4}$ (use $n^4 + 2n^3 + n^2 + 0n + 0$), as long as you divide the remainder by 4.

92. Though not a direct focus of this course, the remainder and factor theorems, as well as synthetic division, *can also be applied using complex numbers*. Use the remainder theorem to show the value given is a zero of $P(x)$.

   a. $P(x) = x^3 - 4x^2 + 9x - 36$; $x = 3i$
   b. $P(x) = x^4 + x^3 + 2x^2 + 4x - 8$; $x = -2i$
   c. $P(x) = -x^3 + x^2 - 3x - 5$; $x = 1 + 2i$
   d. $P(x) = x^3 + 2x^2 + 16x + 32$; $x = -4i$
   e. $P(x) = x^4 + x^3 - 5x^2 + x - 6$; $x = i$
   f. $P(x) = -x^3 + x^2 - 8x - 10$; $x = 1 + 3i$

MAINTAINING YOUR SKILLS

93. (1.1) John and Rick are out orienteering. Rick finds the last marker first and is heading for the finish line, 1275 yd away. John is just seconds behind, and after locating the last marker tries to overtake Rick, who by now has a 250-yd lead. If Rick runs at 4 yd/sec and John runs at 5 yd/sec, will John catch Rick before they reach the finish line?

94. (1.5) Solve for w: $-2(3w^2 + 5) + 3 = -7w + w^2 - 7$

95. (2.3) The profit of a small business increased linearly from $5000 in 2005 to $12,000 in 2010. Find a linear function $G(t)$ modeling the growth of the company's profit (let $t = 0$ correspond to 2005).

96. (2.7) Given $f(x) = x^2 - 4x$, use the average rate of change formula to find $\frac{\Delta y}{\Delta x}$ in the interval $x \in [1.0, 1.1]$.

3.3 | The Zeroes of Polynomial Functions

Learning Objectives

In Section 3.3 you will learn how to:

☐ **A.** Apply the fundamental theorem of algebra and the linear factorization theorem

☐ **B.** Locate zeroes of a polynomial using the intermediate value theorem

☐ **C.** Find rational zeroes of a polynomial using the rational zeroes theorem

☐ **D.** Use Descartes' rule of signs and the upper/lower bounds theorem

☐ **E.** Solve applications of polynomials

This section represents one of the highlights in the college algebra curriculum, because it offers a look at what many call *the big picture*. The ideas presented are the result of a cumulative knowledge base developed over a long period of time, and give a fairly comprehensive view of the study of polynomial functions.

A. The Fundamental Theorem of Algebra

From Section 1.4, we know that real numbers are a subset of the complex numbers: $\mathbb{R} \subset \mathbb{C}$. Because complex numbers are the "larger" set (containing all other number sets), properties and theorems about complex numbers are more powerful and far reaching than theorems about real numbers. In the same way, real polynomials are a subset of the complex polynomials, and the same principle applies.

Complex Polynomial Functions

A complex polynomial of degree n has the form

$$P(x) = a_n x^n + a_{n-1} x^{n-1} + \cdots + a_1 x^1 + a_0,$$

where $a_n, a_{n-1}, \cdots, a_1, a_0$ are complex numbers and $a_n \neq 0$.

Notice that real polynomials have the same form, but here $a_n, a_{n-1}, \ldots, a_1, a_0$ *represent complex numbers.* In 1797, Carl Friedrich Gauss (1777–1855) proved that *all* polynomial functions have zeroes, and that the number of zeroes is equal to the degree of the polynomial. The proof of this statement is based on a theorem that is the bedrock for a complete study of polynomial functions, and has come to be known as the **fundamental theorem of algebra.**

The Fundamental Theorem of Algebra

Every complex polynomial of degree $n \geq 1$ has at least one complex zero.

Although the statement may seem trivial, it allows us to draw two important conclusions. The first is that our search for a solution will not be fruitless or wasted—zeroes for *all* polynomial equations exist. Second, the fundamental theorem combined with the factor theorem allows us to state the **linear factorization theorem.**

The Linear Factorization Theorem

If $p(x)$ is a polynomial function of degree $n \geq 1$, then p has exactly n linear factors and can be written in the form,

$$p(x) = a(x - c_1)(x - c_2) \cdot \cdots \cdot (x - c_n)$$

where $a \neq 0$ and $c_1, c_2, \ldots, c_n$ are (not necessarily distinct) complex numbers.

In other words, every complex polynomial of degree n can be rewritten as the product of a nonzero constant and exactly n linear factors (for a proof of this theorem, see Appendix II).

EXAMPLE 1 ▶ **Writing Polynomials as a Product of Linear Factors**

Rewrite $P(x) = x^4 - 8x^2 - 9$ as a product of linear factors, and find its zeroes.

Solution ▶ From its given form, we know $a = 1$. Since P has degree 4, the factored form must be $P(x) = (x - c_1)(x - c_2)(x - c_3)(x - c_4)$. Noting that P is in quadratic form, we substitute u for x^2 and u^2 for x^4 and attempt to factor:

$$x^4 - 8x^2 - 9 \rightarrow u^2 - 8u - 9 \qquad \text{substitute } u \text{ for } x^2; u^2 \text{ for } x^4$$
$$= (u - 9)(u + 1) \qquad \text{factor in terms of } u$$
$$= (x^2 - 9)(x^2 + 1) \qquad \text{rewrite in terms of } x \text{ (substitute } x^2 \text{ for } u)$$

We know $x^2 - 9$ will factor since it is a difference of squares. From our work with complex numbers (Section 1.4), we know $(a + bi)(a - bi) = a^2 + b^2$, and the factored form of $x^2 + 1$ must be $(x + i)(x - i)$. The completely factored form is

$$P(x) = (x + 3)(x - 3)(x + i)(x - i), \text{ and}$$

the zeroes of P are $-3, 3, -i,$ and i.

Now try Exercises 7 through 10 ▶

EXAMPLE 2 ▶ **Writing Polynomials as a Product of Linear Factors**

Rewrite $P(x) = x^3 + 2x^2 - 4x - 8$ as a product of linear factors and find its zeroes.

Solution ▶ We observe that $a = 1$ and P has degree 3, so the factored form must be $P(x) = (x - c_1)(x - c_2)(x - c_3)$. Noting that $ad = bc$ (Section R.4), we start with factoring by grouping.

$$\begin{aligned} P(x) &= x^3 + 2x^2 - 4x - 8 \qquad &\text{group terms (in color)} \\ &= x^2(x + 2) - 4(x + 2) \qquad &\text{remove common factors (note sign change)} \\ &= (x + 2)(x^2 - 4) \qquad &\text{factor common binomial} \\ &= (x + 2)(x + 2)(x - 2) \qquad &\text{factor difference of squares} \end{aligned}$$

The zeroes of P are -2, -2, and 2.

Now try Exercises 11 through 14 ▶

Note the polynomial in Example 2 has three zeroes, but the zero -2 was repeated two times. In this case we say -2 is a zero of multiplicity two, and a zero of **even multiplicity.** It is also possible for a zero to be repeated three or more times, with those repeated an odd number of times called zeroes of **odd multiplicity** [the factor $(x - 2) = (x - 2)^1$ also gives a zero of odd multiplicity]. In general, repeated factors are written in exponential form and we have

Zeroes of Multiplicity

If p is a polynomial function with degree $n \geq 1$, and $(x - c)$ occurs as a factor of p exactly m times, then c is a zero of multiplicity m.

EXAMPLE 3 ▶ **Identifying the Multiplicity of a Zero**

Factor the given function completely, writing repeated factors in exponential form. Then state the multiplicity of each zero: $P(x) = (x^2 + 8x + 16)(x^2 - x - 20)(x - 5)$

Solution ▶
$$\begin{aligned} P(x) &= (x^2 + 8x + 16)(x^2 - x - 20)(x - 5) \qquad &\text{given polynomial} \\ &= (x + 4)(x + 4)(x - 5)(x + 4)(x - 5) \qquad &\text{trinomial factoring} \\ &= (x + 4)^3(x - 5)^2 \qquad &\text{exponential form} \end{aligned}$$

For function P, -4 is a zero of multiplicity 3 (odd multiplicity), and 5 is a zero of multiplicity 2 (even multiplicity).

Now try Exercises 15 through 18 ▶

WORTHY OF NOTE

When reconstructing a polynomial P having complex zeroes, it is often more efficient to determine the irreducible quadratic factors of P separately, as shown here. For the zeroes $2 \pm \sqrt{3}i$ we have

$$\begin{aligned} x &= 2 \pm i\sqrt{3} \\ x - 2 &= \pm i\sqrt{3} \\ (x - 2)^2 &= (\pm i\sqrt{3})^2 \\ x^2 - 4x + 4 &= -3 \\ x^2 - 4x + 7 &= 0. \end{aligned}$$

The quadratic factor is $(x^2 - 4x + 7)$.

These examples help illustrate three important consequences of the linear factorization theorem. From Example 1, if the coefficients of P are real, the polynomial can be factored into linear and quadratic factors using real numbers only $[(x + 3)(x - 3)(x^2 + 1)]$, where the quadratic factors have no real zeroes. Quadratic factors of this type are said to be **irreducible.**

Corollary I: Irreducible Quadratic Factors

If p is a polynomial with real coefficients, p can be factored into a product of linear factors (which are not necessarily distinct) and irreducible quadratic factors having real coefficients.

Closely related to this corollary and our previous study of quadratic functions, complex zeroes of the irreducible factors must occur in conjugate pairs.

Corollary II: Complex Conjugates

If p is a polynomial with real coefficients, complex zeroes must occur in conjugate pairs. If $a + bi$, $b \neq 0$ is a zero, then $a - bi$ will also be a zero.

Finally, the polynomial in Example 1 has degree 4 with 4 zeroes (two real, two complex), and the polynomial in Example 2 has degree 3 with 3 zeroes (three real, one repeated). While not shown explicitly, the polynomial in Example 3 has degree 5, and there were 5 zeroes (one repeated twice, one repeated three times). This suggests our final corollary.

Corollary III: Number of Zeroes

If p is a polynomial function with degree $n \geq 1$, then p has exactly n zeroes (real or complex), where zeroes of multiplicity m are counted m times.

These corollaries help us gain valuable information about a polynomial, when only partial information is given or known.

EXAMPLE 4 ▶ Constructing a Polynomial from Its Zeroes

A polynomial P of degree 3 with real coefficients has zeroes of -1 and $2 + i\sqrt{3}$. Find the polynomial (assume $a = 1$).

Solution ▶ Using the factor theorem, two of the factors are $(x + 1)$ and $x - (2 + i\sqrt{3})$. From Corollary II, $2 - i\sqrt{3}$ must also be a zero and $x - (2 - i\sqrt{3})$ is also a factor of P. This gives

$$P(x) = (x + 1)[x - (2 + i\sqrt{3})][x - (2 - i\sqrt{3})]$$
$$= (x + 1)[(x - 2) - i\sqrt{3}][(x - 2) + i\sqrt{3}] \quad \text{associative property}$$
$$= (x + 1)[(x^2 - 4x + 4) + 3] \quad (a + bi)(a - bi) = a^2 + b^2$$
$$= (x + 1)(x^2 - 4x + 7) \quad \text{simplify}$$
$$= x^3 - 3x^2 + 3x + 7 \quad \text{result}$$

The polynomial is $P(x) = x^3 - 3x^2 + 3x + 7$, which can be verified using the remainder theorem and any of the original zeroes.

Now try Exercises 19 through 22 ▶

EXAMPLE 5 ▶ Building a Polynomial from Its Zeroes

Find a fourth degree polynomial P with real coefficients, if 3 is the only real zero and $2i$ is also a zero of P.

Solution ▶ Since complex zeroes must occur in conjugate pairs, $-2i$ is also a zero, but this accounts for only three zeroes. Since P has degree 4, 3 must be a *repeated* zero, and the factors of P are $(x - 3)(x - 3)(x - 2i)(x + 2i)$.

☑ **A.** You've just learned how to apply the fundamental theorem of algebra and the linear factorization theorem

$$P(x) = (x - 3)(x - 3)(x - 2i)(x + 2i) \quad \text{factored form}$$
$$= (x^2 - 6x + 9)(x^2 + 4) \quad \text{multiply binomials, } (a + bi)(a - bi) = a^2 + b^2$$
$$= x^4 - 6x^3 + 13x^2 - 24x + 36 \quad \text{result}$$

The polynomial is $P(x) = x^4 - 6x^3 + 13x^2 - 24x + 36$, which can be verified using the remainder theorem and any of the original zeroes.

Now try Exercises 23 through 28 ▶

B. Real Polynomials and the Intermediate Value Theorem

The fundamental theorem of algebra is called an **existence theorem,** as it affirms the *existence* of the zeroes but does not tell us where or how to find them. Because polynomial graphs are continuous (there are no holes or breaks in the graph), the **intermediate value theorem (IVT)** can be used for this purpose.

The Intermediate Value Theorem

Given P is a polynomial with real coefficients, if $P(a)$ and $P(b)$ have opposite signs, there is *at least* one value c between a and b such that $P(c) = 0$.

Figure 3.8

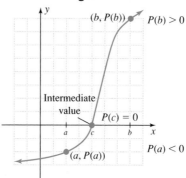

WORTHY OF NOTE

You might recall a similar idea was used in Section 2.5, where we noted the graph of $P(x)$ crosses the x-axis at the zeroes determined by linear factors, with a corresponding change of sign in the function values.

EXAMPLE 6 ▶ **Finding Zeroes Using the Intermediate Value Theorem**

Use the intermediate value theorem to show $P(x) = x^3 - 9x + 6$ has at least one zero in the interval given:

a. $[-4, -3]$ **b.** $[0, 1]$

Solution ▶ **a.** Begin by evaluating P at $x = -4$ and $x = -3$.

$$P(-4) = (-4)^3 - 9(-4) + 6 \qquad P(-3) = x^3 - 9x + 6$$
$$= -64 + 36 + 6 \qquad\qquad = -27 + 27 + 6$$
$$= -22 \qquad\qquad\qquad = 6$$

Since $P(-4) < 0$ and $P(-3) > 0$, there must be at least one number c_1 between -4 and -3 where $P(c_1) = 0$. The graph must cross the x-axis in this interval.

b. Evaluate P at $x = 0$ and $x = 1$.

$$P(0) = (0)^3 - 9(0) + 6 \qquad P(1) = (1)^3 - 9(1) + 6$$
$$= 0 - 0 + 6 \qquad\qquad = 1 - 9 + 6$$
$$= 6 \qquad\qquad\qquad = -2$$

Since $P(0) > 0$ and $P(1) < 0$, there must be at least one number c_2 between 0 and 1 where $P(c_2) = 0$.

☑ **B.** You've just learned how to locate zeroes of a real polynomial function using the intermediate value theorem

Now try Exercises 29 through 32 ▶

C. The Rational Zeroes Theorem

The fundamental theorem of algebra tells us that zeroes of a polynomial function *exist*. The intermediate value theorem tells us how to *locate* zeroes within an interval. Our next theorem gives us the information we need to actually *find* certain zeroes of a polynomial. Recall that if c is a zero of P, then $P(c) = 0$, and when $P(x)$ is divided by $x - c$ using synthetic division, the remainder is zero (from the remainder and factor theorems).

To find *divisors that give a remainder of zero,* we make the following observations. To solve $3x^2 - 11x - 20 = 0$ by factoring, a beginner might write out all possible binomial pairs where the **F**irst term in the F-O-I-L process multiplies to $3x^2$ and the **L**ast term multiplies to 20. The six possibilities are shown here:

$$(3x \quad 1)(x \quad 20) \qquad (3x \quad 20)(x \quad 1) \qquad (3x \quad 2)(x \quad 10) \qquad (3x \quad 10)(x \quad 2)$$

$$(3x \quad 4)(x \quad 5) \qquad (3x \quad 5)(x \quad 4)$$

If $3x^2 - 11x - 20$ is factorable using integers, the factors *must be somewhere in this list*. Also, the first coefficient in each binomial must be a factor of the leading coefficient, and the second coefficient must be a factor of the constant term. This means that regardless of which factored form is correct, the solution will be a rational number whose numerator comes from the factors of 20, and whose denominator comes from the factors of 3. The correct factored form is shown here, along with the solution:

$$3x^2 - 11x - 20 = 0$$
$$(3x + 4)(x - 5) = 0$$
$$3x + 4 = 0 \qquad x - 5 = 0$$

$$x = \frac{-4}{3} \quad \begin{array}{l}\leftarrow \text{from the factors of 20} \\ \leftarrow \text{from the factors of 3}\end{array} \qquad x = \frac{5}{1} \quad \begin{array}{l}\leftarrow \text{from the factors of 20} \\ \leftarrow \text{from the factors of 3}\end{array}$$

This same principle also applies to polynomials of higher degree, and these observations suggest the following theorem.

The Rational Zeroes Theorem

Given polynomial P with integer coefficients, and $\frac{p}{q}$ a rational number in lowest terms, the rational zeroes of P (if they exist) must be of the form $\frac{p}{q}$, where p is a factor of the constant term, and q is a factor of the leading coefficient.

Note that if the leading coefficient is 1, the possible rational zeroes are limited to factors of the constant term: $\frac{p}{1} = p$. If the leading coefficient is not "1" and the constant term has a large number of factors, the set of possible rational zeroes becomes rather large. To list these possibilities, it helps to begin with all factor *pairs* of the constant a_0, then divide each of these by the factors of a_n as shown in Example 7.

EXAMPLE 7 ▶ **Identifying the Possible Rational Zeroes of a Polynomial**

List all possible rational zeroes for $3x^4 + 14x^3 - x^2 - 42x - 24 = 0$, but do not solve.

Solution ▶ All rational zeroes must be of the form $\frac{p}{q}$, where p is a factor of $a_0 = -24$ and q is a factor of $a_n = 3$. The factor pairs of -24 are: $\pm 1, \pm 24, \pm 2, \pm 12, \pm 3, \pm 8, \pm 4$ and ± 6. Dividing each by ± 1 and ± 3 (the factor pairs of 3), we note division by ± 1 will not change any of the previous values, while division by ± 3 gives $\pm\frac{1}{3}, \pm\frac{2}{3}, \pm\frac{8}{3}, \pm\frac{4}{3}$ as additional possibilities. Any rational zeroes must be in the set $\{\pm 1, \pm 24, \pm 2, \pm 12, \pm 3, \pm 8, \pm 4, \pm 6, \pm\frac{1}{3}, \pm\frac{2}{3}, \pm\frac{8}{3}, \pm\frac{4}{3}\}$.

Now try Exercises 33 through 40 ▶

WORTHY OF NOTE

To test for -1, only the sign of terms with odd degree must be changed, since $(-1)^{\text{even\#}} = 1$, while $(-1)^{\text{odd\#}} = -1$. The method simply gives a shortcut for evaluating $P(1)$ and $P(-1)$, which often helps to break down a higher degree polynomial.

The actual solutions to the equation in Example 7 are $x = \sqrt{3}, x = -\sqrt{3}, x = -\frac{2}{3}$, and $x = -4$. Although the *rational* zeroes are indeed in the set noted, it's apparent we need a way to narrow down the number of possibilities (we don't want to try all 24 possible zeroes). If we're able to find even one factor easily, we can rewrite the polynomial using this factor and the quotient polynomial, with the hope of factoring further using trinomial factoring or factoring by grouping. Many times testing to see if 1 or -1 are zeroes will help.

Tests to Determine If 1 or −1 is a Zero of P

For any polynomial P with real coefficients,

1. *If the sum of all coefficients is zero, then 1 is a root and $(x - 1)$ is a factor.*
2. *After changing the sign of all terms with odd degree, if the sum of the coefficients is zero, then -1 is a root and $(x + 1)$ is a factor.*

EXAMPLE 8 ▶ **Finding the Rational Zeroes of a Polynomial**

Find all rational zeroes of $P(x) = 3x^4 - x^3 - 8x^2 + 2x + 4$, and use them to write the function in completely factored form. Then use the factored form to name all zeroes of P.

Solution ▶ Instead of listing all possibilities using the rational zeroes theorem, we first test for 1 and -1, then see if we're able to complete the factorization using other means. The sum of the coefficients is: $3 - 1 - 8 + 2 + 4 = 0$, which means 1 is a zero and $x - 1$ is a factor. By changing the sign on terms of odd degree, we have $3x^4 + x^3 - 8x^2 - 2x + 4$ and $3 + 1 - 8 - 2 + 4 = -2$, showing -1 is *not* a zero. Using $x = 1$ and the factor theorem, we have

use 1 as a "divisor"
$$\begin{array}{r|rrrrr} 1 & 3 & -1 & -8 & 2 & 4 \\ & & 3 & 2 & -6 & -4 \\ \hline & 3 & 2 & -6 & -4 & \underline{|0} \end{array}$$

and we write P as $P(x) = (x - 1)(\mathbf{3x^3 + 2x^2 - 6x - 4})$. Noting the quotient polynomial can be factored by grouping ($ad = bc$), we need not continue with synthetic division or the factor theorem.

$$\begin{aligned} P(x) &= (x - 1)(\underline{3x^3 + 2x^2} - \underline{6x - 4}) && \text{group terms} \\ &= (x - 1)[x^2(3x + 2) - 2(3x + 2)] && \text{factor common terms} \\ &= (x - 1)(3x + 2)(x^2 - 2) && \text{factor common binomial} \\ &= (x - 1)(3x + 2)(x + \sqrt{2})(x - \sqrt{2}) && \text{completely factored form} \end{aligned}$$

The zeroes of P are $1, \frac{-2}{3}$, and $\pm\sqrt{2}$.

Now try Exercises 41 through 62 ▶

WORTHY OF NOTE

In the second to last line of Example 8, we factored $x^2 - 2$ as $(x + \sqrt{2})(x - \sqrt{2})$. As discussed in Section R.4, this is an application of factoring the difference of two squares: $a^2 - b^2 = (a + b)(a - b)$. By mentally rewriting $x^2 - 2$ as $x^2 - (\sqrt{2})^2$, we obtain the result shown. Also **see Exercise 113.**

In cases where the quotient polynomial is not easily factored, we continue with synthetic division and other possible zeroes, until the remaining zeroes can be determined.

EXAMPLE 9 ▶ **Finding the Zeroes of a Polynomial**

Find all zeroes of $P(x) = x^5 - 3x^4 + 3x^3 - 5x^2 + 12$.

Solution ▶ Using the rational zeroes theorem, the possibilities are: $\{\pm1, \pm12, \pm2, \pm6, \pm3, \pm4\}$. The test for 1 shows 1 is not a zero. After changing the signs of all terms with odd degree, we have $-1 - 3 - 3 - 5 + 12 = 0$, and find -1 *is* a zero. Using -1 with the factor theorem, we continue our search for additional factors. Noting that P is missing a linear term, we include a place-holder zero:

use -1 as a "divisor"
$$\begin{array}{r|rrrrrr} -1 & 1 & -3 & 3 & -5 & 0 & 12 \\ & & -1 & 4 & -7 & 12 & -12 \\ \hline & 1 & -4 & 7 & -12 & 12 & \underline{|0} \end{array}$$
coefficients of P

coefficients of $q_1(x)$

Here the quotient polynomial $q_1(x) = x^4 - 4x^3 + 7x^2 - 12x + 12$ is not easily factored, so we next try 2, *using the quotient polynomial:*

use 2 as a "divisor" on $q_1(x)$
$$\begin{array}{r|rrrrr} 2 & 1 & -4 & 7 & -12 & 12 \\ & & 2 & -4 & 6 & -12 \\ \hline & 1 & -2 & 3 & -6 & \underline{|0} \end{array}$$
coefficients of $q_1(x)$

coefficients of $q_2(x)$

If you miss the fact that $q_2(x)$ is actually factorable ($ad = bc$), the process would continue using -2 and the current quotient.

use -2 as a "divisor"
$$\begin{array}{r|rrrr} -2 & 1 & -2 & 3 & -6 \\ & & -2 & 8 & -22 \\ \hline & 1 & -4 & 11 & \underline{|-28} \end{array}$$
coefficients of $q_2(x)$

-2 is not a zero

We find -2 is not a zero, and in fact, trying *all other possible zeroes* will show that *none* of them are zeroes. As there must be five zeroes, we are reminded of three things:

1. This process can only find *rational zeros* (the remaining zeroes may be irrational or complex),
2. This process cannot find irreducible quadratic factors (unless they appear as the quotient polynomial), and
3. Some of the zeroes *may have multiplicities greater than 1!*

Testing the zero 2 for a second time using $q_2(x)$ gives

$$\text{use 2 as a "divisor"} \quad \begin{array}{r} 2\rfloor \ \ 1 \quad -2 \quad 3 \quad -6 \\ \quad 2 \quad 0 \quad 6 \\ \hline 1 \quad \ \ 0 \quad 3 \quad \underline{|0} \end{array} \quad \text{coefficients of } q_2(x)$$

2 is a *repeated* zero

✓ **C.** You've just learned how to find rational zeroes of a real polynomial function using the rational zeroes theorem

and we see that 2 is actually a zero of multiplicity two, and the final quotient is the irreducible quadratic factor $x^2 + 3$. Using this information produces the factored form $P(x) = (x + 1)(x - 2)^2(x^2 + 3) = (x + 1)(x - 2)^2(x + i\sqrt{3})(x - i\sqrt{3})$, and the zeroes of P are $-i\sqrt{3}, i\sqrt{3}, -1$, and 2 with multiplicity two.

Now try Exercises 63 through 82 ▶

D. Descartes' Rule of Signs and Upper/Lower Bounds

Testing $x = 1$ and $x = -1$ is one way to reduce the number of possible rational zeroes, but unless we're very lucky, factoring the polynomial can still be a challenge. **Descartes' rule of signs** and the **upper and lower bounds property** offer additional assistance.

Descartes' Rule of Signs

Given the real polynomial equation $P(x) = 0$,

1. The number of positive real zeroes is equal to the number of variations in sign for $P(x)$, or an even number less.
2. The number of negative real zeroes is equal to the number of variations in sign for $P(-x)$, or an even number less.

EXAMPLE 10 ▶ **Finding the Zeroes of a Polynomial**

For $P(x) = 2x^5 - 5x^4 + x^3 + x^2 - x + 6$,

a. Use the rational zeroes theorem to list all possible rational zeroes.
b. Apply Descartes' rule to count the number of possible positive, negative, and complex roots.
c. Use this information and the tools of this section to find all zeroes of P.

Solution ▶ a. The factors of 2 are $\{\pm 1, \pm 2\}$ and the factors of 6 are $\{\pm 1, \pm 6, \pm 2, \pm 3\}$. The possible rational zeroes for P are $\{\pm 1, \pm 6, \pm 2, \pm 3, \pm\frac{1}{2}, \pm\frac{3}{2}\}$.

b. For Descartes' rule, we organize our work in a table. Since P has degree 5, there must be a total of five zeroes. For this illustration, positive terms are in **blue** and negative terms in **red**: $P(x) = 2x^5 - 5x^4 + x^3 + x^2 - x + 6$. The terms change sign a total of four times, meaning there are four, two, or zero positive roots. For the negative roots, recall that $P(-x)$ will change the sign of *all odd-degree terms*, giving $P(-x) = -2x^5 - 5x^4 - x^3 + x^2 + x + 6$. This time there is only one sign change (from negative to positive) showing there is exactly one negative root, a fact that is highlighted in the following table.

possible positive zeroes	known negative zeroes	possibilities for complex roots	total number must be 5
4	1	0	5
2	1	2	5
0	1	4	5

c. Testing 1 and -1 shows $x = 1$ is not a root, but $x = -1$ *is*, and using -1 in synthetic division gives:

use -1 as a "divisor" -1⌋ 2 -5 1   1 -1 6   coefficients of $P(x)$
                                    -2 7 -8 7 -6
                          ───────────────────────────────────
                              2   -7 8 -7 6   0 $q_1(x)$ is not easily factored

Since there is *only one* negative root, we need only check the remaining positive zeroes. The quotient $q_1(x)$ is not easily factored, so we continue with synthetic division using the next larger positive root, $x = 2$.

use 2 as a "divisor"   2⌋ 2 -7 8 -7 6   coefficients of $q_1(x)$
                                 4   -6 4 -6
                       ─────────────────────────────────
                           2   -3 2 -3 0 $q_2(x)$ is easily factored

The partially factored form is $P(x) = (x + 1)(x - 2)(2x^3 - 3x^2 + 2x - 3)$, which we can complete using factoring by grouping. The factored form is

$$
\begin{aligned}
P(x) &= (x + 1)(x - 2)(\underline{2x^3 - 3x^2} + \underline{2x - 3}) && \text{group terms} \\
&= (x + 1)(x - 2)[x^2(2x - 3) + 1(2x - 3)] && \text{factor common terms} \\
&= (x + 1)(x - 2)(2x - 3)(x^2 + 1) && \text{factor out common binomial} \\
&= (x + 1)(x - 2)(2x - 3)(x + i)(x - i) && \text{completely factored form}
\end{aligned}
$$

The zeroes of P are -1, 2, $\frac{3}{2}$, $-i$ and i, with two positive, one negative, and two complex zeroes.

Now try Exercises 83 through 96 ▶

One final idea that helps reduce the number of possible zeroes is the **upper and lower bounds property.** A number b is an **upper bound** on the positive zeroes of a function if no positive zero is greater than b. In the same way, a number a is a **lower bound** on the negative zeroes if no negative zero is less than a.

Upper and Lower Bounds Property

Given $P(x)$ is a polynomial with real coefficients.

1. If $P(x)$ is divided by $x - b$ ($b > 0$) using synthetic division and all coefficients in the quotient row are either positive or zero, then b is an upper bound on the zeroes of P.
2. If $P(x)$ is divided by $x - a$ ($a < 0$) using synthetic division and all coefficients in the quotient row alternate in sign, then a is a lower bound on the zeroes of P.

For both 1 and 2, zero coefficients can be either positive or negative as needed.

☑ **D.** You just learned how to gain more information on the zeroes of real polynomials using Descartes' rule of signs and upper/lower bounds

While this test certainly helps narrow the possibilities, we gain the additional benefit of knowing the property actually places boundaries on *all* real zeroes of the polynomial, both rational and irrational. In Part (c) of Example 10, the quotient row of the first division alternates in sign, showing $x = -1$ is both a zero and a lower bound on the real zeroes of P. For more on the upper and lower bounds property, **see Exercise 111.**

E. Applications of Polynomial Functions

Polynomial functions can be very accurate models of real-world phenomena, though we often must restrict their domain, as illustrated in Example 11.

EXAMPLE 11 ▶ **Using the Remainder Theorem to Solve an Oceanography Application**

As part of an environmental study, scientists use radar to map the ocean floor from the coastline to a distance 12 mi from shore. In this study, ocean trenches appear as negative values and underwater mountains as positive values, as measured from the surrounding ocean floor. The terrain due west of a particular island can be modeled by $h(x) = x^4 - 25x^3 + 200x^2 - 560x + 384$, where $h(x)$ represents the height in feet, x mi from shore ($0 < x \leq 12$).

  a. Use the remainder theorem to find the "height of the ocean floor" 10 mi out.

  b. Use the tools developed in this section to find the number of times the ocean floor has height $h(x) = 0$ in this interval, given this occurs 12 mi out.

Solution ▶   **a.** For part (a) we simply evaluate $h(10)$ using the remainder theorem.

use 10 as a "divisor"

$$
\begin{array}{r|rrrrr}
10] & 1 & -25 & 200 & -560 & 384 \\
   &   & 10 & -150 & 500 & -600 \\
\hline
   & 1 & -15 & 50 & -60 & \underline{|-216} \\
\end{array}
$$

coefficients of $h(x)$

remainder is -216

Ten miles from shore, there is an ocean trench 216 ft deep.

  b. For part (b), we know 12 is zero, so we again use the remainder theorem and work with the quotient polynomial.

use 12 as a "divisor"

$$
\begin{array}{r|rrrrr}
12] & 1 & -25 & 200 & -560 & 384 \\
   &   & 12 & -156 & 528 & -384 \\
\hline
   & 1 & -13 & 44 & -32 & |0 \quad q_1(x) \\
\end{array}
$$

coefficients of $h(x)$

The quotient is $q_1(x) = x^3 - 13x^2 + 44x - 32$. Since $a = 1$, we know the remaining zeroes must be factors of -32: $\{\pm 1, \pm 32, \pm 2, \pm 16, \pm 4, \pm 8\}$. Using $x = 1$ gives

use 1 as a "divisor"

$$
\begin{array}{r|rrrr}
1] & 1 & -13 & 44 & -32 \\
  &   & 1 & -12 & 32 \\
\hline
  & 1 & -12 & 32 & |0 \quad q_2(x) \\
\end{array}
$$

coefficients of $q_1(x)$

☑ **E. You've just learned how to solve an application of polynomial functions**

The function can now be written as $h(x) = (x - 12)(x - 1)(x^2 - 12x + 32)$ and in completely factored form $h(x) = (x - 12)(x - 1)(x - 4)(x - 8)$. The ocean floor has height zero at distances of 1, 4, 8, and 12 mi from shore.

Now try Exercises 99 through 110 ▶

GRAPHICAL SUPPORT

The graph of $h(x)$ is shown here using a window size of $X \in [0, 13]$ and $Y \in [-450, 450]$. The graphs shows a great deal of variation in the ocean floor, but the zeroes occurring at 1, 4, 8, and 12 mi out are clearly evident.

TECHNOLOGY HIGHLIGHT

The Intermediate Value Theorem and Split Screen Viewing

Graphical support for the results of Example 6 is shown in Figure 3.9 using the window $x \in [-5, 5]$ and $y \in [-10, 20]$. The zero of P between 0 and 1 is highlighted, and the zero between $x = -4$ and $x = -3$ is clearly seen. Note there is also a third zero between 2 and 3.

The TI 84 Plus (and other models) offer a useful feature called *split screen viewing*, that enables us to view a table of values and the graph of a function at the same time. To illustrate, enter the function $y = x^3 - 9x + 6$ for Y_1 on the Y= screen. Press the ZOOM **4:ZDecimal** keys to view the graph, then adjust the viewing window as needed to get a comprehensive view. Set up your table in **AUTO** mode with ΔTbl = 1 [use 2nd WINDOW (TBLSET)]. Use the table of values (2nd GRAPH) to locate any real zeroes of f [look for where $f(x)$ changes in sign]. To support this concept we can view *both the graph and table at the same time.* Press the MODE key and notice the second-to-last entry on this screen reads: **Full** (for full screen viewing), **Horiz** for splitting the screen horizontally with the graph above a reduced home screen, and **G-T**, which represents **Graph-Table** and splits the screen vertically. In the **G-T** mode, the graph appears on the left and the table of values on the right. Navigate the cursor to the **G-T** mode and press ENTER . Pressing the GRAPH key at this point should give you a screen similar to Figure 3.10. Use this feature to complete the following exercises.

Figure 3.9

Figure 3.10

Exercise 1: What do the graph, table, and the IVT tell you about the zeroes of this function?

Exercise 2: Go to TBLSET and reset TblStart = −4 and ΔTbl = 0.1. Use 2nd GRAPH to walk through the table values. Does this give you a better idea about where the zeroes are located?

Exercise 3: Press the TRACE key. What happens to the table as you trace through the points on Y_1?

3.3 EXERCISES

▶ CONCEPTS AND VOCABULARY

Fill in each blank with the appropriate word or phrase. Carefully reread the section if needed.

1. A complex polynomial is one where one or more _____ are complex numbers.

2. A polynomial function of degree n will have exactly _____ zeroes, real or _____, where zeroes of multiplicity m are counted m times.

3. If $a + bi$ is a complex zero of polynomial P with real coefficients, then _____ is also a zero.

4. According to Descartes' rule of signs, there are as many _____ real roots as changes in sign from term to term, or an _____ number less.

5. Which of the following values is *not* a possible root of $f(x) = 6x^3 - 2x^2 + 5x - 12$:
   a. $x = \frac{4}{3}$ b. $x = \frac{3}{4}$ c. $x = \frac{1}{2}$

   Discuss/Explain why.

6. Discuss/Explain each of the following:
   (a) irreducible quadratic factors, (b) factors that are complex conjugates, (c) zeroes of multiplicity m, and (d) upper bounds on the zeroes of a polynomial.

▶ **DEVELOPING YOUR SKILLS**

Rewrite each polynomial as a product of linear factors, and find the zeroes of the polynomial.

7. $P(x) = x^4 + 5x^2 - 36$

8. $Q(x) = x^4 + 21x^2 - 100$

9. $Q(x) = x^4 - 16$

10. $P(x) = x^4 - 81$

11. $P(x) = x^3 + x^2 - x - 1$

12. $Q(x) = x^3 - 3x^2 - 9x + 27$

13. $Q(x) = x^3 - 5x^2 - 25x + 125$

14. $P(x) = x^3 + 4x^2 - 16x - 64$

Factor each polynomial completely. Write any repeated factors in exponential form, then name all zeroes and their multiplicity.

15. $p(x) = (x^2 - 10x + 25)(x^2 + 4x - 45)(x + 9)$

16. $q(x) = (x^2 + 12x + 36)(x^2 + 2x - 24)(x - 4)$

17. $P(x) = (x^2 - 5x - 14)(x^2 - 49)(x + 2)$

18. $Q(x) = (x^2 - 9x + 18)(x^2 - 36)(x - 3)$

Find a polynomial $P(x)$ having real coefficients, with the degree and zeroes indicated. Assume the lead coefficient is 1. Recall $(a + bi)(a - bi) = a^2 + b^2$.

19. degree 3, $x = 3, x = 2i$

20. degree 3, $x = -5, x = -3i$

21. degree 4, $x = -1, x = 2, x = i$

22. degree 4, $x = -1, x = 3, x = -2i$

23. degree 4, $x = 3, x = 2i$

24. degree 4, $x = -2, x = -3i$

25. degree 4, $x = -1, x = 1 + 2i$

26. degree 4, $x = -1, x = 1 - 3i$

27. degree 4, $x = -3, x = 1 + i\sqrt{2}$

28. degree 4, $x = -2, x = 1 + i\sqrt{3}$

Use the intermediate value theorem to verify the given polynomial has at least one zero "c_i" in the intervals specified. Do not find the zeroes.

29. $f(x) = x^3 + 2x^2 - 8x - 5$
   a. $[-4, -3]$ b. $[2, 3]$

30. $g(x) = x^4 - 2x^2 + 6x - 3$
   a. $[-3, -2]$ b. $[0, 1]$

31. $h(x) = 2x^3 + 13x^2 + 3x - 36$
   a. $[1, 2]$ b. $[-3, -2]$

32. $H(x) = 2x^4 + 3x^3 - 14x^2 - 9x + 8$
   a. $[-4, -3]$ b. $[-2, -1]$

List all possible rational zeroes for the polynomials given, but do not solve.

33. $f(x) = 4x^3 - 19x - 15$

34. $g(x) = 3x^3 - 2x + 20$

35. $h(x) = 2x^3 - 5x^2 - 28x + 15$

36. $H(x) = 2x^3 - 19x^2 + 37x - 14$

37. $p(x) = 6x^4 - 2x^3 + 5x^2 - 28$

38. $q(x) = 7x^4 + 6x^3 - 49x^2 + 36$

39. $Y_1 = 32t^3 - 52t^2 + 17t + 3$

40. $Y_2 = 24t^3 + 17t^2 - 13t - 6$

Use the rational zeroes theorem to write each function in factored form and find all zeroes. Note $a = 1$.

41. $f(x) = x^3 - 13x + 12$

42. $g(x) = x^3 - 21x + 20$

43. $h(x) = x^3 - 19x - 30$

44. $H(x) = x^3 - 28x - 48$

45. $p(x) = x^3 - 2x^2 - 11x + 12$

46. $q(x) = x^3 - 4x^2 - 7x + 10$

47. $Y_1 = x^3 - 6x^2 - x + 30$

48. $Y_2 = x^3 - 4x^2 - 20x + 48$

49. $Y_3 = x^4 - 15x^2 + 10x + 24$

50. $Y_4 = x^4 - 23x^2 - 18x + 40$

51. $f(x) = x^4 + 7x^3 - 7x^2 - 55x - 42$

52. $g(x) = x^4 + 4x^3 - 17x^2 - 24x + 36$

Find all rational zeroes of the functions given and use them to write the function in factored form. Use the factored form to state *all* zeroes of *f*. Begin by applying the tests for 1 and −1.

53. $f(x) = 4x^3 - 7x + 3$

54. $g(x) = 9x^3 - 7x - 2$

55. $h(x) = 4x^3 + 8x^2 - 3x - 9$

56. $H(x) = 9x^3 + 3x^2 - 8x - 4$

57. $Y_1 = 2x^3 - 3x^2 - 9x + 10$

58. $Y_2 = 3x^3 - 14x^2 + 17x - 6$

59. $p(x) = 2x^4 + 3x^3 - 9x^2 - 15x - 5$

60. $q(x) = 3x^4 + x^3 - 11x^2 - 3x + 6$

61. $r(x) = 3x^4 - 5x^3 + 14x^2 - 20x + 8$

62. $s(x) = 2x^4 - x^3 + 17x^2 - 9x - 9$

Find the zeroes of the polynomials given using any combination of the rational zeroes theorem, testing for 1 and −1, and/or the remainder and factor theorems.

63. $f(x) = 2x^4 - 9x^3 + 4x^2 + 21x - 18$

64. $g(x) = 3x^4 + 4x^3 - 21x^2 - 10x + 24$

65. $h(x) = 3x^4 + 2x^3 - 9x^2 + 4$

66. $H(x) = 7x^4 + 6x^3 - 49x^2 + 36$

67. $p(x) = 2x^4 + 3x^3 - 24x^2 - 68x - 48$

68. $q(x) = 3x^4 - 19x^3 + 6x^2 + 96x - 32$

69. $r(x) = 3x^4 - 20x^3 + 34x^2 + 12x - 45$

70. $s(x) = 4x^4 - 15x^3 + 9x^2 + 16x - 12$

71. $Y_1 = x^5 + 6x^2 - 49x + 42$

72. $Y_2 = x^5 + 2x^2 - 9x + 6$

73. $P(x) = 3x^5 + x^4 + x^3 + 7x^2 - 24x + 12$

74. $P(x) = 2x^5 - x^4 - 3x^3 + 4x^2 - 14x + 12$

75. $Y_1 = x^4 - 5x^3 + 20x - 16$

76. $Y_2 = x^4 - 10x^3 + 90x - 81$

77. $r(x) = x^4 + 2x^3 - 5x^2 - 4x + 6$

78. $s(x) = x^4 + x^3 - 5x^2 - 3x + 6$

79. $p(x) = 2x^4 - x^3 + 3x^2 - 3x - 9$

80. $q(x) = 3x^4 + x^3 + 13x^2 + 5x - 10$

81. $f(x) = 2x^5 - 7x^4 + 13x^3 - 23x^2 + 21x - 6$

82. $g(x) = 4x^5 + 3x^4 + 3x^3 + 11x^2 - 27x + 6$

Gather information on each polynomial using (a) the rational zeroes theorem, (b) testing for 1 and −1, (c) applying Descartes' rule of signs, and (d) using the upper and lower bounds property. Respond explicitly to each.

83. $f(x) = x^4 - 2x^3 + 4x - 8$

84. $g(x) = x^4 + 3x^3 - 7x - 6$

85. $h(x) = x^5 + x^4 - 3x^3 + 5x + 2$

86. $H(x) = x^5 + x^4 - 2x^3 + 4x - 4$

87. $p(x) = x^5 - 3x^4 + 3x^3 - 9x^2 - 4x + 12$

88. $q(x) = x^5 - 2x^4 - 8x^3 + 16x^2 + 7x - 14$

89. $r(x) = 2x^4 + 7x^2 + 11x - 20$

90. $s(x) = 3x^4 - 8x^3 - 13x - 24$

Use Descartes' rule of signs to determine the possible combinations of real and complex zeroes for each polynomial. Then graph the function on the standard window of a graphing calculator and adjust it as needed until you're certain all real zeroes are in clear view. Use this screen and a list of the possible rational zeroes to factor the polynomial and find all zeroes (real and complex).

91. $f(x) = 4x^3 - 16x^2 - 9x + 36$

92. $g(x) = 6x^3 - 41x^2 + 26x + 24$

93. $h(x) = 6x^3 - 73x^2 + 10x + 24$

94. $H(x) = 4x^3 + 60x^2 + 53x - 42$

95. $p(x) = 4x^4 + 40x^3 - 97x^2 - 10x + 24$

96. $q(x) = 4x^4 - 42x^3 - 70x^2 - 21x - 36$

▶ **WORKING WITH FORMULAS**

97. **The absolute value of a complex number**
$z = a + bi$: $|z| = \sqrt{a^2 + b^2}$

The absolute value of a complex number z, denoted $|z|$, represents the distance between the origin and the point (a, b) in the complex plane. Use the formula to find $|z|$ for the complex numbers given (also see Section 1.4, Exercise 69): (a) $3 + 4i$, (b) $-5 + 12i$, and (c) $1 + \sqrt{3}\,i$.

98. **The square root of $z = a + bi$:**
$\sqrt{z} = \frac{\sqrt{2}}{2}\left(\sqrt{|z| + a} \pm i\sqrt{|z| - a}\right)$

The square roots of a complex number are given by the relations shown, where $|z|$ represents the absolute value of z and the sign is chosen to match the sign of b. Use the formula to find the square root of each complex number from Exercise 97, then check your answer by squaring the result (also see Section 1.4, Exercise 82).

▶ APPLICATIONS

99. Maximum and minimum values: To locate the maximum and minimum values of $F(x) = x^4 - 4x^3 - 12x^2 + 32x + 15$ requires finding the zeroes of $f(x) = 4x^3 - 12x^2 - 24x + 32$. Use the rational zeroes theorem and synthetic division to find the zeroes of f, then graph $F(x)$ on a calculator and see if the graph tends to support your calculations—do the maximum and minimum values occur at the zeroes of f?

100. Graphical analysis: Use the rational zeroes theorem and synthetic division to find the zeroes of $F(x) = x^4 - 4x^3 - 12x^2 + 32x + 15$ (see Exercise 99).

101. Maximum and minimum values: To locate the maximum and minimum values of $G(x) = x^4 - 6x^3 + x^2 + 24x - 20$ requires finding the zeroes of $g(x) = 4x^3 - 18x^2 + 2x + 24$. Use the rational zeroes theorem and synthetic division to find the zeroes of g, then graph $G(x)$ on a calculator and see if the graph tends to support your calculations—do the maximum and minimum values occur at the zeroes of g?

102. Graphical analysis: Use the rational zeroes theorem and synthetic division to find the zeroes of $G(x) = x^4 - 6x^3 + x^2 + 24x - 20$ (see Exercise 101).

Geometry: The volume of a cube is $V = x \cdot x \cdot x = x^3$, where x represents the length of the edges. If a slice 1 unit thick is removed from the cube, the remaining volume is $v = x \cdot x \cdot (x - 1) = x^3 - x^2$. Use this information for Exercises 103 and 104.

103. A slice 1 unit in thickness is removed from one side of a cube. Use the rational zeroes theorem and synthetic division to find the original dimensions of the cube, if the remaining volume is (a) 48 cm³ and (b) 100 cm³.

104. A slice 1 unit in thickness is removed from one side of a cube, then a second slice of the same thickness is removed from a different side (not the opposite side). Use the rational zeroes theorem and synthetic division to find the original dimensions of the cube, if the remaining volume is (a) 36 cm³ and (b) 80 cm³.

Geometry: The volume of a rectangular box is $V = LWH$. For the box to satisfy certain requirements, its length must be twice the width, and its height must be two inches less than the width. Use this information for Exercises 105 and 106.

105. Use the rational zeroes theorem and synthetic division to find the dimensions of the box if it must have a volume of 150 in³.

106. Suppose the box must have a volume of 64 in³. Use the rational zeroes theorem and synthetic division to find the dimensions required.

Government deficits: Over a 14-yr period, the balance of payments (deficit versus surplus) for a certain county government was modeled by the function $f(x) = \frac{1}{4}x^4 - 6x^3 + 42x^2 - 72x - 64$, where $x = 0$ corresponds to 1990 and $f(x)$ is the deficit or surplus in tens of thousands of dollars. Use this information for Exercises 107 and 108.

107. Use the rational zeroes theorem and synthetic division to find the years when the county "broke even" (debt = surplus = 0) from 1990 to 2004. How many years did the county run a surplus during this period?

108. The deficit was at the \$84,000 level $[f(x) = -84]$, four times from 1990 to 2004. Given this occurred in 1992 and 2000 ($x = 2$ and $x = 10$), use the rational zeroes theorem, synthetic division, and the remainder theorem to find the other two years the deficit was at \$84,000.

109. Drag resistance on a boat: In a scientific study on the effects of drag against the hull of a sculling boat, some of the factors to consider are displacement, draft, speed, hull shape, and length, among others. If the first four are held constant and we assume a flat, calm water surface, length becomes the sole variable (as length changes, we adjust the beam by a uniform scaling to keep a constant displacement). For a fixed sculling speed of 5.5 knots, the relationship between drag and length can be modeled by $f(x) = -0.4192x^4 + 18.9663x^3 - 319.9714x^2 + 2384.2x - 6615.8$, where $f(x)$ is the efficiency rating of a boat with length x ($8.7 < x < 13.6$). Here, $f(x) = 0$ represents an *average* efficiency rating. (a) Under these conditions, what lengths (to the nearest hundredth) will give the boat an average rating? (b) What length will maximize the efficiency of the boat? What is this rating?

110. Comparing densities: Why is it that when you throw a rock into a lake, it sinks, while a wooden ball will float half submerged, but the bobber on your fishing line floats on the surface? It all depends on the density of the object compared to the density of water ($d = 1$). For uniformity, we'll consider spherical objects of various densities, each with a radius of 5 cm. When placed into water, the depth that the sphere will sink beneath the surface (while still floating) is modeled by the polynomial $p(x) = \frac{\pi}{3}x^3 - 5\pi x^2 + \frac{500\pi}{3}d$, where d is the density of the object and the smallest positive zero of p is the depth of the sphere below the surface (in centimeters). How far submerged is the sphere if it's made of (a) balsa wood, $d = 0.17$;

(b) pine wood, $d = 0.55$; (c) ebony wood, $d = 1.12$; (d) a large bobber made of lightweight plastic, $d = 0.05$?

▶ EXTENDING THE CONCEPT

111. In the figure, $P(x) = 0.02x^3 - 0.24x^2 - 1.04x + 2.68$ is graphed on the standard screen ($-10 \le x \le 10$), which shows two real zeroes. Since P has degree 3, there must be one more real zero but is it negative or positive? Use the upper/lower bounds property (a) to see if -10 is a lower bound and (b) to see if 10 is an upper bound. (c) Then use your calculator to find the remaining zero.

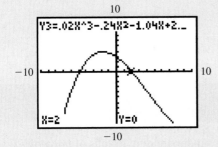

112. From Example 11, (a) what is the significance of the y-intercept? (b) If the domain were extended to include $0 < x \le 13$, what happens when x is approximately 12.8?

113A. It is often said that while the difference of two squares is factorable, $a^2 - b^2 = (a + b)(a - b)$, the sum of two squares is prime. To be 100% correct, we should say the sum of two squares cannot be factored *using real numbers*. If complex numbers are used, $(a^2 + b^2) = (a + bi)(a - bi)$. Use this idea to factor the following binomials.
  a. $p(x) = x^2 + 25$ **b.** $q(x) = x^2 + 9$
  c. $r(x) = x^2 + 7$

113B. It is often said that while $x^2 - 16$ is factorable as a difference of squares, $a^2 - b^2 = (a + b)(a - b)$, $x^2 - 17$ is not. To be 100% correct, we should say that $x^2 - 17$ is not factorable *using integers*. Since $(\sqrt{17})^2 = 17$, it can actually be factored in the same way: $x^2 - 17 = (x + \sqrt{17})(x - \sqrt{17})$. Use this idea to solve the following equations.
  a. $x^2 - 7 = 0$ **b.** $x^2 - 12 = 0$ **c.** $x^2 - 18 = 0$

114. Every general cubic equation $aw^3 + bw^2 + cw + d = 0$ can be written in the form $x^3 + px + q = 0$ (where the squared term has been "depressed"), using the transformation $w = x - \frac{b}{3}$. Use this transformation to solve the following equations.

  a. $w^3 - 3w^2 + 6w - 4 = 0$
  b. $w^3 - 6w^2 + 21w - 26 = 0$

Note: It is actually very rare that the transformation produces a value of $q = 0$ for the "depressed" cubic $x^3 + px + q = 0$, and general solutions must be found using what has become known as *Cardano's formula*. For a complete treatment of cubic equations and their solutions, visit our website at www.mhhe.com/coburn. Here we'll focus on the primary root of selected cubics.

115. For each of the following complex polynomials, one of its zeroes is given. Use this zero to help write the polynomial in completely factored form. (*Hint:* Synthetic division and the quadratic formula can be applied to *all polynomials,* even those with complex coefficients.)

a. $C(z) = z^3 + (1 - 4i)z^2 + (-6 - 4i)z + 24i$;
$z = 4i$

b. $C(z) = z^3 + (5 - 9i)z^2 + (4 - 45i)z - 36i$;
$z = 9i$

c. $C(z) = z^3 + (-2 - 3i)z^2 + (5 + 6i)z - 15i$;
$z = 3i$

d. $C(z) = z^3 + (-4 - i)z^2 + (29 + 4i)z - 29i$;
$z = i$

e. $C(z) = z^3 + (-2 - 6i)z^2 + (4 + 12i)z - 24i$;
$z = 6i$

f. $C(z) = z^3 + (-6 + 4i)z^2 + (11 - 24i)z + 44i$;
$z = -4i$

g. $C(z) = z^3 + (-2 - i)z^2 + (5 + 4i)z + (-6 + 3i)$;
$z = 2 - i$

h. $C(z) = z^3 - 2z^2 + (19 + 6i)z + (-20 + 30i)$;
$z = 2 - 3i$

▶ **MAINTAINING YOUR SKILLS**

116. (2.6) Graph the piecewise-defined function and
find the value of $f(-3), f(2),$ and $f(5)$.

$$f(x) = \begin{cases} 2 & x \le -1 \\ |x - 1| & -1 < x < 5 \\ 4 & x \ge 5 \end{cases}$$

117. (3.1) For a county fair, officials need to fence off a
large rectangular area, then subdivide it into three
equal (rectangular) areas. If the county provides
1200 ft of fencing, (a) what dimensions will
maximize the area of the larger (outer) rectangle?
(b) What is the area of each smaller rectangle?

118. (2.7) Use the graph given to
(a) state intervals where
$f(x) \ge 0$, (b) locate local
maximum and minimum
values, and (c) state intervals
where $f(x)\uparrow$ and $f(x)\downarrow$.

119. (2.5) Write the equation of the function shown.

3.4 | Graphing Polynomial Functions

Learning Objectives

In Section 3.4 you will learn how to:

☐ **A.** Identify the graph of a
polynomial function and
determine its degree

☐ **B.** Describe the end
behavior of a polynomial
graph

☐ **C.** Discuss the attributes of
a polynomial graph with
zeroes of multiplicity

☐ **D.** Graph polynomial
functions in standard
form

☐ **E.** Solve applications of
polynomials

As with linear and quadratic functions, understanding graphs of *polynomial* functions
will help us apply them more effectively as mathematical models. Since all real polynomials can be written in terms of their linear and quadratic factors (Section 3.3), these
functions provide the basis for our continuing study.

A. Identifying the Graph of a Polynomial Function

Consider the graphs of $f(x) = x + 2$ and $g(x) = (x - 1)^2$, which we know are smooth, continuous
curves. The graph of f is a straight line with positive
slope, that crosses the x-axis at -2. The graph of g is
a parabola, opening upward, shifted 1 unit to the right,
and touching the x-axis at $x = 1$. When f and g are "combined" into the single function $P(x) = (x + 2)(x - 1)^2$,
the behavior of the graph at these zeroes is still evident.
In Figure 3.11, the graph of P crosses the x-axis at
$x = -2$, "bounces" off the x-axis at $x = 1$, and is still
a smooth, continuous curve. This observation could be

Figure 3.11

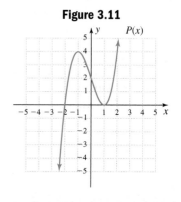

WORTHY OF NOTE

While defined more precisely in a future course, we will take "smooth" to mean the graph has no sharp turns or jagged edges, and "continuous" to mean the entire graph can be drawn without lifting your pencil.

extended to include additional linear or quadratic factors, and helps affirm that the graph of a polynomial function is a *smooth, continuous curve*.

Further, after the graph of P crosses the axis at $x = -2$, it must "turn around" at some point to reach the zero at $x = 1$, then turn again as it touches the x-axis without crossing. By combining this observation with our work in Section 3.3, we can state the following:

Polynomial Graphs and Turning Points

1. If $P(x)$ is a polynomial function of degree n, then the graph of P has at most $n - 1$ turning points.
2. If the graph of a function P has $n - 1$ turning points, then the degree of $P(x)$ is at least n.

EXAMPLE 1 ▶ **Identifying Polynomial Graphs**

Determine whether each graph could be the graph of a polynomial. If not, discuss why. If so, use the number of turning points and zeroes to identify the least possible degree of the function.

a.

b.

c.

d.

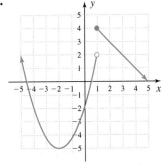

Solution ▶ **a.** This is not a polynomial graph, as it has a sharp turn (called a **cusp**) at (1, 3). A polynomial graph is always smooth.

b. This graph is smooth and continuous, and could be that of a polynomial. With two turning points and three zeroes, the function is at least degree 3.

c. This graph is smooth and continuous, and could be that of a polynomial. With three turning points and two zeroes, the function is at least degree 4.

d. This is not a polynomial graph, as it has a break (discontinuity) at $x = 1$. A polynomial graph is always continuous.

☑ **A.** You've just learned how to identify the graph of a polynomial function and determine its degree

Now try Exercises 7 through 12 ▶

B. The End Behavior of a Polynomial Graph

Once the graph of a function has "made its last turn" and crossed or touched its last real zero, it will continue to increase or decrease without bound as $|x|$ becomes large. As before, we refer to this as the **end behavior** of the graph. In previous sections we noted that quadratic functions (degree 2) with a positive leading coefficient ($a > 0$), had the end behavior "up on the left" and "up on the right (up/up)." If the leading coefficient was negative ($a < 0$), end behavior was "down on the left" and "down on the right (down/down)." These descriptions were also applied to the graph of a linear function $y = mx + b$ (degree 1). A positive leading coefficient ($m > 0$) indicates the graph will be down on the left, up on the right (down/up), and so on. All polynomial graphs exhibit some form of end behavior, which can be likewise described.

EXAMPLE 2 ▶   Identifying the End Behavior of a Graph

State the end behavior of each graph shown:

a. $f(x) = x^3 - 4x + 1$ **b.** $g(x) = -2x^5 + 7x^3 - 4x$ **c.** $h(x) = -2x^4 + 5x^2 + x - 1$

  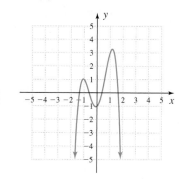

Solution ▶   **a.** down on the left, up on the right

b. up on the left, down on the right

c. down on the left, down on the right

Now try Exercises 13 through 16 ▶

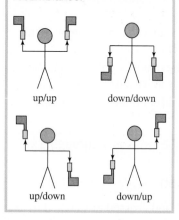
The leading term ax^n of a polynomial function is said to be the **dominant term,** because for large values of |x|, the value of ax^n is much larger than all other terms combined. This means that like linear and quadratic graphs, polynomial end behavior can be predicted in advance by analyzing this term alone.

1. For ax^n when n is even, any nonzero number raised to an even power is positive, so the ends of the graph must point in the same direction. If $a > 0$, both point upward. If $a < 0$, both point downward.
2. For ax^n when n is odd, any number raised to an odd power has the same sign as the input value, so the ends of the graph must point in opposite directions. If $a > 0$, end behavior is down on the left, up on the right. If $a < 0$, end behavior is up on the left, down on the right.

From this we find that end behavior depends on two things: *the degree of the function* (even or odd) and the *sign of the leading coefficient* (positive or negative). In more formal terms, this is described in terms of how the graph "behaves" for large values of x. For end behavior that is "up on the right," we mean that as x becomes a large positive number, y becomes a large positive number. This is indicated using the notation: as $x \rightarrow \infty$, $y \rightarrow \infty$. Similar notation is used for the other possibilities. These facts are summarized in Table 3.1. The interior portion of each graph is dashed since the actual number of turning points may vary, although a polynomial of odd degree will have an even number of turning points, and a polynomial of even degree will have an odd number of turning points.

Table 3.1
Polynomial End Behavior

Note the end behavior of $y = mx$ can be used as a representative of all odd degree functions, and the end behavior of $y = ax^2$ as a representative of all even degree functions.

The End Behavior of a Polynomial Graph

Given a polynomial $P(x)$ with leading term ax^n and $n \geq 1$.
If n is **even,** ends will point in the **same direction,**

  1. for $a > 0$: up on the left, up on the right (*as with* $y = x^2$);
            as $x \to -\infty$, $y \to \infty$;   as $x \to \infty$, $y \to \infty$
  2. for $a < 0$: down on the left, down on the right (*as with* $y = -x^2$);
            as $x \to -\infty$, $y \to -\infty$;   as $x \to \infty$, $y \to -\infty$

If n is **odd,** the ends will point in **opposite directions,**

  1. for $a > 0$: down on the left, up on the right (*as with* $y = x$);
            as $x \to -\infty$, $y \to -\infty$;   as $x \to \infty$, $y \to \infty$
  2. for $a < 0$: up on the left, down on the right (*as with* $y = -x$);
            as $x \to -\infty$, $y \to \infty$;   as $x \to \infty$, $y \to -\infty$

EXAMPLE 3 ▶ Identifying the End Behavior of a Function

State the end behavior of each function, without actually graphing.

  a. $f(x) = 0.5x^4 + 3x^3 - 5x + 6$ **b.** $g(x) = -2x^5 - 5x^3 - 3$

Solution ▶ **a.** The function has degree 4 (even), and the ends will point in the same direction. The leading coefficient is positive, so end behavior is up/up.

b. The function has degree 5 (odd), and the ends will point in opposite directions. The leading coefficient is negative, so the end behavior is up/down.

☑ **B. You've just learned how to describe the end behavior of a polynomial graph**

Now try Exercises 17 through 22 ▶

C. Attributes of Polynomial Graphs with Zeroes of Multiplicity

Another important aspect of polynomial functions is the behavior of a graph near its zeroes. In the simplest case, consider the functions $f(x) = x$ and $g(x) = x^3$. Both have odd degree, like end behavior (down/up), and a zero at $x = 0$. But the zero of f has multiplicity 1, while the zero from g has multiplicity 3. Notice the graph of g is vertically compressed near $x = 0$ and seems to approach this zero "more gradually."

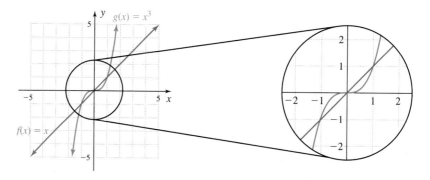

This behavior can be explained by noting that for $x = -1$ and 1, $f(x) = g(x)$. But for $|x| < 1$, the graph of g *will be closer to the x-axis* since the cube of a fractional number is smaller than the fraction itself. We further note that for $|x| > 1$, g increases much faster than f, and $|g(x)| > |f(x)|$. Similar observations can be made regarding $f(x) = x^2$ and $g(x) = x^4$. Both functions have even degree, a zero at $x = 0$, and $f(x) = g(x)$ for $x = -1$ and 1. But for $|x| < 1$, the function with higher degree is once again closer to the x-axis.

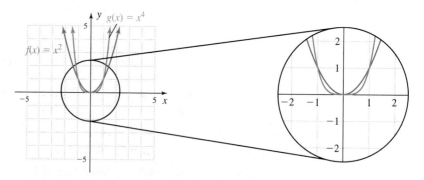

These observations can be generalized and applied to all real zeroes of a function.

Polynomial Graphs and Zeroes of Multiplicity

Given $P(x)$ is a polynomial with factors of the form $(x - c)^m$, with c a real number,

- If m is odd, the graph will cross through the x-axis.
- If m is even, the graph will bounce off the x-axis (touching at just one point).

In each case, the graph will be more compressed (flatter) near c for larger values of m.

To illustrate, compare the graph of $P(x) = (x + 2)(x - 1)^2$ from page 320, with the graph of $p(x) = (x + 2)^3(x - 1)^4$ shown, noting the increased multiplicity of each zero.

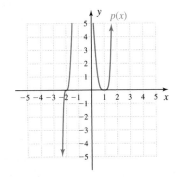

Both graphs show the expected zeroes at $x = -2$ and $x = 1$, but the graph of $p(x)$ is flatter near $x = -2$ and $x = 1$, due to the increased multiplicity of each zero. We lose sight of the graph of $p(x)$ between $x = -2$ and $x = 0$, since the increased multiplicities produce larger values than the original grid could display.

EXAMPLE 4 ▶ **Naming Attributes of a Function from Its Graph**

The graph of a polynomial $f(x)$ is shown.
  a. State whether the degree of f is even or odd.
  b. Use the graph to name the zeroes of f, then state whether their multiplicity is even or odd.
  c. State the minimum possible degree of f.
  d. State the domain and range of f.

Solution ▶   **a.** Since the ends of the graph point in opposite directions, the degree of the function must be odd.

  b. The graph crosses the x-axis at $x = -3$ and is compressed near -3, meaning it must have odd multiplicity with $m > 1$. The graph bounces off the x-axis at $x = 2$ and 2 must be a zero of even multiplicity.

  c. The minimum possible degree of f is 5, as in $f(x) = a(x - 2)^2(x + 3)^3$.

  d. $x \in \mathbb{R}, y \in \mathbb{R}$.

Now try Exercises 23 through 28 ▶

To find the degree of a polynomial from its factored form, add the exponents on all linear factors, then add 2 for each irreducible quadratic factor (the degree of any quadratic factor is 2). The sum gives the degree of the polynomial, from which end behavior can be determined. To find the y-intercept, substitute 0 for x as before, noting this is equivalent to applying the exponent to the constant from each factor.

EXAMPLE 5 ▶ **Naming Attributes of a Function from Its Factored Form**

State the degree of each function, then describe the end behavior and name the y-intercept of each graph.
  a. $f(x) = (x + 2)^3(x - 3)$ **b.** $g(x) = -(x + 1)^2(x^2 + 3)(x - 6)$

Solution ▶   **a.** The degree of f is $3 + 1 = 4$. With even degree and positive leading coefficient, end behavior is up/up. For $f(0) = (2)^3(-3) = -24$, the y-intercept is $(0, -24)$.

  b. The degree of g is $2 + 2 + 1 = 5$. With odd degree and negative leading coefficient, end behavior is up/down. For $g(0) = -1(1)^2(3)(-6) = 18$, the y-intercept is $(0, 18)$.

Now try Exercises 29 through 36 ▶

EXAMPLE 6 ▶ **Matching Graphs to Functions Using Zeroes of Multiplicity**

The following functions all have zeroes at $x = -2, -1$, and 1. Match each function to the corresponding graph *using its degree and the multiplicity of each zero.*

    a. $y = (x + 2)(x + 1)^2(x - 1)^3$ **b.** $y = (x + 2)(x + 1)(x - 1)^3$

    c. $y = (x + 2)^2(x + 1)^2(x - 1)^3$ **d.** $y = (x + 2)^2(x + 1)(x - 1)^3$

Solution ▶ The functions in Figures 3.12 and 3.14 must have even degree due to end behavior, so each corresponds to (a) or (d). At $x = -1$ the graph in Figure 3.12 "crosses," while the graph in Figure 3.14 "bounces." This indicates Figure 3.12 matches equation (d), while Figure 3.14 matches equation (a).

    The graphs in Figures 3.13 and 3.15 must have odd degree due to end behavior, so each corresponds to (b) or (c). Here, one graph "bounces" at $x = -2$, while the other "crosses." The graph in Figure 3.13 matches equation (c), the graph in Figure 3.15 matches equation (b).

Figure 3.12

Figure 3.13

Figure 3.14

Figure 3.15

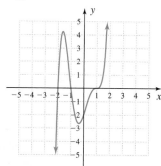

Now try Exercises 37 through 42 ▶

Using the ideas from Examples 5 and 6, we're able to draw a fairly accurate graph given the factored form of a polynomial. Convenient values between two zeroes, called **mid-interval points,** can be used to help complete the graph.

EXAMPLE 7 ▶ **Graphing a Function Given the Factored Form**

Sketch the graph of $f(x) = (x - 2)(x - 1)^2(x + 1)^3$ using end behavior; the x- and y-intercepts, and zeroes of multiplicity.

Solution ▶ Adding the exponents of each factor, we find that f is a function of degree 6 with a positive lead coefficient, so end behavior will be up/up. Since $f(0) = -2$, the y-intercept is $(0, -2)$. The graph will bounce off the x-axis at $x = 1$ (even multiplicity), and cross the axis at $x = -1$ and 2 (odd multiplicities). The graph will "flatten out" near $x = -1$ because of its higher multiplicity. To help "round-out" the graph we evaluate f at $x = 1.5$, giving $(-0.5)^2(0.5)^3(2.5) \approx -1.95$ (note scaling of the x- and y-axes).

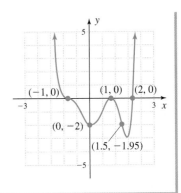

☑ **C.** You've just learned how to discuss the attributes of a polynomial graph with zeroes of multiplicity

Now try Exercises 43 through 56 ▶

D. The Graph of a Polynomial Function

Using the cumulative observations from this and previous sections, a general strategy emerges for the graphing of polynomial functions.

Guidelines for Graphing Polynomial Functions

1. Determine the end behavior of the graph.
2. Find the y-intercept $(0, a_0)$
3. Find the zeroes using any combination of the rational zeroes theorem, the factor and remainder theorems, tests for 1 and -1 (p. 310), factoring, and the quadratic formula.
4. Use the y-intercept, end behavior, the multiplicity of each zero, and midinterval points as needed to sketch a smooth, continuous curve.

   Additional tools include (a) polynomial zeroes theorem, (b) complex conjugates theorem, (c) number of turning points, (d) Descartes' rule of signs, (e) upper and lower bounds, and (f) symmetry.

EXAMPLE 8 ▶ Graphing a Polynomial Function

Sketch the graph of $g(x) = -x^4 + 9x^2 - 4x - 12$.

Solution ▶ 1. End behavior: The function has degree 4 (even) with a negative leading coefficient, so end behavior is *down on the left, down on the right.*

2. Since $g(0) = -12$, the y-intercept is $(0, -12)$.

3. Zeroes: Using the test for $x = 1$ gives $-1 + 9 - 4 - 12 = -8$, showing $x = 1$ is not a zero but $(1, -8)$ is a point on the graph. Using the test for $x = -1$ gives $-1 + 9 + 4 - 12 = 0$, so -1 is a zero and $(x + 1)$ is a factor. Using $x = -1$ with the factor theorem yields

$$
\begin{array}{r|rrrrr}
-1| & -1 & 0 & 9 & -4 & -12 \\
   &   & 1 & -1 & -8 & 12 \\
\hline
   & -1 & 1 & 8 & -12 & 0
\end{array}
$$

The quotient polynomial is not easily factorable so we continue with synthetic division. Using the rational zeroes theorem, the possible rational zeroes are $\{\pm1, \pm12, \pm2, \pm6, \pm3, \pm4\}$, so we try $x = 2$.

use 2 as a "divisor" on
the quotient polynomial

$$
\begin{array}{r|rrrr}
2| & -1 & 1 & 8 & -12 \\
   &   & -2 & -2 & 12 \\
\hline
   & -1 & -1 & 6 & 0
\end{array}
$$

This shows $x = 2$ is a zero, $x - 2$ is a factor, and the function can now be written as

$$g(x) = (x + 1)(x - 2)(-x^2 - x + 6).$$

Factoring -1 from the trinomial gives

$$g(x) = -1(x + 1)(x - 2)(x^2 + x - 6)$$
$$= -1(x + 1)(x - 2)(x + 3)(x - 2)$$
$$= -1(x + 1)(x - 2)^2(x + 3)$$

The zeroes of g are $x = -1$ and -3, both with multiplicity 1, and $x = 2$ with multiplicity 2.

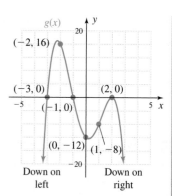

4. To help "round-out" the graph we evaluate the midinterval point $x = -2$ using the remainder theorem, which shows that $(-2, 16)$ is also a point on the graph.

$$
\begin{array}{r|rrrrr}
\text{use } -2 \text{ as a "divisor"} \quad -2 & -1 & 0 & 9 & -4 & -12 \\
 & & 2 & -4 & -10 & 28 \\
\hline
 & -1 & 2 & 5 & -1 & \underline{|16} \\
\end{array}
$$

The final result is the graph shown.

> **Now try Exercises 57 through 72** ▶

⚠ **CAUTION** ▶ Sometimes using a midinterval point to help draw a graph will give the illusion that a maximum or minimum value has been located. This is rarely the case, as demonstrated in the figure in Example 8, where the maximum value in Quadrant II is actually closer to $(-2.22, 16.95)$.

EXAMPLE 9 ▶ **Using the Guidelines to Sketch a Polynomial Graph**

Sketch the graph of $h(x) = x^7 - 4x^6 + 7x^5 - 12x^4 + 12x^3$.

Solution ▶ 1. End behavior: The function has degree 7 (odd) and the ends will point in opposite directions. The leading coefficient is positive and the end behavior will be *down on the left* and *up on the right*.

2. y-intercept: Since $h(0) = 0$, the y-intercept is $(0, 0)$.

3. Zeroes: Testing 1 and -1 shows neither are zeroes but $(1, 4)$ and $(-1, -36)$ are points on the graph. Factoring out x^3 produces $h(x) = x^3(x^4 - 4x^3 + 7x^2 - 12x + 12)$, and we see that $x = 0$ is a zero of multiplicity 3. We next use synthetic division with $x = 2$ on the fourth-degree polynomial:

$$
\begin{array}{r|rrrrr}
\text{use 2 as a "divisor"} \quad 2 & 1 & -4 & 7 & -12 & 12 \\
 & & 2 & -4 & 6 & -12 \\
\hline
 & 1 & -2 & 3 & -6 & \underline{|0} \\
\end{array}
$$

This shows $x = 2$ is a zero and $x - 2$ is a factor. At this stage, it appears the quotient can be factored by grouping. From $h(x) = x^3(x - 2)(x^3 - 2x^2 + 3x - 6)$, we obtain $h(x) = x^3(x - 2)(x^2 + 3)(x - 2)$ after factoring and

$$h(x) = x^3(x - 2)^2(x^2 + 3)$$

as the completely factored form. We find that $x = 2$ is a zero of multiplicity 2, and the remaining two zeroes are complex.

4. Using this information produces the graph shown in the figure.

> **Now try Exercises 73 through 76** ▶

☑ **D.** You've just learned how to graph polynomial functions in standard form

For practice with these ideas using a graphing calculator, **see Exercises 77 through 80.** Similar to our work in previous sections, **Exercises 81 and 82** ask you to reconstruct the complete equation of a polynomial from its given graph.

E. Applications of Polynomials

EXAMPLE 10 ▶ **Modeling the Value of an Investment**

In the year 2000, Marc and his wife Maria decided to invest some money in precious metals. As expected, the value of the investment fluctuated over the years, sometimes being worth more than they paid, other times less. Through 2008, the value of the investment was modeled by $v(t) = t^4 - 11t^3 + 38t^2 - 40t$, where $v(t)$ represents the gain or loss (in hundreds of dollars) in year t ($t = 0 \rightarrow 2000$).

   a. Use the rational zeroes theorem to find the years when their gain/loss was zero.

   b. Sketch the graph of the function.

   c. In what years was the investment worth less than they paid?

   d. What was their gain or loss in 2008?

Solution ▶ **a.** Writing the function as $v(t) = t(t^3 - 11t^2 + 38t - 40)$, we note $t = 0$ shows no gain or loss on purchase, and attempt to find the remaining zeroes. Testing for 1 and -1 shows neither is a zero, but $(1, -12)$ and $(-1, 90)$ are points on the graph of v. Next we try $t = 2$ with the factor theorem and the cubic polynomial.

$$\underline{2|}\quad \begin{array}{rrrr} 1 & -11 & 38 & -40 \\ & 2 & -18 & 40 \\ \hline 1 & -9 & 2 & \underline{|0} \end{array}$$

We find that 2 is a zero and write $v(t) = t(t - 2)(t^2 - 9t + 20)$, then factor to obtain $v(t) = t(t - 2)(t - 4)(t - 5)$. Since $v(t) = 0$ for $t = 0, 2, 4,$ and 5, they "broke even" in years 2000, 2002, 2004, and 2005.

b. With even degree and a positive leading coefficient, the end behavior is up/up. All zeroes have multiplicity 1. As an additional midinterval point we find $v(3) = 6$:

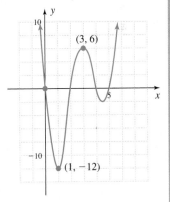

$$\underline{3|}\quad \begin{array}{rrrrr} 1 & -11 & 38 & -40 & 0 \\ & 3 & -24 & 42 & 6 \\ \hline 1 & -8 & 14 & 2 & \underline{|6} \end{array}$$

The complete graph is shown.

c. The investment was worth less than what they paid (outputs are negative) from 2000 to 2002 and 2004 to 2005.

d. In 2008, they were "sitting pretty," as their investment had gained $576.

$$\underline{8|}\quad \begin{array}{rrrrr} 1 & -11 & 38 & -40 & 0 \\ & 8 & -24 & 112 & 576 \\ \hline 1 & -3 & 14 & 72 & \underline{|576} \end{array}$$

WORTHY OF NOTE

Due to the context, the domain of $v(t)$ in Example 10 actually begins at $x = 0$, which we could designate with a point at $(0, 0)$. In addition, note there are three sign changes in the terms of $v(t)$, indicating there will be 3 or 1 positive roots (we found 3).

☑ **E.** You've just learned how to solve an application of polynomials

Now try Exercises 85 through 88 ▶

3.4 EXERCISES

▶ **CONCEPTS AND VOCABULARY**

Fill in each blank with the appropriate word or phrase. Carefully reread the section if needed.

1. For a polynomial with factors of the form $(x - c)^m$, c is called a _____ of multiplicity _____.

2. A polynomial function of degree n has _____ zeroes and at most _____ "turning points."

3. The graphs of $Y_1 = (x - 2)^2$ and $Y_2 = (x - 2)^4$ both _____ at $x = 2$, but the graph of Y_2 is _____ than the graph of Y_1 at this point.

4. Since $x^4 > 0$ for all x, the ends of its graph will always point in the _____ direction. Since

$x^3 > 0$ when $x > 0$ and $x^3 < 0$ when $x < 0$, the ends of its graph will always point in the _____ direction.

5. In your own words, explain/discuss how to find the degree and y-intercept of a function that is given in factored form. Use $f(x) = (x + 1)^3(x - 2)(x + 4)^2$ to illustrate.

6. Name all of the "tools" at your disposal that play a role in the graphing of polynomial functions. Which tools are indispensable and always used? Which tools are used only as the situation merits?

▶ **DEVELOPING YOUR SKILLS**

Determine whether each graph is the graph of a polynomial function. If yes, state the least possible degree of the function. If no, state why.

7.

8.

9.

10.

11.

12.

State the end behavior of the functions given.

13. $f(x)$

14. $g(x)$

15. $H(x)$

16. $h(x)$

State the end behavior and y-intercept of the functions given. Do not graph.

17. $f(x) = x^3 + 6x^2 - 5x - 2$

18. $g(x) = x^4 - 4x^3 - 2x^2 + 16x - 12$

19. $p(x) = -2x^4 + x^3 + 7x^2 - x - 6$

20. $q(x) = -2x^3 - 18x^2 + 7x + 3$

21. $Y_1 = -3x^5 + x^3 + 7x^2 - 6$

22. $Y_2 = -x^6 - 4x^5 + 4x^3 + 16x - 12$

For each polynomial graph, (a) state whether the degree of the function is even or odd; (b) use the graph to name the zeroes of f, then state whether their multiplicity is even or odd; (c) state the minimum possible degree of f and write it in factored form; and (d) estimate the domain and range. Assume all zeroes are real.

23.

24.

25.

26.

27.

28.

State the degree of each function, the end behavior, and y-intercept of its graph.

29. $f(x) = (x - 3)(x + 1)^3(x - 2)^2$

30. $g(x) = (x + 2)^2(x - 4)(x + 1)$

31. $Y_1 = -(x + 1)^2(x - 2)(2x - 3)(x + 4)$

32. $Y_2 = -(x + 1)(x - 2)^3(5x - 3)$

33. $r(x) = (x^2 + 3)(x + 4)^3(x - 1)$

34. $s(x) = (x + 2)^2(x - 1)^2(x^2 + 5)$

35. $h(x) = (x^2 + 2)(x - 1)^2(1 - x)$

36. $H(x) = (x + 2)^2(2 - x)(x^2 + 4)$

Every function in Exercises 37 through 42 has the zeroes $x = -1$, $x = -3$, and $x = 2$. Match each to its corresponding graph using degree, end behavior, and the multiplicity of each zero.

37. $f(x) = (x + 1)^2(x + 3)(x - 2)$

38. $F(x) = (x + 1)(x + 3)^2(x - 2)$

39. $g(x) = (x + 1)(x + 3)(x - 2)^3$

40. $G(x) = (x + 1)^3(x + 3)(x - 2)$

41. $Y_1 = (x + 1)^2(x + 3)(x - 2)^2$

42. $Y_2 = (x + 1)^3(x + 3)(x - 2)^2$

a.

b.

c.

d.

e.

f.

Sketch the graph of each function using the degree, end behavior, x- and y-intercepts, zeroes of multiplicity, and a few midinterval points to round-out the graph. Connect all points with a smooth, continuous curve.

43. $f(x) = (x + 3)(x + 1)(x - 2)$

44. $g(x) = (x + 2)(x - 4)(x - 1)$

45. $p(x) = -(x + 1)^2(x - 3)$

46. $q(x) = -(x + 2)(x - 2)^2$

47. $Y_1 = (x + 1)^2(3x - 2)(x + 3)$

48. $Y_2 = (x + 2)(x - 1)^2(5x - 2)$

49. $r(x) = -(x + 1)^2(x - 2)^2(x - 1)$

50. $s(x) = -(x - 3)(x - 1)^2(x + 1)^2$

51. $f(x) = (2x + 3)(x - 1)^3$

52. $g(x) = (3x - 4)(x + 1)^3$

53. $h(x) = (x + 1)^3(x - 3)(x - 2)$

54. $H(x) = (x + 3)(x + 1)^2(x - 2)^2$

55. $Y_3 = (x + 1)^3(x - 1)^2(x - 2)$

56. $Y_4 = (x - 3)(x - 1)^3(x + 1)^2$

Use the *Guidelines for Graphing Polynomial Functions* to graph the polynomials.

57. $y = x^3 + 3x^2 - 4$

58. $y = x^3 - 13x + 12$

59. $f(x) = x^3 - 3x^2 - 6x + 8$

60. $g(x) = x^3 + 2x^2 - 5x - 6$

61. $h(x) = -x^3 - x^2 + 5x - 3$

62. $H(x) = -x^3 - x^2 + 8x + 12$

63. $p(x) = -x^4 + 10x^2 - 9$

64. $q(x) = -x^4 + 13x^2 - 36$

65. $r(x) = x^4 - 9x^2 - 4x + 12$

66. $s(x) = x^4 - 5x^3 + 20x - 16$

67. $Y_1 = x^4 - 6x^3 + 8x^2 + 6x - 9$

68. $Y_2 = x^4 - 4x^3 - 3x^2 + 10x + 8$

69. $Y_3 = 3x^4 + 2x^3 - 36x^2 + 24x + 32$

70. $Y_4 = 2x^4 - 3x^3 - 15x^2 + 32x - 12$

71. $F(x) = 2x^4 + 3x^3 - 9x^2$

72. $G(x) = 3x^4 + 2x^3 - 8x^2$

73. $f(x) = x^5 + 4x^4 - 16x^2 - 16x$

74. $g(x) = x^5 - 3x^4 + x^3 - 3x^2$

75. $h(x) = x^6 - 2x^5 - 4x^4 + 8x^3$

76. $H(x) = x^6 + 3x^5 - 4x^4$

In preparation for future course work, it becomes helpful to recognize the most common square roots in mathematics: $\sqrt{2} \approx 1.414$, $\sqrt{3} \approx 1.732$, and $\sqrt{6} \approx 2.449$. Graph the following polynomials *on a graphing calculator,* and use the calculator to locate the maximum/minimum values and all zeroes. Use the zeroes to write the polynomial in factored form, then verify the *y*-intercept from the factored form and polynomial form.

77. $h(x) = x^5 + 4x^4 - 9x - 36$

78. $H(x) = x^5 + 5x^4 - 4x - 20$

79. $f(x) = 2x^5 + 5x^4 - 10x^3 - 25x^2 + 12x + 30$

80. $g(x) = 3x^5 + 2x^4 - 24x^3 - 16x^2 + 36x + 24$

Use the graph of each function to construct its equation in factored form and in polynomial form. Be sure to check the *y*-intercept and adjust the lead coefficient if necessary.

81.

82.

▶ **WORKING WITH FORMULAS**

83. Root tests for quartic polynomials:
$ax^4 + bx^3 + cx^2 + dx + e = 0$

If $u, v, w,$ and z represent the roots of a quartic polynomial, then the following relationships are true: (a) $u + v + w + z = -b$, (b) $u(v + z) + v(w + z) + w(u + z) = c$, (c) $u(vw + wz) + v(uz + wz) = -d$, and (d) $u \cdot v \cdot w \cdot z = e$. Use these tests to verify that $x = -3, -1, 2, 4$ are the solutions to $x^4 - 2x^3 - 13x^2 + 14x + 24 = 0$,

then use these zeroes and the factored form to write the equation in polynomial form to confirm results.

84. It is worth noting that the root tests in Exercise 83 still apply when the roots are irrational and/or complex. Use these tests to verify that $x = -\sqrt{3}, \sqrt{3}, 1 + 2i,$ and $1 - 2i$ are the solutions to $x^4 - 2x^3 + 2x^2 + 6x - 15 = 0$, then use these zeroes and the factored form to write the equation in polynomial form to confirm results.

▶ **APPLICATIONS**

85. Traffic volume: Between the hours of 6:00 A.M. and 6.00 P.M., the volume of traffic at a busy intersection can be modeled by the polynomial $v(t) = -t^4 + 25t^3 - 192t^2 + 432t$, where $v(t)$ represents the number of vehicles above/below average, and t is number of hours past 6:00 A.M. (6:00 A.M. → 0). (a) Use the remainder theorem to find the volume of traffic during rush hour (8:00 A.M.), lunch time (12 noon), and the trip home (5:00 P.M.). (b) Use the rational zeroes theorem to find the times when the volume of

traffic is at its average $[v(t) = 0]$. (c) Use this information to graph $v(t)$, then use the graph to estimate the maximum and minimum flow of traffic and the time at which each occurs.

86. Insect population: The population of a certain insect varies dramatically with the weather, with spring-like temperatures causing a population boom and extreme weather (summer heat and winter cold) adversely affecting the population. This phenomena can be modeled by the polynomial $p(m) = -m^4 + 26m^3 - 217m^2 + 588m$, where $p(m)$

represents the number of live insects (in hundreds of thousands) in month m ($m = 1 \rightarrow$ Jan). (a) Use the remainder theorem to find the population of insects during the cool of spring (March) and the fair weather of fall (October). (b) Use the rational zeroes theorem to find the times when the population of insects becomes dormant $[p(m) = 0]$. (c) Use this information to graph $p(m)$, then use the graph to estimate the maximum and minimum population of insects, and the month at which each occurs.

87. **Balance of payments:** The graph shown represents the balance of payments (surplus versus deficit) for a large county over a 9-yr period. Use it to answer the following:

a. What is the minimum possible degree polynomial that can model this graph?

b. How many years did this county run a deficit?

c. Construct an equation model in factored form and in polynomial

form, adjusting the lead coefficient as needed. How large was the deficit in year 8?

88. **Water supply:** The graph shown represents the water level in a reservoir (above and below normal) that supplies water to a metropolitan area, over a 6-month period. Use it to answer the following:

a. What is the minimum possible degree polynomial that can model this graph?

b. How many months was the water level below normal in this 6-month period?

c. At the beginning of this period ($m = 0$), the water level was 36 in. above normal, due to a long period of rain. Use this fact to help construct an equation model in factored form and in polynomial form, adjusting the lead coefficient as needed. Use the equation to determine the water level in months three and five.

▶ EXTENDING THE CONCEPT

89. As discussed in this section, the study of end behavior looks at what happens to the graph of a function as $|x| \rightarrow \infty$. Notice that as $|x| \rightarrow \infty$, both $\frac{1}{x}$ and $\frac{1}{x^2}$ approach zero. This fact can be used to study the end behavior of polynomial graphs.

a. For $f(x) = x^3 + x^2 - 3x + 6$, factoring out x^3 gives the expression

$$f(x) = x^3\left(1 + \frac{1}{x} - \frac{3}{x^2} + \frac{6}{x^3}\right).$$ What happens to the value of the expression as $x \rightarrow \infty$? As $x \rightarrow -\infty$?

b. Factor out x^4 from $g(x) = x^4 + 3x^3 - 4x^2 + 5x - 1$. What happens to the value of the expression as $x \rightarrow \infty$? As $x \rightarrow -\infty$? How does this affirm the end behavior must be up/up?

90. For what value of c will three of the four real roots of $x^4 + 5x^3 + x^2 - 21x + c = 0$ be shared by the polynomial $x^3 + 2x^2 - 5x - 6 = 0$?

Show that the following equations have no rational roots.

91. $x^5 - x^4 - x^3 + x^2 - 2x + 3 = 0$

92. $x^5 - 2x^4 - x^3 + 2x^2 - 3x + 4 = 0$

▶ MAINTAINING YOUR SKILLS

93. **(2.8)** Given $f(x) = x^2 - 2x$ and $g(x) = \frac{1}{x}$, find the compositions $h(x) = (f \circ g)(x)$ and $H(x) = (g \circ f)(x)$, then state the domain of each.

94. **(1.5)** By direct substitution, verify that $x = 1 - 2i$ is a solution to $x^2 - 2x + 5 = 0$ and name the second solution.

95. **(1.1/1.6)** Solve each of the following equations.

a. $-(2x + 5) - (6 - x) + 3 = x - 3(x + 2)$

b. $\sqrt{x + 1} + 3 = \sqrt{2x} + 2$

c. $\dfrac{2}{x - 3} + 5 = \dfrac{21}{x^2 - 9} + 4$

96. **(2.2)** Determine if the relation shown is a function. If not, explain how the definition of a function is violated.

MID-CHAPTER CHECK

1. Compute $(x^3 + 8x^2 + 7x - 14) \div (x + 2)$ using long division and write the result in two ways:
   (a) dividend = (quotient)(divisor) + remainder and
   (b) $\dfrac{\text{dividend}}{\text{divisor}} = (\text{quotient}) + \dfrac{\text{remainder}}{\text{divisor}}$.

2. Given that $x - 2$ is a factor of $f(x) = 2x^4 - x^3 - 8x^2 + x + 6$, use the rational zeroes theorem to write $f(x)$ in completely factored form.

3. Use the remainder theorem to evaluate $f(-2)$, given $f(x) = -3x^4 + 7x^2 - 8x + 11$.

4. Use the factor theorem to find a third-degree polynomial having $x = -2$ and $x = 1 + i$ as roots.

5. Use the intermediate value theorem to show that $g(x) = x^3 - 6x - 4$ has a root in the interval $(2, 3)$.

6. Use the rational zeroes theorem, tests for -1 and 1, synthetic division, and the remainder theorem to write $f(x) = x^4 + 5x^3 - 20x - 16$ in completely factored form.

7. Find all the zeroes of h, real and complex: $h(x) = x^4 + 3x^3 + 10x^2 + 6x - 20$.

8. Sketch the graph of p using its degree, end behavior, y-intercept, zeroes of multiplicity, and any midinterval points needed, given $p(x) = (x + 1)^2(x - 1)(x - 3)$.

9. Use the *Guidelines for Graphing* to draw the graph of $q(x) = x^3 + 5x^2 + 2x - 8$.

10. When fighter pilots train for dogfighting, a "hard-deck" is usually established below which no competitive activity can take place. The polynomial graph given shows Maverick's altitude above and below this hard-deck during a 5-sec interval.

    a. What is the minimum possible degree polynomial that could form this graph? Why?

    b. How many seconds (total) was Maverick below the hard-deck for these 5 sec of the exercise?

    c. At the beginning of this time interval ($t = 0$), Maverick's altitude was 1500 ft above the hard-deck. Use this fact and the graph given to help construct an equation model in factored form and in polynomial form, adjusting the lead coefficient if needed. Use the equation to determine Maverick's altitude in relation to the hard-deck at $t = 2$ and $t = 4$.

REINFORCING BASIC CONCEPTS

Approximating Real Zeroes

Consider the equation $x^4 + x^3 + x - 6 = 0$. Using the rational zeroes theorem, the possible rational zeroes are $\{\pm 1, \pm 6, \pm 2, \pm 3\}$. The tests for 1 and -1 indicate that neither is a zero: $f(1) = -3$ and $f(-1) = -7$. Descartes' rule of signs reveals there must be one positive real zero since the coefficients of $f(x)$ change sign one time: $f(x) = x^4 + x^3 + x - 6$, and one negative real zero since $f(-x)$ also changes sign one time: $f(-x) = x^4 - x^3 - x - 6$. The remaining two zeroes must be complex. Using $x = 2$ with synthetic division shows 2 is not a zero, but the coefficients in the quotient row are all positive, so 2 is an upper bound:

$$
\begin{array}{r|rrrrr}
2| & 1 & 1 & 0 & 1 & -6 \quad \text{coefficients of } f(x) \\
   &   & 2 & 6 & 12 & 26 \\
\hline
   & 1 & 3 & 6 & 13 & 20 \quad q(x)
\end{array}
$$

Using $x = -2$ shows that -2 is a zero *and a lower bound* for all other zeroes (quotient row alternates in sign):

$$
\begin{array}{r|rrrrr}
-2| & 1 & 1 & 0 & 1 & -6 \quad \text{coefficients of } f(x) \\
    &   & -2 & 2 & -4 & 6 \\
\hline
    & 1 & -1 & 2 & -3 & 0 \quad q_1(x)
\end{array}
$$

This means the remaining real zero must be a positive irrational number less than 2 (all other possible rational zeroes were eliminated). The quotient polynomial $q_1(x) = x^3 - x^2 + 2x - 3$ is not factorable, yet we're left with the challenge of finding this final zero. While there are many advanced techniques available for approximating irrational zeroes, at this level either technology or a technique called **bisection** is commonly used. The bisection method combines the intermediate value theorem with successively smaller intervals of the input variable, to narrow down the location of the irrational zero. Although "bisection" implies halving the interval each time, any number within the interval will do. The bisection method may be most efficient using a succession of short input/output tables as shown, with the number of tables increased if greater accuracy is desired. Since $f(1) = -3$ and $f(2) = 20$, the intermediate value theorem tells us the zero must be in the interval $[1, 2]$. We begin our search here, rounding noninteger outputs to the nearest 100th. As a visual aid, positive outputs are in blue, negative outputs in red.

x	$f(x)$	Conclusion
1	-3	$\leftarrow$ Zero is here, use $x = 1.25$ next
1.5	3.94	
2	20	

x	$f(x)$	Conclusion
1	-3	Zero is here, $\leftarrow$ use $x = 1.30$ next
1.25	-0.36	
1.5	3.94	

x	$f(x)$	Conclusion
1.25	-0.36	$\leftarrow$ Zero is here, use $x = 1.275$ next
1.30	0.35	
1.5	3.94	

A reasonable estimate for the zero appears to be $x = 1.275$. Evaluating the function at this point gives $f(1.275) \approx 0.0098$, which is very close to zero.

Naturally, a closer approximation is obtained using the capabilities of a graphing calculator. To seven decimal places the zero is $x \approx 1.2756822$.

Exercise 1: Use the intermediate value theorem to show that $f(x) = x^3 - 3x + 1$ has a zero in the interval $[1, 2]$, then use bisection to locate the zero to three decimal place accuracy.

Exercise 2: The function $f(x) = x^4 + 3x - 15$ has two real zeroes in the interval $[-5, 5]$. Use the intermediate value theorem to locate the zeroes, then use bisection to find the zeroes accurate to three decimal places.

3.5 | Graphing Rational Functions

Learning Objectives

In Section 3.5 you will learn how to:

☐ **A.** Identify horizontal and vertical asymptotes

☐ **B.** Find the domain of a rational function

☐ **C.** Apply the concept of "multiplicity" to rational graphs

☐ **D.** Find the horizontal asymptotes of a rational function

☐ **E.** Graph general rational functions

☐ **F.** Solve applications of rational functions

In this section we introduce an entirely new kind of relation, called a **rational function.** While we've already studied a variety of functions, we still lack the ability to model a large number of important situations. For example, functions that model the amount of medication remaining in the bloodstream over time, the relationship between altitude and weightlessness, and the relationship between predator and prey populations are all rational functions.

A. Rational Functions and Asymptotes

Just as a rational number is the ratio of two integers, a **rational function** is the ratio of two polynomials. In general,

Rational Functions

A rational function $V(x)$ is one of the form

$$V(x) = \frac{p(x)}{d(x)},$$

where p and d are polynomials and $d(x) \neq 0$.
The domain of $V(x)$ is all real numbers, *except the zeroes of d.*

The simplest rational functions are the reciprocal function $y = \frac{1}{x}$ and the reciprocal square function $y = \frac{1}{x^2}$, as both have a constant numerator and a single term in the denominator, with the domain of both excluding $x = 0$.

The Reciprocal Function: $y = \dfrac{1}{x}$

The reciprocal function takes any input (other than zero) and gives its reciprocal as the output. This means large inputs produce small outputs and vice versa. A table of values (Table 3.2) and the resulting graph (Figure 3.16) are shown.

Table 3.2

x	y	x	y
-1000	$-1/1000$	$1/1000$	1000
-5	$-1/5$	$1/3$	3
-4	$-1/4$	$1/2$	2
-3	$-1/3$	1	1
-2	$-1/2$	2	$1/2$
-1	-1	3	$1/3$
$-1/2$	-2	4	$1/4$
$-1/3$	-3	5	$1/5$
$-1/1000$	-1000	1000	$1/1000$
0	undefined		

Figure 3.16

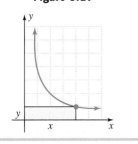

WORTHY OF NOTE

The notation used for graphical behavior always begins by describing what is happening to the *x*-values, and the resulting effect on the *y*-values. Using Figure 3.17, visualize that for a point (x, y) on the graph of $y = \frac{1}{x}$, as *x* gets larger, *y* must become smaller, particularly since their product must always be 1 ($y = \frac{1}{x} \Rightarrow xy = 1$).

Figure 3.17

Table 3.2 and Figure 3.16 reveal some interesting features. First, the graph passes the vertical line test, verifying $y = \frac{1}{x}$ is indeed a function. Second, since division by zero is undefined, there can be no corresponding point on the graph, *creating a break at $x = 0$*. In line with our definition of rational functions, the domain is $x \in (-\infty, 0) \cup (0, \infty)$. Third, this is an odd function, with a "branch" of the graph in the first quadrant and one in the third quadrant, as the reciprocal of any input maintains its sign. Finally, we note in QI that as *x* becomes an infinitely large positive number, *y* gets closer and closer to zero. It seems convenient to symbolize this end behavior using the notation adopted in Section 3.4, and we write as $x \to \infty$, $y \to 0$. Graphically, the curve becomes very close to, or *approaches the x-axis*.

We also note that as *x* approaches zero from the right, *y* becomes an infinitely large positive number: as $x \to 0^+$, $y \to \infty$. Note a superscript $+$ or $-$ sign is used to indicate the *direction of the approach,* meaning *from the positive side* (right) or *from the negative side* (left).

EXAMPLE 1 ▶ **Describing the End Behavior of Rational Functions**

For $y = \frac{1}{x}$ in QIII,

a. Describe the end behavior of the graph.

b. Describe what happens as *x* approaches zero.

Solution ▶ Similar to the graph's behavior in QI, we have

a. In words: As *x* becomes an infinitely large negative number, *y* approaches zero. In notation: As $x \to -\infty$, $y \to 0$.

b. In words: As *x* approaches zero from the left, y becomes an infinitely large negative number. In notation: As $x \to 0^-$, $y \to -\infty$.

Now try Exercises 7 and 8 ▶

The Reciprocal Square Function: $y = \dfrac{1}{x^2}$

From our previous work, we anticipate this graph will also have a break at $x = 0$. But since the square of any negative number is positive, the branches of the **reciprocal square function** are both *above the x-axis*. Note the result is the graph of an even function. See Table 3.3 and Figure 3.18.

Table 3.3

x	y	x	y
-1000	1/1,000,000	1/1000	1,000,000
-5	1/25	1/3	9
-4	1/16	1/2	4
-3	1/9	1	1
-2	1/4	2	1/4
-1	1	3	1/9
$-1/2$	4	4	1/16
$-1/3$	9	5	1/25
$-1/1000$	1,000,000	1000	1/1,000,000
0	undefined		

Figure 3.18

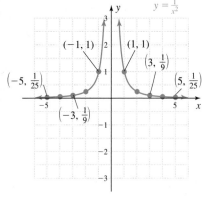

Similar to $y = \frac{1}{x}$, large positive inputs generate small, positive outputs: as $x \to \infty$, $y \to 0$. This is one indication of **asymptotic behavior** in the horizontal direction, and we say the line $y = 0$ is a **horizontal asymptote** for the reciprocal and reciprocal square functions. In general,

Horizontal Asymptotes

Given a constant k, the line $y = k$ is a horizontal asymptote for a function V if as x increases without bound, $V(x)$ approaches k:

$$\text{as } x \to -\infty, V(x) \to k \qquad \text{or} \qquad \text{as } x \to \infty, V(x) \to k$$

Figure 3.19 shows a horizontal asymptote at $y = 1$, which suggests the graph of $f(x)$ is the graph of $y = \frac{1}{x}$ shifted up 1 unit. Figure 3.20 shows a horizontal asymptote at $y = -2$, which suggests the graph of $g(x)$ is the graph of $y = \frac{1}{x^2}$ shifted down 2 units.

Figure 3.19   **Figure 3.20**

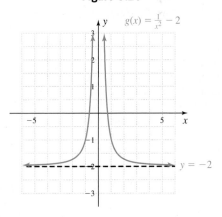

WORTHY OF NOTE

As seen in Figures 3.19 and 3.20, asymptotes appear graphically as dashed lines that seem to "guide" the branches of the graph.

EXAMPLE 2 ▶ **Describing the End Behavior of Rational Functions**

For the graph in Figure 3.20, use mathematical notation to
  a. Describe the end behavior of the graph.
  b. Describe what happens as x approaches zero.

Solution ▶ **a.** as $x \to -\infty$, $g(x) \to -2$ **b.** as $x \to 0^-$, $g(x) \to \infty$
            as $x \to \infty$, $g(x) \to -2$ as $x \to 0^+$, $g(x) \to \infty$

Now try Exercises 9 and 10 ▶

From Example 2b, we note that as x becomes *smaller and close to 0, g* becomes very large and *increases without bound*. This is an indication of asymptotic behavior in the vertical direction, and we say the line $x = 0$ is a **vertical asymptote** for g ($x = 0$ is also a vertical asymptote for f). In general,

Vertical Asymptotes

Given a constant h, the line $x = h$ is a vertical asymptote for a function V if as x approaches h, $V(x)$ increases or decreases without bound:

$$\text{as } x \to h^{+}, V(x) \to \pm\infty \qquad \text{or} \qquad \text{as } x \to h^{-}, V(x) \to \pm\infty$$

Identifying these asymptotes is useful because the graphs of $y = \frac{1}{x}$ and $y = \frac{1}{x^2}$ can be transformed *in exactly the same way as the toolbox functions*. When their graphs shift—the vertical and horizontal asymptotes shift with them and can be used as guides to redraw the graph. In shifted form, $f(x) = \dfrac{a}{(x \pm h)} \pm k$ for the reciprocal function, and $g(x) = \dfrac{a}{(x \pm h)^2} \pm k$ for the reciprocal square function.

EXAMPLE 3 ▶ **Writing the Equation of a Basic Rational Function, Given Its Graph**

Identify the function family for the graph given, then use the graph to write the equation of the function in "shifted form." Assume $|a| = 1$.

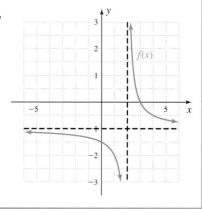

Solution ▶ The graph appears to be from the reciprocal function family, and has been shifted 2 units right (the vertical asymptote is at $x = 2$), and 1 unit down (the horizontal asymptote is at $y = -1$). From $y = \frac{1}{x}$, we obtain $f(x) = \frac{1}{x-2} - 1$ as the shifted form.

☑ **A.** You've just learned how to identify and name horizontal and vertical asymptotes

Now try Exercises 11 through 22 ▶

B. Vertical Asymptotes and the Domain

Much of what we know about these basic functions can be generalized and applied to general rational functions. The graphs in Figures 3.21 through 3.24 show that rational graphs come in many shapes, often in "pieces," and exhibit asymptotic behavior.

WORTHY OF NOTE

In Section 2.7, we studied special cases of $\frac{p(x)}{d(x)}$, where p and d shared a common factor, creating a "hole" in the graph. In this section, we'll assume the functions are given in simplest form (the numerator and denominator have no common factors).

Figure 3.21
$$f(x) = \frac{1}{x + 2}$$

Figure 3.22
$$g(x) = \frac{2x}{x^2 - 1}$$

Figure 3.23
$$w(x) = \frac{3}{x^2 + 1}$$

Figure 3.24
$$H(x) = \frac{x^2}{x^2 - 2x - 3}$$

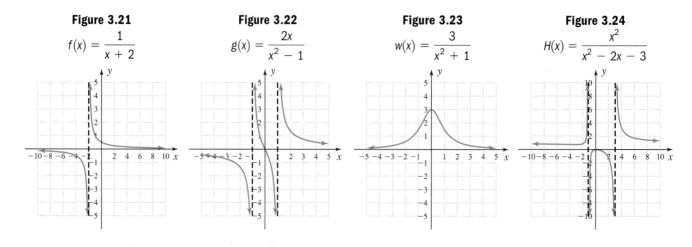

WORTHY OF NOTE

Breaks created by vertical asymptotes are said to be **nonremovable,** because there is no way to repair the break, even if a piecewise-defined function were used. See Example 5, Section 2.7.

For $y = \frac{1}{x}$ and $y = \frac{1}{x^2}$, a vertical asymptote occurred at the zero of the denominator. This actually applies to all rational functions *in simplified form*. For $V(x) = \frac{p(x)}{d(x)}$, if c is a zero of $d(x)$, the function can be evaluated at every point near c, but not *at* c. This creates a **break** or **discontinuity** in the graph, resulting in the asymptotic behavior.

Vertical Asymptotes of a Rational Function

Given $V(x) = \frac{p(x)}{d(x)}$ is a rational function in simplest form, vertical asymptotes will occur at the real zeroes of d.

EXAMPLE 4 ▶ **Finding Vertical Asymptotes**

Locate the vertical asymptote(s) of each function given, then state its domain.

a. $f(x) = \dfrac{2x}{x^2 - 1}$ **b.** $g(x) = \dfrac{3}{x^2 + 1}$ **c.** $v(x) = \dfrac{x^2}{x^2 - 2x - 3}$

Solution ▶ **a.** Setting the denominator equal to zero gives $x^2 - 1 = 0$, so vertical asymptotes will occur at $x = -1$ and $x = 1$. The domain of f is $x \in (-\infty, -1) \cup (-1, 1) \cup (1, \infty)$.

b. Since the equation $x^2 + 1 = 0$ has no real zeroes, there are no vertical asymptotes and the domain of g is unrestricted: $x \in R$.

☑ **B. You've just learned how to find the domain of a rational function**

c. Solving $x^2 - 2x - 3 = 0$ gives $(x + 1)(x - 3) = 0$, with solutions $x = -1$ and $x = 3$. There are vertical asymptotes at $x = -1$ and $x = 3$, and the domain of v is $x \in (-\infty, -1) \cup (-1, 3) \cup (3, \infty)$.

Now try Exercises 23 through 30 ▶

C. Vertical Asymptotes and Multiplicities

The "cross" and "bounce" concept used for polynomial graphs can also be applied to rational graphs, particularly when viewed in terms of sign changes in the dependent variable. As you can see in Figures 3.25 to 3.27, the function $f(x) = \dfrac{1}{x + 2}$ changes sign at the asymptote $x = -2$ (negative on one side, positive on the other), and the denominator has multiplicity 1 (odd). The function $g(x) = \dfrac{1}{(x - 1)^2}$ does not change sign at the asymptote $x = 1$ (positive on both sides), and its denominator has multiplicity 2 (even). As with our earlier study of multiplicities, when these two are combined into the single function $v(x) = \dfrac{1}{(x + 2)(x - 1)^2}$, the function still changes sign at $x = -2$, and does not change sign at $x = 1$.

Figure 3.25

$f(x) = \dfrac{1}{x + 2}$

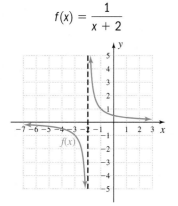

Figure 3.26

$g(x) = \dfrac{1}{(x - 1)^2}$

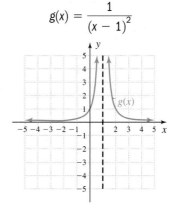

Figure 3.27

$h(x) = \dfrac{1}{(x + 2)(x - 1)^2}$

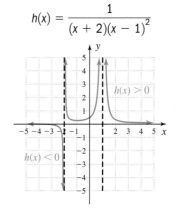

EXAMPLE 5 ▶ **Finding Sign Changes at Vertical Asymptotes**

Locate the vertical asymptotes of each function and state whether the function will change sign from one side of the asymptote(s) to the other.

a. $f(x) = \dfrac{x^2 - 4x + 4}{x^2 - 2x - 3}$ **b.** $g(x) = \dfrac{x^2 + 2}{x^2 + 2x + 1}$

Solution ▶ **a.** Factoring the denominator of f and setting it equal to zero gives $(x + 1)(x - 3) = 0$, and vertical asymptotes will occur at $x = -1$ and $x = 3$ (both multiplicity 1). The function will change sign at each asymptote (see Figure 3.28).

b. Factoring the denominator of g and setting it equal to zero gives $(x + 1)^2 = 0$. There will be a vertical asymptote at $x = -1$, but the function will not change sign since it's a zero of even multiplicity (see Figure 3.29).

Figure 3.28

$f(x) = \dfrac{x^2 - 4x + 4}{x^2 - 2x - 3}$

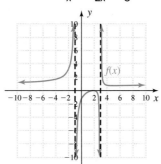

Figure 3.29

$g(x) = \dfrac{x^2 + 2}{x^2 + 2x + 1}$

☑ **C.** You've just learned how to apply the concept of "multiplicity" to rational graphs

Now try Exercises 31 through 36 ▶

D. Finding Horizontal Asymptotes

A study of horizontal asymptotes is closely related to our study of "dominant terms" in Section 3.4. Recall the highest degree term in a polynomial tends to dominate all other terms as $|x| \to \infty$. For $v(x) = \dfrac{2x^2 + 4x + 3}{x^2 + 2x + 1}$, both polynomials *have the same degree*, so $\dfrac{2x^2 + 4x + 3}{x^2 + 2x + 1} \approx \dfrac{2x^2}{x^2} = 2$ for large values of x: as $|x| \to \infty$, $y \to 2$ and $y = 2$ is a horizontal asymptote for v. When the degree of the numerator is *smaller* than the degree of the denominator, our earlier work with $y = \frac{1}{x}$ and $y = \frac{1}{x^2}$ showed there was a horizontal asymptote at $y = 0$ (the x-axis), since as $|x| \to \infty$, $y \to 0$. In general,

LOOKING AHEAD

In Section 3.6 we will explore two additional kinds of asymptotic behavior, (1) oblique (slant) asymptotes and (2) asymptotes that are nonlinear.

Horizontal Asymptotes

Given $V(x) = \frac{p(x)}{d(x)}$ is a rational function in lowest terms, where the leading term of p is ax^n and the leading term of d is bx^m (polynomial p has degree n, polynomial d has degree m).

  I. If $n < m$, there is a horizontal asymptote at $y = 0$ (the x-axis).
 II. If $n = m$, there is a horizontal asymptote at $y = \frac{a}{b}$.
III. If $n > m$, the graph has no horizontal asymptote.

Finally, while the graph of a rational function can never "cross" the vertical asymptote $x = h$ (since the function simply cannot be evaluated at h), it is possible for a graph to cross the horizontal asymptote $y = k$ (some do, others do not). To find out which is the case, we set the function equal to k and solve.

EXAMPLE 6 ▶ **Locating Horizontal Asymptotes**

Locate the horizontal asymptote for each function, if one exists. Then determine if the graph will cross the asymptote.

   a. $f(x) = \dfrac{3x}{x^2 + 2}$ **b.** $g(x) = \dfrac{x^2 - 4}{x^2 - 1}$ **c.** $v(x) = \dfrac{3x^2 - x - 6}{x^2 + x - 6}$

Solution ▶ **a.** For $f(x)$, the degree of the numerator $<$ degree of the denominator, indicating a horizontal asymptote at $y = 0$. Solving $f(x) = 0$, we find $x = 0$ is the only solution and the graph will cross the horizontal asymptote at $(0, 0)$ (see Figure 3.30).

b. For $g(x)$, the degree of the numerator and the denominator are equal. This means $g(x) \approx \dfrac{x^2}{x^2} = 1$ for large values of x, and there is a horizontal asymptote at $y = 1$. Solving $g(x) = 1$ gives

$$\frac{x^2 - 4}{x^2 - 1} = 1 \qquad y = 1 \rightarrow \text{horizontal asymptote}$$
$$x^2 - 4 = x^2 - 1 \qquad \text{multiply by } x^2 - 1$$
$$-4 = -1 \qquad \text{no solution}$$

The graph will not cross the asymptote (see Figure 3.31).

c. For $v(x)$, the degree of the numerator and denominator are once again equal, so $v(x) \approx \dfrac{3x^2}{x^2} = 3$ and there is a horizontal asymptote at $y = 3$.

Solving $v(x) = 3$ gives

$$\frac{3x^2 - x - 6}{x^2 + x - 6} = 3 \qquad y = 3 \rightarrow \text{horizontal asymptote}$$
$$3x^2 - x - 6 = 3(x^2 + x - 6) \qquad \text{multiply by } x^2 + x - 6$$
$$3x^2 - x - 6 = 3x^2 + 3x - 18 \qquad \text{distribute}$$
$$-4x + 12 = 0 \qquad \text{simplify}$$
$$x = 3 \qquad \text{result}$$

☑ **D. You've just learned how to find the horizontal asymptotes of a rational function**

The graph will cross its asymptote at $x = 3$ (see Figure 3.32).

Figure 3.30

$f(x) = \dfrac{3x}{x^2 + 2}$

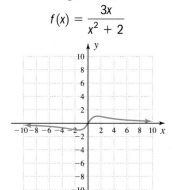

Figure 3.31

$g(x) = \dfrac{x^2 - 4}{x^2 - 1}$

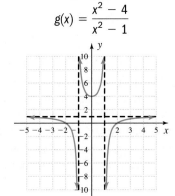

Figure 3.32

$v(x) = \dfrac{3x^2 - x - 6}{x^2 + x - 6}$

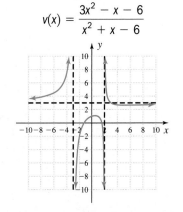

Now try Exercises 37 through 42 ▶

It's helpful to note that all nonvertical asymptotes and whether they cross the graph can actually be found using long division. The quotient $q(x)$ gives the equation of the asymptote, and the zeroes of the remainder $r(x)$ will indicate if or where the two will cross. From Example 6c, long division gives $q(x) = 3$ and $r(x) = -4x + 12$ (verify this), showing there is a horizontal asymptote at $y = 3$, which the graph crosses at $x = 3$ [the zero of $r(x)$].

E. The Graph of a Rational Function

Our observations to this point lead us to this general strategy for graphing rational functions. Not all graphs require every step, but together they provide an effective approach.

Guidelines for Graphing Rational Functions

Given $V(x) = \frac{p(x)}{d(x)}$, $d(x) \neq 0$, is a rational function in lowest terms,

1. Find the y-intercept at $V(0)$.
2. Find vertical asymptotes (if any) at $d(x) = 0$.
3. Find x-intercepts at $p(x) = 0$.
4. Locate the horizontal asymptote (if any).
5. Determine if the graph will cross the horizontal asymptote.
6. If needed, compute "midinterval" points to help complete the graph.
7. Draw the asymptotes, plot the intercepts and additional points, and use intervals where $V(x)$ changes sign to complete the graph.

EXAMPLE 7 ▶ **Graphing Rational Functions**

Graph each function given.

a. $f(x) = \dfrac{x^2 - x - 6}{x^2 + x - 6}$ **b.** $g(x) = \dfrac{2x^2 - 4x + 2}{x^2 - 7}$

Solution ▶ **a.** Begin by writing f in factored form: $f(x) = \dfrac{(x + 2)(x - 3)}{(x + 3)(x - 2)}$.

1. y-intercept: $f(0) = \dfrac{(2)(-3)}{(3)(-2)} = 1$, so the y-intercept is $(0, 1)$.

2. Vertical asymptote(s): Setting the denominator equal to zero gives $(x + 3)(x - 2) = 0$, showing there will be vertical asymptotes at $x = -3, x = 2$.

3. x-intercepts: Setting the numerator equal to zero gives $(x + 2)(x - 3) = 0$, showing the x-intercepts will be $(-2, 0)$ and $(3, 0)$.

4. Horizontal asymptote: Since the degree of the numerator and the degree of the denominator are equal, $y = \dfrac{x^2}{x^2} = 1$ is a horizontal asymptote.

5. Solving $\dfrac{x^2 - x - 6}{x^2 + x - 6} = 1$ $f(x) = 1 \rightarrow$ horizontal asymptote

       $x^2 - x - 6 = x^2 + x - 6$ multiply by $x^2 + x - 6$

             $-2x = 0$ simplify

               $x = 0$ solve

The graph will cross the horizontal asymptote at $(0, 1)$.

     The information from steps 1 through 5 is shown in Figure 3.33, and indicates we have no information about the graph in the interval $(-\infty, -3)$. Since rational functions are defined for all real numbers except the zeroes of d, we know there must be a "piece" of the graph in this interval.

6. Selecting $x = -4$ to compute one additional point, we find $f(-4) = \dfrac{(-2)(-7)}{(-1)(-6)} = \dfrac{14}{6} = \dfrac{7}{3}$. The point is $\left(-4, \dfrac{7}{3}\right)$.

7. All factors of f are linear, so function values will alternate sign in the intervals created by x-intercepts and vertical asymptotes. The y-intercept $(0, 1)$ shows $f(x)$ is positive in the interval containing 0. To meet all necessary conditions, we complete the graph, as shown in Figure 3.34.

Figure 3.33　　　　　　　　　**Figure 3.34**

　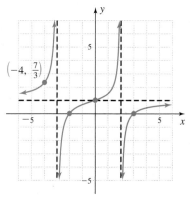

b. Writing g in factored form gives $g(x) = \dfrac{2(x^2 - 2x + 1)}{x^2 - 7} = \dfrac{2(x - 1)^2}{(x + \sqrt{7})(x - \sqrt{7})}$.

1. y-intercept: $g(0) = \dfrac{2(-1)^2}{(\sqrt{7})(-\sqrt{7})} = -\dfrac{2}{7}$. The y-intercept is $(0, -\dfrac{2}{7})$.

2. Vertical asymptote(s): Setting the denominator equal to zero gives $(x + \sqrt{7})(x - \sqrt{7}) = 0$, showing there will be asymptotes at $x = -\sqrt{7}, x = \sqrt{7}$.

3. x-intercept(s): Setting the numerator equal to zero gives $2(x - 1)^2 = 0$, with $x = 1$ a zero of multiplicity 2. The x-intercept is $(1, 0)$.

4. Horizontal asymptote: The degree of the numerator is equal to the degree of denominator, so $y = \dfrac{2x^2}{x^2} = 2$ is a horizontal asymptote.

5. Solve $\dfrac{2x^2 - 4x + 2}{x^2 - 7} = 2$　　　　$g(x) = 2 \rightarrow$ horizontal asymptote

$2x^2 - 4x + 2 = 2x^2 - 14$　　multiply by $x^2 - 7$

$-4x = -16$　　simplify

$x = 4$　　solve

The graph will cross its horizontal asymptote at $(4, 2)$.

The information from steps 1 to 5 is shown in Figure 3.35, and indicates we have no information about the graph in the interval $(-\infty, -\sqrt{7})$.

Figure 3.35

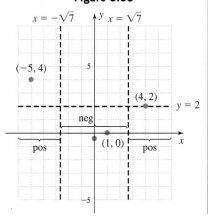

6. Selecting $x = -5$, $g(-5) = \dfrac{2(-5 - 1)^2}{(-5)^2 - 7}$

$= \dfrac{2(-6)^2}{25 - 7}$

$= \dfrac{2(36)}{18}$

$= 4$

The point $(-5, 4)$ is on the graph.

7. Since factors of the denominator have odd multiplicity, function values will alternate sign on either side of the asymptotes. The factor in the numerator has even multiplicity, so the graph will "bounce off" the x-axis at $x = 1$ (no change in sign). The y-intercept $(0, -\frac{2}{7})$ shows the function is negative in the interval containing 0. This information and the completed graph are shown in Figure 3.36.

Figure 3.36

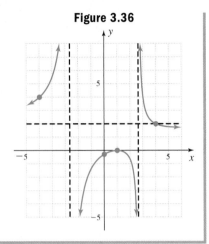

Now try Exercises 43 through 66 ▶

Examples 6 and 7 demonstrate that graphs of rational functions come in a large variety. Once the components of the graph have been found, completing the graph presents an intriguing and puzzle-like challenge as we attempt to sketch a graph that meets all conditions. As we've done with other functions, can you reverse this process? That is, given the <u>graph</u> of a rational function, can you construct its equation?

EXAMPLE 8 ▶ **Finding the Equation of a Rational Function from Its Graph**

Use the graph of $f(x)$ shown to construct its equation.

Solution ▶ The x-intercepts are $(-1, 0)$ and $(4, 0)$, so the numerator must contain the factors $(x + 1)$ and $(x - 4)$. The vertical asymptotes are $x = -2$ and $x = 3$, so the denominator must have the factors $(x + 2)$ and $(x - 3)$. So far we have:

$$f(x) = \frac{a(x + 1)(x - 4)}{(x + 2)(x - 3)}$$

Since $(2, 3)$ is on the graph, we substitute 2 for x and 3 for $f(x)$ to solve for a:

$$3 = \frac{a(2 + 1)(2 - 4)}{(2 + 2)(2 - 3)} \qquad \text{substitute 3 for } f(x) \text{ and 2 for } x$$

$$3 = \frac{3a}{2} \qquad \text{simplify}$$

$$2 = a \qquad \text{solve}$$

☑ **E.** You've just learned how to graph general rational functions

The result is $f(x) = \dfrac{2(x + 1)(x - 4)}{(x + 2)(x - 3)} = \dfrac{2x^2 - 6x - 8}{x^2 - x - 6}$, with a horizontal asymptote at $y = 2$ and a y-intercept of $(0, \frac{4}{3})$, which fits the graph very well.

Now try Exercises 67 through 70 ▶

F. Applications of Rational Functions

In many applications of rational functions, the coefficients can be rather large and the graph should be scaled appropriately.

EXAMPLE 9 ▶ **Modeling the Cost to Remove Chemical Waste**

For a large urban-centered county, the cost to remove chemical waste from a local river is modeled by $C(p) = \frac{180p}{100 - p}$, where $C(p)$ represents the cost (in thousands of dollars) to remove p percent of the pollutants.

a. Find the cost to remove 25%, 50%, and 75% of the pollutants and comment.

b. Graph the function using an appropriate scale.

c. In mathematical notation, state what happens if the county attempts to remove 100% of the pollutants.

Solution ▶ a. Evaluating the function for the values indicated, we find $C(25) = 60$, $C(50) = 180$, and $C(75) = 540$. The cost is escalating rapidly. The change from 25% to 50% brought a $120,000 increase, but the change from 50% to 75% brought *a $360,000 increase!*

b. From $C(p) = \frac{180p}{100 - p}$, we see that C has a y-intercept at $(0, 0)$ and a vertical asymptote at $p = 100$. Since the degree of the numerator and denominator are equal, there is a horizontal asymptote at $y = \frac{180p}{-p} = -180$. From the context we need only graph the portion from $0 \le p < 100$, producing the following graph:

☑ **F. You've just learned how to solve an application of rational functions**

c. As the percentage of the pollutants removed approaches 100%, the cost of the cleanup skyrockets. Notationally, as $p \to 100^-$, $C \to \infty$.

Now try Exercises 73 through 84 ▶

TECHNOLOGY HIGHLIGHT

Rational Functions and Appropriate Domains

In Example 9, portions of the graph were ignored due to the context of the application. To see the full graph, we use the fact that a second branch of C occurs on the opposite side of the vertical and horizontal asymptotes, and set a window size like the one shown in Figure 3.37. After entering $C(p)$ as Y_1 on the [Y=] screen and pressing [GRAPH], the full graph shown in Figure 3.38 appears (the horizontal asymptote was drawn using $Y_2 = -180$).

Figure 3.37

```
WINDOW
 Xmin=0
 Xmax=200
 Xscl=20
 Ymin=-2000
 Ymax=2000
 Yscl=200
 Xres=1
```

Figure 3.38

Exercise 1: Use the [TRACE] feature to verify that as $p \to 100^-$, $C \to \infty$. Approximately how much money must be spent to remove 95% of the pollutants? What happens when you [TRACE] to 100%? Past 100%?

Exercise 2: Calculate the rate of change $\frac{\Delta C}{\Delta p}$ for the intervals [60, 65], [85, 90], and [90, 95] (use the *Technology Extension* from Chapter 3 at www.mhhe.com/coburn if desired). Comment on what you notice.

Exercise 3: Reset the window size changing only Xmax to 100 and Ymin to 0 for a more relevant graph. How closely does it resemble the graph from Example 9?

3.5 EXERCISES

▶ CONCEPTS AND VOCABULARY

Fill in each blank with the appropriate word or phrase. Carefully reread the section if needed.

1. Write the following in direction/approach notation. *As x becomes an infinitely large negative number, y approaches 2.* _____

2. For any constant k, the notation as $|x| \to +\infty$, $y \to k$ is an indication of a _____ asymptote, while $x \to k$, $|y| \to +\infty$ indicates a _____ asymptote.

3. Vertical asymptotes are found by setting the _____ equal to zero. The x-intercepts are found by setting the _____ equal to zero.

4. If the degree of the numerator is equal to the degree of the denominator, a horizontal asymptote occurs at $y = \frac{a}{b}$, where $\frac{a}{b}$ represents the ratio of the _____ _____.

5. Use the function $g(x) = \dfrac{3x^2 - 2x}{2x^2 - 3}$ and a table of values to discuss the concept of horizontal asymptotes. At what positive value of x is the graph of g within 0.01 of its horizontal asymptote?

6. Name all of the "tools" at your disposal that play a role in the graphing of rational functions. Which tools are indispensable and always used? Which are used only as the situation merits?

▶ DEVELOPING YOUR SKILLS

For each graph given, (a) use mathematical notation to describe the end behavior of each graph and (b) describe what happens as x approaches 1.

7. $V(x) = \dfrac{1}{(x-1)} + 2$ 8. $v(x) = \dfrac{1}{(x-1)} - 2$

For each graph given, (a) use mathematical notation to describe the end behavior of each graph, and (b) describe what happens as x approaches −2.

9. $Q(x) = \dfrac{1}{(x+2)^2} + 1$ 10. $q(x) = \dfrac{-1}{(x+2)^2} + 2$

Identify the parent function for each graph given, then use the graph to construct the equation of the function in shifted form. Assume $|a| = 1$.

11.

12.

13.

14.

15.

16.

Use the graph shown to complete each statement using the direction/approach notation.

Exercises 17 through 22

17. As $x \to -\infty$, y _____.

18. As $x \to \infty$, y _____.

19. As $x \to -1^+$, y _____.

20. As $x \to -1^-$, y _____.

21. The line $x = -1$ is a vertical asymptote, since: as $x \to$ _____, $y \to$ _____.

22. The line $y = -2$ is a horizontal asymptote, since: as $x \to$ _____, $y \to$ _____.

▶ DEVELOPING YOUR SKILLS

Give the location of the vertical asymptote(s) if they exist, and state the function's domain.

23. $f(x) = \dfrac{x + 2}{x - 3}$ **24.** $F(x) = \dfrac{4x}{2x - 3}$

25. $g(x) = \dfrac{3x^2}{x^2 - 9}$ **26.** $G(x) = \dfrac{x + 1}{9x^2 - 4}$

27. $h(x) = \dfrac{x^2 - 1}{2x^2 + 3x - 5}$ **28.** $H(x) = \dfrac{x - 5}{2x^2 - x - 3}$

29. $p(x) = \dfrac{2x + 3}{x^2 + x + 1}$ **30.** $q(x) = \dfrac{2x^3}{x^2 + 4}$

Give the location of the vertical asymptote(s) if they exist, and state whether function values will change sign (positive to negative or negative to positive) from one side of the asymptote to the other.

31. $Y_1 = \dfrac{x + 1}{x^2 - x - 6}$ **32.** $Y_2 = \dfrac{2x + 3}{x^2 - x - 20}$

33. $r(x) = \dfrac{x^2 + 3x - 10}{x^2 - 6x + 9}$ **34.** $R(x) = \dfrac{x^2 - 2x - 15}{x^2 - 4x + 4}$

35. $Y_1 = \dfrac{x}{x^3 + 2x^2 - 4x - 8}$

36. $Y_2 = \dfrac{-2x}{x^3 + x^2 - x - 1}$

For the functions given, (a) determine if a horizontal asymptote exists and (b) determine if the graph will cross the asymptote, and if so, where it crosses.

37. $Y_1 = \dfrac{2x - 3}{x^2 + 1}$ **38.** $Y_2 = \dfrac{4x + 3}{2x^2 + 5}$

39. $r(x) = \dfrac{4x^2 - 9}{x^2 - 3x - 18}$ **40.** $R(x) = \dfrac{2x^2 - x - 10}{x^2 + 5}$

41. $p(x) = \dfrac{3x^2 - 5}{x^2 - 1}$ **42.** $P(x) = \dfrac{3x^2 - 5x - 2}{x^2 - 4}$

Give the location of the x- and y-intercepts (if they exist), and discuss the behavior of the function (bounce or cross) at each x-intercept.

43. $f(x) = \dfrac{x^2 - 3x}{x^2 - 5}$ **44.** $F(x) = \dfrac{2x - x^2}{x^2 + 2x - 3}$

45. $g(x) = \dfrac{x^2 + 3x - 4}{x^2 - 1}$ **46.** $G(x) = \dfrac{x^2 + 7x + 6}{x^2 - 2}$

47. $h(x) = \dfrac{x^3 - 6x^2 + 9x}{4 - x^2}$ **48.** $H(x) = \dfrac{4x + 4x^2 + x^3}{x^2 - 1}$

Use the *Guidelines for Graphing Rational Functions* to graph the functions given.

49. $f(x) = \dfrac{x + 3}{x - 1}$ **50.** $g(x) = \dfrac{x - 4}{x + 2}$

51. $F(x) = \dfrac{8x}{x^2 + 4}$ **52.** $G(x) = \dfrac{-12x}{x^2 + 3}$

53. $p(x) = \dfrac{-2x^2}{x^2 - 4}$ **54.** $P(x) = \dfrac{3x^2}{x^2 - 9}$

55. $q(x) = \dfrac{2x - x^2}{x^2 + 4x - 5}$ **56.** $Q(x) = \dfrac{x^2 + 3x}{x^2 - 2x - 3}$

57. $h(x) = \dfrac{-3x}{x^2 - 6x + 9}$ **58.** $H(x) = \dfrac{2x}{x^2 - 2x + 1}$

59. $Y_1 = \dfrac{x - 1}{x^2 - 3x - 4}$ **60.** $Y_2 = \dfrac{1 - x}{x^2 - 2x}$

61. $s(x) = \dfrac{4x^2}{2x^2 + 4}$ **62.** $S(x) = \dfrac{-2x^2}{x^2 + 1}$

63. $Y_1 = \dfrac{x^2 - 4}{x^2 - 1}$ **64.** $Y_2 = \dfrac{x^2 - x - 6}{x^2 + x - 6}$

65. $v(x) = \dfrac{-2x}{x^3 + 2x^2 - 4x - 8}$

66. $V(x) = \dfrac{3x}{x^3 + x^2 - x - 1}$

Use the vertical asymptotes, *x*-intercepts, and their multiplicities to construct an equation that corresponds to each graph. Be sure the *y*-intercept estimated from the graph matches the value given by your equation for $x = 0$. Check work on a graphing calculator.

67.

68.

69.

70.

▶ WORKING WITH FORMULAS

71. Population density: $D(x) = \dfrac{ax}{x^2 + b}$

The population density of urban areas (in people per square mile) can be modeled by the formula shown, where *a* and *b* are constants related to the overall population and sprawl of the area under study, and $D(x)$ is the population density (in hundreds), *x* mi from the center of downtown.

Graph the function for $a = 63$ and $b = 20$ over the interval $x \in [0, 25]$, and then use the graph to answer the following questions.

  a. What is the significance of the *horizontal asymptote* (what does it mean in this context)?

  b. How far from downtown does the population density fall below 525 people per square mile? How far until the density falls below 300 people per square mile?

  c. Use the graph and a table to determine how far from downtown the population density reaches a maximum? What is this maximum?

72. Cost of removing pollutants: $C(x) = \dfrac{kx}{100 - x}$

Some industries resist cleaner air standards because the cost of removing pollutants rises dramatically as higher standards are set. This phenomenon can be modeled by the formula given, where $C(x)$ is the cost (in thousands of dollars) of removing *x*% of the pollutant and *k* is a constant that depends on the type of pollutant and other factors.

Graph the function for $k = 250$ over the interval $x \in [0, 100]$, and then use the graph to answer the following questions.

  a. What is the significance of the *vertical asymptote* (what does it mean in this context)?

  b. If new laws are passed that require 80% of a pollutant to be removed, while the existing law requires only 75%, how much will the new legislation cost the company? Compare the cost of the 5% increase from 75% to 80% with the cost of the 1% increase from 90% to 91%.

  c. What percent of the pollutants can be removed if the company budgets 2250 thousand dollars?

▶ APPLICATIONS

73. For a certain coal-burning power plant, the cost to remove pollutants from plant emissions can be modeled by $C(p) = \dfrac{80p}{100 - p}$, where $C(p)$ represents the cost (in thousands of dollars) to remove *p* percent of the pollutants. (a) Find the cost to remove 20%, 50%, and 80% of the pollutants, then comment on the results; (b) graph the function using an appropriate scale; and (c) use the direction/approach notation to state what happens if the power company attempts to remove 100% of the pollutants.

74. A large city has initiated a new recycling effort, and wants to distribute recycling bins for use in separating various recyclable materials. City planners anticipate the cost of the program can be modeled by the function $C(p) = \dfrac{220p}{100 - p}$, where $C(p)$ represents the cost (in $10,000) to distribute the bins to p percent of the population. (a) Find the cost to distribute bins to 25%, 50%, and 75% of the population, then comment on the results; (b) graph the function using an appropriate scale; and (c) use the direction/approach notation to state what happens if the city attempts to give recycling bins to 100% of the population.

75. The concentration C of a certain medicine in the bloodstream h hours after being injected into the shoulder is given by the function: $C(h) = \dfrac{2h^2 + h}{h^3 + 70}$. Use the given

graph of the function to answer the following questions.

   a. Approximately how many hours after injection did the maximum concentration occur? What was the maximum concentration?

   b. Use $C(h)$ to *compute* the rate of change for the intervals $h = 8$ to $h = 10$ and $h = 20$ to $h = 22$. What do you notice?

   c. Use the direction/approach notation to state what happens to the concentration C as the number of hours becomes infinitely large. What role does the h-axis play for this function?

76. In response to certain market demands, manufacturers will quickly get a product out on the market to take advantage of consumer interest. Once the product is released, it is not uncommon for sales to initially skyrocket, taper off and then gradually decrease as consumer interest wanes. For a certain product, sales can be modeled by the function $S(t) = \dfrac{250t}{t^2 + 150}$, where $S(t)$ represents the daily sales (in $10,000) t days after the product has debuted. Use the given graph of the function to answer the following questions.

   a. Approximately how many days after the product came out did sales reach a maximum? What was the maximum sales?

   b. Use $S(t)$ to compute the rate of change for the intervals $t = 7$ to $t = 8$ and $t = 60$ to $t = 62$. What do you notice?

   c. Use the direction/approach notation to state what happens to the daily sales S as the number of days becomes infinitely large. What role does the t-axis play for this function?

Memory retention: Due to their asymptotic behavior, rational functions are often used to model the mind's ability to retain information over a long period of time—the "use it or lose it" phenomenon.

77. A large group of students is asked to memorize a list of 50 Italian words, a language that is unfamiliar to them. The group is then tested regularly to see how many of the words are retained over a period of time. The average number of words retained is modeled by the function $W(t) = \dfrac{6t + 40}{t}$, where $W(t)$ represents the number of words remembered after t days.

   a. Graph the function over the interval $t \in [0, 40]$. How many days until only half the words are remembered? How many days until only one-fifth of the words are remembered?

   b. After 10 days, what is the average number of words retained? How many days until only 8 words can be recalled?

   c. What is the significance of the horizontal asymptote (what does it mean in this context)?

78. A similar study asked students to memorize 50 Hawaiian words, a language that is both unfamiliar and phonetically foreign to them (see Exercise 77). The average number of words retained is modeled by the function $W(t) = \dfrac{4t + 20}{t}$, where $W(t)$ represents the number of words after t days.

   a. Graph the function over the interval $t \in [0, 40]$. How many days until only half the words are remembered? How does this compare to Exercise 77? How many days until only one-fifth of the words are remembered?

   b. After 7 days, what is the average number of words retained? How many days until only 5 words can be recalled?

   c. What is the significance of the horizontal asymptote (what does it mean in this context)?

Concentration and dilution: When antifreeze is mixed with water, it becomes diluted—less than 100% antifreeze. The more water added, the less concentrated the antifreeze becomes, with this process continuing until a desired concentration is met. This application and many similar to it can be modeled by rational functions.

79. A 400-gal tank currently holds 40 gal of a 25% antifreeze solution. To raise the concentration of the antifreeze in the tank, x gal of a 75% antifreeze solution is pumped in.

   a. Show the formula for the resulting concentration is $C(x) = \dfrac{40 + 3x}{160 + 4x}$ after simplifying, and graph the function over the interval $x \in [0, 360]$.

   b. What is the concentration of the antifreeze in the tank after 10 gal of the new solution are added? After 120 gal have been added? How much liquid is now in the tank?

   c. If the concentration level is now at 65%, how many gallons of the 75% solution have been added? How many gallons of liquid are in the tank now?

   d. What is the maximum antifreeze concentration that can be attained in a tank of this size? What is the maximum concentration that can be attained in a tank of "unlimited" size?

80. A sodium chloride solution has a concentration of 0.2 oz (weight) per gallon. The solution is pumped into an 800-gal tank currently holding 40 gal of pure water, at a rate of 10 gal/min.

   a. Find a function $A(t)$ modeling the amount of liquid in the tank after t min, and a function $S(t)$ for the amount of sodium chloride in the tank after t min.

   b. The concentration $C(t)$ in ounces per gallon is measured by the ratio $\dfrac{S(t)}{A(t)}$, a rational function. Graph the function on the interval $t \in [0, 100]$. What is the concentration level (in ounces per gallon) after 6 min? After 28 min? How many gallons of liquid are in the tank at this time?

   c. If the concentration level is now 0.184 oz/gal, how long have the pumps been running? How many gallons of liquid are in the tank now?

   d. What is the maximum concentration that can be attained in a tank of this size? What is the maximum concentration that can be attained in a tank of "unlimited" size?

Average cost of manufacturing an item: The cost "C" to manufacture an item depends on the relatively fixed costs "K" for remaining in business (utilities, maintenance, transportation, etc.) and the actual cost "c" of manufacturing the item (labor and materials). For x items the cost is $C(x) = K + cx$. The average cost "A" of manufacturing an item is then $A(x) = \dfrac{C(x)}{x}$.

81. A company that manufactures water heaters finds their fixed costs are normally $50,000 per month, while the cost to manufacture each heater is $125. Due to factory size and the current equipment, the company can produce a maximum of 5000 water heaters per month during a good month.

   a. Use the average cost function to find the average cost if 500 water heaters are manufactured each month. What is the average cost if 1000 heaters are made?

   b. What level of production will bring the average cost down to $150 per water heater?

   c. If the average cost is currently $137.50, how many water heaters are being produced that month?

   d. What's the significance of the horizontal asymptote for the average cost function (what does it mean in this context)? Will the company ever break the $130 average cost level? Why or why not?

82. An enterprising company has finally developed a disposable diaper that is biodegradable. The brand becomes wildly popular and production is soaring. The fixed cost of production is $20,000 per month, while the cost of manufacturing is $6.00 per case (48 diapers). Even while working three shifts around-the-clock, the maximum production level is 16,000 cases per month. The company figures it will be profitable if it can bring costs down to an average of $7 per case.

   a. Use the average cost function to find the average cost if 2000 cases are produced each month. What is the average cost if 4000 cases are made?

   b. What level of production will bring the average cost down to $8 per case?

   c. If the average cost is currently $10 per case, how many cases are being produced?

   d. What's the significance of the horizontal asymptote for the average cost function (what does it mean in this context)? Will the company ever reach its goal of $7/case at its maximum production? What level of production would help them meet their goal?

Test averages and grade point averages: To calculate a test average we sum all test points P and divide by the number of tests N: $\dfrac{P}{N}$. To compute

the score or scores needed on future tests to raise the average grade to a desired grade G, we add the number of additional tests n to the denominator, and the number of additional tests times the projected grade g on each test to the numerator:

$G(n) = \dfrac{P + ng}{N + n}$. The result is a rational function with some "eye-opening" results.

83. After four tests, Bobby Lou's test average was an 84. [*Hint: P* = 4(84) = 336.]

a. Assume that she gets a 95 on all remaining tests ($g = 95$). Graph the resulting function on a calculator using the window $n \in [0, 20]$ and $G(n) \in [80$ to $100]$. Use the calculator to determine how many tests are required to lift her grade to a 90 under these conditions.

b. At some colleges, the range for an "A" grade is 93–100. How many tests would Bobby Lou have to score a 95 on, to raise her average to higher than 93? Were you surprised?

c. Describe the significance of the horizontal asymptote of the average grade function. Is a test average of 95 possible for her under these conditions?

d. Assume now that Bobby Lou scores 100 on all remaining tests ($g = 100$). Approximately how many more tests are required to lift her grade average to higher than 93?

84. At most colleges, $A \rightarrow 4$ grade points, $B \rightarrow 3$, $C \rightarrow 2$, and $D \rightarrow 1$. After taking 56 credit hours, Aurelio's GPA is 2.5. [*Hint:* In the formula given, $P = 2.5(56) = 140$.]

a. Assume Aurelio is determined to get A's (4 grade points or $g = 4$), for all remaining credit hours. Graph the resulting function on a calculator using the window $n \in [0, 60]$ and $G(n) \in [2, 4]$. Use the calculator to determine the number of credit hours required to lift his GPA to over 2.75 under these conditions.

b. At some colleges, scholarship money is available only to students with a 3.0 average or higher. How many (perfect 4.0) credit hours would Aurelio have to earn, to raise his GPA to 3.0 or higher? Were you surprised?

c. Describe the significance of the horizontal asymptote of the GPA function. Is a GPA of 4.0 possible for him under these conditions?

▶ **EXTENDING THE CONCEPT**

85. In addition to determining *if* a function has a vertical asymptote, we are often interested in *how fast* the graph approaches the asymptote. As in previous investigations, this involves the function's rate of change over a small interval. Exercise 72 describes the rising cost of removing pollutants from the air. As noted there, the rate of increase in the cost changes as higher requirements are set. To quantify this change, we'll compute the rate of change

$\dfrac{\Delta C}{\Delta x} = \dfrac{C(x_2) - C(x_1)}{x_2 - x_1}$ for $C(x) = \dfrac{250x}{100 - x}$.

a. Find the rate of change of the function in the following intervals:

$x \in [60, 61]$ $x \in [70, 71]$
$x \in [80, 81]$ $x \in [90, 91]$

b. What do you notice? How much did the rate increase from the first interval to the second? From the second to the third? From the third to the fourth?

c. Recompute parts (a) and (b) using the function $C(x) = \dfrac{350x}{100 - x}$. Comment on what you notice.

86. Consider the function $f(x) = \dfrac{ax^2 + k}{bx^2 + h}$, where a, b, k, and h are constants and $a, b > 0$.

a. What can you say about asymptotes and intercepts of this function if $h, k > 0$?

b. Now assume $k < 0$ and $h > 0$. How does this affect the asymptotes? The intercepts?

c. If $b = 1$ and $a > 1$, how does this affect the results from part (b)?

d. How is the graph affected if $k > 0$ and $h < 0$?

e. Find values of a, b, h, and k that create a function with a horizontal asymptote at $y = \frac{3}{2}$, x-intercepts at $(-2, 0)$ and $(2, 0)$, a y-intercept of $(0, -4)$, and no vertical asymptotes.

87. The horizontal asymptotes of a rational function, and whether or not a graph crosses this asymptote, can be found using long division. The quotient polynomial $q(x)$ gives the equation of the asymptote, and the zeroes of the remainder $r(x)$ will indicate if and where the graph crosses it. Use this idea to help graph these functions.

a. $V(x) = \dfrac{3x^2 - 16x - 20}{x^2 - 3x - 10}$

b. $v(x) = \dfrac{-2x^2 + 4x + 13}{x^2 - 2x - 3}$

▶ **MAINTAINING YOUR SKILLS**

88. (R.1/1.4) Describe/Define each set of numbers: complex C, rational Q, and integers Z.

89. (2.3) Find the equation of a line that is perpendicular to $3x - 4y = 12$ and contains the point $(2, -3)$.

90. (1.5) Solve the following equation using the quadratic formula, then write the equation in factored form: $12x^2 + 55x - 48 = 0$.

91. (3.2) Use synthetic division and the remainder theorem to find the value of $f(4)$, $f(\frac{3}{2})$, and $f(2)$: $f(x) = 2x^3 - 7x^2 + 5x + 3$.

3.6 | Additional Insights into Rational Functions

Learning Objectives

In Section 3.6 you will learn how to:

☐ **A.** Graph rational functions with removable discontinuities

☐ **B.** Graph rational functions with oblique or non-linear asymptotes

☐ **C.** Solve applications involving rational functions

In Section 3.5, we saw that rational graphs can have both a horizontal and vertical asymptote. In this section, we'll study functions with asymptotes that are *neither* horizontal nor vertical. In addition, we'll further explore the "break" we saw in graphs of certain piecewise-defined functions, that of a simple "hole" created when the numerator and denominator share a common variable factor.

A. Rational Functions and Removable Discontinuities

In Example 5 of Section 2.7, we graphed the piecewise-defined function $h(x) = \begin{cases} \dfrac{x^2 - 4}{x - 2} & x \neq 2 \\ 1 & x = 2 \end{cases}$.

Figure 3.39

The second piece is simply the point $(2, 1)$. The first piece is a rational function, but instead of a vertical asymptote at $x = 2$ (the zero of the denominator), its graph was actually the line $y = x + 2$ with a "hole" at $(2, 4)$, called a **removable discontinuity** (Figure 3.39). As the name implies, we can *remove* or fix this break by redefining the second piece as $h(x) = 4$, when $x = 2$. This would create a new and continuous function,

$$H(x) = \begin{cases} \dfrac{x^2 - 4}{x - 2} & x \neq 2 \\ 4 & x = 2 \end{cases}$$ (Figure 3.40). It's possible

for a rational graph to have more than one removable discontinuity, or to be nonlinear with a removable discontinuity. For cases where we elect to repair the break, we will adopt the convention of using the corresponding upper case letter to name the new function, as we did here.

Figure 3.40

WORTHY OF NOTE

The graph of $f(x) = \dfrac{1}{x - 2}$ also has a break at $x = 2$, but this time the result is a *vertical asymptote*. The difference is the numerator and denominator of $h(x) = \dfrac{x^2 - 4}{x - 2}$ share a common factor, and canceling these factors leaves $y = x + 2$, which is a continuous function. However, the *original function* is not defined at $x = 2$, so we must remove the single point $(2, 4)$ from the domain of $y = x + 2$ (Figure 3.39).

EXAMPLE 1 ▶   Graphing Rational Functions with Removable Discontinuities

Graph the function $t(x) = \dfrac{x^3 + 8}{x + 2}$. If there is a removable discontinuity, repair the break using an appropriate piecewise-defined function.

Solution ▶ Note the domain of t does not include $x = -2$. We begin by factoring as before to identify zeroes and asymptotes, but find the numerator and denominator share a common factor, which we remove.

$$t(x) = \frac{x^3 + 8}{x + 2}$$

$$= \frac{(\cancel{x + 2})(x^2 - 2x + 4)}{\cancel{x + 2}}$$

$$= x^2 - 2x + 4; \text{ where } x \neq -2$$

The graph of t will be the same as $y = x^2 - 2x + 4$ *for all values except* $x = -2$. Here we have a parabola, opening upward, with y-intercept $(0, 4)$. From the vertex formula, the x-coordinate of the vertex will be $\dfrac{-b}{2a} = \dfrac{-(-2)}{2(1)} = 1$, giving $y = 3$ after substitution. The vertex is $(1, 3)$. Evaluating $t(-1)$ we find $(-1, 7)$ is on the graph, giving the point $(3, 7)$ using the axis of symmetry. We draw a parabola through these points, noting the original function is not defined at -2, and there will be a "hole" in the graph at $(-2, y)$. The value of y is found by substituting -2 for x in the simplified form: $(-2)^2 - 2(-2) + 4 = 12$. This information produces the graph shown. We can repair the break using the function

$$T(x) = \begin{cases} \dfrac{x^3 + 8}{x + 2} & x \neq -2 \\ 12 & x = -2 \end{cases}$$

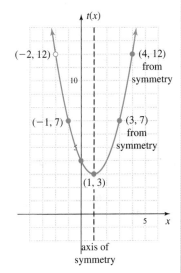

WORTHY OF NOTE

For more on removable discontinuities, see the *Technology Highlight* feature on page 370.

✓ **A.** You've just learned how to graph rational functions with removable discontinuities

Now try Exercises 7 through 18 ▶

B. Rational Functions with Oblique and Nonlinear Asymptotes

In Section 3.5, we found that for $V(x) = \dfrac{p(x)}{d(x)}$, the location of nonvertical asymptotes was determined by comparing the degree of p with the degree of d. As review, for $p(x)$ with leading term ax^n and $d(x)$ with leading term degree bx^m,

• If $n < m$, the line $y = 0$ is a horizontal asymptote.
• If $n = m$, the line $y = \frac{a}{b}$ is a horizontal asymptote.

But what happens if the degree of the numerator is *greater than* the degree of the denominator? To investigate, consider the functions f, g, and h in Figures 3.41 to 3.43, whose only difference is the degree of the numerator.

Figure 3.41

$$f(x) = \frac{2x}{x^2 + 1}$$

Figure 3.42

$$g(x) = \frac{2x^2}{x^2 + 1}$$

Figure 3.43

$$h(x) = \frac{2x^3}{x^2 + 1}$$

The graph of f has a horizontal asymptote at $y = 0$ since the denominator is of larger degree (as $|x| \to \infty$, $y \to 0$). As we might have anticipated, the horizontal asymptote for g is $y = 2$, the ratio of leading coefficients (as $|x| \to \infty$, $y \to 2$). The graph of h has no horizontal asymptote, yet appears to be asymptotic to some slanted line. The table in Figure 3.44 suggests that as $|x| \to \infty$, $y \to 2x = Y_2$. To see why, note the function $h(x) = \dfrac{2x^3}{x^2 + 1}$ can be considered an "improper fraction," similar to how we apply this designation to the fraction $\frac{3}{2}$. To write h in "proper" form, we use long division, writing the dividend as $2x^3 + 0x^2 + 0x + 0$, and the divisor as $x^2 + 0x + 1$.

Figure 3.44

Y$_2$ = 2X

The ratio $\dfrac{2x^3 \text{ from dividend}}{x^2 \text{ from divisor}}$ shows $2x$ will be our first multiplier.

$$
\begin{array}{r}
2x \\
x^2 + 0x + 1 \overline{)\, 2x^3 + 0x^2 + 0x + 0} \\
\underline{-(2x^3 + 0x^2 + 2x)} \\
-2x
\end{array}
$$

multiply $2x(x^2 + 0x + 1)$

subtract, next term is 0

The result shows $h(x) = 2x + \dfrac{-2x}{x^2 + 1}$. Note as $|x| \to \infty$, the term $\dfrac{-2x}{x^2 + 1}$ becomes very small and closer to zero, so $h(x) \approx 2x$ for large x. This is an example of an **oblique asymptote.** In general,

---

Oblique and Nonlinear Asymptotes

Given $V(x) = \frac{p(x)}{d(x)}$ is a rational function in simplest form, where the degree of p is greater than the degree of d, the graph will have an oblique or nonlinear asymptote as determined by $q(x)$, where $q(x)$ is the quotient polynomial after division.

We conclude that an oblique or slant asymptote occurs when the degree of the numerator is one more than the degree of the denominator, and a nonlinear asymptote occurs when its degree is larger by two or more.

EXAMPLE 2 ▶ **Graphing a Rational Function with an Oblique Asymptote**

Graph the function $f(x) = \dfrac{x^2 - 1}{x}$.

Solution ▶ Using the *Guidelines*, we find $f(x) = \dfrac{(x + 1)(x - 1)}{x}$ and proceed:

1. *y*-intercept: The graph has no *y*-intercept.
2. Vertical asymptote(s): $x = 0$ with multiplicity 1. The function will change sign at $x = 0$.
3. *x*-intercepts: From $(x + 1)(x - 1) = 0$, the *x*-intercepts are $(-1, 0)$ and $(1, 0)$. Since both have multiplicity 1, the graph will cross the *x*-axis and the function will change sign at these points
4. Horizontal/oblique asymptote: Since the degree of numerator $>$ the degree of denominator, we rewrite *f* using division. Using term-by-term division (the denominator is a monomial) produces $f(x) = \dfrac{x^2 - 1}{x} = \dfrac{x^2}{x} - \dfrac{1}{x} = x - \dfrac{1}{x}$. The quotient polynomial is $q(x) = x$ and the graph has the oblique asymptote $y = x$.
5. To determine if the function will cross the asymptote, we solve

$$\dfrac{x^2 - 1}{x} = x \qquad \text{\small } q(x) = x \text{ is the slant asymptote}$$

$$x^2 - 1 = x^2 \qquad \text{\small multiply by } x$$

$$-1 = 0 \qquad \text{\small no solutions possible}$$

The graph will not cross the oblique asymptote.

The information from steps 1 through 5 is displayed in Figure 3.45. While this is sufficient to complete the graph, we select $x = -4$ and 4 to compute additional points and find $f(-4) = -\frac{15}{4}$ and $f(4) = \frac{15}{4}$. To meet all necessary conditions, we complete the graph as shown in Figure 3.46.

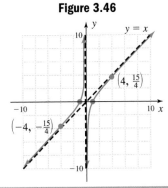

Figure 3.45

Figure 3.46

Now try Exercises 19 through 24 ▶

EXAMPLE 3 ▶ **Graphing a Rational Function with an Oblique Asymptote**

Graph the function: $h(x) = \dfrac{x^2}{x - 1}$

Solution ▶ The function is already in "factored form."

1. *y*-intercept: Since $h(0) = 0$, the *y*-intercept is $(0, 0)$.
2. Vertical asymptote: Solving $x - 1 = 0$ gives $x = 1$ with multiplicity one. There is a vertical asymptote at $x = 1$ and the function will change sign here.

3. *x*-intercept: (0, 0); From, $x^2 = 0$, we have $x = 0$ with multiplicity two. The *x*-intercept is (0, 0) and the function will not change sign here.

4. Horizontal/oblique asymptote: Since the degree of numerator > the degree of denominator, we rewrite *h* using division. The denominator is linear so we use synthetic division:

$$\begin{array}{r|rrr} \text{use 1 as a ``divisor''} \quad 1 & 1 & 0 & 0 \quad \text{coefficients of dividend} \\ & \downarrow & 1 & 1 \\ \hline & 1 & 1 & 1 \quad \text{quotient and remainder} \end{array}$$

Since $q(x) = x + 1$ the graph has an oblique asymptote at $y = x + 1$.

5. To determine if the function crosses the asymptote, we solve

$$\frac{x^2}{x - 1} = x + 1 \qquad q(x) = x + 1 \text{ is the slant asymptote}$$
$$x^2 = x^2 - 1 \qquad \text{cross multiply}$$
$$0 = -1 \qquad \text{no solutions possible}$$

The graph will not cross the slant asymptote.

The information gathered in steps 1 through 5 is shown Figure 3.47, and is actually sufficient to complete the graph. If you feel a little unsure about how to "puzzle" out the graph, find additional points in the first and third quadrants: $h(2) = 4$ and $h(-2) = -\frac{4}{3}$. Since the graph will "bounce" at $x = 0$ and output values must change sign at $x = 1$, all conditions are met with the graph shown in Figure 3.48.

Figure 3.47

Figure 3.48

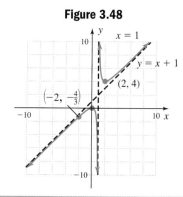

Now try Exercises 25 through 46 ▶

Finally, it would be a mistake to think that all asymptotes are linear. In fact, when the degree of the numerator is two more than the degree of the denominator, a parabolic asymptote results. Functions of this type often occur in applications of rational functions, and are used to minimize cost, materials, distances, or other considerations of great importance to business and industry. For $f(x) = \dfrac{x^4 + 1}{x^2}$, term-by-term division

Figure 3.49

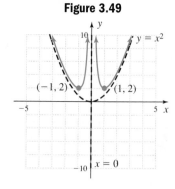

☑ **B.** You've just learned how to graph rational functions with oblique or nonlinear asymptotes

gives $x^2 + \dfrac{1}{x^2}$ and the quotient $q(x) = x^2$ is a nonlinear, parabolic asymptote (see Figure 3.49). For more on nonlinear asymptotes, **see Exercises 47 through 50.**

C. Applications of Rational Functions

Rational functions have applications in a wide variety of fields, including environmental studies, manufacturing, and various branches of medicine. In most practical applications, only the values from Quadrant I have meaning since inputs and outputs must often be positive **(see Exercises 51 and 52).** Here we investigate an application involving manufacturing and average cost.

EXAMPLE 4 ▶ Solving an Application of Rational Functions

Suppose the cost (in thousands of dollars) of manufacturing x thousand of a given item is modeled by the function $C(x) = x^2 + 4x + 3$. The *average cost* of each item would then be expressed by

$$A(x) = \frac{x^2 + 4x + 3}{x} = \frac{\text{total cost}}{\text{number of items}}$$

a. Graph the function $A(x)$.

b. Find how many thousand items are manufactured when the average cost is $8.

c. Determine how many thousand items should be manufactured to minimize the average cost (use the graph to estimate this minimum average cost).

Solution ▶ a. The function is already in simplest form.

1. y-intercept: none [$A(0)$ is undefined]

2. Vertical asymptote: $x = 0$, multiplicity one; the function will change sign at $x = 0$.

3. x-intercept(s): After factoring we obtain $(x + 3)(x + 1) = 0$, and the zeroes of the numerator are $x = -1$ and $x = -3$, both with multiplicity one. The graph will cross the x-axis at each intercept.

4. Horizontal/oblique asymptote: The degree of numerator > the degree of denominator, so we divide using term-by-term division:

$$\frac{x^2 + 4x + 3}{x} = \frac{x^2}{x} + \frac{4x}{x} + \frac{3}{x}$$

$$= x + 4 + \frac{3}{x}$$

The line $q(x) = x + 4$ is an oblique asymptote.

5. Solve

$$\frac{x^2 + 4x + 3}{x} = x + 4 \qquad q(x) = x + 4 \text{ is a slant asymptote}$$

$$x^2 + 4x + 3 = x^2 + 4x \qquad \text{cross multiply}$$

$$3 = 0 \qquad \text{no solutions possible}$$

The graph will not cross the slant asymptote.

The function changes sign at both x-intercepts and at the asymptote $x = 0$. The information from steps 1 through 5 is shown in Figure 3.50 and perhaps an additional point in Quadrant I would help to complete the graph: $A(1) = 8$. The point $(1, 8)$ is on the graph, showing A is positive in the interval containing 1. Since output values will alternate in sign as stipulated above, all conditions are met with the graph shown in Figure 3.51.

Figure 3.50

Figure 3.51

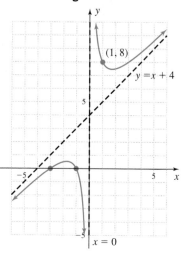

b. To find the number of items manufactured when average cost is $8, we replace $A(x)$ with 8 and solve: $\dfrac{x^2 + 4x + 3}{x} = 8$:

$$x^2 + 4x + 3 = 8x$$
$$x^2 - 4x + 3 = 0$$
$$(x - 1)(x - 3) = 0$$
$$x = 1 \quad \text{or} \quad x = 3$$

The average cost is $8 when 1000 items or 3000 items are manufactured.

c. From the graph, it appears that the minimum average cost is close to $7.50, when approximately 1500 to 1800 items are manufactured.

> **Now try Exercises 55 and 56 ▶**

GRAPHICAL SUPPORT

In the *Technology Highlight* from Section 2.5, we saw how a graphing calculator can be used to locate the extreme values of a function. Applying this technology to the graph from Example 4 we find that the minimum average cost is approximately $7.46, when about 1732 items are manufactured.

In some applications, the functions we use are initially defined in *two variables* rather than just one, as in $H(x, y) = (x - 50)(y - 80)$. However, in the solution process a substitution is used to rewrite the relationship as a rational function in one variable and we can proceed as before.

EXAMPLE 5 ▶ **Using a Rational Function to Solve a Layout Application**

The building codes in a new subdivision require that a rectangular home be built at least 20 ft from the street, 40 ft from the neighboring lots, and 30 ft from the house to the rear fence line.

a. Find a function $A(x, y)$ for the area of the lot, and a function $H(x, y)$ for the area of the home (the inner rectangle).

b. If a new home is to have a floor area of 2000 ft², $H(x, y) = 2000$. Substitute 2000 for $H(x, y)$ and solve for y, then substitute the result in $A(x, y)$ to write the area A as a function of x alone (simplify the result).

c. Graph $A(x)$ on a calculator, using the window $X \in [-50, 150]$; $Y \in [-30{,}000, 30{,}000]$. Then graph $y = 80x + 2000$ on the same screen. How are these two graphs related?

d. Use the graph of $A(x)$ in Quadrant I to determine the minimum dimensions of a lot that satisfies the subdivision's requirements (to the nearest tenth of a foot). Also state the dimensions of the house.

Solution ▶ **a.** The area of the lot is simply width times length, so $A(x, y) = xy$. For the house, these dimensions are decreased by 50 ft and 80 ft, respectively, so $H(x, y) = (x - 50)(y - 80)$.

b. Given $H(x, y) = 2000$ produces the equation $2000 = (x - 50)(y - 80)$, and solving for y gives

$$2000 = (x - 50)(y - 80) \quad \text{given equation}$$

$$\frac{2000}{x - 50} = y - 80 \quad \text{divide by } x - 50$$

$$\frac{2000}{x - 50} + 80 = y \quad \text{add 80}$$

$$\frac{2000}{x - 50} + \frac{80(x - 50)}{x - 50} = y \quad \text{find LCD}$$

$$\frac{80x - 2000}{x - 50} = y \quad \text{combine terms}$$

Substituting this expression for y in $A(x, y) = xy$ produces

$$A(x) = x\left(\frac{80x - 2000}{x - 50}\right) \quad \text{substitute } \frac{80x - 2000}{x - 50} \text{ for } y$$

$$= \frac{80x^2 - 2000x}{x - 50} \quad \text{multiply}$$

c. The graph of $Y1 = A(x)$ appears in Figure 3.52 using the prescribed window. $Y_2 = 80x + 2000$ appears to be an oblique asymptote for A, which can be verified using synthetic division.

Figure 3.52

d. Using the $\boxed{\text{2nd}}$ $\boxed{\text{TRACE}}$ (CALC) **3:minimum** feature of a calculator, the minimum width is $x \approx 85.4$ ft. Substituting 85.4 for x in

$$y = \frac{80x - 2000}{x - 50},\ \text{gives the length}$$

$y \approx 136.5$ ft. The dimensions of the house must be $85.4 - 50 = 35.4$ ft, by $136.5 - 80 = 56.5$ ft (see Figure 3.53).

As expected, the area of the house will be $(35.4)(56.5) \approx 2000\ \text{ft}^2$.

☑ **C.** You've just learned how to solve applications involving rational functions

Figure 3.53

Now try Exercises 57 through 60 ▶

TECHNOLOGY HIGHLIGHT

Removable Discontinuities

Graphing calculators offer both numerical and visual representations of removable discontinuities. For instance, enter the function $r(x) = \dfrac{x^2 - 4x + 3}{x - 1}$ on the $\boxed{\text{Y =}}$ screen, then use the $\boxed{\text{TBLSET}}$ feature to set up the table as shown in Figure 3.54. Pressing $\boxed{\text{2nd}}$ $\boxed{\text{GRAPH}}$ displays the expected table, which shows the function cannot be evaluated at $x = 1$ (see Figure 3.55). Now change the $\boxed{\text{TBLSET}}$ screen so that **ΔTbl = 0.01**. Note again that the function is defined for all values except $x = 1$. Reset the table to **ΔTbl = 0.001** and investigate further.

Figure 3.54

We can actually see the gap or hole in the graph using a "friendly window." Since the screen of the TI-84 Plus is 95 pixels wide and 63 pixels high, multiples of 4.7 for Xmin and Xmax, and multiples of 3.1 for Ymin and Ymax, display what happens at integer (and other) values (see Figure 3.56). Pressing $\boxed{\text{GRAPH}}$ gives Figure 3.57, which shows a noticeable gap at $(1, -2)$. With the $\boxed{\text{TRACE}}$ feature, move the cursor over to the gap and notice what happens.

Use these ideas to view the discontinuities in the following rational functions. State the ordered pair location of each discontinuity.

Figure 3.55

Figure 3.56

Figure 3.57

Exercise 1: $r(x) = \dfrac{x^2 - 4}{x + 2}$

Exercise 2: $f(x) = \dfrac{x^2 - 2x - 3}{x + 1}$

Exercise 3: $r(x) = \dfrac{x^3 + 1}{x + 1}$

Exercise 4: $f(x) = \dfrac{x^3 - 7x + 6}{x^2 + x - 6}$

3.6 EXERCISES

▶ CONCEPTS AND VOCABULARY

Fill in each blank with the appropriate word or phrase. Carefully reread the section if needed.

1. The discontinuity in the graph of $y = \dfrac{1}{(x + 3)^2}$ is

   called a _____ discontinuity, since it cannot be "repaired."

2. If the degree of the numerator is greater than the degree of the denominator, the graph will have an _____ or _____ asymptote.

3. If the degree of the numerator is _____ more than the degree of the denominator, the graph will have a parabolic asymptote.

4. If the denominator is a _____, use term by term division to find the quotient. Otherwise _____ or long division must be used.

5. Discuss/Explain how you would create a function with a parabolic asymptote and two vertical asymptotes.

6. Complete Exercise 7 in expository form. That is, work this exercise out completely, discussing each step of the process as you go.

▶ DEVELOPING YOUR SKILLS

Graph each function. If there is a removable discontinuity, repair the break using an appropriate piecewise-defined function.

7. $f(x) = \dfrac{x^2 - 4}{x + 2}$

8. $f(x) = \dfrac{x^2 - 9}{x + 3}$

9. $g(x) = \dfrac{x^2 - 2x - 3}{x + 1}$

10. $g(x) = \dfrac{x^2 - 3x - 10}{x - 5}$

11. $h(x) = \dfrac{3x - 2x^2}{2x - 3}$

12. $h(x) = \dfrac{4x - 5x^2}{5x - 4}$

13. $p(x) = \dfrac{x^3 - 8}{x - 2}$

14. $p(x) = \dfrac{8x^3 - 1}{2x - 1}$

15. $q(x) = \dfrac{x^3 - 7x - 6}{x + 1}$

16. $q(x) = \dfrac{x^3 - 3x + 2}{x + 2}$

17. $r(x) = \dfrac{x^3 + 3x^2 - x - 3}{x^2 + 2x - 3}$

18. $r(x) = \dfrac{x^3 - 2x^2 - 4x + 8}{x^2 - 4}$

Graph each function using the *Guidelines for Graphing Rational Functions,* which is simply modified to include nonlinear asymptotes. Clearly label all intercepts and asymptotes and any additional points used to sketch the graph.

19. $Y_1 = \dfrac{x^2 - 4}{x}$

20. $Y_2 = \dfrac{x^2 - x - 6}{x}$

21. $v(x) = \dfrac{3 - x^2}{x}$

22. $V(x) = \dfrac{7 - x^2}{x}$

23. $w(x) = \dfrac{x^2 + 1}{x}$

24. $W(x) = \dfrac{x^2 + 4}{2x}$

25. $h(x) = \dfrac{x^3 - 2x^2 + 3}{x^2}$

26. $H(x) = \dfrac{x^3 + x^2 - 2}{x^2}$

27. $Y_1 = \dfrac{x^3 + 3x^2 - 4}{x^2}$

28. $Y_2 = \dfrac{x^3 - 3x^2 + 4}{x^2}$

29. $f(x) = \dfrac{x^3 - 3x + 2}{x^2}$

30. $F(x) = \dfrac{x^3 - 12x - 16}{x^2}$

31. $Y_3 = \dfrac{x^3 - 5x^2 + 4}{x^2}$

32. $Y_4 = \dfrac{x^3 + 5x^2 - 6}{x^2}$

33. $r(x) = \dfrac{x^3 - x^2 - 4x + 4}{x^2}$

34. $R(x) = \dfrac{x^3 - 2x^2 - 9x + 18}{x^2}$

35. $g(x) = \dfrac{x^2 + 4x + 4}{x + 3}$ **36.** $G(x) = \dfrac{x^2 - 2x + 1}{x - 2}$

37. $f(x) = \dfrac{x^2 + 1}{x + 1}$ **38.** $F(x) = \dfrac{x^2 + x + 1}{x - 1}$

39. $Y_3 = \dfrac{x^2 - 4}{x + 1}$ **40.** $Y_4 = \dfrac{x^2 - x - 6}{x - 1}$

41. $v(x) = \dfrac{x^3 - 4x}{x^2 - 1}$ **42.** $V(x) = \dfrac{9x - x^3}{x^2 - 4}$

43. $w(x) = \dfrac{16x - x^3}{x^2 + 4}$ **44.** $W(x) = \dfrac{x^3 - 7x + 6}{2 + x^2}$

45. $Y_1 = \dfrac{x^3 - 3x + 2}{x^2 - 9}$ **46.** $Y_2 = \dfrac{x^3 - x^2 - 12x}{x^2 - 7}$

47. $p(x) = \dfrac{x^4 + 4}{x^2 + 1}$ **48.** $P(x) = \dfrac{x^4 - 5x^2 + 4}{x^2 + 2}$

49. $q(x) = \dfrac{10 + 9x^2 - x^4}{x^2 + 5}$ **50.** $Q(x) = \dfrac{x^4 - 2x^2 + 3}{x^2}$

Graph each function and its nonlinear asymptote on the same screen, using the window specified. Then locate the minimum value of f in the first quadrant.

51. $f(x) = \dfrac{x^3 + 500}{x}$;

$x \in [-24, 24], y \in [-500, 500]$

52. $f(x) = \dfrac{2\pi x^3 + 750}{x}$;

$x \in [-12, 12], y \in [-750, 750]$

▶ WORKING WITH FORMULAS

53. Area of a first quadrant triangle:

$$A(a) = \frac{1}{2}\left(\frac{ka^2}{a - h}\right)$$

The area of a right triangle in the first quadrant, formed by a line with negative slope through the point (h, k) and legs that lie along the positive axes is given by the formula shown, where a represents the x-intercept of

the resulting line ($h < a$). The area of the triangle varies with the slope of the line. Assume the line contains the point $(5, 6)$.

a. Find the equation of the vertical and slant asymptotes.

b. Find the area of the triangle if it has an x-intercept of $(11, 0)$.

c. Use a graphing calculator to graph the function on an appropriate window. Does the shape of the graph look familiar? Use the calculator to find the value of a that minimizes $A(a)$. That is, find the x-intercept that results in a triangle with the smallest possible area.

54. Surface area of a cylinder with fixed volume:

$$S = \frac{2\pi r^3 + 2V}{r}$$

It's possible to construct many different cylinders that will hold a specified volume, by changing the radius and height. This is critically important to producers who want to minimize the cost of packing canned goods and marketers who want to present an attractive product. The surface area of the cylinder can be found using the formula shown, where the radius is r and $V = \pi r^2 h$ is known. Assume the fixed volume is 750 cm^3.

a. Find the equation of the vertical asymptote. How would you describe the nonlinear asymptote?

b. If the radius of the cylinder is 2 cm, what is its surface area?

c. Use a graphing calculator to graph the function on an appropriate window, and use it to find the value of r that minimizes $S(r)$. That is, find the radius that results in a cylinder with the smallest possible area, while still holding a volume of 750 cm^3.

▶ APPLICATIONS

Costs of manufacturing: As in Example 4, the cost $C(x)$ of manufacturing is sometimes nonlinear and can increase dramatically with each item. For the average cost function $A(x) = \dfrac{C(x)}{x}$, consider the following.

55. Assume the monthly cost of manufacturing custom-crafted storage sheds is modeled by the function $C(x) = 4x^2 + 53x + 250$.

a. Write the average cost function and state the equation of the vertical and oblique asymptotes.

b. Enter the cost function $C(x)$ as Y_1 on a graphing calculator, and the average cost function $A(x)$ as Y_2. Using the TABLE feature, find the cost and average cost of making 1, 2, and 3 sheds.

c. Scroll down the table to where it appears that average cost is a minimum. According to the table, how many sheds should be made each month to minimize costs? What is the minimum cost?

d. Graph the average cost function and its asymptotes, using a window that shows the entire function. Use the graph to confirm the result from part (c).

56. Assume the monthly cost of manufacturing playground equipment that combines a play house, slides, and swings is modeled by the function $C(x) = 5x^2 + 94x + 576$. The company has projected that they will be profitable if they can bring their average cost down to $200 per set of playground equipment.

a. Write the average cost function and state the equation of the vertical and oblique asymptotes.

b. Enter the cost function $C(x)$ as Y_1 on a graphing calculator, and the average cost function $A(x)$ as Y_2. Using the TABLE feature, find the cost and average cost of making 1, 2, and 3 playground equipment combinations. Why would the average cost fall so dramatically early on?

c. Scroll down the table to where it appears that average cost is a minimum. According to the table, how many sets of equipment should be made each month to minimize costs? What is the minimum cost? Will the company be profitable under these conditions?

d. Graph the average cost function and its asymptotes, using a window that shows the entire function. Use the graph to confirm the result from part (c).

Minimum cost of packaging: Similar to Exercise 54, manufacturers can minimize their costs by shipping merchandise in packages that use a minimum amount of material. After all, rectangular boxes come in different sizes and there are many combinations of length, width, and height that will hold a specified volume.

57. A clothing manufacturer wishes to ship lots of 12 ft³ of clothing in boxes with square ends and rectangular sides.

a. Find a function $S(x, y)$ for the surface area of the box, and a function $V(x, y)$ for the volume of the box.

b. Solve for y in $V(x, y) = 12$ (volume is 12 ft³) and use the result to write the surface area as a function $S(x)$ in terms of x alone (simplify the result).

c. On a graphing calculator, graph the function $S(x)$ using the window $x \in [-8, 8]$; $y \in [-100, 100]$. Then graph $y = 2x^2$ on the same screen. How are these two graphs related?

d. Use the graph of $S(x)$ in Quadrant I to determine the dimensions that will minimize the surface area of the box, yet still hold 12 ft³ of clothing. Clearly state the values of x and y, *in terms of feet and inches*, rounded to the nearest $\frac{1}{2}$ in.

58. A maker of packaging materials needs to ship 36 ft³ of foam "peanuts" to his customers across the country, using boxes with the dimensions shown.

a. Find a function $S(x, y)$ for the surface area of the box, and a function $V(x, y)$ for the volume of the box.

b. Solve for y in $V(x, y) = 36$ (volume is 36 ft³), and use the result to write the surface area as a function $S(x)$ in terms of x alone (simplify the result).

c. On a graphing calculator, graph the function $S(x)$ using the window $x \in [-10, 10]$; $y \in [-200, 200]$. Then graph $y = 2x^2 + 4x$ on the same screen. How are these two graphs related?

d. Use the graph of $S(x)$ in Quadrant I to determine the dimensions that will minimize the surface area of the box, yet still hold the foam peanuts. Clearly state the values of x and y, *in terms of feet and inches*, rounded to the nearest $\frac{1}{2}$ in.

Printing and publishing: In the design of magazine pages, posters, and other published materials, an effort is made to maximize the usable area of the page while maintaining an attractive border, or minimizing the page size that will hold a certain amount of print or art work.

59. An editor has a story that requires 60 in² of print. Company standards require a 1-in. border at the top and bottom of a page, and 1.25-in. borders along both sides.

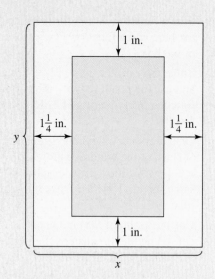

a. Find a function $A(x, y)$ for the area of the page, and a function $R(x, y)$ for the area of the inner rectangle (the printed portion).

b. Solve for y in $R(x, y) = 60$, and use the result to write the area from part (a) as a function $A(x)$ in terms of x alone (simplify the result).

c. On a graphing calculator, graph the function $A(x)$ using the window $x \in [-30, 30]$; $y \in [-100, 200]$. Then graph $y = 2x + 60$ on the same screen. How are these two graphs related?

d. Use the graph of $A(x)$ in Quadrant I to determine the page of minimum size that satisfies these border requirements and holds the necessary print. Clearly state the values of x and y, rounded to the nearest hundredth of an inch.

60. *The Poster Shoppe* creates posters, handbills, billboards, and other advertising for business customers. An order comes in for a poster with 500 in² of usable area, with margins of 2 in. across the top, 3 in. across the bottom, and 2.5 in. on each side.

a. Find a function $A(x, y)$ for the area of the page, and a function $R(x, y)$ for the area of the inner rectangle (the usable area).

b. Solve for y in $R(x, y) = 500$, and use the result to write the area from part (a) as a function $A(x)$ in terms of x alone (simplify the result).

c. On a graphing calculator, graph $A(x)$ using the window $x \in [-100, 100]$; $y \in [-800, 1600]$. Then graph $y = 5x + 500$ on the same screen. How are these two graphs related?

d. Use the graph of $A(x)$ in Quadrant I to determine the poster of minimum size that satisfies these border requirements and has the necessary usable area. Clearly state the values of x and y, rounded to the nearest hundredth of an inch.

61. The formula from Exercise 54 has an interesting derivation. The volume of a cylinder is $V = \pi r^2 h$, while the surface area is given by $S = 2\pi r^2 + 2\pi r h$ (the circular top and bottom + the area of the side).

a. Solve the volume formula for the variable h.

b. Substitute the resulting expression for h into the surface area formula and simplify.

c. Combine the resulting two terms using the least common denominator, and the result is the formula from Exercise 54.

d. Assume the volume of a can must be 1200 cm³. Use a calculator to graph the function S using an appropriate window, then use it to find the radius r and height h that will result in a cylinder with the smallest possible area, while still holding a volume of 1200 cm³. Also see Exercise 62.

62. The surface area of a spherical cap is given by $S = 2\pi r h$, where r is the radius of the sphere and h is the perpendicular distance from the sphere's surface to the plane intersecting the sphere, forming the cap. The volume of the cap is $V = \frac{1}{3}\pi h^2(3r - h)$. Similar to Exercise 61, a formula can be found that will minimize the area of a cap that holds a specified volume.

a. Solve the volume formula for the variable r.

b. Substitute the resulting expression for r into the surface area formula and simplify. The result is a formula for surface area given solely in terms of the volume V and the height h.

c. Assume the volume of the spherical cap is 500 cm³. Use a graphing calculator to graph the resulting function on an appropriate window, and use the graph to find the height h that will result in a spherical cap with the

smallest possible area, while still holding a volume of 500 cm³.

d. Use this value of h and $V = 500$ cm³ to find the radius of the sphere.

▶ EXTENDING THE CONCEPT

63. Consider rational functions of the form $f(x) = \dfrac{x^2 - a}{x - b}$. Use a graphing calculator to explore cases where $a = b^2 + 1$, $a = b^2$, and $a = b^2 - 1$. What do you notice? Explain/Discuss why the graphs differ. It's helpful to note that when graphing functions of this form, the "center" of the graph will be at $(b, b^2 - a)$, and the window size can be set accordingly for an optimal view. Do some investigation on this function and determine/explain *why* the "center" of the graph is at $(b, b^2 - a)$.

64. The formula from Exercise 53 also has an interesting derivation, and the process involves this sequence:

a. Use the points $(a, 0)$ and (h, k) to find the slope of the line, and the point-slope formula to find the equation of the line in terms of y.

b. Use this equation to find the x- and y-intercepts of the line in terms of a, k, and h.

c. Complete the derivation using these intercepts and the triangle formula $A = \frac{1}{2}BH$.

d. If the lines goes through $(4, 4)$ the area formula becomes $A = \dfrac{1}{2}\left(\dfrac{4a^2}{a - 4}\right)$. Find the minimum value of this rational function. What can you say about the triangle with minimum area through (h, k), where $h = k$? Verify using the points $(5, 5)$, and $(6, 6)$.

65. Referring to Exercises 54 and 61, suppose that instead of a closed cylinder, with both a top and bottom, we needed to manufacture *open cylinders,* like tennis ball cans that use a lid made from a different material. Derive the formula that will minimize the surface area of an open cylinder, and use it to find the cylinder with minimum surface area that will hold 90 in³ of material.

▶ MAINTAINING YOUR SKILLS

66. (1.4) Compute the quotient $\dfrac{5i}{1 + 2i}$, then check your answer using multiplication.

67. (2.3) Write the equation of the line in slope intercept form and state the slope and y-intercept: $-3x + 4y = -16$.

68. (1.5) Given $f(x) = ax^2 + bx + c$, for what real values of a, b, and c will the function have: (a) two, real/rational roots, (b) two, real/irrational roots, (c) one real and rational root, (d) one real/irrational root, (e) one complex root, and (f) two complex roots?

69. (R.2/1.5) For triangle ABC as shown, (a) find the perimeter; (b) find the length of $\overline{CD}$, given $(\overline{CB})^2 = \overline{AB} \cdot \overline{DB}$; (c) find the area; and (d) find the area of the two smaller triangles.

Learning Objectives

In Section 3.7 you will learn how to:

☐ **A.** Solve quadratic inequalities

☐ **B.** Solve polynomial inequalities

☐ **C.** Solve rational inequalities

☐ **D.** Use interval tests to solve inequalities

☐ **E.** Solve applications of inequalities

The study of polynomial and rational inequalities is simply an extension of our earlier work in analyzing functions (Section 2.5). While we've developed the ability to graph a variety of new functions, solution sets will still be determined by analyzing the behavior of the function at its zeroes, and in the case of rational functions, on either side of any vertical asymptotes. The key idea is to recognize the following statements are synonymous:

1. $f(x) > 0$. **2.** Outputs are positive. **3.** The graph is *above the x-axis*.

Similar statements can be made using the other inequality symbols.

A. Quadratic Inequalities

Solving a quadratic inequality only requires that we (a) locate any real zeroes of the function and (b) determine whether the graph opens upward or downward. If there are no x-intercepts, the graph is entirely above the x-axis (output values are positive), or entirely below the x-axis (output values are negative), making the solution either all real numbers or the empty set.

EXAMPLE 1 ▶ **Solving a Quadratic Inequality**

For $f(x) = x^2 + x - 6$, solve $f(x) > 0$.

Solution ▶ The graph of f will open upward since $a > 0$. Factoring gives $f(x) = (x + 3)(x - 2)$, with zeroes at -3 and 2. Using a the x-axis alone (since graphing the function is not our focus), we plot $(-3, 0)$ and $(2, 0)$ and visualize a parabola opening upward through these points (Figure 3.58).

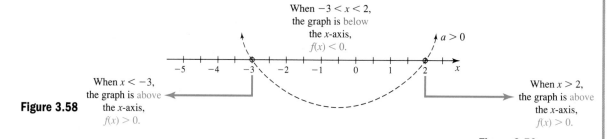

When $-3 < x < 2$, the graph is below the x-axis, $f(x) < 0$.

When $x < -3$, the graph is above the x-axis, $f(x) > 0$.

$a > 0$

When $x > 2$, the graph is above the x-axis, $f(x) > 0$.

Figure 3.58

Figure 3.59

The diagram clearly shows the graph is *above* the x-axis (outputs are positive) when $x < -3$ or when $x > 2$. The solution is $x \in (-\infty, -3) \cup (2, \infty)$. For reference only, the complete graph is given in Figure 3.59.

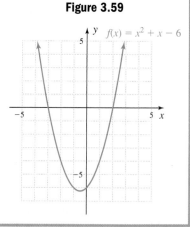

$f(x) = x^2 + x - 6$

Now try Exercises 7 through 18 ▶

When solving general inequalities, zeroes of multiplicity continue to play a role. In Example 1, the zeroes of f were both of multiplicity 1, and the graph crossed the x-axis at these points. In other cases, the zeroes may have even multiplicity.

EXAMPLE 2 ▶ **Solving a Quadratic Inequality**

Solve the inequality $-x^2 + 6x \leq 9$.

Solution ▶ Begin by writing the inequality in standard form: $-x^2 + 6x - 9 \leq 0$. Note this is equivalent to $g(x) \leq 0$ for $g(x) = -x^2 + 6x - 9$. Since $a < 0$, the graph of g will open downward. The factored form is $g(x) = -(x - 3)^2$, showing 3 is a zero with multiplicity 2. Using the x-axis, we plot the point $(3, 0)$ and visualize a parabola opening downward through this point.

Figure 3.60

$a < 0$

Figure 3.61

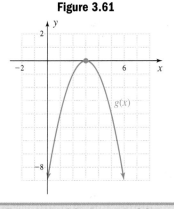

WORTHY OF NOTE

Since $x = 3$ was a zero of multiplicity 2, the graph "bounced off" the x-axis at this point, with no change of sign for g. The graph is entirely below the x-axis, except at the vertex $(3, 0)$.

Figure 3.60 shows the graph is *below* the x-axis (outputs are negative) for *all values* of x except $x = 3$. But since this is a less than *or equal to* inequality, the solution is $x \in \mathbb{R}$. For reference only, the complete graph is given in Figure 3.61.

☑ **A.** You've just learned how to solve quadratic inequalities

Now try Exercises 19 through 36 ▶

B. Polynomial Inequalities

The reasoning in Examples 1 and 2 transfers seamlessly to inequalities involving higher degree polynomials. After writing the polynomial in standard form, find the zeroes, plot them on the x-axis, and determine the solution set using end behavior and the behavior at each zero (cross—sign change; or bounce—no change in sign). In this process, any irreducible quadratic factors can be ignored, as they have no effect on the solution set. In summary,

Solving Polynomial Inequalities

Given $f(x)$ is a polynomial in standard form,
   1. Write f in completely factored form.
   2. Plot real zeroes on the x-axis, noting their multiplicity.
      • If the multiplicity is odd the function will **change** sign.
      • If the multiplicity is even, there will be **no change** in sign.
   3. Use the end behavior to determine the sign of f in the outermost intervals, then label the other intervals as $f(x) < 0$ or $f(x) > 0$ by analyzing the multiplicity of neighboring zeroes.
   4. State the solution in interval notation.

EXAMPLE 3 ▶ Solving a Polynomial Inequality

Solve the inequality $x^3 - 18 < -4x^2 + 3x$.

Solution ▶ In standard form we have $x^3 + 4x^2 - 3x - 18 < 0$, which is equivalent to $f(x) < 0$ where $f(x) = x^3 + 4x^2 - 3x - 18$. The polynomial cannot be factored by grouping and testing 1 and -1 shows neither is a zero. Using $x = 2$ and synthetic division gives

use 2 as a "divisor"

$$
\begin{array}{r|rrrr}
2 & 1 & 4 & -3 & -18 \\
  & \downarrow & 2 & 12 & 18 \\
\hline
  & 1 & 6 & 9 & 0,
\end{array}
$$

with a quotient of $x^2 + 6x + 9$ and a remainder of zero.

1. The factored form is $f(x) = (x - 2)(x^2 + 6x + 9) = (x - 2)(x + 3)^2$.

2. The graph will bounce off the x-axis at $x = -3$ (f will not change sign), and cross the x-axis at $x = 2$ (f will change sign). This is illustrated in Figure 3.62, which uses open dots due to the strict inequality.

Figure 3.62

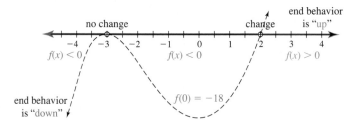

3. The polynomial has odd degree with a positive lead coefficient, so end behavior is down/up, which we note in the outermost intervals. Working from the left, f will not change sign at $x = -3$, showing $f(x) < 0$ in the left and middle intervals. This is supported by the y-intercept $(0, -18)$. See Figure 3.63.

Figure 3.63

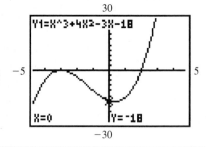

4. From the diagram, we see that $f(x) < 0$ for $x \in (-\infty, -3) \cup (-3, 2)$, which must also be the solution interval for $x^3 - 18 < -4x^2 + 3x$.

Now try Exercises 37 through 48 ▶

GRAPHICAL SUPPORT

The results from Example 3 can easily be verified using a graphing calculator. The graph shown here is displayed using a window of $X \in [-5, 5]$ and $Y \in [-30, 30]$, and definitely shows the graph is below the x-axis $[f(x) < 0]$ from $-\infty$ to 2, except at $x = -3$ where the graph touches the x-axis without crossing.

EXAMPLE 4 ▶ **Solving a Polynomial Inequality**

Solve the inequality $x^4 + 4x \le 9x^2 - 12$.

Solution ▶ Writing the polynomial in standard form gives $x^4 - 9x^2 + 4x + 12 \le 0$. The equivalent inequality is $f(x) \le 0$. Testing 1 and -1 shows $x = 1$ is not a zero, but $x = -1$ is. Using synthetic division with $x = -1$ gives

use −1 as a "divisor" $\underline{-1|}$ 1   0   −9   4   12
                                    ↓   −1   1   8   −12
                                    ——————————————————————————
                                        1   −1   −8   12 $\lfloor 0$

with a quotient of $q_1(x) = x^3 - x^2 - 8x + 12$ and a remainder of zero. As $q_1(x)$ is not easily factored, we continue with synthetic division using $x = 2$.

use 2 as a "divisor" $\underline{2|}$ 1   −1   −8   12
                                   ↓   2   2   −12
                                   ——————————————————————————
                                       1   1   −6 $\lfloor 0$

The result is $q_2(x) = x^2 + x - 6$ with a remainder of zero.

1. The factored form is
   $f(x) = (x + 1)(x - 2)(x^2 + x - 6) = (x + 1)(x - 2)^2(x + 3)$.
2. The graph will "cross" at $x = -1$ and -3, and f will change sign. The graph will bounce at $x = 2$ and f will not change sign. This is illustrated in Figure 3.65 which uses closed dots since $f(x)$ can be equal to zero. See Figure 3.64.

Figure 3.64

3. With even degree and positive lead coefficient, the end behavior is up/up. Working from the leftmost interval, $f(x) > 0$, the function must change sign at $x = -3$ (going below the x-axis), and again at $x = -1$ (going above the x-axis). This is supported by the y-intercept $(0, 12)$. The graph then "bounces" at $x = 2$, remaining above the x-axis (no sign change). This produces the sketch shown in Figure 3.65.

End behavior is "up"   End behavior is "up"

$f(x) > 0$ $f(x) < 0$ $f(x) > 0$ $f(x) > 0$

$f(0) = 12$

Figure 3.65

☑ **B.** You've just learned how to solve polynomial inequalities

4. From the diagram, we see that $f(x) \le 0$ for $x \in [-3, -1]$, and at the single point $x = 2$. This shows the solution for $x^4 + 4x \le 9x^2 - 12$ is $x \in [-3, -1] \cup \{2\}$.

Now try Exercises 49 through 54 ▶

GRAPHICAL SUPPORT

As with Example 3, the results from Example 4 can be confirmed using a graphing calculator. The graph shown here is displayed using $X \in [-5, 5]$ and $Y \in [-20, 20]$. The graph is below or touching the x-axis $[f(x) < 0]$ from -3 to -1 and at $x = 2$.

C. Rational Inequalities

In general, the solution process for polynomial and rational inequalities is virtually identical, once we recognize that vertical asymptotes also break the x-axis into intervals where function values may change sign. However, for rational functions it's more efficient to begin the analysis using the y-intercept or a test point, rather than end behavior, although either will do.

EXAMPLE 5 ▶ **Solving a Rational Inequality**

Solve $\dfrac{x^2 - 9}{x^3 - x^2 - x + 1} \leq 0$.

Solution ▶ In function form, $v(x) = \dfrac{x^2 - 9}{x^3 - x^2 - x + 1}$ and we want the solution for $v(x) \leq 0$.
The numerator and denominator are in standard form. The numerator factors easily, and the denominator can be factored by grouping.

1. The factored form is $v(x) = \dfrac{(x - 3)(x + 3)}{(x - 1)^2(x + 1)}$.

2. $v(x)$ will change sign at $x = 3$, -3, and -1 as all have odd multiplicity, but will not change sign at $x = 1$ (even multiplicity). Note that zeroes of the denominator will always be indicated by open dots (Figure 3.66) as they are excluded from any solution set.

WORTHY OF NOTE

End behavior can also be used to analyze rational inequalities, although using the y-intercept may be more efficient. For the function $v(x)$ from Example 5 we have
$\dfrac{x^2 - 9}{x^3 - x^2 - x + 1} \approx \dfrac{x^2}{x^3} = \dfrac{1}{x}$ for
large values of x, indicating $v(x) > 0$ to the far right and $v(x) < 0$ to the far left. The analysis of each interval can then begin from either side.

Figure 3.66

3. The y-intercept is $(0, -9)$, indicating that function values will be negative in the interval containing zero. Working outward from this interval using the "change/no change" approach, gives the solution indicated in Figure 3.67.

Figure 3.67

4. For $v(x) \leq 0$, the solution is $x \in (-\infty, -3] \cup (-1, 1) \cup (1, 3]$.

Now try Exercises 55 through 66 ▶

GRAPHICAL SUPPORT

Sometimes finding a window that clearly displays all features of rational function can be difficult. In these cases, we can investigate each piece separately to confirm solutions. For Example 5, most of the features of $v(x)$ can be seen using a window $X \in [-5, 5]$ and $Y \in [-20, 10]$, and we note the graph displayed strongly tends to support our solution.

If the rational inequality is not given in function form or is composed of more than one term, start by writing the inequality with zero on one side, then combine terms into a single expression.

EXAMPLE 6 ▶ **Solving a Rational Inequality**

Solve $\dfrac{x-2}{x-3} \le \dfrac{1}{x+3}$.

Solution ▶ Rewrite the inequality with zero on one side: $\dfrac{x-2}{x-3} - \dfrac{1}{x+3} \le 0$. This is

equivalent to $v(x) \le 0$, where $v(x) = \dfrac{x-2}{x-3} - \dfrac{1}{x+3}$. Combining the expressions

on the right, we have

$$v(x) = \dfrac{(x-2)(x+3) - 1(x-3)}{(x+3)(x-3)} \qquad \text{LCD is } (x+3)(x-3)$$

$$= \dfrac{x^2 + x - 6 - x + 3}{(x+3)(x-3)} \qquad \text{multiply}$$

$$= \dfrac{x^2 - 3}{(x+3)(x-3)} \qquad \text{simplify}$$

1. The factored form is $v(x) = \dfrac{(x+\sqrt{3})(x-\sqrt{3})}{(x+3)(x-3)}$. $\quad x^2 - k = (x + \sqrt{k})(x - \sqrt{k})$

2. $v(x)$ will change sign at $x = -\sqrt{3}, \sqrt{3}, -3$, and 3, as all have odd
   multiplicity (Figure 3.68).

Figure 3.68

3. Since $v(0) = \frac{1}{3}$ (verify this), function values will be positive in the interval
   containing zero. Working outward from this interval produces the diagram
   shown in Figure 3.69.

Figure 3.69

4. The solution for $\dfrac{x-2}{x-3} \le \dfrac{1}{x+3}$ is $x \in (-3, -\sqrt{3}] \cup [\sqrt{3}, 3)$.

☑ **C.** You've just learned how
to solve rational inequalities

Now try Exercises 67 through 82 ▶

GRAPHICAL SUPPORT

To check the solutions to $\dfrac{x-2}{x-3} \le \dfrac{1}{x+3}$, we
subtract $\dfrac{1}{x+3}$ and graph $Y_1 = \dfrac{x-2}{x-3} - \dfrac{1}{x+3}$
to look for intervals where the graph is below
the x-axis. The graph is shown here using the
window $X \in [-5, 5]$ and $y \in [-10, 10]$, and
verifies our solution.

D. Solving Function Inequalities Using Interval Tests

As an alternative to the "zeroes method," an **interval test method** can be used to solve polynomial and rational inequalities. The x-intercepts and vertical asymptotes (in the case of rational functions) are noted on the x-axis, then a test number is selected from each interval. Since polynomial and rational functions are continuous over their entire domain, the sign of the function at these test values will be the sign of the function for all values of x in the chosen interval.

EXAMPLE 7 ▶   **Solving a Polynomial Inequality**

Solve the inequality $x^3 + 8 \leq 5x^2 - 2x$.

Solution ▶   Writing the relationship in function form gives $p(x) = x^3 - 5x^2 + 2x + 8$, with solutions needed to $p(x) \leq 0$. The tests for 1 and -1 show $x = -1$ is a root, and using -1 with synthetic division gives

$$\text{use } -1 \text{ as a "divisor"} \quad \underline{-1|} \begin{array}{rrrr} 1 & -5 & 2 & 8 \\ \downarrow & -1 & 6 & -8 \\ \hline 1 & -6 & 8 & \underline{|0} \end{array}$$

The quotient is $q(x) = x^2 - 6x + 8$, with a remainder of 0.

The factored form is $p(x) = (x + 1)(x^2 - 6x + 8) = (x + 1)(x - 2)(x - 4)$. The x-intercepts are $(-1, 0)$, $(2, 0)$, and $(4, 0)$. Plotting these intercepts creates four intervals on the x-axis (Figure 3.70).

When evaluating a function using the interval test method, it's usually easier to use the factored form instead of the polynomial form, since all you really need is whether the result will be positive or negative. For instance, you could likely tell $p(3) = (3 + 1)(3 - 2)(3 - 4)$ is going to be negative, more quickly than $p(3) = (3)^3 - 5(3)^2 + 2(3) + 8$.

Figure 3.70

Selecting a test value from each interval gives Figure 3.71.

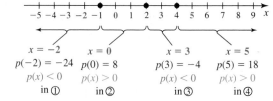

Figure 3.71

The interval tests show $x^3 + 8 \leq 5x^2 - 2x$ for $x \in (-\infty, -1] \cup [2, 4]$.

☑ D. You've just learned how to use interval tests to solve inequalities

Now try Exercises 83 through 90 ▶

E. Applications of Inequalities

Applications of inequalities come in many varieties. In addition to stating the solution algebraically, these exercises often compel us to consider the *context of each application* as we state the solution set.

EXAMPLE 8 ▶   **Solving Applications of Inequalities**

The velocity of a particle (in feet per second) as it floats through air turbulence is given by $V(t) = t^5 - 10t^4 + 35t^3 - 50t^2 + 24t$, where t is the time in seconds and $0 < t < 4.5$. During what intervals of time is the particle moving in the positive direction $[V(t) > 0]$?

Solution ▶   Begin by writing V in factored form. Testing 1 and -1 shows $t = 1$ is a root. Factoring out t gives $V(t) = t(t^4 - 10t^3 + 35t^2 - 50t + 24)$, and using $t = 1$ with synthetic division yields

$$\text{use 1 as a "divisor"} \quad \underline{1|} \begin{array}{rrrrr} 1 & -10 & 35 & -50 & 24 \\ \downarrow & 1 & -9 & 26 & -24 \\ \hline 1 & -9 & 26 & -24 & \underline{|0} \end{array}$$

The quotient is $q_1(t) = t^3 - 9t^2 + 26t - 24$. Using $t = 2$, we continue with the division on $q_1(t)$ which gives:

use 2 as a "divisor" $2|$

$$
\begin{array}{r|rrrr}
2 & 1 & -9 & 26 & -24 \\
  & \downarrow & 2 & -14 & 24 \\
\hline
  & 1 & -7 & 12 & \underline{|0} \\
\end{array}
$$

This shows $V(t) = t(t - 1)(t - 2)(t^2 - 7t + 12)$.

1. The completely factored form is $V(t) = t(t - 1)(t - 2)(t - 3)(t - 4)$.
2. All zeroes have odd multiplicity and function values will change sign.
3. With odd degree and a positive leading coefficient, end behavior is down/up.

Function values will be negative in the far left interval and alternate in sign thereafter. The solution diagram is shown in the figure.

Since end behavior is down/up, function values are negative in this interval, and will alternate thereafter.

change   change   change   change   change

-1 0 1 2 3 4 t

$V(t) < 0$ $V(t) > 0$ $V(t) < 0$ $V(t) > 0$ $V(t) < 0$ $V(t) > 0$

☑ **E.** You've just learned how to solve applications of inequalities

4. For $V(t) > 0$, the solution is $t \in (0, 1) \cup (2, 3) \cup (4, 4.5)$. The particle is moving in the positive direction in these time intervals.

Now try Exercises 93 through 100 ▶

GRAPHICAL SUPPORT

To verify our analysis of Example 8, we graph $V(t)$ using the window $X \in [-1, 5]$ and $Y \in [-5, 5]$. As the graph shows, function values are positive (graph is above the x-axis) when $t \in (0, 1) \cup (2, 3) \cup (4, 4.5)$. Also see the *Technology Highlight* for this section.

TECHNOLOGY HIGHLIGHT

Polynomial and Rational Inequalities

Consider the results from Example 8, where we solved the inequality $V(t) > 0$ for $V(t) = t^5 - 10t^4 + 35t^3 - 50t^2 + 24t$. To emphasize that we are seeking intervals where the function is above the x-axis (the horizontal line $y = 0$), we can have the calculator *shade these areas*. Begin by entering $V(t)$ as Y_1 on the [Y=] screen, and the line $y = 0$ as Y_2. Using $x \in [0, 4.7]$ and $y \in [-5, 5]$ (a "friendly" window) and [GRAPH]ing the functions produces Figure 3.72. To shade all portions of the graph that are above the x-axis, go to the home screen, and press [2nd] [PRGM] (DRAW) 7:Shade. This feature requires six arguments, all separated by commas. These are (in order): *lower function, upper function, left endpoint, right endpoint, pattern choice,* and *density*. The calculator will then shade the area between the lower and upper functions,

Figure 3.72

between the left and right endpoints, using the pattern and density
chosen. The patterns are (1) vertical lines, (2) horizontal lines,
(3) lines with negative slope, and (4) lines with positive slope. There
are eight density settings, from every pixel (1), to every eight pixels
(8). Figure 3.73 shows the options we've selected, with the
resulting graph shown in Figure 3.74. The friendly window makes it
easy to investigate the inequality further using the TRACE feature.

Use these ideas to visually study and explore the solution to
the following inequality.

Figure 3.73

Exercise 1: Use window size $x \in [-4.7, 4.7]$; $y \in [-10, 20]$,
(DRAW) 7:Shade, and TRACE to solve $P(x) < 0$ for
$P(x) = x^4 + 1.1x^3 - 9.37x^2 - 4.523x + 16.4424$.

Figure 3.74

3.7 EXERCISES

▶ CONCEPTS AND VOCABULARY

**Fill in each blank with the appropriate word or phrase.
Carefully reread the section if needed.**

1. To solve a polynomial or rational inequality, begin
by plotting the location of all zeroes and _____
asymptotes (if they exist), then consider the
_____ of each.

2. For strict inequalities, the zeroes are _____
from the solution set. For nonstrict inequalities,
zeroes are _____. The values at which vertical
asymptotes occur are always _____.

3. If the graph of a quadratic function $g(x)$ opens
downward with a vertex at $(5, -1)$, the solution set
for $g(x) > 0$ is _____.

4. To solve a polynomial/rational inequality, it helps
to find the sign of f in some interval. This can
quickly be done using the _____ _____ or
_____ of the function.

5. Compare/Contrast the process for solving
$x^2 - 3x - 4 \geq 0$ with $\dfrac{1}{x^2 - 3x - 4} \geq 0$. Are there
similarities? What are the differences?

6. Compare/Contrast the process for solving
$(x + 1)(x - 3)(x^2 + 1) > 0$ with
$(x + 1)(x - 3) > 0$. Are there similarities? What
are the differences?

▶ DEVELOPING YOUR SKILLS

**Solve each quadratic inequality by locating the
x-intercept(s) (if they exist), and noting the end behavior
of the graph. Begin by writing the inequality in function
form as needed.**

7. $f(x) = -x^2 + 4x$; $f(x) > 0$

8. $g(x) = x^2 - 5x$; $g(x) < 0$

9. $h(x) = x^2 + 4x - 5$; $h(x) \geq 0$

10. $p(x) = -x^2 + 3x + 10$; $p(x) \leq 0$

11. $q(x) = 2x^2 - 5x - 7$; $q(x) < 0$

12. $r(x) = -2x^2 - 3x + 5$; $r(x) > 0$

13. $7 \geq x^2$ 14. $x^2 \leq 13$

15. $x^2 + 3x \leq 6$ 16. $x^2 - 2 \leq 5x$

17. $3x^2 \geq -2x + 5$ 18. $4x^2 \geq 3x + 7$

19. $s(x) = x^2 - 8x + 16; s(x) \geq 0$

20. $t(x) = x^2 - 6x + 9; t(x) \geq 0$

21. $r(x) = 4x^2 + 12x + 9; r(x) < 0$

22. $f(x) = 9x^2 - 6x + 1; f(x) < 0$

23. $g(x) = -x^2 + 10x - 25; g(x) < 0$

24. $h(x) = -x^2 + 14x - 49; h(x) < 0$

25. $-x^2 > 2$ 26. $x^2 < -4$

27. $x^2 - 2x > -5$ 28. $-x^2 + 3x < 3$

29. $2x^2 \geq 6x - 9$ 30. $5x^2 \geq 4x - 4$

Recall that for a square root expression to represent a real number, the radicand must be greater than or equal to zero. Applying this idea results in an inequality that can be solved using the skills from this section. Determine the domain of the following radical functions.

31. $h(x) = \sqrt{x^2 - 25}$ 32. $p(x) = \sqrt{25 - x^2}$

33. $q(x) = \sqrt{x^2 - 5x}$ 34. $r(x) = \sqrt{6x - x^2}$

35. $t(x) = \sqrt{-x^2 + 3x - 4}$

36. $Y_1 = \sqrt{x^2 - 6x + 9}$

Solve the inequality indicated using a number line and the behavior of the graph at each zero. Write all answers in interval notation.

37. $(x + 3)(x - 5) < 0$ 38. $(x - 2)(x + 7) < 0$

39. $(x + 1)^2(x - 4) \geq 0$ 40. $(x + 6)(x - 1)^2 \leq 0$

41. $(x + 2)^3(x - 2)^2(x - 4) \geq 0$

42. $(x - 1)^3(x + 2)^2(x - 3) \leq 0$

43. $x^2 + 4x + 1 < 0$ 44. $x^2 - 6x + 4 > 0$

45. $x^3 + x^2 - 5x + 3 \leq 0$

46. $x^3 + x^2 - 8x - 12 \geq 0$

47. $x^3 - 7x + 6 > 0$ 48. $x^3 - 13x + 12 > 0$

49. $x^4 - 10x^2 > -9$ 50. $x^4 + 36 < 13x^2$

51. $x^4 - 9x^2 > 4x - 12$ 52. $x^4 - 16 > 5x^3 - 20x$

53. $x^4 - 6x^3 \leq -8x^2 - 6x + 9$

54. $x^4 - 3x^2 + 8 \leq 4x^3 - 10x$

55. $f(x) = \dfrac{x + 3}{x - 2}; f(x) \leq 0$

56. $F(x) = \dfrac{x - 4}{x + 1}; F(x) \geq 0$

57. $g(x) = \dfrac{x + 1}{x^2 + 4x + 4}; g(x) < 0$

58. $G(x) = \dfrac{x - 3}{x^2 - 2x + 1}; G(x) > 0$

59. $\dfrac{2 - x}{x^2 - x - 6} \geq 0$ 60. $\dfrac{1 - x}{x^2 - 2x - 8} \leq 0$

61. $\dfrac{2x - x^2}{x^2 + 4x - 5} < 0$ 62. $\dfrac{x^2 + 3x}{x^2 - 2x - 3} > 0$

63. $\dfrac{x^2 - 4}{x^3 - 13x + 12} \geq 0$ 64. $\dfrac{x^2 + x - 6}{x^3 - 7x + 6} \leq 0$

65. $\dfrac{x^2 + 5x - 14}{x^3 + x^2 - 5x + 3} > 0$

66. $\dfrac{x^2 + 2x - 8}{x^3 + 5x^2 + 3x - 9} < 0$

67. $\dfrac{2}{x - 2} \leq \dfrac{1}{x}$ 68. $\dfrac{5}{x + 3} \geq \dfrac{3}{x}$

69. $\dfrac{x - 3}{x + 17} > \dfrac{1}{x - 1}$ 70. $\dfrac{1}{x + 5} < \dfrac{x - 2}{x - 7}$

71. $\dfrac{x + 1}{x - 2} \geq \dfrac{x + 2}{x + 3}$ 72. $\dfrac{x - 3}{x - 6} \leq \dfrac{x + 1}{x + 4}$

73. $\dfrac{x + 2}{x^2 + 9} > 0$ 74. $\dfrac{x^2 + 4}{x - 3} < 0$

75. $\dfrac{x^3 + 1}{x^2 + 1} > 0$ 76. $\dfrac{x^2 + 4}{x^3 - 8} < 0$

77. $\dfrac{x^4 - 5x^2 - 36}{x^2 - 2x + 1} > 0$ 78. $\dfrac{x^4 - 3x^2 - 4}{x^2 - x - 20} < 0$

79. $x^2 - 2x \geq 15$ 80. $x^2 + 3x \geq 18$

81. $x^3 \geq 9x$ 82. $x^3 \leq 4x$

83. $-4x + 12 < -x^3 + 3x^2$

84. $x^3 + 8 < 5x^2 - 2x$ 85. $\dfrac{x^2 - x - 6}{x^2 - 1} \geq 0$

86. $\dfrac{x^2 - 4x - 21}{x - 3} < 0$

Match the correct solution with the inequality and graph given.

87. $f(x) < 0$

a. $x \in (-5, -2) \cup (3, 5)$

b. $x \in (-\infty, -2) \cup (-2, 1) \cup (3, \infty)$

c. $x \in (-\infty, -2) \cup (3, \infty)$

d. $x \in (-\infty, -2) \cup (-2, 1] \cup [3, \infty)$

e. none of these

88. $g(x) \geq 0$

   a. $x \in (-4, -0.5) \cup (4, \infty)$

   b. $x \in [-0.5, 4] \cup [4, 5]$

   c. $x \in (-\infty, -4) \cup (-0.5, 4)$

   d. $x \in [-4, -0.5] \cup [4, \infty)$

   e. none of these

89. $r(x) \geq 0$

a. $x \in (-\infty, -2) \cup [-1, 1] \cup [3, \infty)$

b. $x \in (-2, -1] \cup [1, 2) \cup (2, 3]$

c. $x \in (-\infty, -2) \cup (2, \infty)$

d. $x \in (-2, -1) \cup (1, 2) \cup (2, 3]$

e. none of these

90. $R(x) \leq 0$

   a. $x \in (-\infty, -1) \cup (0, 2)$

   b. $x \in [0, 1] \cup (2, \infty)$

   c. $x \in [-5, -1] \cup [2, 5]$

   d. $x \in (-\infty, -1) \cup [0, 2)$

   e. none of these

▶ WORKING WITH FORMULAS

91. Discriminant of the reduced cubic
$x^3 + px + q = 0: D = -(4p^3 + 27q^2)$

The discriminant of a cubic equation is less well known than that of the quadratic, but serves the same purpose. The discriminant of the reduced cubic is given by the formula shown, where p is the linear coefficient and q is the constant term. If $D > 0$, there will be three real and distinct roots. If $D = 0$, there are still three real roots, but one is a repeated root (multiplicity two). If $D < 0$, there are one real and two complex roots. Suppose we wish to study the family of cubic equations where $q = p + 1$.

   a. Verify the resulting discriminant is
$D = -(4p^3 + 27p^2 + 54p + 27)$.

   b. Determine the values of p and q for which this family of equations has a repeated real root. In other words, solve the equation
$-(4p^3 + 27p^2 + 54p + 27) = 0$ using the rational zeroes theorem and synthetic division to write D in completely factored form.

   c. Use the factored form from part (b) to determine the values of p and q for which this family of equations has three real and distinct roots. In other words, solve $D > 0$.

   d. Verify the results of parts (b) and (c) on a graphing calculator.

92. Coordinates for the folium of Descartes:

$$\begin{cases} a = \dfrac{3kx}{1 + x^3} \\ b = \dfrac{3kx^2}{1 + x^3} \end{cases}$$

The interesting relation shown here is called the folium (leaf) of Descartes. The folium is most often graphed using what are called *parametric equations,* in which the coordinates a and b are expressed in terms of the parameter x

Folium of Descartes

("k" is a constant that affects the size of the leaf). Since each is an individual function, the x- and y-coordinates can be investigated individually in rectangular coordinates using $F(x) = \dfrac{3x}{1 + x^3}$ and

$G(x) = \dfrac{3x^2}{1 + x^3}$ (assume $k = 1$ for now).

   a. Graph each function using the techniques from this section.

   b. According to your graph, for what values of x will the x-coordinate of the folium be positive? In other words, solve $F(x) = \dfrac{3x}{1 + x^3} > 0$.

   c. For what values of x will the y-coordinate of the folium be positive? Solve
$G(x) = \dfrac{3x^2}{1 + x^3} > 0$.

   d. Will $F(x)$ ever be equal to $G(x)$? If so, for what values of x?

▶ **APPLICATIONS**

Deflection of a beam: The amount of deflection in a rectangular wooden beam of length L ft can be approximated by $d(x) = k(x^3 - 3L^2x + 2L^3)$, where k is a constant that depends on the characteristics of the wood and the force applied, and x is the *distance from the unsupported end* of the beam $(x < L)$.

Weight

Deflection

93. Find the equation for a beam 8 ft long and use it for the following:

   a. For what distances x is the quantity $\dfrac{d(x)}{k}$ less than 189 units?

   b. What is the amount of deflection 4 ft from the unsupported end $(x = 4)$?

   c. For what distances x is the quantity $\dfrac{d(x)}{k}$ greater than 475 units?

   d. If safety concerns prohibit a deflection of more than 648 units, what is the shortest distance from the end of the beam that the force can be applied?

94. Find the equation for a beam 9 ft long and use it for the following:

   a. For what distances x is the quantity $\dfrac{d(x)}{k}$ less than 216 units?

   b. What is the amount of deflection 4 ft from the unsupported end $(x = 4)$?

   c. For what distances x is the quantity $\dfrac{d(x)}{k}$ greater than 550 units?

    d. Compare the answer to 93b with the answer to 94b. What can you conclude?

Average speed for a round-trip: Surprisingly, the average speed of a round-trip is *not* the sum of the average speed in each direction divided by two. For a fixed distance D, consider rate r_1 in time t_1 for one direction, and rate r_2 in time t_2 for the other, giving $r_1 = \dfrac{D}{t_1}$ and $r_2 = \dfrac{D}{t_2}$. The average speed for the round-trip is $R = \dfrac{2D}{t_1 + t_2}$.

95. The distance from St. Louis, Missouri, to Springfield, Illinois, is approximately 80 mi. Suppose that Sione, due to the age of his vehicle, made the round-trip with an average speed of 40 mph.

   a. Use the relationships stated to verify that $r_2 = \dfrac{20r_1}{r_1 - 20}$.

   b. Discuss the meaning of the horizontal and vertical asymptotes in this context.

   c. Verify algebraically the speed returning would be greater than the speed going for $20 < r_1 < 40$. In other words, solve the inequality $\dfrac{20r_1}{r_1 - 20} > r_1$ using the ideas from this section.

96. The distance from Boston, Massachusetts, to Hartford, Connecticut, is approximately 100 mi. Suppose that Stella, due to excellent driving conditions, made the round-trip with an average speed of 60 mph.

   a. Use the relationships above to verify that $r_2 = \dfrac{30r_1}{r_1 - 30}$.

   b. Discuss the meaning of the horizontal and vertical asymptotes in this context.

   c. Verify algebraically the speed returning would be greater than the speed going for $30 < r_1 < 60$. In other words, solve the inequality $\dfrac{30r_1}{r_1 - 30} > r_1$ using the ideas from this section.

Electrical resistance and temperature: The amount of electrical resistance R in a medium depends on the temperature, and for certain materials can be modeled by the equation $R(t) = 0.01t^2 + 0.1t + k$, where $R(t)$ is the resistance (in ohms Ω) at temperature t $(t \geq 0°)$ in degrees Celsius, and k is the resistance at $t = 0°C$.

97. Suppose $k = 30$ for a certain medium. Write the resistance equation and use it to answer the following.

   a. For what temperatures is the resistance less than 42 Ω?

   b. For what temperatures is the resistance greater than 36 Ω?

   c. If it becomes uneconomical to run electricity through the medium for resistances greater than 60 Ω, for what temperatures should the electricity generator be shut down?

98. Suppose $k = 20$. Write the resistance equation and solve the following.

    a. For what temperatures is the resistance less than 26 Ω?

    b. For what temperatures is the resistance greater than 40 Ω?

    c. If it becomes uneconomical to run electricity through the medium for resistances greater than 50 Ω, for what temperatures should the electricity generator be shut down?

99. Sum of consecutive squares: The sum of the first n squares $1^2 + 2^2 + 3^2 + \cdots + n^2$ is given by the formula $S(n) = \dfrac{2n^3 + 3n^2 + n}{6}$. Use the equation to solve the following inequalities.

    a. For what number of consecutive squares is $S(n) \geq 30$?

b. For what number of consecutive squares is $S(n) \leq 285$?

 c. What is the maximum number of consecutive squares that can be summed without the result exceeding three digits?

100. Sum of consecutive cubes: The sum of the first n cubes $1^3 + 2^3 + 3^3 + \cdots + n^3$ is given by the formula $S(n) = \dfrac{n^4 + 2n^3 + n^2}{4}$. Use the equation to solve the following inequalities.

    a. For what number of consecutive cubes is $S(n) \geq 100$?

    b. For what number of consecutive cubes is $S(n) \leq 784$?

     c. What is the maximum number of consecutive cubes that can be summed without the result exceeding three digits?

▶ **EXTENDING THE CONCEPT**

101. (a) Is it possible for the solution set of a polynomial inequality to be all real numbers? If not, discuss why. If so, provide an example. (b) Is it possible for the solution set of a rational inequality to be all real numbers? If not, discuss why. If so, provide an example.

102. The domain of radical functions: As in Exercises 31–36, if n is an even number, the expression $\sqrt[n]{A}$ represents a real number only if $A \geq 0$. Use this idea to find the domain of the following functions.

    a. $f(x) = \sqrt{2x^3 - x^2 - 16x + 15}$

    b. $g(x) = \sqrt[4]{2x^3 + x^2 - 22x + 24}$

    c. $p(x) = \sqrt[4]{\dfrac{x + 2}{x^2 - 2x - 35}}$

    d. $q(x) = \sqrt{\dfrac{x^2 - 1}{x^2 - x - 6}}$

103. Find one polynomial inequality and one rational inequality that have the solution $x \in (-\infty, -2) \cup (0, 1) \cup (1, \infty)$.

104. Using the tools of calculus, it can be shown that $f(x) = x^4 - 4x^3 - 12x^2 + 32x + 39$ is increasing in the intervals where $F(x) = x^3 - 3x^2 - 6x + 8$ is positive. Solve the inequality $F(x) > 0$ using the ideas from this section, then verify $f(x)\!\uparrow$ in these intervals by graphing f on a graphing calculator and using the TRACE feature.

105. Using the tools of calculus, it can be shown that $r(x) = \dfrac{x^2 - 3x - 4}{x - 8}$ is decreasing in the intervals where $R(x) = \dfrac{x^2 - 16x + 28}{(x - 8)^2}$ is negative. Solve the inequality $R(x) < 0$ using the ideas from this section, then verify $r(x)\!\downarrow$ in these intervals by graphing r on a graphing calculator and using the TRACE feature.

▶ **MAINTAINING YOUR SKILLS**

106. (2.5) Use the graph of $f(x)$ given to sketch the graph of $y = f(x + 2) - 3$.

Exercise 106

107. (3.5) Graph the function $f(x) = \dfrac{x^2 + 2x - 8}{x + 4}$. If there is a removable discontinuity,

repair the break using an appropriate piecewise-defined function.

108. (1.6) Solve the equation $\dfrac{1}{2}\sqrt{16 - x} - \dfrac{x}{2} = 2$. Check solutions in the original equation.

109. (1.2/3.7) Graph the solution set for the relation: $3x + 1 < 10$ *and* $x^2 - 3 < 1$.

Learning Objectives

In Section 3.8 you will learn how to:

- ☐ **A.** Solve direct variations
- ☐ **B.** Solve inverse variations
- ☐ **C.** Solve joint variations

A study of direct and inverse variation offers perhaps our clearest view of how mathematics is used to model real-world phenomena. While the basis of our study is elementary, involving only the toolbox functions, the applications are at the same time elegant, powerful, and far reaching. In addition, these applications unite some of the most important ideas in algebra, including functions, transformations, rates of change, and graphical analysis, to name a few.

A. Toolbox Functions and Direct Variation

If a car gets 24 miles per gallon (mpg) of gas, we could express the distance d it could travel as $d = 24g$. Table 3.4 verifies the distance traveled by the car changes in *direct* or *constant proportion* to the number of gallons used, and here we say, "distance traveled *varies directly* with gallons used." The equation $d = 24g$ is called a **direct variation,** and the coefficient 24 is called the **constant of variation.**

Table 3.4

g	d
1	24
2	48
3	72
4	96

Using the rate of change notation, $\dfrac{\Delta \text{distance}}{\Delta \text{gallons}} = \dfrac{\Delta d}{\Delta g} = \dfrac{24}{1}$, and we note this is actually a *linear equation* with slope $m = 24$. When working with variations, the constant k is preferred over m, and in general we have the following:

Direct Variation

y varies directly with x, or *y is directly proportional to x*,
if there is a nonzero constant k such that

$$y = kx.$$

k is called the *constant of variation.*

EXAMPLE 1 ▶ **Writing a Variation Equation**

Write the variation equation for these statements:

- **a.** Wages earned varies directly with the number of hours worked.
- **b.** The value of an office machine varies directly with time.
- **c.** The circumference of a circle varies directly with the length of the diameter.

Solution ▶
- **a.** **W**ages varies directly with **h**ours worked: $W = kh$
- **b.** The **V**alue of an office machine varies directly with **t**ime: $V = kt$
- **c.** The **C**ircumference varies directly with the **d**iameter: $C = kd$

Now try Exercises 7 through 10 ▶

Once we determine the relationship between two variables is a direct variation, we try to find the value of k and develop a general equation model for the relationship indicated. Note that "varies directly" indicates that one value is a constant multiple of the other. In Example 1(c), you may have realized that for $C = kd$, $k = \pi$ since $\pi = \frac{C}{d}$ and the formula for a circle's circumference is $C = \pi d$. The connection helps illustrate the procedure for finding k, as it shows that only *one known relationship is needed!* This suggests the following procedure:

> **Solving Applications of Variation**
>
> 1. Write the information given as an equation, using k as the constant multiple.
> 2. Substitute the first relationship (pair of values) given and solve for k.
> 3. Substitute this value for k in the original equation to obtain the variation equation.
> 4. Use the variation equation to complete the application.

EXAMPLE 2 ▶ **Solving an Application of Direct Variation**

The weight of an astronaut on the surface of another planet **varies directly** with their weight on Earth. An astronaut weighing 140 lb on Earth weighs only 53.2 lb on Mars. How much would a 170-lb astronaut weigh on Mars?

Solution ▶
1. $M = kE$ "Mars weight **varies directly** with Earth weight"
2. $53.2 = k(140)$ substitute 53.2 for M and 140 for E
    $k = 0.38$ solve for k (constant of variation)

Substitute this value of k in the original equation to obtain the variation equation, then find the weight of a 170-lb astronaut that landed on Mars.

3. $M = 0.38E$ variation equation
4. $\quad = 0.38(170)$ substitute 170 for E
    $\quad = 64.6$ result

An astronaut weighing 170 lb on Earth weighs only 64.6 lb on Mars.

Now try Exercises 11 through 14 ▶

The toolbox function from Example 2 was a line with slope $k = 0.38$, or $k = \frac{19}{50}$ as a fraction in simplest form. As a rate of change, $k = \frac{\Delta M}{\Delta E} = \frac{19}{50}$, and we see that for every 50 additional pounds on Earth, the weight of an astronaut would increase by only 19 lb on Mars.

EXAMPLE 3 ▶ **Making Estimates from the Graph of a Variation**

The scientists at NASA are planning to send additional probes to the red planet (Mars), that will weigh from 250 to 450 lb. Graph the variation equation from Example 2, then *use the graph* to estimate the corresponding range of weights on Mars. Check your estimate using the variation equation.

Solution ▶ After selecting an appropriate scale, begin at $(0, 0)$ and count off the slope $k = \frac{\Delta M}{\Delta E} = \frac{19}{50}$. This gives the points $(50, 19)$, $(100, 38)$, $(200, 76)$, and so on. From the graph (see dashed arrows), it appears the weights corresponding to 250 lb and 450 lb on Earth are near 95 lb and 170 lb on Mars. Using the equation gives

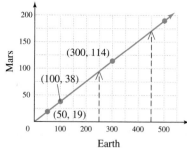

$M = 0.38E$ variation equation
$\quad = 0.38(250)$ substitute 250 for E
$\quad = 95$, and
$M = 0.38E$ variation equation
$\quad = 0.38(450)$ substitute 450 for E
$\quad = 171$, very close to our estimate from the graph.

Now try Exercises 15 and 16 ▶

When a toolbox function is used to model a variation, our knowledge of their graphs and defining characteristics strengthens a contextual understanding of the application. Consider Examples 4 and 5, where the squaring function is used.

EXAMPLE 4 ▶ **Writing Variation Equations**

Write the variation equation for these statements:

a. In free fall, the distance traveled by an object varies directly with the square of the time.

b. The area of a circle varies directly with the square of its radius.

Solution ▶ **a.** Distance varies directly with the square of the time: $D = kt^2$.

b. Area varies directly with the square of the radius: $A = kr^2$.

Now try Exercises 17 through 20 ▶

Both variations in Example 4 use the squaring function, where k represents the amount of stretch or compression applied, and whether the graph will open upward or downward. However, regardless of the function used, the four-step solution process remains the same.

EXAMPLE 5 ▶ **Solving an Application of Direct Variation**

The range of a projectile varies directly with the square of its initial velocity. As part of a circus act, Bailey the Human Bullet is shot out of a cannon with an initial velocity of 80 feet per second (ft/sec), into a net 200 ft away.

a. Find the constant of variation and write the variation equation.

b. Graph the equation and *use the graph* to estimate how far away the net should be placed if initial velocity is increased to 95 ft/sec.

c. Determine the accuracy of the estimate from (b) using the variation equation.

Solution ▶ **a. 1.** $R = kv^2$ "Range varies directly with the square of the velocity"

2. $200 = k(80)^2$ substitute 200 for R and 80 for v

$k = 0.03125$ solve for k (constant of variation)

3. $R = 0.03125v^2$ variation equation (substitute 0.03125 for k)

b. Since velocity and distance are positive, we again use only QI. The graph is a parabola that opens upward, with the vertex at $(0, 0)$. Selecting velocities from 50 to 100 ft/s, we have:

$R = 0.03125v^2$ variation equation

$= 0.03125(50)^2$ substitute 50 for v

$= 78.125$ result

Likewise substituting 100 for v gives $R = 312.5$ ft. Scaling the axes and using $(0, 0)$, $(50, 78)$, and $(100, 313)$ produces the graph shown. At 95 ft/s (dashed lines), it appears the net should be placed about 280 ft away.

c. Using the variation equation gives:

4. $R = 0.03125v^2$ variation equation

$= 0.03125(95)^2$ substitute 95 for v

$R = 282.03125$ result

☑ **A.** You've just learned how to solve direct variations

Our estimate was off by about 2 ft. The net should be placed about 282 ft away.

Now try Exercises 21 through 26 ▶

Note: For Examples 6 to 8, the four steps of the solution process are used in sequence, but are not numbered.

B. Inverse Variation

Table 3.5

Price (dollars)	Demand (1000s)
8	288
9	144
10	96
11	72
12	57.6

Numerous studies have been done that relate the price of a commodity to the demand—the willingness of a consumer to pay that price. For instance, if there is a sudden increase in the price of a popular tool, hardware stores know there will be a corresponding decrease in the demand for that tool. The question remains, "What is this rate of decrease?" Can it be modeled by a linear function with a negative slope? A parabola that opens downward? Some other function? Table 3.5 shows some (simulated) data regarding price versus demand. It appears that a linear function is not appropriate because the rate of change in the number of tools sold is not constant. Likewise a quadratic model seems inappropriate, since we don't expect demand to suddenly start rising again as the price continues to increase. This phenomenon is actually an example of an **inverse variation**, modeled by a transformation of the reciprocal function $y = \frac{k}{x}$. We will often rewrite the equation as $y = k\left(\frac{1}{x}\right)$ to clearly see the inverse relationship. In the case at hand, we might write $D = k\left(\frac{1}{P}\right)$, where k is the constant of variation, D represents the demand for the product, and P the price of the product. In words, we say that "demand *varies inversely* as the price." In other applications of inverse variation, one quantity may vary inversely as the *square* of another, and in general we have

---

Inverse Variation

y varies inversely with x, or *y is inversely proportional to x*, if there is a nonzero constant k such that

$$y = k\left(\frac{1}{x}\right).$$

k is called the *constant of variation*.

EXAMPLE 6 ▶   **Writing Inverse Variation Equations**

Write the variation equation for these statements:

a. In a closed container, pressure varies inversely with the volume of gas.
b. The intensity of light varies inversely with the square of the distance from the source.

Solution ▶   a. Pressure varies inversely with the *Volume* of gas: $P = k\left(\frac{1}{V}\right)$.
b. Intensity of light varies inversely with the square of the distance: $I = k\frac{1}{(d^2)}$.

Now try Exercises 27 through 30 ▶

EXAMPLE 7 ▶   **Solving an Application of Inverse Variation**

Boyle's law tells us that in a closed container with constant temperature, the pressure of a gas varies inversely with its volume (see illustration on page 393). Suppose the air pressure in a closed cylinder is 60 pounds per square inch (psi) when the volume of the cylinder is 50 in³.

a. Find the constant of variation and write the variation equation.
b. Use the equation to find the pressure, if volume is compressed to 30 in³.

Illustration of Boyle's Law

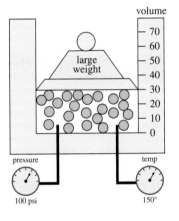

✅ **B. You've just learned how to solve inverse variations**

Solution ▶ **a.** $P = k\left(\dfrac{1}{V}\right)$ "Pressure varies inversely with the volume"

$60 = k\left(\dfrac{1}{50}\right)$ substitute 60 for *P* and 50 for *V*.

$k = 3000$ constant of variation

$P = 3000\left(\dfrac{1}{V}\right)$ variation equation (substitute 3000 for *k*)

b. Using the variation equation we have:

$P = 3000\left(\dfrac{1}{V}\right)$ variation equation

$= 3000\left(\dfrac{1}{30}\right)$ substitute 30 for *V*

$= 100$ result

When the volume is decreased to 30 in³, the pressure increases to 100 psi.

Now try Exercises 31 through 34 ▶

C. Joint or Combined Variations

Just as some decisions might be based on many considerations, often the relationship between two variables depends on a combination of factors. Imagine a wooden plank laid across the banks of a stream for hikers to cross the streambed (see Figure 3.75). The amount of weight the plank will support depends on the type of wood, the width and height of the plank's cross section, and the distance between the supported ends **(see Exercises 59 and 60).** This is an example of a **joint variation,** which can combine

Figure 3.75

any number of variables in different ways. Two general possibilities are: (1) *y varies jointly with the product of x and p: y = kxp;* and (2) *y varies jointly with the product of x and p, and inversely with the square of q:* $y = kxp\left(\dfrac{1}{q^2}\right)$. For practice writing joint variations as an equation model, **see Exercises 35 through 40.**

EXAMPLE 8 ▶ **Solving an Application of Joint Variation**

The amount of fuel used by a ship traveling at a uniform speed varies jointly with the distance it travels and the square of the velocity. If 200 barrels of fuel are used to travel 10 mi at 20 nautical miles per hour, how far does the ship travel on 500 barrels of fuel at 30 nautical miles per hour?

Solution ▶

$F = kdv^2$ "fuel use *varies jointly* with distance and velocity squared"

$200 = k(10)(20)^2$ substitute 200 for *F*, 10 for *d*, and 20 for *v*

$200 = 4000k$ simplify and solve for *k*

$0.05 = k$ constant of variation

$F = 0.05dv^2$ equation of variation

To find the distance traveled at 30 nautical miles per hour using 500 barrels of fuel, substitute 500 for F and 30 for v:

$$F = 0.05dv^2 \qquad \text{equation of variation}$$
$$500 = 0.05d(30)^2 \qquad \text{substitute 500 for } F \text{ and 30 for } v$$
$$500 = 45d \qquad \text{simplify}$$
$$11.\overline{1} = d \qquad \text{result}$$

If 500 barrels of fuel are consumed while traveling 30 nautical miles per hour, the ship covers a distance of just over 11 mi.

Now try Exercises 41 through 44 ▶

It's interesting to note that the ship covers just over one additional mile, but consumes over 2.5 times the amount of fuel. The additional speed requires a great deal more fuel.
There is a variety of additional applications in the Exercise Set. **See Exercises 47 through 55.**

☑ **C. You've just learned how to solve joint variations**

3.8 EXERCISES

▶ **CONCEPTS AND VOCABULARY**

Fill in each blank with the appropriate word or phrase. Carefully reread the section if needed.

1. The phrase "y varies directly with x" is written $y = kx$, where k is called the _____ of variation.

2. If more than two quantities are related in a variation equation, the result is called a _____ variation.

3. The statement "y varies inversely with the square of x" is written _____.

4. For $y = kx$, $y = kx^2$, $y = kx^3$, and $y = k\sqrt{x}$, it is true that as $x \to \infty$, $y \to \infty$ (functions increase). One important difference among the functions is $\frac{\Delta y}{\Delta x}$, or their _____ of _____.

5. Discuss/Explain the general procedure for solving applications of variation. Include references to keywords, and illustrate using an example.

6. The basic percent formula is *amount equals percent times base,* or $A = PB$. In words, write this out as a direct variation with B as the constant of variation, then as an inverse variation with the amount A as the constant of variation.

▶ **DEVELOPING YOUR SKILLS**

Write the variation equation for each statement.

7. distance traveled varies directly with rate of speed

8. cost varies directly with the quantity purchased

9. force varies directly with acceleration

10. length of a spring varies directly with attached weight

For Exercises 11 and 12, find the constant of variation and write the variation equation. Then use the equation to complete the table.

11. y varies directly with x; $y = 0.6$ when $x = 24$.

x	y
500	
	16.25
750	

12. w varies directly with v; $w = \frac{1}{3}$ when $v = 5$.

v	w
291	
	21.8
339	

13. Wages earned varies directly with the number of hours worked. Last week I worked 37.5 hr and my gross pay was $344.25. Write the variation equation and determine how much I will gross this week if I work 35 hr. What does the value of k represent in this case?

14. The thickness of a paperback book varies directly as the number of pages. A book 3.2 cm thick has 750 pages. Write the variation equation and approximate the thickness of *Roget's 21st Century Thesaurus* (paperback—2nd edition), which has 957 pages.

15. The number of stairs in the stairwell of tall buildings and other structures varies directly as the height of the structure. The base and pedestal for the Statue of Liberty are 47 m tall, with 192 stairs from ground level to the observation deck at the top of the pedestal (at the statue's feet). (a) Find the constant of variation and write the variation equation, (b) graph the variation equation, (c) use the graph to estimate the number of stairs from ground level to the observation deck in the statue's crown 81 m above ground level, and (d) use the equation to check this estimate. Was it close?

16. The height of a projected image varies directly as the distance of the projector from the screen. At a distance of 48 in., the image on the screen is 16 in. high. (a) Find the constant of variation and write the variation equation, (b) graph the variation equation, (c) use the graph to estimate the height of the image if the projector is placed at a distance of 5 ft 3 in., and (d) use the equation to check this estimate. Was it close?

Write the variation equation for each statement.

17. Surface area of a cube varies directly with the square of a side.

18. Potential energy in a spring varies directly with the square of the distance the spring is compressed.

19. Electric power varies directly with the square of the current (amperes).

20. Manufacturing cost varies directly as the square of the number of items made.

For Exercises 21 through 26, find the constant of variation and write the variation equation. Then use the equation to complete the table or solve the application.

21. p varies directly with the square of q; $p = 280$ when $q = 50$

q	p
45	
	338.8
70	

22. n varies directly with m squared; $n = 24.75$ when $m = 30$

m	n
40	
	99
88	

23. The surface area of a cube varies directly as the square of one edge. A cube with edges of $14\sqrt{3}$ cm has a surface area of 3528 cm^2. Find the surface area in square meters of the spaceships used by the Borg Collective in *Star Trek—The Next Generation,* cubical spacecraft with edges of 3036 m.

24. The area of an equilateral triangle varies directly as the square of one side. A triangle with sides of 50 yd has an area of 1082.5 yd^2. Find the area in mi^2 of the region bounded by straight lines connecting the cities of Cincinnati, Ohio, Washington, D.C., and Columbia, South Carolina, which are each approximately 400 mi apart.

25. The distance an object falls varies directly as the square of the time it has been falling. The cannonballs dropped by Galileo from the Leaning Tower of Pisa fell about 169 ft in 3.25 sec. (a) Find the constant of variation and write the variation equation, (b) graph the variation equation, (c) use the graph to estimate how long it would take a hammer, accidentally dropped from a height of 196 ft by a bridge repair crew, to splash into the water below, and (d) use the equation to check this estimate. Was it close? (e) According to the equation, if a camera accidentally fell out of the *News 4 Eye-in-the-Sky* helicopter from a height of 121 ft, how long until it strikes the ground?

26. When a child blows small soap bubbles, they come out in the form of a sphere because the surface tension in the soap seeks to minimize the surface area. The surface area of any sphere varies directly with the square of its radius. A soap bubble with a $\frac{3}{4}$ in. radius has a surface area of approximately 7.07 in^2. (a) Find the constant of variation and write the variation equation, (b) graph the variation equation, (c) use the graph to estimate the radius of a seventeenth-century cannonball that has a surface area of 113.1 in^2, and (d) use the equation to check this estimate. Was it close? (e) According to the equation, what is the surface area of an orange with a radius of $1\frac{1}{2}$ in.?

Write the variation equation for each statement.

27. The force of gravity varies inversely as the square of the distance between objects.

28. Pressure varies inversely as the area over which it is applied.

29. The safe load of a beam supported at both ends varies inversely as its length.

30. The intensity of sound varies inversely as the square of its distance from the source.

For Exercises 31 through 34, find the constant of variation and write the variation equation. Then use the equation to complete the table or solve the application.

31. Y varies inversely as the square of Z; $Y = 1369$ when $Z = 3$

Z	Y
37	
	2.25
111	

32. A varies inversely with B; $A = 2450$ when $B = 0.8$

B	A
140	
	6.125
560	

33. The effect of Earth's gravity on an object (its weight) varies inversely as the square of its distance from the center of the planet (assume the Earth's radius is 6400 km). If the weight of an astronaut is 75 kg on Earth (when $r = 6400$), what would this weight be at an altitude of 1600 km *above the surface* of the Earth?

34. The demand for a popular new running shoe varies inversely with the cost of the shoes. When the wholesale price is set at $45, the manufacturer ships 5500 orders per week to retail outlets. Based on this information, how many orders would be shipped if the wholesale price rose to $55?

Write the variation equation for each statement.

35. Interest earned varies jointly with the rate of interest and the length of time on deposit.

36. Horsepower varies jointly as the number of cylinders in the engine and the square of the cylinder's diameter.

37. The area of a trapezoid varies jointly with its height and the sum of the bases.

38. The area of a triangle varies jointly with its height and the length of the base.

39. The volume of metal in a circular coin varies directly with the thickness of the coin and the square of its radius.

40. The electrical resistance in a wire varies directly with its length and inversely as the cross-sectional area of the wire.

For Exercises 41–44, find the constant of variation and write the related variation equation. Then use the equation to complete the table or solve the application.

41. C varies directly with R and inversely with S squared, and $C = 21$ when $R = 7$ and $S = 1.5$.

R	S	C
120		22.5
200	12.5	
	15	10.5

42. J varies directly with P and inversely with the square root of Q, and $J = 19$ when $P = 4$ and $Q = 25$.

P	Q	J
47.5		118.75
112	31.36	
	44.89	66.5

43. **Kinetic energy:** Kinetic energy (energy attributed to motion) varies jointly with the mass of the object and the square of its velocity. Assuming a unit mass of $m = 1$, an object with a velocity of 20 m per sec (m/s) has kinetic energy of 200 J. How much energy is produced if the velocity is increased to 35 m/s?

44. **Safe load:** The load that a horizontal beam can support varies jointly as the width of the beam, the square of its height, and inversely as the length of the beam. A beam 4 in. wide and 8 in. tall can safely support a load of 1 ton when the beam has a length of 12 ft. How much could a similar beam 10 in. tall safely support?

► WORKING WITH FORMULAS

45. Required interest rate: $R(A) = \sqrt[3]{A} - 1$

To determine the simple interest rate R that would be required for each dollar ($1) left on deposit for 3 yr to grow to an amount A, the formula $R(A) = \sqrt[3]{A} - 1$ can be applied. To what function family does this formula belong? Complete the table using a calculator, then use the table to estimate the interest rate required for each $1 to grow to $1.17. Compare your estimate to the value you get by evaluating $R(1.17)$.

Amount A	Rate R
1.0	
1.05	
1.10	
1.15	
1.20	
1.25	

 46. Force between charged particles: $F = k\dfrac{Q_1 Q_2}{d^2}$

The force between two charged particles is given by the formula shown, where F is the force (in joules—J), Q_1 and Q_2 represent the electrical charge on each particle (in coulombs—C), and d is the distance between them (in meters). If the particles have a like charge, the force is repulsive; if the charges are unlike, the force is attractive. (a) Write the variation equation in words. (b) Solve for k and use the formula to find the electrical constant k, given $F = 0.36$ J, $Q_1 = 2 \times 10^{-6}$ C, $Q_2 = 4 \times 10^{-6}$ C, and $d = 0.2$ m. Express the result in scientific notation.

► APPLICATIONS

Find the constant of variation "k" and write the variation equation, then use the equation to solve.

47. Cleanup time: The time required to pick up the trash along a stretch of highway varies inversely as the number of volunteers who are working. If 12 volunteers can do the cleanup in 4 hr, how many volunteers are needed to complete the cleanup in just 1.5 hr?

48. Wind power: The wind farms in southern California contain wind generators whose power production varies directly with the cube of the wind's speed. If one such generator produces 1000 W of power in a 25 mph wind, find the power it generates in a 35 mph wind.

49. Pull of gravity: The weight of an object on the moon varies directly with the weight of the object on Earth. A 96-kg object on Earth would weigh only 16 kg on the moon. How much would a fully suited 250-kg astronaut weigh on the moon?

50. Period of a pendulum: The time that it takes for a simple pendulum to complete one period (swing over and back) varies directly as the square root of its length. If a pendulum 20 ft long has a period of 5 sec, find the period of a pendulum 30 ft long.

51. Stopping distance: The stopping distance of an automobile varies directly as the square root of its speed when the brakes are applied. If a car requires 108 ft to stop from a speed of 25 mph, estimate the stopping distance if the brakes were applied when the car was traveling 45 mph.

52. Supply and demand: A chain of hardware stores finds that the demand for a special power tool varies inversely with the advertised price of the tool. If the price is advertised at $85, there is a monthly demand for 10,000 units at all participating stores. Find the projected demand if the price were lowered to $70.83.

53. Cost of copper tubing: The cost of copper tubing varies jointly with the length and the diameter of the tube. If a 36-ft spool of $\frac{1}{4}$-in.-diameter tubing costs $76.50, how much does a 24-ft spool of $\frac{3}{8}$-in.-diameter tubing cost?

54. Electrical resistance: The electrical resistance of a copper wire varies directly with its length and inversely with the square of the diameter of the wire. If a wire 30 m long with a diameter of 3 mm has a resistance of 25 Ω, find the resistance of a wire 40 m long with a diameter of 3.5 mm.

55. Volume of phone calls: The number of phone calls per day between two cities varies directly as the product of their populations and inversely as the square of the distance between them. The city of Tampa, Florida (pop. 300,000), is 430 mi from the city of Atlanta, Georgia (pop. 420,000).

Telecommunications experts estimate there are about 300 calls per day between the two cities. Use this information to estimate the number of daily phone calls between Amarillo, Texas (pop. 170,000), and Denver, Colorado (pop. 550,000), which are also separated by a distance of about 430 mi. Note: Population figures are for the year 2000 and rounded to the nearest ten-thousand.

Source: 2005 World Almanac, p. 626.

56. **Internet commerce:** The likelihood of an eBay® item being sold for its "Buy it Now®" price P, varies directly with the feedback rating of the seller, and inversely with the cube of $\frac{P}{MSRP}$, where MSRP represents the manufacturer's suggested retail price. A power eBay® seller with a feedback rating of 99.6%, knows she has a 60% likelihood of selling an item at 90% of the MSRP. What is the likelihood a seller with a 95.3% feedback rating can sell the same item at 95% of the MSRP?

57. **Volume of an egg:** The volume of an egg laid by an average chicken varies jointly with its length and the square of its width. An egg measuring 2.50 cm wide and 3.75 cm long has a volume of 12.27 cm^3. A Barret's Blue Ribbon hen can lay an egg measuring 3.10 cm wide and 4.65 cm long. (a) What is the volume of this egg? (b) As a percentage, how much greater is this volume than that of an average chicken?

58. **Athletic performance:** Researchers have estimated that a sprinter's time in the 100-m dash varies directly as the square root of her age and inversely as the number of hours spent training each week. At 20 yr old, Gail trains 10 hr per week (hr/wk) and has an average time of 11 sec. Assuming she continues to train 10 hr/wk, (a) what will her average time be at 30 yr old? (b) If she wants to keep her average time at 11 sec, how many hours per week should she train?

59. **Maximum safe-load:** The maximum safe load M that can be placed on a uniform horizontal beam supported at both ends varies directly as the width w and the square of the height h of the beam's cross section, and inversely as its length L (width and height are assumed to be in inches, and length in feet). (a) Write the variation equation. (b) If a beam 18 in. wide, 2 in. high, and 8 ft long can safely support 270 lb, what is the safe load for a beam of like dimensions with a length of 12 ft?

60. **Maximum safe load:** Suppose a 10-ft wooden beam with dimensions 4 in. by 6 in. is made from the same material as the beam in Exercise 59 (the same k value can be used). (a) What is the maximum safe load if the beam is placed so that width is 6 in. and height is 4 in.? (b) What is the maximum safe load if the beam is placed so that width is 4 in. and height is 6 in.?

▶ **EXTENDING THE CONCEPT**

61. In function form, the variations $Y_1 = k\frac{1}{x}$ and $Y_2 = k\frac{1}{x^2}$ become $f(x) = k\frac{1}{x}$ and $g(x) = k\frac{1}{x^2}$. Both graphs appear similar in Quadrant I and both may "fit" a scatter-plot fairly well, but there is a big difference between them—they decrease as x gets larger, but *they decrease at very different rates*. Assume $k = 1$ and use the ideas from Section 2.5 to compute the rate of change for f and g for the interval from $x = 0.5$ to $x = 0.6$. Were you surprised? In the interval $x = 0.7$ to $x = 0.8$, will the rate of decrease for each function be greater or less than in the interval $x = 0.5$ to $x = 0.6$? Why?

62. The gravitational force F between two celestial bodies varies jointly as the product of their masses and inversely as the square of the distance d between them. The relationship is modeled by Newton's law of universal gravitation: $F = k\frac{m_1 m_2}{d^2}$.

Given that $k = 6.67 \times 10^{-11}$, what is the gravitational force exerted by a 1000-kg sphere on another identical sphere that is 10 m away?

63. The intensity of light and sound both vary inversely as the square of their distance from the source.

a. Suppose you're relaxing one evening with a copy of *Twelfth Night* (Shakespeare), and the reading light is placed 5 ft from the surface of the book. At what distance would the intensity of the light be twice as great?

b. *Tamino's Aria* (*The Magic Flute*—Mozart) is playing in the background, with the speakers 12 ft away. At what distance from the speakers would the intensity of sound be three times as great?

▶ **MAINTAINING YOUR SKILLS**

64. (R.3) Evaluate: $\left(\dfrac{2x^4}{3x^3y}\right)^{-2}$

65. (1.5) Find all zeroes, real and complex:
$x^3 + 4x^2 + 8x = 0$.

66. (2.2) State the domains of f and g given here:

   a. $f(x) = \dfrac{x-3}{x^2-16}$ **b.** $g(x) = \dfrac{x-3}{\sqrt{x^2-16}}$

67. (2.5) Graph by using transformations of the parent function and plotting a minimum number of points:
$f(x) = -2|x-3| + 5$.

SUMMARY AND CONCEPT REVIEW

SECTION 3.1   Quadratic Functions and Applications

KEY CONCEPTS

- A quadratic function is one of the form $f(x) = ax^2 + bx + c;\ a \neq 0$. The simplest quadratic is the squaring function $f(x) = x^2$, where $a = 1$ and $b, c = 0$.
- The graph of a quadratic function is a parabola. Parabolas have three distinctive features: (1) like end behavior on the left and right, (2) an axis of symmetry, (3) a highest or lowest point called the vertex.
- For a quadratic function in the standard form $y = ax^2 + bx + c$,
  - End behavior: graph opens upward if $a > 0$, opens downward if $a < 0$
  - Zeroes/x-intercepts: substitute 0 for y and solve for x (if they exist)
  - y-intercept: substitute 0 for $x \rightarrow (0, c)$
  - Vertex: (h, k), where $h = \dfrac{-b}{2a},\ k = f\left(\dfrac{-b}{2a}\right)$
  - Maximum value: If the parabola opens downward, $y = k$ is the maximum value of f.
  - Minimum value: If the parabola opens upward, $y = k$ is the minimum value of f.
  - Line of symmetry: $x = h$ is the line (or axis) of symmetry [if $(h + c, y)$ is on the graph, then $(h - c, y)$ is also on the graph].
- By completing the square, $f(x) = ax^2 + bx + c$ can be written as the transformation $f(x) = a(x + h)^2 \pm k$, and graphed using transformations of $y = x^2$.

EXERCISES

Graph $f(x)$ by completing the square and using transformations of the parent function. Graph $g(x)$ and $h(x)$ using the vertex formula and y-intercept. Find the x-intercepts (if they exist) for all functions.

  1. $f(x) = x^2 + 8x + 15$ **2.** $g(x) = -x^2 + 4x - 5$ **3.** $h(x) = 4x^2 - 12x + 3$

 4. Height of a superball: A teenager tries to see how high she can bounce her superball by throwing it downward on her driveway. The height of the ball (in feet) at time t (in seconds) is given by $h(t) = -16t^2 + 96t$. (a) How high is the ball at $t = 0$? (b) How high is the ball after 1.5 sec? (c) How long until the ball is 135 ft high? (d) What is the maximum height attained by the ball? At what time t did this occur?

SECTION 3.2   Synthetic Division; the Remainder and Factor Theorems

KEY CONCEPTS

- Synthetic division is an abbreviated form of long division. Only the coefficients of the dividend are used, since "standard form" ensures like place values are aligned. Zero placeholders are used for "missing" terms. The "divisor" must be linear with leading coefficient 1.
- To divide a polynomial by $x - c$, use c in the synthetic division; to divide by $x + c$, use $-c$.
- After setting up the synthetic division template, drop the leading coefficient of the dividend into place, then multiply in the diagonal direction, place the product in the next column, and add in the vertical direction, continuing to the last column.
- The final sum is the remainder r, the numbers preceding it are the coefficients of $q(x)$.
- Remainder theorem: If $p(x)$ is divided by $x - c$, the remainder is equal to $p(c)$. The theorem can be used to evaluate polynomials at $x = c$.
- Factor theorem: If $p(c) = 0$, then c is a zero of p and $(x - c)$ is a factor. Conversely, if $(x - c)$ is a factor of p, then $p(c) = 0$. The theorem can be used to factor a polynomial or build a polynomial from its zeroes.
- The remainder and factor theorems also apply when c is a complex number.

EXERCISES

Divide using long division and clearly identify the quotient and remainder:

5. $\dfrac{x^3 + 4x^2 - 5x - 6}{x - 2}$

6. $\dfrac{x^3 + 2x - 4}{x^2 - x}$

7. Use the factor theorem to show that $x + 7$ is a factor of $2x^4 + 13x^3 - 6x^2 + 9x + 14$.

8. Complete the division and write $h(x)$ as $h(x) = d(x)q(x) + r(x)$, given $\dfrac{h(x)}{d(x)} = \dfrac{x^3 - 4x + 5}{x - 2}$.

9. Use the factor theorem to help factor $p(x) = x^3 + 2x^2 - 11x - 12$ completely.

10. Use the factor and remainder theorems to factor h, given $x = 4$ is a zero: $h(x) = x^4 - 3x^3 - 4x^2 - 2x + 8$.

Use the remainder theorem:

11. Show $x = \frac{1}{2}$ is a zero of V: $V(x) = 4x^3 + 8x^2 - 3x - 1$.

12. Show $x = 3i$ is a zero of W: $W(x) = x^3 - 2x^2 + 9x - 18$.

13. Find $h(-7)$ given $h(x) = x^3 + 9x^2 + 13x - 10$.

Use the factor theorem:

14. Find a degree 3 polynomial in standard form with zeroes $x = 1$, $x = -\sqrt{5}$, and $x = \sqrt{5}$.

15. Find a fourth-degree polynomial in standard form with one real zero, given $x = 1$ and $x = -2i$ are zeroes.

16. Use synthetic division and the remainder theorem to answer: At a busy shopping mall, customers are constantly coming and going. One summer afternoon during the hours from 12 o'clock noon to 6 in the evening, the number of customers in the mall could be modeled by $C(t) = 3t^3 - 28t^2 + 66t + 35$, where $C(t)$ is the number of customers (in tens), t hours after 12 noon. (a) How many customers were in the mall at noon? (b) Were more customers in the mall at 2:00 or at 3:00 P.M.? How many more? (c) Was the mall busier at 1:00 P.M. (after lunch) or 6:00 P.M. (around dinner time)?

SECTION 3.3   Zeroes of Polynomial Functions

KEY CONCEPTS

- Fundamental theorem of algebra: Every complex polynomial of degree $n \geq 1$ has at least one complex zero.
- Linear factorization theorem: Every complex polynomial of degree $n \geq 1$ has exactly n linear factors, and can be written in the form $p(x) = a(x - c_1)(x - c_2) \ldots (x - c_n)$, where $a \neq 0$ and $c_1, c_2, \ldots, c_n$ are (not necessarily distinct) complex numbers.

- For a polynomial p in factored form with repeated factors $(x - c)^m$, c is a zero of multiplicity m. If m is odd, c is a zero of odd multiplicity; if m is even, c is a zero of even multiplicity.
- Corollaries to the linear factorization theorem:
  I. If p is a polynomial with real coefficients, p can be factored into linear factors (not necessarily distinct) and irreducible quadratic factors having real coefficients.
  II. If p is a polynomial with real coefficients, the complex zeroes of p must occur in conjugate pairs. If $a + bi$ $(b \neq 0)$, is a zero, then $a - bi$ is also a zero.
  III. If p is a polynomial with degree $n \geq 1$, then p will have exactly n zeroes (real or complex), where zeroes of multiplicity m are counted m times.
- Intermediate value theorem: If p is a polynomial with real coefficients where $p(a)$ and $p(b)$ have opposite signs, then there is at least one c between a and b such that $p(c) = 0$.
- Rational zeroes theorem: If a real polynomial has integer coefficients, rational zeroes must be of the form $\frac{p}{q}$, where p is a factor of the constant term and q is a factor of the leading coefficient.
- Descartes' rule of signs, upper and lower bounds property, tests for -1 and 1, and graphing technology can all be used with the rational zeroes theorem to factor, solve, and graph polynomial functions.

EXERCISES

Using the tools from this section,

17. List all possible rational zeroes of
$p(x) = 4x^3 - 16x^2 + 11x + 10$.

18. Find all rational zeroes of
$p(x) = 4x^3 - 16x^2 + 11x + 10$.

19. Write $P(x) = 2x^3 - 3x^2 - 17x - 12$
in completely factored form.

20. Prove that $h(x) = x^4 - 7x^2 - 2x + 3$
has no rational zeroes.

21. Identify two intervals (of those given) that contain a zero of $P(x) = x^4 - 3x^3 - 8x^2 + 12x + 6$: $[-2, -1]$, $[1, 2]$, $[2, 3]$, $[4, 5]$. Then verify your answer using a graphing calculator.

22. Discuss the number of possible positive, negative, and complex zeroes for $g(x) = x^4 + 3x^3 - 2x^2 - x - 30$. Then identify which combination is correct using a graphing calculator.

SECTION 3.4 Graphing Polynomial Functions

KEY CONCEPTS

- All polynomial graphs are smooth, continuous curves.
- A polynomial of degree n has *at most* $n - 1$ turning points. The precise location of these turning points are the local maximums or local minimums of the function.
- If the degree of a polynomial is odd, the ends of its graph will point in opposite directions (like $y = mx$). If the degree is even, the ends will point in the same direction (like $y = ax^2$). The sign of the lead coefficient determines the actual behavior.
- The "behavior" of a polynomial graph near its zeroes is determined by the multiplicity of the zero. For any factor $(x - c)^m$, the graph will "cross through" the x-axis if m is odd and "bounce off" the x-axis (touching at just one point) if m is even. The larger the value of m, the flatter (more compressed) the graph will be near c.
- To "round-out" a graph, additional *midinterval points* can be found between known zeroes.
- These ideas help to establish the *Guidelines for Graphing Polynomial Functions*. See page 327.

EXERCISES

State the degree, end behavior, and y-intercept, but do not graph.

23. $f(x) = -3x^5 + 2x^4 + 9x - 4$

24. $g(x) = (x - 1)(x + 2)^2(x - 2)$

Graph using the *Guidelines for Graphing Polynomials.*

25. $p(x) = (x + 1)^3(x - 2)^2$ **26.** $q(x) = 2x^3 - 3x^2 - 9x + 10$ **27.** $h(x) = x^4 - 6x^3 + 8x^2 + 6x - 9$

28. For the graph of $P(x)$ shown, (a) state whether the degree of P is even or odd, (b) use the graph to locate the zeroes of P and state whether their multiplicity is even or odd, and (c) find the minimum possible degree of P and write it in factored form. Assume all zeroes are real.

SECTION 3.5   Graphing Rational Functions

KEY CONCEPTS

- A rational function is one of the form $V(x) = \dfrac{p(x)}{d(x)}$, where p and d are polynomials and $d(x) \neq 0$.

- The domain of V is all real numbers, except the zeroes of d.
- If zero is in the domain of V, substitute 0 for x to find the y-intercept.
- The zeroes of V (if they exist), are solutions to $p(x) = 0$.
- The line $y = k$ is a horizontal asymptote of V if as $|x|$ increases without bound, $V(x)$ approaches k.
- If $\dfrac{p(x)}{d(x)}$ is in simplest form, vertical asymptotes will occur at the zeroes of d.
- The line $x = h$ is a vertical asymptote of V if as x approaches h, $V(x)$ increases/decreases without bound.
- If the degree of p is less than the degree of d, $y = 0$ (the x-axis) is a horizontal asymptote. If the degree of p is equal to the degree of d, $y = \dfrac{a}{b}$ is a horizontal asymptote, where a is the leading coefficient of p, and b is the leading coefficient of d.
- The *Guidelines for Graphing Rational Functions* can be found on page 342.

EXERCISES

29. For the function $V(x) = \dfrac{x^2 - 9}{x^2 - 3x - 4}$, state the following but do not graph: (a) domain (in set notation), (b) equations of the horizontal and vertical asymptotes, (c) the x- and y-intercept(s), and (d) the value of $V(1)$.

30. For $v(x) = \dfrac{(x + 1)^2}{x + 2}$, will the function change sign at $x = -1$? Will the function change sign at $x = -2$? Justify your responses.

Graph using the *Guidelines for Graphing Rational Functions*.

31. $v(x) = \dfrac{x^2 - 4x}{x^2 - 4}$

32. $t(x) = \dfrac{2x^2}{x^2 - 5}$

33. Use the vertical asymptotes, x-intercepts, and their multiplicities to construct an equation that corresponds to the given graph. Be sure the y-intercept on the graph matches the value given by your equation. Assume these features are integer-valued. Check your work on a graphing calculator.

34. The average cost of producing a popular board game is given by the function $A(x) = \dfrac{5000 + 15x}{x}$; $x \geq 1000$. (a) Identify the horizontal asymptote of the function and explain its meaning in this context. (b) To be profitable, management believes the average cost must be below $17.50. What levels of production will make the company profitable?

SECTION 3.6   Additional Insights into Rational Functions

KEY CONCEPTS

- If $V = \dfrac{p(x)}{d(x)}$ is not in simplest form, with p and d sharing factors of the form $x - c$, the graph will have a removable discontinuity (a hole or gap) at $x = c$. The discontinuity can be "removed" (repaired) by redefining V using a piecewise-defined function.

- If $V = \dfrac{p(x)}{d(x)}$ is in simplest form, and the degree of p is greater than the degree of d, the graph will have an oblique or nonlinear asymptote, as determined by the quotient polynomial after division. If the degree of p is greater by 1, the result is a slant (oblique) asymptote. If the degree of p is greater by 2, the result is a parabolic asymptote.

EXERCISES

35. Determine if the graph of h will have a vertical asymptote or a removable discontinuity, then graph the function

$$h(x) = \frac{x^3 - 2x^2 - 9x + 18}{x - 2}.$$

36. Sketch the graph of $h(x) = \dfrac{x^2 - 3x - 4}{x + 1}$. If there is a removable discontinuity, repair the break by redefining h using an appropriate piecewise-defined function.

Graph the functions using the *Guidelines for Graphing Rational Functions.*

37. $h(x) = \dfrac{x^2 - 2x}{x - 3}$ **38.** $t(x) = \dfrac{x^3 - 7x + 6}{x^2}$

39. The cost to make x thousand party favors is given by $C(x) = x^2 - 2x + 6$, where $x \geq 1$ and C is in thousands of dollars. For the average cost of production $A(x) = \dfrac{x^2 - 2x + 6}{x}$, (a) graph the function, (b) use the graph to estimate the level of production that will make average cost a minimum, and (c) state the average cost of a single party favor at this level of production.

SECTION 3.7   Polynomial and Rational Inequalities

KEY CONCEPTS

- To solve polynomial inequalities, write $P(x)$ in factored form and note the multiplicity of each real zero.
- Plot real zeroes on a number line. The graph will cross the x-axis at zeroes of odd multiplicity (P will change sign), and bounce off the axis at zeroes of even multiplicity (P will not change sign).
- Use the end behavior, y-intercept, or a test point to determine the sign of P in a given interval, then label all other intervals as $P(x) > 0$ or $P(x) < 0$ by analyzing the multiplicity of neighboring zeroes. Use the resulting diagram to state the solution.
- The solution process for rational inequalities and polynomial inequalities is virtually identical, considering that vertical asymptotes also create intervals where function values may change sign, depending on their multiplicity.
- Polynomial and rational inequalities can also be solved using an interval test method. Since polynomials and rational functions are continuous on their domains, the sign of the function at any one point in an interval will be the same as for all other points in that interval.

EXERCISES

Solve each inequality indicated using a number line and the behavior of the graph at each zero.

40. $x^3 + x^2 > 10x - 8$ **41.** $\dfrac{x^2 - 3x - 10}{x - 2} \geq 0$ **42.** $\dfrac{x}{x - 2} \leq \dfrac{-1}{x}$

SECTION 3.8   Variation: Function Models in Action

KEY CONCEPTS

- *Direct variation:* If there is a nonzero constant k such that $y = kx$, we say, "y varies directly with x" or "y is directly proportional to x" (k is called the constant of variation).

- *Inverse variation:* If there is a nonzero constant k such that $y = k\left(\dfrac{1}{x}\right)$ we say, "y varies inversely with x" or y is inversely proportional to x.

- In some cases, direct and inverse variations work simultaneously to form a *joint variation.*
- The process for solving variation equations can be found on page 380.

EXERCISES

Find the constant of variation and write the equation model, then use this model to complete the table.

43. y varies directly as the cube root of x; $y = 52.5$ when $x = 27$.

x	y
216	
	12.25
729	

44. z varies directly as v and inversely as the square of w; $z = 1.62$ when $w = 8$ and $v = 144$.

v	w	z
196	7	
	1.25	17.856
24		48

45. Given t varies jointly with u and v, and inversely as w, if $t = 30$ when $u = 2$, $v = 3$, and $w = 5$, find t when $u = 8$, $v = 12$, and $w = 15$.

46. The time that it takes for a simple pendulum to complete one period (swing over and back) is directly proportional to the square root of its length. If a pendulum 16 ft long has a period of 3 sec, find the time it takes for a 36-ft pendulum to complete one period.

MIXED REVIEW

1. Find the equation of the function whose graph is shown.

2. Complete the square to write each function as a transformation. Then graph each function, clearly labeling the vertex and all intercepts (if they exist).

  a. $f(x) = 2x^2 + 8x + 3$ **b.** $g(x) = -x^2 - 4x$

3. A computer components manufacturer produces external 2.5″ hard drives. Their sizes range from 20 GB to 200 GB. The cost of producing a hard drive can be modeled by the function $C(s) = \dfrac{1}{180}s^2 - \dfrac{8}{9}s + \dfrac{680}{9}$, where s is the size of the hard drive, in gigabytes. Find the hard drive size that has the lowest cost of production. What is the cost of production?

4. Divide using long division and name the quotient and remainder: $\dfrac{x^3 + 3x^2 - 5x - 7}{x + 3}$.

5. Divide using synthetic division and name the quotient and remainder: $\dfrac{x^4 - 3x^2 + 5x - 1}{x + 2}$.

Use synthetic division and the remainder theorem to complete Exercises 6 and 7.

6. State which of the following *are not factors* of $x^3 - 9x^2 + 2x + 48$: (a) $(x + 6)$, (b) $(x - 8)$, (c) $(x - 12)$, (d) $(x - 4)$, (e) $(x + 2)$.

7. Given $P(x) = 6x^3 - 23x^2 - 40x + 31$, find (a) $P(-1)$, (b) $P(1)$, and (c) $P(5)$.

8. Use the factor theorem.

  a. Find a real polynomial of degree 3 with roots $x = 3$ and $x = -5i$.

  b. Find a real polynomial of degree 2 with $x = 2 - 3i$ as one of the roots.

9. Use the rational zeroes theorem.

   a. Which of the following *cannot be* roots of
$6x^3 + x^2 - 20x - 12 = 0$?
$x = 9$ $x = -3$ $x = \frac{3}{2}$ $x = \frac{8}{3}$ $x = -\frac{2}{3}$

   b. Write P in completely factored form. Then state all zeroes of P, real and complex.
$P(x) = x^4 - x^3 + 7x^2 - 9x - 18$.

10. Graph using the *Guidelines for Graphing Polynomials*.

   a. $f(x) = x^3 - 13x + 12$

   b. $g(x) = x^4 - 10x^2 + 9$

   c. $h(x) = (x - 1)^3 (x + 2)^2(x + 1)$

Graph using the *Guidelines for Graphing Rational Functions.*

11. $p(x) = \dfrac{x^2 - 2x}{x^2 - 2x + 1}$ **12.** $q(x) = \dfrac{x^2 - 4}{x^2 - 3x - 4}$

13. $r(x) = \dfrac{x^3 - 13x + 12}{x^2}$ **14.** $y = \dfrac{x^2 - 4x}{x - 3}$

(see Exercise 10a)

Solve each inequality.

15. $x^3 - 4x < 12 - 3x^2$ **16.** $\dfrac{4}{x + 2} \geq \dfrac{3}{x}$

17. An open, rectangular box is to be made from a 24-in. by 16-in. piece of sheet metal, by cutting a square from each corner and folding up the sides.

   a. Show that the resulting volume is given by
$V(x) = 4x^3 - 80x^2 + 384x$.

24 in.

16 in.

   b. Show that for a desired volume of 512 in³, the height "x" of the box can be found by solving $x^3 - 20x^2 + 96x - 128 = 0$.

   c. According to the rational roots theorem *and the context of this application,* what are the possible rational zeroes for this equation?

   d. Find the *rational zero x* (the height) that gives the box a volume of 512 in³.

   e. Use the zero from part (d) and synthetic division to help find the *irrational zero x* that also gives the box a volume of 512 in³. Round the solution to hundredths.

Write the variation equation for each statement.

18. The volume of metal in a circular coin varies directly with the thickness of the coin and the square of its radius.

19. The electrical resistance in a wire varies directly with its length and inversely as the cross-sectional area of the wire.

20. Cost of copper tubing: The cost of copper tubing varies jointly with the length and the diameter of the tube. If a 36-ft spool of $\frac{1}{4}$-in. diameter tubing costs $76.50, how much does a 24-ft spool of $\frac{3}{8}$-in. diameter tubing cost?

PRACTICE TEST

1. Complete the square to write each function as a transformation. Then graph each function and label the vertex and all intercepts (if they exist).

   a. $f(x) = -x^2 + 10x - 16$

   b. $g(x) = \dfrac{1}{2}x^2 + 4x + 16$

2. The graph of a quadratic function has a vertex of $(-1, -2)$, and passes through the origin. Find the other intercept, and the equation of the graph in standard form.

3. Suppose the function $d(t) = t^2 - 14t$ models the depth of a scuba diver at time t, as she dives

underwater from a steep shoreline, reaches a certain depth, and swims back to the surface.

   a. What is her depth after 4 sec? After 6 sec?

   b. What was the maximum depth of the dive?

   c. How many seconds was the diver beneath the surface?

4. Compute the quotient using long division:
$\dfrac{x^3 - 3x^2 + 5x - 2}{x^2 + 2x + 1}$.

5. Find the quotient and remainder using synthetic division: $\dfrac{x^3 + 4x^2 - 5x - 20}{x + 2}$.

6. Use the remainder theorem to show $(x + 3)$ is a factor of $x^4 - 15x^2 - 10x + 24$.

7. Given $f(x) = 2x^3 + 4x^2 - 5x + 2$, find the value of $f(-3)$ using synthetic division and the remainder theorem.

8. Given $x = 2$ and $x = 3i$ are two zeroes of a real polynomial $P(x)$ with degree 3. Use the factor theorem to find $P(x)$.

9. Factor the polynomial and state the multiplicity of each zero: $Q(x) = (x^2 - 3x + 2)(x^3 - 2x^2 - x + 2)$.

10. Given $C(x) = x^4 + x^3 + 7x^2 + 9x - 18$, (a) use the rational zeroes theorem to list all possible rational zeroes; (b) apply Descartes' rule of signs to count the number of possible positive, negative, and complex zeroes; and (c) use this information along with the tests for 1 and -1, synthetic division, and the factor theorem to factor C completely.

11. Over a 10-yr period, the balance of payments (deficit versus surplus) for a small county was modeled by the function $f(x) = \frac{1}{2}x^3 - 7x^2 + 28x - 32$, where $x = 0$ corresponds to 1990 and $f(x)$ is the deficit or surplus in millions of dollars. (a) Use the rational roots theorem and synthetic division to find the years the county "broke even" (debt = surplus = 0) from 1990 to 2000. (b) How many years did the county run a surplus during this period? (c) What was the surplus/deficit in 1993?

12. Sketch the graph of $f(x) = (x - 3)(x + 1)^3(x + 2)^2$ using the degree, end behavior, x- and y-intercepts, zeroes of multiplicity, and a few "midinterval" points.

13. Use the *Guidelines for Graphing Polynomials* to graph $g(x) = x^4 - 9x^2 - 4x + 12$.

14. Use the *Guidelines for Graphing Rational Functions* to graph $h(x) = \dfrac{x - 2}{x^2 - 3x - 4}$.

15. Suppose the cost of cleaning contaminated soil from a dump site is modeled by $C(x) = \dfrac{300x}{100 - x}$, where $C(x)$ is the cost (in \$1000s) to remove $x\%$ of the contaminants. Graph using $x \in [0, 100]$, and use the graph to answer the following questions.

    a. What is the significance of the *vertical asymptote* (what does it mean in this context)?

    b. If EPA regulations are changed so that 85% of the contaminants must be removed, instead of the 80% previously required, how much additional cost will the new regulations add? Compare the cost of the 5% increase from 80% to 85% with the cost of the 5% increase from 90% to 95%. What do you notice?

    c. What percent of the pollutants can be removed if the company budgets \$2,200,000?

16. Graph using the *Guidelines for Graphing Rational Functions*.

    a. $r(x) = \dfrac{x^3 - x^2 - 9x + 9}{x^2}$

    b. $R(x) = \dfrac{x^3 + 7x - 6}{x^2 - 4}$

17. Find the level of production that will minimize the average cost of an item, if production costs are modeled by $C(x) = 2x^2 + 25x + 128$, where $C(x)$ is the cost to manufacture x hundred items.

18. Solve each inequality

    a. $x^3 - 13x \le 12$ **b.** $\dfrac{3}{x - 2} < \dfrac{2}{x}$

19. Suppose the concentration of a chemical in the bloodstream of a large animal h hr after injection into muscle tissue is modeled by the formula $C(h) = \dfrac{2h^2 + 5h}{h^3 + 55}$.

    a. Sketch a graph of the function for the intervals $x \in [-5, 20]$, $y \in [0, 1]$.

    b. Where is the vertical asymptote? Does it play a role in this context?

    c. What is the concentration after 2 hr? After 8 hr?

    d. How long does it take the concentration to fall below 20% $[C(h) < 0.2]$?

    e. When does the maximum concentration of the chemical occur? What is this maximum?

    f. Describe the significance of the horizontal asymptote in this context.

20. The maximum load that can be supported by a rectangular beam varies jointly with its width and its height squared and inversely with its length. If a beam 10 ft long, 3 in. wide, and 4 in. high can support 624 lb, how many pounds could a beam support with the same dimensions but 12 ft long?

CALCULATOR EXPLORATION AND DISCOVERY

Complex Zeroes, Repeated Zeroes, and Inequalities

This *Calculator Exploration and Discovery* will explore the relationship between the solution of a polynomial (or rational) inequality and the complex zeroes and repeated zeroes of the related function. After all, if complex zeroes can never create an x-intercept, how do they affect the function? And if a zero of even multiplicity never crosses the x-axis (always bounces), can it still affect a nonstrict (*less than or equal to* or *greater than or equal to*) inequality? These are interesting and important questions, with numerous avenues of exploration. To begin, consider the function $Y_1 = (x + 3)^2(x^3 - 1)$. In completely factored form $Y_1 = (x + 3)^2(x - 1)(x^2 + x + 1)$. This is a polynomial function of degree 5 with two real zeroes (one repeated), two complex zeroes (the quadratic factor is irreducible), and after viewing the graph on Figure 3.76, four turning points. From the graph (or by analysis), we have $Y_1 \leq 0$ for $x \leq 1$. Now let's consider $Y_2 = (x + 3)^2(x - 1)$, the same function as Y_1, less the quadratic factor. Since complex zeroes never "cross the x-axis" anyway, the removal of this factor *cannot affect the solution set of the inequality!* But how does it affect the function? Y_2 is now a function of degree three, with three real zeroes (one repeated) and only two turning points (Figure 3.77). But even so, the solution to $Y_2 \leq 0$ is the same as for $Y_1 \leq 0$: $x \leq 1$. Finally, let's look at $Y_3 = x - 1$, the same function as Y_2 but with the repeated zero removed. The key here is to notice that since $(x - 3)^2$

will be nonnegative for any value of x, it too does not change the solution set of the "less than or equal to inequality," only the shape of the graph. Y_3 is a function of degree 1, with one real zero and no turning points, *but the solution interval for $Y_3 \leq 0$ is the same solution interval as Y_2 and Y_1: $x \leq 1$* (see Figure 3.78).

Explore these relationships further using the following exercises and a "greater than or equal to" inequality. Begin by writing Y_1 in completely factored form.

Exercise 1: $Y_1 = (x^3 - 6x^2 + 32)(x^2 + 1)$
$\qquad\qquad Y_2 = x^3 - 6x^2 + 32$
$\qquad\qquad Y_3 = x + 2$

Exercise 2: $Y_1 = (x + 3)^2(x^3 - 2x^2 + x - 2)$
$\qquad\qquad Y_2 = (x + 3)^2(x - 2)$
$\qquad\qquad Y_3 = x - 2$

Exercise 3: Based on what you've noticed, comment on how the irreducible quadratic factors of a polynomial affect its graph. What role do they play in the solution of inequalities?

Exercise 4: How do zeroes of even multiplicity affect the solution set of nonstrict inequalities (less/greater than or equal to)?

For more on these ideas, see the *Strengthening Core Skills* feature from this chapter.

Figure 3.76

Figure 3.77

Figure 3.78

STRENGTHENING CORE SKILLS

Solving Inequalities Using the Push Principle

The most common method for solving polynomial inequalities involves finding the zeroes of the function and checking the sign of the function in the intervals between these zeroes. In Section 3.7, we relied on the end behavior of the graph, the sign of the function at the y-intercept, and the multiplicity of the zeroes to determine the solution. There is a third method that is more conceptual in nature,

but in many cases highly efficient. It is based on two very simple ideas, the first involving only order relations and the number line:

A. Given any number x and constant $k > 0$: $x > x - k$ and $x < x + k$.

This statement simply reinforces the idea that if a is left of b on the number line, then $a < b$. As shown in the diagram, $x - 4 < x$ and $x < x + 3$, from which $x - 4 < x + 3$ for any x.

B. The second idea reiterates well-known ideas regarding the multiplication of signed numbers. For any number of factors:

if there are an even number of negative factors, the result is positive;

if there are an odd number of negative factors, the result is negative.

These two ideas work together to solve inequalities using what we'll call the *push principle*. Consider the inequality $x^2 - x - 12 > 0$. The factored form is $(x - 4)(x + 3) > 0$ and we want the product of these two factors to be positive. From (A), both factors will be positive if $(x - 4)$ is positive, since it's the smaller of the two; and both factors will be negative if $x + 3 < 0$, since it's the larger. The solution set is found by solving these two simple inequalities: $x - 4 > 0$ gives $x > 4$ and $x + 3 < 0$ gives $x < -3$. If the inequality were $(x - 4)(x + 3) < 0$ instead, we require one negative factor and one positive factor. Due to order relations and the number line, the larger factor must be the positive one: $x + 3 > 0$ so $x > -3$. The smaller factor must be the negative one: $x - 4 < 0$ and $x < 4$. This gives the solution $-3 < x < 4$ as can be verified using any alternative method. Solutions to all other polynomial and rational inequalities are an extension of these two cases.

Illustration 1 ▶ Solve $x^3 - 7x + 6 < 0$ using the push principle.

Solution ▶ The polynomial can be factored using the tests for 1 and -1 and synthetic division. The factors are $(x - 2)(x - 1)(x + 3) < 0$, which we've conveniently written in increasing order. For the product of three factors to be negative we require: (1) three negative factors or (2) one negative and two positive factors. The first condition is met

by simply making the largest factor negative, as it will ensure the smaller factors are also negative: $x + 3 < 0$ so $x < -3$. The second condition is met by making the smaller factor negative and the "middle" factor positive: $x - 2 < 0$ *and* $x - 1 > 0$. The second solution interval is $x < 2$ and $x > 1$, or $1 < x < 2$.

Note the push principle does not require the testing of intervals between the zeroes, nor the "cross/bounce" analysis at the zeroes and vertical asymptotes (of rational functions). In addition, irreducible quadratic factors can still be ignored as they contribute nothing to the solution of real inequalities, and factors of even multiplicity can be overlooked precisely because there is no sign change at these roots.

Illustration 2 ▶ Solve $(x^2 + 1)(x - 2)^2(x + 3) \geq 0$ using the push principle.

Solution ▶ Since the factor $(x^2 + 1)$ does not affect the solution set, this inequality will have the same solution as $(x - 2)^2(x + 3) \geq 0$. Further, since $(x - 2)^2$ will be non-negative for all x, the original inequality *has the same solution set as* $(x + 3) \geq 0$! The solution is $x \geq -3$.

With some practice, the push principle can be a very effective tool. Use it to solve the following exercises. Check all solutions by graphing the function on a graphing calculator.

Exercise 1: $x^3 - 3x - 18 \leq 0$

Exercise 2: $\dfrac{x + 1}{x^2 - 4} > 0$

Exercise 3: $x^3 - 13x + 12 < 0$

Exercise 4: $x^3 - 3x + 2 \geq 0$

Exercise 5: $x^4 - x^2 - 12 > 0$

Exercise 6: $(x^2 + 5)(x^2 - 9)(x + 2)^2(x - 1) \geq 0$

CUMULATIVE REVIEW CHAPTERS R–3

1. Solve for R: $\dfrac{1}{R} = \dfrac{1}{R_1} + \dfrac{1}{R_2}$

2. Solve for x: $\dfrac{2}{x + 1} + 1 = \dfrac{5}{x^2 - 1}$

3. Factor the expressions:
   a. $x^3 - 1$ **b.** $x^3 - 3x^2 - 4x + 12$

4. Solve using the quadratic formula. Write answers in both exact and approximate form:
   $2x^2 + 4x + 1 = 0$.

5. Solve the following inequality: $x + 3 < 5$ *or* $5 - x < 4$.

6. Name the eight toolbox functions, give their equations, then draw a sketch of each.

7. Use substitution to verify that $x = 2 - 3i$ is a solution to $x^2 - 4x + 13 = 0$.

8. Solve the rational inequality:
   $\dfrac{x + 4}{x - 2} < 3$.

9. As part of a study on traffic conditions, the mayor of a small city tracks her driving time to work each day for six months and finds a linear and increasing relationship. On day 1, her drive time was 17 min. By day 61 the drive time had increased to 28 min. Find a linear function that models the drive time and use it to estimate the drive time on day 121, if the trend continues. Explain what the slope of the line means in this context.

10. Does the relation shown represent a function? If not, discuss/explain why not.

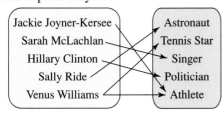

11. The data given shows the profit of a new company for the first 6 months of business, and is closely modeled by the function $p(m)$ $= 1.18x^2 - 10.99x + 4.6$, where $p(m)$ is the profit earned in month m. Assuming this trend continues, use this function to find the first month a profit will be earned $(p > 0)$.

Exercise 11

Month	Profit (1000s)
1	−5
2	−13
3	−18
4	−20
5	−21
6	−19

12. Graph the function
$$g(x) = \frac{-1}{(x + 2)^2} + 3 \text{ using}$$
transformations of a basic function.

13. Find $f^{-1}(x)$, given $f(x) = \sqrt[3]{2x - 3}$, then use composition to verify your inverse is correct.

14. Graph $f(x) = x^2 - 4x + 7$ by completing the square, then state intervals where:
   a. $f(x) \geq 0$ **b.** $f(x)\uparrow$

15. Given the graph of a general function $f(x)$, graph $F(x) = -f(x + 1) + 2$.

Exercise 15

16. Graph the piecewise-defined function given:
$$f(x) = \begin{cases} -3 & x < -1 \\ x & -1 \leq x \leq 1 \\ 3x & x > 1 \end{cases}$$

17. Y varies directly with X and inversely with the square of Z. If $Y = 10$ when $X = 32$ and $Z = 4$, find X when $Z = 15$ and $Y = 1.4$.

18. Use the rational zeroes theorem and synthetic division to find all zeroes (real and complex) of $f(x) = x^4 - 2x^2 + 16x - 15$.

19. Sketch the graph of $f(x) = x^3 - 3x^2 - 6x + 8$.

20. Sketch the graph of $h(x) = \dfrac{x - 1}{x^2 - 4}$ and use the zeroes and vertical asymptotes to solve $h(x) \geq 0$.

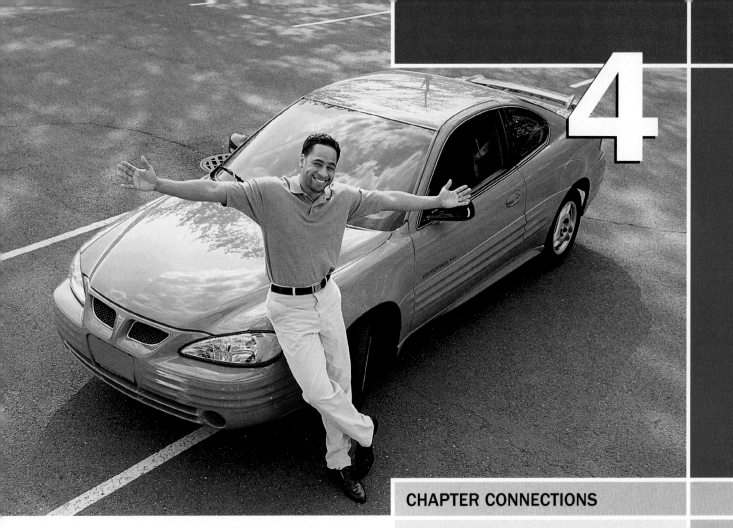

Exponential and Logarithmic Functions

CHAPTER OUTLINE

CHAPTER CONNECTIONS

The largest purchase that most individuals will make in their lifetime is that of a car or home. The monthly payment $\mathcal{P}$ required to amortize (pay off) the loan can be calculated using the formula

$$\mathcal{P} = \frac{AR}{1 - (1 - R)^{-12t}}$$

where A is the amount financed; t is the time in years; and $R = \dfrac{r}{12}$, where r is the annual rate of interest. This study of exponential and logarithmic functions will help you become a more knowledgeable consumer. This application appears as Exercise 53 in Section 4.5.

Check out these other real-world connections:

▶ Calculating the Effects of Inflation (Section 4.2, Exercises 87 and 88)
▶ Calculating the Intensity of Sound (Section 4.3, Exercises 87 to 90)
▶ Calculating the Proper Ventilation of a Home (Section 4.3, Exercise 97)
▶ Calculating Freezing Time for Water Puddles (Section 4.4, Exercise 120)

411

Learning Objectives

In Section 4.1 you will learn how to:

- ☐ **A.** Identify one-to-one functions
- ☐ **B.** Explore inverse functions using ordered pairs
- ☐ **C.** Find inverse functions using an algebraic method
- ☐ **D.** Graph a function and its inverse
- ☐ **E.** Solve applications of inverse functions

Consider the function $f(x) = 2x - 3$. If $f(x) = 7$, the equation becomes $2x - 3 = 7$, and the corresponding value of x can be found using *inverse operations*. In this section, we introduce the concept of an *inverse function,* which can be viewed as a formula for finding x-values that correspond to *any* given value of $f(x)$.

A. Identifying One-to-One Functions

The graphs of $y = 2x$ and $y = x^2$ are shown in Figures 4.1 and 4.2. The dashed, vertical lines clearly indicate both are functions, with each x-value corresponding to only one y. But the points on $y = 2x$ have one characteristic those from $y = x^2$ do not—*each y-value also corresponds to only one x* (for $y = x^2$, 4 corresponds to both -2 and 2). If each element from the range of a function corresponds to only one element of the domain, the function is said to be **one-to-one.**

Figure 4.1

Figure 4.2

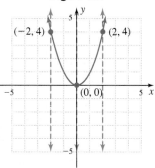

One-to-One Functions

A function f is one-to-one if every element in the range corresponds to only one element of the domain.

In symbols,   if $f(x_1) = f(x_2)$ then $x_1 = x_2$, or
if $x_1 \neq x_2$, then $f(x_1) \neq f(x_2)$.

From this definition we note the graph of a one-to-one function must not only pass a vertical line test (to show each x corresponds to only one y), but also pass a **horizontal line test** (to show each y corresponds to only one x).

Horizontal Line Test

If every horizontal line intersects the graph of a function in at most one point, the function is one-to-one.

Notice the graph of $y = 2x$ (Figure 4.3) passes the horizontal line test, while the graph of $y = x^2$ (Figure 4.4) does not.

Figure 4.3

Figure 4.4

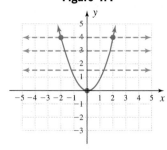

EXAMPLE 1 ▶ Identifying One-to-One Functions

Use the horizontal line test to determine whether each graph is the graph of a one-to-one function.

a.

b.

c.

d.
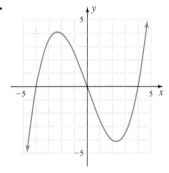

Solution ▶ A careful inspection shows all four graphs depict a function, since each passes the vertical line test. Only (a) and (b) pass the horizontal line test and are *one-to-one* functions.

Now try Exercises 7 through 28 ▶

☑ **A.** You've just learned how to identify a one-to-one function

If the function is given in ordered pair form, we simply check to see that no given second coordinate is paired with more than one first coordinate.

B. Inverse Functions and Ordered Pairs

Consider the function $f(x) = 2x - 3$ and the solutions shown in Table 4.1. Figure 4.5 shows this function in diagram form (in blue), and illustrates that for each element of the domain, we *multiply by 2, then subtract 3*. An **inverse function** for f is one that takes the result of these operations (elements of the range), and returns the original domain element. Figure 4.6 shows that function F achieves this by "undoing" the operations in reverse order: *add 3, then divide by 2* (in red). A table of values for $F(x)$ is shown (Table 4.2).

Table 4.1

x	$f(x)$
-3	-9
0	-3
2	1
5	7
8	13

Table 4.2

x	$F(x)$
-9	-3
-3	0
1	2
7	5
13	8

Figure 4.5

Figure 4.6

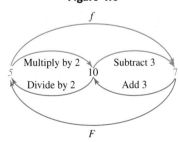

From this illustration we make the following observations regarding an inverse function, which we actually denote as $f^{-1}(x)$.

Inverse Functions

If f is a one-to-one function with ordered pairs (a, b),
1. $f^{-1}(x)$ is a one-to-one function with ordered pairs (b, a).
2. The range of f will be the domain of $f^{-1}(x)$.
3. The domain of f will be the range of $f^{-1}(x)$.

⚠ **CAUTION** ▶ The notation $f^{-1}(x)$ is simply a way of denoting an inverse function and has nothing to do with exponential properties. In particular, $f^{-1}(x)$ does *not* mean $\frac{1}{f(x)}$.

EXAMPLE 2 ▶ **Finding the Inverse of a Function**

Find the inverse of each one-to-one function given:
 a. $f(x) = \{(-4, 13), (-1, 7), (0, 5), (2, 1), (5, -5), (8, -11)\}$
 b. $p(x) = -3x + 2$

Solution ▶ a. When a function is defined as a set of ordered pairs, the inverse function is found by simply interchanging the x- and y-coordinates:
$f^{-1}(x) = \{(13, -4), (7, -1), (5, 0), (1, 2), (-5, 5), (-11, 8)\}$.
 b. Using diagrams similar to Figures 4.5 and 4.6, we reason that $p^{-1}(x)$ will subtract 2, then divide the result by -3: $p^{-1}(x) = \dfrac{x - 2}{-3}$. As a test, we find that $(-2, 8)$, $(0, 2)$, and $(3, -7)$ are solutions to $p(x)$, and note that $(8, -2)$, $(2, 0)$, and $(-7, 3)$ are indeed solutions to $p^{-1}(x)$.

☑ **B.** You've just learned how to explore inverse functions using ordered pairs

Now try Exercises 29 through 40 ▶

C. Finding Inverse Functions Using an Algebraic Method

The fact that interchanging x- and y-values helps determine an inverse function can be generalized to develop an **algebraic method** for finding inverses. Instead of interchanging *specific* x- and y-values, we actually interchange the x- and y-*variables*, then solve the equation for y. The process is summarized here.

WORTHY OF NOTE

If a function is *not* one-to-one, no inverse function exists since interchanging the x- and y-coordinates will result in a nonfunction. For instance, interchanging the coordinates of $(-2, 4)$ and $(2, 4)$ from $y = x^2$ results in $(4, -2)$ and $(4, 2)$, and we have one x-value being mapped to two y-values, in violation of the function definition.

Finding an Inverse Function

1. Use y instead of $f(x)$.
2. Interchange x and y.
3. Solve the equation for y.
4. The result gives the inverse function: substitute $f^{-1}(x)$ for y.

In this process, it might seem like we're using the *same* y to represent two different functions. To see why there is actually no contradiction, **see Exercise 103.**

EXAMPLE 3 ▶ **Finding Inverse Functions Algebraically**

Use the algebraic method to find the inverse function for

 a. $f(x) = \sqrt[3]{x + 5}$ b. $g(x) = \dfrac{2x}{x + 1}$

Solution ▶ **a.** $f(x) = \sqrt[3]{x+5}$ given function

$y = \sqrt[3]{x+5}$ use y instead of $f(x)$

$x = \sqrt[3]{y+5}$ interchange x and y

$x^3 = y + 5$ cube both sides

$x^3 - 5 = y$ solve for y

$x^3 - 5 = f^{-1}(x)$ the result is $f^{-1}(x)$

For $f(x) = \sqrt[3]{x+5}$, $f^{-1}(x) = x^3 - 5$.

b. $g(x) = \dfrac{2x}{x+1}$ given function

$y = \dfrac{2x}{x+1}$ use y instead of $f(x)$

$x = \dfrac{2y}{y+1}$ interchange x and y

$xy + x = 2y$ multiply by $y + 1$ and distribute

$x = 2y - xy$ gather terms with y

$x = y(2 - x)$ factor

$\dfrac{x}{2-x} = y$ solve for y

$\dfrac{x}{2-x} = g^{-1}(x)$ the result is $g^{-1}(x)$

For $g(x) = \dfrac{2x}{x+1}$, $g^{-1}(x) = \dfrac{x}{2-x}$.

Now try Exercises 41 through 48 ▶

In cases where a given function is *not* one-to-one, we can sometimes restrict the domain to create a function that *is,* and then determine an inverse. The restriction we use is arbitrary, and only requires that the result still produce all possible range values. For the most part, we simply choose a limited domain that seems convenient or reasonable.

EXAMPLE 4 ▶   Restricting the Domain to Create a One-to-One Function

Given $f(x) = (x - 4)^2$, restrict the domain to create a one-to-one function, then find $f^{-1}(x)$. State the domain and range of both resulting functions.

Solution ▶ The graph of f is a parabola, opening upward with the vertex at $(4, 0)$. Restricting the domain to $x \geq 4$ (see figure) leaves only the "right branch" of the parabola, creating a one-to-one function without affecting the range, $y \in [0, \infty)$. For $f(x) = (x - 4)^2$, $x \geq 4$, we have

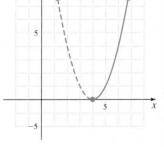

$f(x) = (x - 4)^2$ given function

$y = (x - 4)^2$ use y instead of $f(x)$

$x = (y - 4)^2$ interchange x and y

$\pm\sqrt{x} = y - 4$ take square roots

$\sqrt{x} + 4 = y$ solve for y, use $\sqrt{x}$ since $x \geq 4$

The result shows $f^{-1}(x) = \sqrt{x} + 4$, with domain $x \in [0, \infty)$ and range $y \in [4, \infty)$ (the domain of f becomes the range of f^{-1}, and the range of f becomes the domain of f^{-1}).

Now try Exercises 49 through 54 ▶

While we now have the ability to *find* the inverse of a function, we still lack a definitive method of *verifying* the inverse is correct. Actually, the diagrams in Figures 4.5 and 4.6 suggest just such a method. If we use the function *f* itself as an input for f^{-1}, or the function f^{-1} as an input for *f*, the end result should simply be *x*, as each function "undoes" the operations of the other. From Section 2.8 this is called a composition of functions and using the notation for composition we have,

---

Verifying Inverse Functions

If *f* is a one-to-one function, then the function f^{-1} exists, where

$$(f \circ f^{-1})(x) = x \quad \text{and} \quad (f^{-1} \circ f)(x) = x$$

EXAMPLE 5 ▶ **Finding and Verifying an Inverse Function**

Use the algebraic method to find the inverse function for $f(x) = \sqrt{x + 2}$. Then verify the inverse you found is correct.

Solution ▶ Since the graph of *f* is the graph of $y = \sqrt{x}$ shifted 2 units left, we know *f* is one-to-one with domain $x \in [-2, \infty)$ and range $y \in [0, \infty)$. This is important since the *domain and range values will be interchanged for the inverse function.* The domain of f^{-1} will be $x \in [0, \infty)$ and its range $y \in [-2, \infty)$.

$f(x) = \sqrt{x + 2}$	given function; $x \geq -2$
$y = \sqrt{x + 2}$	use *y* instead of $f(x)$
$x = \sqrt{y + 2}$	interchange *x* and *y*
$x^2 = y + 2$	solve for *y* (square both sides)
$x^2 - 2 = y$	subtract 2
$f^{-1}(x) = x^2 - 2$	the result is $f^{-1}(x)$; $D: x \in [0, \infty)$, $R: y \in [-2, \infty)$

Verify ▶

$(f \circ f^{-1})(x) = f[f^{-1}(x)]$	$f^{-1}(x)$ is an input for *f*
$\quad = \sqrt{f^{-1}(x) + 2}$	*f* adds 2 to inputs, then takes the square root
$\quad = \sqrt{(x^2 - 2) + 2}$	substitute $x^2 - 2$ for $f^{-1}(x)$
$\quad = \sqrt{x^2}$	simplify
$\quad = x$ ✓	since the domain of $f^{-1}(x)$ is $x \in [0, \infty)$

Verify ▶

$(f^{-1} \circ f)(x) = f^{-1}[f(x)]$	$f(x)$ is an input for f^{-1}
$\quad = [f(x)]^2 - 2$	f^{-1} squares inputs, then subtracts 2
$\quad = [\sqrt{x + 2}]^2 - 2$	substitute $\sqrt{x + 2}$ for $f(x)$
$\quad = x + 2 - 2$	simplify
$\quad = x$ ✓	result

☑ **C. You've just learned how to find inverse functions using an algebraic method**

Now try Exercises 55 through 80 ▶

D. The Graph of a Function and Its Inverse

Graphing a function and its inverse on the same axes reveals an interesting and useful relationship—the graphs are reflections across the line $y = x$ (the identity function).

Consider the function $f(x) = 2x + 3$, and its inverse $f^{-1}(x) = \dfrac{x - 3}{2} = \dfrac{1}{2}x - \dfrac{3}{2}$. In Figure 4.7, the points $(1, 5)$, $(0, 3)$, $(-\frac{3}{2}, 0)$, and $(-4, -5)$ from *f* (see Table 4.3) are graphed in blue, with the points $(5, 1)$, $(3, 0)$, $(0, -\frac{3}{2})$, and $(-5, -4)$ (see Table 4.4) from f^{-1} graphed in red (note the x- and y-values are reversed). Graphing both lines illustrates this symmetry (Figure 4.8).

Table 4.3

x	$f(x)$
1	5
0	3
$-\dfrac{3}{2}$	0
-4	-5

Figure 4.7

Figure 4.8

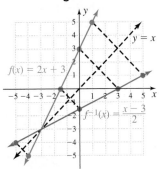

Table 4.4

x	$f^{-1}(x)$
5	1
3	0
0	$-\dfrac{3}{2}$
-5	-4

EXAMPLE 6 ▶ **Graphing a Function and Its Inverse**

In Example 5, we found the inverse function for $f(x) = \sqrt{x + 2}$ was $f^{-1}(x) = x^2 - 2, x \geq 0$. Graph these functions on the same axes and comment on how the graphs are related.

Solution ▶ The graph of f is a square root function with initial point $(-2, 0)$, a y-intercept of $(0, \sqrt{2})$, and an x-intercept of $(-2, 0)$ (Figure 4.9 in blue). The graph of $x^2 - 2, x \geq 0$ is the right-hand branch of a parabola, with y-intercept at $(0, -2)$ and an x-intercept at $(\sqrt{2}, 0)$ (Figure 4.9 in red).

Figure 4.9

Figure 4.10

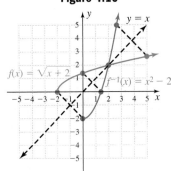

Connecting these points with a smooth curve indeed shows their graphs are symmetric to the line $y = x$ (Figure 4.10).

Now try Exercises 81 through 88 ▶

EXAMPLE 7 ▶ **Graphing a Function and Its Inverse**

Given the graph shown in Figure 4.11, use the grid in Figure 4.12 to draw a graph of the inverse function.

Figure 4.11

Figure 4.12

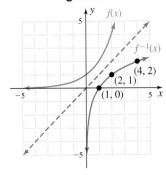

Solution ▶ From the graph, the domain of f appears to be $x \in \mathbb{R}$ and the range is $y \in (0, \infty)$. This means the domain of f^{-1} will be $x \in (0, \infty)$ and the range will be $y \in \mathbb{R}$. To sketch f^{-1}, draw the line $y = x$, interchange the x- and y-coordinates of the selected points, then plot these points and draw a smooth curve using the domain and range boundaries as a guide.

Now try Exercises 89 through 94 ▶

A summary of important points is given here followed by their application in Example 8.

Functions and Inverse Functions

D. You've just learned how to graph a function and its inverse

1. If the graph of a function passes the horizontal line test, the function is one-to-one.
2. If a function f is one-to-one, the function f^{-1} exists.
3. The domain of f is the range of f^{-1}, and the range of f is the domain of f^{-1}.
4. For a function f and its inverse f^{-1}, $(f \circ f^{-1})(x) = x$ and $(f^{-1} \circ f)(x) = x$.
5. The graphs of f and f^{-1} are symmetric with respect to the line $y = x$.

E. Applications of Inverse Functions

Our final example illustrates one of the many ways that inverse functions can be applied.

EXAMPLE 8 ▶   **Using Volume to Understand Inverse Functions**

The volume of an equipoise cylinder (height equal to diameter) is given by $v(x) = 2\pi x^3$ (since $h = d = 2r$), where $v(x)$ represents the volume in units cubed and x represents the radius of the cylinder.

    a. Find the volume of such a cylinder if $x = 10$ ft.
    b. Find $v^{-1}(x)$, and discuss what the input and output variables represent.
    c. If a volume of 1024π ft^3 is required, which formula would be easier to use to find the radius? What is this radius?

Solution ▶   **a.** $v(x) = 2\pi x^3$ given function
           $v(10) = 2\pi(10)^3$ substitute 10 for x
                 $= 2000\pi$ $10^3 = 1000$, exact form

With a radius of 10 ft, the volume of the cylinder would be 2000π ft^3.

    b. $v(x) = 2\pi x^3$ given function
              $y = 2\pi x^3$ use y instead of $v(x)$
              $x = 2\pi y^3$ interchange x and y

$$\frac{x}{2\pi} = y^3 \qquad \text{solve for } y$$

$$\sqrt[3]{\frac{x}{2\pi}} = y \qquad \text{result}$$

The inverse function is $v^{-1}(x) = \sqrt[3]{\dfrac{x}{2\pi}}$. In this case, the input x is a given volume, the output $v^{-1}(x)$ is the radius of an equipoise cylinder that will hold this volume.

c. Since the volume is known and we need the radius, using $v^{-1}(x) = \sqrt[3]{\dfrac{x}{2\pi}}$ would be more efficient.

$$v^{-1}(1024\pi) = \sqrt[3]{\dfrac{1024\pi}{2\pi}} \qquad \text{substitute } 1024\pi \text{ for } x \text{ in } v^{-1}(x)$$
$$= \sqrt[3]{512} \qquad \dfrac{2\pi}{2\pi} = 1$$
$$= 8 \qquad \text{result}$$

☑ **E.** You've just learned how to solve an application of inverse functions

The radius of the cylinder would be 8 ft.

Now try Exercises 97 through 102 ▶

TECHNOLOGY HIGHLIGHT

Investigating Inverse Functions

Many important ideas from this section can be illustrated using a graphing calculator. To begin, enter the function

$Y_1 = 2\sqrt[3]{x - 2}$ and $Y_2 = \dfrac{x^3}{8} + 2$ (which appear to be inverse

functions) on the [Y=] screen, then press [ZOOM] **5:ZSquare**. The graphs seem to be reflections across the line $y = x$ (Figure 4.13). To verify, use the [TABLE] feature with inputs $x = -2, -1, 0, 1, 2$, and 3. As shown in Figure 4.14, the points $(1, -2)$, $(2, 0)$, and $(3, 2)$ are on Y_1, and the points $(-2, 1)$, $(0, 2)$, and $(2, 3)$ are all on Y_2. While this seems convincing (the x- and y-coordinates are interchanged), the technology can actually *compose the two functions* to verify an inverse relationship. Function names Y_1 and Y_2 can be accessed using the [VARS] and [▶] keys, then pressing [ENTER]. After entering $Y_3 = Y_1(Y_2)$ and $Y_4 = Y_2(Y_1)$ on the [Y=] screen, we observe whether one function "undoes" the other using the TABLE feature (Figure 4.15).

Figure 4.13

For the functions given, (a) find $f^{-1}(x)$, then use your calculator to verify they are inverses by (b) using ordered pairs, (c) composing the functions, and (d) showing their graphs are symmetric to $y = x$.

Figure 4.14

X	Y₁	Y₂
-2	-3.175	1
-1	-2.884	1.875
0	-2.52	2
1	-2	2.125
2	0	3
3	2	5.375

X=

Figure 4.15

X	Y₃	Y₄
-2	-2	-2
-1	-1	-1
0	0	0
1	1	1
2	2	2
3	3	3

X=

Exercise 1: $f(x) = 2x + 1$ **Exercise 2:** $g(x) = x^2 + 1; x \geq 0$

Exercise 3: $h(x) = \dfrac{x}{x + 1}$

4.1 EXERCISES

▶ CONCEPTS AND VOCABULARY

Fill in each blank with the appropriate word or phrase. Carefully reread the section if needed.

1. A function is one-to-one if each _____ coordinate corresponds to exactly _____ first coordinate.

2. If every _____ line intersects the graph of a function in at most _____ point, the function is one-to-one.

3. A certain function is defined by the ordered pairs $(-2, -11)$, $(0, -5)$, $(2, 1)$, and $(4, 19)$. The inverse function is _____ .

4. To find f^{-1} using the algebraic method, we (1) use _____ instead of $f(x)$, (2) _____ x and y, (3) _____ for y and replace y with $f^{-1}(x)$.

5. State true or false and explain why: *To show that g is the inverse function for f, simply show that* $(f \circ g)(x) = x$. Include an example in your response.

6. Discuss/Explain why no inverse function exists for $f(x) = (x + 3)^2$ and $g(x) = \sqrt{4 - x^2}$. How would the domain of each function have to be restricted to allow for an inverse function?

▶ DEVELOPING YOUR SKILLS

Determine whether each graph given is the graph of a one-to-one function. If not, give examples of how the definition of one-to-oneness is violated.

7.

8.

9.

10.

11.

12.

13.

14.

15.

Determine whether the functions given are one-to-one. If not, state why.

16. $\{(-7, 4), (-1, 9), (0, 5), (-2, 1), (5, -5)\}$

17. $\{(9, 1), (-2, 7), (7, 4), (3, 9), (2, 7)\}$

18. $\{(-6, 1), (4, -9), (0, 11), (-2, 7), (-4, 5), (8, 1)\}$

19. $\{(-6, 2), (-3, 7), (8, 0), (12, -1), (2, -3), (1, 3)\}$

Determine if the functions given are one-to-one by noting the function family to which each belongs and mentally picturing the shape of the graph. If a function is not one-to-one, discuss how the definition of one-to-oneness is violated.

20. $f(x) = 3x - 5$

21. $g(x) = (x + 2)^3 - 1$

22. $h(x) = -|x - 4| + 3$

23. $p(t) = 3t^2 + 5$

24. $s(t) = \sqrt{2t - 1} + 5$

25. $r(t) = \sqrt[3]{t + 1} - 2$

26. $y = 3$

27. $y = -2x$

28. $y = x$

For Exercises 29 to 32, find the inverse function of the one-to-one functions given.

29. $f(x) = \{(-2, 1), (-1, 4), (0, 5), (2, 9), (5, 15)\}$

30. $g(x) = \{(-2, 30), (-1, 11), (0, 4), (1, 3), (2, 2)\}$

31. $v(x)$ is defined by the ordered pairs shown.

X	Y1	
-4	3	
-3	2	
0	1	
5	0	
12	-1	
21	-2	
▮	-3	
X=32		

32. $w(x)$ is defined by the ordered pairs shown.

X	Y1	
-6	4	
-5	2.5	
-2	-2	
0	-5	
3	-9.5	
4	-11	
▮	-15.5	
X=7		

Find the inverse function using diagrams similar to those illustrated in Example 2. Check the result using three test points.

33. $f(x) = x + 5$

34. $g(x) = x - 4$

35. $p(x) = -\dfrac{4}{5}x$

36. $r(x) = \dfrac{3}{4}x$

37. $f(x) = 4x + 3$

38. $g(x) = 5x - 2$

39. $Y_1 = \sqrt[3]{x - 4}$

40. $Y_2 = \sqrt[3]{x + 2}$

Find each function $f(x)$ given, (a) find any three ordered pair solutions (a, b), then (b) algebraically compute $f^{-1}(x)$, and (c) verify the ordered pairs (a, b) satisfy $f^{-1}(x)$.

41. $f(x) = \sqrt[3]{x - 2}$

42. $f(x) = \sqrt[3]{x + 3}$

43. $f(x) = x^3 + 1$

44. $f(x) = x^3 - 2$

45. $f(x) = \dfrac{8}{x + 2}$

46. $f(x) = \dfrac{12}{x - 1}$

47. $f(x) = \dfrac{x}{x + 1}$

48. $f(x) = \dfrac{x + 2}{1 - x}$

The functions given in Exercises 49 through 54 are not one-to-one. (a) Determine a domain restriction that preserves all range values, then state this domain and range. (b) Find the inverse function and state its domain and range.

49. $f(x) = (x + 5)^2$

50. $g(x) = x^2 + 3$

51. $v(x) = \dfrac{8}{(x - 3)^2}$

52. $V(x) = \dfrac{4}{x^2} + 2$

53. $p(x) = (x + 4)^2 - 2$

54. $q(x) = \dfrac{4}{(x - 2)^2} + 1$

For each function $f(x)$ given, prove (using a composition) that $g(x) = f^{-1}(x)$.

55. $f(x) = -2x + 5$, $g(x) = \dfrac{x - 5}{-2}$

56. $f(x) = 3x - 4$, $g(x) = \dfrac{x + 4}{3}$

57. $f(x) = \sqrt[3]{x} + 5$, $g(x) = x^3 - 5$

58. $f(x) = \sqrt[3]{x} - 4$, $g(x) = x^3 + 4$

59. $f(x) = \frac{2}{3}x - 6$, $g(x) = \frac{3}{2}x + 9$

60. $f(x) = \frac{4}{5}x + 6$, $g(x) = \frac{5}{4}x - \frac{15}{2}$

61. $f(x) = x^2 - 3; x \geq 0$, $g(x) = \sqrt{x + 3}$

62. $f(x) = x^2 + 8; x \geq 0$, $g(x) = \sqrt{x - 8}$

Find the inverse of each function $f(x)$ given, then prove (by composition) your inverse function is correct. Note the domain of f is all real numbers.

63. $f(x) = 3x - 5$

64. $f(x) = 5x + 4$

65. $f(x) = \dfrac{x - 5}{2}$

66. $f(x) = \dfrac{x + 4}{3}$

67. $f(x) = \frac{1}{2}x - 3$

68. $f(x) = \frac{2}{3}x + 1$

69. $f(x) = x^3 + 3$

70. $f(x) = x^3 - 4$

71. $f(x) = \sqrt[3]{2x + 1}$

72. $f(x) = \sqrt[3]{3x - 2}$

73. $f(x) = \dfrac{(x - 1)^3}{8}$

74. $f(x) = \dfrac{(x + 3)^3}{-27}$

Find the inverse of each function, then prove (by composition) your inverse function is correct. State the implied domain and range as you begin, and use these to state the domain and range of the inverse function.

75. $f(x) = \sqrt{3x + 2}$

76. $g(x) = \sqrt{2x - 5}$

77. $p(x) = 2\sqrt{x - 3}$

78. $q(x) = 4\sqrt{x + 1}$

79. $v(x) = x^2 + 3; x \geq 0$

80. $w(x) = x^2 - 1; x \geq 0$

Graph each function $f(x)$ and its inverse $f^{-1}(x)$ on the same grid and "dash-in" the line $y = x$. Note how the graphs are related. Then verify the "inverse function" relationship using a composition.

81. $f(x) = 4x + 1; f^{-1}(x) = \dfrac{x - 1}{4}$

82. $f(x) = 2x - 7; f^{-1}(x) = \dfrac{x + 7}{2}$

83. $f(x) = \sqrt[3]{x + 2}; f^{-1}(x) = x^3 - 2$

84. $f(x) = \sqrt[3]{x - 7}; f^{-1}(x) = x^3 + 7$

85. $f(x) = 0.2x + 1; f^{-1}(x) = 5x - 5$

86. $f(x) = \dfrac{2}{9}x + 4; f^{-1}(x) = \dfrac{9}{2}x - 18$

87. $f(x) = (x + 2)^2; x \geq -2; f^{-1}(x) = \sqrt{x} - 2$

88. $f(x) = (x - 3)^2; x \geq 3; f^{-1}(x) = \sqrt{x} + 3$

Determine the domain and range for each function whose graph is given, and use this information to state the domain and range of the inverse function. Then sketch in the line $y = x$, estimate the location of two or

more points on the graph, and use these to graph $f^{-1}(x)$ on the same grid.

89.

90.

91.

92.

93.

94.

▶ WORKING WITH FORMULAS

95. **The height of a projected image:** $f(x) = \frac{1}{2}x - 8.5$

The height of an image projected on a screen by a projector is given by the formula shown, where $f(x)$ represents the actual height of the image on the projector (in centimeters) and x is the distance of the projector from the screen (in centimeters). (a) When the projector is 80 cm from the screen, how large is the image? (b) Show that the inverse function is $f^{-1}(x) = 2x + 17$, then input your answer from part (a) and comment on the result. What information does the inverse function give?

96. **The radius of a sphere:** $r(x) = \sqrt[3]{\dfrac{3x}{4\pi}}$

In generic form, the radius of a sphere is given by the formula shown, where $r(x)$ represents the radius and x represents the volume of the sphere in cubic units. (a) If a weather balloon that is roughly spherical holds 14,130 in³ of air, what is the radius of the balloon (use $\pi \approx 3.14$)? (b) Show that the inverse function is $r^{-1}(x) = \frac{4}{3}\pi x^3$, then input your answer from part (a) and comment on the result. What information does the inverse function give?

▶ APPLICATIONS

97. **Temperature and altitude:** The temperature (in degrees Fahrenheit) at a given altitude can be approximated by the function $f(x) = -\frac{7}{2}x + 59$, where $f(x)$ represents the temperature and x represents the altitude in thousands of feet. (a) What is the approximate temperature at an altitude of 35,000 ft (normal cruising altitude for commercial airliners)? (b) Find $f^{-1}(x)$, and state what the independent and dependent variables represent. (c) If the temperature outside a weather

balloon is $-18°$F, what is the approximate altitude of the balloon?

98. **Fines for speeding:** In some localities, there is a set formula to determine the amount of a fine for exceeding posted speed limits. Suppose the amount of the fine for exceeding a 50 mph speed limit was given by the function $f(x) = 12x - 560 \, (x > 50)$ where $f(x)$ represents the fine in dollars for a speed of x mph. (a) What is the fine for traveling 65 mph through this speed zone? (b) Find $f^{-1}(x)$, and state

what the independent and dependent variables represent. (c) If a fine of $172 were assessed, how fast was the driver going through this speed zone?

99. Effect of gravity: Due to the effect of gravity, the distance an object has fallen after being dropped is given by the function $f(x) = 16x^2$; $x \geq 0$, where $f(x)$ represents the distance in feet after x sec. (a) How far has the object fallen 3 sec after it has been dropped? (b) Find $f^{-1}(x)$, and state what the independent and dependent variables represent. (c) If the object is dropped from a height of 784 ft, how many seconds until it hits the ground (stops falling)?

100. Area and radius: In generic form, the area of a circle is given by $f(x) = \pi x^2$, where $f(x)$ represents the area in square units for a circle with radius x. (a) A pet dog is tethered to a stake in the backyard. If the tether is 10 ft long, how much area does the dog have to roam (use $\pi \approx 3.14$)? (b) Find $f^{-1}(x)$, and state what the independent and dependent variables represent. (c) If the owners want to allow the dog 1256 ft^2 of area to live and roam, how long a tether should be used?

101. Volume of a cone: In generic form, the volume of an equipoise cone (height equal to radius) is given by $f(x) = \frac{1}{3}\pi x^3$, where $f(x)$ represents the volume

in units^3 and x represents the height of the cone. (a) Find the volume of such a cone if $r = 30$ ft (use $\pi \approx 3.14$). (b) Find $f^{-1}(x)$, and state what the independent and dependent variables represent. (c) If the volume of water in the cone is 763.02 ft^3, how deep is the water at its deepest point?

102. Wind power: The power delivered by a certain wind-powered generator can be modeled by the function $f(x) = \dfrac{x^3}{2500}$, where $f(x)$ is the horsepower (hp) delivered by the generator and x represents the speed of the wind in miles per hour. (a) Use the model to determine how much horsepower is generated by a 30 mph wind. (b) The person monitoring the output of the generators (wind generators are usually erected in large numbers) would like a function that gives the wind speed based on the horsepower readings on the gauges in the monitoring station. For this purpose, find $f^{-1}(x)$ and state what the independent and dependent variables represent. (c) If gauges show 25.6 hp is being generated, how fast is the wind blowing?

▶ **EXTENDING THE CONCEPT**

103. For a deeper understanding of the algebraic method for finding an inverse, suppose a function f is defined as $f(x)$: $\{(x, y)\,|\,y = 3x - 6\}$. We can then define the inverse as f^{-1}: $\{(x, y)\,|\,x = 3y - 6\}$, having interchanged x and y in the equation portion. The equation for f^{-1} is not in standard form, but (x, y) still represents all ordered pairs satisfying either equation. Solving for y gives f^{-1}: $\left\{(x, y)\,\middle|\,y = \dfrac{x}{3} + 2\right\}$, and demonstrates the role of steps 2, 3, and 4 of the method. (a) Find five ordered pairs that satisfy the equation for f, then (b) interchange their coordinates and show they satisfy the equation for f^{-1}.

104. The function $f(x) = \dfrac{1}{x}$ is one of the few functions that is its own inverse. This means the ordered pairs (a, b) and (b, a) must satisfy both f and f^{-1}. (a) Find f^{-1} using the algebraic method to verify that $f(x) = f^{-1}(x) = \dfrac{1}{x}$. (b) Graph the function $f(x) = \dfrac{1}{x}$ using a table of integers from -4 to 4. Note that for any ordered pair (a, b) on f, the

ordered pair (b, a) is also on f. (c) State where the graph of $y = x$ will intersect the graph of this function and discuss why.

105. By inspection, which of the following is the inverse function for $f(x) = \dfrac{2}{3}\left(x - \dfrac{1}{2}\right)^5 + \dfrac{4}{5}$?

   a. $f^{-1}(x) = \sqrt[5]{\dfrac{1}{2}\left(x - \dfrac{2}{3}\right)} - \dfrac{4}{5}$

   b. $f^{-1}(x) = \dfrac{3}{2}\sqrt[5]{(x - 2)} - \dfrac{5}{4}$

   c. $f^{-1}(x) = \dfrac{3}{2}\sqrt[5]{\left(x + \dfrac{1}{2}\right)} - \dfrac{5}{4}$

   d. $f^{-1}(x) = \sqrt[5]{\dfrac{3}{2}\left(x - \dfrac{4}{5}\right)} + \dfrac{1}{2}$

106. Suppose a function is defined as $f(x) = $ *the exponent that goes on 9 to obtain x*. For example, $f(81) = 2$ since 2 is the exponent that goes on 9 to obtain 81, and $f(3) = \frac{1}{2}$ since $\frac{1}{2}$ is the exponent that goes on 9 to obtain 3. Determine the value of each of the following:

   a. $f(1)$ **b.** $f(729)$ **c.** $f^{-1}(2)$ **d.** $f^{-1}\left(\dfrac{1}{2}\right)$

▶ **MAINTAINING YOUR SKILLS**

107. (2.5) Given $f(x) = x^2 - x - 2$, solve the inequality $f(x) \leq 0$ using the x-intercepts and end behavior of the graph.

108. (2.4) For the function $y = 2\sqrt{x + 3}$, find the average rate of change between $x = 1$ and $x = 2$, and between $x = 4$ and $x = 5$. Which is greater? Why?

109. (R.7) Write as many of the following formulas as you can from memory:

  a. perimeter of a rectangle

  b. area of a circle

  c. volume of a cylinder

  d. volume of a cone

  e. circumference of a circle

  f. area of a triangle

  g. area of a trapezoid

  h. volume of a sphere

  i. Pythagorean theorem

110. (1.3) Solve the following cubic equations by factoring:

  a. $x^3 - 5x = 0$

  b. $x^3 - 7x^2 - 4x + 28 = 0$

  c. $x^3 - 3x^2 = 0$

  d. $x^3 - 3x^2 - 4x = 0$

4.2 | Exponential Functions

Learning Objectives

In Section 4.2 you will learn how to:

  ☐ **A.** Evaluate an exponential function

  ☐ **B.** Graph general exponential functions

  ☐ **C.** Graph base-*e* exponential functions

  ☐ **D.** Solve exponential equations and applications

Demographics is the statistical study of human populations. In this section, we introduce the family of *exponential functions,* which are widely used to model population growth or decline with additional applications in science, engineering, and many other fields. As with other functions, we begin with a study of the graph and its characteristics.

A. Evaluating Exponential Functions

In the boomtowns of the old west, it was not uncommon for a town to double in size every year (at least for a time) as the lure of gold drew more and more people westward. When this type of growth is modeled using mathematics, exponents play a lead role. Suppose the town of Goldsboro had 1000 residents when gold was first discovered. After 1 yr the population doubled to 2000 residents. The next year it doubled again to 4000, then

again to 8000, then to 16,000 and so on. You probably recognize the digits in blue as powers of two (indicating the population is *doubling*), with each one multiplied by 1000 (the initial population). This suggests we can model the relationship using

$$P(x) = 1000 \cdot 2^x$$

where $P(x)$ is the population after x yr. Further, we can evaluate this function, called an **exponential function,** for *fractional parts of a year* using rational exponents. The population of Goldsboro one-and-a-half years after the gold rush was

$$P\left(\frac{3}{2}\right) = 1000 \cdot 2^{\frac{3}{2}}$$

$$= 1000 \cdot (\sqrt{2})^3$$

$$\approx 2828 \text{ people}$$

In general, exponential functions are defined as follows.

Exponential Functions

For $b > 0$, $b \neq 1$, and all real numbers x,

$$f(x) = b^x$$

defines the base b exponential function.

Limiting b to positive values ensures that outputs will be real numbers, and the restriction $b \neq 1$ is needed since $y = 1^x$ is a constant function (1 raised to *any* power is still 1). Specifically note the domain of an exponential function is *all real numbers*, and that all of the familiar properties of exponents still hold. A summary of these properties follows. For a complete review, see Section R.3.

Exponential Properties

For real numbers a, b, m, and n, with $a, b > 0$,

$$b^m \cdot b^n = b^{m+n} \qquad \frac{b^m}{b^n} = b^{m-n} \qquad (b^m)^n = b^{mn}$$

$$(ab)^n = a^n \cdot b^n \qquad b^{-n} = \frac{1}{b^n} \qquad \left(\frac{b}{a}\right)^{-n} = \left(\frac{a}{b}\right)^n$$

EXAMPLE 1 ▶ **Evaluating Exponential Functions**

Evaluate each exponential function for $x = 2$, $x = -1$, $x = \frac{1}{2}$, and $x = \pi$. Use a calculator for $x = \pi$, rounding to five decimal places.

a. $f(x) = 4^x$ **b.** $g(x) = \left(\dfrac{4}{9}\right)^x$

Solution ▶ **a.** For $f(x) = 4^x$,

$$f(2) = 4^2 = 16$$

$$f(-1) = 4^{-1} = \frac{1}{4}$$

$$f\left(\frac{1}{2}\right) = 4^{\frac{1}{2}} = \sqrt{4} = 2$$

$$f(\pi) = 4^{\pi} \approx 77.88023$$

☑ **A.** You've just learned how to evaluate an exponential function

b. For $g(x) = \left(\dfrac{4}{9}\right)^x$,

$$g(2) = \left(\frac{4}{9}\right)^2 = \frac{16}{81}$$

$$g(-1) = \left(\frac{4}{9}\right)^{-1} = \frac{9}{4}$$

$$g\left(\frac{1}{2}\right) = \left(\frac{4}{9}\right)^{\frac{1}{2}} = \sqrt{\frac{4}{9}} = \frac{2}{3}$$

$$g(\pi) = \left(\frac{4}{9}\right)^{\pi} \approx 0.07827$$

Now try Exercises 7 through 12 ▶

B. Graphing Exponential Functions

To gain a better understanding of exponential functions, we'll graph examples of $y = b^x$ and note some of the characteristic features. Since $b \neq 1$, it seems reasonable that we graph one exponential function where $b > 1$ and one where $0 < b < 1$.

EXAMPLE 2 ▶ **Graphing Exponential Functions with $b > 1$**

Graph $y = 2^x$ using a table of values.

Solution ▶ To get an idea of the graph's shape we'll use integer values from -3 to 3 in our table, then draw the graph as a continuous curve, since the function is defined for all real numbers.

x	$y = 2^x$
-3	$2^{-3} = \frac{1}{8}$
-2	$2^{-2} = \frac{1}{4}$
-1	$2^{-1} = \frac{1}{2}$
0	$2^0 = 1$
1	$2^1 = 2$
2	$2^2 = 4$
3	$2^3 = 8$

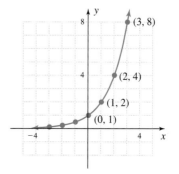

Now try Exercises 13 and 14 ▶

WORTHY OF NOTE

As in Example 2, functions that are increasing for all $x \in D$ are said to be **monotonically increasing** or simply **monotonic functions.** The function in Example 3 is monotonically decreasing.

Several important observations can now be made. First note the x-axis (the line $y = 0$) is a horizontal asymptote for the function, because as $x \to -\infty$, $y \to 0$. Second, the function is increasing over its entire domain, giving the function a range of $y \in (0, \infty)$.

EXAMPLE 3 ▶ **Graphing Exponential Functions with $0 < b < 1$**

Graph $y = \left(\frac{1}{2}\right)^x$ using a table of values.

Solution ▶ Using properties of exponents, we can write $\left(\frac{1}{2}\right)^x$ as $\left(\frac{2}{1}\right)^{-x} = 2^{-x}$. Again using integers from -3 to 3, we plot the ordered pairs and draw a continuous curve.

x	$y = 2^{-x}$
-3	$2^{-(-3)} = 2^3 = 8$
-2	$2^{-(-2)} = 2^2 = 4$
-1	$2^{-(-1)} = 2^1 = 2$
0	$2^0 = 1$
1	$2^{-1} = \frac{1}{2}$
2	$2^{-2} = \frac{1}{4}$
3	$2^{-3} = \frac{1}{8}$

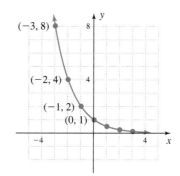

Now try Exercises 15 and 16 ▶

We note this graph is also asymptotic to the x-axis, but *decreasing on its domain.* In addition, both $y = 2^x$ and $y = 2^{-x} = \left(\frac{1}{2}\right)^x$ are one-to-one, and have a y-intercept of $(0, 1)$—which we expect since any base to the zero power is 1. Finally, observe that $y = b^{-x}$ is a reflection of $y = b^x$ across the y-axis, a property that suggests these basic graphs might also be transformed in other ways, as were the toolbox functions. The characteristics of exponential functions are summarized here:

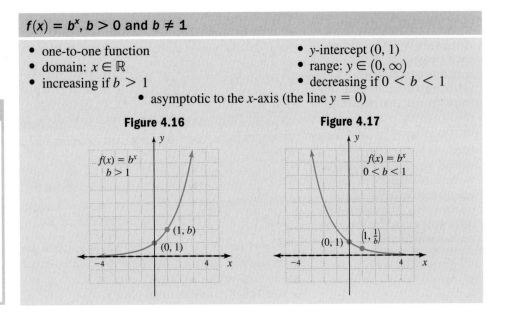

$f(x) = b^x, b > 0$ and $b \neq 1$

- one-to-one function
- domain: $x \in \mathbb{R}$
- increasing if $b > 1$
- y-intercept $(0, 1)$
- range: $y \in (0, \infty)$
- decreasing if $0 < b < 1$
- asymptotic to the x-axis (the line $y = 0$)

Figure 4.16

Figure 4.17

WORTHY OF NOTE

When an exponential function is increasing, it can be referred to as a "growth function." When decreasing, it is often called a "decay function." Each of the graphs shown in Figures 4.16 and 4.17 should now be added to your repertoire of basic functions, to be sketched from memory and analyzed or used as needed.

Just as the graph of a quadratic function maintains its parabolic shape regardless of the transformations applied, exponential functions will also maintain their general shape and features. Any sum or difference applied to the basic function ($y = b^x \pm k$ vs. $y = b^x$) will cause a vertical shift in the same direction as the sign, and any change to input values ($y = b^{x+h}$ vs. $y = b^x$) will cause a horizontal shift in a direction opposite the sign.

EXAMPLE 4 ▶ **Graphing Exponential Functions Using Transformations**

Graph $F(x) = 2^{x-1} + 2$ using transformations of the basic function (not by simply plotting points). Clearly name the parent function and state what transformations are applied.

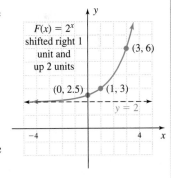

Solution ▶ The graph of F is that of the basic function $y = 2^x$ shifted 1 unit right and 2 units up. With this in mind the horizontal asymptote shifts from $y = 0$ to $y = 2$ and $(0, 1)$ shifts to $(1, 3)$. The y-intercept of F is at $(0, 2.5)$ and to help sketch a more accurate graph, the additional point $(3, 6)$ is used: $F(3) = 6$.

☑ **B.** You've just learned how to graph general exponential functions

Now try Exercises 17 through 38 ▶

C. The Base-e Exponential Function: $f(x) = e^x$

In nature, exponential growth occurs when the rate of change in a population's growth, is in constant proportion to its current size. Using the rate of change notation, $\dfrac{\Delta P}{\Delta t} = kP$, where k is a constant. For the city of Goldsboro, we know the population at time t is given by $P(t) = 1000 \cdot 2^t$, but have no information on this value of k (**see Exercise 96**). We can actually rewrite this function, and other exponential functions, using a base that gives the value of k directly and without having to apply the difference quotient. This new base is an irrational number, symbolized by the letter e and defined as follows.

The Number e

For $x > 0$,

$$\text{as } x \to \infty, \left(1 + \frac{1}{x}\right)^x \to e$$

In words, e is the number that $\left(1 + \dfrac{1}{x}\right)^x$ approaches as x becomes infinitely large.

It has been proven that as x grows without bound, $\left(1 + \dfrac{1}{x}\right)^x$ indeed approaches the unique, irrational number that we have named e (**also see Exercise 97**). Table 4.5 gives approximate values of the expression for selected values of x, and shows $e \approx 2.71828$ to five decimal places.

The result is the base-e **exponential function:** $f(x) = e^x$, also called the **natural exponential function.** Instead of having to enter a decimal approximation when computing with e, most calculators have an "e^x" key, usually as the 2nd function for the key marked LN. To find the value of e^2, use the keystrokes 2nd LN 2) ENTER, and the calculator display should read 7.389056099. Note the calculator supplies the left parenthesis for the exponent, and you must supply the right.

Table 4.5

x	$(1 + \frac{1}{x})^x$
1	2
10	2.59
100	2.705
1000	2.7169
10,000	2.71815
100,000	2.718268
1,000,000	2.7182804
10,000,000	2.71828169

EXAMPLE 5 ▶ **Evaluating the Natural Exponential Function**

Use a calculator to evaluate $f(x) = e^x$ for the values of x given. Round to six decimal places.

a. $f(3)$ **b.** $f(1)$ **c.** $f(0)$ **d.** $f(\frac{1}{2})$

Solution ▶   **a.** $f(3) = e^3 \approx 20.085537$ **b.** $f(1) = e^1 \approx 2.718282$

c. $f(0) = e^0 = 1$ (exactly)   **d.** $f(\frac{1}{2}) = e^{\frac{1}{2}} \approx 1.648721$

Now try Exercises 39 through 46 ▶

Although e is an irrational number, the graph of $y = e^x$ behaves in exactly the same way and has the same characteristics as other exponential graphs. Figure 4.18 shows this graph on the same grid as $y = 2^x$ and $y = 3^x$. As we might expect, all three graphs are increasing, have an asymptote at $y = 0$, and contain the point $(0, 1)$, with the graph of $y = e^x$ "between" the other two. The domain for all three functions, as with all basic exponential functions, is $x \in (-\infty, \infty)$ with range $y \in (0, \infty)$. The same transformations applied earlier can also be applied to the graph of $y = e^x$.

Figure 4.18

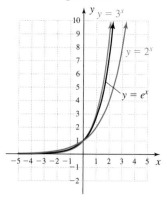

EXAMPLE 6 ▶ **Graphing Exponential Functions Using a Transformation**

Graph $f(x) = e^{x+1} - 2$ using transformations of $y = e^x$. Clearly state the transformations applied.

Solution ▶ The graph of f is the same as $y = e^x$, shifted 1 unit left and 2 units down. The point $(0, 1)$ becomes $(-1, -1)$, and the horizontal asymptote becomes $y = -2$. As the basic shape of the graph is known, we compute $f(1) \approx 5.4$, and complete the graph as shown.

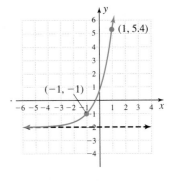

☑ **C.** You've just learned how to graph base-e exponential functions

Now try Exercises 47 through 52 ▶

D. Solving Exponential Equations Using the Uniqueness Property

Since exponential functions are one-to-one, we can solve equations where each side is an exponential term with the identical base. This is because one-to-oneness guarantees a unique solution to the equation.

WORTHY OF NOTE

Exponential functions are very different from the power functions studied earlier. For power functions, the base is variable and the exponent is constant: $y = x^b$, while for exponential functions the *exponent is a variable* and the *base is constant*: $y = b^x$.

Exponential Equations and the Uniqueness Property

For all real numbers m, n, and b, where $b > 0$ and $b \neq 1$,

$$\text{If } b^m = b^n,$$
$$\text{then } m = n.$$

Equal bases imply exponents are equal.

The equation $2^x = 32$ can be written as $2^x = 2^5$, and we note $x = 5$ is a solution. Although $3^x = 32$ can be written as $3^x = 2^5$, the bases are not alike and the solution to this equation must wait until additional tools are developed in Section 4.4.

EXAMPLE 7 ▶ **Solving Exponential Equations**

Solve the exponential equations using the uniqueness property.

a. $3^{2x-1} = 81$ **b.** $25^{-2x} = 125^{x+7}$

c. $\left(\frac{1}{6}\right)^{-3x-2} = 36^{x+1}$ **d.** $e^x e^2 = \dfrac{e^4}{e^{x+1}}$

Solution ▶ **a.** $3^{2x-1} = 81$ given
                  $3^{2x-1} = 3^4$ rewrite using base 3
             $\Rightarrow 2x - 1 = 4$ uniqueness property
                  $x = \dfrac{5}{2}$ solve for x

Check ▶ $3^{2x-1} = 81$ given
               $3^{2\left(\frac{5}{2}\right)-1} = 81$ substitute $\frac{5}{2}$ for x
               $3^{5-1} = 81$ simplify
               $3^4 = 81$ result checks
               $81 = 81$

The remaining checks are left to the student.

b.
$$25^{-2x} = 125^{x+7} \qquad \text{given}$$
$$(5^2)^{-2x} = (5^3)^{x+7} \qquad \text{rewrite using base 5}$$
$$5^{-4x} = 5^{3x+21} \qquad \text{power property of exponents}$$
$$\Rightarrow -4x = 3x + 21 \qquad \text{uniqueness property}$$
$$x = -3 \qquad \text{solve for } x$$

c.
$$\left(\frac{1}{6}\right)^{-3x-2} = 36^{x+1} \qquad \text{given}$$
$$(6^{-1})^{-3x-2} = (6^2)^{x+1} \qquad \text{rewrite using base 6}$$
$$6^{3x+2} = 6^{2x+2} \qquad \text{power property of exponents}$$
$$\Rightarrow 3x + 2 = 2x + 2 \qquad \text{uniqueness property}$$
$$x = 0 \qquad \text{solve for } x$$

d.
$$e^x e^2 = \frac{e^4}{e^{x+1}} \qquad \text{given}$$
$$e^{x+2} = e^{4-(x+1)} \qquad \text{product property; quotient property}$$
$$e^{x+2} = e^{3-x} \qquad \text{simplify}$$
$$\Rightarrow x + 2 = 3 - x \qquad \text{uniqueness property}$$
$$2x = 1 \qquad \text{add } x, \text{ subtract 2}$$
$$x = \frac{1}{2} \qquad \text{solve for } x$$

Now try Exercises 53 through 72 ▶

One very practical application of the natural exponential function involves **Newton's law of cooling.** This law or formula models the temperature of an object as it cools down, as when a pizza is removed from the oven and placed on the kitchen counter. The function model is

$$T(x) = T_R + (T_0 - T_R)e^{kx}, \ k < 0$$

where T_0 represents the initial temperature of the object, T_R represents the temperature of the room or surrounding medium, $T(x)$ is the temperature of the object x min later, and k is the cooling rate as determined by the nature and physical properties of the object.

EXAMPLE 8 ▶ Applying an Exponential Function—Newton's Law of Cooling

A pizza is taken from a 425°F oven and placed on the counter to cool. If the temperature in the kitchen is 75°F, and the cooling rate for this type of pizza is $k = -0.35$,

a. What is the temperature (to the nearest degree) of the pizza 2 min later?

b. To the nearest minute, how long until the pizza has cooled to a temperature below 90°F?

c. If Zack and Raef like to eat their pizza at a temperature of about 110°F, how many minutes should they wait to "dig in"?

Solution ▶ Begin by substituting the given values to obtain the equation model:

$$T(x) = T_R + (T_0 - T_R)e^{kx} \qquad \text{general equation model}$$
$$= 75 + (425 - 75)e^{-0.35x} \qquad \text{substitute 75 for } T_R, \text{ 425 for } T_0 \text{ and } -0.35 \text{ for } k$$
$$= 75 + 350e^{-0.35x} \qquad \text{simplify}$$

For part (a) we simply find $T(2)$:

a. $T(2) = 75 + 350e^{-0.35(2)} \qquad \text{substitute 2 for } x$
$\approx 249 \qquad \text{result}$

Two minutes later, the temperature of the pizza is near 249°.

b. Using the [TABLE] feature of a graphing calculator shows the pizza reaches a temperature of just under 90° after 9 min: $T(9) \approx 90°F$.

c. We elect to use the intersection of graphs method (see the *Technology Highlight* on page 432). After setting an appropriate window, we enter $Y_1 = 75 + 350e^{-0.35x}$ and $Y_2 = 110$, then press [2nd] [CALC] option **5: intersect**. After pressing [ENTER] three times, the coordinates of the point of intersection appear at the bottom of the screen: $x \approx 6.6$, $y = 110$. It appears the boys should wait about $6\frac{1}{2}$ min for the pizza to cool.

Now try Exercises 75 and 76 ▶

EXAMPLE 9 ▶   Applications of Exponential Functions—Depreciation

For insurance purposes, it is estimated that large household appliances lose $\frac{1}{5}$ of their value each year. The current value can then be modeled by the function $V(t) = V_0(\frac{4}{5})^t$, where V_0 is the initial value and $V(t)$ represents the value after t years. How many years does it take a washing machine that cost \$625 new, to depreciate to a value of \$256?

Solution ▶   For this exercise, $V_0 = \$625$ and $V(t) = \$256$. The formula yields

$$V(t) = V_0\left(\frac{4}{5}\right)^t \qquad \text{given}$$

$$256 = 625\left(\frac{4}{5}\right)^t \qquad \text{substitute known values}$$

$$\frac{256}{625} = \left(\frac{4}{5}\right)^t \qquad \text{divide by 625}$$

$$\left(\frac{4}{5}\right)^4 = \left(\frac{4}{5}\right)^t \qquad \text{equate bases } \frac{256}{625} = \left(\frac{4}{5}\right)^4$$

$$\Rightarrow 4 = t \qquad \text{Uniqueness Property}$$

☑ **D.** You've just learned how to solve exponential equations and applications

After 4 yr, the washing machine's value has dropped to \$256.

Now try Exercises 77 through 90 ▶

TECHNOLOGY HIGHLIGHT

Solving Exponential Equations Graphically

In this section, we showed that the exponential function $f(x) = b^x$ was defined for all real numbers. This is important because it establishes that equations like $2^x = 7$ must have a solution, even if x is not rational. In fact, since $2^2 = 4$ and $2^3 = 8$, the following inequalities indicate the solution must be between 2 and 3

$$4 < 7 < 8 \qquad \text{7 is between 4 and 8}$$

$$2^2 < 2^x < 2^3 \qquad \text{replace 4 with } 2^2, \text{ 8 with } 2^3$$

$$2 < x < 3 \qquad x \text{ must be between 2 and 3}$$

—continued

Until we develop an inverse for exponential functions, we are unable to solve many of these equations in exact form. We can, however, get a very close approximation using a graphing calculator. For the equation $2^x = 7$, enter $Y_1 = 2^x$ and $Y_2 = 7$ on the [Y =] screen. Then press [ZOOM] 6 to graph both functions (see Figure 4.19). To find the point of intersection, press [2nd] [TRACE] (CALC) and select option **5: intersect** and press [ENTER] *three* times (to identify the intersecting functions and bypass "Guess"). The x- and y-coordinates of the point of intersection will appear at the bottom of the screen, with the x-coordinate being the solution. As you can see, x is indeed between 2 and 3. Solve the following equations. First estimate the answer by bounding it between two integers, then solve the equation graphically. Adjust the viewing window as needed.

Figure 4.19

Intersection
X=2.8073549 Y=7

Exercise 1: $3^x = 22$

Exercise 2: $2^x = 0.125$

Exercise 3: $e^{x-1} = 9$

Exercise 4: $e^{0.5x} = 0.1x^3$

4.2 EXERCISES

▶ CONCEPTS AND VOCABULARY

Fill in each blank with the appropriate word or phrase. Carefully reread the section if needed.

1. An exponential function is one of the form $y = $ _____, where _____ > 0, _____ $\neq 1$, and _____ is any real number.

2. The domain of $y = b^x$ is all ___ _____, and the range is $y \in$ _____. Further, as $x \to -\infty$, y _____.

3. For exponential functions of the form $y = ab^x$, the y-intercept is (0, _____), since $b^0 = $ _____ for any real number b.

4. If each side of an equation can be written as an exponential term with the same base, the equation can be solved using the _____ _____.

5. State true or false and explain why: $y = b^x$ is always increasing if $0 < b < 1$.

6. Discuss/Explain the statement, "For $k > 0$, the y-intercept of $y = ab^x + k$ is $(0, a + k)$."

▶ DEVELOPING YOUR SKILLS

 Use a calculator (as needed) to evaluate each function as indicated. Round answers to thousandths.

7. $P(t) = 2500 \cdot 4^t$;
   $t = 2, t = \frac{1}{2}, t = \frac{3}{2}$,
   $t = \sqrt{3}$

8. $Q(t) = 5000 \cdot 8^t$;
   $t = 2, t = \frac{1}{3}, t = \frac{5}{3}$,
   $t = 5$

9. $f(x) = 0.5 \cdot 10^x$;
   $x = 3, x = \frac{1}{2}, x = \frac{2}{3}$,
   $x = \sqrt{7}$

10. $g(x) = 0.8 \cdot 5^x$;
    $x = 4, x = \frac{1}{4}, x = \frac{4}{5}$,
    $x = \pi$

11. $V(n) = 10{,}000(\frac{2}{3})^n$;
    $n = 0, n = 4, n = 4.7$,
    $n = 5$

12. $W(m) = 3300(\frac{4}{5})^m$;
    $m = 0, m = 5, m = 7.2$,
    $m = 10$

Graph each function using a table of values and integer inputs between -3 and 3. Clearly label the y-intercept and one additional point, then indicate whether the function is increasing or decreasing.

13. $y = 3^x$ **14.** $y = 4^x$

15. $y = \left(\frac{1}{3}\right)^x$ **16.** $y = \left(\frac{1}{4}\right)^x$

Graph each of the following functions by *translating the basic function* $y = b^x$, sketching the asymptote, and strategically plotting a few points to round out the graph. Clearly state the basic function and what shifts are applied.

17. $y = 3^x + 2$ **18.** $y = 3^x - 3$

19. $y = 3^{x+3}$ **20.** $y = 3^{x-2}$

21. $y = 2^{-x}$ **22.** $y = 3^{-x}$

23. $y = 2^{-x} + 3$ **24.** $y = 3^{-x} - 2$

25. $y = 2^{x+1} - 3$ **26.** $y = 3^{x-2} + 1$

27. $y = \left(\frac{1}{3}\right)^x + 1$ **28.** $y = \left(\frac{1}{3}\right)^x - 4$

29. $y = \left(\frac{1}{3}\right)^{x-2}$ **30.** $y = \left(\frac{1}{3}\right)^{x+2}$

31. $f(x) = \left(\frac{1}{3}\right)^x - 2$ **32.** $g(x) = \left(\frac{1}{3}\right)^x + 2$

Match each graph to the correct exponential equation.

33. $y = 5^{-x}$ **34.** $y = 4^{-x}$

35. $y = 3^{-x+1}$ **36.** $y = 3^{-x} + 1$

37. $y = 2^{x+1} - 2$ **38.** $y = 2^{x+2} - 1$

a.

b.

c.

d.

e.

f.

Use a calculator to evaluate each expression, rounded to six decimal places.

39. e^1 **40.** e^0

41. e^2 **42.** e^5

43. $e^{1.5}$ **44.** $e^{-3.2}$

45. $e^{\sqrt{2}}$ **46.** e^{π}

Graph each exponential function.

47. $f(x) = e^{x+3} - 2$ **48.** $g(x) = e^{x-2} + 1$

49. $r(t) = -e^t + 2$ **50.** $s(t) = -e^{t+2}$

51. $p(x) = e^{-x+2} - 1$ **52.** $q(x) = e^{-x-1} + 2$

Solve each exponential equation and check your answer by substituting into the original equation.

53. $10^x = 1000$ **54.** $144 = 12^x$

55. $25^x = 125$ **56.** $81 = 27^x$

57. $8^{x+2} = 32$ **58.** $9^{x-1} = 27$

59. $32^x = 16^{x+1}$ **60.** $100^{x+2} = 1000^x$

61. $\left(\frac{1}{5}\right)^x = 125$ **62.** $\left(\frac{1}{4}\right)^x = 64$

63. $\left(\frac{1}{3}\right)^{2x} = 9^{x-6}$ **64.** $\left(\frac{1}{2}\right)^{3x} = 8^{x-2}$

65. $\left(\frac{1}{9}\right)^{x-5} = 3^{3x}$ **66.** $2^{-2x} = \left(\frac{1}{32}\right)^{x-3}$

67. $25^{3x} = 125^{x-2}$ **68.** $27^{2x+4} = 9^{4x}$

69. $\dfrac{e^4}{e^{2-x}} = e^3 e$ **70.** $e^x(e^x + e) = \dfrac{e^x + e^{3x}}{e^{-x}}$

71. $(e^{2x-4})^3 = \dfrac{e^{x+5}}{e^2}$ **72.** $e^x e^{x+3} = (e^{x+2})^3$

▶ WORKING WITH FORMULAS

73. The growth of a bacteria population: $P(t) = 1000 \cdot 3^t$

If the initial population of a common bacterium is 1000 and the population triples every day, its population is given by the formula shown, where $P(t)$ is the total population after t days. (a) Find the total population 12 hr, 1 day, $1\frac{1}{2}$ days, and 2 days later. (b) Do the outputs show the population is tripling every 24 hr (1 day)? (c) Explain why this is an increasing function. (d) Graph the function using an appropriate scale.

74. Games involving a spinner with numbers 1 through 4: $P(x) = (\frac{1}{4})^x$

Games that involve moving pieces around a board using a fair spinner are fairly common. If the spinner has the numbers 1 through 4, the probability that any one number is spun repeatedly is given by the formula shown, where x represents the number of spins and $P(x)$ represents the probability the same number results x times. (a) What is the probability that the first player spins a 2? (b) What is the probability that all four players spin a 2? (c) Explain why this is a decreasing function.

▶ APPLICATIONS

Use Newton's law of cooling to complete Exercises 75 and 76: $T(x) = T_R + (T_0 - T_R)e^{kx}$.

75. Cold party drinks: Janae was late getting ready for the party, and the liters of soft drinks she bought were still at room temperature (73°F) with guests due to arrive in 15 min. If she puts these in her freezer at −10°F, will the drinks be cold enough (35°F) for her guests? Assume $k \approx -0.031$.

76. Warm party drinks: Newton's law of cooling applies equally well if the "cooling is negative," meaning the object is taken from a colder medium and placed in a warmer one. If a can of soft drink is taken from a 35°F cooler and placed in a room where the temperature is 75°F, how long will it take the drink to warm to 65°F? Assume $k \approx -0.031$.

77. Depreciation:
The financial analyst for a large construction firm estimates that its heavy equipment loses one-fifth of its value each year. The current value of the equipment is then modeled by the function $V(t) = V_0(\frac{4}{5})^t$, where V_0 represents the initial value, t is in years, and $V(t)$ represents the value after t years. (a) How much is a large earthmover worth after 1 yr if it cost $125 thousand new? (b) How many years does it take for the earthmover to depreciate to a value of $64 thousand?

78. Depreciation: Photocopiers have become a critical part of the operation of many businesses, and due to their heavy use they can depreciate in value very quickly. If a copier loses $\frac{3}{8}$ of its value each year,

the current value of the copier can be modeled by the function $V(t) = V_0(\frac{5}{8})^t$, where V_0 represents the initial value, t is in years, and $V(t)$ represents the value after t yr. (a) How much is this copier worth after one year if it cost $64 thousand new? (b) How many years does it take for the copier to depreciate to a value of $25 thousand?

79. Depreciation: Margaret Madison, DDS, estimates that her dental equipment loses one-sixth of its value each year. (a) Determine the value of an x-ray machine after 5 yr if it cost $216 thousand new, and (b) determine how long until the machine is worth less than $125 thousand.

80. Exponential decay: The groundskeeper of a local high school estimates that due to heavy usage by the baseball and softball teams, the pitcher's mound loses one-fifth of its height every month. (a) Determine the height of the mound after 3 months if it was 25 cm to begin, and (b) determine how long until the pitcher's mound is less than 16 cm high (meaning it must be rebuilt).

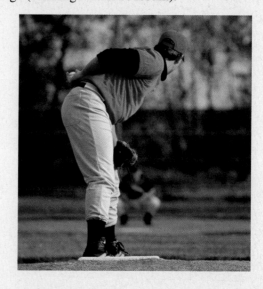

81. **Exponential growth:** Similar to a small town doubling in size after a discovery of gold, a business that develops a product in high demand has the potential for doubling its revenue each year for a number of years. The revenue would be modeled by the function $R(t) = R_0 2^t$, where R_0 represents the initial revenue, and $R(t)$ represents the revenue after t years. (a) How much revenue is being generated after 4 yr, if the company's initial revenue was $2.5 million? (b) How many years does it take for the business to be generating $320 million in revenue?

82. **Exponential growth:** If a company's revenue grows at a rate of 150% per year (rather than doubling as in Exercise 81), the revenue would be modeled by the function $R(t) = R_0 (\frac{3}{2})^t$, where R_0 represents the initial revenue, and $R(t)$ represents the revenue after t years. (a) How much revenue is being generated after 3 yr, if the company's initial revenue was $256 thousand? (b) How long until the business is generating $1944 thousand in revenue? (*Hint:* Reduce the fraction.)

Photochromatic sunglasses: Sunglasses that darken in sunlight (photochromatic sunglasses) contain millions of molecules of a substance known as *silver halide*. The molecules are transparent indoors in the absence of ultraviolent (UV) light. Outdoors, UV light from the sun causes the molecules to change shape, darkening the lenses in response to the intensity of the UV light. For certain lenses, the function $T(x) = 0.85^x$ models the transparency of the lenses (as a percentage) based on a UV index x. Find the transparency (to the nearest percent), if the lenses are exposed to

83. sunlight with a UV index of 7 (a high exposure).

84. sunlight with a UV index of 5.5 (a moderate exposure).

85. Given that a UV index of 11 is very high and most individuals should stay indoors, what is the minimum transparency percentage for these lenses?

86. Use trial-and-error to determine the UV index when the lenses are 50% transparent.

Modeling inflation: Assuming the rate of inflation is 5% per year, the predicted price of an item can be modeled by the function $P(t) = P_0 (1.05)^t$, where P_0 represents the initial price of the item and t is in years. Use this information to solve Exercises 87 and 88.

87. What will the price of a new car be in the year 2010, if it cost $20,000 in the year 2000?

88. What will the price of a gallon of milk be in the year 2010, if it cost $2.95 in the year 2000? Round to the nearest cent.

Modeling radioactive decay: The half-life of a radioactive substance is the time required for half an initial amount of the substance to disappear through decay. The amount of the substance remaining is given by the formula $Q(t) = Q_0 (\frac{1}{2})^{\frac{t}{h}}$, where h is the half-life, t represents the elapsed time, and $Q(t)$ represents the amount that remains (t and h must have the same unit of time). Use this information to solve Exercises 89 and 90.

89. Some isotopes of the substance known as thorium have a half-life of only 8 min. (a) If 64 grams are initially present, how many grams (g) of the substance remain after 24 min? (b) How many minutes until only 1 gram (g) of the substance remains?

90. Some isotopes of sodium have a half-life of about 16 hr. (a) If 128 g are initially present, how many grams of the substance remain after 2 days (48 hr)? (b) How many hours until only 1 g of the substance remains?

▶ **EXTENDING THE CONCEPT**

91. The formula $f(x) = (\frac{1}{2})^x$ gives the probability that "x" number of flips result in heads (or tails). First determine the probability that 20 flips results in *20 heads in a row*. Then use the Internet or some other resource to determine the probability of winning a state lottery (expressed as a decimal). Which has the greater probability? Were you surprised?

92. If $10^{2x} = 25$, what is the value of 10^{-x}?

93. If $5^{3x} = 27$, what is the value of 5^{2x}?

94. If $3^{0.5x} = 5$, what is the value of 3^{x+1}?

95. If $\left(\frac{1}{2}\right)^{x+1} = \frac{1}{3}$, what is the value of $\left(\frac{1}{2}\right)^{-x}$?

The growth rate constant that governs an exponential function was introduced on page 427.

96. In later sections, we will easily be able to find the growth constant k for Goldsboro, where

$P(t) = 1000 \cdot 2^t$. For now we'll approximate its value using the rate of change formula on a very small interval of the domain. From the definition of an exponential function, $\dfrac{\Delta P}{\Delta t} = kP(t)$. Since k is constant, we can choose any value of t, say $t = 4$. For $h = 0.0001$, we have

$$\frac{1000 \cdot 2^{4+0.0001} - 1000 \cdot 2^4}{0.0001} = k \cdot P(4)$$

(a) Use the equation shown to solve for k (round to thousandths). (b) Show that k is constant by completing the same exercise for $t = 2$ and $t = 6$. (c) Verify that $P(t) = 1000 \cdot 2^t$ and $P(t) = 1000e^{kt}$ give approximately the same results.

97. As we analyze the expression $\left(1 + \dfrac{1}{x}\right)^x$, we notice a battle (of sorts) takes place between the base $\left(1 + \dfrac{1}{x}\right)$ and the exponent x. As $x \to \infty$, $\dfrac{1}{x}$ becomes infinitely small, but the exponent becomes

infinitely large. So what happens? The answer is best understood by computing a series of *average rates of change*, using the intervals given here. Using the tools of Calculus, it can be shown that this rate of change becomes infinitely small, and that the "battle" ends at the irrational number e. In other words, e is an upper bound on the value of this expression, regardless of how large x becomes.

a. Use a calculator to find the average rate of change for $y = \left(1 + \dfrac{1}{x}\right)^x$ in these intervals: [1, 1.01], [4, 4.01], [10, 10.01], and [20, 20.01]. What do you notice?

b. What is the smallest integer value for x that gives the value of e correct to four decimal places?

c. Use a graphing calculator to graph this function on a window size of $x \in [0, 25]$ and $y \in [0, 3]$. Does the graph seem to support the statements above?

▶ MAINTAINING YOUR SKILLS

98. (2.2) Given $f(x) = 2x^2 - 3x$, determine:

$f(-1), \quad f(\tfrac{1}{3}), \quad f(a), \quad f(a + h)$

99. (3.3) Graph $g(x) = \sqrt{x + 2} - 1$ using a shift of the parent function. Then state the domain and range of g.

100. (1.3) Solve the following equations:

a. $-2\sqrt{x - 3} + 7 = 21$

b. $\dfrac{9}{x + 3} + 3 = \dfrac{12}{x - 3}$

101. (R.7) Identify each formula:

a. $\tfrac{4}{3}\pi r^3$

b. $\tfrac{1}{2}bh$

c. lwh

d. $a^2 + b^2 = c^2$

4.3 | Logarithms and Logarithmic Functions

Learning Objectives

In Section 4.3 you will learn how to:

☐ **A.** Write exponential equations in logarithmic form

☐ **B.** Find common logarithms and natural logarithms

☐ **C.** Graph logarithmic functions

☐ **D.** Find the domain of a logarithmic function

☐ **E.** Solve applications of logarithmic functions

A **transcendental function** is one whose solutions are beyond or *transcend* the methods applied to polynomial functions. The exponential function and its inverse, called the logarithmic function, are transcendental functions. In this section, we'll use the concept of an inverse to develop an understanding of the logarithmic function, which has numerous applications that include measuring pH levels, sound and earthquake intensities, barometric pressure, and other natural phenomena.

A. Exponential Equations and Logarithmic Form

While exponential functions have a large number of significant applications, we can't appreciate their full value until we develop the inverse function. Without it, we're unable to solve all but the simplest equations, of the type encountered in Section 4.2. Using the fact that $f(x) = b^x$ is one-to-one, we have the following:

1. The function $f^{-1}(x)$ must exist.
2. We can graph $f^{-1}(x)$ by interchanging the x- and y-coordinates of points from $f(x)$.
3. The domain of $f(x)$ will become the range of $f^{-1}(x)$.
4. The range of $f(x)$ will become the domain of $f^{-1}(x)$.
5. The graph of $f^{-1}(x)$ will be a reflection of $f(x)$ across the line $y = x$.

Table 4.6 contains selected values for $f(x) = 2^x$. The values for $f^{-1}(x)$ in Table 4.7 were found by interchanging x- and y-coordinates. Both functions were then graphed using these values.

Table 4.6

$f(x): y = 2^x$

x	y
-3	$\frac{1}{8}$
-2	$\frac{1}{4}$
-1	$\frac{1}{2}$
0	1
1	2
2	4
3	8

Table 4.7

$f^{-1}(x): x = 2^y$

x	(x)
$\frac{1}{8}$	-3
$\frac{1}{4}$	-2
$\frac{1}{2}$	-1
1	0
2	1
4	2
8	3

Figure 4.20

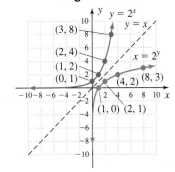

The interchange of x and y and the graphs in Figure 4.20 show that $f^{-1}(x)$ has an x-intercept of $(1, 0)$, a vertical asymptote at $x = 0$, a domain of $x \in (0, \infty)$, and a range of $y \in (-\infty, \infty)$. To find *an equation* for $f^{-1}(x)$, we'll attempt to use the algebraic approach employed previously. For $f(x) = 2^x$,

1. use y instead of $f(x)$: $y = 2^x$.
2. interchange x and y: $x = 2^y$.

At this point we have an *implicit* equation for the inverse function, but no algebraic operations that enable us to solve *explicitly* for y in terms of x. Instead, we write $x = 2^y$ in function form by noting that "y is the exponent that goes on base 2 to obtain x."

In the language of mathematics, this phrase is represented by $y = \log_2 x$ and is called a **logarithmic function** with base 2. For $y = b^x$, $x = b^y \rightarrow y = \log_b x$ is the inverse function, and is read, "y is the logarithm base b of x." For this new function, we must always keep in mind what y *represents*—y is an exponent. In fact, y *is the exponent that goes on base b to obtain x*: $y = \log_b x$.

Logarithmic Functions

For positive numbers x and b, with $b \neq 1$,

$$y = \log_b x \text{ if and only if } x = b^y$$

The function $f(x) = \log_b x$ is a logarithmic function with base b. The expression $\log_b x$ is simply called a logarithm, and represents the exponent on b that yields x.

Finally, note the equations $x = b^y$ and $y = \log_b x$ are equivalent. We say that $x = b^y$ is the **exponential form** of the equation, whereas $y = \log_b x$ is written in **logarithmic form**.

EXAMPLE 1 ▶ **Converting from Logarithmic Form to Exponential Form**

Write each equation in words, then in exponential form.

a. $3 = \log_2 8$ **b.** $1 = \log_{10} 10$ **c.** $0 = \log_e 1$ **d.** $-2 = \log_3(\frac{1}{9})$

Solution ▶   **a.** $3 = \log_2 8 \to 3$ is the exponent on base 2 for 8: $2^3 = 8$.

b. $1 = \log_{10} 10 \to 1$ is the exponent on base 10 for 10: $10^1 = 10$.

c. $0 = \log_e 1 \to 0$ is the exponent on base e for 1: $e^0 = 1$.

d. $-2 = \log_3(\frac{1}{9}) \to -2$ is the exponent on base 3 for $\frac{1}{9}$: $3^{-2} = \frac{1}{9}$.

Now try Exercises 7 through 22 ▶

To convert from exponential form to logarithmic form, note the exponent on the base and read from there. For $5^3 = 125$, "3 is the exponent that goes on base 5 for 125," or *3 is the logarithm base 5 of 125*: $3 = \log_5 125$.

EXAMPLE 2 ▶ **Converting from Exponential Form to Logarithmic Form**

Write each equation in words, then in logarithmic form.

a. $10^3 = 1000$ **b.** $2^{-1} = \frac{1}{2}$ **c.** $e^2 \approx 7.389$ **d.** $9^{\frac{3}{2}} = 27$

Solution ▶   **a.** $10^3 = 1000 \to 3$ is the exponent on base 10 for 1000, or
3 is the logarithm base 10 of 1000: $3 = \log_{10} 1000$.

b. $2^{-1} = \frac{1}{2} \to -1$ is the exponent on base 2 for $\frac{1}{2}$, or
-1 is the logarithm base 2 of $\frac{1}{2}$: $-1 = \log_2(\frac{1}{2})$.

c. $e^2 \approx 7.389 \to 2$ is the exponent on base e for 7.389, or
2 is the logarithm base e of 7.389: $2 \approx \log_e 7.389$.

d. $9^{\frac{3}{2}} = 27 \to \frac{3}{2}$ is the exponent on base 9 for 27, or
$\frac{3}{2}$ is the logarithm base 9 of 27: $\frac{3}{2} = \log_9 27$.

☑ **A.** You've just learned how to write exponential equations in logarithmic form

Now try Exercises 23 through 38 ▶

B. Finding Common Logarithms and Natural Logarithms

Of all possible bases for $\log_b x$, the most common are base 10 (likely due to our base-10 number system), and base e *(due to the advantages it offers in advanced courses).* The expression $\log_{10} x$ is called a **common logarithm,** and we simply write $\log x$ for $\log_{10} x$. The expression $\log_e x$ is called a **natural logarithm,** and is written in abbreviated form as $\ln x$.

Some logarithms are easy to evaluate. For example, $\log 100 = 2$ since $10^2 = 100$, and $\log \frac{1}{100} = -2$ since $10^{-2} = \frac{1}{100}$. But what about the expressions $\log 850$ and $\ln 4$? Because logarithmic functions are continuous on their domains, a value exists for $\log 850$ and the equation $10^x = 850$ must have a solution. Further, the inequalities

$$\log 100 < \log 850 < \log 1000$$
$$2 < \log 850 < 3$$

tell us that $\log 850$ must be between 2 and 3. Fortunately, modern calculators can compute base-10 and base-e logarithms instantly, often with nine-decimal-place accuracy. For $\log 850$, press **LOG**, then input 850 and press **ENTER**. The display should read 2.929418926. We can also use the calculator to verify $10^{2.929418926} = 850$ (see Figure 4.21). For $\ln 4$, press the **LN** key, then input 4 and press **ENTER** to obtain 1.386294361. Figure 4.22 verifies that $e^{1.386294361} = 4$.

WORTHY OF NOTE

We do something similar with square roots. Technically, the "square root of x" should be written $\sqrt[2]{x}$. However, square roots are so common we often leave off the two, assuming that if no index is written, an index of two is intended.

Figure 4.21

```
log(850)
            2.929418926
10^Ans
                     850
```

Figure 4.22

```
ln(4)
            1.386294361
e^(Ans)
                       4
```

EXAMPLE 3 ▶ Finding the Value of a Logarithm

Determine the value of each logarithm without using a calculator:

a. $\log_2 8$ **b.** $\log_5(\frac{1}{25})$ **c.** $\log_e e$ **d.** $\log_{10}\sqrt{10}$

Solution ▶ **a.** $\log_2 8$ represents the exponent on 2 for 8: $\log_2 8 = 3$, since $2^3 = 8$.
b. $\log_5(\frac{1}{25})$ represents the exponent on 5 for $\frac{1}{25}$: $\log_5\frac{1}{25} = -2$, since $5^{-2} = \frac{1}{25}$.
c. $\log_e e$ represents the exponent on e for e: $\log_e e = 1$, since $e^1 = e$.
d. $\log_{10}\sqrt{10}$ represents the exponent on 10 for $\sqrt{10}$: $\log_{10}\sqrt{10} = \frac{1}{2}$, since $10^{\frac{1}{2}} = \sqrt{10}$.

Now try Exercises 39 through 50 ▶

EXAMPLE 4 ▶ Using a Calculator to Find Logarithms

Use a calculator to evaluate each logarithmic expression. Verify the result.

a. log 1857   **b.** log 0.258   **c.** ln 3.592

Solution ▶ **a.** $\log 1857 = 3.268811904$,
$10^{3.268811904} = 1857$ ✓

b. $\log 0.258 = -0.588380294$,
$10^{-0.588380294} = 0.258$ ✓

☑ **B.** You've just learned how to find common logarithms and natural logarithms

c. $\ln 3.592 \approx 1.27870915$
$e^{1.27870915} \approx 3.592$ ✓

Now try Exercises 51 through 58 ▶

C. Graphing Logarithmic Functions

For convenience and ease of calculation, our first examples of logarithmic graphs are done using base-2 logarithms. However, the basic shape of a logarithmic graph remains unchanged regardless of the base used, and transformations can be applied to $y = \log_b(x)$ for any value of b. For $y = a\log(x \pm h) \pm k$, a continues to govern stretches, compressions, and vertical reflections, the graph will shift horizontally h units opposite the sign, and shift k units vertically in the same direction as the sign. Our earlier graph of $y = \log_2 x$ was completed using $x = 2^y$ as the inverse function for $y = 2^x$ (Figure 4.20). For reference, the graph is repeated in Figure 4.23.

Figure 4.23

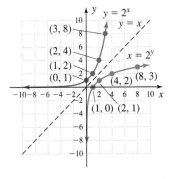

EXAMPLE 5 ▶ Graphing Logarithmic Functions Using Transformations

Graph $f(x) = \log_2(x - 3) + 1$ using transformations of $y = \log_2 x$ (not by simply plotting points). Clearly state what transformations are applied.

Solution ▶ The graph of f is the same as that of $y = \log_2 x$, shifted 3 units right and 1 unit up. The vertical asymptote will be at $x = 3$ and the point $(1, 0)$ from the basic graph becomes $(1 + 3, 0 + 1) = (4, 1)$. Knowing the graph's basic shape, we compute one additional point using $x = 7$:

$$f(7) = \log_2(7 - 3) + 1$$
$$= \log_2 4 + 1$$
$$= 2 + 1$$
$$= 3$$

The point $(7, 3)$ is on the graph, shown in Figure 4.24.

Figure 4.24

Now try Exercises 59 through 62 ▶

As with the exponential functions, much can be learned from graphs of logarithmic functions and a summary of important characteristics is given here.

$f(x) = \log_b x,\ b > 0$ and $b \neq 1$

- one-to-one function
- domain: $x \in (0, \infty)$
- increasing if $b > 1$
- x-intercept $(1, 0)$
- range: $y \in \mathbb{R}$
- decreasing if $0 < b < 1$
- asymptotic to the y-axis (the line $x = 0$)

EXAMPLE 6 ▶ Graphing Logarithmic Functions Using Transformations

Graph $g(x) = -\ln(x + 2)$ using transformations of $y = \ln x$ (not by simply plotting points). Clearly state what transformations are applied.

Solution ▶ The graph of g is the same as $y = \ln x$, shifted 2 units left, then reflected across the x-axis. The vertical asymptote will be at $x = -2$, and the point $(1, 0)$ from the basic function becomes $(-1, 0)$. To complete the graph we compute $f(6)$:

$$f(6) = -\ln(6 + 2)$$
$$= -\ln 8$$
$$\approx -2.1 \text{ (using a calculator)}.$$

The point $(6, -2.1)$ is on the graph shown in the figure.

WORTHY OF NOTE

Accurate graphs can actually be drawn for logarithms of *any base* using what is called the base-change formula, introduced in Section 4.4.

☑ **C.** You've just learned how to graph logarithmic functions

D. Finding the Domain of a Logarithmic Function

Examples 5 and 6 illustrate how the domain of a logarithmic function can change when certain transformations are applied. Since the domain consists of *positive* real numbers, the argument of a logarithmic function must be greater than zero. This means finding the domain often consists of solving various inequalities, which can be done using the skills acquired in Sections 2.5 and 3.7.

EXAMPLE 7 ▶ **Finding the Domain of a Logarithmic Function**

Determine the domain of each function.

a. $p(x) = \log_2(2x + 3)$ **b.** $q(x) = \log_5(x^2 - 2x)$

c. $r(x) = \log\left(\dfrac{3 - x}{x + 3}\right)$ **d.** $f(x) = \ln|x - 2|$

Solution ▶ Begin by writing the argument of each logarithmic function as a greater than inequality.

a. Solving $2x + 3 > 0$ for x gives $x > -\frac{3}{2}$, and the domain of p is $x \in (-\frac{3}{2}, \infty)$.

b. For $x^2 - 2x > 0$, we note $y = x^2 - 2x$ is a parabola, opening upward, with zeroes at $x = 0$ and $x = 2$ (see Figure 4.25). This means $x^2 - 2x$ will be positive for $x < 0$ and $x > 2$. The domain of q is $x \in (-\infty, 0) \cup (2, \infty)$.

Figure 4.25

c. For $\dfrac{3 - x}{x + 3} > 0$, we note $y = \dfrac{3 - x}{x + 3}$ has a zero at $x = 3$, and a vertical asymptote at $x = -3$. Outputs are positive when $x = 0$ (see Figure 4.26), so y is positive in the interval $(-3, 3)$ and negative elsewhere. The domain of r is $x \in (-3, 3)$.

Figure 4.26

When $-3 < x < 3$,
$y > 0$
(interval test)

When $x < -3, y < 0$ When $x > 3, y < 0$

☑ **D. You've just learned how to find the domain of a logarithmic function**

d. For $|x - 2| > 0$, we note $y = |x - 2|$ is the graph of $y = |x|$ shifted 2 units right, with its vertex at $(2, 0)$. The graph is positive for all x, except at $x = 2$. The domain of f is $x \in (-\infty, 2) \cup (2, \infty)$.

Now try Exercises 73 through 78 ▶

GRAPHICAL SUPPORT

The domain for $r(x) = \log_{10}\left(\dfrac{3 - x}{x + 3}\right)$ from Example 6c can be confirmed using the [LOG] key on a graphing calculator. Use the key to enter the equation as Y_1 on the [Y=] screen, then graph the function using the [ZOOM] **4:ZDecimal** option. Both the graph and TABLE feature help to confirm the domain is $x \in (-3, 3)$.

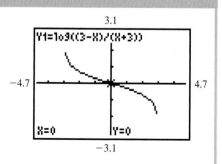

E. Applications of Logarithms

As we use mathematics to model the real world, there are times when the range of outcomes is so large that using a linear scale would be hard to manage. For example, compared to a whisper—the scream of a jet engine may be up to *ten billion times* louder. Similar ranges exist in the measurement of earthquakes, light, acidity, and voltage. In lieu of a linear scale, logarithms are used where each whole number increase in magnitude represents a tenfold increase in the intensity. For earthquake intensities (a measure of the wave energy produced by the quake), units called **magnitudes** (or **Richter values**) are used. Earthquakes with a magnitude of 6.5 or more often cause significant damage, while the slightest earthquakes have magnitudes near 1 and are barely perceptible. The magnitude of the intensity $M(I)$ is given by $M(I) = \log\left(\dfrac{I}{I_0}\right)$ where I is the measured intensity and I_0 represents a minimum or **reference intensity.** The value of I is often given as a multiple of this reference intensity.

> **WORTHY OF NOTE**
>
> The **decibel** (dB) is the reference unit for sound, and is based on the faintest sound a person can hear, called the **threshold of audibility.** It is a base-10 logarithmic scale, meaning a sound 10 times more intense is one decibel louder.

EXAMPLE 8A ▶ Finding the Magnitude of an Earthquake

Find the magnitude of an earthquake (rounded to hundredths) with the intensities given.

  a. $I = 4000I_0$ **b.** $I = 8{,}252{,}000I_0$

Solution ▶ **a.**

$$M(I) = \log\left(\frac{I}{I_0}\right) \qquad \text{magnitude equation}$$

$$M(4000I_0) = \log\left(\frac{4000I_0}{I_0}\right) \qquad \text{substitute } 4000I_0 \text{ for } I$$

$$= \log 4000 \qquad \text{simplify}$$

$$\approx 3.60 \qquad \text{result}$$

The earthquake had a magnitude of 3.6.

  b.

$$M(I) = \log\left(\frac{I}{I_0}\right) \qquad \text{magnitude equation}$$

$$M(8{,}252{,}000I_0) = \log\left(\frac{8252000I_0}{I_0}\right) \qquad \text{substitute } 8{,}252{,}000\, I_0 \text{ for } I$$

$$= \log 8{,}252{,}000 \qquad \text{simplify}$$

$$\approx 6.92 \qquad \text{result}$$

The earthquake had a magnitude of about 6.92.

EXAMPLE 8B ▶ Comparing Earthquake Intensity to the Reference Intensity

How many times more intense than the reference intensity I_0 is an earthquake with a magnitude of 6.7?

Solution ▶

$$M(I) = \log\left(\frac{I}{I_0}\right) \qquad \text{magnitude equation}$$

$$6.7 = \log\left(\frac{I}{I_0}\right) \qquad \text{substitute 6.7 for } M(I)$$

$$10^{6.7} = \left(\frac{I}{I_0}\right) \qquad \text{exponential form}$$

$$I = 10^{6.7}I_0 \qquad \text{solve for } I$$

$$I = 5{,}011{,}872I_0 \qquad 10^{6.7} \approx 5{,}011{,}872$$

An earthquake of magnitude 6.7 is over 5 million times more intense than the reference intensity.

EXAMPLE 8C ▶ **Comparing Earthquake Intensities**

The Great San Francisco Earthquake of 1906 left over 800 dead, did $80,000,000 in damage (see photo), and had an estimated magnitude of 7.7. The 2004 Indian Ocean earthquake, which had a magnitude of approximately 9.2, triggered a series of deadly tsunamis and was responsible for nearly 300,000 casualties. How much more intense was the 2004 quake?

Solution ▶ To find the intensity of each quake, substitute the given magnitude for $M(I)$ and solve for I:

$$M(I) = \log\left(\frac{I}{I_0}\right) \quad \text{magnitude equation} \qquad M(I) = \log\left(\frac{I}{I_0}\right)$$

$$7.7 = \log\left(\frac{I}{I_0}\right) \quad \text{substitute for } M(I) \qquad 9.2 = \log\left(\frac{I}{I_0}\right)$$

$$10^{7.7} = \left(\frac{I}{I_0}\right) \quad \text{exponential form} \qquad 10^{9.2} = \left(\frac{I}{I_0}\right)$$

$$10^{7.7}I_0 = I \quad \text{solve for } I \qquad 10^{9.2}I_0 = I$$

Using these intensities, we find that the Indian Ocean quake was $\frac{10^{9.2}}{10^{7.7}} = 10^{1.5} \approx 31.6$ times more intense.

Now try Exercises 81 through 90 ▶

A second application of logarithmic functions involves the relationship between altitude and barometric pressure. The altitude or height above sea level can be determined by the formula $H = (30T + 8000)\ln\left(\frac{P_0}{P}\right)$, where H is the altitude in meters for a temperature T in degrees Celsius, P is the barometric pressure at a given altitude in units called **centimeters of mercury** (cmHg), and P_0 is the barometric pressure at sea level: 76 cmHg.

EXAMPLE 9 ▶ **Using Logarithms to Determine Altitude**

Hikers at the summit of Mt. Shasta in northern California take a pressure reading of 45.1 cmHg at a temperature of 9°C. How high is Mt. Shasta?

Solution ▶ For this exercise, $P_0 = 76$, $P = 45.1$, and $T = 9$. The formula yields

$$H = (30T + 8000)\ln\left(\frac{P_0}{P}\right) \quad \text{given formula}$$

$$= [30(9) + 8000]\ln\left(\frac{76}{45.1}\right) \quad \text{substitute given values}$$

$$= 8270\ln\left(\frac{76}{45.1}\right) \quad \text{simplify}$$

$$\approx 4316 \quad \text{result}$$

Mt. Shasta is about 4316 m high.

Now try Exercises 91 through 94 ▶

Our final application shows the versatility of logarithmic functions, and their value as a real-world model. Large advertising agencies are well aware that after a new ad campaign, sales will increase rapidly as more people become aware of the product. Continued advertising will give the new product additional market share, but once the "newness" wears off and the competition begins responding, sales tend to taper off—regardless of any additional amount spent on ads. This phenomenon can be modeled by the function

$$S(d) = k + a \ln d,$$

where $S(d)$ is the number of expected sales after d dollars are spent, and a and k are constants related to product type and market size (**see Exercises 95 and 96**).

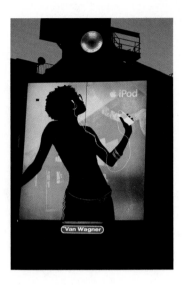

Van Wagner

EXAMPLE 10 ▶ **Using Logarithms for Marketing Strategies**

Market research has shown that sales of the MusicMaster, a new system for downloading and playing music, can be approximated by the equation $S(d) = 2500 + 250 \ln d$, where $S(d)$ is the number of sales after d thousand dollars is spent on advertising.

  a. What sales volume is expected if the advertising budget is \$40,000?

  b. If the company needs to sell 3500 units to begin making a profit, how much should be spent on advertising?

  c. To gain a firm hold on market share, the company is willing to continue spending on advertising up to a point where only 3 additional sales are gained for each \$1000 spent, in other words, $\frac{\Delta S}{\Delta d} = \frac{3}{1}$. Verify that spending between \$83,200 and \$83,300 puts them very close to this goal.

Solution ▶   **a.** For sales volume, we simply evaluate the function for $d = 40$ (d in thousands):

$$
\begin{aligned}
S(d) &= 2500 + 250 \ln d && \text{given equation} \\
S(40) &= 2500 + 250 \ln 40 && \text{substitute 40 for } d \\
&\approx 2500 + 922 && 250 \ln 40 \approx 922 \\
&= 3422
\end{aligned}
$$

Spending \$40,000 on advertising will generate approximately 3422 sales.

  b. To find the advertising budget needed, we substitute number of sales and solve for d.

$$
\begin{aligned}
S(d) &= 2500 + 250 \ln d && \text{given equation} \\
3500 &= 2500 + 250 \ln d && \text{substitute 2500 for } S(d) \\
1000 &= 250 \ln d && \text{subtract 2500} \\
4 &= \ln d && \text{divide by 250} \\
e^4 &= d && \text{exponential form} \\
54.598 &\approx d && e^4 \approx 54.598
\end{aligned}
$$

About \$54,600 should be spent in order to sell 3500 units.

c. To verify, we calculate the average rate of change on the interval [83.2, 83.3].

$$\frac{\Delta S}{\Delta d} = \frac{S(d_2) - S(d_1)}{d_2 - d_1} \qquad \text{formula for average rate of change}$$

$$= \frac{S(83.3) - S(83.2)}{83.3 - 83.2} \qquad \text{substitute 83.3 for } d_2 \text{ and 83.2 for } d_1$$

$$\approx \frac{3605.6 - 3605.3}{0.1} \qquad \text{evaluate } S(83.3) \text{ and } S(83.2)$$

$$= 3$$

☑ **E.** You've just learned how to solve applications of logarithmic functions

The average rate of change in this interval is very close to $\frac{3}{1}$.

Now try Exercises 95 and 96 ▶

4.3 EXERCISES

▶ **CONCEPTS AND VOCABULARY**

Fill in each blank with the appropriate word or phrase. Carefully reread the section if needed.

1. A logarithmic function is of the form $y =$ _____, where _____ > 0, _____ $\neq 1$ and inputs are _____ than zero.

2. The range of $y = \log_b x$ is all _____ _____, and the domain is $x \in$ _____. Further, as $x \to 0$, $y \to$ ___.

3. For logarithmic functions of the form $y = \log_b x$, the x-intercept is _____, since $\log_b 1 =$ _____.

4. The function $y = \log_b x$ is an increasing function if _____, and a decreasing function if _____.

5. What number does the expression $\log_2 32$ represent? Discuss/Explain how $\log_2 32 = \log_2 2^5$ justifies this fact.

6. Explain how the graph of $Y = \log_b(x - 3)$ can be obtained from $y = \log_b x$. Where is the "new" x-intercept? Where is the new asymptote?

▶ **DEVELOPING YOUR SKILLS**

Write each equation in exponential form.

7. $3 = \log_2 8$

8. $2 = \log_3 9$

9. $-1 = \log_7 \frac{1}{7}$

10. $-3 = \log_e \frac{1}{e^3}$

11. $0 = \log_9 1$

12. $0 = \log_e 1$

13. $\frac{1}{3} = \log_8 2$

14. $\frac{1}{2} = \log_{81} 9$

15. $1 = \log_2 2$

16. $1 = \log_e e$

17. $\log_7 49 = 2$

18. $\log_4 16 = 2$

19. $\log_{10} 100 = 2$

20. $\log_{10} 10{,}000 = 4$

21. $\log_e(54.598) \approx 4$

22. $\log_{10} 0.001 = -3$

Write each equation in logarithmic form.

23. $4^3 = 64$

24. $e^3 \approx 20.086$

25. $3^{-2} = \frac{1}{9}$

26. $2^{-3} = \frac{1}{8}$

27. $e^0 = 1$

28. $8^0 = 1$

29. $\left(\frac{1}{3}\right)^{-3} = 27$

30. $\left(\frac{1}{5}\right)^{-2} = 25$

31. $10^3 = 1000$

32. $e^1 = e$

33. $10^{-2} = \frac{1}{100}$

34. $10^{-5} = \frac{1}{100{,}000}$

35. $4^{\frac{3}{2}} = 8$

36. $e^{\frac{3}{4}} \approx 2.117$

37. $4^{\frac{-3}{2}} = \frac{1}{8}$

38. $27^{\frac{-2}{3}} = \frac{1}{9}$

Determine the value of each logarithm without using a calculator.

39. $\log_4 4$

40. $\log_9 9$

41. $\log_{11} 121$

42. $\log_{12} 144$

43. $\log_e e$

44. $\log_e e^2$

45. $\log_4 2$

46. $\log_{81} 9$

47. $\log_7 \frac{1}{49}$

48. $\log_9 \frac{1}{81}$

49. $\log_e \frac{1}{e^2}$

50. $\log_e \frac{1}{\sqrt{e}}$

Use a calculator to evaluate each expression, rounded to four decimal places.

51. $\log 50$

52. $\log 47$

53. $\ln 1.6$

54. $\ln 0.75$

55. $\ln 225$

56. $\ln 381$

57. $\log \sqrt{37}$

58. $\log 4\pi$

Graph each function *using transformations* of $y = \log_b x$ and strategically plotting a few points. Clearly state the transformations applied.

59. $f(x) = \log_2 x + 3$

60. $g(x) = \log_2(x - 2)$

61. $h(x) = \log_2(x - 2) + 3$

62. $p(x) = \log_3 x - 2$

63. $q(x) = \ln(x + 1)$

64. $r(x) = \ln(x + 1) - 2$

65. $Y_1 = -\ln(x + 1)$

66. $Y_2 = -\ln x + 2$

Use the transformation equation $y = af(x \pm h) \pm k$ and the asymptotes and intercept(s) of the parent function to match each equation to one of the graphs given.

67. $y = \log_b(x + 2)$

68. $y = 2\log_b x$

69. $y = 1 - \log_b x$

70. $y = \log_b x - 1$

71. $y = \log_b x + 2$

72. $y = -\log_b x$

I.

II.

III.

IV.

V.

VI.

Determine the domain of the following functions.

73. $y = \log_6\left(\dfrac{x + 1}{x - 3}\right)$

74. $y = \ln\left(\dfrac{x - 2}{x + 3}\right)$

75. $y = \log_5 \sqrt{2x - 3}$

76. $y = \ln \sqrt{5 - 3x}$

77. $y = \log(9 - x^2)$

78. $y = \ln(9x - x^2)$

▶ **WORKING THE FORMULAS**

79. pH level: $f(x) = -\log_{10} x$

The pH level of a solution indicates the concentration of hydrogen (H^+) ions in a unit called *moles per liter*. The pH level $f(x)$ is given by the formula shown, where x is the ion concentration (given in scientific notation). A solution with pH < 7 is called an acid (lemon juice: pH ≈ 2), and a solution with pH > 7 is called a base (household ammonia: pH ≈ 11). Use the formula to determine the pH level of tomato juice if $x = 7.94 \times 10^{-5}$ moles per liter. Is this an acid or base solution?

80. Time required for an investment to double:

$$T(r) = \frac{\log 2}{\log(1 + r)}$$

The time required for an investment to double in value is given by the formula shown, where r represents the interest rate (expressed as a decimal) and $T(r)$ gives the years required. How long would it take an investment to double if the interest rate were (a) 5%, (b) 8%, (c) 12%?

▶ **APPLICATIONS**

Earthquake intensity: Use the information provided in Example 8 to answer the following.

81. Find the value of $M(I)$ given

   a. $I = 50,000I_0$ and   **b.** $I = 75,000I_0$.

82. Find the intensity I of the earthquake given

   a. $M(I) = 3.2$ and   **b.** $M(I) = 8.1$.

83. **Earthquake intensity:** On June 25, 1989, an earthquake with magnitude 6.2 shook the southeast side of the Island of Hawaii (near Kalapana), causing some $1,000,000 in damage. On October 15, 2006, an earthquake measuring 6.7 on the Richter scale shook the northwest side of the island, causing over $100,000,000 in damage. How much more intense was the 2006 quake?

84. **Earthquake intensity:** The most intense earthquake of the modern era occurred in Chile on May 22, 1960, and measured 9.5 on the Richter scale. How many times more intense was this earthquake, than the quake that hit Northern Sumatra (Indonesia) on March 28, 2005, and measured 8.7?

Brightness of a star: The brightness or intensity I of a star as perceived by the naked eye is measured in units called *magnitudes*. The brightest stars have magnitude 1 $[M(I) = 1]$ and the dimmest have magnitude 6 $[M(I) = 6]$. The magnitude of a star is given by the equation $M(I) = 6 - 2.5 \cdot \log\left(\dfrac{I}{I_0}\right)$, where I is the actual intensity of light from the star and I_0 is the faintest light visible to the human eye, called the reference intensity. The intensity I is often given as a multiple of this reference intensity.

85. Find the value of $M(I)$ given

   a. $I = 27I_0$ and   **b.** $I = 85I_0$.

86. Find the intensity I of a star given

   a. $M(I) = 1.6$ and   **b.** $M(I) = 5.2$.

Intensity of sound: The intensity of sound as perceived by the human ear is measured in units called decibels (dB). The loudest sounds that can be withstood without damage to the eardrum are in the 120- to 130-dB range, while a whisper may measure in the 15- to 20-dB range. Decibel measure is given by the equation $D(I) = 10 \log\left(\dfrac{I}{I_0}\right)$, where I is the actual intensity of the sound and I_0 is the

faintest sound perceptible by the human ear—called the reference intensity. The intensity I is often given as a multiple of this reference intensity, but often the constant 10^{-16} (watts per cm²; W/cm²) is used as the threshold of audibility.

87. Find the value of $D(I)$ given

   a. $I = 10^{-14}$ and   **b.** $I = 10^{-4}$.

88. Find the intensity I of the sound given

   a. $D(I) = 83$ and   **b.** $D(I) = 125$.

89. **Sound intensity of a hair dryer:** Every morning (it seems), Jose is awakened by the mind-jarring, ear-jamming sound of his daughter's hair dryer (75 dB). He knew he was exaggerating, but told her (many times) of how it reminded him of his railroad days, when the air compressor for the pneumatic tools was running (110 dB). In fact, how many times more intense was the sound of the air compressor compared to the sound of the hair dryer?

90. **Sound intensity of a busy street:** The decibel level of noisy, downtown traffic has been estimated at 87 dB, while the laughter and banter at a loud party might be in the 60 dB range. How many times more intense is the sound of the downtown traffic?

The *barometric equation* $H = (30T + 8000)\ln\left(\dfrac{P_0}{P}\right)$ was discussed in Example 9.

91. **Temperature and atmospheric pressure:** Determine the height of Mount McKinley (Alaska), if the temperature at the summit is $-10°C$, with a barometric reading of 34 cmHg.

92. **Temperature and atmospheric pressure:** A large passenger plane is flying cross-country. The instruments on board show an air temperature of 3°C, with a barometric pressure of 22 cmHg. What is the altitude of the plane?

93. **Altitude and atmospheric pressure:** By definition, a mountain pass is a low point between two mountains. Passes may be very short with steep slopes, or as large as a valley between two peaks. Perhaps the highest drivable pass in the world is the Semo La pass in central Tibet. At its highest elevation, a temperature reading of 8°C was taken, along with a barometer reading of 39.3 cmHg. (a) Approximately how high is the Semo La pass? (b) While traveling up to this pass, an elevation marker is seen. If the barometer reading was 47.1 cmHg at a temperature of 12°C, what height did the marker give?

94. Altitude and atmospheric pressure: Hikers on Mt. Everest take successive readings of 35 cmHg at 5°C and 30 cmHg at −10°C. (a) How far up the mountain are they at each reading? (b) Approximate the height of Mt. Everest if the temperature at the summit is −27°C and the barometric pressure is 22.2 cmHg.

95. Marketing budgets: An advertising agency has determined the number of items sold by a certain client is modeled by the equation $N(A) = 1500 + 315 \ln A$, where $N(A)$ represents the number of sales after spending A thousands of dollars on advertising. Determine the approximate number of items sold on an advertising budget of (a) $10,000; (b) $50,000. (c) Use the TABLE feature of a calculator to estimate how large a budget is needed (to the nearest $500 dollars) to sell 3000 items. (d) This company is willing to continue advertising as long as eight additional sales are gained for every $1000 spent: $\dfrac{\Delta N}{\Delta A} = \dfrac{8}{1}$. Show this occurs by spending between $39,300 to $39,400.

96. Sports promotions: The accountants for a major boxing promoter have determined that the number of pay-per-view subscriptions sold to their championship bouts can be modeled by the function $N(d) = 15,000 + 5850 \ln d$, where $N(d)$ represents the number of subscriptions sold after spending d thousand dollars on promotional activities. Determine the number of subscriptions sold if (a) $50,000 and (b) $100,000 is spent. (c) Use the TABLE feature of a calculator to estimate how much should be spent (to the nearest $1000 dollars) to sell over 50,000 subscriptions. (d) This promoter is willing to continue promotional spending as long as 14 additional subscriptions are sold for every $1000 spent: $\dfrac{\Delta N}{\Delta d} = \dfrac{14}{1}$. Show this occurs by spending between $417,800 and $417,900.

97. Home ventilation: In the construction of new housing, there is considerable emphasis placed on correct ventilation. If too little outdoor air enters a home, pollutants can sometimes accumulate to levels that pose a health risk. For homes of various sizes, ventilation requirements have been established and are based on floor area and the number of bedrooms. For a three-bedroom home, the relationship can be modeled by the function $C(x) = 42 \ln x - 270$, where $C(x)$ represents the number of cubic feet of air per minute (cfm) that should be exchanged with outside air in a home with floor area x (in square feet). (a) How many cfm of exchanged air are needed for a three-bedroom home with a floor area of 2500 ft²? (b) If a three-bedroom home is being mechanically ventilated by a system with 40 cfm capacity, what is the square footage of the home, assuming it is built to code?

98. Runway takeoff distance: Many will remember the August 27, 2006, crash of a commuter jet at Lexington's Blue Grass Airport, that was mistakenly trying to take off on a runway that was just too short. Forty-nine lives were lost. The minimum required length of a runway depends on the maximum allowable takeoff weight (mtw) of a specific plane. This relationship can be approximated by the function

$L(x) = 2085 \ln x - 14,900$, where $L(x)$ represents the required length of a runway in feet, for a plane with x mtw in pounds.

  a. The Airbus-320 has a 169,750 lb mtw. What minimum runway length is required for takeoff?

  b. A Learjet 30 model requires a runway of 5550 ft to takeoff safely. What is its mtw?

Memory retention: Under certain conditions, a person's retention of random facts can be modeled by the equation $P(x) = 95 - 14 \log_2 x$, where $P(x)$ is the percentage of those facts retained after x number of days. Find the percentage of facts a person might retain after:

99. a. 1 day   **b.** 4 days   **c.** 16 days

100. a. 32 days   **b.** 64 days   **c.** 78 days
                                                             ≈7%

101. pH level: Use the formula given in Exercise 79 to determine the pH level of black coffee if $x = 5.1 \times 10^{-5}$ moles per liter. Is black coffee considered an acid or base solution?

102. The length of time required for an amount of money to *triple* is given by the formula $T(r) = \dfrac{\log 3}{\log(1 + r)}$ (refer to Exercise 80). Construct a table of values to help estimate what interest rate is needed for an investment to triple in nine years.

▶ **EXTENDING THE CONCEPT**

103. Many texts and reference books give estimates of the noise level (in decibels dB) of common sounds. Through reading and research, try to locate or approximate where the following sounds would fall along this scale. In addition, determine at what point pain or ear damage begins to occur.

    a. threshold of audibility   **b.** lawn mower

    c. whisper   **d.** loud rock concert

    e. lively party   **f.** jet engine

104. Determine the value of x that makes the equation true: $\log_3[\log_3(\log_3 x)] = 0$.

105. Find the value of each expression without using a calculator.

    a. $\log_{64}\frac{1}{16}$ **b.** $\log_{\frac{4}{9}}\frac{27}{8}$ **c.** $\log_{0.25}32$

106. Suppose you and I represent two different numbers. Is the following cryptogram true or false? *The log of me base me is one and the log of you base you is one, but the log of you base me is equal to the log of me base you turned upside down.*

▶ **MAINTAINING YOUR SKILLS**

107. (3.3) Graph $g(x) = \sqrt[3]{x + 2} - 1$ by shifting the parent function. Then state the domain and range of g.

108. (R.4) Factor the following expressions:

    a. $x^3 - 8$ **b.** $a^2 - 49$

    c. $n^2 - 10n + 25$ **d.** $2b^2 - 7b + 6$

109. (3.4/3.7) For the graph shown, write the solution set for $f(x) < 0$. Then write the equation of the graph in factored form and in polynomial form.

110. (2.2) A function $f(x)$ is defined by the ordered pairs shown in the table. Is the function (a) linear? (b) increasing? Justify your answers.

x	y
-10	0
-9	-2
-8	-8
-6	-18
-5	-50
-4	-72

MID-CHAPTER CHECK

1. Write the following in logarithmic form.

    a. $27^{\frac{2}{3}} = 9$ **b.** $81^{\frac{5}{4}} = 243$

2. Write the following in exponential form.

    a. $\log_8 32 = \frac{5}{3}$ **b.** $\log_{1296} 6 = 0.25$

3. Solve each equation for the unknown:

    a. $4^{2x} = 32^{x-1}$ **b.** $\left(\frac{1}{3}\right)^{4b} = 9^{2b-5}$

4. Solve each equation for the unknown:

    a. $\log_{27} x = \frac{1}{3}$ **b.** $\log_b 125 = 3$

 5. The homes in a popular neighborhood are growing in value according to the formula $V(t) = V_0\left(\frac{9}{8}\right)^t$, where t is the time in years, V_0 is the purchase price of the home, and $V(t)$ is the current value of the home. (a) In 3 yr, how much will a $50,000 home be worth? (b) Use the TABLE feature of your calculator to estimate how many years (to the nearest year) until the home doubles in value.

6. The graph of the function $f(x) = 5^x$ has been shifted right 3 units, up 2 units, and stretched by a factor of 4. What is the equation of the resulting function?

7. State the domain and range for $f(x) = \sqrt{x - 3} + 1$, then find $f^{-1}(x)$ and state its domain and range. Verify the inverse relationship using composition.

8. Write the following equations in logarithmic form, then verify the result on a calculator.

    a. $81 = 3^4$ **b.** $e^4 \approx 54.598$

9. Write the following equations in exponential form, then verify the result on a calculator.

   a. $\dfrac{2}{3} = \log_{27} 9$ b. $1.4 \approx \ln 4.0552$

10. On August 15, 2007, an earthquake measuring 8.0 on the Richter scale struck coastal Peru. On October 17, 1989, right before Game 3 of the World Series between the Oakland A's and the San Francisco Giants, the Loma Prieta earthquake, measuring 7.1 on the Richter scale, struck the San Francisco Bay area. How much more intense was the Peruvian earthquake?

REINFORCING BASIC CONCEPTS

Linear and Logarithm Scales

The use of logarithmic scales as a tool of measurement is primarily due to the range of values for the phenomenon being measured. For instance, time is generally measured on a linear scale, and for short periods a linear scale is appropriate. For the time line in Figure 4.27, each tick-mark *represents 1 unit,* and the time line can display a period of 10 yr. However, the scale would be useless in a study of world history or geology. If we scale the number line logarithmically, each tick-mark *represents a power of 10* (Figure 4.28) and a scale of the same length can now display a time period of 10 billion years.

Figure 4.27

Figure 4.28

In much the same way, logarithmic measures are needed in a study of sound and earthquake intensity, as the scream of a jet engine is over 1 billion times more intense than the threshold of hearing, and the most destructive earthquakes are billions of times stronger than the slightest earth movement that can be felt. Figures 4.29 and 4.30 show logarithmic scales for measuring sound in decibels (1 bel = 10 decibels) and earthquake intensity in Richter values (or magnitudes).

Figure 4.29

Figure 4.30

As you view these scales, remember that each unit increase represents a power of 10. For instance, the 1906 San Francisco earthquake was $8.1 - 5.5 = 2.6$ magnitudes greater than the San Jose quake of 1992, meaning it was $10^{2.6} \approx$ *398 times more intense.* Use this information to complete the following exercises. Determine how many times more intense the first sound is compared to the second.

Exercise 1: jet engine: 14 bels
              rock concert: 11.8 bels

Exercise 2: pneumatic hammer: 11.2 bels
              heavy lawn mower: 8.5 bels

Exercise 3: train horn: 7.5 bels
              soft music: 3.4 bels

Determine how many times more intense the first quake was compared to the second.

Exercise 4: Great Chilean quake (1960): magnitude 9.5
              Kobe, Japan, quake (1995): magnitude 6.9

Exercise 5: Northern Sumatra (2004): magnitude 9.1
              Southern, Greece (2008): magnitude 4.5

**Properties of Logarithms; Solving Exponential/
Logarithmic Equations**

Learning Objectives

In Section 4.4 you will learn how to:

☐ **A.** Solve logarithmic equations using the fundamental properties of logarithms

☐ **B.** Apply the product, quotient, and power properties of logarithms

☐ **C.** Solve general logarithmic and exponential equations

☐ **D.** Solve applications involving logistic, exponential, and logarithmic functions

In this section, we develop the ability to solve logarithmic and exponential equations of any base. A **logarithmic equation** has at least one term that involves the logarithm of a variable. Likewise, an **exponential equation** has at least one term that involves a variable exponent. In the same way that we might *square both sides* or *divide both sides* of an equation in the solution process, we'll show that we can also *exponentiate both sides* or *take logarithms of both sides* to help obtain a solution.

A. Solving Equations Using the Fundamental Properties of Logarithms

In Section 4.3, we converted expressions from exponential form to logarithmic form using the basic definition: $x = b^y \Leftrightarrow y = \log_b x$. This relationship reveals the following four properties:

Fundamental Properties of Logarithms

For any base $b > 0$, $b \neq 1$,

I. $\log_b b = 1$, since $b^1 = b$

II. $\log_b 1 = 0$, since $b^0 = 1$

III. $\log_b b^x = x$, since $b^x = b^x$ (exponential form)

IV. $b^{\log_b x} = x$, since $\log_b x = \log_b x$ (logarithmic form)

To see the verification of Property IV more clearly, again note that for $y = \log_b x$, $b^y = x$ is the exponential form, and substituting $\log_b x$ for y yields $b^{\log_b x} = x$. Also note that Properties III and IV demonstrate that $y = \log_b x$ and $y = b^x$ are inverse functions. In common language, "a base-b logarithm *undoes* a base-b exponential," and "a base-b exponential *undoes* a base-b logarithm." For $f(x) = \log_b x$ and $f^{-1}(x) = b^x$, using a composition verifies the inverse relationship:

$$(f \circ f^{-1})(x) = f[f^{-1}(x)] \qquad (f^{-1} \circ f)(x) = f^{-1}[f(x)]$$
$$= \log_b b^x \qquad\qquad\qquad = b^{\log_b x}$$
$$= x \qquad\qquad\qquad\qquad = x$$

These properties can be used to solve basic equations involving logarithms and exponentials.

EXAMPLE 1 ▶ **Solving Basic Logarithmic Equations**

Solve each equation by applying fundamental properties. Answer in exact form and approximate form using a calculator (round to 1000ths).

a. $\ln x = 2$ **b.** $-0.52 = \log x$

Solution ▶
a. $\ln x = 2$ given
$e^{\ln x} = e^2$ exponentiate both sides
$x = e^2$ Property IV
≈ 7.389 result

b. $-0.52 = \log x$ given
$10^{-0.52} = 10^{\log x}$ exponentiate both sides
$10^{-0.52} = x$ Property IV
$0.302 \approx x$ result

Now try Exercises 7 through 10 ▶

Note that exponentiating both sides of the equation produced the same result as simply writing the original equation in exponential form, and either approach can be used.

EXAMPLE 2 ▶ **Solving Basic Exponential Equations**

Solve each equation by applying fundamental properties. Answer in exact form and approximate form using a calculator (round to 1000ths).

a. $e^x = 167$ **b.** $10^x = 8.223$

Solution ▶ **a.**
$$e^x = 167 \quad \text{given}$$
$$\ln e^x = \ln 167 \quad \text{take natural log of both sides}$$
$$x = \ln 167 \quad \text{Property III}$$
$$x \approx 5.118 \quad \text{result}$$

b.
$$10^x = 8.223 \quad \text{given}$$
$$\log 10^x = \log 8.223 \quad \text{take common log of both sides}$$
$$x = \log 8.223 \quad \text{Property III}$$
$$x \approx 0.915 \quad \text{result}$$

Now try Exercises 11 through 14 ▶

Similar to our previous observation, taking the logarithm of both sides produced the same result as writing the original equation in logarithmic form, and either approach can be used.

If an equation has a single logarithmic or exponential term (base 10 or base e), the equation can be solved by isolating this term and applying one of the fundamental properties.

EXAMPLE 3 ▶ **Solving Exponential Equations**

Solve each equation. Write answers in exact form and approximate form to four decimal places.

a. $10^x - 29 = 51$ **b.** $3e^{x+1} - 5 = 7$

Solution ▶ **a.**
$$10^x - 29 = 51 \quad \text{given}$$
$$10^x = 80 \quad \text{add 29}$$

Since the left-hand side is base 10, we apply a common logarithm.

$$\log 10^x = \log 80 \quad \text{take the common log of both sides}$$
$$x = \log 80 \quad \text{Property III (exact form)}$$
$$\approx 1.9031 \quad \text{approximate form}$$

WORTHY OF NOTE

To check solutions using a calculator, we can STO▶ (store) the exact result in storage location $\boxed{x, T, \theta, n}$ (the function variable x) and simply enter the original equation on the home screen. The figure shows this verification for Example 3b.

```
ln(4)-1→X
          .3862943611
3e^(X+1)-5
                      7
```

b.
$$3e^{x+1} - 5 = 7 \quad \text{given}$$
$$3e^{x+1} = 12 \quad \text{add 5}$$
$$e^{x+1} = 4 \quad \text{divide by 3}$$

Since the left-hand side is base e, we apply a natural logarithm.

$$\ln e^{x+1} = \ln 4 \quad \text{take the natural log of both sides}$$
$$x + 1 = \ln 4 \quad \text{Property III}$$
$$x = \ln 4 - 1 \quad \text{solve for } x \text{ (exact form)}$$
$$\approx 0.3863 \quad \text{approximate form}$$

Now try Exercises 15 through 20 ▶

EXAMPLE 4 ▶ **Solving Logarithmic Equations**

Solve each equation. Write answers in exact form and approximate form to four decimal places.

a. $2 \log(7x) + 1 = 4$　　　**b.** $-4 \ln(x + 1) - 5 = 7$

Solution ▶ **a.**
$$2 \log(7x) + 1 = 4 \qquad \text{given}$$
$$2 \log(7x) = 3 \qquad \text{subtract 1}$$
$$\log(7x) = \frac{3}{2} \qquad \text{divide by 2}$$
$$7x = 10^{\frac{3}{2}} \qquad \text{exponential form}$$
$$x = \frac{10^{\frac{3}{2}}}{7} \qquad \text{divide by 7 (exact form)}$$
$$\approx 4.5175 \qquad \text{approximate form}$$

b.
$$-4 \ln(x + 1) - 5 = 7 \qquad \text{given}$$
$$-4 \ln(x + 1) = 12 \qquad \text{add 5}$$
$$\ln(x + 1) = -3 \qquad \text{divide by } -4$$
$$x + 1 = e^{-3} \qquad \text{exponential form}$$
$$x = e^{-3} - 1 \qquad \text{subtract 1 (exact form)}$$
$$\approx -0.9502 \qquad \text{approximate form}$$

☑ **A.** You've just learned how to solve logarithmic equations using the fundamental properties of logarithms

Now try Exercises 21 through 26 ▶

GRAPHICAL SUPPORT

Solutions can be also checked using the intersection of graphs method (Technology Highlight, page 432). For Example 4a, enter $Y_1 = 2 \log(7x) + 1$ and $Y_2 = 4$ on the [Y=] screen. From the domain of the function and the expected answer, we set a window that includes only Quadrant I. Use the keystrokes [2nd] [TRACE] (CALC) 5:intersect, and identify each graph by pressing [ENTER] 3 times. The calculator will find the point of intersection and display it at the bottom of the screen.

B. The Product, Quotient, and Power Properties of Logarithms

Generally speaking, equation solving involves simplifying the equation, isolating a variable term on one side, and applying an inverse to solve for the unknown. For logarithmic equations such as $\log x + \log(x + 3) = 1$, we must find a way to combine the terms on the left, before we can work toward a solution. This requires a further exploration of logarithmic properties.

Due to the close connection between exponents and logarithms, their properties are very similar. To illustrate, we'll use terms that can all be written in the form 2^x, and write the equations $8 \cdot 4 = 32$, $\frac{8}{4} = 2$, and $8^2 = 64$ in both exponential form and logarithmic form.

The exponents from a product are added:　　exponential form:　　$2^3 \cdot 2^2 = 2^{3+2}$

　　　　　　　　　　　　　　　logarithmic form:　　$\log_2(8 \cdot 4) = \log_2 8 + \log_2 4$

The exponents from a quotient are subtracted:

exponential form: $\dfrac{2^3}{2^2} = 2^{3-2}$

logarithmic form: $\log_2\left(\dfrac{8}{4}\right) = \log_2 8 - \log_2 4$

The exponents from a power are multiplied:

exponential form: $(2^3)^2 = 2^{3\cdot 2}$

logarithmic form: $(\log_2 8)^2 = 2 \cdot \log_2 8$

Each illustration can be generalized and applied with any base *b*.

Properties of Logarithms

Give *M*, *N*, and *b* ≠ 1 are *positive* real numbers, and *any* real number *p*.

Product Property	**Quotient Property**	**Power Property**
$\log_b(MN) = \log_b M + \log_b N$	$\log_b\left(\dfrac{M}{N}\right) = \log_b M - \log_b N$	$\log_b M^p = p\log_b M$
The log of a product is a sum of logarithms.	The log of a quotient is a difference of logarithms.	The log of a quantity to a power is the power times the log of the quantity.

For a detailed verification of these properties, see Appendix I.

⚠ **CAUTION** ▶ It's very important that you read and understand these properties correctly. In particular, note that $\log_b(M + N) \neq \log_b M + \log_b N$, and $\log_b\left(\dfrac{M}{N}\right) \neq \dfrac{\log_b M}{\log_b N}$. In the first case, it might help to compare the statement with $f(x + 3)$, which represents a horizontal shift of the graph 3 units left, and in particular, $f(x + 3) \neq f(x) + f(3)$.

In many cases, these properties are applied to consolidate logarithmic terms in preparation for equation solving.

EXAMPLE 5 ▶ **Rewriting Expressions Using Logarithmic Properties**

Use the properties of logarithms to write each expression as a single term.

   a. $\log_2 7 + \log_2 5$ **b.** $2\ln x + \ln(x + 6)$ **c.** $\ln(x + 2) - \ln x$

Solution ▶   **a.** $\log_2 7 + \log_2 5 = \log_2(7 \cdot 5)$ product property

                          $= \log_2 35$ simplify

   b. $2\ln x + \ln(x + 6) = \ln x^2 + \ln(x + 6)$ power property

                         $= \ln[x^2(x + 6)]$ product property

                         $= \ln[x^3 + 6x^2]$ simplify

   c. $\ln(x + 2) - \ln x = \ln\left(\dfrac{x + 2}{x}\right)$ quotient property

Now try Exercises 27 through 42 ▶

EXAMPLE 6 ▶ **Rewriting Logarithmic Expressions Using the Power Property**

Use the power property of logarithms to rewrite each term as a product.

   a. $\ln 5^x$ **b.** $\log 32^{x+2}$ **c.** $\log \sqrt{x}$

Solution ▶   **a.** $\ln 5^x = x \ln 5$ power property

      b. $\log 32^{x+2} = (x+2)\log 32$ power property (note use of parentheses)

      c. $\log \sqrt{x} = \log x^{\frac{1}{2}}$ write radical using a rational exponent

         $= \dfrac{1}{2} \log x$ power property

Now try Exercises 43 through 50 ▶

⚠ **CAUTION** ▶ Note from Example 6b that parentheses *must be used* whenever the exponent is a sum or difference. There is a huge difference between $(x+2)\log 32$ and $x + 2\log 32$.

In other cases, these properties help rewrite an expression so that certain procedures can be applied more easily. Example 7 actually lays the foundation for more advanced work in mathematics.

EXAMPLE 7 ▶   **Rewriting Expressions Using Logarithmic Properties**

Use the properties of logarithms to write the following expressions as a sum or difference of simple logarithmic terms.

    a. $\log(x^2 z)$ **b.** $\ln \sqrt{\dfrac{x}{x+5}}$ **c.** $\ln\left[\dfrac{e\sqrt{x^2+1}}{(2x+5)^3}\right]$

Solution ▶   **a.** $\log(x^2 z) = \log x^2 + \log z$ product property

             $= 2\log x + \log z$ power property

      b. $\ln \sqrt{\dfrac{x}{x+5}} = \ln\left(\dfrac{x}{x+5}\right)^{\frac{1}{2}}$ write radical using a rational exponent

             $= \dfrac{1}{2}\ln\left(\dfrac{x}{x+5}\right)$ power property

             $= \dfrac{1}{2}[\ln x - \ln(x+5)]$ quotient property

      c. $\ln\left[\dfrac{e\sqrt{x^2+1}}{(2x+5)^3}\right] = \ln\left[\dfrac{e(x^2+1)^{\frac{1}{2}}}{(2x+5)^3}\right]$ write radical using a rational exponent

             $= \ln[e(x^2+1)^{\frac{1}{2}}] - \ln(2x+5)^3$ quotient property

             $= \ln e + \ln(x^2+1)^{\frac{1}{2}} - \ln(2x+3)^3$ product property

             $= 1 + \dfrac{1}{2}\ln(x^2+1) - 3\ln(2x+3)$ power property

Now try Exercises 51 through 60 ▶

Although base-10 and base-*e* logarithms dominate the mathematical landscape, there are many practical applications that use other bases. Fortunately, a formula exists that will convert any given base into either base 10 or base *e*. It's called the **change-of-base formula.**

Change-of-Base Formula

For the positive real numbers M, a, and b, with $a, b \neq 1$,

$$\log_b M = \frac{\log M}{\log b} \qquad \qquad \log_b M = \frac{\ln M}{\ln b} \qquad \qquad \log_b M = \frac{\log_a M}{\log_a b}$$

$$\text{base 10} \qquad \qquad \qquad \qquad \text{base } e \qquad \qquad \qquad \qquad \text{arbitrary base } a$$

Proof of the Change-of-Base Formula:

For $y = \log_b M$, we have $b^y = M$ in exponential form. It follows that

$$\log_a(b^y) = \log_a M \qquad \text{take base-}a \text{ logarithm of both sides}$$

$$y \log_a b = \log_a M \qquad \text{power property of logarithms}$$

$$y = \frac{\log_a M}{\log_a b} \qquad \text{divide by } \log_a b$$

$$\log_b M = \frac{\log_a M}{\log_a b} \qquad \text{substitute } \log_b M \text{ for } y$$

EXAMPLE 8 ▶ **Using the Change-of-Base Formula to Evaluate Expressions**

Find the value of each expression using the change-of-base formula. Answer in exact form and approximate form using nine digits, then *verify the result* using the original base.

a. $\log_3 29$ **b.** $\log_5 3.6$

Solution ▶ **a.** $\log_3 29 = \dfrac{\log 29}{\log 3}$ **b.** $\log_5 3.6 = \dfrac{\log 3.6}{\log 5}$

$= 3.065044752$ $= 0.795888947$

Check: $3^{3.065044752} = 29$ ✓   **Check:** $5^{0.795888947} = 3.6$ ✓

Now try Exercises 61 through 72 ▶

☑ **B.** You've just learned how to apply the product, quotient, and power properties of logarithms

The change-of-base formula can also be used to study and graph logarithmic functions of *any* base. For $y = \log_b x$, the right-hand expression is simply rewritten using the formula and the equivalent function is $y = \dfrac{\log x}{\log b}$. The new function can then be evaluated as in Example 8, or used to study the graph of $y = \log_b x$ for any base b.

C. Solving Logarithmic Equations

One of the most common mistakes in solving exponential and logarithmic equations is to apply the inverse function too early—before the equation has been simplified. In addition, since the domain of $y = \log_b x$ is $x > 0$, logarithmic equations can sometimes produce **extraneous roots,** and checking all answers is a good practice. We'll illustrate by solving the equation mentioned earlier: $\log x + \log(x + 3) = 1$.

EXAMPLE 9 ▶ **Solving a Logarithmic Equation**

Solve for x and check your answer: $\log x + \log(x + 3) = 1$.

Solution ▶ $\log x + \log (x + 3) = 1$ original equation

$\log [x(x + 3)] = 1$ product property

$x^2 + 3x = 10^1$ exponential form, distribute x

$x^2 + 3x - 10 = 0$ set equal to 0

$(x + 5)(x - 2) = 0$ factor

$x = -5$ or $x = 2$ result

Check: The "solution" $x = -5$ is outside the domain and is discarded. For $x = 2$,

$$\log x + \log (x + 3) = 1 \quad \text{original equation}$$
$$\log 2 + \log (2 + 3) = 1 \quad \text{substitute 2 for } x$$
$$\log 2 + \log 5 = 1 \quad \text{simplify}$$
$$\log (2 \cdot 5) = 1 \quad \text{product property}$$
$$\log 10 = 1 \quad \text{Property I}$$

Now try Exercises 73 through 80 ▶

As an alternative check, you could also use a calculator to verify $\log 2 + \log 5 = 1$ directly.

If the simplified form of an equation yields a logarithmic term on both sides, the **uniqueness property of logarithms** provides an efficient way to work toward a solution. Since logarithmic functions are one-to-one, we have

The Uniqueness Property of Logarithms

For positive real numbers m, n, and $b \neq 1$,

$$\text{If } \log_b m = \log_b n, \quad \text{then } m = n$$

Equal bases imply equal arguments.

EXAMPLE 10 ▶ Solving Logarithmic Equations Using the Uniqueness Property

Solve each equation using the uniqueness property.

a. $\log (x + 2) = \log 7 + \log x$ **b.** $\ln 87 - \ln x = \ln 29$

Solution ▶ **a.** $\log (x + 2) = \log 7 + \log x$ **b.** $\ln 87 - \ln x = \ln 29$

$\log (x + 2) = \log 7x$ properties of logarithms $\ln \left(\dfrac{87}{x} \right) = \ln 29$

WORTHY OF NOTE

The uniqueness property can also be viewed as exponentiating both sides using the appropriate base, then applying Property IV.

$x + 2 = 7x$ uniqueness property $\dfrac{87}{x} = 29$

$2 = 6x$ solve for x $87 = 29x$

$\dfrac{1}{3} = x$ result $3 = x$

The checks are left to the student.

Now try Exercises 81 through 86 ▶

Often the solution may depend on using a variety of algebraic skills in addition to logarithmic or exponential properties.

EXAMPLE 11 ▶ Solving Logarithmic Equations

Solve each equation and check your answers.

a. $\ln (x + 7) - 2 \ln 5 = 0.9$ **b.** $\log (x + 12) - \log x = \log (x + 9)$

Solution ▶   **a.** $\ln(x+7) - 2\ln 5 = 0.9$ given

$\qquad\quad \ln(x+7) - \ln 5^2 = 0.9$ power property

$\qquad\quad \ln\left(\dfrac{x+7}{25}\right) = 0.9$ quotient property

$\qquad\quad \dfrac{x+7}{25} = e^{0.9}$ exponential form

$\qquad\quad x+7 = 25e^{0.9}$ clear denominator

$\qquad\quad x = 25e^{0.9} - 7$ solve for x (exact form)

$\qquad\quad\quad \approx 54.49$ approximate form (to 100ths)

 Check: $\ln(x+7) - 2\ln 5 = 0.9$ original equation

$\qquad\quad \ln(54.49 + 7) - 2\ln 5 \approx 0.9$ substitute 54.49 for x

$\qquad\quad\quad \ln 61.49 - 2\ln 5 \approx 0.9$ simplify

$\qquad\quad\qquad\qquad 0.9 \approx 0.9$ ✓   result checks

b. $\log(x+12) - \log x = \log(x+9)$ given equation

$\qquad\quad \log\left(\dfrac{x+12}{x}\right) = \log(x+9)$ quotient property

$\qquad\quad \dfrac{x+12}{x} = x+9$ uniqueness property

$\qquad\quad x+12 = x^2 + 9x$ clear denominator

$\qquad\quad 0 = x^2 + 8x - 12$ set equal to 0

The equation is not factorable, and the quadratic formula must be used.

$x = \dfrac{-b \pm \sqrt{b^2 - 4ac}}{2a}$ quadratic formula

$ = \dfrac{-8 \pm \sqrt{(8)^2 - 4(1)(-12)}}{2(1)}$ substitute 1 for a, 8 for b, -12 for c

$ = \dfrac{-8 \pm \sqrt{112}}{2} = \dfrac{-8 \pm 4\sqrt{7}}{2}$ simplify

$ = -4 \pm 2\sqrt{7}$ result

Substitution shows $x = -4 + 2\sqrt{7}$ ($x \approx 1.29150$) checks, but substituting $-4 - 2\sqrt{7}$ for x gives $\log(2.7085) - \log(-9.2915) = \log(-0.2915)$ and two of the three terms do not represent real numbers ($x = -4 - 2\sqrt{7}$ is an extraneous root).

Now try Exercises 87 through 102 ▶

> **WORTHY OF NOTE**
>
> If all digits given by your calculator are used in the check, a calculator will generally produce "exact" answers. Try using the solution $x = 54.49007778$ in Example 11a by substituting directly, or by storing the result of the original computation and using your home screen.

GRAPHICAL SUPPORT

Logarithmic equations can also be checked using the intersection of graphs method. For Example 11b, we first enter $\log(x+12) - \log x$ as Y_1 and $\log(x+9)$ as Y_2 on the [Y =] screen. Using [2nd] [TRACE] **(CALC) 5:intersect**, we find the graphs intersect at $x = 1.2915026$, and that *this is the only solution* (knowing the graph's basic shape, we conclude they cannot intersect again).

> ⚠ **CAUTION** ▶ Be careful not to dismiss or discard a possible solution simply because it's negative. For the equation $\log(-6 - x) = 1$, $x = -16$ is the solution (the domain here allows negative numbers: $-6 - x > 0$ yields $x < -6$ as the domain). In general, when a logarithmic equation has multiple solutions, all solutions should be checked.

Solving exponential equations likewise involves isolating an exponential term on one side, or writing the equation where exponential terms of like base occur on each side. The latter case can be solved using the uniqueness property. If the exponential base is neither 10 nor e, logarithms of either base can be used along with the Power Property to solve the equation.

EXAMPLE 12 ▶ **Solving an Exponential Equation Using Base 10 or Base e**

Solve the exponential equation. Answer in both exact form, and approximate form to four decimal places: $4^{3x} - 1 = 8$

Solution ▶ $4^{3x} - 1 = 8$ given equation

$4^{3x} = 9$ add 1

WORTHY OF NOTE

The equation $\log 4^{3x} = \log 9$ from Example 12, can actually be solved using the change-of-base property, by taking logarithms base 4 of both sides.

$\log_4 4^{3x} = \log_4 9$ logarithms base 4

$3x = \dfrac{\log 9}{\log 4}$ Property III; change-of-base property

$x = \dfrac{\log 9}{3 \log 4}$ divide by 3

The left-hand side is neither base 10 or base e, so the choice is arbitrary. Here we chose base 10 to solve.

$\log 4^{3x} = \log 9$ take logarithm base 10 of both sides

$3x \log 4 = \log 9$ power property

$x = \dfrac{\log 9}{3 \log 4}$ divide by 3 log 4 (exact form)

$x \approx 0.5283$ approximate form

Now try Exercises 103 through 106 ▶

In some cases, two exponential terms with *unlike* bases may be involved. Here again, either common logs or natural logs can be used, but be sure to distinguish between constant terms like *ln 5* and variable terms like *x ln 5*. As with all equations, the goal is to isolate the *variable terms* on one side, with all constant terms on the other.

EXAMPLE 13 ▶ **Solving an Exponential Equation with Unlike Bases**

Solve the exponential equation $5^{x+1} = 6^{2x}$.

Solution ▶ $5^{x+1} = 6^{2x}$ original equation

Begin by taking the natural log of both sides:

$\ln(5^{x+1}) = \ln(6^{2x})$ apply base-e logarithms

$(x + 1)\ln 5 = 2x \ln 6$ power property

$x \ln 5 + \ln 5 = 2x \ln 6$ distribute

$\ln 5 = 2x \ln 6 - x \ln 5$ variable terms to one side

$\ln 5 = x(2 \ln 6 - \ln 5)$ factor out x

$\dfrac{\ln 5}{2 \ln 6 - \ln 5} = x$ solve for x (exact form)

$0.8153 \approx x$ approximate form

☑ **C. You've just learned how to solve general logarithmic and exponential equations**

The solution can be checked on a calculator.

Now try Exercises 107 through 110 ▶

D. Applications of Logistic, Exponential, and Logarithmic Functions

Applications of exponential and logarithmic functions take many different forms and it would be impossible to illustrate them all. As you work through the exercises, try to adopt a "big picture" approach, applying the general principles illustrated here to other applications. Some may have been introduced in previous sections. The difference here is that we can now *solve for the independent variable,* instead of simply evaluating the relationships.

In applications involving the **logistic growth** of animal populations, the initial stage of growth is virtually exponential, but due to limitations on food, space, or other resources, growth slows and at some point it reaches a limit. In business, the same principle applies to the logistic growth of sales or profits, due to market saturation. In these cases, the exponential term appears in the denominator of a quotient, and we "clear denominators" to begin the solution process.

EXAMPLE 14 ▶ Solving a Logistics Equation

A small business makes a new discovery and begins an aggressive advertising campaign, confident they can capture 66% of the market in a short period of time. They anticipate their market share will be modeled by the function

$M(t) = \dfrac{66}{1 + 10e^{-0.05t}}$, where $M(t)$ represents the percentage after t days. Use this function to answer the following.

a. What was the company's initial market share ($t = 0$)? What was their market share 30 days later?

b. How long will it take the company to reach a 60% market share?

Solution ▶ **a.** $M(t) = \dfrac{66}{1 + 10e^{-0.05t}}$ given

$M(0) = \dfrac{66}{1 + 10e^{-0.05(0)}}$ substitute 0 for t

$= \dfrac{66}{11}$ simplify

$= 6$ result

The company originally had only a 6% market share.

$M(30) = \dfrac{66}{1 + 10e^{-0.05(30)}}$ substitute 30 for t

$= \dfrac{66}{1 + 10e^{-1.5}}$ simplify

≈ 20.4 result

After 30 days, they held a 20.4% market share.

b. For Part b, we replace $M(t)$ with 60 and solve for t.

$60 = \dfrac{66}{1 + 10e^{-0.05t}}$ given

$60(1 + 10e^{-0.05t}) = 66$ multiply by $1 + 10e^{-0.05t}$

$1 + 10e^{-0.05t} = 1.1$ divide by 60

$$10e^{-0.05t} = 0.1 \qquad \text{subtract 1}$$
$$e^{-0.05t} = 0.01 \qquad \text{divide by 10}$$
$$\ln e^{-0.05t} = \ln 0.01 \qquad \text{apply base-}e\text{ logarithms}$$
$$-0.05t = \ln 0.01 \qquad \text{Property III}$$
$$t = \frac{\ln 0.01}{-0.05} \qquad \text{solve for } t \text{ (exact form)}$$
$$\approx 92 \qquad \text{approximate form}$$

According to this model, the company will reach a 60% market share in about 92 days.

Now try Exercises 111 through 116 ▶

Earlier we used the barometric equation $H = (30T + 8000)\ln\left(\dfrac{P_0}{P}\right)$ to find an altitude H, given a temperature and the atmospheric (barometric) pressure in centimeters of mercury (cmHg). Using the tools from this section, we are now able to find the atmospheric pressure for a given altitude and temperature.

EXAMPLE 15 ▶ Using Logarithms to Determine Atmospheric Pressure

Suppose a group of climbers has just scaled Mt. Rainier, the highest mountain of the Cascade Range in western Washington State. If the mountain is about 4395 m high and the temperature at the summit is $-22.5°C$, what is the atmospheric pressure at this altitude? The pressure at sea level is $P_0 = 76$ cmHg.

Solution ▶

$$H = (30T + 8000)\ln\left(\frac{P_0}{P}\right) \qquad \text{given}$$

$$4395 = [30(-22.5) + 8000]\ln\left(\frac{76}{P}\right) \qquad \text{substitute 4395 for } H, \text{ 76 for } P_0, \text{ and } -22.5 \text{ for } T$$

$$4395 = 7325\ln\left(\frac{76}{P}\right) \qquad \text{simplify}$$

$$0.6 = \ln\left(\frac{76}{P}\right) \qquad \text{divide by 7325}$$

$$e^{0.6} = \frac{76}{P} \qquad \text{exponential form}$$

$$Pe^{0.6} = 76 \qquad \text{multiply by } P$$

☑ D. You've just learned how to solve applications involving logistic, exponential, and logarithmic functions

$$P = \frac{76}{e^{0.6}} \qquad \text{divide by } e^{0.6} \text{ (exact form)}$$

$$\approx 41.7 \qquad \text{approximate form}$$

Under these conditions and at this altitude, the atmospheric pressure would be 41.7 cmHg.

Now try Exercises 117 through 120 ▶

4.4 EXERCISES

▶ CONCEPTS AND VOCABULARY

Fill in each blank with the appropriate word or phrase. Carefully reread the section if needed.

1. For $e^{-0.02x+1} = 10$, the solution process is most efficient if we apply a base _____ logarithm to both sides.

2. To solve $3 \ln x - \ln(x + 3) = 0$, we can combine terms using the _____ property, or add $\ln(x + 3)$ to both sides and use the _____ property.

3. Since logarithmic functions are not defined for all real numbers, we should check all "solutions" for _____ roots.

4. The statement $\log_e 10 = \dfrac{\log 10}{\log e}$ is an example of the _____ -of- _____ property.

5. Solve the equation here, giving a step-by-step discussion of the solution process: $\ln(4x + 3) + \ln(2) = 3.2$

6. Describe the difference between *evaluating* the equation below given $x = 9.7$ and *solving* the equation given $y = 9.7$: $y = 3 \log_2(x - 1.7) - 2.3$.

▶ DEVELOPING YOUR SKILLS

Solve each equation by applying fundamental properties. Round to thousandths.

7. $\ln x = 3.4$ 8. $\ln x = \frac{1}{2}$

9. $\log x = \frac{1}{4}$ 10. $\log x = 1.6$

11. $e^x = 9.025$ 12. $e^x = 0.343$

13. $10^x = 18.197$ 14. $10^x = 0.024$

Solve each equation. Write answers in exact form and in approximate form to four decimal places.

15. $4e^{x-2} + 5 = 70$ 16. $2 - 3e^{0.4x} = -7$

17. $10^{x+5} - 228 = -150$ 18. $10^{2x} + 27 = 190$

19. $-150 = 290.8 - 190e^{-0.75x}$

20. $250e^{0.05x+1} + 175 = 1175$

Solve each equation. Write answers in exact form and in approximate form to four decimal places.

21. $3 \ln(x + 4) - 5 = 3$ 22. $-15 = -8 \ln(3x) + 7$

23. $-1.5 = 2 \log(5 - x) - 4$

24. $-4 \log(2x) + 9 = 3.6$

25. $\frac{1}{2} \ln(2x + 5) + 3 = 3.2$

26. $\frac{3}{4} \ln(4x) - 6.9 = -5.1$

Use properties of logarithms to write each expression as a single term.

27. $\ln(2x) + \ln(x - 7)$ 28. $\ln(x + 2) + \ln(3x)$

29. $\log(x + 1) + \log(x - 1)$

30. $\log(x - 3) + \log(x + 3)$

31. $\log_3 28 - \log_3 7$ 32. $\log_6 30 - \log_6 10$

33. $\log x - \log(x + 1)$ 34. $\log(x - 2) - \log x$

35. $\ln(x - 5) - \ln x$ 36. $\ln(x + 3) - \ln(x - 1)$

37. $\ln(x^2 - 4) - \ln(x + 2)$

38. $\ln(x^2 - 25) - \ln(x + 5)$

39. $\log_2 7 + \log_2 6$ 40. $\log_9 2 + \log_9 15$

41. $\log_5(x^2 - 2x) + \log_5 x^{-1}$

42. $\log_3(3x^2 + 5x) - \log_3 x$

Use the power property of logarithms to rewrite each term as the product of a constant and a logarithmic term.

43. $\log 8^{x+2}$ 44. $\log 15^{x-3}$

45. $\ln 5^{2x-1}$ 46. $\ln 10^{3x+2}$

47. $\log \sqrt{22}$ 48. $\log \sqrt[3]{34}$

49. $\log_5 81$ 50. $\log_7 121$

Use the properties of logarithms to write the following expressions as a sum or difference of simple logarithmic terms.

51. $\log(a^3 b)$

52. $\log(m^2 n)$

53. $\ln(x \sqrt[4]{y})$

54. $\ln(\sqrt[3]{pq})$

55. $\ln\left(\dfrac{x^2}{y}\right)$

56. $\ln\left(\dfrac{m^2}{n^3}\right)$

57. $\log\left(\sqrt{\dfrac{x-2}{x}}\right)$

58. $\log\left(\sqrt[3]{\dfrac{3-v}{2v}}\right)$

59. $\ln\left(\dfrac{7x\sqrt{3-4x}}{2(x-1)^3}\right)$

60. $\ln\left(\dfrac{x^4\sqrt{x^2-4}}{\sqrt[3]{x^2+5}}\right)$

Evaluate each expression using the change-of-base formula and either base 10 or base e. Answer in exact form and in approximate form using nine decimal places, then verify the result using the original base.

61. $\log_7 60$

62. $\log_8 92$

63. $\log_5 152$

64. $\log_6 200$

65. $\log_3 1.73205$

66. $\log_2 1.41421$

67. $\log_{0.5} 0.125$

68. $\log_{0.2} 0.008$

Use the change-of-base formula to write an equivalent function, then evaluate the function as indicated (round to four decimal places). Investigate and discuss any patterns you notice in the output values, then determine the next input that will continue the pattern.

69. $f(x) = \log_3 x; f(5), f(15), f(45)$

70. $g(x) = \log_2 x; g(5), g(10), g(20)$

71. $h(x) = \log_9 x; h(2), h(4), h(8)$

72. $H(x) = \log_\pi x; H(\sqrt{2}), H(2), H(\sqrt{2^3})$

Solve each equation and check your answers.

73. $\log 4 + \log(x-7) = 2$

74. $\log 5 + \log(x-9) = 1$

75. $\log(2x-5) - \log 78 = -1$

76. $\log(4-3x) - \log 145 = -2$

77. $\log(x-15) - 2 = -\log x$

78. $\log x - 1 = -\log(x-9)$

79. $\log(2x+1) = 1 - \log x$

80. $\log(3x-13) = 2 - \log x$

Solve each equation using the uniqueness property of logarithms.

81. $\log(5x+2) = \log 2$

82. $\log(2x-3) = \log 3$

83. $\log_4(x+2) - \log_4 3 = \log_4(x-1)$

84. $\log_3(x+6) - \log_3 x = \log_3 5$

85. $\ln(8x-4) = \ln 2 + \ln x$

86. $\ln(x-1) + \ln 6 = \ln(3x)$

Solve each logarithmic equation using any appropriate method. Clearly identify any extraneous roots. If there are no solutions, so state.

87. $\log(2x-1) + \log 5 = 1$

88. $\log(x-7) + \log 3 = 2$

89. $\log_2(9) + \log_2(x+3) = 3$

90. $\log_3(x-4) + \log_3(7) = 2$

91. $\ln(x+7) + \ln 9 = 2$

92. $\ln 5 + \ln(x-2) = 1$

93. $\log(x+8) + \log x = \log(x+18)$

94. $\log(x+14) - \log x = \log(x+6)$

95. $\ln(2x+1) = 3 + \ln 6$

96. $\ln 21 = 1 + \ln(x-2)$

97. $\log(-x-1) = \log(5x) - \log x$

98. $\log(1-x) + \log x = \log(x+4)$

99. $\ln(2t+7) = \ln 3 - \ln(t+1)$

100. $\ln 6 - \ln(5-r) = \ln(r+2)$

101. $\log(x-1) - \log x = \log(x-3)$

102. $\ln x + \ln(x-2) = \ln 4$

103. $7^{x+2} = 231$ **104.** $6^{x+2} = 3589$

105. $5^{3x-2} = 128{,}965$ **106.** $9^{5x-3} = 78{,}462$

107. $2^{x+1} = 3^x$ **108.** $7^x = 4^{2x-1}$

109. $5^{2x+1} = 9^{x+1}$ **110.** $\left(\dfrac{1}{5}\right)^{x-1} = \left(\dfrac{1}{2}\right)^{3-x}$

111. $\dfrac{250}{1+4e^{-0.06x}} = 200$ **112.** $\dfrac{80}{1+15e^{-0.06x}} = 50$

▶ **WORKING WITH FORMULAS**

113. Logistic growth: $P(t) = \dfrac{C}{1 + ae^{-kt}}$

For populations that exhibit logistic growth, the population at time t is modeled by the function shown, where C is the carrying capacity of the population (the maximum population that can be supported over a long period of time), k is the growth constant, and $a = \dfrac{C - P(0)}{P(0)}$. Solve the formula for t, then use the result to find the value of t given $C = 450$, $a = 8$, $P = 400$, and $k = 0.075$.

114. Forensics—estimating time of death:

$$h = -3.9 \cdot \ln\left(\frac{T - T_R}{T_0 - T_R}\right)$$

Using the formula shown, a forensic expert can compute the approximate time of death for a person found recently expired, where T is the body temperature when it was found, T_R is the (constant) temperature of the room, T_0 is the body temperature at the time of death ($T_0 = 98.6°F$), and h is the number of hours since death. If the body was discovered at 9:00 A.M. with a temperature of 86.2°F, in a room at 73°F, at approximately what time did the person expire? (Note this formula is a version of Newton's law of cooling.)

▶ **APPLICATIONS**

115. Stocking a lake: A farmer wants to stock a private lake on his property with catfish. A specialist studies the area and depth of the lake, along with other factors, and determines it can support a maximum population of around 750 fish, with growth modeled by the function $P(t) = \dfrac{750}{1 + 24e^{-0.075t}}$, where $P(t)$ gives the current population after t months.
(a) How many catfish did the farmer initially put in the lake? (b) How many months until the population reaches 300 fish?

116. Increasing sales: After expanding their area of operations, a manufacturer of small storage buildings believes the larger area can support sales of 40 units per month. After increasing the advertising budget and enlarging the sales force, sales are expected to grow according to the model $S(t) = \dfrac{40}{1 + 1.5e^{-0.08t}}$, where $S(t)$ is the expected number of sales after t months. (a) How many sales were being made each month, prior to the expansion? (b) How many months until sales reach 25 units per month?

Use the *barometric equation* $H = (30T + 8000)\ln\left(\dfrac{P_0}{P}\right)$ for exercises 117 and 118. Recall that $P_0 = 76$ cmHg.

117. Altitude and temperature: A sophisticated spy plane is cruising at an altitude of 18,250 m. If the temperature at this altitude is −75°C, what is the barometric pressure?

118. Altitude and temperature: A large weather balloon is released and takes altitude, pressure, and temperature readings as it climbs, and radios the information back to Earth. What is the pressure reading at an altitude of 5000 m, given the temperature is −18°C?

Use *Newton's law of cooling* $T = T_R + (T_0 - T_R)e^{kh}$ to complete Exercises 119 and 120. Recall that water freezes at 32°F and use $k = -0.012$. Refer to Section 4.2, page 430 as needed.

119. Making popsicles: On a hot summer day, Sean and his friends mix some Kool-Aid® and decide to freeze it in an ice tray to make popsicles. If the water used for the Kool-Aid® was 75°F and the freezer has a temperature of −20°F, how long will they have to wait to enjoy the treat?

120. Freezing time: Suppose the current temperature in Esconabe, Michigan, was 47°F when a 5°F arctic cold front moved over the state. How long would it take a puddle of water to freeze over?

Depreciation/appreciation: As time passes, the value of certain items decrease (appliances, automobiles, etc.), while the value of other items increase (collectibles, real estate, etc.). The time T in years for an item to reach a future value can be modeled by the formula $T = k \ln\left(\dfrac{V_n}{V_f}\right)$, where V_n is the purchase price when new, V_f is its future value, and k is a constant that depends on the item.

121. Automobile depreciation: If a new car is purchased for $28,500, find its value 3 yr later if $k = 5$.

122. Home appreciation: If a new home in an "upscale" neighborhood is purchased for $130,000, find its value 12 yr later if $k = -16$.

Drug absorption: The time required for a certain percentage of a drug to be *absorbed* by the body depends on the drug's absorption rate. This can be modeled by the function $T(p) = \dfrac{-\ln p}{k}$, where p represents the percent of the drug that *remains unabsorbed* (expressed as a decimal), k is the absorption rate of the drug, and $T(p)$ represents the elapsed time.

123. For a drug with an absorption rate of 7.2%, (a) find the time required (to the nearest hour) for the body to *absorb* 35% of the drug, and (b) find the percent of this drug (to the nearest half percent) that remains unabsorbed after 24 hr.

124. For a drug with an absorption rate of 5.7%, (a) find the time required (to the nearest hour) for the body to *absorb* 50% of the drug, and (b) find the percent of this drug (to the nearest half percent) that remains unabsorbed after 24 hr.

Spaceship velocity: In space travel, the change in the velocity of a spaceship V_s (in km/sec) depends on the mass of the ship M_s (in tons), the mass of the fuel which has been burned M_f (in tons) and the escape velocity of the exhaust V_e (in km/sec). Disregarding frictional forces, these are related by the equation
$$V_s = V_e \ln\left(\frac{M_s}{M_s - M_f}\right).$$

125. For the Jupiter VII rocket, find the mass of the fuel M_f that has been burned if $V_s = 6$ km/sec when $V_e = 8$ km/sec, and the ship's mass is 100 tons.

126. For the Neptune X satellite booster, find the mass of the ship M_s if $M_f = 75$ tons of fuel has been burned when $V_s = 8$ km/sec and $V_e = 10$ km/sec.

Learning curve: The job performance of a new employee when learning a repetitive task (as on an assembly line) improves very quickly at first, then grows more slowly over time. This can be modeled by the function $P(t) = a + b \ln t$, where a and b are constants that depend on the type of task and the training of the employee.

127. The number of toy planes an employee can assemble from its component parts depends on the length of time the employee has been working. This output is modeled by $P(t) = 5.9 + 12.6 \ln t$, where $P(t)$ is the number of planes assembled daily after working t days. (a) How many planes is an employee making after 5 days on the job? (b) How many days until the employee is able to assemble 34 planes per day?

128. The number of circuit boards an associate can assemble from its component parts depends on the length of time the associate has been working. This output is modeled by $B(t) = 1 + 2.3 \ln t$, where $B(t)$ is the number of boards assembled daily after working t days. (a) How many boards is an employee completing after 9 days on the job? (b) How long will it take until the employee is able to complete 10 boards per day?

▶ **EXTENDING THE CONCEPT**

Use prime factors, properties of logs, and the values given to evaluate each expression without a calculator. Check each result using the change-of-base formula:

129. $\log_3 4 = 1.2619$ and $\log_3 5 = 1.4649$:

    a. $\log_3 20$

    b. $\log_3 \dfrac{4}{5}$

    c. $\log_3 25$

130. $\log_5 2 \approx 0.4307$ and $\log_5 3 \approx 0.6826$:

    a. $\log_5 \dfrac{9}{2}$

    b. $\log_5 216$

    c. $\log_5 \sqrt[3]{6}$

131. Match each equation with the most appropriate solution strategy, and justify/discuss why.

    a. $e^{x+1} = 25$ ____   apply base-10 logarithm to both sides

    b. $\log(2x + 3) = \log 53$ ____   rewrite and apply uniqueness property for exponentials

    c. $\log(x^2 - 3x) = 2$ ____   apply uniqueness property for logarithms

    d. $10^{2x} = 97$ ____   apply either base-10 or base-e logarithm

    e. $2^{5x-3} = 32$ ____   apply base-e logarithm

    f. $7^{x+2} = 23$ ____   write in exponential form

Solve the following equations. Note that equations Exercises 132 and 133 are in quadratic form.

132. $2e^{2x} - 7e^x = 15$

133. $3e^{2x} - 4e^x - 7 = -3$

134. $\log_2(x + 5) = \log_4(21x + 1)$

135. Show that $g(x) = f^{-1}(x)$ by composing the functions.

    a. $f(x) = 3^{x-2}$; $g(x) = \log_3 x + 2$

    b. $f(x) = e^{x-1}$; $g(x) = \ln x + 1$

136. Use the algebraic method to find the inverse function.

    a. $f(x) = 2^{x+1}$ **b.** $y = 2\ln(x - 3)$

137. Use properties of logarithms and/or exponents to show

    a. $y = 2^x$ is equivalent to $y = e^{x \ln 2}$.

    b. $y = b^x$ is equivalent to $y = e^{rx}$, where $r = \ln b$.

138. To understand the formula for the half-life of radioactive material, consider that for each time increment, a constant proportion of mass m is lost. In symbols; $m(t + 1) - m(t) = -km(t)$. (a) Solve for $m(t + 1)$ and factor the right-hand side. (b) Evaluate the new equation for $t = 0, 1, 2,$ and 3, to show that $m(t) = m(0)(1 - k)^t$. (c) For any half-life h, we have $m(h) = m(0)(1 - k)^h = \frac{1}{2}m(0)$. Solve for $1 - k$, raise both sides to the power t, and substitute to show $m(t) = m(0)(\frac{1}{2})^{\frac{t}{h}}$.

139. Use test values for p and q to demonstrate that the following relationships are *false*, then state the correct property and use the same test value to verify the property.

    a. $\ln(pq) = \ln p \ln q$

    b. $\ln\left(\dfrac{p}{q}\right) = \dfrac{\ln p}{\ln q}$

    c. $\ln p + \ln q = \ln(p + q)$

140. Verify that $\ln x = (\ln 10)(\log x)$, and discuss *why* they're equal. Then use the relationship to find the value of $\ln e$, $\ln 10$, and $\ln 2$.

► **MAINTAINING YOUR SKILLS**

141. (2.4) Match the graph shown with its correct equation, without actually graphing the function.

    a. $y = x^2 + 4x - 5$

    b. $y = -x^2 - 4x + 5$

    c. $y = -x^2 + 4x + 5$

    d. $y = x^2 - 4x - 5$

142. (3.3) State the domain and range of the functions.

    a. $y = \sqrt{2x + 3}$ **b.** $y = |x + 2| - 3$

143. (4.6) Graph the function $r(x) = \dfrac{x^2 - 4}{x - 1}$. Label all intercepts and asymptotes.

144. (3.6) Suppose the maximum load (in tons) that can be supported by a cylindrical post varies directly with its diameter raised to the fourth power and inversely as the square of its height. A post 8 ft high and 2 ft in diameter can support 6 tons. How many tons can be supported by a post 12 ft high and 3 ft in diameter?

4.5 | Applications from Business, Finance, and Science

Learning Objectives

In Section 4.5 you will learn how to:

☐ **A.** Calculate simple interest and compound interest

☐ **B.** Calculate interest compounded continuously

☐ **C.** Solve applications of annuities and amortization

☐ **D.** Solve applications of exponential growth and decay

Would you pay $750,000 for a home worth only $250,000? Surprisingly, when a conventional mortgage is repaid over 30 years, this is not at all rare. Over time, the accumulated interest on the mortgage is easily more than two or three times the original value of the house. In this section we explore how interest is paid or charged, and look at other applications of exponential and logarithmic functions from business, finance, as well as the physical and social sciences.

A. Simple and Compound Interest

Simple interest is an amount of interest that is computed only once during the lifetime of an investment (or loan). In the world of finance, the initial deposit or base amount is referred to as the **principal p,** the **interest rate r** is given as a percentage and stated as an annual rate, with the term of the investment or loan most often given as *time t* in years. Simple interest is merely an application of the basic percent equation, with the additional element of time coming into play: *interest = principal × rate × time*, or $I = prt$. To find the total amount A that has accumulated (for deposits) or is due (for loans) after t years, we merely add the accumulated interest to the initial principal: $A = p + prt$.

> **WORTHY OF NOTE**
>
> If a loan is kept for only a certain number of months, weeks, or days, the time t should be stated as a fractional part of a year so the time period for the rate (years) matches the time period over which the loan is repaid.

Simple Interest Formula

If principal p is deposited or borrowed at interest rate r for a period of t years, the simple interest on this account will be

$$I = prt$$

The total amount A accumulated or due after this period will be:

$$A = p + prt \qquad \text{or} \qquad A = p(1 + rt)$$

EXAMPLE 1 ▶ **Solving an Application of Simple Interest**

Many finance companies offer what have become known as *PayDay Loans*—a small $50 loan to help people get by until payday, usually no longer than 2 weeks. If the cost of this service is $12.50, determine the annual rate of interest charged by these companies.

Solution ▶ The interest charge is $12.50, the initial principal is $50.00, and the time period is 2 weeks or $\frac{2}{52} = \frac{1}{26}$ of a year. The simple interest formula yields

$$I = prt \qquad \text{simple interest formula}$$

$$12.50 = 50r\left(\frac{1}{26}\right) \qquad \text{substitute \$12.50 for } I, \text{\$50.00 for } p, \text{ and } \tfrac{1}{26} \text{ for } t$$

$$6.5 = r \qquad \text{solve for } r$$

The annual interest rate on these loans is a whopping 650%!

Now try Exercises 7 through 16 ▶

Compound Interest

Many financial institutions pay **compound interest** on deposits they receive, which is interest paid on previously accumulated interest. The most common compounding periods are yearly, semiannually (two times per year), quarterly (four times per year), monthly (12 times per year), and daily (365 times per year). Applications of compound interest typically involve exponential functions. For convenience, consider $1000 in principal, deposited at 8% for 3 yr. The simple interest calculation shows $240 in interest is earned and there will be $1240 in the account: $A = 1000[1 + (0.08)(3)] = \$1240$. If the interest is *compounded each year* $(t = 1)$ instead of once at the start of the 3-yr period, the interest calculation shows

$A_1 = 1000(1 + 0.08) = 1080$ in the account at the end of year 1,
$A_2 = 1080(1 + 0.08) = 1166.40$ in the account at the end of year 2,
$A_3 = 1166.40(1 + 0.08) \approx 1259.71$ in the account at the end of year 3.

The account has earned an additional $19.71 interest. More importantly, notice that we're multiplying by $(1 + 0.08)$ each compounding period, meaning results can be computed more efficiently by simply applying the factor $(1 + 0.08)^t$ to the initial principal p. For example,

$$A_3 = 1000(1 + 0.08)^3 \approx \$1259.71.$$

In general, for interest compounded yearly the **accumulated value** is $A = p(1 + r)^t$. Notice that solving this equation for p will tell us the amount we need to deposit *now*, in order to accumulate A dollars in t years: $p = \frac{A}{(1 + r)^t}$. This is called the **present value equation.**

Interest Compounded Annually

If a principal p is deposited at interest rate r and compounded yearly for a period of t yr, the ***accumulated value*** is

$$A = p(1 + r)^t$$

If an accumulated value A is desired after t yr, and the money is deposited at interest rate r and compounded yearly, the *present value* is

$$p = \frac{A}{(1 + r)^t}$$

EXAMPLE 2 ▶   Finding the Doubling Time of an Investment

An initial deposit of $1000 is made into an account paying 6% compounded yearly. How long will it take for the money to double?

Solution ▶   Using the formula for interest compounded yearly we have

$A = p(1 + r)^t$	given
$2000 = 1000(1 + 0.06)^t$	substitute 2000 for A, 1000 for p, and 0.06 for r
$2 = 1.06^t$	isolate variable term
$\ln 2 = t \ln 1.06$	apply base-e logarithms; power property
$\dfrac{\ln 2}{\ln 1.06} = t$	solve for t
$11.9 \approx t$	approximate form

The money will double in just under 12 yr.

Now try Exercises 17 through 22 ▶

If interest is compounded monthly (12 times each year), the bank will divide the interest rate by 12 (the number of compoundings), but then pay you interest 12 times per year (interest is *compounded*). The net effect is an increased gain in the interest you earn, and the final compound interest formula takes this form:

$$\text{total amount} = \text{principal}\left(1 + \frac{\text{interest rate}}{\text{compoundings per year}}\right)^{(\text{years} \times \text{compoundings per year})}$$

Compounded Interest Formula

If principal p is deposited at interest rate r and compounded n times per year for a period of t yr, the *accumulated value* will be:

$$A = p\left(1 + \frac{r}{n}\right)^{nt}$$

EXAMPLE 3 ▶ **Solving an Application of Compound Interest**

Macalyn won $150,000 in the Missouri lottery and decides to invest the money for retirement in 20 yr. Of all the options available here, which one will produce the most money for retirement?

  a. A certificate of deposit paying 5.4% compounded yearly.
  b. A money market certificate paying 5.35% compounded semiannually.
  c. A bank account paying 5.25% compounded quarterly.
  d. A bond issue paying 5.2% compounded daily.

Solution ▶ **a.** $A = \$150{,}000\left(1 + \dfrac{0.054}{1}\right)^{(20 \times 1)}$

$\approx \$429{,}440.97$

  b. $A = \$150{,}000\left(1 + \dfrac{0.0535}{2}\right)^{(20 \times 2)}$

$\approx \$431{,}200.96$

  c. $A = \$150{,}000\left(1 + \dfrac{0.0525}{4}\right)^{(20 \times 4)}$

$\approx \$425{,}729.59$

  d. $A = \$150{,}000\left(1 + \dfrac{0.052}{365}\right)^{(20 \times 365)}$

☑ **A.** You've just learned how to calculate simple interest and compound interest

$\approx \$424{,}351.12$

The best choice is (b), semiannual compounding at 5.35% for 20 yr.

Now try Exercises 23 through 30 ▶

B. Interest Compounded Continuously

It seems natural to wonder what happens to the interest accumulation as n (the number of compounding periods) becomes very large. It appears the interest rate becomes very small (because we're dividing by n), but the exponent becomes very large (since we're multiplying by n). To see the result of this interplay more clearly, it will help to rewrite the compound interest formula $A = p(1 + \frac{r}{n})^{nt}$ using the substitution $n = xr$. This gives $\frac{r}{n} = \frac{1}{x}$, and by direct substitution (xr for n and $\frac{1}{x}$ for $\frac{r}{n}$) we obtain the form

$$A = p\left[\left(1 + \frac{1}{x}\right)^{x}\right]^{rt}$$

by regrouping. This allows for a more careful study of the "denominator versus exponent" relationship using $(1 + \frac{1}{x})^{x}$, *the same expression we used in Section 4.2 to define the number e* (also **see Section 4.2 Exercise 97**). Once again, note what happens as $x \to \infty$ (meaning the number of compounding periods increase without bound).

x	1	10	100	1000	10,000	100,000	1,000,000
$\left(1 + \dfrac{1}{x}\right)^x$	2	2.56374	2.70481	2.71692	2.71815	2.71827	2.71828

As before, we have, as $x \to \infty$, $(1 + \frac{1}{x})^x \to e$. The net result of this investigation is a formula for **interest compounded continuously,** derived by replacing $(1 + \frac{1}{x})^x$ with the number e in the formula for compound interest, where

$$A = p\left[\left(1 + \frac{1}{x}\right)^x\right]^{rt} = pe^{rt}$$

Interest Compounded Continuously

If a principal p is deposited at interest rate r and compounded continuously for a period of t years, the *accumulated value* will be

$$A = pe^{rt}$$

EXAMPLE 4 ▶ **Solving an Application of Interest Compounded Continuously**

Jaimin has $10,000 to invest and wants to have at least $25,000 in the account in 10 yr for his daughter's college education fund. If the account pays interest compounded continuously, what interest rate is required?

Solution ▶ In this case, $P = \$10,000$, $A = \$25,000$, and $t = 10$.

$A = pe^{rt}$ — given

$25,000 = 10,000e^{10r}$ — substitute 25,000 for A, 10,000 for p, and 10 for t

$2.5 = e^{10r}$ — isolate variable term

$\ln 2.5 = 10r \ln e$ — apply base-e logarithms ($\ln e = 1$); power property

$\dfrac{\ln 2.5}{10} = r$ — solve for r

$0.092 \approx r$ — approximate form

Jaimin will need an interest rate of about 9.2% to meet his goal.

Now try Exercises 31 through 40 ▶

☑ **B.** You've just learned how to calculate interest compounded continuously

GRAPHICAL SUPPORT

To check the result from Example 4, use $Y_1 = 10{,}000e^{10x}$ and $Y_2 = 25{,}000$, then look for their point of intersection. We need only set an appropriate window size to ensure the answer will appear in the viewing window. Since 25,000 is the goal, $y \in [0, 30{,}000]$ seems reasonable for y. Although 12% interest ($x = 0.12$) is too good to be true, $x \in [0, 0.12]$ leaves a nice frame for the x-values. Verify that the calculator's answer is equal to $\frac{\ln 2.5}{10}$.

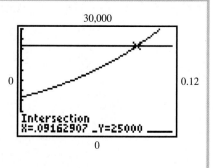

C. Applications Involving Annuities and Amortization

Our previous calculations for simple and compound interest involved a single (lump) deposit (the principal) that accumulated interest over time. Many savings and investment plans involve a regular schedule of deposits (monthly, quarterly, or annual deposits) over the life of the investment. Such an investment plan is called an **annuity.**

Suppose that for 4 yr, $100 is deposited annually into an account paying 8% compounded yearly. Using the compound interest formula we can track the accumulated value A in the account:

$$A = 100 + 100(1.08)^1 + 100(1.08)^2 + 100(1.08)^3$$

To develop an annuity formula, we multiply the annuity equation by 1.08, then subtract the original equation. This leaves only the first and last terms, since the other (interior) terms add to zero:

$$1.08A = 100(1.08) + 100(1.08)^2 + 100(1.08)^3 + 100(1.08)^4 \quad \text{multiply by 1.08}$$
$$-A = -[100 + 100(1.08)^1 + 100(1.08)^2 + 100(1.08)^3] \quad \text{original equation}$$
$$1.08A - A = 100(1.08)^4 - 100 \quad \text{subtract ("interior terms" sum to zero)}$$
$$0.08A = 100[(1.08)^4 - 1] \quad \text{factor out 100}$$
$$A = \frac{100[(1.08)^4 - 1]}{0.08} \quad \text{solve for } A$$

This result can be generalized for any periodic payment $\mathcal{P}$, interest rate r, number of compounding periods n, and number of years t. This would give

$$A = \frac{\mathcal{P}\left[\left(1 + \dfrac{r}{n}\right)^{nt} - 1\right]}{\dfrac{r}{n}}$$

The formula can be made less formidable using $R = \frac{r}{n}$, where R is the interest rate per compounding period.

Accumulated Value of an Annuity

If a periodic payment $\mathcal{P}$ is deposited n times per year at an *annual interest rate r* with interest compounded n times per year for t years, the accumulated value is given by

$$A = \frac{\mathcal{P}}{R}[(1 + R)^{nt} - 1], \text{ where } R = \frac{r}{n}$$

This is also referred to as the **future value** of the account.

EXAMPLE 5 ▶ **Solving an Application of Annuities**

Since he was a young child, Fitisemanu's parents have been depositing $50 each month into an annuity that pays 6% annually and is compounded monthly. If the account is now worth $9875, how long has it been open?

Solution ▶ In this case $\mathcal{P} = 50$, $r = 0.06$, $n = 12$, $R = 0.005$, and $A = 9875$. The formula gives

$$A = \frac{\mathcal{P}}{R}[(1 + R)^{nt} - 1] \qquad \text{future value formula}$$

$$9875 = \frac{50}{0.005}[(1.005)^{(12)(t)} - 1] \qquad \text{substitute 9875 for } A, \text{ 50 for } \mathcal{P}, \text{ 0.005 for } R, \text{ and 12 for } n$$

$$1.9875 = 1.005^{12t} \qquad \text{simplify and isolate variable term}$$

$$\ln(1.9875) = 12t(\ln 1.005) \quad \text{apply base-}e\text{ logarithms; power property}$$

$$\frac{\ln(1.9875)}{12 \ln(1.005)} = t \quad \text{solve for } t \text{ (exact form)}$$

$$11.5 \approx t \quad \text{approximate form}$$

The account has been open approximately 11.5 yr.

Now try Exercises 41 through 44 ▶

The periodic payment required to meet a future goal or obligation can be computed by solving for $\mathcal{P}$ in the future value formula: $\mathcal{P} = \dfrac{AR}{\left[(1 + R)^{nt} - 1\right]}$. In this form, $\mathcal{P}$ is referred to as a **sinking fund.**

EXAMPLE 6 ▶   Solving an Application of Sinking Funds

Sheila is determined to stay out of debt and decides to save $20,000 to pay cash for a new car in 4 yr. The best investment vehicle she can find pays 9% compounded monthly. If $300 is the most she can invest each month, can she meet her "4-yr" goal?

Solution ▶   Here we have $\mathcal{P} = 300$, $A = 20{,}000$, $r = 0.09$, $n = 12$, and $R = 0.0075$. The sinking fund formula gives

$$\mathcal{P} = \frac{AR}{\left[(1 + R)^{nt} - 1\right]} \quad \text{sinking fund}$$

$$300 = \frac{(20{,}000)(0.0075)}{(1.0075)^{12t} - 1} \quad \begin{array}{l}\text{substitute 300 for } \mathcal{P}, 20{,}000 \text{ for } A, \\ 0.0075 \text{ for } R, \text{ and 12 for } n\end{array}$$

$$300(1.0075^{12t} - 1) = 150 \quad \text{multiply in numerator, clear denominators}$$

$$1.0075^{12t} = 1.5 \quad \text{isolate variable term}$$

$$12t \ln(1.0075) = \ln 1.5 \quad \text{apply base-}e\text{ logarithms; power property}$$

$$t = \frac{\ln(1.5)}{12 \ln(1.0075)} \quad \text{solve for } t \text{ (exact form)}$$

$$\approx 4.5 \quad \text{approximate form}$$

No. She is close, but misses her original 4-yr goal.

Now try Exercises 45 and 46 ▶

☑ **C. You've just learned how to solve applications of annuities and amortization**

For Example 6, we could have substituted 4 for t and left $\mathcal{P}$ unknown, to see if a payment of $300 per month would be sufficient. You can verify the result would be $\mathcal{P} \approx \$347.70$, which is what Sheila would need to invest to meet her 4-yr goal exactly.

For additional practice with the formulas for interest earned or paid, the *Working with Formulas* portion of this Exercise Set has been expanded. See **Exercises 47 through 54.**

WORTHY OF NOTE

Notice the formula for exponential growth is virtually identical to the formula for interest compounded continuously. In fact, both are based on the same principles. If we let $A(t)$ represent the amount in an account after t years and A_0 represent the initial deposit (instead of P), we have: $A(t) = A_0 e^{rt}$ versus $Q(t) = Q_0 e^{rt}$ and the two cannot be distinguished.

D. Applications Involving Exponential Growth and Decay

Closely related to interest compounded continuously are applications of **exponential growth** and **exponential decay**. If Q (quantity) and t (time) are variables, then Q grows exponentially as a function of t if $Q(t) = Q_0 e^{rt}$ for positive constants Q_0 and r. Careful studies have shown that population growth, whether it be humans, bats, or bacteria, can be modeled by these "base-e" exponential growth functions. If $Q(t) = Q_0 e^{-rt}$, then we say Q decreases or **decays exponentially** over time. The constant r determines how rapidly a quantity grows or decays and is known as the **growth rate** or **decay rate** constant.

EXAMPLE 7 ▶ **Solving an Application of Exponential Growth**

Because fruit flies multiply very quickly, they are often used in a study of genetics. Given the necessary space and food supply, a certain population of fruit flies is known to double every 12 days. If there were 100 flies to begin, find (a) the growth rate r and (b) the number of days until the population reaches 2000 flies.

Solution ▶

a. Using the formula for exponential growth with $Q_0 = 100$, $t = 12$, and $Q(t) = 200$, we can solve for the growth rate r.

$$Q(t) = Q_0 e^{rt}$$ exponential growth function

$$200 = 100 e^{12r}$$ substitute 200 for $Q(t)$ 100 for Q_0, and 12 for t

$$2 = e^{12r}$$ isolate variable term

$$\ln 2 = 12r \ln e$$ apply base-e logarithms; power property

$$\frac{\ln 2}{12} = r$$ solve for r (exact form)

$$0.05776 \approx r$$ approximate form

The growth rate is approximately 5.78%.

b. To find the number of days until the fly population reaches 2000, we substitute 0.05776 for r in the exponential growth function.

$$Q(t) = Q_0 e^{rt}$$ exponential growth function

$$2000 = 100 e^{0.05776t}$$ substitute 2000 for $Q(t)$, 100 for Q_0, and 0.05776 for r

$$20 = e^{0.05776t}$$ isolate variable term

$$\ln 20 = 0.05776t \ln e$$ apply base-e logarithms; power property

$$\frac{\ln 20}{0.05776} = t$$ solve for t (exact form)

$$51.87 \approx t$$ approximate form

The fruit fly population will reach 2000 on day 51.

Now try Exercises 55 and 56 ▶

WORTHY OF NOTE

Many population growth models assume an unlimited supply of resources, nutrients, and room for growth. When this is not the case, a logistic growth model often results. See the *Modeling with Technology* feature following this chapter.

Perhaps the best-known examples of exponential decay involve radioactivity. Ever since the end of World War II, common citizens have been aware of the existence of **radioactive elements** and the power of atomic energy. Today, hundreds of additional applications have been found for these materials, from areas as diverse as biological research, radiology, medicine, and archeology. Radioactive elements decay of their own accord by emitting radiation. The rate of decay is measured using the **half-life** of the substance, which is the time required for a mass of radioactive material to decay until only one-half of its original mass remains. This half-life is used to find the rate of decay r, first mentioned in Section 4.4. In general, if h represents the half-life of the substance, one-half the initial amount remains when $t = h$.

$$Q(t) = Q_0 e^{-rt}$$ exponential decay function

$$\frac{1}{2} Q_0 = Q_0 e^{-rh}$$ substitute $\frac{1}{2} Q_0$ for $Q(t)$, h for t

$$\frac{1}{2} = \frac{1}{e^{rh}}$$ divide by Q_0; rewrite expression

$$2 = e^{rh}$$ property of ratios

$$\ln 2 = rh \ln e$$ apply base-e logarithms; power property

$$\frac{\ln 2}{h} = r$$ solve for r

Radioactive Rate of Decay

If h represents the half-life of a radioactive substance per unit time, the nominal rate of decay per a like unit of time is given by

$$r = \frac{\ln 2}{h}$$

The rate of decay for known radioactive elements varies greatly. For example, the element carbon-14 has a half-life of about 5730 yr, while the element lead-211 has a half-life of only about 3.5 min. Radioactive elements can be detected in extremely small amounts. If a drug is "labeled" (mixed with) a radioactive element and injected into a living organism, its passage through the organism can be traced and information on the health of internal organs can be obtained.

EXAMPLE 8 ▶ **Solving a Rate of Decay Application**

The radioactive element potassium-42 is often used in biological experiments, since it has a half-life of only about 12.4 hr. How much of a 2-g sample will remain after 18 hr and 45 min?

Solution ▶ To begin we must find the nominal rate of decay r and use this value in the exponential decay function.

$$r = \frac{\ln 2}{h} \qquad \text{radioactive rate of decay}$$

$$r = \frac{\ln 2}{12.4} \qquad \text{substitute 12.4 for } h$$

$$r \approx 0.056 \qquad \text{result}$$

The rate of decay is approximately 5.6%. To determine how much of the sample remains after 18.75 hr, we use $r = 0.056$ in the decay function and evaluate it at $t = 18.75$.

$$Q(t) = Q_0 e^{-rt} \qquad \text{exponential decay function}$$

$$Q(18.75) = 2e^{(-0.056)(18.75)} \qquad \text{substitute 2 for } Q_0, \text{ 0.056 for } r, \text{ and 18.75 for } t$$

$$Q(18.75) \approx 0.7 \qquad \text{evaluate}$$

☑ **D.** You've just learned how to solve applications of exponential growth and decay

After 18 hr and 45 min, only 0.7 g of potassium-42 will remain.

Now try Exercises 57 through 62 ▶

TECHNOLOGY HIGHLIGHT

Exploring Compound Interest

The graphing calculator is an excellent tool for exploring mathematical relationships, particularly when many variables work simultaneously to produce a single result. For example, the formula $A = P \left(1 + \frac{r}{n}\right)^{nt}$ has five different unknowns. In Example 2, we asked how long it would take $1000 to double if it were compounded yearly at 6% ($n = 1$, $r = 0.06$). What if we deposited $5000 instead of $1000? Compounded daily instead of quarterly? Or invested at 12% rather than 10%? There are many ways a graphing calculator can be used to answer such questions. In this exercise, we make use of the calculator's "alpha constants." Most graphing calculators can use any of the 26 letters of the English alphabet (and even a few other symbols) to store constant values. We can use them to write a formula

—continued

Figure 4.31   **Figure 4.32**   **Figure 4.33**

on the Y = screen, then change any constant on the home screen to see how other values are affected. On the TI-84 Plus, these alpha constants are shown in green and accessed by pressing the ALPHA key. Suppose we wanted to study the relationship between an interest rate r and the time t required for a deposit to double. Using Y$_1$ in place of A as output variable, and x in place of t, enter $A = P \left(1 + \frac{r}{n}\right)^{nt}$ as Y$_1$ on the Y = screen (Figure 4.31). Let's start with a deposit of $1000 at 7% interest compounded monthly. The keystrokes are: 1000 STO → ALPHA 8 ENTER , 0.07 STO → ALPHA × ENTER , and 12 STO → ALPHA LOG ENTER (Figure 4.32). After setting an appropriate window size (perhaps Xmax = 15 and Ymax = 3000), and entering Y$_2$ = 2000 we can graph both functions and use the intersection of graphs method to find the doubling time. This produces the result in Figure 4.33, where we note it will take about 9.9 yr. Return to the home screen (2nd MODE), change the interest rate to 10%, and graph the functions again. This time the point of intersection is just less than 7 (yr). Experiment with other rates and compounding periods.

Exercise 1: With $P = \$1000$, and $r = 0.08$, investigate the "doubling time" for interest compounded quarterly, monthly, daily, and hourly.

Exercise 2: With $P = \$1000$, investigate "doubling time" for rates of 6%, 8%, 10%, and 12%, and $n = 4$, $n = 12$, and $n = 365$. Which had a more significant impact, more compounding periods, or a greater interest rate?

Exercise 3: Will a larger principal cause the money to double faster? Investigate and respond.

4.5 EXERCISES

▶ CONCEPTS AND VOCABULARY

Fill in each blank with the appropriate word or phrase. Carefully reread the section if needed.

1. _____ interest is interest paid to you on previously accumulated interest.

2. The formula for interest compounded _____ is $A = pe^{rt}$, where e is approximately _____.

3. Given constants Q_0 and r, and that Q decays exponentially as a function of t, the equation model is $Q(t) = $ _____.

4. Investment plans calling for regularly scheduled deposits are called _____. The annuity formula gives the _____ value of the account.

5. Explain/Describe the difference between the future value and present value of an annuity. Include an example.

6. Describe/Explain how you would find the rate of growth r, given that a population of ants grew from 250 to 3000 in 6 weeks.

▶ DEVELOPING YOUR SKILLS

For simple interest accounts, the interest earned or due depends on the principal p, interest rate r, and the time t in years according to the formula $I = prt$.

7. Find p given $I = \$229.50$, $r = 6.25\%$, and $t = 9$ months.

8. Find r given $I = \$1928.75$, $p = \$8500$, and $t = 3.75$ yr.

9. Larry came up a little short one month at bill-paying time and had to take out a title loan on his car at Check Casher's, Inc. He borrowed $260, and 3 weeks later he paid off the note for $297.50. What was the annual interest rate on this title loan? (*Hint:* How much *interest* was charged?)

10. Angela has $750 in a passbook savings account that pays 2.5% simple interest. How long will it take the account balance to hit the $1000 mark at this rate of interest, if she makes no further deposits? (*Hint:* How much *interest* will be paid?)

For simple interest accounts, the amount A accumulated or due depends on the principal p, interest rate r, and the time t in years according to the formula $A = p(1 + rt)$.

11. Find p given $A = \$2500$, $r = 6.25\%$, and $t = 31$ months.

12. Find r given $A = \$15,800$, $p = \$10,000$, and $t = 3.75$ yr.

13. Olivette Custom Auto Service borrowed $120,000 at 4.75% simple interest to expand their facility from three service bays to four. If they repaid $149,925, what was the term of the loan?

14. Healthy U sells nutritional supplements and borrows $50,000 to expand their product line. When the note is due 3 yr later, they repay the lender $62,500. If it was a simple interest note, what was the annual interest rate?

15. **Simple interest:** The owner of Paul's Pawn Shop loans Larry $200.00 using his Toro riding mower as collateral. Thirteen weeks later Larry comes back to get his mower out of pawn and pays Paul $240.00. What was the annual simple interest rate on this loan?

16. **Simple interest:** To open business in a new strip mall, Laurie's Custom Card Shoppe borrows $50,000 from a group of investors at 4.55% simple interest. Business booms and blossoms, enabling Laurie to repay the loan fairly quickly. If Laurie repays $62,500, how long did it take?

For accounts where interest is compounded annually, the amount A accumulated or due depends on the principal p, interest rate r, and the time t in years according to the formula $A = p(1 + r)^t$.

17. Find t given $A = \$48,428$, $p = \$38,000$, and $r = 6.25\%$.

18. Find p given $A = \$30,146$, $r = 5.3\%$, and $t = 7$ yr.

19. How long would it take $1525 to triple if invested at 7.1%?

20. What interest rate will ensure a $747.26 deposit will be worth $1000 in 5 yr?

For accounts where interest is compounded annually, the principal P needed to ensure an amount A has been accumulated in the time period t when deposited at interest rate r is given by the formula $P = \dfrac{A}{(1 + r)^t}$.

21. The Stringers need to make a $10,000 balloon payment in 5 yr. How much should be invested now at 5.75%, so that the money will be available?

22. Morgan is 8 yr old. If her mother wants to have $25,000 for Morgan's first year of college (in 10 yr), how much should be invested now if the account pays a 6.375% fixed rate?

For compound interest accounts, the amount A accumulated or due depends on the principal p, interest rate r, number of compoundings per year n, and the time t in years according to the formula $A = p\left(1 + \frac{r}{n}\right)^{nt}$.

23. Find t given $A = \$129,500$, $p = \$90,000$, and $r = 7.125\%$ compounded weekly.

24. Find r given $A = \$95,375$, $p = \$65,750$, and $t = 15$ yr with interest compounded monthly.

25. How long would it take a $5000 deposit to double, if invested at a 9.25% rate and compounded daily?

26. What principal should be deposited at 8.375% compounded monthly to ensure the account will be worth $20,000 in 10 yr?

27. **Compound interest:** As a curiosity, David decides to invest $10 in an account paying 10% interest compounded 10 times per year for 10 yr. Is that enough time for the $10 to triple in value?

28. **Compound interest:** As a follow-up experiment (see Exercise 27), David invests $10 in an account paying 12% interest compounded 10 times per year

for 10 yr, and another $10 in an account paying 10% interest compounded 12 times per year for 10 yr. Which produces the better investment—more compounding periods or a higher interest rate?

29. Compound interest: Due to demand, Donovan's Dairy (Wisconsin, USA) plans to double its size in 4 yr and will need $250,000 to begin development. If they invest $175,000 in an account that pays 8.75% compounded semiannually, (a) will there be sufficient funds to break ground in 4 yr? (b) If not, find the *minimum interest rate* that will allow the dairy to meet its 4-yr goal.

30. Compound interest: To celebrate the birth of a new daughter, Helyn invests 6000 Swiss francs in a college savings plan to pay for her daughter's first year of college in 18 yr. She estimates that 25,000 francs will be needed. If the account pays 7.2% compounded daily, (a) will she meet her investment goal? (b) If not, find the *minimum rate of interest* that will enable her to meet this 18-yr goal.

For accounts where interest is compounded continuously, the amount A accumulated or due depends on the principal p, interest rate r, and the time t in years according to the formula $A = pe^{rt}$.

31. Find t given $A = \$2500$, $p = \$1750$, and $r = 4.5\%$.

32. Find r given $A = \$325,000$, $p = \$250,000$, and $t = 10$ yr.

33. How long would it take $5000 to double if it is invested at 9.25%? Compare the result to Exercise 25.

34. What principal should be deposited at 8.375% to ensure the account will be worth $20,000 in 10 yr? Compare the result to Exercise 26.

35. Interest compounded continuously: Valance wants to build an addition to his home outside Madrid (Spain) so he can watch over and help his parents in their old age. He hopes to have 20,000 euros put aside for this purpose within 5 yr. If he invests 12,500 euros in an account paying 8.6% interest compounded continuously, (a) will he meet his investment goal? (b) If not, find the *minimum rate of interest* that will enable him to meet this 5-yr goal.

36. Interest compounded continuously: Minh-Ho just inherited her father's farm near Mito (Japan), which badly needs a new barn. The estimated cost of the barn is 8,465,000 yen and she would like to begin construction in 4 yr. If she invests 6,250,000 yen in

an account paying 6.5% interest compounded continuously, (a) will she meet her investment goal? (b) If not, find the *minimum rate of interest* that will enable her to meet this 4-yr goal.

37. Interest compounded continuously: William and Mary buy a small cottage in Dovershire (England), where they hope to move after retiring in 7 yr. The cottage needs about 20,000 euros worth of improvements to make it the retirement home they desire. If they invest 12,000 euros in an account paying 5.5% interest compounded continuously, (a) will they have enough to make the repairs? (b) If not, find the *minimum amount they need to deposit* that will enable them to meet this goal in 7 yr.

38. Interest compounded continuously: After living in Oslo (Norway) for 20 years, Zirkcyt and Shybrt decide to move inland to help operate the family ski resort. They hope to make the move in 6 yr, after they have put aside 140,000 kroner. If they invest 85,000 kroner in an account paying 6.9% interest compounded continuously, (a) will they meet their 140,000 kroner goal? (b) If not, find the *minimum amount they need to deposit* that will allow them to meet this goal in 6 yr.

The length of time T (in years) required for an initial principal P to grow to an amount A at a given interest rate r is given by $T = \frac{1}{r}\ln\left(\frac{A}{P}\right)$.

39. Investment growth: A small business is planning to build a new $350,000 facility in 8 yr. If they deposit $200,000 in an account that pays 5% interest compounded continuously, will they have enough for the new facility in 8 yr? If not, what amount should be invested on these terms to meet the goal?

40. Investment growth: After the twins were born, Sasan deposited $25,000 in an account paying 7.5% compounded continuously, with the goal of having $120,000 available for their college education 20 yr later. Will Sasan meet the 20-yr goal? If not, what amount should be invested on these terms to meet the goal?

Ordinary annuities: If a periodic payment $\mathcal{P}$ is deposited n times per year, with annual interest rate r also compounded n times per year for t years, the future value of the account is given by $A = \frac{\mathcal{P}\left[(1 + R)^{nt} - 1\right]}{R}$, where $R = \frac{r}{n}$ (if the rate is 9% compounded monthly, $R = \frac{0.09}{12} = 0.0075$).

41. Saving for a rainy day: How long would it take Jasmine to save $10,000 if she deposits $90/month at an annual rate of 7.75% compounded monthly?

42. Saving for a sunny day: What quarterly investment amount is required to ensure that Larry can save $4700 in 4 yr at an annual rate of 8.5% compounded quarterly?

43. Saving for college: At the birth of their first child, Latasha and Terrance opened an annuity account and have been depositing $50 per month in the account ever since. If the account is now worth $30,000 and the interest on the account is 6.2% compounded monthly, how old is the child?

44. Saving for a bequest: When Cherie (Brandon's first granddaughter) was born, he purchased an annuity account for her and stipulated that she should receive the funds (in trust, if necessary) upon his death. The quarterly annuity payments were $250 and interest on the account was 7.6% compounded quarterly. The account balance of $17,500 was recently given to Cherie. How much longer did Brandon live?

45. Saving for a down payment: Tae-Hon is tired of renting and decides that within the next 5 yr he must save $22,500 for the down payment on a home. He finds an investment company that offers 8.5% interest compounded monthly and begins depositing $250 each month in the account. (a) Is this monthly amount sufficient to help him meet his 5 yr goal? (b) If not, find the *minimum amount he needs to deposit each month* that will allow him to meet his goal in 5 yr.

46. Saving to open a business: Madeline feels trapped in her current job and decides to save $75,000 over the next 7 yr to open up a Harley Davidson franchise. To this end, she invests $145 every week in an account paying $7\frac{1}{2}$% interest compounded weekly. (a) Is this weekly amount sufficient to help her meet the seven-year goal? (b) If not, find the *minimum amount she needs to deposit each week* that will allow her to meet this goal in 7 yr?

▶ WORKING WITH FORMULAS

Solve for the indicated unknowns.

47. $A = p + prt$
   a. solve for t
   b. solve for p

48. $A = p(1 + r)^t$
   a. solve for t
   b. solve for r

49. $A = P\left(1 + \dfrac{r}{n}\right)^{nt}$
   a. solve for r
   b. solve for t

50. $A = pe^{rt}$
   a. solve for p
   b. solve for r

51. $Q(t) = Q_0 e^{rt}$
   a. solve for Q_0
   b. solve for t

52. $p = \dfrac{AR}{[(1 + R)^{nt} - 1]}$
   a. solve for A
   b. solve for n

 53. Amount of a mortgage payment:

$$\mathcal{P} = \frac{AR}{1 - (1 + R)^{-nt}}$$

The mortgage payment required to pay off (or amortize) a loan is given by the formula shown, where $\mathcal{P}$ is the payment amount, A is the original amount of the loan, t is the time in years, r is the annual interest rate, n is the number of payments per year, and $R = \frac{r}{n}$. Find the *monthly payment* required to amortize a $125,000 home, if the interest rate is 5.5%/year and the home is financed over 30 yr.

54. Total interest paid on a home mortgage:

$$I = \left[\frac{prt}{1 - \left(\dfrac{1}{1 + 0.08\overline{3}r}\right)^{12t}}\right] - p$$

The total interest I paid in t years on a home mortgage of p dollars is given by the formula shown, where r is the interest rate on the loan (note that $0.08\overline{3} = \frac{1}{12}$). If the original mortgage was $198,000 at an interest rate of 6.5%, (a) how much interest has been paid in 10 yr? (b) Use a table of values to determine how many years it will take for the interest paid to exceed the amount of the original mortgage.

▶ APPLICATIONS

55. Exponential growth: As part of a lab experiment, Luamata needs to grow a culture of 200,000 bacteria, which are known to double in number in 12 hr. If he begins with 1000 bacteria, (a) find the growth rate r and (b) find how many hours it takes for the culture to produce the 200,000 bacteria.

56. Exponential growth: After the wolf population was decimated due to overhunting, the rabbit population in the Boluhti Game Reserve began to double every 6 months. If there were an estimated 120 rabbits to begin, (a) find the growth rate r and (b) find the number of months required for the population to reach 2500.

57. Radioactive decay: The radioactive element iodine-131 has a half-life of 8 days and is often used to help diagnose patients with thyroid problems. If a certain thyroid procedure requires 0.5 g and is scheduled to take place in 3 days, what is the minimum amount that must be on hand now (to the nearest hundredth of a gram)?

58. Radioactive decay: The radioactive element sodium-24 has a half-life of 15 hr and is used to help locate obstructions in blood flow. If the procedure requires 0.75 g and is scheduled to take place in 2 days (48 hr), what minimum amount must be on hand *now* (to the nearest hundredth of a gram)?

59. Radioactive decay: The radioactive element americium-241 has a half-life of 432 yr and although extremely small amounts are used (about 0.0002 g), it is the most vital component of standard household smoke detectors. How many years will it take a 10-g mass of americium-241 to decay to 2.7 g?

60. Radioactive decay: Carbon-14 is a radioactive compound that occurs naturally in all living organisms, with the amount in the organism constantly renewed. After death, no new carbon-14 is acquired and the amount in the organism begins to decay exponentially. If the half-life of carbon-14 is 5730 yr, how old is a mummy having only 30% of the normal amount of carbon-14?

Carbon-14 dating: If the percentage p of carbon-14 that remains in a fossil can be determined, the formula $T = -8267 \ln p$ can be used to estimate the number of years T since the organism died.

61. Dating the Lascaux Cave Dwellers: Bits of charcoal from Lascaux Cave (home of the prehistoric Lascaux Cave Paintings) were used to estimate that the fire had burned some 17,255 yr ago. What percent of the original amount of carbon-14 remained in the bits of charcoal?

62. Dating Stonehenge: Using organic fragments found near Stonehenge (England), scientists were able to determine that the organism that produced the fragments lived about 3925 yr ago. What percent of the original amount of carbon-14 remained in the organism?

▶ EXTENDING THE CONCEPT

63. Many claim that inheritance taxes are put in place simply to prevent a massive accumulation of wealth by a select few. Suppose that in 1890, your great-grandfather deposited $10,000 in an account paying 6.2% compounded continuously. If the account were to pass to you untaxed, what would it be worth in 2010? Do some research on the inheritance tax laws in your state. In particular, what amounts can be inherited untaxed (i.e., before the inheritance tax kicks in)?

64. In Section 4.2, we noted that one important characteristic of exponential functions is their rate of growth is in constant proportion to the population at time t: $\frac{\Delta P}{\Delta t} = kP$. This rate of growth can also be applied to finance and biological models, as well as the growth of tumors, and is of great value in studying these applications. In Exercise 96 of Section 4.2, we computed the value of k for the Goldsboro model ($P = 1000 \cdot 2^t$) using the difference quotient. If we rewrite this model in terms of base e ($P = 1000 \cdot e^{kt}$), the value of k is given directly. The following sequence shows how

this is done, and you are asked to supply the reason or justification for each step.

$$
\begin{array}{ll}
P = b^t & \text{base-}b \text{ exponential} \\
\ln P = \ln b^t & \underline{\hspace{3cm}} \\
\ln P = t \ln b & \underline{\hspace{3cm}} \\
P = e^{t \ln b} & \underline{\hspace{3cm}} \\
P = e^{kt} & k = \ln b
\end{array}
$$

The last step shows the growth rate constant is equal to the natural log of the given base b: $k = \ln b$.

a. Use this result to verify the growth rate constant for Goldsboro is 0.6931472.

b. After the Great Oklahoma Land Run of 1890, the population of the state grew rapidly for the next 2 decades. For this time period, population growth could be approximated by $P = 260(1.10^t)$. Find the growth rate constant for this model, and use it to write the base-e population equation. Use the TABLE feature of a graphing calculator to verify that the equations are equivalent.

65. If you have not already completed Exercise 30, please do so. For *this* exercise, *solve the compound interest equation for r* to find the exact rate of interest that will allow Helyn to meet her 18-yr goal.

66. If you have not already completed Exercise 43, please do so. Suppose the final balance of the account was $35,100 with interest again being compounded monthly. For *this* exercise, use a graphing calculator to find *r*, the exact rate of interest the account would have been earning.

▶ **MAINTAINING YOUR SKILLS**

67. (2.1) In an effort to boost tourism, a trolley car is being built to carry sightseers from a strip mall to the top of Mt. Vernon, 1580-m high. Approximately how long will the trolley cables be?

2000 m

68. (2.2) Is the following relation a function? If not, state how the definition of a function is violated.

69. (4.3) A polynomial is known to have the zeroes $x = 3$, $x = -1$, and $x = 1 + 2i$. Find the equation of the polynomial, given it has degree 4 and a y-intercept of $(0, -15)$.

70. (2.2/3.8) Name the toolbox functions that are (a) one-to-one, (b) even, (c) increasing for $x \in R$, and (d) asymptotic.

SUMMARY AND CONCEPT REVIEW

SECTION 4.1   One-to-One and Inverse Functions

KEY CONCEPTS

* A function is one-to-one if each element of the range corresponds to a unique element of the domain.
* If every horizontal line intersects the graph of a function in at most one point, the function is one-to-one.
* If f is a one-to-one function with ordered pairs (a, b), then the inverse of f exists and is that one-to-one function f^{-1} with ordered pairs of the form (b, a).
* The range of f becomes the domain of f^{-1}, and the domain of f becomes the range of f^{-1}.
* To find f^{-1} using the algebraic method:
  1. Use y instead of $f(x)$.   2. Interchange x and y.
  3. Solve the equation for y.   4. Substitute $f^{-1}(x)$ for y.
* If f is a one-to-one function, the inverse f^{-1} exists, where $(f \circ f^{-1})(x) = x$ and $(f^{-1} \circ f)(x) = x$.
* The graphs of f and f^{-1} are symmetric with respect to the identity function $y = x$.

EXERCISES

Determine whether the functions given are one-to-one by noting the function family to which each belongs and mentally picturing the shape of the graph.

1. $h(x) = -|x - 2| + 3$ **2.** $p(x) = 2x^2 + 7$ **3.** $s(x) = \sqrt{x - 1} + 5$

Find the inverse of each function given. Then show using composition that your inverse function is correct. State any necessary restrictions.

4. $f(x) = -3x + 2$ **5.** $f(x) = x^2 - 2, x \geq 0$ **6.** $f(x) = \sqrt{x - 1}$

Determine the domain and range for each function whose graph is given, and use this information to state the domain and range of the inverse function. Then sketch in the line $y = x$, estimate the location of three points on the graph, and use these to graph $f^{-1}(x)$ on the same grid.

7. **8.** **9.**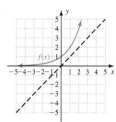

10. **Fines for overdue material:** Some libraries have set fees and penalties to discourage patrons from holding borrowed materials for an extended period. Suppose the fine for overdue DVDs is given by the function $f(t) = 0.15t + 2$, where $f(t)$ is the amount of the fine t days after it is due. (a) What is the fine for keeping a DVD seven (7) extra days? (b) Find $f^{-1}(t)$, then input your answer from part (a) and comment on the result. (c) If a fine of \$3.80 was assessed, how many days was the DVD overdue?

SECTION 4.2 Exponential Functions

KEY CONCEPTS
- An exponential function is defined as $f(x) = b^x$, where $b > 0$, $b \neq 1$, and b, x are real numbers.
- The natural exponential function is $f(x) = e^x$, where $e \approx 2.71828182846$.
- For exponential functions, we have
  - one-to-one function
  - y-intercept $(0, 1)$
  - domain: $x \in \mathbb{R}$
  - range: $y \in (0, \infty)$
  - increasing if $b > 1$
  - decreasing if $0 < b < 1$
  - asymptotic to x-axis
- The graph of $y = b^{x \pm h} \pm k$ is a translation of the basic graph of $y = b^x$, horizontally h units opposite the sign and vertically k units in the same direction as the sign.
- If an equation can be written with like bases on each side, we solve it using the uniqueness property: If $b^m = b^n$, then $m = n$ (equal bases imply equal exponents).
- All previous properties of exponents also apply to exponential functions.

EXERCISES
Graph each function using *transformations of the basic function,* then strategically plot a few points to check your work and round out the graph. Draw and label the asymptote.

11. $y = 2^x + 3$ **12.** $y = 2^{-x} - 1$ **13.** $y = -e^{x+1} - 2$

Solve using the uniqueness property.

14. $3^{2x-1} = 27$ **15.** $4^x = \frac{1}{16}$ **16.** $e^x \cdot e^{x+1} = e^6$

17. A ballast machine is purchased new for \$142,000 by the AT & SF Railroad. The machine loses 15% of its value each year and must be replaced when its value drops below \$20,000. How many years will the machine be in service?

SECTION 4.3 Logarithms and Logarithmic Functions

KEY CONCEPTS
- A logarithm is an exponent. For $x, b > 0$, and $b \neq 1$, the expression $\log_b x$ represents the exponent that goes on base b to obtain x: If $y = \log_b x$, then $b^y = x \Rightarrow b^{\log_b x} = x$ (by substitution).

- The equations $x = b^y$ and $y = \log_b x$ are equivalent. We say $x = b^y$ is the *exponential* form and $y = \log_b x$ is the *logarithmic* form of the equation.
- The value of $\log_b x$ can sometimes be determined by writing the expression in exponential form. If $b = 10$ or $b = e$, the value of $\log_b x$ can be found directly using a calculator.
- A logarithmic *function* is defined as $f(x) = \log_b x$, where $x, b > 0$, and $b \neq 1$.
  - $y = \log_{10} x = \log x$ is called a *common* logarithmic function.
  - $y = \log_e x = \ln x$ is called a *natural* logarithmic function.
- For $f(x) = \log_b x$ as defined we have
  - one-to-one function
  - x-intercept $(1, 0)$
  - domain: $x \in (0, \infty)$
  - range: $y \in \mathbb{R}$
  - increasing if $b > 1$
  - decreasing if $0 < b < 1$
  - asymptotic to y-axis
- The graph of $y = \log_b(x \pm h) \pm k$ is a translation of the graph of $y = \log_b x$, horizontally h units opposite the sign and vertically k units in the same direction as the sign.

EXERCISES

Write each expression in *exponential* form.

18. $\log_3 9 = 2$ **19.** $\log_5 \frac{1}{125} = -3$ **20.** $\ln 43 \approx 3.7612$

Write each expression in *logarithmic* form.

21. $5^2 = 25$ **22.** $e^{-0.25} \approx 0.7788$ **23.** $3^4 = 81$

Find the value of each expression without using a calculator.

24. $\log_2 32$ **25.** $\ln\left(\frac{1}{e}\right)$ **26.** $\log_9 3$

Graph each function using *transformations of the basic function,* then strategically plot a few points to check your work and round out the graph. Draw and label the asymptote.

27. $f(x) = \log_2 x$ **28.** $f(x) = \log_2(x + 3)$ **29.** $f(x) = 2 + \ln(x - 1)$

Find the domain of the following functions.

30. $f(x) = \ln(x^2 - 6x)$ **31.** $g(x) = \log \sqrt{2x + 3}$

32. The magnitude of an earthquake is given by $M(I) = \log\frac{I}{I_0}$, where I is the intensity and I_0 is the reference intensity. (a) Find $M(I)$ given $I = 62{,}000I_0$ and (b) find the intensity I given $M(I) = 7.3$.

SECTION 4.4   Properties of Logarithms; Solving Exponential and Logarithmic Equations

KEY CONCEPTS

- The basic definition of a logarithm gives rise to the following properties: For any base $b > 0$, $b \neq 1$,
  1. $\log_b b = 1$ (since $b^1 = b$)   2. $\log_b 1 = 0$ (since $b^0 = 1$)
  3. $\log_b b^x = x$ (since $b^x = b^x$)   4. $b^{\log_b x} = x$
- Since a logarithm is an exponent, they have properties that parallel those of exponents.

Product Property	**Quotient Property**	**Power Property**
like base and multiplication, add exponents:	like base and division, subtract exponents:	exponent raised to a power, multiply exponents:
$\log_b(MN) = \log_b M + \log_b N$	$\log_b\left(\dfrac{M}{N}\right) = \log_b M - \log_b N$	$\log_b M^p = p\log_b M$

- The logarithmic properties can be used to expand an expression: $\log(2x) = \log 2 + \log x$.
- The logarithmic properties can be used to contract an expression: $\ln(2x) - \ln(x + 3) = \ln\left(\dfrac{2x}{x + 3}\right)$.

- To evaluate logarithms with bases other than 10 or e, use the change-of-base formula:

$$\log_b M = \frac{\log M}{\log b} = \frac{\ln M}{\ln b}$$

- If an equation can be written with like bases on each side, we solve it using the uniqueness property: if $\log_b m = \log_b n$, then $m = n$ (equal bases imply equal arguments).
- If a single exponential or logarithmic term can be isolated on one side, then for any base b:

$$\text{If } b^x = k, \text{ then } x = \frac{\log k}{\log b} \qquad\qquad \text{If } \log_b x = k, \text{ then } x = b^k.$$

EXERCISES

33. Solve each equation by applying fundamental properties.

  a. $\ln x = 32$ **b.** $\log x = 2.38$ **c.** $e^x = 9.8$ **d.** $10^x = \sqrt{7}$

34. Solve each equation. Write answers in exact form and in approximate form to four decimal places.

  a. $15 = 7 + 2e^{0.5x}$ **b.** $10^{0.2x} = 19$ **c.** $-2\log(3x) + 1 = -5$ **d.** $-2\ln x + 1 = 6.5$

35. Use the product or quotient property of logarithms to write each sum or difference as a single term.

  a. $\ln 7 + \ln 6$ **b.** $\log_9 2 + \log_9 15$ **c.** $\ln(x + 3) - \ln(x - 1)$ **d.** $\log x + \log(x + 1)$

36. Use the power property of logarithms to rewrite each term as a product.

  a. $\log_5 9^2$ **b.** $\log_7 4^2$ **c.** $\ln 5^{2x-1}$ **d.** $\ln 10^{3x+2}$

37. Use the properties of logarithms to write the following expressions as a sum or difference of simple logarithmic terms.

  a. $\ln(x\sqrt[4]{y})$ **b.** $\ln(\sqrt[3]{pq})$ **c.** $\log\left(\dfrac{\sqrt[3]{x^5} \cdot y^4}{\sqrt{x^5 y^3}}\right)$ **d.** $\log\left(\dfrac{4\sqrt[3]{p^5 q^4}}{\sqrt{p^3 q^2}}\right)$

38. Evaluate using a change-of-base formula. Answer in exact form and approximate form to thousandths.

  a. $\log_6 45$ **b.** $\log_3 128$ **c.** $\ln_2 124$ **d.** $\ln_5 0.42$

Solve each equation.

39. $2^x = 7$ **40.** $3^{x+1} = 5$ **41.** $e^{x-2} = 3^x$

42. $\ln(x + 1) = 2$ **43.** $\log x + \log(x - 3) = 1$ **44.** $\log_{25}(x + 2) - \log_{25}(x - 3) = \frac{1}{2}$

45. The rate of decay for radioactive material is related to its half-life by the formula $R(h) = \frac{\ln 2}{h}$, where h represents the half-life of the material and $R(h)$ is the rate of decay expressed as a decimal. The element radon-222 has a half-life of approximately 3.9 days. (a) Find its rate of decay to the nearest hundredth of a percent. (b) Find the half-life of thorium-234 if its rate of decay is 2.89% per day.

46. The *barometric equation* $H = (30T + 8000)\ln\left(\frac{P_0}{P}\right)$ relates the altitude H to atmospheric pressure P, where $P_0 = 76$ cmHg. Find the atmospheric pressure at the summit of Mount Pico de Orizaba (Mexico), whose summit is at 5657 m. Assume the temperature at the summit is $T = 12°C$.

SECTION 4.5   Applications from Investment, Finance, and Physical Science

KEY CONCEPTS

- Simple interest: $I = prt$; p is the initial principal, r is the interest rate per year, and t is the time in years.
- Amount in an account after t years: $A = p + prt$ or $A = p(1 + rt)$.
- Interest compounded n times per year: $A = p\left(1 + \dfrac{r}{n}\right)^{nt}$; p is the initial principal, r is the interest rate per year, t is the time in years, and n is the times per year interest is compounded.
- Interest compounded continuously: $A = pe^{rt}$; p is the initial principal, r is the interest rate per year, and t is the time in years.

- If a loan or savings plan calls for a regular schedule of deposits, the plan is called an annuity.
- For periodic payment $\mathcal{P}$, deposited or paid n times per year, at annual interest rate r, with interest compounded or calculated n times per year for t years, and $R = \dfrac{r}{n}$:

  - The accumulated value of the account is $A = \dfrac{\mathcal{P}}{R}[(1 + R)^{nt} - 1]$.

  - The payment required to meet a future goal is $\mathcal{P} = \dfrac{AR}{[(1 + R)^{nt} - 1]}$.

  - The payment required to amortize an amount A is $\mathcal{P} = \dfrac{AR}{1 - (1 + R)^{-nt}}$.

  - The general formulas for exponential growth and decay are $Q(t) = Q_0 e^{rt}$ and $Q(t) = Q_0 e^{-rt}$, respectively.

EXERCISES

Solve each application.

47. Jeffery borrows $600.00 from his dad, who decides it's best to charge him interest. Three months later Jeff repays the loan plus interest, a total of $627.75. What was the annual interest rate on the loan?

48. To save money for her first car, Cheryl invests the $7500 she inherited in an account paying 7.8% interest compounded monthly. She hopes to buy the car in 6 yr and needs $12,000. Is this possible?

49. To save up for the vacation of a lifetime, Al-Harwi decides to save $15,000 over the next 4 yr. For this purpose he invests $260 every month in an account paying $7\frac{1}{2}\%$ interest compounded monthly. (a) Is this monthly amount sufficient to meet the four-year goal? (b) If not, find the *minimum amount he needs to deposit each month* that will allow him to meet this goal in 4 yr.

50. Eighty prairie dogs are released in a wilderness area in an effort to repopulate the species. Five years later a statistical survey reveals the population has reached 1250 dogs. Assuming the growth was exponential, approximate the growth rate to the nearest tenth of a percent.

MIXED REVIEW

1. Evaluate each expression using the change-of-base formula.
  a. $\log_2 30$ **b.** $\log_{0.25} 8$
  c. $\log_8 2$

2. Solve each equation using the uniqueness property.
  a. $10^{4x-5} = 1000$ **b.** $5^{3x-1} = \sqrt{5}$
  c. $2^x \cdot 2^{0.5x} = 64$

3. Use the power property of logarithms to rewrite each expression as a product.
  a. $\log_{10} 20^2$ **b.** $\log 10^{0.05x}$
  c. $\ln 2^{x-3}$

Graph each of the following functions by shifting the basic function, then strategically plotting a few points to check your work and round out the graph. Graph and label the asymptote.

4. $y = -e^x + 15$ **5.** $y = 5 \cdot 2^{-x}$
6. $y = \ln(x + 5) + 7$ **7.** $y = \log_2(-x) - 4$

8. Use the properties of logarithms to write the following expressions as a sum or difference of simple logarithmic terms.
  a. $\ln\left(\dfrac{x^3}{2y}\right)$ **b.** $\log(10a\sqrt[3]{a^2 b})$
  c. $\log_2\left(\dfrac{8x^4 \sqrt{x}}{3\sqrt{y}}\right)$

9. Write the following expressions in exponential form.
  a. $\log_5 625 = 4$ **b.** $\ln 0.15x = 0.45$
  c. $\log(0.1 \times 10^8) = 7$

10. Write the following expressions in logarithmic form.
  a. $343^{1/3} = 7$ **b.** $256^{3/4} = 64$
  c. $2^{-3} = \frac{1}{8}$

11. For $g(x) = \sqrt{x-1} + 2$, (a) state the domain and range, (b) find $g^{-1}(x)$ and state its domain and range, and (c) compute at least three ordered pairs (a, b) for g and show the order pairs (b, a) are solutions to g^{-1}.

Solve the following equations. State answers in exact form.

12. $\log_5(4x + 7) = 0$ **13.** $10^{x-4} = 200$

14. $e^{x+1} = 3^x$

15. $\log_2(2x - 5) + \log_2(x - 2) = 4$

16. $\log(3x - 4) - \log(x - 2) = 1$

Solve each application.

17. The magnitude of an earthquake is given by $M(I) = \log\left(\dfrac{I}{I_0}\right)$, where I is the intensity of the quake and I_0 is the reference intensity 2×10^{11} (energy released from the smallest detectable quake). On October 23, 2004, the Niigata region of Japan was hit by an earthquake that registered 6.5 on the Richter scale. Find the intensity of this earthquake by solving the following equation for I: $6.5 = \log\left(\dfrac{I}{2 \times 10^{11}}\right)$.

18. Serene is planning to buy a house. She has $6500 to invest in a certificate of deposit that compounds interest quarterly at an annual rate of 4.4%. (a) Find how long it will take for this account to grow to the $12,500 she will need for a 10% down payment for a $125,000 house. Round to the nearest tenth of a year. (b) Suppose instead of investing an initial $6500, Serene deposits $500 a quarter in an account paying 4% each quarter. Find how long it will take for this account to grow to $12,500. Round to the nearest tenth of a year.

19. British artist Simon Thomas designs sculptures he calls hypercones. These sculptures involve rings of exponentially decreasing radii rotated through space. For one sculpture, the radii follow the model $r(n) = 2(0.8)^n$, where n counts the rings (outer-most first) and $r(n)$ is radii in meters. Find the radii of the six largest rings in the sculpture. Round to the nearest hundredth of a meter.

Source: http://www.plus.maths.org/issue8/features/art/

20. Ms. Chan-Chiu works for MediaMax, a small business that helps other companies purchase advertising in publications. Her model for the benefits of advertising is $P(a) = 1000(1.07)^a$, where P represents the number of potential customers reached when a dollars (in thousands) are invested in advertising.

 a. Use this model to predict (to the nearest thousand) how many potential customers will be reached when $50,000 is invested in advertising.

 b. Use this model to determine how much money a company should expect to invest in advertising (to the nearest thousand), if it wants to reach 100,000 potential customers.

PRACTICE TEST

1. Write the expression $\log_3 81 = 4$ in exponential form.

2. Write the expression $25^{1/2} = 5$ in logarithmic form.

3. Write the expression $\log_b\left(\dfrac{\sqrt{x^5 y^3}}{z}\right)$ as a sum or difference of logarithmic terms.

4. Write the expression $\log_b m + \left(\frac{3}{2}\right)\log_b n - \frac{1}{2}\log_b p$ as a single logarithm.

Solve for x using the uniqueness property.

5. $5^{x-7} = 125$ **6.** $2 \cdot 4^{3x} = \dfrac{8^x}{16}$

Given $\log_a 3 \approx 0.48$ and $\log_a 5 \approx 1.72$, evaluate the following without the use of a calculator:

7. $\log_a 45$ **8.** $\log_a 0.6$

Graph using transformations of the parent function. Verify answers using a graphing calculator.

9. $g(x) = -2^{x-1} + 3$ **10.** $h(x) = \log_2(x - 2) + 1$

11. Use the change-of-base formula to evaluate. Verify results using a calculator.

 a. $\log_3 100$ **b.** $\log_6 0.235$

12. State the domain and range of $f(x) = (x - 2)^2 - 3$ and determine if f is a one-to-one function. If so, find its inverse. If not, restrict the domain of f to create a one-to-one function, then find the inverse of this new function, including the domain and range.

Solve each equation.

13. $3^{x-1} = 89$

14. $\log_5 x + \log_5(x + 4) = 1$

15. A copier is purchased new for $8000. The machine loses 18% of its value each year and must be replaced when its value drops below $3000. How many years will the machine be in service?

16. How long would it take $1000 to double if invested at 8% annual interest compounded daily?

17. The number of ounces of unrefined platinum drawn from a mine is modeled by $Q(t) = -2600 + 1900 \ln(t)$, where $Q(t)$ represents the number of ounces mined in t months. How many months did it take for the number of ounces mined to exceed 3000?

18. Septashi can invest his savings in an account paying 7% compounded semi-annually, or in an account paying 6.8% compounded daily. Which is the better investment?

19. Jacob decides to save $4000 over the next 5 yr so that he can present his wife with a new diamond ring for their 20th anniversary. He invests $50 every month in an account paying $8\frac{1}{4}\%$ interest compounded monthly. (a) Is this amount sufficient to meet the 5-yr goal? (b) If not, find the *minimum amount he needs to save monthly* that will enable him to meet this goal.

20. Chaucer is a typical Welsh Corgi puppy. During his first year of life, his weight very closely follows the model $W(t) = 6.79 \ln t - 11.97$, where $W(t)$ is his weight in pounds after t weeks and $8 \leq t \leq 52$.

a. How much will Chaucer weigh when he is 6 months old (to the nearest one-tenth pound)?

b. To the nearest week, how old is Chaucer when he weighs 8 lb?

CALCULATOR EXPLORATION AND DISCOVERY

Investigating Logistic Equations

As we saw in Section 4.4, logistics models have the form $P(t) = \dfrac{c}{1 + ae^{-bt}}$, where a, b, and c are constants and $P(t)$ represents the population at time t. For populations modeled by a logistics curve (sometimes called an "S" curve) growth is very rapid at first (like an exponential function), but this growth begins to slow down and level off due to various factors. This *Calculator Exploration and Discovery* is designed to investigate the effects that a, b, and c have on the resulting graph.

I. From our earlier observation, as t becomes larger and larger, the term ae^{-bt} becomes smaller and smaller (approaching 0) because it is a decreasing function: as $t \rightarrow \infty$, $ae^{-bt} \rightarrow 0$. If we allow that the term eventually becomes so small it can be disregarded, what remains is $P(t) = \dfrac{c}{1}$ or c. This is why c is called the capacity constant and the population can get no larger than c. In Figure 4.34, the graph of

Figure 4.34

$P(t) = \dfrac{1000}{1 + 50e^{-1x}}$ $(a = 50, b = 1, \text{and } c = 1000)$ is shown using a lighter line, while the graph of $P(t) = \dfrac{750}{1 + 50e^{-1x}}$ $(a = 50, b = 1, \text{and } c = 750)$, is given in bold. The window size is indicated in Figure 4.35.

Figure 4.35

```
WINDOW
 Xmin=-1
 Xmax=10
 Xscl=1
 Ymin=-150
 Ymax=1100
 Yscl=100
 Xres=1
```

Also note that if a is held constant, smaller values of c cause the "interior" of the S curve to grow at a slower rate than larger values, a concept studied in some detail in a Calculus I class.

II. If $t = 0$, $ae^{-bt} = ae^0 = a$, and we note the ratio $P(0) = \dfrac{c}{1 + a}$ represents the *initial population*. This also means for constant values of c, larger values of a make the ratio $\dfrac{c}{1 + a}$ smaller; while smaller values of a make the ratio $\dfrac{c}{1 + a}$ larger. From this we conclude that a primarily affects the initial population. For the screens shown next, $P(t) = \dfrac{1000}{1 + 50e^{-1x}}$ (from I) is graphed using a lighter line. For comparison, the graph of $P(t) = \dfrac{1000}{1 + 5e^{-1x}}$ ($a = 5$, $b = 1$, and $c = 1000$) is shown in bold in Figure 4.36, while the graph of $P(t) = \dfrac{1000}{1 + 500e^{-1x}}$ ($a = 500$, $b = 1$, and $c = 1000$) is shown in bold in Figure 4.37.

Figure 4.36

Figure 4.37

Note that changes in a appear to have no effect on the rate of growth in the interior of the S curve.

III. As for the value of b, we might expect that it affects the rate of growth in much the same way as the growth rate r does for exponential functions $Q(t) = Q_0 e^{-rt}$. Sure enough, we note from the graphs shown that b has no effect on the initial value or the eventual capacity, but causes the population to approach this capacity more quickly for larger values of b, and more slowly for smaller values of b. For the screens shown, $P(t) = \dfrac{1000}{1 + 50e^{-1x}}$ ($a = 50$, $b = 1$, and $c = 1000$) is graphed using a lighter line. For comparison, the graph of $P(t) = \dfrac{1000}{1 + 50e^{-1.2x}}$ ($a = 50$, $b = 1.2$, and $c = 1000$) is shown in bold in Figure 4.38, while the graph of $P(t) =$

Figure 4.38

Figure 4.39

$\dfrac{1000}{1 + 50e^{-0.8x}}$ ($a = 50$, $b = 0.8$, and $c = 1000$) is shown in bold in Figure 4.39.

The following exercises are based on the population of an ant colony, modeled by the logistic function $P(t) = \dfrac{2500}{1 + 25e^{-0.5x}}$. Respond to Exercises 1 through 6 without the use of a calculator.

Exercise 1: Identify the values of a, b, and c for this logistics curve.

Exercise 2: What was the approximate initial population of the colony?

Exercise 3: Which gives a larger initial population: (a) $c = 2500$ and $a = 25$ or (b) $c = 3000$ and $a = 15$?

Exercise 4: What is the maximum population capacity for this colony?

Exercise 5: Would the population of the colony surpass 2000 more quickly if $b = 0.6$ or if $b = 0.4$?

Exercise 6: Which causes a slower population growth: (a) $c = 2000$ and $a = 25$ or (b) $c = 3000$ and $a = 25$?

Exercise 7: Verify your responses to Exercises 2 through 6 using a graphing calculator.

STRENGTHENING CORE SKILLS

Understanding Properties of Logarithms

To effectively use the properties of logarithms as a mathematical tool, a student must attain some degree of comfort and fluency in their application. Otherwise we are resigned to using them as a template or formula, leaving little room for growth or insight. This feature is divided into two parts. The first is designed to promote an understanding of the product and quotient properties of logarithms, which play a role in the solution of logarithmic and exponential equations.

We begin by looking at some logarithmic expressions that are obviously true:

$$\log_2 2 = 1 \qquad \log_2 4 = 2 \qquad \log_2 8 = 3$$
$$\log_2 16 = 4 \qquad \log_2 32 = 5 \qquad \log_2 64 = 6$$

Next, we view the same expressions with their value *understood mentally,* illustrated by the numbers in the background, rather than expressly written.

$$\log_2 2 \quad \log_2 4 \quad \log_2 8 \quad \log_2 16 \quad \log_2 32 \quad \log_2 64$$

This will make the product and quotient properties of equality much easier to "see." Recall the product property states: $\log_b M + \log_b N = \log_b(MN)$ and the quotient property states: $\log_b M - \log_b N = \log_b\left(\dfrac{M}{N}\right)$. Consider the following.

$$\log_2 4 + \log_2 8 = \log_2 32 \qquad \log_2 64 - \log_2 32 = \log_2 2$$

which is the same as saying

$$\log_2 4 + \log_2 8 = \log_2(4 \cdot 8)$$
(since $4 \cdot 8 = 32$)

which is the same as saying

$$\log_2 64 - \log_2 32 = \log_2\left(\tfrac{64}{32}\right)$$
(since $\tfrac{64}{32} = 2$)

$$\log_b M + \log_b N = \log_b(MN) \qquad \log_b M - \log_b N = \log_b\left(\frac{M}{N}\right)$$

Exercise 1: Repeat this exercise using logarithms of base 3 and various sums and differences.

Exercise 2: Use the basic concept behind these exercises to combine these expressions: (a) $\log(x) + \log(x + 3)$, (b) $\ln(x + 2) + \ln(x - 2)$, and (c) $\log(x) - \log(x + 3)$.

The second part is similar to the first, but highlights the power property: $\log_b M^x = x \log_b M$. For instance, knowing that $\log_2 64 = 6$, $\log_2 8 = 3$, and $\log_2 2 = 1$, consider the following:

$\log_2 8$ can be written as $\log_2 2^3$ (since $2^3 = 8$). Applying the power property gives $3 \cdot \log_2 2 = 3$.

$\log_2 64$ can be written as $\log_2 2^6$ (since $2^6 = 64$). Applying the power property gives $6 \cdot \log_2 2 = 6$.

$$\log_b M^x = x \log_b M$$

Exercise 3: Repeat this exercise using logarithms of base 3 and various powers.

Exercise 4: Use the basic concept behind these exercises to rewrite each expression as a product: (a) $\log 3^x$, (b) $\ln x^5$, and (c) $\ln 2^{3x-1}$.

CUMULATIVE REVIEW CHAPTERS 1–4

Use the quadratic formula to solve for x.

1. $x^2 - 4x + 53 = 0$ **2.** $6x^2 + 19x = 36$

3. Use substitution to show that $4 + 5i$ is a zero of $f(x) = x^2 - 8x + 41$.

4. Graph using transformations of a basic function: $y = 2\sqrt{x + 2} - 3$.

5. Find $(f \circ g)(x)$ and $(g \circ f)(x)$ and comment on what you notice: $f(x) = x^3 - 2$; $g(x) = \sqrt[3]{x + 2}$.

6. State the domain of $h(x)$ in interval notation:
$$h(x) = \frac{\sqrt{x + 3}}{x^2 + 6x + 8}.$$

7. According to the 2002 *National Vital Statistics Report* (Vol. 50, No. 5, page 19) there were 3100 sets of triplets born in the United States in 1991, and 6740 sets of triplets born in 1999. Assuming the relationship (year, sets of triplets) is linear: (a) find the equation of the line, (b) explain the meaning of the slope in this context, and (c) use the equation to estimate the number of sets born in 1996, and to project the number of sets that will be born in 2007 if this trend continues.

8. State the following geometric formulas:
   a. area of a circle
   b. Pythagorean theorem
   c. perimeter of a rectangle
   d. area of a trapezoid

9. Graph the following piecewise-defined function and state its domain, range, and intervals where it is increasing and decreasing.

$$h(x) = \begin{cases} -4 & -10 \le x < -2 \\ -x^2 & -2 \le x < 3 \\ 3x - 18 & x \ge 3 \end{cases}$$

10. Solve the inequality and write the solution in interval notation: $\dfrac{2x + 1}{x - 3} \ge 0$.

11. Use the rational roots theorem to find all zeroes of $f(x) = x^4 - 3x^3 - 12x^2 + 52x - 48$.

12. Given $f(c) = \dfrac{9}{5}c + 32$, find k, where $k = f(25)$.

Then find the inverse function using the algebraic method, and verify that $f^{-1}(k) = 25$.

13. Solve the formula $V = \dfrac{1}{2}\pi b^2 a$ (the

volume of a paraboloid) for the variable b.

14. Use the *Guidelines for Graphing* to graph

   a. $p(x) = x^3 - 4x^2 + x + 6$.

   b. $r(x) = \dfrac{5x^2}{x^2 + 4}$.

15. For $f(x) = \dfrac{2x + 3}{5}$, (a) find f^{-1}, (b) graph both

functions and verify they are symmetric to the line $y = x$, and (c) show they are inverses using composition.

16. Solve for x: $10 = -2e^{-0.05x} + 25$.

17. Solve for x: $\ln(x + 3) + \ln(x - 2) = \ln(24)$.

18. Once in orbit, satellites are often powered by radioactive isotopes. From the natural process of radioactive decay, the power output declines over a period of time. For an initial amount of 50 g, suppose the power output is modeled by the function $p(t) = 50e^{-0.002t}$, where $p(t)$ is the power output in watts, t days after the satellite has been put into service. (a) Approximately how much power remains 6 *months* later? (b) How many *years* until only one-fourth of the original power remains?

19. Simon and Christine own a sport wagon and a minivan. The sport wagon has a power curve that is closely modeled by $H(r) = 123 \ln r - 897$, where $H(r)$ is the horsepower at r rpm, with $2200 \le r \le 5600$. The power curve for the minivan is $h(r) = 193 \ln r - 1464$, for $2600 < r < 5800$.

   a. How much horsepower is generated by each engine at 3000 rpm?

   b. At what rpm are the engines generating the same horsepower?

   c. If Christine wants the maximum horsepower available, which vehicle should she drive? What is the maximum horsepower?

20. Wilson's disease is a hereditary disease that causes the body to retain copper. Radioactive copper, ^{64}Cu, has been used extensively to study and understand this disease. ^{64}Cu has a relatively short half-life of 12.7 hr. How many hours will it take for a 5-g mass of ^{64}Cu to decay to 1 g?

Modeling with Technology II — Exponential, Logarithmic, and Other Regression Models

Learning Objectives

In this feature you will learn how to:

☐ **A.** Choose an appropriate form of regression for a set of data

☐ **B.** Use a calculator to obtain exponential and logarithmic regression models

☐ **C.** Determine when a logistics model is appropriate and apply logistics models to a set of data

☐ **D.** Use a regression model to answer questions and solve applications

WORTHY OF NOTE

For more information on the use of residuals, see the *Calculator Exploration and Discovery* feature on Residuals at www.mhhe.com/coburn

The basic concepts involved in calculating a regression equation were presented in *Modeling with Technology I*. In this section, we extend these concepts to data sets that are best modeled by power, exponential, logarithmic, or logistic functions. All data sets, while contextual and accurate, *have been carefully chosen* to provide a maximum focus on regression fundamentals and related mathematical concepts. In reality, data sets are often not so "well-behaved" and many require sophisticated statistical tests before any conclusions can be drawn.

A. Choosing an Appropriate Form of Regression

Most graphing calculators have the ability to perform several forms of regression, and selecting which of these to use is a critical issue. When various forms are applied to a given data set, some are easily discounted due to a poor fit. Others may fit very well for only a portion of the data, while still others may compete for being the "best-fit" equation. In a statistical study of regression, an in-depth look at the correlation coefficient (r), the coefficient of determination (r^2 or R^2), and a study of **residuals** are used to help make an appropriate choice. For our purposes, the correct or best choice will generally depend on two things: (1) how well the graph appears to fit the scatter-plot, and (2) the context or situation that generated the data, coupled with a dose of common sense.

As we've noted previously, the final choice of regression can rarely be based on the scatter-plot alone, although relying on the basic characteristics and end behavior of certain graphs can be helpful **(see Exercise 58).** With an awareness of the toolbox functions, polynomial graphs, and applications of exponential and logarithmic functions, the context of the data can aid a decision.

EXAMPLE 1 ▶ **Choosing an Appropriate Form of Regression**

Suppose a set of data is generated from each context given. Use common sense, previous experience, or your own knowledge base to state whether a linear, quadratic, logarithmic, exponential, or power regression might be most appropriate. Justify your answers.

  a. population growth of the United States since 1800

  b. the distance covered by a jogger running at a constant speed

  c. height of a baseball t seconds after it's thrown

  d. the time it takes for a cup of hot coffee to cool to room temperature

Solution ▶   **a.** From examples in Section 4.5 and elsewhere, we've seen that animal and human populations tend to grow exponentially over time. Here, an exponential model is likely most appropriate.

  b. Since the jogger is moving at a constant speed, the rate-of-change $\dfrac{\Delta \text{distance}}{\Delta \text{time}}$ is constant and a linear model would be most appropriate.

  c. As seen in numerous places throughout the text, the height of a projectile is modeled by the equation $h(t) = -16t^2 + vt + k$, where $h(t)$ is the height after t seconds. Here, a quadratic model would be most appropriate.

  d. Many have had the experience of pouring a cup of hot chocolate, coffee, or tea, only to leave it on the counter as they turn their attention to other things. The hot drink seems to cool quickly at first, then slowly approach room temperature. This experience, perhaps coupled with our awareness of *Newton's law of cooling*, shows a logarithmic or exponential model might be appropriate here.

☑ **A. You've just learned how to choose an appropriate form of regression for a set of data**

Now try Exercises 1 through 14 ▶

B. Exponential and Logarithmic Regression Models

We now focus our attention on regression models that involve exponential and logarithmic functions. Recall the process of developing a regression equation involves these five stages: (1) clearing old data, (2) entering new data, (3) displaying the data, (4) calculating the regression equation, and (5) displaying and using the regression graph and equation.

EXAMPLE 2 ▶ **Calculating an Exponential Regression Model**

The number of centenarians (people who are 100 years of age or older) has been climbing steadily over the last half century. The table shows the number of centenarians (per million population) for selected years. Use the data and a graphing calculator to draw the scatter-plot, then use the scatter-plot and context to decide on an appropriate form of regression.

Source: Data from 2004 *Statistical Abstract of the United States*, Table 14; various other years

Year "t" (1950 → 0)	Number "N" (per million)
0	16
10	18
20	25
30	74
40	115
50	262

Solution ▶ After clearing any existing data in the data lists, enter the input values (years since 1950) in L1 and the output values (number of centenarians per million population) in L2 (Figure MWT II.1). For the viewing window, scale the *x*-axis (years since 1950) from −10 to 70 and the *y*-axis (number per million) from −50 to 300 to comfortably fit the data and allow room for the coordinates to be shown at the bottom of the screen (Figure MWT II.2). The scatter-plot rules out a linear model. While a quadratic model may fit the data, we expect that the correct model should exhibit asymptotic behavior since extremely few people lived to be 100 years of age prior to dramatic advances in hygiene, diet, and medical care. This would lead us toward an exponential equation model. The keystrokes [STAT] [▶] brings up the **CALC** menu, with **ExpReg** (exponential regression) being option "0." The option can be selected by simply pressing "0," or by using the up arrow [▲] or down arrow [▼] to scroll to **0:ExpReg** then pressing [ENTER]. The exponential model seems to fit the data very well (Figures MWT II.3 and MWT II.4). To four decimal places the equation model is $y = (11.5090)1.0607^x$.

Figure MWT II.1

Figure MWT II.2

Figure MWT II.3

Figure MWT II.4

EXAMPLE 3 ▶ Calculating a Logarithmic Regression Model

One measure used in studies related to infant growth, nutrition, and development, is the relation between the circumference of a child's head and their age. The table to the right shows the average circumference of a female child's head for ages 0 to 36 months. Use the data and a graphing calculator to draw the scatter-plot, then use the scatter-plot and context to decide on an appropriate form of regression.

Source: *National Center for Health Statistics*

Age a (months)	Circumference C (cm)
0	34.8
6	43.0
12	45.2
18	46.5
24	47.5
30	48.2
36	48.6

Solution ▶ After clearing any existing data, enter the child's age (in months) as L1 and the circumference of the head (in cm) as L2. For the viewing window, scale the *x*-axis from −5 to 50 and the *y*-axis from 25 to 60 to comfortably fit the data (Figure MWT II.5). The scatter-plot again rules out a linear model, and the context rules out a polynomial model due to end-behavior. As we expect the circumference of the head to continue increasing slightly for many more months, it

WORTHY OF NOTE

For applications involving exponential growth and logarithmic functions, it helps to remember that while both basic functions are increasing, a logarithmic function increases at a much slower rate.

Figure MWT II.5

appears a logarithmic model may be the best fit. Note that since ln (0) is undefined, $a = 0.1$ was used to represent the age at birth (rather than $a = 0$), prior to running the regression. The **LnReg** (logarithmic regression) option is option 9, and the keystrokes [STAT] [▶] (CALC) **9** [ENTER] gives the equation shown in Figure MWT II.6, which fits the data very well (Figure MWT II.7).

Figure MWT II.6   **Figure MWT II.7**

☑ **B.** You've just learned how to use a calculator to obtain exponential and logarithmic regression models

Now try Exercises 17 and 18 ▶

C. Logistics Equations and Regression Models

Many population growth models assume an unlimited supply of resources, nutrients, and room for growth, resulting in an exponential growth model. When resources become scarce or room for further expansion is limited, the result is often a **logistic growth model.** At first, growth is very rapid (like an exponential function), but this growth begins to taper off and slow down as nutrients are used up, living space becomes restricted, or due to other factors. Surprisingly, this type of growth can take many forms, including population growth, the spread of a disease, the growth of a tumor, or the spread of a stain in fabric. Specific logistic equations were encountered in Section 4.4. The general equation model for logistic growth is

Logistic Growth

Given constants a, b, and c, the logistic growth $P(t)$ of a population depends on time t according to the model

$$P(t) = \frac{c}{1 + ae^{-bt}}$$

The constant c is called the **carrying capacity** of the population, in that as $t \to \infty$, $P(t) \to c$. In words, as the elapsed time becomes very large, the population will approach (but not exceed) c.

EXAMPLE 4 ▶ Calculating a Logistic Regression Model

Yeast cultures have a number of applications that are a great benefit to civilization and have been an object of study for centuries. A certain strain of yeast is grown in a lab, with its population checked at 2-hr intervals, and the data gathered are given in the table. Use the data and a graphing calculator to draw a scatter-plot, and decide on an appropriate form of regression. If a logistic regression is the best model, attempt to estimate the capacity coefficient c prior to using your calculator to find the regression equation. How close were you to the actual value?

Elapsed Time (hours)	Population (100s)
2	20
4	50
6	122
8	260
10	450
12	570
14	630
16	650

Solution ▶ After clearing the data lists, enter the input values (elapsed time) in L1 and the output values (population) in L2. For the viewing window, scale the t-axis from 0 to 20 and the P-axis from 0 to 700 to comfortably fit the data. From the context and scatter-plot, it's apparent the data are best modeled by a logistic function. Noting that Ymax = 700 and the data seem to level off near the top of the window, a good estimate for c would be about 675. Using logistic regression on the home screen (option **B:Logistic**), we obtain the equation $Y_1 = \dfrac{663}{1 + 123.9e^{-0.553x}}$ (rounded).

WORTHY OF NOTE

Notice that calculating a logistic regression model takes a few seconds longer than for other forms.

☑ **C.** You've just learned how to determine when a logistic model is appropriate and apply logistics models to a set of data

Now try Exercises 19 and 20 ▶

When a regression equation is used to gather information, many of the equation solving skills from prior sections are employed. **Exercises 21 through 28** offer a variety of these equations for practice and warm-up.

D. Applications of Regression

Once the equation model for a data set has been obtained, it can be used to **interpolate** or approximate values that might occur *between* those given in the data set. It can also be used to **extrapolate** or predict future values. In this case, the investigation extends *beyond* the values from the data set, and is based on the assumption that projected trends will continue for an extended period of time.

Regardless of the regression applied, interpolation and extrapolation involve substituting a given or known value, then solving for the remaining unknown. We'll demonstrate here using the regression model from Example 3. The exercise set offers a large variety of regression applications, including some power regressions and additional applications of linear and quadratic regression.

EXAMPLE 5 ▶ **Using a Regression Equation to Interpolate or Extrapolate Information**

Use the regression equation from Example 3 to answer the following questions:

  a. What is the average circumference of a female child's head, if the child is 21 months old?

  b. According to the equation model, what will the average circumference be when the child turns $3\frac{1}{2}$ years old?

  c. If the circumference of the child's head is 46.9 cm, about how old is the child?

Solution ▶   **a.** Using function notation we have $C(a) \approx 39.8171 + 2.3344 \ln(a)$. Substituting 21 for a gives:

$$C(21) \approx 39.8171 + 2.3344 \ln(21) \quad \text{substitute 21 for } a$$
$$\approx 46.9 \quad \text{result}$$

The circumference is approximately 46.9 cm.

  b. Substituting 3.5 yr $\times$ 12 = 42 months for a gives:

$$C(42) \approx 39.8171 + 2.3344 \ln(42) \quad \text{substitute 42 for } a$$
$$\approx 48.5 \quad \text{result}$$

The circumference will be approximately 48.5 cm.

  c. For part (c) we're given the circumference C and are asked to find the age a in which this circumference (46.9) occurs. Substituting 46.9 for $C(a)$ we obtain:

$$46.9 = 39.8171 + 2.3344 \ln(a) \quad \text{substitute 46.9 for } C(a)$$

$$\frac{7.0829}{2.3344} = \ln(a) \quad \text{subtract 39.8171, then divide by 2.3344}$$

$$e^{\frac{7.0859}{2.3344}} = a \quad \text{write in exponential form}$$

$$20.8 \approx a \quad \text{result}$$

The child must be about 21 months old.

 D. You've just learned how to use a regression model to answer questions and solve applications

WORTHY OF NOTE

When extrapolating from a set of data, care and common sense must be used or results can be very misleading. For example, while the Olympic record for the 100-m dash has been steadily declining since the first Olympic Games, it would be foolish to think it will ever be run in 0 sec.

Now try Exercises 29 through 32 ▶

MODELING WITH TECHNOLOGY EXERCISES

▶ **DEVELOPING YOUR SKILLS**

Match each scatter-plot given with one of the following: **(a)** likely linear, **(b)** likely quadratic, **(c)** likely exponential, **(d)** likely logarithmic, **(e)** likely logistic, or **(f)** none of these.

3.   **4.**

5.   **6.**

For Exercises 7 to 14, suppose a set of data is generated from the context indicated. Use common sense, previous experience, or your own knowledge base to state whether a linear, quadratic, logarithmic, exponential, power, or logistic regression might be most appropriate. Justify your answers.

7. total revenue and number of units sold

8. page count in a book and total number of words

9. years on the job and annual salary

10. population growth with unlimited resources

11. population growth with limited resources

12. elapsed time and the height of a projectile

13. the cost of a gallon of milk over time

14. elapsed time and radioactive decay

Discuss why an exponential model could be an appropriate form of regression for each data set, then find the regression equation.

15. Radioactive Studies

Time in Hours	Grams of Material
0.1	1.0
1	0.6
2	0.3
3	0.2
4	0.1
5	0.06

16. Rabbit Population

Month	Population (in hundreds)
0	2.5
3	5.0
6	6.1
9	12.3
12	17.8
15	30.2

Discuss why a logarithmic model could be an appropriate form of regression for each data set, then find the regression equation.

17. Total number of sales compared to the amount spent on advertising

Advertising Costs ($1000s)	Total Number of Sales
1	125
5	437
10	652
15	710
20	770
25	848
30	858
35	864

18. Cumulative weight of diamonds extracted from a diamond mine

Time (months)	Weight (carats)
1	500
3	1748
6	2263
9	2610
12	3158
15	3501
18	3689
21	3810

19. Spread of disease: Estimates of the cumulative number of SARS (sudden acute respiratory syndrome) cases reported in Hong-Kong during the spring of 2003 are shown in the table, with day 0 corresponding to February 20. (a) Use the data to draw a scatter-plot, then use the context and scatter-plot to decide on the best form of regression. (b) If a logistic model seems best, attempt to estimate the carrying capacity c, then (c) use your calculator to find the regression equation.

Days After Outbreak	Cumulative Total
0	100
14	560
21	870
35	1390
56	1660
70	1710
84	1750

Source: Center for Disease Control @ www.cdc.gov/ncidod/EID/vol9no12.

20. Cable television subscribers: The percentage of American households having cable television is given in the table for select years from 1976 to 2004. (a) Use the data to draw a scatter-plot, then use the context and scatter-plot to decide on the best form of regression. (b) If a logistic model seems best, attempt to estimate the carrying capacity c, then (c) use your calculator to find the regression equation (use 1976 → 0).

Year 1976 → 0	Percentage with Cable TV
0	16
4	22.6
8	43.7
12	53.8
16	61.5
20	66.7
24	68
28	70

Source: Data pooled from the 2001 *New York Times Almanac*, p. 393; 2004 *Statistical Abstract of the United States*, Table 1120; various other years.

The applications in this section require solving equations similar to those that follow. Solve each equation.

21. $96.35 = (9.4)1.6^x$ **22.** $(3.7)2.9^x = 1253.93$

23. $4.8x^{2.5} = 468.75$ **24.** $4375 = 1.4x^{-1.25}$

25. $52 = 63.9 - 6.8\ln x$

26. $498.53 + 18.2\ln x = 595.9$

27. $52 = \dfrac{67}{1 + 20e^{-0.62x}}$ **28.** $\dfrac{975}{1 + 82.3e^{-0.423x}} = 890$

▶ APPLICATIONS

Answer the questions using the given data and the related regression equation. All extrapolations assume the mathematical model will continue to represent future trends.

29. Weight loss: Harold needed to lose weight and started on a new diet and exercise regimen. The number of pounds he's lost since the diet began is given in the table. Draw the scatter-plot, decide on an appropriate form of regression, and find an equation that models the data.

Time (days)	Pounds Lost
10	2
20	14
30	20
40	23
50	25.5
60	27.6
70	29.2
80	30.7

   a. What was Harold's total weight loss after 15 days?

   b. Approximately how many days did it take to lose a total of 18 pounds?

   c. According to the model, what is the projected weight loss for 100 days?

30. Depletion of resources: The longer an area is mined for gold, the more difficult and expensive it gets to obtain. The cumulative total of the ounces produced by a particular mine is shown in the table. Draw the scatter-plot, use the scatter-plot and context to determine whether an exponential or logarithmic model is more appropriate, then find an equation that models the data.

Time (months)	Ounces Mined
5	275
10	1890
15	2610
20	3158
25	3501
30	3789
35	4109
40	4309

   a. What was his total number of ounces mined after 18 months?

   b. About how many months did it take to mine a total of 4000 oz?

   c. According to the model, what is the projected total after 50 months?

31. Number of U.S. post offices: Due in large part to the ease of travel and increased use of telephones, e-mail and instant messaging, the number of post offices in the United States has been on the decline since the twentieth century. The data given show number of post offices (in thousands) for selected years. Use the data to draw a scatter-plot, then use the context and scatter-plot to find the regression equation (use 1900 → 0).

Year (1900 → 0)	Offices (1000s)
1	77
20	52
40	43
60	37
80	32
100	28

Source: Statistical Abstract of the United States; The First Measured Century

   a. Approximately how many post offices were there in 1915?

   b. In what year did the number of post offices drop below 34,000?

   c. According to the model, how many post offices will there be in the year 2010?

32. Automobile value: While it is well known that most cars decrease in value over time, what is the equation model for this decline? Use the data given to draw a scatter-plot, then use the context and scatter-plot to find the regression equation.

Age of Car	Value of Car
1	19,500
2	16,950
4	12,420
6	11,350
8	8,375
10	7,935
12	6,900

   a. What was the car's value after 7.5 years?

   b. About how old is the car if its current value is $8150?

   c. Using the model, how old is the car when value ≤ $3000?

33. Female physicians: The number of females practicing medicine as MDs is given in the table for selected years. Use the data to draw a scatter-plot, then use the context and scatter-plot to find the regression equation.

Year (1980 → 0)	Number (in 1000s)
0	48.7
5	74.8
10	96.1
13	117.2
14	124.9
15	140.1
16	148.3

Source: Statistical Abstract of the United States.

a. What was the approximate number of female MDs in 1988?

b. Approximately how many female MDs will there be in 2005?

c. In what year did the number of female MDs exceed 100,000?

34. Telephone use: The number of telephone calls per capita has been rising dramatically since the invention of the telephone in 1876. The table shows the number of phone calls per capita per year for selected years. Use the data to draw a scatter-plot, then use the context and scatter-plot to find the regression equation.

Year (1900 → 0)	Number (per capita/ per year)
0	38
20	180
40	260
60	590
80	1250
97	2325

Source: The First Measured Century by Theodore Caplow, Louis Hicks, and Ben J. Wattenberg, The AEI Press, Washington, D.C., 2001.

a. What was the approximate number of calls per capita in 1970?

b. Approximately how many calls per capita will there be in 2005?

c. In what year did the number of calls per capita exceed 1800?

35. Milk production: Since 1980, the number of family farms with milk cows for commercial production has been decreasing. Use the data from the table given to draw a scatter-plot, then use the context and scatter-plot to find the regression equation.

Year (1980 → 0)	Number (in 1000s)
0	334
5	269
10	193
15	140
17	124
18	117
19	111

Source: Statistical Abstract of the United States, 2000.

a. What was the approximate number of farms with milk cows in 1993?

b. Approximately how many farms will have milk cows in 2004?

c. In what year did this number of farms drop below 150 thousand?

36. Froth height—carbonated beverages: The height of the froth on carbonated drinks and other beverages can be manipulated by the ingredients used in making the beverage and lends itself very well to the modeling process. The data in the table given show the froth height of a certain beverage as a function of time, after the froth has reached a maximum height. Use the data to draw a scatter-plot, then use the context and scatter-plot to find the regression equation.

Time (seconds)	Height of Froth (in.)
0	0.90
2	0.65
4	0.40
6	0.21
8	0.15
10	0.12
12	0.08

a. What was the approximate height of the froth after 6.5 sec?

b. How long does it take for the height of the froth to reach one-half of its maximum height?

c. According to the model, how many seconds until the froth height is 0.02 in.?

37. Chicken production: In 1980, the production of chickens in the United States was about 392 million. In the next decade, the demand for chicken first dropped, then rose dramatically. The number of chickens produced is given in the table to the right for selected years. Use the data to draw a scatter-plot, then use the context and scatter-plot to find the regression equation.

Year (1980 → 0)	Number (millions)
0	392
5	370
9	356
14	386
16	393
17	410
18	424

Source: Statistical Abstract of the United States, 2000.

a. What was the approximate number of chickens produced in 1987?

b. Approximately how many chickens will be produced in 2004?

c. According to the model, for what years was the production of chickens below 365 million?

38. Veterans in civilian life: The number of military veterans in civilian life fluctuates with the number of persons inducted into the military (higher in times of war) and the passing of time. The number of living veterans is given in the table for selected years from 1950 to 1999. Use the data to draw a scatter-plot, then use the context and scatter-plot to find the regression equation.

Year (1950 → 0)	Number (millions)
0	19.1
10	22.5
20	27.6
30	28.6
40	27
48	25.1
49	24.6

Source: Statistical Abstract of the United States, 2000.

a. What was the approximate number of living military veterans in 1995?

b. Approximately how many living veterans will there be in 2006?

c. According to the model, in what years did the number of veterans exceed 26 million?

39. Use of debit cards:
Since 1990, the use of debit cards to obtain cash and pay for purchases has become very common. The number of debit cards nationwide is given in the table for selected years. Use the data to draw a scatter-plot, then use the context and scatter-plot to find the regression equation.

Year (1990 → 0)	Number of Cards (millions)
0	164
5	201
8	217
10	230

Source: Statistical Abstract of the United States, 2000.

a. Approximately how many debit cards were there in 1999?

b. Approximately how many debit cards will there be in 2005?

c. In what year did the number of debit cards exceed 300 million?

40. Quiz grade versus study time: To determine the value of doing homework, a student in college algebra records the time spent by classmates in preparation for a quiz the next day. Then she records their scores, which are shown in the table. Use the data to draw a scatter-plot, then use the context and scatter-plot to find the regression equation. According to the model, what grade can I expect if I study for 120 min?

x (min study)	y (score)
45	70
30	63
10	59
20	67
60	73
70	85
90	82
75	90

41. Population of coastal areas: The percentage of the U.S. population that can be categorized as living in *Pacific coastal areas* (minimum of 15% of the state's land area is a coastal watershed) has been growing steadily for decades, as indicated by the data given for selected years. Use the

Year	Percentage
1970	22.8
1980	27.0
1990	33.2
1995	35.2
2000	37.8
2001	38.5
2002	38.9
2003	39.4

data to draw a scatter-plot, then use the context and scatter-plot to find the regression equation. According to the model, what is the predicted percentage of the population living in Pacific coastal areas in 2005 and 2010?

Source: 2004 Statistical Abstract of the United States, Table 23.

42. Water depth and pressure:
As anyone who's been swimming knows, the deeper you dive, the more pressure you feel on your body and eardrums. This pressure (in pounds per square inch or psi) is shown in the table for selected depths. Use the data to draw a scatter-plot, then use the context and scatter-plot to find the regression equation. According to the model, what pressure can be expected at a depth of 100 ft?

Depth (ft)	Pressure (psi)
15	6.94
25	11.85
35	15.64
45	19.58
55	24.35
65	28.27
75	32.68

43. Personal debt-load: The data given tracks the total amount of debt carried by a family over a 6-month period. Use the data to draw a scatter-plot, then use the context and scatter-plot to find the regression equation. According to the model, how much debt will the family have by the end of December? When will their debt-load exceed $10,000?

Month (x)	Debt (y)
(start)	$0
Jan	471
Feb	1105
March	1513
April	1921
May	2498
June	3129

44. Use of debit cards:
Since 1990, the dollar volume of business transacted using debit cards has been growing. The volume of business nationwide is given in the table to the right for selected years. Use the

Year (1990 → 0)	Volume (billions)
0	12
5	62
8	239
10	423

data to draw a scatter-plot, then use the context and scatter-plot to find the regression equation.

Source: Statistical Abstract of the United States, 2004.

a. In 1993, what was the approximate dollar volume of business transacted with debit cards?

b. Approximately how much dollar volume of business was transacted in 1997?

c. In what year did the volume of business transacted using debit cards exceed 1000 billion?

45. Musical notes: The table shown gives the frequency (vibrations per second for each of the twelve notes in a selected octave) from the standard chromatic scale. Use the data to draw a scatter-plot, then use the context and scatter-plot to find the regression equation.

#	Note	Frequency
1	A	110.00
2	A#	116.54
3	B	123.48
4	C	130.82
5	C#	138.60
6	D	146.84
7	D#	155.56
8	E	164.82
9	F	174.62
10	F#	185.00
11	G	196.00
12	G#	207.66

a. What is the frequency of the "A" note that is an octave higher than the one shown? [*Hint:* The names repeat every 12 notes (one octave), so this would be the 13th note in this sequence.]

b. If the frequency is 370.00 what note is being played?

c. What pattern do you notice for the F#'s in each octave (the 10th, 22nd, 34th, and 46th notes in sequence)? Does the pattern hold for all notes?

46. Basketball salaries: In 1970, the average player salary for a professional basketball player was about $43,000. Since that time player salaries have risen dramatically. The average player salary for a professional player is given in the table to the right for selected years. Use the data to draw a scatter-plot, then use the context and scatter-plot to find the regression equation.

Year (1970 → 0)	Salary ($1000s)
0	43
10	260
15	325
20	750
25	1900
27	2200
28	2600

Source: Wall Street Journal Almanac.

a. What was the approximate salary for a player in 1993?

b. Approximately how much will the average salary be in 2005?

c. In what year did the average salary exceed $5,000,000?

47. Cost of cable service: The average monthly cost of cable TV has been rising steadily since it became very popular in the early 1980s. The data given shows the average monthly rate for selected years (1980 → 0). Use the data to draw a scatter-plot, then use the context and

Year (1980 → 0)	Monthly Charge
0	$7.69
5	$9.73
10	$16.78
20	$23.07
25	$30.70

scatter-plot to find the regression equation. According to the model, what will be the cost of cable service in 2010? 2015?

Source: 2004–2005 Statistical Abstract of the United States, page 725, Table 1138.

48. Research and development expenditures: The development of new products, improved health care, greater scientific achievement, and other advances is fueled by huge investments in research and development (R & D). Since 1960, total R & D expenditures in the United States have shown a distinct pattern of growth, and the data are given in the table for selected years from 1960 to 1999. Use the data to draw a scatter-plot, then use the context and scatter-plot to find the regression equation. According to the model, what was spent on R & D in 1992? In what year did expenditures for R & D exceed 450 billion?

Year (1960 → 0)	R & D (billion $)
0	13.7
5	20.3
10	26.3
15	35.7
20	63.3
25	114.7
30	152.0
35	183.2
39	247.0

49. Business start-up costs: As many new businesses open, they experience a period where little or no profit is realized due to start-up expenses, equipment purchases, and so on. The data given shows the profit of a new company for the first 6 months of business. Use the data to draw a scatter-plot, then use the context and scatter-plot to find the regression equation. According to the model, what is the first month that a profit will be earned?

Month	Profit ($1000s)
1	−5
2	−13
3	−18
4	−20
5	−21
6	−19

50. Low birth weight: For many years, the association between low birth weight (less than 2500 g or about 5.5 lb) and a mother's age has been well documented. The data given are grouped by age and give the percent of total births with low birth weight.

Source: National Vital Statistics Report, Vol. 50, No. 5, February 12, 2002.

Ages	Percent
15–19	8.5
20–24	6.5
25–29	5.2
30–34	5
35–39	6
40–44	8
45–54	10

a. Using the median age of each group, use the data to draw a scatter-plot and decide on an appropriate form of regression.

b. Find a regression equation that models the data. According to the model, what percent of births will have a low birth weight if the mother was 58 years old?

51. Growth of cell phone use: The tremendous surge in cell phone use that began in the early nineties has continued unabated into the new century. The total number of subscriptions is shown in the table for selected years, with 1990 → 0 and the number of subscriptions in millions. Use the data to draw a scatter-plot. Does the data seem to follow an exponential or logistic pattern? Find the regression equation. According to the model, how many subscriptions were there in 1997? How many subscriptions does your model project for 2005? 2010? In what year will the subscriptions exceed 220 million?

Year (1990 → 0)	Subscriptions (millions)
0	5.3
3	16.0
6	44.0
8	69.2
12	140.0
13	158.7

Source: 2000/2004 Statistical Abstracts of the United States, Tables 919/1144.

52. Absorption rates of fabric: Using time lapse photography, the spread of a liquid is tracked in one-fifth of a second intervals, as a small amount of liquid is dropped on a piece of fabric. Use the data to draw a scatter-plot, then use the context and scatter-plot to find the regression equation. To the nearest hundredth of a second, how long will it take the stain to reach a size of 15 mm?

Time (sec)	Size (mm)
0.2	0.39
0.4	1.27
0.6	3.90
0.8	10.60
1.0	21.50
1.2	31.30
1.4	36.30
1.6	38.10
1.8	39.00

53. Planetary orbits: The table shown gives the time required for the first five planets to make one complete revolution around the Sun (in years), along with the average orbital radius of the planet in astronomical units (1 AU = 92.96 million miles). Use a graphing calculator to draw the scatter-plot, then use the scatter-plot, the context, and any previous experience to decide whether a polynomial, exponential, logarithmic,

Planet	Years	Radius
Mercury	0.24	0.39
Venus	0.62	0.72
Earth	1.00	1.00
Mars	1.88	1.52
Jupiter	11.86	5.20

or power regression is most appropriate. Then (a) find the regression equation and use it to estimate the average orbital radius of Saturn, given it orbits the Sun every 29.46 yr, and (b) estimate how many years it takes Uranus to orbit the Sun, given it has an average orbital radius of 19.2 AU.

54. Ocean temperatures: The temperature of ocean water depends on several factors, including salinity, latitude, depth, and density. However, between depths of 125 m and 2000 m, ocean temperatures are relatively predictable, as indicated by the data shown for tropical oceans in the table. Use a graphing calculator to draw the scatter-plot, then use the scatter-plot, the context, and any previous experience to decide whether a polynomial, exponential, logarithmic, or power regression is most appropriate (end behavior rules out linear and quadratic models as possibilities).

Depth (meters)	Temp (°C)
125	13.0
250	9.0
500	6.0
750	5.0
1000	4.4
1250	3.8
1500	3.1
1750	2.8
2000	2.5

Source: UCLA at www.msc.ucla.oceanglobe/pdf/ thermo_plot_lab

a. Find the regression equation and use it to estimate the water temperature at a depth of 2850 m.

b. If the model were still valid at greater depths, what is the ocean temperature at the bottom of the Marianas Trench, some 10,900 m below sea level?

55. Predator/prey model: In the wild, some rodent populations vary inversely with the number of predators in the area. Over a period of time, a conservation team does an extensive study on this relationship and gathers the data shown. Draw a scatter-plot of the data and (a) find a regression equation that models the data. According to the model, (b) if there are 150 predators in the area, what is the rodent population? (c) How many predators are in the area if studies show a rodent population of 3000 animals?

Predators	Rodents
10	5100
20	2500
30	1600
40	1200
50	950
60	775
70	660
80	575
90	500
100	450

56. Children and AIDS:
Largely due to research, education, prevention, and better health care, estimates of the number of AIDS (acquired immune deficiency syndrome) cases diagnosed in children less than 13 yr of age have been declining. Data for the years 1995 through 2002 is given in the table.

Source: National Center for Disease Control and Prevention.

Years Since 1990	Cases
5	686
6	518
7	328
8	238
9	183
10	118
11	110
12	92

a. Use the data to draw a scatter-plot and decide on an appropriate form of regression.

b. Find a regression equation that models the data. According to the model, how many cases of AIDS in children are projected for 2010?

c. In what year did the number of cases fall below 50?

57. Growth rates of children: After reading a report from The National Center for Health Statistics regarding the growth of children from age 0 to 36 months, Maryann decides to track the relationships (length in inches, weight in pounds) and (age in months, circumference of head in centimeters) for her newborn child, a beautiful baby girl—Morgan.

a. Use the (length, weight) data to draw a scatter-plot, then use the context and scatter-plot to find the regression equation. According to the model, how much will Morgan weigh when she reaches a height (length) of 39 in.? What will her length be when she weighs 28 lb?

b. Use the (age, circumference) data to draw a scatter-plot, then use the context and

scatter-plot to find the regression equation. According to the model, what is the circumference of Morgan's head when she is 27 months old? How old will she be when the circumference of her head is 50 cm?

Exercise 57a

Length (in.)	Weight (lb)
17.5	5.50
21	10.75
25.5	16.25
28.5	19.00
33	25.25

Exercise 57b

Age (months)	Circumference (cm)
1	38.0
6	44.0
12	46.5
18	48.0
21	48.3

58. Correlation coefficients: Although correlation coefficients can be very helpful, other factors must also be considered when selecting the most appropriate equation model for a set of data. To see why, use the data given to (a) find a linear regression equation and note its correlation coefficient, and (b) find an exponential

regression equation and note its correlation coefficient. What do you notice? Without knowing the context of the data, would you be able to tell which model might be more suitable? (c) Use your calculator to graph the scatter-plot and both functions. Which function appears to be a better fit?

5

Systems of Equations and Inequalities

CHAPTER OUTLINE

5.1 Linear Systems in Two Variables with Applications 504

5.2 Linear Systems in Three Variables with Applications 516

5.3 Nonlinear Systems of Equations and Inequalities 529

5.4 Systems of Inequalities and Linear Programming 536

CHAPTER CONNECTIONS

The disposal of hazardous waste is a growing concern for today's communities, and with many budgets stretched to the breaking point, there is a cost/benefit analysis involved. One major hauler uses trucks with a carrying capacity of 800 ft^3, and can transport at most 10 tons. A full container of liquid waste weighs 800 lb and has a volume of 20 ft^3, while a full container of solid waste weighs 600 lb and has a volume of 30 ft^3. If the hauler makes \$300 for disposing of liquid waste and \$400 for disposing of solid waste, what is the maximum revenue that can be generated per truck? Chapter 5 outlines a systematic process for answering this question. This application appears as Exercise 58 in Section 5.4.

Check out these other real-world connections:

▶ Appropriate Measurements in Dietetics (Section 5.1, Exercise 64)

▶ Allocating Winnings to Different Investments (Section 5.2, Exercise 54)

▶ Minimizing Shipping Costs (Section 5.4, Exercise 61)

▶ Market Pricing for Organic Produce (Section 5.3, Exercise 56)

503

Learning Objectives

In Section 5.1 you will learn how to:

☐ **A.** Verify ordered pair solutions

☐ **B.** Solve linear systems by graphing

☐ **C.** Solve linear systems by substitution

☐ **D.** Solve linear systems by elimination

☐ **E.** Recognize inconsistent systems and dependent systems

☐ **F.** Use a system of equations to model and solve applications

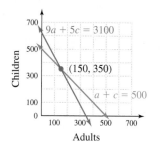

In earlier chapters, we used linear equations in two variables to model a number of real-world situations. Graphing these equations gave us a visual image of how the variables were related, and helped us better understand this relationship. In many applications, two different measures of the independent variable must be considered simultaneously, leading to a **system of two linear equations in two unknowns.** Here, a graphical presentation once again supports a better understanding, as we explore systems and their many applications.

A. Solutions to a System of Equations

A **system of equations** is a set of two or more equations for which a common solution is sought. Systems are widely used to model and solve applications when the information given enables the relationship between variables to be stated in different ways. For example, consider an amusement park that brought in $3100 in revenue by charging $9.00 for adults and $5.00 for children, while selling 500 tickets. Using a for adult and c for children, we could write one equation modeling the number of tickets sold: $a + c = 500$, and a second modeling the amount of revenue brought in: $9a + 5c = 3100$. To show that we're considering both equations simultaneously, a large "left brace" is used and the result is called a **system of two equations in two variables:**

$$\begin{cases} a + c = 500 & \text{number of tickets} \\ 9a + 5c = 3100 & \text{amount of revenue} \end{cases}$$

We note that both equations are linear and will have different slope values, so their graphs must intersect at some point. Since every point on a line satisfies the equation of that line, this point of intersection must satisfy *both* equations simultaneously and is the solution to the system. The figure that accompanies Example 1 shows the point of intersecion for this system is (150, 350).

EXAMPLE 1 ▶ **Verifying Solutions to a System**

Verify that (150, 350) is a solution to $\begin{cases} a + c = 500 \\ 9a + 5c = 3100 \end{cases}$.

Solution ▶ Substitute the **150** for a and **350** for c in each equation.

$$a + c = 500 \quad \text{first equation} \qquad\qquad 9a + 5c = 3100 \quad \text{second equation}$$
$$(150) + (350) = 500 \qquad\qquad\qquad 9(150) + 5(350) = 3100$$
$$500 = 500 \checkmark \qquad\qquad\qquad\qquad 3100 = 3100 \checkmark$$

☑ **A.** You've just learned how to verify ordered pair solutions

Since (150, 350) satisfies both equations, it is the solution to the system and we find the park sold 150 adult tickets and 350 tickets for children.

Now try Exercises 7 through 18 ▶

B. Solving Systems Graphically

To **solve a system of equations** means we apply various methods in an attempt to find ordered pair solutions. As Example 1 suggests, one method for finding solutions is to graph the system. Any method for graphing the lines can be employed, but to keep important concepts fresh, the slope-intercept method is used here.

EXAMPLE 2 ▶ **Solving a System Graphically**

Solve the system by graphing: $\begin{cases} 4x - 3y = 9 \\ -2x + y = -5 \end{cases}$.

Solution ▶ First write each equation in slope-intercept form (solve for y):

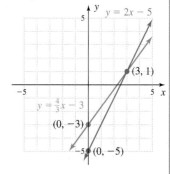

$$\begin{cases} 4x - 3y = 9 \\ -2x + y = -5 \end{cases} \rightarrow \begin{cases} y = \dfrac{4}{3}x - 3 \\ y = 2x - 5 \end{cases}$$

For the first line, $\frac{\Delta y}{\Delta x} = \frac{4}{3}$ with y-intercept $(0, -3)$. The second equation yields $\frac{\Delta y}{\Delta x} = \frac{2}{1}$ with $(0, -5)$ as the y-intercept. Both are then graphed on the grid as shown. The point of intersection appears to be $(3, 1)$, and checking this point in both equations gives

$$4x - 3y = 9 \qquad\qquad -2x + y = -5$$
$$4(3) - 3(1) = 9 \quad \text{substitute 3} \quad -2(3) + (1) = -5$$
$$9 = 9 \checkmark \quad \text{for } x \text{ and 1 for } y \qquad -5 = -5 \checkmark$$

☑ **B. You've just learned how to solve linear systems by graphing**

This verifies that $(3, 1)$ is the solution to the system.

Now try Exercises 19 through 22 ▶

C. Solving Systems by Substitution

While a graphical approach best illustrates *why* the solution must be an ordered pair, it does have one obvious drawback—noninteger solutions are difficult to spot. The ordered pair $\left(\frac{2}{5}, \frac{12}{5}\right)$ is the solution to $\begin{cases} 4x + y = 4 \\ y = x + 2 \end{cases}$, but this would be difficult to "pinpoint" as a precise location on a hand-drawn graph. To overcome this limitation, we next consider a method known as **substitution.** The method involves converting a system of two equations in two variables into a single equation in one variable by using an appropriate substitution. For $\begin{cases} 4x + y = 4 \\ y = x + 2 \end{cases}$, the second equation says "y is two more than x." We reason that *all* points on this line are related this way, *including the point where this line intersects the other.* For this reason, we can substitute $x + 2$ for y in *the first equation,* obtaining a single equation in x.

EXAMPLE 3 ▶ **Solving a System Using Substitution**

Solve using substitution: $\begin{cases} 4x + y = 4 \\ y = x + 2 \end{cases}$.

Solution ▶ Since $y = x + 2$, we can replace y with $x + 2$ in the first equation.

$$4x + y = 4 \qquad \text{first equation}$$
$$4x + (x + 2) = 4 \qquad \text{substitute } x + 2 \text{ for } y$$
$$5x + 2 = 4 \qquad \text{simplify}$$
$$x = \frac{2}{5} \qquad \text{result}$$

The x-coordinate is $\frac{2}{5}$. To find the y-coordinate, substitute $\frac{2}{5}$ for x into either of the original equations. Substituting in the second equation gives

$$y = x + 2 \qquad \text{second equation}$$
$$= \frac{2}{5} + 2 \qquad \text{substitute } \frac{2}{5} \text{ for } x$$
$$= \frac{12}{5} \qquad \frac{2}{1} = \frac{10}{5}, \frac{10}{5} + \frac{2}{5} = \frac{12}{5}$$

The solution to the system is $\left(\frac{2}{5}, \frac{12}{5}\right)$. Verify by substituting $\frac{2}{5}$ for x and $\frac{12}{5}$ for y into both equations.

Now try Exercises 23 through 32 ▶

If neither equation allows an immediate substitution, we first solve for one of the variables, either x or y, and *then* substitute. The method is summarized here, and can actually be used with either like variables or like variable expressions. **See Exercises 57 to 60.**

Solving Systems Using Substitution

1. Solve one of the equations for x in terms of y or y in terms of x.
2. Substitute for the appropriate variable in the *other* equation and solve for the variable that remains.
3. Substitute the value from step 2 into either of the original equations and solve for the other unknown.
4. Write the answer as an ordered pair and check the solution in both original equations.

☑ **C.** You've just learned how to solve linear systems by substitution

D. Solving Systems Using Elimination

Now consider the system $\begin{cases} -2x + 5y = 13 \\ 2x - 3y = -7 \end{cases}$, where solving for any one of the variables will result in fractional values. The substitution method can still be used, but often the **elimination method** is more efficient. The method takes its name from what happens when you add certain equations in a system (by adding the like terms from each). If the coefficients of either x or y are additive inverses—they sum to zero and are *eliminated*. For the system shown, adding the equations produces $2y = 6$, giving $y = 3$, then $x = 1$ using back-substitution (verify).

When neither variable term meets this condition, we can multiply one or both equations by a nonzero constant to "match up" the coefficients, so an elimination will take place. In doing so, we create an **equivalent system of equations,** meaning one that has the same solution as the original system. For $\begin{cases} 7x - 4y = 16 \\ -3x + 2y = -6 \end{cases}$, multiplying the second equation by 2 produces $\begin{cases} 7x - 4y = 16 \\ -6x + 4y = -12 \end{cases}$, giving $x = 4$ after "adding the equations." Note the three systems produced are equivalent, and have the solution $(4, 3)$ ($y = 3$ was found using back-substitution).

$$1. \begin{cases} 7x - 4y = 16 \\ -3x + 2y = -6 \end{cases} \qquad 2. \begin{cases} 7x - 4y = 16 \\ -6x + 4y = -12 \end{cases} \qquad 3. \begin{cases} 7x - 4y = 16 \\ \quad\quad x = 4 \end{cases}$$

In summary,

Operations that Produce an Equivalent System

1. Changing the order of the equations.
2. Replacing an equation by a nonzero constant multiple of that equation.
3. Replacing an equation with the sum of two equations from the system.

Before beginning a solution using elimination, check to make sure the equations are written in the **standard form** $Ax + By = C$, so that like terms will appear above/below each other. Throughout this chapter, we will use R1 to represent the equation in *row 1* of the system, R2 to represent the equation in *row 2*, and so on. These designations are used to help describe and document the steps being used to solve a system, as in Example 4 where 2R1 + R2 indicates the first equation has been multiplied by two, with the result added to the second equation.

EXAMPLE 4 ▶ **Solving a System by Elimination**

Solve using elimination: $\begin{cases} 2x - 3y = 7 \\ 6y + 5x = 4 \end{cases}$

Solution ▶ The second equation is not in standard form, so we re-write the system as $\begin{cases} 2x - 3y = 7 \\ 5x + 6y = 4 \end{cases}$. If we "add the equations" now, we would get $7x + 3y = 11$, with neither variable eliminated. However, if we multiply *both sides* of the first equation by 2, the y-coefficients will be additive inverses. The sum then results in an equation with x as the only unknown.

$$
\begin{array}{r}
2\text{R1} \\
+ \\
\text{R2} \\
\hline
\text{sum}
\end{array}
\left\{
\begin{array}{rcr}
4x - 6y &=& 14 \\
5x + 6y &=& 4 \\
\hline
9x + 0y &=& 18
\end{array}
\right. \quad \text{add}
$$

$$9x = 18$$
$$x = 2 \quad \text{solve for } x$$

Substituting 2 for x back into either of the original equations yields $y = -1$. The ordered pair solution is $(2, -1)$. Verify using the original equations.

Now try Exercises 33 through 38 ▶

The elimination method is summarized here. If either equation has fraction or decimal coefficients, we can "clear" them using an appropriate constant multiplier.

> **Solving Systems Using Elimination**
>
> **1.** Write each equation in standard form: $Ax + By = C$.
> **2.** Multiply one or both equations by a constant that will create coefficients of x (or y) that are additive inverses.
> **3.** Combine the two equations using vertical addition and solve for the variable that remains.
> **4.** Substitute the value from step 3 into either of the original equations and solve for the other unknown.
> **5.** Write the answer as an ordered pair and check the solution in both original equations.

EXAMPLE 5 ▶ **Solving a System Using Elimination**

Solve using elimination: $\begin{cases} \frac{5}{8}x - \frac{3}{4}y = \frac{1}{4} \\ \frac{1}{2}x - \frac{2}{3}y = 1 \end{cases}$.

Solution ▶ Multiplying the first equation by 8(8R1) and the second equation by 6(6R2) will clear the fractions from each.

$$
\begin{array}{c}
8\text{R1} \\
6\text{R2}
\end{array}
\left\{
\begin{array}{l}
\frac{8}{1}\left(\frac{5}{8}\right)x - \frac{8}{1}\left(\frac{3}{4}\right)y = \frac{8}{1}\left(\frac{1}{4}\right) \\
\frac{6}{1}\left(\frac{1}{2}\right)x - \frac{6}{1}\left(\frac{2}{3}\right)y = 6(1)
\end{array}
\right.
\rightarrow
\left\{
\begin{array}{l}
5x - 6y = 2 \\
3x - 4y = 6
\end{array}
\right.
$$

The x-terms can now be eliminated if we use $3\text{R1} + (-5\text{R2})$.

$$
\begin{array}{r}
3\text{R1} \\
+ \\
-5\text{R2} \\
\hline
\text{sum}
\end{array}
\left\{
\begin{array}{rcr}
15x - 18y &=& 6 \\
-15x + 20y &=& -30 \\
\hline
0x + 2y &=& -24
\end{array}
\right. \quad \text{add}
$$

$$y = -12 \quad \text{solve for } y$$

Substituting $y = -12$ in either of the original equations yields $x = -14$, and the solution is $(-14, -12)$. Verify by substituting in both equations.

☑ **D.** You've just learned how to solve linear systems by elimination

Now try Exercises 39 through 44 ▶

⚠ **CAUTION** ▶ Be sure to multiply *all* terms (on both sides) of the equation when using a constant multiplier. Also, note that for Example 5, we could have eliminated the *y*-terms using 2R1 with −3R2.

E. Inconsistent and Dependent Systems

A system having *at least one* solution is called a **consistent system.** As seen in Example 2, if the lines have different slopes, they intersect at a single point and the system has exactly one solution. Here, the lines are *independent* of each other and the system is called an **independent system.** If the lines have equal slopes *and* the same *y*-intercept, they are identical or **coincident lines.** Since one is right atop the other, they *intersect at all points,* and the system has an infinite number of solutions. Here, one line *depends* on the other and the system is called a **dependent system.** Using substitution or elimination on a dependent system results in the elimination of all variable terms and leaves a statement that is *always true,* such as 0 = 0 or some other simple identity.

EXAMPLE 6 ▶ **Solving a Dependent System**

Solve using elimination: $\begin{cases} 3x + 4y = 12 \\ 6x = 24 - 8y \end{cases}$.

Solution ▶ Writing the system in standard form gives $\begin{cases} 3x + 4y = 12 \\ 6x + 8y = 24 \end{cases}$. By applying −2R1, we can eliminate the variable *x*:

$$\begin{array}{r} -2R1 \\ + \\ \underline{R2} \\ \text{sum} \end{array} \begin{cases} -6x - 8y = -24 \\ 6x + 8y = 24 \\ \overline{0x + 0y = 0} \end{cases} \begin{array}{l} \\ \\ \text{add} \\ \text{variables are eliminated} \end{array}$$

$$0 = 0 \quad \text{true statement}$$

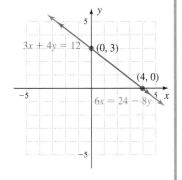
$3x + 4y = 12$ (0, 3)
(4, 0)
$6x = 24 - 8y$

Although we didn't expect it, both variables were eliminated and the final statement is true (0 = 0). This indicates the system is dependent, which the graph verifies (the lines are coincident). Writing both equations in slope-intercept form shows they represent the same line.

$$\begin{cases} 3x + 4y = 12 \\ 6x + 8y = 24 \end{cases} \longrightarrow \begin{cases} 4y = -3x + 12 \\ 8y = -6x + 24 \end{cases} \longrightarrow \begin{cases} y = -\dfrac{3}{4}x + 3 \\ y = -\dfrac{3}{4}x + 3 \end{cases}$$

WORTHY OF NOTE

When writing the solution to a dependent system using a parameter, the solution can be written in many different ways. For instance, if we let $p = 4b$ for the first coordinate of the solution to Example 6, we have $\dfrac{-3(4b)}{4} + 3 = -3b + 3$ as the second coordinate, and the solution becomes $(4b, -3b + 3)$ for any constant *b*.

The solutions of a dependent system are often written in set notation as the set of ordered pairs (*x*, *y*), where *y* is a specified function of *x*. For Example 6 the solution would be $\{(x, y) | y = -\frac{3}{4}x + 3\}$. Using an ordered pair with an arbitrary variable, called a **parameter,** is also common: $\left(p, \dfrac{-3p}{4} + 3 \right)$.

Now try Exercises 45 through 56 ▶

✓ **E.** You've just learned how to recognize inconsistent systems and dependent systems

Finally, if the lines have equal slopes and *different y-intercepts,* they are parallel and the system will have no solution. A system with no solutions is called an **inconsistent system.** An "inconsistent system" produces an "inconsistent answer," such as 12 = 0 or some other false statement when substitution or elimination is applied. In other words, all variable terms are once again eliminated, but the remaining statement is *false.* A summary of the three possibilities is shown here for arbitrary slope *m* and *y*-intercept (0, *b*).

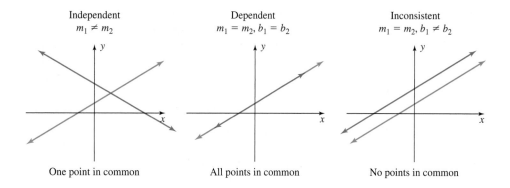

Independent $m_1 \neq m_2$	Dependent $m_1 = m_2, b_1 = b_2$	Inconsistent $m_1 = m_2, b_1 \neq b_2$
One point in common	All points in common	No points in common

F. Systems and Modeling

In previous chapters, we solved numerous real-world applications by writing all given relationships in terms of a single variable. Many situations are easier to model using a system of equations with each relationship modeled independently using *two* variables. We begin here with a **mixture** application. Although they appear in many different forms (coin problems, metal alloys, investments, merchandising, and so on), mixture problems all have a similar theme. Generally one equation is related to *quantity* (how much of each item is being combined) and one equation is related to *value* (what is the value of each item being combined).

EXAMPLE 7 ▶ **Solving a Mixture Application**

A jeweler is commissioned to create a piece of artwork that will weigh 14 oz and consist of 75% gold. She has on hand two alloys that are 60% and 80% gold, respectively. How much of each should she use?

Solution ▶ Let x represent ounces of the 60% alloy and y represent ounces of the 80% alloy. The first equation must be $x + y = 14$, since the piece of art must weigh exactly 14 oz (this is the *quantity* equation). The x ounces are 60% gold, the y ounces are 80% gold, and the 14 oz will be 75% gold. This gives the *value* equation: $0.6x + 0.8y = 0.75(14)$. The system is $\begin{cases} x + y = 14 \\ 6x + 8y = 105 \end{cases}$ (after clearing decimals). Solving for y in the first equation gives $y = 14 - x$. Substituting $14 - x$ for y in the second equation gives

$$6x + 8y = 105 \quad \text{second equation}$$
$$6x + 8(14 - x) = 105 \quad \text{substitute } 14 - x \text{ for } y$$
$$-2x + 112 = 105 \quad \text{simplify}$$
$$x = \frac{7}{2} \quad \text{solve for } x$$

Substituting $\frac{7}{2}$ for x in the first equation gives $y = \frac{21}{2}$. She should use 3.5 oz of the 60% alloy and 10.5 oz of the 80% alloy.

WORTHY OF NOTE

As an estimation tool, note that if equal amounts of the 60% and 80% alloys were used (7 oz each), the result would be a 70% alloy (halfway in between). Since a 75% alloy is needed, more of the 80% gold will be used.

Now try Exercises 63 through 70 ▶

Systems of equations also play a significant role in *cost-based pricing* in the business world. The costs involved in running a business can broadly be understood as either a **fixed cost k** or a **variable cost v.** Fixed costs might include the monthly rent paid for facilities, which remains the same regardless of how many items are produced and sold. Variable costs would include the cost of materials needed to produce the item, which depends on the number of items made. The total cost can then be modeled by

$C(x) = vx + k$ for x number of items. Once a **selling price p** has been determined, the revenue equation is simply $R(x) = px$ (price times number of items sold). We can now set up and solve a system of equations that will determine how many items must be sold to break even, performing what is called a **break-even analysis.**

EXAMPLE 8 ▶ Solving an Application of Systems: Break-Even Analysis

In home businesses that produce items to sell on Ebay®, fixed costs are easily determined by rent and utilities, and variable costs by the price of materials needed to produce the item. Karen's home business makes large, decorative candles for all occasions. The cost of materials is $3.50 per candle, and her rent and utilities average $900 per month. If her candles sell for $9.50, how many candles must be sold each month to break even?

Solution ▶ Let x represent the number of candles sold. Her total cost is $C(x) = 3.5x + 900$ (variable cost plus fixed cost), and projected revenue is $R(x) = 9.5x$. This gives the system $\begin{cases} C(x) = 3.5x + 900 \\ R(x) = 9.5x \end{cases}$. To break even, Cost = Revenue which gives

$$9.5x = 3.5x + 900$$
$$6x = 900$$
$$x = 150$$

WORTHY OF NOTE

This break-even concept can also be applied in studies of supply and demand, as well as in the decision to buy a new car or appliance that will enable you to break even over time due to energy and efficiency savings.

The analysis shows that Karen must sell 150 candles each month to break even.

Now try Exercises 71 through 74 ▶

Our final example involves an application of uniform motion (distance = rate · time), and explores concepts of great importance to the navigation of ships and airplanes. As a simple illustration, if you've ever walked at your normal rate r on the "moving walkways" at an airport, you likely noticed an increase in your total speed. This is because the resulting speed combines your walking rate r with the speed w of the walkway: *total speed = $r + w$*. If you walk in the *opposite direction* of the walkway, your total speed is much slower, as now *total speed = $r - w$*.

This same phenomenon is observed when an airplane is flying with or against the wind, or a ship is sailing with or against the current.

EXAMPLE 9 ▶ Solving an Application of Systems — Uniform Motion

An airplane flying due south from St. Louis, Missouri, to Baton Rouge, Louisiana, uses a strong, steady tailwind to complete the trip in only 2.5 hr. On the return trip, the same wind slows the flight and it takes 3 hr to get back. If the flight distance between these cities is 912 km, what is the cruising speed of the airplane (speed with no wind)? How fast is the wind blowing?

Solution ▶ Let r represent the rate of the plane and w the rate of the wind. Since $D = RT$, the flight to Baton Rouge can be modeled by $912 = (r + w)(2.5)$, and the return flight by $912 = (r - w)(3)$. This produces the system $\begin{cases} 912 = 2.5r + 2.5w \\ 912 = 3r - 3w \end{cases}$. Using $\dfrac{R1}{2.5}$ and $\dfrac{R2}{3}$ gives the equivalent system $\begin{cases} 364.8 = r + w \\ 304 = r - w \end{cases}$, which is easily solved using elimination with R1 + R2.

$$668.8 = 2r \qquad \text{R1 + R2}$$
$$334.4 = r \qquad \text{divide by 2}$$

☑ **F. You've just learned how to use a system of equations to model and solve applications**

The cruising speed of the plane (with no wind) is 334.4 kph. Using $r - w = 304$ shows the wind is blowing at 30.4 kph.

Now try Exercises 75 through 78 ▶

TECHNOLOGY HIGHLIGHT

Solving Systems Graphically

When used with care, graphing calculators offer an accurate way to solve linear systems and to check solution(s) obtained by hand. We'll illustrate using the system from Example 3: $\begin{cases} 4x + y = 4 \\ y = x + 2 \end{cases}$, where we found the solution was $(\frac{2}{5}, \frac{12}{5})$.

Figure 5.1

1. Solve for y in both equations:

$$\begin{cases} y = -4x + 4 \\ y = \quad x + 2 \end{cases}$$

3. Graph using ZOOM 6

$Y_1 = -4x + 4$
$Y_2 = x + 2$

2. Enter the equations as

$Y_1 = -4x + 4$
$Y_2 = x + 2$

4. Press 2nd TRACE (CALC) 5 ENTER ENTER ENTER to have the calculator compute the point of intersection.

The coordinates of the intersection appear as decimal fractions at the bottom of the screen (Figure 5.1). In step 4, The first ENTER selects Y_1, the second ENTER selects Y_2 and the third ENTER bypasses the GUESS option (this option is most often used if the graphs intersect at more than one point). The calculator automatically registers the x-coordinate as its most recent entry, and from the home screen, converting it to a standard fraction (using MATH 1: ▶ Frac ENTER) shows $x = \frac{2}{5}$. You can also get an *approximate solution* by tracing along either line towards the point of intersection using the TRACE key and the left or right arrows.

Solve each system graphically, using a graphing calculator.

Exercise 1: $\begin{cases} 3x - y = -7 \\ y + 5x = -1 \end{cases}$

Exercise 2: $\begin{cases} 2x - 3y = 3 \\ 6 = 8x - 3y \end{cases}$

5.1 EXERCISES

▶ CONCEPTS AND VOCABULARY

Fill in the blank with the appropriate word or phrase. Carefully reread the section if needed.

1. Systems that have no solution are called _____ systems.

2. Systems having at least one solution are called _____ systems.

3. If the lines in a system intersect at a single point, the system is said to be _____ and _____.

4. If the lines in a system are coincident, the system is referred to as _____ and _____.

5. The given systems are equivalent. How do we obtain the second system from the first?

$$\begin{cases} \frac{2}{3}x + \frac{1}{2}y = \frac{5}{3} \\ 0.2x + 0.4y = 1 \end{cases} \quad \begin{cases} 4x + 3y = 10 \\ 2x + 4y = 10 \end{cases}$$

6. For $\begin{cases} 2x + 5y = 8 \\ 3x + 4y = 5 \end{cases}$, which solution method would be more efficient, substitution or elimination? Discuss/Explain why.

▶ DEVELOPING YOUR SKILLS

Show the lines in each system would intersect in a single point by writing the equations in slope-intercept form.

7. $\begin{cases} 7x - 4y = 24 \\ 4x + 3y = 15 \end{cases}$

8. $\begin{cases} 0.3x - 0.4y = 2 \\ 0.5x + 0.2y = -4 \end{cases}$

An ordered pair is a solution to an equation if it makes the equation true. Given the graph shown here, determine which equation(s) have the indicated point as a solution. If the point satisfies more than one equation, write the system for which it is a solution.

9. *A*   10. *B*

11. *C*   12. *D*

13. *E*   14. *F*

Substitute the *x*- and *y*-values indicated by the ordered pair to determine if it solves the system.

15. $\begin{cases} 3x + y = 11 \\ -5x + y = -13; \end{cases} (3, 2)$

16. $\begin{cases} 3x + 7y = -4 \\ 7x + 8y = -21; \end{cases} (-6, 2)$

17. $\begin{cases} 8x - 24y = -17 \\ 12x + 30y = 2; \end{cases} \left(-\dfrac{7}{8}, \dfrac{5}{12}\right)$

18. $\begin{cases} 4x + 15y = 7 \\ 8x + 21y = 11; \end{cases} \left(\dfrac{1}{2}, \dfrac{1}{3}\right)$

Solve each system by *graphing*. If the coordinates do not appear to be integers, estimate the solution to the nearest tenth (indicate that your solution is an estimate).

19. $\begin{cases} 3x + 2y = 12 \\ x - y = 9 \end{cases}$ 20. $\begin{cases} 5x + 2y = -2 \\ -3x + y = 10 \end{cases}$

21. $\begin{cases} 5x - 2y = 4 \\ x + 3y = -15 \end{cases}$ 22. $\begin{cases} 3x + y = 2 \\ 5x + 3y = 12 \end{cases}$

Solve each system using *substitution*. Write solutions as an ordered pair.

23. $\begin{cases} x = 5y - 9 \\ x - 2y = -6 \end{cases}$ 24. $\begin{cases} 4x - 5y = 7 \\ 2x - 5 = y \end{cases}$

25. $\begin{cases} y = \frac{2}{3}x - 7 \\ 3x - 2y = 19 \end{cases}$ 26. $\begin{cases} 2x - y = 6 \\ y = \frac{3}{4}x - 1 \end{cases}$

Identify the equation and variable that makes the substitution method easiest to use. Then solve the system.

27. $\begin{cases} 3x - 4y = 24 \\ 5x + y = 17 \end{cases}$ 28. $\begin{cases} 3x + 2y = 19 \\ x - 4y = -3 \end{cases}$

29. $\begin{cases} 0.7x + 2y = 5 \\ x - 1.4y = 11.4 \end{cases}$ 30. $\begin{cases} 0.8x + y = 7.4 \\ 0.6x + 1.5y = 9.3 \end{cases}$

31. $\begin{cases} 5x - 6y = 2 \\ x + 2y = 6 \end{cases}$ 32. $\begin{cases} 2x + 5y = 5 \\ 8x - y = 6 \end{cases}$

Solve using *elimination*. In some cases, the system must first be written in standard form.

33. $\begin{cases} 2x - 4y = 10 \\ 3x + 4y = 5 \end{cases}$ 34. $\begin{cases} -x + 5y = 8 \\ x + 2y = 6 \end{cases}$

35. $\begin{cases} 4x - 3y = 1 \\ 3y = -5x - 19 \end{cases}$ 36. $\begin{cases} 5y - 3x = -5 \\ 3x + 2y = 19 \end{cases}$

37. $\begin{cases} 2x = -3y + 17 \\ 4x - 5y = 12 \end{cases}$ 38. $\begin{cases} 2y = 5x + 2 \\ -4x = 17 - 6y \end{cases}$

39. $\begin{cases} 0.5x + 0.4y = 0.2 \\ 0.3y = 1.3 + 0.2x \end{cases}$ 40. $\begin{cases} 0.2x + 0.3y = 0.8 \\ 0.3x + 0.4y = 1.3 \end{cases}$

41. $\begin{cases} 0.32m - 0.12n = -1.44 \\ -0.24m + 0.08n = 1.04 \end{cases}$

42. $\begin{cases} 0.06g - 0.35h = -0.67 \\ -0.12g + 0.25h = 0.44 \end{cases}$

43. $\begin{cases} -\frac{1}{6}u + \frac{1}{4}v = 4 \\ \frac{1}{2}u - \frac{2}{3}v = -11 \end{cases}$ 44. $\begin{cases} \frac{3}{4}x + \frac{1}{3}y = -2 \\ \frac{3}{2}x + \frac{1}{5}y = 3 \end{cases}$

Solve using any method and identify the system as consistent, inconsistent, or dependent.

45. $\begin{cases} 4x + \frac{3}{4}y = 14 \\ -9x + \frac{5}{8}y = -13 \end{cases}$ 46. $\begin{cases} \frac{2}{3}x + y = 2 \\ 2y = \frac{5}{6}x - 9 \end{cases}$

47. $\begin{cases} 0.2y = 0.3x + 4 \\ 0.6x - 0.4y = -1 \end{cases}$ 48. $\begin{cases} 1.2x + 0.4y = 5 \\ 0.5y = -1.5x + 2 \end{cases}$

49. $\begin{cases} 6x - 22 = -y \\ 3x + \frac{1}{2}y = 11 \end{cases}$ 50. $\begin{cases} 15 - 5y = -9x \\ -3x + \frac{5}{3}y = 5 \end{cases}$

51. $\begin{cases} -10x + 35y = -5 \\ y = 0.25x \end{cases}$ 52. $\begin{cases} 2x + 3y = 4 \\ x = -2.5y \end{cases}$

53. $\begin{cases} 7a + b = -25 \\ 2a - 5b = 14 \end{cases}$ 54. $\begin{cases} -2m + 3n = -1 \\ 5m - 6n = 4 \end{cases}$

55. $\begin{cases} 4a = 2 - 3b \\ 6b + 2a = 7 \end{cases}$ **56.** $\begin{cases} 3p - 2q = 4 \\ 9p + 4q = -3 \end{cases}$

57. $\begin{cases} 2x + 4y = 6 \\ x + 12 = 4y \end{cases}$ **58.** $\begin{cases} 8x = 3y + 24 \\ 8x - 5y = 36 \end{cases}$

59. $\begin{cases} 5x - 11y = 21 \\ 11y = 5 - 8x \end{cases}$ **60.** $\begin{cases} -6x = 5y - 16 \\ 5y - 6x = 4 \end{cases}$

The substitution method can be used for like variables *or for like expressions.* Solve the following systems, *using the expression* common to both equations (do not solve for *x* or *y* alone).

▶ **WORKING WITH FORMULAS**

61. Uniform motion with current: $\begin{cases} (R + C)T_1 = D_1 \\ (R - C)T_2 = D_2 \end{cases}$

The formula shown can be used to solve uniform motion problems involving a *current,* where *D* represents distance traveled, *R* is the rate of the object with no current, *C* is the speed of the current, and *T* is the time. Chan-Li rows 9 mi up river (against the current) in 3 hr. It only took him 1 hr to row 5 mi downstream (with the current). How fast was the current? How fast can he row in still water?

62. Fahrenheit and Celsius temperatures:
$\begin{cases} y = \frac{9}{5}x + 32 & °F \\ y = \frac{5}{9}(x - 32) & °C \end{cases}$

Many people are familiar with temperature measurement in degrees Celsius and degrees Fahrenheit, but few realize that the equations are linear and there is one temperature at which the two scales agree. Solve the system using the method of your choice and find this temperature.

▶ **APPLICATIONS**

Solve each application by modeling the situation with a linear system. Be sure to clearly indicate what each variable represents.

Mixture

63. Theater productions: At a recent production of *A Comedy of Errors,* the Community Theater brought in a total of $30,495 in revenue. If adult tickets were $9 and children's tickets were $6.50, how many tickets of each type were sold if 3800 tickets in all were sold?

64. Milk-fat requirements: A dietician needs to mix 10 gal of milk that is $2\frac{1}{2}\%$ milk fat for the day's rounds. He has some milk that is 4% milk fat and some that is $1\frac{1}{2}\%$ milk fat. How much of each should be used?

65. Filling the family cars: Cherokee just filled both of the family vehicles at a service station. The total cost for 20 gal of regular unleaded and 17 gal of premium unleaded was $144.89. The premium gas was $0.10 more per gallon than the regular gas. Find the price per gallon for each type of gasoline.

66. Household cleaners: As a cleaning agent, a solution that is 24% vinegar is often used. How much pure (100%) vinegar and 5% vinegar must be mixed to obtain 50 oz of a 24% solution?

67. Alumni contributions: A wealthy alumnus donated $10,000 to his alma mater. The college used the funds to make a loan to a science major at 7% interest and a loan to a nursing student at 6% interest. That year the college earned $635 in interest. How much was loaned to each student?

68. Investing in bonds: A total of $12,000 is invested in two municipal bonds, one paying 10.5% and the other 12% simple interest. Last year the annual interest earned on the two investments was $1335. How much was invested at each rate?

69. Saving money: Bryan has been doing odd jobs around the house, trying to earn enough money to buy a new Dirt-Surfer©. He saves all quarters and dimes in his piggy bank, while he places all nickels and pennies in a drawer to spend. So far, he has 225 coins in the piggy bank, worth a total of $45.00. How many of the coins are quarters? How many are dimes?

70. Coin investments: In 1990, Molly attended a coin auction and purchased some rare "Flowing Hair" fifty-cent pieces, and a number of very rare two-cent pieces from the Civil War Era. If she bought 47 coins with a face value of $10.06, how many of each denomination did she buy?

71. **Lawn service:** Dave and his sons run a lawn service, which includes mowing, edging, trimming, and aerating a lawn. His fixed cost includes insurance, his salary, and monthly payments on equipment, and amounts to $4000/mo. The variable costs include gas, oil, hourly wages for his employees, and miscellaneous expenses, which run about $75 per lawn. The average charge for full-service lawn care is $115 per visit. Do a break-even analysis to (a) determine how many lawns Dave must service each month to break even and (b) the revenue required to break even.

72. **Production of mini-microwave ovens:** Due to high market demand, a manufacturer decides to introduce a new line of mini-microwave ovens for personal and office use. By using existing factory space and retraining some employees, fixed costs are estimated at $8400/mo. The components to assemble and test each microwave are expected to run $45 per unit. If market research shows consumers are willing to pay at least $69 for this product, find (a) how many units must be made and sold each month to break even and (b) the revenue required to break even.

In a market economy, the availability of goods is closely related to the market price. Suppliers are willing to produce more of the item at a higher price (the supply), with consumers willing to buy more of the item at a lower price (the demand). This is called the law of supply and demand. When supply and demand are equal, both the buyer and seller are satisfied with the current price and we have *market equilibrium*.

73. **Farm commodities:** One area where the law of supply and demand is clearly at work is farm commodities. Both growers and consumers watch this relationship closely, and use data collected by government agencies to track the relationship and make adjustments, as when a farmer decides to convert a large portion of her farmland from corn to soybeans to improve profits. Suppose that for x billion bushels of soybeans, supply is modeled by $y = 1.5x + 3$, where y is the current market price (in dollars per bushel). The related demand equation might be $y = -2.20x + 12$. (a) How many billion bushels will be supplied at a market price of $5.40? What will the demand be at this price? Is supply less than demand? (b) How many billion bushels will be supplied at a market price of $7.05? What will the demand be at this price? Is demand less than supply? (c) To the nearest cent, at what price does the market reach equilibrium? How many bushels are being supplied/demanded?

74. **Digital music:** Market research has indicated that by 2010, sales of MP3 portables will mushroom into a $70 billion dollar market. With a market this large, competition is often fierce—with suppliers fighting to earn and hold market shares. For x million MP3 players sold, supply is modeled by $y = 10.5x + 25$, where y is the current market price (in dollars). The related demand equation might be $y = -5.20x + 140$. (a) How many million MP3 players will be supplied at a market price of $88? What will the demand be at this price? Is supply less than demand? (b) How many million MP3 players will be supplied at a market price of $114? What will the demand be at this price? Is demand less than supply? (c) To the nearest cent, at what price does the market reach equilibrium? How many units are being supplied/demanded?

Uniform Motion

75. **Canoeing on a stream:** On a recent camping trip, it took Molly and Sharon 2 hr to row 4 mi upstream from the drop in point to the campsite. After a leisurely weekend of camping, fishing, and relaxation, they rowed back downstream to the drop in point in just 30 min. Use this information to find (a) the speed of the current and (b) the speed Sharon and Molly would be rowing in still water.

76. **Taking a luxury cruise:** A luxury ship is taking a Caribbean cruise from Caracas, Venezuela, to just off the coast of Belize City on the Yucatan Peninsula, a distance of 1435 mi. En route they encounter the Caribbean Current, which flows to the northwest, parallel to the coastline. From Caracas to the Belize coast, the trip took 70 hr. After a few days of fun in the sun, the ship leaves for Caracas, with the return trip taking 82 hr. Use

this information to find (a) the speed of the Caribbean Current and (b) the cruising speed of the ship.

77. **Airport walkways:** As part of an algebra field trip, Jason takes his class to the airport to use their moving walkways for a demonstration. The class measures the longest walkway, which turns out to be 256 ft long. Using a stop watch, Jason shows it takes him just 32 sec to complete the walk going in the same direction as the walkway. Walking in a direction opposite the walkway, it takes him 320 sec—10 times as long! The next day in class, Jason hands out a two-question quiz: (1) What was the speed of the walkway in feet per second? (2) What is my (Jason's) normal walking speed? Create the answer key for this quiz.

78. **Racing pigeons:** The American Racing Pigeon Union often sponsors opportunities for owners to fly their birds in friendly competitions. During a recent competition, Steve's birds were liberated in Topeka, Kansas, and headed almost due north to their loft in Sioux Falls, South Dakota, a distance of 308 mi. During the flight, they encountered a steady wind from the north and the trip took 4.4 hr. The next month, Steve took his birds to a competition in Grand Forks, North Dakota, with the birds heading almost due south to home, also a distance of 308 mi. This time the birds were aided by the same wind from the north, and the trip took only 3.5 hr. Use this information to (a) find the racing speed of Steve's birds and (b) find the speed of the wind.

Descriptive Translation

79. **Important dates in U.S. history:** If you sum the year that the Declaration of Independence was signed and the year that the Civil War ended, you get 3641. There are 89 yr that separate the two events. What year was the Declaration signed? What year did the Civil War end?

80. **Architectual wonders:** When it was first constructed in 1889, the Eiffel Tower in Paris, France, was the tallest structure in the world. In 1975, the CN Tower in Toronto, Canada, became the world's tallest structure. The CN Tower is 153 ft less than twice the height of the Eiffel Tower, and the sum of their heights is 2799 ft. How tall is each tower?

81. **Pacific islands land area:** In the South Pacific, the island nations of Tahiti and Tonga have a combined land area of 692 mi^2. Tahiti's land area is 112 mi^2 more than Tonga's. What is the land area of each island group?

82. **Card games:** On a cold winter night, in the lobby of a beautiful hotel in Sante Fe, New Mexico, Marc and Klay just barely beat John and Steve in a close game of Trumps. If the sum of the team scores was 990 points, and there was a 12-point margin of victory, what was the final score?

▶ **EXTENDING THE CONCEPT**

83. Answer using observations only—no calculations. Is the given system consistent/independent, consistent/dependent, or inconsistent? Explain/Discuss your answer. $\begin{cases} y = 5x + 2 \\ y = 5.01x + 1.9 \end{cases}$

84. Federal income tax reform has been a hot political topic for many years. Suppose tax plan A calls for a flat tax of 20% tax on all income (no deductions or loopholes). Tax plan B requires taxpayers to pay $5000 plus 10% of all income. For what income level do both plans require the same tax?

85. Suppose a certain amount of money was invested at 6% per year, and another amount at 8.5% per year, with a total return of $1250. If the amounts invested at each rate were switched, the yearly income would have been $1375. To the nearest whole dollar, how much was invested at each rate?

▶ **MAINTAINING YOUR SKILLS**

86. (2.6) Given the parent function $f(x) = |x|$, sketch the graph of $F(x) = -|x + 3| - 2$.

87. (3.3) Use the RRT to write the polynomial in completely factored form: $3x^4 - 19x^3 + 15x^2 + 27x - 10 = 0$.

88. (4.4) Solve for x (rounded to the nearest thousandth): $33 = 77.5e^{-0.0052x} - 8.37$.

89. (3.1) Graph $y = x^2 - 6x - 16$ by completing the square and state the interval where $f(x) \leq 0$.

Learning Objectives

In Section 5.2 you will learn how to:

☐ **A.** Visualize a solution in three dimensions

☐ **B.** Check ordered triple solutions

☐ **C.** Solve linear systems in three variables

☐ **D.** Recognize inconsistent and dependent systems

☐ **E.** Use a system of three equations in three variables to solve applications

The transition to systems of three equations in three variables requires a fair amount of "visual gymnastics" along with good organizational skills. Although the techniques used are identical and similar results are obtained, the third equation and variable give us more to track, and we must work more carefully toward the solution.

A. Visualizing Solutions in Three Dimensions

The solution to an equation in one variable is the single number that satisfies the equation. For $x + 1 = 3$, the solution is $x = 2$ and its graph is a single *point* on the number line, a **one-dimensional graph.** The solution to an equation in two variables, such as $x + y = 3$, is an ordered pair (x, y) that satisfies the equation. When we graph this solution set, the result is a *line* on the xy-coordinate grid, a **two-dimensional graph.** The solutions to an equation in three variables, such as $x + y + z = 6$, are the **ordered triples** (x, y, z) that satisfy the equation. When we graph this solution set, the result is a **plane** in **space**, a *graph in three dimensions.* Recall a plane is a flat surface having infinite length and width, but no depth. We can graph this plane using the intercept method and the result is shown in Figure 5.2. For graphs in three dimensions, the xy-plane is parallel to the ground (the y-axis points to the right) and z is the **vertical axis.** To find an additional point on this plane, we use any three numbers whose sum is 6, such as $(2, 3, 1)$. Move 2 units along the x-axis, 3 units parallel to the y-axis, and 1 unit parallel to the z-axis, as shown in Figure 5.3.

WORTHY OF NOTE

We can visualize the location of a point in space by considering a large rectangular box 2 ft long × 3 ft wide × 1 ft tall, placed snugly in the corner of a room. The floor is the xy-plane, one wall is the xz-plane, and the other wall is the yz-plane. The z-axis is formed where the two walls meet and the corner of the room is the origin $(0, 0, 0)$. To find the corner of the box located at $(2, 3, 1)$, first locate the point $(2, 3)$ in the xy-plane (the floor), then move up 1 ft.

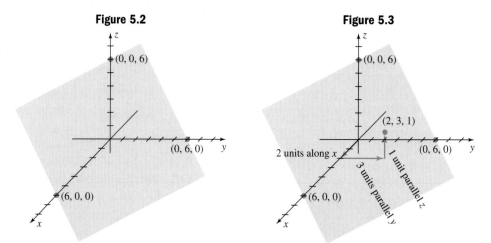

Figure 5.2   **Figure 5.3**

EXAMPLE 1 ▶ **Finding Solutions to an Equation in Three Variables**

Use a guess-and-check method to find four additional points on the plane determined by $x + y + z = 6$.

Solution ▶ We can begin by letting $x = 0$, then use any combination of y and z that sum to 6. Two examples are $(0, 2, 4)$ and $(0, 5, 1)$. We could also select any two values for x and y, then determine a value for z that results in a sum of 6. Two examples are $(-2, 9, -1)$ and $(8, -3, 1)$.

☑ **A.** You've just learned how to visualize a solution in three dimensions

Now try Exercises 7 through 10 ▶

B. Solutions to a System of Three Equations in Three Variables

When solving a system of three equations in three variables, remember each equation represents a plane in space. These planes can intersect in various ways, creating different possibilities for a solution set (see Figures 5.4 to 5.7). The system could have a **unique solution** (a, b, c), if the planes intersect at a single point (Figure 5.4) (the point satisfies all three equations simultaneously). If the planes intersect in a line (Figure 5.5), the system is **linearly dependent** and there are an infinite number of solutions. Unlike the two-dimensional case, the equation of a line in three dimensions is somewhat complex, and the coordinates of all points on this line are usually *represented* by a specialized ordered triple, which we use to state the solution set. If the planes intersect at all points, the system has **coincident dependence** (see Figure 5.6). This indicates the equations of the system differ by only a constant multiple—they are all "disguised forms" of the *same equation*. The solution set is any ordered triple (a, b, c) satisfying this equation. Finally, the system may have no solutions. This can happen a number of different ways, most notably if the planes intersect as shown in Figure 5.7 (other possibilities are discussed in the exercises). In the case of "no solutions," an ordered triple may satisfy none of the equations, only one of the equations, only two of the equations, but not all three equations.

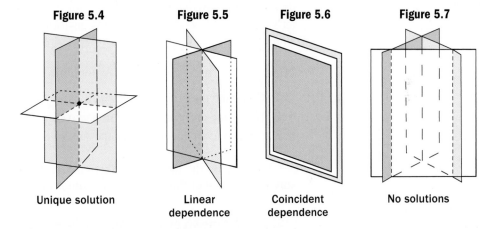

Figure 5.4	Figure 5.5	Figure 5.6	Figure 5.7
Unique solution	Linear dependence	Coincident dependence	No solutions

---

EXAMPLE 2 ▶ **Determining If an Ordered Triple Is a Solution**

Determine if the ordered triple $(1, -2, 3)$ is a solution to the systems shown.

a. $\begin{cases} x + 4y - z = -10 \\ 2x + 5y + 8z = 4 \\ x - 2y - 3z = -4 \end{cases}$ **b.** $\begin{cases} 3x + 2y - z = -4 \\ 2x - 3y - 2z = 2 \\ x - y + 2z = 9 \end{cases}$

Solution ▶ Substitute 1 for x, -2 for y, and 3 for z in the first system.

a. $\begin{cases} x + 4y - z = -10 \\ 2x + 5y + 8z = 4 \\ x - 2y - 3z = -4 \end{cases}$ → $\begin{cases} (1) + 4(-2) - (3) = -10 \\ 2(1) + 5(-2) + 8(3) = 4 \\ (1) - 2(-2) - 3(3) = -4 \end{cases}$ → $\begin{cases} -10 = -10 \text{ true} \\ 16 = 4 \text{ false} \\ -4 = -4 \text{ true} \end{cases}$

No, the ordered triple $(1, -2, 3)$ is not a solution to the first system. Now use the same substitutions in the second system.

b. $\begin{cases} 3x + 2y - z = -4 \\ 2x - 3y - 2z = 2 \\ x - y + 2z = 9 \end{cases}$ → $\begin{cases} 3(1) + 2(-2) - (3) = -4 \\ 2(1) - 3(-2) - 2(3) = 2 \\ (1) - (-2) + 2(3) = 9 \end{cases}$ → $\begin{cases} -4 = -4 \text{ true} \\ 2 = 2 \text{ true} \\ 9 = 9 \text{ true} \end{cases}$

☑ **B.** You've just learned how to check ordered triple solutions

The ordered triple $(1, -2, 3)$ is a solution to the second system only.

Now try Exercises 11 and 12 ▶

C. Solving Systems of Three Equations in Three Variables Using Elimination

From Section 5.1, we know that two systems of equations are **equivalent** if they have the same solution set. The systems

$$\begin{cases} 2x + y - 2z = -7 \\ x + y + z = -1 \\ -2y - z = -3 \end{cases} \text{ and } \begin{cases} 2x + y - 2z = -7 \\ y + 4z = 5 \\ z = 1 \end{cases}$$

are equivalent, as both have the unique solution $(-3, 1, 1)$. In addition, it is evident that the second system can be solved more easily, since R2 and R3 have fewer variables than the first system. In the simpler system, mentally substituting 1 for z into R2 immediately gives $y = 1$, and these values can be back-substituted into the first equation to find that $x = -3$. This observation guides us to a general approach for solving larger systems—we would like to *eliminate variables in the second and third equations, until we obtain an equivalent system that can easily be solved by back-substitution.* To begin, let's review the three operations that "transform" a given system, and produce an equivalent system.

Operations That Produce an Equivalent System

1. Changing the order of the equations.
2. Replacing an equation by a nonzero constant multiple of that equation.
3. Replacing an equation with the sum of two equations from the system.

Building on the ideas from Section 5.1, we develop the following approach for solving a system of three equations in three variables.

Solving a System of Three Equations in Three Variables

1. Write each equation in standard form: $Ax + By + Cz = D$.
2. If the "x" term in any equation has a coefficient of 1, interchange equations (if necessary) so this equation becomes R1.
3. Use the x-term in R1 to eliminate the x-terms from R2 and R3. The original R1, with the new R2 and R3, form an equivalent system that contains a smaller "subsystem" of two equations in two variables.
4. Solve the subsystem and keep the result as the new R3. The result is an equivalent system that can be solved using back-substitution.

We'll begin by solving the system $\begin{cases} 2x + y - 2z = -7 \\ x + y + z = -1 \\ -2y - z = -3 \end{cases}$ using the elimination method and the procedure outlined. In Example 3, the notation $-2R1 + R2 \rightarrow R2$ indicates the equation in row 1 has been multiplied by -2 and added to the equation in row 2, with the result placed in the system as the new row 2.

EXAMPLE 3 ▶ **Solving a System of Three Equations in Three Variables**

Solve using elimination: $\begin{cases} 2x + y - 2z = -7 \\ x + y + z = -1. \\ -2y - z = -3 \end{cases}$

Solution ▶ 1. The system is in standard form.

2. If the x-term in any equation has a coefficient of 1, interchange equations so this equation becomes R1.

$$\begin{cases} 2x + y - 2z = -7 \\ x + y + z = -1 \\ -2y - z = -3 \end{cases} \xrightarrow{\text{R2} \leftrightarrow \text{R1}} \begin{cases} x + y + z = -1 \\ 2x + y - 2z = -7 \\ -2y - z = -3 \end{cases}$$

3. Use R1 to eliminate the x-term in R2 and R3. Since R3 has no x-term, the only elimination needed is the x-term from R2. Using $-2\text{R1} + \text{R2}$ will eliminate this term:

$$\begin{array}{rl} -2\text{R1} & -2x - 2y - 2z = 2 \\ + & \\ \underline{\text{R2}} & \underline{2x + y - 2z = -7} \\ & 0x - 1y - 4z = -5 \quad \text{sum} \\ & y + 4z = 5 \quad \text{simplify} \end{array}$$

The new R2 is $y + 4z = 5$. The original R1 and R3, along with the new R2 form an equivalent system that contains a smaller **subsystem**

$$\begin{cases} x + y + z = -1 \\ 2x + y - 2z = -7 \\ -2y - z = -3 \end{cases} \xrightarrow[\text{R3} \to \text{R3}]{-2\text{R1} + \text{R2} \to \text{R2}} \begin{cases} x + y + z = -1 \quad \text{new} \\ y + 4z = 5 \quad \text{equivalent} \\ -2y - z = -3 \quad \text{system} \end{cases}$$

4. Solve the subsystem for either y or z, and keep the result as a *new* R3. We choose to eliminate y using $2\text{R2} + \text{R3}$:

$$\begin{array}{rl} 2\text{R2} & 2y + 8z = 10 \\ + & \\ \underline{\text{R3}} & \underline{-2y - z = -3} \\ & 0y + 7z = 7 \quad \text{sum} \\ & z = 1 \quad \text{simplify} \end{array}$$

The new R3 is $z = 1$.

$$\begin{cases} x + y + z = -1 \\ y + 4z = 5 \\ -2y - z = -3 \end{cases} \xrightarrow{2\text{R2} + \text{R3} \to \text{R3}} \begin{cases} x + y + z = -1 \quad \text{new} \\ y + 4z = 5 \quad \text{equivalent} \\ z = 1 \quad \text{system} \end{cases}$$

The new R3, along with the original R1 and R2 from step 3, form an equivalent system that can be solved using back-substitution. Substituting 1 for z in R2 yields $y = 1$. Substituting 1 for z and 1 for y in R1 yields $x = -3$. The solution is $(-3, 1, 1)$.

Now try Exercises 13 through 18 ▶

While not absolutely needed for the elimination process, there are two reasons for wanting the coefficient of x to be "1" in R1. First, it makes the elimination method more efficient since we can more easily see what to use as a multiplier. Second, it lays the foundation for developing other methods of solving larger systems. If no equation has an x-coefficient of 1, we simply use the y- or z-variable instead (see Example 7). Since solutions to larger systems generally are worked out in stages, we will sometimes track the transformations used by writing them *between* the original system and the equivalent system, rather than to the left as we did in Section 5.1.

Here is an additional example illustrating the elimination process, but in *abbreviated form.* Verify the calculations indicated using a separate sheet.

EXAMPLE 4 ▶ **Solving a System of Three Equations in Three Variables**

Solve using elimination:
$$\begin{cases} -5y + 2x - z = -8 \\ -x + 3z + 2y = 13 \\ -z + 3y + x = 5 \end{cases}$$

Solution ▶ 1. Write the equations in standard form:
$$\begin{cases} 2x - 5y - z = -8 \\ -x + 2y + 3z = 13 \\ x + 3y - z = 5 \end{cases}$$

2. $\begin{cases} 2x - 5y - z = -8 \\ -x + 2y + 3z = 13 \\ x + 3y - z = 5 \end{cases}$ $\xrightarrow{R3 \leftrightarrow R1}$ $\begin{cases} x + 3y - z = 5 \\ -x + 2y + 3z = 13 \quad \text{equivalent} \\ 2x - 5y - z = -8 \quad \text{system} \end{cases}$

3. Using R1 + R2 will eliminate the x-term from R2, yielding $5y + 2z = 18$.
   Using -2R1 + R3 eliminates the x-term from R3, yielding $-11y + z = -18$.

$\begin{cases} x + 3y - z = 5 \\ -x + 2y + 3z = 13 \\ 2x - 5y - z = -8 \end{cases}$ $\xrightarrow[-2R1 + R3 \to R3]{R1 + R2 \to R2}$ $\begin{cases} x + 3y - z = 5 \\ 5y + 2z = 18 \quad \text{equivalent} \\ -11y + z = -18 \quad \text{system} \end{cases}$

4. Using -2R3 + R2 will eliminate z from the subsystem, leaving $27y = 54$.

$\begin{cases} x + 3y - z = 5 \\ 5y + 2z = 18 \\ -11y + z = -18 \end{cases}$ $\xrightarrow{-2R3 + R2 \to R3}$ $\begin{cases} x + 3y - z = 5 \\ 5y + 2z = 18 \quad \text{equivalent} \\ 27y = 54 \quad \text{system} \end{cases}$

☑ **C.** You've learned just how to solve linear systems in three variables

Solving for y in R3 shows $y = 2$. Substituting 2 for y in R2 yields $z = 4$, and substituting 2 for y and 4 for z in R1 shows $x = 3$. The solution is $(3, 2, 4)$.

Now try Exercises 19 through 24 ▶

D. Inconsistent and Dependent Systems

As mentioned, it is possible for larger systems to have no solutions or an infinite number of solutions. As with our work in Section 5.1, an inconsistent system (no solutions) will produce inconsistent results, ending with a statement such as $0 = -3$ or some other **contradiction.**

EXAMPLE 5 ▶ **Attempting to Solve an Inconsistent System**

Solve using elimination:
$$\begin{cases} 2x + y - 3z = -3 \\ 3x - 2y + 4z = 2 \\ 4x + 2y - 6z = -7 \end{cases}$$

Solution ▶ 1. This system has no equation where the coefficient of x is 1.

2. We can still use R1 to begin the solution process, but this time we'll use the variable y since it *does* have coefficient 1.

Using 2R1 + R2 eliminates the y-term from R2, leaving $7x - 2z = -4$. But using -2R1 + R3 to eliminate the y-term from R3 results in a contradiction:

$\begin{array}{r} 2R1 \quad 4x + 2y - 6z = -6 \\ + \\ \underline{R2 \quad 3x - 2y + 4z = 2} \\ 7x \qquad - 2z = -4 \end{array}$ $\qquad$ $\begin{array}{r} -2R1 \quad -4x - 2y + 6z = 6 \\ + \\ \underline{R3 \quad 4x + 2y - 6z = -7} \\ 0x + 0y + 0z = -1 \end{array}$

$0 = -1 \quad \text{contradiction}$

We conclude the system is inconsistent. The answer is the empty set $\emptyset$, and we need work no further.

Now try Exercises 25 and 26 ▶

Unlike our work with systems having only two variables, systems in three variables can have two forms of dependence—*linear dependence* or *coincident dependence*. To help understand linear dependence, consider a system of two equations in three variables: $\begin{cases} -2x + 3y - z = 5 \\ x - 3y + 2z = -1 \end{cases}$. Each of these equations represents a plane, and unless the planes are parallel, their intersection will be a line (see Figure 5.5). As in Section 5.1, we can state solutions to a dependent system using set notation with two of the variables written in terms of the third, or as an ordered triple using a parameter. The relationships named can then be used to generate specific solutions to the system.

Systems with two equations and two variables or three equations and three variables are called **square systems,** meaning there are exactly as many equations as there are variables. A system of linear equations cannot have a unique solution unless there are at least as many equations as there are variables in the system.

EXAMPLE 6 ▶ Solving a Dependent System

Solve using elimination: $\begin{cases} -2x + 3y - z = 5 \\ x - 3y + 2z = -1 \end{cases}$.

Solution ▶ Using R1 + R2 eliminates the y-term from R2, yielding $-x + z = 4$. This means (x, y, z) will satisfy both equations only when $x = z - 4$ (the x-coordinate must be 4 less than the z-coordinate). Since x is written in terms of z, we substitute $z - 4$ for x *in either equation* to find how y is related to z. Using R2 we have: $(z - 4) - 3y + 2z = -1$, which yields $y = z - 1$ (verify). This means the y-coordinate of the solution must be 1 less than z. In set notation the solution is $\{(x, y, z,) \mid x = z - 4, y = z - 1, z \in \mathbb{R}\}$. For $z = -2, 0$, and 3, the solutions would be $(-6, -3, -2), (-4, -1, 0)$, and $(-1, 2, 3)$, respectively. Verify that these satisfy both equations. Using p as our parameter, the solution could be written $(p - 4, p - 1, p)$ in parameterized form.

Now try Exercises 27 through 30 ▶

The system in Example 6 was nonsquare, and we knew ahead of time the system would be dependent. The system in Example 7 *is* square, but only by applying the elimination process can we determine the nature of its solution(s).

EXAMPLE 7 ▶ Solving a Dependent System

Solve using elimination: $\begin{cases} 3x - 2y + z = -1 \\ 2x + y - z = 5 \\ 10x - 2y = 8 \end{cases}$.

Solution ▶ This system has no equation where the coefficient of x is 1. We will still use R1, but we'll try to eliminate z in R2 (there is no z-term in R3).
Using R1 + R2 eliminates the z-term from R2, yielding $5x - y = 4$.

$$\begin{cases} 3x - 2y + z = -1 \\ 2x + y - z = 5 \\ 10x - 2y = 8 \end{cases} \quad \begin{array}{c} R1 + R2 \to R2 \\ \underline{R3 \to R3} \\ \end{array} \quad \begin{cases} 3x - 2y + z = -1 \\ 5x - y = 4 \\ 10x - 2y = 8 \end{cases}$$

We next solve the subsystem. Using -2R2 + R3 eliminates the y-term in R3, but also all other terms:

$$\begin{array}{rl} -2\text{R2} & -10x + 2y = -8 \\ + & \\ \underline{\text{R3}} & \underline{10x - 2y = 8} \\ & 0x + 0y = 0 \;\text{\scriptsize sum} \\ & \, 0 = 0 \;\text{\scriptsize result} \end{array}$$

Since R3 is the same as 2R2, the system is linearly dependent and equivalent to $\begin{cases} 3x - 2y + z = -1 \\ 5x - y \quad\quad = 4 \end{cases}$. We can solve for y in R2 to write y in terms of x: $y = 5x - 4$.

Substituting $5x - 4$ for y in R1 enables us to also write z in terms of x:

$$3x - 2y \quad\quad + z = -1 \quad\quad \text{R1}$$
$$3x - 2(5x - 4) + z = -1 \quad\quad \text{substitute } 5x - 4 \text{ for } y$$
$$3x - 10x + 8 + z = -1 \quad\quad \text{distribute}$$
$$-7x + z = -9 \quad\quad \text{simplify}$$
$$z = 7x - 9 \quad\quad \text{solve for } z$$

The solution set is $\{(x, y, z) | x \in \mathbb{R}, y = 5x - 4, z = 7x - 9)\}$. Three of the infinite number of solutions are $(0, -4, -9)$ for $x = 0$, $(2, 6, 5)$ for $x = 2$, and $(-1, -9, -16)$ for $x = -1$. Verify these triples satisfy all three equations. Again using the parameter p, the solution could be written as $(p, 5p - 4, 7p - 9)$ in parameterized form.

☑ **D.** You've just learned how to recognize inconsistent and dependent systems

Now try Exercises 31 through 34 ▶

Solutions to linearly dependent systems can actually be written in terms of either x, y, or z, depending on which variable is eliminated in the first step and the variable we elect to solve for afterward.

For **coincident dependence** the equations in a system differ by only a constant multiple. After applying the elimination process—all variables are eliminated from the other equations, leaving statements that are always true (such as $2 = 2$ or some other). See **Exercises 35 and 36.** For additional practice solving various kinds of systems, see **Exercises 37 to 51.**

E. Applications

Applications of larger systems are simply an extension of our work with systems of two equations in two variables. Once again, the applications come in a variety of forms and from many fields. In the world of business and finance, systems can be used to diversify investments or spread out liabilities, a financial strategy hinted at in Example 8.

EXAMPLE 8 ▶ **Modeling the Finances of a Business**

A small business borrowed $225,000 from three different lenders to expand their product line. The interest rates were 5%, 6%, and 7%. Find how much was borrowed at each rate if the annual interest came to $13,000 and twice as much was borrowed at the 5% rate than was borrowed at the 7% rate.

Solution ▶ Let x, y, and z represent the amount borrowed at 5%, 6%, and 7%, respectively. This means our first equation is $x + y + z = 225$ (in thousands). The second equation is determined by the total interest paid, which was $13,000: $0.05x + 0.06y + 0.07z = 13$. The third is found by carefully reading the problem.

"twice as much was borrowed at the 5% rate than was borrowed at the 7% rate", or $x = 2z$.

These equations form the system: $\begin{cases} x + y + z = 225 \\ 0.05x + 0.06y + 0.07z = 13 \\ x = 2z \end{cases}$. The x-term of the first equation has a coefficient of 1. Written in standard form we have:

$$\begin{cases} x + y + z = 225 & \text{R1} \\ 5x + 6y + 7z = 1300 & \text{R2} \quad \text{(multiplied by 100)} \\ x - 2z = 0 & \text{R3} \end{cases}$$

Using $-5\text{R1} + \text{R2}$ will eliminate the x term in R2, while $-\text{R1} + \text{R3}$ will eliminate the x-term in R3.

$$\begin{array}{ll} -5\text{R1} & -5x - 5y - 5z = -1125 \\ + & \\ \underline{\text{R2}} & \underline{5x + 6y + 7z = 1300} \\ & y + 2z = 175 \end{array} \qquad \begin{array}{ll} -\text{R1} & -x - y - z = -225 \\ + & \\ \underline{\text{R3}} & \underline{x - 2z = 0} \\ & -y - 3z = -225 \end{array}$$

The new R2 is $y + 2z = 175$, and the new R3 (after multiplying by -1) is

$y + 3z = 225$, yielding the equivalent system $\begin{cases} x + y + z = 225 \\ y + 2z = 175. \\ y + 3z = 225 \end{cases}$

✓ **E.** You've just learned how to use a system of three equations in three variables to solve applications

Solving the 2×2 subsystem using $-\text{R2} + \text{R3}$ yields $z = 50$. Back-substitution shows $y = 75$ and $x = 100$, yielding the solution $(100, 75, 50)$. This means $50,000 was borrowed at the 7% rate, $75,000 was borrowed at 6%, and $100,000 at 5%.

Now try Exercises 54 through 63 ▶

TECHNOLOGY HIGHLIGHT

More on Parameterized Solutions

For linearly dependent systems, a graphing calculator can be used to both find and check possible solutions using the parameters Y_1, Y_2, and Y_3. This is done by assigning the chosen parameter to Y_1, then using Y_2 and Y_3 to form the other coordinates of the solution. We can then build the equations in the system using Y_1, Y_2, and Y_3 in place of x, y, and z. The system from Example 7 is

$$\begin{cases} 3x - 2y + z = -1 \\ 2x + y - z = 5 \\ 10x - 2y = 8 \end{cases}$$, which we found had solutions of the form $(x, 5x - 4, 7x - 9)$. We first form the

solution using $Y_1 = X$, $Y_2 = 5Y_1 - 4$ (for y), and $Y_3 = 7Y_1 - 9$ (for z). Then we form the equations in the system using $Y_4 = 3Y_1 - 2Y_2 + Y_3$, $Y_5 = 2Y_1 + Y_2 - Y_3$, and $Y_6 = 10Y_1 - 2Y_2$ (see Figure 5.8). After setting up the table (set on **AUTO**), solutions can be found by enabling only Y_1, Y_2, and Y_3, which gives values of x, y, and z, respectively (see Figure 5.9—use the right arrow ▶ to view Y_3). By enabling Y_4, Y_5, and Y_6 you can verify that for any value of the parameter, the first equation is equal to -1, the second is equal to 5, and the third is equal to 8 (see Figure 5.10—use the right arrow ▶ to view Y_6).

Figure 5.8

```
Plot1 Plot2 Plot3
\Y1=X
\Y2=5Y1-4
\Y3=7Y1-9
\Y4=3Y1-2Y2+Y3
\Y5=2Y1+Y2-Y3
\Y6=10Y1-2Y2
\Y7=
```

Figure 5.9

X	Y1	Y2
-3	-3	-19
-2	-2	-14
-1	-1	-9
0	0	-4
1	1	1
2	2	6
3	3	11
X= -3		

Figure 5.10

X	Y4	Y5
-3	-1	5
-2	-1	5
-1	-1	5
0	-1	5
1	-1	5
2	-1	5
3	-1	5
X= -3		

Exercise 1: Use the ideas from this Technology Highlight to (a) find four specific solutions to Example 6, (b) check multiple variations of the solution given, and (c) determine if $(-9, -6, -5)$, $(-2, 1, 2)$, and $(6, 2, 4)$ are solutions.

5.2 EXERCISES

▶ CONCEPTS AND VOCABULARY

Fill in the blank with the appropriate word or phrase. Carefully reread the section if needed.

1. The solution to an equation in three variables is an ordered _____.

2. The graph of the solutions to an equation in three variables is a(n) _____.

3. Systems that have the same solution set are called _____ _____.

4. If a 3 × 3 system is linearly dependent, the ordered triple solutions can be written in terms of a single variable called a(n) _____.

5. Find a value of z that makes the ordered triple $(2, -5, z)$ a solution to $2x + y + z = 4$. Discuss/Explain how this is accomplished.

6. Explain the difference between linear dependence and coincident dependence, and describe how the equations are related.

▶ DEVELOPING YOUR SKILLS

Find any four ordered triples that satisfy the equation given.

7. $x + 2y + z = 9$

8. $3x + y - z = 8$

9. $-x + y + 2z = -6$

10. $2x - y + 3z = -12$

Determine if the given ordered triples are solutions to the system.

11. $\begin{cases} x + y - 2z = -1 \\ 4x - y + 3z = 3 \quad ; \quad (0, 3, 2) \\ 3x + 2y - z = 4 \qquad (-3, 4, 1) \end{cases}$

12. $\begin{cases} 2x + 3y + z = 9 \\ 5x - 2y - z = -32; \quad (-4, 5, 2) \\ x - y - 2z = -13 \quad (5, -4, 11) \end{cases}$

Solve each system using elimination and back-substitution.

13. $\begin{cases} x - y - 2z = -10 \\ x - \qquad z = 1 \\ \qquad z = 4 \end{cases}$

14. $\begin{cases} x + y + 2z = -1 \\ 4x - y \qquad = 3 \\ 3x \qquad = 6 \end{cases}$

15. $\begin{cases} x + 3y + 2z = 16 \\ -2y + 3z = 1 \\ 8y - 13z = -7 \end{cases}$

16. $\begin{cases} -x + y + 5z = 1 \\ 4x + y \qquad = 1 \\ -3x - 2y \qquad = 8 \end{cases}$

17. $\begin{cases} 2x - y + 4z = -7 \\ x + 2y - 5z = 13 \\ y - 4z = 9 \end{cases}$

18. $\begin{cases} 2x + 3y + 4z = -18 \\ x - 2y + z = 4 \\ 4x + \qquad z = -19 \end{cases}$

19. $\begin{cases} -x + y + 2z = -10 \\ x + y - z = 7 \\ 2x + y + z = 5 \end{cases}$

20. $\begin{cases} x + y - 2z = -1 \\ 4x - y + 3z = 3 \\ 3x + 2y - z = 4 \end{cases}$

21. $\begin{cases} 3x + y - 2z = 3 \\ x - 2y + 3z = 10 \\ 4x - 8y + 5z = 5 \end{cases}$

22. $\begin{cases} 2x - 3y + 2z = 0 \\ 3x - 4y + z = -20 \\ x + 2y - z = 16 \end{cases}$

23. $\begin{cases} 3x - y + z = 6 \\ 2x + 2y - z = 5 \\ 2x - y + z = 5 \end{cases}$

24. $\begin{cases} 2x - 3y - 2z = 7 \\ x - y + 2z = -5 \\ 2x - 2y + 3z = -7 \end{cases}$

Solve using the elimination method. If a system is inconsistent or dependent, so state. For systems with linear dependence, write solutions in set notation and as an ordered triple in terms of a parameter.

25. $\begin{cases} 3x + y + 2z = 3 \\ x - 2y + 3z = 1 \\ 4x - 8y + 12z = 7 \end{cases}$

26. $\begin{cases} 2x - y + 3z = 8 \\ 3x - 4y + z = 4 \\ -4x + 2y - 6z = 5 \end{cases}$

27. $\begin{cases} 4x + y + 3z = 8 \\ x - 2y + 3z = 2 \end{cases}$

28. $\begin{cases} 4x - y + 2z = 9 \\ 3x + y + 5z = 5 \end{cases}$

29. $\begin{cases} 6x - 3y + 7z = 2 \\ 3x - 4y + z = 6 \end{cases}$

30. $\begin{cases} 2x - 4y + 5z = -2 \\ 3x - 2y + 3z = 7 \end{cases}$

Solve using elimination. If the system is linearly dependent, state the general solution in terms of a parameter. Different forms of the solution are possible.

31. $\begin{cases} 3x - 4y + 5z = 5 \\ -x + 2y - 3z = -3 \\ 3x - 2y + z = 1 \end{cases}$

32. $\begin{cases} 5x - 3y + 2z = 4 \\ -9x + 5y - 4z = -12 \\ -3x + y - 2z = -12 \end{cases}$

33. $\begin{cases} x + 2y - 3z = 1 \\ 3x + 5y - 8z = 7 \\ x + y - 2z = 5 \end{cases}$

34. $\begin{cases} -2x + 3y - 5z = 3 \\ 5x - 7y + 12z = -8 \\ x - y + 2z = -2 \end{cases}$

Solve using elimination. If the system has coincident dependence, state the solution in set notation.

35. $\begin{cases} -0.2x + 1.2y - 2.4z = -1 \\ 0.5x - 3y + 6z = 2.5 \\ x - 6y + 12z = 5 \end{cases}$

36. $\begin{cases} 6x - 3y + 9z = 21 \\ 4x - 2y + 6z = 14 \\ -2x + y - 3z = -7 \end{cases}$

Solve using the elimination method. If a system is inconsistent or dependent, so state. For systems with linear dependence, write the answer in terms of a parameter. For coincident dependence, state the solution in set notation.

37. $\begin{cases} x + 2y - z = 1 \\ x + z = 3 \\ 2x - y + z = 3 \end{cases}$
38. $\begin{cases} 3x + 5y - z = 11 \\ 2x + y - 3z = 12 \\ y + 2z = -4 \end{cases}$

39. $\begin{cases} 2x - 5y - 4z = 6 \\ x - 2.5y - 2z = 3 \\ -3x + 7.5y + 6z = -9 \end{cases}$

40. $\begin{cases} x - 2y + 2z = 6 \\ 2x - 6y + 3z = 13 \\ 3x + 4y - z = -11 \end{cases}$

41. $\begin{cases} 4x - 5y - 6z = 5 \\ 2x - 3y + 3z = 0 \\ x + 2y - 3z = 5 \end{cases}$

42. $\begin{cases} x - 5y - 4z = 3 \\ 2x - 9y - 7z = 2 \\ 3x - 14y - 11z = 5 \end{cases}$

43. $\begin{cases} 2x + 3y - 5z = 4 \\ x + y - 2z = 3 \\ x + 3y - 4z = -1 \end{cases}$

44. $\begin{cases} \dfrac{1}{6}x + \dfrac{1}{3}y - \dfrac{1}{2}z = 2 \\ \dfrac{3}{4}x - \dfrac{1}{3}y + \dfrac{1}{2}z = 9 \\ \dfrac{1}{2}x - y + \dfrac{1}{2}z = 2 \end{cases}$
45. $\begin{cases} \dfrac{x}{2} + \dfrac{y}{3} - \dfrac{z}{2} = 2 \\ \dfrac{2x}{3} - y - z = 8 \\ \dfrac{x}{6} + 2y + \dfrac{3z}{2} = 6 \end{cases}$

Some applications of systems lead to systems similar to those that follow. Solve using elimination.

46. $\begin{cases} -2A - B - 3C = 21 \\ B - C = 1 \\ A + B = -4 \end{cases}$

47. $\begin{cases} -A + 3B + 2C = 11 \\ 2B + C = 9 \\ B + 2C = 8 \end{cases}$

48. $\begin{cases} A + 2C = 7 \\ 2A - 3B = 8 \\ 3A + 6B - 8C = -33 \end{cases}$

49. $\begin{cases} A - 2B = 5 \\ B + 3C = 7 \\ 2A - B - C = 1 \end{cases}$

50. $\begin{cases} C = -2 \\ 5A - 2C = 5 \\ -4B - 9C = 16 \end{cases}$

51. $\begin{cases} C = 3 \\ 2A + 3C = 10 \\ 3B - 4C = -11 \end{cases}$

▶ WORKING WITH FORMULAS

52. Dimensions of a rectangular solid:

$\begin{cases} 2w + 2h = P_1 \\ 2l + 2w = P_2 \\ 2l + 2h = P_3 \end{cases}$

$P_2 = 16$ cm (top)
$P_1 = 14$ cm (small side)
$P_3 = 18$ cm (large side)

Using the formula shown, the dimensions of a rectangular solid can be found if the perimeters of the three distinct faces are known. Find the dimensions of the solid shown.

53. Distance from a point (x, y, z) to the plane

$Ax + By + Cz = D$: $\left| \dfrac{Ax + By + Cz - D}{\sqrt{A^2 + B^2 + C^2}} \right|$

The perpendicular distance from a given point (x, y, z) to the plane defined by $Ax + By + Cz = D$ is given by the formula shown. Consider the plane given in Figure 5.2 ($x + y + z = 6$). What is the distance from this plane to the point $(3, 4, 5)$?

▶ APPLICATIONS

Solve the following applications by setting up and solving a system of three equations in three variables. Note that some equations may have only two of the three variables used to create the system.

Investment/Finance and Simple Interest Problems

54. **Investing the winnings:** After winning $280,000 in the lottery, Maurika decided to place the money in three different investments: a certificate of deposit paying 4%, a money market certificate paying 5%, and some Aa bonds paying 7%. After 1 yr she earned $15,400 in interest. Find how much was invested at each rate if $20,000 more was invested at 7% than at 5%.

55. **Purchase at auction:** At an auction, a wealthy collector paid $7,000,000 for three paintings: a Monet, a Picasso, and a van Gogh. The Monet cost $800,000 more than the Picasso. The price of the van Gogh was $200,000 more than twice the price of the Monet. What was the price of each painting?

Descriptive Translation

56. **Major wars:** The United States has fought three major wars in modern times: World War II, the Korean War, and the Vietnam War. If you sum the years that each conflict ended, the result is 5871. The Vietnam War ended 20 years after the Korean War and 28 years after World War II. In what year did each end?

57. **Animal gestation periods:** The average gestation period (in days) of an elephant, rhinoceros, and camel sum to 1520 days. The gestation period of a rhino is 58 days longer than that of a camel. Twice the camel's gestation period decreased by 162 gives the gestation period of an elephant. What is the gestation period of each?

58. **Moments in U.S. history:** If you sum the year the Declaration of Independence was signed, the year the 13th Amendment to the Constitution abolished slavery, and the year the Civil Rights Act was signed, the total would be 5605. Ninety-nine years separate the 13th Amendment and the Civil Rights Act. The Civil Rights Act was signed 188 years after the Declaration of Independence. What year was each signed?

59. **Aviary wingspan:** If you combine the wingspan of the California Condor, the Wandering Albatross (see photo), and the prehistoric Quetzalcoatlus, you get an astonishing 18.6 m (over 60 ft). If the wingspan of the Quetzalcoatlus is equal to five times that of the Wandering Albatross minus twice

that of the California Condor, and six times the wingspan of the Condor is equal to five times the wingspan of the Albatross, what is the wingspan of each?

Mixtures

60. **Chemical mixtures:** A chemist mixes three different solutions with concentrations of 20%, 30%, and 45% glucose to obtain 10 L of a 38% glucose solution. If the amount of 30% solution used is 1 L more than twice the amount of 20% solution used, find the amount of each solution used.

61. **Value of gold coins:** As part of a promotion, a local bank invites its customers to view a large sack full of $5, $10, and $20 gold pieces, promising to give the sack to the first person able to state the number of coins for each denomination. Customers are told there are exactly 250 coins, with a total face value of $1875. If there are also seven times as many $5 gold pieces as $20 gold pieces, how many of each denomination are there?

62. **Rewriting a rational function:** It can be shown that the rational function $V(x) = \dfrac{3x + 11}{x^3 - 3x^2 + x - 3}$ can be written as a sum of the terms $\dfrac{A}{x - 3} + \dfrac{Bx + C}{x^2 + 1}$, where the coefficients A, B, and C are solutions to $\begin{cases} A + B & = 0 \\ -3B + C = 3 \\ A - 3C = 11 \end{cases}$. Find the missing coefficients and verify your answer by adding the terms.

63. **Rewriting a rational function:** It can be shown that the rational function $V(x) = \dfrac{x - 9}{x^3 - 6x^2 + 9x}$ can be written as a sum of the terms $\dfrac{A}{x} + \dfrac{B}{x - 3} + \dfrac{C}{(x - 3)^2}$, where the coefficients A, B, and C are solutions to $\begin{cases} A + B & = 0 \\ -6A - 3B + C = 1 \\ 9A & = -9 \end{cases}$. Find the missing coefficients and verify your answer by adding the terms.

▶ EXTENDING THE CONCEPT

64. The system $\begin{cases} x - 2y - z = 2 \\ x - 2y + kz = 5 \\ 2x - 4y + 4z = 10 \end{cases}$ is inconsistent if

$k = $ _____, and dependent if $k = $ ____.

65. One form of the equation of a circle is $x^2 + y^2 + Dx + Ey + F = 0$. Use a system to find the equation of the circle through the points $(2, -1), (4, -3),$ and $(2, -5)$.

66. The lengths of each side of the squares $A, B, C, D,$ $E, F, G, H,$ and I (the smallest square) shown are whole numbers. Square B has sides of 15 cm and square G has sides of 7 cm. What are the dimensions of square D?

a. 9 cm   **b.** 10 cm   **c.** 11 cm

d. 12 cm   **e.** 13 cm

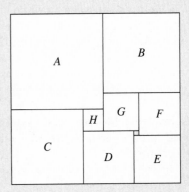

▶ MAINTAINING YOUR SKILLS

67. (3.7) If $p(x) = 2x^2 - x - 3$, in what intervals is $p(x) \leq 0$?

68. (3.4) Graph the polynomial defined by $f(x) = x^4 - 5x^2 + 4$.

69. (4.4) Solve the logarithmic equation: $\log(x + 2) + \log x = \log 3$

70. (2.5) Analyze the graph of g shown. Clearly state the domain and range, the zeroes of g, intervals

where $g(x) > 0$, intervals where $g(x) < 0$, local maximums or minimums, and intervals where the function is increasing or decreasing. Assume each tick mark is one unit and estimate endpoints to the nearest tenths.

MID-CHAPTER CHECK

1. Solve using the substitution method. State whether the system is consistent, inconsistent, or dependent.
$\begin{cases} x - 3y = -2 \\ 2x + y = 3 \end{cases}$

2. Solve the system using elimination. State whether the system is consistent, inconsistent, or dependent.
$\begin{cases} x - 3y = -4 \\ 2x + y = 13 \end{cases}$

3. Solve using a system of linear equations and any method you choose: How many ounces of a 40% acid, should be mixed with 10 oz of a 64% acid, to obtain a 48% acid solution?

4. Determine whether the ordered triple is a solution to the system.
$\begin{cases} 5x + 2y - 4z = 22 \\ 2x - 3y + z = -1 \quad (2, 0, -3) \\ 3x - 6y + z = 2 \end{cases}$

5. The system given is a dependent system. Without solving, state why.
$\begin{cases} x + 2y - 3z = 3 \\ 2x + 4y - 6z = 6 \\ x - 2y + 5z = -1 \end{cases}$

6. Solve the system of equations:
$\begin{cases} x + 2y - 3z = -4 \\ 2y + z = 7 \\ 5y - 2z = 4 \end{cases}$

7. Solve using elimination:
$\begin{cases} 2x + 3y - 4z = -4 \\ x - 2y + z = 0 \\ -3x - 2y + 2z = -1 \end{cases}$

8. Solve the following system and write the solution as an ordered triple in terms of the parameter p.

$$\begin{cases} 2x - y + z = 1 \\ -5x + 2y - 3z = 2 \end{cases}$$

9. If you add Mozart's age when he wrote his first symphony, with the age of American chess player Paul Morphy when he began dominating the international chess scene, and the age of Blaise Pascal when he formulated his well-known *Essai pour les coniques* (Essay on Conics), the sum is 37. At the time of each event, Paul Morphy's age was 3 yr less than twice Mozart's, and Pascal was 3 yr older than Morphy. Set up a system of equations and find the age of each.

10. The *William Tell Overture* (Gioachino Rossini, 1829) is one of the most famous, and best-loved overtures known. It is played in four movements: a prelude, the storm (often used in animations with great clashes of thunder and a driving rain), the sunrise (actually, *A call to the dairy cows . . .*), and the finale (better known as the Lone Ranger theme song). The prelude takes 2.75 min. Depending on how fast the finale is played, the total playing time is about 11 min. The playing time for the prelude and finale is 1 min longer than the playing time of the storm and the sunrise. Also, the playtime of the storm plus twice the playtime of the sunrise is 1 min longer than twice the finale. Find the playtime for each movement.

REINFORCING BASIC CONCEPTS

Window Size and Graphing Technology

Since most substantial applications involve noninteger values, technology can play an important role in applying mathematical models. However, with its use comes a heavy responsibility to use it carefully. A very real effort must be made to determine the best approach and to secure a reasonable estimate. This is the only way to guard against (the inevitable) keystroke errors, or ensure a window size that properly displays the results.

Rationale

On October 1, 1999, the newspaper *USA TODAY* ran an article titled, "Bad Math added up to Doomed Mars Craft." The article told of how a $125,000,000.00 spacecraft was lost, apparently because the team of scientists that *plotted the course* for the craft used U.S. units of measurement, while the team of scientists *guiding* the craft were using metric units. NASA's space chief was later quoted, "The problem here was not the error, it was the failure of . . . the checks and balances in our process to detect the error."

No matter how powerful the technology, always try to begin your problem-solving efforts with an estimate. Begin by exploring the **context** of the problem, asking questions about the range of possibilities: How fast can a human run? How much does a new car cost? What is a reasonable price for a ticket? What is the total available to invest? There is no calculating involved in these estimates, they simply rely on "horse sense" and human experience. In many applied problems, the input and output values must be positive — which means the solution will appear in the first quadrant, narrowing the possibilities considerably.

This information will be used to set the viewing window of your graphing calculator, in preparation for solving the problem using a system and graphing technology.

Illustration 1 ▶ Erin just filled both her boat and Blazer with gas, at a total cost of $211.14. She purchased 35.7 gallons of premium for her boat and 15.3 gal of regular for her Blazer. Premium gasoline cost $0.10 per gallon more than regular. What was the cost per gallon of each grade of gasoline?

Solution ▶ Asking how much *you* paid for gas the last time you filled up should serve as a fair estimate. Certainly (in 2008) a cost of $6.00 or more per gallon in the United States is too high, and a cost of $2.50 per gallon or less would be too low. Also, we can estimate a solution by assuming that both kinds of gasoline cost the same. This would mean 51 gal were purchased for about $211, and a quick division would place the estimate at near $\frac{211}{51} \approx \$4.14$ per gallon. A good viewing window would be restricted to the first quadrant (since cost > 0) with maximum values of Xmax = 6 and Ymax = 6.

Exercise 1: Solve Illustration 1 using graphing technology.

Exercise 2: Re-solve Exercises 63 and 64 from Section 5.1 using graphing technology. Verify results are identical.

Learning Objectives

In Section 5.3 you will learn how to:

☐ **A.** Visualize possible solutions

☐ **B.** Solve nonlinear systems using substitution

☐ **C.** Solve nonlinear systems using elimination

☐ **D.** Solve nonlinear systems of inequalities

☐ **E.** Solve applications of nonlinear systems

Equations where the variables have exponents other than 1 or that are transcendental (like logarithmic and exponential equations) are all nonlinear equations. A nonlinear system of equations has at least one nonlinear equation, and these occur in a great variety.

A. Possible Solutions for a Nonlinear System

When solving nonlinear systems, it is often helpful to *visualize* the graphs of each equation in the system. This can help determine the number of possible intersections and further assist the solution process.

EXAMPLE 1 ▶ **Sketching Graphs to Visualize the Number of Possible Solutions**

Identify each equation in the system as the equation of a line, parabola, circle, or one of the toolbox functions. Then determine the number of solutions possible by considering the different ways the graphs might intersect: $\begin{cases} x^2 + y^2 = 25 \\ x - y = 1 \end{cases}$.

Finally, solve the system by graphing.

Solution ▶ The first equation contains a sum of second-degree terms with equal coefficients, which we recognize as the equation of a circle. The second equation is obviously linear. This means the system may have no solution, one solution, or two solutions, as shown in Figure 5.11. The graph of the system is shown in Figure 5.12 and the two points of intersection appear to be $(-3, -4)$ and $(4, 3)$. After checking these in the original equations we find that both are solutions to the system.

☑ **A.** You've just learned how to visualize possible solutions

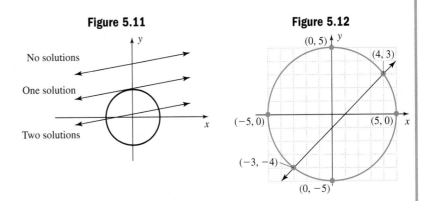

Figure 5.11

No solutions

One solution

Two solutions

Figure 5.12

$(0, 5)$ $(4, 3)$

$(-5, 0)$ $(5, 0)$

$(-3, -4)$

$(0, -5)$

Now try Exercises 7 through 12 ▶

B. Solving Nonlinear Systems by Substitution

Since graphical methods at best offer an estimate for the solution (points of intersection may not have integer values), we more often turn to the algebraic methods just developed. Recall the substitution method involves solving one of the equations for a variable or expression that can be substituted in the other equation to eliminate one of the variables.

EXAMPLE 2 ▶ **Solving a Nonlinear System Using Substitution**

Solve the system using substitution: $\begin{cases} y = x^2 - 2x - 3 \\ 2x - y = 7 \end{cases}$.

Solution ▶ The first equation is the equation of a parabola. The second equation is linear. Since the first equation is already written with y in terms of x, we can substitute $x^2 - 2x - 3$ for y in the second equation to solve.

$$2x - y = 7 \qquad \text{second equation}$$
$$2x - (x^2 - 2x - 3) = 7 \qquad \text{substitute } x^2 - 2x - 3 \text{ for } y$$
$$2x - x^2 + 2x + 3 = 7 \qquad \text{distribute}$$
$$-x^2 + 4x + 3 = 7 \qquad \text{simplify}$$
$$x^2 - 4x + 4 = 0 \qquad \text{set equal to zero}$$
$$(x - 2)^2 = 0 \qquad \text{factor}$$

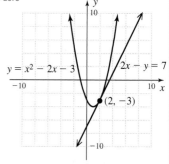

We find that $x = 2$ is a repeated root.

Since the second equation is simpler than the first, we substitute 2 for x in this equation and find $y = -3$. The system has only one (repeated) solution at $(2, -3)$, as shown in the figure.

☑ **B** You've just learned how to solve nonlinear systems using substitution

Now try Exercises 13 through 18 ▶

C. Solving Nonlinear Systems by Elimination

When both equations in the system have second-degree terms with like variables, it is generally easier to use the elimination method, rather than substitution. Remember to watch for systems that have no solutions.

EXAMPLE 3 ▶ **Solving a Nonlinear System Using Elimination**

Solve the system using elimination: $\begin{cases} y - \frac{1}{2}x^2 = -3 \\ x^2 + y^2 = 41 \end{cases}$.

Solution ▶ The first equation can be rewritten as $y = \frac{1}{2}x^2 - 3$ and is a parabola opening upward with vertex $(0, -3)$. The second equation represents a circle with center at $(0, 0)$ and radius $r = \sqrt{41} \approx 6.4$. Mentally visualizing these graphs indicates there will be two solutions (see figure). After writing the system with x- and y-terms in the same order, we find that using $2R1 + R2$ will eliminate the variable x:

WORTHY OF NOTE

Note that the x-terms sum to zero, and the y-terms cannot be combined as they are not like terms.

$$\begin{array}{r} 2R1 \\ + \\ R2 \end{array} \begin{cases} -x^2 + 2y = -6 \qquad \text{rewrite first equation; multiply by 2} \\ \underline{x^2 + y^2 = 41} \qquad \text{second equation} \\ y^2 + 2y = 35 \qquad \text{add} \end{cases}$$

To find solutions, we set the equation equal to zero and factor or use the quadratic formula if needed.

$$y^2 + 2y - 35 = 0 \quad \text{standard form}$$
$$(y + 7)(y - 5) = 0 \quad \text{factored form}$$
$$y = -7 \text{ or } y = 5 \quad \text{result}$$

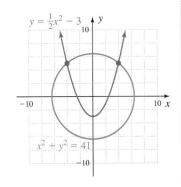

The solution $y = -7$ is extraneous, due to the radius of the circle. Using $y = 5$ in the second equation gives the following:

$$x^2 + y^2 = 41 \quad \text{equation 2}$$
$$x^2 + (5)^2 = 41 \quad \text{substitute 5 for } y$$
$$x^2 + 25 = 41 \quad 5^2 = 25$$
$$x^2 = 16 \quad \text{subtract 25}$$
$$x = \pm 4 \quad \text{square root property}$$

The solutions are $(-4, 5)$ and $(4, 5)$, which is supported by the graph shown.

> **Now try Exercises 19 through 24 ▶**

Nonlinear systems may involve other relations as well, including power, polynomial, logarithmic, or exponential functions. These are solved using the same methods.

EXAMPLE 4 ▶ **Solving a System of Logarithmic Equations**

Solve the system using the method of your choice: $\begin{cases} y = -\log(x + 7) + 2 \\ y = \log(x + 4) + 1 \end{cases}$.

Solution ▶ Since both equations have y written in terms of x, substitution appears to be the better choice. The result is a logarithmic equation, which we can solve using the techniques from Chapter 4.

$$\log(x + 4) + 1 = -\log(x + 7) + 2 \quad \text{substitute } \log(x + 4) + 1 \text{ for } y \text{ in first equation}$$
$$\log(x + 4) + \log(x + 7) = 1 \quad \text{add } \log(x + 7); \text{ subtract 1}$$
$$\log(x + 4)(x + 7) = 1 \quad \text{product property of logarithms}$$
$$(x + 4)(x + 7) = 10^1 \quad \text{exponential form}$$
$$x^2 + 11x + 18 = 0 \quad \text{eliminate parentheses and set equal to zero}$$
$$(x + 9)(x + 2) = 0 \quad \text{factor}$$
$$x + 9 = 0 \quad \text{or} \quad x + 2 = 0 \quad \text{zero factor theorem}$$
$$x = -9 \quad \text{or} \quad x = -2 \quad \text{possible solutions}$$

By inspection, we see that $x = -9$ is not a solution, since $\log(-9 + 4)$ and $-\log(-9 + 7)$ are not real numbers. Substituting -2 for x in the second equation we find one form of the (exact) solution is $(-2, \log 2 + 1)$. If we substitute -2 for x in the first equation the exact solution is $(-2, -\log 5 + 2)$. Use a calculator to verify the answers are equivalent and approximately $(-2, 1.3)$.

> **Now try Exercises 25 through 36 ▶**

☑ **C.** You've just learned how to solve nonlinear systems using elimination

For practice solving more complex systems using a graphing calculator, see **Exercises 37 to 42.**

Figure 5.13

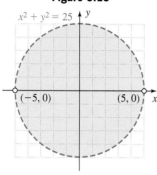

$x^2 + y^2 = 25$

$(-5, 0)$ $(5, 0)$ x

D. Solving Systems of Nonlinear Inequalities

Nonlinear inequalities can be solved by graphing the boundary given by the related equation, and checking the regions that result using a test point. For example, the inequality $x^2 + y^2 < 25$ is solved by first graphing $x^2 + y^2 = 25$, a circle with radius 5, and deciding if the boundary is included or excluded (in this case it is not). We then use a test point from either "outside" or "inside" the region formed. The test point $(0, 0)$ results in a true statement since $(0)^2 + (0)^2 < 25$, so the inside of the circle is shaded to indicate the solution region (Figure 5.13). For a *system* of nonlinear inequalities, we identify regions where the solution set for both inequalities overlap, paying special attention to points of intersection.

EXAMPLE 5 ▶ **Solving Systems of Nonlinear Inequalities**

Solve the system: $\begin{cases} x^2 + y^2 < 25 \\ 2y - x \geq 5 \end{cases}$.

Solution ▶ We recognize the first inequality from Figure 5.13, a circle with radius 5, and a solution region in the interior. The second inequality is linear and after solving for x we'll use a substitution to find points of intersection (if they exist). From $2y - x = 5$, we obtain $x = 2y - 5$.

$$x^2 + y^2 = 25 \qquad \text{given}$$
$$(2y - 5)^2 + y^2 = 25 \qquad \text{substitute } 2y - 5 \text{ for } x$$
$$5y^2 - 20y + 25 = 25 \qquad \text{expand and simplify}$$
$$y^2 - 4y = 0 \qquad \text{subtract 25; divide by 5}$$
$$y(y - 4) = 0 \qquad \text{factor}$$
$$y = 0 \quad \text{or} \quad y = 4 \qquad \text{result}$$

Back-substitution shows the graphs intersect at $(-5, 0)$ and $(3, 4)$. Graphing a line through these points and using $(0, 0)$ as a test point shows the upper half plane is the solution region for the linear inequality [$2(0) - 0 \geq 5$ is *false*]. The overlapping (solution) region for *both* inequalities is the circular section shown. Note the points of intersection are graphed using "open dots" (see figure), since points on the graph of the circle are excluded from the solution set.

y $-x + 2y = 5$

$(3, 4)$

$(-5, 0)$ x

$x^2 + y^2 = 25$

☑ **D** You've just learned how to solve nonlinear systems of inequalities

Now try Exercises 43 through 50 ▶

E. Applications of Nonlinear Systems

In the business world, a fast growing company can often reduce the average price of its products using what are called the **economies of scale.** These would include the ability to buy necessary materials in larger quantities, integrating new technology into the production process, and other means. However, there are also countering forces called the **diseconomies of scale,** which may include the need to hire additional employees, rent more production space, and the like.

EXAMPLE 6 ▶ **Solving an Application of Nonlinear Systems**

Suppose the cost to produce a new and inexpensive shoe made from molded plastic is modeled by the function $C(x) = x^2 - 5x + 18$, where $C(x)$ represents the cost to produce x thousand of these shoes. The revenue from the sales of these shoes is modeled by $R(x) = -x^2 + 10x - 4$. Use a break-even analysis to find the quantity of sales that will cause the company to break even.

Solution ▶ Essentially we are asked to solve the system formed by the two equations:
$$\begin{cases} C(x) = x^2 - 5x + 18 \\ R(x) = -x^2 + 10x - 4 \end{cases}$$. Since we want to know the point where the company breaks even, we set $C(x) = R(x)$ and solve.

$$C(x) = R(x)$$
$$x^2 - 5x + 18 = -x^2 + 10x - 4 \quad \text{substitute for } C(x) \text{ and } R(x)$$
$$2x^2 - 15x + 22 = 0 \quad \text{set equal to zero}$$
$$(2x - 11)(x - 2) = 0 \quad \text{factored form}$$
$$x = \frac{11}{2} \text{ or } x = 2 \quad \text{result}$$

☑ **E You've just learned how to solve applications of nonlinear systems**

With x in thousands, it appears the company will break even if either 2000 shoes or 5500 shoes are made and sold.

Now try Exercises 53 and 54 ▶

5.3 EXERCISES

▶ CONCEPTS AND VOCABULARY

1. Draw sketches showing the different ways each pair of relations can intersect and give one, two, three, and/or four points of intersection. If a given number of intersections is not possible, so state.

   a. circle and line

   b. parabola and line

   c. circle and parabola

   d. circle and absolute value function

   e. absolute value function and line

   f. absolute value function and parabola

2. By inspection only, identify the systems having *no solutions* and justify your choices.

   a. $\begin{cases} y = |x| - 6 \\ x^2 + y^2 = 9 \end{cases}$ b. $\begin{cases} y = x^2 + 4 \\ x^2 + y^2 = 4 \end{cases}$

   c. $\begin{cases} y = x + 1 \\ x^2 + y^2 = 12 \end{cases}$

3. The solution to a system of nonlinear inequalities is a(n) _____ of the plane where the _____ for each individual inequality overlap.

4. When both equations in the system have at least one _____ -degree term, it is generally easier to use the _____ method to find a solution.

5. Suppose a nonlinear system contained a central hyperbola and an exponential function. Are three solutions possible? Are four solutions possible? Explain/Discuss.

6. Solve the system twice, once using elimination, then again using substitution. Compare/contrast each process and comment on which is more efficient in this case: $\begin{cases} x^2 + y^2 = 25 \\ x^2 + y = 5 \end{cases}$.

▶ DEVELOPING YOUR SKILLS

Identify each equation in the system as that of a line, parabola, circle, or absolute value function, then solve the system by graphing.

7. $\begin{cases} x^2 + y = 6 \\ x + y = 4 \end{cases}$ 8. $\begin{cases} -x + y = 4 \\ x^2 + y^2 = 16 \end{cases}$

9. $\begin{cases} y^2 + x^2 = 100 \\ y = |x - 2| \end{cases}$ 10. $\begin{cases} x^2 + y^2 = 25 \\ x^2 + y = 13 \end{cases}$

11. $\begin{cases} -(x - 1)^2 + 2 = y \\ y - x^2 = -3 \end{cases}$ 12. $\begin{cases} y - 4 = -x^2 \\ y = -|x - 1| + 3 \end{cases}$

Solve using substitution. In Exercises 17 and 18, solve for x^2 or y^2 and use the result as a substitution.

13. $\begin{cases} x^2 + y^2 = 25 \\ y - x = 1 \end{cases}$ 14. $\begin{cases} x + 7y = 50 \\ x^2 + y^2 = 100 \end{cases}$

15. $\begin{cases} x^2 + y = 9 \\ -2x + y = 1 \end{cases}$ 16. $\begin{cases} x^2 - y = 8 \\ x + y = 4 \end{cases}$

17. $\begin{cases} x^2 + y = 13 \\ x^2 + y^2 = 25 \end{cases}$ 18. $\begin{cases} y^2 + (x - 3)^2 = 25 \\ y^2 + (x + 1)^2 = 9 \end{cases}$

Solve each system.

19. $\begin{cases} x^2 + y^2 = 25 \\ \frac{1}{4}x^2 + y = 1 \end{cases}$ 20. $\begin{cases} y - \frac{1}{2}x^2 = -1 \\ x^2 + y^2 = 65 \end{cases}$

21. $\begin{cases} x^2 + y^2 = 4 \\ y + x^2 = 5 \end{cases}$ 22. $\begin{cases} y + x^2 = 6x \\ y - 11 = (x - 3)^2 \end{cases}$

23. $\begin{cases} x^2 + y^2 = 65 \\ y = 3x + 25 \end{cases}$ 24. $\begin{cases} y - 2x = 5 \\ x^2 + y^2 = 85 \end{cases}$

Solve using the method of your choice.

25. $\begin{cases} y - 5 = \log x \\ y = 6 - \log(x - 3) \end{cases}$

26. $\begin{cases} y = \log(x + 4) + 1 \\ y - 2 = -\log(x + 7) \end{cases}$

27. $\begin{cases} y = \ln(x^2) + 1 \\ y - 1 = \ln(x + 12) \end{cases}$

28. $\begin{cases} \log(x + 1.1) = y + 3 \\ y + 4 = \log(x^2) \end{cases}$ 29. $\begin{cases} y - 9 = e^{2x} \\ 3 = y - 7e^x \end{cases}$

30. $\begin{cases} y - 2e^{2x} = 5 \\ y - 1 = 6e^x \end{cases}$ 31. $\begin{cases} y = 4^{x+3} \\ y - 2^{x^2+3x} = 0 \end{cases}$

32. $\begin{cases} y - 3^{x^2+2x} = 0 \\ y = 9^{x+2} \end{cases}$ 33. $\begin{cases} x^3 - y = 2x \\ y - 5x = -6 \end{cases}$

34. $\begin{cases} y - x^3 = -2 \\ y + 4 = 3x \end{cases}$ 35. $\begin{cases} x^2 - 6x = y - 4 \\ y - 2x = -8 \end{cases}$

36. $\begin{cases} y + x = -2 \\ y + 4x = x^2 \end{cases}$

Solve each system using a graphing calculator. Round solutions to hundredths (as needed).

37. $\begin{cases} x^2 + y^2 = 34 \\ y^2 + (x - 3)^2 = 25 \end{cases}$ 38. $\begin{cases} 5x^2 + 5y^2 = 40 \\ y + 2x = x^2 - 6 \end{cases}$

39. $\begin{cases} y = 2^x - 3 \\ y + 2x^2 = 9 \end{cases}$ 40. $\begin{cases} y = -2\log(x + 8) \\ y + x^3 = 4x - 2 \end{cases}$

41. $\begin{cases} y = \dfrac{1}{(x - 3)^2} + 2 \\ (x - 3)^2 + y^2 = 10 \end{cases}$ 42. $\begin{cases} y^2 + x^2 = 5 \\ y = \dfrac{1}{x - 1} - 2 \end{cases}$

Solve each system of inequalities.

43. $\begin{cases} y - x^2 \ge 1 \\ x + y \le 3 \end{cases}$ 44. $\begin{cases} x^2 + y^2 \le 25 \\ x + 2y \le 5 \end{cases}$

45. $\begin{cases} x^2 + y^2 > 16 \\ x^2 + y^2 \le 64 \end{cases}$ 46. $\begin{cases} y + 4 \ge x^2 \\ x^2 + y^2 \le 34 \end{cases}$

47. $\begin{cases} y - x^2 \le -16 \\ y^2 + x^2 < 9 \end{cases}$ 48. $\begin{cases} x^2 + y^2 \le 16 \\ x + 2y > 10 \end{cases}$

49. $\begin{cases} y^2 + x^2 \le 25 \\ |x| - 1 > -y \end{cases}$ 50. $\begin{cases} y^2 + x^2 \le 4 \\ x + y < 4 \end{cases}$

▶ **WORKING WITH FORMULAS**

51. **Tunnel clearance:** $h = \sqrt{r^2 - d^2}$

The maximum rectangular clearance allowed by a circular tunnel can be found using the formula shown, where $x^2 + y^2 = r^2$ models the tunnel's circular cross section and h is the height of the tunnel at a distance d from the center. If $r = 50$ ft, find the maximum clearance at distances of $d = 20$, 30, and 40 ft from center.

52. **Manufacturing cylindrical vents:** $\begin{cases} A = 2\pi rh \\ V = \pi r^2 h \end{cases}$

In the manufacture of cylindrical vents, a rectangular piece of sheet metal is rolled, riveted, and sealed to form the vent. The radius and height required to form a vent with a specified volume, using a piece of sheet metal with a given area, can be found by solving the system shown. Use the system to find the radius and height if the volume required is 4071 cm³ and the area of the rectangular piece is 2714 cm².

▶ APPLICATIONS

Market equilibrium: In a free-enterprise (supply and demand) economy, the amount buyers are willing to pay for an item and the number of these items manufacturers are willing to produce depend on the price of the item. As the price increases, demand for the item decreases since buyers are less willing to pay the higher price. On the other hand, an increase in price increases the supply of the item since manufacturers are now more willing to supply it. When the supply and demand curves are graphed, their point of intersection is called the market equilibrium for the item.

Solve the following applications of economies of scale.

53. **World's most inexpensive car:** Early in 2008, the Tata Company (India) unveiled the new Tata Nano, the world's most inexpensive car. With its low price and 54 miles per gallon, the car may prove to be very popular. *Assume* the cost to produce these cars is modeled by the function $C(x) = 2.5x^2 - 120x + 3500$, where $C(x)$ represents the cost to produce x-thousand cars. Suppose the revenue from the sale of these cars is modeled by $R(x) = -2x^2 + 180x - 500$. Use a break-even analysis to find the quantity of sales (to the nearest hundred) that will cause the company to break even.

54. **Document reproduction:** In a world of technology, document reproduction has become a billion dollar business. With very stiff competition, the price of a single black and white copy has varied greatly in recent years. Suppose the cost to produce these copies is modeled by the function $C(x) = 0.1x^2 - 1.2x + 7$, where $C(x)$ represents the cost to produce x hundred thousand copies. If the revenue from the sale of these copies is modeled by $R(x) = -0.1x^2 + 1.8x - 2$, use a break-even analysis to find the quantity of copies that will cause the company to break even.

55. Suppose the monthly market demand D (in ten-thousands of gallons) for a new synthetic oil is related to the price P in dollars by the equation $10P^2 + 6D = 144$. For the market price P, assume the amount D that manufacturers are willing to supply is modeled by $8P^2 - 8P - 4D = 12$. (a) What is the minimum price at which manufacturers are willing to begin supplying the oil? (b) Use this information to create a system of nonlinear equations, then solve the system to find the market equilibrium price (per gallon) and the quantity of oil supplied and sold at this price.

56. The weekly demand D for organically grown carrots (in thousands of pounds) is related to the price per pound P by the equation $8P^2 + 4D = 84$. At this market price, the amount that growers are willing to supply is modeled by the equation $8P^2 + 6P - 2D = 48$. (a) What is the minimum price at which growers are willing to supply the organically grown carrots? (b) Use this information to create a system of nonlinear equations, then solve the system to find the market equilibrium price (per pound) and the quantity of carrots supplied and sold at this price.

Solve by setting up and solving a system of nonlinear equations.

57. **Dimensions of a flag:** A large American flag has an area of 85 m^2 and a perimeter of 37 m. Find the dimensions of the flag.

58. **Dimensions of a sail:** The sail on a boat is a right triangle with a perimeter of 36 ft and a hypotenuse of 15 ft. Find the height and width of the sail.

59. **Dimensions of a tract:** The area of a rectangular tract of land is 45 km^2. The length of a diagonal is $\sqrt{106}$ km. Find the dimensions of the tract.

60. **Dimensions of a deck:** A rectangular deck has an area of 192 ft^2 and the length of the diagonal is 20 ft. Find the dimensions of the deck.

61. **Dimensions of a trailer:** The surface area of a rectangular trailer with square ends is 928 ft^2. If the sum of all edges of the trailer is 164 ft, find its dimensions.

62. **Dimensions of a cylindrical tank:** The surface area of a closed cylindrical tank is 192π m^2. Find the dimensions of the tank if the volume is 320π m^3 and the radius is as small as possible.

▶ **EXTENDING THE CONCEPT**

63. The area of a vertical parabolic segment is given by $A = \frac{2}{3}BH$, where B is the length of the horizontal base of the segment and H is the height from the base to the vertex. Investigate how this formula can be used to find the *area* of the solution region for the general system of inequalities shown.

$$\begin{cases} y \geq x^2 - bx + c \\ y \leq c + bx - x^2 \end{cases}$$

(*Hint:* Begin by investigating with $b = 6$ and $c = 8$, then use other values and try to generalize what you find.)

64. Find the area of the trapezoid formed by joining the points where the parabola $y = \frac{1}{2}x^2 - 26$ and the circle $x^2 + y^2 = 100$ intersect.

65. A rectangular fish tank has a bottom and four sides made out of glass. Use a system of equations to help find the dimensions of the tank if the height is 18 in., surface area is 4806 in^2, the tank must hold 108 gal (1 gal = 231 in^3), and all three dimensions are integers.

▶ **MAINTAINING YOUR SKILLS**

66. (1.5) Solve by factoring:
    a. $2x^2 + 5x - 63 = 0$
    b. $4x^2 - 121 = 0$
    c. $2x^3 - 3x^2 - 8x + 12 = 0$

67. (1.6) Solve each equation:
    a. $3x^2 + 4x - 12 = 0$
    b. $\sqrt{3x + 1} - \sqrt{2x} = 1$
    c. $\dfrac{1}{x + 2} + \dfrac{3}{x^2 + 5x + 6} = \dfrac{2}{x + 3}$

68. (5.2) Solve using any method. As an investment for retirement, Donovan bought three properties for a total of $250,000. Ten years later, the first property had doubled in value, the second property had tripled in value, and the third property was worth $10,000 less than when he bought it, for a current value of $485,000. Find the original purchase price if he paid $20,000 more for the first property than he did for the second.

69. (2.3) In 2001, a small business purchased a copier for $4500. In 2004, the value of the copier had decreased to $3300. Assuming the depreciation is linear: (a) find the rate-of-change $m = \dfrac{\Delta\text{value}}{\Delta\text{time}}$ and discuss its meaning in this context; (b) find the depreciation equation; and (c) use the equation to predict the copier's value in 2008. (d) If the copier is traded in for a new model when its value is less than $700, how long will the company use this copier?

5.4 | Systems of Inequalities and Linear Programming

Learning Objectives

In Section 5.4 you will learn how to:

☐ **A.** Solve a linear inequality in two variables

☐ **B.** Solve a system of linear inequalities

☐ **C.** Solve applications using a system of linear inequalities

☐ **D.** Solve applications using linear programming

In this section, we'll build on many of the ideas from Section 5.3, with a more direct focus on systems of linear inequalities. While systems of linear *equations* have an unlimited number of applications, there are many situations that can only be modeled using linear *inequalities*. For example, many decisions in business and industry are based on a large number of limitations or constraints, with many different ways these constraints can be satisfied.

A. Linear Inequalities in Two Variables

A linear equation in two variables is any equation that can be written in the form $Ax + By = C$, where A and B are real numbers, not simultaneously equal to zero. A **linear inequality** in two variables is similarly defined, with the " $=$ " sign replaced by the " $<$," " $>$," " $\leq$," or " $\geq$ " symbol:

$$\begin{array}{ll} Ax + By < C & \quad Ax + By > C \\ Ax + By \leq C & \quad Ax + By \geq C \end{array}$$

Solving a linear inequality in two variables has many similarities with the one variable case. For one variable, we graph the *boundary point* on a number line, decide whether the endpoint is *included* or *excluded,* and *shade the appropriate half line.* For $x + 1 \leq 3$, we have the solution $x \leq 2$ with the endpoint included and the line shaded to the left (Figure 5.14):

Figure 5.14

$$-\infty \qquad\qquad\qquad\qquad\qquad\qquad\qquad\qquad \infty$$

$$\xleftarrow{\hspace{2cm}}\!\!\!\!\mid\!\!\!\!-\!\!-\!\!\!\mid\!\!\!-\!\!-\!\!\!\mid\!\!\!-\!\!-\!\!\!\mid\!\!\!-\!\!-\!\!\!\mid\!\!\!-\!\!-\!\!\!\mid\!\!\!-\!\!-\!\!\!\mid\!\!-\!\!\!\rightarrow$$

$$-3 \quad -2 \quad -1 \quad\; 0 \quad\;\; 1 \quad\;\; 2 \quad\;\; 3$$

Interval notation: $x \in (-\infty, 2]$

Figure 5.15

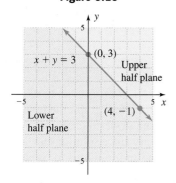

For linear inequalities in two variables, we graph a *boundary line,* decide whether the boundary line is *included* or *excluded,* and *shade the appropriate half plane.* For $x + y \leq 3$, the boundary line $x + y = 3$ is graphed in Figure 5.15. Note it divides the coordinate plane into two regions called **half planes,** and it forms the **boundary** between the two regions. If the boundary is **included** in the solution set, we graph it using a *solid line.* If the boundary is **excluded,** a *dashed line* is used. Recall that solutions to a linear equation are ordered pairs that make the equation true. We use a similar idea to find or verify solutions to linear inequalities. If any one point in a half plane makes the inequality true, all points in that half plane will satisfy the inequality.

EXAMPLE 1 ▶ **Checking Solutions to an Inequality in Two Variables**

Determine whether the given ordered pairs are solutions to $-x + 2y < 2$:

a. $(4, -3)$ **b.** $(-2, 1)$

Solution ▶ **a.** Substitute 4 for x and -3 for y: $-(4) + 2(-3) < 2$ substitute 4 for x, −3 for y

$$-10 < 2 \qquad \text{true}$$

$(4, -3)$ is a solution.

b. Substitute -2 for x and 1 for y: $-(-2) + 2(1) < 2$ substitute −2 for x, 1 for y

$$4 < 2 \qquad \text{false}$$

$(-2, 1)$ is not a solution.

Now try Exercises 7 through 10 ▶

WORTHY OF NOTE

This relationship is often called the **trichotomy axiom** or the *"three-part truth."* Given any two quantities, they are either equal to each other, or the first is less than the second, or the first is greater than the second.

Earlier we graphed linear equations by plotting a small number of ordered pairs or by solving for y and using the slope-intercept method. The line represented all ordered pairs that made the equation true, meaning *the left-hand expression was equal to the right-hand expression.* To graph linear inequalities, we reason that if the line represents all ordered pairs that make the expressions *equal,* then any point *not on that line* must make the expressions *unequal*—either greater than or less than. These ordered pair solutions must lie in one of the half planes formed by the line, which we shade to indicate the **solution region.** Note this implies the boundary line for any inequality *is determined by the related equation,* temporarily replacing the inequality symbol with an "=" sign.

EXAMPLE 2 ▶ **Solving an Inequality in Two Variables**

Solve the inequality $-x + 2y \leq 2$.

Solution ▶ The related equation and boundary line is $-x + 2y = 2$. Since the inequality is inclusive (less than *or equal to*), we graph a solid line. Using the intercepts, we graph the line through $(0, 1)$ and $(-2, 0)$ shown in Figure 5.16. To determine the solution region and which side to shade, we select $(0, 0)$ as a test point, which results in a true statement: $-(0) + 2(0) \leq 2✓$. Since $(0, 0)$ is in the "lower" half plane, we shade this side of the boundary (see Figure 5.17).

Figure 5.16

Figure 5.17

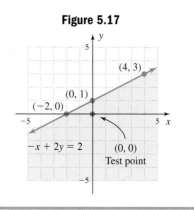

Now try Exercises 11 through 14 ▶

☑ **A.** You've just learned how to solve a linear inequality in two variables

The same solution would be obtained if we first solve for y and graph the boundary line using the slope-intercept method. However, using the slope-intercept method offers a distinct advantage — test points are no longer necessary since solutions to "less than" inequalities will always appear *below* the boundary line and solutions to "greater than" inequalities appear *above* the line. Written in slope-intercept form, the inequality from Example 2 is $y \leq \frac{1}{2}x + 1$. Note that $(0, 0)$ still results in a true statement, but the "less than or equal to" symbol now indicates directly that solutions will be found in the lower half plane. This observation leads to our general approach for solving linear inequalities:

Solving a Linear Inequality

1. Graph the boundary line by solving for y and using the slope-intercept form.
   - Use a solid line if the boundary is included in the solution set.
   - Use a dashed line if the boundary is excluded from the solution set.
2. For "greater than" inequalities shade the upper half plane. For "less than" inequalities shade the lower half plane.

B. Solving Systems of Linear Inequalities

To solve a **system of inequalities,** we apply the procedure outlined above to all inequalities in the system, and note the ordered pairs that satisfy *all inequalities simultaneously.* In other words, we find *the intersection of all solution regions* (where they overlap), which then represents the solution for the system. In the case of vertical boundary lines, the designations *"above"* or *"below" the line* cannot be applied, and instead we simply note that for any vertical line $x = k$, points with x-coordinates larger than k will occur to the right.

EXAMPLE 3 ▶   Solving a System of Linear Inequalities

Solve the system of inequalities: $\begin{cases} 2x + y \geq 4 \\ x - y < 2 \end{cases}$.

Solution ▶   Solving for y, we obtain $y \geq -2x + 4$ and $y > x - 2$. The line $y = -2x + 4$ will be a solid boundary line (included), while $y = x - 2$ will be dashed (not included). Both inequalities are "greater than" and so we shade the upper half plane for each. The regions overlap and form the solution region (the lavender region shown). This sequence of events is illustrated here:

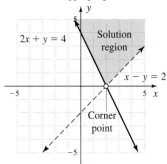

Shade above $y = -2x + 4$ (in blue)

Shade above $y = x - 2$ (in pink)

Overlapping region

The solutions are all ordered pairs found in this region and its included boundaries. To verify the result, test the point $(2, 3)$ from inside the region, $(5, -2)$ from outside the region (the point $(2, 0)$ is not a solution since $x - y < 2$).

Now try Exercises 15 through 42 ▶

✓ **B.** You've just learned how to solve a system of linear inequalities

For further reference, the point of intersection $(2, 0)$ is called a **corner point** or **vertex** of the solution region. If the point of intersection is not easily found from the graph, we can find it by solving a linear system using the two lines. For Example 3, the system is

$$\begin{cases} 2x + y = 4 \\ x - y = 2 \end{cases}$$

and solving by elimination gives $3x = 6$, $x = 2$, and $(2, 0)$ as the point of intersection.

C. Applications of Systems of Linear Inequalities

Systems of inequalities give us a way to model the decision-making process when certain *constraints* must be satisfied. A constraint is a fact or consideration that somehow limits or governs possible solutions, like the number of acres a farmer plants—which may be limited by time, size of land, government regulation, and so on.

EXAMPLE 4 ▶ **Solving Applications of Linear Inequalities**

As part of their retirement planning, James and Lily decide to invest up to $30,000 in two separate investment vehicles. The first is a bond issue paying 9% and the second is a money market certificate paying 5%. A financial adviser suggests they invest at least $10,000 in the certificate and not more than $15,000 in bonds. What various amounts can be invested in each?

Solution ▶ Consider the ordered pairs (B, C) where B represents the money invested in bonds and C the money invested in the certificate. Since they plan to invest no more than $30,000, the investment constraint would be $B + C \leq 30$ (in thousands). Following the adviser's recommendations, the constraints on each investment would be $B \leq 15$ and $C \geq 10$. Since they cannot invest less than zero dollars, the last two constraints are $B \geq 0$ and $C \geq 0$.

$$\begin{cases} B + C \leq 30 \\ B \leq 15 \\ C \geq 10 \\ B \geq 0 \\ C \geq 0 \end{cases}$$

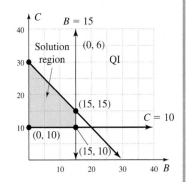

The resulting system is shown in the figure, and indicates solutions will be in the first quadrant.

There is a vertical boundary line at $B = 15$ with shading to the left (less than) and a horizontal boundary line at $C = 10$ with shading above (greater than). After graphing $C = 30 - B$, we see the solution region is a quadrilateral with vertices at $(0, 10)$, $(0, 30)$, $(15, 10)$, and $(15, 15)$, as shown.

Now try Exercises 53 and 54 ▶

D. Linear Programming

To become as profitable as possible, corporations look for ways to maximize their revenue and minimize their costs, while keeping up with delivery schedules and product demand. To operate at peak efficiency, plant managers must find ways to maximize productivity, while minimizing related costs and considering employee welfare, union agreements, and other factors. Problems where the goal is to **maximize** or **minimize** the value of a given quantity under certain **constraints** or restrictions are called programming problems. The quantity we seek to maximize or minimize is called the **objective function.** For situations where *linear* programming is used, the objective function is given as a linear function in two variables and is denoted $f(x, y)$. A function in two variables is evaluated in much the same way as a single variable function. To evaluate $f(x, y) = 2x + 3y$ at the point $(4, 5)$, we substitute 4 for x and 5 for y: $f(4, 5) = 2(4) + 3(5) = 23$.

EXAMPLE 5 ▶ **Determining Maximum Values**

Determine which of the following ordered pairs maximizes the value of $f(x, y) = 5x + 4y$: $(0, 6)$, $(5, 0)$, $(0, 0)$, or $(4, 2)$.

Solution ▶ Organizing our work in table form gives

Given Point	Evaluate $f(x, y) = 5x + 4y$
$(0, 6)$	$f(0, 6) = 5(0) + 4(6) = 24$
$(5, 0)$	$f(5, 0) = 5(5) + 4(0) = 25$
$(0, 0)$	$f(0, 0) = 5(0) + 4(0) = 0$
$(4, 2)$	$f(4, 2) = 5(4) + 4(2) = 28$

The function $f(x, y) = 5x + 4y$ is maximized at $(4, 2)$.

Now try Exercises 43 through 46 ▶

Figure 5.18

Convex   Not convex

When the objective is stated as a linear function in two variables and the constraints are expressed as a system of linear inequalities, we have what is called a **linear programming** problem. The systems of inequalities solved earlier produced a solution region that was either **bounded** (as in Example 4) or **unbounded** (as in Example 3). We interpret the word *bounded* to mean we can enclose the solution region within a circle of appropriate size. If we cannot draw a circle around the region because it extends indefinitely in some direction, the region is said to be *unbounded*. In this study, we will consider only situations that produce a bounded solution region, meaning the region will have three or more vertices. The regions we study will also be **convex,** meaning that for any two points in the feasible region, the line segment between them is also in the region (Figure 5.18). Under these conditions, it can be shown that the optimal solution(s) *must occur at one of the corner points of the solution region,* also called the **feasible region.**

EXAMPLE 6 ▶   **Finding the Maximum of an Objective Function**

Find the maximum value of the objective function $f(x, y) = 2x + y$ given the

constraints shown: $\begin{cases} x + y \le 4 \\ 3x + y \le 6 \\ x \ge 0 \\ y \ge 0 \end{cases}$.

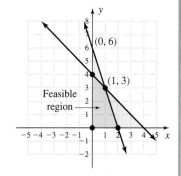

Solution ▶   Begin by noting that the solutions must be in QI, since $x \ge 0$ and $y \ge 0$. Graph the boundary lines $y = -x + 4$ and $y = -3x + 6$, shading the lower half plane in each case since they are "less than" inequalities. This produces the feasible region shown in lavender. There are four corner points to this region: $(0, 0)$, $(0, 4)$, $(2, 0)$, and $(1, 3)$. Three of these points are intercepts and can be found quickly. The point $(1, 3)$ was found by solving the system $\begin{cases} x + y = 4 \\ 3x + y = 6 \end{cases}$. Knowing that the objective function will be maximized at one of the corner points, we test them in the objective function, using a table to organize our work.

Corner Point	Objective Function $f(x, y) = 2x + y$
$(0, 0)$	$f(0, 0) = 2(0) + (0) = 0$
$(0, 4)$	$f(0, 4) = 2(0) + (4) = 4$
$(2, 0)$	$f(2, 0) = 2(2) + (0) = 4$
$(1, 3)$	$f(1, 3) = 2(1) + (3) = 5$

The objective function $f(x, y) = 2x + y$ is maximized at $(1, 3)$.

Now try Exercises 47 through 50 ▶

Figure 5.19

To help understand why solutions must occur at a vertex, note the objective function $f(x, y)$ is maximized using only (x, y) ordered pairs from the feasible region. If we let K represent this maximum value, the function from Example 6 becomes $K = 2x + y$ or $y = -2x + K$, which is a line with slope -2 and y-intercept K. The table in Example 6 suggests that K should range from 0 to 5 and graphing $y = -2x + K$ for $K = 1$, $K = 3$, and $K = 5$ produces the family of parallel lines shown in Figure 5.19. Note that values of K larger than 5 will cause the line to miss the solution region, and the maximum value of 5 occurs where the line intersects the feasible region at the vertex $(1, 3)$. These observations lead to the following principles, which we offer without a formal proof.

Linear Programming Solutions

1. If the feasible region is convex and bounded, a maximum and a minimum value exist.
2. If a unique solution exists, it will occur at a vertex of the feasible region.
3. If more than one solution exists, at least one of them occurs at a vertex of the feasible region with others on a boundary line.
4. If the feasible region is unbounded, a linear programming problem may have no solutions.

Solving linear programming problems depends in large part on two things: (1) identifying the **objective** and the **decision variables** (what each variable represents in context), and (2) using the decision variables to write the *objective function* and **constraint inequalities.** This brings us to our five-step approach for solving linear programming applications.

Solving Linear Programming Applications

1. Identify the main objective and the decision variables (descriptive variables may help) and write the objective function in terms of these variables.
2. Organize all information in a table, with the *decision variables* and *constraints* heading up the columns, and their *components* leading each row.
3. Complete the table using the information given, and write the constraint inequalities using the decision variables, constraints, and the domain.
4. Graph the constraint inequalities, determine the feasible region, and identify all corner points.
5. Test these points in the objective function to determine the optimal solution(s).

EXAMPLE 7 ▶ **Solving an Application of Linear Programming**

The owner of a snack food business wants to create two nut mixes for the holiday season. The regular mix will have 14 oz of peanuts and 4 oz of cashews, while the deluxe mix will have 12 oz of peanuts and 6 oz of cashews. The owner estimates he will make a profit of $3 on the regular mixes and $4 on the deluxe mixes. How many of each should be made in order to maximize profit, if only 840 oz of peanuts and 348 oz of cashews are available?

Solution ▶ Our *objective* is to maximize profit, and the *decision variables* could be r to represent the regular mixes sold, and d for the number of deluxe mixes. This gives $P(r, d) = \$3r + \$4d$ as our *objective function*. The information is organized in Table 5.1, using the variables r, d, and the constraints to head each column. Since the mixes are composed of peanuts and cashews, these lead the rows in the table.

Table 5.1

$$P(r, d) = \$3r \quad + \quad \$4d$$

	Regular r	Deluxe d	Constraints: Total Ounces Available
Peanuts	14	12	840
Cashews	4	6	348

After filling in the appropriate values, reading the table from left to right along the "peanut" row and the "cashew" row, gives the constraint inequalities $14r + 12d \le 840$ and $4r + 6d \le 348$. Realizing we won't be making a negative number of mixes, the remaining constraints are $r \ge 0$ and $d \ge 0$. The complete system is

$$\begin{cases} 14r + 12d \le 840 \\ 4r + 6d \le 348 \\ r \ge 0 \\ d \ge 0 \end{cases}$$

Note once again that the solutions must be in QI, since $r \ge 0$ and $d \ge 0$. Graphing the first two inequalities using slope-intercept form gives $d \le -\frac{7}{6}r + 70$ and $d \le -\frac{2}{3}r + 58$ producing the feasible region shown in lavender. The four corner points are (0, 0), (60, 0), (0, 58), and (24, 42). Three of these points are intercepts and can be read from a table of values or the graph itself. The point (24, 42) was found by solving the system $\begin{cases} 14r + 12d = 840 \\ 4r + 6d = 348 \end{cases}$.

Knowing the objective function will be maximized at one of these points, we test them in the objective function (Table 5.2).

Table 5.2

Corner Point	Objective Function $P(r, d) = \$3r + \$4d$
(0, 0)	$P(0, 0) = \$3(0) + \$4(0) = 0$
(60, 0)	$P(60, 0) = \$3(60) + \$4(0) = \$180$
(0, 58)	$P(0, 58) = \$3(0) + \$4(58) = \$232$
(24, 42)	$P(24, 42) = \$3(24) + \$4(42) = \$240$

Profit will be maximized if 24 boxes of the regular mix and 42 boxes of the deluxe mix are made and sold.

Now try Exercises 55 through 60 ▶

Linear programming can also be used to minimize an objective function, as in Example 8.

EXAMPLE 8 ▶   Minimizing Costs Using Linear Programming

A beverage producer needs to minimize shipping costs from its two primary plants in Kansas City (KC) and St. Louis (STL). All wholesale orders within the state are shipped from one of these plants. An outlet in Macon orders 200 cases of soft drinks on the same day an order for 240 cases comes from Springfield. The plant in KC has 300 cases ready to ship and the plant in STL has 200 cases. The cost of shipping each case to Macon is $0.50 from KC, and $0.70 from STL. The cost of shipping each case to Springfield is $0.60 from KC, and $0.65 from STL. How many cases should be shipped from each warehouse to minimize costs?

Solution ▶   Our *objective* is to minimize costs, which depends on the number of cases shipped from each plant. To begin we use the following assignments:

$$A \rightarrow \text{cases shipped from KC to Macon}$$
$$B \rightarrow \text{cases shipped from KC to Springfield}$$
$$C \rightarrow \text{cases shipped from STL to Macon}$$
$$D \rightarrow \text{cases shipped from STL to Springfield}$$

From this information, the equation for total cost T is

$$T = 0.5A + 0.6B + 0.7C + 0.65D,$$

an equation in *four* variables. To make the cost equation more manageable, note since Macon ordered 200 cases, $A + C = 200$. Similarly, Springfield ordered 240 cases, so $B + D = 240$. After solving for C and D, respectively, these equations enable us to substitute for C and D, resulting in an equation with just two variables. For $C = 200 - A$ and $D = 240 - B$ we have

$$T(A, B) = 0.5A + 0.6B + 0.7(200 - A) + 0.65(240 - B)$$
$$= 0.5A + 0.6B + 140 - 0.7A + 156 - 0.65B$$
$$= 296 - 0.2A - 0.05B$$

The constraints involving the KC plant are $A + B \leq 300$ with $A \geq 0$, $B \geq 0$. The constraints for the STL plant are $C + D \leq 200$ with $C \geq 0$, $D \geq 0$. Since we want a system in terms of A and B only, we again substitute $C = 200 - A$ and $D = 240 - B$ in all the STL inequalities:

$C + D \leq 200$	STL inequalities	$C \geq 0$	$D \geq 0$
$(200 - A) + (240 - B) \leq 200$	substitute $200 - A$	$200 - A \geq 0$	$240 - B \geq 0$
	for C, $240 - B$ for D	$200 \geq A$	$240 \geq B$
$440 - A - B \leq 200$	simplify		
$240 \leq A + B$	result		

Combining the new STL constraints with those from KC produces the following system and solution. All points of intersection were read from the graph or located using the related system of equations.

$$\begin{cases} A + B \leq 300 \\ A + B \geq 240 \\ A \leq 200 \\ B \leq 240 \\ A \geq 0 \\ B \geq 0 \end{cases}$$

To find the minimum cost, we check each vertex in the objective function.

Vertices	Objective Function $T(A, B) = 296 - 0.2A - 0.05B$
$(0, 240)$	$P(0, 240) = 296 - 0.2(0) - 0.05(240) = \$284$
$(60, 240)$	$P(60, 240) = 296 - 0.2(60) - 0.05(240) = \$272$
$(200, 100)$	$P(200, 100) = 296 - 0.2(200) - 0.05(100) = \$251$
$(200, 40)$	$P(200, 40) = 296 - 0.2(200) - 0.05(40) = \$254$

The minimum cost occurs when $A = 200$ and $B = 100$, meaning the producer should ship the following quantities:

$A \rightarrow$ cases shipped from KC to Macon $= 200$
$B \rightarrow$ cases shipped from KC to Springfield $= 100$
$C \rightarrow$ cases shipped from STL to Macon $= 0$
$D \rightarrow$ cases shipped from STL to Springfield $= 140$

☑ **D.** You've just learned how to solve applications using linear programming

Now try Exercises 61 and 62 ▶

TECHNOLOGY HIGHLIGHT

Systems of Linear Inequalities

Solving systems of linear inequalities on the TI-84 Plus involves three steps, which are performed on both equations: (1) enter the related equations in Y_1 and Y_2 (solve for y in each equation) to create the boundary lines, (2) graph both lines and test the resulting half planes, and (3) shade the appropriate half plane. Since many real-world applications of linear inequalities preclude the use of negative numbers, we **set Xmin = 0 and Ymin = 0 for the WINDOW size.** Xmax and Ymax will depend on the equations given. We illustrate by solving the system $\begin{cases} 3x + 2y < 14 \\ x + 2y < 8 \end{cases}$.

—continued

1. *Enter the related equations.* For $3x + 2y = 14$, we have $y = -1.5 + 7$. For $x + 2y = 8$, we have $y = -0.5x + 4$. Enter these as Y_1 and Y_2 on the Y= screen.

2. *Graph the boundary lines.* Note the x- and y-intercepts of both lines are less than 10, so we can graph them using a friendly window where $x \in [0, 9.4]$ and $y \in [0, 6.2]$. After setting the window, press GRAPH to graph the lines.

3. *Shade the appropriate half plane.* Since both equations are in slope-intercept form, we shade *below* both lines for the less than inequalities, using the "◣" feature located to the far left of Y_1 and Y_2. Simply overlay the diagonal line and press ENTER repeatedly until the symbol appears (Figure 5.20). After pressing the GRAPH key, the calculator draws both lines and shades the appropriate regions (Figure 5.21). Note the calculator uses two different kinds of shading. This makes it easy to identify the solution region—it will be the "checkerboard area" where the horizontal and vertical lines cross. As a final check, you could navigate the position marker into the solution region and test a few points in both equations.

Figure 5.20

Figure 5.21

Use these ideas to solve the following systems of linear inequalities. Assume all solutions lie in Quadrant I.

Exercise 1: $\begin{cases} y + 2x < 8 \\ y + x < 6 \end{cases}$ **Exercise 2:** $\begin{cases} 3x + y < 8 \\ x + y < 4 \end{cases}$ **Exercise 3:** $\begin{cases} -4x - y > -9 \\ -3x - y > -7 \end{cases}$

5.4 EXERCISES

▶ CONCEPTS AND VOCABULARY

Fill in the blank with the appropriate word or phrase. Carefully reread the section if needed.

1. Any line $y = mx + b$ drawn in the coordinate plane divides the plane into two regions called _____ _____.

2. For the line $y = mx + b$ drawn in the coordinate plane, solutions to $y > mx + b$ are found in the region _____ the line.

3. The overlapping region of two or more linear inequalities in a system is called the _____ region.

4. If a linear programming problem has a unique solution (x, y), it must be a _____ of the feasible region.

5. Suppose two boundary lines in a system of linear inequalities intersect, but the point of intersection is not a vertex of the feasible region. Describe how this is possible.

6. Describe the conditions necessary for a linear programming problem to have multiple solutions. (*Hint:* Consider the diagram in Figure 5.19, and the slope of the line from the objective function.)

▶ DEVELOPING YOUR SKILLS

Determine whether the ordered pairs given are solutions.

7. $2x + y > 3$; $(0, 0), (3, -5), (-3, -4), (-3, 9)$

8. $3x - y > 5$; $(0, 0), (4, -1), (-1, -5), (1, -2)$

9. $4x - 2y \leq -8$; $(0, 0), (-3, 5), (-3, -2), (-1, 1)$

10. $3x + 5y \geq 15$; $(0, 0), (3, 5), (-1, 6), (7, -3)$

Solve the linear inequalities by shading the appropriate half plane.

11. $x + 2y < 8$

12. $x - 3y > 6$

13. $2x - 3y \geq 9$

14. $4x + 5y \geq 15$

Determine whether the ordered pairs given are solutions to the accompanying system.

15. $\begin{cases} 5y - x \geq 10 \\ 5y + 2x \leq -5 \end{cases}$;
$(-2, 1), (-5, -4), (-6, 2), (-8, 2.2)$

16. $\begin{cases} 8y + 7x \geq 56 \\ 3y - 4x \geq -12 \\ y \geq 4 \end{cases}$; $(1, 5), (4, 6), (8, 5), (5, 3)$

Solve each system of inequalities by graphing the solution region. Verify the solution using a test point.

17. $\begin{cases} x + 2y \geq 1 \\ 2x - y \leq -2 \end{cases}$

18. $\begin{cases} -x + 5y < 5 \\ x + 2y \geq 1 \end{cases}$

19. $\begin{cases} 3x + y > 4 \\ x > 2y \end{cases}$

20. $\begin{cases} 3x \leq 2y \\ y \geq 4x + 3 \end{cases}$

21. $\begin{cases} 2x + y < 4 \\ 2y > 3x + 6 \end{cases}$

22. $\begin{cases} x - 2y < -7 \\ 2x + y > 5 \end{cases}$

23. $\begin{cases} x > -3y - 2 \\ x + 3y \leq 6 \end{cases}$

24. $\begin{cases} 2x - 5y < 15 \\ 3x - 2y > 6 \end{cases}$

25. $\begin{cases} 5x + 4y \geq 20 \\ x - 1 \geq y \end{cases}$

26. $\begin{cases} 10x - 4y \leq 20 \\ 5x - 2y > -1 \end{cases}$

27. $\begin{cases} 0.2x > -0.3y - 1 \\ 0.3x + 0.5y \leq 0.6 \end{cases}$

28. $\begin{cases} x > -0.4y - 2.2 \\ x + 0.9y \leq -1.2 \end{cases}$

29. $\begin{cases} y \leq \dfrac{3}{2}x \\ 4y \geq 6x - 12 \end{cases}$

30. $\begin{cases} 3x + 4y > 12 \\ y < \dfrac{2}{3}x \end{cases}$

31. $\begin{cases} \dfrac{-2}{3}x + \dfrac{3}{4}y \leq 1 \\ \dfrac{1}{2}x + 2y \geq 3 \end{cases}$

32. $\begin{cases} \dfrac{1}{2}x + \dfrac{2}{5}y \leq 5 \\ \dfrac{5}{6}x - 2y \geq -5 \end{cases}$

33. $\begin{cases} x - y \geq -4 \\ 2x + y \leq 4 \\ x \geq 1, y \geq 0 \end{cases}$

34. $\begin{cases} 2x - y \leq 5 \\ x + 3y \leq 6 \\ x \geq 1 \end{cases}$

35. $\begin{cases} y \leq x + 3 \\ x + 2y \leq 4 \\ y \geq 0 \end{cases}$

36. $\begin{cases} 4y < 3x + 12 \\ x \geq 0 \\ y \leq x + 1 \end{cases}$

37. $\begin{cases} 2x + 3y \leq 18 \\ x \geq 0 \\ y \geq 0 \end{cases}$

38. $\begin{cases} 8x + 5y \leq 40 \\ x \geq 0 \\ y \geq 0 \end{cases}$

Use the equations given to write the system of linear inequalities represented by each graph.

39.

40.

41.

42.

Determine which of the ordered pairs given produces the maximum value of $f(x, y)$.

43. $f(x, y) = 12x + 10y$; $(0, 0), (0, 8.5), (7, 0), (5, 3)$

44. $f(x, y) = 50x + 45y$; $(0, 0), (0, 21), (15, 0), (7.5, 12.5)$

Determine which of the ordered pairs given produces the minimum value of $f(x, y)$.

45. $f(x, y) = 8x + 15y$; $(0, 20), (35, 0), (5, 15), (12, 11)$

46. $f(x, y) = 75x + 80y$; $(0, 9), (10, 0), (4, 5), (5, 4)$

For Exercises 47 and 48, find the *maximum* value of the objective function $f(x, y) = 8x + 5y$ given the constraints shown.

47. $\begin{cases} x + 2y \leq 6 \\ 3x + y \leq 8 \\ x \geq 0 \\ y \geq 0 \end{cases}$

48. $\begin{cases} 2x + y \leq 7 \\ x + 2y \leq 5 \\ x \geq 0 \\ y \geq 0 \end{cases}$

For Exercises 49 and 50, find the *minimum* value of the objective function $f(x, y) = 36x + 40y$ given the constraints shown.

49. $\begin{cases} 3x + 2y \geq 18 \\ 3x + 4y \geq 24 \\ x \geq 0 \\ y \geq 0 \end{cases}$

50. $\begin{cases} 2x + y \geq 10 \\ x + 4y \geq 3 \\ x \geq 2 \\ y \geq 0 \end{cases}$

▶ WORKING WITH FORMULAS

Area Formulas

51. The area of a triangle is usually given as $A = \frac{1}{2}BH$, where B and H represent the base and height, respectively. The area of a rectangle can be stated as $A = BH$. If the base of both the triangle and rectangle is equal to 20 in., what are the possible values for H if the triangle must have an area *greater than* 50 in^2 and the rectangle must have an area *less than* 200 in^2?

Volume Formulas

52. The volume of a cone is $V = \frac{1}{3}\pi r^2 h$, where r is the radius of the base and h is the height. The volume of a cylinder is $V = \pi r^2 h$. If the radius of both the cone and cylinder is equal to 10 cm, what are the possible values for h if the cone must have a volume *greater than* 200 cm^3 and the volume of the cylinder must be *less than* 850 cm^3?

▶ APPLICATIONS

Write a system of linear inequalities that models the information given, then solve.

53. Gifts to grandchildren: Grandpa Augustus is considering how to divide a $50,000 gift between his two grandchildren, Julius and Anthony. After weighing their respective positions in life and family responsibilities, he decides he must bequeath at least $20,000 to Julius, but no more than $25,000 to Anthony. Determine the possible ways that Grandpa can divide the $50,000.

54. Guns versus butter: Every year, governments around the world have to make the decision as to how much of their revenue must be spent on national defense and domestic improvements (guns versus butter). Suppose total revenue for these two needs was $120 billion, and a government decides they need to spend at least $42 billion on butter and no more than $80 billion on defense. Determine the possible amounts that can go toward each need.

Solve the following linear programming problems.

55. Land/crop allocation: A farmer has 500 acres of land to plant corn and soybeans. During the last few years, market prices have been stable and the farmer anticipates a profit of $900 per acre on the corn harvest and $800 per acre on the soybeans. The farmer must take into account the time it takes to plant and harvest each crop, which is 3 hr/acre for corn and 2 hr/acre for soybeans. If the farmer has at most 1300 hr to plant, care for, and harvest each crop, how many acres of each crop should be planted in order to maximize profits?

56. Coffee blends: The owner of a coffee shop has decided to introduce two new blends of coffee in order to attract new customers—a *Deluxe Blend* and a *Savory Blend.* Each pound of the deluxe blend contains 30% Colombian and 20% Arabian coffee, while each pound of the savory blend contains 35% Colombian and 15% Arabian coffee (the remainder of each is made up of cheap and plentiful domestic varieties). The profit on the deluxe blend will be $1.25 per pound, while the profit on the savory blend will be $1.40 per pound. How many pounds of each should the owner make in order to maximize profit, if only 455 lb of Colombian coffee and 250 lb of Arabian coffee are currently available?

57. Manufacturing screws: A machine shop manufactures two types of screws—sheet metal screws and wood screws, using three different machines. Machine Moe can make a sheet metal screw in 20 sec and a wood screw in 5 sec. Machine Larry can make a sheet metal screw in 5 sec and a wood screw in 20 sec. Machine Curly, the newest machine (nyuk, nyuk) can make a sheet metal screw in 15 sec and a wood screw in 15 sec. (Shemp couldn't get a job because he failed the math portion of the employment exam.) Each machine can operate for only 3 hr each day before shutting down for maintenance. If sheet metal screws sell for 10 cents and wood screws sell for 12 cents, how many of each type should the machines be programmed to make in order to maximize revenue? (*Hint:* Standardize time units.)

58. Hauling hazardous waste: A waste disposal company is contracted to haul away some hazardous waste material. A full container of liquid waste weighs 800 lb and has a volume of 20 ft^3. A full container of solid waste weighs 600 lb and has a volume of 30 ft^3. The trucks used can carry at most 10 tons (20,000 lb) and have a carrying volume of 800 ft^3. If the trucking company makes $300 for disposing of liquid waste and $400 for disposing of solid waste, what is the maximum revenue per truck that can be generated?

59. Maximizing profit—food service: P. Barrett & Justin, Inc., is starting up a fast-food restaurant specializing in peanut butter and jelly sandwiches. Some of the peanut butter varieties are smooth, crunchy, reduced fat, and reduced sugar. The jellies will include those expected and common, as well as some exotic varieties such as kiwi and mango. Independent research has determined the two most popular sandwiches will be the traditional P&J (smooth peanut butter and grape jelly), and the Double-T (three slices of bread). A traditional P&J uses 2 oz of peanut butter and 3 oz of jelly. The Double-T uses 4 oz of peanut butter and 5 oz of jelly. The traditional sandwich will be priced at $2.00, and a Double-T at $3.50. If the restaurant has 250 oz of smooth peanut butter and 345 oz of grape jelly on hand for opening day, how many of each should they make and sell to maximize revenue?

60. Maximizing profit—construction materials: Mooney and Sons produces and sells two varieties of concrete mixes. The mixes are packaged in 50-lb bags. Type A is appropriate for finish work, and contains 20 lb of cement and 30 lb of sand. Type B is appropriate for foundation and footing work, and contains 10 lb of cement and 20 lb of sand. The remaining weight comes from gravel aggregate. The profit on type A is $1.20/bag, while the profit on type B is $0.90/bag. How many bags of each should the company make to maximize profit, if 2750 lb of cement and 4500 lb of sand are currently available?

61. Minimizing shipping costs: An oil company is trying to minimize shipping costs from its two primary refineries in Tulsa, Oklahoma, and Houston, Texas. All orders within the region are shipped from one of these two refineries. An order for 220,000 gal comes in from a location in Colorado, and another for 250,000 gal from a location in Mississippi. The Tulsa refinery has 320,000 gal ready to ship, while the Houston refinery has 240,000 gal. The cost of transporting each gallon to Colorado is $0.05 from Tulsa and $0.075 from Houston. The cost of transporting each gallon to Mississippi is $0.06 from Tulsa and $0.065 from Houston. How many gallons should be distributed from each refinery to minimize the cost of filling both orders?

62. Minimizing transportation costs: Robert's Las Vegas Tours needs to drive 375 people and 19,450 lb of luggage from Salt Lake City, Utah, to Las Vegas, Nevada, and can charter buses from two companies. The buses from company X carry 45 passengers and 2750 lb of luggage at a cost of $1250 per trip. Company Y offers buses that carry 60 passengers and 2800 lb of luggage at a cost of $1350 per trip. How many buses should be chartered from each company in order for Robert to minimize the cost?

▶ **EXTENDING THE CONCEPT**

63. Graph the feasible region formed by the system
$$\begin{cases} x \geq 0 \\ y \geq 0 \\ y \leq 3 \\ x \leq 3 \end{cases}.$$ How would you describe this region?

Select random points within the region or on any boundary line and evaluate the objective function $f(x, y) = 4.5x + 7.2y$. At what point (x, y) will this function be maximized? How does this relate to optimal solutions to a linear programing problem?

64. Find the maximum value of the objective function $f(x, y) = 22x + 15y$ given the constraints
$$\begin{cases} 2x + 5y \leq 24 \\ 3x + 4y \leq 29 \\ x + 6y \leq 26 \\ x \geq 0 \\ y \geq 0 \end{cases}.$$

▶ **MAINTAINING YOUR SKILLS**

65. (1.4/1.5) Find all solutions (real and complex) by factoring: $x^3 - 5x^2 + 3x - 15 = 0$.

66. (3.7) Solve the rational inequality. Write your answer in interval notation. $\dfrac{x + 2}{x^2 - 9} > 0$

67. (3.8) The resistance to current flow in copper wire varies directly as its length and inversely as the square of its diameter. A wire 8 m long with a 0.004-m diameter has a resistance of 1500 Ω. Find the resistance in a wire of like material that is 2.7 m long with a 0.005-m diameter.

68. (4.4) Solve for x: $-350 = 211e^{-0.025x} - 450$.

SUMMARY AND CONCEPT REVIEW

SECTION 5.1   Linear Systems in Two Variables with Applications

KEY CONCEPTS

- A *solution* to a linear system in two variables is an ordered pair (x, y) that makes all equations in the system true.
- Since every point on the graph of a line satisfies the equation of that line, a point where two lines intersect must satisfy both equations and is a solution of the system.
- A system with at least one solution is called a *consistent system.*
- If the lines have different slopes, there is a unique solution to the system (they intersect at a single point). The system is called a *consistent* and *independent system.*
- If the lines have equal slopes and the same y-intercept, they form identical or *coincident* lines. Since one line is right atop the other, they intersect at all points with an infinite number of solutions. The system is called a *consistent* and *dependent system.*
- If the lines have equal slopes but different y-intercepts, they will never intersect. The system has no solution and is called an *inconsistent system.*

EXERCISES

Solve each system by graphing. If the solution does not have integer values indicate your solution is an estimate. If the system is inconsistent or dependent, so state.

1. $\begin{cases} 3x - 2y = 4 \\ -x + 3y = 8 \end{cases}$

2. $\begin{cases} 0.2x + 0.5y = -1.4 \\ x - 0.3y = 1.4 \end{cases}$

3. $\begin{cases} 2x + y = 2 \\ x - 2y = 4 \end{cases}$

Solve using substitution. Indicate whether each system is consistent, inconsistent, or dependent. Write unique solutions as an ordered pair.

4. $\begin{cases} y = 5 - x \\ 2x + 2y = 13 \end{cases}$

5. $\begin{cases} x + y = 4 \\ 0.4x + 0.3y = 1.7 \end{cases}$

6. $\begin{cases} x - 2y = 3 \\ x - 4y = -1 \end{cases}$

Solve using elimination. Indicate whether each system is consistent, inconsistent, or dependent. Write unique solutions as an ordered pair.

7. $\begin{cases} 2x - 4y = 10 \\ 3x + 4y = 5 \end{cases}$

8. $\begin{cases} -x + 5y = 8 \\ x + 2y = 6 \end{cases}$

9. $\begin{cases} 2x = 3y + 6 \\ 2.4x + 3.6y = 6 \end{cases}$

10. When it was first constructed in 1968, the John Hancock building in Chicago, Illinois, was the tallest structure in the world. In 1985, the Sears Tower in Chicago became the world's tallest structure. The Sears Tower is 323 ft taller than the John Hancock Building, and the sum of their heights is 2577 ft. How tall is each structure?

SECTION 5.2   Linear Systems in Three Variables with Applications

KEY CONCEPTS

- The graph of a linear equation in three variables is a *plane.*
- Systems in three variables can be solved using substitution and elimination.
- A linear system in three variables has the following possible solution sets:
  - If the planes intersect at a point, the system has one *unique solution* (x, y, z).
  - If the planes intersect at a line, the system has *linear dependence* and the solution (x, y, z) can be written as linear combinations of a single variable *(a parameter).*
  - If the planes are *coincident,* the equations in the system differ by a constant multiple, meaning they are all "disguised forms" of the *same equation.* The solutions have *coincident dependence,* and the solution set can be represented by any one of the equations.
  - In all other cases, the system has *no solutions* and is an inconsistent system.

EXERCISES

Solve using elimination. If a system is inconsistent or dependent, so state. For systems with linear dependence, give the answer as an ordered triple using a parameter.

11. $\begin{cases} x + y - 2z = -1 \\ 4x - y + 3z = 3 \\ 3x + 2y - z = 4 \end{cases}$ **12.** $\begin{cases} -x + y + 2z = 2 \\ x + y - z = 1 \\ 2x + y + z = 4 \end{cases}$ **13.** $\begin{cases} 3x + y + 2z = 3 \\ x - 2y + 3z = 1 \\ 4x - 8y + 12z = 7 \end{cases}$

Solve using a system of three equations in three variables.

14. In one version of the card game Gin Rummy, numbered cards (*N*) 2 through 9 are worth 5 points, the 10s and all face cards (*F*) are worth 10 points, and aces (*A*) are worth 20 points. At the moment his opponent said "Gin!" Kenan had 12 cards in his hand, worth a total value of 125 points. If the value of his aces and face cards was equal to four times the value of his numbered cards, how many aces, face cards, and numbered cards was he holding?

15. A vending machine accepts nickels, dimes, and quarters. At the end of a week, there is a total of $536 in the machine. The number of nickels and dimes combined is 360 more than the number of quarters. The number of quarters is 110 more than twice the number of nickels. How many of each type of coin are in the machine?

SECTION 5.3   Nonlinear Systems of Equations and Inequalities

KEY CONCEPTS

- Nonlinear systems of equations can be solved using substitution or elimination.
- First identify the graphs of the equations in the system to help determine the number of solutions possible.
- For nonlinear systems of inequalities, graph the related equation for each inequality given, then use a test point to decide what region to shade as the solution.
- The solution for the system is the overlapping region (if it exists) created by solutions to the individual inequalities.
- If the boundary is included, graph it using a solid line; if the boundary is not included use a dashed line.

EXERCISES

Solve Exercises 16–21 using substitution or elimination. Identify the graph of each relation before you begin.

16. $\begin{cases} x^2 + y^2 = 25 \\ y - x = -1 \end{cases}$ **17.** $\begin{cases} x = y^2 - 1 \\ x + 4y = -5 \end{cases}$ **18.** $\begin{cases} -x^2 + y = -1 \\ x^2 + y^2 = 7 \end{cases}$

19. $\begin{cases} x^2 + y^2 = 10 \\ y - 3x^2 = 0 \end{cases}$ **20.** $\begin{cases} y \le x^2 - 2 \\ x^2 + y^2 \le 16 \end{cases}$ **21.** $\begin{cases} x^2 + y^2 > 9 \\ x^2 + y \le -3 \end{cases}$

SECTION 5.4   Systems of Linear Inequalities and Linear Programming

KEY CONCEPTS

- As in Section 5.3, to solve a *system of linear inequalities,* we find the intersecting or *overlapping areas* of the solution regions from the individual inequalities. The common area is called the *feasible region.*
- The process known as *linear programming* seeks to *maximize* or *minimize* the value of a given quantity under certain *constraints* or restrictions.
- The quantity we attempt to maximize or minimize is called the *objective function.*
- The solution(s) to a linear programming problem *occur at one of the corner points of the feasible region.*
- The process of solving a linear programming application contains these six steps:
  - Identify the main objective and the decision variables.
  - Write the objective function in terms of these variables.
  - Organize all information in a table, using the decision variables and constraints.
  - Fill in the table with the information given and write the constraint inequalities.
  - Graph the constraint inequalities and determine the feasible region.
  - Identify all corner points of the feasible region and test these points in the objective function.

EXERCISES

Graph the solution region for each system of linear inequalities and verify the solution using a test point.

22. $\begin{cases} -x - y > -2 \\ -x + y < -4 \end{cases}$ **23.** $\begin{cases} x - 4y \le 5 \\ -x + 2y \le 0 \end{cases}$ **24.** $\begin{cases} x + 2y \ge 1 \\ 2x - y \le -2 \end{cases}$

25. Carefully graph the feasible region for the system of inequalities shown, then maximize the objective function: $f(x, y) = 30x + 45y$ $\begin{cases} x + y \le 7 \\ 2x + y \le 10 \\ 2x + 3y \le 18 \\ x \ge 0, y \ge 0 \end{cases}$

26. After retiring, Oliver and Lisa Douglas buy and work a small farm (near Hooterville) that consists mostly of milk cows and egg-laying chickens. Although the price of a commodity is rarely stable, suppose that milk sales bring in an average of $85 per cow and egg sales an average of $50 per chicken over a period of time. During this time period, the new ranchers estimate that care and feeding of the animals took about 3 hr per cow and 2 hr per chicken, while maintaining the related equipment took 2 hr per cow and 1 hr per chicken. How many animals of each type should be maintained in order to maximize profits, if at most 1000 hr can be spent on care and feeding, and at most 525 hr on equipment maintenance?

MIXED REVIEW

1. Write the equations in each system in slope-intercept form, then state whether the system is consistent/independent, consistent/dependent, or inconsistent. Do not solve.

a. $\begin{cases} -3x + 5y = 10 \\ 6x + 20 = 10y \end{cases}$ **b.** $\begin{cases} 4x - 3y = 9 \\ -2x + 5y = -10 \end{cases}$

c. $\begin{cases} x - 3y = 9 \\ -6y + 2x = 10 \end{cases}$

2. Solve by graphing.
$\begin{cases} x - 2y = 6 \\ -2x + y = -9 \end{cases}$

3. Solve using a substitution.
$\begin{cases} 2x + 3y = 5 \\ -x + 5y = 17 \end{cases}$

4. Solve using elimination.
$\begin{cases} 7x - 4y = -5 \\ 3x + 2y = 9 \end{cases}$

5. A burrito stand sells a veggie burrito for $2.45 and a beef burrito for $2.95. If the stand sold 54 burritos in one day, for a total revenue of $148.80, how many of each did they sell?

Solve using elimination.

6. $\begin{cases} x + 2y - 3z = -4 \\ -3x + 4y + z = 1 \\ 2x - 6y + z = 1 \end{cases}$

7. $\begin{cases} 0.1x - 0.2y + z = 1.7 \\ 0.3x + y - 0.1z = 3.6 \\ -0.2x - 0.1y + 0.2z = -1.7 \end{cases}$

Solve using elimination. If the system has coincident dependence, state the solution set using set notation.

8. $\begin{cases} x - 2y + 3z = 4 \\ 2x + y - z = 1 \\ 5x + z = 2 \end{cases}$ **9.** $\begin{cases} x - 2y + 3z = 4 \\ 2x + y - z = 1 \\ 5x + z = 6 \end{cases}$

10. It's the end of another big day at the circus, and the clowns are putting away their riding equipment—a motley collection of unicycles, bicycles, and tricycles. As she loads them into the storage shed, Trixie counts 21 cycles in all with a total of 40 wheels. In addition, she notes the number of bicycles is one fewer than twice the number of tricycles. How many cycles of each type do the clowns use?

Solve each system of inequalities by graphing the solution region.

11. $\begin{cases} 2x + y \le 4 \\ x - 3y > 6 \end{cases}$ **12.** $\begin{cases} 2x + y < 3 \\ 2x + y > -3 \end{cases}$

13. $\begin{cases} x - 2y \ge 5 \\ x \le 2y \end{cases}$

14. Graph the solution region for the system of inequalities. $\begin{cases} 4x + 2y \le 14 \\ 2x + 3y \le 15 \\ y \ge 0 \\ x \ge 0 \end{cases}$

15. Maximize   given

$$\begin{cases} x + y \le 8 \\ x + 2y \le 14 \\ 4x + 3y \le 30 \\ x, \ y \ge 0 \end{cases}$$

16. Solve the system by substitution.

$$\begin{cases} x^2 + y^2 = 1 \\ x - y = -1 \end{cases}$$

17. Solve using elimination: $\begin{cases} 4x^2 - y^2 = -9 \\ x^2 + 3y^2 = 79 \end{cases}$

18. Solve using the method of your choice.

$$\begin{cases} y + 1 = x^2 \\ x^2 + y = 7 \end{cases}$$

Solve each system of inequalities.

19. $\begin{cases} x + y > 1 \\ x^2 + y^2 \ge 16 \end{cases}$ **20.** $\begin{cases} x^2 + y^2 < 4 \\ x^2 + y < 0 \end{cases}$

PRACTICE TEST

Solve each system and state whether the system is consistent, inconsistent, or dependent.

1. Solve graphically:

$$\begin{cases} 3x + 2y = 12 \\ -x + 4y = 10 \end{cases}$$

2. Solve using substitution:

$$\begin{cases} 3x - y = 2 \\ -7x + 4y = -6 \end{cases}$$

3. Solve using elimination:

$$\begin{cases} 5x + 8y = 1 \\ 3x + 7y = 5 \end{cases}$$

4. Solve using elimination:

$$\begin{cases} x + 2y - z = -4 \\ 2x - 3y + 5z = 27 \\ -5x + y - 4z = -27 \end{cases}$$

5. Solve using elimination:

$$\begin{cases} 2x - y + z = 4 \\ -x + 2z = 1 \\ x - 2y + 8z = 11 \end{cases}$$

6. Find values of a and b such that $(2, -1)$ is a solution of the system.

$$\begin{cases} ax - by = 12 \\ bx + ay = -1 \end{cases}$$

Create a system of equations to model each exercise, then solve using the method of your choice.

7. The perimeter of a "legal-size" paper is 114.3 cm. The length of the paper is 7.62 cm less than twice the width. Find the dimensions of a legal-size sheet of paper.

8. The island nations of Tahiti and Tonga have a combined land area of 692 mi². Tahiti's land area is 112 mi² more than Tonga's. What is the land area of each island group?

9. Many years ago, two cans of corn (C), 3 cans of green beans (B), and 1 can of peas (P) cost $1.39. Three cans of C, 2 of B, and 2 of P cost $1.73. One

can of C, 4 of B, and 3 of P cost $1.92. What was the price of a single can of C, B, and P?

10. After inheriting $30,000 from a rich aunt, David decides to place the money in three different investments: a savings account paying 5%, a bond account paying 7%, and a stock account paying 9%. After 1 yr he earned $2080 in interest. Find how much was invested at each rate if $8000 less was invested at 9% than at 7%.

11. Solve the system of inequalities by graphing.

$$\begin{cases} x - y \le 2 \\ x + 2y \ge 8 \end{cases}$$

12. Maximize the objective function: $P = 50x - 12y$

$$\begin{cases} x + 2y \le 8 \\ 8x + 5y \ge 40 \\ x, \quad y \ge 0 \end{cases}$$

Solve the linear programming problem.

13. A company manufactures two types of T-shirts, a plain T-shirt and a deluxe monogrammed T-shirt. To produce a plain shirt requires 1 hr of working time on machine A and 2 hr on machine B. To produce a deluxe shirt requires 1 hr on machine A and 3 hr on machine B. Machine A is available for at most 50 hr/week, while machine B is available for at most 120 hr/week. If a plain shirt can be sold at a profit of $4.25 each and a deluxe shirt can be sold at a profit of $5.00 each, how many of each should be manufactured to maximize the profit?

Solve each nonlinear system using the technique of your choice.

14. $\begin{cases} x^2 + y^2 = 16 \\ y - x = 2 \end{cases}$ **15.** $\begin{cases} 4y - x^2 = 1 \\ y^2 + x^2 = 4 \end{cases}$

16. A support bracket on the frame of a large ship is a steel right triangle with a hypotenuse of 25 ft and a perimeter of 60 ft. Find the lengths of the other sides using a system of nonlinear equations.

17. Solve $\begin{cases} x^2 - y \leq 2 \\ x - y^2 \geq -2 \end{cases}$.

18. Write a system of inequalities that describes all the points with positive y-values that are less than 3 units away from the origin.

19. Solve the system of inequalities.
$$\begin{cases} 2x - y \leq -1 \\ 3x + 2y \geq 2 \\ x - 3y \geq -3 \end{cases}$$

20. Write a system of four inequalities that describes the location of the dart on the dartboard shown.

Exercise 20

CALCULATOR EXPLORATION AND DISCOVERY

Optimal Solutions and Linear Programming

In this exercise, we'll use a graphing calculator to explore various areas of the feasible region, repeatedly evaluating the objective function to see where the maximal values (optimal solutions) seem to "congregate." If all goes as expected, ordered pairs nearest to a vertex should give relatively larger values. To demonstrate, we'll use Example 6 from Section 5.4, stated below.

Example 6 ▶ Find the maximum value of the objective function $f(x, y) = 2x + y$ given the constraints shown:
$$\begin{cases} x + y \leq 4 \\ 3x + y \leq 6 \\ x \geq 0 \\ y \geq 0 \end{cases}$$

Solution ▶ The feasible region is shown in lavender. There are four corner points to this region: $(0, 0)$, $(0, 4)$, $(2, 0)$, and $(1, 3)$, and we found $f(x, y)$ was maximized at $(1, 3)$: $f(1, 3) = 5$.

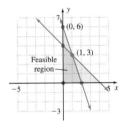

To explore this feasible region in terms of the objective function $f(x, y) = 2x + y$, enter the boundary lines $Y_1 = -x + 4$ and $Y_2 = -3x + 6$ on the $\boxed{Y=}$ screen. However, instead of shading below the lines to show the feasible region (using the ◣ feature to the extreme left), we shade above both lines (using the ◥ feature) so that the feasible region remains clear. Setting the window size at $x \in [0, 3]$ and $y \in [-1.5, 4]$ produces Figure 5.22. Using YMin $= -1.5$ will leave a blank area just below QI that enables you to explore the feasible region as the x- and y-values are displayed. Next we place the calculator in "split-screen" mode so that we can view the graph and the home screen simultaneously. Press the $\boxed{\text{MODE}}$ key

and notice the second-to-last line reads **Full Horiz G-T.** The **Full** (screen) mode is the default operating mode. The **Horiz** mode splits the screen horizontally, placing the graph directly above a shorter home screen. Highlight **Horiz,** then press $\boxed{\text{ENTER}}$ and $\boxed{\text{GRAPH}}$ to have the calculator reset the screen in this mode. The TI-84 Plus has a free-moving cursor that is brought into view by pressing the left $\boxed{◄}$ or right $\boxed{►}$ arrow (Figure 5.23). A useful feature of this cursor is that it automatically stores the current X value as the variable X ($\boxed{X,T,\theta,n}$ or $\boxed{\text{ALPHA}}$ $\boxed{\text{STO} →}$) and the current Y value as the variable Y ($\boxed{\text{ALPHA}}$ 1), which allows us to evaluate the objective function $f(x, y) = 2x + y$ right on the home screen. To access the graph and free-moving cursor you must press $\boxed{\text{GRAPH}}$ each time, and to access the home screen you must press $\boxed{\text{2nd}}$ $\boxed{\text{MODE}}$ (QUIT) each time. Begin by moving the cursor to the upper-left corner of the region, near the y-intercept [we stopped at ($\sim$0.0957, 3.26)]. Once you have the cursor "tucked up into the corner," press $\boxed{\text{2nd}}$ $\boxed{\text{MODE}}$ (QUIT) to get to the home screen, then enter the objective function: 2X + Y. Pressing $\boxed{\text{ENTER}}$ evaluates the function for the values indicated by the cursor's location

Figure 5.22

Figure 5.23

(Figure 5.24). It appears the value of the objective function for points (x, y) in this corner are close to 4, and it's no accident that at the corner point $(0, 4)$ the maximum value is in fact 4. Repeating this

Figure 5.24

3.458156029

procedure for the lower-right corner suggests the maximum value near $(2, 0)$ is also 4. Finally, press **GRAPH** to explore the region in the upper-right corner, where the lines intersect. Move the cursor to this vicinity, locate it very near the point of intersection [we stopped at $(\sim 0.957, 2.71\overline{6})$] and return to the home screen and evaluate (Figure 5.25). The value of the objective function is

near 5 in this corner of the region, and at the corner point $(1, 3)$ the maximum value is 5.

Exercise 1: The feasible region for the system given to the right has four corner points. Use the ideas here to explore the area near each corner point of the feasible region to determine which point is the likely candidate to produce the *minimum* value of the objective function $f(x, y) = 2x + 4y$. Then solve the linear programming problem to verify your guess.

Figure 5.25

4.631560284

$$\begin{cases} 2x + 2y \le 15 \\ x + y \ge 6 \\ x + 4y \ge 9 \\ x, y \ge 0 \end{cases}$$

STRENGTHENING CORE SKILLS

Understanding Why Elimination and Substitution "Work"

When asked to solve a system of two equations in two variables, we first select an appropriate method. In Section 5.1, we learned three basic techniques: graphing, substitution, and elimination. In this feature, we'll explore how these methods are related using Example 2 from Section 5.1 where we were asked to solve the system $\begin{cases} 4x - 3y = 9 \\ -2x + y = -5 \end{cases}$ by graphing. The resulting graph, shown here in Figure 5.26, clearly indicates the solution is $(3, 1)$.

Figure 5.26

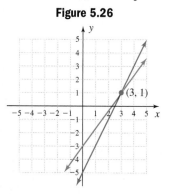

As for the elimination method, either x or y can be easily eliminated. If the second equation is multiplied by 2, the x-coefficients will be additive inverses, and the sum results in an equation with y as the only unknown.

R1 $\quad 4x - 3y = 9$
+
2R2 $\quad -4x + 2y = -10$
sum $\quad -y = -1$

The result is $y = 1$ but remember, this is a system of *linear equations,* and $y = 1$ is still the equation of a (horizontal) line. Since the system $\begin{cases} 4x - 3y = 9 \\ y = 1 \end{cases}$ is equivalent to the original, it will have the same solution set. In Figure 5.27, we note the point of intersection for the new system is still $(3, 1)$. If we eliminate the y-terms instead

(using R1 + 3R2), the result is $x = 3$, which is also the equation of a (vertical) line. Creating another equivalent system using this line produces $\begin{cases} x = 3 \\ y = 1 \end{cases}$ and the graph shown in Figure 5.28, where the vertical and horizontal lines intersect at $(3, 1)$, making the solution trivial.

Figure 5.27

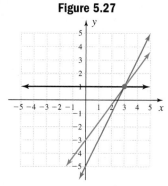

Note: Here we see a close connection to solving general equations, in that the goal is to write a series of equivalent yet simpler equations, continuing until the solution is obvious.

As for the substitution method, consider the second equation written as $y = 2x - 5$. This equation represents every point (x, y) on its graph, meaning the relationship for the ordered pair solutions can also be written $(x, 2x - 5)$. The same thing can be said for the line $4x - 3y = 9$, with its ordered pair solutions represented by $(x, \frac{4}{3}x - 3)$. At the point of intersection the y-coordinates must be identical, giving $2x - 5 = \frac{4}{3}x - 3$. In other words, we can substitute $2x - 5$ for y in the first equation, or $\frac{4}{3}x - 3$ for y in the second equation, with both yielding the

Figure 5.28

correct solution. Substituting $2x - 5$ for y in the first equation gives

$$4x - 3(2x - 5) = 9 \quad \text{substitute } 2x - 5 \text{ for } y$$
$$4x - 6x + 15 = 9 \quad \text{expand}$$
$$-2x = -6 \quad \text{simplify}$$
$$x = 3$$

and the solution $(x, 2x - 5)$ becomes $(3, 2(3) - 5)$ or $(3, 1)$.

All three methods will produce the same solution, and the best method to use at the time often depends on the nature of the system given, or even personal preference.

Exercise 1: Solve the system by (a) graphing, (b) elimination, and (c) substitution. Which method was most efficient for solving this system? $\begin{cases} 2x + y = 2 \\ 4x + 3y = 8 \end{cases}$

CUMULATIVE REVIEW CHAPTERS 1–5

Graph each of the following. Include x- and y-intercepts and other important features of each graph.

1. $y = \frac{2}{3}x + 2$

2. $f(x) = |x - 2| + 3$

3. $g(x) = \sqrt{x - 3} + 1$

4. $h(x) = \dfrac{1}{x - 1} + 2$

5. $g(x) = (x - 3)(x + 1)(x + 4)$

6. $y = 2^x + 3$

7. Determine the following for the graph shown to the right. Write all answers in interval notation:

  a. domain   **b.** range

  c. interval(s) where $f(x)$ is increasing or decreasing

  d. interval(s) where $f(x)$ is constant

  e. location of any maximum or minimum value(s)

  f. interval(s) where $f(x)$ is positive or negative

  g. the average rate of change using $(-4, 0)$ and $(-2, 3.5)$.

8. Suppose the cost of making a rubber ball is given by $C(x) = 3x + 10$, where x is the number of balls in hundreds. If the revenue from the sale of these balls is given by $R(x) = -x^2 + 123x - 1990$, find the profit function (Profit = Revenue − Cost). How many balls should be produced and sold to obtain the maximum profit? What is this maximum profit?

9. Find all zeroes (real or complex): $g(v) = v^3 - 9v^2 + 2v - 18$.

10. A polynomial has roots $x = 2$ and $x = 2 \pm 3i$. Find the polynomial and write it in standard form.

Given $f(x) = 2x - 5$ and $g(x) = 3x^2 + 2x$ find:

11. $(g - f)(x)$ **12.** $(fg)(-2)$ **13.** $(g \circ f)(2)$

14. Calculate the difference quotient for $f(x) = x^2 - 3x$.

15. Use the rational roots theorem to factor the polynomial completely: $x^4 - 6x^3 - 13x^2 + 24x + 36$.

Solve each inequality. Write your answer using interval notation.

16. $x^2 - 3x - 10 < 0$

17. $\dfrac{x - 2}{x + 3} \le 3$

18. Solve each equation.

  a. $\sqrt{x - 2} = \sqrt{3x + 4}$

  b. $x^{\frac{3}{2}} + 8 = 0$ **c.** $2|n + 4| + 3 = 13$

  d. $x^2 - 6x + 13 = 0$ **e.** $x^{-2} - 3x^{-1} - 40 = 0$

  f. $4 \cdot 2^{x+1} = \dfrac{1}{8}$ **g.** $3^{x-2} = 7$

  h. $\log_3 81 = x$ **i.** $\log_3 x + \log_3(x - 2) = 1$

19. If a person invests $5000 at 9% compounded continuously, how long until the account grows to $12,000?

20. Graph the piecewise function shown and state its domain and range.

$$f(x) = \begin{cases} |x + 2| & -5 \le x \le 0 \\ 2(x - 1)^2 & 0 < x \le 5 \end{cases}$$

21. Graph $h(x) = \dfrac{9 - x^2}{x^2 - 4}$. Give the coordinates of all intercepts and the equations of all asymptotes.

22. Solve the system of equations graphically.

$$\begin{cases} x - 3y = 1 \\ 2x + y = -5 \end{cases}$$

Exercise 22

23. A truck is delivering a shipment of arrows, bowling balls, and cricket bats to a sporting goods store. There is a total of 120 items in the shipment. There are twice as many arrows as balls and bats combined. There are 10 more balls than bats. How many of each item are there?

24. Sketch the solution region.
$$\begin{cases} y < 2^x \\ x + 2y > 0 \\ 3x + y < 5 \end{cases}$$

25. Solve the system of equations.
$$\begin{cases} y = \log x + 4 \\ y = 5 - \log(x - 3) \end{cases}$$

6

Matrices and Matrix Applications

CHAPTER OUTLINE

CHAPTER CONNECTIONS

From pediatric and geriatric care, to the training of a modern athlete, dietetic applications have become increasingly effective. In the latter case, athletes generally need high levels of carbohydrates and protein, but only moderate levels of fat. Suppose a physical trainer wants to supply one of her clients with 24 g of fat, 244 g of "carbs," and 40 g of protein for the noontime meal. Knowing the amount of these nutrients contained in certain foods, the trainer can recommend a variety of foods and the amount of each that should be eaten. The matrix operations in this chapter demonstrate how to do this effectively. This application occurs as Exercise 78 in Section 6.3.

Check out these other real-world connections:

Learning Objectives

In Section 6.1 you will learn how to:

☐ **A.** State the size of a matrix and identify its entries

☐ **B.** Form the augmented matrix of a system of equations

☐ **C.** Solve a system of equations using row operations

☐ **D.** Recognize inconsistent and dependent systems

☐ **E.** Solve applications using linear systems

Just as synthetic division streamlines the process of polynomial division, matrices and row operations streamline the process of solving systems using elimination. With the equations of the system in standard form, the location of the variable terms and constant terms are set, and we simply apply the elimination process on the coefficients and constants.

A. Introduction to Matrices

In general terms, a **matrix** is simply a rectangular arrangement of numbers, called the **entries** of the matrix. **Matrices** (plural of matrix) are denoted by enclosing the entries between a left and right bracket, and named using a capital letter, such as

$$A = \begin{bmatrix} 1 & -3 & 2 \\ 5 & 1 & -1 \end{bmatrix} \text{ and } B = \begin{bmatrix} 2 & -1 & 3 \\ 4 & 6 & -2 \\ 1 & 0 & -1 \end{bmatrix}. \text{ They occur in many different sizes}$$

as defined by the number of **rows** and **columns** each has, with the number of rows always given first. Matrix A is said to be a 2×3 (two by three) matrix, since it has two rows and three columns. Matrix B is a 3×3 (three by three) matrix.

EXAMPLE 1A ▶ **Identifying the Size and Entries of a Matrix**

Determine the size of each matrix and identify the entry located in the second row and first column.

a. $C = \begin{bmatrix} 3 & -2 \\ 1 & 5 \\ -4 & 3 \end{bmatrix}$ **b.** $D = \begin{bmatrix} 0.2 & -0.5 & 0.7 & 3.3 \\ -0.4 & 0.3 & 1 & 2 \\ 2.1 & -0.1 & 0.6 & 4.1 \end{bmatrix}$

Solution ▶ **a.** Matrix C is 3×2. The row 2, column 1 entry is 1.

b. Matrix D is 3×4. The row 2, column 1 entry is -0.4.

If a matrix has the same number of rows and columns, it's called a **square matrix.** Matrix B above is a square matrix, while matrix A is not. For square matrices, the values on a diagonal line *from the upper left to the lower right* are called the **diagonal entries** and are said to be **on the diagonal** of the matrix. When solving systems using matrices, much of our focus is on these diagonal entries.

EXAMPLE 1B ▶ **Identifying the Diagonal Entries of a Square Matrix**

Name the diagonal entries of each matrix.

a. $E = \begin{bmatrix} 1 & 4 \\ -2 & -3 \end{bmatrix}$ **b.** $F = \begin{bmatrix} 0.2 & -0.5 & 0.7 \\ -0.4 & 0.3 & 1 \\ 2.1 & -0.1 & 0.6 \end{bmatrix}$

Solution ▶ **a.** The diagonal entries of matrix E are 1 and -3.

b. For matrix F, the diagonal entries are 0.2, 0.3, and 0.6.

☑ **A.** You've just learned how to state the size of a matrix and identify its entries

Now try Exercises 7 through 9 ▶

B. The Augmented Matrix of a System of Equations

A system of equations can be written in matrix form by augmenting or joining the **coefficient matrix,** formed by the variable coefficients, with the **matrix of constants.**

The coefficient matrix for the system $\begin{cases} 2x + 3y - z = 1 \\ x + z = 2 \\ x - 3y + 4z = 5 \end{cases}$ is $\begin{bmatrix} 2 & 3 & -1 \\ 1 & 0 & 1 \\ 1 & -3 & 4 \end{bmatrix}$ with

column 1 for the coefficients of x, column 2 for the coefficients of y, and so on. The

matrix of constants is $\begin{bmatrix} 1 \\ 2 \\ 5 \end{bmatrix}$. These two are joined to form the **augmented matrix,**

with a dotted line often used to separate the two as shown here: $\begin{bmatrix} 2 & 3 & -1 & \vdots & 1 \\ 1 & 0 & 1 & \vdots & 2 \\ 1 & -3 & 4 & \vdots & 5 \end{bmatrix}$.

It's important to note the use of a zero placeholder for the y-variable in the second row of the matrix, signifying there is no y-variable in the corresponding equation.

EXAMPLE 2 ▶ Forming Augmented Matrices

Form the augmented matrix for each system, and name the diagonal entries of each coefficient matrix.

a. $\begin{cases} 2x + y = 11 \\ -x + 3y = -2 \end{cases}$ b. $\begin{cases} x + 4y - z = -10 \\ 2x + 5y + 8z = 4 \\ x - 2y - 3z = -7 \end{cases}$ c. $\begin{cases} \frac{1}{2}x + y = -7 \\ x + \frac{2}{3}y + \frac{5}{6}z = \frac{11}{12} \\ -2y - z = -3 \end{cases}$

Solution ▶ a. $\begin{cases} 2x + y = 11 \\ -x + 3y = -2 \end{cases} \longrightarrow \begin{bmatrix} 2 & 1 & \vdots & 11 \\ -1 & 3 & \vdots & -2 \end{bmatrix}$

Diagonal entries: 2 and 3.

b. $\begin{cases} x + 4y - z = -10 \\ 2x + 5y + 8z = 4 \\ x - 2y - 3z = -7 \end{cases} \longrightarrow \begin{bmatrix} 1 & 4 & -1 & \vdots & -10 \\ 2 & 5 & 8 & \vdots & 4 \\ 1 & -2 & -3 & \vdots & -7 \end{bmatrix}$

Diagonal entries: 1, 5, and -3.

c. $\begin{cases} \frac{1}{2}x + y = -7 \\ x + \frac{2}{3}y + \frac{5}{6}z = \frac{11}{12} \\ -2y - z = -3 \end{cases} \longrightarrow \begin{bmatrix} \frac{1}{2} & 1 & 0 & \vdots & -7 \\ 1 & \frac{2}{3} & \frac{5}{6} & \vdots & \frac{11}{12} \\ 0 & -2 & -1 & \vdots & -3 \end{bmatrix}$

Diagonal entries: $\frac{1}{2}$, $\frac{2}{3}$, and -1.

Now try Exercises 10 through 12 ▶

This process can easily be reversed to write a system of equations from a given augmented matrix.

EXAMPLE 3 ▶ Writing the System Corresponding to an Augmented Matrix

Write the system of equations corresponding to each matrix.

a. $\begin{bmatrix} 3 & -5 & \vdots & -14 \\ 0 & 1 & \vdots & 4 \end{bmatrix}$ b. $\begin{bmatrix} 1 & 4 & -1 & \vdots & -10 \\ 0 & -3 & 10 & \vdots & 7 \\ 0 & 0 & 1 & \vdots & 1 \end{bmatrix}$

Solution ▶ a. $\begin{bmatrix} 3 & -5 & \vdots & -14 \\ 0 & 1 & \vdots & 4 \end{bmatrix} \longrightarrow \begin{cases} 3x - 5y = -14 \\ 0x + 1y = 4 \end{cases}$

☑ **B.** You've just learned how to form the augmented matrix of a system of equations

b. $\begin{bmatrix} 1 & 4 & -1 & \vdots & -10 \\ 0 & -3 & 10 & \vdots & 7 \\ 0 & 0 & 1 & \vdots & 1 \end{bmatrix} \longrightarrow \begin{cases} 1x + 4y - 1z = -10 \\ 0x - 3y + 10z = 7 \\ 0x + 0y + 1z = 1 \end{cases}$

Now try Exercises 13 through 18 ▶

C. Solving a System Using Matrices

When a system of equations is written in augmented matrix form, we can solve the system by applying the same operations to each row of the matrix, that would be applied to the equations in the system. In this context, the operations are referred to as **elementary row operations.**

Elementary Row Operations

1. Any two rows in a matrix can be interchanged.
2. The elements of any row can be multiplied by a nonzero constant.
3. Any two rows can be added together, and the sum used to replace one of the rows.

In this section, we'll use these operations to **triangularize the augmented matrix,** employing a solution method known as **Gaussian elimination.** A matrix is said to be in **triangular form** when all of the entries below the diagonal are zero. For example,

the matrix $\begin{bmatrix} 1 & 4 & -1 & | & -10 \\ 0 & -3 & 10 & | & 7 \\ 0 & 0 & 1 & | & 1 \end{bmatrix}$ is in triangular form: $\begin{bmatrix} 1 & 4 & -1 & | & -10 \\ 0 & -3 & 10 & | & 7 \\ 0 & 0 & 1 & | & 1 \end{bmatrix}$.

In system form we have $\begin{cases} x + 4y - z = -10 \\ \quad\;\; -3y + 10z = 7 \\ \qquad\qquad\;\; z = 1 \end{cases}$, meaning a matrix written in triangular form can be used to solve the system using back-substitution. We'll illustrate by

solving $\begin{cases} 1x + 4y - 1z = 4 \\ 2x + 5y + 8z = 15 \\ 1x + 3y - 3z = 1 \end{cases}$, using elimination to the left, and *row operations on the*

augmented matrix to the right. As before, R1 represents the first equation in the system and the first row of the matrix, R2 represents equation 2 and row 2, and so on. The calculations involved are shown for the first stage only and are designed to offer a careful comparison. In actual practice, the format shown in Example 4 is used.

Elimination (System of Equations)	Row Operations (Augmented Matrix)
$\begin{cases} 1x + 4y - 1z = 4 \\ 2x + 5y + 8z = 15 \\ 1x + 3y - 3z = 1 \end{cases}$	$\begin{bmatrix} 1 & 4 & -1 & | & 4 \\ 2 & 5 & 8 & | & 15 \\ 1 & 3 & -3 & | & 1 \end{bmatrix}$

To eliminate the x-term in R2, we use $-2\text{R1} + \text{R2} \rightarrow \text{R2}$. For R3 the operations would be $-1\text{R1} + \text{R3} \rightarrow \text{R3}$. Identical operations are performed on the matrix, which begins the process of triangularizing the matrix.

System Form		**Matrix Form**	
-2R1	$-2x - 8y + 2z = -8$	-2R1	$-2 \quad -8 \quad 2 \quad -8$
$+$		$+$	
R2	$2x + 5y + 8z = 15$	R2	$\underline{2 \quad\;\; 5 \quad\;\; 8 \quad\;\; 15}$
New R2	$-3y + 10z = 7$	New R2	$0 \quad -3 \quad 10 \quad\;\; 7$
-1R1	$-1x - 4y + 1z = -4$	-1R1	$-1 \quad -4 \quad 1 \quad -4$
$+$		$+$	
R3	$1x + 3y - 3z = 1$	R3	$\underline{1 \quad\;\; 3 \quad -3 \quad\;\; 1}$
New R3	$-1y - 2z = -3$	New R3	$0 \quad -1 \quad -2 \quad -3$

As always, we should look for opportunities to simplify any equation in the system (and any row in the matrix). Note that $-1R3$ will make the coefficients and related matrix entries positive. Here is the new system and matrix.

New System   **New Matrix**

$$\begin{cases} 1x + 4y - 1z = 4 \\ -3y + 10z = 7 \\ 1y + 2z = 3 \end{cases} \qquad \begin{bmatrix} 1 & 4 & -1 & | & 4 \\ 0 & -3 & 10 & | & 7 \\ 0 & 1 & 2 & | & 3 \end{bmatrix}$$

On the left, we would finish by solving the 2×2 subsystem using $R2 + 3R3 \to R3$. In matrix form, we eliminate the corresponding entry (third row, second column) to triangularize the matrix.

$$\begin{bmatrix} 1 & 4 & -1 & | & 4 \\ 0 & -3 & 10 & | & 7 \\ 0 & 1 & 2 & | & 3 \end{bmatrix} \xrightarrow{R2 + 3R3 \to R3} \begin{bmatrix} 1 & 4 & -1 & | & 4 \\ 0 & -3 & 10 & | & 7 \\ 0 & 0 & 16 & | & 16 \end{bmatrix}$$

Dividing R3 by 16 gives $z = 1$ in the system, and entries of $0 \; 0 \; 1 \; 1$ in the augmented matrix. Completing the solution by back-substitution in the system gives the ordered triple $(1, 1, 1)$. **See Exercises 19 through 27.**

The general solution process is summarized here.

> ## Solving Systems by Triangularizing the Augmented Matrix
>
> 1. Write the system as an augmented matrix.
> 2. Use row operations to obtain zeroes below the first diagonal entry.
> 3. Use row operations to obtain zeroes below the second diagonal entry.
> 4. Continue until the matrix is triangularized (entries below diagonal are zero).
> 5. Divide to obtain a "1" in the last diagonal entry (if it is nonzero), then convert to equation form and solve using back-substitution.
>
> *Note:* At each stage, look for opportunities to simplify row entries using multiplication or division. Also, to begin the process any equation with an x-coefficient of 1 can be made R1 by interchanging the equations.

WORTHY OF NOTE

The procedure outlined for solving systems using matrices is virtually identical to that for solving systems by elimination. Using a 3×3 system for illustration, the "zeroes below the first diagonal entry" indicates we've eliminated the x-term from R2 and R3, the "zeroes below the second entry" indicates we've eliminated the y-term from the subsystem, and the division "to obtain a '1' in the final entry" indicates we have just solved for z.

EXAMPLE 4 ▶ **Solving Systems Using the Augmented Matrix**

Solve by triangularizing the augmented matrix: $\begin{cases} 2x + y - 2z = -7 \\ x + y + z = -1 \\ -2y - z = -3 \end{cases}$

Solution ▶ $\begin{cases} 2x + y - 2z = -7 \\ x + y + z = -1 \\ -2y - z = -3 \end{cases}$ $\xrightarrow{\text{matrix form}}$ $\begin{bmatrix} 2 & 1 & -2 & | & -7 \\ 1 & 1 & 1 & | & -1 \\ 0 & -2 & -1 & | & -3 \end{bmatrix}$ $\xrightarrow{R1 \leftrightarrow R2}$ $\begin{bmatrix} 1 & 1 & 1 & | & -1 \\ 2 & 1 & -2 & | & -7 \\ 0 & -2 & -1 & | & -3 \end{bmatrix}$

$\begin{bmatrix} 1 & 1 & 1 & | & -1 \\ 2 & 1 & -2 & | & -7 \\ 0 & -2 & -1 & | & -3 \end{bmatrix}$ $\xrightarrow{-2R1 + R2 \to R2}$ $\begin{bmatrix} 1 & 1 & 1 & | & -1 \\ 0 & -1 & -4 & | & -5 \\ 0 & -2 & -1 & | & -3 \end{bmatrix}$ $\xrightarrow{-1R2 \to R2}$ $\begin{bmatrix} 1 & 1 & 1 & | & -1 \\ 0 & 1 & 4 & | & 5 \\ 0 & -2 & -1 & | & -3 \end{bmatrix}$

$\begin{bmatrix} 1 & 1 & 1 & | & -1 \\ 0 & 1 & 4 & | & 5 \\ 0 & -2 & -1 & | & -3 \end{bmatrix}$ $\xrightarrow{2R2 + R3 \to R3}$ $\begin{bmatrix} 1 & 1 & 1 & | & -1 \\ 0 & 1 & 4 & | & 5 \\ 0 & 0 & 7 & | & 7 \end{bmatrix}$ $\xrightarrow{\frac{R3}{7} \to R3}$ $\begin{bmatrix} 1 & 1 & 1 & | & -1 \\ 0 & 1 & 4 & | & 5 \\ 0 & 0 & 1 & | & 1 \end{bmatrix}$

Converting the augmented matrix back into equation form yields $\begin{cases} x + y + z = -1 \\ \quad\quad y + 4z = 5 \\ \quad\quad\quad\quad z = 1 \end{cases}$.

Back-substitution shows the solution is $(-3, 1, 1)$.

Now try Exercises 28 through 32 ▶

The process used in Example 4 is also called **Gaussian elimination** (Carl Friedrich Gauss, 1777–1855), with the last matrix written in **row-echelon form.** It's possible to solve a system entirely using only the augmented matrix, by continuing to use row operations until the diagonal entries are 1's, with 0's for all other entries:

$\begin{bmatrix} 1 & 0 & 0 & | & a \\ 0 & 1 & 0 & | & b \\ 0 & 0 & 1 & | & c \end{bmatrix}$. The process is then called **Gauss-Jordan elimination** (Wilhelm Jordan, 1842–1899), with the final matrix written in **reduced row-echelon form** (see Appendix II).

☑ C. You've just learned how to solve a system of equations using row operations

Note that with Gauss-Jordan elimination, our *initial* focus is less on getting 1's along the diagonal, and more on obtaining zeroes for all entries *other than* the diagonal entries. This will enable us to work with integer values in the solution process.

EXAMPLE 5 ▶ Solving a System Using Gauss-Jordan Elimination

Solve using Gauss-Jordan elimination $\begin{cases} 2x + 5z - 15 = 2y \\ 2x + 3y = -1 + z \\ \quad\quad 4y + z = -7 \end{cases}$.

Solution ▶

standard form

$\begin{cases} 2x - 2y + 5z = 15 \\ 2x + 3y - 1z = -1 \\ 0x + 4y + 1z = -7 \end{cases}$ matrix form →
$\begin{bmatrix} 2 & -2 & 5 & | & 15 \\ 2 & 3 & -1 & | & -1 \\ 0 & 4 & 1 & | & -7 \end{bmatrix}$
$-R1 + R2 \rightarrow R2$
$\begin{bmatrix} 2 & -2 & 5 & | & 15 \\ 0 & 5 & -6 & | & -16 \\ 0 & 4 & 1 & | & -7 \end{bmatrix}$

$\begin{bmatrix} 2 & -2 & 5 & | & 15 \\ 0 & 5 & -6 & | & -16 \\ 0 & 4 & 1 & | & -7 \end{bmatrix}$
$\begin{matrix} 2R2 + 5R1 \rightarrow R1 \\ \\ -4R2 + 5R3 \rightarrow R3 \end{matrix}$
$\begin{bmatrix} 10 & 0 & 13 & | & 43 \\ 0 & 5 & -6 & | & -16 \\ 0 & 0 & 29 & | & 29 \end{bmatrix}$
$\begin{matrix} \frac{R3}{29} \rightarrow R3 \\ \\ \frac{R1}{10} \rightarrow R1 \end{matrix}$
$\begin{bmatrix} 10 & 0 & 13 & | & 43 \\ 0 & 5 & -6 & | & -16 \\ 0 & 0 & 1 & | & 1 \end{bmatrix}$

$\begin{bmatrix} 10 & 0 & 13 & | & 43 \\ 0 & 5 & -6 & | & -16 \\ 0 & 0 & 1 & | & 1 \end{bmatrix}$
$\begin{matrix} -13R3 + R1 \rightarrow R1 \\ \\ 6R3 + R2 \rightarrow R2 \end{matrix}$
$\begin{bmatrix} 10 & 0 & 0 & | & 30 \\ 0 & 5 & 0 & | & -10 \\ 0 & 0 & 1 & | & 1 \end{bmatrix}$
$\begin{matrix} \frac{R1}{10} \rightarrow R1 \\ \\ \frac{R2}{5} \rightarrow R2 \end{matrix}$
$\begin{bmatrix} 1 & 0 & 0 & | & 3 \\ 0 & 1 & 0 & | & -2 \\ 0 & 0 & 1 & | & 1 \end{bmatrix}$

The final matrix shows the solution is $(3, -2, 1)$.

Now try Exercises 33 through 36 ▶

D. Inconsistent and Dependent Systems

Due to the strong link between a linear system and its augmented matrix, inconsistent and dependent systems can be recognized just as in Sections 5.1 and 5.2. An inconsistent system will yield an inconsistent or contradictory statement such as $0 = -12$, meaning all entries in a row of the matrix of coefficients are zero, but the constant is not. A linearly dependent system will yield an identity statement such as $0 = 0$, meaning all entries in one row of the matrix are zero. If the system has coincident dependence, there will be only one nonzero row of coefficients.

EXAMPLE 6 ▶ **Solving a Dependent System**

Solve the system using Gauss-Jordan elimination: $\begin{cases} x + y - 5z = 3 \\ -x + 2z = -1 \\ 2x - y - z = 0 \end{cases}$

Solution ▶

$\begin{cases} x + y - 5z = 3 \\ -x + 2z = -1 \\ 2x - y - z = 0 \end{cases}$ standard form → $\begin{cases} x + y - 5z = 3 \\ -x + 0y + 2z = -1 \\ 2x - y - z = 0 \end{cases}$ matrix form → $\begin{bmatrix} 1 & 1 & -5 & | & 3 \\ -1 & 0 & 2 & | & -1 \\ 2 & -1 & -1 & | & 0 \end{bmatrix}$

$\begin{bmatrix} 1 & 1 & -5 & | & 3 \\ -1 & 0 & 2 & | & -1 \\ 2 & -1 & -1 & | & 0 \end{bmatrix}$ $\begin{array}{l} R1 + R2 \to R2 \\ \\ -2R1 + R3 \to R3 \end{array}$ $\begin{bmatrix} 1 & 1 & -5 & | & 3 \\ 0 & 1 & -3 & | & 2 \\ 0 & -3 & 9 & | & -6 \end{bmatrix}$ $\begin{array}{l} -1R2 + R1 \to R1 \\ \\ 3R2 + R3 \to R3 \end{array}$ $\begin{bmatrix} 1 & 0 & -2 & | & 1 \\ 0 & 1 & -3 & | & 2 \\ 0 & 0 & 0 & | & 0 \end{bmatrix}$

Since all entries in the last row are zeroes and it's the only row of zeroes, we conclude the system is linearly dependent and equivalent to $\begin{cases} x \phantom{{}+ y} - 2z = 1 \\ y - 3z = 2 \end{cases}$. As in Chapter 5, we demonstrate this dependence by writing the (x, y, z) solution in terms of a parameter. Solving for y in R2 gives y in terms of z: $y = 3z + 2$. Solving for x in R1 gives x in terms of z: $x = 2z + 1$. As written, the solutions all depend on z: $x = 2z + 1$, $y = 3z + 2$, and $z = z$. Selecting p as the parameter (or some other "neutral" variable), we write the solution as $(2p + 1, 3p + 2, p)$. Two of the infinite number of solutions would be $(1, 2, 0)$ for $p = 0$, and $(-1, -1, -1)$ for $p = -1$. Test these triples in the original equations.

Now try Exercises 37 through 45 ▶

☑ D. You've just learned how to recognize inconsistent and dependent systems

E. Solving Applications Using Matrices

As in other areas, solving applications using systems relies heavily on the ability to mathematically model information given verbally or in context. As you work through the exercises, read each problem carefully. Look for relationships that yield a system of two equations in two variables, three equations in three variables and so on.

EXAMPLE 7 ▶ **Determining the Original Value of Collector's Items**

A museum purchases a famous painting, a ruby tiara, and a rare coin for its collection, spending a total of $30,000. One year later, the painting has tripled in value, while the tiara and the coin have doubled in value. The items now have a total value of $75,000. Find the purchase price of each if the original price of the painting was $1000 more than twice the coin.

Solution ▶ Let P represent the price of the painting, T the tiara, and C the coin.

Total spent was $30,000:   → $P + T + C = 30,000$

One year later: → $3P + 2T + 2C = 75,000$

Value of painting versus coin: → $P = 2C + 1000$

$\begin{cases} P + T + C = 30000 \\ 3P + 2T + 2C = 75000 \\ P = 2C + 1000 \end{cases}$ standard form → $\begin{cases} 1P + 1T + 1C = 30000 \\ 3P + 2T + 2C = 75000 \\ 1P + 0T - 2C = 1000 \end{cases}$ matrix form → $\begin{bmatrix} 1 & 1 & 1 & | & 30000 \\ 3 & 2 & 2 & | & 75000 \\ 1 & 0 & -2 & | & 1000 \end{bmatrix}$

$$\begin{bmatrix} 1 & 1 & 1 & \vdots & 30000 \\ 3 & 2 & 2 & \vdots & 75000 \\ 1 & 0 & -2 & \vdots & 1000 \end{bmatrix} \begin{array}{l} -3R1 + R2 \to R2 \\ \\ -1R1 + R3 \to R3 \end{array} \begin{bmatrix} 1 & 1 & 1 & \vdots & 30000 \\ 0 & -1 & -1 & \vdots & -15000 \\ 0 & -1 & -3 & \vdots & -29000 \end{bmatrix} \begin{array}{l} -1R2 \to R2 \\ \\ -1R3 \to R3 \end{array} \begin{bmatrix} 1 & 1 & 1 & \vdots & 30000 \\ 0 & 1 & 1 & \vdots & 15000 \\ 0 & 1 & 3 & \vdots & 29000 \end{bmatrix}$$

$$\begin{bmatrix} 1 & 1 & 1 & \vdots & 30000 \\ 0 & 1 & 1 & \vdots & 15000 \\ 0 & 1 & 3 & \vdots & 29000 \end{bmatrix} \begin{array}{l} -1R2 + R3 \to R3 \end{array} \begin{bmatrix} 1 & 1 & 1 & \vdots & 30000 \\ 0 & 1 & 1 & \vdots & 15000 \\ 0 & 0 & 2 & \vdots & 14000 \end{bmatrix} \begin{array}{l} \dfrac{R3}{2} \to R3 \end{array} \begin{bmatrix} 1 & 1 & 1 & \vdots & 30000 \\ 0 & 1 & 1 & \vdots & 15000 \\ 0 & 0 & 1 & \vdots & 7000 \end{bmatrix}$$

From R3 of the triangularized form, $C = \$7000$ directly. Since R2 represents $T + C = 15{,}000$, we find the tiara was purchased for $T = \$8000$. Substituting these values into the first equation shows the painting was purchased for $\$15{,}000$. The solution is (15,000, 8000, 7000).

Now try Exercises 48 through 55 ▶

 E. You've just learned how to solve applications using linear systems

---

TECHNOLOGY HIGHLIGHT

Solving Systems Using Matrices and Calculating Technology

Graphing calculators offer a very efficient way to solve systems using matrices. Once the system has been written in matrix form, it can easily be entered and solved by asking the calculator to instantly perform the row operations needed. Pressing [2nd] [X⁻¹] (**MATRIX**) gives a screen similar to the one shown in Figure 6.1, where we begin by selecting the **EDIT** option (push the right arrow [▶] twice). Pressing [ENTER] places you on a screen where you can EDIT matrix A, changing the size as needed. Using the 3×4 matrix from Example 4, we press 3 and [ENTER], then 4 and [ENTER], giving the screen shown in Figure 6.2. The dash marks to the right indicate that there is a fourth column that cannot be seen, but that comes into view as you enter the elements of the matrix. Begin entering the first row of the matrix resulting from $R_1 \leftrightarrow R_2$, which has entries {1, 1, 1, −1}. Press [ENTER] after each entry and the cursor automatically goes to the next position in the matrix (note that the TI-84 Plus automatically shifts left and right to allow all four columns to be entered). After entering the second row {2, 1, −2, −7} and the third row {0, −2, −1, −3}, the completed matrix should look like the one shown in Figure 6.3 (the matrix is currently shifted to the right, showing the fourth column). To write this matrix in reduced row-echelon form **(rref)** we return to the home screen by pressing [2nd] [MODE] **(QUIT)**. Press the [CLEAR] key for a clean home screen. To access the **rref** function, press [2nd] [X⁻¹] (**MATRIX**) and select the **MATH** option, then scroll upward (or downward) until you get to **B:rref.** Pressing [ENTER] places this function on the home screen, where we must tell it to perform the **rref** operation on matrix [A]. Press [2nd] [X⁻¹] (**MATRIX**) to select a matrix (notice that matrix **NAMES** is automatically highlighted. Press [ENTER] to select matrix

Figure 6.1

Figure 6.2

Figure 6.3

[A] as the object of the **rref** function. After pressing ENTER the calculator quickly computes the reduced row-echelon form and displays it on the screen as in Figure 6.4. The solution is easily read as $x = -3$, $y = 1$, and $z = 1$, as we found in Example 4. Use these ideas to complete the following.

Figure 6.4

Exercise 1:   Use this method to solve the 2×2 system from Exercise 30.

Exercise 2:   Use this method to solve the 3×3 system from Exercise 32.

6.1 EXERCISES

▶ CONCEPTS AND VOCABULARY

Fill in the blank with the appropriate word or phrase. Carefully reread the section if needed.

1. A matrix with the same number of rows and columns is called a(n) _____ matrix.

2. When the coefficient matrix is used with the matrix of constants, the result is a(n) _____ matrix.

3. Matrix $A = \begin{bmatrix} 2 & 4 & -3 \\ 1 & -2 & 1 \end{bmatrix}$ is a ___ by ___ matrix. The entry in the second row and third column is ___.

4. Given matrix B shown here, the diagonal entries are ___, ___, and ___.
$$B = \begin{bmatrix} 1 & 4 & 3 \\ -1 & 5 & 2 \\ 3 & -2 & 1 \end{bmatrix}$$

5. The notation $-2R1 + R2 \rightarrow R2$ indicates that an equivalent matrix is formed by performing what operations/replacements?

6. Describe how to tell an inconsistent system apart from a dependent system when solving using matrix methods (row reduction).

▶ DEVELOPING YOUR SKILLS

Determine the size (order) of each matrix and identify the third row and second column entry. If the matrix given is a square matrix, identify the diagonal entries.

7. $\begin{bmatrix} 1 & 0 \\ 2.1 & 1 \\ -3 & 5.8 \end{bmatrix}$

8. $\begin{bmatrix} 1 & 0 & 4 \\ 1 & 3 & -7 \\ 5 & -1 & 2 \end{bmatrix}$

9. $\begin{bmatrix} 1 & 0 & 4 \\ 1 & 3 & -7 \\ 5 & -1 & 2 \\ 2 & -3 & 9 \end{bmatrix}$

Form the augmented matrix, then name the diagonal entries of the coefficient matrix.

10. $\begin{cases} 2x - 3y - 2z = 7 \\ x - y + 2z = -5 \\ 3x + 2y - z = 11 \end{cases}$

11. $\begin{cases} x + 2y - z = 1 \\ x + z = 3 \\ 2x - y + z = 3 \end{cases}$

12. $\begin{cases} 2x + 3y + z = 5 \\ 2y - z = 7 \\ x - y - 2z = 5 \end{cases}$

Write the system of equations for each matrix. Then use back-substitution to find its solution.

13. $\begin{bmatrix} 1 & 4 & | & 5 \\ 0 & 1 & | & \frac{1}{2} \end{bmatrix}$

14. $\begin{bmatrix} 1 & -5 & | & -15 \\ 0 & -1 & | & -2 \end{bmatrix}$

15. $\begin{bmatrix} 1 & 2 & -1 & | & 0 \\ 0 & 1 & 2 & | & 2 \\ 0 & 0 & 1 & | & 3 \end{bmatrix}$

16. $\begin{bmatrix} 1 & 0 & 7 & | & -5 \\ 0 & 1 & -5 & | & 15 \\ 0 & 0 & 1 & | & -26 \end{bmatrix}$

17. $\begin{bmatrix} 1 & 3 & -4 & | & 29 \\ 0 & 1 & -\frac{3}{2} & | & \frac{21}{2} \\ 0 & 0 & 1 & | & 3 \end{bmatrix}$

18. $\begin{bmatrix} 1 & 2 & -1 & | & 3 \\ 0 & 1 & \frac{1}{6} & | & \frac{2}{3} \\ 0 & 0 & 1 & | & \frac{22}{7} \end{bmatrix}$

Perform the indicated row operation(s) and write the new matrix.

19. $\begin{bmatrix} \frac{1}{2} & -3 & | & -1 \\ -5 & 2 & | & 4 \end{bmatrix}$ $\begin{matrix} 2R1 \rightarrow R1, \\ 5R1 + R2 \rightarrow R2 \end{matrix}$

20. $\begin{bmatrix} 7 & 4 & | & 3 \\ 4 & -8 & | & 12 \end{bmatrix}$ $\begin{matrix} \frac{1}{4}R2 \rightarrow R2, \\ R1 \leftrightarrow R2 \end{matrix}$

21. $\begin{bmatrix} -2 & 1 & 0 & | & 4 \\ 5 & 8 & 3 & | & -5 \\ 1 & -3 & 3 & | & 2 \end{bmatrix}$ $\begin{matrix} R1 \leftrightarrow R3, \\ -5R1 + R2 \rightarrow R2 \end{matrix}$

22. $\begin{bmatrix} -3 & 2 & 0 & | & 0 \\ 1 & 1 & 2 & | & 6 \\ 4 & 1 & -3 & | & 2 \end{bmatrix}$ $\begin{matrix} R1 \leftrightarrow R2, \\ -4R1 + R3 \rightarrow R3 \end{matrix}$

23. $\begin{bmatrix} 3 & 1 & 1 & | & 8 \\ 6 & -1 & -1 & | & 10 \\ 4 & -2 & -3 & | & 22 \end{bmatrix}$ $\begin{matrix} -2R1 + R2 \rightarrow R2, \\ -4R1 + 3R3 \rightarrow R3 \end{matrix}$

24. $\begin{bmatrix} 2 & 1 & -1 & | & -3 \\ 3 & 1 & 1 & | & 0 \\ 4 & 3 & 2 & | & 3 \end{bmatrix}$ $\begin{matrix} -3R1 + 2R2 \rightarrow R2, \\ -2R1 + R3 \rightarrow R3 \end{matrix}$

What row operations would produce zeroes beneath the first entry in the diagonal?

25. $\begin{bmatrix} 1 & 3 & 0 & | & 2 \\ -2 & 4 & 1 & | & 1 \\ 3 & -1 & -2 & | & 9 \end{bmatrix}$

26. $\begin{bmatrix} 1 & 1 & -4 & | & -3 \\ 3 & 0 & 1 & | & 5 \\ -5 & 3 & 2 & | & 3 \end{bmatrix}$

27. $\begin{bmatrix} 1 & 2 & 0 & | & 10 \\ 5 & 1 & 2 & | & 6 \\ -4 & 3 & -3 & | & 2 \end{bmatrix}$

Solve each system by triangularizing the augmented matrix and using back-substitution. Simplify by clearing fractions or decimals before beginning.

28. $\begin{cases} 2y = 5x + 4 \\ -5x = 2 - 4y \end{cases}$

29. $\begin{cases} 0.15g - 0.35h = -0.5 \\ -0.12g + 0.25h = 0.1 \end{cases}$

30. $\begin{cases} -\frac{1}{5}u + \frac{1}{4}v = 1 \\ \frac{1}{10}u + \frac{1}{2}v = 7 \end{cases}$

31. $\begin{cases} x - 2y + 2z = 7 \\ 2x + 2y - z = 5 \\ 3x - y + z = 6 \end{cases}$

32. $\begin{cases} 2x - 3y - 2z = 7 \\ x - y + 2z = -5 \\ 3x + 2y - z = 11 \end{cases}$

33. $\begin{cases} x + 2y - z = 1 \\ x + z = 3 \\ 2x - y + z = 3 \end{cases}$

34. $\begin{cases} 2x + 3y + z = 5 \\ 2y - z = 7 \\ x - y - 2z = 5 \end{cases}$

35. $\begin{cases} -x + y + 2z = 2 \\ x + y - z = 1 \\ 2x + y + z = 4 \end{cases}$

36. $\begin{cases} x + y - 2z = -1 \\ 4x - y + 3z = 3 \\ 3x + 2y - z = 4 \end{cases}$

Solve each system by triangularizing the augmented matrix and using back-substitution. If the system is linearly dependent, give the solution in terms of a parameter. If the system has coincident dependence, answer in set notation as in Chapter 5.

37. $\begin{cases} 4x - 8y + 8z = 24 \\ 2x - 6y + 3z = 13 \\ 3x + 4y - z = -11 \end{cases}$

38. $\begin{cases} 3x + y + z = -2 \\ x - 2y + 3z = 1 \\ 2x - 3y + 5z = 3 \end{cases}$

39. $\begin{cases} x + 3y + 5z = 20 \\ 2x + 3y + 4z = 16 \\ x + 2y + 3z = 12 \end{cases}$

40. $\begin{cases} -x + 2y + 3z = -6 \\ x - y + 2z = -4 \\ 3x - 6y - 9z = 18 \end{cases}$

41. $\begin{cases} 3x - 4y + 2z = -2 \\ \frac{3}{2}x - 2y + z = -1 \\ -6x + 8y - 4z = 4 \end{cases}$

42. $\begin{cases} 2x - y + 3z = 1 \\ 4x - 2y + 6z = 2 \\ 10x - 5y + 15z = 5 \end{cases}$

43. $\begin{cases} 2x - y + 3z = 1 \\ 2y + 6z = 2 \\ x - \frac{1}{2}y + \frac{3}{2}z = 5 \end{cases}$

44. $\begin{cases} x + 2y + z = 4 \\ 3x - 4y + z = 4 \\ 6x - 8y + 2z = 8 \end{cases}$

45. $\begin{cases} -2x + 4y - 3z = 4 \\ 5x - 6y + 7z = -12 \\ x + 2y + z = -4 \end{cases}$

▶ WORKING WITH FORMULAS

Area of a triangle in the plane:

$$A = \pm\frac{1}{2}(x_1y_2 - x_2y_1 + x_2y_3 - x_3y_2 + x_3y_1 - x_1y_3)$$

The area of a triangle in the plane is given by the formula shown, where the vertices of the triangle are located at the points (x_1, y_1), (x_2, y_2), and (x_3, y_3), and the sign is chosen to ensure a positive value.

46. Find the area of a triangle whose vertices are $(-1, -3), (5, 2),$ and $(1, 8)$.

47. Find the area of a triangle whose vertices are $(6, -2), (-5, 4),$ and $(-1, 7)$.

▶ APPLICATIONS

Model each problem using a system of linear equations. Then solve using the augmented matrix.

Descriptive Translation

48. The distance (via air travel) from Los Angeles (LA), California, to Saint Louis (STL), Missouri, to Cincinnati (CIN), Ohio, to New York City (NYC), New York, is approximately 2480 mi. Find the distances between each city if the distance from LA to STL is 50 mi more than five times the distance between STL and CIN and 110 mi less than three times the distance from CIN to NYC.

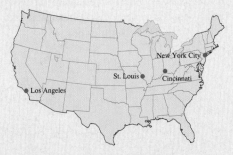

49. In the 2006 NBA Championship Series, Dwayne Wade of the Miami Heat carried his team to the title after the first two games were lost to the Dallas Mavericks. If 187 points were scored in the

title game and the Heat won by 3 points, what was the final score?

50. Moe is lecturing Larry and Curly once again (Moe, Larry, and Curly of *The Three Stooges* fame) claiming he is twice as smart as Larry and three times as smart as Curly. If he is correct and the sum of their IQs is 165, what is the IQ of each stooge?

51. A collector of rare books buys a handwritten, autographed copy of Edgar Allan Poe's *Annabel Lee,* an original advance copy of L. Frank Baum's *The Wonderful Wizard of Oz,* and a first print copy of *The Caine Mutiny* by Herman Wouk, paying a total of $100,000. Find the cost of each one, given that the cost of *Annabel Lee* and twice the cost of *The Caine Mutiny* sum to the price paid for *The Wonderful Wizard of Oz,* and *The Caine Mutiny* cost twice as much as *Annabel Lee.*

Geometry

52. A right triangle has a hypotenuse of 39 m. If the perimeter is 90 m, and the longer leg is 6 m longer than twice the shorter leg, find the dimensions of the triangle.

53. In triangle ABC, the sum of angles A and C is equal to three times angle B. Angle C is 10 degrees more than twice angle B. Find the measure of each angle.

Investment and Finance

54. Suppose $10,000 is invested in three different investment vehicles paying 5%, 7%, and 9% annual interest. Find the amount invested at each rate if the interest earned after 1 yr is $760 and the amount invested at 9% is equal to the sum of the amounts invested at 5% and 7%.

55. The trustee of a union's pension fund has invested the funds in three ways: a savings fund paying 4% annual interest, a money market fund paying 7%, and government bonds paying 8%. Find the amount invested in each if the interest earned after one year is $0.178 million and the amount in government bonds is $0.3 million more than twice the amount in money market funds. The total amount invested is $2.5 million dollars.

▶ EXTENDING THE CONCEPT

56. In previous sections, we noted that one condition for a 3×3 system to be dependent was for the third equation to be a linear combination of the other two. To test this, write any two (different) equations using the same three variables, then form a third equation by performing some combination of elementary row operations. Solve the resulting 3×3 system. What do you notice?

57. Given the drawing shown, use a system of equations and the matrix method to find the measure of the angles labeled as x and y. Recall that vertical angles are equal and that the sum of the angles in a triangle is $180°$.

58. The system given here has a solution of $(1, -2, 3)$. Find the value of a and b.

$$\begin{bmatrix} 1 & a & b & \vdots & 1 \\ 2b & 2a & 5 & \vdots & 13 \\ 2a & 7 & 3b & \vdots & -8 \end{bmatrix}$$

▶ MAINTAINING YOUR SKILLS

59. (2.8) Given $f(x) = x^3 - 8$ and $g(x) = x - 2$, find $f + g, f - g, fg,$ and $\dfrac{f}{g}$.

60. (3.3) Given $x = 2$ is a zero of $h(x) = x^4 - x^2 - 12$, find all zeroes of h, real and complex.

61. (4.4) Since 2005, cable installations for an Internet company have been modeled by the function $C(t) = 15 \ln (t + 1)$, where $C(t)$ represents cable installations in thousands, t yr after 2005. In what year will the number of installations be greater than 30,000?

62. (4.5) If a set amount of money p is deposited regularly (daily, weekly, monthly, etc.) n times per year at a fixed interest rate r, the amount of money A accumulated in t years is given by the formula shown. If a parent deposits $250 per month for 18 yr at 4.6% beginning when her first child was born, how much has been accumulated to help pay for college expenses?

$$A = \dfrac{p\left[\left(1 + \dfrac{r}{n}\right)^{nt} - 1\right]}{\dfrac{r}{n}}$$

Learning Objectives

In Section 6.2 you will learn how to:

☐ **A.** Determine if two matrices are equal

☐ **B.** Add and subtract matrices

☐ **C.** Compute the product of two matrices

Matrices serve a much wider purpose than just a convenient method for solving systems. To understand their broader application, we need to know more about matrix theory, the various ways matrices can be combined, and some of their more practical uses. The common operations of addition, subtraction, multiplication, and division are all defined for matrices, as are other operations. Practical applications of matrix theory can be found in the social sciences, inventory management, genetics, operations research, engineering, and many other fields.

A. Equality of Matrices

To effectively study matrix algebra, we first give matrices a more general definition. For the *general* matrix A, all entries will be denoted using the lowercase letter "a," with their position in the matrix designated by the dual subscript a_{ij}. The letter "i" gives the *row* and the letter "j" gives the *column* of the entry's location. The general $m \times n$ matrix A is written

$$
\begin{array}{cccccc}
 & \text{col 1} & \text{col 2} & \text{col 3} & \text{col } j & \text{col } n \\
\text{row 1} \rightarrow & a_{11} & a_{12} & a_{13} & \cdots & a_{1j} & \cdots & a_{1n} \\
\text{row 2} \rightarrow & a_{21} & a_{22} & a_{23} & \cdots & a_{2j} & \cdots & a_{2n} \\
\text{row 3} \rightarrow & a_{31} & a_{32} & a_{33} & \cdots & a_{3j} & \cdots & a_{3n} \\
 & \vdots & \vdots & \vdots & & \vdots & & \vdots \\
\text{row } i \rightarrow & a_{i1} & a_{i2} & a_{i3} & \cdots & a_{ij} & \cdots & a_{in} \\
 & \vdots & \vdots & \vdots & & \vdots & & \vdots \\
\text{row } m \rightarrow & a_{m1} & a_{m2} & a_{m3} & \cdots & a_{mj} & \cdots & a_{mn}
\end{array}
$$

a_{ij} is a general matrix element

The size of a matrix is also referred to as its **order**, and we say the order of general matrix A is $m \times n$. Note that diagonal entries have the same row and column number, a_{ij}, where $i = j$. Also, where the general entry of matrix A is a_{ij}, the general entry of matrix B is b_{ij}, of matrix C is c_{ij}, and so on.

EXAMPLE 1 ▶ **Identifying the Order and Entries of a Matrix**

State the order of each matrix and name the entries corresponding to a_{22}, a_{31}; b_{22}, b_{31}; and c_{22}, c_{31}.

a. $A = \begin{bmatrix} 1 & 4 \\ -2 & -3 \end{bmatrix}$ **b.** $B = \begin{bmatrix} 3 & -2 \\ 1 & 5 \\ -4 & 3 \end{bmatrix}$ **c.** $C = \begin{bmatrix} 0.2 & -0.5 & 0.7 \\ -1 & 0.3 & 1 \\ 2.1 & -0.1 & 0.6 \end{bmatrix}$

Solution ▶ **a.** matrix A: order 2×2. Entry $a_{22} = -3$ (the row 2, column 2 entry is -3). There is no a_{31} entry (A is only 2×2).

b. matrix B: order 3×2. Entry $b_{22} = 5$, entry $b_{31} = -4$.

c. matrix C: order 3×3. Entry $c_{22} = 0.3$, entry $c_{31} = 2.1$.

Now try Exercises 7 through 12 ▶

Equality of Matrices

Two matrices are equal if they have the same order and their corresponding entries are equal. In symbols, this means that $A = B$ if $a_{ij} = b_{ij}$ for all i and j.

EXAMPLE 2 ▶ **Determining If Two Matrices Are Equal**

Determine whether the following statements are true, false, or conditional. If false, explain why. If conditional, find values that will make the statement true.

a. $\begin{bmatrix} 1 & 4 \\ -2 & -3 \end{bmatrix} = \begin{bmatrix} -3 & -2 \\ 4 & 1 \end{bmatrix}$ b. $\begin{bmatrix} 3 & -2 \\ 1 & 5 \\ -4 & 3 \end{bmatrix} = \begin{bmatrix} 3 & -2 & 1 \\ 5 & -4 & 3 \end{bmatrix}$

c. $\begin{bmatrix} 1 & 4 \\ -2 & -3 \end{bmatrix} = \begin{bmatrix} a - 2 & 2b \\ c & -3 \end{bmatrix}$

Solution ▶ a. $\begin{bmatrix} 1 & 4 \\ -2 & -3 \end{bmatrix} = \begin{bmatrix} -3 & -2 \\ 4 & 1 \end{bmatrix}$ is false. The matrices have the same order and entries, but corresponding entries are not equal.

b. $\begin{bmatrix} 3 & -2 \\ 1 & 5 \\ -4 & 3 \end{bmatrix} = \begin{bmatrix} 3 & -2 & 1 \\ 5 & -4 & 3 \end{bmatrix}$ is false. Their orders are not equal.

☑ **A.** You've just learned how to determine if two matrices are equal

c. $\begin{bmatrix} 1 & 4 \\ -2 & -3 \end{bmatrix} = \begin{bmatrix} a - 2 & 2b \\ c & -3 \end{bmatrix}$ is conditional. The statement is true when $a - 2 = 1$ $(a = 3)$, $2b = 4$ $(b = 2)$, $c = -2$, and is false otherwise.

Now try Exercises 13 through 16 ▶

B. Addition and Subtraction of Matrices

A sum or difference of matrices is found by combining the corresponding entries. This limits the operations to matrices of like orders, so that every entry in one matrix has a "corresponding entry" in the other. This also means the result is a new matrix of *like order,* whose entries are the corresponding sums or differences. Note that since a_{ij} represents a general entry of matrix A, $[a_{ij}]$ represents the entire matrix.

Addition and Subtraction of Matrices

Given matrices A, B, and C having like orders.

The sum $A + B = C$,   The difference $A - B = D$,
where $[a_{ij} + b_{ij}] = [c_{ij}]$.   where $[a_{ij} - b_{ij}] = [d_{ij}]$.

EXAMPLE 3 ▶ **Adding and Subtracting Matrices**

Compute the sum or difference of the matrices indicated.

$$A = \begin{bmatrix} 2 & 6 \\ 1 & 0 \\ 1 & -3 \end{bmatrix} \qquad B = \begin{bmatrix} -3 & 2 & -1 \\ -5 & 4 & 3 \end{bmatrix} \qquad C = \begin{bmatrix} 3 & -2 \\ 1 & 5 \\ -4 & 3 \end{bmatrix}$$

a. $A + C$ b. $A + B$ c. $C - A$

Solution ▶ a. $A + C = \begin{bmatrix} 2 & 6 \\ 1 & 0 \\ 1 & -3 \end{bmatrix} + \begin{bmatrix} 3 & -2 \\ 1 & 5 \\ -4 & 3 \end{bmatrix}$ sum of A and C

$= \begin{bmatrix} 2 + 3 & 6 + (-2) \\ 1 + 1 & 0 + 5 \\ 1 + (-4) & -3 + 3 \end{bmatrix} = \begin{bmatrix} 5 & 4 \\ 2 & 5 \\ -3 & 0 \end{bmatrix}$ add corresponding entries

b. $A + B = \begin{bmatrix} 2 & 6 \\ 1 & 0 \\ 1 & -3 \end{bmatrix} + \begin{bmatrix} -3 & 2 & -1 \\ -5 & 4 & 3 \end{bmatrix}$ Addition and subtraction are not defined for matrices of unlike order.

c. $C - A = \begin{bmatrix} 3 & -2 \\ 1 & 5 \\ -4 & 3 \end{bmatrix} - \begin{bmatrix} 2 & 6 \\ 1 & 0 \\ 1 & -3 \end{bmatrix}$ difference of C and A

☑ **B.** You've just learned how to add and subtract matrices

$= \begin{bmatrix} 3 - 2 & -2 - 6 \\ 1 - 1 & 5 - 0 \\ -4 - 1 & 3 - (-3) \end{bmatrix} = \begin{bmatrix} 1 & -8 \\ 0 & 5 \\ -5 & 6 \end{bmatrix}$ subtract corresponding entries

Now try Exercises 17 through 20 ▶

Since the addition of two matrices is defined as the sum of corresponding entries, we find the properties of matrix addition closely resemble those of real number addition.

Properties of Matrix Addition

Given matrices A, B, C, and Z are $m \times n$ matrices, with Z the zero matrix. Then,

I. $A + B = B + A$ matrix addition is commutative
II. $(A + B) + C = A + (B + C)$ matrix addition is associative
III. $A + Z = Z + A = A$ Z is the additive identity
IV. $A + (-A) = (-A) + A = Z$ $-A$ is the additive inverse of A

C. Matrices and Multiplication

The algebraic terms $2a$ and ab have counterparts in matrix algebra. The product $2A$ represents a constant times a matrix and is called **scalar multiplication.** The product AB represents the product of two matrices.

Scalar Multiplication

Scalar multiplication is defined by taking the product of the constant with *each entry* in the matrix, forming a new matrix of like size. In symbols, for any real number k and matrix A, $kA = [ka_{ij}]$. Similar to standard algebraic properties, $-A$ represents the scalar product $-1 \cdot A$ and any subtraction can be rewritten as an algebraic sum: $A - B = A + (-B)$. As noted in the properties box, for any matrix A, the sum $A + (-A)$ will yield the **zero matrix Z,** a matrix of like size whose entries are all zeroes. Also note that matrix $-A$ is the **additive inverse** for A, while Z is the **additive identity.**

EXAMPLE 4 ▶ Computing Operations on Matrices

Given $A = \begin{bmatrix} 4 & 3 \\ \frac{1}{2} & 1 \\ 0 & -3 \end{bmatrix}$ and $B = \begin{bmatrix} 3 & -2 \\ 0 & 6 \\ -4 & 0.4 \end{bmatrix}$, compute the following:

a. $\frac{1}{2}B$ **b.** $-4A - \frac{1}{2}B$

Solution ▶ **a.** $\frac{1}{2}B = \left(\frac{1}{2}\right)\begin{bmatrix} 3 & -2 \\ 0 & 6 \\ -4 & 0.4 \end{bmatrix}$

$= \begin{bmatrix} \left(\frac{1}{2}\right)(3) & \left(\frac{1}{2}\right)(-2) \\ \left(\frac{1}{2}\right)(0) & \left(\frac{1}{2}\right)(6) \\ \left(\frac{1}{2}\right)(-4) & \left(\frac{1}{2}\right)(0.4) \end{bmatrix} = \begin{bmatrix} \frac{3}{2} & -1 \\ 0 & 3 \\ -2 & 0.2 \end{bmatrix}$

b. $-4A - \dfrac{1}{2}B = -4A + \left(-\dfrac{1}{2}\right)B$ rewrite using algebraic addition

$$= \begin{bmatrix} (-4)(4) & (-4)(3) \\ (-4)(\frac{1}{2}) & (-4)(1) \\ (-4)(0) & (-4)(-3) \end{bmatrix} + \begin{bmatrix} (-\frac{1}{2})(3) & (-\frac{1}{2})(-2) \\ (-\frac{1}{2})(0) & (-\frac{1}{2})(6) \\ (-\frac{1}{2})(-4) & (-\frac{1}{2})(0.4) \end{bmatrix}$$

$$= \begin{bmatrix} -16 & -12 \\ -2 & -4 \\ 0 & 12 \end{bmatrix} + \begin{bmatrix} -\frac{3}{2} & 1 \\ 0 & -3 \\ 2 & -0.2 \end{bmatrix} \quad \text{simplify}$$

$$= \begin{bmatrix} -16 + (-\frac{3}{2}) & -12 + 1 \\ -2 + 0 & -4 + (-3) \\ 0 + 2 & 12 + (-0.2) \end{bmatrix} = \begin{bmatrix} -\frac{35}{2} & -11 \\ -2 & -7 \\ 2 & 11.8 \end{bmatrix} \quad \text{result}$$

Now try Exercises 21 through 24 ▶

Matrix Multiplication

Consider a cable company offering three different levels of Internet service: Bronze—fast, Silver—very fast, and Gold—lightning fast. Table 6.1 shows the number and types of programs sold to households and businesses for the week. Each program has an incentive package consisting of a rebate and a certain number of free weeks, as shown in Table 6.2.

Table 6.1 Matrix A

	Bronze	Silver	Gold
Homes	40	20	**25**
Businesses	10	15	45

Table 6.2 Matrix B

	Rebate	Free Weeks
Bronze	$15	2
Silver	$25	4
Gold	**$35**	6

To compute the amount of rebate money the cable company paid to households for the week, we would take the first row (R1) in Table 6.1 and multiply by the corresponding entries (bronze with bronze, silver with silver, and so on) in the first column (C1) of Table 6.2 and add these products. In matrix form, we have

$$\begin{bmatrix} 40 & 20 & \mathbf{25} \end{bmatrix} \cdot \begin{bmatrix} 15 \\ 25 \\ \mathbf{35} \end{bmatrix} = 40 \cdot 15 + 20 \cdot 25 + \mathbf{25} \cdot \mathbf{35} = \mathbf{\$1975}.$$ Using R1 of Table 6.1

with C2 from Table 6.2 gives the number of free weeks awarded to homes:

$$\begin{bmatrix} 40 & 20 & \mathbf{25} \end{bmatrix} \cdot \begin{bmatrix} 2 \\ 4 \\ \mathbf{6} \end{bmatrix} = 40 \cdot 2 + 20 \cdot 4 + \mathbf{25} \cdot \mathbf{6} = 310.$$ Using the second row (R2) of

Table 6.1 with the two columns from Table 6.2 will give the amount of rebate money and the number of free weeks, respectively, awarded to business customers. When all computations are complete, the result is a product matrix P with order 2×2. This is because the product of R1 from matrix A, with C1 from matrix B, *gives the entry in*

position P_{11} *of the product matrix:* $\begin{bmatrix} 40 & 20 & 25 \\ 10 & 15 & 45 \end{bmatrix} \cdot \begin{bmatrix} 15 & 2 \\ 25 & 4 \\ 35 & 6 \end{bmatrix} = \begin{bmatrix} 1975 & 310 \\ 2100 & 350 \end{bmatrix}.$

Likewise, the product R1 · C2 will give entry P_{12} (310), the product of R2 with C1 will give P_{21} (2100), and so on. This "row × column" multiplication can be generalized, and leads to the following. Given $m \times n$ matrix A and $s \times t$ matrix B,

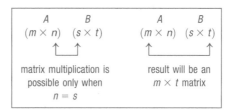

A	B	A	B
$(m \times n)$	$(s \times t)$	$(m \times n)$	$(s \times t)$

matrix multiplication is possible only when $n = s$

result will be an $m \times t$ matrix

In more formal terms, we have the following definition of matrix multiplication.

Matrix Multiplication

Given the $m \times n$ matrix $A = [a_{ij}]$ and the $s \times t$ matrix $B = [b_{ij}]$. If $n = s$, then matrix multiplication is possible and the product AB is an $m \times t$ matrix $P = [p_{ij}]$, where p_{ij} is product of the ith row of A with the jth column of B.

In less formal terms, matrix multiplication involves multiplying the row entries of the first matrix with the corresponding column entries of the second, and adding them together.

In Example 5, two of the matrix products [parts (a) and (b)] are shown in full detail, with the first entry of the product matrix color-coded.

EXAMPLE 5 ▶ Multiplying Matrices

Given the matrices A through E shown here, compute the following products:

a. AB **b.** CD **c.** DC **d.** AE **e.** EA

$$A = \begin{bmatrix} -2 & 1 \\ 3 & 4 \end{bmatrix} \quad B = \begin{bmatrix} 4 & 3 \\ 6 & 1 \end{bmatrix} \quad C = \begin{bmatrix} -2 & 1 & 3 \\ 1 & 0 & 2 \\ 4 & 1 & -1 \end{bmatrix} \quad D = \begin{bmatrix} 2 & 5 & 1 \\ 4 & -1 & 1 \\ 0 & 3 & -2 \end{bmatrix} \quad E = \begin{bmatrix} -2 & -1 \\ 3 & 0 \\ 1 & 2 \end{bmatrix}$$

Solution ▶ a. $AB = \begin{bmatrix} -2 & 1 \\ 3 & 4 \end{bmatrix}\begin{bmatrix} 4 & 3 \\ 6 & 1 \end{bmatrix} = \begin{bmatrix} -2 & -5 \\ 36 & 13 \end{bmatrix}$

Computation: $\begin{bmatrix} (-2)(4) + (1)(6) & (-2)(3) + (1)(1) \\ (3)\,4 + (4)(6) & (3)(3) + (4)(1) \end{bmatrix}$

A	B	A	B
(2×2)	(2×2)	(2×2)	(2×2)

multiplication is possible since 2 = 2

result will be a 2 × 2 matrix

b. $CD = \begin{bmatrix} -2 & 1 & 3 \\ 1 & 0 & 2 \\ 4 & 1 & -1 \end{bmatrix}\begin{bmatrix} 2 & 5 & 1 \\ 4 & -1 & 1 \\ 0 & 3 & -2 \end{bmatrix} = \begin{bmatrix} 0 & -2 & -7 \\ 2 & 11 & -3 \\ 12 & 16 & 7 \end{bmatrix}$

C	D	C	D
(3×3)	(3×3)	(3×3)	(3×3)

multiplication is possible since 3 = 3

result will be a 3 × 3 matrix

Computation: $\begin{bmatrix} (-2)(2) + (1)(4) + (3)(0) & (-2)(5) + (1)(-1) + (3)(3) & (-2)(1) + (1)(1) + (3)(-2) \\ (1) + (0)(4) + (2)(0) & (1)(5) + (0)(-1) + (2)(3) & (1)(1) + (0)(1) + (2)(-2) \\ (4) + (1)(4) + (-1)(0) & (4)(5) + (1)(-1) + (-1)(3) & (4)(1) + (1)(1) + (-1)(-2) \end{bmatrix}$

c. $DC = \begin{bmatrix} 2 & 5 & 1 \\ 4 & -1 & 1 \\ 0 & 3 & -2 \end{bmatrix}\begin{bmatrix} -2 & 1 & 3 \\ 1 & 0 & 2 \\ 4 & 1 & -1 \end{bmatrix} = \begin{bmatrix} 5 & 3 & 15 \\ -5 & 5 & 9 \\ -5 & -2 & 8 \end{bmatrix}$

D	C	D	C
(3×3)	(3×3)	(3×3)	(3×3)

multiplication is possible since 3 = 3

result will be a 3 × 3 matrix

d. $AE = \begin{bmatrix} -2 & 1 \\ 3 & 4 \end{bmatrix} \begin{bmatrix} -2 & -1 \\ 3 & 0 \\ 1 & 2 \end{bmatrix}$

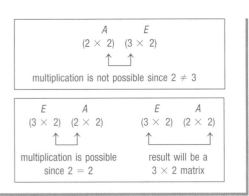

e. $EA = \begin{bmatrix} -2 & -1 \\ 3 & 0 \\ 1 & 2 \end{bmatrix} \begin{bmatrix} -2 & 1 \\ 3 & 4 \end{bmatrix} = \begin{bmatrix} 1 & -6 \\ -6 & 3 \\ 4 & 9 \end{bmatrix}$

Now try Exercises 25 through 36 ▶

Example 5 shows that in general, matrix multiplication is not commutative. Parts (b) and (c) show $CD \neq DC$ since we get different results, and parts (d) and (e) show $AE \neq EA$, since AE is not defined while EA is.

Operations on matrices can be a laborious process for larger matrices and for matrices with noninteger or large entries. For these, we can turn to available technology for assistance. This shifts our focus from a meticulous computation of entries, to carefully entering each matrix into the calculator, double-checking each entry, and appraising results to see if they're reasonable.

EXAMPLE 6 ▶ Using Technology for Matrix Operations

Use a calculator to compute the difference $A - B$ for the matrices given.

$$A = \begin{bmatrix} \frac{2}{11} & -0.5 & \frac{6}{5} \\ 0.9 & \frac{3}{4} & -4 \\ 0 & 6 & -\frac{5}{12} \end{bmatrix} \qquad B = \begin{bmatrix} \frac{1}{6} & \frac{-7}{10} & 0.75 \\ \frac{11}{25} & 0 & -5 \\ -4 & \frac{-5}{9} & \frac{-5}{12} \end{bmatrix}$$

Solution ▶ The entries for matrix A are shown in Figure 6.5. After entering matrix B, exit to the home screen [**2nd** **MODE** (**QUIT**)], call up matrix A, press the **–** (subtract) key, then call up matrix B and press **ENTER** . The calculator quickly finds the difference and displays the results shown in Figure 6.6. The last line on the screen shows the result can be stored for future use in a new matrix C by pressing the **STO ▸** key, calling up matrix C, and pressing **ENTER** .

Figure 6.5

Figure 6.6

Now try Exercises 37 through 40 ▶

Figure 6.7

In Figure 6.6 the dots to the right on the calculator screen indicate there are additional digits or matrix columns that can't fit on the display, as often happens with larger matrices or decimal numbers. Sometimes, converting entries to fraction form will provide a display that's easier to read. Here, this is done by calling up the matrix C, and using the **MATH** **1: ▸ Frac** option. After pressing **ENTER** , all entries are converted to fractions in simplest form (where possible), as in Figure 6.7. The third column can be viewed by pressing the right arrow.

EXAMPLE 7 ▶ Using Technology for Matrix Operations

Use a calculator to compute the product AB.

$$A = \begin{bmatrix} 2 & -3 & 0 \\ -1 & 5 & 4 \\ 6 & 0 & 2 \\ 3 & 2 & -1 \end{bmatrix} \quad B = \begin{bmatrix} \frac{1}{2} & -0.7 & 1 \\ 0.5 & 3.2 & -3 \\ -2 & \frac{3}{4} & 4 \end{bmatrix}$$

Solution ▶ Carefully enter matrices A and B into the calculator, then press [2nd] [MODE] (**QUIT**) to get to the home screen. Use [A][B] [ENTER], and the calculator finds the product shown in the figure.

$$AB = \begin{bmatrix} 2 & -3 & 0 \\ -1 & 5 & 4 \\ 6 & 0 & 2 \\ 3 & 2 & -1 \end{bmatrix} \begin{bmatrix} \frac{1}{2} & -0.7 & 1 \\ 0.5 & 3.2 & -3 \\ -2 & \frac{3}{4} & 4 \end{bmatrix}$$

```
[A] [B]
[[-.5  -11    11]
 [-6   19.7   0 ]
 [-1   -2.7  14]
 [4.5  3.55  -7]]
```

Now try Exercises 41 through 52 ▶

Properties of Matrix Multiplication

Earlier, Example 5 demonstrated that matrix multiplication is not commutative. Here is a group of properties that *do* hold for matrices. You are asked to check these properties in the exercise set using various matrices. See **Exercises 53 through 56.**

Properties of Matrix Multiplication	
Given matrices A, B, and C for which the products are defined:	
I. $A(BC) = (AB)C$	matrix multiplication is associative
II. $A(B + C) = AB + AC$	matrix multiplication is distributive from the left
III. $(B + C)A = BA + CA$	matrix multiplication is distributive from the right
IV. $k(A + B) = kA + kB$	a constant k can be distributed over addition

We close this section with an application of matrix multiplication. There are many other interesting applications in the exercise set.

EXAMPLE 8 ▶ Using Matrix Multiplication to Track Volunteer Enlistments

In a certain country, the number of males and females that will join the military depends on their age. This information is stored in matrix A (Table 6.3). The likelihood a volunteer will join a *particular branch* of the military also depends on their age, with this information stored in matrix B (Table 6.4). (a) Compute the product $P = AB$ and discuss/interpret what is indicated by the entries P_{11}, P_{13}, and P_{24} of the product matrix. (b) How many males are expected to join the Navy this year?

Table 6.3 Matrix A

A	Age Groups		
Sex	18–19	20–21	22–23
Female	1000	1500	500
Male	2500	3000	2000

Table 6.4 Matrix B

B	Likelihood of Joining			
Age Group	Army	Navy	Air Force	Marines
18–19	0.42	0.28	0.17	0.13
20–21	0.38	0.26	0.27	0.09
22–23	0.33	0.25	0.35	0.07

Solution ▶ **a.** Matrix A has order 2×3 and matrix B has order 3×4. The product matrix P can be found and is a 2×4 matrix. Carefully enter the matrices in your calculator. Figure 6.8 shows the entries of matrix B. Using $[A][B]$ **ENTER**, the calculator finds the product matrix shown in Figure 6.9. Pressing the right arrow shows the complete product matrix is

$$P = \begin{bmatrix} 1155 & 795 & 750 & 300 \\ 2850 & 1980 & 1935 & 735 \end{bmatrix}.$$

The entry P_{11} is the product of R1 from A and C1 from B, and indicates that for the year, 1155 females are projected to join the Army. In like manner, entry P_{13} shows that 750 females are projected to join the Air Force. Entry P_{24} indicates that 735 males are projected to join the Marines.

C. You've just learned how to compute the product of two matrices

Figure 6.8

Figure 6.9

b. The product R2 (males) · C2 (Navy) gives $P_{22} = 1980$, meaning 1980 males are expected to join the Navy.

Now try Exercise 59 through 66 ▶

6.2 EXERCISES

CONCEPTS AND VOCABULARY

Fill in the blank with the appropriate word or phrase. Carefully reread the section if needed.

1. Two matrices are equal if they are like size and the corresponding entries are equal. In symbols, $A = B$ if _____ = _____.

2. The sum of two matrices (of like size) is found by adding the corresponding entries. In symbols, $A + B =$ _____.

3. The product of a constant times a matrix is called _____ multiplication.

4. The size of a matrix is also referred to as its _____. The order of $A = \begin{bmatrix} 1 & 2 & 3 \\ 4 & 5 & 6 \end{bmatrix}$ is _____.

5. Give two reasons why matrix multiplication is generally not commutative. Include several examples using matrices of various sizes.

6. Discuss the conditions under which matrix multiplication is defined. Include several examples using matrices of various sizes.

▶ DEVELOPING YOUR SKILLS

State the order of each matrix and name the entries in positions a_{12} and a_{23} if they exist. Then name the position a_{ij} of the 5 in each.

7. $\begin{bmatrix} 1 & -3 \\ 5 & -7 \end{bmatrix}$ **8.** $\begin{bmatrix} 19 \\ -11 \\ 5 \end{bmatrix}$

9. $\begin{bmatrix} 2 & -3 & 0.5 \\ 0 & 5 & 6 \end{bmatrix}$ **10.** $\begin{bmatrix} 2 & 0.4 \\ -0.1 & 5 \\ 0.3 & -3 \end{bmatrix}$

11. $\begin{bmatrix} -2 & 1 & -7 \\ 0 & 8 & 1 \\ 5 & -1 & 4 \end{bmatrix}$ **12.** $\begin{bmatrix} 89 & 55 & 34 & 21 \\ 13 & 8 & 5 & 3 \\ 2 & 1 & 1 & 0 \end{bmatrix}$

Determine if the following statements are true, false, or conditional. If false, explain why. If conditional, find values of $a, b, c, p, q,$ and r that will make the statement true.

13. $\begin{bmatrix} \sqrt{1} & \sqrt{4} & \sqrt{8} \\ \sqrt{16} & \sqrt{32} & \sqrt{64} \end{bmatrix} = \begin{bmatrix} 1 & 2 & 2\sqrt{2} \\ 4 & 4\sqrt{2} & 8 \end{bmatrix}$

14. $\begin{bmatrix} \dfrac{3}{2} & \dfrac{-7}{5} & \dfrac{13}{10} \\ \dfrac{-1}{2} & \dfrac{-2}{5} & \dfrac{1}{3} \end{bmatrix} = \begin{bmatrix} 1.5 & -1.4 & 1.3 \\ -0.5 & -0.4 & 0.\overline{3} \end{bmatrix}$

15. $\begin{bmatrix} -2 & 3 & a \\ 2b & -5 & 4 \\ 0 & -9 & 3c \end{bmatrix} = \begin{bmatrix} c & 3 & -4 \\ 6 & -5 & -a \\ 0 & -3b & -6 \end{bmatrix}$

16. $\begin{bmatrix} 2p+1 & -5 & 9 \\ 1 & 12 & 0 \\ q+5 & 9 & -2r \end{bmatrix} = \begin{bmatrix} 7 & -5 & 2-q \\ 1 & 3r & 0 \\ -2 & 3p & -8 \end{bmatrix}$

For matrices A through H as given, perform the indicated operation(s), if possible. Do not use a calculator. If an operation cannot be completed, state why.

$A = \begin{bmatrix} 2 & 3 \\ 5 & 8 \end{bmatrix}$ $B = \begin{bmatrix} 2 \\ 1 \\ -3 \end{bmatrix}$

$C = \begin{bmatrix} 2 & 0.5 \\ 0.2 & 5 \\ -1 & 3 \end{bmatrix}$ $D = \begin{bmatrix} 1 & 0 & 0 \\ 0 & 1 & 0 \\ 0 & 0 & 1 \end{bmatrix}$

$E = \begin{bmatrix} 1 & -2 & 0 \\ 0 & -1 & 2 \\ 4 & 3 & -6 \end{bmatrix}$ $F = \begin{bmatrix} 6 & -3 & 9 \\ 12 & 0 & -6 \end{bmatrix}$

$G = \begin{bmatrix} -1 & 2 & 0 \\ 0 & 1 & -2 \\ -4 & -3 & 6 \end{bmatrix}$ $H = \begin{bmatrix} 8 & -3 \\ -5 & 2 \end{bmatrix}$

17. $A + H$ **18.** $E + G$

19. $F + H$ **20.** $G + D$

21. $3H - 2A$ **22.** $2E + 3G$

23. $\dfrac{1}{2}E - 3D$ **24.** $F - \dfrac{2}{3}F$

25. ED **26.** DE

27. AH **28.** HA

29. FD **30.** FH

31. HF **32.** EB

33. H^2 **34.** F^2

35. FE **36.** EF

For matrices A through H as given, use a calculator to perform the indicated operation(s), if possible. If an operation cannot be completed, state why.

$A = \begin{bmatrix} -5 & 4 \\ 3 & 9 \end{bmatrix}$ $B = \begin{bmatrix} 1 & 0 \\ 0 & 1 \end{bmatrix}$

$C = \begin{bmatrix} \dfrac{\sqrt{3}}{2} & \dfrac{\sqrt{3}}{3} \\ \sqrt{3} & 2\sqrt{3} \end{bmatrix}$ $D = \begin{bmatrix} 1 & 0 & 0 \\ 0 & 1 & 0 \\ 0 & 0 & 1 \end{bmatrix}$

$E = \begin{bmatrix} 1 & -2 & 0 \\ 0 & -1 & 2 \\ 4 & 3 & -6 \end{bmatrix}$ $F = \begin{bmatrix} -0.52 & 0.002 & 1.032 \\ 1.021 & -1.27 & 0.019 \end{bmatrix}$

$G = \begin{bmatrix} 0 & \dfrac{3}{4} & \dfrac{1}{4} \\ \dfrac{-1}{2} & \dfrac{3}{8} & \dfrac{1}{8} \\ \dfrac{-1}{4} & \dfrac{11}{16} & \dfrac{1}{16} \end{bmatrix}$ $H = \begin{bmatrix} \dfrac{-3}{19} & \dfrac{4}{57} \\ \dfrac{1}{19} & \dfrac{5}{57} \end{bmatrix}$

37. $C + H$ **38.** $A - H$

39. $E + G$ **40.** $G - E$

41. AH **42.** HA

43. EG **44.** GE

45. HB **46.** BH

47. DG **48.** GD

49. C^2 **50.** E^2

51. FG **52.** AF

For Exercises 53 through 56, use a calculator and matrices A, B, and C to verify each statement.

$$A = \begin{bmatrix} -1 & 3 & 5 \\ 2 & 7 & -1 \\ 4 & 0 & 6 \end{bmatrix} \quad B = \begin{bmatrix} 0.3 & -0.4 & 1.2 \\ -2.5 & 2 & 0.9 \\ 1 & -0.5 & 0.2 \end{bmatrix}$$

$$C = \begin{bmatrix} 45 & -1 & 3 \\ -6 & 10 & -15 \\ 21 & -28 & 36 \end{bmatrix}$$

53. Matrix multiplication is not generally commutative: (a) $AB \neq BA$, (b) $AC \neq CA$, and (c) $BC \neq CB$.

54. Matrix multiplication is distributive from the left: $A(B + C) = AB + AC$.

55. Matrix multiplication is distributive from the right: $(B + C)A = BA + CA$.

56. Matrix multiplication is associative: $(AB)C = A(BC)$.

▶ **WORKING WITH FORMULAS**

$$\begin{bmatrix} 2 & 2 \\ W & 0 \end{bmatrix} \cdot \begin{bmatrix} L \\ W \end{bmatrix} = \begin{bmatrix} \text{Perimeter} \\ \text{Area} \end{bmatrix}$$

The perimeter and area of a rectangle can be simultaneously calculated using the matrix formula shown, where L represents the length and W represents the width of the rectangle. Use the matrix formula and your calculator to find the perimeter and area of the

rectangles shown, then check the results using $P = 2L + 2W$ and $A = LW$.

57. 6.374 cm, 4.35 cm

58. 5.02 cm, 3.75 cm

▶ **APPLICATIONS**

59. Custom T's designs and sells specialty T-shirts and sweatshirts, with plants in Verdi and Minsk. The company offers this apparel in three quality levels: standard, deluxe, and premium. Last fall the Verdi office produced 3820 standard, 2460 deluxe, and 1540 premium T-shirts, along with 1960 standard, 1240 deluxe, and 920 premium sweatshirts. The Minsk office produced 4220 standard, 2960 deluxe, and 1640 premium T-shirts, along with 2960 standard, 3240 deluxe, and 820 premium sweatshirts in the same time period.

a. Write a 3 × 2 "production matrix" for each plant [$V \to$ Verdi, $M \to$ Minsk], with a *T-shirt* column, a *sweatshirt* column, and three rows showing how many of the different types of apparel were manufactured.

b. Use the matrices from Part (a) to determine how many more or less articles of clothing were produced by Minsk than Verdi.

c. Use scalar multiplication to find how many shirts of each type will be made at Verdi and Minsk next fall, if each is expecting a 4% increase in business.

d. What will be Custom T's total production next fall (from both plants), for each type of apparel?

60. Terry's Tire Store sells automobile and truck tires through three retail outlets. Sales at the Cahokia store for the months of January, February, and March

broke down as follows: 350, 420, and 530 auto tires and 220, 180, and 140 truck tires. The Shady Oak branch sold 430, 560, and 690 auto tires and 280, 320, and 220 truck tires during the same 3 months. Sales figures for the downtown store were 864, 980, and 1236 auto tires and 535, 542, and 332 truck tires.

a. Write a 2 × 3 "sales matrix" for each store [$C \to$ Cahokia, $S \to$ Shady Oak, $D \to$ Downtown], with *January, February,* and *March* columns, and two rows showing the sales of auto and truck tires respectively.

b. Use the matrices from Part (a) to determine how many more or fewer tires of each type the downtown store sold (each month) over the other two stores combined.

c. Market trends indicate that for the same three months in the following year, the Cahokia store will likely experience a 10% increase in sales, the Shady Oak store a 3% decrease, with sales at the downtown store remaining level (no change). What will be the combined monthly sales from all three stores next year, for each type of tire?

61. Home improvements: Dream-Makers Home Improvements specializes in replacement windows, replacement doors, and new siding. During the peak season, the number of contracts that came from various parts of the city (North, South, East, and West) are shown in matrix C. The average profit per contract is shown in matrix *P*. Compute the product *PC* and discuss what each entry of the product matrix represents.

$$\begin{array}{c} \phantom{\text{Windows}} \begin{array}{cccc} \text{N} & \text{S} & \text{E} & \text{W} \end{array} \\ \begin{array}{c} \text{Windows} \\ \text{Doors} \\ \text{Siding} \end{array} \left[\begin{array}{cccc} 9 & 6 & 5 & 4 \\ 7 & 5 & 7 & 6 \\ 2 & 3 & 5 & 2 \end{array}\right] = C \end{array}$$

$$\begin{array}{ccc} \text{Windows} & \text{Doors} & \text{Siding} \\ \left[\begin{array}{ccc} 1500 & 500 & 2500 \end{array}\right] = P \end{array}$$

62. Classical music: Station 90.7—*The Home of Classical Music*—is having their annual fund drive. Being a loyal listener, Mitchell decides that for the next 3 days he will donate money according to his favorite composers, by the number of times their music comes on the air: $3 for every piece by Mozart (*M*), $2.50 for every piece by Beethoven (*B*), and $2 for every piece by Vivaldi (*V*). This information is displayed in matrix *D*. The number of pieces he heard from each composer is displayed in matrix *C*. Compute the product *DC* and discuss what each entry of the product matrix represents.

$$\begin{array}{c} \begin{array}{ccc} \text{Mon.} & \text{Tue.} & \text{Wed.} \end{array} \\ \begin{array}{c} M \\ B \\ V \end{array} \left[\begin{array}{ccc} 4 & 3 & 5 \\ 3 & 2 & 4 \\ 2 & 3 & 3 \end{array}\right] = C \end{array}$$

$$\begin{array}{ccc} M & B & V \\ \left[\begin{array}{ccc} 3 & 2.5 & 2 \end{array}\right] = D \end{array}$$

63. Pizza and salad: The science department and math department of a local college are at a pre-semester retreat, and decide to have pizza, salads, and soft drinks for lunch. The quantity of food ordered by each department is shown in matrix *Q*. The cost of the food item at each restaurant is shown in matrix *C* using the published prices from three popular restaurants: Pizza Home (PH), Papa Jeff's (PJ), and Dynamos (D).

a. What is the total cost to the math department if the food is ordered from Pizza Home?

b. What is the total cost to the science department if the food is ordered from Papa Jeff's?

c. Compute the product *QC* and discuss the meaning of each entry in the product matrix.

$$\begin{array}{c} \phantom{\text{Science}} \begin{array}{ccc} \text{Pizza} & \text{Salad} & \text{Drink} \end{array} \\ \begin{array}{c} \text{Science} \\ \text{Math} \end{array} \left[\begin{array}{ccc} 8 & 12 & 20 \\ 10 & 8 & 18 \end{array}\right] = Q \end{array}$$

$$\begin{array}{c} \phantom{\text{Pizza}} \begin{array}{ccc} \text{PH} & \text{PJ} & \text{D} \end{array} \\ \begin{array}{c} \text{Pizza} \\ \text{Salad} \\ \text{Drink} \end{array} \left[\begin{array}{ccc} 8 & 7.5 & 10 \\ 1.5 & 1.75 & 2 \\ 0.90 & 1 & 0.75 \end{array}\right] = C \end{array}$$

64. Manufacturing pool tables: Cue Ball Incorporated makes three types of pool tables, for homes, commercial use, and professional use. The amount of time required to pack, load, and install each is summarized in matrix *T*, with all times in hours. The cost of these components in dollars per hour, is summarized in matrix *C* for two of its warehouses, one on the west coast and the other in the midwest.

a. What is the cost to package, load, and install a commercial pool table from the coastal warehouse?

b. What is the cost to package, load, and install a commercial pool table from the warehouse in the midwest?

c. Compute the product *TC* and discuss the meaning of each entry in the product matrix.

$$\begin{array}{c} \phantom{\text{Home}} \begin{array}{ccc} \text{Pack} & \text{Load} & \text{Install} \end{array} \\ \begin{array}{c} \text{Home} \\ \text{Comm} \\ \text{Prof} \end{array} \left[\begin{array}{ccc} 1 & 0.2 & 1.5 \\ 1.5 & 0.5 & 2.2 \\ 1.75 & 0.75 & 2.5 \end{array}\right] = T \end{array}$$

$$\begin{array}{c} \phantom{\text{Pack}} \begin{array}{cc} \text{Coast} & \text{Midwest} \end{array} \\ \begin{array}{c} \text{Pack} \\ \text{Load} \\ \text{Install} \end{array} \left[\begin{array}{cc} 10 & 8 \\ 12 & 10.5 \\ 13.5 & 12.5 \end{array}\right] = C \end{array}$$

65. Joining a club: Each school year, among the students planning to join a club, the likelihood a student joins a particular club depends on their class standing. This information is stored in matrix *C*. The number of males and females from each class that are projected to join a club each year is stored in matrix *J*. Compute the product *JC* and use the result to answer the following:

a. Approximately how many females joined the chess club?

b. Approximately how many males joined the writing club?

c. What does the entry P_{13} of the product matrix tells us?

$$\begin{array}{c} \phantom{\text{Female}} \begin{array}{ccc} \text{Fresh} & \text{Soph} & \text{Junior} \end{array} \\ \begin{array}{c} \text{Female} \\ \text{Male} \end{array} \left[\begin{array}{ccc} 25 & 18 & 21 \\ 22 & 19 & 18 \end{array}\right] = J \end{array}$$

$$\begin{array}{c} \phantom{\text{Fresh}} \begin{array}{ccc} \text{Spanish} & \text{Chess} & \text{Writing} \end{array} \\ \begin{array}{c} \text{Fresh} \\ \text{Soph} \\ \text{Junior} \end{array} \left[\begin{array}{ccc} 0.6 & 0.1 & 0.3 \\ 0.5 & 0.2 & 0.3 \\ 0.4 & 0.2 & 0.4 \end{array}\right] = C \end{array}$$

66. Designer shirts: The SweatShirt Shoppe sells three types of designs on its products: stenciled (S), embossed (E), and applique (A). The quantity of each size sold is shown in matrix *Q*. The retail price of each sweatshirt depends on its size and whether it was finished by hand or machine. Retail prices are shown in matrix C. Assuming all stock is sold,

   a. How much revenue was generated by the large sweatshirts?

   b. How much revenue was generated by the extra-large sweatshirts?

   c. What does the entry P_{11} of the product matrix QC tell us?

$$\begin{array}{c} \\ \text{med} \\ \text{large} \\ \text{x-large} \end{array}\begin{array}{ccc} S & E & A \\ \left[\begin{array}{ccc} 30 & 30 & 15 \\ 60 & 50 & 20 \\ 50 & 40 & 30 \end{array}\right] \end{array} = Q$$

$$\begin{array}{c} \\ S \\ E \\ A \end{array}\begin{array}{cc} \text{Hand} & \text{Machine} \\ \left[\begin{array}{cc} 40 & 25 \\ 60 & 40 \\ 90 & 60 \end{array}\right] \end{array} = C$$

▶ EXTENDING THE CONCEPT

67. For the matrix A shown, use your calculator to compute A^2, A^3, A^4, and A^5. Do you notice a pattern? Try to write a "matrix formula" for A^n, where n is a positive integer, then use your formula to find A^6. Check results using a calculator.

$$A = \begin{bmatrix} 1 & 0 & 1 \\ 1 & 1 & 1 \\ 1 & 0 & 1 \end{bmatrix}$$

68. The matrix $M = \begin{bmatrix} 2 & 1 \\ -3 & -2 \end{bmatrix}$ has some very interesting properties. Compute the powers M^2,

M^3, M^4, and M^5, then discuss what you find. Try to find/create another 2×2 matrix that has similar properties.

69. For the "matrix equation"
$$\begin{bmatrix} 2 & 1 \\ -3 & -2 \end{bmatrix} \cdot \begin{bmatrix} a & b \\ c & d \end{bmatrix} = \begin{bmatrix} 1 & 0 \\ 0 & 1 \end{bmatrix},$$ use matrix multiplication and two systems of equations to find the entries a, b, c, and d that make the equation true.

▶ MAINTAINING YOUR SKILLS

70. (5.2) Solve the system using elimination.

$$\begin{cases} x + 2y - z = 3 \\ -2x - y + 3z = -5 \\ 5x + 3y - 2z = 2 \end{cases}$$

71. (3.7) Use the graph of $f(x)$ given to solve $f(x) \geq 0$.

72. (4.3) Evaluate using the change-of-base formula, then check using exponentiation.

$$\log_2 21$$

73. (3.2) Find the quotient using synthetic division, then check using multiplication.

$$\frac{x^3 - 9x + 10}{x - 2}$$

MID-CHAPTER CHECK

State the size of each matrix and identify the entry in second row, third column.

1. $A = \begin{bmatrix} 0.4 & 1.1 & 0.2 \\ -0.2 & 0.1 & -0.9 \\ 0.7 & 0.4 & 0.8 \end{bmatrix}$

2. $\begin{bmatrix} -2 & 1 & \frac{1}{2} & 5 \\ 4 & \frac{3}{4} & 0 & -3 \end{bmatrix}$

Write each system in matrix form and solve using row operations to triangularize the matrix. If the system is linearly dependent, write the solution using a parameter.

3. $\begin{cases} 2x + 3y = -5 \\ -5x - 4y = 2 \end{cases}$

4. $\begin{cases} -x + y - 5z = 23 \\ 2x + 4y - z = 9 \\ 3x - 5y + z = 1 \end{cases}$

5. $\begin{cases} x + y - 3z = 11 \\ 4x - y - 2z = -4 \\ 3x - 2y + z = 7 \end{cases}$

6. For matrices A and B given, compute:

$$A = \begin{bmatrix} -3 & -2 \\ 5 & 4 \end{bmatrix} \quad B = \begin{bmatrix} 10 & 15 \\ -30 & -5 \end{bmatrix}$$

a. $A - B$ **b.** $\dfrac{2}{5}B$ **c.** $5A + B$

7. For matrices C and D given, use a calculator to find:

$$C = \begin{bmatrix} -0.2 & 0 & 0.2 \\ 0.4 & 0.8 & 0 \\ 0.1 & -0.2 & -0.1 \end{bmatrix} \quad D = \begin{bmatrix} 5 & 2.5 & 10 \\ -2.5 & 0 & -5 \\ 10 & 2.5 & 10 \end{bmatrix}$$

a. $C + \dfrac{1}{5}D$ **b.** $-0.6D$ **c.** CD

8. For the matrices A, B, C, and D given, compute the products indicated (if possible):

$$A = \begin{bmatrix} 4 & -1 \\ 0 & -5 \end{bmatrix} \qquad B = \begin{bmatrix} 6 & -2 \\ 0 & 1 \\ 4 & 7 \end{bmatrix}$$

$$C = \begin{bmatrix} 4 & -8 & -3 \\ -1 & 0 & 1 \end{bmatrix} \qquad D = \begin{bmatrix} 2 & 0 & -6 \\ -1 & -3 & 0 \\ 1 & 5 & -4 \end{bmatrix}$$

a. AC **b.** $-2CD$ **c.** BA **d.** $CB - 4A$

9. Create a system of equations to model this exercise, then write the system in matrix form and solve. The campus bookstore offers both new and used texts to students. In a recent biology class with 24 students, 14 bought used texts and 10 bought new texts, with the class as a whole paying $2370. Of the 6 premed students in class, 2 bought used texts, and 4 bought new texts, with the group paying a total of $660. How much does a used text cost? How much does a new text cost?

10. Matrix A shown gives the number and type of extended warranties sold to individual car owners and to business fleets. Matrix B shows the promotions offered to those making the purchase. Compute the product matrix $P = AB$, and state what each entry of the product matrix represents.

Matrix A

Extended Warranties	80,000 mi	100,000 mi	120,000 mi
Individuals	30	25	10
Businesses	20	12	5

Matrix B

Promotions	Rebate	Free AAA Membership
80,000 mi	$50	1 yr
100,000 mi	$75	2 yr
120,000 mi	$100	3 yr

REINFORCING BASIC CONCEPTS

More on Matrix Multiplication

To help understand and master the concept of matrix multiplication, it helps to take a closer look at the entries of the product matrix. Recall for the product $AB = P$, the entry P_{11} in the product matrix is the result of multiplying the 1st row of A with the 1st column of B, the entry P_{12} is the result of multiplying 1st row of A, with the 2nd column of B, and so on.

Exercise 1: The product of the 3rd row of A with the 2nd column of B, gives what entry in P?

Exercise 2: The entry P_{13} is the result of what product? The entry P_{22} is the result of what product?

Exercise 3: If P_{33} is the last entry of the product matrix, what are the possible sizes of A and B?

Exercise 4: Of the eight matrices shown here, only two produce the product matrix P shown. Use the ideas highlighted above to determine which two.

$$A = \begin{bmatrix} 3 & -1 \\ 2 & 1 \\ 4 & 1 \end{bmatrix} \qquad B = \begin{bmatrix} 1 & 1 \\ 2 & 0 \end{bmatrix}$$

$$C = \begin{bmatrix} 2 & 0 & -6 \\ -1 & -3 & 0 \\ 1 & 5 & -4 \end{bmatrix} \qquad D = \begin{bmatrix} 2 & 4 \\ -1 & 0 \\ 3 & 5 \end{bmatrix}$$

$$E = \begin{bmatrix} 1 & 2 \\ 3 & -1 \end{bmatrix} \qquad F = \begin{bmatrix} 1 & 0 \\ 0 & 2 \\ 4 & 1 \end{bmatrix}$$

$$G = \begin{bmatrix} 1 \\ 4 \\ 6 \end{bmatrix} \qquad H = \begin{bmatrix} 1 & 3 \end{bmatrix}$$

$$P = \begin{bmatrix} 1 & 2 \\ 6 & -2 \\ 7 & 7 \end{bmatrix}$$

Learning Objectives

In Section 6.3 you will learn how to:

☐ **A.** Recognize the identity matrix for multiplication

☐ **B.** Find the inverse of a square matrix

☐ **C.** Solve systems using matrix equations

☐ **D.** Use determinants to find whether a matrix is invertible

While using matrices and row operations offers a degree of efficiency in solving systems, we are still required to solve for each variable *individually*. Using matrix multiplication we can actually rewrite a given system as a single *matrix equation*, in which the solutions are computed *simultaneously*. As with other kinds of equations, the use of identities and inverses are involved, which we now develop in the context of matrices.

A. Multiplication and Identity Matrices

From the properties of real numbers, 1 is the identity for multiplication since $n \cdot 1 = 1 \cdot n = n$. A similar identity exists for matrix multiplication. Consider the 2×2 matrix $A = \begin{bmatrix} 1 & 4 \\ -2 & 3 \end{bmatrix}$. While matrix multiplication is not *generally* commutative, if we can find a matrix B where $AB = BA = A$, then B is a prime candidate for the identity matrix, which is denoted I. For the products AB and BA to be possible and have the same order as A, we note B must also be a 2×2 matrix. Using the arbitrary matrix $B = \begin{bmatrix} a & b \\ c & d \end{bmatrix}$, we have the following.

EXAMPLE 1A ▶ **Solving $AB = A$ to Find the Identity Matrix**

For $\begin{bmatrix} 1 & 4 \\ -2 & 3 \end{bmatrix} \begin{bmatrix} a & b \\ c & d \end{bmatrix} = \begin{bmatrix} 1 & 4 \\ -2 & 3 \end{bmatrix}$, use matrix multiplication, the equality of matrices, and systems of equations to find the value of a, b, c, and d.

Solution ▶ The product on the left gives $\begin{bmatrix} a + 4c & b + 4d \\ -2a + 3c & -2b + 3d \end{bmatrix} = \begin{bmatrix} 1 & 4 \\ -2 & 3 \end{bmatrix}$.

Since corresponding entries must be equal (shown by matching colors), we can find a, b, c, and d by solving the systems $\begin{cases} a + 4c = 1 \\ -2a + 3c = -2 \end{cases}$ and $\begin{cases} b + 4d = 4 \\ -2b + 3d = 3 \end{cases}$. For the first system, $2R1 + R2$ shows $a = 1$ and $c = 0$. Using $2R1 + R2$ for the second shows $b = 0$ and $d = 1$. It appears $\begin{bmatrix} 1 & 0 \\ 0 & 1 \end{bmatrix}$ is a candidate for the identity matrix.

▶

Before we name B as the identity matrix, we must show that $AB = BA = A$.

EXAMPLE 1B ▶ **Verifying $AB = BA = A$**

Given $A = \begin{bmatrix} 1 & 4 \\ -2 & 3 \end{bmatrix}$ and $B = \begin{bmatrix} 1 & 0 \\ 0 & 1 \end{bmatrix}$, determine if $AB = A$ and $BA = A$.

Solution ▶ $AB = \begin{bmatrix} 1 & 4 \\ -2 & 3 \end{bmatrix} \begin{bmatrix} 1 & 0 \\ 0 & 1 \end{bmatrix}$ $\qquad\qquad BA = \begin{bmatrix} 1 & 0 \\ 0 & 1 \end{bmatrix} \begin{bmatrix} 1 & 4 \\ -2 & 3 \end{bmatrix}$

$= \begin{bmatrix} 1(1) + 4(0) & 1(0) + 4(1) \\ -2(1) + 3(0) & -2(0) + 3(1) \end{bmatrix}$ $\qquad = \begin{bmatrix} 1(1) + 0(-2) & 1(4) + 0(3) \\ 0(1) + 1(-2) & 0(4) + 1(3) \end{bmatrix}$

$= \begin{bmatrix} 1 & 4 \\ -2 & 3 \end{bmatrix} = A \checkmark$ $\qquad\qquad\qquad = \begin{bmatrix} 1 & 4 \\ -2 & 3 \end{bmatrix} = A \checkmark$

Since $AB = A = BA$, B is the identity matrix I.

Now try Exercises 7 through 10 ▶

By replacing the entries of $A = \begin{bmatrix} 1 & 4 \\ -2 & -3 \end{bmatrix}$ with those of the general matrix $\begin{bmatrix} a_{11} & a_{12} \\ a_{21} & a_{22} \end{bmatrix}$, we can show that $I = \begin{bmatrix} 1 & 0 \\ 0 & 1 \end{bmatrix}$ is the identity for *all* 2 × 2 matrices. In considering the identity for larger matrices, we find that only *square matrices* have inverses, since $AI = IA$ is the primary requirement (the multiplication must be possible in both directions). This is commonly referred to as *multiplication from the right* and *multiplication from the left.* Using the same procedure as before we can show $\begin{bmatrix} 1 & 0 & 0 \\ 0 & 1 & 0 \\ 0 & 0 & 1 \end{bmatrix}$ is the identity for 3 × 3 matrices (denoted I_3). The $n \times n$ identity

☑ A. You've just learned how to recognize the identity matrix for multiplication

matrix I_n consists of 1's down the main diagonal and 0's for all other entries. Also, the identity I_n for a square matrix is unique.

As in Section 6.2, a graphing calculator can be used to investigate operations on matrices and matrix properties. For the 3 × 3 matrix $A = \begin{bmatrix} 2 & 5 & 1 \\ 4 & -1 & 1 \\ 0 & 3 & -2 \end{bmatrix}$ and

Figure 6.10

$I_3 = \begin{bmatrix} 1 & 0 & 0 \\ 0 & 1 & 0 \\ 0 & 0 & 1 \end{bmatrix}$, a calculator will confirm that $AI_3 = A = I_3 A$. Carefully enter A into your calculator as matrix A, and I_3 as matrix B. Figure 6.10 shows $AB = A$ and after pressing **ENTER**, the calculator will verify $BA = A$, although the screen cannot display the result without scrolling. **See Exercises 11 through 14.**

B. The Inverse of a Matrix

Again from the properties of real numbers, we know the multiplicative inverse for a is $a^{-1} = \dfrac{1}{a} (a \neq 0)$, since the products $a \cdot a^{-1}$ and $a^{-1} \cdot a$ yield the identity 1. To show that a similar inverse exists for matrices, consider the square matrix $A = \begin{bmatrix} 6 & 5 \\ 2 & 2 \end{bmatrix}$ and an arbitrary matrix $B = \begin{bmatrix} a & b \\ c & d \end{bmatrix}$. If we can find a matrix B, where $AB = BA = I$, then B is a prime candidate for the inverse matrix of A, which is denoted A^{-1}. Proceeding as in Examples 1A and 1B gives the result shown in Example 2.

EXAMPLE 2A ▶ **Solving $AB = I$ to find A^{-1}**

For $\begin{bmatrix} 6 & 5 \\ 2 & 2 \end{bmatrix} \begin{bmatrix} a & b \\ c & d \end{bmatrix} = \begin{bmatrix} 1 & 0 \\ 0 & 1 \end{bmatrix}$, use matrix multiplication, the equality of matrices, and systems of equations to find the entries of B.

Solution ▶ The product on the left gives $\begin{bmatrix} 6a + 5c & 6b + 5d \\ 2a + 2c & 2b + 2d \end{bmatrix} = \begin{bmatrix} 1 & 0 \\ 0 & 1 \end{bmatrix}$. Since corresponding entries must be equal (shown by matching colors), we find the values of a, b, c, and d by solving the systems $\begin{cases} 6a + 5c = 1 \\ 2a + 2c = 0 \end{cases}$ and $\begin{cases} 6b + 5d = 0 \\ 2b + 2d = 1 \end{cases}$. Using $-3R2 + R1$ for the first system shows $a = 1$ and $c = -1$, while $-3R2 + R1$ for the second system shows $b = -2.5$ and $d = 3$. Matrix $B = \begin{bmatrix} a & b \\ c & d \end{bmatrix} = \begin{bmatrix} 1 & -2.5 \\ -1 & 3 \end{bmatrix}$ is the prime candidate for A^{-1}.

To determine if A^{-1} has truly been found, we check to see if multiplication from the right and multiplication from the left yields the matrix I: $AB = BA = I$.

EXAMPLE 2B ▶ **Verifying $B = A^{-1}$**

For the matrices $A = \begin{bmatrix} 6 & 5 \\ 2 & 2 \end{bmatrix}$ and $B = \begin{bmatrix} 1 & -2.5 \\ -1 & 3 \end{bmatrix}$ from Example 2A, determine if $AB = BA = I$.

Solution ▶
$$AB = \begin{bmatrix} 6 & 5 \\ 2 & 2 \end{bmatrix}\begin{bmatrix} 1 & -2.5 \\ -1 & 3 \end{bmatrix} \qquad BA = \begin{bmatrix} 1 & -2.5 \\ -1 & 3 \end{bmatrix}\begin{bmatrix} 6 & 5 \\ 2 & 2 \end{bmatrix}$$
$$= \begin{bmatrix} 6(1)+5(-1) & 6(-2.5)+5(3) \\ 2(1)+2(-1) & 2(-2.5)+2(3) \end{bmatrix} \qquad = \begin{bmatrix} 1(6)+(-2.5)(2) & 1(5)+(-2.5)(2) \\ -1(6)+3(2) & -1(5)+3(2) \end{bmatrix}$$
$$= \begin{bmatrix} 1 & 0 \\ 0 & 1 \end{bmatrix}\checkmark \qquad = \begin{bmatrix} 1 & 0 \\ 0 & 1 \end{bmatrix}\checkmark$$

Since $AB = BA = I$, we conclude $B = A^{-1}$.

Now try Exercises 15 through 22 ▶

These observations guide us to the following definition of an inverse matrix.

The Inverse of a Matrix

Given an $n \times n$ matrix A, if there exists an $n \times n$ matrix A^{-1} such that $AA^{-1} = A^{-1}A = I_n$, then A^{-1} is the inverse of matrix A.

We will soon discover that while only square matrices have inverses, not every square matrix has an inverse. If an inverse exists, the matrix is said to be **invertible.** For 2×2 matrices that are invertible, a simple formula exists for computing the inverse. The formula is derived in the *Strengthening Core Skills* feature at the end of Chapter 6.

The Inverse of a 2 × 2 Matrix

If $A = \begin{bmatrix} a & b \\ c & d \end{bmatrix}$, then $A^{-1} = \dfrac{1}{ad-bc}\begin{bmatrix} d & -b \\ -c & a \end{bmatrix}$ provided $ad - bc \neq 0$

To "test" the formula, again consider the matrix $A = \begin{bmatrix} 6 & 5 \\ 2 & 2 \end{bmatrix}$, where $a = 6$, $b = 5$, $c = 2$, and $d = 2$:

☑ **B. You've just learned how to find the inverse of a square matrix**

$$A^{-1} = \frac{1}{(6)(2)-(5)(2)}\begin{bmatrix} 2 & -5 \\ -2 & 6 \end{bmatrix}$$
$$= \frac{1}{2}\begin{bmatrix} 2 & -5 \\ -2 & 6 \end{bmatrix} = \begin{bmatrix} 1 & -2.5 \\ -1 & 3 \end{bmatrix}\checkmark$$

See Exercises 63 through 66 for more practice with this formula.

Almost without exception, real-world applications involve much larger matrices, with entries that are not integer-valued. Although the *equality of matrices* method from Example 2 can be extended to find the inverse of larger matrices, the process becomes very tedious and too time consuming to be useful. As an alternative, the **augmented matrix method** can be used. This process is discussed in the *Strengthening Core Skills* feature at the end Chapter 6 (see page 614). For practical reasons, we will rely on a calculator to produce these larger inverse matrices. This is done by (1) carefully entering a square matrix A into the calculator, (2) returning to the home screen and (3) calling up matrix A and pressing the $\boxed{x^{-1}}$ key and $\boxed{\text{ENTER}}$ to find A^{-1}. In the context of matrices, calculators are programmed to compute an inverse matrix, rather than to somehow find a reciprocal. **See Exercises 23 through 26.**

C. Solving Systems Using Matrix Equations

One reason matrix multiplication has its row $\times$ column definition is to assist in writing a linear system of equations as a single matrix equation. The equation consists of the matrix of constants B on the right, and a product of the coefficient matrix A with the matrix of variables X on the left: $AX = B$. For $\begin{cases} x + 4y - z = 10 \\ 2x + 5y - 3z = 7, \text{ the matrix} \\ 8x + y - 2z = 11 \end{cases}$

equation is $\begin{bmatrix} 1 & 4 & -1 \\ 2 & 5 & -3 \\ 8 & 1 & -2 \end{bmatrix} \begin{bmatrix} x \\ y \\ z \end{bmatrix} = \begin{bmatrix} 10 \\ 7 \\ 11 \end{bmatrix}$. Note that computing the product on the left

will yield the original system.

Once written as a matrix equation, the system can be solved using an inverse matrix and the following sequence. If A represents the matrix of coefficients, X the matrix of variables, B the matrix of constants, and I the appropriate identity, the sequence is

(1) $\qquad AX = B$ matrix equation

(2) $\quad A^{-1}(AX) = A^{-1}B$ multiply from the left by the inverse of A

(3) $\quad (A^{-1}A)X = A^{-1}B$ associative property

(4) $\qquad IX = A^{-1}B$ $A^{-1}A = I$

(5) $\qquad X = A^{-1}B$ $IX = X$

Lines 1 through 5 illustrate the steps that make the method work. In actual practice, after carefully entering the matrices, only step 5 is used when solving matrix equations using technology. Once matrix A is entered, the calculator will automatically *find* and *use* A^{-1} as we enter $A^{-1}B$.

EXAMPLE 3 ▶ **Using Technology to Solve a Matrix Equation**

Use a calculator and a matrix equation to solve the system

$$\begin{cases} x + 4y - z = 10 \\ 2x + 5y - 3z = 7. \\ 8x + y - 2z = 11 \end{cases}$$

Solution ▶ As before, the matrix equation is $\begin{bmatrix} 1 & 4 & -1 \\ 2 & 5 & -3 \\ 8 & 1 & -2 \end{bmatrix} \begin{bmatrix} x \\ y \\ z \end{bmatrix} = \begin{bmatrix} 10 \\ 7 \\ 11 \end{bmatrix}$.

Carefully enter (and double-check) the matrix of coefficients as matrix A in your calculator, and the matrix of constants as matrix B. The product $A^{-1}B$ shows the solution is $x = 2, y = 3, z = 4$. Verify by substitution.

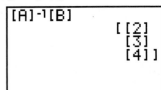

Now try Exercises 27 through 44

Figure 6.11

Figure 6.11

The matrix equation method does have a few shortcomings. Consider the system whose corresponding matrix equation is $\begin{bmatrix} 4 & -10 \\ -2 & 5 \end{bmatrix}\begin{bmatrix} x \\ y \end{bmatrix} = \begin{bmatrix} -8 \\ 13 \end{bmatrix}$. After entering the matrix of coefficients A and matrix of constants B, attempting to compute $A^{-1}B$ results in the error message shown in Figure 6.11. The calculator is unable to return a solution due to something called a **"singular matrix."** To investigate further, we attempt to find A^{-1} for $\begin{bmatrix} 4 & -10 \\ -2 & 5 \end{bmatrix}$ using the formula for a 2×2 matrix. With $a = 4, b = -10, c = -2$, and $d = 5$, we have

$$A^{-1} = \frac{1}{ad-bc}\begin{bmatrix} d & -b \\ -c & a \end{bmatrix} = \frac{1}{(4)(5)-(-10)(-2)}\begin{bmatrix} 5 & 10 \\ 2 & 4 \end{bmatrix}$$

$$= \frac{1}{0}\begin{bmatrix} 5 & 10 \\ 2 & 4 \end{bmatrix}$$

☑ **C. You've just learned how to solve systems using matrix equations**

Since division by zero is undefined, we conclude that matrix A has no inverse. A matrix having no inverse is said to be **singular** or **noninvertible.** Solving systems using matrix equations is only possible when the matrix of coefficients is **nonsingular.**

D. Determinants and Singular Matrices

As a practical matter, it becomes important to know ahead of time whether a particular matrix has an inverse. To help with this, we introduce one additional operation on a square matrix, that of calculating its **determinant.** For a 1×1 matrix the determinant is the entry itself. For a 2×2 matrix $A = \begin{bmatrix} a_{11} & a_{12} \\ a_{21} & a_{22} \end{bmatrix}$, the determinant of A, written as $\det(A)$ or denoted with vertical bars as $|A|$, is computed as *a difference of diagonal products* beginning with the upper-left entry:

2nd diagonal
product

$$\det(A) = \begin{vmatrix} a_{11} & a_{12} \\ a_{21} & a_{22} \end{vmatrix} = a_{11}a_{22} - a_{21}a_{12}$$

1st diagonal
product

The Determinant of a 2 × 2 Matrix

Given any 2×2 matrix $A = \begin{bmatrix} a_{11} & a_{12} \\ a_{21} & a_{22} \end{bmatrix}$,

$$\det(A) = |A| = a_{11}a_{22} - a_{21}a_{12}$$

EXAMPLE 4 ▶ **Calculating Determinants**

Compute the determinant of each matrix given.

a. $B = \begin{bmatrix} 3 & 2 \\ 1 & -6 \end{bmatrix}$

b. $C = \begin{bmatrix} 5 & 2 & 1 \\ -1 & -3 & 4 \end{bmatrix}$

c. $D = \begin{bmatrix} 4 & -10 \\ -2 & 5 \end{bmatrix}$

Solution ▶   **a.** $\det(B) = \begin{vmatrix} 3 & 2 \\ 1 & -6 \end{vmatrix} = (3)(-6) - (1)(2) = -20$

      b. Determinants are only defined for square matrices.

      c. $\det(D) = \begin{vmatrix} 4 & -10 \\ -2 & 5 \end{vmatrix} = (4)(5) - (-2)(-10) = 20 - 20 = 0$

Now try Exercises 45 through 48 ▶

Notice from Example 4c, the determinant of $\begin{bmatrix} 4 & -10 \\ -2 & 5 \end{bmatrix}$ was zero, and this is the same matrix we earlier found had no inverse. This observation can be extended to larger matrices and offers the connection we seek between a given matrix, its inverse, and matrix equations.

Singular Matrices

> If A is a square matrix and $\det(A) = 0$, the inverse matrix *does not exist* and A is said to be *singular* or *noninvertible*.

In summary, inverses exist only for square matrices, but not every square matrix has an inverse. If the determinant of a square matrix is zero, an inverse does not exist and the method of matrix equations cannot be used to solve the system.

To use the determinant test for a 3 × 3 system, we need to compute a 3 × 3 determinant. At first glance, our experience with 2 × 2 determinants appears to be of little help. However, every entry in a 3 × 3 matrix is associated with a smaller 2 × 2 matrix, formed by *deleting the row and column* of that entry and using the entries that remain. These 2 × 2's are called the **associated minor matrices** or simply the **minors.** Using a general matrix of coefficients, we'll identify the minors associated with the entries in the first row.

WORTHY OF NOTE

For the determinant of a general $n \times n$ matrix using **cofactors,** see Appendix II.

$$\begin{bmatrix} \boxed{a_{11}} & a_{12} & a_{13} \\ a_{21} & a_{22} & a_{23} \\ a_{31} & a_{32} & a_{33} \end{bmatrix} \qquad \begin{bmatrix} a_{11} & \boxed{a_{12}} & a_{13} \\ a_{21} & a_{22} & a_{23} \\ a_{31} & a_{32} & a_{33} \end{bmatrix} \qquad \begin{bmatrix} a_{11} & a_{12} & \boxed{a_{13}} \\ a_{21} & a_{22} & a_{23} \\ a_{31} & a_{32} & a_{33} \end{bmatrix}$$

     Entry: a_{11}   **Entry: a_{12}**   **Entry: a_{13}**
   associated minor   **associated minor**   **associated minor**

$$\begin{bmatrix} a_{22} & a_{23} \\ a_{32} & a_{33} \end{bmatrix} \qquad\qquad \begin{bmatrix} a_{21} & a_{23} \\ a_{31} & a_{33} \end{bmatrix} \qquad\qquad \begin{bmatrix} a_{21} & a_{22} \\ a_{31} & a_{32} \end{bmatrix}$$

To illustrate, consider the system shown, and (1) form the matrix of coefficients, (2) identify the minor matrices associated with the entries in the first row, and (3) compute the determinant of each *minor*.

$$\begin{cases} 2x + 3y - z = 1 \\ x - 4y + 2z = -3 \\ 3x + y = -1 \end{cases} \qquad \text{(1) Matrix of coefficients} \begin{bmatrix} 2 & 3 & -1 \\ 1 & -4 & 2 \\ 3 & 1 & 0 \end{bmatrix}$$

$$\text{(2)} \begin{bmatrix} \boxed{2} & 3 & -1 \\ 1 & -4 & 2 \\ 3 & 1 & 0 \end{bmatrix} \qquad \begin{bmatrix} 2 & \boxed{3} & -1 \\ 1 & -4 & 2 \\ 3 & 1 & 0 \end{bmatrix} \qquad \begin{bmatrix} 2 & 3 & \boxed{-1} \\ 1 & -4 & 2 \\ 3 & 1 & 0 \end{bmatrix}$$

    Entry a_{11}: 2   **Entry a_{12}: 3**   **Entry a_{13}: -1**
   associated minor   **associated minor**   **associated minor**

$$\begin{bmatrix} -4 & 2 \\ 1 & 0 \end{bmatrix} \qquad\qquad \begin{bmatrix} 1 & 2 \\ 3 & 0 \end{bmatrix} \qquad\qquad \begin{bmatrix} 1 & -4 \\ 3 & 1 \end{bmatrix}$$

(3) **Determinant** of minor	**Determinant of** minor	**Determinant of** minor
$(-4)(0) - (1)(2) = -2$	$(1)(0) - (3)(2) = -6$	$(1)(1) - (3)(-4) = 13$

For computing a 3×3 determinant, we illustrate a technique called **expansion by minors.**

The Determinant of a 3×3 Matrix—Expansion by Minors

For the matrix M shown, $\det(M)$ is the unique number computed as follows:

matrix M

$$\begin{bmatrix} a_{11} & a_{12} & a_{13} \\ a_{21} & a_{22} & a_{23} \\ a_{31} & a_{32} & a_{33} \end{bmatrix}$$

1. Select any row or column and form the product of each entry with its minor matrix. The illustration here uses the entries in row 1:

$$\det(M) = +a_{11}\begin{vmatrix} a_{22} & a_{23} \\ a_{32} & a_{33} \end{vmatrix} - a_{12}\begin{vmatrix} a_{21} & a_{23} \\ a_{31} & a_{33} \end{vmatrix} + a_{13}\begin{vmatrix} a_{21} & a_{22} \\ a_{31} & a_{32} \end{vmatrix}$$

2. The *signs used between terms* of the expansion depends on the row or column chosen, according to the *sign chart* shown.

Sign Chart

$$\begin{bmatrix} + & - & + \\ - & + & - \\ + & - & + \end{bmatrix}$$

The determinant of a matrix is unique and *any* row or column can be used. For this reason, it's helpful to select the row or column having the most zero, positive, and/or smaller entries.

EXAMPLE 5 ▶ **Calculating a 3×3 Determinant**

Compute the determinant of $M = \begin{bmatrix} 2 & 1 & -3 \\ 1 & -1 & 0 \\ -2 & 1 & 4 \end{bmatrix}$.

Solution ▶ Since the second row has the "smallest" entries as well as a zero entry, we compute the determinant using this row. According to the sign chart, the signs of the terms will be negative–positive–negative, giving

$$\det(M) = -(1)\begin{vmatrix} 1 & -3 \\ 1 & 4 \end{vmatrix} + (-1)\begin{vmatrix} 2 & -3 \\ -2 & 4 \end{vmatrix} - (0)\begin{vmatrix} 2 & 1 \\ -2 & 1 \end{vmatrix}$$
$$= -1(4 + 3) + (-1)(8 - 6) - (0)(2 + 2)$$
$$= \qquad -7 \quad + \qquad (-2) - \qquad 0$$
$$= -9 \rightarrow \text{The value of } \det(M) \text{ is } -9.$$

Now try Exercises 49 through 54 ▶

Try computing the determinant of M two more times, using a different row or column each time. Since the determinant is unique, you should obtain the same result.

There are actually other alternatives for computing a 3×3 determinant. The first is called **determinants by column rotation,** and takes advantage of patterns generated from the expansion of minors. This method is applied to the matrix shown, which uses alphabetical entries for simplicity.

$$\det\begin{bmatrix} a & b & c \\ d & e & f \\ g & h & i \end{bmatrix} = a(ei - fh) - b(di - fg) + c(dh - eg) \qquad \text{expansion using R1}$$
$$= aei - afh - bdi + bfg + cdh - ceg \qquad \text{distribute}$$
$$= aei + bfg + cdh - afh - bdi - ceg \qquad \text{rewrite result}$$

Although history is unsure of who should be credited, notice that if you repeat the first two columns to the right of the given matrix ("rotation of columns"), identical products are obtained using the six diagonals formed—three in the downward direction using addition, three in the upward direction using subtraction.

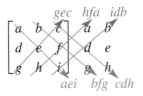

Adding the products in **blue** (regardless of sign) and subtracting the products in **red** (regardless of sign) gives the determinant. This method is more efficient than expansion by minors, *but can only be used for* 3×3 *matrices!*

EXAMPLE 6 ▶ **Calculating det(A) Using Column Rotation**

Use the column rotation method to find the determinant of $A = \begin{bmatrix} 1 & 5 & 3 \\ -2 & -8 & 0 \\ -3 & -11 & 1 \end{bmatrix}$.

Solution ▶ Rotate columns 1 and 2 to the right as above, and compute the diagonal products.

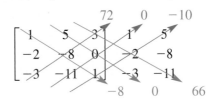

Adding the products in **blue** (regardless of sign) and subtracting the products in **red** (regardless of sign) shows $\det(A) = -4$:

$$-8 + 0 + 66 - 72 - 0 - (-10) = -4.$$

Now try Exercises 55 through 58 ▶

The final method is presented in the *Extending the Concept* feature of the Exercise Set, and shows that if certain conditions are met, the determinant of a matrix can be found using its triangularized form.

EXAMPLE 7 ▶ **Solving a System after Verifying A is Invertible**

Given the system shown here, (1) form the matrix equation $AX = B$; (2) compute the determinant of the coefficient matrix and determine if you can proceed; and (3) if so, solve the system using a matrix equation.

$$\begin{cases} 2x + 1y - 3z = 11 \\ 1x - 1y \qquad = 1 \\ -2x + 1y + 4z = -8 \end{cases}$$

Solution ▶ **1.** Form the matrix equation $AX = B$:

$$\begin{bmatrix} 2 & 1 & -3 \\ 1 & -1 & 0 \\ -2 & 1 & 4 \end{bmatrix} \begin{bmatrix} x \\ y \\ z \end{bmatrix} = \begin{bmatrix} 11 \\ 1 \\ -8 \end{bmatrix}$$

2. Since $\det(A)$ is nonzero (from Example 5), we can proceed.

3. For $X = A^{-1}B$, input $A^{-1}B$ on your calculator and press ENTER .

$$X = \begin{bmatrix} \frac{4}{9} & \frac{7}{9} & \frac{1}{3} \\ \frac{4}{9} & -\frac{2}{9} & \frac{1}{3} \\ \frac{1}{9} & \frac{4}{9} & \frac{1}{3} \end{bmatrix} \begin{bmatrix} 11 \\ 1 \\ -8 \end{bmatrix} = \begin{bmatrix} 3 \\ 2 \\ -1 \end{bmatrix}$$ calculator computes and uses A^{-1} in one step

The solution is the ordered triple $(3, 2, -1)$

Now try Exercises 59 through 62 ▶

We close this section with an application involving a 4×4 system.

EXAMPLE 8 ▶   **Solving an Application Using Technology and Matrix Equations**

A local theater sells four sizes of soft drinks: 32 oz @ \$2.25; 24 oz @ \$1.90; 16 oz @ \$1.50; and 12 oz @ \$1.20/each. As part of a "free guest pass" promotion, the manager asks employees to try and determine the number of each size sold, given the following information: (1) the total revenue from soft drinks was \$719.80; (2) there were 9096 oz of soft drink sold; (3) there was a total of 394 soft drinks sold; and (4) the number of 24-oz and 12-oz drinks sold was 12 more than the number of 32-oz and 16-oz drinks sold. Write a system of equations that models this information, then solve the system using a matrix equation.

Solution ▶   If we let x, l, m, and s represent the number of 32-oz, 24-oz, 16-oz, and 12-oz soft drinks sold, the following system is produced:

$$\begin{array}{rl} \text{revenue:} & \left\{ \begin{array}{l} 2.25x + 1.90l + 1.50m + 1.20s = 719.8 \\ \text{ounces sold:} \quad 32x + 24l + 16m + 12s = 9096 \\ \text{quantity sold:} \quad x + l + m + s = 394 \\ l + s = x + m + 12 \end{array} \right. \end{array}$$

☑ **D.** You've just learned how to use determinants to find whether a matrix is invertible

relationship between amounts sold:

When written as a matrix equation the system becomes:

$$\begin{bmatrix} 2.25 & 1.9 & 1.5 & 1.2 \\ 32 & 24 & 16 & 12 \\ 1 & 1 & 1 & 1 \\ -1 & 1 & -1 & 1 \end{bmatrix} \begin{bmatrix} x \\ l \\ m \\ s \end{bmatrix} = \begin{bmatrix} 719.8 \\ 9096 \\ 394 \\ 12 \end{bmatrix}$$

To solve, carefully enter the matrix of coefficients as matrix A, and the matrix of constants as matrix B, then compute $A^{-1}B = X$ [verify $\det(A) \neq 0$]. This gives a solution of $(x, l, m, s) =$ (112, 151, 79, 52).

Now try Exercises 67 to 78 ▶

6.3 EXERCISES

▶ CONCEPTS AND VOCABULARY

Fill in the blank with the appropriate word or phrase. Carefully reread the section if needed.

1. The $n \times n$ identity matrix I_n consists of 1's down the _____ and _____ for all other entries.

2. Given square matrices A and B of like size, B is the inverse of A if __ = __ = ___. Notationally we write $B =$ ____.

3. The product of a square matrix A and its inverse A^{-1} yields the _____ matrix.

4. If the determinant of a matrix is zero, the matrix is said to be _____ or _____, meaning no inverse exists.

5. Explain why inverses exist only for square matrices, then discuss why some square matrices do not have an inverse. Illustrate each point with an example.

6. What is the connection between the determinant of a 2×2 matrix and the formula for finding its inverse? Use the connection to create a 2×2 matrix that is invertible, and another that is not.

▶ DEVELOPING YOUR SKILLS

Use matrix multiplication, equality of matrices, and the arbitrary matrix given to show that $\begin{bmatrix} a & b \\ c & d \end{bmatrix} = \begin{bmatrix} 1 & 0 \\ 0 & 1 \end{bmatrix}$.

7. $A = \begin{bmatrix} 2 & 5 \\ -3 & -7 \end{bmatrix}\begin{bmatrix} a & b \\ c & d \end{bmatrix} = \begin{bmatrix} 2 & 5 \\ -3 & -7 \end{bmatrix}$

8. $A = \begin{bmatrix} 9 & -7 \\ -5 & 4 \end{bmatrix}\begin{bmatrix} a & b \\ c & d \end{bmatrix} = \begin{bmatrix} 9 & -7 \\ -5 & 4 \end{bmatrix}$

9. $A = \begin{bmatrix} 0.4 & 0.6 \\ 0.3 & 0.2 \end{bmatrix}\begin{bmatrix} a & b \\ c & d \end{bmatrix} = \begin{bmatrix} 0.4 & 0.6 \\ 0.3 & 0.2 \end{bmatrix}$

10. $A = \begin{bmatrix} \frac{1}{2} & \frac{1}{4} \\ \frac{1}{3} & \frac{1}{8} \end{bmatrix}\begin{bmatrix} a & b \\ c & d \end{bmatrix} = \begin{bmatrix} \frac{1}{2} & \frac{1}{4} \\ \frac{1}{3} & \frac{1}{8} \end{bmatrix}$

For $I_2 = \begin{bmatrix} 1 & 0 \\ 0 & 1 \end{bmatrix}, I_3 = \begin{bmatrix} 1 & 0 & 0 \\ 0 & 1 & 0 \\ 0 & 0 & 1 \end{bmatrix}$, **and**

$I_4 = \begin{bmatrix} 1 & 0 & 0 & 0 \\ 0 & 1 & 0 & 0 \\ 0 & 0 & 1 & 0 \\ 0 & 0 & 0 & 1 \end{bmatrix}$, **show** $AI = IA = A$ **for the**

matrices of like size. Use a calculator for Exercise 14.

11. $\begin{bmatrix} -3 & 8 \\ -4 & 10 \end{bmatrix}$ 12. $\begin{bmatrix} 0.5 & -0.2 \\ -0.7 & 0.3 \end{bmatrix}$

13. $\begin{bmatrix} -4 & 1 & 6 \\ 9 & 5 & 3 \\ 0 & -2 & 1 \end{bmatrix}$ 14. $\begin{bmatrix} 9 & 1 & 3 & -1 \\ 2 & 0 & -5 & 3 \\ 4 & 6 & 1 & 0 \\ 0 & -2 & 4 & 1 \end{bmatrix}$

Find the inverse of each 2×2 matrix using matrix multiplication, equality of matrices, and a system of equations.

15. $\begin{bmatrix} 5 & -4 \\ 2 & 2 \end{bmatrix}$ 16. $\begin{bmatrix} 1 & -5 \\ 0 & -4 \end{bmatrix}$

17. $\begin{bmatrix} 1 & -3 \\ 4 & -10 \end{bmatrix}$ 18. $\begin{bmatrix} -2 & 0.4 \\ 1 & 0.8 \end{bmatrix}$

Demonstrate that $B = A^{-1}$, by showing $AB = BA = I$. Do not use a calculator.

19. $A = \begin{bmatrix} 1 & 5 \\ -2 & -9 \end{bmatrix}$ 20. $A = \begin{bmatrix} -2 & -6 \\ 4 & 11 \end{bmatrix}$

$B = \begin{bmatrix} -9 & -5 \\ 2 & 1 \end{bmatrix}$ $B = \begin{bmatrix} 5.5 & 3 \\ -2 & -1 \end{bmatrix}$

21. $A = \begin{bmatrix} 4 & -5 \\ 0 & 2 \end{bmatrix}$ 22. $A = \begin{bmatrix} -2 & 5 \\ 3 & -4 \end{bmatrix}$

$B = \begin{bmatrix} \frac{1}{4} & \frac{5}{8} \\ 0 & \frac{1}{2} \end{bmatrix}$ $B = \begin{bmatrix} \frac{4}{7} & \frac{5}{7} \\ \frac{3}{7} & \frac{2}{7} \end{bmatrix}$

Use a calculator to find $A^{-1} = B$, then confirm the inverse by showing $AB = BA = I$.

23. $A = \begin{bmatrix} -2 & 3 & 1 \\ 5 & 2 & 4 \\ 2 & 0 & -1 \end{bmatrix}$

24. $A = \begin{bmatrix} 0.5 & 0.2 & 0.1 \\ 0 & 0.3 & 0.6 \\ 1 & 0.4 & -0.3 \end{bmatrix}$

25. $A = \begin{bmatrix} -7 & 5 & -3 \\ 1 & 9 & 0 \\ 2 & -2 & -5 \end{bmatrix}$

26. $A = \dfrac{1}{12} \begin{bmatrix} 12 & -6 & 3 & 0 \\ 0 & -4 & 8 & -12 \\ 12 & -12 & 0 & 0 \\ 0 & 12 & 0 & -12 \end{bmatrix}$

Write each system in the form of a matrix equation. Do not solve.

27. $\begin{cases} 2x - 3y = 9 \\ -5x + 7y = 8 \end{cases}$

28. $\begin{cases} 0.5x - 0.6y = 0.6 \\ -0.7x + 0.4y = -0.375 \end{cases}$

29. $\begin{cases} x + 2y - z = 1 \\ x + z = 3 \\ 2x - y + z = 3 \end{cases}$

30. $\begin{cases} 2x - 3y - 2z = 4 \\ \frac{1}{4}x - \frac{2}{5}y + \frac{3}{4}z = \frac{-1}{3} \\ -2x + 1.3y - 3z = 5 \end{cases}$

31. $\begin{cases} -2w + x - 4y + 5z = -3 \\ 2w - 5x + y - 3z = 4 \\ -3w + x + 6y + z = 1 \\ w + 4x - 5y + z = -9 \end{cases}$

32. $\begin{cases} 1.5w + 2.1x - 0.4y + z = 1 \\ 0.2w - 2.6x + y = 5.8 \\ 3.2x + z = 2.7 \\ 1.6w + 4x - 5y + 2.6z = -1.8 \end{cases}$

Write each system as a matrix equation and solve (if possible) using inverse matrices and your calculator. If the coefficient matrix is singular, write *no solution*.

33. $\begin{cases} 0.05x - 3.2y = -15.8 \\ 0.02x + 2.4y = 12.08 \end{cases}$

34. $\begin{cases} 0.3x + 1.1y = 3.5 \\ -0.5x - 2.9y = -10.1 \end{cases}$

35. $\begin{cases} \frac{-1}{6}u + \frac{1}{4}v = 1 \\ \frac{1}{2}u - \frac{2}{3}v = -2 \end{cases}$

36. $\begin{cases} \sqrt{2}a + \sqrt{3}b = \sqrt{5} \\ \sqrt{6}a + 3b = \sqrt{7} \end{cases}$

37. $\begin{cases} \frac{-1}{8}a + \frac{3}{5}b = \frac{5}{6} \\ \frac{5}{16}a - \frac{3}{2}b = \frac{-4}{5} \end{cases}$

38. $\begin{cases} 3\sqrt{2}a + 2\sqrt{3}b = 12 \\ 5\sqrt{2}a - 3\sqrt{3}b = 1 \end{cases}$

39. $\begin{cases} 0.2x - 1.6y + 2z = -1.9 \\ -0.4x - y + 0.6z = -1 \\ 0.8x + 3.2y - 0.4z = 0.2 \end{cases}$

40. $\begin{cases} 1.7x + 2.3y - 2z = 41.5 \\ 1.4x - 0.9y + 1.6z = -10 \\ -0.8x + 1.8y - 0.5z = 16.5 \end{cases}$

41. $\begin{cases} x - 2y + 2z = 6 \\ 2x - 1.5y + 1.8z = 2.8 \\ \frac{-2}{3}x + \frac{1}{2}y - \frac{3}{5}z = -\frac{11}{30} \end{cases}$

42. $\begin{cases} 4x - 5y - 6z = 5 \\ \frac{1}{8}x - \frac{3}{5}y + \frac{5}{4}z = \frac{-2}{3} \\ -0.5x + 2.4y - 5z = 5 \end{cases}$

43. $\begin{cases} -2w + 3x - 4y + 5z = -3 \\ 0.2w - 2.6x + y - 0.4z = 2.4 \\ -3w + 3.2x + 2.8y + z = 6.1 \\ 1.6w + 4x - 5y + 2.6z = -9.8 \end{cases}$

44. $\begin{cases} 2w - 5x + 3y - 4z = 7 \\ 1.6w + 4.2y - 1.8z = 5.4 \\ 3w + 6.7x - 9y + 4z = -8.5 \\ 0.7x - 0.9z = 0.9 \end{cases}$

Compute the determinant of each matrix and state whether an inverse matrix exists. Do not use a calculator.

45. $\begin{bmatrix} 4 & -7 \\ 3 & -5 \end{bmatrix}$

46. $\begin{bmatrix} 0.6 & 0.3 \\ 0.4 & 0.5 \end{bmatrix}$

47. $\begin{bmatrix} 1.2 & -0.8 \\ 0.3 & -0.2 \end{bmatrix}$

48. $\begin{bmatrix} -2 & 6 \\ -3 & 9 \end{bmatrix}$

Compute the determinant of each matrix without using a calculator. If the determinant is zero, write *singular matrix*.

49. $A = \begin{bmatrix} 1 & 0 & -2 \\ 0 & -1 & -1 \\ 2 & 1 & -4 \end{bmatrix}$

50. $B = \begin{bmatrix} -2 & 2 & 1 \\ 0 & -1 & 2 \\ 4 & -4 & 0 \end{bmatrix}$

51. $C = \begin{bmatrix} -2 & 3 & 4 \\ 0 & 6 & 2 \\ 1 & -1.5 & -2 \end{bmatrix}$

52. $D = \begin{bmatrix} 1 & 2 & -0.8 \\ 2.5 & 5 & -2 \\ 3 & 0 & -2.5 \end{bmatrix}$

Use a calculator to compute the determinant of each matrix. If the determinant is zero, write *singular matrix*. If the determinant is nonzero, find A^{-1} and store the result as matrix B (STO→ 2nd X⁻¹ 2: [B] ENTER). Then verify each inverse by showing $AB = BA = I$.

53. $A = \begin{bmatrix} 1 & 0 & 3 & -4 \\ 2 & 5 & 0 & 1 \\ 8 & 15 & 6 & -5 \\ 0 & 8 & -4 & 1 \end{bmatrix}$

54. $M = \begin{bmatrix} 1 & 2 & 1 & 1 \\ 0 & 1 & -3 & 2 \\ -1 & 0 & 2 & -3 \\ 2 & -1 & 1 & 4 \end{bmatrix}$

Compute the determinant of each matrix using the column rotation method.

55. $\begin{bmatrix} 2 & -3 & 1 \\ 4 & -1 & 5 \\ 1 & 0 & -2 \end{bmatrix}$ **56.** $\begin{bmatrix} -3 & 2 & 4 \\ -1 & -2 & 0 \\ 3 & 1 & 5 \end{bmatrix}$

57. $\begin{bmatrix} 1 & -1 & 2 \\ 3 & -2 & 4 \\ 4 & 3 & 1 \end{bmatrix}$ **58.** $\begin{bmatrix} 5 & 6 & 2 \\ -2 & 1 & -2 \\ 3 & 4 & -1 \end{bmatrix}$

For each system shown, form the matrix equation $AX = B$; compute the determinant of the coefficient matrix and determine if you can proceed; and if possible, solve the system using the matrix equation.

59. $\begin{cases} x - 2y + 2z = 7 \\ 2x + 2y - z = 5 \\ 3x - y + z = 6 \end{cases}$ **60.** $\begin{cases} 2x - 3y - 2z = 7 \\ x - y + 2z = -5 \\ 3x + 2y - z = 11 \end{cases}$

61. $\begin{cases} x - 3y + 4z = -1 \\ 4x - y + 5z = 7 \\ 3x + 2y + z = -3 \end{cases}$ **62.** $\begin{cases} 5x - 2y + z = 1 \\ 3x - 4y + 9z = -2 \\ 4x - 3y + 5z = 6 \end{cases}$

▶ WORKING WITH FORMULAS

The inverse of a 2 × 2 matrix:

$$A = \begin{bmatrix} a & b \\ c & d \end{bmatrix} \rightarrow A^{-1} = \frac{1}{ad - bc} \cdot \begin{bmatrix} d & -b \\ -c & a \end{bmatrix}$$

The inverse of a 2 × 2 matrix can be found using the formula shown, as long as $ad - bc \neq 0$. Use the formula to find inverses for the matrices here (if possible), then verify by showing $A \cdot A^{-1} = A \cdot A^{-1} = I$.

63. $A = \begin{bmatrix} 3 & -5 \\ 2 & 1 \end{bmatrix}$ **64.** $B = \begin{bmatrix} 2 & 3 \\ -5 & -4 \end{bmatrix}$

65. $C = \begin{bmatrix} 0.3 & -0.4 \\ -0.6 & 0.8 \end{bmatrix}$ **66.** $\begin{bmatrix} 0.2 & 0.3 \\ -0.4 & -0.6 \end{bmatrix}$

▶ APPLICATIONS

Solve each application using a matrix equation.

Descriptive Translation

67. Convenience store sales: The local Moto-Mart sells four different sizes of Slushies—behemoth, 60 oz @ \$2.59; gargantuan, 48 oz @ \$2.29; mammoth, 36 oz @ \$1.99; and jumbo, 24 oz @ \$1.59. As part of a promotion, the owner offers free gas to any customer who can tell how many of each size were sold last week, given the following information: (1) The total revenue for the Slushies was \$402.29; (2) 7884 ounces were sold; (3) a total of 191 Slushies were sold; and (4) the number of behemoth Slushies sold was one more than the number of jumbo. How many of each size were sold?

68. Cartoon characters: In America, four of the most beloved cartoon characters are Foghorn Leghorn, Elmer Fudd, Bugs Bunny, and Tweety Bird. Suppose that Bugs Bunny is four times as tall as Tweety Bird. Elmer Fudd is as tall as the combined height of Bugs Bunny and Tweety Bird. Foghorn Leghorn is 20 cm taller than the combined height of Elmer Fudd and Tweety Bird. The combined height of all four characters is 500 cm. How tall is each one?

69. Rolling Stones music: One of the most prolific and popular rock-and-roll bands of all time is the Rolling Stones. Four of their many great hits include: *Jumpin' Jack Flash*, *Tumbling Dice*, *You Can't Always Get What You Want*, and *Wild Horses*. The total playing time of all four songs is 20.75

min. The combined playing time of *Jumpin' Jack Flash* and *Tumbling Dice* equals that of *You Can't Always Get What You Want*. *Wild Horses* is 2 min longer than *Jumpin' Jack Flash,* and *You Can't Always Get What You Want* is twice as long as *Tumbling Dice.* Find the playing time of each song.

70. Mozart's arias: Mozart wrote some of vocal music's most memorable arias in his operas, including *Tamino's Aria, Papageno's Aria,* the *Champagne Aria,* and the *Catalogue Aria.* The total playing time of all four arias is 14.3 min. *Papageno's Aria* is 3 min shorter than the *Catalogue Aria.* The *Champagne Aria* is 2.7 min shorter than *Tamino's Aria.* The combined time of *Tamino's Aria* and *Papageno's Aria* is five times that of the *Champagne Aria.* Find the playing time of all four arias.

Manufacturing

71. Resource allocation: Time Pieces Inc. manufactures four different types of grandfather clocks. Each clock requires these four stages: (1) assembly, (2) installing the clockworks, (3) inspection and testing, and (4) packaging for delivery. The time required for each stage is shown in the table, for each of the four clock types. At the end of a busy week, the owner determines that personnel on the assembly line worked for 262 hours, the installation crews for 160 hours, the testing department for 29 hours, and the packaging department for 68 hours. How many clocks of each type were made?

Dept.	Clock A	Clock B	Clock C	Clock D
Assemble	2.2	2.5	2.75	3
Install	1.2	1.4	1.8	2
Test	0.2	0.25	0.3	0.5
Pack	0.5	0.55	0.75	1.0

72. Resource allocation: Figurines Inc. makes and sells four sizes of metal figurines, mostly historical figures and celebrities. Each figurine goes through four stages of development: (1) casting, (2) trimming, (3) polishing, and (4) painting. The time required for each stage is shown in the table, for each of the four sizes. At the end of a busy week, the manager finds that the casting department put in 62 hr, and the trimming department worked for 93.5 hr, with the polishing and painting departments logging 138 hr and 358 hr respectively. How many figurines of each type were made?

Dept.	Small	Medium	Large	X-Large
Casting	0.5	0.6	0.75	1
Trimming	0.8	0.9	1.1	1.5
Polishing	1.2	1.4	1.7	2
Painting	2.5	3.5	4.5	6

73. Thermal conductivity: In lab experiments designed to measure the heat conductivity of a square metal plate of uniform density, the edges are held at four different (constant) temperatures. The *mean-value principle* from physics tells us that the temperature at a given point p_i on the plate, is equal to the average temperature of nearby points. Use this information to form a system of four equations in four variables, and determine the temperature at interior points p_1, p_2, p_3, and p_4 on the plate shown. (*Hint:* Use the temperature of the four points closest to each.)

74. Thermal conductivity: Repeat Exercise 73 if (a) the temperatures at the top and bottom of the plate were *increased* by 10°, with the temperatures at the left and right edges *decreased* by 10° (what do you notice?); (b) the temperature at the top and the temperature to the left were *decreased* by 10°, with the temperatures at the bottom and right held at their original temperature.

Curve Fitting

75. Cubic fit: Find a cubic function of the form $y = ax^3 + bx^2 + cx + d$ such that $(-4, -6)$, $(-1, 0)$, $(1, -16)$, and $(3, 8)$ are on the graph of the function.

76. Cubic fit: Find a cubic function of the form $y = ax^3 + bx^2 + cx + d$ such that $(-2, 5)$, $(0, 1)$, $(2, -3)$, and $(3, 25)$ are on the graph of the function.

Nutrition

77. Animal diets: A zoo dietician needs to create a specialized diet that regulates an animal's intake of fat, carbohydrates, and protein during a meal. The table given shows three different foods and the amount of these nutrients (in grams) that each ounce of food provides. How many ounces of each should the dietician recommend to supply 20 g of fat, 30 g of carbohydrates, and 44 g of protein?

Nutrient	Food I	Food II	Food III
Fat	2	4	3
Carb.	4	2	5
Protein	5	6	7

78. Training diet: A physical trainer is designing a workout diet for one of her clients, and wants to supply him with 24 g of fat, 244 g of carbohydrates, and 40 g of protein for the noontime meal. The table given shows three different foods and the amount of these nutrients (in grams) that each ounce of food provides. How many ounces of each should the trainer recommend?

Nutrient	Food I	Food II	Food III
Fat	2	5	0
Carb.	10	15	18
Protein	2	10	0.75

▶ EXTENDING THE CONCEPT

79. Some matrix applications require that you solve a matrix equation of the form $AX + B = C$, where A, B, and C are matrices with the appropriate number of rows and columns and A^{-1} exists. Investigate the solution process for such equations using $A = \begin{bmatrix} 2 & 3 \\ -5 & -4 \end{bmatrix}$, $B = \begin{bmatrix} 4 \\ 9 \end{bmatrix}$, $C = \begin{bmatrix} 12 \\ -4 \end{bmatrix}$, and $X = \begin{bmatrix} x \\ y \end{bmatrix}$, then solve $AX + B = C$ for X symbolically (using A^{-1}, I, and so on).

80. It is possible for the matrix of coefficients to be singular, yet for solutions to exist. If the system is dependent instead of inconsistent, there may be infinitely many solutions and the solution set must be written using a parameter or the set notation seen previously. Try solving the exercise given here using matrix equations. If this is not possible, discuss why, then solve using elimination. If the system is dependent, find at least *two* sets of three fractions that fit the criteria. The sum of the two smaller fractions equals the larger, the larger less the smaller equals the "middle" fraction, and four times the smaller fraction equals the sum of the other two.

81. Another alternative for finding determinants uses the triangularized form of a matrix and is offered without proof: *If nonsingular matrix A is written in triangularized form without exchanging any rows and without using the operations kR_i to replace any row (k a constant), then det(A) is equal to the product of resulting diagonal entries.* Compute the determinant of each matrix using this method. Be careful not to interchange rows and do not replace any row by a multiple of that row in the process.

a. $\begin{bmatrix} 1 & -2 & 3 \\ -4 & 5 & -6 \\ 2 & 5 & 3 \end{bmatrix}$ $^{-45}$ **b.** $\begin{bmatrix} 2 & 5 & -1 \\ -2 & -3 & 4 \\ 4 & 6 & 5 \end{bmatrix}$

c. $\begin{bmatrix} -2 & 4 & 1 \\ 5 & 7 & -2 \\ 3 & -8 & -1 \end{bmatrix}$ **d.** $\begin{bmatrix} 3 & -1 & 4 \\ 0 & -2 & 6 \\ -2 & 1 & -3 \end{bmatrix}$

82. Find 2×2 nonzero matrices A and B whose product gives the zero matrix $\begin{bmatrix} 0 & 0 \\ 0 & 0 \end{bmatrix}$.

▶ MAINTAINING YOUR SKILLS

83. (3.3) Solve using the rational zeroes theorem:
$x^3 - 7x^2 = -36$

84. (2.5/4.3) Match each equation to its related graph. Justify your answers.

$y = \log_2(x - 2)$ $y = \log_2 x - 2$

a.

b.

85. (1.3) Solve the absolute value inequality:
$-3|2x + 5| - 7 \leq -19$.

86. (3.8) A coin collector believes that the value of a coin varies inversely as the number of coins still in circulation. If 4 million coins are in circulation, the coin has a value of $25, (a) find the variation equation and (b) determine how many coins are in circulation if the value of the coin is $6.25.

Learning Objectives

In Section 6.4 you will learn how to:

☐ **A.** Solve a system using determinants and Cramer's rule

☐ **B.** Decompose a rational expression into partial fractions

☐ **C.** Use determinants in applications involving geometry in the coordinate plane

In addition to solving systems, matrices can be used to accomplish such diverse things as finding the volume of a three-dimensional solid or establishing certain geometrical relationships in the coordinate plane. Numerous uses are also found in higher mathematics, such as checking whether solutions to a differential equation are linearly independent.

A. Solving Systems Using Determinants and Cramer's Rule

In addition to identifying singular matrices, determinants can actually be used to *develop a formula* for the solution of a system. Consider the following solution to a *general* 2×2 system, which parallels the solution to a specific 2×2 system. With a view toward a solution involving determinants, the coefficients of x are written as a_{11} and a_{21} in the general system, and the coefficients of y are a_{12} and a_{22}.

Specific System

$$\begin{cases} 2x + 5y = 9 \\ 3x + 4y = 10 \end{cases}$$

eliminate the *x*-term
$$-3R1 + 2R2$$

sums to zero
$$\begin{cases} \boxed{-3 \cdot 2x} \ - 3 \cdot 5y = -3 \cdot 9 \\ \boxed{2 \cdot 3x} \ + 2 \cdot 4y = 2 \cdot 10 \end{cases}$$
$$2 \cdot 4y - 3 \cdot 5y = 2 \cdot 10 - 3 \cdot 9$$

General System

$$\begin{cases} a_{11}x + a_{12}y = c_1 \\ a_{21}x + a_{22}y = c_2 \end{cases}$$

eliminate the *x*-term
$$-a_{21}R1 + a_{11}R2$$

sums to zero
$$\begin{cases} \boxed{-a_{21}a_{11}x} \ - a_{21}a_{12}y = -a_{21}c_1 \\ \boxed{a_{11}a_{21}x} \ + a_{11}a_{22}y = a_{11}c_2 \end{cases}$$
$$a_{11}a_{22}y - a_{21}a_{12}y = a_{11}c_2 - a_{21}c_1$$

Notice the *x*-terms sum to zero in both systems. We are deliberately leaving the solution on the left unsimplified to show the pattern developing on the right. Next we solve for *y*.

Factor Out *y*

$$(2 \cdot 4 - 3 \cdot 5)y = 2 \cdot 10 - 3 \cdot 9$$

$$y = \frac{2 \cdot 10 - 3 \cdot 9}{2 \cdot 4 - 3 \cdot 5} \quad \text{divide}$$

Factor Out *y*

$$(a_{11}a_{22} - a_{21}a_{12})y = a_{11}c_2 - a_{21}c_1$$

$$\text{divide} \quad y = \frac{a_{11}c_2 - a_{21}c_1}{a_{11}a_{22} - a_{21}a_{12}}$$

On the left we find $y = \frac{-7}{-7} = 1$ and back-substitution shows $x = 2$. But more importantly, on the right we obtain a formula for the *y*-value of *any* 2×2 system: $y = \dfrac{a_{11}c_2 - a_{21}c_1}{a_{11}a_{22} - a_{21}a_{12}}$. If we had chosen to solve for *x*, the solution would be $x = \dfrac{a_{22}c_1 - a_{12}c_2}{a_{11}a_{22} - a_{21}a_{12}}$. Note these formulas are defined only if $a_{11}a_{22} - a_{21}a_{12} \neq 0$. You may have already noticed, but this denominator is the *determinant of the matrix of coefficients* $\begin{bmatrix} a_{11} & a_{12} \\ a_{21} & a_{22} \end{bmatrix}$ from the previous section! Since the numerator is also a difference of two products, we investigate the possibility that it too can be expressed as a determinant. Working backward, we're able to reconstruct the numerator for *x* in determinant form as $\begin{bmatrix} c_1 & a_{12} \\ c_2 & a_{22} \end{bmatrix}$, where it is apparent this matrix was formed by *replacing the coefficients of the x-variables with the constant terms.*

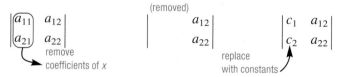

It is also apparent the numerator for y can be also written in determinant form as $\begin{vmatrix} a_{11} & c_1 \\ a_{21} & c_2 \end{vmatrix}$, or the determinant formed by *replacing the coefficients of the y-variables with the constant terms*:

$$\begin{vmatrix} a_{11} & \boxed{a_{12}} \\ a_{21} & \boxed{a_{22}} \end{vmatrix} \qquad \begin{vmatrix} a_{11} & \overset{\text{(removed)}}{} \\ a_{21} & \end{vmatrix} \qquad \begin{vmatrix} a_{11} & c_1 \\ a_{21} & c_2 \end{vmatrix}$$

remove coefficients of y replace with constants

If we use the notation D_y for this determinant, D_x for the determinant where x coefficients were replaced by the constants, and D as the determinant for the matrix of coefficients—the solutions can be written as shown next, with the result known as **Cramer's rule.**

Cramer's Rule for 2 × 2 Systems

Given a 2 × 2 system of linear equations

$$\begin{cases} a_{11}x + a_{12}y = c_1 \\ a_{21}x + a_{22}y = c_2 \end{cases}$$

the solution is the ordered pair (x, y), where

$$x = \frac{D_x}{D} = \frac{\begin{vmatrix} c_1 & a_{12} \\ c_2 & a_{22} \end{vmatrix}}{\begin{vmatrix} a_{11} & a_{12} \\ a_{21} & a_{22} \end{vmatrix}} \quad \text{and} \quad y = \frac{D_y}{D} = \frac{\begin{vmatrix} a_{11} & c_1 \\ a_{21} & c_2 \end{vmatrix}}{\begin{vmatrix} a_{11} & a_{12} \\ a_{21} & a_{22} \end{vmatrix}}$$

provided $D \neq 0$.

EXAMPLE 1 ▶ Solving a System Using Cramer's Rule

Use Cramer's rule to solve the system $\begin{cases} 2x - 5y = 9 \\ -3x + 4y = -10 \end{cases}$.

Solution ▶ Begin by finding the value of D, D_x, and D_y.

$$D = \begin{vmatrix} 2 & -5 \\ -3 & 4 \end{vmatrix} \qquad D_x = \begin{vmatrix} 9 & -5 \\ -10 & 4 \end{vmatrix} \qquad D_y = \begin{vmatrix} 2 & 9 \\ -3 & -10 \end{vmatrix}$$

$$(2)(4) - (-3)(-5) \qquad (9)(4) - (-10)(-5) \qquad (2)(-10) - (-3)(9)$$

$$= -7 \qquad\qquad\qquad = -14 \qquad\qquad\qquad = 7$$

This gives $x = \dfrac{D_x}{D} = \dfrac{-14}{-7} = 2$ and $y = \dfrac{D_y}{D} = \dfrac{7}{-7} = -1$. The solution is $(2, -1)$. Check by substituting these values into the original equations.

Now try Exercises 7 through 14 ▶

Regardless of the method used to solve a system, always be aware that a consistent, inconsistent, or dependent system is possible. The system $\begin{cases} y - 2x = -3 \\ 4x + 6 = 2y \end{cases}$ yields

$\begin{cases} -2x + y = -3 \\ 4x - 2y = -6 \end{cases}$ in standard form, with $D = \begin{vmatrix} -2 & 1 \\ 4 & -2 \end{vmatrix} = (-2)(-2) - (4)(1) = 0$.

Since $\det(D) = 0$, Cramer's rule cannot be applied, and the system is either inconsistent or dependent. To find out which, we write the equations in function form (solve for y). The result is $\begin{cases} y = 2x - 3 \\ y = 2x + 3 \end{cases}$, showing the system consists of two parallel lines and has no solutions.

Cramer's Rule for 3 × 3 Systems

Cramer's rule can be extended to a 3 × 3 system of linear equations, using the same pattern as for 2 × 2 systems. Given the general 3 × 3 system

$$\begin{cases} a_{11}x + a_{12}y + a_{13}z = c_1 \\ a_{21}x + a_{22}y + a_{23}z = c_2 \\ a_{31}x + a_{32}y + a_{33}z = c_3 \end{cases}$$

the solutions are $x = \dfrac{D_x}{D}$, $y = \dfrac{D_y}{D}$, and $z = \dfrac{D_z}{D}$, where D_x, D_y, and D_z are again formed by replacing the coefficients of the indicated variable with the constants, and D is the determinant of the matrix of coefficients ($D \neq 0$).

Cramer's Rule Applied to 3 × 3 Systems

Given a 3 × 3 system of linear equations

$$\begin{cases} a_{11}x + a_{12}y + a_{13}z = c_1 \\ a_{21}x + a_{22}y + a_{23}z = c_2 \\ a_{31}x + a_{32}y + a_{33}z = c_3 \end{cases}$$

The solution is an ordered triple (x, y, z), where

$$x = \frac{\begin{vmatrix} c_1 & a_{12} & a_{13} \\ c_2 & a_{22} & a_{23} \\ c_3 & a_{32} & a_{33} \end{vmatrix}}{\begin{vmatrix} a_{11} & a_{12} & a_{13} \\ a_{21} & a_{22} & a_{23} \\ a_{31} & a_{32} & a_{33} \end{vmatrix}} \quad y = \frac{\begin{vmatrix} a_{11} & c_1 & a_{13} \\ a_{21} & c_2 & a_{23} \\ a_{31} & c_3 & a_{33} \end{vmatrix}}{\begin{vmatrix} a_{11} & a_{12} & a_{13} \\ a_{21} & a_{22} & a_{23} \\ a_{31} & a_{32} & a_{33} \end{vmatrix}} \quad z = \frac{\begin{vmatrix} a_{11} & a_{12} & c_1 \\ a_{21} & a_{22} & c_2 \\ a_{31} & a_{32} & c_3 \end{vmatrix}}{\begin{vmatrix} a_{11} & a_{12} & a_{13} \\ a_{21} & a_{22} & a_{23} \\ a_{31} & a_{32} & a_{33} \end{vmatrix}},$$

☑ **A.** You've just learned how to solve a system using determinants and Cramer's rule

provided $D \neq 0$.

EXAMPLE 2 ▶ Solving a 3 × 3 System Using Cramer's Rule

Solve using Cramer's rule: $\begin{cases} x - 2y + 3z = -1 \\ -2x + y - 5z = 1 \\ 3x + 3y + 4z = 2 \end{cases}$

Solution ▶ Begin by computing the determinant of the matrix of coefficients, to ensure that Cramer's rule can be applied. Using the third row, we have

$$D = \begin{vmatrix} 1 & -2 & 3 \\ -2 & 1 & -5 \\ 3 & 3 & 4 \end{vmatrix} = +3\begin{vmatrix} -2 & 3 \\ 1 & -5 \end{vmatrix} - 3\begin{vmatrix} 1 & 3 \\ -2 & -5 \end{vmatrix} + 4\begin{vmatrix} 1 & -2 \\ -2 & 1 \end{vmatrix}$$

$$= 3(7) - 3(1) + 4(-3) = 6$$

Since $D \neq 0$ we continue, electing to compute the remaining determinants using a calculator.

$$D_x = \begin{vmatrix} -1 & -2 & 3 \\ 1 & 1 & -5 \\ 2 & 3 & 4 \end{vmatrix} = 12 \quad D_y = \begin{vmatrix} 1 & -1 & 3 \\ -2 & 1 & -5 \\ 3 & 2 & 4 \end{vmatrix} = 0 \quad D_z = \begin{vmatrix} 1 & -2 & -1 \\ -2 & 1 & 1 \\ 3 & 3 & 2 \end{vmatrix} = -6$$

The solution is $x = \dfrac{D_x}{D} = \dfrac{12}{6} = 2$, $y = \dfrac{D_y}{D} = \dfrac{0}{6} = 0$, and $z = \dfrac{D_z}{D} = \dfrac{-6}{6} = -1$, or $(2, 0, -1)$ in triple form. Check this solution in the original equations.

Now try Exercises 15 through 22 ▶

B. Rational Expressions and Partial Fractions

Recall that a rational expression is one of the form $\dfrac{P(x)}{Q(x)}$, where P and Q are polynomials and $Q(x) \neq 0$. The addition of rational expressions is widely taught in courses prior to college algebra, and involves combining two rational expressions into a single term using a common denominator. In some applications of higher mathematics, we seek to reverse this process and *decompose* a rational expression into a sum of its **partial fractions**. To begin, we make the following observations:

1. Consider the sum $\dfrac{7}{x+2} + \dfrac{5}{x-3}$, noting both terms are proper fractions (the degree of the numerator is less than the degree of the denominator) and have distinct linear denominators.

$$\dfrac{7}{x+2} + \dfrac{5}{x-3} = \dfrac{7(x-3)}{(x+2)(x-3)} + \dfrac{5(x+2)}{(x-3)(x+2)} \qquad \text{common denominators}$$

$$= \dfrac{7(x-3) + 5(x+2)}{(x+2)(x-3)} \qquad \text{combine numerators}$$

$$= \dfrac{12x - 11}{(x+2)(x-3)} \qquad \text{result}$$

Assuming we didn't have the original sum to look at, reversing the process would require us to begin with a **decomposition template** such as

$$\dfrac{12x - 11}{(x+2)(x-3)} = \dfrac{A}{x+2} + \dfrac{B}{x-3}$$

and solve for the *constants* A and B. We know the numerators must be constant, else the fraction(s) would be improper while the original expression is not.

2. Consider the sum $\dfrac{3}{x-1} + \dfrac{5}{x^2 - 2x + 1}$, again noting both terms are proper fractions.

$$\dfrac{3}{x-1} + \dfrac{5}{x^2 - 2x + 1} = \dfrac{3}{x-1} + \dfrac{5}{(x-1)(x-1)} \qquad \text{factor denominators}$$

$$= \dfrac{3(x-1)}{(x-1)(x-1)} + \dfrac{5}{(x-1)(x-1)} \qquad \text{common denominators}$$

$$= \dfrac{(3x-3) + 5}{(x-1)(x-1)} \qquad \text{combine numerators}$$

$$= \dfrac{3x + 2}{(x-1)^2} \qquad \text{result}$$

Note that while the new denominator is the repeated factor $(x-1)^2$, *both* $(x-1)$ and $(x-1)^2$ were denominators in the original sum. Assuming we didn't know the original sum, reversing the process would require us to begin with the template

$$\dfrac{3x + 2}{(x-1)^2} = \dfrac{A}{x-1} + \dfrac{B}{(x-1)^2}$$

and solve for the constants A and B. As with observation 1, we know the numerator of the first term must be constant. While the second term would still be a proper fraction if the numerator were linear (degree 1), the denominator is a *repeated* linear factor and using a single constant in the numerator of *all such fractions* will ensure we obtain unique values for A and B. In the end, for any repeated linear factor $(ax + b)^n$ in the original denominator, terms of the form $\dfrac{A_1}{ax + b} + \dfrac{A_2}{(ax + b)^2} + \cdots + \dfrac{A_{n-1}}{(ax + b)^{n-1}} +$ $\dfrac{A_n}{(ax + b)^n}$ must appear in the decomposition template, although some of these numerators may turn out to be zero.

EXAMPLE 3 ▶ **Writing the Decomposition Template for Unique and Repeated Linear Factors**

Write the decomposition template for

a. $\dfrac{x - 8}{2x^2 + 5x + 3}$ b. $\dfrac{x + 1}{x^2 - 6x + 9}$

Solution ▶ a. Factoring the denominator gives $\dfrac{x - 8}{(2x + 3)(x + 1)}$. With two distinct linear factors in the denominator, the decomposition template is

$$\frac{x - 8}{(2x + 3)(x + 1)} = \frac{A}{2x + 3} + \frac{B}{x + 1} \qquad \text{decomposition template}$$

b. After factoring we have $\dfrac{x + 1}{(x - 3)^2}$, and the denominator is a repeated linear factor. Using our previous observations the template would be

$$\frac{x + 1}{(x - 3)^2} = \frac{A}{x - 3} + \frac{B}{(x - 3)^2} \qquad \text{decomposition template}$$

Now try Exercises 23 through 28 ▶

When both distinct and repeated linear factors are present in the denominator, the decomposition template maintains the elements illustrated in both observations 1 and 2.

EXAMPLE 4 ▶ **Writing the Decomposition Template for Unique and Repeated Linear Factors**

Write the decomposition template for $\dfrac{x^2 - 4x - 15}{x^3 - 2x^2 + x}$.

Solution ▶ Factoring the denominator gives $\dfrac{x^2 - 4x - 15}{x(x^2 - 2x + 1)}$ or $\dfrac{x^2 - 4x - 15}{x(x - 1)^2}$ after factoring completely. With a distinct linear factor of x, and the repeated linear factor $(x - 1)^2$, the decomposition template becomes

$$\frac{x^2 - 4x - 15}{x(x - 1)^2} = \frac{A}{x} + \frac{B}{x - 1} + \frac{C}{(x - 1)^2} \qquad \text{decomposition template}$$

Now try Exercises 29 and 30 ▶

To continue our observations,

3. Consider the sum $\dfrac{4}{x} + \dfrac{2x + 3}{x^2 + 1}$, noting the denominator of the first term is linear, while the denominator of the second is an irreducible quadratic.

$$\begin{aligned}
\frac{4}{x} + \frac{2x + 3}{x^2 + 1} &= \frac{4(x^2 + 1)}{x(x^2 + 1)} + \frac{(2x + 3)x}{(x^2 + 1)x} \qquad && \text{find common denominator} \\
&= \frac{(4x^2 + 4) + (2x^2 + 3x)}{x(x^2 + 1)} \qquad && \text{combine numerators} \\
&= \frac{6x^2 + 3x + 4}{x(x^2 + 1)} \qquad && \text{result}
\end{aligned}$$

Here, reversing the process would require us to begin with the template

$$\frac{6x^2 + 3x + 4}{x(x^2 + 1)} = \frac{A}{x} + \frac{Bx + C}{x^2 + 1},$$

allowing that the numerator of the second term might be linear since the denominator is quadratic but *not due to a repeated linear factor.*

4. Finally, consider the sum $\dfrac{1}{x^2 + 3} + \dfrac{x - 2}{(x^2 + 3)^2}$, where the denominator of the first term is an irreducible quadratic, with the second being *the same factor* with multiplicity two.

$$\frac{1}{x^2 + 3} + \frac{x - 2}{(x^2 + 3)^2} = \frac{1(x^2 + 3)}{(x^2 + 3)(x^2 + 3)} + \frac{x - 2}{(x^2 + 3)(x^2 + 3)} \quad \text{common denominators}$$

$$= \frac{(x^2 + 3) + (x - 2)}{(x^2 + 3)(x^2 + 3)} \quad \text{combine numerators}$$

$$= \frac{x^2 + x + 1}{(x^2 + 3)^2} \quad \text{result after simplifying}$$

Reversing the process would require us to begin with the template

$$\frac{x^2 + x + 1}{(x^2 + 3)^2} = \frac{Ax + B}{x^2 + 3} + \frac{Cx + D}{(x^2 + 3)^2}$$

allowing that the numerator of either term might be nonconstant for the reasons in observation 3. Similar to our reasoning in observation 2, all powers of a repeated quadratic factor must be present in the template.

> **WORTHY OF NOTE**
>
> Note that the second term in the decomposition template would still be a proper fraction if the numerator were quadratic or cubic, but since the denominator is a *repeated* quadratic factor, using only a linear form ensures we obtain unique values for all coefficients.

EXAMPLE 5 ▶ **Writing the Decomposition Template for Linear and Quadratic Factors**

Write the decomposition template for

a. $\dfrac{x^2 + 10x + 1}{(x + 1)(x^2 + 3x + 1)}$ **b.** $\dfrac{x^2}{(x^2 + 2)^3}$

Solution ▶ **a.** One factor of the denominator is a distinct linear factor, and the other is an irreducible quadratic. The decomposition template is

$$\frac{x^2 + 10x + 1}{(x + 1)(x^2 + 3x + 1)} = \frac{A}{x + 1} + \frac{Bx + C}{x^2 + 3x + 1} \quad \text{decomposition template}$$

b. The denominator consists of a repeated, irreducible quadratic factor. Using our previous observations the template would be

$$\frac{x^2}{(x^2 + 2)^3} = \frac{Ax + B}{x^2 + 2} + \frac{Cx + D}{(x^2 + 2)^2} + \frac{Ex + F}{(x^2 + 2)^3} \quad \text{decomposition template}$$

Now try Exercises 31 and 32 ▶

When both distinct and repeated factors are present in the denominator, the decomposition template maintains the essential elements determined by observations 1 through 4. Using these observations, we can formulate a general approach to the decomposition template.

Decomposition Template for Rational Expressions

For the rational expression $\dfrac{P(x)}{Q(x)}$ in lowest terms and constants $A, B, C, D, \ldots$

1. Factor Q completely into linear factors and irreducible quadratic factors.
2. For the linear factors, each distinct linear factor and each power of a repeated linear factor must appear in the decomposition template with a constant numerator.
3. For the irreducible quadratic factors, each distinct quadratic factor and each power of a repeated quadratic factor must appear in the decomposition template with a linear numerator.
4. If the degree of P is greater than or equal to the degree of Q, find the quotient and remainder using polynomial division. Only the remainder portion need be decomposed into partial fractions.

Once the template is obtained, we multiply both sides of the equation by the factored form of the original denominator and simplify. The resulting equation is an identity — a true statement for all real numbers x, and in many cases the constants A, B, C, and so on can be identified using a choice of **convenient values** for x, as in Example 6.

EXAMPLE 6 ▶ **Decomposing a Rational Expression with Linear Factors**

Decompose the expression $\dfrac{4x + 11}{x^2 + 7x + 10}$ into partial fractions.

Solution ▶ Factoring the denominator gives $\dfrac{4x + 11}{(x + 5)(x + 2)}$, with two distinct linear factors in the denominator. The required template is

$$\frac{4x + 11}{(x + 5)(x + 2)} = \frac{A}{x + 5} + \frac{B}{x + 2} \qquad \text{decomposition template}$$

Multiplying both sides by $(x + 5)(x + 2)$ clears all denominators and yields

$$4x + 11 = A(x + 2) + B(x + 5) \qquad \text{clear denominators}$$

Since the equation must be true for all x, using $x = -5$ will *conveniently* eliminate the term with B, and enable us to solve for A directly:

$$4(-5) + 11 = A(-5 + 2) + B(-5 + 5) \qquad \text{substitute } -5 \text{ for } x$$
$$-20 + 11 = -3A + B(0) \qquad \text{simplify}$$
$$-9 = -3A \qquad \text{term with } B \text{ is eliminated}$$
$$3 = A \qquad \text{solve for } A$$

To find B, we repeat this procedure, using an x-value that *conveniently* eliminates the term with A, namely, $x = -2$.

$$4x + 11 = A(x + 2) + B(x + 5) \qquad \text{original equation}$$
$$4(-2) + 11 = A(-2 + 2) + B(-2 + 5) \qquad \text{substitute } -2 \text{ for } x$$
$$-8 + 11 = A(0) + 3B \qquad \text{simplify}$$
$$3 = 3B \qquad \text{term with } A \text{ is eliminated}$$
$$1 = B \qquad \text{solve for } B$$

With $A = 3$ and $B = 1$, the complete decomposition is

$$\frac{4x + 11}{(x + 5)(x + 2)} = \frac{3}{x + 5} + \frac{1}{x + 2}$$

which can be checked by adding the fractions on the right.

Now try Exercises 33 through 38 ▶

EXAMPLE 7 ▶ **Decomposing a Rational Expression with Repeated Linear Factors**

Decompose the expression $\dfrac{9}{(x + 5)(x^2 + 7x + 10)}$ into partial fractions.

Solution ▶ Factoring the denominator gives $\dfrac{9}{(x + 5)(x + 2)(x + 5)} = \dfrac{9}{(x + 2)(x + 5)^2}$,
(one distinct linear factor, one repeated linear factor). The decomposition template
is $\dfrac{9}{(x + 2)(x + 5)^2} = \dfrac{A}{x + 2} + \dfrac{B}{x + 5} + \dfrac{C}{(x + 5)^2}$. Multiplying both sides by
$(x + 2)(x + 5)^2$ clears all denominators and yields

$$9 = A(x + 5)^2 + B(x + 2)(x + 5) + C(x + 2).$$

Using $x = -5$ will eliminate the terms with A and B, giving

$$9 = A(-5 + 5)^2 + B(-5 + 2)(-5 + 5) + C(-5 + 2) \quad \text{substitute } -5 \text{ for } x$$
$$9 = A(0) + B(-3)(0) - 3C \quad \text{simplify}$$
$$9 = -3C \quad \text{terms with } A \text{ and } B \text{ are eliminated}$$
$$-3 = C \quad \text{solve for } C$$

Using $x = -2$ will eliminate the terms with B and C, and we have

$$9 = A(x + 5)^2 + B(x + 2)(x + 5) + C(x + 2) \quad \text{original equation}$$
$$9 = A(-2 + 5)^2 + B(-2 + 2)(-2 + 5) + C(-2 + 2) \quad \text{substitute } -2 \text{ for } x$$
$$9 = A(3)^2 + B(0)(3) + C(0) \quad \text{simplify}$$
$$9 = 9A \quad \text{terms with } B \text{ and } C \text{ are eliminated}$$
$$1 = A \quad \text{solve for } A$$

To find B, we substitute $A = 1$ and $C = -3$ into the original equation, *with
any value of x that does not eliminate B*. For efficiency, we'll often use $x = 0$ or
$x = 1$ for this purpose (if possible).

$$9 = A(x + 5)^2 + B(x + 2)(x + 5) + C(x + 2) \quad \text{original equation}$$
$$9 = 1(0 + 5)^2 + B(0 + 2)(0 + 5) - 3(0 + 2) \quad \text{substitute 1 for } A, -3 \text{ for } C, 0 \text{ for } x$$
$$9 = 25 + 10B - 6 \quad \text{simplify}$$
$$-1 = B \quad \text{solve for } B$$

With $A = 1$, $B = -1$, and $C = -3$ the complete decomposition is

$$\frac{9}{(x + 2)(x + 5)^2} = \frac{1}{x + 2} + \frac{-1}{x + 5} + \frac{-3}{(x + 5)^2}$$
$$= \frac{1}{x + 2} - \frac{1}{x + 5} - \frac{3}{(x + 5)^2}$$

Now try Exercises 39 and 40 ▶

As an alternative to using convenient values, a system of equations can be set up
by multiplying out the right-hand side (after clearing fractions) and equating coefficients of the terms with like degrees.

EXAMPLE 8 ▶ **Decomposing a Rational Expression with Linear and Quadratic Factors**

Decompose the given expression into partial fractions: $\dfrac{3x^2 - x - 11}{x^3 - 3x^2 + 4x - 12}$.

Solution ▶ A careful inspection indicates the denominator will factor by grouping, giving
$x^3 - 3x^2 + 4x - 12 = x^2(x - 3) + 4(x - 3) = (x - 3)(x^2 + 4)$. With one linear factor and one irreducible quadratic factor, the required template is

$$\frac{3x^2 - x - 11}{(x - 3)(x^2 + 4)} = \frac{A}{x - 3} + \frac{Bx + C}{x^2 + 4} \qquad \text{decomposition template}$$

$$3x^2 - x - 11 = A(x^2 + 4) + (Bx + C)(x - 3) \qquad \substack{\text{multiply by } (x - 3)(x^2 + 4) \\ \text{(clear denominators)}}$$

$$= Ax^2 + 4A + Bx^2 - 3Bx + Cx - 3C \qquad \text{distribute/F-O-I-L}$$

$$= (A + B)x^2 + (C - 3B)x + 4A - 3C \qquad \text{group and factor}$$

For the left side to equal the right, we must equate coefficients of terms with like degree: $A + B = 3$, $C - 3B = -1$, and $4A - 3C = -11$. This gives the 3×3 system

$$\begin{bmatrix} 1 & 1 & 0 & | & 3 \\ 0 & -3 & 1 & | & -1 \\ 4 & 0 & -3 & | & -11 \end{bmatrix} \text{ in matrix form (verify this). Using the matrix of}$$

coefficients we find that $D = 13$, and we complete the solution using Cramer's rule.

$$D_A = \begin{vmatrix} 3 & 1 & 0 \\ -1 & -3 & 1 \\ -11 & 0 & -3 \end{vmatrix} = 13 \quad D_B = \begin{vmatrix} 1 & 3 & 0 \\ 0 & -1 & 1 \\ 4 & -11 & -3 \end{vmatrix} = 26 \quad D_C = \begin{vmatrix} 1 & 1 & 3 \\ 0 & -3 & -1 \\ 4 & 0 & -11 \end{vmatrix} = 65$$

The result is $A = \frac{13}{13} = 1$, $B = \frac{26}{13} = 2$, and $C = \frac{65}{13} = 5$, giving the decomposition

$$\frac{3x^2 - x - 11}{(x - 3)(x^2 + 4)} = \frac{1}{x - 3} + \frac{2x + 5}{x^2 + 4}.$$

Now try Exercises 41 through 46 ▶

In some cases, the "convenient values" method cannot be applied and a system of equations is our *only* option. Also, if the decomposition template produces a large or cumbersome system, a graphing calculator can be used to assist the solution process.

EXAMPLE 9 ▶ **Decomposing a Rational Expression Using Technology**

Use matrix equations and a graphing calculator to decompose the given expression into partial fractions: $\dfrac{12x^3 + 62x^2 + 102x + 56}{x(x + 2)^3}$.

Solution ▶ The decomposition template will be

$$\frac{12x^3 + 62x^2 + 102x + 56}{x(x + 2)^3} = \frac{A}{x} + \frac{B}{x + 2} + \frac{C}{(x + 2)^2} + \frac{D}{(x + 2)^3}.$$

Clearing denominators and simplifying yields

$$12x^3 + 62x^2 + 102x + 56 = A(x + 2)^3 + Bx(x + 2)^2$$
$$+ Cx(x + 2) + Dx \qquad \text{clear denominators}$$

After expanding the powers on the right and factoring we obtain

$$12x^3 + 62x^2 + 102x + 56 = (A + B)x^3 + (6A + 4B + C)x^2$$
$$+ (12A + 4B + 2C + D)x + 8A.$$

By equating the coefficients of like terms, the following system and matrix equation are obtained:

$$\begin{cases} A + B = 12 \\ 6A + 4B + C = 62 \\ 12A + 4B + 2C + D = 102 \\ 8A = 56 \end{cases} \rightarrow \begin{bmatrix} 1 & 1 & 0 & 0 \\ 6 & 4 & 1 & 0 \\ 12 & 4 & 2 & 1 \\ 8 & 0 & 0 & 0 \end{bmatrix} \begin{bmatrix} A \\ B \\ C \\ D \end{bmatrix} = \begin{bmatrix} 12 \\ 62 \\ 102 \\ 56 \end{bmatrix}.$$

After carefully entering the matrices F (coefficients) and G (constants), we obtain the solution $A = 7$, $B = 5$, $C = 0$, and $D = -2$ as shown in the figure. The decomposed form is

```
[F]⁻¹[G]
           [[7 ]
            [5 ]
            [0 ]
            [-2]]
```

$$\frac{12x^3 + 62x^2 + 102x + 56}{x(x + 2)^3} = \frac{7}{x} + \frac{5}{x + 2} + \frac{-2}{(x + 2)^3}.$$

Now try Exercises 47 and 48 ▶

☑ **B. You've just learned how to decompose a rational expression into partial fractions**

As a final reminder, if the degree of the numerator is *greater than* the degree of the denominator, divide using long division and apply the preceding methods to the remainder polynomial. For instance, you can check that $\dfrac{3x^3 + 6x^2 + 5x - 7}{x^2 + 2x + 1} = 3x + \dfrac{2x - 7}{(x + 1)^2}$, and decomposing the remainder polynomial gives a final result of $3x + \dfrac{2}{x + 1} - \dfrac{9}{(x + 1)^2}$.

C. Determinants, Geometry, and the Coordinate Plane

As mentioned in the introduction, the use of determinants extends far beyond solving systems of equations. Here, we'll demonstrate how determinants can be used to find the area of a triangle whose vertices are given as three points in the coordinate plane.

The Area of a Triangle in the *xy*-Plane

Given a triangle with vertices at (x_1, y_1), (x_2, y_2), and (x_3, y_3),

$$\text{Area} = \left| \frac{\det(T)}{2} \right| \quad \text{where } T = \begin{bmatrix} x_1 & y_1 & 1 \\ x_2 & y_2 & 1 \\ x_3 & y_3 & 1 \end{bmatrix}$$

EXAMPLE 10 ▶   **Finding the Area of a Triangle Using Determinants**

Find the area of a triangle with vertices at $(3, 1)$, $(-2, 3)$, and $(1, 7)$.

Solution ▶   Begin by forming matrix T and computing $\det(T)$:

$$\det(T) = \begin{vmatrix} x_1 & y_1 & 1 \\ x_2 & y_2 & 1 \\ x_3 & y_3 & 1 \end{vmatrix} = \begin{vmatrix} 3 & 1 & 1 \\ -2 & 3 & 1 \\ 1 & 7 & 1 \end{vmatrix}$$

$$= 3(3 - 7) - 1(-2 - 1) + 1(-14 - 3)$$

$$= -12 + 3 + (-17) = -26$$

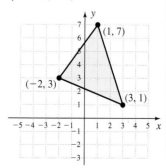

Compute the area: $A = \left| \dfrac{\det(T)}{2} \right| = \left| \dfrac{-26}{2} \right|$

$$= 13$$

The area of this triangle is 13 units^2.

Now try Exercises 51 through 56 ▶

As an extension of this formula, what if the three points were collinear? After a moment, it may occur to you that the formula would give an area of 0 units^2, since no triangle could be formed. This gives rise to a **test for collinear points.**

Test for Collinear Points

Three points (x_1, y_1), (x_2, y_2), and (x_3, y_3) are collinear if

$$\det(A) = \begin{vmatrix} x_1 & y_1 & 1 \\ x_2 & y_2 & 1 \\ x_3 & y_3 & 1 \end{vmatrix} = 0.$$

☑ **C.** You've just learned how to use determinants in applications involving geometry in the coordinate plane

See Exercises 57 through 62.

6.4 EXERCISES

▶ **CONCEPTS AND VOCABULARY**

Fill in the blank with the appropriate word or phrase. Carefully reread the section if needed.

1. The determinant $\begin{vmatrix} a_{11} & a_{12} \\ a_{21} & a_{22} \end{vmatrix}$ is evaluated as:_____.

2. _____ rule uses a ratio of determinants to solve for the unknowns in a system.

3. Given the matrix of coefficients D, the matrix D_x is formed by replacing the coefficients of x with the _____ terms.

4. The three points (x_1, y_1), (x_2, y_2), and (x_3, y_3) are collinear if $|T| = \begin{vmatrix} x_1 & y_1 & 1 \\ x_2 & y_2 & 1 \\ x_3 & y_3 & 1 \end{vmatrix}$ has a value of ___.

5. Discuss/Explain the process of writing $\dfrac{8x - 3}{x^2 - x}$ as a sum of partial fractions.

6. Discuss/Explain why Cramer's rule cannot be applied if $D = 0$. Use an example to illustrate.

▶ **DEVELOPING YOUR SKILLS**

Write the determinants $D, D_x,$ and D_y for the systems given. Do not solve.

7. $\begin{cases} 2x + 5y = 7 \\ -3x + 4y = 1 \end{cases}$

8. $\begin{cases} -x + 5y = 12 \\ 3x - 2y = -8 \end{cases}$

Solve each system of equations using Cramer's rule, if possible. Do not use a calculator.

9. $\begin{cases} 4x + y = -11 \\ 3x - 5y = -60 \end{cases}$

10. $\begin{cases} x = -2y - 11 \\ y = 2x - 13 \end{cases}$

11. $\begin{cases} \dfrac{x}{8} + \dfrac{y}{4} = 1 \\ \dfrac{y}{5} = \dfrac{x}{2} + 6 \end{cases}$ **12.** $\begin{cases} \dfrac{2}{3}x - \dfrac{3}{8}y = \dfrac{7}{5} \\ \dfrac{5}{6}x + \dfrac{3}{4}y = \dfrac{11}{10} \end{cases}$

13. $\begin{cases} 0.6x - 0.3y = 8 \\ 0.8x - 0.4y = -3 \end{cases}$ **14.** $\begin{cases} -2.5x + 6y = -1.5 \\ 0.5x - 1.2y = 3.6 \end{cases}$

Write the determinants D, D_x, D_y, and D_z for the systems given, then determine if a solution using Cramer's rule is possible by computing the value of D without the use of a calculator (do not solve the system). Try to determine how the system from part (a) is related to the system in part (b).

15. a. $\begin{cases} 4x - y + 2z = -5 \\ -3x + 2y - z = 8 \\ x - 5y + 3z = -3 \end{cases}$ **b.** $\begin{cases} 4x - y + 2z = -5 \\ -3x + 2y - z = 8 \\ x + y + z = -3 \end{cases}$

16. a. $\begin{cases} 2x + 3z = -2 \\ -x + 5y + z = 12 \\ 3x - 2y + z = -8 \end{cases}$ **b.** $\begin{cases} 2x + 3z = -2 \\ -x + 5y + z = 12 \\ x + 5y + 4z = -8 \end{cases}$

Use Cramer's rule to solve each system of equations.

17. $\begin{cases} x + 2y + 5z = 10 \\ 3x + 4y - z = 10 \\ x - y - z = -2 \end{cases}$ **18.** $\begin{cases} x + 3y + 5z = 6 \\ 2x - 4y + 6z = 14 \\ 9x - 6y + 3z = 3 \end{cases}$

19. $\begin{cases} y + 2z = 1 \\ 4x - 5y + 8z = -8 \\ 8x - 9z = 9 \end{cases}$ **20.** $\begin{cases} x + 2y + 5z = 10 \\ 3x - z = 8 \\ -y - z = -3 \end{cases}$

21. $\begin{cases} w + 2x - 3y = -8 \\ x - 3y + 5z = -22 \\ 4w - 5x = 5 \\ -y + 3z = -11 \end{cases}$

22. $\begin{cases} w - 2x + 3y - z = 11 \\ 3w - 2y + 6z = -13 \\ 2x + 4y - 5z = 16 \\ 3x - 4z = 5 \end{cases}$

▶ DECOMPOSITION OF RATIONAL EXPRESSIONS

Exercises 23 through 32 are designed solely to reinforce the various possibilities for decomposing a rational expression. All are proper fractions whose denominators are completely factored. Set up the partial fraction decomposition using appropriate numerators, but *do not solve*.

23. $\dfrac{3x + 2}{(x + 3)(x - 2)}$ **24.** $\dfrac{-4x + 1}{(x - 2)(x - 5)}$

25. $\dfrac{3x^2 - 2x + 5}{(x - 1)(x + 2)(x - 3)}$ **26.** $\dfrac{-2x^2 + 3x - 4}{(x + 3)(x + 1)(x - 2)}$

27. $\dfrac{x^2 + 5}{x(x - 3)(x + 1)}$ **28.** $\dfrac{x^2 - 7}{(x + 4)(x - 2)x}$

29. $\dfrac{x^2 + x - 1}{x^2(x + 2)}$ **30.** $\dfrac{x^2 - 3x + 5}{(x - 3)(x + 2)^2}$

31. $\dfrac{x^3 + 3x - 2}{(x + 1)(x^2 + 2)^2}$ **32.** $\dfrac{2x^3 + 3x^2 - 4x + 1}{x(x^2 + 3)^2}$

Decompose each rational expression into partial fractions.

33. $\dfrac{4 - x}{x^2 + x}$ **34.** $\dfrac{3x + 13}{x^2 + 5x + 6}$

35. $\dfrac{2x - 27}{2x^2 + x - 15}$ **36.** $\dfrac{-11x + 6}{5x^2 - 4x - 12}$

37. $\dfrac{8x^2 - 3x - 7}{x^3 - x}$ **38.** $\dfrac{x^2 + 24x - 12}{x^3 - 4x}$

39. $\dfrac{3x^2 + 7x - 1}{x^3 + 2x^2 + x}$ **40.** $\dfrac{-2x^2 - 7x + 28}{x^3 - 4x^2 + 4x}$

41. $\dfrac{3x^2 + 10x + 4}{8 - x^3}$ **42.** $\dfrac{3x^2 + 4x - 1}{x^3 - 1}$

43. $\dfrac{6x^2 + x + 13}{x^3 + 2x^2 + 3x + 6}$ **44.** $\dfrac{2x^2 - 14x - 7}{x^3 - 2x^2 + 5x - 10}$

45. $\dfrac{x^4 - x^2 - 2x + 1}{x^5 + 2x^3 + x}$

46. $\dfrac{-3x^4 + 13x^2 + x - 12}{x^5 + 4x^3 + 4x}$

47. $\dfrac{x^3 - 17x^2 + 76x - 98}{(x^2 - 6x + 9)(x^2 - 2x - 3)}$

48. $\dfrac{16x^3 - 66x^2 + 98x - 54}{(2x^2 - 3x)(4x^2 - 12x + 9)}$

▶ WORKING WITH FORMULAS

Area of a Norman window: $A = \begin{vmatrix} L & r^2 \\ -\dfrac{\pi}{2} & W \end{vmatrix}$. The determinant shown can be used to find the area of a Norman

window (rectangle + half-circle) with length L, width W, and radius $r = \dfrac{W}{2}$. Find the area of the following windows.

49.

16 in.
20 in.

50.

58 cm
32 cm

▶ APPLICATIONS

Geometric Applications

Find the area of the triangle with the vertices given. Assume units are in centimeters.

51. $(2, 1), (3, 7),$ and $(5, 3)$

52. $(-2, 3), (-3, -4),$ and $(-6, 1)$

Find the area of the parallelogram with vertices given. Assume units are in feet.

53. $(-4, 2), (-6, -1), (3, -1),$ and $(5, 2)$

54. $(-5, -6), (5, 0), (5, 4),$ and $(-5, -2)$

The volume of a triangular pyramid is given by the formula $V = \frac{1}{3}Bh$, where B represents the area of the triangular base and h is the height of the pyramid. Find the volume of a triangular pyramid whose height is given and whose base has the coordinates shown. Assume units are in meters.

55. $h = 6$ m; vertices $(3, 5), (-4, 2),$ and $(-1, 6)$

56. $h = 7.5$ m; vertices $(-2, 3), (-3, -4),$ and $(-6, 1)$

Determine if the following sets of points are collinear.

57. $(1, 5), (-2, -1),$ and $(4, 11)$

58. $(1, 1), (3, -5),$ and $(-2, 9)$

59. $(-2.5, 5.2), (1.2, -5.6),$ and $(2.2, -8.5)$

60. $(-0.5, 1.25), (-2.8, 3.75),$ and $(3, 6.25)$

For each linear equation given, substitute the first two points to verify they are solutions. Then use the test for collinear points to determine if the third point is also a solution.

61. $2x - 3y = 7$; $(2, -1), (-1.3, -3.2), (-3.1, -4.4)$

62. $5x + 2y = 4$; $(2, -3), (3.5, -6.75), (-2.7, 8.75)$

Write a linear system that models each application. Then solve using Cramer's rule.

63. **Return on investments:** If $15,000 is invested at a certain interest rate and $25,000 is invested at another interest rate, the total return was $2900. If the investments were reversed the return would be $2700. What was the interest rate paid on each investment?

64. **Cost of fruit:** Many years ago, two pounds of apples, 2 lb of kiwi, and 10 lb of pears cost $3.26. Three pounds of apples, 2 lb of kiwi, and 7 lb of pears cost $2.98. Two pounds of apples, 3 lb of kiwi, and 6 lb of pears cost $2.89. Find the cost of a pound of each fruit.

▶ EXTENDING THE CONCEPT

65. Solve the given system four different ways: (1) elimination, (2) row reduction, (3) Cramer's rule, and (4) using a matrix equation. Which method seems to be the least error-prone? Which method seems most efficient (takes the least time)? Discuss the advantages and drawbacks of each method.

$$\begin{cases} x + 3y + 5z = 6 \\ 2x - 4y + 6z = 14 \\ 9x - 6y + 3z = 3 \end{cases}$$

66. Find the area of the pentagon whose vertices are: $(-5, -5), (5, -5), (8, 6), (-8, 6),$ and $(0, 12.5)$.

67. The polynomial form for the equation of a circle is $x^2 + y^2 + Dx + Ey + F = 0$. Find the equation of the circle that contains the points $(-1, 7), (2, 8),$ and $(5, -1)$.

▶ **MAINTAINING YOUR SKILLS**

68. (3.4) Graph the polynomial using information about end behavior, y-intercept, x-intercept(s), and midinterval points: $f(x) = x^3 - 2x^2 - 7x + 6$.

69. (2.5) Which is the graph (left or right) of $g(x) = -|x + 1| + 3$? Justify your answer.

70. (4.2/4.5) Solve the equation $3^{2x-1} = 9^{2-x}$ two ways. First using logarithms, then by equating the bases and using properties of equality.

71. (3.4) Which is the graph (left or right) of a degree 3 polynomial? Justify your answer.

SUMMARY AND CONCEPT REVIEW

SECTION 6.1   Solving Linear Systems Using Matrices and Row Operations

KEY CONCEPTS

- A *matrix* is a rectangular arrangement of numbers. An $m \times n$ matrix has m rows and n columns.
- An *augmented matrix* is derived from a system of linear equations by augmenting the *coefficient matrix* (formed by the variable coefficients) with the *matrix of constants*.
- One matrix method for solving systems of equations is by triangularizing the augmented matrix.
- An inconsistent system with no solutions will yield a contradictory statement such as $0 = 1$. A dependent system with infinitely many solutions will yield an identity statement such as $0 = 0$.

EXERCISES

1. Write an example of the following matrices:

     a. 2×3 **b.** 3×2 **c.** 3×4, in triangular form

Solve by triangularizing the matrix. Use a calculator for Exercise 4.

2. $\begin{cases} x - 2y = 6 \\ 4x - 3y = 4 \end{cases}$

3. $\begin{cases} x - 2y + 2z = 7 \\ 2x + 2y - z = 5 \\ 3x - y + z = 6 \end{cases}$

4. $\begin{cases} 2w + x + 2y - 3z = -19 \\ w - 2x - y + 4z = 15 \\ x + 2y - z = 1 \\ 3w - 2x - 5z = -60 \end{cases}$

5. $\begin{cases} 2x - y + 2z = -1 \\ x + 2y + 2z = -3 \\ 3x - 4y + 2z = 1 \end{cases}$

SECTION 6.2   The Algebra of Matrices

KEY CONCEPTS

- The entries of a matrix are denoted a_{ij}, where i gives the row and j gives the column of its location.
- Two matrices A and B of equal size (or *order*) are equal if corresponding entries are equal.
- The sum or difference of two matrices of equal order is found by combining corresponding entries: $A + B = [a_{ij} + b_{ij}]$
- The *identity matrix for addition* is an $m \times n$ matrix whose entries are all zeroes.
- To perform *scalar multiplication,* take the product of the constant with each entry in the matrix, forming a new matrix of like size. For matrix A: $kA = [ka_{ij}]$.

- *Matrix multiplication* is performed as row entry × column entry. For an $m \times n$ matrix $A = [a_{ij}]$ and an $s \times t$ matrix $B = [b_{ij}]$, AB is possible if $n = s$. The result will be an $m \times t$ matrix $P = [p_{ij}]$, where p_{ij} is the product of the ith row of A with the jth column of B.
- When technology is used to perform operations on matrices, carefully enter each matrix into the calculator. Then double check that each entry is correct and appraise the results to see if they are reasonable.

EXERCISES

Compute the operations indicated below (if possible), using the following matrices.

$$A = \begin{bmatrix} \frac{-1}{4} & \frac{-3}{4} \\ \frac{-1}{8} & \frac{-7}{8} \end{bmatrix} \qquad B = \begin{bmatrix} -7 & 6 \\ 1 & -2 \end{bmatrix} \qquad C = \begin{bmatrix} -1 & 3 & 4 \\ 5 & -2 & 0 \\ 6 & -3 & 2 \end{bmatrix} \qquad D = \begin{bmatrix} 2 & -3 & 0 \\ 0.5 & 1 & -1 \\ 4 & 0.1 & 5 \end{bmatrix}$$

6. $A + B$ **7.** $B - A$ **8.** $C - B$ **9.** $8A$ **10.** BA

11. $C + D$ **12.** $D - C$ **13.** BC **14.** $-4D$ **15.** CD

SECTION 6.3   Solving Linear Systems Using Matrix Equations

KEY CONCEPTS

- I, the *identity matrix for multiplication*, has 1's on the main diagonal and 0's for all other entries. For any $n \times n$ matrix A, the identity matrix is also an $n \times n$ matrix, is called I_n, and $AI_n = I_nA = A$.
- For an $n \times n$ (square) matrix A, the *inverse matrix* for multiplication is a matrix B such that $AB = BA = I_n$. For matrix A the inverse is denoted A^{-1}. Only square matrices have an inverse.
- Any $n \times n$ system of equations can be written as a matrix equation and solved (if a unique solution exists) using an inverse matrix. The system

$$\begin{cases} 2x + 3y = 7 \\ x - 4y = -2 \end{cases} \text{ is written as } \begin{bmatrix} 2 & 3 \\ 1 & -4 \end{bmatrix}\begin{bmatrix} x \\ y \end{bmatrix} = \begin{bmatrix} 7 \\ -2 \end{bmatrix}.$$

- Every square matrix has a real number associated with it, called its *determinant*. For 2×2 matrix $A = \begin{bmatrix} a_{11} & a_{12} \\ a_{21} & a_{22} \end{bmatrix}$, $\det(A) = a_{11}a_{22} - a_{21}a_{12}$.
- If the determinant of a matrix is zero, the matrix is said to be *singular* or *noninvertible*. If the coefficient matrix of a matrix equation is noninvertible, the system is either inconsistent or dependent.

EXERCISES

Complete Exercises 16 through 18 using the following matrices:

$$A = \begin{bmatrix} 1 & 0 \\ 0 & 1 \end{bmatrix} \qquad B = \begin{bmatrix} 0.2 & 0.2 \\ -0.6 & 0.4 \end{bmatrix} \qquad C = \begin{bmatrix} 2 & -1 \\ 3 & 1 \end{bmatrix} \qquad D = \begin{bmatrix} 10 & -6 \\ -15 & 9 \end{bmatrix}$$

16. Exactly one of the matrices given is singular. Compute each determinant to identify it.

17. Show that $AB = BA = B$. What can you conclude about the matrix A?

18. Show that $BC = CB = I$. What can you conclude about the matrix C?

Complete Exercises 19 through 21 using the matrices given:

$$E = \begin{bmatrix} 1 & -2 & 3 \\ -2 & 1 & -5 \\ -1 & -1 & -2 \end{bmatrix} \qquad F = \begin{bmatrix} 1 & -1 & 1 \\ 0 & 1 & 0 \\ -2 & 1 & -1 \end{bmatrix} \qquad G = \begin{bmatrix} -1 & 0 & -1 \\ 0 & 1 & 0 \\ 2 & 1 & 1 \end{bmatrix}$$

19. Exactly one of the matrices is singular. Use a calculator to determine which.

20. Compute the products FG and GF. What can you conclude about matrix G?

21. Verify that $EG \neq GE$ and $EF \neq FE$. What can you conclude?

Solve manually using a matrix equation.

22. $\begin{cases} 2x - 5y = 14 \\ -3y + 4x = -14 \end{cases}$

Solve using a matrix equation and your calculator.

23. $\begin{cases} 0.5x - 2.2y + 3z = -8 \\ -0.6x - y + 2z = -7.2 \\ x + 1.5y - 0.2z = 2.6 \end{cases}$

SECTION 6.4 Applications of Matrices and Determinants: Cramer's Rule, Partial Fractions, and More

KEY CONCEPTS

• Cramer's rule uses a ratio of determinants to solve systems of equations (if solutions exist).

• The determinant of the 2×2 matrix $\begin{vmatrix} a_{11} & a_{12} \\ a_{21} & a_{22} \end{vmatrix}$ is $a_{11}a_{22} - a_{21}a_{12}$.

• To compute the value of 3×3 and larger determinants, a calculator is generally used.

• Determinants can be used to find the area of a triangle in the plane if the vertices of the triangle are known, and as a test to see if three points are collinear.

• A system of equations can be used to write a rational expression as a sum of its partial fractions.

EXERCISES

Solve using Cramer's rule.

24. $\begin{cases} 5x + 6y = 8 \\ 10x - 2y = -9 \end{cases}$

25. $\begin{cases} 2x + y = -2 \\ -x + y + 5z = 12 \\ 3x - 2y + z = -8 \end{cases}$

26. $\begin{cases} 2x + y - z = -1 \\ x - 2y + z = 5 \\ 3x - y + 2z = 8 \end{cases}$

27. Find the area of a triangle whose vertices have the coordinates $(6, 1)$, $(-1, -6)$, and $(-6, 2)$.

28. Find the partial fraction decomposition for $\dfrac{7x^2 - 5x + 17}{x^3 - 2x^2 + 3x - 6}$.

 MIXED REVIEW

Solve using row operations to triangularize the matrix.

1. $\begin{cases} \dfrac{1}{2}x + \dfrac{2}{3}y = 3 \\ \dfrac{-2}{5}x - \dfrac{1}{4}y = 1 \end{cases}$

2. $\begin{cases} 2x - 5y = 5 \\ y = 0.4x + 1.2 \end{cases}$

3. $\begin{cases} 3x - 4y + 5z = 5 \\ -x + 2y - 3z = -3 \\ 3x - 2y + z = 1 \end{cases}$

4. $\begin{cases} -2x + y - 4z = -11 \\ x + 3y - z = -4 \\ 3x - 2y + z = 7 \end{cases}$

Compute as indicated for

$A = \begin{bmatrix} 2 & -1 \\ 0 & 3 \end{bmatrix}$ $\qquad B = \begin{bmatrix} 1 & -2 \\ 3 & 0 \\ 2 & 4 \end{bmatrix}$

$C = \begin{bmatrix} 1 & -4 & 2 \\ -2 & 0 & -1 \end{bmatrix}$ $\qquad D = \begin{bmatrix} 3 & 0 & 1 \\ -1 & 2 & 0 \\ 1 & 1 & -4 \end{bmatrix}$

5. a. $-2AC$ **b.** CD

6. a. BA **b.** $CB - 4A$

7. Find: **a.** a_{22} **b.** b_{21} **c.** c_{12} **d.** d_{32}

8. Find values of x, y, and z so that $A = B$, given:

$A = \begin{bmatrix} 2x & y - z \\ y + z & 4 \end{bmatrix}$ and $B = \begin{bmatrix} 5 & -1 \\ 5 & 4 \end{bmatrix}$

9. Solve using a matrix equation:

$\begin{cases} -x - 2z = 5 \\ 2y + z = -4 \\ -x + 2y = 3 \end{cases}$

 10. Use a matrix equation and a calculator to solve:

$\begin{cases} w + \dfrac{1}{2}x + \dfrac{2}{3}y - z = -3 \\ \dfrac{3}{4}x - y + \dfrac{5}{8}z = \dfrac{41}{8} \\ \dfrac{2}{5}w - x - \dfrac{3}{10}z = -\dfrac{27}{10} \\ w + 2x - 3y + 4z = 16 \end{cases}$

11. Find the inverse of $A = \begin{bmatrix} 2 & 3 \\ 5 & 8 \end{bmatrix}$.

12. Compute the determinant of the following matrix without the use of a calculator, and state whether an inverse exists. $A = \begin{bmatrix} 1 & 3 & 2 \\ 3 & 7 & 4 \\ 0 & -1 & -1 \end{bmatrix}$

13. Given $A = \begin{bmatrix} 1 & 1 & 2 \\ 2 & 0 & -1 \\ -1 & 1 & 1 \end{bmatrix}$, $X = \begin{bmatrix} x \\ y \\ z \end{bmatrix}$, and $B = \begin{bmatrix} 1 \\ 3 \\ 3 \end{bmatrix}$, write the system of equations represented by the matrix equation $AX = B$.

14. Solve using Cramer's rule: $\begin{cases} -2x + y = -8 \\ x - 2y = 7 \end{cases}$

15. Solve using Cramer's rule:
$$\begin{cases} -x + 5y - 2z = 1 \\ 2x + 3y - z = 3 \\ 3x - y + 3z = -2 \end{cases}$$

16. Decompose the expression into partial fractions:
$$\frac{x^2 + 1}{x^3 - 3x + 2}$$

17. Find the area of a triangle with vertices at $(-1, 2)$, $(3, 1)$, and $(0, 4)$.

18. Determine if the following points are collinear:
  a. $(1, -1)$, $(2, 1)$, and $(-2, -7)$
  b. $(2, 0)$, $(-3, 2)$, $(0, 1)$

19. The ancient Gaelic game of Hurling is one of the most demanding and skillful field sports being played today. The field (or pitch) is a rectangle whose perimeter is 438 m. Its length is 55 m longer than its width. Use a system of linear equations and the augmented matrix to find the dimensions of the pitch.

20. A local fitness center is offering incentives in an effort to boost membership. If you buy a year's membership (Y), you receive a $50 rebate and six tickets to a St. Louis Cardinals home game. For a half-year membership (H), you receive a $30 rebate and four tickets to a Cardinals home game. For a monthly trial membership (M), you receive a $10 rebate and two tickets to a Cardinals home game. During the last month, male clients purchased 40 one-year, 52 half-year, and 70 monthly memberships, while female clients purchased 50 one-year, 44 half-year, and 60 monthly memberships. Write the number of sales of each type to males and females as a 2×3 matrix, and the amount of the rebates and number of Cards tickets awarded per type of membership as a 3×2 matrix. Use these matrices to determine (a) the total amount of rebate money paid to males and (b) the number of Cardinals tickets awarded to females.

PRACTICE TEST

Solve each system by triangularizing the augmented matrix and using back-substitution.

1. $\begin{cases} 3x + 8y = -5 \\ x + 10y = 2 \end{cases}$

2. $\begin{cases} 3x - y + 5z = 1 \\ 3x + y + 4z = 4 \\ x + y + z = \frac{7}{3} \end{cases}$

3. $\begin{cases} 4x - 5y - 6z = 5 \\ 2x - 3y + 3z = 0 \\ x + 2y - 3z = 5 \end{cases}$

4. Given matrices A and B, compute:
  a. $A - B$ **b.** $\frac{2}{5}B$ **c.** AB **d.** A^{-1} **e.** $|A|$

$A = \begin{bmatrix} -3 & -2 \\ 5 & 4 \end{bmatrix}$ $B = \begin{bmatrix} 3 & 3 \\ -3 & -5 \end{bmatrix}$

5. Given matrices C and D, use a calculator to find:
  a. $C - D$ **b.** $-0.6D$ **c.** DC **d.** D^{-1} **e.** $|D|$

$C = \begin{bmatrix} 0.5 & 0 & 0.2 \\ 0.4 & -0.5 & 0 \\ 0.1 & -0.4 & -0.1 \end{bmatrix}$

$D = \begin{bmatrix} 0.5 & 0.1 & 0.2 \\ -0.1 & 0.1 & 0 \\ 0.3 & 0.4 & 0.8 \end{bmatrix}$

6. Use matrices to find three different solutions of the dependent system:
$$\begin{cases} 2x - y + z = 4 \\ 3x - 2y + 4z = 9 \\ x - 2y + 8z = 11 \end{cases}$$

7. Solve using Cramer's rule: $\begin{cases} 2x - 3y = 2 \\ x - 6y = -2 \end{cases}$

8. Solve using a calculator and Cramer's rule:
$$\begin{cases} 2x + 3y + z = 3 \\ x - 2y - z = 4 \\ x - y - 2z = -1 \end{cases}$$

9. Solve using a matrix equation and your calculator:
$$\begin{cases} 2x - 5y = 11 \\ 4x + 7y = 4 \end{cases}$$

10. Solve using a matrix equation and your calculator:
$$\begin{cases} x - 2y + 2z = 7 \\ 2x + 2y - z = 5 \\ 3x - y + z = 6 \end{cases}$$

11. Find values of x, y, and z so that $A = B$, given
$$A = \begin{bmatrix} 2x + y & 3 \\ x + z & 3x + 2z \end{bmatrix} \text{ and }$$
$$B = \begin{bmatrix} z - 1 & 3 \\ 2y + 5 & y + 8 \end{bmatrix}$$

12. Given matrix X is a solution to $AX = B$ for the matrix A given. Find matrix B.
$$X = \begin{bmatrix} -1 \\ \frac{-3}{2} \\ 2 \end{bmatrix} \quad A = \begin{bmatrix} 1 & -2 & 2 \\ 2 & -6 & 3 \\ 3 & 4 & -1 \end{bmatrix}$$

13. Use matrices to determine which three of the following four points are collinear: $(-1, 4)$, $(1, 3)$, $(2, 1)$, $(4, -1)$

14. A farmer plants a triangular field with wheat. The first vertex of the triangular field is 1 mi east and 1 mi north of his house. The second vertex is 3 mi east and 1 mi south of his house. The third vertex is 1 mi west and 2 mi south of his house. What is the area of the field?

15. For $A = \begin{bmatrix} r & 2 \\ 3 & s \end{bmatrix}$ and $A^2 = \begin{bmatrix} 10 & -2 \\ -3 & 7 \end{bmatrix}$ given, find r and s.

Create a system of equations to model each exercise, then solve using any matrix method.

16. Dr. Brown and Dr. Stamper graduate from medical school with $155,000 worth of student loans. Due to her state's tuition reimbursement plan, Dr. Brown owes one fourth of what Dr. Stamper owes. How much does each doctor owe?

17. Justin is rehabbing two old houses simultaneously. He calculates that last week he spent 23 hr working on these houses. If he spent 8 more hours on one of the houses, how many hours did he spend on each house?

18. In his first month as assistant principal of Washington High School, Mr. Johnson gave out 20 detentions. They were either for 1 day, 2 days, or 5 days. He recorded a total of 38 days of detention served. He also noted that there were twice as many 2-day detentions as 5-day detentions. How many of each type of detention did Mr. Johnson give out?

19. The city of Cherrywood has approved a $1,800,000 plan to renovate its historic commercial district. The money will be coming from three separate sources. The first is a federal program that charges a low 2% interest annually. The second is a municipal bond offering that will cost 5% annually. The third is a standard loan from a neighborhood bank, but it will cost 8.5% annually. In the first year, the city will not make any repayment on these loans and will accrue $94,500 more debt. The federal program and bank loan together are responsible for $29,500 of this interest. How much money was originally provided by each source?

20. Decompose the expression into partial fractions:
$$\frac{4x^2 - 4x + 3}{x^3 - 27}$$

CALCULATOR EXPLORATION AND DISCOVERY

Cramer's Rule

In Section 6.4, we saw that one interesting application of matrices is Cramer's rule. You may have noticed that when technology is used with Cramer's rule, the chances of making an error are fairly high, as we need to input the entries for numerous matrices. However, as we mentioned in the chapter introduction, one of the advantages of matrices is that they *are easily programmable,* and we can actually write a very simple program that will make Cramer's rule a more efficient method.

To begin, press the **PRGM** key, and then the right arrow ▶ twice to enter a name for our program. At the prompt, we'll enter **CRAMER2.** As we write the program, note that the needed commands (**ClrHome, Disp, Prompt, Stop**) are all located in the submenus of the **PRGM** key, and the **=** sign is found under the **TEST** menu, accessed using the **2nd** **MATH** keys (the arrows "→" are used to indicate the **STO▶** key.

This program asks for the coefficients of two linear equations in two variables, written in standard form. Then, using Cramer's rule and the formula for the determinant of a 2 × 2 matrix, it returns the solutions of the system.

:**PROGRAM:CRAMER2**
:**ClrHome**
:**Disp "2×2 SYSTEM"**
:**Disp "AX+BY = C"**
:**Disp "DX+EY = F"**
:**Prompt A,B,C,D,E,F**
:**(CE–BF)/(AE–BD)→X**
:**(AF–DC)/(AE–BD)→Y**
:**Disp "X =",X,"Y = ",Y**
:**Stop**

Exercise 1: Create 2 × 2 systems of your own that are (a) consistent, (b) inconsistent, and (c) dependent. Then verify results using the program.

Exercise 2: Use the formula from page 598 of Section 6.4 to write a similar program for 3 × 3 systems. Call the program CRAMER3, and repeat parts (a), (b), and (c) from Exercise 1.

STRENGTHENING CORE SKILLS

Augmented Matrices and Matrix Inverses

The formula for finding the inverse of a 2 × 2 matrix has its roots in the more general method of computing the inverse of an $n \times n$ matrix. This involves augmenting a square matrix M with its corresponding identity I_n on the right (forming an $n \times 2n$ matrix), and using row operations to *transform M into the identity*. In some sense, as the original matrix is transformed, the "identity part" keeps track of the operations we used to convert M and we can use the results to "get back home," so to speak. We'll illustrate with the 2 × 2 matrix from Section 6.3, Example 2B, where we found that $\begin{bmatrix} 1 & -2.5 \\ -1 & 3 \end{bmatrix}$ was the inverse matrix for $\begin{bmatrix} 6 & 5 \\ 2 & 2 \end{bmatrix}$. We begin by augmenting $\begin{bmatrix} 6 & 5 \\ 2 & 2 \end{bmatrix}$ with the 2 × 2 identity.

$$\begin{bmatrix} 6 & 5 & | & 1 & 0 \\ 2 & 2 & | & 0 & 1 \end{bmatrix} -2R1 + 6R2 \rightarrow R2 \begin{bmatrix} 6 & 5 & | & 1 & 0 \\ 0 & 2 & | & -2 & 6 \end{bmatrix} \frac{R2}{2} \rightarrow R2 \begin{bmatrix} 6 & 5 & | & 1 & 0 \\ 0 & 1 & | & -1 & 3 \end{bmatrix}$$

$$\begin{bmatrix} 6 & 5 & | & 1 & 0 \\ 0 & 1 & | & -1 & 3 \end{bmatrix} -5R2 + R1 \rightarrow R1 \begin{bmatrix} 6 & 0 & | & 6 & -15 \\ 0 & 1 & | & -1 & 3 \end{bmatrix} \frac{R1}{6} \rightarrow R1 \begin{bmatrix} 1 & 0 & | & 1 & -2.5 \\ 0 & 1 & | & -1 & 3 \end{bmatrix}$$

As you can see, the identity is automatically transformed into the inverse matrix when this method is applied. Performing similar row operations on the general matrix $\begin{bmatrix} a & b \\ c & d \end{bmatrix}$ results in the formula given earlier.

$$\begin{bmatrix} a & b & | & 1 & 0 \\ c & d & | & 0 & 1 \end{bmatrix} -cR1 + aR2 \rightarrow R2 \begin{bmatrix} a & b & | & 1 & 0 \\ 0 & ad-bc & | & -c & a \end{bmatrix}$$

$$\begin{bmatrix} a & b & | & 1 & 0 \\ 0 & ad-bc & | & c & a \end{bmatrix} \frac{R2}{ad-bc} \rightarrow R2 \begin{bmatrix} a & b & | & 1 & 0 \\ 0 & 1 & | & \dfrac{-c}{ad-bc} & \dfrac{a}{ad-bc} \end{bmatrix}$$

$$\begin{bmatrix} a & b & | & 1 & 0 \\ 0 & 1 & | & \dfrac{-c}{ad-bc} & \dfrac{a}{ad-bc} \end{bmatrix} -bR2 + R1 \rightarrow R1 \begin{bmatrix} a & 0 & | & \dfrac{bc}{ad-bc}+1 & \dfrac{-ba}{ad-bc} \\ 0 & 1 & | & \dfrac{-c}{ad-bc} & \dfrac{a}{ad-bc} \end{bmatrix}$$

Finding a common denominator for $\dfrac{bc}{ad-bc} + 1$ and combining like terms gives $\dfrac{bc}{ad-bc} + \dfrac{ad-bc}{ad-bc} = \dfrac{ad}{ad-bc}$, and making this replacement we then have

$$\begin{bmatrix} a & 0 & \bigg| & \dfrac{ad}{ad-bc} & \dfrac{-ba}{ad-bc} \\ 0 & 1 & \bigg| & \dfrac{-c}{ad-bc} & \dfrac{a}{ad-bc} \end{bmatrix} \xrightarrow{\;\frac{R1}{a}\to R1\;} \begin{bmatrix} 1 & 0 & \bigg| & \dfrac{d}{ad-bc} & \dfrac{-b}{ad-bc} \\ 0 & 1 & \bigg| & \dfrac{-c}{ad-bc} & \dfrac{a}{ad-bc} \end{bmatrix}$$. This shows

$$A^{-1} = \begin{bmatrix} \dfrac{d}{ad-bc} & \dfrac{-b}{ad-bc} \\ \dfrac{-c}{ad-bc} & \dfrac{a}{ad-bc} \end{bmatrix}$$ and factoring out $\dfrac{1}{ad-bc}$ produces the familiar formula.

As you might imagine, attempting this on a general 3×3 matrix is problematic at best, and instead we simply apply the augmented matrix method to find A^{-1} for the 3×3 matrix shown in blue.

$$\begin{bmatrix} 2 & 1 & 0 & \big| & 1 & 0 & 0 \\ -1 & 3 & -2 & \big| & 0 & 1 & 0 \\ 3 & -1 & 2 & \big| & 0 & 0 & 1 \end{bmatrix} \xrightarrow[\;-3R1+2R3\to R3\;]{\;R1+2R2\to R2\;} \begin{bmatrix} 2 & 1 & 0 & \big| & 1 & 0 & 0 \\ 0 & 7 & -4 & \big| & 1 & 2 & 0 \\ 0 & -5 & 4 & \big| & -3 & 0 & 2 \end{bmatrix}$$

$$\xrightarrow[\;5R2+7R3\to R3\;]{\;R2-7R1\to R1\;} \begin{bmatrix} -14 & 0 & -4 & \big| & -6 & 2 & 0 \\ 0 & 7 & -4 & \big| & 1 & 2 & 0 \\ 0 & 0 & 8 & \big| & -16 & 10 & 14 \end{bmatrix}$$

$$\xrightarrow{\;\frac{R3}{8}\to R3\;} \begin{bmatrix} -14 & 0 & -4 & \big| & -6 & 2 & 0 \\ 0 & 7 & -4 & \big| & 1 & 2 & 0 \\ 0 & 0 & 1 & \big| & -2 & 1.25 & 1.75 \end{bmatrix}$$

$$\xrightarrow[\;4R3+R1\to R1\;]{\;4R3+R2\to R2\;} \begin{bmatrix} -14 & 0 & 0 & \big| & -14 & 7 & 7 \\ 0 & 7 & 0 & \big| & -7 & 7 & 7 \\ 0 & 0 & 1 & \big| & -2 & 1.25 & 1.75 \end{bmatrix}$$

Using $\dfrac{R1}{-14} \to R1$ and $\dfrac{R2}{7} \to R2$ produces the inverse matrix $A^{-1} = \begin{bmatrix} 1 & -0.5 & -0.5 \\ -1 & 1 & 1 \\ -2 & 1.25 & 1.75 \end{bmatrix}$.

To verify, we show $AA^{-1} = I$: $\begin{bmatrix} 2 & 1 & 0 \\ -1 & 3 & -2 \\ 3 & -1 & 2 \end{bmatrix} \begin{bmatrix} 1 & -0.5 & -0.5 \\ -1 & 1 & 1 \\ -2 & 1.25 & 1.75 \end{bmatrix} = \begin{bmatrix} 1 & 0 & 0 \\ 0 & 1 & 0 \\ 0 & 0 & 1 \end{bmatrix}$ ✓ ($A^{-1}A$ is also true).

Exercise 1: Use the preceding inverse and a matrix equation to solve the system
$$\begin{cases} 2x + y = -2 \\ -x + 3y - 2z = -15. \\ 3x - y + 2z = 9 \end{cases}$$

CUMULATIVE REVIEW CHAPTERS 1–6

1. Perform the operations indicated.
   a. $(3 - 2i)(2 + i)$
   b. $(5 - 3i)^2$
   c. $\dfrac{8 - i}{2 + i}$
   d. i^{49}

2. Solve $S = 2\pi rh + 2\pi r^2$ for h.

Solve the following equations

3. $2x - 4(3x + 1) = 5 - 4x$

4. $\dfrac{x + 6}{x + 2} - \dfrac{1}{x} = \dfrac{12}{x^2 + 2x}$

5. $9x^2 + 1 = 6x$

6. $\sqrt{2x + 11} - x = 6$

7. Find the equation of a line that passes through the points $(2, -2)$ and $(-3, 5)$.

8. Find the equation of the line perpendicular to the line shown, with the same y-intercept.

9. Given $f(x) = 3 - 4x - x^2$ and $g(x) = |4 - \sqrt{x + 2}|$, find

  a. $f(3)$ **b.** $g(23)$

  c. $f(-2)$ **d.** $g(-3)$

Graph the following by using transformations of the toolbox functions.

10. $y = (x - 3)^2$ **11.** $y = \frac{1}{2}|x| - 3$

12. $y = -(x + 2)^3$ **13.** $y = \sqrt{-x} - 2$

Solve the following inequalities. Express your answer in interval notation.

14. $2(x - 2) + 3 \le 8$ **15.** $x^2 + 5 > 6x$

16. A long-distance calling card advertises one of the lowest rates available. It claims to charge $3¢$ per minute for a long distance call, after a $15¢$ connection fee. The cost of a call is modeled by the ceiling function $c(m) = 9\left\lceil \frac{1}{3}m \right\rceil + 15$, where m is the length of the call in minutes and c is the cost of the call in cents. Graph this function and explain why this card may not be as inexpensive as advertised.

Name interval(s) where the following functions are increasing, decreasing, or constant. Write answers using interval notation.

17. $y = f(x)$ **18.** $g(x) = \dfrac{1}{(x + 3)} + 3$

19. Compute the quotient $(4y^2 - 3y + 10) \div (2y + 1)$.

20. Use the remainder theorem to find $P(-3)$ for $P(x) = x^3 + 4x^2 - 15$.

21. Find a cubic polynomial with real coefficients having the roots $x = -2$ and $x = 3 - i$. Assume a lead coefficient of 1.

22. State the end behavior and y-intercept of the following functions.

  a. $f(x) = 3x^3 - 16$

  b. $g(x) = -5x^6 + 2x^3 - 3x + 7$

  c. $h(x) = 10 - 2x^5$

23. Solve the system of equations by any method.

$$\begin{cases} 5x + 6y = -30 \\ -3x + 2y = 4 \end{cases}$$

24. Graph the solution of the system and find the coordinates of the vertices.

$$\begin{cases} y > x^2 - 1 \\ x^2 + y^2 \le 13 \end{cases}$$

25. Simon Legré, a notorious French criminal, is planning to embezzle money from the Prefect of Police, and believes he can remove 12% of their $5,000,000 budget without being detected. He plans to use his ill-gotten gain to buy a legitimate business, and purchase a condominium and a delivery truck for the business. The business will cost $20,000 more than twice the delivery truck, and the condominium will cost $200,000 more than the truck and business combined. How much will each cost?

Modeling With Technology III   Matrix Applications

Learning Objectives

In this section, you will learn how to:

☐ **A.** Find market equilibrium graphically

☐ **B.** Use matrix equations to solve static systems

☐ **C.** Use matrices for encryption/decryption

Most of the skills needed for this study have been presented in previous sections. Here we'll use various types of regression, combined with systems of equations, to solve practical applications from business and industry.

A. Supply/Demand Curves and Market Equilibrium

In a "free-market" economy, also referred to as a "supply-and-demand" economy, there are naturally occurring forces that invariably come into play if no outside forces act on the producers (suppliers) and consumers (demanders). Generally speaking, the higher the price of an item, the lower the demand. A good advertising campaign can increase the demand, but the increasing demand brings an increase in price, which moderates the demand—and so it goes until a balance is reached. There are limitations to this model, and this interplay can be affected by the number of available consumers, production limits, "shelf-life" issues, and so on, but at any given moment in the life cycle of a product, consumer demand responds to price in this way. On the other hand, producers also respond to price in a very predictable way. The higher the price of an item, the more producers are willing to supply, but as the price of the item decreases, the producers' willingness to supply the item also decreases. These free-market forces ebb and flow until **market equilibrium** occurs, at the specific price where the supply and demand are equal. In the mathematical model for these market forces, it seems reasonable that *price* is the independent variable, with quantity demanded and quantity produced as the dependent variables.

EXAMPLE 1 ▶ Finding Market Equilibrium Graphically

An electronics company manufactures mp3 players. The monthly demand for their player can be modeled by the function $D(p) = -95p + 11{,}900$, where $D(p)$ represents the number of players *purchased* (demanded) at price p. The supply function for these players is $S(p) = 130p - 1600$, where $S(p)$ represents the number of players manufactured (supplied) at price p. Use a graphing calculator to help find the price at which market equilibrium will occur.

Solution ▶ Enter the demand equation as Y_1 and the supply equation as Y_2, as shown in Figure MWT III.1. To set an appropriate viewing window, note that negative values do not fit the context, so we restrict our attention to QI. Since the y-intercepts are 11,900 and 600, respectively, appropriate settings for the y-values would be Ymin $= 0$ and Ymax $= 12{,}000$. The x-intercepts are approximately 12 and 125, so appropriate settings for the x-values would be Xmin $= 0$ and Xmax $= 130$ to allow for a frame around the viewing window. Pressing the [GRAPH] key at this point should produce the graph shown in Figure MWT III.2. To find the equilibrium point (where supply and demand meet), we'll use the [2nd] [TRACE] (**CALC**), **5:intersect** option. As before, we press [ENTER] three times to identify the first curve,

Figure MWT III.1

```
Plot1  Plot2  Plot3
\Y1 ▤ -95X+11900
\Y2 ▤ 130X-1600
\Y3 =
\Y4 =
\Y5 =
\Y6 =
\Y7 =
```

Figure MWT III.2

the second curve, and to bypass the "Guess" option. The result shows (60, 6200) is the point of intersection, meaning equilibrium will occur at a price of $60/player, and 6200 units will be demanded and supplied at this price.

Now try Exercises 1 through 4 ▶

As a follow-up to Example 1, a person could reasonably ask, "Where did the equations come from?" In this instance, they were artificially constructed to yield a "nice" solution. In actual practice, the equations and coefficients are not so "well behaved" and are based on the collection and interpretation of real data. While market analysts have sophisticated programs and numerous models to help develop these equations, in this study we'll use our experience with regression to develop the supply and demand curves.

EXAMPLE 2 ▶   **Using Technology to Find Market Equilibrium**

The company from Exercise 1 hired a consulting firm to do market research on their "next generation" mp3 player. Over a 10-week period, the firm collected the data shown for the mp3 player market (data includes mp3 players sold and expected to sell).

Price (dollars)	Demand	Supply (Inventory)
107.10	6,900	12,200
85.50	7,900	9,900
64.80	13,200	8,000
52.20	13,500	7,900
108.00	6,700	14,000
91.80	7,600	12,000
77.40	9,200	9,400
46.80	13,800	6,100
74.70	10,600	8,800
68.40	12,800	8,600

a. Use a graphing calculator to simultaneously display the demand and supply scatter-plots.

b. Calculate a line of best fit for each and graph them with the scatter-plots (identify each curve).

c. Find the equilibrium point.

Solution ▶   a. Begin by clearing all lists. This can be done manually, or by pressing [2nd] [+] (MEM) and selecting option **4:ClrAllLists** (the command appears on the home screen). Pressing [ENTER] will execute the command, and the word DONE will appear. Carefully input price in L1, demand in L2, and supply in L3 (see Figure MWT III.3). With the window settings given in Figure MWT III.4, pressing [GRAPH] will display the price/demand and price/supply scatter-plots shown. If this is not the case, use [2nd] [Y=] (STAT PLOT) to be sure that "On" is highlighted in Plot1 and Plot2, and that Plot1 uses L1 and L2, while Plot2 uses L1 and L3 (Figure MWT III.5). Note we've chosen a different mark to indicate the data points in Plot2.

b. Calculate the linear regression equation for L1 and L2 (demand), and paste it in Y_1: **LinReg (ax + b) L1, L2, Y_1** [ENTER]. Next, calculate the linear regression for L1 and L3 (supply) and paste it in

Figure MWT III.3

L1	L2	L3	1
107.1	6900	12200	
85.5	7900	9900	
64.8	13200	8000	
52.2	13500	7900	
108	6700	14000	
91.8	7600	12000	
77.4	9200	9400	

L1(1)=107.1

Figure MWT III.4

Figure MWT III.5   **Figure MWT III.6**

Y_2: **LinReg(ax + b) L1, L3, Y_2** ENTER
(recall that Y_1 and Y_2 are accessed
using the VARS key). The resulting
equations and graphs are shown in
Figures MWT III.6 and MWT III.7.

Figure MWT III.7

c. Once again we use 2nd TRACE (CALC)
5:intersect to find the equilibrium point,
which is approximately (80, 9931).
Supply and demand for this mp3 player
model are approximately equal at a price
of about $80, with 9931 mp3 players bought and sold.

Now try Exercises 5 through 8 ▶

☑ **A.** You've just learned how
to find market equilibrium
graphically

B. Solving Static Systems with Varying Constraints

When the considerations of a business or industry involve more than two variables,
solutions using matrix methods have a distinct advantage over other methods. Companies often have to perform calculations using basic systems weekly, daily, or even
hourly, to keep up with trends, market changes, changes in cost of raw materials, and
so on. In many situations, the basic requirements remain the same, but the frequently
changing inputs require a recalculation each time they change.

EXAMPLE 3 ▶ **Determining Supply Inventories Using Matrices**

BNN Soft Drinks receives new orders daily for its most popular drink, Saratoga
Cola. It can deliver the carbonated beverage in a twelve-pack of 12-ounce (oz)
cans, a six-pack of 20-oz bottles, or in a 2-L bottle. The ingredients required to
produce a twelve-pack include 1 gallon (gal) of carbonated water, 1.25 pounds (lb)
of sugar, 2 cups (c) of flavoring, and 0.5 grams (g) of caffeine. For the six-pack,
0.8 gal of carbonated water, 1 lb of sugar, 1.6 c of flavoring, and 0.4 g of caffeine
are needed. The 2-L bottle contains 0.47 gal of carbonated water, 0.59 lb of sugar,
0.94 c of flavoring, and 0.24 g of caffeine. How much of each ingredient must be
on hand for Monday's order of 300 twelve-packs, 200 six-packs, and 500
2-L bottles? What quantities must be on hand for Tuesday's order: 410 twelve-
packs, 320 six-packs, and 275 2-L bottles?

Solution ▶ Begin by setting up a general system of equations, letting x represent the number
of twelve-packs, y the number of six-packs, and z the number of 2-L bottles:

$$\begin{cases} 1x + 0.8y + 0.47z = \text{gallons of carbonated water} \\ 1.25x + 1y + 0.59z = \text{pounds of sugar} \\ 2x + 1.6y + 0.94z = \text{cups of flavoring} \\ 0.5x + 0.4y + 0.24z = \text{grams of caffeine} \end{cases}$$

As a matrix equation we have

$$\begin{bmatrix} 1 & 0.8 & 0.47 \\ 1.25 & 1 & 0.59 \\ 2 & 1.6 & 0.94 \\ 0.5 & 0.4 & 0.24 \end{bmatrix} \begin{bmatrix} x \\ y \\ z \end{bmatrix} = \begin{bmatrix} w \\ s \\ f \\ c \end{bmatrix}$$

Enter the 4×3 matrix as matrix A, and the size of the order as matrix as B. Using a calculator, we find

$$AB = \begin{bmatrix} 1 & 0.8 & 0.47 \\ 1.25 & 1 & 0.59 \\ 2 & 1.6 & 0.94 \\ 0.5 & 0.4 & 0.24 \end{bmatrix} \begin{bmatrix} 300 \\ 200 \\ 500 \end{bmatrix} = \begin{bmatrix} 695 \\ 870 \\ 1390 \\ 350 \end{bmatrix},$$

```
[A][B]
        [[695  ]
         [870  ]
         [1390 ]
         [350  ]]
```

and BNN Soft Drinks will need 695 gal of carbonated water, 870 lb of sugar, 1390 c of flavoring, and 350 g of caffeine for Monday's order. After entering $C = \begin{bmatrix} 410 \\ 320 \\ 275 \end{bmatrix}$

for Tuesday's orders, computing the product AC shows 795.25 gal of carbonated water, 994.75 lb of sugar, 1590.5 c of flavoring, and 399 g of caffeine are needed for Tuesday.

Now try Exercises 9 through 14 ▶

Example 3 showed how the creation of a static matrix can help track and control inventory requirements. In Example 4, we use a static matrix to solve a system that will identify the amount of data traffic used by a company during various hours of the day.

EXAMPLE 4 ▶ Identifying the Source of Data Traffic Using Matrices

Mariño Imports is a medium-size company that is considering upgrading from a 1.544 megabytes per sec (Mbps) T1 Internet line to a fractional T3 line with a bandwidth of 7.72 Mbps. They currently use their bandwidth for phone traffic, office data, and Internet commerce. The IT (Internet Technology) director devises a plan to monitor how much data traffic each resource uses on an hourly basis. Because of the physical arrangement of the hardware, she cannot monitor each resource individually. The table shows the information she collected for the first 3 hr. Determine how many gigabytes (GB) each resource used individually during these 3 hrs.

	Phone, Data, and Commerce	Phone and Data	Data and Commerce
9–10:00 A.M.	5.4 GB	4.0 GB	4.2 GB
10–11:00 A.M.	5.3 GB	3.8 GB	4.2 GB
11–12:00 P.M.	5.1 GB	3.5 GB	3.6 GB

Solution ▶ Using p to represent the phone traffic, d for office data, and c for Internet commerce, we create the system shown, which models data use for the 9:00 o'clock hour. Since we actually need to solve two more systems whose only difference is the constant terms (for the 10 and 11 o'clock hours), using a matrix equation to solve the system (Section 6.3)

$$\begin{cases} p + d + c = 5.4 \\ p + d \quad\;\; = 4.0 \\ \quad\;\; d + c = 4.2 \end{cases}$$

will be most convenient. Begin by writing the related matrix equation for this system:

$$AX = B: \begin{bmatrix} 1 & 1 & 1 \\ 1 & 1 & 0 \\ 0 & 1 & 1 \end{bmatrix} \begin{bmatrix} p \\ d \\ c \end{bmatrix} = \begin{bmatrix} 5.4 \\ 4.0 \\ 4.2 \end{bmatrix}$$

```
[A]⁻¹[B]
            [[1.2]
             [2.8]
             [1.4]]
```

Using $X = A^{-1}B$ to solve the system (see figure), we find there was 1.2 GB of phone traffic, 2.8 GB of office data, and 1.4 GB of Internet commerce during this hour. Note that the IT director may make this calculation 10 or more times a day (once for every hour of business). While we *could* solve for the 10 and 11 o'clock hours using

$$C = \begin{bmatrix} 5.3 \\ 3.8 \\ 4.2 \end{bmatrix} \text{ and } D = \begin{bmatrix} 5.1 \\ 3.5 \\ 3.6 \end{bmatrix}, \text{ then calculate } A^{-1}C \text{ and } A^{-1}D, \text{ these calculations}$$

can all be performed simultaneously by combining the matrices B, C, and D into one 3×3 matrix and multiplying by A^{-1}. Due to the properties of matrix multiplication, each column of the product will represent the data information for a given hour, as shown here:

$$[A^{-1}] \begin{bmatrix} 5.4 & 5.3 & 5.1 \\ 4.0 & 3.8 & 3.5 \\ 4.2 & 4.2 & 3.6 \end{bmatrix} = \begin{bmatrix} 1.2 & 1.1 & 1.5 \\ 2.8 & 2.7 & 2.0 \\ 1.4 & 1.5 & 1.6 \end{bmatrix}$$

☑ **B.** You've just learned how to use matrix equations to solve static systems

In the second hour, there was 1.1 GB of phone traffic, 2.7 GB of office data, and 1.5 GB of Internet commerce. In the third hour, 1.5 GB of phone traffic, 2.0 GB of office data, and 1.6 GB of Internet commerce bandwidth was used.

Now try Exercises 15 through 18 ▶

C. Using Matrices to Encode Messages

In *The Gold-Bug,* by Edgar Allan Poe, the narrator deciphers the secret message on a treasure map by accounting for the frequencies of specific letters and words. The coding used was a simple substitution cipher, which today can be broken with relative ease. In modern times where information wields great power, a simple cipher like the one used here will not suffice. When you pay your tuition, register for classes, make on-line purchases, and so on, you are publicly transmitting very private data, which needs to be protected. While there are many complex encryption methods available (including the now famous symmetric-key and public-key techniques), we will use a matrix-based technique. The nature of matrix multiplication makes it very difficult to determine the exact matrices that yield a given product, and we'll use this fact to our advantage. Beginning with a fixed, invertible matrix A, we will develop a matrix B such that the product AB is possible, and our secret message is encrypted in AB. At the receiving end, they will need to know A^{-1} to decipher the message, since $A^{-1}(AB) = (A^{-1}A)B = B$, which is the original message. Note that in case an intruder were to find matrix A (perhaps purchasing the information from a disgruntled employee), we must be able to change it easily.

This means we should develop a method for generating a matrix A, with integer entries, where A is invertible and A^{-1} also has integer entries.

EXAMPLE 5 ▶ Finding an Invertible Matrix A Where Both A and A^{-1} Have Integer Entries

Find an invertible 3×3 matrix A as just described, and its inverse A^{-1}.

Solution ▶ Begin with any 3×3 matrix that has only 1s or -1s on its main diagonal, and 0s below the diagonal. The upper triangle can consist of any integer values you choose, as in

$$\begin{bmatrix} -1 & 5 & -1 \\ 0 & 1 & 8 \\ 0 & 0 & 1 \end{bmatrix}$$

Now, use **any** of the elementary row operations to make the matrix more complex. For instance, we'll using a calculator to create a new matrix by (1) using $R1 + R2 \rightarrow R2$, and $R1 + R3 \rightarrow R3$ to create matrix [C], then (2) using $R2 + R3 \rightarrow R2$ to create matrix [D], then (3) using $R3 + R2 \rightarrow R2$ to create matrix [E], and finally (4) $-2\,R1 + R3 \rightarrow R3$ to obtain our final matrix [A]. To begin, enter the initial matrix as matrix B. For (1) $R1 + R2 \rightarrow R2$, go to the (MATRIX) **MATH** submenu, select option **D:row+(** and press **ENTER** to bring this option to the home screen (Figure MWT III.8). This feature requires us to name the matrix we're using, and to indicate what rows to add, so we enter **D:row+([B], 1, 2).** The screen shown in Figure MWT III.9 indicates we've placed the result in matrix [C]. For $R1 + R3 \rightarrow R3$, recall **D:row+([B], 1, 2)** using **2nd** **ENTER** and change it to **D:row+([C], 1, 3)** **STO ►** [D]. Repeat this process for (3) $R3 + R2 \rightarrow R2$ to create matrix [E]: **D:row+([D], 3, 2)** **STO ►** [E] (Figure MWT III.10). Finally, we compute (4) $-2R1 + R3 \rightarrow R3$ using the (new) option **F:*row+(−2, [E], 1, 3)** **STO ►** [A] and the process is complete. The entries of matrix A are all integers, and A^{-1} exists and also has integer entries (see Figures MWT III.11 and MWT III.12). This will always be the case for matrices created in this way.

Figure MWT III.8

```
row+([B],1,2)→[C
]
       [[-1 5 -1]
        [-1 6 7 ]
        [0  0 1 ]]
```

Figure MWT III.9

```
row+([C],1,3)→[D
]
       [[-1 5 -1]
        [-1 6 7 ]
        [-1 5 0 ]]
```

Figure MWT III.10

```
row+([D],3,2)→[E
]
       [[-1 5  -1]
        [-2 11 7 ]
        [-1 5  0 ]]
```

Figure MWT III.11

```
*row+(-2,[E],1,3
)→[A]
       [[-1 5  -1]
        [-2 11 7 ]
        [1  -5 2 ]]
```

Figure MWT III.12

```
[A]-1
       [[-57 5 -46]
        [-11 1 -9 ]
        [1   0 1  ]]
```

Now try Exercises 19 through 24 ▶

EXAMPLE 6 ▶ **Using Matrices to Encode Messages**

Set up a substitution cipher to encode the message MATH IS SWEET, and then use the matrix A from Example 5 to encrypt it.

Solution ▶ For the cipher, we will associate a unique number to every letter in the alphabet. This can be done randomly or using a systematic approach. Here we choose to associate 0 with a blank space, and assign 1 to A, -1 to B, 2 to C, -2 to D, and so on.

Blank	A	B	C	D	E	F	G	H	I	J	K	L	M
0	1	-1	2	-2	3	-3	4	-4	5	-5	6	-6	7

N	O	P	Q	R	S	T	U	V	W	X	Y	Z
-7	8	-8	9	-9	10	-10	11	-11	12	-12	13	-13

Now encode the secret message as shown:

M	A	T	H		I	S		S	W	E	E	T
7	1	-10	-4	0	5	10	0	10	12	3	3	-10

We next enter the coded message into a new matrix B, by entering it letter by letter *into the columns* of B. Note that since the encrypting matrix A is 3×3, B must have 3 rows for multiplication to be possible. The result is

$$B = \begin{bmatrix} M & H & S & W & T \\ A & * & * & E & * \\ T & I & S & E & * \end{bmatrix} = \begin{bmatrix} 7 & -4 & 10 & 12 & -10 \\ 1 & 0 & 0 & 3 & 0 \\ -10 & 5 & 10 & 3 & 0 \end{bmatrix}$$

If the message is too short to fill matrix B, we use blank spaces to complete the final column. Computing the product AB encrypts the message, and only someone with access to A^{-1} will be able to read it:

$$AB = \begin{bmatrix} -1 & 5 & -1 \\ -2 & 11 & 7 \\ 1 & -5 & 2 \end{bmatrix}\begin{bmatrix} 7 & -4 & 10 & 12 & -10 \\ 1 & 0 & 0 & 3 & 0 \\ -10 & 5 & 10 & 3 & 0 \end{bmatrix}$$

$$= \begin{bmatrix} 8 & -1 & -20 & 0 & 10 \\ -73 & 43 & 50 & 30 & 20 \\ -18 & 6 & 30 & 3 & -10 \end{bmatrix}$$

The coded message is $8, -73, -18, -1, 43, 6, -20, 50, 30, 0, 30, 3, 10, 20, -10$.

Now try Exercises 25 through 30 ▶

EXAMPLE 7 ▶ **Deciphering Coded Messages Using an Inverse Matrix**

Decipher the coded message from Example 6, using A^{-1} from Example 5.

Solution ▶ The received message is $8, -73, -18, -1, 43, 6, -20, 50, 30, 0, 30, 3, 10, 20, -10$, and is the result of the product AB. To find matrix B, we apply A^{-1} since $A^{-1}(AB) = (A^{-1}A)B = B$. Writing the received message in matrix form we have

$$AB = \begin{bmatrix} 8 & -1 & -20 & 0 & 10 \\ -73 & 43 & 50 & 30 & 20 \\ -18 & 6 & 30 & 3 & -10 \end{bmatrix}$$

Next multiply AB by A^{-1} on the left, to determine matrix B:

$$A^{-1}(AB) = \begin{bmatrix} -57 & 5 & -46 \\ -11 & 1 & -9 \\ 1 & 0 & 1 \end{bmatrix} \begin{bmatrix} 8 & -1 & -20 & 0 & 10 \\ -73 & 43 & 50 & 30 & 20 \\ -18 & 6 & 30 & 3 & -10 \end{bmatrix}$$

$$= \begin{bmatrix} 7 & -4 & 10 & 12 & -10 \\ 1 & 0 & 0 & 3 & 0 \\ -10 & 5 & 10 & 3 & 0 \end{bmatrix} = B$$

 C. You've just learned how to use matrices for encryption/decryption

Writing matrix B in sentence form gives 7, 1, −10, −4, 0, 5, 10, 0, 10, 12, 3, 3, −10, 0, 0, and using the substitution cipher to replace numbers with letters, reveals the message MATH IS SWEET.

Now try Exercises 31 through 36 ▶

MODELING WITH TECHNOLOGY EXERCISES

1. A water sports company manufactures high-end wakeboards. The monthly demand for their boards is modeled by $D(p) = -2.25p + 1125$, where $D(p)$ represents the number of boards bought at price p. The supply function is $S(p) = 0.75p - 75$, where $S(p)$ represents the number of boards supplied at price p. Use a graphing calculator to find the market equilibrium for this product.

2. A local metal shop manufactures rabbit cages. The monthly demand for their cage is related to its price. The function $D(p) = -\frac{1}{2}p + \frac{45}{2}$ models this demand, where $D(p)$ represents the number of cages bought at price p. The supply function is $S(p) = p - 15$, where $S(p)$ represents the number of cages supplied at price p. Use a graphing calculator to find the market equilibrium for this product.

3. The Bureau of Labor and Statistics (BLS) keeps track of important statistics for many different markets. In studying the supply and demand for refined gasoline in Atlanta, Georgia, for the month of April, the BLS stated that monthly demand was modeled by the function D shown, where $D(p)$ represents the number of gallons bought at price p. The monthly supply was modeled by the function S shown, where $S(p)$ represents the number of gallons supplied at price p. Use a graphing calculator to find the market equilibrium for this product:
$D(p) = (-5.000 \times 10^7)p + (2.435 \times 10^8)$,
$S(p) = (5.000 \times 10^7)p - (6.350 \times 10^7)$

4. The Bureau of Labor and Statistics has been closely following the energy-efficient fluorescent lightbulb market over the past year. The yearly demand for these "green" lightbulbs is related to its price. The demand function is

$D(p) = (-4.500 \times 10^6)p + (5.191 \times 10^7)$, where $D(p)$ represents the number of 30-W bulbs bought at price p. The supply function is $S(p) = (8.5 \times 10^6)p - (6.483 \times 10^7)$, where $S(p)$ represents the number of 30-watt bulbs supplied at price p. Use a graphing calculator to find the market equilibrium for this product.

5. The company from Exercise 1 has hired an outside consulting firm to do some market research on their wakeboard. This consulting firm collected the following supply and demand data for this and comparable wakeboards over a 10-week period. Find the equilibrium point. Round your answer to the nearest integer and dollar.

Average Price (in U.S. dollars)	Quantity Demanded	Available Inventory
424.85	175	232
445.25	166	247
389.55	291	215
349.98	391	201
402.22	218	226
413.87	200	222
481.73	139	251
419.45	177	235
397.05	220	219
361.90	317	212

6. The metal shop from Exercise 2 has collected some data on sales and production of its rabbit cages over the past 8 weeks. The following table shows the supply and demand data for this cage. Find the equilibrium point (round to the nearest cent and whole cage).

Average Price (in U.S. dollars)	Quantity Sold (Demand)	Production (Supply)
22.99	12	7
21.49	14	6
23.99	11	7
26.99	9	11
25.99	8	10
27.99	8	13
24.49	10	9
26.49	9	11

7. The Bureau of Labor and Statistics from Exercise 3 collected the following supply-and-demand data for refined gasoline for the month of May. Data were collected every Tuesday and Friday. Find the equilibrium point, rounding your answer to the nearest hundred thousand gallons and whole cent.

Average Price (in U.S. dollars)	Quantity Demanded (1×10^7 gal)	Available Inventory (1×10^7 gal)
3.17	8.82	9.10
3.12	8.87	9.05
3.04	9.08	8.97
2.84	9.22	8.91
3.11	8.92	9.02
3.15	8.76	9.08
3.10	9.01	8.99
3.11	8.94	9.01
2.93	9.13	8.93

8. The Bureau of Labor and Statistics from Exercise 4 collected the following supply and demand data for the energy-efficient fluorescent lightbulbs sold each month for the past year. Find the equilibrium point, rounding your answer to the nearest ten thousand lightbulbs and whole cent. What is the yearly demand at the equilibrium point?

Average Price (in U.S. dollars)	Quantity Demanded (in millions)	Available Inventory (in millions)
9.40	0.84	1.23
8.51	1.17	0.95
8.78	1.05	1.11
10.82	0.68	1.29
6.77	1.47	0.77
9.33	0.91	1.21
8.34	1.25	0.88
10.37	0.76	1.27
8.62	1.09	1.02
8.44	1.21	0.92
8.58	1.18	0.97
8.96	1.01	1.17

9. Slammin' Drums manufactures several different types of drums. Its most popular drums are the 22″ bass drum, the 12″ tom, and the 14″ snare drum. The 22″ bass drum requires 7 ft² of skin, 8.5 ft² of wood veneer, 8 tension rods, and 11.5 ft of hoop. The 12″ tom requires 2 ft² of skin, 3 ft² of wood veneer, 6 tension rods, and 6.5 ft of hoop. The 14″ snare requires 2.5 ft² of skin, 1.5 ft² of wood veneer, 10 tension rods, and 7 ft of hoop. In February, Slammin' Drums received orders for 15 bass drums, 21 toms, and 27 snares. Use your calculator and a matrix equation to determine how much of each raw material they need to have on hand, to fill these orders.

10. In March, Slammin' Drum's orders consisted of 19 bass drums, 19 toms, and 25 snares. Use your calculator and a matrix equation to determine how much of each raw material they need to have on hand, to fill their orders. (See Exercise 9.)

11. The following table represents Slammin's orders for the months of April through July. Use your calculator and a matrix equation to determine how much of each raw material they need to have on hand to fill these orders. (See Exercise 9.) (*Hint:* Using a clever 4×3 and 3×1 matrix can reduce this problem to a single step.)

	April	May	June	July
Bass drum	23	21	17	14
Tom	20	18	15	17
Snare drum	29	35	27	25

12. The following table represents Slammin's orders for the months of August through November. Use your calculator and a matrix equation to determine how much of each raw material they need to have on hand to fill their orders. (See Exercise 9.) (*Hint:* Using a clever 4 × 3 and 3 × 1 matrix can reduce this problem to a single step.)

	August	September	October	November
Bass drum	17	22	16	12
Tom	15	14	13	11
Snare drum	32	28	27	21

13. Midwest Petroleum (MP) produces three types of combustibles using common refined gasoline and vegetable products. The first is E10 (also known as gasohol), the second is E85, and the third is biodiesel. One gallon of E10 requires 0.90 gal of gasoline, 2 lb of corn, 1 oz of yeast, and 0.5 gal of water. One gallon of E85 requires 0.15 gal of gasoline, 17 lb of corn, 8.5 oz of yeast, and 4.25 gal of water. One gallon of biodiesel requires 20 lb of corn and 3 gal of water. One week's production at MP consisted of 100,000 gal of E10, 15,000 gal of E85, and 7000 gal of biodiesel. Use your calculator and a matrix equation to determine how much of each raw material they used to fill their orders.

14. The following table represents Midwest Petroleum's production for the next 3 weeks. Use your calculator and a matrix equation to determine the total amount of raw material they used to fill their orders. See Exercise 13.

	Week 2	Week 3	Week 4
E10	110,000	95,000	105,000
E85	17,000	18,000	20,000
Biodiesel	6,000	8,000	10,000

15. Roll-X Watches makes some of the finest wristwatches in the world. Their most popular model is the Clam. It comes in three versions: Silver, Gold, and Platinum. Management thinks there might be a thief in the production line, so they decide to closely monitor the precious metal consumption. A Silver Clam contains 1.2 oz of silver and 0.2 oz of gold. A Gold Clam contains 0.5 oz of silver, 0.8 oz of gold, and 0.1 oz of platinum. A Platinum Clam contains 0.2 oz of silver, 0.5 oz of gold, and 0.7 oz of platinum. During the first week of monitoring, the production team used 10.9 oz of silver, 9.2 oz of gold, and 2.3 oz of platinum. Use your calculator and a matrix equation to determine the number of each type of watch that should have been produced.

16. The following table contains the precious metal consumption of the Roll-X Watch production line during the next five weeks (see Exercise 15). Use your graphing calculator to determine the number of each type of watch that should have been produced each week. For which week does the data seem to indicate a possible theft of precious metal?

Ounces	Week 1	Week 2	Week 3	Week 4	Week 5
Silver	13.1	9	12.9	11.9	11.2
Gold	11	7.7	8.6	8.4	9.5
Platinum	2.5	1.5	0.9	2.8	1.7

17. There are three classes of grain, of which three bundles from the first class, two from the second, and one from the third make 39 measures. Two of the first, three of the second, and one of the third make 34 measures. And one of the first, two of the second, and three of the third make 26 measures. How many measures of grain are contained in one bundle of each class? (*This is the historic problem from the Chiu chang suan shu.*)

18. During a given week, the measures of grain that make up the bundles in Exercise 17 can vary slightly. Three local Chinese bakeries always buy the same amount of bundles, as outlined in Exercise 17. That is to say, bakery 1 buys three bundles of the first class, two of the second, and one of the third. Bakery 2 buys two of the first, three of the second, and one of the third. And finally, bakery 3 buys one of the first, two of the second, and three of the third. The following table outlines how many measures of grain each bakery received each day. How many measures of grain were contained in one bundle of each class, on each day?

	Mon	Tues	Wed	Thurs	Fri
Bakery 1 (measures)	39	38	38	37.75	39.75
Bakery 2 (measures)	34	33	33.5	32.5	35
Bakery 3 (measures)	26	26	27	26.25	27.25

For Exercises 19–24, use the criteria indicated to find 3×3 matrices A and A^{-1}, where the entries of both are all integers.

19. The lower triangle is all zeroes.

20. The upper triangle is all zeroes.

21. $a_{2,1} = 5$

22. $a_{3,2} = -2$

23. $a_{3,1} = 1$ and $a_{2,3} = 2$

24. $a_{2,1} = -3$ and $a_{1,3} = 1$

25. Use the matrix A you created in Exercise 19 and the substitution cipher from Example 6 to encrypt your full name.

26. Use the matrix A you created in Exercise 20 and the substitution cipher from Example 6 to encrypt your school's name.

27. Design your own substitution cipher. Then use it and the matrix A you created in Exercise 21 to encrypt the title of your favorite movie.

28. Design your own substitution cipher. Then use it and the matrix A you created in Exercise 22 to encrypt the title of your favorite snack food.

29. Design your own substitution cipher. Then use it and the matrix A you created in Exercise 23 to encrypt the White House switchboard phone number, 202-456-1414.

30. Design your own substitution cipher. Then use it and the matrix A you created in Exercise 24 to encrypt the Casa Rosada switchboard phone number 54-11-4344-3600. The Casa Rosada, or Pink House, consists of the offices of the president of Argentina.

31. Use the matrix A^{-1} from Exercise 19, and the appropriate substitution cipher to decrypt the message from Exercise 25.

32. Use the matrix A^{-1} from Exercise 20, and the appropriate substitution cipher to decrypt the message from Exercise 26.

33. Use the matrix A^{-1} from Exercise 21, and the appropriate substitution cipher to decrypt the message from Exercise 27.

34. Use the matrix A^{-1} from Exercise 22, and the appropriate substitution cipher to decrypt the message from Exercise 28.

35. Use the matrix A^{-1} from Exercise 23, and the appropriate substitution cipher to decrypt the message from Exercise 29.

36. Use the matrix A^{-1} from Exercise 24, and the appropriate substitution cipher to decrypt the message from Exercise 30.

7

Analytic Geometry and the Conic Sections

CHAPTER OUTLINE

CHAPTER CONNECTIONS

One of the most breathtaking, daredevil stunts performed at air shows is the *power dive*. In some cases, as the pilot dives toward the ground and pulls out of the dive just in time, the plane flys along a path that can be modeled by a hyperpola (the third member of the family of conic sections). If we consider a given point on the ground and under the grandstand as the origin (0, 0), we can use the equation that models the hyperbolic path of the plane to determine its minimum altitude as it passes over the stands. This application appears as Exercise 81 in Section 7.3.

Check out these other real-world connections:

▶ Designing an Elliptical Garden with Fountains (Section 7.2, Exercise 66)

▶ The Design of a Lithotripter for treating kidney stones with shockwaves (Section 7.2, Exercise 68)

▶ Locating a Ship Using Radar (Section 7.3, Exercise 85)

▶ Parabolic Shape of a Solar Furnace (Section 7.4, Exercise 90)

Learning Objectives

In Section 7.1 you will learn how to:

☐ **A.** Verify theorems from basic geometry involving the distance between two points

☐ **B.** Verify that points (x, y) are an equal distance from a given point and a given line

☐ **C.** Use the defining characteristics of a conic section to find its equation

Generally speaking, **analytical geometry** is a study of geometry using the tools of algebra and a coordinate system. These tools include the midpoint and distance formulas; the algebra of parallel, perpendicular, and intersecting lines; and other tools that help establish geometric concepts. In this section, we'll use these tools to verify certain relationships, then use these relationships to introduce a family of curves known as the **conic sections.**

A. Verifying Relationships from Plane Geometry

For the most part, the algebraic tools used in this study were introduced in previous chapters. As the midpoint and distance formulas play a central role, they are restated here for convenience.

Algebraic Tools Used in Analytical Geometry

Given two points $P_1 = (x_1, y_1)$ and $P_2 = (x_2, y_2)$ in the xy-plane.

Midpoint Formula
The midpoint of line segment P_1P_2 is

$$(x, y) = \left(\frac{x_1 + x_2}{2}, \frac{y_1 + y_2}{2} \right)$$

Distance Formula
The distance from P_1 to P_2 is

$$d = \sqrt{(x_2 - x_1)^2 + (y_2 - y_1)^2}$$

These formulas can be used to verify the conclusion of many theorems from Euclidean geometry, while providing important links to an understanding of the conic sections.

EXAMPLE 1 ▶ **Verifying a Theorem from Basic Geometry**

A theorem from basic geometry states: *The midpoint of the hypotenuse of a right triangle is an equal distance from all three vertices.* Verify this statement for the right triangle formed by $(-4, -2)$, $(4, -2)$, and $(4, 4)$.

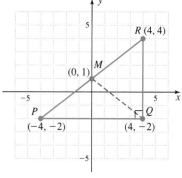

Solution ▶ After the plotting points and drawing a triangle, we note the hypotenuse has endpoints $(-4, -2)$ and $(4, 4)$, with midpoint

$$\left(\frac{4 + (-4)}{2}, \frac{4 + (-2)}{2} \right) = (0, 1).$$ Using the

distance formula to find the distance from $(0, 1)$ to $(4, 4)$ gives

$$\begin{aligned} d &= \sqrt{(0 - 4)^2 + (1 - 4)^2} \\ &= \sqrt{(-4)^2 + (-3)^2} \\ &= \sqrt{25} \\ &= 5 \end{aligned}$$

From the definition of midpoint, $(0, 1)$ is also 5 units from $(-4, -2)$. Checking the distance from $(0, 1)$ to the vertex $(4, -2)$ gives

$$\begin{aligned} d &= \sqrt{(4 - 0)^2 + (-2 - 1)^2} \\ &= \sqrt{4^2 + (-3)^2} \\ &= \sqrt{25} \\ &= 5 \end{aligned}$$

The midpoint of the hypotenuse *is* an equal distance from all three vertices (see the figure).

Now try Exercises 7 through 12 ▶

Recall from Section 2.1 that a circle is the set of all points that are an equal distance (called the radius) from a given point (called the center). If all three vertices of a triangle lie on the circumference of a circle, we say the circle **circumscribes** the triangle. Based on our earlier work, it appears we could also state the theorem in Example 1 as *For any circle in the xy-plane whose center (h, k) is the midpoint of the hypotenuse of a right triangle, all vertices lie on the circle defined by* $(x - h)^2 + (y - k)^2 = (\frac{d}{2})^2$, *where d is the length of the hypotenuse.* See Figure 7.1 and **Exercises 13 through 20.**

Figure 7.1

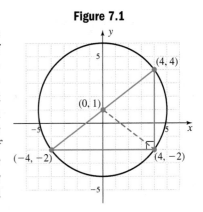

☑ **A.** You've just learned how to verify theorems from basic geometry involving the distance between two points

B. The Distance between a Point and a Line

In a study of analytical geometry, we are also interested in the distance *d* between a point and a *line*. This is always defined as the **perpendicular distance,** or the length of a line segment perpendicular to the given line, with the given point and the point of intersection as endpoints (see Figure 7.2).

Figure 7.2

EXAMPLE 2 ▶ **Locating Points That Are an Equal Distance from a Given Point and Line**

In Figure 7.3, the origin (0, 0) is seen to be an equal distance from the point (0, 2) and the line $y = -2$. Show that the following points are also an equal distance from (0, 2) and $y = -2$:

     a. $(2, \frac{1}{2})$ **b.** (4, 2)   **c.** (8, 8)

Solution ▶ Since the given line is horizontal, the perpendicular distance from the line to each point can be found by vertically counting the units. It remains to show that this is also the distance from the given point to (0, 2) (see Figure 7.4).

     a. The distance from $(2, \frac{1}{2})$ to $y = -2$ is **2.5 units**. The distance from $(2, \frac{1}{2})$ to (0, 2) is

Figure 7.3

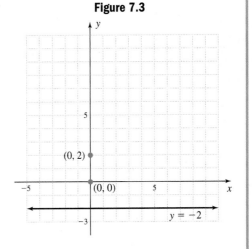

$$d = \sqrt{(0 - 2)^2 + (2 - 0.5)^2}$$
$$= \sqrt{(-2)^2 + (1.5)^2}$$
$$= \sqrt{6.25}$$
$$= 2.5 \ \checkmark$$

     b. The distance from (4, 2) to $y = -2$ is **4 units**. The distance from (4, 2) to (0, 2) is

$$d = \sqrt{(0 - 4)^2 + (2 - 2)^2}$$
$$= \sqrt{(-4)^2 + (0)^2}$$
$$= \sqrt{16}$$
$$= 4 \ \checkmark$$

c. The distance from $(8, 8)$ to $y = -2$ is **10 units**. The distance from $(8, 8)$ to $(0, 2)$ is

$$d = \sqrt{(0 - 8)^2 + (2 - 8)^2}$$
$$= \sqrt{(-8)^2 + (-6)^2}$$
$$= \sqrt{100}$$
$$= 10 \checkmark$$

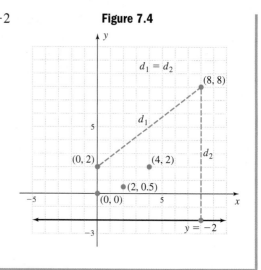

Figure 7.4

☑ **B.** You've just learned how to verify that points (x, y) are an equal distance from a given point and a given line

Now try Exercises 23 through 26 ▶

C. Characteristics of the Conic Sections

Examples 1 and 2 bring us one step closer to the wider application of these ideas in a study of the conic sections. But before the connection is clearly made, we'll introduce some background on this family of curves. In common use, a cone might bring to mind the conical paper cups found at a water cooler. The point of the cone is called the **vertex** and the sheet of paper forming the sides is called a **nappe.** In mathematical terms, a cone has two nappes, formed by rotating a nonvertical line (called the generator), about a vertical line (called the axis), at their point of intersection—the vertex (see Figure 7.5). The conic sections are so named because all curves in the family can be formed by a *section* of the *cone,* or more precisely the intersection of a plane and a cone. Figure 7.6 shows that if the plane does not go through the vertex, the intersection will produce a circle, ellipse, parabola, or hyperbola.

Figure 7.5

Axis

Nappe

Generator

Vertex

Nappe

WORTHY OF NOTE

If the plane *does* go through the vertex, the result is a single point, a single line (if the plane contains the generator), or a pair of intersecting lines (if the plane contains the axis).

Figure 7.6

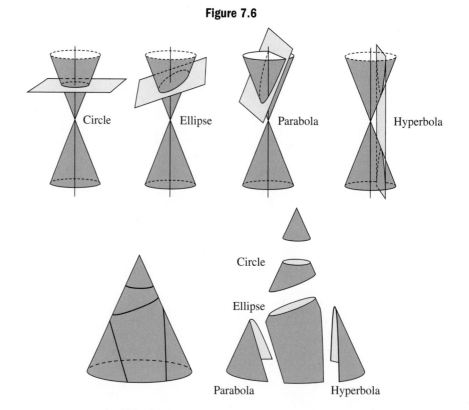

Circle   Ellipse   Parabola   Hyperbola

Circle

Ellipse

Parabola   Hyperbola

The connection we seek to make is that each conic section can be defined in terms of the distance between points in the plane, as in Example 1, or the distance between a given point and a line, as in Example 2. In Example 1, we noted the points $(-4, -2)$, $(4, -2)$, and $(4, 4)$ were all on a circle of radius 5 with center $(0, 1)$, in line with the analytic definition of a circle: *A circle is the set of all points that are an equal distance (called the radius) from a given point (called the center).*

In Example 2, you may have noticed that the points seemed to form the right branch of a parabola (see Figure 7.7), and in fact, this example illustrates the analytic definition of a parabola: *A parabola is the set of all points that are an equal distance from a given point (called the **focus**), and a given line (called the **directrix**).*

Figure 7.7

The focus and directrix are not actually part of the graph, they are simply used to locate points on the graph. For this reason all foci (plural of focus) will be represented by a "*" symbol rather than a point.

EXAMPLE 3 ▶ **Finding an Equation for All Points That Form a Certain Parabola**

With Example 2 as a pattern, use the analytic definition to find a formula (equation) for the set of all points that form the parabola.

Solution ▶ Use the ordered pair (x, y) to represent an arbitrary point on the parabola. Since any point on the line $y = -2$ has coordinates $(x, -2)$, we set the distance from $(x, -2)$ to (x, y) equal to the distance from $(0, 2)$ to (x, y). The result is

$$\sqrt{(x - x)^2 + [y - (-2)]^2} = \sqrt{(x - 0)^2 + (y - 2)^2} \qquad \text{distances are equal}$$
$$\sqrt{(y + 2)^2} = \sqrt{x^2 + (y - 2)^2} \qquad \text{simplify}$$
$$(y + 2)^2 = x^2 + (y - 2)^2 \qquad \text{power property}$$
$$y^2 + 4y + 4 = x^2 + y^2 - 4y + 4 \qquad \text{expand binomials}$$
$$8y = x^2 \qquad \text{simplify}$$
$$y = \frac{1}{8}x^2 \qquad \text{result}$$

All points satisfying these conditions are on the parabola defined by $y = \frac{1}{8}x^2$.

Now try Exercises 27 and 28 ▶

At this point, it seems reasonable to ask what happens when the distance from the focus to (x, y) is *less than* the distance from the directrix to (x, y). For example, what if the distance is only two-thirds as long? As you might guess, the result is one of the other conic sections, in this case an ellipse. If the distance from the focus to a point (x, y) is *greater than* the distance from the directrix to (x, y), one branch of a hyperbola is formed. While we will defer a development of their general equations until later in the chapter, the following diagrams serve to illustrate this relationship for the ellipse, and show why we refer to the conic sections as a *family of curves*. In Figure 7.8, the line segment from the focus to each point on the graph (shown in blue), is exactly two-thirds the length of the line segment from the directrix to the same point (shown in red). Note the graph of these points forms the right half of an ellipse. In Figure 7.9, the lines and points forming the first half are moved to the background to more clearly show the remaining points that form the complete graph.

Figure 7.8

Figure 7.9

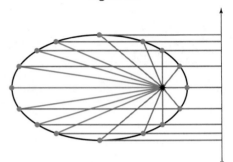

EXAMPLE 4 ▶ **Finding an Equation for All Points That Form a Certain Ellipse**

Suppose we arbitrarily select the point $(1, 0)$ as a focus and the (vertical) line $x = 4$ as the directrix. Use these to find an equation for the set of all points where the distance from the focus to a point (x, y) is $\frac{1}{2}$ the distance from the directrix to (x, y).

Solution ▶ Since any point on the line $x = 4$ has coordinates $(4, y)$, we have

$$\text{Distance from } (1, 0) \text{ to } (x, y) = \frac{1}{2} \left[\text{distance from } (4, y) \text{ to } (x, y) \right] \quad \text{in words}$$

$$\sqrt{(x - 1)^2 + [y - (0)]^2} = \frac{1}{2}\sqrt{(x - 4)^2 + (y - y)^2} \quad \text{resulting equation}$$

$$\sqrt{(x - 1)^2 + y^2} = \frac{1}{2}\sqrt{(x - 4)^2} \quad \text{simplify}$$

$$(x - 1)^2 + y^2 = \frac{1}{4}(x - 4)^2 \quad \text{power property}$$

$$x^2 - 2x + 1 + y^2 = \frac{1}{4}(x^2 - 8x + 16) \quad \text{expand binomials}$$

$$x^2 - 2x + 1 + y^2 = \frac{1}{4}x^2 - 2x + 4 \quad \text{distribute}$$

$$\frac{3}{4}x^2 + y^2 = 3 \quad \text{simplify: } 1x^2 - \frac{1}{4}x^2 = \frac{3}{4}x^2$$

$$3x^2 + 4y^2 = 12 \quad \text{polynomial form}$$

All points satisfying these conditions are on the ellipse defined by $3x^2 + 4y^2 = 12$.

Now try Exercises 29 and 30 ▶

Actually, any given ellipse has two foci (see Figure 7.10) and the equation from Example 4 could also have been developed using the left focus (with the directrix also on the left). This symmetrical relationship leads us to an *alternative definition* for the ellipse, which we will explore further in Section 7.2.

Figure 7.10

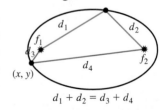

For foci f_1 and f_2, an ellipse is the set of all points (x, y) where the sum of the distances from f_1 to (x, y) and f_2 to (x, y) is constant.

See Figure 7.11 and **Exercises 31 and 32.** Both the focus/directrix definition and the two foci definition have merit, and simply tend to call out different characteristics and applications of the ellipse. The hyperbola also has a focus/directrix definition and a two foci definition. **See Exercises 33 and 34.**

Figure 7.11

$d_1 + d_2 = d_3 + d_4$

☑ **C.** You've just learned how to use the defining characteristics of a conic section to find its equation

7.1 EXERCISES

▶ CONCEPTS AND VOCABULARY

Fill in the blank with the appropriate word or phrase. Carefully reread the section if needed.

1. Analytical geometry is a study of _____ using the tools of _____.

2. The distance formula is $d =$ _____; the midpoint formula is $M =$ _____.

3. The distance between a point and a line always refers to the _____ distance.

4. The conic sections are formed by the intersection of a _____ and a _____.

5. If a plane intersects a cone at its vertex, the result is a _____, a line, or a pair of _____ lines.

6. A circle is defined relative to an equal distance between two _____. A parabola is defined relative to an equal distance between a _____ and a _____.

▶ DEVELOPING YOUR SKILLS

The three points given form a right triangle. Find the midpoint of the hypotenuse and verify that the midpoint is an equal distance from all three vertices.

7. $P_1 = (-5, 2)$
    $P_2 = (1, 2)$
    $P_3 = (-5, -6)$

8. $P_1 = (3, 2)$
    $P_2 = (3, 14)$
    $P_3 = (8, 2)$

9. $P_1 = (-2, 1)$
    $P_2 = (6, -5)$
    $P_3 = (2, -7)$

10. $P_1 = (0, -5)$
    $P_2 = (-6, 4)$
    $P_3 = (6, -1)$

11. $P_1 = (10, -21)$
    $P_2 = (-6, -9)$
    $P_3 = (3, 3)$

12. $P_1 = (6, -6)$
    $P_2 = (-12, 18)$
    $P_3 = (20, 42)$

13. Find the equation of the circle that circumscribes the triangle in Exercise 7.

14. Find the equation of the circle that circumscribes the triangle in Exercise 8.

15. Find the equation of the circle that circumscribes the triangle in Exercise 9.

16. Find the equation of the circle that circumscribes the triangle in Exercise 10.

17. Find the equation of the circle that circumscribes the triangle in Exercise 11.

18. Find the equation of the circle that circumscribes the triangle in Exercise 12.

19. Of the following six points, four are an equal distance from the point $A(2, 3)$ and two are not. (a) Identify which four, and (b) find any two additional points that are this same (nonvertical, nonhorizontal) distance from $(2, 3)$:

    $B(7, 15)$ $C(-10, 8)$ $D(9, 14)$ $E(-3, -9)$
    $F(5, 4 + 3\sqrt{10})$ $G(2 - 2\sqrt{30}, 10)$

20. Of the following six points, four are an equal distance from the point $P(-1, 4)$ and two are not. (a) Identify which four, and (b) Find any two additional points that are the same (nonvertical, nonhorizontal) distance from $(-1, 4)$.

    $Q(-9, 10)$ $R(5, 12)$ $S(-7, 11)$ $T(4, 4 + 5\sqrt{3})$
    $U(-1 + 4\sqrt{6}, 6)$ $V(-7, 4 + \sqrt{51})$

▶ WORKING WITH FORMULAS

The Perpendicular Distance from a Point to a Line:
$$d = \left| \frac{Ax_1 + By_1 + C}{\sqrt{A^2 + B^2}} \right|$$

The perpendicular distance from a point (x_1, y_1) to a given line can be found using the formula shown, where $Ax + By + C = 0$ is the equation of the line in standard form (A, B, and C are integers).

21. Use the formula to verify that $P(-6, 2)$ and $Q(6, 4)$ are an equal distance from the line $y = -\frac{1}{2}x + 3$.

Exercise 21

22. Find the value(s) for y that ensure $(1, y)$ is this same distance from $y = -\frac{1}{2}x + 3$.

▶ APPLICATIONS

23. Of the following four points, three are an equal distance from the point $A(0, 1)$ and the line $y = -1$. (a) Identify which three, and (b) find any two additional points that satisfy these conditions.

$B(-6, 9)$ $C(4, 4)$ $D(-2\sqrt{2}, 6)$ $E(4\sqrt{2}, 8)$

24. Of the following four points, three are an equal distance from the point $P(2, 4)$ and the line $y = -4$. (a) Identify which three, and (b) find any two additional points that satisfy these conditions.

$Q(-10, 9)$ $R(2 + 4\sqrt{2}, 3)$ $S(10, 4)$
$T(2 - 4\sqrt{5}, 5)$

25. Consider a fixed *point* $(0, -4)$ and a fixed *line* $y = 4$. Verify that the distance from each point to $(0, -4)$, is equal to the distance from the point to the line $y = 4$.

$A(4, -1)$ $B\left(10, -\dfrac{25}{4}\right)$ $C(4\sqrt{2}, -2)$

$D(8\sqrt{5}, -20)$

26. Consider a fixed *point* $(0, -2)$ and a fixed *line* $y = 2$. Verify that the distance from each point to $(0, -2)$, is equal to the distance from the point to the line $y = 2$.

$P(12, -18)$ $Q\left(6, -\dfrac{9}{2}\right)$ $R(4\sqrt{5}, -10)$

$S(4\sqrt{6}, -12)$

27. The points from Exercise 25 are on the graph of a parabola. Find the equation of the parabola.

28. The points from Exercise 26 are on the graph of a parabola. Find the equation of the parabola.

29. Using $(0, -2)$ as the focus and the horizontal line $y = -8$ as the directrix, find an equation for the set of all points (x, y) where the distance from the focus to (x, y) is one-half the distance from the directrix to (x, y).

30. Using $(4, 0)$ as the focus and the vertical line $x = 9$ as the directrix, find an equation for the set of all points (x, y) where the distance from the focus to (x, y) is two-thirds the distance from the directrix to (x, y).

31. From Exercise 29, verify the points $(-3, 2)$ and $(\sqrt{12}, 0)$ are on the ellipse defined by $4x^2 + 3y^2 = 48$. Then verify that $d_1 + d_2 = d_3 + d_4$.

Exercise 31

32. From Exercise 30, verify the points $(4, \frac{10}{3})$ and $(-3, -\sqrt{15})$ are on the ellipse defined by $5x^2 + 9y^2 = 180$. Then verify that $d_1 + d_2 = d_3 + d_4$.

Exercise 32

33. From the focus/directrix definition of a hyperbola: *If the distance from the focus to a point (x, y) is* ***greater than*** *the distance from the directrix to (x, y), one branch of a hyperbola is formed.* Using $(2, 0)$ as the focus and the vertical line $x = \frac{1}{2}$ as the directrix, find an equation for the set of all points (x, y) where the distance from the focus to (x, y), is twice the distance from the directrix to (x, y).

34. From the two foci definition of a hyperbola: *For foci f_1 and f_2, a hyperbola is the set of all points (x, y) where the difference of the distances from f_1 to (x, y) and f_2 to (x, y) is constant.* Verify the points $(2, 3)$ and $(-3, -2\sqrt{6})$ are on the graph of the hyperbola from Exercise 33. Then verify $d_1 - d_2 = d_3 - d_4$.

▶ EXTENDING THE CONCEPT

35. Do some reading or research on the **orthocenter** of a triangle, and the **centroid** of a triangle. How are they found? What are their properties? Use the ideas and skills from this section to find the (a) orthocenter and (b) centroid of the triangle formed by the points $A(-8, 2)$, $B(-2, -6)$, and $C(4, 0)$.

36. **Properties of a circle:** A theorem from elementary geometry states: *If a radius is perpendicular to a chord, it bisects the chord.* Verify this is true for the circle, radii, and chords shown.

37. Verify that points $C(-2, 3)$ and $D(2\sqrt{2}, \sqrt{6})$ are points on the ellipse with foci at $A(-2, 0)$ and $B(2, 0)$, by verifying $d(AC) + d(BC) = d(AD) + d(BD)$. The expression that results has the form

$\sqrt{A + B} + \sqrt{A - B}$, which prior to the common use of technology had to be simplified using the formula $\sqrt{A + B} + \sqrt{A - B} = \sqrt{a} + \sqrt{b}$, where $a = 2A$ and $b = 4(A^2 - B^2)$. Use this relationship to verify the equation above.

▶ **MAINTAINING YOUR SKILLS**

38. (4.5) $5,000 is deposited at 4% compounded continuously. How many years will it take for the account to exceed $8000?

39. (4.4) Solve for x in both exact and approximate form:

   a. $5 = \dfrac{10}{1 + 9e^{-0.5x}}$

   b. $345 = 5e^{0.4x} + 75$

40. (3.4) Use the rational zeroes theorem and other tools to factor $f(x)$ and sketch its graph:

   $f(x) = x^4 - 3x^3 - 3x^2 + 11x - 6.$

41. (3.5) Sketch a complete graph of $h(x) = \dfrac{x^2 - 9}{x^2 - 4}$.

   Clearly label all intercepts and asymptotes.

7.2 | The Circle and the Ellipse

Learning Objectives

In Section 7.2 you will learn how to:

☐ **A.** Use the characteristics of a circle and its graph to understand the equation of an ellipse

☐ **B.** Use the equation of an ellipse to graph central and noncentral ellipses

☐ **C.** Locate the foci of an ellipse and use the foci and other features to write the equation

☐ **D.** Solve applications involving the foci

In Section 7.1, we introduced the equation of an ellipse using analytical geometry and the focus-directrix definition. Here we'll take a different approach, and use the equation of a circle to demonstrate that a circle is simply a special ellipse. In doing so, we'll establish a relationship between the foci and vertices of the ellipse, that enables us to apply these characteristics in context.

A. The Equation and Graph of a Circle

Recall that the equation of a circle with radius r and center at (h, k) is

$$(x - h)^2 + (y - k)^2 = r^2.$$

As in Section 2.1, the standard form can be used to construct the equation of the circle given the center and radius as in Example 1, or to graph the circle as in Example 2.

EXAMPLE 1 ▶ **Determining the Equation of a Circle Given Its Center and Radius**

Find the equation of a circle with radius 5 and center at $(2, -1)$.

Solution ▶ With a center of $(2, -1)$, we have $h = 2$, $k = -1$, and $r = 5$. Making the corresponding substitutions into the standard form we obtain

$$(x - h)^2 + (y - k)^2 = r^2 \quad \text{standard form}$$
$$(x - 2)^2 + [y - (-1)]^2 = 5^2 \quad \text{substitute 2 for } h, -1 \text{ for } k, \text{ and 5 for } r$$
$$(x - 2)^2 + (y + 1)^2 = 25 \quad \text{simplify}$$

The equation of this circle is $(x - 2)^2 + (y + 1)^2 = 25$.

Now try Exercises 7 through 12 ▶

If the equation is given in polynomial form, recall that we first complete the square in x and y to identify the center and radius.

EXAMPLE 2 ▶   **Completing the Square to Graph a Circle**

Find the center and radius of the circle whose equation is given, then sketch its graph: $x^2 + y^2 - 6x + 4y - 3 = 0$.

Solution ▶   Begin by completing the square in both x and y.

$$(x^2 - 6x + \underline{\quad}) + (y^2 + 4y + \underline{\quad}) = 3 \qquad \text{group } x\text{- and } y\text{-terms; add 3}$$

$$(x^2 - 6x + 9) + (y^2 + 4y + 4) = 3 + 9 + 4 \qquad \text{complete the square}$$

$$\underbrace{\qquad}_{\text{adds 9 to left side}} \quad \underbrace{\qquad}_{\text{adds 4 to left side}} \quad \underbrace{\qquad}_{\text{add } 9 + 4 \text{ to right side}}$$

$$(x - 3)^2 + (y + 2)^2 = 16 \qquad \text{factor and simplify}$$

The center is at $(3, -2)$, with radius $r = \sqrt{16} = 4$.

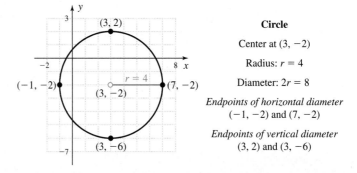

Circle

Center at $(3, -2)$

Radius: $r = 4$

Diameter: $2r = 8$

Endpoints of horizontal diameter
$(-1, -2)$ and $(7, -2)$

Endpoints of vertical diameter
$(3, 2)$ and $(3, -6)$

Now try Exercises 13 through 18 ▶

The equation of a circle in **standard form** provides a useful link to some of the other conic sections, and is obtained by *setting the equation equal to 1.* In the case of a circle, this means we simply divide by r^2.

$$(x - h)^2 + (y - k)^2 = r^2 \qquad \text{standard form}$$

$$\frac{(x - h)^2}{r^2} + \frac{(y - k)^2}{r^2} = 1 \qquad \text{divide by } r^2$$

In this form, the value of r in each denominator gives the *horizontal* and *vertical* distances, respectively, from the center to the graph. This is not so important in the case of a circle, since this distance is the same in *any* direction. But for other conics, these horizontal and vertical distances are *not* the same, making the new form a valuable tool for graphing. To distinguish the horizontal from the vertical distance, r^2 is replaced by a^2 in the "x-term" (horizontal distance), and by b^2 in the "y-term" (vertical distance).

☑ **A. You've just learned how to use the characteristics of a circle and its graph to understand the equation of an ellipse**

B. The Equation of an Ellipse

It then seems reasonable to ask, "What happens to the graph when $a \neq b$?" To answer, consider the equation from Example 2. We have $\dfrac{(x - 3)^2}{4^2} + \dfrac{(y + 2)^2}{4^2} = 1$ (after

dividing by 16), which we now compare to $\dfrac{(x - 3)^2}{4^2} + \dfrac{(y + 2)^2}{3^2} = 1$, where $a = 4$

and $b = 3$. The center of the graph is still at $(3, -2)$, since $h = 3$ and $k = -2$ remain unchanged. Substituting $y = -2$ to find additional points, eliminates the y-term and gives two values for x:

$$\frac{(x - 3)^2}{4^2} + \frac{(-2 + 2)^2}{3^2} = 1 \qquad \text{substitute } -2 \text{ for } y$$

$$\frac{(x - 3)^2}{4^2} + 0 = 1 \qquad \text{simplify}$$

$$(x - 3)^2 = 16 \qquad \text{multiply by } 4^2 = 16$$

$$x - 3 = \pm 4 \qquad \text{property of square roots}$$

$$x = 3 \pm 4 \qquad \text{add } 3$$

$$x = 7 \text{ and } x = -1$$

This shows the horizontal distance from the center to the graph is still $a = 4$, and the points $(-1, -2)$ and $(7, -2)$ are on the graph (see Figure 7.12). Similarly, for $x = 3$ we have $(y + 2)^2 = 9$, giving $y = -5$ and $y = 1$, and showing the vertical distance from the center to the graph is $b = 3$, with points $(3, 1)$ and $(3, -5)$ also on the graph. Using this information to sketch the curve reveals the "circle" is elongated and has become an **ellipse.**

For an ellipse, the longest distance across the graph is called the **major axis,** with the endpoints of the major axis called **vertices.** The segment perpendicular to and bisecting the major axis (with its endpoints on the ellipse) is called the **minor axis,** as shown in Figure 7.13.

Figure 7.12

Figure 7.13

The case where $a > b$

- If $a > b$, the major axis is horizontal (parallel to the x-axis) with length $2a$, and the minor axis is vertical with length $2b$ (see Example 3).
- If $b > a$ the major axis is vertical (parallel to the y-axis) with length $2b$, and the minor axis is horizontal with length $2a$ (see Example 4).

Generalizing this observation we obtain the equation of an ellipse in standard form.

The Equation of an Ellipse in Standard Form

Given $\dfrac{(x - h)^2}{a^2} + \dfrac{(y - k)^2}{b^2} = 1$.

If $a \neq b$ the equation represents the graph of an ellipse with center at (h, k).

- $|a|$ gives the horizontal distance from center to graph.
- $|b|$ gives the vertical distance from center to graph.

EXAMPLE 3 ▶ **Graphing a Horizontal Ellipse**

Sketch the graph of
$$\frac{(x-2)^2}{25} + \frac{(y+1)^2}{9} = 1.$$

Solution ▶ Noting $a \neq b$, we have an ellipse with center $(h, k) = (2, -1)$. The horizontal distance from the center to the graph is $a = 5$, and the vertical distance from the center to the graph is $b = 3$. After plotting the corresponding points and connecting them with a smooth curve, we obtain the graph shown.

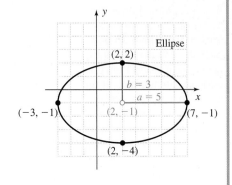

Now try Exercises 19 through 24 ▶

As with the circle, the equation of an ellipse can be given in polynomial form, and here our knowledge of circles is helpful. For the equation $25x^2 + 4y^2 = 100$, we know the graph cannot be a circle since the coefficients are unequal, and the center of the graph must be at the origin since $h = k = 0$. To actually draw the graph, we convert the equation to standard form.

EXAMPLE 4 ▶ **Graphing a Vertical Ellipse**

For $25x^2 + 4y^2 = 100$, (a) write the equation in standard form and identify the center and the values of a and b, (b) identify the major and minor axes and name the vertices, and (c) sketch the graph.

Solution ▶ The coefficients of x^2 and y^2 are unequal, and 25, 4, and 100 have like signs. The equation represents an ellipse with center at $(0, 0)$. To obtain standard form:

a. $25x^2 + 4y^2 = 100$ given equation

$\dfrac{25x^2}{100} + \dfrac{4y^2}{100} = 1$ divide by 100

$\dfrac{x^2}{4} + \dfrac{y^2}{25} = 1$ standard form

$\dfrac{x^2}{2^2} + \dfrac{y^2}{5^2} = 1$ write denominators in squared form; $a = 2, b = 5$

b. The result shows $a = 2$ and $b = 5$, indicating the major axis will be vertical and the minor axis will be horizontal. With the center at the origin, the x-intercepts will be $(2, 0)$ and $(-2, 0)$, with the vertices (and y-intercepts) at $(0, 5)$ and $(0, -5)$.

c. Plotting these intercepts and sketching the ellipse results in the graph shown.

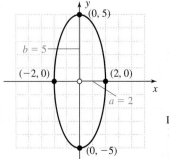

Vertical ellipse

Center at $(0, 0)$

Endpoints of major axis (vertices)
$(0, -5)$ and $(0, 5)$

Endpoints of minor axis
$(-2, 0)$ and $(2, 0)$

Length of major axis $2b$: $2(5) = 10$
Length of minor axis $2a$: $2(2) = 4$

Now try Exercises 25 through 36 ▶

After writing the equation in standard form, it is possible to end up with a constant that is zero or negative. In the first case, the graph is a single point. In the second case, no graph is possible since roots of the equation will be complex numbers. These are called *degenerate cases*. See **Exercise 78.**

If the center of the ellipse is not at the origin, the polynomial form has additional linear terms and we must first complete the square in x and y, then write the equation in standard form to sketch the graph (see the Reinforcing Basic Concepts feature for more on completing the square). Figure 7.14 illustrates how the central ellipse and the shifted ellipse are related.

Figure 7.14

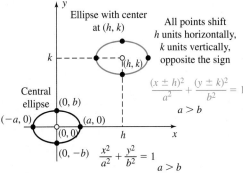

Ellipse with center at (h, k)

All points shift h units horizontally, k units vertically, opposite the sign

$$\frac{(x \pm h)^2}{a^2} + \frac{(y \pm k)^2}{b^2} = 1$$

$a > b$

Central ellipse

$$\frac{x^2}{a^2} + \frac{y^2}{b^2} = 1$$

$a > b$

EXAMPLE 5 ▶ **Completing the Square to Graph an Ellipse**

Sketch the graph of $25x^2 + 4y^2 + 150x - 16y + 141 = 0$.

Solution ▶ The coefficients of x^2 and y^2 are unequal and have like signs, and we assume the equation represents an ellipse but wait until we have the factored form to be certain.

$25x^2 + 4y^2 + 150x - 16y + 141 = 0$ given equation (polynomial form)

$25x^2 + 150x + 4y^2 - 16y = -141$ group like terms; subtract 141

$25(x^2 + 6x + \_\_) + 4(y^2 - 4y + \_\_) = -141$ factor out leading coefficient from each group

$25(x^2 + 6x + 9) + 4(y^2 - 4y + 4) = -141 + 225 + 16$ complete the square

add $225 + 16$ to right

adds $25(9) = 225$ adds $4(4) = 16$

$25(x + 3)^2 + 4(y - 2)^2 = 100$ factor

$$\frac{25(x + 3)^2}{100} + \frac{4(y - 2)^2}{100} = \frac{100}{100}$$ divide both sides by 100

$$\frac{(x + 3)^2}{4} + \frac{(y - 2)^2}{25} = 1$$ simplify (standard form)

$$\frac{(x + 3)^2}{2^2} + \frac{(y - 2)^2}{5^2} = 1$$ write denominators in squared form

The result is a vertical ellipse with center at $(-3, 2)$, with $a = 2$ and $b = 5$. The vertices are a vertical distance of 5 units from center, and the endpoints of the minor axis are a horizontal distance of 2 units from center.

Note this is the same ellipse as in Example 4, but shifted 3 units left and 2 up.

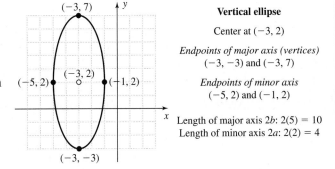

Vertical ellipse

Center at $(-3, 2)$

Endpoints of major axis (vertices)
$(-3, -3)$ and $(-3, 7)$

Endpoints of minor axis
$(-5, 2)$ and $(-1, 2)$

Length of major axis $2b$: $2(5) = 10$
Length of minor axis $2a$: $2(2) = 4$

Now try Exercises 37 through 44 ▶

☑ **B. You've just learned how to use the equation of an ellipse to graph central and noncentral ellipses**

C. The Foci of an Ellipse

In Section 7.1, we noted that an ellipse could also be defined in terms of two special points called the **foci.** The Museum of Science and Industry in Chicago, Illinois (http://www.msichicago.org), has a permanent exhibit called the *Whispering Gallery*. The construction of the room is based on some of the reflective properties of an ellipse. If two people stand at designated points in the room and one of them whispers very softly, the other person can hear the whisper quite clearly—even though they are over 40 ft apart! The point where each person stands is a **focus** of the ellipse. This reflective property also applies to light and radiation, giving the ellipse some powerful applications in science, medicine, acoustics, and other areas. To understand and appreciate these applications, we introduce the analytic definition of an ellipse.

WORTHY OF NOTE

You can easily draw an ellipse that satisfies the definition. Press two pushpins (these form the foci of the ellipse) halfway down into a piece of heavy cardboard about 6 in. apart. Take an 8-in. piece of string and loop each end around the pins. Use a pencil to draw the string taut and keep it taut as you move the pencil in a circular motion—and the result is an ellipse! A different length of string or a different distance between the foci will produce a different ellipse.

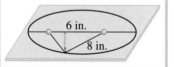

Definition of an Ellipse

Given two fixed points f_1 and f_2 in a plane, an ellipse is the set of all points (x, y) where the distance from f_1 to (x, y) added to the distance from f_2 to (x, y) remains constant.

$$d_1 + d_2 = k$$

The fixed points f_1 and f_2 are called the *foci* of the ellipse, and the points $P(x, y)$ are on the graph of the ellipse.

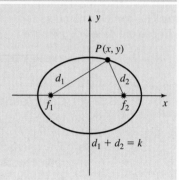

To find the equation of an ellipse in terms of a and b we combine the definition just given with the distance formula. Consider the ellipse shown in Figure 7.15 (for calculating ease we use a central ellipse). Note the vertices have coordinates $(-a, 0)$ and $(a, 0)$, and the endpoints of the minor axis have coordinates $(0, -b)$ and $(0, b)$ as before. It is customary to assign foci the coordinates $f_1 \rightarrow (-c, 0)$ and $f_2 \rightarrow (c, 0)$. We can calculate the distance between $(c, 0)$ and any point $P(x, y)$ on the ellipse using the distance formula:

Figure 7.15

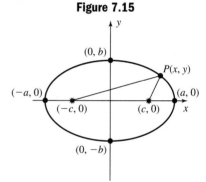

$$\sqrt{(x - c)^2 + (y - 0)^2}$$

Likewise the distance between $(-c, 0)$ and any point (x, y) is

$$\sqrt{(x + c)^2 + (y - 0)^2}$$

According to the definition, the sum must be constant:

$$\sqrt{(x - c)^2 + y^2} + \sqrt{(x + c)^2 + y^2} = k$$

EXAMPLE 6 ▶ **Finding the Value of k from the Definition of an Ellipse**

Use the definition of an ellipse and the diagram given to determine the constant k used for this ellipse (also see the following *Worthy of Note*). Note that $a = 5, b = 3$, and $c = 4$.

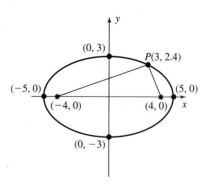

Solution ▶

$$\sqrt{(x - c)^2 + (y - 0)^2} + \sqrt{(x + c)^2 + (y - 0)^2} = k \quad \text{given}$$

$$\sqrt{(3 - 4)^2 + (2.4 - 0)^2} + \sqrt{(3 + 4)^2 + (2.4 - 0)^2} = k \quad \text{substitute}$$

$$\sqrt{(-1)^2 + 2.4^2} + \sqrt{7^2 + 2.4^2} = k \quad \text{add}$$

$$\sqrt{6.76} + \sqrt{54.76} = k \quad \text{simplify radicals}$$

$$2.6 + 7.4 = k \quad \text{compute square roots}$$

$$10 = k \quad \text{result}$$

The constant used for this ellipse is 10 units.

> **WORTHY OF NOTE**
>
> Note that if the foci are coincident (both at the origin) the "ellipse" will actually be a circle with radius $\frac{k}{2}$; $\sqrt{x^2 + y^2} + \sqrt{x^2 + y^2} = k$ leads to $x^2 + y^2 = \frac{k^2}{4}$. In Example 1 we found $k = 10$, giving $\frac{10}{2} = 5$, and if we used the "string" to draw the circle, the pencil would be 5 units from the center, creating a circle of radius 5.

Now try Exercises 45 through 48 ▶

Figure 7.16

These two segments are equal: $d_1 + d_2 = 2a$

In Example 6, the sum of the distances could also be found by moving the point (x, y) to the location of a vertex $(a, 0)$, then using the symmetry of the ellipse. The sum is identical to the length of the major axis, since the overlapping part of the string from $(c, 0)$ to $(a, 0)$ is the same length as from $(-a, 0)$ to $(-c, 0)$ (see Figure 7.16). This shows the constant k is equal to 2a *regardless of the distance between foci.*

As we noted, the result is

$$\sqrt{(x - c)^2 + y^2} + \sqrt{(x + c)^2 + y^2} = 2a \quad \text{substitute } 2a \text{ for } k$$

The details for simplifying this expression are given in Appendix III, and the result is very close to the standard form seen previously:

$$\frac{x^2}{a^2} + \frac{y^2}{a^2 - c^2} = 1$$

By comparing the standard form $\frac{x^2}{a^2} + \frac{y^2}{b^2} = 1$ with $\frac{x^2}{a^2} + \frac{y^2}{a^2 - c^2} = 1$, we might suspect that $b^2 = a^2 - c^2$, and this is indeed the case. Note from Example 6 the relationship yields

$$b^2 = a^2 - c^2$$
$$3^2 = 5^2 - 4^2$$
$$9 = 25 - 16$$

Additionally, when we consider that $(0, b)$ is a point on the ellipse, the distance from $(0, b)$ to $(c, 0)$ must be equal to a due to symmetry (the "constant distance" used to form the ellipse is always 2a). We then see in Figure 7.17, that $b^2 + c^2 = a^2$ (Pythagorean Theorem), yielding $b^2 = a^2 - c^2$ as above.

With this development, we now have the ability to *locate the foci of any ellipse*—an important step toward using the ellipse in practical applications. Because we're often asked to find the location of the foci, it's best to rewrite the relationship in terms of c^2, using absolute value bars to allow for a major axis that is vertical: $c^2 = |a^2 - b^2|$.

Figure 7.17

EXAMPLE 7 ▶ **Completing the Square to Graph an Ellipse and Locate the Foci**

For the ellipse defined by $25x^2 + 9y^2 - 100x - 54y - 44 = 0$, find the coordinates of the center, vertices, foci, and endpoints of the minor axis. Then sketch the graph.

Solution ▶

$$25x^2 + 9y^2 - 100x - 54y - 44 = 0 \qquad \text{given}$$

$$25x^2 - 100x + 9y^2 - 54y = 44 \qquad \text{group terms; add 44}$$

$$25(x^2 - 4x + \underline{\quad}) + 9(y^2 - 6y + \underline{\quad}) = 44 \qquad \text{factor out lead coefficients}$$

$$25(x^2 - 4x + 4) + 9(y^2 - 6y + 9) = 44 + 100 + 81 \qquad \text{add } 100 + 81 \text{ to right-hand side}$$

adds $25(4) = 100$ adds $9(9) = 81$

$$25(x - 2)^2 + 9(y - 3)^3 = 225 \qquad \text{factored form}$$

$$\frac{25(x - 2)^2}{225} + \frac{9(y - 3)^2}{225} = \frac{225}{225} \qquad \text{divide by 225}$$

$$\frac{(x - 2)^2}{9} + \frac{(y - 3)^2}{25} = 1 \qquad \text{simplify (standard form)}$$

$$\frac{(x - 2)^2}{3^2} + \frac{(y - 3)^2}{5^2} = 1 \qquad \text{write denominators in squared form}$$

The result shows a vertical ellipse with $a = 3$ and $b = 5$. The center of the ellipse is at $(2, 3)$. The vertices are a vertical distance of $b = 5$ units from center at $(2, 8)$ and $(2, -2)$. The endpoints of the minor axis are a horizontal distance of $a = 3$ units from center at $(-1, 3)$ and $(5, 3)$. To locate the foci, we use the foci formula for an ellipse: $c^2 = |a^2 - b^2|$, giving $c^2 = |3^2 - 5^2| = 16$. This shows the foci "✸" are located a vertical distance of 4 units from center at $(2, 7)$ and $(2, -1)$.

Vertical ellipse

Center at $(2, 3)$

Endpoints of major axis (vertices)
$(2, 8)$ and $(2, -2)$

Endpoints of minor axis
$(-1, 3)$ and $(5, 3)$

Location of foci
$(2, 7)$ and $(2, -1)$

Length of major axis: $2b = 2(5) = 10$

Length of minor axis: $2a = 2(3) = 6$

Now try Exercises 49 through 54 ▶

For future reference, remember the foci of an ellipse always occur on the major axis, with $a > c$ and $a^2 > c^2$ for a horizontal ellipse. This makes it easier to remember the **foci formula** for ellipses: $c^2 = |a^2 - b^2|$. Since a^2 is larger, it must be decreased by b^2 to equal c^2.

If any two of the values for a, b, and c are known, the relationship between them can be used to construct the equation of the ellipse.

EXAMPLE 8 ▶ **Finding the Equation of an Ellipse**

Find the equation of the ellipse (in standard form) that has foci at $(0, -2)$ and $(0, 2)$, with a minor axis 6 units in length.

Solution ▶ Since the foci must be on the major axis, we know this is a vertical and central ellipse with $c = 2$ and $c^2 = 4$. The minor axis has a length of $2a = 6$ units, meaning $a = 3$ and $a^2 = 9$. To find b^2, use the foci equation and solve.

Now try Exercises 55 through 64 ▶

<table>
<tr><td>

LOOKING AHEAD

For the hyperbola, we'll find that $c > a$, and the formula for the foci of a hyperbola will be $c^2 = a^2 + b^2$.

</td></tr>
</table>

☑ **C.** You've just learned how to locate the foci of an ellipse and use the foci and other features to write the equation

$$c^2 = |a^2 - b^2| \qquad \text{foci equation (ellipse)}$$
$$4 = |9 - b^2| \qquad \text{substitute}$$
$$-4 = 9 - b^2 \qquad 4 = 9 - b^2 \qquad \text{solve}$$
$$b^2 = 13 \qquad b^2 = 5 \qquad \text{result}$$

Since we know b^2 must be greater than a^2 (the major axis is always longer), $b^2 = 5$ can be discarded. The standard form is $\dfrac{x^2}{3^2} + \dfrac{y^2}{(\sqrt{13})^2} = 1$.

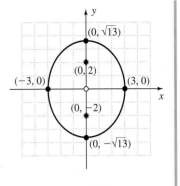

D. Applications Involving Foci

Applications involving the foci of a conic section can take various forms. In many cases, only partial information about the conic section is available and the ideas from Example 8 must be used to "fill in the gaps." In other applications, we must rewrite a known or given equation to find information related to the values of a, b, and c.

EXAMPLE 9 ▶ **Solving Applications Using the Characteristics of an Ellipse**

In Washington, D.C., there is a park called the *Ellipse* located between the White House and the Washington Monument. The park is surrounded by a path that forms an ellipse with the length of the major axis being about 1502 ft and the minor axis having a length of 1280 ft. Suppose the park manager wants to install water fountains at the location of the foci. Find the distance between the fountains rounded to the nearest foot.

Solution ▶ Since the major axis has length $2a = 1502$, we know $a = 751$ and $a^2 = 564,001$. The minor axis has length $2b = 1280$, meaning $b = 640$ and $b^2 = 409,600$. To find c, use the foci equation:

$$c^2 = a^2 - b^2$$
$$= 564,001 - 409,600$$
$$= 154,401$$
$$c \approx -393 \text{ and } c \approx 393$$

☑ **D.** You've just learned how to solve applications involving the foci

The distance between the water fountains would be $2(393) = 786$ ft.

Now try Exercises 65 through 76 ▶

7.2 EXERCISES

▶ CONCEPTS AND VOCABULARY

Fill in the blank with the appropriate word or phrase. Carefully reread the section if needed.

1. For an ellipse, the relationship between a, b, and c is given by the foci equation _____, since $c < a$ or $c < b$.

2. The greatest distance across an ellipse is called the _____ _____ and the endpoints are called _____.

3. For a vertical ellipse, the length of the minor axis is _____ and the length of the major axis is _____.

4. To write the equation $2x^2 + y^2 - 6x = 7$ in standard form, _____ the _____ in x.

5. Explain/Discuss how the relations $a > b$, $a = b$ and $a < b$ affect the graph of a conic section with equation $\dfrac{(x - h)^2}{a^2} + \dfrac{(y - k)^2}{b^2} = 1$.

6. Suppose foci are located at $(-3, 2)$ and $(5, 2)$. Discuss/Explain the conditions necessary for the graph to be an ellipse.

▶ **DEVELOPING YOUR SKILLS**

Find the equation of a circle satisfying the conditions given.

7. center $(0, 0)$, radius 7

8. center $(0, 0)$, radius 9

9. center $(5, 0)$, radius $\sqrt{3}$

10. center $(0, 4)$, radius $\sqrt{5}$

11. diameter has endpoints $(4, 9)$ and $(-2, 1)$

12. diameter has endpoints $(-2, -3)$, and $(3, 9)$

Identify the center and radius of each circle, then sketch its graph.

13. $x^2 + y^2 - 12x - 10y + 52 = 0$

14. $x^2 + y^2 + 8x - 6y - 11 = 0$

15. $x^2 + y^2 - 4x + 10y + 4 = 0$

16. $x^2 + y^2 + 4x + 6y - 3 = 0$

17. $x^2 + y^2 + 6x - 5 = 0$

18. $x^2 + y^2 - 8y - 5 = 0$

Sketch the graph of each ellipse.

19. $\dfrac{(x - 1)^2}{9} + \dfrac{(y - 2)^2}{16} = 1$

20. $\dfrac{(x - 3)^2}{4} + \dfrac{(y - 1)^2}{25} = 1$

21. $\dfrac{(x - 2)^2}{25} + \dfrac{(y + 3)^2}{4} = 1$

22. $\dfrac{(x + 5)^2}{1} + \dfrac{(y - 2)^2}{16} = 1$

23. $\dfrac{(x + 1)^2}{16} + \dfrac{(y + 2)^2}{9} = 1$

24. $\dfrac{(x + 1)^2}{36} + \dfrac{(y + 3)^2}{9} = 1$

For each exercise, (a) write the equation in standard form, then identify the center and the values of a and b, (b) state the coordinates of the vertices and the coordinates of the endpoints of the minor axis, and (c) sketch the graph.

25. $x^2 + 4y^2 = 16$

26. $9x^2 + y^2 = 36$

27. $16x^2 + 9y^2 = 144$

28. $25x^2 + 9y^2 = 225$

29. $2x^2 + 5y^2 = 10$

30. $3x^2 + 7y^2 = 21$

Identify each equation as that of an ellipse or circle, then sketch its graph.

31. $(x + 1)^2 + 4(y - 2)^2 = 16$

32. $9(x - 2)^2 + (y + 3)^2 = 36$

33. $2(x - 2)^2 + 2(y + 4)^2 = 18$

34. $(x - 6)^2 + y^2 = 49$

35. $4(x - 1)^2 + 9(y - 4)^2 = 36$

36. $25(x - 3)^2 + 4(y + 2)^2 = 100$

Complete the square in both x and y to write each equation in standard form. Then draw a complete graph of the relation and identify all important features.

37. $4x^2 + y^2 + 6y + 5 = 0$

38. $x^2 + 3y^2 + 8x + 7 = 0$

39. $x^2 + 4y^2 - 8y + 4x - 8 = 0$

40. $3x^2 + y^2 - 8y + 12x - 8 = 0$

41. $5x^2 + 2y^2 + 20y - 30x + 75 = 0$

42. $4x^2 + 9y^2 - 16x + 18y - 11 = 0$

43. $2x^2 + 5y^2 - 12x + 20y - 12 = 0$

44. $6x^2 + 3y^2 - 24x + 18y - 3 = 0$

Use the definition of an ellipse to find the constant k used for each ellipse (figures are not drawn to scale).

45.

46.

47.

48.

Find the coordinates of the (a) center, (b) vertices, (c) foci, and (d) endpoints of the minor axis. Then (e) sketch the graph.

49. $4x^2 + 25y^2 - 16x - 50y - 59 = 0$

50. $9x^2 + 16y^2 - 54x - 64y + 1 = 0$

51. $25x^2 + 16y^2 - 200x + 96y + 144 = 0$

52. $49x^2 + 4y^2 + 196x - 40y + 100 = 0$

53. $6x^2 + 24x + 9y^2 + 36y + 6 = 0$

54. $5x^2 - 50x + 2y^2 - 12y + 93 = 0$

Find the equation of an ellipse (in standard form) that satisfies the following conditions:

55. vertices at $(-6, 0)$ and $(6, 0)$; foci at $(-4, 0)$ and $(4, 0)$

56. vertices at $(-8, 0)$ and $(8, 0)$; foci at $(-5, 0)$ and $(5, 0)$

57. foci at $(3, -6)$ and $(3, 2)$; length of minor axis: 6 units

58. foci at $(-4, -3)$ and $(8, -3)$; length of minor axis: 8 units

Use the characteristics of an ellipse and the graph given to write the related equation and find the location of the foci.

59.

60.

61.

62.

▶ **WORKING WITH FORMULAS**

63. Area of an Ellipse: $A = \pi ab$
The area of an ellipse is given by the formula shown, where a is the distance from the center to the graph in the horizontal direction and b is the distance from center to graph in the vertical direction. Find the area of the ellipse defined by $16x^2 + 9y^2 = 144$.

64. The Perimeter of an Ellipse: $P = 2\pi\sqrt{\dfrac{a^2 + b^2}{2}}$
The perimeter of an ellipse can be *approximated* by the formula shown, where a represents the length of the semimajor axis and b represents the length of the semiminor axis. Find the perimeter of the ellipse defined by the equation $\dfrac{x^2}{49} + \dfrac{y^2}{4} = 1$.

▶ **APPLICATIONS**

65. Decorative fireplaces: A bricklayer intends to build an elliptical fireplace 3 ft high and 8 ft wide, with two glass doors that open at the middle. The hinges to these doors are to be screwed onto a spine that is perpendicular to the hearth and goes through the foci of the ellipse. How far from center will the spines be located? How tall will each spine be?

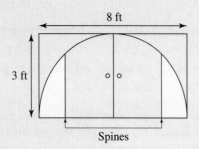

66. Decorative gardens: A retired math teacher decides to present her husband with a beautiful elliptical garden to help celebrate their 50th anniversary. The ellipse is to be 8 m long and 5 m across, with decorative fountains located at the foci. How far from the center of the ellipse should the fountains be located (round to the nearest 100th of a meter)? How far apart are the fountains?

67. Attracting attention to art: As part of an art show, a gallery owner asks a student from the local university to design a unique exhibit that will highlight one of the more significant pieces in the collection, an ancient sculpture. The student decides to create an elliptical showroom with reflective walls, with a rotating laser light on a stand at one focus, and the sculpture placed at the other focus on a stand of equal height. The laser light then points continually at the sculpture as it rotates. If the elliptical room is 24 ft long and 16 ft wide, how far from the center of the ellipse should the stands be located (round to the nearest 10th of a foot)? How far apart are the stands?

68. Medical procedures: The medical procedure called *lithotripsy* is a noninvasive medical procedure that is used to break up kidney and bladder stones in the body. A machine called a *lithotripter* uses its three-dimensional semielliptical shape and the foci properties of an ellipse to concentrate shock waves generated at one focus, on a kidney stone located at the other focus (see diagram — not drawn to scale). If the lithotripter has a length (semimajor axis) of 16 cm and a radius (semiminor axis) of 10 cm, how far from the vertex should a kidney stone be located

for the best result? Round to the nearest hundredth.

Exercise 68

69. Elliptical arches: In some situations, bridges are built using uniform elliptical archways as shown in the figure given. Find the equation of the ellipse forming each arch if it has a total width of 30 ft and a maximum center height (above level ground) of 8 ft. What is the height of a point 9 ft to the right of the center of each arch?

Exercise 69

70. Elliptical arches: An elliptical arch bridge is built across a one lane highway. The arch is 20 ft across and has a maximum center height of 12 ft. Will a farm truck hauling a load 10 ft wide with a clearance height of 11 ft be able to go through the bridge without damage? (*Hint:* See Exercise 69.)

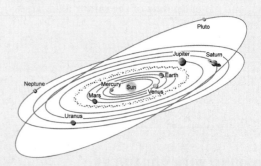

As a planet orbits around the Sun, it traces out an ellipse. If the center of the ellipse were placed at (0, 0) on a coordinate grid, the Sun would be actually off-centered (located at the *focus* of the ellipse). Use this information and the graphs provided to complete Exercises 71 through 74.

71. Orbit of Mercury: The approximate orbit of the planet Mercury is shown in the figure given. Find an equation that models this orbit.

Exercise 71

72. **Orbit of Pluto:** The approximate orbit of the dwarf planet Pluto is shown in the figure given. Find an equation that models this orbit.

Exercise 72

73. **Planetary orbits:** Except for small variations, a planet's orbit around the Sun is elliptical with the Sun at one focus. The aphelion (maximum distance from the Sun) of the planet Mars is approximately 156 million miles, while the perihelion (minimum distance from the Sun) of Mars is about 128 million miles. Use this information to find the lengths of the semimajor and semiminor axes, rounded to the nearest million. If Mars has an orbital velocity of 54,000 miles per hour

(1.296 million miles per day), how many days does it take Mars to orbit the Sun? (*Hint:* Use the formula from Exercise 64).

74. **Planetary orbits:** The aphelion (maximum distance from the Sun) of the planet Saturn is approximately 940 million miles, while the perihelion (minimum distance from the Sun) of Saturn is about 840 million miles. Use this information to find the lengths of the semimajor and semiminor axes, rounded to the nearest million. If Saturn has an orbital velocity of 21,650 miles per hour (about 0.52 million miles per day), how many days does it take Saturn to orbit the Sun? How many years?

75. **Area of a race track**: Suppose the *Toronado 500* is a car race that is run on an elliptical track. The track is bounded by two ellipses with equations of $4x^2 + 9y^2 = 900$ and $9x^2 + 25y^2 = 900$, where x any y are in hundreds of yards. Use the formula given in Exercise 63 to find the area of the race track.

76. **Area of a border**: The table cloth for a large oval table is elliptical in shape. It is designed with two concentric ellipses (one within the other) as shown in the figure. The equation of the outer ellipse is $9x^2 + 25y^2 = 225$, and the equation of the inner ellipse is $4x^2 + 16y^2 = 64$ with x and y in feet. Use the formula given in Exercise 63 to find the area of the border of the tablecloth.

Exercise 76

▶ **EXTENDING THE THOUGHT**

77. When graphing the conic sections, it is often helpful to use what is called a **focal chord,** as it gives additional points on the graph with very little effort. A focal chord is a line segment through a focus (perpendicular to the major or transverse axis), with the endpoints on the graph. For an ellipse, the length of the focal chord is given by $L = \frac{2m^2}{n}$, where m is the length of the semiminor axis, and n is the length of the semimajor axis. The focus will always be the midpoint of this line segment. Find the length of the focal chord for the ellipse

Exercise 77

Focal chords

$\frac{x^2}{81} + \frac{y^2}{36} = 1$ and the coordinates of the endpoints. Verify (by substituting into the equation) that these endpoints are indeed points on the graph, then use them to help complete the graph.

78. For $6x^2 + 36x + 3y^2 - 24y + 74 = -28$, does the equation appear to be that of a circle, ellipse, or parabola? Write the equation in factored form. What do you notice? What can you say about the graph of this equation?

79. Verify that for the ellipse $\frac{x^2}{a^2} + \frac{y^2}{b^2} = 1$, the length of the focal chord is $\frac{2b^2}{a}$.

▶ **MAINTAINING YOUR SKILLS**

80. (4.4) Evaluate the expression using the change-of-base formula: $\log_3 20$.

81. (1.4) Compute the product $z_1 z_2$ and quotient $\dfrac{z_1}{z_2}$ of: $z_1 = 2\sqrt{3} + 2\sqrt{3}i$; $z_2 = 5\sqrt{3} - 5i$

82. (1.3) Solve the absolute value inequality (a) graphically and (b) analytically: $-2|x - 3| + 10 > 4$.

83. (3.8) The resistance R to current flow in an electrical wire varies directly as the length L of the wire and inversely as the square of its diameter d. (a) Write the equation of variation; (b) find the constant of variation if a wire 2 m long with diameter $d = 0.005$ m has a resistance of 240 ohms (Ω); and (c) find the resistance in a similar wire 3 m long and 0.006 m in diameter.

MID-CHAPTER CHECK

Sketch the graph of each conic section.

1. $(x - 4)^2 + (y + 3)^2 = 9$

2. $x^2 + y^2 - 10x + 4y + 4 = 0$

3. $\dfrac{(x - 2)^2}{16} + \dfrac{(y + 3)^2}{1} = 1$

4. $9x^2 + 4y^2 + 18x - 24y + 9 = 0$

5. $\dfrac{(x + 3)^2}{9} + \dfrac{(y - 4)^2}{4} = 1$

6. $9x^2 + 4y^2 - 36x + 24y - 72 = 0$

7. Find the equation for all points located an equal distance from the point (0, 3) and the line $y = -3$.

8. Find the equation of each relation and state its domain and range.

a.

b.

9. Find the equation of the ellipse having foci at (0, 13) and (0, -13), with a minor axis of length 10 units.

10. Find the equation of the ellipse (in standard form) if the vertices are (-4, 0) and (4, 0) and the distance between the foci is $4\sqrt{3}$ units.

REINFORCING BASIC CONCEPTS

More on Completing the Square

From our work so far in Chapter 7, we realize the process of *completing the square* has much greater use than simply as a tool for working with quadratic equations. It is a valuable tool in the application of the conic sections, as well as other areas. The purpose of this *Reinforcing Basic Concepts* is to strengthen the ability and confidence needed to apply the process correctly. This is important because in some cases the values of a and b are rational or irrational numbers. No matter what the context,

1. The process begins *with a coefficient of 1*. For $20x^2 + 180x + 27y^2 - 54y + 192 = 0$, we recognize the equation of an ellipse, since the coefficients of the squared terms are positive and unequal. To study or graph this ellipse, we'll use the standard form to identify the values of a, b, and c. Grouping the like-variable terms gives

$$(20x^2 + 120x \quad) + (27y^2 - 54y \quad) + 192 = 0$$

and to complete the square, we factor out the lead coefficient of each group (to get a coefficient of 1):

$$20(x^2 + 6x \quad) + 27(y^2 - 2y \quad) + 192 = 0$$

This brings us to the fundamental step for completing the square.

2. **The quantity** $\left(\dfrac{1}{2} \cdot linear\ cofficient\right)^2$ **will complete a trinomial square.** For this example we obtain

$$\left(\frac{1}{2} \cdot 6\right)^2 = 9 \text{ for } x, \text{ and } \left(\frac{1}{2} \cdot -2\right)^2 = 1 \text{ for } y, \text{ with these numbers inserted in the appropriate group:}$$

$$20(x^2 + 6x + 9) + 27(y^2 - 2y + 1) + 192 = 0 \quad \text{\small complete the square}$$

Due to the distributive property, we have in effect added $20 \cdot 9 = 180$ and $27 \cdot 1 = 27$ (for a total of 207) to the left side of the equation:

$$\underset{\substack{\text{adds } 20 \cdot 9 = 180 \\ \text{to left side}}}{20(x^2 + 6x + 9)} + \underset{\substack{\text{adds } 27 \cdot 1 = 27 \\ \text{to left side}}}{27(y^2 - 2y + 1)} + 192 = 0$$

This brings us to the final step.

3. **Keep the equation in balance.** Since the left side was increased by 207, we also increased the right side by 207, after subtracting 192:

$$\underset{\substack{\text{adds } 20 \cdot 9 = 180 \\ \text{to left side}}}{20(x^2 + 6x + 9)} + \underset{\substack{\text{adds } 27 \cdot 1 = 27 \\ \text{to left side}}}{27(y^2 - 2y + 1)} = \underset{\substack{\text{adds } 180 + 27 = 207 \\ \text{to right side}}}{-192 + 207}$$

The quantities in parentheses factor, giving $20(x + 3)^2 + 27(y - 1)^2 = 15$. We then divide by 15 and simplify, obtaining the standard form $\dfrac{4(x + 3)^2}{3} + \dfrac{9(y - 1)^2}{5} = 1$. Note the coefficient of each binomial square is not 1, even after setting the equation equal to 1. In the *Strengthening Core Skills* feature of this chapter, we'll look at how to write equations of this type in standard form to obtain the values of a and b. For now, practice completing the square using these exercises.

Exercise 1: $100x^2 - 400x - 18y^2 - 108y + 230 = 0$

Exercise 2: $28x^2 - 56x + 48y^2 + 192y + 195 = 0$

7.3 | The Hyperbola

Learning Objectives

In Section 7.3 you will learn how to:

☐ **A.** Use the equation of a hyperbola to graph central and noncentral hyperbolas

☐ **B.** Distinguish between the equations of a circle, ellipse, and hyperbola

☐ **C.** Locate the foci of a hyperbola and use the foci and other features to write its equation

☐ **D.** Solve applications involving foci

As seen in Section 7.1 (see Figure 7.18), a hyperbola is a conic section formed by a plane that cuts both nappes of a right circular cone. A hyperbola has two symmetric parts called **branches,** which open in opposite directions. Although the branches appear to resemble parabolas, we will soon discover they are actually a very different curve.

A. The Equation of a Hyperbola

In Section 7.2, we noted that for the equation $Ax^2 + By^2 = F$, if $A = B$, the equation is that of a circle, if $A \neq B$, the equation represents an ellipse. Both cases contain a *sum* of second-degree terms. Perhaps driven by curiosity, we might wonder what happens if the equation has a *difference* of second-degree terms. Consider the equation $9x^2 - 16y^2 = 144$. It appears the graph will be centered at $(0, 0)$ since no shifts are applied (h and k are both zero). Using the intercept method to graph this equation reveals an entirely new curve, called a *hyperbola*.

Figure 7.18

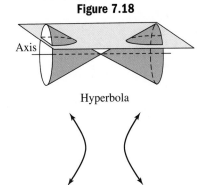

EXAMPLE 1 ▶ **Graphing a Central Hyperbola**

Graph the equation $9x^2 - 16y^2 = 144$ using intercepts and additional points as needed.

Solution ▶

$$9x^2 - 16y^2 = 144 \quad \text{given}$$
$$9(0)^2 - 16y^2 = 144 \quad \text{substitute 0 for } x$$
$$-16y^2 = 144 \quad \text{simplify}$$
$$y^2 = -9 \quad \text{divide by } -16$$

Since y^2 can never be negative, we conclude that the graph has *no y-intercepts*. Substituting $y = 0$ to find the *x*-intercepts gives

$$9x^2 - 16y^2 = 144 \quad \text{given}$$
$$9x^2 - 16(0)^2 = 144 \quad \text{substitute 0 for } y$$
$$9x^2 = 144 \quad \text{simplify}$$
$$x^2 = 16 \quad \text{divide by 9}$$
$$x = \sqrt{16} \quad \text{and} \quad x = -\sqrt{16} \quad \text{square root property}$$
$$x = 4 \quad \text{and} \quad x = -4 \quad \text{simplify}$$
$$(4, 0) \quad \text{and} \quad (-4, 0) \quad \text{x-intercepts}$$

Knowing the graph has no *y*-intercepts, we select inputs greater than 4 and less than −4 to help sketch the graph. Using $x = 5$ and $x = -5$ yields

$9x^2 - 16y^2 = 144$	given	$9x^2 - 16y^2 = 144$
$9(5)^2 - 16y^2 = 144$	substitute for x	$9(-5)^2 - 16y^2 = 144$
$9(25) - 16y^2 = 144$	$5^2 = (-5)^2 = 25$	$9(25) - 16y^2 = 144$
$225 - 16y^2 = 144$	simplify	$225 - 16y^2 = 144$
$-16y^2 = -81$	subtract 225	$-16y^2 = -81$
$y^2 = \dfrac{81}{16}$	divide by -16	$y^2 = \dfrac{81}{16}$
$y = \dfrac{9}{4} \quad y = -\dfrac{9}{4}$	square root property	$y = \dfrac{9}{4} \quad y = -\dfrac{9}{4}$
$y = 2.25 \quad y = -2.25$	decimal form	$y = 2.25 \quad y = -2.25$
$(5, 2.25) \quad (5, -2.25)$	ordered pairs	$(-5, 2.25) \quad (-5, -2.25)$

Plotting these points and connecting them with a smooth curve, while *knowing there are no y-intercepts,* produces the graph in the figure. The point at the origin (in blue) is not a part of the graph, and is given only to indicate the "center" of the hyperbola.

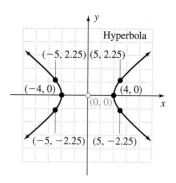

Now try Exercises 7 through 22 ▶

Since the hyperbola crosses a horizontal line of symmetry, it is referred to as a **horizontal hyperbola.** The points $(-4, 0)$ and $(4, 0)$ are called **vertices,** and the **center** of the hyperbola is always the point halfway between them. If the center is at the origin, we have a **central hyperbola.** The line passing through the center and both vertices is called the **transverse axis** (vertices are always on the transverse axis), and the line passing through the center and perpendicular to this axis is called the **conjugate axis** (see Figure 7.19).

In Example 1, the coefficient of x^2 was positive and we were subtracting $16y^2$: $9x^2 - 16y^2 = 144$. The result was a horizontal hyperbola. If the y^2-term is positive and we subtract the term containing x^2, the result is a **vertical hyperbola** (Figure 7.20).

Figure 7.19

Horizontal hyperbola

Figure 7.20

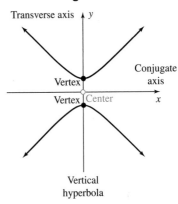

Vertical hyperbola

EXAMPLE 2 ▶ **Identifying the Axes, Vertices, and Center of a Hyperbola from Its Graph**

For the hyperbola shown, state the location of the vertices and the equation of the transverse axis. Then identify the location of the center and the equation of the conjugate axis.

Solution ▶ By inspection we locate the vertices at (0, 0) and (0, 4). The equation of the transverse axis is $x = 0$. The center is halfway between the vertices at (0, 2), meaning the equation of the conjugate axis is $y = 2$.

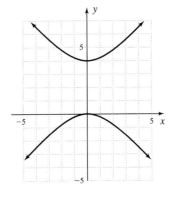

Now try Exercises 23 through 26 ▶

Standard Form

As with the ellipse, the polynomial form of the equation is helpful for *identifying* hyperbolas, but not very helpful when it comes to *graphing* a hyperbola (since we still must go through the laborious process of finding additional points). For graphing, standard form is once again preferred. Consider the hyperbola $9x^2 - 16y^2 = 144$ from Example 1. To write the equation in standard form, we divide by 144 and obtain $\dfrac{x^2}{4^2} - \dfrac{y^2}{3^2} = 1$. By comparing the standard form to the graph, we note $a = 4$ represents the distance from center to vertices, similar to the way we used a previously. But since the graph has no y-intercepts, what could $b = 3$ represent? The answer lies in the fact that branches of a hyperbola are **asymptotic,** meaning they will approach and become very close to imaginary lines that can be used to sketch the graph.

For a central hyperbola, the slopes of the asymptotic lines are given by the ratios $\dfrac{b}{a}$ and

$-\dfrac{b}{a}$, with the related equations being $y = \dfrac{b}{a}x$ and $y = -\dfrac{b}{a}x$. The graph from Example 1 is repeated in Figure 7.21, with the asymptotes drawn. For a clearer understanding of how the equations for the asymptotes were determined, see **Exercise 88.**

A second method of drawing the asymptotes involves drawing a **central rectangle** with dimensions $2a$ by $2b$, as shown in Figure 7.22. The asymptotes will be the *extended diagonals* of this rectangle. This brings us to the equation of a hyperbola in standard form.

Figure 7.21   **Figure 7.22**

Slope method

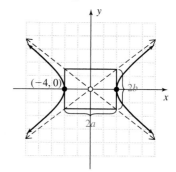

Central rectangle method

The Equation of a Hyperbola in Standard Form

The equation

$$\frac{(x - h)^2}{a^2} - \frac{(y - k)^2}{b^2} = 1$$

represents a *horizontal* hyperbola with center (h, k)
- *transverse* axis $y = k$
- *conjugate* axis $x = h$
- $|a|$ gives the distance from center to vertices.

The equation

$$\frac{(y - k)^2}{b^2} - \frac{(x - h)^2}{a^2} = 1$$

represents a *vertical* hyperbola with center (h, k)
- *transverse* axis $x = h$
- *conjugate* axis $y = k$
- $|b|$ gives the distance from center to vertices.

- Asymptotes can be drawn by starting at (h, k) and using slopes $m = \pm\dfrac{b}{a}$.

EXAMPLE 3 ▶ **Graphing a Hyperbola Using Its Equation in Standard Form**

Sketch the graph of $16(x - 2)^2 - 9(y - 1)^2 = 144$. Label the center, vertices, and asymptotes.

Solution ▶ Begin by noting a difference of the second-degree terms, with the x^2-term occurring first. This means we'll be graphing a horizontal hyperbola whose center is at $(2, 1)$. Continue by writing the equation in standard form.

$$16(x - 2)^2 - 9(y - 1)^2 = 144 \qquad \text{given equation}$$

$$\frac{16(x - 2)^2}{144} - \frac{9(y - 1)^2}{144} = \frac{144}{144} \qquad \text{divide by 144}$$

$$\frac{(x - 2)^2}{9} - \frac{(y - 1)^2}{16} = 1 \qquad \text{simplify}$$

$$\frac{(x - 2)^2}{3^2} - \frac{(y - 1)^2}{4^2} = 1 \qquad \text{write denominators in squared form}$$

Since $a = 3$ the vertices are a horizontal distance of 3 units from the center $(2, 1)$, giving $(2 + 3, 1) \rightarrow (5, 1)$ and $(2 - 3, 1) \rightarrow (-1, 1)$. After plotting the center and vertices, we can begin at the center and count off slopes of $m = \pm\dfrac{b}{a} = \pm\dfrac{4}{3}$, or

draw a rectangle centered at $(2, 1)$ with dimensions $2(3) = 6$ (horizontal dimension) by $2(4) = 8$ (vertical dimension) to sketch the asymptotes. The complete graph is shown here.

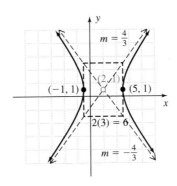

Horizontal hyperbola

Center at $(2, 1)$

Vertices at $(-1, 1)$ and $(5, 1)$

Transverse axis: $y = 1$
Conjugate axis: $x = 2$

Width of rectangle
$\begin{pmatrix} \text{horizontal dimension and} \\ \text{distance between vertices} \end{pmatrix}$
$2a = 2(3) = 6$

Length of rectangle
(vertical dimension)
$2b = 2(4) = 8$

Now try Exercises 27 through 38 ▶

Polynomial Form

If the equation is given as a polynomial in expanded form, complete the square in x and y, then write the equation in standard form.

EXAMPLE 4 ▶ **Graphing a Hyperbola by Completing the Square**

Graph the equation $9y^2 - x^2 + 54y + 4x + 68 = 0$.

Solution ▶ Since the y^2-term occurs first, we assume the equation represents a vertical hyperbola, but wait for the factored form to be sure (see **Exercise 87**).

$$9y^2 - x^2 + 54y + 4x + 68 = 0 \qquad \text{given}$$

$$9y^2 + 54y - x^2 + 4x = -68 \qquad \text{collect like-variable terms; subtract 68}$$

$$9(y^2 + 6y + \underline{\quad}) - 1(x^2 - 4x + \underline{\quad}) = -68 \qquad \text{factor out 9 from } y\text{-terms and } -1 \text{ from } x\text{-terms}$$

$$9(\underbrace{y^2 + 6y + 9}_{\text{adds } 9(9) = 81}) - 1(\underbrace{x^2 - 4x + 4}_{\text{adds } -1(4) = -4}) = -68 + 81 + (-4) \qquad \begin{array}{l} \text{complete the square} \\ \text{add } 81 + (-4) \text{ to right} \end{array}$$

$$9(y + 3)^2 - 1(x - 2)^2 = 9 \qquad \text{factor} \rightarrow \text{vertical hyperbola}$$

$$\frac{(y + 3)^2}{1} - \frac{(x - 2)^2}{9} = 1 \qquad \text{divide by 9 (standard form)}$$

$$\frac{(y + 3)^2}{1^2} - \frac{(x - 2)^2}{3^2} = 1 \qquad \text{write denominators in squared form}$$

The center of the hyperbola is $(2, -3)$ with $a = 3$, $b = 1$, and a transverse axis of $x = 2$. The vertices are at $(2, -3 + 1)$ and $(2, -3 - 1) \rightarrow (2, -2)$ and $(2, -4)$. After plotting the center and vertices, we draw a rectangle centered at $(2, -3)$ with a horizontal "width" of $2(3) = 6$ and a vertical "length" of $2(1) = 2$ to sketch the asymptotes. The completed graph is given in the figure.

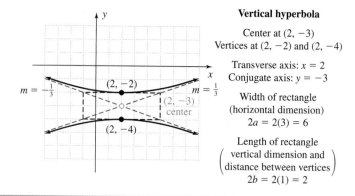

Vertical hyperbola

Center at $(2, -3)$
Vertices at $(2, -2)$ and $(2, -4)$

Transverse axis: $x = 2$
Conjugate axis: $y = -3$

Width of rectangle
(horizontal dimension)
$2a = 2(3) = 6$

Length of rectangle
$\left(\begin{array}{c}\text{vertical dimension and}\\\text{distance between vertices}\end{array}\right)$
$2b = 2(1) = 2$

☑ **A.** You've just learned how to use the equation of a hyperbola to graph central and noncentral hyperbolas

Now try Exercises 39 through 48 ▶

B. Distinguishing between the Equations of a Circle, Ellipse, and Hyperbola

So far we've explored numerous graphs of circles, ellipses, and hyperbolas. In Example 5 we'll attempt to identify a given conic section from its equation alone (without graphing the equation). As you've seen, the corresponding equations have unique characteristics that can help distinguish one from the other.

EXAMPLE 5 ▶   **Identifying a Conic Section from Its Equation**

Identify each equation as that of a circle, ellipse, or hyperbola. Justify your choice and name the center, but do not draw the graphs.

a. $y^2 = 36 + 9x^2$ **b.** $4x^2 = 16 - 4y^2$

c. $x^2 = 225 - 25y^2$ **d.** $25x^2 = 100 + 4y^2$

e. $3(x - 2)^2 + 4(y + 3)^2 = 12$ **f.** $4(x + 5)^2 = 36 + 9(y - 4)^2$

Solution ▶   **a.** Writing the equation in factored form gives $y^2 - 9x^2 = 36$ ($h = 0, k = 0$). Since the equation contains a difference of second-degree terms, it is the equation of a (vertical) hyperbola. The center is at $(0, 0)$.

b. Rewriting the equation as $4x^2 + 4y^2 = 16$ and dividing by 4 gives $x^2 + y^2 = 4$. The equation represents a circle of radius 2, with the center at $(0, 0)$.

c. Writing the equation as $x^2 + 25y^2 = 225$ we note a sum of second-degree terms with unequal coefficients. The equation is that of an ellipse, with the center at $(0, 0)$.

d. Rewriting the equation as $25x^2 - 4y^2 = 100$ we note the equation contains a difference of second-degree terms. The equation represents a central (horizontal) hyperbola, whose center is at $(0, 0)$.

e. The equation is in factored form and contains a sum of second-degree terms with unequal coefficients. This is the equation of an ellipse with the center at $(2, -3)$.

☑ **B.** You've just learned how to distinguish between the equations of a circle, ellipse, and hyperbola

f. Rewriting the equation as $4(x + 5)^2 - 9(y - 4)^2 = 36$ we note a difference of second-degree terms. The equation represents a horizontal hyperbola with center $(-5, 4)$.

Now try Exercises 49 through 60 ▶

C. The Foci of a Hyperbola

Like the ellipse, the foci of a hyperbola play an important part in their application. A long distance radio navigation system (called LORAN for short), can be used to determine the location of ships and airplanes and is based on the characteristics of a hyperbola (see **Exercises 85 and 86**). Hyperbolic mirrors are also used in some telescopes, and have the property that a beam of light directed at one focus will be reflected to the second focus. To understand and appreciate these applications, we use the analytic definition of a hyperbola:

Definition of a Hyperbola

Given two fixed points f_1 and f_2 in a plane, a hyperbola is the set of all points (x, y) such that the distance from f_2 to (x, y) subtracted from the distance from f_1 to (x, y) is a positive constant. In symbols,

$$d_1 - d_2 = k, k > 0$$

The fixed points f_1 and f_2 are called the foci of the hyperbola, and the points (x, y) are on the graph of the hyperbola.

Figure 7.23

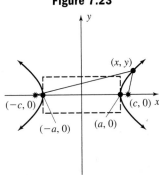

As with the analytic definition of the ellipse, it can be shown that the constant k is again equal to $2a$ (for horizontal hyperbolas). To find the equation of a hyperbola in terms of a and b, we use an approach similar to that of the ellipse (see Appendix III), and the result is identical to that seen earlier: $\dfrac{x^2}{a^2} - \dfrac{y^2}{b^2} = 1$ where $b^2 = c^2 - a^2$ (see Figure 7.23).

We now have the ability to *find the foci of any hyperbola*—and can use this information in many significant applications. Since the location of the foci play such an important role, it is best to remember the relationship as $c^2 = a^2 + b^2$ (called the **foci formula** for hyperbolas), noting that for a hyperbola, $c > a$ and $c^2 > a^2$ (also $c > b$ and $c^2 > b^2$).

EXAMPLE 6 ▶ **Graphing a Hyperbola and Identifying Its Foci by Completing the Square.**

For the hyperbola defined by $7x^2 - 9y^2 - 14x + 72y - 200 = 0$, find the coordinates of the center, vertices, foci, and the dimensions of the central rectangle. Then sketch the graph.

Solution ▶

$$7x^2 - 9y^2 - 14x + 72y - 200 = 0 \qquad \text{given}$$

$$7x^2 - 14x - 9y^2 + 72y = 200 \qquad \text{group terms; add 200}$$

$$7(x^2 - 2x + \underline{}) - 9(y^2 - 8y + \underline{}) = 200 \qquad \text{factor out leading coefficients}$$

$$7(x^2 - 2x + 1) - 9(y^2 - 8y + 16) = 200 + 7 + (-144) \qquad \text{complete the square}$$
$$\rightarrow \text{add } 7 + (-144)$$
$$\text{to right-hand side}$$

adds $7(1) = 7$ \qquad adds $-9(16) = -144$

$$7(x - 1)^2 - 9(y - 4)^2 = 63 \qquad \text{factored form}$$

$$\frac{(x - 1)^2}{9} - \frac{(y - 4)^2}{7} = 1 \qquad \text{divide by 63 and simplify}$$

$$\frac{(x - 1)^2}{3^2} - \frac{(y - 4)^2}{(\sqrt{7})^2} = 1 \qquad \text{write denominators in squared form}$$

This is a horizontal hyperbola with $a = 3$ ($a^2 = 9$) and $b = \sqrt{7}$ ($b^2 = 7$). The center is at $(1, 4)$, with vertices $(-2, 4)$ and $(4, 4)$. Using the foci formula $c^2 = a^2 + b^2$ yields $c^2 = 9 + 7 = 16$, showing the foci are $(-3, 4)$ and $(5, 4)$ (4 units from center). The central rectangle is $2\sqrt{7} \approx 5.29$ by $2(3) = 6$.

Drawing the rectangle and sketching the asymptotes to complete the graph, results in the graph shown.

Horizontal hyperbola

Center at (1, 4)
Vertices at (−2, 4) and (4, 4)

Transverse axis: $y = 4$
Conjugate axis: $x = 1$
Location of foci: (−3, 4) and (5, 4)

Width of rectangle
$\left(\begin{array}{c}\text{horizontal dimension and}\\ \text{distance between vertices}\end{array}\right)$
$2a = 2(3) = 6$

Length of rectangle
(vertical dimension)
$2b = 2(\sqrt{7}) \approx 5.29$

☑ C. You've just learned how to locate the foci of a hyperbola and use the foci and other features to write its equation

Now try Exercises 61 through 70 ▶

As with the ellipse, if any two of the values for a, b, and c are known, the relationship between them can be used to construct the equation of the hyperbola. See **Exercises 71 through 78.**

D. Applications Involving Foci

Applications involving the foci of a conic section can take many forms. As before, only partial information about the hyperbola may be available, and we'll determine a solution by manipulating a given equation, or constructing an equation from given facts.

EXAMPLE 7 ▶ **Applying the Properties of a Hyperbola—The Path of a Comet**

Comets with a high velocity cannot be captured by the Sun's gravity, and are slung around the Sun in a hyperbolic path with the Sun at one focus. If the path illustrated by the graph shown is modeled by the equation $2116x^2 - 400y^2 = 846{,}400$, how close did the comet get to the Sun? Assume units are in millions of miles and round to the nearest million.

Solution ▶ We are essentially asked to find the distance between a vertex and focus. Begin by writing the equation in standard form:

$2116x^2 - 400y^2 = 846{,}400$ given

$\dfrac{x^2}{400} - \dfrac{y^2}{2116} = 1$ divide by 846,400

$\dfrac{x^2}{20^2} - \dfrac{y^2}{46^2} = 1$ write denominators in squared form

This is a horizontal hyperbola with $a = 20$ ($a^2 = 400$) and $b = 46$ ($b^2 = 2116$). Use the foci formula to find c^2 and c.

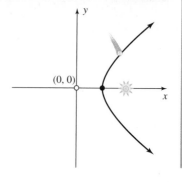

$$c^2 = a^2 + b^2$$
$$c^2 = 400 + 2116$$
$$c^2 = 2516$$
$$c \approx -50 \text{ and } c \approx 50$$

Since $a = 20$ and $|c| \approx 50$, the comet came within $50 - 20 = 30$ million miles of the Sun.

Now try Exercises 81 through 84 ▶

EXAMPLE 8 ▶ **Applying the Properties of a Hyperbola—The Location of a Storm**

Two amateur meteorologists, living 4 km apart (4000 m), see a storm approaching. The one farthest from the storm hears a loud clap of thunder 9 sec after the one nearest. Assuming the speed of sound is 340 m/sec, determine an equation that models possible locations for the storm at this time.

Solution ▶ Let M_1 represent the meteorologist nearest the storm and M_2 the farthest. Since M_2 heard the thunder 9 sec after M_1, M_2 must be $9 \cdot 340 = 3060$ m farther away from the storm S. In other words, $|M_2S| - |M_1S| = 3060$. The set of all points that satisfy this description fit the definition of a hyperbola, and we'll use this fact to develop an equation model for possible locations of the storm. Let's place the information on a coordinate grid. For convenience, we'll use the straight line distance between M_1 and M_2 as the x-axis, with the origin an equal distance from each. With the constant difference equal to 3060, we have $2a = 3060$, $a = 1530$ from the definition of a hyperbola,

giving $\dfrac{x^2}{1530^2} - \dfrac{y^2}{b^2} = 1$. With $c = 2000$ m (the distance from the origin to M_1 or M_2), we find the value of b using the equation $c^2 = a^2 + b^2$: $2000^2 = 1530^2 + b^2$ or $b^2 = (2000)^2 - (1530)^2 = 1,659,100 \approx 1288^2$. The equation that models possible locations of the storm is

☑ **D.** You've just learned how to solve applications involving foci

$$\dfrac{x^2}{1530^2} - \dfrac{y^2}{1288^2} \approx 1.$$

Now try Exercises 85 and 86 ▶

TECHNOLOGY HIGHLIGHT

Studying Hyperbolas

As with the circle and ellipse, the hyperbola must also be defined in two pieces in order to use a graphing calculator to study its graph. Consider the *relation* $4x^2 - 9y^2 = 36$. From our work in this section, we know this is the equation of a horizontal hyperbola centered at (0, 0). Solving for *y* gives

$$4x^2 - 9y^2 = 36 \qquad \text{original equation}$$

$$-9y^2 = 36 - 4x^2 \qquad \text{isolate } y^2\text{-term}$$

$$y^2 = \frac{36 - 4x^2}{-9} \qquad \text{divide by } -9$$

$$y = \pm\sqrt{\frac{36 - 4x^2}{-9}} \qquad \text{solve for } y$$

Figure 7.24

We can again separate this result into two parts: $Y_1 = \sqrt{\dfrac{36 - 4x^2}{-9}}$ gives the "upper half" of the hyperbola, and $Y_2 = -\sqrt{\dfrac{36 - 4x^2}{-9}}$ gives the "lower half." In Figure 7.24, note the use of parentheses on the
Y= screen to ensure we're taking the square root of the entire expression.

Entering these on the Y= screen, graphing them with the window shown, and pressing the TRACE key gives the graph shown in Figure 7.25. Note the location of the cursor at $x = 0$, but no y-value is displayed. This is because the hyperbola is not defined at $x = 0$. Press the right arrow key ▶ and walk the cursor to the right until the y-values begin appearing. In fact, they begin to appear at (3, 0), which is one of the vertices of the hyperbola. We could also graph the asymptotes ($y = \pm\frac{2}{3}x$) by entering the lines as Y_3 and Y_4 on the Y= screen. The resulting graph is shown in Figure 7.26 using the standard window (the TRACE key has been pushed and the down arrow used to highlight Y_2). Use these ideas to complete the following exercises.

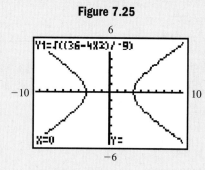

Figure 7.25

Figure 7.26

Exercise 1: Graph the hyperbola $25y^2 - 4x^2 = 100$ using a friendly window. What are the coordinates of the vertices? Use the TRACE feature to find the value(s) of y when $x = 4$. Determine (from the graph) the value(s) of y when $x = -4$, then verify your response using the TABLE feature.

Exercise 2: Graph the hyperbola $9x^2 - 16y^2 = 144$ using the standard window. Then determine the equations of the asymptotes and graph these as well. Why do the asymptotes intersect at the origin? When will the asymptotes *not* intersect at the origin?

7.3 EXERCISES

▶ CONCEPTS AND VOCABULARY

Fill in the blank with the appropriate word or phrase. Carefully reread the section if needed.

1. The line that passes through the vertices of a hyperbola is called the _____ axis.

2. The conjugate axis is _____ to the _____ axis and contains the _____ of the hyperbola.

3. The center of a hyperbola is located _____ between the vertices.

4. The center of the hyperbola defined by
$$\frac{(x-2)^2}{4^2} - \frac{(y-3)^2}{5^2} = 1 \text{ is at } _____.$$

5. Compare/Contrast the two methods used to find the asymptotes of a hyperbola. Include an example illustrating both methods.

6. Explore/Explain why $A(x-h)^2 - B(y-k)^2 = F$ results in a hyperbola regardless of whether $A = B$ or $A \neq B$. Illustrate with an example.

▶ DEVELOPING YOUR SKILLS

Graph each hyperbola. Label the center, vertices, and any additional points used.

7. $\dfrac{x^2}{9} - \dfrac{y^2}{4} = 1$

8. $\dfrac{x^2}{16} - \dfrac{y^2}{9} = 1$

9. $\dfrac{x^2}{4} - \dfrac{y^2}{9} = 1$

10. $\dfrac{x^2}{25} - \dfrac{y^2}{16} = 1$

11. $\dfrac{x^2}{49} - \dfrac{y^2}{16} = 1$

12. $\dfrac{x^2}{25} - \dfrac{y^2}{9} = 1$

13. $\dfrac{x^2}{36} - \dfrac{y^2}{16} = 1$

14. $\dfrac{x^2}{81} - \dfrac{y^2}{16} = 1$

15. $\dfrac{y^2}{9} - \dfrac{x^2}{1} = 1$

16. $\dfrac{y^2}{1} - \dfrac{x^2}{4} = 1$

17. $\dfrac{y^2}{12} - \dfrac{x^2}{4} = 1$

18. $\dfrac{y^2}{9} - \dfrac{x^2}{18} = 1$

19. $\dfrac{y^2}{9} - \dfrac{x^2}{9} = 1$

20. $\dfrac{y^2}{4} - \dfrac{x^2}{4} = 1$

21. $\dfrac{y^2}{36} - \dfrac{x^2}{25} = 1$

22. $\dfrac{y^2}{16} - \dfrac{x^2}{4} = 1$

For the graphs given, state the location of the vertices and the equation of the transverse axis. Then identify the location of the center and the equation of the conjugate axis. Note the scale used on each axis.

23.

24.

25.

26.

Sketch a complete graph of each equation, including the asymptotes. Be sure to identify the center and vertices.

27. $\dfrac{(y+1)^2}{4} - \dfrac{x^2}{25} = 1$

28. $\dfrac{y^2}{4} - \dfrac{(x-2)^2}{9} = 1$

29. $\dfrac{(x-3)^2}{36} - \dfrac{(y+2)^2}{49} = 1$

30. $\dfrac{(x-2)^2}{9} - \dfrac{(y-1)^2}{4} = 1$

31. $\dfrac{(y + 1)^2}{7} - \dfrac{(x + 5)^2}{9} = 1$

32. $\dfrac{(y - 3)^2}{16} - \dfrac{(x + 2)^2}{5} = 1$

33. $(x - 2)^2 - 4(y + 1)^2 = 16$

34. $9(x + 1)^2 - (y - 3)^2 = 81$

35. $2(y + 3)^2 - 5(x - 1)^2 = 50$

36. $9(y - 4)^2 - 5(x - 3)^2 = 45$

37. $12(x - 4)^2 - 5(y - 3)^2 = 60$

38. $8(x - 4)^2 - 3(y - 3)^2 = 24$

39. $16x^2 - 9y^2 = 144$

40. $16x^2 - 25y^2 = 400$

41. $9y^2 - 4x^2 = 36$

42. $25y^2 - 4x^2 = 100$

43. $12x^2 - 9y^2 = 72$

44. $36x^2 - 20y^2 = 180$

45. $4x^2 - y^2 + 40x - 4y + 60 = 0$

46. $x^2 - 4y^2 - 12x - 16y + 16 = 0$

47. $x^2 - 4y^2 - 24y - 4x - 36 = 0$

48. $-9x^2 + 4y^2 - 18x - 24y - 9 = 0$

Classify each equation as that of a circle, ellipse, or hyperbola. Justify your response.

49. $-4x^2 - 4y^2 = -24$

50. $9y^2 = -4x^2 + 36$

51. $x^2 + y^2 = 2x + 4y + 4$

52. $x^2 = y^2 + 6y - 7$

53. $2x^2 - 4y^2 = 8$

54. $36x^2 + 25y^2 = 900$

55. $x^2 + 5 = 2y^2$

56. $x + y^2 = 3x^2 + 9$

57. $2x^2 = -2y^2 + x + 20$

58. $2y^2 + 3 = 6x^2 + 8$

59. $16x^2 + 5y^2 - 3x + 4y = 538$

60. $9x^2 + 9y^2 - 9x + 12y + 4 = 0$

Use the definition of a hyperbola to find the distance between the vertices and the dimensions of the rectangle centered at (h, k). Figures are not drawn to scale. Note that Exercises 63 and 64 are *vertical hyperbolas*.

61.

62.

63.

64.

Find and list the coordinates of the (a) center, (b) vertices, (c) foci, and (d) dimensions of the central rectangle. Then (e) sketch the graph, including the asymptotes.

65. $4x^2 - 9y^2 - 24x + 72y - 144 = 0$

66. $4x^2 - 36y^2 - 40x + 144y - 188 = 0$

67. $16x^2 - 4y^2 + 24y - 100 = 0$

68. $81x^2 - 162x - 4y^2 - 243 = 0$

69. $9x^2 - 3y^2 - 54x - 12y + 33 = 0$

70. $10x^2 + 60x - 5y^2 + 20y - 20 = 0$

Find the equation of the hyperbola (in standard form) that satisfies the following conditions:

71. vertices at $(-6, 0)$ and $(6, 0)$; foci at $(-8, 0)$ and $(8, 0)$

72. vertices at $(-4, 0)$ and $(4, 0)$; foci at $(-6, 0)$ and $(6, 0)$

73. foci at $(-2, -3\sqrt{2})$ and $(-2, 3\sqrt{2})$; length of conjugate axis: 6 units

74. foci at $(-5, 2)$ and $(7, 2)$; length of conjugate axis: 8 units

Use the characteristics of a hyperbola and the graph given to write the related equation and state the location of the foci (75 and 76) or the dimensions of the central rectangle (77 and 78).

75.

76.

77.

78.

▶ WORKING WITH FORMULAS

79. Equation of a semi-hyperbola: $y = \sqrt{\dfrac{36 - 4x^2}{-9}}$

The "upper half" of a certain hyperbola is given by the equation shown. (a) Simplify the radicand, (b) state the domain of the expression, and (c) enter the expression as Y_1 on a graphing calculator and graph. What is the equation for the "lower half" of this hyperbola?

80. Focal chord of a hyperbola: $L = \dfrac{2b^2}{a}$

The focal chords of a hyperbola are line segments parallel to the conjugate axis with endpoints on the hyperbola, and containing certain points f_1 and f_2 called the *foci* (see grid). The length of the chord is given by the formula shown. Use it to find the length of the focal chord for the hyperbola indicated, then compare the calculated value with the length estimated from the given graph:

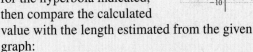

$$\frac{(x - 2)^2}{4} - \frac{(y - 1)^2}{5} = 1.$$

▶ APPLICATIONS

81. Stunt pilots: At an air show, a stunt plane dives along a hyperbolic path whose vertex is directly over the grandstands. If the plane's flight path can be modeled by the hyperbola $25y^2 - 1600x^2 = 40,000$, what is the minimum altitude of the plane as it passes over the stands? Assume x and y are in yards.

82. Flying clubs: To test their skill as pilots, the members of a flight club attempt to drop sandbags on a target placed in an open field, by diving along a hyperbolic path whose vertex is directly over the target area. If the flight path of the plane flown by the club's president is modeled by $9y^2 - 16x^2 = 14,400$, what is the minimum altitude of her plane as it passes over the target? Assume x and y are in feet.

83. Nuclear cooling towers: The natural draft cooling towers for nuclear power stations are called *hyperboloids of one sheet.* The

perpendicular cross sections of these hyperboloids form two branches of a hyperbola. Suppose the central cross section of one such tower is modeled by the hyperbola $1600x^2 - 400(y - 50)^2 = 640,000$. What is the minimum distance between the sides of the tower? Assume x and y are in feet.

84. Charged particles: It has been shown that when like particles with a common charge are hurled at each other, they deflect and travel along paths that are hyperbolic. Suppose the path of two such particles is modeled by the hyperbola $x^2 - 9y^2 = 36$. What is the minimum distance between the particles as they approach each other? Assume x and y are in microns.

85. Locating a ship using radar: Under certain conditions, the properties of a hyperbola can be used to help locate the position of a ship. Suppose two radio stations are located 100 km apart along a straight shoreline. A ship is sailing parallel to the shore and is 60 km out to sea. The ship sends out a distress call that is picked up by the closer station in 0.4 milliseconds (msec—one-thousandth of a second), while it takes 0.5 msec to reach the station that is farther away. Radio waves travel at a speed of approximately 300 km/msec. Use this information to find the equation of a hyperbola that will help you find the location of the ship, then find the coordinates of the ship. (*Hint:* Draw the hyperbola on a coordinate system with the radio stations on the *x*-axis at the foci, then use the definition of a hyperbola.)

86. **Locating a plane using radar:** Two radio stations are located 80 km apart along a straight shoreline, when a "mayday" call (a plea for immediate help) is received from a plane that is about to ditch in the ocean (attempt a water landing). The plane was flying at low altitude, parallel to the shoreline, and 20 km out when it ran into trouble. The plane's distress call is picked up by the closer station in 0.1 msec, while it takes 0.3 msec to reach the other. Use this information to construct the equation of a hyperbola that will help you find the location of the ditched plane, then find the coordinates of the plane. Also see Exercise 85.

87. It is possible for the plane to intersect only the vertex of the cone or to be tangent to the sides. These are called **degenerate cases** of a conic section. Many times we're unable to tell if the equation represents a degenerate case until it's written in standard form. Write the following equations in standard form and comment.

a. $4x^2 - 32x - y^2 + 4y + 60 = 0$

b. $x^2 - 4x + 5y^2 - 40y + 84 = 0$

88. For a greater understanding as to *why* the branches of a hyperbola are asymptotic, solve $\dfrac{x^2}{a^2} - \dfrac{y^2}{b^2} = 1$ for *y*, then consider what happens as $x \to \infty$ (note that $x^2 - k \approx x^2$ for large *x*).

89. Which has a greater area: (a) The central rectangle of the hyperbola given by $(x - 5)^2 - (y + 4)^2 = 57$, (b) the circle given by $(x - 5)^2 + (y + 4)^2 = 57$, or (c) the ellipse given by $9(x - 5)^2 + 10(y + 4)^2 = 570$?

90. Find the equation of the circle shown, given the equation of the hyperbola: $9(x - 2)^2 - 25(y - 3)^2 = 225$

91. Find the equation of the ellipse shown, given the equation of the hyperbola and (2, 0) is on the graph of the ellipse. The hyperbola and ellipse share the same foci: $9(x - 2)^2 - 25(y - 3)^2 = 225$

92. Verify that for a central hyperbola, a circle that circumscribes the central rectangle must also go through both foci.

▶ **MAINTAINING YOUR SKILLS**

93. (2.7) Graph the piecewise-defined function:

$$f(x) = \begin{cases} 4 - x^2 & -2 \le x < 3 \\ 5 & x \ge 3 \end{cases}$$

94. (3.2) Use synthetic division and the remainder theorem to determine if $x = 2$ is a zero of $g(x) = x^5 - 5x^4 + 4x^3 + 16x^2 - 32x + 16$. If yes, find its multiplicity.

95. (1.4) The number $z = 1 + i\sqrt{2}$ is a solution to two out of the three equations given. Which two?

a. $x^4 + 4 = 0$

b. $x^3 - 6x^2 + 11x - 12 = 0$

c. $x^2 - 2x + 3 = 0$

96. (5.4) A government-approved company is licensed to haul toxic waste. Each container of solid waste weighs 800 lb and has a volume of 100 ft³. Each container of liquid waste weighs 1000 lb and is 60 ft³ in volume. The revenue from hauling solid waste is $300 per container, while the revenue from liquid waste is $375 per container. The truck used by this company has a weight capacity of 39.8 tons and a volume capacity of 6960 ft³. What combination of solid and liquid weight containers will produce the maximum revenue?

Learning Objectives

In Section 7.4 you will learn how to:

- ☐ **A.** Graph parabolas with a horizontal axis of symmetry
- ☐ **B.** Identify and use the focus-directrix form of the equation of a parabola
- ☐ **C.** Solve nonlinear systems involving the conic sections
- ☐ **D.** Solve Applications of the Analytic parabola

In previous coursework, you likely learned that the graph of a quadratic function was a parabola. Parabolas are actually the fourth and final member of the family of conic sections, and as we saw in Section 7.1, the graph can be obtained by observing the intersection of a plane and a cone. If the plane is parallel to the generator of the cone (shown as a dark line in Figure 7.27), the intersection of the plane with one nappe forms a parabola. In this section we develop the general equation of a parabola from its analytic definition, opening a new realm of applications that extends far beyond those involving only zeroes and extreme values.

Figure 7.27

A. Parabolas with a Horizontal Axis

An introductory study of parabolas generally involves those with a vertical axis, defined by the equation $y = ax^2 + bx + c$. Unlike the previous conic sections, this equation has *only one second-degree (squared) term in x* and defines a function. As a review, the primary characteristics are listed here and illustrated in Figure 7.28.

Figure 7.28

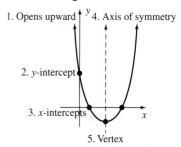

1. Opens upward
2. y-intercept
3. x-intercepts
4. Axis of symmetry
5. Vertex

Vertical Parabolas

For a second-degree equation of the form $y = ax^2 + bx + c$, the graph is a vertical parabola with these characteristics:

1. opens upward if $a > 0$, downward if $a < 0$.
2. axis of symmetry: $x = \dfrac{-b}{2a}$, vertex: $\left(\dfrac{-b}{2a}, y\right)$
3. y-intercept: $(0, c)$ (substitute 0 for x)
4. x-intercept(s): substitute 0 for y and solve.

See **Exercises 7 through 12.**

Horizontal Parabolas

Similar to our study of horizontal and vertical hyperbolas, the graph of a parabola can open *to the right or left,* as well as up or down. After interchanging the variables x and y in the standard equation, we obtain the parabola $x = ay^2 + by + c$, noting the resulting graph will be a reflection about the line $y = k$. Here, the axis of symmetry is a horizontal line and factoring or the quadratic formula is used to find the *y-intercepts* (if they exist). Note that although the graph is still a parabola—*it is not the graph of a function.*

Horizontal Parabolas

For a second-degree equation of the form $x = ay^2 + by + c$, the graph is a horizontal parabola with these characteristics:

1. opens right if $a > 0$, left if $a < 0$.
2. axis of symmetry: $y = \dfrac{-b}{2a}$, vertex: $\left(x, \dfrac{-b}{2a}\right)$
3. x-intercept: $(c, 0)$ (substitute 0 for y)
4. y-intercepts(s): substitute 0 for x and solve.

EXAMPLE 1 ▶ **Graphing a Horizontal Parabola**

Graph the relation whose equation is $x = y^2 + 3y - 4$, then state the domain and range of the relation.

Solution ▶ Since the equation has a single squared term in y, the graph will be a horizontal parabola. With $a > 0$ $(a = 1)$, the parabola opens to the right. The x-intercept is $(-4, 0)$. Factoring shows the y-intercepts are $y = -4$ and $y = 1$. The axis of symmetry is $y = \frac{-3}{2} = -1.5$, and substituting this value into the original equation gives $x = -6.25$. The coordinates of the vertex are $(-6.25, -1.5)$. Using horizontal and vertical boundary lines we find the domain for this relation is $x \in [-6.25, \infty)$ and the range is $y \in (-\infty, \infty)$. The graph is shown.

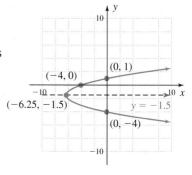

Now try Exercises 13 through 18 ▶

As with the vertical parabola, the equation of a horizontal parabola can be written as a transformation: $x = a(y \pm k)^2 \pm h$ by completing the square. Note that in this case, the vertical shift is k units *opposite the sign,* with a horizontal shift of h units in the same direction as the sign.

EXAMPLE 2 ▶ **Graphing a Horizontal Parabola by Completing the Square**

Graph by completing the square: $x = -2y^2 - 8y - 9$.

Solution ▶ Using the original equation, we note the graph will be a horizontal parabola opening to the left $(a = -2)$ and have an x-intercept of $(-9, 0)$. Completing the square gives $x = -2(y^2 + 4y + 4) - 9 + 8$, so $x = -2(y + 2)^2 - 1$. The vertex is at $(-1, -2)$ and $y = -2$ is the axis of symmetry. This means there are no y-intercepts, a fact that comes to light when we attempt to solve the equation after substituting 0 for x:

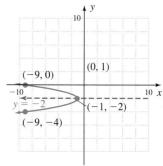

$$-2(y + 2)^2 - 1 = 0 \qquad \text{substitute 0 for } x$$

$$(y + 2)^2 = -\frac{1}{2} \qquad \text{isolate squared term}$$

☑ **A.** You've just learned how to graph parabolas with a horizontal axis of symmetry

The equation has no real roots. Using symmetry, the point $(-9, -4)$ is also on the graph. After plotting these points we obtain the graph shown.

Now try Exercises 19 through 36 ▶

B. The Focus-Directrix Form of the Equation of a Parabola

As with the ellipse and hyperbola, many significant applications of the parabola rely on its analytical definition rather than its algebraic form. From the construction of radio telescopes to the manufacture of flashlights, the location of the focus of a parabola is critical. To understand these and other applications, we use the analytic definition of a parabola first introduced in Section 7.1.

Definition of a Parabola

Given a fixed point f and fixed line D in the plane, a parabola is the set of all points (x, y) such that the distance from f to (x, y) is equal to the distance from line D to (x, y). The fixed point f is the **focus** of the parabola, and the fixed line is the **directrix.**

Figure 7.29

The general equation of a parabola can be obtained by combining this definition with the distance formula. With no loss of generality, we can assume the parabola shown in the definition box is oriented in the plane with the vertex at $(0, 0)$ and the focus at $(0, p)$. As the diagram in Figure 7.29 indicates, this gives the directrix an equation of $y = -p$ with all points on D having coordinates of $(x, -p)$.

Using $d_1 = d_2$ the distance formula yields

$$\sqrt{(x - 0)^2 + (y - p)^2} = \sqrt{(x - x)^2 + (y + p)^2} \quad \text{from definition}$$
$$(x - 0)^2 + (y - p)^2 = (x - x)^2 + (y + p)^2 \quad \text{square both sides}$$
$$x^2 + y^2 - 2py + p^2 = 0 + y^2 + 2py + p^2 \quad \text{simplify; expand binomials}$$
$$x^2 - 2py = 2py \quad \text{subtract } p^2 \text{ and } y^2$$
$$x^2 = 4py \quad \text{isolate } x^2$$

The resulting equation is called the **focus-directrix form** of a *vertical parabola* with center at $(0, 0)$. If we had begun by orienting the parabola so it opened to the right, we would have obtained the equation of a *horizontal parabola* with center $(0, 0)$: $y^2 = 4px$.

The Equation of a Parabola in Focus-Directrix Form

Vertical Parabola	Horizontal Parabola
$x^2 = 4py$	$y^2 = 4px$
focus $(0, p)$, directrix: $y = -p$	focus at $(p, 0)$ directrix: $x = -p$
If $p > 0$, opens upward.	If $p > 0$, opens to the right.
If $p < 0$, opens downward.	If $p < 0$, opens to the left.

For a parabola, note there is only one second-degree term.

EXAMPLE 3 ▶ **Locating the Focus and Directrix of a Parabola**

Find the vertex, focus, and directrix for the parabola defined by $x^2 = -12y$. Then sketch the graph, including the focus and directrix.

Solution ▶ Since the x-term is squared and no shifts have been applied, the graph will be a vertical parabola with a vertex of $(0, 0)$. Use a direct comparison between the given equation and the focus-directrix form to determine the value of p:

$$x^2 = -12y \quad \text{given equation}$$
$$\downarrow$$
$$x^2 = 4py \quad \text{focus-directrix form}$$

This shows:

$$4p = -12$$
$$p = -3$$

Since $p = -3$ ($p < 0$), the parabola opens downward, with the focus at $(0, -3)$ and directrix $y = 3$. To complete the graph we need a few additional points. Since 36 $(6)^2$ is divisible by 12, we can use inputs of $x = 6$ and $x = -6$, giving the points $(6, -3)$ and $(-6, -3)$. Note the axis of symmetry is $x = 0$. The graph is shown.

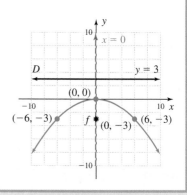

Now try Exercises 37 through 48 ▶

Figure 7.30

As an alternative to calculating additional points to sketch the graph, we can use what is called the **focal chord** of the parabola. Similar to the ellipse and hyperbola, the focal chord is a line segment that contains the focus, is parallel to the directrix, and has its endpoints on the graph. Using the definition of a parabola and the diagram in Figure 7.30, we see the horizontal distance from f to (x, y) is $2p$. Since $d_1 = d_2$ a line segment parallel to the directrix from the focus to the graph will also have a length of $|2p|$, and the focal chord of any parabola has a total length of $|4p|$. Note that in Example 3, the points we happened to choose were actually the endpoints of the focal chord.

Finally, if the vertex of a vertical parabola is shifted to (h, k), the equation will have the form $(x \pm h)^2 = 4p(y \pm k)$. As with the other conic sections, both the horizontal and vertical shifts are "opposite the sign."

EXAMPLE 4 ▶ **Locating the Focus and Directrix of a Parabola**

Find the vertex, focus, and directrix for the parabola whose equation is given, then sketch the graph, including the focus and directrix: $x^2 - 6x + 12y - 15 = 0$.

Solution ▶ Since only the x-term is squared, the graph will be a vertical parabola. To find the concavity, vertex, focus, and directrix, we complete the square in x and use a direct comparison between the shifted form and the focus-directrix form:

$$x^2 - 6x + 12y - 15 = 0 \qquad \text{given equation}$$
$$x^2 - 6x + \underline{} = -12y + 15 \qquad \text{complete the square in } x$$
$$x^2 - 6x + 9 = -12y + 24 \qquad \text{add 9}$$
$$(x - 3)^2 = -12(y - 2) \qquad \text{factor}$$

Notice the parabola has been shifted 3 right and 2 up, so *all features of the parabola will likewise be shifted*. Since we have $4p = -12$ (the coefficient of the linear term), we know $p = -3$ ($p < 0$) and the parabola opens downward. If the parabola were in standard position, the vertex would be at $(0, 0)$, the focus at $(0, -3)$ and the directrix a horizontal line at $y = 3$. But since the parabola is shifted 3 right and 2 up, we add 3 to all x-values and 2 to all y-values to locate the features of the shifted parabola. The vertex is at

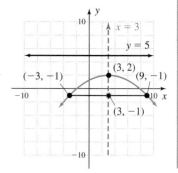

$(0 + 3, 0 + 2) = (3, 2)$. The focus is $(0 + 3, -3 + 2) = (3, -1)$ and the directrix is $y = 3 + 2 = 5$. Finally, the horizontal distance from the focus to the graph is $|2p| = 6$ units (since $|4p| = 12$), giving us the additional points $(-3, -1)$ and $(9, -1)$. The graph is shown.

Now try Exercises 49 through 60 ▶

In many cases, we need to construct the equation of the parabola when only partial information in known, as illustrated in Example 5.

EXAMPLE 5 ▶   Constructing the Equation of a Parabola

Find the equation of the parabola with vertex (4, 4) and focus (4, 1). Then graph the parabola using the equation and focal chord.

Solution ▶   As the vertex and focus are on a vertical line, we have a vertical parabola with general equation $(x \pm h)^2 = 4p(y \pm k)$. The distance p from vertex to focus is 3 units, and with the focus below the vertex, the parabola opens downward so $p = -3$. Using the focal chord, the horizontal distance from (4, 1) to the graph is $|2p| = |2(-3)| = 6$, giving points $(-2, 1)$ and $(10, 1)$. The vertex is shifted 4 units right and 4 units up from (0, 0), showing $h = 4$ and $k = 4$, and the equation of the parabola must be $(x - 4)^2 = -12(y - 4)$, with directrix $y = 7$. The graph is shown.

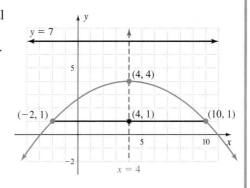

☑ **B. You've just learned how to identify and use the focus-directrix form of the equation of a parabola**

Now try Exercises 61 through 76 ▶

C. Nonlinear Systems and the Conic Sections

Similar to our work with nonlinear systems in Section 5.3, the graphing, substitution, or elimination method can still be used when the system involves a conic section. When both equations in the system have at least one second degree term, it is generally easier to use the elimination method.

EXAMPLE 6 ▶   Solving a System of Nonlinear Equations

Solve the system using elimination: $\begin{cases} 2y^2 - 5x^2 = 13 \\ 3x^2 + 4y^2 = 39 \end{cases}$

Solution ▶   The first equation represents a vertical and central hyperbola, while the second represents a horizontal and central ellipse. After writing the system with the x- and y-terms in the same order, we obtain $\begin{cases} -5x^2 + 2y^2 = 13 \\ 3x^2 + 4y^2 = 39 \end{cases}$. Using $-2R1 + R2$ will eliminate the y-term.

$$\begin{cases} 10x^2 - 4y^2 = -26 \qquad \text{\small{−2R1}} \\ \underline{\ 3x^2 + 4y^2 = 39} \qquad \text{\small{+R2}} \\ \ \ 13x^2 + 0 = 13 \qquad \text{\small{sum}} \\ \qquad\quad x^2 = 1 \qquad \text{\small{divide by 13}} \end{cases}$$

$$x = -1 \quad \text{or} \quad x = 1 \qquad \text{\small{square root property}}$$

Substituting $x = 1$ and $x = -1$ into the second equation we obtain

$$3(1)^2 + 4y^2 = 39 \qquad\qquad 3(-1)^2 + 4y^2 = 39$$
$$3 + 4y^2 = 39 \qquad\qquad\qquad 3 + 4y^2 = 39$$
$$4y^2 = 36 \qquad\qquad\qquad\qquad 4y^2 = 36$$
$$y^2 = 9 \qquad\qquad\qquad\qquad\quad y^2 = 9$$
$$y = -3 \quad \text{or} \quad y = 3 \qquad\qquad y = -3 \quad \text{or} \quad y = 3$$

Since -1 and 1 each generated *two outputs*, there are a total of four ordered pair solutions: $(1, -3)$, $(1, 3)$, $(-1, -3)$, and $(-1, 3)$. The graph is shown and supports our results.

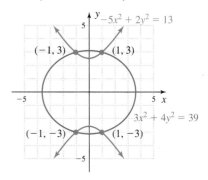

☑ **C.** You've just learned how to solve nonlinear systems involving the conic sections

> **Now try Exercises 77 through 82** ▶

D. Application of the Analytic Parabola

Here is just one of the many ways the analytic definition of a parabola can be applied. There are several others in the exercise set.

EXAMPLE 7 ▶ **Locating the Focus of a Parabolic Receiver**

The diagram shows the cross section of a radio antenna dish. Engineers have located a point on the cross section that is 0.75 m above and 6 m to the right of the vertex. At what coordinates should the engineers build the focus of the antenna?

Solution ▶ By inspection we see this is a vertical parabola with center at $(0, 0)$. This means its equation must be of the form $x^2 = 4py$. Because we know $(6, 0.75)$ is a point on this graph, we can substitute $(6, 0.75)$ in this equation and solve for p:

$$x^2 = 4py \qquad \text{equation for vertical parabola, vertex at (0, 0)}$$
$$(6)^2 = 4p(0.75) \qquad \text{substitute 6 for } x \text{ and 0.75 for } y$$
$$36 = 3p \qquad \text{simplify}$$
$$p = 12 \qquad \text{result}$$

With $p = 12$, we see that the focus must be located at $(0, 12)$, or 12 m directly above the vertex.

> **Now try Exercises 85 through 92**

☑ **D.** You've just learned how to solve applications of the analytic parabola

Note that in many cases, the focus of a parabolic dish may be taller than the rim of the dish.

7.4 EXERCISES

▶ CONCEPTS AND VOCABULARY

**Fill in the blank with the appropriate word or phrase.
Carefully reread the section if needed.**

1. The equation $x = ay^2 + by + c$ is that of a(n)
   _____ parabola, opening to the _____ if $a > 0$
   and to the left if _____.

2. If point P is on the graph of a parabola with
   directrix D, the distance from P to line D is equal
   to the distance between P and the _____ of the
   parabola.

3. Given $y^2 = 4px$, the focus is at _____ and the
   equation of the directrix is _____.

4. Given $x^2 = -16y$, the value of p is _____ and
   the coordinates of the focus are _____.

5. Discuss/Explain how to find the vertex,
   directrix, and focus from the equation
   $(x - h)^2 = 4p(y - k)$.

6. If a horizontal parabola has a vertex of $(2, -3)$
   with $a > 0$, what can you say about the
   y-intercepts? Will the graph always have an
   x-intercept? Explain.

▶ DEVELOPING YOUR SKILLS

**Find the x- and y-intercepts (if they exist) and the
vertex of the parabola. Then sketch the graph by using
symmetry and a few additional points or completing the
square and shifting a parent function. Scale the axes as
needed to comfortably fit the graph and state the
domain and range.**

7. $y = x^2 - 2x - 3$ 8. $y = x^2 + 6x + 5$

9. $y = 2x^2 - 8x - 10$ 10. $y = 3x^2 + 12x - 15$

11. $y = 2x^2 + 5x - 7$ 12. $y = 2x^2 - 7x + 3$

**Find the x- and y-intercepts (if they exist) and the vertex
of the graph. Then sketch the graph using symmetry
and a few additional points (scale the axes as needed).
Finally, state the domain and range of the relation.**

13. $x = y^2 - 2y - 3$ 14. $x = y^2 - 4y - 12$

15. $x = -y^2 + 6y + 7$ 16. $x = -y^2 + 8y - 12$

17. $x = -y^2 + 8y - 16$ 18. $x = -y^2 + 6y - 9$

**Sketch using symmetry and shifts of a basic function. Be
sure to find the x- and y-intercepts (if they exist) and the
vertex of the graph, then state the domain and range of
the relation.**

19. $x = y^2 - 6y$ 20. $x = y^2 - 8y$

21. $x = y^2 - 4$ 22. $x = y^2 - 9$

23. $x = -y^2 + 2y - 1$ 24. $x = -y^2 + 4y - 4$

25. $x = y^2 + y - 6$ 26. $x = y^2 + 4y - 5$

27. $x = y^2 - 10y + 4$ 28. $x = y^2 + 12y - 5$

29. $x = 3 - 8y - 2y^2$ 30. $x = 2 - 12y + 3y^2$

31. $y = (x - 2)^2 + 3$ 32. $y = (x + 2)^2 - 4$

33. $x = (y - 3)^2 + 2$ 34. $x = (y + 1)^2 - 4$

35. $x = 2(y - 3)^2 + 1$ 36. $x = -2(y + 3)^2 - 5$

**Find the vertex, focus, and directrix for the parabolas
defined by the equations given, then use this information
to sketch a complete graph (illustrate and name these
features). For Exercises 43 to 60, also include the focal
chord.**

37. $x^2 = 8y$ 38. $x^2 = 16y$

39. $x^2 = -24y$ 40. $x^2 = -20y$

41. $x^2 = 6y$ 42. $x^2 = 18y$

43. $y^2 = -4x$ 44. $y^2 = -12x$

45. $y^2 = 18x$ 46. $y^2 = 20x$

47. $y^2 = -10x$ 48. $y^2 = -14x$

49. $x^2 - 8x - 8y + 16 = 0$

50. $x^2 - 10x - 12y + 25 = 0$

51. $x^2 - 14x - 24y + 1 = 0$

52. $x^2 - 10x - 12y + 1 = 0$

53. $3x^2 - 24x - 12y + 12 = 0$

54. $2x^2 - 8x - 16y - 24 = 0$

55. $y^2 - 12y - 20x + 36 = 0$

56. $y^2 - 6y - 16x + 9 = 0$

57. $y^2 - 6y + 4x + 1 = 0$

58. $y^2 - 2y + 8x + 9 = 0$

59. $2y^2 - 20y + 8x + 2 = 0$

60. $3y^2 - 18y + 12x + 3 = 0$

For Exercises 61–72, find the equation of the parabola in standard form that satisfies the conditions given.

61. focus: $(0, 2)$
    directrix: $y = -2$

62. focus: $(0, -3)$
    directrix: $y = 3$

63. focus: $(4, 0)$
    directrix: $x = -4$

64. focus: $(-3, 0)$
    directrix: $x = 3$

65. focus: $(0, -5)$
    directrix: $y = 5$

66. focus: $(5, 0)$
    directrix: $x = -5$

67. vertex: $(2, -2)$
    focus: $(-1, -2)$

68. vertex: $(4, 1)$
    focus: $(1, 1)$

69. vertex: $(4, -7)$
    focus: $(4, -4)$

70. vertex: $(-3, -4)$
    focus: $(-3, -1)$

71. focus: $(3, 4)$
    directrix: $y = 0$

72. focus: $(-1, 2)$
    directrix: $x = -5$

For the graphs in Exercises 73–76, only two of the following four features are displayed: vertex, focus, directrix, and endpoints of the focal chord. Find the remaining two features and the equation of the parabola.

73.

74.

75.

76.

Solve using substitution or elimination, then graph the system.

77. $\begin{cases} x^2 + y^2 = 25 \\ 2x^2 - 3y^2 = 5 \end{cases}$

78. $\begin{cases} y^2 - x^2 = 12 \\ x^2 + y^2 = 20 \end{cases}$

79. $\begin{cases} x^2 - y = 4 \\ y^2 - x^2 = 16 \end{cases}$

80. $\begin{cases} 2x^2 - 3y^2 = 38 \\ x^2 + 5y = 35 \end{cases}$

81. $\begin{cases} 5x^2 - 2y^2 = 75 \\ 2x^2 + 3y^2 = 125 \end{cases}$

82. $\begin{cases} 3x^2 - 7y^2 = 20 \\ 4x^2 + 9y^2 = 45 \end{cases}$

▶ WORKING WITH FORMULAS

83. The area of a right parabolic segment: $A = \frac{2}{3}ab$

A *right parabolic segment* is that part of a parabola formed by a line perpendicular to its axis, which cuts the parabola. The area of this segment is given by the formula shown, where b is the length of the chord cutting the parabola

and a is the perpendicular distance from the vertex to this chord. What is the area of the parabolic segment shown in the figure?

84. The arc length of a right parabolic segment:

$$\frac{1}{2}\sqrt{b^2 + 16a^2} + \frac{b^2}{8a}\ln\left(\frac{4a + \sqrt{b^2 + 16a^2}}{b}\right)$$

Although a fairly simple concept, finding the length of the parabolic arc traversed by a projectile requires a good deal of computation. To find the length of the arc ABC shown, we use the formula given where a is the maximum height attained by the projectile, b is the horizontal distance it traveled, and

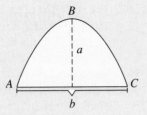

"ln" represents the natural log function. Suppose a baseball thrown from centerfield reaches a maximum height of 20 ft and traverses an arc length of 340 ft. Will the ball reach the catcher 310 ft away without bouncing?

▶ **APPLICATIONS**

85. **Parabolic car headlights:** The cross section of a typical car headlight can be modeled by an equation similar to $25x = 16y^2$, where x and y are in inches and $x \in [0, 4]$. Use this information to graph the relation for the indicated domain.

86. **Parabolic flashlights:** The cross section of a typical flashlight reflector can be modeled by an equation similar to $4x = y^2$, where x and y are in centimeters and $x \in [0, 2.25]$. Use this information to graph the relation for the indicated domain.

87. **Parabolic sound receivers:** Sound technicians at professional sports events often use parabolic receivers as they move along the sidelines. If a two-dimensional cross section of the receiver is modeled by the equation $y^2 = 54x$, and is 36 in. in *diameter,* how deep is the parabolic receiver? What is the location of the focus? [*Hint:* Graph the parabola on the coordinate grid (scale the axes).]

Exercise 87

88. **Parabolic sound receivers:** Private investigators will often use a smaller and less expensive

parabolic receiver (see Exercise 87) to gather information for their clients. If a two-dimensional cross section of the receiver is modeled by the equation $y^2 = 24x$, and the receiver is 12 in. in *diameter,* how deep is the parabolic dish? What is the location of the focus?

89. **Parabolic radio wave receivers:** The program known as S.E.T.I. (Search for Extra-Terrestrial Intelligence) identifies a group of scientists using radio telescopes to look for radio signals from possible intelligent species in outer

space. The radio telescopes are actually parabolic dishes that vary in size from a few feet to hundreds of feet in diameter. If a particular radio telescope is 100 ft in diameter and has a cross section modeled by the equation $x^2 = 167y$, how deep is the parabolic dish? What is the location of the focus? [*Hint:* Graph the parabola on the coordinate grid (scale the axes).]

90. **Solar furnace:** Another form of technology that uses a parabolic dish is called a solar furnace. In general, the rays of the Sun are reflected by the dish and concentrated at the focus, producing extremely high temperatures. Suppose the dish of one of these parabolic reflectors had a 30-ft diameter and a cross section modeled by the equation $x^2 = 50y$. How deep is the parabolic dish? What is the location of the focus?

91. The reflector of a large, commercial flashlight has the shape of a parabolic dish, with a diameter of 10 cm and a depth of 5 cm. What equation will the engineers and technicians use for the manufacture of the dish? How far from the vertex (the lowest point of the dish) will the bulb be placed? (*Hint:* Analyze the information using a coordinate system.)

92. The reflector of an industrial spotlight has the shape of a parabolic dish with a diameter of 120 cm. What is the depth of the dish if the correct placement of the bulb is 11.25 cm above the vertex (the lowest point of the dish)? What equation will the engineers and technicians use for the manufacture of the dish? (*Hint:* Analyze the information using a coordinate system.)

▶ EXTENDING THE CONCEPT

93. In a study of quadratic graphs from the equation $y = ax^2 + bx + c$, no mention is made of a parabola's focus and directrix. Generally, when $a \geq 1$, the focus of a parabola is very near its vertex. Complete the square of the function $y = 2x^2 - 8x$ and write the result in the form $(x - h)^2 = 4p(y - k)$. What is the value of p? What are the coordinates of the vertex?

94. Like the ellipse and hyperbola, the focal chord of a parabola (also called the **latus rectum**) can be used to help sketch its graph. From our earlier work, we know the endpoints of the focal chord are $2p$ units from the focus. Write the equation $-12y + 15 = x^2 - 6x$ in the form $4p(y \pm k) = (x \pm h)^2$, and use the endpoints of the focal chord to help graph the parabola.

Exercise 94

95. In Exercise 83, a formula was given for the area of a right parabolic segment. The area of an *oblique* parabolic segment (the line segment cutting the parabola is *not* perpendicular to the axis) is more complex, as it involves locating the point where a line parallel to this segment is tangent (touches at only one point) to the parabola. The formula is $A = \frac{4}{3}T$, where T represents the area of the triangle formed by the endpoints of the segment and this point of tangency. What is the area of the parabolic segment shown (assuming the lines are parallel)? See Section 6.1, Exercises 46 and 47 and Section 6.4, Example 3.

Exercise 95

▶ MAINTAINING YOUR SKILLS

96. **(6.3)** Construct a system of three equations in three variables using the general equation $y = ax^2 + bx + c$ and the points $(-3, 3)$, $(0, 6)$, and $(1, -1)$. Then use a matrix equation and your calculator to find the equation of the parabola containing these points.

97. **(3.3/3.4)** Use the function
$f(x) = x^5 + 2x^4 + 17x^3 + 34x^2 - 18x - 36$
to comment and give illustrations of the tools available for working with polynomials: (a) synthetic division, (b) rational roots theorem, (c) the remainder and factor theorems, (d) the test for $x = -1$ and $x = 1$, (e) the upper/lower bounds

property, (f) Descartes' rule of signs, and (g) roots of multiplicity (bounces, cuts, alternating intervals).

98. **(1.6)** Find all roots (real and complex) to the equation $x^6 - 64 = 0$. (*Hint:* Begin by factoring the expression as the difference of two perfect squares.)

99. **(2.5)** What are the characteristics of an *even* function? What are the characteristics of an *odd* function?

SUMMARY AND CONCEPT REVIEW

SECTION 7.1 A Brief Introduction to Analytical Geometry

KEY CONCEPTS

- The midpoint and distance formulas play an important role in the study of analytical geometry:

$$\text{midpoint: } (x, y) = \left(\frac{x_2 + x_1}{2}, \frac{y_2 + y_1}{2}\right) \quad \text{distance: } d = \sqrt{(x_2 - x_1)^2 + (y_2 - y_1)^2}$$

- The perpendicular distance from a point to a line is the length of a line segment perpendicular to a given line with the given point and the point of intersection as endpoints.
- Using these tools, we can verify or construct relationships between points, lines, and curves in the plane; verify properties of geometric figures; prove theorems from Euclidean geometry; and construct relationships that define the conic sections.

EXERCISES

1. Verify the closed figure with vertices $(-3, -4)$, $(-5, 4)$, $(3, 6)$, and $(5, -2)$ is a square.

2. Find the equation of the circle that circumscribes the square in Exercise 1.

3. A theorem from Euclidean geometry states: *If any two points are equidistant from the endpoints of a line segment, they are on the perpendicular bisector of the segment.* Determine if the line through $(-3, 6)$ and $(6, -9)$ is a perpendicular bisector of the segment through $(-5, -2)$ and $(5, 4)$.

4. Four points are given below. Verify that the distance from each point to the line $y = -1$ is the same as the distance from the given point to the fixed point $(0, 1)$: $(-6, 9)$, $(-2, 1)$, $(4, 4)$, and $(8, 16)$.

SECTION 7.2 The Circle and the Ellipse

KEY CONCEPTS

- The equation of a circle centered at (h, k) with radius r is $(x - h)^2 + (y - k)^2 = r^2$.

- Dividing both sides by r^2, we obtain the standard form $\dfrac{(x - h)^2}{r^2} + \dfrac{(y - k)^2}{r^2} = 1$, showing the horizontal and vertical distance from center to graph is r.

- The equation of an ellipse in standard form is $\dfrac{(x - h)^2}{a^2} + \dfrac{(y - k)^2}{b^2} = 1$. The center of the ellipse is (h, k), with horizontal distance a and vertical distance b from center to graph.

- Given two fixed points f_1 and f_2 in a plane (called the foci), an ellipse is the set of all points (x, y) such that the distance from the first focus to (x, y), plus the distance from the second focus to (x, y), remains constant.

- For an ellipse, the distance a from center to vertex is *greater than* the distance c from center to one focus.

- To find the foci of an ellipse: $a^2 = b^2 + c^2$ (since $a > c$).

EXERCISES

Sketch the graph of each equation in Exercises 5 through 9.

5. $x^2 + y^2 = 16$

6. $x^2 + 4y^2 = 36$

7. $9x^2 + y^2 - 18x - 27 = 0$

8. $x^2 + y^2 + 6x + 4y + 12 = 0$

9. $\dfrac{(x + 3)^2}{16} + \dfrac{(y - 2)^2}{9} = 1$

10. Find the equation of the ellipse with minor axis of length 6 and foci at $(-4, 0)$ and $(4, 0)$.

11. Find the equation of the ellipse with vertices at (a) $(-13, 0)$ and $(13, 0)$, foci at $(-12, 0)$ and $(12, 0)$; (b) foci at $(0, -16)$ and $(0, 16)$, major axis: 40 units.

12. Write the equation in standard form and sketch the graph, noting all of the characteristic features of the ellipse. $4x^2 + 25y^2 - 16x - 50y - 59 = 0$

SECTION 7.3   The Hyperbola

KEY CONCEPTS

- The equation of a *horizontal* hyperbola in standard form is $\dfrac{(x - h)^2}{a^2} - \dfrac{(y - k)^2}{b^2} = 1$. The center of the hyperbola is (h, k) with horizontal distance a from center to vertices and vertical distance b from center to the midpoint of one side of the central rectangle.

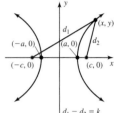

- Given two fixed points f_1 and f_2 in a plane (called the foci), a hyperbola is the set of all points (x, y) such that the distance from the first focus to point (x, y), less the distance from the second focus to (x, y), remains constant.
- For a hyperbola, the distance from center to one of the vertices is *less than* the distance from center to one focus.
- To find the foci of a hyperbola: $c^2 = a^2 + b^2$ (since $c > a$).

EXERCISES

Sketch the graph of each equation, indicating the center, vertices, and asymptotes. For Exercise 18, also give the equation of the hyperbola in standard form.

13. $4y^2 - 25x^2 = 100$ **14.** $\dfrac{(y - 3)^2}{16} - \dfrac{(x + 2)^2}{9} = 1$ **15.** $\dfrac{(x + 2)^2}{9} - \dfrac{(y - 1)^2}{4} = 1$

16. $9y^2 - x^2 - 18y - 72 = 0$ **17.** $x^2 - 4y^2 - 12x - 8y + 16 = 0$

18. vertices at $(-3, 0)$ and $(3, 0)$, asymptotes of $y = \pm\dfrac{4}{3}x$

19. Find the equation of the hyperbola with (a) vertices at $(\pm 15, 0)$, foci at $(\pm 17, 0)$, and (b) foci at $(0, \pm 5)$ with vertical dimension of central rectangle 8 units.

20. Write the equation in standard form and sketch the graph, noting all of the characteristic features of the hyperbola. $4x^2 - 9y^2 - 40x + 36y + 28 = 0$

SECTION 7.4   The Analytic Parabola

KEY CONCEPTS

- Horizontal parabolas have equations of the form $x = ay^2 + by + c; a \neq 0$.
- A horizontal parabola will open to the right if $a > 0$, and to the left if $a < 0$. The axis of symmetry is $y = \dfrac{-b}{2a}$, with the vertex (h, k) found by evaluating at $y = \dfrac{-b}{2a}$ or by completing the square and writing the equation in shifted form: $x = a(y - k)^2 + h$.

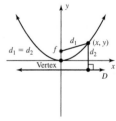

- Given a fixed point f (called the focus) and fixed line D in the plane, a parabola is the set of all points (x, y) such that the distance from f to (x, y) is equal to the distance from (x, y) to line D.
- The equation $x^2 = 4py$ describes a vertical parabola, opening upward if $p > 0$, and opening downward if $p < 0$.
- The equation $y^2 = 4px$ describes a horizontal parabola, opening to the right if $p > 0$, and opening to the left if $p < 0$.

- The focal chord of a parabola is a line segment that contains the focus and is parallel the directrix, with its endpoints on the graph. It has a total length of $|4p|$, meaning the distance from the focus to a point of the graph is $|2p|$. It is commonly used to assist in drawing a graph of the parabola.

EXERCISES

For Exercises 21 and 22, find the vertex and x- and y-intercepts if they exist. Then sketch the graph using symmetry and a few points or by completing the square and shifting a parent function.

21. $x = y^2 - 4$ **22.** $x = y^2 + y - 6$

For Exercises 23 and 24, find the vertex, focus, and directrix for each parabola. Then sketch the graph using this information and the focal chord. Also graph the directrix.

23. $x^2 = -20y$ **24.** $x^2 - 8x - 8y + 16 = 0$

MIXED REVIEW

For Exercises 1 through 21, graph the conic section and label/draw the center, vertices, directrix, foci, focal chords, asymptotes, and other important features as these apply to a particular equation and conic section.

1. $9x^2 + 9y^2 = 54$

2. $16x^2 + 25y^2 = 400$

3. $9y^2 - 25x^2 = 225$

4. $\dfrac{(x - 3)^2}{9} + \dfrac{(y + 1)^2}{25} = 1$

5. $\dfrac{(y - 2)^2}{25} - \dfrac{(x + 3)^2}{16} = 1$

6. $4(x - 1)^2 - 36(y + 2)^2 = 144$

7. $16(x + 2)^2 + 4(y - 1)^2 = 64$

8. $49(x + 2)^2 + (y - 3)^2 = 49$

9. $x^2 + y^2 - 8x + 12y + 16 = 0$

10. $y = 2x^2 - 10x + 15$

11. $y = 2x^2 - 8x - 10$

12. $x = -y^2 - 8y - 11$

13. $x = -y^2 + 2y + 3$

14. $x = (y + 2)^2 - 3$

15. $x = y^2 - 9$

16. $x = y^2 - y - 6$

17. $x^2 = -24y$

18. $x^2 - 8x - 4y + 16 = 0$

19. $4x^2 + 16y^2 - 12x - 48y - 19 = 0$

20. $4x^2 - 25y^2 - 24x + 150y - 289 = 0$

21. $x^2 + y^2 + 8x + 12y + 2 = 0$

22. Solve using elimination: $\begin{cases} 4x^2 - y^2 = -9 \\ x^2 + 3y^2 = 79 \end{cases}$

23. Verify the closed figure with vertices $P_1(-4, -9)$, $P_2(6, 1)$, and $P_3(-6, 3)$ is an isosceles triangle.

24. A theorem from Euclidean geometry states: *A line drawn from the vertex of an isosceles triangle to the midpoint of the base is the perpendicular bisector of the base.* Verify the theorem is true for the isosceles triangle given in Exercise 23.

25. Planetary orbits: Except for small variations, a planet's orbit around the Sun is elliptical, with the Sun at one focus. The *perihelion* or minimum distance from the planet Mercury to the Sun is about 46 million kilometers. Its *aphelion* or maximum distance from the Sun is approximately 70 million kilometers. Assume this orbit is a central ellipse on the coordinate grid. (a) Find the coordinates of the Sun; (b) find the length $2a$ of the major axis, and the length $2b$ of the minor axis; and (c) determine the equation model for the orbit of Mercury in standard form.

PRACTICE TEST

By inspection only (no graphing, completing the square, etc.), match each equation to its correct description.

1. $x^2 + y^2 - 6x + 4y + 9 = 0$ _____

2. $4y^2 + x^2 - 4x + 8y + 20 = 0$ _____

3. $y - x^2 - 4x + 20 = 0$ _____

4. $x^2 - 4y^2 - 4x + 12y + 20 = 0$ _____

    a. Parabola **b.** Hyperbola **c.** Circle **d.** Ellipse

Graph each conic section, and label the center, vertices, foci, focal chords, asymptotes, and other important features where applicable.

5. $(x - 4)^2 + (y + 3)^2 = 9$

6. $\dfrac{(x - 2)^2}{16} + \dfrac{(y + 3)^2}{1} = 1$

7. $\dfrac{(x + 3)^2}{9} - \dfrac{(y - 4)^2}{4} = 1$

8. $x^2 + y^2 - 10x + 4y + 4 = 0$

9. $9x^2 + 4y^2 + 18x - 24y + 9 = 0$

10. $9x^2 - 4y^2 + 18x - 24y - 63 = 0$

11. $x = (y + 3)^2 - 2$

12. $y^2 - 6y - 12x - 15 = 0$

Solve each nonlinear system using the technique of your choice.

13. a. $\begin{cases} 4x^2 - y^2 = 16 \\ y - x = 2 \end{cases}$ **b.** $\begin{cases} 4y^2 - x^2 = 4 \\ x^2 + y^2 = 4 \end{cases}$

14. A support bracket on the frame of a large ship is a steel right triangle with a hypotenuse of 25 ft and a perimeter of 60 ft. Find the lengths of the other sides using a system of nonlinear equations.

15. Find an equation for the circle whose center is at $(-2, 5)$ and whose graph goes through the point $(0, 3)$.

16. Find the equation of the ellipse (in standard form) with vertices at $(-4, 0)$ and $(4, 0)$ with foci located at $(-2, 0)$ and $(2, 0)$.

17. The orbit of Mars around the Sun is elliptical, with the Sun at one focus. When the orbit is expressed as a central ellipse on the coordinate grid, its equation is $\dfrac{x^2}{(141.65)^2} + \dfrac{y^2}{(141.03)^2} = 1$, with a and b in millions of miles. Use this information to find the *aphelion* of Mars (distance from the Sun at its farthest point), and the *perihelion* of Mars (distance from the Sun at its closest point).

Determine the equation of each relation and state its domain and range. For the parabola and the ellipse, also give the location of the foci.

18.

19.

20.

CALCULATOR EXPLORATION AND DISCOVERY

Elongation and Eccentricity

Technically speaking, a circle is an ellipse with both foci at the center. As the distance between foci increases, the ellipse becomes more elongated. We saw other instances of elongation in stretches and compressions of parabolic graphs, and in hyperbolic graphs where the asymptotic slopes varied depending on the values a and b. The measure used to quantify this elongation is called the *eccentricity* e, and is determined by the ratio $e = \dfrac{c}{a}$. For this *Exploration and Discovery*, we'll use the **repeat**

graph feature of a graphing calculator to explore the eccentricity of the graph of a conic. The "repeat graph" feature enables you to graph a family of curves by enclosing changes in a parameter in braces "{ }." For instance, entering $y = \{-2, -1, 0, 1, 2\}x + 3$ as Y_1 on the Y= screen will automatically graph these five lines:

$$y = -2x + 3 \quad y = -x + 3 \quad y = 3$$
$$y = x + 3 \quad y = 2x + 3$$

We'll use this feature to graph a family of ellipses, observing the result and calculating the eccentricity for each curve in the family. The standard form is $\dfrac{x^2}{a^2} + \dfrac{y^2}{b^2} = 1$, which we solve for y and enter as Y_1. After simplification the result is $y = b\sqrt{1 - \dfrac{x^2}{a^2}}$, but for this investigation we'll use the constant $b = 2$ and vary the parameter a using the values $a = 2, 4, 6$, and 8. The result is $y = 2\sqrt{1 - \dfrac{x^2}{\{4, 16, 36, 64\}}}$.

Note from Figure 7.31 that we've set $Y_2 = -Y_1$ to graph the lower half of the ellipse. Using the "friendly window" shown (Figure 7.32) gives the result shown in Figure 7.33, where we see the ellipse is increasingly elongated in the horizontal direction (note when $a = 2$ the result is a circle since $a = b$). Using $a = 2, 4, 6$, and 8 with $b = 2$ in the foci formula $c = \sqrt{a^2 - b^2}$ gives $c = 0, 2\sqrt{3}, 4\sqrt{2}$, and $2\sqrt{15}$, respectively,

Figure 7.31

```
Plot1  Plot2  Plot3
\Y1■2√(1-(X²/{4,
16,36,64))
\Y2■-Y1
\Y3=
\Y4=
\Y5=
\Y6=
```

Figure 7.32

```
WINDOW
Xmin=-9.4
Xmax=9.4
Xscl=1
Ymin=-6.2
Ymax=6.2
Yscl=1
Xres=1
```

with these eccentricities:

$$e = \frac{0}{2}, \frac{2\sqrt{3}}{4}, \frac{4\sqrt{2}}{6},$$

and $\dfrac{2\sqrt{15}}{8}$. While difficult to see in radical form, we find that the eccentricity of an ellipse always satisfies the inequality $0 < e < 1$ (excluding the circle = ellipse case). To two decimal places, the values are $e = 0, 0.87, 0.94$, and 0.97, respectively.

Figure 7.33

Exercise 1: Perform a similar exploration using a family of *hyperbolas*. What do you notice about the eccentricity?

Exercise 2: Perform a similar exploration using a family of *parabolas*. What do you notice about the eccentricity?

WORTHY OF NOTE

It's interesting to note how the $e = \dfrac{c}{a}$ definition of eccentricity relates to our everyday use of the word "eccentric." A normal or "noneccentric" person is thought to be well-rounded, and sure enough $e = 0$ produces a well-rounded figure—a circle. A person who is highly eccentric is thought to be far from the norm, deviating greatly from the center, and greater values of e produce very elongated ellipses.

STRENGTHENING CORE SKILLS

Ellipses and Hyperbolas with Rational/Irrational Values of *a* and *b*

Using the process known as completing the square, we were able to convert from the polynomial form of a conic section to the standard form. However, for some equations, values of a and b are somewhat difficult to identify, since the coefficients are not factors. Consider the equation $20x^2 + 120x + 27y^2 - 54y + 192 = 0$, the equation of an ellipse.

$$20x^2 + 120x + 27y^2 - 54y + 192 = 0 \qquad \text{original equation}$$

$$20(x^2 + 6x + \underline{\quad}) + 27(y^2 - 2y + \underline{\quad}) = -192 \qquad \text{subtract 192, begin process}$$

$$20(x^2 + 6x + 9) + 27(y^2 - 2y + 1) = -192 + 27 + 180 \qquad \text{complete the square in } x \text{ and } y$$

$$20(x + 3)^2 + 27(y - 1)^2 = 15 \qquad \text{factor and simplify}$$

$$\frac{4(x + 3)^2}{3} + \frac{9(y - 1)^2}{5} = 1 \qquad \text{standard form}$$

Unfortunately, we cannot easily identify the values of a and b, since the coefficients of each binomial square were not "1." In these cases, we can write the equation in standard form by using a simple property of fractions—the numerator and denominator of any fraction can be divided by the same quantity to obtain an equivalent fraction. Although the result may look odd, it can nevertheless be applied here, giving a result of $\dfrac{(x + 3)^2}{\dfrac{3}{4}} + \dfrac{(y - 1)^2}{\dfrac{5}{9}} = 1$.

We can now identify a and b by writing these denominators in squared form, which gives the following expression: $\dfrac{(x + 3)^2}{\left(\dfrac{\sqrt{3}}{2}\right)^2} + \dfrac{(y - 1)^2}{\left(\dfrac{\sqrt{5}}{3}\right)^2} = 1$. The values of a and b are now easily seen as $a = \dfrac{\sqrt{3}}{2} \approx 0.866$ and $b = \dfrac{\sqrt{5}}{3} \approx 0.745$.

Use this idea to complete the following exercises.

Exercise 1: Identify the values of a and b by writing the equation $100x^2 - 400x - 18y^2 - 108y + 230 = 0$ in standard form.

Exercise 2: Identify the values of a and b by writing the equation $28x^2 - 56x + 48y^2 + 192y + 195 = 0$ in standard form.

Exercise 3: Write the equation in standard form, then identify the values of a and b and use them to graph the ellipse.

$$\frac{4(x + 3)^2}{49} + \frac{25(y - 1)^2}{36} = 1$$

Exercise 4: Write the equation in standard form, then identify the values of a and b and use them to graph the hyperbola.

$$\frac{9(x + 3)^2}{80} - \frac{4(y - 1)^2}{81} = 1$$

CUMULATIVE REVIEW CHAPTERS 1–7

Solve each equation.

1. $x^3 - 2x^2 + 4x - 8 = 0$

2. $2|n + 4| + 3 = 13$

3. $\sqrt{x + 2} - 2 = \sqrt{3x + 4}$

4. $x^{\frac{3}{2}} + 8 = 0$ 5. $x^2 - 6x + 13 = 0$

6. $4 \cdot 2^{x+1} = \dfrac{1}{8}$ 7. $3^{x-2} = 7$

8. $\log_3 81 = x$

9. $\log_3 x + \log_3(x - 2) = 1$

Graph each function. Include vertices, x- and y-intercepts, asymptotes, and other features.

10. $y = \dfrac{2}{3}x + 2$ 11. $y = |x - 2| + 3$

12. $y = \dfrac{1}{x - 1} + 2$ 13. $y = \sqrt{x - 3} + 1$

14. **a.** $g(x) = (x - 3)(x + 1)(x + 4)$
    b. $f(x) = x^4 + x^3 - 13x^2 - x + 12$

15. $h(x) = \dfrac{x - 2}{x^2 - 9}$ 16. $q(x) = 2^x + 3$

17. $f(x) = \log_2(x + 1)$ 18. $x = y^2 + 4y + 7$

19. $x^2 + y^2 + 10x - 4y + 20 = 0$

20. $4(x - 1)^2 - 36(y + 2)^2 = 144$

21. Determine the following for the indicated graph (write all answers in interval notation): (a) the domain, (b) the range, (c) interval(s) where $f(x)$ is increasing or decreasing, (d) interval(s) where $f(x)$ is constant, (e) location of any maximum or minimum value(s), (f) interval(s) where $f(x)$ is positive, and (g) interval(s) where $f(x)$ is negative.

Solve each system of equations.

22. $\begin{cases} 4x + 3y = 13 \\ -9y + 5z = 19 \\ x - 4z = -4 \end{cases}$ 23. $\begin{cases} x^2 + y^2 = 25 \\ 64x^2 + 12y^2 = 768 \end{cases}$

24. If a person invests \$5000 at 9% compounded quarterly, how long would it take for the money to grow to \$12,000?

25. A radiator contains 10 L of liquid that is 40% antifreeze. How much should be drained off and replaced with pure antifreeze for a 60% mixture?

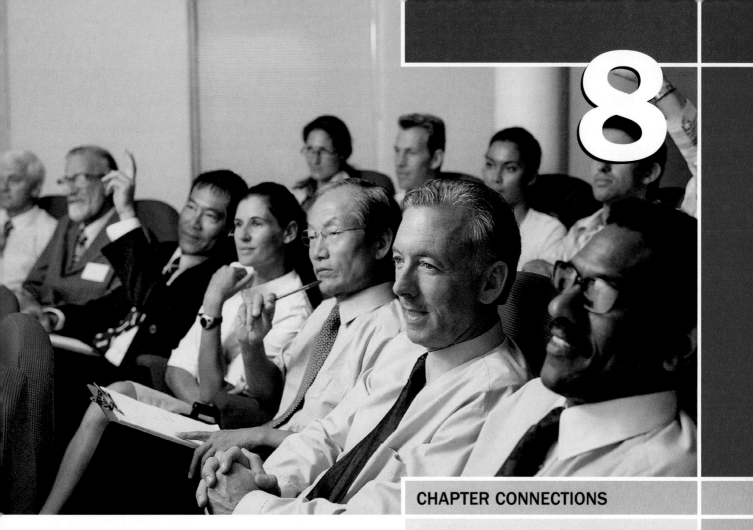

8

Additional Topics in Algebra

CHAPTER OUTLINE

CHAPTER CONNECTIONS

For a corporation of any size, decisions made by upper management often depend on a large number of factors, with the desired outcome attainable in many different ways. For instance, consider a legal firm that specializes in family law, with a support staff of 15 employees—6 paralegals and 9 legal assistants. Due to recent changes in the law, the firm wants to send some combination of five support staff to a conference dedicated to the new changes. In Chapter 8, we'll see how counting techniques and probability can be used to determine the various ways such a group can be randomly formed, even if certain constraints are imposed. This application appears as Exercise 34 in Section 8.6.

Check out these other real-world connections:

▶ Determining the Effects of Inflation (Section 8.1, Exercise 86)

▶ Counting the Number of Possible Area Codes and Phone Numbers (Section 8.5, Exercise 84)

▶ Calculating Possible Movements of a Computer Animation (Section 8.2, Exercise 73)

▶ Tracking and Improving Customer Service Using Probability (Section 8.6, Exercise 53)

Learning Objectives

In Section 8.1 you will learn how to:

☐ **A.** Write out the terms of a sequence given the general or *n*th term

☐ **B.** Work with recursive sequences and sequences involving a factorial

☐ **C.** Find the partial sum of a series

☐ **D.** Use summation notation to write and evaluate series

☐ **E.** Use sequences to solve applied problems

A *sequence* can be thought of as a pattern of numbers listed in a prescribed order. A *series* is the sum of the numbers in a sequence. Sequences and series come in countless varieties, and we'll introduce some general forms here. In following sections we'll focus on two special types: arithmetic and geometric sequences. These are used in a number of different fields, with a wide variety of significant applications.

A. Finding the Terms of a Sequence Given the General Term

Suppose a person had $10,000 to invest, and decided to place the money in government bonds that guarantee an annual return of 7%. From our work in Chapter 4, we know the amount of money in the account after x years can be modeled by the function $f(x) = 10,000(1.07)^x$. If you reinvest your earnings each year, the amount in the account would be (rounded to the nearest dollar):

Year:	$f(1)$	$f(2)$	$f(3)$	$f(4)$	$f(5)\ldots$
	$\downarrow$	$\downarrow$	$\downarrow$	$\downarrow$	$\downarrow$
Value:	$10,700	$11,449	$12,250	$13,108	$14,026\ldots$

Note the relationship (year, value) is a function that pairs 1 with $10,700, 2 with $11,449, 3 with $12,250 and so on. This is an example of a **sequence.** To distinguish sequences from other algebraic functions, we commonly name the functions a instead of f, use the variable n instead of x, and employ a subscript notation. The function $f(x) = 10,000(1.07)^x$ would then be written $a_n = 10,000(1.07)^n$. Using this notation $a_1 = 10,700$, $a_2 = 11,449$, and so on.

The values $a_1, a_2, a_3, a_4, \ldots$ are called the **terms** of the sequence. If the account were closed after a certain number of years (for example, after the fifth year) we have a **finite sequence.** If we let the investment grow indefinitely, the result is called an **infinite sequence.** The expression a_n that defines the sequence is called the **general** or **nth term** and the terms immediately preceding it are called the $(n - 1)$st term, the $(n - 2)$nd term, and so on.

Sequences

A *finite sequence* is a function a_n whose domain is the set of natural numbers from 1 to n. The terms of the sequence are labeled

$$a_1, a_2, a_3, \ldots, a_k, a_{k+1}, \ldots, a_{n-1}, a_n$$

where a_k represents an arbitrary "interior" term and a_n also represents the last term of the sequence.

An *infinite sequence* is a function a_n whose domain is the set of <u>all</u> natural numbers.

EXAMPLE 1A ▶ Computing Specified Terms of a Sequence

For $a_n = \dfrac{n + 1}{n^2}$, find a_1, a_3, a_6, and a_7.

Solution ▶

$a_1 = \dfrac{1 + 1}{1^2} = 2$ $a_3 = \dfrac{3 + 1}{3^2} = \dfrac{4}{9}$

$a_6 = \dfrac{6 + 1}{6^2} = \dfrac{7}{36}$ $a_7 = \dfrac{7 + 1}{7^2} = \dfrac{8}{49}$

EXAMPLE 1B ▶ Computing the First *k* Terms of a Sequence

Find the first four terms of the sequence $a_n = (-1)^n 2^n$. Write the terms of the sequence as a list.

Solution ▶ $a_1 = (-1)^1 2^1 = -2$ $a_2 = (-1)^2 2^2 = 4$

$a_3 = (-1)^3 2^3 = -8$ $a_4 = (-1)^4 2^4 = 16$

The sequence can be written $-2, 4, -8, 16, \ldots$, or more generally as $-2, 4, -8, 16, \ldots, (-1)^n 2^n, \ldots$ to show how each term was generated.

Now try Exercises 7 through 32 ▶

☑ **A.** You've just learned how
to write out the terms of a
sequence given the general or
*n*th term

B. Recursive Sequences and Factorial Notation

Sometimes the formula defining a sequence uses the preceding term or terms to generate those that follow. These are called **recursive sequences** and are particularly useful in writing computer programs. Because of how they are defined, recursive sequences must give an inaugural term or **seed element,** to begin the recursion process.

    Perhaps the most famous recursive sequence is associated with the work of Leonardo of Pisa (A.D. 1180–1250), better known to history as *Fibonacci.* In fact, it is commonly called the Fibonacci sequence in which each successive term is the sum of the previous two, beginning with $1, 1, \ldots$.

EXAMPLE 2 ▶ **Computing the Terms of a Recursive Sequence**

Write out the first eight terms of the recursive (Fibonacci) sequence defined by $c_1 = 1$, $c_2 = 1$, and $c_n = c_{n-1} + c_{n-2}$.

Solution ▶   The first two terms are given, so we begin with $n = 3$.

$$
\begin{array}{lll}
c_3 = c_{3-1} + c_{3-2} & c_4 = c_{4-1} + c_{4-2} & c_5 = c_{5-1} + c_{5-2} \\
\quad = c_2 + c_1 & \quad = c_3 + c_2 & \quad = c_4 + c_3 \\
\quad = 1 + 1 & \quad = 2 + 1 & \quad = 3 + 2 \\
\quad = 2 & \quad = 3 & \quad = 5
\end{array}
$$

At this point we can simply use the fact that each successive term is simply the sum of the preceding two, and find that $c_6 = 3 + 5 = 8$, $c_7 = 5 + 8 = 13$, and $c_8 = 13 + 8 = 21$. The first eight terms are 1, 1, 2, 3, 5, 8, 13, and 21.

Now try Exercises 33 through 38 ▶

    Sequences can also be defined using a **factorial,** which is the product of a given natural number with all those that precede it. The expression **5!** is read, "five factorial," and is evaluated as: $5! = 5 \cdot 4 \cdot 3 \cdot 2 \cdot 1 = 120$.

Factorials

For any natural number n,

$$n! = n \cdot (n-1) \cdot (n-2) \cdots 3 \cdot 2 \cdot 1$$

    Rewriting a factorial in equivalent forms often makes it easier to simplify certain expressions. For example, we can rewrite $5! = 5 \cdot 4!$ or $5! = 5 \cdot 4 \cdot 3!$. Consider Example 3.

EXAMPLE 3 ▶ **Simplifying Expressions Using Factorial Notation**

Simplify by writing the numerator in an equivalent form.

     a. $\dfrac{9!}{7!}$ **b.** $\dfrac{11!}{8!2!}$ **c.** $\dfrac{6!}{3!5!}$

Solution ▶   **a.** $\dfrac{9!}{7!} = \dfrac{9 \cdot 8 \cdot \cancel{7!}}{\cancel{7!}}$ **b.** $\dfrac{11!}{8!2!} = \dfrac{11 \cdot 10 \cdot 9 \cdot \cancel{8!}}{\cancel{8!}2!}$ **c.** $\dfrac{6!}{3!5!} = \dfrac{6 \cdot \cancel{5!}}{3!\cancel{5!}}$

$= 9 \cdot 8$ $= \dfrac{990}{2}$ $= \dfrac{6}{6}$

$= 72$ $= 495$ $= 1$

> **WORTHY OF NOTE**
>
> Most calculators have a factorial option or key. On the TI-84 Plus it is located on a submenu of the `MATH` key:
>
> `MATH` **PRB** (option) **4: !**

Now try Exercises 39 through 44 ▶

EXAMPLE 4 ▶   **Computing a Specified Term from a Sequence Defined Using Factorials**

Find the third term of each sequence.

a. $a_n = \dfrac{n!}{2^n}$ **b.** $c_n = \dfrac{(-1)^n (2n - 1)!}{n!}$

Solution ▶   **a.** $a_3 = \dfrac{3!}{2^3}$ **b.** $c_3 = \dfrac{(-1)^3 [2(3) - 1]!}{3!}$

☑ **B.** You've just learned how to work with recursive sequences and sequences involving a factorial

$= \dfrac{6}{8} = \dfrac{3}{4}$ $= \dfrac{(-1)(5)!}{3!} = \dfrac{(-1)[5 \cdot 4 \cdot \cancel{3!}]}{\cancel{3!}}$

$= -20$

Now try Exercises 45 through 50 ▶

Figure 8.1

C. Series and Partial Sums

Sometimes the terms of a sequence are dictated by context rather than a formula. Consider the stacking of large pipes in a storage yard. If there are 10 pipes in the bottom row, then 9 pipes, then 8 (see Figure 8.1), how many pipes are in the stack if there is a single pipe at the top? The sequence generated is 10, 9, 8, . . . , 3, 2, 1 and to answer the question we would have to *compute the sum of all terms in the sequence.* When the terms of a finite sequence are added, the result is called a **finite series.**

> **Finite Series**
>
> Given the sequence $a_1, a_2, a_3, a_4, \ldots, a_n$, the sum of the terms is called a **finite series** or **partial sum** and is denoted S_n:
>
> $$S_n = a_1 + a_2 + a_3 + a_4 + \cdots + a_n$$

EXAMPLE 5 ▶   **Computing a Partial Sum**

Given $a_n = 2n$, find the value of

a. S_4 and   **b.** S_7.

Solution ▶   Since we eventually need the sum of the first seven terms (for Part b), begin by writing out these terms: 2, 4, 6, 8, 10, 12, and 14.

a. $S_4 = a_1 + a_2 + a_3 + a_4$

$= 2 + 4 + 6 + 8$

$= 20$

b. $S_7 = a_1 + a_2 + a_3 + a_4 + a_5 + a_6 + a_7$

☑ **C.** You've just learned how to find the partial sum of a series

$= 2 + 4 + 6 + 8 + 10 + 12 + 14$

$= 56$

Now try Exercises 51 through 56 ▶

Handwritten in top margin: $8 + 12 + 16 + 20$ $+ 24 + 28 = 108$

D. Summation Notation

When the general term of a sequence is known, the Greek letter *sigma* Σ can be used to write the related series as a formula. For instance, to indicate the sum of the first four terms of $a_n = 3n + 2$, we write $\sum_{i=1}^{4}(3i + 2)$. This result is called **summation** or **sigma notation** and the letter i is called the **index of summation.** The letters j, k, l, and m are also used as index numbers, and the summation need not start at 1.

EXAMPLE 6 ▶ **Computing a Partial Sum**

Compute each sum:

    a. $\displaystyle\sum_{i=1}^{4}(3i + 2)$ **b.** $\displaystyle\sum_{j=1}^{5}\frac{1}{j}$ **c.** $\displaystyle\sum_{k=3}^{6}(-1)^{k}k^2$

Solution ▶   **a.** $\displaystyle\sum_{i=1}^{4}(3i + 2) = (3\cdot 1 + 2) + (3\cdot 2 + 2) + (3\cdot 3 + 2) + (3\cdot 4 + 2)$

              $= 5 + 8 + 11 + 14 = 38$

  b. $\displaystyle\sum_{j=1}^{5}\frac{1}{j} = \frac{1}{1} + \frac{1}{2} + \frac{1}{3} + \frac{1}{4} + \frac{1}{5}$

       $= \dfrac{60}{60} + \dfrac{30}{60} + \dfrac{20}{60} + \dfrac{15}{60} + \dfrac{12}{60} = \dfrac{137}{60}$

  c. $\displaystyle\sum_{k=3}^{6}(-1)^{k}k^2 = (-1)^{3}\cdot 3^2 + (-1)^{4}\cdot 4^2 + (-1)^{5}\cdot 5^2 + (-1)^{6}\cdot 6^2$

             $= -9 + 16 + -25 + 36 = 18$

Now try Exercises 57 through 68 ▶

If a definite pattern is noted in a given series expansion, this process can be reversed, with the expanded form being expressed in summation notation using the *n*th term.

EXAMPLE 7 ▶ **Writing a Sum in Sigma Notation**

Write each of the following sums in summation (sigma) notation.
    a. $1 + 3 + 5 + 7 + 9$ **b.** $6 + 9 + 12 + \cdots$

Solution ▶   **a.** The series has five terms and each term is an odd number, or 1 less than a multiple of 2. The general term is $a_n = 2n - 1$, and the series is $\displaystyle\sum_{n=1}^{5}(2n - 1)$.

  b. This is an infinite sum whose terms are multiples of 3. The general term is $a_n = 3n$, but the series starts at 2 and not 1. The series is $\displaystyle\sum_{j=2}^{\infty}3j$.

Now try Exercises 69 through 78 ▶

WORTHY OF NOTE

By varying the function given and/or where the sum begins, more than one acceptable form is possible.

For example 7(b) $\displaystyle\sum_{n=1}^{\infty}(3 + 3n)$ also works.

Since the commutative and associative laws hold for the addition of real numbers, summations have the following properties:

Properties of Summation

Given any real number c and natural number n,

(I) $\displaystyle\sum_{i=1}^{n} c = cn$

If you add a constant c "n" times the result is cn.

(II) $\displaystyle\sum_{i=1}^{n} ca_i = c\sum_{i=1}^{n} a_i$

A constant can be factored out of a sum.

(III) $\displaystyle\sum_{i=1}^{n} (a_i \pm b_i) = \sum_{i=1}^{n} a_i \pm \sum_{i=1}^{n} b_i$

A summation can be distributed to two (or more) sequences.

(IV) $\displaystyle\sum_{i=1}^{m} a_i + \sum_{i=m+1}^{n} a_i = \sum_{i=1}^{n} a_i; \; 1 \le m < n$

A summation is cumulative and can be written as a sum of smaller parts.

The verification of property II depends solely on the distributive property.

Proof: $\displaystyle\sum_{i=1}^{n} ca_i = ca_1 + ca_2 + ca_3 + \cdots + ca_n$ expand sum

$= c(a_1 + a_2 + a_3 + \cdots + a_n)$ factor out c

$= c\displaystyle\sum_{i=1}^{n} a_i$ write series in summation form

The verification of properties III and IV simply uses the commutative and associative properties. You are asked to prove property III in **Exercise 91.**

EXAMPLE 8 ▶ Computing a Sum Using Summation Properties

Recompute the sum $\displaystyle\sum_{i=1}^{4} (3i + 2)$ from Example 6(a) using summation properties.

Solution ▶ $\displaystyle\sum_{i=1}^{4} (3i+2) = \sum_{i=1}^{4} 3i + \sum_{i=1}^{4} 2$ property III

$= 3\displaystyle\sum_{i=1}^{4} i + \sum_{i=1}^{4} 2$ property II

$= 3(10) + 2(4)$ $1+2+3+4 = 10$; property I

$= 38$ result

☑ **D.** You've just learned how to use summation notation to write and evaluate series

Now try Exercises 79 through 82 ▶

E. Applications of Sequences

To solve applications of sequences, (1) identify where the sequence begins (the initial term) and (2) write out the first few terms to help identify the nth term.

EXAMPLE 9 ▶ **Solving an Application — Accumulation of Stock**

Hydra already owned 1420 shares of stock when her company began offering employees the opportunity to purchase 175 discounted shares per year. If she made no purchases other than these discounted shares each year, how many shares will she have 9 yr later? If this continued for the 25 yr she will work for the company, how many shares will she have at retirement?

Solution ▶ To begin, it helps to simply write out the first few terms of the sequence. Since she already had 1420 shares before the company made this offer, we let $a_0 = 1420$ be the inaugural element, showing $a_1 = 1595$ (after 1 yr, she owns 1595 shares). The first few terms are 1595, 1770, 1945, 2120, and so on. This supports a general term of $a_n = 1595 + 175(n - 1)$.

After 9 years

$$a_9 = 1595 + 175(8)$$
$$= 2995$$

After 25 years

$$a_{25} = 1595 + 175(24)$$
$$= 5795$$

☑ **E. You've just learned to use sequences to solve applied problems**

After 9 yr she would have 2995 shares. Upon retirement she would own 5795 shares of company stock.

Now try Exercises 85 through 90 ▶

TECHNOLOGY HIGHLIGHT

Studying Sequences and Series

To support a study of sequences and series, we can use a graphing calculator to generate the desired terms. This can be done either on the home screen or directly into the LIST feature of the calculator. On the TI-84 Plus this is accomplished using the **"seq("** and **"sum("** commands, which are accessed using the keystrokes [2nd] [STAT] (LIST) and the screen shown in Figure 8.2. The **"seq("** feature is option 5 under the **OPS** submenu (press [▶] 5) and the **"sum("** feature is option 5 under the **MATH** submenu (press [◀] 5).

To generate the first four terms of the sequence $a_n = n^2 + 1$, and to find the sum, [CLEAR] the home screen and press [2nd] [STAT] [▶] 5 to place **"seq("** on the home screen. This command requires four inputs: a_n (the nth term), variable used (the calculator can work with any letter), initial term and the last term. For this example the screen reads "**seq**($x^2 + 1, x, 1, 4$)," with the result being the four terms shown in Figure 8.3. To find the *sum* of these terms, we simply precede the **seq**($x^2 + 1, x, 1, 4$) command by "**sum(,**" and two methods are shown in Figure 8.4.

Each of the following sequences have some interesting properties or mathematical connections. Use your graphing calculator to generate the first 10 terms of each sequence and the sum of the first 10 terms. Next, generate the first 20 terms of each sequence and the sum of these terms. What conclusion (if any) can you reach about the sum of each sequence?

Figure 8.2

```
NAMES OPS MATH
1:SortA(
2:SortD(
3:dim(
4:Fill(
5 seq(
6:cumSum(
7↓ΔList(
```

Figure 8.3

```
seq(X²+1,X,1,4)
      {2 5 10 17}
```

Figure 8.4

```
seq(X²+1,X,1,4)
      {2 5 10 17}
sum(Ans)
                34
sum(seq(X²+1,X,1
,4)
                34
```

Exercise 1: $a_n = \dfrac{1}{3^n}$ **Exercise 2:** $a_n = \dfrac{2}{n(n + 1)}$ **Exercise 3:** $a_n = \dfrac{1}{(2n - 1)(2n + 1)}$

8.1 EXERCISES

▶ **CONCEPTS AND VOCABULARY**

Fill in the blank with the appropriate word or phrase. Carefully reread the section if needed.

1. A sequence is a(n) _____ of numbers listed in a specific _____.

2. A series is the _____ of the numbers from a given sequence.

3. When each term of a sequence is larger than the preceding term, the sequence is said to be _____.

4. When each term of a sequence is smaller than the preceding term, the sequence is said to be _____.

5. Describe the characteristics of a recursive sequence and give one example.

6. Describe the characteristics of an alternating sequence and give one example.

▶ **DEVELOPING YOUR SKILLS**

Find the first four terms, then find the 8th and 12th term for each nth term given.

7. $a_n = 2n - 1$

8. $a_n = 2n + 3$

9. $a_n = 3n^2 - 3$

10. $a_n = 2n^3 - 12$

11. $a_n = (-1)^n n$

12. $a_n = \dfrac{(-1)^n}{n}$

13. $a_n = \dfrac{n}{n + 1}$

14. $a_n = \left(1 + \dfrac{1}{n}\right)^n$

15. $a_n = \left(\dfrac{1}{2}\right)^n$

16. $a_n = \left(\dfrac{2}{3}\right)^n$

17. $a_n = \dfrac{1}{n}$

18. $a_n = \dfrac{1}{n^2}$

19. $a_n = \dfrac{(-1)^n}{n(n + 1)}$

20. $a_n = \dfrac{(-1)^{n+1}}{2n^2 - 1}$

21. $a_n = (-1)^n 2^n$

22. $a_n = (-1)^n 2^{-n}$

Find the indicated term for each sequence.

23. $a_n = n^2 - 2;\ a_9$

24. $a_n = (n - 2)^2;\ a_9$

25. $a_n = \dfrac{(-1)^{n+1}}{n};\ a_5$

26. $a_n = \dfrac{(-1)^{n+1}}{2n - 1};\ a_5$

27. $a_n = 2\left(\dfrac{1}{2}\right)^{n-1};\ a_7$

28. $a_n = 3\left(\dfrac{1}{3}\right)^{n-1};\ a_7$

29. $a_n = \left(1 + \dfrac{1}{n}\right)^n;\ a_{10}$ 30. $a_n = \left(n + \dfrac{1}{n}\right)^n;\ a_9$

31. $a_n = \dfrac{1}{n(2n + 1)};\ a_4$

32. $a_n = \dfrac{1}{(2n - 1)(2n + 1)};\ a_5$

Find the first five terms of each recursive sequence.

33. $\begin{cases} a_1 = 2 \\ a_n = 5a_{n-1} - 3 \end{cases}$

34. $\begin{cases} a_1 = 3 \\ a_n = 2a_{n-1} - 3 \end{cases}$

35. $\begin{cases} a_1 = -1 \\ a_n = (a_{n-1})^2 + 3 \end{cases}$

36. $\begin{cases} a_1 = -2 \\ a_n = a_{n-1} - 16 \end{cases}$

37. $\begin{cases} c_1 = 64,\ c_2 = 32 \\ c_n = \dfrac{c_{n-2} - c_{n-1}}{2} \end{cases}$

38. $\begin{cases} c_1 = 1,\ c_2 = 2 \\ c_n = c_{n-1} + (c_{n-2})^2 \end{cases}$

Simplify each factorial expression.

39. $\dfrac{8!}{5!}$ **40.** $\dfrac{12!}{10!}$ **41.** $\dfrac{9!}{7!2!}$

42. $\dfrac{6!}{3!3!}$ **43.** $\dfrac{8!}{2!6!}$ **44.** $\dfrac{10!}{3!7!}$

Write out the first four terms in each sequence.

45. $a_n = \dfrac{n!}{(n+1)!}$ **46.** $a_n = \dfrac{n!}{(n+3)!}$

47. $a_n = \dfrac{(n+1)!}{(3n)!}$ **48.** $a_n = \dfrac{(n+3)!}{(2n)!}$

49. $a_n = \dfrac{n^n}{n!}$ **50.** $a_n = \dfrac{2^n}{n!}$

Find the indicated partial sum for each sequence.

51. $a_n = n; S_5$ **52.** $a_n = n^2; S_7$

53. $a_n = 2n - 1; S_8$ **54.** $a_n = 3n - 1; S_6$

55. $a_n = \dfrac{1}{n}; S_5$ **56.** $a_n = \dfrac{n}{n+1}; S_4$

Expand and evaluate each series.

57. $\displaystyle\sum_{i=1}^{4} (3i - 5)$ **58.** $\displaystyle\sum_{i=1}^{5} (2i - 3)$

59. $\displaystyle\sum_{k=1}^{5} (2k^2 - 3)$ **60.** $\displaystyle\sum_{k=1}^{5} (k^2 + 1)$

61. $\displaystyle\sum_{k=1}^{7} (-1)^k k$ **62.** $\displaystyle\sum_{k=1}^{5} (-1)^k 2^k$

63. $\displaystyle\sum_{i=1}^{4} \dfrac{i^2}{2}$ **64.** $\displaystyle\sum_{i=2}^{4} i^2$

65. $\displaystyle\sum_{j=3}^{7} 2j$ **66.** $\displaystyle\sum_{j=3}^{7} \dfrac{j}{2^j}$

67. $\displaystyle\sum_{k=3}^{8} \dfrac{(-1)^k}{k(k-2)}$ **68.** $\displaystyle\sum_{k=2}^{6} \dfrac{(-1)^{k+1}}{k^2 - 1}$

Write each sum using sigma notation. Answers are not necessarily unique.

69. $4 + 8 + 12 + 16 + 20$

70. $5 + 10 + 15 + 20 + 25$

71. $-1 + 4 - 9 + 16 - 25 + 36$

72. $1 - 8 + 27 - 64 + 125 - 216$

For the given general term a_n, write the indicated sum using sigma notation.

73. $a_n = n + 3; S_5$

74. $a_n = \dfrac{n^2 + 1}{n + 1}; S_4$

75. $a_n = \dfrac{n^2}{3}$; third partial sum

76. $a_n = 2n - 1$; sixth partial sum

77. $a_n = \dfrac{n}{2^n}$; sum for $n = 3$ to 7

78. $a_n = n^2$; sum for $n = 2$ to 6

Compute each sum by applying properties of summation.

79. $\displaystyle\sum_{i=1}^{5} (4i - 5)$ **80.** $\displaystyle\sum_{i=1}^{6} (3 + 2i)$

81. $\displaystyle\sum_{k=1}^{4} (3k^2 + k)$ **82.** $\displaystyle\sum_{k=1}^{4} (2k^3 + 5)$

▶ **WORKING WITH FORMULAS**

---

83. Sum of $a_n = 3n - 2$: $S_n = \dfrac{n(3n - 1)}{2}$

The sum of the first n terms of the sequence defined by $a_n = 3n - 2 = 1, 4, 7, 10, \ldots,$ $(3n - 2), \ldots$ is given by the formula shown. Find S_5 using the formula, then verify by direct calculation.

84. Sum of $a_n = 3n - 1$: $S_n = \dfrac{n(3n + 1)}{2}$

The sum of the first n terms of the sequence defined by $a_n = 3n - 1 = 2, 5, 8, 11, \ldots, (3n - 1), \ldots$ is given by the formula shown. Find S_8 using the formula, then verify by direct calculation. Observing the results of Exercises 83 and 84, can you now state the sum formula for $a_n = 3n - 0$?

▶ **APPLICATIONS**

Use the information given in each exercise to determine the nth term a_n for the sequence described. Then use the nth term to list the specified number of terms.

85. **Blue-book value:** Steve's car has a blue-book value of $6000. Each year it loses 20% of its value (its value each year is 80% of the year before). List the value of Steve's car for the next 5 yr. (*Hint:* For $a_1 = 6000$, we need the *next* five terms.)

86. **Effects of inflation:** Suppose inflation (an increase in value) will average 4% for the next 5 yr. List the growing cost (year by year) of a DVD that costs $15 right now. (*Hint:* For $a_1 = 15$, we need the *next* five terms.)

87. **Wage increases:** Latisha gets $5.20 an hour for filling candy machines for Archtown Vending. Each year she receives a $0.50 hourly raise. List Latisha's wage for the first 5 yr. How much will she make in the fifth year if she works 8 hr per day for 240 working days?

88. **Average birth weight:** The average birth weight of a certain animal species is 900 g, with the baby gaining 125 g each day for the first 10 days. List the infant's weight for the first 10 days. How much does the infant weigh on the 10th day?

89. **Stocking a lake:** A local fishery stocks a large lake with 1500 bass and then adds an additional 100 mature bass per month until the lake nears maximum capacity. If the bass population grows at a rate of 5% per month through natural reproduction, the number of bass in the pond after n months is given by the recursive sequence $b_0 = 1500$, $b_n = 1.05b_{n-1} + 100$. How many bass will be in the lake after 6 months?

90. **Species preservation:** The Interior Department introduces 50 wolves (male and female) into a large wildlife area in an effort to preserve the species. Each year about 12 additional adult wolves are added from capture and relocation programs. If the wolf population grows at a rate of 10% per year through natural reproduction, the number of wolves in the area after n years is given by the recursive sequence $w_0 = 50$, $w_n = 1.10w_{n-1} + 12$. How many wolves are in the wildlife area after 6 years?

▶ **EXTENDING THE CONCEPT**

91. Verify that a summation may be distributed to two (or more) sequences. That is, verify that the following statement is true:

$$\sum_{i=1}^{n} (a_i \pm b_i) = \sum_{i=1}^{n} a_i \pm \sum_{i=1}^{n} b_i.$$

Surprisingly, some of the most celebrated numbers in mathematics can be represented or approximated by a series expansion. Use your calculator to find the partial sums for $n = 4$, $n = 8$, and $n = 12$ for the summations given, and attempt to name the number the summation approximates:

92. $\displaystyle\sum_{k=0}^{n} \frac{1}{k!}$

93. $\displaystyle\sum_{k=1}^{n} \frac{1}{3^k}$

94. $\displaystyle\sum_{k=1}^{n} \frac{1}{2^k}$

▶ **MAINTAINING YOUR SKILLS**

95. (4.2) Write $\log_{\frac{1}{81}} = -x$ in exponential form, then solve by equating bases.

96. (2.5) Set up the difference quotient for $f(x) = \sqrt{x}$, then rationalize the numerator.

97. (5.3) Solve the nonlinear system. $\begin{cases} x^2 + y^2 = 9 \\ 9y^2 - 4x^2 = 16 \end{cases}$

98. (6.3) Solve the system using a matrix equation. $\begin{cases} 25x + y - 2z = -14 \\ 2x - y + z = 40 \\ -7x + 3y - z = -13 \end{cases}$

Learning Objectives

In Section 8.2 you will learn how to:

☐ **A.** Identify an arithmetic sequence and its common difference

☐ **B.** Find the *n*th term of an arithmetic sequence

☐ **C.** Find the *n*th partial sum of an arithmetic sequence

☐ **D.** Solve applications involving arithmetic sequences

Similar to the way polynomials fall into certain groups or families (linear, quadratic, cubic, etc.), sequences and series with common characteristics are likewise grouped. In this section, we focus on sequences where each successive term is generated by adding a constant value, as in the sequence 1, 8, 15, 22, 29, . . . , where 7 is added to a given term in order to produce the next term.

A. Identifying an Arithmetic Sequence and Finding the Common Difference

An **arithmetic sequence** is one where each successive term is found by adding a fixed constant to the preceding term. For instance 3, 7, 11, 15, . . . is an arithmetic sequence, since adding 4 to any given term produces the next term. This also means if you take the difference of any two consecutive terms, the result will be 4 and in fact, 4 is called the **common difference** *d* for this sequence. Using the notation developed earlier, we can write $d = a_{k+1} - a_k$, where a_k represents any term of the sequence and a_{k+1} represents the term that follows a_k.

Arithmetic Sequences

Given a sequence $a_1, a_2, a_3, \ldots, a_k, a_{k+1}, \ldots, a_n$, where $k, n \in \mathbb{N}$ and $k < n$,

if there exists a common difference *d* such that $a_{k+1} - a_k = d$,

then the sequence is an *arithmetic sequence*.

The difference of successive terms can be rewritten as $a_{k+1} = a_k + d$ (for $k \geq 1$) to highlight that each following term is found by adding *d* to the previous term.

EXAMPLE 1 ▶ **Identifying an Arithmetic Sequence**

Determine if the given sequence is arithmetic.

a. 2, 5, 8, 11, . . .   **b.** $\frac{1}{2}, \frac{5}{6}, \frac{4}{3}, \frac{7}{6}, \cdots$

Solution ▶ **a.** Begin by looking for a common difference $d = a_{k+1} - a_k$. Checking each pair of consecutive terms we have

$$5 - 2 = 3 \quad 8 - 5 = 3 \quad 11 - 8 = 3 \quad \text{and so on.}$$

This is an arithmetic sequence with common difference $d = 3$.

b. Checking each pair of consecutive terms yields

$$\frac{5}{6} - \frac{1}{2} = \frac{5}{6} - \frac{3}{6} \qquad\qquad \frac{4}{3} - \frac{5}{6} = \frac{8}{6} - \frac{5}{6}$$

$$= \frac{2}{6} = \frac{1}{3} \qquad\qquad\qquad = \frac{3}{6} = \frac{1}{2}$$

Since the difference is not constant, this is not an arithmetic sequence.

Now try Exercises 7 through 18 ▶

EXAMPLE 2 ▶ **Writing the First *k* Terms of an Arithmetic Sequence**

Write the first five terms of the arithmetic sequence, given the first term a_1 and the common difference *d*.

a. $a_1 = 12$ and $d = -4$ **b.** $a_1 = \frac{1}{2}$ and $d = \frac{1}{3}$

a. $a_1 = 12$ and $d = -4$. Starting at $a_1 = 12$, add -4 to each new term to generate the sequence: 12, 8, 4, 0, -4

b. $a_1 = \frac{1}{2}$ and $d = \frac{1}{3}$. Starting at $a_1 = \frac{1}{2}$ and adding $\frac{1}{3}$ to each new term will generate the sequence: $\frac{1}{2}, \frac{5}{6}, \frac{7}{6}, \frac{3}{2}, \frac{11}{6}$

Now try Exercises 19 through 30 ▶

B. Finding the *n*th Term of an Arithmetic Sequence

If the values a_1 and d from an arithmetic sequence are known, we could generate the terms of the sequence by adding *multiples of d to the first term,* instead of adding d to each new term. For example, we can generate the sequence 3, 8, 13, 18, 23 by adding multiples of 5 to the first term $a_1 = 3$:

$$3 = 3 + (0)5 \qquad\qquad a_1 = a_1 + 0d$$
$$8 = 3 + (1)5 \qquad\qquad a_2 = a_1 + 1d$$
$$13 = 3 + (2)5 \qquad\qquad a_3 = a_1 + 2d$$
$$18 = 3 + (3)5 \qquad\qquad a_4 = a_1 + 3d$$
$$23 = 3 + (4)5 \qquad\qquad a_5 = a_1 + 4d$$

current term ⟶   initial ⟵ coefficient of common
term   difference

It's helpful to note the coefficient of d is 1 less than the subscript of the current term (as shown): $5 - 1 = 4$. This observation leads us to a formula for the *n*th term.

> **The *n*th Term of an Arithmetic Sequence**
>
> The *n*th term of an *arithmetic sequence* is given by
> $$a_n = a_1 + (n - 1)d$$
> where d is the common difference.

EXAMPLE 3 ▶ **Finding a Specified Term in an Arithmetic Sequence**

Find the 24th term of the sequence 0.1, 0.4, 0.7, 1, ...

Solution ▶ Instead of creating all terms up to the 24th, we determine the constant d and use the *n*th term formula. By inspection we note $a_1 = 0.1$ and $d = 0.3$.

$$a_n = a_1 + (n - 1)d \qquad \text{\small nth term formula}$$
$$= 0.1 + (n - 1)0.3 \qquad \text{\small substitute 0.1 for } a_1 \text{ and 0.3 for } d$$
$$= 0.1 + 0.3n - 0.3 \qquad \text{\small eliminate parentheses}$$
$$= 0.3n - 0.2 \qquad \text{\small simplify}$$

To find the 24th term we substitute 24 for n:

$$a_{24} = 0.3(24) - 0.2 \qquad \text{\small substitute 24 for } n$$
$$= 7.0 \qquad \text{\small result}$$

Now try Exercises 31 through 42 ▶

EXAMPLE 4 ▶ **Finding the Number of Terms in an Arithmetic Sequence**

Find the number of terms in the arithmetic sequence 2, -5, -12, -19, . . . , -411.

Solution ▶ By inspection we see that $a_1 = 2$ and $d = -7$. As before,

$$
\begin{aligned}
a_n &= a_1 + (n-1)d && \text{\textit{n}th term formula} \\
&= 2 + (n-1)(-7) && \text{substitute 2 for } a_1 \text{ and } -7 \text{ for } d \\
&= -7n + 9 && \text{simplify}
\end{aligned}
$$

Although we don't know the number of terms in the sequence, we *do* know the last or *n*th term is -411. Substituting -411 for a_n gives

$$
\begin{aligned}
-411 &= -7n + 9 && \text{substitute } -411 \text{ for } a_n \\
60 &= n && \text{solve for } n
\end{aligned}
$$

There are 60 terms in this sequence.

Now try Exercises 43 through 50 ▶

If the term a_1 is unknown but a term a_k is given, the *n*th term can be written

$$a_n = a_k + (n-k)d \text{ since } n = k + (n-k)$$

(the subscript of the term a_k and coefficient of d sum to n).

EXAMPLE 5 ▶ **Finding the First Term of an Arithmetic Sequence**

Given an arithmetic sequence where $a_6 = 0.55$ and $a_{13} = 0.9$, find the common difference d and the value of a_1.

Solution ▶ At first it seems that not enough information is given, but recall we can express a_{13} as the sum of any earlier term and the appropriate multiple of d. Since a_6 is known, we write $a_{13} = a_6 + 7d$ (note $13 = 6 + 7$ as required).

$$
\begin{aligned}
a_{13} &= a_6 + 7d && a_1 \text{ is unknown} \\
0.9 &= 0.55 + 7d && \text{substitute 0.9 for } a_{13} \text{ and 0.55 for } a_6 \\
0.35 &= 7d && \text{subtract 0.55} \\
d &= 0.05 && \text{solve for } d
\end{aligned}
$$

Having found d, we can now solve for a_1.

$$
\begin{aligned}
a_{13} &= a_1 + 12d && \text{\textit{n}th term formula for } n = 13 \\
0.9 &= a_1 + 12(0.05) && \text{substitute 0.9 for } a_{13} \text{ and 0.05 for } d \\
0.9 &= a_1 + 0.6 && \text{simplify} \\
a_1 &= 0.3 && \text{solve for } a_1
\end{aligned}
$$

The first term is $a_1 = 0.3$ and the common difference is $d = 0.05$.

☑ **B. You've just learned how to find the *n*th term of an arithmetic sequence**

Now try Exercises 51 through 56 ▶

C. Finding the *n*th Partial Sum of an Arithmetic Sequence

Using sequences and series to solve applications often requires computing the sum of a given number of terms. Consider the sequence $a_1, a_2, a_3, a_4, \ldots, a_n$ with common difference d. Use S_n to represent the sum of the first n terms and write the original series, then the series in reverse order underneath. Since one row increases at the same rate the other decreases, the sum of each column remains constant, and for simplicity's sake we choose $a_1 + a_n$ to represent this sum.

$$
\begin{array}{llllllll}
S_n = & a_1 & + & a_2 & + & a_3 & + \cdots + & a_{n-2} & + & a_{n-1} & + & a_n & \text{add} \\
S_n = & a_n & + & a_{n-1} & + & a_{n-2} & + \cdots + & a_3 & + & a_2 & + & a_1 & \text{columns} \\
\hline
2S_n = & (a_1 + a_n) & + & (a_1 + a_n) & + & (a_1 + a_n) & + \cdots + & (a_1 + a_n) & + & (a_1 + a_n) & + & (a_1 + a_n) & \text{vertically}
\end{array}
$$

To understand why each column adds to $a_1 + a_n$, consider the sum in the second column: $a_2 + a_{n-1}$. From $a_2 = a_1 + d$ and $a_{n-1} = a_n - d$, we obtain $a_2 + a_{n-1} = (a_1 + d) + (a_n - d)$ by direct substitution, which gives a result of $a_1 + a_n$. Since there are n columns, we end up with $2S_n = n(a_1 + a_n)$, and solving for S_n gives the formula for the first n terms of an arithmetic sequence.

The *n*th Partial Sum of an Arithmetic Sequence

Given an arithmetic sequence with first term a_1, the nth partial sum is given by

$$S_n = n\left(\frac{a_1 + a_n}{2}\right).$$

In words: The sum of an arithmetic sequence is the number of terms times the average of the first and last term.

EXAMPLE 6 ▶   **Computing the Sum of an Arithmetic Sequence**

Find the sum of the first 75 positive, odd integers: $\displaystyle\sum_{k=1}^{75} (2k - 1)$.

Solution ▶   The initial terms of the sequence are 1, 3, 5, . . . and we note $a_1 = 1$, $d = 2$, and $n = 75$. To use the sum formula, we need the value of $a_n = a_{75}$. The nth term formula shows $a_{75} = a_1 + 74d = 1 + 74(2)$, so $a_{75} = 149$.

$$S_n = \frac{n(a_1 + a_n)}{2} \qquad \text{sum formula}$$

$$S_{75} = \frac{75(a_1 + a_{75})}{2} \qquad \text{substitute 75 for } n$$

$$= \frac{75(1 + 149)}{2} \qquad \text{substitute 1 for } a_1,\ 149 \text{ for } a_{75}$$

$$= 5625 \qquad \text{result}$$

☑ **C. You've just learned how to find the *n*th partial sum of an arithmetic sequence**

The sum of the first 75 positive, odd integers is 5625.

Now try Exercises 57 through 62 ▶

Figure 8.5

spiral fern

Figure 8.6

By substituting the nth term formula directly into the formula for partial sums, we're able to find a partial sum without actually having to find the nth term:

$$S_n = \frac{n(a_1 + a_n)}{2} \qquad \text{sum formula}$$

$$= \frac{n(a_1 + [a_1 + (n-1)d])}{2} \qquad \text{substitute } a_1 + (n-1)d \text{ for } a_n$$

$$= \frac{n}{2}[2a_1 + (n-1)d] \qquad \textit{alternative formula for the nth partial sum}$$

See **Exercises 63 through 68** for more on this alternative formula.

nautilus

D. Applications

In the evolution of certain plants and shelled animals, sequences and series seem to have been one of nature's favorite tools (see Figures 8.5 and 8.6). Sequences and series also provide a good mathematical model for a variety of other situations as well.

EXAMPLE 7 ▶ **Solving an Application of Arithmetic Sequences: Seating Capacity**

Cox Auditorium is an amphitheater that has 40 seats in the first row, 42 seats in the second row, 44 in the third, and so on. If there are 75 rows in the auditorium, what is the auditorium's seating capacity?

Solution ▶ The number of seats in each row gives the terms of an arithmetic sequence with $a_1 = 40$, $d = 2$, and $n = 75$. To find the seating capacity, we need to find the total number of seats, which is the sum of this arithmetic sequence. Since the value of a_{75} is unknown, we opt for the alternative formula $S_n = \dfrac{n}{2}[2a_1 + (n-1)d]$.

$$S_n = \frac{n}{2}[2a_1 + (n-1)d] \qquad \text{sum formula}$$

$$S_{75} = \frac{75}{2}[2(40) + (75-1)(2)] \qquad \text{substitute 40 for } a_1\text{, 2 for } d \text{ and 75 for } n$$

$$= \frac{75}{2}(228) \qquad \text{simplify}$$

$$= 8550 \qquad \text{result}$$

☑ **D.** You've just learned how to solve applications involving arithmetic sequences

The seating capacity for Cox Auditorium is 8550.

Now try Exercises 71 through 76 ▶

8.2 EXERCISES

▶ **CONCEPTS AND VOCABULARY**

Fill in the blank with the appropriate word or phrase. Carefully reread the section if needed.

1. Consecutive terms in an arithmetic sequence differ by a constant called the _____ _____.

2. The sum of the first n terms of an arithmetic sequence is called the nth _____ _____.

3. The formula for the nth partial sum of an arithmetic sequence is $s_n =$ _____, where a_n is the _____ term.

4. The nth term formula for an arithmetic sequence is $a_n =$ _____, where a_1 is the _____ term and d is the _____ _____.

5. Discuss how the terms of an arithmetic sequence can be written in various ways using the relationship $a_n = a_k + (n-k)d$.

6. Describe how the formula for the nth partial sum was derived, and illustrate its application using a sequence from the exercise set.

▶ **DEVELOPING YOUR SKILLS**

Determine if the sequence given is arithmetic. If yes, name the common difference. If not, try to determine the pattern that forms the sequence.

7. $-5, -2, 1, 4, 7, 10, \ldots$

8. $1, -2, -5, -8, -11, -14, \ldots$

9. $0.5, 3, 5.5, 8, 10.5, \ldots$

10. $1.2, 3.5, 5.8, 8.1, 10.4, \ldots$

11. $2, 3, 5, 7, 11, 13, 17, \ldots$

12. $1, 4, 8, 13, 19, 26, 34, \ldots$

13. $\frac{1}{24}, \frac{1}{12}, \frac{1}{8}, \frac{1}{6}, \frac{5}{24}, \ldots$ 14. $\frac{1}{12}, \frac{1}{15}, \frac{1}{20}, \frac{1}{30}, \frac{1}{60}, \ldots$

15. 1, 4, 9, 16, 25, 36, . . .

16. $-125, -64, -27, -8, -1, . . .$

17. $\pi, \dfrac{5\pi}{6}, \dfrac{2\pi}{3}, \dfrac{\pi}{2}, \dfrac{\pi}{3}, \dfrac{\pi}{6}, . . .$ **18.** $\pi, \dfrac{7\pi}{8}, \dfrac{3\pi}{4}, \dfrac{5\pi}{8}, \dfrac{\pi}{2}, . . .$

Write the first four terms of the arithmetic sequence with the given first term and common difference.

19. $a_1 = 2, d = 3$

20. $a_1 = 8, d = 3$

21. $a_1 = 7, d = -2$

22. $a_1 = 60, d = -12$

23. $a_1 = 0.3, d = 0.03$

24. $a_1 = 0.5, d = 0.25$

25. $a_1 = \frac{3}{2}, d = \frac{1}{2}$

26. $a_1 = \frac{1}{5}, d = \frac{1}{10}$

27. $a_1 = \frac{3}{4}, d = -\frac{1}{8}$

28. $a_1 = \frac{1}{6}, d = -\frac{1}{3}$

29. $a_1 = -2, d = -3$

30. $a_1 = -4, d = -4$

Identify the first term and the common difference, then write the expression for the general term a_n and use it to find the 6th, 10th, and 12th terms of the sequence.

31. 2, 7, 12, 17, . . .

32. 7, 4, 1, $-2, -5, . . .$

33. 5.10, 5.25, 5.40, . . .

34. 9.75, 9.40, 9.05, . . .

35. $\frac{3}{2}, \frac{9}{4}, 3, \frac{15}{4}, . . .$

36. $\frac{5}{7}, \frac{3}{14}, -\frac{2}{7}, -\frac{11}{14}, . . .$

Find the indicated term using the information given.

37. $a_1 = 5, d = 4$; find a_{15}

38. $a_1 = 9, d = -2$; find a_{17}

39. $a_1 = \frac{3}{2}, d = -\frac{1}{12}$; find a_7

40. $a_1 = \frac{12}{25}, d = -\frac{1}{10}$; find a_9

41. $a_1 = -0.025, d = 0.05$; find a_{50}

42. $a_1 = 3.125, d = -0.25$; find a_{20}

Find the number of terms in each sequence.

43. $a_1 = 2, a_n = -22, d = -3$

44. $a_1 = 4, a_n = 42, d = 2$

45. $a_1 = 0.4, a_n = 10.9, d = 0.25$

46. $a_1 = -0.3, a_n = -36, d = -2.1$

47. $-3, -0.5, 2, 4.5, 7, . . . , 47$

48. $-3.4, -1.1, 1.2, 3.5, . . . , 38$

49. $\frac{1}{12}, \frac{1}{8}, \frac{1}{6}, \frac{5}{24}, \frac{1}{4}, . . . , \frac{9}{8}$ **50.** $\frac{1}{12}, \frac{1}{15}, \frac{1}{20}, \frac{1}{30}, . . . , -\frac{1}{4}$

Find the common difference d and the value of a_1 using the information given.

51. $a_3 = 7, a_7 = 19$

52. $a_5 = -17, a_{11} = -2$

53. $a_2 = 1.025, a_{26} = 10.025$

54. $a_6 = -12.9, a_{30} = 1.5$

55. $a_{10} = \frac{13}{18}, a_{24} = \frac{27}{2}$

56. $a_4 = \frac{5}{4}, a_8 = \frac{9}{4}$

Evaluate each sum. For Exercises 61 and 62, use the summation properties from Section 8.1.

57. $\displaystyle\sum_{n=1}^{30} (3n - 4)$

58. $\displaystyle\sum_{n=1}^{29} (4n - 1)$

59. $\displaystyle\sum_{n=1}^{37} \left(\frac{3}{4}n + 2\right)$

60. $\displaystyle\sum_{n=1}^{20} \left(\frac{5}{2}n - 3\right)$

61. $\displaystyle\sum_{n=4}^{15} (3 - 5n)$

62. $\displaystyle\sum_{n=7}^{20} (7 - 2n)$

Use the alternative formula for the nth partial sum to compute the sums indicated.

63. The sum S_{15} for the sequence
$-12 + (-9.5) + (-7) + (-4.5) + \cdots$

64. The sum S_{20} for the sequence $\frac{9}{2} + \frac{7}{2} + \frac{5}{2} + \frac{3}{2} + \cdots$

65. The sum S_{30} for the sequence
$0.003 + 0.173 + 0.343 + 0.513 + \cdots$

66. The sum S_{50} for the sequence
$(-2) + (-7) + (-12) + (-17) + \cdots$

67. The sum S_{20} for the sequence
$\sqrt{2} + 2\sqrt{2} + 3\sqrt{2} + 4\sqrt{2} + \cdots$

68. The sum S_{10} for the sequence
$12\sqrt{3} + 10\sqrt{3} + 8\sqrt{3} + 6\sqrt{3} + \cdots$

► **WORKING WITH FORMULAS**

69. Sum of the first n natural numbers: $S_n = \dfrac{n(n + 1)}{2}$

The sum of the first n natural numbers can be found using the formula shown, where n represents the number of terms in the sum. Verify the formula by adding the first six natural numbers by hand, and then evaluating S_6. Then find the sum of the first 75 natural numbers.

70. Sum of the squares of the first n natural numbers: $S_n = \dfrac{n(n + 1)(2n + 1)}{6}$

If the first n natural numbers are squared, the sum of these squares can be found using the formula shown, where n represents the number of terms in the sum. Verify the formula by computing the sum of the squares of the first six natural numbers by hand, and then evaluating S_6. Then find the sum of the squares of the first 20 natural numbers: $(1^2 + 2^2 + 3^2 + \cdots + 20^2)$.

▶ **APPLICATIONS**

71. Temperature fluctuation: At 5 P.M. in Coldwater, the temperature was a chilly 36°F. If the temperature decreased by 3°F every half-hour for the next 7 hr, at what time did the temperature hit 0°F?

72. Arc of a baby swing: When Mackenzie's baby swing is started, the first swing (one way) is a 30-in. arc. As the swing slows down, each successive arc is $\frac{3}{2}$ in. less than the previous one. Find (a) the length of the tenth swing and (b) how far Mackenzie has traveled during the 10 swings.

73. Computer animations: The animation on a new computer game initially allows the hero of the game to jump a (screen) distance of 10 in. over booby traps and obstacles. Each successive jump is limited to $\frac{3}{4}$ in. less than the previous one. Find (a) the length of the seventh jump and (b) the total distance covered after seven jumps.

74. Seating capacity: The Fox Theater creates a "theater in the round" when it shows any of Shakespeare's plays. The first row has 80 seats, the second row has 88, the third row has 96, and so on. How many seats are in the 10th row? If there is room for 25 rows, how many chairs will be needed to set up the theater?

75. Sales goals: *At the time that I was newly hired, 100 sales per month was what I required. Each following month—the last plus 20 more, as I work for the goal of top sales award. When 2500 sales are thusly made, it's Tahiti, Hawaii, and pina coladas in the shade.* How many sales were made by this person in the seventh month? What were the total sales after the 12th month? Was the goal of 2500 total sales met after the 12th month?

76. Bequests to charity: *At the time our mother left this Earth, she gave $9000 to her children of birth. This we kept and each year added $3000 more, as a lasting memorial from the children she bore. When $42,000 is thusly attained, all goes to charity that her memory be maintained.* What was the balance in the sixth year? In what year was the goal of $42,000 met?

▶ **EXTENDING THE THOUGHT**

77. From a study of numerical analysis, a function is known to be linear if its "first differences" (differences between each output) are constant. Likewise, a function is known to be quadratic if its "first differences" form an *arithmetic sequence*. Use this information to determine if the following sets of output come from a linear or quadratic function:

   a. 19, 11.8, 4.6, −2.6, −9.8, −17, −24.2, . . .
   b. −10.31, −10.94, −11.99, −13.46, −15.35, . . .

78. From elementary geometry it is known that the interior angles of a triangle sum to 180°, the interior angles of a quadrilateral sum to 360°, the interior angles of a pentagon sum to 540°, and so on. Use the pattern created by the relationship between the number of sides to the number of angles to develop a formula for the sum of the interior angles of an *n*-sided polygon. The interior angles of a decagon (10 sides) sum to how many degrees?

▶ **MAINTAINING YOUR SKILLS**

79. (4.4) Solve for *t*: $2530 = 500e^{0.45t}$

80. (3.1) Graph by completing the square. Label all important features: $y = x^2 - 2x - 3$.

81. (2.3) In 2000, the deer population was 972. By 2005 it had grown to 1217. Assuming the growth is linear, find the function that models this data and use it to estimate the deer population in 2008.

82. (3.8) Given *y* varies inversely with *x* and jointly with *w*. If $y = 4$ when $x = 15$ and $w = 52.5$, find the value of *y* if $x = 32$ and $w = 208$.

Learning Objectives

In Section 8.3 you will learn how to:

☐ **A.** Identify a geometric sequence and its common ratio

☐ **B.** Find the *n*th term of a geometric sequence

☐ **C.** Find the *n*th partial sum of a geometric sequence

☐ **D.** Find the sum of an infinite geometric series

☐ **E.** Solve application problems involving geometric sequences and series

Recall that arithmetic sequences are those where each term is found by *adding* a constant value to the preceding term. In this section, we consider **geometric sequences,** where each term is found by *multiplying* the preceding term by a constant value. Geometric sequences have many interesting applications, as do **geometric series.**

A. Geometric Sequences

A geometric sequence is one where each successive term is found by multiplying the preceding term by a fixed constant. Consider growth of a bacteria population, where a single cell splits in two every hour over a 24-hr period. Beginning with a single bacterium ($a_0 = 1$), after 1 hr there are 2, after 2 hr there are 4, and so on. Writing the number of bacteria as a sequence we have:

hours:	a_1	a_2	a_3	a_4	a_5	. . .
	↓	↓	↓	↓	↓	
bacteria:	2	4	8	16	32	. . .

The sequence 2, 4, 8, 16, 32, . . . is a geometric sequence since each term is found by multiplying the previous term by the constant factor 2. This also means that the ratio of any two consecutive terms must be 2 and in fact, 2 is called the **common ratio *r*** for this sequence. Using the notation from Section 8.1 we can write $r = \dfrac{a_{k+1}}{a_k}$, where a_k represents any term of the sequence and a_{k+1} represents the term that follows a_k.

EXAMPLE 1 ▶ **Testing a Sequence for a Common Ratio**

Determine if the given sequence is geometric.

   a. 1, 0.5, 0.25, 0.125, . . .   **b.** $\frac{1}{8}, \frac{1}{4}, \frac{3}{4}, 3, 15, \ldots$

Solution ▶ Apply the definition to check for a common ratio $r = \dfrac{a_{k+1}}{a_k}$.

   a. For 1, 0.5, 0.25, 0.125, . . . , the ratio of consecutive terms gives

$$\frac{0.5}{1} = 0.5, \qquad \frac{0.25}{0.5} = 0.5, \qquad \frac{0.125}{0.25} = 0.5, \qquad \text{and so on.}$$

   This is a geometric sequence with common ratio $r = 0.5$.

   b. For $\frac{1}{8}, \frac{1}{4}, \frac{3}{4}, 3, 15, \ldots$, we have:

$$\frac{1}{4} \div \frac{1}{8} = \frac{1}{4} \cdot \frac{8}{1} \qquad \frac{3}{4} \div \frac{1}{4} = \frac{3}{4} \cdot \frac{4}{1} \qquad 3 \div \frac{3}{4} = \frac{3}{1} \cdot \frac{4}{3} \qquad \text{and so on.}$$
$$= 2 \qquad\qquad\qquad = 3 \qquad\qquad\qquad = 4$$

   Since the ratio is not constant, this is not a geometric sequence.

Now try Exercises 7 through 24 ▶

EXAMPLE 2 ▶ **Writing the Terms of a Geometric Sequence**

Write the first five terms of the geometric sequence, given the first term $a_1 = -16$ and the common ratio $r = 0.25$.

Solution ▶ Given $a_1 = -16$ and $r = 0.25$. Starting at $a_1 = -16$, multiply each term by 0.25 to generate the sequence.

$$a_2 = -16 \cdot 0.25 = -4 \qquad a_3 = -4 \cdot 0.25 = -1$$
$$a_4 = -1 \cdot 0.25 = -0.25 \qquad a_5 = -0.25 \cdot 0.25 = -0.0625$$

The first five terms of this sequence are $-16, -4, -1, -0.25,$ and -0.0625.

☑ **A. You've just learned how to identify a geometric sequence and its common ratio**

Now try Exercises 25 through 38 ▶

B. Find the *n*th Term of a Geometric Sequence

If the values a_1 and r from a geometric sequence are known, we could generate the terms of the sequence by applying *additional factors of r to the first term,* instead of multiplying each new term by *r.* If $a_1 = 3$ and $r = 2$, we simply begin at a_1, and continue applying additional factors of *r* for each successive term.

$$3 = 3 \cdot 2^0 \qquad a_1 = a_1 r^0$$
$$6 = 3 \cdot 2^1 \qquad a_2 = a_1 r^1$$
$$12 = 3 \cdot 2^2 \qquad a_3 = a_1 r^2$$
$$24 = 3 \cdot 2^3 \qquad a_4 = a_1 r^3$$
$$48 = 3 \cdot 2^4 \qquad a_5 = a_1 r^4$$

current term ⟶ ↑ initial term   ⟵ exponent on common ratio

From this pattern, we note the exponent on *r* is always 1 less than the subscript of the current term: $5 - 1 = 4$, which leads us to the formula for the *n*th term of a geometric sequence.

The *n*th Term of a Geometric Sequence

The *n*th term of a *geometric sequence* is given by

$$a_n = a_1 r^{n-1}$$

where *r* is the common ratio.

EXAMPLE 3 ▶ **Finding a Specific Term in a Sequence**

Find the 10th term of the sequence $3, -6, 12, -24, \ldots$.

Solution ▶ Instead of writing out all 10 terms, we determine the constant ratio *r* and use the *n*th term formula. By inspection we note that $a_1 = 3$ and $r = -2$.

$$a_n = a_1 r^{n-1} \qquad \text{\textit{n}th term formula}$$
$$= 3(-2)^{n-1} \qquad \text{substitute 3 for } a_1 \text{ and } -2 \text{ for } r$$

To find the 10th term we substitute $n = 10$:

$$a_{10} = 3(-2)^{10-1} \qquad \text{substitute 10 for } n$$
$$= 3(-2)^9 = -1536 \qquad \text{simplify}$$

Now try Exercises 39 through 46 ▶

EXAMPLE 4 ▶ **Determining the Number of Terms in a Geometric Sequence**

Find the number of terms in the geometric sequence $4, 2, 1, \ldots, \frac{1}{64}$.

Solution ▶ Observing that $a_1 = 4$ and $r = \frac{1}{2}$, we have

$$a_n = a_1 r^{n-1} \qquad \text{\textit{n}th term formula}$$
$$= 4\left(\frac{1}{2}\right)^{n-1} \qquad \text{substitute 4 for } a_1 \text{ and } \frac{1}{2} \text{ for } r$$

Although we don't know the number of terms in the sequence, we *do* know the last or *n*th term is $\frac{1}{64}$. Substituting $a_n = \frac{1}{64}$ gives

$$\frac{1}{64} = 4\left(\frac{1}{2}\right)^{n-1} \qquad \text{substitute } \frac{1}{64} \text{ for } a_n$$
$$\frac{1}{256} = \left(\frac{1}{2}\right)^{n-1} \qquad \text{divide by 4} \left(\text{multiply by } \frac{1}{4}\right)$$

From our work in Chapter 4, we attempt to write both sides as exponentials with a like base, or apply logarithms. Since $256 = 2^8$, we equate bases.

$$\left(\frac{1}{2}\right)^8 = \left(\frac{1}{2}\right)^{n-1} \qquad \text{write } \frac{1}{256} \text{ as } \left(\frac{1}{2}\right)^8$$

$$\rightarrow 8 = n - 1 \qquad \text{like bases imply exponents must be equal}$$

$$9 = n \qquad \text{solve for } n$$

This shows there are nine terms in the sequence.

Now try Exercises 47 through 58 ▶

If the term a_1 is unknown but a term a_k is given, the nth term can be written

$$a_n = a_k r^{n-k}, \text{ since } n = k + (n - k)$$

(the subscript on the term a_k and the exponent on r sum to n).

EXAMPLE 5 ▶   **Finding the First Term of a Geometric Sequence**

Given a geometric sequence where $a_4 = 0.075$ and $a_7 = 0.009375$, find the common ratio r and the value of a_1.

Solution ▶   Since a_1 is not known, we express a_7 as the product of a known term and the appropriate number of common ratios: $a_7 = a_4 r^3$ ($7 - 4 = 3$, as required).

$$a_7 = a_4 \cdot r^3 \qquad a_1 \text{ is unknown}$$

$$0.009375 = 0.075 r^3 \qquad \text{substitute } 0.009375 \text{ for } a_7 \text{ and } 0.075 \text{ for } a_4$$

$$0.125 = r^3 \qquad \text{divide by } 0.075$$

$$r = 0.5 \qquad \text{solve for } r$$

Having found r, we can now solve for a_1

$$a_7 = a_1 r^6 \qquad n\text{th term formula}$$

$$0.009375 = a_1(0.5)^6 \qquad \text{substitute } 0.009375 \text{ for } a_7 \text{ and } 0.5 \text{ for } r$$

$$0.009375 = a_1(0.015625) \qquad \text{simplify}$$

$$a_1 = 0.6 \qquad \text{solve for } a_1$$

☑ **B. You've just learned how to find the nth term of a geometric sequence**

The first term is $a_1 = 0.6$ and the common ratio is $r = 0.5$.

Now try Exercises 59 through 64 ▶

C. Find the nth Partial Sum of a Geometric Sequence

As with arithmetic series, applications of geometric series often involve computing a sum of consecutive terms. We can adapt the method for finding the sum of an arithmetic sequence to develop a formula for adding the first n terms of a geometric sequence.

For the nth term $a_n = a_1 r^{n-1}$, we have $S_n = a_1 + a_1 r + a_1 r^2 + a_1 r^3 + \cdots + a_1 r^{n-1}$. If we multiply S_n by $-r$ then add the original series, the "interior terms" sum to zero.

$$- rS_n = \quad -a_1 r + (-a_1 r^2) + (-a_1 r^3) + \cdots + (-a_1 r^{n-1}) + (-a_1 r^n)$$

$$+ \; S_n = \quad a_1 + \quad a_1 r \quad + \quad a_1 r^2 \quad + \cdots + \quad a_1 r^{n-2} \quad + a_1 r^{n-1}$$

$$\overline{S_n - rS_n = \quad a_1 + \quad 0 \quad + \quad 0 \quad + 0 + \quad 0 \quad + \quad 0 \quad + (-a_1 r^n)}$$

We then have $S_n - rS_n = a_1 - a_1 r^n$, and can now solve for S_n:

$$S_n - rS_n = a_1 - a_1 r^n \qquad \text{difference of } S_n \text{ and } rS_n$$

$$S_n(1 - r) = a_1 - a_1 r^n \qquad \text{factor out } S_n$$

$$S_n = \frac{a_1 - a_1 r^n}{1 - r} \qquad \text{solve for } S_n$$

The result is a formula for the nth partial sum of a geometric sequence.

The nth Partial Sum of a Geometric Sequence

Given a geometric sequence with first term a_1 and common ratio r, the nth partial sum (the sum of the first n terms) is

$$S_n = \frac{a_1 - a_1 r^n}{1 - r} = \frac{a_1(1 - r^n)}{1 - r}, r \neq 1$$

In words: The sum of a geometric sequence is the difference of the first and $(n + 1)$st term, divided by 1 minus the common ratio.

EXAMPLE 6 ▶ **Computing a Partial Sum**

Find the sum: $\displaystyle\sum_{i=1}^{9} 3^i$ (the first nine powers of 3).

Solution ▶ The initial terms of this series are $3 + 9 + 27 + \cdots$, and we note $a_1 = 3$, $r = 3$, and $n = 9$. We could find the first nine terms and add, but using the partial sum formula gives

$$S_n = \frac{a_1(1 - r^n)}{1 - r} \qquad \text{sum formula}$$

$$S_9 = \frac{3(1 - 3^9)}{1 - 3} \qquad \text{substitute 3 for } a_1, 9 \text{ for } n, \text{ and 3 for } r$$

$$= \frac{3(-19,682)}{-2} \qquad \text{simplify}$$

$$= 29,523 \qquad \text{result}$$

☑ **C.** You've just learned how to find the nth partial sum of a geometric sequence

Now try Exercises 65 through 88 ▶

D. The Sum of an Infinite Geometric Series

To this point we've considered only partial sums of a geometric series. While it is impossible to add an infinite number of these terms, some of these "infinite sums" appear to have a limiting value. The sum appears to get ever closer to this value but never exceeds it—much like the asymptotic behavior of some graphs. We will define the sum of this **infinite geometric series** to be this limiting value, if it exists. Consider the illustration in Figure 8.7, where a standard sheet of typing paper is cut in half. One of the halves is again cut in half and the process is continued indefinitely, as shown. Notice the "halves" create an infinite sequence $\frac{1}{2}, \frac{1}{4}, \frac{1}{8}, \frac{1}{16}, \frac{1}{32}, \cdots$ with $a_1 = \frac{1}{2}$ and $r = \frac{1}{2}$. The corresponding infinite series is $\frac{1}{2} + \frac{1}{4} + \frac{1}{8} + \frac{1}{16} + \frac{1}{32} + \cdots + \frac{1}{2^n} + \cdots$.

Figure 8.7

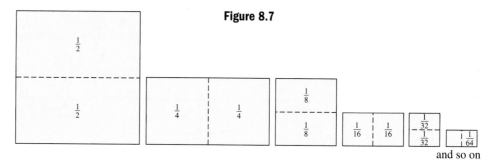

and so on

Figure 8.8

If we arrange one of the halves from each stage as shown in Figure 8.8, we would be rebuilding the original sheet of paper. As we add more and more of these halves together, we get closer and closer to the size of the original sheet. We gain an intuitive sense that this series must add to 1, because the *pieces* of the original sheet of paper must add to 1 whole sheet. To explore this idea further, consider what happens to $\left(\frac{1}{2}\right)^n$ as n becomes large.

$$n = 4: \left(\frac{1}{2}\right)^4 = 0.0625 \qquad n = 8: \left(\frac{1}{2}\right)^8 \approx 0.004 \qquad n = 12: \left(\frac{1}{2}\right)^{12} \approx 0.0002$$

Further exploration with a calculator seems to support the idea that as $n \to \infty$, $\left(\frac{1}{2}\right)^n \to 0$, although a definitive proof is left for a future course. In fact, it can be shown that for any $|r| < 1$, r^n becomes very close to zero as n becomes large. In symbols: as $n \to \infty$, $r^n \to 0$. For $S_n = \dfrac{a_1 - a_1 r^n}{1 - r} = \dfrac{a_1}{1 - r} - \dfrac{a_1 r^n}{1 - r}$, note that if $|r| < 1$ and "we sum an infinite number of terms," the second term becomes zero, leaving only the first term. In other words, the limiting value (represented by S_∞) is

$$S_\infty = \frac{a_1}{1 - r}.$$

WORTHY OF NOTE

The formula for the sum of an infinite geometric series can also be derived by noting that $S_\infty = a_1 + a_1 r + a_1 r^2 + a_1 r^3 + \cdots$ can be rewritten as $S_\infty = a_1 + r(a_1 + a_1 r + a_1 r^2 + a_1 r^3 + \cdots) = a_1 + r S_\infty$.

$$S_\infty - r S_\infty = a_1$$
$$S_\infty(1 - r) = a_1$$
$$S_\infty = \frac{a_1}{1 - r}.$$

Infinite Geometric Series

Given a geometric sequence with first term a_1 and $|r| < 1$, the sum of the related infinite series is given by

$$S_\infty = \frac{a_1}{1 - r}; r \neq 1$$

If $|r| > 1$, no finite sum exists.

EXAMPLE 7 ▶ **Computing an Infinite Sum**

Find the limiting value of each infinite geometric series (if it exists).
a. $1 + 2 + 4 + 8 + \cdots$ **b.** $3 + 2 + \frac{4}{3} + \frac{8}{9} + \cdots$
c. $0.185 + 0.000185 + 0.000000185 + \cdots$

Solution ▶ Begin by determining if the infinite series is geometric. If so, use $S_\infty = \dfrac{a_1}{1 - r}$.

a. Since $r = 2$ (by inspection), a finite sum does not exist.
b. Using the ratio of consecutive terms we find $r = \frac{2}{3}$ and the infinite sum exists. With $a_1 = 3$, we have

$$S_\infty = \frac{3}{1 - \frac{2}{3}} = \frac{3}{\frac{1}{3}} = 9$$

c. This series is equivalent to the repeating decimal $0.185185185\ldots = 0.\overline{185}$. The common ratio is $r = \frac{0.000185}{0.185} = 0.001$ and the infinite sum exists:

$$S_\infty = \frac{0.185}{1 - 0.001} = \frac{5}{27}$$

☑ **D.** You've just learned how to find the sum of an infinite geometric series

Now try Exercises 89 through 104 ▶

E. Applications Involving Geometric Sequences and Series

Here are a few of the ways these ideas can be put to use.

EXAMPLE 8 ▶ **Solving an Application of Geometric Sequences: Pendulums**

A pendulum is any object attached to a fixed point and allowed to swing freely under the influence of gravity. Suppose each swing is 0.9 the length of the previous one. Gradually the swings become shorter and shorter and at some point the pendulum will appear to have stopped (although *theoretically* it never does).

  a. How far does the pendulum travel on its eighth swing, if the first swing was 2 m?

  b. What is the total distance traveled by the pendulum for these eight swings?

  c. How many swings until the length of each swing falls below 0.5 m?

  d. What total distance does the pendulum travel before coming to rest?

Solution ▶ **a.** The lengths of each swing form the terms of a geometric sequence with $a_1 = 2$ and $r = 0.9$. The first few terms are 2, 1.8, 1.62, 1.458, and so on. For the 8th term we have:

$$a_n = a_1 r^{n-1} \qquad \text{\textit{n}th term formula}$$
$$a_8 = 2(0.9)^{8-1} \qquad \text{substitute 8 for \textit{n}, 2 for } a_1, \text{ and 0.9 for } r$$
$$\approx 0.956$$

The pendulum travels about 0.956 m on its 8th swing.

  b. For the total distance traveled after eight swings, we compute the value of S_8.

$$S_n = \frac{a_1(1 - r^n)}{1 - r} \qquad \text{\textit{n}th partial sum formula}$$
$$S_8 = \frac{2(1 - 0.9^8)}{1 - 0.9} \qquad \text{substitute 2 for } a_1, \text{ 0.9 for } r, \text{ and 8 for } n$$
$$\approx 11.4$$

The pendulum has traveled about 11.4 m by the end of the 8th swing.

  c. To find the number of swings until the length of each swing is less than 0.5 m, we solve for n in the equation $0.5 = 2(0.9)^{n-1}$. This yields

$$0.25 = (0.9)^{n-1} \qquad \text{divide by 2}$$
$$\ln 0.25 = (n - 1)\ln 0.9 \qquad \text{take the natural log, apply power property}$$
$$\frac{\ln 0.25}{\ln 0.9} + 1 = n \qquad \text{solve for \textit{n} (exact form)}$$
$$14.16 \approx n \qquad \text{solve for \textit{n} (approximate form)}$$

After the 14th swing, each successive swing will be less than 0.5 m.

  d. For the total distance traveled before coming to rest, we consider the related infinite geometric series, with $a_1 = 2$ and $r = 0.9$.

$$S_\infty = \frac{a_1}{1 - r} \qquad \text{infinite sum formula}$$
$$S_\infty = \frac{2}{1 - 0.9} \qquad \text{substitute 2 for } a_1 \text{ and 0.9 for } r$$
$$= 20 \qquad \text{result}$$

The pendulum would travel 20 m before coming to rest.

☑ **E. You've just learned how to solve application problems involving geometric sequences and series**

Now try Exercises 107 through 119 ▶

8.3 EXERCISES

▶ CONCEPTS AND VOCABULARY

Fill in the blank with the appropriate word or phrase. Carefully reread the section if needed.

1. In a geometric sequence, each successive term is found by _____ the preceding term by a fixed value r.

2. In a geometric sequence, the common ratio r can be found by computing the _____ of any two consecutive terms.

3. The nth term of a geometric sequence is given by $a_n = $ _____, for any $n \geq 1$.

4. For the general sequence $a_1, a_2, a_3, \ldots, a_k, \ldots$, the fifth partial sum is given by $S_5 = $ _____.

5. Describe/Discuss how the formula for the nth partial sum is related to the formula for the sum of an infinite geometric series.

6. Describe the difference(s) between an arithmetic and a geometric sequence. How can a student prevent confusion between the formulas?

▶ DEVELOPING YOUR SKILLS

Determine if the sequence given is geometric. If yes, name the common ratio. If not, try to determine the pattern that forms the sequence.

7. 4, 8, 16, 32, …

8. 2, 6, 18, 54, 162, …

9. 3, −6, 12, −24, 48, …

10. 128, −32, 8, −2, …

11. 2, 5, 10, 17, 26, …

12. −13, −9, −5, −1, 3, …

13. 3, 0.3, 0.03, 0.003, …

14. 12, 0.12, 0.0012, 0.000012, …

15. −1, 3, −12, 60, −360, …

16. $-\frac{2}{3}$, 2, −8, 40, −240, …

17. 25, 10, 4, $\frac{8}{5}$, …

18. −36, 24, −16, $\frac{32}{3}$, …

19. $\frac{1}{2}, \frac{1}{4}, \frac{1}{8}, \frac{1}{16}, \ldots$

20. $\frac{2}{3}, \frac{4}{9}, \frac{8}{27}, \frac{16}{81}, \ldots$

21. $3, \dfrac{12}{x}, \dfrac{48}{x^2}, \dfrac{192}{x^3}, \ldots$

22. $5, \dfrac{10}{a}, \dfrac{20}{a^2}, \dfrac{40}{a^3}, \ldots$

23. 240, 120, 40, 10, 2, …

24. −120, −60, −20, −5, −1, …

Write the first four terms of the sequence, given a_1 and r.

25. $a_1 = 5, r = 2$

26. $a_1 = 2, r = -4$

27. $a_1 = -6, r = -\frac{1}{2}$

28. $a_1 = \frac{2}{3}, r = \frac{1}{5}$

29. $a_1 = 4, r = \sqrt{3}$

30. $a_1 = \sqrt{5}, r = \sqrt{5}$

31. $a_1 = 0.1, r = 0.1$

32. $a_1 = 0.024, r = 0.01$

Find the indicated term for each sequence.

33. $a_1 = -24, r = \frac{1}{2}$; find a_7

34. $a_1 = 48, r = -\frac{1}{3}$; find a_6

35. $a_1 = -\frac{1}{20}, r = -5$; find a_4

36. $a_1 = \frac{3}{20}, r = 4$; find a_5

37. $a_1 = 2, r = \sqrt{2}$; find a_7

38. $a_1 = \sqrt{3}, r = \sqrt{3}$; find a_8

Identify a_1 and r, then write the expression for the nth term $a_n = a_1 r^{n-1}$ and use it to find a_6, a_{10}, and a_{12}.

39. $\frac{1}{27}, -\frac{1}{9}, \frac{1}{3}, -1, 3, \ldots$

40. $-\frac{7}{8}, \frac{7}{4}, -\frac{7}{2}, 7, -14, \ldots$

41. 729, 243, 81, 27, 9, …

42. 625, 125, 25, 5, 1, …

43. $\frac{1}{2}, \frac{\sqrt{2}}{2}, 1, \sqrt{2}, 2, \ldots$

44. $36\sqrt{3}, 36, 12\sqrt{3}, 12, 4\sqrt{3}, \ldots$

45. $0.2, 0.08, 0.032, 0.0128, \ldots$

46. $0.5, -0.35, 0.245, -0.1715, \ldots$

Find the number of terms in each sequence.

47. $a_1 = 9, a_n = 729, r = 3$

48. $a_1 = 1, a_n = -128, r = -2$

49. $a_1 = 16, a_n = \frac{1}{64}, r = \frac{1}{2}$

50. $a_1 = 4, a_n = \frac{1}{512}, r = \frac{1}{2}$

51. $a_1 = -1, a_n = -1296, r = \sqrt{6}$

52. $a_1 = 2, a_n = 1458, r = -\sqrt{3}$

53. $2, -6, 18, -54, \ldots, -4374$

54. $3, -6, 12, -24, \ldots, -6144$

55. $64, 32\sqrt{2}, 32, 16\sqrt{2}, \ldots, 1$

56. $243, 81\sqrt{3}, 81, 27\sqrt{3}, \ldots, 1$

57. $\frac{3}{8}, -\frac{3}{4}, \frac{3}{2}, -3, \ldots, 96$

58. $-\frac{5}{27}, \frac{5}{9}, -\frac{5}{3}, -5, \ldots, -135$

Find the common ratio r and the value of a_1 using the information given (assume $r > 0$).

59. $a_3 = 324, a_7 = 64$ **60.** $a_5 = 6, a_9 = 486$

61. $a_4 = \frac{4}{9}, a_8 = \frac{9}{4}$ **62.** $a_2 = \frac{16}{81}, a_5 = \frac{2}{3}$

63. $a_4 = \frac{32}{3}, a_8 = 54$ **64.** $a_3 = \frac{16}{25}, a_7 = 25$

Find the indicated sum. For Exercises 81 and 82, use the summation properties from Section 8.1.

65. $a_1 = 8, r = -2$; find S_{12}

66. $a_1 = 2, r = -3$; find S_8

67. $a_1 = 96, r = \frac{1}{3}$; find S_5

68. $a_1 = 12, r = \frac{1}{2}$; find S_8

69. $a_1 = 8, r = \frac{3}{2}$; find S_7

70. $a_1 = -1, r = -\frac{3}{2}$; find S_{10}

71. $2 + 6 + 18 + \cdots$; find S_6

72. $2 + 8 + 32 + \cdots$; find S_7

73. $16 - 8 + 4 - \cdots$; find S_8

74. $4 - 12 + 36 - \cdots$; find S_8

75. $\frac{4}{3} + \frac{2}{9} + \frac{1}{27} + \cdots$; find S_9

76. $\frac{1}{18} - \frac{1}{6} + \frac{1}{2} - \cdots$; find S_7

77. $\displaystyle\sum_{j=1}^{5} 4^j$ **78.** $\displaystyle\sum_{k=1}^{10} 2^k$

79. $\displaystyle\sum_{k=1}^{8} 5\left(\frac{2}{3}\right)^{k-1}$ **80.** $\displaystyle\sum_{j=1}^{7} 3\left(\frac{1}{5}\right)^{j-1}$

81. $\displaystyle\sum_{i=4}^{10} 9\left(-\frac{1}{2}\right)^{i-1}$ **82.** $\displaystyle\sum_{i=3}^{8} 5\left(-\frac{1}{4}\right)^{i-1}$

Find the indicated partial sum using the information given. Write all results in simplest form.

83. $a_2 = -5, a_5 = \frac{1}{25}$; find S_5

84. $a_3 = 1, a_6 = -27$; find S_6

85. $a_3 = \frac{4}{9}, a_7 = \frac{9}{64}$; find S_6

86. $a_2 = \frac{16}{81}, a_5 = \frac{2}{3}$; find S_8

87. $a_3 = 2\sqrt{2}, a_6 = 8$; find S_7

88. $a_2 = 3, a_5 = 9\sqrt{3}$; find S_7

Determine whether the infinite geometric series has a finite sum. If so, find the limiting value.

89. $3 + 6 + 12 + 24 + \cdots$

90. $4 + 8 + 16 + 32 + \cdots$

91. $9 + 3 + 1 + \cdots$

92. $36 + 24 + 16 + \cdots$

93. $25 + 10 + 4 + \frac{8}{5} + \cdots$

94. $10 + 2 + \frac{2}{5} + \frac{2}{25} + \cdots$

95. $6 + 3 + \frac{3}{2} + \frac{3}{4} + \cdots$

96. $-49 + (-7) + \left(-\frac{1}{7}\right) + \cdots$

97. $6 - 3 + \frac{3}{2} - \frac{3}{4} + \cdots$

98. $10 - 5 + \frac{5}{2} - \frac{5}{4} + \cdots$

99. $0.3 + 0.03 + 0.003 + \cdots$

100. $0.63 + 0.0063 + 0.000063 + \cdots$

101. $\displaystyle\sum_{k=1}^{\infty} \frac{3}{4}\left(\frac{2}{3}\right)^k$ **102.** $\displaystyle\sum_{i=1}^{\infty} 5\left(\frac{1}{2}\right)^i$

103. $\displaystyle\sum_{j=1}^{\infty} 9\left(-\frac{2}{3}\right)^j$ **104.** $\displaystyle\sum_{k=1}^{\infty} 12\left(\frac{4}{3}\right)^k$

► WORKING WITH FORMULAS

105. Sum of the cubes of the first n natural numbers:

$$S_n = \frac{n^2(n+1)^2}{4}$$

Compute $1^3 + 2^3 + 3^3 + \cdots + 8^3$ using the formula given. Then confirm the result by direct calculation.

106. Student loan payment: $A_n = P(1 + r)^n$

If P dollars is borrowed at an annual interest rate r with interest compounded annually, the amount of money to be paid back after n years is given by the indicated formula. Find the total amount of money that the student must repay to clear the loan, if $8000 is borrowed at 4.5% interest and the loan is paid back in 10 yr.

► APPLICATIONS

107. Pendulum movement: On each swing, a pendulum travels only 80% as far as it did on the previous swing. If the first swing is 24 ft, how far does the pendulum travel on the 7th swing? What total distance is traveled before the pendulum comes to rest?

108. Pendulum movement: Ernesto is swinging to and fro on his backyard tire swing. Using his legs and body, he pumps each swing until reaching a maximum height, then suddenly relaxes until the swing comes to a stop. With each swing, Ernesto travels 75% as far as he did on the previous swing. If the first arc (or swing) is 30 ft, find the distance Ernesto travels on the 5th arc. What total distance will he travel before coming to rest?

109. Depreciation: A certain new SUV depreciates in value about 20% per year (meaning it holds 80% of its value each year). If the SUV is purchased for $46,000, how much is it worth 4 yr later? How many years until its value is less than $5000?

110. Depreciation: A new photocopier under heavy use will depreciate about 25% per year (meaning it holds 75% of its value each year). If the copier is purchased for $7000, how much is it worth 4 yr later? How many years until its value is less than $1246?

111. Equipment aging: Tests have shown that the pumping power of a heavy-duty oil pump decreases by 3% per month. If the pump can move 160 gallons per minute (gpm) new, how many gpm can the pump move 8 months later? If the pumping rate falls below 118 gpm, the pump must be replaced. How many months until this pump is replaced?

112. Equipment aging: At the local mill, a certain type of saw blade can saw approximately 2 log-feet/sec when it is new. As time goes on, the blade becomes worn, and loses 6% of its cutting speed each week. How many log-feet/sec can the saw blade cut after 6 weeks? If the cutting speed falls below 1.2 log-feet/sec, the blade must be replaced. During what week of operation will this blade be replaced?

113. Population growth: At the beginning of the year 2000, the population of the United States was approximately 277 million. If the population is growing at a rate of 2.3% per year, what will the population be in 2010, 10 yr later?

114. Population growth: The population of the Zeta Colony on Mars is 1000 people. Determine the population of the Colony 20 yr from now, if the population is growing at a constant rate of 5% per year.

115. Population growth: A biologist finds that the population of a certain type of bacteria doubles *each half-hour*. If an initial culture has 50 bacteria, what is the population after 5 hr? How long will it take for the number of bacteria to reach 204,800?

116. Population growth: Suppose the population of a "boom town" in the old west doubled *every 2 months* after gold was discovered. If the initial population was 219, what was the population 8 months later? How many months until the population exceeds 28,000?

117. **Elastic rebound:** Megan discovers that a rubber ball dropped from a height of 2 m rebounds four-fifths of the distance it has previously fallen. How high does it rebound on the 7th bounce? How far does the ball travel before coming to rest?

118. **Elastic rebound:** The screen saver on my computer is programmed to send a colored ball vertically down the middle of the screen so that it rebounds 95% of the distance it last traversed. If the ball always begins at the top and the screen is 36 cm tall,

how high does the ball bounce after its 8th rebound? How far does the ball travel before coming to rest (and a new screen saver starts)?

119. **Creating a vacuum:** To create a vacuum, a hand pump is used to remove the air from an air-tight cube with a volume of 462 in^3. With each stroke of the pump, two-fifths of the air that remains in the cube is removed. How much air remains inside after the 5th stroke? How many strokes are required to remove all but 12.9 in^3 of the air?

▶ EXTENDING THE CONCEPT

120. As part of a science experiment, identical rubber balls are dropped from a certain height on these surfaces: slate, cement, and asphalt. When dropped on slate, the ball rebounds 80% of the height from which it last fell. On cement the figure is 75% and on asphalt the figure is 70%. The ball is dropped from 130 m on the slate, 175 m on the cement, and 200 m on the asphalt. Which ball has traveled the shortest total distance at the time of the fourth bounce? Which ball will travel farthest before coming to rest?

121. Consider the following situation. A person is hired at a salary of $40,000 per year, with a guaranteed raise of $1750 per year. At the same time, inflation is running about 4% per year. How many years until this person's salary is overtaken and eaten up by the actual cost of living?

122. A standard piece of typing paper is approximately 0.001 in. thick. Suppose you were able to fold this

piece of paper in half 26 times. How thick would the result be? (a) As tall as a hare, (b) as tall as a hen, (c) as tall as a horse, (d) as tall as a house, or (e) over 1 mi high? Find the actual height by computing the 27th term of a geometric sequence. Discuss what you find.

123. Find an alternative formula for the sum
$$S_n = \sum_{k=1}^{n} \log k, \text{ that does not use the sigma notation.}$$

124. Verify the following statements:

   a. If $a_1, a_2, a_3, \ldots, a_n$ is a geometric sequence with r and a_1 greater than zero, then $\log a_1$, $\log a_2, \log a_3, \ldots, \log a_n$ is an arithmetic sequence.

   b. If $a_1, a_2, a_3, \ldots, a_n$ is an arithmetic sequence, then $10^{a_1}, 10^{a_2}, \ldots, 10^{a_n}$, is a geometric sequence.

▶ MAINTAINING YOUR SKILLS

125. (1.5) Find the zeroes of f using the quadratic formula: $f(x) = x^2 + 5x + 9$.

126. (1.3) Solve for x: $\dfrac{3}{x^2 - 3x - 10} - \dfrac{4}{x - 5} = \dfrac{1}{x + 2}$

127. (4.6) Graph the rational function:
$$h(x) = \frac{x^2}{x - 1}$$

128. (5.6) Given the logistics function shown, find $p(50)$, $p(75)$, $p(100)$, and $p(150)$:
$$p(t) = \frac{4200}{1 + 10e^{-0.055t}}$$

8.4 Mathematical Induction

Since middle school (or even before) we have accepted that, "The product of two negative numbers is a positive number." But have you ever been asked to *prove* it? It's not as easy as it seems. We may think of several patterns that yield the result, analogies that indicate its truth, or even number line illustrations that lead us to believe the statement. But most of us have never seen a *proof* (see www.mhhe.com/coburn). In this section, we introduce one of mathematics' most powerful tools for proving a statement, called **proof by induction.**

A. Subscript Notation and Composition of Functions

One of the challenges in understanding a proof by induction is working with the notation. Earlier in the chapter, we introduced subscript notation as an alternative to function notation, since it is more commonly used when the functions are defined by a sequence. But regardless of the notation used, the functions can still be simplified, evaluated, composed, and even graphed. Consider the function $f(x) = 3x^2 - 1$ and the sequence defined by $a_n = 3n^2 - 1$. Both can be evaluated and graphed, with the only difference being that $f(x)$ is continuous with domain $x \in \mathbb{R}$, while a_n is discrete (made up of distinct points) with domain $n \in \mathbb{N}$.

EXAMPLE 1 ▶ **Using Subscript Notation for a Composition**

For $f(x) = 3x^2 - 1$ and $a_n = 3n^2 - 1$, find $f(k + 1)$ and a_{k+1}.

Solution ▶

$$f(k + 1) = 3(k + 1)^2 - 1 \qquad\qquad a_{k+1} = 3(k + 1)^2 - 1$$
$$= 3(k^2 + 2k + 1) - 1 \qquad\qquad = 3(k^2 + 2k + 1) - 1$$
$$= 3k^2 + 6k + 2 \qquad\qquad = 3k^2 + 6k + 2$$

Now try Exercises 7 through 18 ▶

☑ **A.** You've just learned how to use subscript notation to evaluate and compose functions

No matter which notation is used, every occurrence of the input variable is replaced by the new value or expression indicated by the composition.

B. Mathematical Induction Applied to Sums

Consider the sum of odd numbers $1 + 3 + 5 + 7 + 9 + 11 + 13 + \cdots$. The sum of the first four terms is $1 + 3 + 5 + 7 = 16$, or $S_4 = 16$. If we now add a_5 (the next term in line), would we get the same answer as if we had simply computed S_5? Common sense would say, "Yes!" since $S_5 = 1 + 3 + 5 + 7 + 9 = 25$ and $S_4 + a_5 = 16 + 9 = 25\checkmark$. In diagram form, we have

add next term $a_5 = 9$ to S_4

$$1 + 3 + 5 + 7 + 9 + 11 + 13 + 15 + \cdots$$

S_4 ____ sum of 4 terms

S_5 _____ sum of 5 terms

Our goal is to develop this same degree of clarity in the *notational scheme* of things. For a given series, if we find the kth partial sum S_k (shown next) and then add the next term a_{k+1}, would we get the same answer if we had simply computed S_{k+1}? In other words, is $S_k + a_{k+1} = S_{k+1}$ true?

add next term a_{k+1}

$$a_1 \ + \ a_2 \ + \ a_3 \ + \ \cdots \ + \ a_{k-1} \ + \ a_k \ + \ a_{k+1} \ + \ \cdots \ + \ a_{n-1} \ + \ a_n$$

S_k —————— sum of k terms

S_{k+1} ————————— sum of $k+1$ terms

Now, let's return to the sum $1 + 3 + 5 + 7 + \cdots + 2n - 1$. This is an arithmetic series with $a_1 = 1$, $d = 2$, and nth term $a_n = 2n - 1$. Using the sum formula for an arithmetic sequence, an alternative formula for *this sum* can be established.

$$S_n = \frac{n(a_1 + a_n)}{2} \qquad \text{summation formula for an arithmetic sequence}$$

$$= \frac{n(1 + \mathbf{2n - 1})}{2} \qquad \text{substitute 1 for } a_1 \text{ and } 2n-1 \text{ for } a_n$$

$$= \frac{n(2n)}{2} \qquad \text{simplify}$$

$$= n^2 \qquad \text{result}$$

This shows that the sum of the first n positive odd integers is given by $S_n = n^2$. As a check we compute $S_5 = 1 + 3 + 5 + 7 + 9 = 25$ and compare to $S_5 = 5^2 = 25\checkmark$. We also note $S_6 = 6^2 = 36$, and $S_5 + a_6 = 25 + 11 = 36$, showing $S_6 = S_5 + a_6$. For more on this relationship, **see Exercises 19 through 24.**

While it may seem simplistic now, showing $S_5 + a_6 = S_6$ and $S_k + a_{k+1} = S_{k+1}$ (in general) is a critical component of a proof by induction. Unfortunately, general summation formulas for many sequences cannot be established from known formulas. In addition, just because a formula works for the first few values of n, we cannot assume that it will hold true for *all* values of n (there are infinitely many). As an illustration, the formula $a_n = n^2 - n + 41$ yields a prime number for *every natural number n from 1 to 40*, but fails to yield a prime for $n = 41$. This helps demonstrate the need for a more conclusive proof, particularly when a relationship appears to be true, and can be "verified" in a finite number of cases, but whether it is true in *all* cases remains in question.

Proof by induction is based on a relatively simple idea. To help understand how it works, consider n relay stations that are used to transport electricity from a generating plant to a distant city. If we know the generating plant is operating, and if we assume that the kth relay station (any station in the series) is making the transfer to the $(k + 1)$st station (the next station in the series), then we're sure the city will have electricity.

Generating plant kth $(k + 1)$st
                                               relay   relay

This idea can be applied mathematically as follows. Consider the statement, "The sum of the first n positive, even integers is $n^2 + n$." In other words, $2 + 4 + 6 + 8 + \cdots + 2n = n^2 + n$. We can certainly verify the statement for the first few even numbers:

The first even number is 2 and . . . $(1)^2 + 1 = 2$

The sum of the first *two* even numbers is $2 + 4 = 6$ and . . . $(2)^2 + 2 = 6$

The sum of the first *three* even numbers is
   $2 + 4 + 6 = 12$ and . . . $(3)^2 + 3 = 12$

The sum of the first *four* even numbers is
   $2 + 4 + 6 + 8 = 20$ and . . . $(4)^2 + 4 = 20$

While we could continue this process for a very long time (or even use a computer), *no finite number of checks can prove a statement is universally true.* To prove the statement true for *all* positive integers, we use a reasoning similar to that applied in the relay stations example. If we are sure the formula works for $n = 1$ (the generating station is operating), and assume that if the formula is true for $n = k$, it must also be true for $n = k + 1$ [the kth relay station is transferring electricity to the $(k + 1)$st station], then the statement is true for all n (the city will get its electricity). The case where $n = 1$ is called the **base case** of an inductive proof, and the assumption that the formula is true for $n = k$ is called the **induction hypothesis.** When the induction hypothesis is applied to a sum formula, we attempt to show that $S_k + a_{k+1} = S_{k+1}$. Since k and $k + 1$ are arbitrary, the statement must be true for all n.

Mathematical Induction Applied to Sums

Let S_n be a sum formula involving positive integers.

If   **1.** S_1 is true, and

2. the truth of S_k implies that S_{k+1} is true,

then S_n must be true for all positive integers n.

WORTHY OF NOTE

To satisfy our finite minds, it might help to show that S_n is true for the first few cases, prior to extending the ideas to the infinite case.

Both parts 1 and 2 must be verified for the proof to be complete. Since the process requires the terms S_k, a_{k+1}, and S_{k+1}, we will usually compute these first.

EXAMPLE 2 ▶ **Proving a Statement Using Mathematical Induction**

Use induction to prove that *the sum of the first n perfect squares is given by*

$$1 + 4 + 9 + 16 + 25 + \cdots + n^2 = \frac{n(n + 1)(2n + 1)}{6}.$$

Solution ▶ Given $a_n = n^2$ and $S_n = \dfrac{n(n + 1)(2n + 1)}{6}$, the needed components are . . .

For $a_n = n^2$: $a_k = k^2$ and $a_{k+1} = (k + 1)^2$

For $S_n = \dfrac{n(n + 1)(2n + 1)}{6}$: $S_k = \dfrac{k(k + 1)(2k + 1)}{6}$ and $S_{k+1} = \dfrac{(k + 1)(k + 2)(2k + 3)}{6}$

1. Show S_n is true for $n = 1$.

$$S_n = \frac{n(n + 1)(2n + 1)}{6} \qquad \text{sum formula}$$

$$S_1 = \frac{1(2)(3)}{6} \qquad \text{base case: } n = 1$$

$$= 1 \checkmark \qquad \text{result checks, the first term is 1}$$

2. Assume S_k is true,

$$1 + 4 + 9 + 16 + \cdots + k^2 = \frac{k(k + 1)(2k + 1)}{6} \qquad \text{induction hypothesis: } S_k \text{ is true}$$

and use it to show the truth of S_{k+1} follows. That is,

$$\underbrace{1 + 4 + 9 + 16 + \cdots + k^2}_{S_k} + \underbrace{(k + 1)^2}_{a_{k+1}} = \underbrace{\frac{(k + 1)(k + 2)(2k + 3)}{6}}_{S_{k+1}}$$

Working with the left-hand side, we have

$$\underbrace{1 + 4 + 9 + 16 + \cdots + k^2} + (k + 1)^2$$

$$= \frac{k(k + 1)(2k + 1)}{6} + (k + 1)^2 \qquad \text{use the induction hypothesis: substitute}$$
$$\frac{k(k+1)(2k+1)}{6} \text{ for } 1 + 4 + 9 + 16 + 25 + \cdots + k^2$$

$$= \frac{k(k + 1)(2k + 1) + 6(k + 1)^2}{6} \qquad \text{common denominator}$$

$$= \frac{(k + 1)[k(2k + 1) + 6(k + 1)]}{6} \qquad \text{factor out } k + 1$$

☑ **B.** You've just learned how to apply the principle of mathematical induction to sum formulas involving natural numbers

$$= \frac{(k + 1)[2k^2 + 7k + 6]}{6} \qquad \text{multiply and combine terms}$$

$$= \frac{(k + 1)(k + 2)(2k + 3)}{6} \qquad \text{factor the trinomial, result is } S_{k+1}\checkmark$$

Since the truth of S_{k+1} follows from S_k, the formula is true for all n.

Now try Exercises 27 through 38 ▶

C. The General Principle of Mathematical Induction

Proof by induction can be used to verify many other kinds of relationships involving a natural number n. In this regard, the basic principles remain the same but are stated more broadly. Rather than having S_n represent a sum, we take it to represent *any statement or relationship* we might wish to verify. This broadens the scope of the proof and makes it more widely applicable, while maintaining its value to the sum formulas verified earlier.

The General Principle of Mathematical Induction

Let S_n be a statement involving natural numbers.

If   **1.** S_1 is true, and
      2. the truth of S_k implies that S_{k+1} is also true

then S_n must be true for all natural numbers n.

EXAMPLE 3 ▶   **Proving a Statement Using the General Principle of Mathematical Induction**

Use the general principle of mathematical induction to show the statement S_n is true for all natural numbers n. S_n: $2^n \geq n + 1$

Solution ▶   The statement S_n is defined as $2^n \geq n + 1$. This means that S_k is represented by $2^k \geq k + 1$ and S_{k+1} by $2^{k+1} \geq k + 2$.

1. Show S_n is true for $n = 1$:

$$S_n: \quad 2^n \geq n + 1 \qquad \text{given statement}$$
$$S_1: \quad 2^1 \geq 1 + 1 \qquad \text{base case: } n = 1$$
$$2 \geq 2\checkmark \qquad \text{true}$$

Although not a part of the formal proof, a table of values can help to illustrate the relationship we're trying to establish. It *appears* that the statement is true.

n	1	2	3	4	5
2^n	2	4	8	16	32
$n+1$	2	3	4	5	6

2. Assume that S_k is true,

$$S_k: \quad 2^k \geq k + 1 \quad \text{induction hypothesis}$$

and use it to show that the truth of S_{k+1}. That is,

$$S_{k+1}: \quad 2^{k+1} \geq k + 2.$$

Begin by working with the left-hand side of the inequality, 2^{k+1}.

$$
\begin{aligned}
2^{k+1} &= 2(2^k) && \text{properties of exponents} \\
&\geq 2(\boldsymbol{k+1}) && \textit{induction hypothesis:} \text{ substitute } k+1 \text{ for } 2^k \\
& && \text{(symbol changes since } k+1 \text{ is less than } 2^k) \\
&\geq 2k + 2 && \text{distribute}
\end{aligned}
$$

Since k is a positive integer, $2k + 2 \geq k + 2$,

showing $2^{k+1} \geq k + 2$.

Since the truth of S_{k+1} follows from S_k, the formula is true for all n.

> **WORTHY OF NOTE**
>
> Note there is no reference to a_n, a_k, or a_{k+1} in the statement of the general principle of mathematical induction.

Now try Exercises 39 through 42 ▶

EXAMPLE 4 ▶   **Proving Divisibility Using Mathematical Induction**

Let S_n be the statement, "$4^n - 1$ *is divisible by 3 for all positive integers n.*" Use mathematical induction to prove that S_n is true.

Solution ▶   If a number is evenly divisible by three, it can be written as the product of 3 and some positive integer we will call p.

1. Show S_n is true for $n = 1$:

$$
\begin{aligned}
S_n: \quad & 4^n - 1 = 3p && 4^n - 1 = 3p, \; p \in \mathbb{Z} \\
S_1: \quad & 4^{(1)} - 1 = 3p && \text{substitute 1 for } n \\
& \quad\quad\;\; 3 = 3p\checkmark && \text{statement is true for } n = 1
\end{aligned}
$$

2. Assume that S_k is true ...

$$
\begin{aligned}
S_k: \quad & 4^k - 1 = 3p && \text{induction hypothesis} \\
& 4^k = 3p + 1
\end{aligned}
$$

and use it to show the truth of S_{k+1}. That is,

$$S_{k+1}: \quad 4^{k+1} - 1 = 3q \text{ for } q \in \mathbb{Z} \text{ is also true.}$$

Beginning with the left-hand side we have:

$$
\begin{aligned}
4^{k+1} - 1 &= 4 \cdot 4^k - 1 && \text{properties of exponents} \\
&= 4 \cdot (\boldsymbol{3p + 1}) - 1 && \textit{induction hypothesis:} \text{ substitute } 3p + 1 \text{ for } 4^k \\
&= 12p + 3 && \text{distribute and simplify} \\
&= 3(4p + 1) = 3q && \text{factor}
\end{aligned}
$$

The last step shows $4^{k+1} - 1$ is divisible by 3. Since the original statement is true for $n = 1$, and the truth of S_k implies the truth of S_{k+1}, the statement, "$4^n - 1$ is divisible by 3" is true for all positive integers n.

Now try Exercises 43 through 47 ▶

We close this section with some final notes. Although the base step of a proof by induction seems trivial, both the base step and the induction hypothesis are necessary parts of the proof. For example, the statement $\dfrac{1}{3^n} < \dfrac{1}{3n}$ is false for $n = 1$, but true for all other positive integers. Finally, for a fixed natural number p, some statements are false for all $n < p$, but true for all $n \geq p$. By modifying the base case to begin at p, we can use the induction hypothesis to prove the statement is true for all n greater than p. For example, $n < \frac{1}{3} n^2$ is false for $n < 4$, but true for all $n \geq 4$.

☑ **C.** You've just learned how to apply the principle of mathematical induction to general statements involving natural numbers

---

8.4 EXERCISES

▶ **CONCEPTS AND VOCABULARY**

Fill in the blank with the appropriate word or phrase. Carefully reread the section if needed.

1. No _____ number of verifications can prove a statement _____ true.

2. Showing a statement is true for $n = 1$ is called the _____ _____ of an inductive proof.

3. Assuming that a statement/formula is true for $n = k$ is called the _____ _____.

4. The graph of a sequence is _____, meaning it is made up of distinct points.

5. Explain the equation $S_k + a_{k+1} = S_{k+1}$. Begin by saying, "Since the kth term is arbitrary . . ." (continue from here).

6. Discuss the similarities and differences between mathematical induction applied to sums and the general principle of mathematical induction.

▶ **DEVELOPING YOUR SKILLS**

For the given nth term a_n, find a_4, a_5, a_k, and a_{k+1}.

7. $a_n = 10n - 6$ 8. $a_n = 6n - 4$

9. $a_n = n$ 10. $a_n = 7n$

11. $a_n = 2^{n-1}$ 12. $a_n = 2(3^{n-1})$

For the given sum formula S_n, find S_4, S_5, S_k, and S_{k+1}.

13. $S_n = n(5n - 1)$ 14. $S_n = n(3n - 1)$

15. $S_n = \dfrac{n(n + 1)}{2}$ 16. $S_n = \dfrac{7n(n + 1)}{2}$

17. $S_n = 2^n - 1$ 18. $S_n = 3^n - 1$

Verify that $S_4 + a_5 = S_5$ for each exercise. These are identical to Exercises 13 through 18.

19. $a_n = 10n - 6$; $S_n = n(5n - 1)$

20. $a_n = 6n - 4$; $S_n = n(3n - 1)$

21. $a_n = n$; $S_n = \dfrac{n(n + 1)}{2}$

22. $a_n = 7n$; $S_n = \dfrac{7n(n + 1)}{2}$

23. $a_n = 2^{n-1}$; $S_n = 2^n - 1$

24. $a_n = 2(3^{n-1})$; $S_n = 3^n - 1$

▶ **WORKING WITH FORMULAS**

25. Sum of the first n cubes (alternative form): $(1 + 2 + 3 + 4 + \cdots + n)^2$

Earlier we noted the formula for the sum of the first n cubes was $\dfrac{n^2(n + 1)^2}{4}$. An alternative is given by the formula shown.

   a. Verify the formula for $n = 1, 5$, and 9.

   b. Verify the formula using

$$1 + 2 + 3 + \cdots + n = \frac{n(n + 1)}{2}.$$

26. Powers of the imaginary unit: $i^{n+4} = i^n$, where $i = \sqrt{-1}$

Use a proof by induction to prove that powers of the imaginary unit are cyclic. That is, that they cycle through the numbers i, -1, $-i$, and 1 for consecutive powers.

▶ **APPLICATIONS**

Use mathematical induction to prove the indicated sum formula is true for all natural numbers n.

27. $2 + 4 + 6 + 8 + 10 + \cdots + 2n$;
$a_n = 2n, S_n = n(n + 1)$

28. $3 + 7 + 11 + 15 + 19 + \cdots + (4n - 1)$;
$a_n = 4n - 1, S_n = n(2n + 1)$

29. $5 + 10 + 15 + 20 + 25 + \cdots + 5n$;
$a_n = 5n, S_n = \dfrac{5n(n + 1)}{2}$

30. $1 + 4 + 7 + 10 + 13 + \cdots + (3n - 2)$;
$a_n = 3n - 2, S_n = \dfrac{n(3n - 1)}{2}$

31. $5 + 9 + 13 + 17 + \cdots + (4n + 1)$;
$a_n = 4n + 1, S_n = n(2n + 3)$

32. $4 + 12 + 20 + 28 + 36 + \cdots + (8n - 4)$;
$a_n = 8n - 4, S_n = 4n^2$

33. $3 + 9 + 27 + 81 + 243 + \cdots + 3^n$;
$a_n = 3^n, S_n = \dfrac{3(3^n - 1)}{2}$

34. $5 + 25 + 125 + 625 + \cdots + 5^n$;
$a_n = 5^n, S_n = \dfrac{5(5^n - 1)}{4}$

35. $2 + 4 + 8 + 16 + 32 + 64 + \cdots + 2^n$;
$a_n = 2^n, S_n = 2^{n+1} - 2$

36. $1 + 8 + 27 + 64 + 125 + 216 + \cdots + n^3$;
$a_n = n^3, S_n = \dfrac{n^2(n + 1)^2}{4}$

37. $\dfrac{1}{1(3)} + \dfrac{1}{3(5)} + \dfrac{1}{5(7)} + \cdots + \dfrac{1}{(2n - 1)(2n + 1)}$;
$a_n = \dfrac{1}{(2n - 1)(2n + 1)}, S_n = \dfrac{n}{2n + 1}$

38. $\dfrac{1}{1(2)} + \dfrac{1}{2(3)} + \dfrac{1}{3(4)} + \cdots + \dfrac{1}{n(n + 1)}$;
$a_n = \dfrac{1}{n(n + 1)}, S_n = \dfrac{n}{n + 1}$

Use the principle of mathematical induction to prove that each statement is true for all natural numbers n.

39. $3^n \geq 2n + 1$ **40.** $2^n \geq n + 1$

41. $3 \cdot 4^{n-1} \leq 4^n - 1$ **42.** $4 \cdot 5^{n-1} \leq 5^n - 1$

43. $n^2 - 7n$ is divisible by 2

44. $n^3 - n + 3$ is divisible by 3

45. $n^3 + 3n^2 + 2n$ is divisible by 3

46. $5^n - 1$ is divisible by 4

47. $6^n - 1$ is divisible by 5

▶ **EXTENDING THE THOUGHT**

 48. You may have noticed that the sum formula for the first n integers was *quadratic,* and the formula for the first n integer squares was *cubic.* Is the formula for the first n integer cubes, if it exists, a quartic (degree four) function? Use your calculator to run a quartic regression on the first five perfect cubes (enter 1

through 5 in L_1 and the cumulative sums in L_2). What did you find? Use proof by induction to show that the sum of the first n cubes is:

$$1 + 8 + 27 + \cdots + n^3 = \frac{n^4 + 2n^3 + n^2}{4} = \frac{n^2(n + 1)^2}{4}.$$

49. Use mathematical induction to prove that
$$\frac{x^n - 1}{x - 1} = (1 + x + x^2 + x^3 + \cdots + x^{n-1}).$$

50. Use mathematical induction to prove that for $1^4 + 2^4 + 3^4 + \cdots + n^4$, where $a_n = n^4$,
$$S_n = \frac{n(n + 1)(2n + 1)(3n^2 + 3n - 1)}{30}.$$

▶ MAINTAINING YOUR SKILLS

51. (6.2) Given the matrices $A = \begin{bmatrix} -1 & 2 \\ 3 & 1 \end{bmatrix}$ and

$B = \begin{bmatrix} 2 & -1 \\ 4 & 3 \end{bmatrix}$, find $A + B, A - B, 2A - 3B$,

AB, BA, and B^{-1}.

52. (2.7) State the domain and range of the piecewise function shown here.

53. (2.1) State the equation of the circle whose graph is shown here.

54. (4.4) Solve: $3e^{(2x-1)} + 5 = 17$. Answer in exact form.

MID-CHAPTER CHECK

In Exercises 1 to 3, the nth term is given. Write the first three terms of each sequence and find a_9.

1. $a_n = 7n - 4$ **2.** $a_n = n^2 + 3$

3. $a_n = (-1)^n(2n - 1)$

4. Evaluate the sum $\sum_{n=1}^{4} 3^{n+1}$

5. Rewrite using sigma notation.
$1 + 4 + 7 + 10 + 13 + 16$

Match each formula to its correct description.

6. $S_n = \dfrac{n(a_1 + a_n)}{2}$

7. $a_n = a_1 r^{n-1}$

8. $S_\infty = \dfrac{a_1}{1 - r}$

9. $a_n = a_1 + (n - 1)d$

10. $S_n = \dfrac{a_1(1 - r^n)}{1 - r}$

    a. sum of an infinite geometric series

    b. nth term formula for an arithmetic series

    c. sum of a finite geometric series

    d. summation formula for an arithmetic series

    e. nth term formula for a geometric series

11. Identify a_1 and the common difference d. Then find an expression for the general term a_n.

    a. 2, 5, 8, 11, ...   **b.** $\frac{3}{2}, \frac{9}{4}, 3, \frac{15}{4}, \ldots$

Find the number of terms in each series and then find the sum.

12. $2 + 5 + 8 + 11 + \cdots + 74$

13. $\frac{1}{2} + \frac{3}{2} + \frac{5}{2} + \frac{7}{2} + \cdots + \frac{31}{2}$

14. For an arithmetic series, $a_3 = -8$ and $a_7 = 4$. Find S_{10}.

15. For a geometric series, $a_3 = -81$ and $a_7 = -1$. Find S_{10}.

16. Identify a_1 and the common ratio r. Then find an expression for the general term a_n.

    a. 2, 6, 18, 54, ...   **b.** $\frac{1}{2}, \frac{1}{4}, \frac{1}{8}, \frac{1}{16}, \ldots$

17. Find the number of terms in the series then compute the sum.

    $\frac{1}{54} + \frac{1}{18} + \frac{1}{6} + \cdots + \frac{81}{2}$

18. Find the infinite sum (if it exists).

$$-49 + (-7) + (-1) + (-\tfrac{1}{7}) + \cdots$$

19. Barrels of toxic waste are stacked at a storage facility in pyramid form, with 60 barrels in the first row, 59 in the second row, and so on, until there are 10 barrels in the top row. How many barrels are in the storage facility?

20. As part of a conditioning regimen, a drill sergeant orders her platoon to do 25 continuous standing broad jumps. The best of these recruits was able to jump 96% of the distance from the previous jump, with a first jump distance of 8 ft. Use a sequence/series to determine the distance the recruit jumped on the 15th try, and the total distance traveled by the recruit after all 25 jumps.

REINFORCING BASIC CONCEPTS

Applications of Summation

The properties of summation play a large role in the development of key ideas in a first semester calculus course, and the following summation formulas are an integral part of these ideas. The first three formulas were verified in Section 8.4, while proof of the fourth was part of Exercise 48 on page 714.

(1) $\displaystyle\sum_{i=1}^{n} c = cn$

(2) $\displaystyle\sum_{i=1}^{n} i = \frac{n(n+1)}{2}$

(3) $\displaystyle\sum_{i=1}^{n} i^2 = \frac{n(n+1)(2n+1)}{6}$

(4) $\displaystyle\sum_{i=1}^{n} i^3 = \frac{n^2(n+1)^2}{4}$

To see the various ways they can be applied consider the following.

Illustration 1 ▶ Over several years, the owner of Morgan's LawnCare has noticed that the company's monthly profits (in thousands) can be approximated by the sequence $a_n = 0.0625n^3 - 1.25n^2 + 6n$, with the points plotted in Figure 8.9 (the continuous graph is shown for effect only). Find the company's approximate annual profit.

Figure 8.9

```
              12
    P1:L1,L2
0  |_____| 12
   X=2              Y=7.5
              0
```

Solution ▶ The most obvious approach would be to simply compute terms a_1 through a_{12} (January through December) and find their sum: **sum(seq(Y1, X, 1, 12)** (see Section 8.1 Technology Highlight), which gives a result of 35.75 or $35,750.

As an alternative, we could add the amount of profit earned by the company in the first 8 months, then add the amount the company lost (or broke even) during the last 4 months. In other words, we could apply summation property

IV: $\displaystyle\sum_{i=1}^{12} a_n = \sum_{i=1}^{8} a_n + \sum_{i=9}^{12} a_n$ (see Figure 8.10), which gives the same result: $42 + (-6.25) = 35.75$ or $35,750.

Figure 8.10

```
sum(seq(Y1,X,1,8
)
                42
sum(seq(Y1,X,9,1
2)
            -6.25
```

As a third option, we could use summation properties along with the appropriate summation formulas, and compute the result manually. Note the function is now written in terms of "i." *Distribute summations and factor out constants* (properties II and III):

$$\sum_{i=1}^{12} (0.0625i^3 - 1.25i^2 + 6i) =$$

$$0.0625\sum_{i=1}^{12} i^3 - 1.25\sum_{i=1}^{12} i^2 + 6\sum_{i=1}^{12} i$$

Replace each summation with the appropriate summation formula, substituting 12 for n:

$$= 0.0625\left[\frac{n^2(n+1)^2}{4}\right] - 1.25\left[\frac{n(n+1)(2n+1)}{6}\right]$$
$$+ 6\left[\frac{n(n+1)}{2}\right]$$

$$= 0.0625\left[\frac{(12)^2(13)^2}{4}\right] - 1.25\left[\frac{(12)(13)(25)}{6}\right]$$
$$+ 6\left[\frac{(12)(13)}{2}\right]$$

$$= 0.0625(6084) - 1.25(650) + 6(78) \text{ or } 35.75$$

As we expected, the result shows profit was $35,750. While some approaches seem "easier" than others, all have great value, are applied in different ways at different times, and are necessary to adequately develop key concepts in future classes.

Exercise 1: Repeat Illustration 1 if the profit sequence is $a_n = 0.125x^3 - 2.5x^2 + 12x$.

8.5 | Counting Techniques

Learning Objectives

In Section 8.5 you will learn how to:

☐ **A.** Count possibilities using lists and tree diagrams

☐ **B.** Count possibilities using the fundamental principle of counting

☐ **C.** Quick-count distinguishable permutations

☐ **D.** Quick-count nondistinguishable permutations

☐ **E.** Quick-count using combinations

How long would it take to estimate the number of fans sitting shoulder-to-shoulder at a sold-out basketball game? Well, it depends. You could actually begin counting 1, 2, 3, 4, 5, . . . , which would take a very long time, or you could try to simplify the process by counting the number of fans in the first row and multiplying by the number of rows. Techniques for "quick-counting" the objects in a set or various subsets of a large set play an important role in a study of probability.

A. Counting by Listing and Tree Diagrams

Consider the simple spinner shown in Figure 8.11, which is divided into three equal parts. What are the different possible outcomes for two spins, spin 1 followed by spin 2? We might begin by organizing the possibilities using a **tree diagram.** As the name implies, each choice or possibility appears as the branch of a tree, with the total possibilities being equal to the number of (unique) paths from the beginning point to the end of a branch. Figure 8.12 shows how the spinner exercise would appear (possibilities for two spins). Moving from top to bottom we can trace nine possible paths: *AA, AB, AC, BA, BB, BC, CA, CB,* and *CC.*

Figure 8.11

Figure 8.12

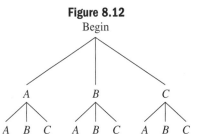

EXAMPLE 1 ▶ **Listing Possibilities Using a Tree Diagram**

A basketball player is fouled and awarded three free throws. Let H represent the possibility of a hit (basket is made), and M the possibility of a miss. Determine the possible outcomes for the three shots using a tree diagram.

Solution ▶ Each shot has two possibilities, hit (H) or miss (M), so the tree will branch in two directions at each level. As illustrated in the figure, there are a total of eight possibilities: HHH, HHM, HMH, HMM, MHH, MHM, MMH, and MMM.

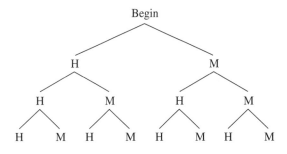

Now try Exercises 7 through 10 ▶

To assist our discussion, an **experiment** is any task that can be done repeatedly and has a well-defined set of possible outcomes. Each repetition of the experiment is called a **trial.** A **sample outcome** is any potential outcome of a trial, and a **sample space** is a set of all possible outcomes.

In our first illustration, the *experiment* was spinning a spinner, there were *three sample outcomes* (*A*, *B*, or *C*), the experiment had *two trials* (spin 1 and spin 2), and there were *nine* elements in the *sample space*. Note that after the first trial, each of the three sample outcomes will again have three possibilities (A, B, and C). For two trials we have $3^2 = 9$ possibilities, while three trials would yield a sample space with $3^3 = 27$ possibilities. In general, we have

---

A "Quick-Counting" Formula for a Sample Space

If an experiment has N sample outcomes that are equally likely and the experiment is repeated t times, the number of elements in the sample space is N^t.

EXAMPLE 2 ▶ **Counting the Outcomes in a Sample Space**

Many combination locks have the digits 0 through 39 arranged along a circular dial. Opening the lock requires stopping at a sequence of three numbers within this range, going counterclockwise to the first number, clockwise to the second, and counterclockwise to the third. How many three-number combinations are possible?

Solution ▶ There are 40 sample outcomes ($N = 40$) in this experiment, and three trials ($t = 3$). The number of possible combinations is identical to the number of elements in the sample space. The quick-counting formula gives $40^3 = 64,000$ possible combinations.

☑ **A.** You've just learned how to count possibilities using lists and tree diagrams

Now try Exercises 11 and 12 ▶

B. Fundamental Principle of Counting

The number of possible outcomes may differ depending on how the event is defined. For example, some security systems, license plates, and telephone numbers exclude certain numbers. For example, phone numbers cannot begin with 0 or 1 because these are reserved for operator assistance, long distance, and international calls. Constructing a three-digit area code is like filling in three blanks $\underset{\text{digit}}{\underline{\quad}}\ \underset{\text{digit}}{\underline{\quad}}\ \underset{\text{digit}}{\underline{\quad}}$ with three digits. Since the area code must start with a number between 2 and 9, there are eight choices for the first blank. Since there are 10 choices for the second digit and 10 choices for the third, there are $8 \cdot 10 \cdot 10 = 800$ possibilities in the sample space.

EXAMPLE 3 ▶ **Counting Possibilities for a Four-Digit Security Code**

A digital security system requires that you enter a four-digit PIN (personal identification number), using only the digits 1 through 9. How many codes are possible if

a. Repetition of digits is allowed?
b. Repetition is not allowed?
c. The first digit must be even and repetitions are not allowed?

Solution ▶ **a.** Consider filling in the four blanks $\underset{\text{digit}}{\underline{\quad}}\ \underset{\text{digit}}{\underline{\quad}}\ \underset{\text{digit}}{\underline{\quad}}\ \underset{\text{digit}}{\underline{\quad}}$ with the number of ways the digit can be chosen. If repetition is allowed, the experiment is similar to that of Example 2 and there are $N^t = 9^4 = 6561$ possible PINs.

b. If repetition is not allowed, there are only eight possible choices for the second digit of the PIN, then seven for the third, and six for the fourth. The number of possible PIN numbers decreases to $9 \cdot 8 \cdot 7 \cdot 6 = 3024$.

c. There are four choices for the first digit (2, 4, 6, 8). Once this choice has been made there are eight choices for the second digit, seven for the third, and six for the last: $4 \cdot 8 \cdot 7 \cdot 6 = 1344$ possible codes.

Now try Exercises 13 through 20 ▶

Given *any* experiment involving a sequence of tasks, if the first task can be completed in p possible ways, the second task has q possibilities, and the third task has r possibilities, a tree diagram will show that the number of possibilities in the sample space for task$_1$–task$_2$–task$_3$ is $p \cdot q \cdot r$. Even though the examples we've considered to this point have varied a great deal, this idea was fundamental to counting all possibilities in a sample space and is, in fact, known as the **fundamental principle of counting** (FPC).

Fundamental Principle of Counting (Applied to Three Tasks)

Given any experiment with three defined tasks, if there are p possibilities for the first task, q possibilities for the second, and r possibilities for the third, the total number of ways the experiment can be completed is $p \cdot q \cdot r$.

This fundamental principle can be extended to include any number of tasks.

EXAMPLE 4 ▶ **Counting Possibilities for Seating Arrangements**

Adrienne, Bob, Carol, Dax, Earlene, and Fabian bought tickets to see *The Marriage of Figaro*. Assuming they sat together in a row of six seats, how many different seating arrangements are possible if

a. Bob and Carol are sweethearts and must sit together?

b. Bob and Carol are enemies and must not sit together?

Solution ▶

a. Since a restriction has been placed on the seating arrangement, it will help to divide the experiment into a sequence of tasks: *task 1:* they sit together; *task 2:* either Bob is on the left or Bob is on the right; and *task 3:* the other four are seated. Bob and Carol can sit together in five different ways, as shown in Figure 8.13, so there are five possibilities for task 1. There are two ways they can be side-by-side: Bob on the left and Carol on the right, as shown, or Carol on the left and Bob on the right. The remaining four people can be seated randomly, so task 3 has 4! = 24 possibilities. Under these conditions they can be seated $5 \cdot 2 \cdot 4! = 240$ ways.

b. This is similar to Part (a), but now we have to count the number of ways they can be separated by *at least one seat: task 1:* Bob and Carol are in nonadjacent seats; *task 2:* either Bob is on the left or Bob is on the right; and *task 3:* the other four are seated. For task 1, be careful to note there is no multiplication involved, just a simple counting. If Bob sits in seat 1, there are four nonadjacent seats. If Bob sits in seat 2, there are three nonadjacent seats, and so on. This gives $4 + 3 + 2 + 1 = 10$ possibilities for Bob and Carol not sitting together. Task 2 and task 3 have the same number of possibilities as in Part (a), giving $10 \cdot 2 \cdot 4! = 480$ possible seating arrangements.

Figure 8.13

Bob	Carol				
1	2	3	4	5	6

	Bob	Carol			
1	2	3	4	5	6

		Bob	Carol		
1	2	3	4	5	6

			Bob	Carol	
1	2	3	4	5	6

				Bob	Carol
1	2	3	4	5	6

Now try Exercises 21 through 28 ▶

☑ **B.** You've just learned how to count possibilities using the fundamental principle of counting

C. Distinguishable Permutations

In the game of Scrabble® (Milton Bradley), players attempt to form words by rearranging letters. Suppose a player has the letters P, S, T, and O at the end of the game. These letters could be rearranged or *permuted* to form the words POTS, SPOT, TOPS, OPTS, POST, or STOP. These arrangements are called permutations of the four letters. A permutation is any new arrangement, listing, or sequence of objects obtained by changing an existing order. A **distinguishable permutation** is a permutation that produces a result different from the original. For example, a distinguishable permutation of the digits in the number 1989 is 8199.

Example 4 considered six people, six seats, and the various ways they could be seated. But what if there were fewer seats than people? By the FPC, with six people and four seats there could be $6 \cdot 5 \cdot 4 \cdot 3 = 360$ different arrangements, with six people and three seats there are $6 \cdot 5 \cdot 4 = 120$ different arrangements, and so on. These rearrangements are called distinguishable permutations. You may have noticed that for six people and six seats, we used all six factors of 6!, while for six people and four seats we used the first four, six people and three seats required only the first three, and so on. Generally, for n people and r seats, the first r factors of $n!$ will be used. The notation and formula for *distinguishable permutations of n objects taken r at a time* is $_nP_r = \dfrac{n!}{(n-r)!}$. By defining $0! = 1$, the formula includes the case where all n objects are selected, which of course results in $_nP_n = \dfrac{n!}{(n-n)!} = \dfrac{n!}{0!} = \dfrac{n!}{1} = n!$.

Distinguishable Permutations: Unique Elements

If r objects are selected from a set containing n unique elements ($r \leq n$) and placed in an ordered arrangement, the number of distinguishable permutations is

$$_nP_r = \frac{n!}{(n-r)!} \quad \text{or} \quad _nP_r = n(n-1)(n-2)\cdots(n-r+1)$$

EXAMPLE 5 ▶ **Computing a Permutation**

Compute each value of $_nP_r$ using the methods just described.

   a. $_7P_4$ **b.** $_{10}P_3$

Solution ▶ Begin by evaluating each expression using the formula $_nP_r = \dfrac{n!}{(n-r)!}$, noting the third line (in bold) gives the first r factors of $n!$.

   a. $_7P_4 = \dfrac{7!}{(7-4)!}$ **b.** $_{10}P_3 = \dfrac{10!}{(10-3)!}$

$\qquad\qquad = \dfrac{7 \cdot 6 \cdot 5 \cdot 4 \cdot \cancel{3!}}{\cancel{3!}}$ $\qquad\qquad\qquad = \dfrac{10 \cdot 9 \cdot 8 \cdot \cancel{7!}}{\cancel{7!}}$

$\qquad\qquad = \mathbf{7 \cdot 6 \cdot 5 \cdot 4}$ $\qquad\qquad\qquad\quad = \mathbf{10 \cdot 9 \cdot 8}$

$\qquad\qquad = 840$ $\qquad\qquad\qquad\qquad\qquad = 720$

Now try Exercises 29 through 36 ▶

EXAMPLE 6 ▶ **Counting the Possibilities for Finishing a Race**

As part of a sorority's initiation process, the nine new inductees must participate in a 1-mi race. Assuming there are no ties, how many first- through fifth-place finishes are possible if it is well known that Mediocre Mary will finish fifth and Lightning Louise will finish first?

Solution ▶ To help understand the situation, we can diagram the possibilities for finishing first through fifth. Since Louise will finish first, this slot can be filled in only one way, by Louise herself. The same goes for Mary and her fifth-place finish: $\dfrac{\text{Louise}}{\text{1st}}$

$\dfrac{\quad\quad}{\text{2nd}} \dfrac{\quad\quad}{\text{3rd}} \dfrac{\quad\quad}{\text{4th}} \dfrac{\text{Mary}}{\text{5th}}$. The remaining three slots can be filled in $_7P_3 = 7 \cdot 6 \cdot 5$ different ways, indicating that under these conditions, there are $1 \cdot 7 \cdot 6 \cdot 5 \cdot 1 = 210$ different ways to finish.

☑ **C.** You've just learned how to quick-count distinguishable permutations

Now try Exercises 37 through 42 ▶

D. Nondistinguishable Permutations

As the name implies, certain permutations are nondistinguishable, meaning you cannot tell one apart from another. Such is the case when the original set contains elements or sample outcomes that are identical. Consider a family with four children, Lyddell, Morgan, Michael, and Mitchell, who are at the photo studio for a family picture. Michael and Mitchell are identical twins and cannot be told apart. In how many ways can they be lined up for the picture? Since this is an ordered arrangement of four children taken from a group of four, there are $_4P_4 = 24$ ways to line them up. A few of them are

Lyddell	Morgan	Michael	Mitchell	Lyddell	Morgan	Mitchell	Michael
Lyddell	Michael	Morgan	Mitchell	Lyddell	Mitchell	Morgan	Michael
Michael	Lyddell	Morgan	Mitchell	Mitchell	Lyddell	Morgan	Michael

But of these six arrangements, half will appear to be the same picture, since the difference between Michael and Mitchell cannot be distinguished. In fact, of the 24 total permutations, every picture where Michael and Mitchell have switched places will be nondistinguishable. To find the distinguishable permutations, we need to take the total permutations ($_4P_4$) *and divide by* 2!, *the number of ways the twins can be permuted:* $\dfrac{_4P_4}{(2)!} = \dfrac{24}{2} = 12$ distinguishable pictures.

These ideas can be generalized and stated in the following way.

Nondistinguishable Permutations: Nonunique Elements

In a set containing n elements where one element is repeated p times, another is repeated q times, and another is repeated r times ($p + q + r = n$), the number of nondistinguishable permutations is

$$\frac{_nP_n}{p!q!r!} = \frac{n!}{p!q!r!}$$

The idea can be extended to include any number of repeated elements.

EXAMPLE 7 ▶ Counting Distinguishable Permutations

A Scrabble player has the seven letters S, A, O, O, T, T, and T in his rack. How many distinguishable arrangements can be formed as he attempts to play a word?

Solution ▶ Essentially the exercise asks for the number of distinguishable permutations of the seven letters, given T is repeated three times and O is repeated twice. There are $\dfrac{_7P_7}{3!2!} = 420$ distinguishable permutations.

☑ **D.** You've just learned how to quick-count nondistinguishable permutations

Now try Exercises 43 through 54 ▶

E. Combinations

Similar to nondistinguishable permutations, there are other times the total number of permutations must be reduced to quick-count the elements of a desired subset. Consider a vending machine that offers a variety of 40¢ candies. If you have a quarter (Q), dime (D), and nickel (N), the machine wouldn't care about the order the coins were deposited. Even though QDN, QND, DQN, DNQ, NQD, and NDQ give the $_3P_3 = 6$ possible permutations, the machine considers them as equal and will vend your snack. Using sets, this is similar to saying the set $A = \{X, Y, Z\}$ has only one subset with three elements, since $\{X, Z, Y\}$, $\{Y, X, Z\}$, $\{Y, Z, X\}$, and so on, all represent the same set. Similarly, there are six, two-letter permutations of X, Y, and Z ($_3P_2 = 6$): XY, XZ, YX,

YZ, *ZX*, and *ZY*, but only three two-letter subsets: {*X*, *Y*}, {*X*, *Z*} and {*Y*, *Z*}. When permutations having the same elements are considered identical, the result is the number of possible **combinations** and is denoted $_nC_r$. Since the *r* objects can be selected in *r*! ways, we divide $_nP_r$ by *r*! to "quick-count" the number of possibilities: $_nC_r = \dfrac{_nP_r}{r!}$, which can be thought of as *the first r factors of n!, divided by r!*. By substituting $\dfrac{n!}{(n-r)!}$ for $_nP_r$ in this formula, we find an alternative method for computing $_nC_r$ is $\dfrac{n!}{r!(n-r)!}$. Take special note that when *r* objects are selected from a set with *n* elements and the order they're listed is unimportant (because you end up with the same subset), the result is a *combination*, not a permutation.

Combinations

The number of combinations of *n* objects taken *r* at a time is given by

$$_nC_r = \frac{_nP_r}{r!} \qquad \text{or} \qquad _nC_r = \frac{n!}{r!(n-r)!}$$

EXAMPLE 8 ▶ **Computing Combinations Using a Formula**

Compute each value of $_nC_r$ given.

a. $_7C_4$ **b.** $_8C_3$ **c.** $_5C_2$

Solution ▶ **a.** $_7C_4 = \dfrac{7\cdot6\cdot5\cdot4}{4!}$ **b.** $_8C_3 = \dfrac{8\cdot7\cdot6}{3!}$ **c.** $_5C_2 = \dfrac{5\cdot4}{2!}$

= 35 = 56 = 10

Now try Exercises 55 through 64 ▶

EXAMPLE 9 ▶ **Applications of Combinations-Lottery Results**

A small city is getting ready to draw five Ping-Pong balls of the nine they have numbered 1 through 9 to determine the winner(s) for its annual raffle. If a ticket holder has the same five numbers, they win. In how many ways can the winning numbers be drawn?

Solution ▶ Since the winning numbers can be drawn in any order, we have a combination of 9 things taken 5 at a time. The five numbers can be drawn in $_9C_5 = \dfrac{9\cdot8\cdot7\cdot6\cdot5}{5!} = 126$ ways.

Now try Exercises 65 and 66 ▶

Somewhat surprisingly, there are many situations where the order things are listed is not important. Such situations include

• The formation of committees, since the order people volunteer is unimportant
• Card games with a standard deck, since the order cards are dealt is unimportant
• Playing BINGO, since the order the numbers are called is unimportant

When the order in which people or objects are selected from a group is unimportant, the number of possibilities is a *combination,* not a permutation.

Another way to tell the difference between permutations and combinations is the following memory device: *Permutations have Priority or Precedence; in other* words, the *Position* of each element matters. By contrast, a *Combination* is like a *Committee* of *Colleagues* or *Collection* of *Commoners*; all members have equal rank. For permutations, *a-b-c* is different from *b-a-c*. For combinations, *a-b-c* is the same as *b-a-c*.

EXAMPLE 10 ▶ **Applications of Quick-Counting—Committees and Government**

The Sociology Department of Lakeside Community College has 12 dedicated faculty members. (a) In how many ways can a three-member textbook selection committee be formed? (b) If the department is in need of a Department Chair, Curriculum Chair, and Technology Chair, in how many ways can the positions be filled?

Solution ▶ **a.** Since textbook selection depends on a *Committee* of *Colleagues*, the order members are chosen is not important. This is a *Combination* of 12 people taken 3 at a time, and there are $_{12}C_3 = 220$ ways the committee can be formed.

☑ **E.** You've just learned how to quick-count using combinations

b. Since those selected will have *Position* or *Priority*, this is a *Permutation* of 12 people taken 3 at a time, giving $_{12}P_3 = 1320$ ways the positions can be filled.

> **Now try Exercises 67 through 78** ▶

The Exercise Set contains a wide variety of additional applications. See **Exercises 81 through 107.**

TECHNOLOGY HIGHLIGHT

Calculating Permutations and Combinations

Both the $_nP_r$ and $_nC_r$ functions are accessed using the [MATH] key and the **PRB** submenu (see Figure 8.14). To compute the permutations of 12 objects taken 9 at a time ($_{12}P_9$), clear the home screen and enter a 12, then press [MATH] [◀] 2:$_nP_r$ to access the $_nP_r$ operation, which is automatically pasted on the home screen after the 12. Now enter a 9, press [ENTER] and a result of 79833600 is displayed (Figure 8.15). Repeat the sequence to compute the value of $_{12}C_9$ ([MATH] [◀] 3:$_nC_r$). Note that the value of $_{12}P_9$ is much larger than $_{12}C_9$ and that they differ by a factor of 9! since $_nC_r = \dfrac{_nP_r}{r!}$.

Figure 8.14

```
MATH NUM CPX PRB
1:rand
2:nPr
3:nCr
4:!
5:randInt(
6:randNorm(
7:randBin(
```

Figure 8.15

```
12 nPr 9
            79833600
12 nCr 9
                 220
220*9!
            79833600
```

Exercise 1: The Department of Humanities has nine faculty members who must serve on at least one committee per semester. How many different committees can be formed that have (a) two members, (b) three members, (c) four members, and (d) five members?

Exercise 2: A certain state places 45 Ping-Pong balls numbered 1 through 45 in a container, then draws out five to form the winning lottery numbers. How many different ways can the five numbers be picked?

Exercise 3: Dairy King maintains six different toppings at a self-service counter, so that customers can top their ice cream sundaes with as many as they like. How many different sundaes can be created if a customer were to select any three ingredients?

8.5 EXERCISES

▶ **CONCEPTS AND VOCABULARY**

Fill in the blank with the appropriate word or phrase. Carefully reread the section if needed.

1. A(n) _____ is any task that can be repeated and has a(n) _____ set of possible outcomes.

2. If an experiment has N equally likely outcomes and is repeated t times, the number of elements in the sample space is given by ____.

3. When unique elements of a set are rearranged, the result is called a(n) _____ permutation.

4. If some elements of a group are identical, certain rearrangements are identical and the result is a(n) _____ permutation.

5. A three-digit number is formed from digits 1 to 9. Explain how forming the number with repetition differs from forming it without repetition.

6. Discuss/Explain the difference between a permutation and a combination. Try to think of new ways to help remember the distinction.

▶ **DEVELOPING YOUR SKILLS**

7. For the spinner shown here, (a) draw a tree diagram illustrating all possible outcomes for two spins and (b) create an ordered list showing all possible outcomes for two spins.

8. For the fair coin shown here, (a) draw a tree diagram illustrating all possible outcomes for four flips and (b) create an ordered list showing the possible outcomes for four flips.

9. A fair coin is flipped five times. If you extend the tree diagram from Exercise 8, how many elements are in the sample space?

10. A spinner has the two equally likely outcomes A or B and is spun four times. How is this experiment related to the one in Exercise 8? How many elements are in the sample space?

11. An inexpensive lock uses the numbers 0 to 24 for a three-number combination. How many different combinations are possible?

12. Grades at a local college consist of A, B, C, D, F, and W. If four classes are taken, how many different report cards are possible?

License plates. In a certain (English-speaking) country, license plates for automobiles consist of two letters followed by one of four symbols (■, ◆, ○, or ●), followed by three digits. How many license plates are possible if

13. Repetition is allowed?

14. Repetition is not allowed?

15. A remote access door opener requires a five-digit (1–9) sequence. How many sequences are possible if (a) repetition is allowed? (b) repetition is not allowed?

16. An instructor is qualified to teach Math 020, 030, 140, and 160. How many different four-course schedules are possible if (a) repetition is allowed? (b) repetition is not allowed?

Use the fundamental principle of counting and other quick-counting techniques to respond.

17. **Menu items:** At Joe's Diner, the manager is offering a dinner special that consists of one choice of entree (chicken, beef, soy meat, or pork), two vegetable servings (corn, carrots, green beans, peas, broccoli, or okra), and one choice of pasta, rice, or potatoes. How many different meals are possible?

18. **Getting dressed:** A frugal businessman has five shirts, seven ties, four pairs of dress pants, and three pairs of dress shoes. Assuming that all possible arrangements are appealing, how many different shirt-tie-pants-shoes outfits are possible?

19. **Number combinations:** How many four-digit numbers can be formed using the even digits 0, 2, 4, 6, 8, if (a) no repetitions are allowed; (b) repetitions are allowed; (c) repetitions are not allowed and the number must be less than 6000 and divisible by 10.

20. **Number combinations:** If I was born in March, April, or May, after the 19th but before the 30th,

and after 1949 but before 1981, how many different MM–DD–YYYY dates are possible for my birthday?

Seating arrangements: William, Xayden, York, and Zelda decide to sit together at the movies. How many ways can they be seated if

21. They sit in random order?

22. York must sit next to Zelda?

23. York and Zelda must be on the outside?

24. William must have the aisle seat?

Course schedule: A college student is trying to set her schedule for the next semester and is planning to take five classes: English, art, math, fitness, and science. How many different schedules are possible if

25. The classes can be taken in any order.

26. She wants her science class to immediately follow her math class.

27. She wants her English class to be first and her fitness class to be last.

28. She can't decide on the best order and simply takes the classes in alphabetical order.

Find the value of $_nP_r$ in two ways: (a) compute r factors of $n!$ and (b) use the formula $_nP_r = \dfrac{n!}{(n-r)!}$.

29. $_{10}P_3$ 30. $_{12}P_2$ 31. $_9P_4$

32. $_5P_3$ 33. $_8P_7$ 34. $_8P_1$

Determine the number of three-letter permutations of the letters given, then use an organized list to write them all out. How many of them are actually words or common names?

35. T, R, and A   36. P, M, and A

37. The regional manager for an office supply store needs to replace the manager and assistant manager at the downtown store. In how many ways can this be done if she selects the personnel from a group of 10 qualified applicants?

38. The local chapter of Mu Alpha Theta will soon be electing a president, vice-president, and treasurer. In how many ways can the positions be filled if the chapter has 15 members?

39. The local school board is going to select a principal, vice-principal, and assistant vice-principal from a pool of eight qualified candidates. In how many ways can this be done?

40. From a pool of 32 applicants, a board of directors must select a president, vice-president, labor relations liaison, and a director of personnel for the company's day-to-day operations. Assuming all applicants are qualified and willing to take on any of these positions, how many ways can this be done?

41. A hugely popular chess tournament now has six finalists. Assuming there are no ties, (a) in how many ways can the finalists place in the final round? (b) In how many ways can they finish first, second, and third? (c) In how many ways can they finish if it's sure that Roberta Fischer is going to win the tournament and that Geraldine Kasparov will come in sixth?

42. A field of 10 horses has just left the paddock area and is heading for the gate. Assuming there are no ties in the big race, (a) in how many ways can the horses place in the race? (b) In how many ways can they finish in the win, place, or show positions? (c) In how many ways can they finish if it's sure that John Henry III is going to win, Seattle Slew III will come in second (place), and either Dumb Luck II or Calamity Jane I will come in tenth?

Assuming all multiple births are identical and the children cannot be told apart, how many distinguishable photographs can be taken of a family of six, if they stand in a single row and there is

43. one set of twins

44. one set of triplets

45. one set of twins and one set of triplets

46. one set of quadruplets

47. How many distinguishable numbers can be made by rearranging the digits of 105,001?

48. How many distinguishable numbers can be made by rearranging the digits in the palindrome 1,234,321?

How many distinguishable permutations can be formed from the letters of the given word?

49. logic   50. leave

51. lotto   52. levee

A Scrabble player (see Example 7) has the six letters shown remaining in her rack. How many distinguishable, six-letter permutations can be formed? (If all six letters are played, what was the word?)

53. A, A, A, N, N, B

54. D, D, D, N, A, E

Find the value of $_nC_r$: (a) using $_nC_r = \dfrac{_nP_r}{r!}$ (r factors of $n!$ over $r!$) and (b) using $_nC_r = \dfrac{n!}{r!(n-r)!}$.

55. $_9C_4$ **56.** $_{10}C_3$ **57.** $_8C_5$

58. $_6C_3$ **59.** $_6C_6$ **60.** $_6C_0$

Use a calculator to verify that each pair of combinations is equal.

61. $_9C_4$, $_9C_5$ **62.** $_{10}C_3$, $_{10}C_7$

63. $_8C_5$, $_8C_3$ **64.** $_7C_2$, $_7C_5$

65. A platoon leader needs to send four soldiers to do some reconnaissance work. There are 12 soldiers in the platoon and each soldier is assigned a number between 1 and 12. The numbers 1 through 12 are placed in a helmet and drawn randomly. If a soldier's number is drawn, then that soldier goes on the mission. In how many ways can the reconnaissance team be chosen?

66. Seven colored balls (red, indigo, violet, yellow, green, blue, and orange) are placed in a bag and three are then withdrawn. In how many ways can the three colored balls be drawn?

67. When the company's switchboard operators went on strike, the company president asked for three volunteers from among the managerial ranks to temporarily take their place. In how many ways can the three volunteers "step forward," if there are 14 managers and assistant managers in all?

68. Becky has identified 12 books she wants to read this year and decides to take four with her to read while on vacation. She chooses *Pastwatch* by Orson Scott Card for sure, then decides to randomly choose any three of the remaining books. In how many ways can she select the four books she'll end up taking?

69. A new garage band has built up their repertoire to 10 excellent songs that really rock. Next month they'll be playing in a *Battle of the Bands* contest, with the winner getting some guaranteed gigs at the city's most popular hot spots. In how many ways can the band select 5 of their 10 songs to play at the contest?

70. Pierre de Guirré is an award-winning chef and has just developed 12 delectable, new main-course recipes for his restaurant. In how many ways can he select three of the recipes to be entered in an international culinary competition?

For each exercise, determine whether a permutation, a combination, counting principles, or a determination of the number of subsets is the most appropriate tool for obtaining a solution, then solve. Some exercises can be completed using more than one method.

71. In how many ways can eight second-grade children line up for lunch?

72. If you flip a fair coin five times, how many different outcomes are possible?

73. Eight sprinters are competing for the gold, silver, and bronze medals. In how many ways can the medals be awarded?

74. Motorcycle license plates are made using two letters followed by three numbers. How many plates can be made if repetition of letters (only) is allowed?

75. A committee of five students is chosen from a class of 20 to attend a seminar. How many different ways can this be done?

76. If onions, cheese, pickles, and tomatoes are available to dress a hamburger, how many different hamburgers can be made?

77. A caterer offers eight kinds of fruit to make various fruit trays. How many different trays can be made using four different fruits?

78. Eighteen females try out for the basketball team, but the coach can only place 15 on her roster. How many different teams can be formed?

▶ WORKING WITH FORMULAS

79. Stirling's Formula: $n! \approx \sqrt{2\pi} \cdot (n^{n+0.5}) \cdot e^{-n}$
Values of $n!$ grow very quickly as n gets larger (13! is already in the billions). For some applications, scientists find it useful to use the approximation for $n!$ shown, called Stirling's Formula.

 a. Compute the value of 7! on your calculator, then use Stirling's Formula with $n = 7$. By what percent does the approximate value differ from the true value?

 b. Compute the value of 10! on your calculator, then use Stirling's Formula with $n = 10$. By

what percent does the approximate value differ from the true value?

80. Factorial formulas: For $n, k \in \mathbb{W}$, where $n > k$,
$$\frac{n!}{(n-k)!} = n(n-1)(n-2)\cdots(n-k+1)$$

 a. Verify the formula for $n = 7$ and $k = 5$.

 b. Verify the formula for $n = 9$ and $k = 6$.

▶ **APPLICATIONS**

81. **Yahtzee:** In the game of "Yahtzee"® (Milton Bradley) five dice are rolled simultaneously on the first turn in an attempt to obtain various arrangements (worth various point values). How many different arrangements are possible?

82. **Twister:** In the game of "Twister"® (Milton Bradley) a simple spinner is divided into four quadrants designated Left Foot (LF), Right Hand (RH), Right Foot (RF), and Left Hand (LH), with four different color possibilities in each quadrant (red, green, yellow, blue). Determine the number of possible outcomes for three spins.

83. **Clue:** In the game of "Clue"® (Parker Brothers) a crime is committed in one of nine rooms, with one of six implements, by one of six people. In how many different ways can the crime be committed?

Phone numbers in North America have 10 digits: a three-digit area code, a three-digit exchange number, and the four final digits that make each phone number unique. Neither area codes nor exchange numbers can start with 0 or 1. Prior to 1994 the second digit of the area code *had to be* a 0 or 1. Sixteen area codes are reserved for special services (such as 911 and 411). In 1994, the last area code was used up and the rules were changed to allow the digits 2 through 9 as the middle digit in area codes.

84. How many different area codes were possible prior to 1994?

85. How many different exchange numbers were possible prior to 1994?

86. How many different phone numbers were possible *prior to* 1994?

87. How many different phone numbers were possible *after* 1994?

Aircraft N-numbers: In the United States, private aircraft are identified by an "N-Number," which is generally the letter "N" followed by five characters and includes these restrictions: (1) the N-Number can consist of five digits, four digits followed by one letter, or three digits followed by two letters; (2) the first digit cannot be a zero; (3) to avoid confusion with the numbers zero and one, the letters O and I cannot be used; and (4) repetition of digits and letters is allowed. How many unique N-Numbers can be formed

88. that have four digits and one letter?

89. that have three digits and two letters?

90. that have five digits?

91. that have three digits, two letters with no repetitions of any kind allowed?

Seating arrangements: Eight people would like to be seated. Assuming some will have to stand, in how many ways can the seats be filled if the number of seats available is

92. eight   93. five

94. three   95. one

Seating arrangements: In how many different ways can eight people (six students and two teachers) sit in a row of eight seats if

96. the teachers must sit on the ends

97. the teachers must sit together

Television station programming: A television station needs to fill eight half-hour slots for its Tuesday evening schedule with eight programs. In how many ways can this be done if

98. there are no constraints

99. *Seinfeld* must have the 8:00 P.M. slot

100. *Seinfeld* must have the 8:00 P.M. slot and *The Drew Carey Show* must be shown at 6:00 P.M.

101. *Friends* can be aired at 7:00 or 9:00 P.M. and *Everybody Loves Raymond* can be aired at 6:00 or 8:00 P.M.

Scholarship awards: Fifteen students at Roosevelt Community College have applied for six available scholarship awards. How many ways can the awards be given if

102. there are six different awards given to six different students

103. there are six identical awards given to six different students

Committee composition: The local city council has 10 members and is trying to decide if they want to be governed by a committee of three people or by a president, vice-president, and secretary.

104. If they are to be governed by committee, how many unique committees can be formed?

105. How many different president, vice-president, and secretary possibilities are there?

106. Team rosters: A soccer team has three goalies, eight defensive players, and eight forwards on its roster. How many different starting line-ups can be formed (one goalie, three defensive players, and three forwards)?

107. e-mail addresses: A business wants to standardize the e-mail addresses of its employees. To make them easier to remember and use, they consist of two letters and two digits (followed by @esmtb.com), with zero being excluded from use as the first digit and no repetition of letters or digits allowed. Will this provide enough unique addresses for their 53,000 employees worldwide?

▶ EXTENDING THE CONCEPT

108. In Exercise 79, we learned that an approximation for $n!$ can be found using Stirling's Formula: $n! \approx \sqrt{2\pi}(n^{n+0.5})e^{-n}$. As with other approximations, mathematicians are very interested in whether the approximation gets better or worse for larger values of n (does their ratio get closer to 1 or farther from 1). Use your calculator to investigate and answer the question.

109. Verify that the following equations are true, then generalize the patterns and relationships noted to create your own equation. Afterward, write each of the four factors from Part (a) (the two combinations on each side) in expanded form and discuss/explain why the two sides are equal.

    a. $_{10}C_3 \cdot _7C_2 = _{10}C_2 \cdot _8C_5$

    b. $_9C_3 \cdot _6C_2 = _9C_2 \cdot _7C_4$

    c. $_{11}C_4 \cdot _7C_5 = _{11}C_5 \cdot _6C_4$

    d. $_8C_3 \cdot _5C_2 = _8C_2 \cdot _6C_3$

110. Tic-Tac-Toe: In the game *Tic-Tac-Toe,* players alternately write an "X" or an "O" in one of nine squares on a 3×3 grid. If either player gets three in a row horizontally, vertically, or diagonally, that player wins. If all nine squares are played with neither person winning, the game is a draw. Assuming "X" always goes first,

    a. How many different "boards" are possible if the game ends after five plays?

    b. How many different "boards" are possible if the game ends after six plays?

▶ MAINTAINING YOUR SKILLS

111. (5.4) Solve the given system of linear inequalities by graphing. Shade the feasible region.

$$\begin{cases} 2x + y < 6 \\ x + 2y < 6 \\ x \geq 0 \\ y \geq 0 \end{cases}$$

112. (8.2) For the series $1 + 5 + 9 + 13 + \cdots + 197$, state the nth term formula then find the 35th term and the sum of the first 35 terms.

113. (6.2/6.3) Given matrices A and B shown, use a calculator to find $A + B$, AB, and A^{-1}.

$$A = \begin{bmatrix} 1 & 0 & 3 \\ -2 & 5 & 1 \\ 2 & 1 & 4 \end{bmatrix} \qquad B = \begin{bmatrix} 0.5 & 0.2 & -7 \\ -9 & 0.1 & 8 \\ 1.2 & 0 & 6 \end{bmatrix}$$

114. (7.3) Graph the hyperbola that is defined by $\dfrac{(x-2)^2}{4} - \dfrac{(y+3)^2}{9} = 1$.

Learning Objectives

In Section 8.6 you will learn how to:

☐ **A.** Define an event on a sample space

☐ **B.** Compute elementary probabilities

☐ **C.** Use certain properties of probability

☐ **D.** Compute probabilities using quick-counting techniques

☐ **E.** Compute probabilities involving nonexclusive events

There are few areas of mathematics that give us a better view of the world than **probability** and **statistics.** Unlike statistics, which seeks to analyze and interpret data, probability (for our purposes) attempts to use observations and data to make statements concerning the likelihood of future events. Such predictions of what *might* happen have found widespread application in such diverse fields as politics, manufacturing, gambling, opinion polls, product life, and many others. In this section, we develop the basic elements of probability.

A. Defining an Event

In Section 8.5 we defined the following terms: experiment and sample outcome. Flipping a coin twice in succession is an *experiment,* and two sample outcomes are HH and HT. An **event** *E* is *any designated set of sample outcomes,* and is a subset of the sample space. One event might be E_1: (two heads occur), another possibility is E_2: (at least one tail occurs).

EXAMPLE 1 ▶ **Stating a Sample Space and Defining an Event**

Consider the experiment of rolling one standard, six-sided die (plural is dice). State the sample space *S* and define any two events relative to *S.*

Solution ▶ *S* is the set of all possible outcomes, so $S = \{1, 2, 3, 4, 5, 6\}$. Two possible events are E_1: (a 5 is rolled) and E_2: (an even number is rolled).

☑ **A.** You've just learned how to define an event on a sample space

Now try Exercises 7 through 10 ▶

B. Elementary Probability

When rolling the die, we know the result can be any of the six equally likely outcomes in the sample space, so the chance of E_1:(a five is rolled) is $\frac{1}{6}$. Since three of the elements in *S* are even numbers, the chance of E_2:(an even number is rolled) is $\frac{3}{6} = \frac{1}{2}$. This suggests the following definition.

> **The Probability of an Event *E***
>
> Given *S* is a sample space of equally likely events and *E* is an event relative to *S*, the probability of *E*, written *P(E)*, is computed as
>
> $$P(E) = \frac{n(E)}{n(S)}$$
>
> where $n(E)$ represents the number of elements in *E*, and $n(S)$ represents the number of elements in *S*.

A standard deck of playing cards consists of 52 cards divided in four groups or *suits.* There are 13 hearts (♥), 13 diamonds (♦), 13 spades (♠), and 13 clubs (♣). As you can see in the illustration, each of the 13 cards in a suit is labeled 2, 3, 4, 5, 6, 7, 8, 9, 10, J, Q, K, and A. Also notice that 26 of the cards are red (hearts and diamonds), 26 are black (spades and clubs) and 12 of the cards are "face cards" (J, Q, K of each suit).

EXAMPLE 2 ▶ **Stating a Sample Space and the Probability of a Single Outcome**

A single card is drawn from a well-shuffled deck. Define S and state the probability of any single outcome. Then define E as *a King is drawn* and find $P(E)$.

Solution ▶ Sample space: $S = \{$the 52 cards$\}$. There are 52 equally likely outcomes, so the probability of any one outcome is $\frac{1}{52}$. Since S has four Kings,

$$P(E) = \frac{n(E)}{n(S)} = \frac{4}{52} \text{ or about } 0.077.$$

> **Now try Exercises 11 through 14 ▶**

EXAMPLE 3 ▶ **Stating a Sample Space and the Probability of a Single Outcome**

A family of five has two girls and three boys named Sophie, Maria, Albert, Isaac, and Pythagoras. Their ages are 21, 19, 15, 13, and 9, respectively. One is to be selected randomly. Find the probability a teenager is chosen.

Solution ▶ The sample space is $S = \{9, 13, 15, 19, 21\}$. Three of the five are teenagers, meaning the probability is $\frac{3}{5}$, 0.6, or 60%.

☑ **B.** You've just learned how to compute elementary probabilities

> **Now try Exercises 15 and 16 ▶**

C. Properties of Probability

A study of probability necessarily includes recognizing some basic and fundamental properties. For example, when a fair die is rolled, what is $P(E)$ if E is defined as a *1, 2, 3, 4, 5,* or *6* is rolled? The event E will occur 100% of the time, since 1, 2, 3, 4, 5, 6 are the only possibilities. In symbols we write P(outcome is in the sample space) or simply $P(S) = 1$ (100%).

What percent of the time will a result *not* in the sample space occur? Since the die has only the six sides numbered 1 through 6, the probability of rolling something else is zero. In symbols, P(outcome is not in sample space) $= 0$ or simply $P(\sim S) = 0$.

WORTHY OF NOTE

In probability studies, the tilde "~" acts as a negation symbol. For any event E defined on the sample space, $\sim E$ means the event does not occur.

Properties of Probability		
Given sample space S and any event E defined relative to S.		
1. $P(S) = 1$	**2.** $P(\sim S) = 0$	**3.** $0 \le P(E) \le 1$

EXAMPLE 4 ▶ **Determining the Probability of an Event**

A game is played using a spinner like the one shown. Determine the probability of the following events:

E_1: A nine is spun. E_2: An integer greater than 0 and less than 9 is spun.

Solution ▶ The sample space consists of eight equally likely outcomes.

$$P(E_1) = \frac{0}{8} = 0 \qquad P(E_2) = \frac{8}{8} = 1.$$

Technically, E_1: A nine is spun is not an "event," since it is not in the sample space and cannot occur, while E_2 contains the entire sample space and must occur.

> **Now try Exercises 17 and 18 ▶**

Because we know $P(S) = 1$ and all sample outcomes are equally likely, the probabilities of all single events defined on the sample space must sum to 1. For the experiment of rolling a fair die, the sample space has six outcomes that are equally likely. Note that $P(1) = P(2) = P(3) = P(4) = P(5) = P(6) = \frac{1}{6}$, and $\frac{1}{6} + \frac{1}{6} + \frac{1}{6} + \frac{1}{6} + \frac{1}{6} + \frac{1}{6} = 1$.

Probability and Sample Outcomes

Given a sample space S with n equally likely sample outcomes $s_1, s_2, s_3, \ldots, s_n$.

$$\sum_{i=1}^{n} P(s_i) = P(s_1) + P(s_2) + P(s_3) + \cdots + P(s_n) = 1$$

The **complement** of an event E is the set of sample outcomes in S not contained in E. Symbolically, $\sim E$ is the complement of E.

Probability and Complementary Events

Given sample space S and any event E defined relative to S, the complement of E, written $\sim E$, is the set of all outcomes not in E and:

1. $P(E) = 1 - P(\sim E)$ **2.** $P(E) + P(\sim E) = 1$

EXAMPLE 5 ▶ **Stating a Probability Using Complements**

Use complementary events to answer the following questions:

a. A single card is drawn from a well-shuffled deck. What is the probability that it is not a diamond?

b. A single letter is picked at random from the letters in the word "divisibility." What is the probability it is not an "i"?

Solution ▶ **a.** Since there are 13 diamonds in a standard 52-card deck, there are 39 nondiamonds: $P(\sim D) = 1 - P(D) = 1 - \frac{13}{52} = \frac{39}{52} = 0.75$.

WORTHY OF NOTE

Probabilities can be written in fraction form, decimal form, or as a percent. For $P(E_2)$ from Example 1, the probability is $\frac{3}{4}$, 0.75, or 75%.

b. Of the 12 letters in d-i-v-i-s-i-b-i-l-i-t-y, 5 are "i's." This means $P(\sim i) = 1 - P(i)$, or $1 - \frac{5}{12} = \frac{7}{12}$. The probability of choosing a letter other than i is $0.58\overline{3}$.

Now try Exercises 19 through 22 ▶

EXAMPLE 6 ▶ **Stating a Probability Using Complements**

Inter-Island Waterways has just opened hydrofoil service between several islands. The hydrofoil is powered by two engines, one forward and one aft, and will operate if either of its two engines is functioning. Due to testing and past experience, the company knows the probability of the aft engine failing is $P(\text{aft engine fails}) = 0.05$, the probability of the forward engine failing is $P(\text{forward engine fails}) = 0.03$, and the probability that both fail is $P(\text{both engines simultaneously fail}) = 0.012$. What is the probability the hydrofoil completes its next trip?

Solution ▶ Although the answer may *seem* complicated, note that $P(\text{trip is completed})$ and $P(\text{both engines simultaneously fail})$ are complements.

$$P(\text{trip is completed}) = 1 - P(\text{both engines simultaneously fail})$$
$$= 1 - 0.012$$
$$= 0.988$$

There is close to a 99% probability the trip will be completed.

Now try Exercises 23 and 24 ▶

The chart in Figure 8.16 shows all 36 possible outcomes (the sample space) from the experiment of rolling two fair dice.

Figure 8.16

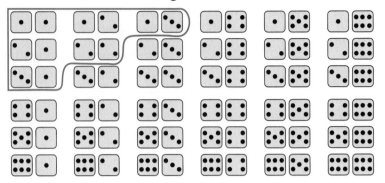

EXAMPLE 7 ▶ **Stating a Probability Using Complements**

Two fair dice are rolled. What is the probability the sum of both dice is greater than or equal to 5, $P(\text{sum} \geq 5)$?

Solution ▶ See Figure 8.16. For $P(\text{sum} \geq 5)$ it may be easier to use complements as there are far fewer possibilities: $P(\text{sum} \geq 5) = 1 - P(\text{sum} < 5)$, which gives

☑ **C. You've just learned how to use certain properties of probability**

$$1 - \frac{6}{36} = 1 - \frac{1}{6} = \frac{5}{6} = 0.8\overline{3}.$$

Now try Exercises 25 and 26 ▶

D. Probability and Quick-Counting

Quick-counting techniques were introduced earlier to help count the number of elements in a large or more complex sample space, and the number of sample outcomes in an event.

EXAMPLE 8A ▶ **Stating a Probability Using Combinations**

Five cards are drawn from a shuffled 52-card deck. Calculate the probability of E_1:(*all five cards are face cards*) or E_2:(*all five cards are hearts*)?

Solution ▶ The sample space for both events consists of all five-card groups that can be formed from the 52 cards or $_{52}C_5$. For E_1 we are to select five face cards from the 12 that are available (three from each suit), or $_{12}C_5$. The probability of five face

WORTHY OF NOTE

It seems reasonable that the probability of 5 hearts is slightly higher, as 13 of the 52 cards are hearts, while only 12 are face cards.

cards is $\dfrac{n(E)}{n(S)} = \dfrac{_{12}C_5}{_{52}C_5}$, which gives $\dfrac{792}{2,598,960} \approx 0.0003$. For E_2 we are to select five

hearts from the 13 available, or $_{13}C_5$. The probability of five hearts is $\dfrac{n(E)}{n(S)} = \dfrac{_{13}C_5}{_{52}C_5}$,

which is $\dfrac{1287}{2,598,960} \approx 0.0005$.

EXAMPLE 8B ▶ **Stating a Probability Using Combinations and the Fundamental Principle of Counting**

Of the 42 seniors at Jacoby High School, 23 are female and 19 are male. A group of five students is to be selected at random to attend a conference in Reno, Nevada. What is the probability the group will have exactly three females?

Solution ▶ The sample space consists of all five-person groups that can be formed from the 42 seniors or $_{42}C_5$. The event consists of selecting 3 females from the 23 available ($_{23}C_3$) and 2 males from the 19 available ($_{19}C_2$). Using the fundamental principle of counting $n(E) = {}_{23}C_3 \cdot {}_{19}C_2$ and the probability the group has 3 females is

$$\frac{n(E)}{n(S)} = \frac{{}_{23}C_3 \cdot {}_{19}C_2}{{}_{42}C_5}, \text{ which gives } \frac{302{,}841}{850{,}668} \approx 0.356. \text{ There is approximately a}$$

35.6% probability the group will have exactly 3 females.

☑ **D.** You've just learned how to compute probabilities using quick-counting techniques

> **Now try Exercises 27 through 34** ▶

E. Probability and Nonexclusive Events

Figure 8.17

Sometimes the way events are defined causes them to share sample outcomes. Using a standard deck of playing cards once again, if we define the events E_1:(a club is drawn) and E_2:(a face card is drawn), they share the outcomes J♣, Q♣, and K♣ as shown in Figure 8.17. This overlapping region is the intersection of the events, or $E_1 \cap E_2$. If we compute $n(E_1 \cup E_2)$ as $n(E_1) + n(E_2)$ as before, this intersecting region gets counted *twice*!

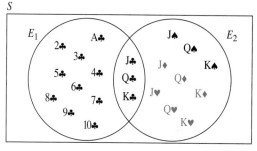

In cases where the events are **nonexclusive** (not mutually exclusive), we maintain the correct count by subtracting one of the two intersections, obtaining $n(E_1 \cup E_2) = n(E_1) + n(E_2) - n(E_1 \cap E_2)$. This leads to the following calculation for the probability of nonexclusive events:

$$P(E_1 \cup E_2) = \frac{n(E_1) + n(E_2) - n(E_1 \cap E_2)}{n(S)} \quad \text{definition of probability}$$

$$= \frac{n(E_1)}{n(S)} + \frac{n(E_1)}{n(S)} - \frac{n(E_1 \cap E_2)}{n(S)} \quad \text{property of rational expressions}$$

$$= P(E_1) + P(E_2) - P(E_1 \cap E_2) \quad \text{definition of probability}$$

Probability and Nonexclusive Events

Given sample space S and *nonexclusive events* E_1 and E_2 defined relative to S, the probability of E_1 *or* E_2 is given by
$$P(E_1 \cup E_2) = P(E_1) + P(E_2) - P(E_1 \cap E_2)$$

EXAMPLE 9A ▶ **Stating the Probability of Nonexclusive Events**

What is the probability that a club or a face card is drawn from a standard deck of 52 well-shuffled cards?

Solution ▶ As before, define the events E_1:(a club is drawn) and E_2:(a face card is drawn). Since there are 13 clubs and 12 face cards, $P(E_1) = \frac{13}{52}$ and $P(E_2) = \frac{12}{52}$. But three of the face cards are clubs, so $P(E_1 \cap E_2) = \frac{3}{52}$. This leads to

$$P(E_1 \cup E_2) = P(E_1) + P(E_2) - P(E_1 \cap E_2) \quad \text{nonexclusive events}$$

$$= \frac{13}{52} + \frac{12}{52} - \frac{3}{52} \quad \text{substitute}$$

$$= \frac{22}{52} \approx 0.423 \quad \text{combine terms}$$

There is about a 42% probability that a club or face card is drawn.

EXAMPLE 9B ▶ Stating the Probability of Nonexclusive Events

A survey of 100 voters was taken to gather information on critical issues and the demographic information collected is shown in the table. One out of the 100 voters is to be drawn at random to be interviewed on the 5 O'Clock News. What is the probability the person is a woman (W) or a Republican (R)?

	Women	Men	Totals
Republican	17	20	37
Democrat	22	17	39
Independent	8	7	15
Green Party	4	1	5
Tax Reform	2	2	4
Totals	53	47	100

Solution ▶ Since there are 53 women and 37 Republicans, $P(W) = 0.53$ and $P(R) = 0.37$. The table shows 17 people are both female and Republican so $P(W \cap R) = 0.17$.

$$P(W \cup R) = P(W) + P(R) - P(W \cap R) \qquad \text{nonexclusive events}$$
$$= 0.53 + 0.37 - 0.17 \qquad \text{substitute}$$
$$= 0.73 \qquad \text{combine}$$

☑ **E.** You've just learned how to compute probabilities involving nonexclusive events

There is a 73% probability the person is a woman or a Republican.

Now try Exercises 35 through 48 ▶

Two events that have no common outcomes are called **mutually exclusive** events (one excludes the other and vice versa). For example, in rolling one die, E_1:(*a 2 is rolled*) and E_2:(*an odd number is rolled*) are mutually exclusive, since 2 is not an odd number. For the probability of E_3:(*a 2 is rolled* or *an odd number is rolled*), we note that $n(E_1 \cap E_2) = 0$ and the previous formula simply reduces to $P(E_1) + P(E_2)$. See **Exercises 49 and 50.**

There is a large variety of additional applications in the Exercise Set. See **Exercises 53 through 68.**

TECHNOLOGY HIGHLIGHT

Principles of Quick-Counting, Combinations, and Probability

At this point you are likely using the [Y=] screen and tables ([TBLSET], [2nd] [GRAPH] **TABLE**, and so on) with relative ease. When probability calculations require a repeated use of permutations and combinations, these features can make the work more efficient and help to explore the patterns they generate. For choosing r children from a group of six children ($n = 6$), set the [TBLSET] to **AUTO**, then press [Y=] and enter 6 $_nC_r$ X as Y_1 (Figure 8.18). Access the **TABLE** ([2nd] [GRAPH]) and note that the calculator has automatically computed the value of $_6C_0, {}_6C_1, {}_6C_2, \ldots, {}_6C_6$ (Figure 8.19) and the pattern of outputs is symmetric. For calculations similar to those required in Example 8B ($_{23}C_3 \cdot {}_{19}C_2$), enter

Figure 8.18

```
Plot1 Plot2 Plot3
\Y1■6 nCr X
\Y2=
\Y3=
\Y4=
\Y5=
\Y6=
\Y7=
```

Figure 8.19

X	Y1	
0	1	
1	6	
2	15	
3	20	
4	15	
5	6	
6	1	

X=0

$Y_1 = 23 \, _nC_r \, X$, $Y_2 = 19 \, _nC_r \, (X - 1)$, and $Y_3 = Y_1 \cdot Y_2$, or any variation of these. Use these ideas to work the following exercises.

Exercise 1: Use your calculator to display the values of $_5C_0$, $_5C_1$, . . . , $_5C_5$. Is the result a pattern similar to that for $_6C_0$, $_6C_1$, $_6C_2$, . . . , $_6C_6$? Repeat for $_7C_r$. Why are the "middle values" repeated for $n = 7$ and $n = 5$, but not for $n = 6$?

Exercise 2: A committee consists of 10 Republicans and eight Democrats. In how many ways can a committee of four Republicans and three Democrats be formed?

8.6 EXERCISES

▶ CONCEPTS AND VOCABULARY

Fill in the blank with the appropriate word or phrase. Carefully reread the section if needed.

1. Given a sample space S and an event E defined relative to S, $P(E) = \dfrac{}{n(S)}$.

2. In elementary probability, we consider all events in the sample space to be _____ likely.

3. Given a sample space S and an event E defined relative to S: ___ $\leq P(E) \leq$ ___, $P(S) =$ ___, and $P(\sim S) =$ ___.

4. The _____ of an event E is the set of sample outcomes in S which are not contained in E.

5. Discuss/Explain the difference between mutually exclusive events and nonmutually exclusive events. Give an example of each.

6. A single die is rolled. With no calculations, explain why the probability of rolling an even number is greater than rolling a number greater than four.

▶ DEVELOPING YOUR SKILLS

State the sample space S and the probability of a single outcome. Then define any two events E relative to S (many answers possible).

7. Two fair coins (heads and tails) are flipped.

8. The simple spinner shown is spun.

    Exercise 8

9. The head coaches for six little league teams (the Patriots, Cougars, Angels, Sharks, Eagles, and Stars) have gathered to discuss new changes in the rule book. One of them is randomly chosen to ask the first question.

10. Experts on the planets Mercury, Venus, Mars, Jupiter, Saturn, Uranus, Neptune, and the dwarf planet Pluto have gathered at a space exploration conference. One group of experts is selected at random to speak first.

Find $P(E)$ for the events defined.

11. Nine index cards numbered 1 through 9 are shuffled and placed in an envelope, then one of the cards is randomly drawn. Define event E as *the number drawn is even.*

12. Eight flash cards used for studying basic geometric shapes are shuffled and one of the cards is drawn at random. The eight cards include information on circles, squares, rectangles, kites, trapezoids, parallelograms, pentagons, and triangles. Define event E as *a quadrilateral is drawn.*

13. One card is drawn at random from a standard deck of 52 cards. What is the probability of

    a. drawing a Jack   b. drawing a spade

    c. drawing a black card   d. drawing a red three

14. Pinochle is a card game played with a deck of 48 cards consisting of 2 Aces, 2 Kings, 2 Queens, 2 Jacks, 2 Tens, and 2 Nines in each of the four

standard suits [hearts (♥), diamonds (♦), spades (♠), and clubs (♣)]. If one card is drawn at random from this deck, what is the probability of

a. drawing an Ace   **b.** drawing a club

c. drawing a red card

d. drawing a face card (Jack, Queen, King)

15. A group of finalists on a game show consists of three males and five females. Hank has a score of 520 points, with Harry and Hester having 490 and 475 points, respectively. Madeline has 532 points, with Mackenzie, Morgan, Maggie, and Melanie having 495, 480, 472, and 470 points, respectively. One of the contestants is randomly selected to start the final round. Define E_1 as *Hester is chosen,* E_2 as *a female is chosen,* and E_3 as *a contestant with less than 500 points is chosen.* Find the probability of each event.

16. Soccer coach Maddox needs to fill the last spot on his starting roster for the opening day of the season and has to choose between three forwards and five defenders. The forwards have jersey numbers 5, 12, and 17, while the defenders have jersey numbers 7, 10, 11, 14, and 18. Define E_1 as *a forward is chosen,* E_2 as *a defender is chosen,* and E_3 as *a player whose jersey number is greater than 10 is chosen.* Find the probability of each event.

17. A game is played using a spinner like the one shown. For each spin,

    a. What is the probability the arrow lands in a shaded region?

    b. What is the probability your spin is less than 5?

    c. What is the probability you spin a 2?

    d. What is the probability the arrow lands on a prime number?

18. A game is played using a spinner like the one shown here. For each spin,

    a. What is the probability the arrow lands in a lightly shaded region?

    b. What is the probability your spin is greater than 2?

    c. What is the probability the arrow lands in a shaded region?

    d. What is the probability you spin a 5?

Use the complementary events to complete Exercises 19 through 22.

19. One card is drawn from a standard deck of 52. What is the probability it is not a club?

20. Four standard dice are rolled. What is the probability the sum is less than 24?

21. A single digit is randomly selected from among the digits of 10!. What is the probability the digit is not a 2?

22. A corporation will be moving its offices to Los Angeles, Miami, Atlanta, Dallas, or Phoenix. If the site is randomly selected, what is the probability Dallas is not chosen?

23. A large manufacturing plant can remain at full production as long as one of its two generators is functioning. Due to past experience and the age difference between the systems, the plant manager estimates the probability of the main generator failing is 0.05, the probability of the secondary generator failing is 0.01, and the probability of both failing is 0.009. What is the probability the plant remains in full production today?

24. A fire station gets an emergency call from a shopping mall in the mid-afternoon. From a study of traffic patterns, Chief Nozawa knows the probability the most direct route is clogged with traffic is 0.07, while the probability of the secondary route being clogged is 0.05. The probability both are clogged is 0.02. What is the probability they can respond to the call unimpeded using one of these routes?

25. Two fair dice are rolled (see Figure 8.16). What is the probability of

    a. a sum less than four

    b. a sum less than eleven

    c. the sum is not nine

    d. a roll is not a "double" (both dice the same)

"Double-six" dominos is a game played with the 28 numbered tiles shown in the diagram.

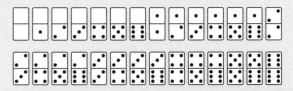

26. The 28 dominos are placed in a bag, shuffled, and then one domino is randomly drawn. What is the probability the total number of dots on the domino

    a. is three or less

    b. is greater than three

    c. does not have a blank half

    d. is not a "double" (both sides the same)

Find $P(E)$ given the values for $n(E)$ and $n(S)$ shown.

27. $n(E) = {}_6C_3 \cdot {}_4C_2$; $n(S) = {}_{10}C_5$

28. $n(E) = {}_{12}C_9 \cdot {}_8C_7$; $n(S) = {}_{20}C_{16}$

29. $n(E) = {}_9C_6 \cdot {}_5C_3$; $n(S) = {}_{14}C_9$

30. $n(E) = {}_7C_6 \cdot {}_3C_2$; $n(S) = {}_{10}C_8$

31. Five cards are drawn from a well-shuffled, standard deck of 52 cards. Which has the greater probability: (a) all five cards are red or (b) all five cards are numbered cards? How much greater?

32. Five cards are drawn from a well-shuffled pinochle deck of 48 cards (see Exercise 14). Which has the greater probability (a) all five cards are face cards (King, Queen, or Jack) or (b) all five cards are black? How much greater?

33. A dietetics class has 24 students. Of these, 9 are vegetarians and 15 are not. The instructor receives enough funding to send six students to a conference. If the students are selected randomly, what is the probability the group will have

    a. exactly two vegetarians

    b. exactly four nonvegetarians

    c. at least three vegetarians

34. A large law firm has a support staff of 15 employees: six paralegals and nine legal assistants. Due to recent changes in the law, the firm wants to send five of them to a forum on the new changes. If the selection is done randomly, what is the probability the group will have

    a. exactly three paralegals

    b. exactly two legal assistants

    c. at least two paralegals

Find the probability indicated using the information given.

35. Given $P(E_1) = 0.7$, $P(E_2) = 0.5$, and $P(E_1 \cap E_2) = 0.3$, compute $P(E_1 \cup E_2)$.

36. Given $P(E_1) = 0.6$, $P(E_2) = 0.3$, and $P(E_1 \cap E_2) = 0.2$, compute $P(E_1 \cup E_2)$.

37. Given $P(E_1) = \frac{3}{8}$, $P(E_2) = \frac{3}{4}$, and $P(E_1 \cup E_2) = \frac{15}{18}$; compute $P(E_1 \cap E_2)$.

38. Given $P(E_1) = \frac{1}{2}$, $P(E_2) = \frac{3}{5}$, and $P(E_1 \cup E_2) = \frac{17}{20}$; compute $P(E_1 \cap E_2)$.

39. Given $P(E_1 \cup E_2) = 0.72$, $P(E_2) = 0.56$, and $P(E_1 \cap E_2) = 0.43$; compute $P(E_1)$.

40. Given $P(E_1 \cup E_2) = 0.85$, $P(E_1) = 0.4$, and $P(E_1 \cap E_2) = 0.21$; compute $P(E_2)$.

41. Two fair dice are rolled. What is the probability the sum of the dice is

    a. a multiple of 3 and an odd number

    b. a sum greater than 5 and a 3 on one die

    c. an even number and a number greater than 9

    d. an odd number and a number less than 10

42. *Eight Ball* is a game played on a pool table with 15 balls numbered 1 through 15 and a cue ball that is solid white. Of the 15 numbered balls, 8 are a solid (nonwhite) color and numbered 1 through 8, and seven are striped balls numbered 9 through 15. The fifteen numbered pool balls (no cueball) are placed in a large bowl and mixed, then one is drawn out. What is the probability of drawing

    a. the eight ball

    b. a number greater than fifteen

    c. an even number

    d. a multiple of three

    e. a solid color and an even number

    f. a striped ball and an odd number

    g. an even number and a number divisible by three

    h. an odd number and a number divisible by 4

43. A survey of 50 veterans was taken to gather information on their service career and what life is like out of the military. A breakdown of those surveyed is shown in the table. One out of the 50 will be selected at random for an interview and a biographical sketch. What is the probability the person chosen is

	Women	Men	Totals
Private	6	9	15
Corporal	10	8	18
Sergeant	4	5	9
Lieutenant	2	1	3
Captain	2	3	5
Totals	24	26	50

    a. a woman and a sergeant

    b. a man and a private

    c. a private and a sergeant

    d. a woman and an officer

    e. a person in the military

44. Referring to Exercise 43, what is the probability the person chosen is

    a. a woman or a sergeant

    b. a man or a private

    c. a woman or a man

    d. a woman or an officer

    e. a captain or a lieutenant

A computer is asked to randomly generate a three-digit number. What is the probability the

45. ten's digit is odd or the one's digit is even

46. first digit is prime and the number is a multiple of 10

A computer is asked to randomly generate a four-digit number. What is the probability the number is

47. at least 4000 or a multiple of 5

48. less than 7000 and an odd number

49. Two fair dice are rolled. What is the probability of rolling

   a. boxcars (a sum of 12) or snake eyes (a sum of 2)

   b. a sum of 7 or a sum of 11

   c. an even-numbered sum or a prime sum

   d. an odd-numbered sum or a sum that is a multiple of 4

   e. a sum of 15 or a multiple of 12

   f. a sum that is a prime number

50. Suppose all 16 balls from a game of pool (see Exercise 42) are placed in a large leather bag and mixed, then one is drawn out. Consider the cue ball as "0." What is the probability of drawing

   a. a striped ball

   b. a solid-colored ball

   c. a polka-dotted ball

   d. the cue ball

   e. the cue ball or the eight ball

   f. a striped ball or a number less than five

   g. a solid color or a number greater than 12

   h. an odd number or a number divisible by 4

▶ WORKING WITH FORMULAS

51. **Games involving a fair spinner (with numbers 1 through 4): $P(n) = \left(\frac{1}{4}\right)^n$**

   Games that involve moving pieces around a board using a fair spinner are fairly common. If a fair spinner has the numbers 1 through 4, the probability that any one number is spun n times in succession is given by the formula shown, where n represents the number of spins. What is the probability (a) the first player spins a two? (b) all four players spin a two? (c) Discuss the graph of $P(n)$ and explain the connection between the graph and the probability of consistently spinning a two.

52. **Games involving a fair coin (heads and tails): $P(n) = \left(\frac{1}{2}\right)^n$**

   When a fair coin is flipped, the probability that heads (or tails) is flipped n times in a row is given by the formula shown, where n represents the number of flips. What is the probability (a) the first flip is heads? (b) the first four flips are heads? (c) Discuss the graph of $P(n)$ and explain the connection between the graph and the probability of consistently flipping heads.

▶ APPLICATIONS

53. To improve customer service, a company tracks the number of minutes a caller is "on hold" and waiting for a customer service representative. The table shows the probability that a caller will wait m minutes. Based on the table, what is the probability a caller waits

   a. at least 2 minutes

   b. less than 2 minutes

   c. 4 minutes or less

   d. over 4 minutes

   e. less than 2 or more than 4 minutes

   f. 3 or more minutes

Wait Time (minutes m)	Probability
0	0.07
$0 < m < 1$	0.28
$1 \leq m < 2$	0.32
$2 \leq m < 3$	0.25
$3 \leq m < 4$	0.08

54. To study the impact of technology on American families, a researcher first determines the probability that a family has n computers at home. Based on the table, what is the probability a home

a. has at least one computer

b. has two or more computers

c. has less than four computers

d. has five computers

e. has one, two, or three computers

f. does not have two computers

Number of Computers	Probability
0	9%
1	51%
2	28%
3	9%
4	3%

55. Jolene is an experienced markswoman and is able to hit a 10 in. by 20 in. target 100% of the time at a range of 100 yd. Assuming the probability she hits a target is related to its area, what is the probability she hits the shaded portions shown?

10 in.

20 in.

a. isosceles triangle

b. right triangle

c. equilateral triangle

56. a. square

b. circle

c. isosceles trapezoid with $b = \frac{B}{2}$

57. A circular dartboard has a total radius of 8 in., with circular bands that are 2 in. wide, as shown. You are skilled enough to hit this board 100% of the time so you always score at least two points each

2
4
6
8

time you throw a dart. Assuming the probabilities are related to area, on the next dart that you throw what is the probability you

a. score at least a 4?   b. score at least a 6?

c. hit the bull's-eye?   d. score exactly 4 points?

58. Three red balls, six blue balls, and four white balls are placed in a bag. What is the probability the first ball you draw out is

a. red   b. blue

c. not white

d. purple

e. red or white

f. red and white

59. Three red balls, six blue balls, and four white balls are placed in a bag, then two are drawn out and placed in a rack. What is the probability the balls drawn are

a. first red, second blue

b. first blue, second red

c. both white

d. first blue, second not red

e. first white, second not blue

f. first not red, second not blue

60. Consider the 210 discrete points found in the first and second quadrants where $-10 \leq x \leq 10$, $1 \leq y \leq 10$, and x and y are integers. The coordinates of each point is written on a slip of paper and placed in a box. One of the slips is then randomly drawn. What is the probability the point (x, y) drawn

a. is on the graph of $y = |x|$

b. is on the graph of $y = 2|x|$

c. is on the graph of $y = 0.5|x|$

d. has coordinates $(x, y > -2)$

e. has coordinates $(x \leq 5, y > -2)$

f. is between the branches of $y = x^2$

61. Your instructor surprises you with a True/False quiz for which you are totally unprepared and must guess randomly. What is the probability you pass the quiz with an 80% or better if there are

a. three questions   b. four questions

c. five questions

62. A robot is sent out to disarm a timed explosive device by randomly changing some switches from a neutral position to a *positive flow* or *negative flow* position. The problem is, the switches are independent and unmarked, and it is unknown which direction is positive and which direction is negative. The bomb is harmless if a majority of the switches yield a positive flow. All switches must be thrown. What is the probability the device is disarmed if there are

a. three switches   **b.** four switches

c. five switches

63. A survey of 100 retirees was taken to gather information concerning how they viewed the Vietnam War back in the early 1970s. A breakdown of those surveyed is shown in the table. One out of the hundred will be selected at random for a personal, taped interview. What is the probability the person chosen had a

a. career of any kind and opposed the war

b. medical career and supported the war

c. military career and opposed the war

d. legal or business career and opposed the war

e. academic or medical career and supported the war

Career	Support	Opposed	T
Military	9	3	12
Medical	8	16	24
Legal	15	12	27
Business	18	6	24
Academics	3	10	13
Totals	53	47	100

64. Referring to Exercise 63, what is the probability the person chosen

a. had a career of any kind or opposed the war

b. had a medical career or supported the war

c. supported the war or had a military career

d. had a medical or a legal career

e. supported or opposed the war

▶ **EXTENDING THE CONCEPT**

69. The function $f(x) = (\frac{1}{2})^x$ gives the probability that x number of flips will all result in heads (or tails). Compute the probability that 20 flips results in *20 heads in a row,* then use the Internet or some other resource to find the probability of winning a state lottery. Which is more likely to happen (which has the greater probability)? Were you surprised?

70. Recall that a function is a relation in which each element of the domain is paired with only one element from the range. Is the relation defined by $C(x) = {}_nC_x$ (n is a constant) a function? To

65. The Board of Directors for a large hospital has 15 members. There are six doctors of nephrology (kidneys), five doctors of gastroenterology (stomach and intestines), and four doctors of endocrinology (hormones and glands). Eight of them will be selected to visit the nation's premier hospitals on a 3-week, expenses-paid tour. What is the probability the group of eight selected consists of exactly

a. four nephrologists and four gastroenterologists

b. three endocrinologists and five nephrologists

66. A support group for hodophobics (an irrational fear of travel) has 32 members. There are 15 aviophobics (fear of air travel), eight siderodrophobics (fear of train travel), and nine thalassophobics (fear of ocean travel) in the group. Twelve of them will be randomly selected to participate in a new therapy. What is the probability the group of 12 selected consists of exactly

a. two aviophobics, six siderodrophobics, and four thalassophobics

b. five thalassophobics, four aviophobics, and three siderodrophobics

67. A trained chimpanzee is given a box containing eight wooden cubes with the letters p, a, r, a, l, l, e, l printed on them (one letter per block). Assuming the chimp can't read or spell, what is the probability he draws the eight blocks in order and actually forms the word "parallel"?

68. A number is called a "perfect number" if the sum of its proper factors is equal to the number itself. Six is the first perfect number since the sum of its proper factors is six: $1 + 2 + 3 = 6$. Twenty-eight is the second since: $1 + 2 + 4 + 7 + 14 = 28$. A young child is given a box containing eight wooden blocks with the following numbers (one per block) printed on them: four 3's, two 5's, one 0, and one 6. What is the probability she draws the eight blocks in order and forms the fifth perfect number: 33,550,336?

investigate, plot the points generated by $C(x) = {}_6C_x$ for $x = 0$ to $x = 6$ and answer the following questions:

a. Is the resulting graph continuous or discrete (made up of distinct points)?

b. Does the resulting graph pass the vertical line test?

c. Discuss the features of the relation and its graph, including the domain, range, maximum or minimum values, and symmetries observed.

► MAINTAINING YOUR SKILLS

71. (6.1) Solve the system using matrices and

row reduction: $\begin{cases} x - 2y + 3z = 10 \\ 2x + y - z = 18 \\ 3x - 2y + z = 26 \end{cases}$

72. (4.4) Complete the following logarithmic properties:

$\log_b b = \underline{}$ $\log_b 1 = \underline{}$

$\log_b b^n = \underline{}$ $b^{\log_b n} = \underline{}$

73. (3.6/3.7) Solve the inequality by graphing the function and labeling the appropriate interval(s):

$\dfrac{x^2 - 1}{x} \geq 0.$

74. (8.3) A rubber ball is dropped from a height of 25 ft onto a hard surface. With each bounce, it rebounds 60% of the height from which it last fell. Use sequences/series to find (a) the height of the sixth bounce, (b) the total distance traveled up to the sixth bounce, and (c) the distance the ball will travel before coming to rest.

8.7 | The Binomial Theorem

Learning Objectives

In Section 8.7 you will learn how to:

☐ **A.** Use Pascal's triangle to find $(a + b)^n$

☐ **B.** Find binomial coefficients using $\binom{n}{k}$ notation

☐ **C.** Use the binomial theorem to find $(a + b)^n$

☐ **D.** Find a specific term of a binomial expansion

Strictly speaking, a binomial is a polynomial with two terms. This limits us to terms with real number coefficients and whole number powers on variables. In this section, we will loosely regard a binomial as the sum or difference of *any* two terms. Hence $3x^2 - y^4$, $\sqrt{x} + 4$, $x + \dfrac{1}{x}$, and $-\dfrac{1}{2} + \dfrac{\sqrt{3}}{2}i$ are all "binomials." Our goal is to develop an ability to raise a binomial to any natural number power, with the results having important applications in genetics, probability, polynomial theory, and other areas. The tool used for this purpose is called the *binomial theorem*.

A. Binomial Powers and Pascal's Triangle

Much of our mathematical understanding comes from a study of patterns. One area where the study of patterns has been particularly fruitful is **Pascal's triangle** (Figure 8.20), named after the French scientist Blaise Pascal (although the triangle was well known before his time). It begins with a "1" at the vertex of the triangle, with 1's extending diagonally downward to the left and right as shown. The entries on the interior of the triangle are found by adding the two entries directly above and to the left and right of each new position.

There are a variety of patterns hidden within the triangle. In this section, we'll use the *horizontal rows* of the triangle to help us raise a binomial to various powers. To begin, recall that $(a + b)^0 = 1$ and $(a + b)^1 = 1a + 1b$ (unit coefficients are included for emphasis). In our earlier work, we saw that a binomial square (a binomial raised to the second power) always

Figure 8.20

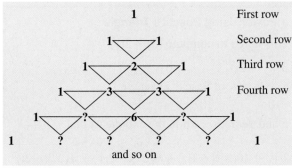

First row
Second row
Third row
Fourth row

and so on

followed the pattern $(a + b)^2 = 1a^2 + 2ab + 1b^2$. Observe the overall pattern that is developing as we include $(a + b)^3$:

$(a + b)^0$	1	row 1
$(a + b)^1$	$1a + 1b$	row 2
$(a + b)^2$	$1a^2 + 2ab + 1b^2$	row 3
$(a + b)^3$	$1a^3 + 3a^2b + 3ab^2 + 1b^3$	row 4

Apparently the coefficients of $(a + b)^n$ will occur in row $n + 1$ of Pascal's triangle. Also observe that in each term of the expansion, the exponent of the first term a *decreases by 1* as the exponent on the second term b *increases by 1*, keeping the degree of each term constant (recall the degree of a term with more than one variable is the sum of the exponents).

$$1a^3b^0 + 3a^2b^1 + 3a^1b^2 + 1a^0b^3$$

$$\begin{array}{cccc} 3+0 & 2+1 & 1+2 & 0+3 \\ \text{degree 3} & \text{degree 3} & \text{degree 3} & \text{degree 3} \end{array}$$

These observations help us to quickly expand a binomial power.

EXAMPLE 1 ▶ **Expanding a Binomial Using Pascal's Triangle**

Use Pascal's triangle and the patterns noted to expand $(x + \frac{1}{2})^4$.

Solution ▶ Working step-by-step we have
1. The coefficients will be in the fifth row of Pascal's triangle.
$$\begin{array}{ccccc} 1 & 4 & 6 & 4 & 1 \end{array}$$
2. The exponents on x begin at 4 and *decrease,* while the exponents on $\frac{1}{2}$ begin at 0 and *increase.*

$$1x^4\left(\frac{1}{2}\right)^0 + 4x^3\left(\frac{1}{2}\right)^1 + 6x^2\left(\frac{1}{2}\right)^2 + 4x^1\left(\frac{1}{2}\right)^3 + 1x^0\left(\frac{1}{2}\right)^4$$

3. Simplify each term.

The result is $x^4 + 2x^3 + \frac{3}{2}x^2 + \frac{1}{2}x + \frac{1}{16}$.

Now try Exercises 7 through 10 ▶

If the exercise involves a difference rather than a sum, we simply rewrite the expression using algebraic addition and proceed as before.

EXAMPLE 2 ▶ **Raising a Complex Number to a Power Using Pascal's Triangle**

Use Pascal's triangle and the patterns noted to compute $(3 - 2i)^5$.

Solution ▶ Begin by rewriting $(3 - 2i)^5$ as $[3 + (-2i)]^5$.
1. The coefficients will be in the sixth row of Pascal's triangle.
$$\begin{array}{cccccc} 1 & 5 & 10 & 10 & 5 & 1 \end{array}$$
2. The exponents on 3 begin at 5 and *decrease,* while the exponents on $(-2i)$ begin at 0 and *increase.*

$$\mathbf{1}(3^5)(-2i)^0 + \mathbf{5}(3^4)(-2i)^1 + \mathbf{10}(3^3)(-2i)^2 + \mathbf{10}(3^2)(-2i)^3 + \mathbf{5}(3^1)(-2i)^4 + \mathbf{1}(3^0)(-2i)^5$$

3. Simplify each term.

$$243 - 810i - 1080 + 720i + 240 - 32i$$

The result is $-597 - 122i$.

☑ **A.** You've just learned how to use Pascal's triangle to find $(a + b)^n$

Now try Exercises 11 and 12 ▶

> **Expanding Binomial Powers $(a + b)^n$**
>
> **1.** The coefficients will be in row $n + 1$ of Pascal's triangle.
> **2.** The exponents on *the first term* begin at n and *decrease,* while the exponents on *the second term* begin at 0 and *increase.*
> **3.** For any binomial difference $(a - b)^n$, rewrite the base as $[a + (-b)]^n$ using algebraic addition and proceed as before, then simplify each term.

B. Binomial Coefficients and Factorials

Pascal's triangle can easily be used to find the coefficients of $(a + b)^n$, as long as the exponent is relatively small. If we needed to expand $(a + b)^{25}$, writing out the first 26 rows of the triangle would be rather tedious. To overcome this limitation, we introduce a *formula* for the binomial coefficients that enables us to find the coefficients of any expansion.

> **The Binomial Coefficients**
>
> For natural numbers n and r where $n \geq r$, the expression $\dbinom{n}{r}$, read "n choose r," is called the **binomial coefficient** and evaluated as:
> $$\binom{n}{r} = \frac{n!}{r!(n - r)!}$$

In Example 1, we found the coefficients of $(a + b)^4$ using the fifth or $(n + 1)$st row of Pascal's triangle. In Example 3, these coefficients are found using the formula for binomial coefficients.

EXAMPLE 3 ▶ **Computing Binomial Coefficients**

Evaluate $\dbinom{n}{r} = \dfrac{n!}{r!(n - r)!}$ as indicated:

a. $\dbinom{4}{1}$ **b.** $\dbinom{4}{2}$ **c.** $\dbinom{4}{3}$

Solution ▶   **a.** $\dbinom{4}{1} = \dfrac{4!}{1!(4 - 1)!} = \dfrac{4 \cdot 3!}{1!3!} = 4$

       b. $\dbinom{4}{2} = \dfrac{4!}{2!(4 - 2)!} = \dfrac{4 \cdot 3 \cdot 2!}{2!2!} = \dfrac{4 \cdot 3}{2} = 6$

       c. $\dbinom{4}{3} = \dfrac{4!}{3!(4 - 3)!} = \dfrac{4 \cdot 3!}{3!1!} = 4$

Now try Exercises 13 through 20 ▶

Note $\dbinom{4}{1} = 4$, $\dbinom{4}{2} = 6$, and $\dbinom{4}{3} = 4$ give the *interior entries* in the fifth row of Pascal's triangle: 1 4 6 4 1. For consistency and symmetry, we define $0! = 1$, which enables the formula to generate all entries of the triangle, including the "1's."

$$\binom{4}{0} = \frac{4!}{0!(4 - 0)!} \quad \text{apply formula} \qquad \binom{4}{4} = \frac{4!}{4!(4 - 4)!} = \frac{4!}{4! \cdot 0!} \quad \text{apply formula}$$

$$= \frac{4!}{1 \cdot 4!} = 1 \quad \text{0! = 1} \qquad\qquad\quad = \frac{4!}{4! \cdot 1} = 1 \quad \text{0! = 1}$$

The formula for $\binom{n}{r}$ with $0 \le r \le n$ now gives all coefficients in the $(n+1)$st row. For $n = 5$, we have

$$\binom{5}{0} \quad \binom{5}{1} \quad \binom{5}{2} \quad \binom{5}{3} \quad \binom{5}{4} \quad \binom{5}{5}$$
$$1 \qquad 5 \qquad 10 \qquad 10 \qquad 5 \qquad 1$$

EXAMPLE 4 ▶ **Computing Binomial Coefficients**

Compute the binomial coefficients:

a. $\binom{9}{0}$ b. $\binom{9}{1}$ c. $\binom{6}{5}$ d. $\binom{6}{6}$

Solution ▶

a. $\binom{9}{0} = \dfrac{9!}{0!(9-0)!}$ b. $\binom{9}{1} = \dfrac{9!}{1!(9-1)!}$

$= \dfrac{9!}{9!} = 1$ $= \dfrac{9!}{8!} = 9$

c. $\binom{6}{5} = \dfrac{6!}{5!(6-5)!}$ d. $\binom{6}{6} = \dfrac{6!}{6!(6-6)!}$

$= \dfrac{6!}{5!} = 6$ $= \dfrac{6!}{6!} = 1$

Now try Exercises 21 through 24 ▶

You may have noticed that the formula for $\binom{n}{r}$ is identical to that of $_nC_r$, and both yield like results for given values of n and r. For future use, it will help to commit the general results from Example 4 to memory: $\binom{n}{0} = 1$, $\binom{n}{1} = n$, $\binom{n}{n-1} = n$, and $\binom{n}{n} = 1$.

☑ **B.** You've just learned how to find binomial coefficients using $\binom{n}{k}$ notation

C. The Binomial Theorem

Using $\binom{n}{r}$ notation and the observations made regarding binomial powers, we can now state the **binomial theorem**.

Binomial Theorem

For any binomial $(a + b)$ and natural number n,

$$(a + b)^n = \binom{n}{0}a^n b^0 + \binom{n}{1}a^{n-1}b^1 + \binom{n}{2}a^{n-2}b^2 + \cdots$$

$$+ \binom{n}{n-1}a^1 b^{n-1} + \binom{n}{n}a^0 b^n$$

The theorem can also be stated in summation form as

$$(a + b)^n = \sum_{r=0}^{n} \binom{n}{r} a^{n-r} b^r$$

The expansion actually looks overly impressive in this form, and it helps to summarize the process in words, as we did earlier. The exponents on the first term a begin at n and decrease, while the exponents on the second term b begin at 0 and increase, keeping the degree of each term constant. The $\binom{n}{r}$ notation simply gives the coefficients of each term. As a final note, observe that the r in $\binom{n}{r}$ gives the exponent on b.

EXAMPLE 5 ▶ Expanding a Binomial Using the Binomial Theorem

Expand $(a + b)^6$ using the binomial theorem.

Solution ▶ $(a + b)^6 = \binom{6}{0}a^6b^0 + \binom{6}{1}a^5b^1 + \binom{6}{2}a^4b^2 + \binom{6}{3}a^3b^3 + \binom{6}{4}a^2b^4 + \binom{6}{5}a^1b^5 + \binom{6}{6}a^0b^6$

$= \dfrac{6!}{0!6!}a^6 + \dfrac{6!}{1!5!}a^5b^1 + \dfrac{6!}{2!4!}a^4b^2 + \dfrac{6!}{3!3!}a^3b^3 + \dfrac{6!}{4!2!}a^2b^4 + \dfrac{6!}{5!1!}a^1b^5 + \dfrac{6!}{6!0!}b^6$

$= 1a^6 + 6a^5b + 15a^4b^2 + 20a^3b^3 + 15a^2b^4 + 6ab^5 + 1b^6$

Now try Exercises 25 through 32 ▶

EXAMPLE 6 ▶ Using the Binomial Theorem to Find the Initial Terms of an Expansion

Find the first three terms of $(2x + y^2)^{10}$.

Solution ▶ Use the binomial theorem with $a = 2x$, $b = y^2$, and $n = 10$.

$(2x + y^2)^{10} = \binom{10}{0}(2x)^{10}(y^2)^0 + \binom{10}{1}(2x)^9(y^2)^1 + \binom{10}{2}(2x)^8(y^2)^2 + \cdots$ first three terms

$= (1)1024x^{10} + (10)512x^9y^2 + \dfrac{10!}{2!8!}256x^8y^4 + \cdots$ $\binom{10}{0} = 1, \binom{10}{1} = 10$

✓ **C.** You've just learned how to use the binomial theorem to find $(a + b)^n$

$= 1024x^{10} + 5120x^9y^2 + (45)256x^8y^4 + \cdots$ $\dfrac{10!}{2!8!} = 45$

$= 1024x^{10} + 5120x^9y^2 + 11{,}520x^8y^4 + \cdots$ result

Now try Exercises 33 through 36 ▶

D. Finding a Specific Term of the Binomial Expansion

In some applications of the binomial theorem, our main interest is a *specific term* of the expansion, rather than the expansion as a whole. To find a specified term, it helps to consider that the expansion of $(a + b)^n$ has $n + 1$ terms: $(a + b)^0$ has one term, $(a + b)^1$ has two terms, $(a + b)^2$ has three terms, and so on. Because the notation $\binom{n}{r}$ always begins at $r = 0$ for the first term, the value of r will be *1 less than the term we are seeking*. In other words, for the seventh term of $(a + b)^9$, we use $r = 6$.

The kth Term of a Binomial Expansion

For the binomial expansion $(a + b)^n$, the kth term is given by

$$\binom{n}{r}a^{n-r}b^r, \text{ where } r = k - 1.$$

EXAMPLE 7 ▶ Finding a Specific Term of a Binomial Expansion

Find the eighth term in the expansion of $(x + 2y)^{12}$.

Solution ▶ By comparing $(x + 2y)^{12}$ to $(a + b)^n$ we have $a = x$, $b = 2y$, and $n = 12$. Since we want the eighth term, $k = 8 \rightarrow r = 7$. The eighth term of the expansion is

$$\binom{12}{7} x^5 (2y)^7 = \frac{12!}{7!5!} 128 x^5 y^7 \qquad 2^7 = 128$$

$$= (792)(128 x^5 y^7) \qquad \binom{12}{7} = 792$$

$$= 101{,}376 x^5 y^7 \qquad \text{result}$$

Now try Exercises 37 through 42 ▶

One application of the binomial theorem involves a **binomial experiment** and **binomial probability.** For binomial probabilities, the following must be true: (1) The experiment must have only two possible outcomes, typically called success and failure, and (2) if the experiment has n trials, the probability of success must be constant for all n trials. If the probability of success for each trial is p, the formula $\binom{n}{k}(1 - p)^{n-k} p^k$ gives the probability that exactly k trials will be successful.

Binomial Probability

Given a binomial experiment with n trials, where the probability for success in each trial is p. The probability that exactly k trials are successful is given by

$$\binom{n}{k}(1 - p)^{n-k} p^k.$$

EXAMPLE 8 ▶ Applying the Binomial Theorem–Binomial Probability

Paula Rodrigues has a free-throw shooting average of 85%. On the last play of the game, with her team behind by three points, she is fouled at the three-point line, and is awarded two additional free throws via technical fouls on the opposing coach (a total of five free-throws). What is the probability she makes *at least three* (meaning they at least tie the game)?

Solution ▶ Here we have $p = 0.85$, $1 - p = 0.15$, and $n = 5$. The key idea is to recognize the phrase *at least three* means "3 or 4 or 5." So $P(\text{at least 3}) = P(3 \cup 4 \cup 5)$.

$$P(\text{at least 3}) = P(3 \cup 4 \cup 5) \qquad \text{"or" implies a union}$$

$$= P(3) + P(4) + P(5) \qquad \text{sum of probabilities (mutually exclusive events)}$$

$$= \binom{5}{3}(0.15)^2(0.85)^3 + \binom{5}{4}(0.15)^1(0.85)^4 + \binom{5}{5}(0.15)^0(0.85)^5$$

$$\approx 0.1382 + 0.3915 + 0.4437$$

$$= 0.9734$$

Paula's team has an excellent chance ($\approx 97.3\%$) of at least tying the game.

☑ D. You've just learned how to find a specific term of a binomial expansion

New try Exercises 45 and 46 ▶

8.7 EXERCISES

▶ **CONCEPTS AND VOCABULARY**

**Fill in the blank with the appropriate word or phrase.
Carefully reread the section if needed.**

1. In any binomial expansion, there is always _____ more term than the power applied.

2. In all terms in the expanded form of $(a + b)^n$, the exponents on a and b must sum to _____.

3. To expand a binomial *difference* such as $(a - 2b)^5$, we rewrite the binomial as _____ and proceed as before.

4. In a binomial experiment with n trials, the probability there are exactly k successes is given by the formula _____.

5. Discuss why the expansion of $(a + b)^n$ has $n + 1$ terms.

6. For any defined binomial experiment, discuss the relationships between the phrases, "exactly k success," and "at least k successes."

▶ **DEVELOPING YOUR SKILLS**

Use *Pascal's triangle* and the patterns explored to write each expansion.

7. $(x + y)^5$ 8. $(a + b)^6$ 9. $(2x + 3)^4$

10. $(x^2 + \frac{1}{3})^3$ 11. $(1 - 2i)^5$ 12. $(2 - 5i)^4$

Evaluate each of the following

13. $\binom{7}{4}$ 14. $\binom{8}{2}$ 15. $\binom{5}{3}$

16. $\binom{9}{5}$ 17. $\binom{20}{17}$ 18. $\binom{30}{26}$

19. $\binom{40}{3}$ 20. $\binom{45}{3}$ 21. $\binom{6}{0}$

22. $\binom{5}{0}$ 23. $\binom{15}{15}$ 24. $\binom{10}{10}$

Use the *binomial theorem* to expand each expression. Write the general form first, then simplify.

25. $(c + d)^5$ 26. $(v + w)^4$ 27. $(a - b)^6$

28. $(x - y)^7$ 29. $(2x - 3)^4$ 30. $(a - 2b)^5$

31. $(1 - 2i)^3$ 32. $(2 + \sqrt{3}i)^5$

Use the *binomial theorem* to write the first three terms.

33. $(x + 2y)^9$ 34. $(3p + q)^8$ 35. $(v^2 - \frac{1}{2}w)^{12}$

36. $(\frac{1}{2}a - b^2)^{10}$

Find the indicated term for each binomial expansion.

37. $(x + y)^7$; 4th term   38. $(m + n)^6$; 5th term

39. $(p - 2)^8$; 7th term   40. $(a - 3)^{14}$; 10th term

41. $(2x + y)^{12}$; 11th term   42. $(3n + m)^9$; 6th term

▶ **WORKING WITH FORMULAS**

43. **Binomial probability:** $P(k) = \binom{n}{k}\left(\frac{1}{2}\right)^k\left(\frac{1}{2}\right)^{n-k}$

    The theoretical probability of getting exactly k heads in n flips of a fair coin is given by the formula above. What is the probability that you would get 5 heads in 10 flips of the coin?

44. **Binomial probability:** $P(k) = \binom{n}{k}\left(\frac{1}{5}\right)^k\left(\frac{4}{5}\right)^{n-k}$

    A multiple choice test has five options per question. The probability of guessing correctly k times out of n questions is found using the formula shown. What is the probability a person scores a 70% by guessing randomly (7 out of 10 questions correct)?

▶ **APPLICATIONS**

45. Batting averages: Tony Gwynn (San Diego Padres) had a lifetime batting average of 0.347, ranking him as one of the greatest hitters of all time. Suppose he came to bat five times in any given game.

   a. What is the probability that he will get exactly three hits?

   b. What is the probability that he will get at least three hits?

46. Pollution testing: Erin suspects that a nearby iron smelter is contaminating the drinking water over a large area. A statistical study reveals that 83% of the wells in this area are likely contaminated. If the figure is accurate, find the probability that if another 10 wells are tested

   a. exactly 8 are contaminated

   b. at least 8 are contaminated

47. Late rental returns: The manager of Victor's DVD Rentals knows that 6% of all DVDs rented are returned *late*. Of the eight videos rented in the last hour, what is the probability that

   a. exactly five are returned on time

   b. exactly six are returned on time

   c. at least six are returned on time

   d. none of them will be returned late

48. Opinion polls: From past experience, a research firm knows that 20% of telephone respondents will agree to answer an opinion poll. If 20 people are contacted by phone, what is the probability that

   a. exactly 18 refuse to be polled

   b. exactly 19 refuse to be polled

   c. at least 18 refuse to be polled

   d. none of them agree to be polled

▶ **EXTENDING THE CONCEPT**

49. Prior to calculators and computers, the binomial theorem was frequently used to approximate the value of compound interest given by the expression $\left(1 + \dfrac{r}{n}\right)^{nt}$ by expanding the first three terms. For example, if the interest rate were 8% ($r = 0.08$) and the interest was compounded quarterly ($n = 4$) for 5 yr ($t = 5$), we have $\left(1 + \frac{0.08}{4}\right)^{(4)(5)} = (1 + 0.02)^{20}$. The first three terms of the expansion give a value of: $1 + 20(0.02) + 190(0.0004) = 1.476$.

   a. Calculate the percent error:
$$\%\text{error} = \frac{\text{approximate value}}{\text{actual value}}$$

   b. What is the percent error if only two terms are used.

50. If you sum the entries in each row of Pascal's triangle, a pattern emerges. Find a formula that generalizes the result for any row of the triangle, and use it to find the sum of the entries in the 12th row of the triangle.

51. Show that $\binom{n}{k} = \binom{n}{n-k}$ for $n = 6$ and $k \le 6$.

52. The *derived polynomial* of $f(x)$ is $f(x + h)$ or the original polynomial evaluated at $x + h$. Use Pascal's triangle or the binomial theorem to find the derived polynomial for $f(x) = x^3 + 3x^2 + 5x - 11$. Simplify the result completely.

▶ **MAINTAINING YOUR SKILLS**

53. (2.7) Graph the function shown and find
$$f(3): f(x) = \begin{cases} x + 2 & x \le 2 \\ (x - 4)^2 & x > 2 \end{cases}$$

54. (3.3) Show that $x = -1 + i$ is a solution to $x^4 + 2x^3 - x^2 - 6x - 6 = 0$.

55. (3.4) Graph the function $g(x) = x^3 - x^2 - 6x$. Clearly indicate all intercepts and intervals where $g(x) > 0$.

56. (4.5) If $2500 is deposited at 6% compounded continuously, how much would be in the account 10 years later?

SUMMARY AND CONCEPT REVIEW

SECTION 8.1   Sequences and Series

KEY CONCEPTS

- A *finite sequence* is a function a_n whose domain is the set of natural numbers from 1 to n.
- The terms of the sequence are labeled $a_1, a_2, a_3, \ldots, a_{k-1}, a_k, a_{k+1}, \ldots, a_{n-2}, a_{n-1}, a_n$.
- The expression a_n, which defines the sequence (generates the terms in order), is called the nth term.
- An *infinite sequence* is a function whose domain is the set of natural numbers.
- When each term of a sequence is larger than the preceding term, it is called an increasing sequence.
- When each term of a sequence is smaller than the preceding term, it is called a decreasing sequence.
- When successive terms of a sequence alternate in sign, it is called an alternating sequence.
- When the terms of a sequence are generated using previous term(s), it is called a recursive sequence.
- Sequences are sometimes defined using factorials, which are the product of a given natural number with all natural numbers that precede it: $n! = n \cdot (n-1) \cdot (n-2) \cdot \cdots \cdot 3 \cdot 2 \cdot 1$.
- Given the sequence $a_1, a_2, a_3, a_4, \ldots, a_n$ the sum is called a finite series and is denoted S_n.
- $S_n = a_1 + a_2 + a_3 + a_4 + \cdots + a_n$. The sum of the first n terms is called a partial sum.
- In sigma notation, the expression $\displaystyle\sum_{i=1}^{k} a_i = a_1 + a_2 + \cdots + a_k$ represents a finite series.
- When sigma notation is used, the letter "i" is called the index of summation.

EXERCISES

Write the first four terms that are defined and the value of a_{10}.

1. $a_n = 5n - 4$

2. $a_n = \dfrac{n+1}{n^2+1}$

Find the general term a_n for each sequence, and the value of a_6.

3. $1, 16, 81, 256, \ldots$

4. $-17, -14, -11, -8, \ldots$

Find the eighth partial sum (S_8).

5. $\frac{1}{2}, \frac{1}{4}, \frac{1}{8}, \ldots$

6. $-21, -19, -17, \ldots$

Evaluate each sum.

7. $\displaystyle\sum_{n=1}^{7} n^2$

8. $\displaystyle\sum_{n=1}^{5} (3n - 2)$

Write the first five terms that are defined.

9. $a_n = \dfrac{n!}{(n-2)!}$

10. $\begin{cases} a_1 = \frac{1}{2} \\ a_{n+1} = 2a_n - \frac{1}{4} \end{cases}$

Write as a single summation and evaluate.

11. $\displaystyle\sum_{n=1}^{7} n^2 + \sum_{n=1}^{7} (3n - 2)$

SECTION 8.2   Arithmetic Sequences

KEY CONCEPTS

- In an arithmetic sequence, successive terms are found by adding a fixed constant to the preceding term.
- In a sequence, if there exists a number d, called the common difference, such that $a_{k+1} - a_k = d$, then the sequence is arithmetic. Alternatively, $a_{k+1} = a_k + d$ for $k \geq 1$.

- The nth term n of an arithmetic sequence is given by $a_n = a_1 + (n - 1)d$, where a_1 is the first term and d is the common difference.
- If the initial term is unknown or is not a_1 the nth term can be written $a_n = a_k + (n - k)d$, where the subscript of the term a_k and the coefficient of d sum to n.
- For an arithmetic sequence with first term a_1, the nth partial sum (the sum of the first n terms) is given by

$$S_n = \frac{n(a_1 + a_n)}{2}.$$

EXERCISES

Find the general term (a_n) for each arithmetic sequence. Then find the indicated term.

12. $2, 5, 8, 11, \ldots$; find a_{40}

13. $3, 1, -1, -3, \ldots$; find a_{35}

Find the sum of each series.

14. $-1 + 3 + 7 + 11 + \cdots + 75$

15. $1 + 4 + 7 + 10 + \cdots + 88$

16. $3 + 6 + 9 + 12 + \cdots$; S_{20}

17. $1 + \frac{3}{4} + \frac{1}{2} + \frac{1}{4} + \cdots$; S_{15}

18. $\displaystyle\sum_{n=1}^{25} (3n - 4)$

19. $\displaystyle\sum_{n=1}^{40} (4n - 1)$

SECTION 8.3 Geometric Sequences

KEY CONCEPTS

- In a geometric sequence, successive terms are found by multiplying the preceding term by a nonzero constant.
- In other words, if there exists a number r, called the common ratio, such that $\dfrac{a_{k+1}}{a_k} = r$, then the sequence is geometric. Alternatively, we can write $a_{k+1} = a_k r$ for $k \geq 1$.
- The nth term a_n of a geometric sequence is given by $a_n = a_1 r^{n-1}$, where a_1 is the first term and a_n represents the general term of a finite sequence.
- If the initial term is unknown or is not a_1, the nth term can be written $a_n = a_k r^{n-k}$, where the subscript of the term a_k and the exponent on r sum to n.
- The nth partial sum of a geometric sequence is $S_n = \dfrac{a_1(1 - r^n)}{1 - r}$.
- If $|r| < 1$, the sum of an infinite geometric series is $S_\infty = \dfrac{a_1}{1 - r}$.

EXERCISES

Find the indicated term for each geometric sequence.

20. $a_1 = 5, r = 3$; find a_7

21. $a_1 = 4, r = \sqrt{2}$; find a_7

22. $a_1 = \sqrt{7}, r = \sqrt{7}$; find a_8

Find the indicated sum, if it exists.

23. $16 - 8 + 4 - \cdots$ find S_7

24. $2 + 6 + 18 + \cdots$; find S_8

25. $\frac{4}{5} + \frac{2}{5} + \frac{1}{5} + \frac{1}{10} + \cdots$; find S_{12}

26. $4 + 8 + 12 + 24 + \cdots$

27. $5 + 0.5 + 0.05 + 0.005 + \cdots$

28. $6 - 3 + \frac{3}{2} - \frac{3}{4} + \cdots$

29. $\displaystyle\sum_{n=1}^{8} 5\left(\frac{2}{3}\right)^n$

30. $\displaystyle\sum_{n=1}^{\infty} 12\left(\frac{4}{3}\right)^n$

31. $\displaystyle\sum_{n=1}^{\infty} 5\left(\frac{1}{2}\right)^n$

32. Charlene began to work for Grayson Natural Gas in January of 1990 with an annual salary of $26,000. Her contract calls for a $1220 raise each year. Use a sequence/series to compute her salary after nine years, and her total earnings up to and including that year. (*Hint:* For $a_1 = 26,000$, her salary after *9 yrs* will be what term of the sequence?)

33. Sumpter reservoir contains 121,500 ft^3 of water and is being drained in the following way. Each day one-third of the water is *drained* (and not replaced). Use a sequence/series to compute how much water *remains in the pond after 7 days.*

34. Credit-hours taught at Cody Community College have been increasing at 7% per year since it opened in 2000 and taught 1225 credit-hours. For the new faculty, the college needs to predict the number of credit-hours that will be taught in 2009. Use a sequence/series to compute the credit-hours for 2009 and to find the total number of credit hours taught through the 2009 school year.

SECTION 8.4   Mathematical Induction

KEY CONCEPTS

- Functions written in subscript notation can be evaluated, graphed, and composed with other functions.
- A sum formula involving only natural numbers n as inputs can be proven valid using a proof by induction. Given that S_n represents a sum formula involving natural numbers, if (1) S_1 is true and (2) $S_k + a_{k+1} = S_{k+1}$, then S_n must be true for all natural numbers.
- Proof by induction can also be used to validate other relationships, using a more general statement of the principle. Let S_n be a statement involving the natural numbers n. If (1) S_1 is true (S_n for $n = 1$) and (2) the truth of S_k implies that S_{k+1} is also true, then S_n must be true for all natural numbers n.

EXERCISES

Use the principle of mathematical induction to prove the indicated sum formula is true for all natural numbers n.

35. $1 + 2 + 3 + 4 + 5 + \cdots + n$;

$a_n = n$ and $S_n = \dfrac{n(n+1)}{2}$.

36. $1 + 4 + 9 + 16 + 25 + 36 + \cdots + n^2$;

$a_n = n^2$ and $S_n = \dfrac{n(n+1)(2n+1)}{6}$.

Use the principle of mathematical induction to prove that each statement is true for all natural numbers n.

37. $4^n \geq 3n + 1$ **38.** $6 \cdot 7^{n-1} \leq 7^n - 1$ **39.** $3^n - 1$ is divisible by 2

SECTION 8.5   Counting Techniques

KEY CONCEPTS

- An experiment is any task that can be repeated and has a well-defined set of possible outcomes.
- Each repetition of an experiment is called a trial.
- Any potential outcome of an experiment is called a sample outcome.
- The set of all sample outcomes is called the sample space.
- An experiment with N (equally likely) sample outcomes that is repeated t times, has a sample space with N^t elements.
- If a sample outcome can be used more than once, the counting is said to be with repetition. If a sample outcome can be used only once the counting is said to be without repetition.
- The fundamental principle of counting states: If there are p possibilities for a first task, q possibilities for the second, and r possibilities for the third, the total number of ways the experiment can be completed is pqr. This fundamental principle can be extended to include any number of tasks.
- If the elements of a sample space have precedence or priority (order or rank is important), the number of elements is counted using a permutation, denoted $_nP_r$ and read, "the distinguishable permutations of n objects taken r at a time."
- To expand $_nP_r$, we can write out the first r factors of $n!$ or use the formula $_nP_r = \dfrac{n!}{(n-r)!}$.
- If any of the sample outcomes are identical, certain permutations will be nondistinguishable. In a set containing n elements where one element is repeated p times, another is repeated q times, and another r times ($p + q + r \leq n$), the number of distinguishable permutations is given by $\dfrac{_nP_n}{p!q!r!} = \dfrac{n!}{p!q!r!}$.

• If the elements of a set have no rank, order, or precedence (as in a committee of colleagues) permutations with the same elements are considered identical. The result is the number of combinations, $_nC_r = \dfrac{n!}{r!(n-r)!}$.

EXERCISES

40. Three slips of paper with the letters A, B, and C are placed in a box and randomly drawn one at a time. Show all possible ways they can be drawn using a tree diagram.

41. The combination for a certain bicycle lock consists of three digits. How many combinations are possible if (a) repetition of digits is not allowed and (b) repetition of digits is allowed.

42. Jethro has three work shirts, four pairs of work pants, and two pairs of work shoes. How many different ways can he dress himself (shirt, pants, shoes) for a day's work?

43. From a field of 12 contestants in a pet show, three cats are chosen at random to be photographed for a publicity poster. In how many different ways can the cats be chosen?

44. How many subsets can be formed from the elements of this set: { ■, ·, □, ■, ▣ }?

45. Compute the following values by hand, showing all work:

    a. $7!$ **b.** $_7P_4$ **c.** $_7C_4$

46. Six horses are competing in a race at the McClintock Ranch. Assuming there are no ties, (a) how many different ways can the horses finish the race? (b) How many different ways can the horses finish first, second, and third place? (c) How many finishes are possible if it is well known that Nellie-the-Nag will finish last and Sea Biscuit will finish first?

47. How many distinguishable permutations can be formed from the letters in the word "tomorrow"?

48. Quality Construction Company has 12 equally talented employees. (a) How many ways can a three-member crew be formed to complete a small job? (b) If the company is in need of a Foreman, Assistant Foreman, and Crew Chief, in how many ways can the positions be filled?

SECTION 8.6 Introduction to Probability

KEY CONCEPTS

• An event E is any designated set of sample outcomes.
• Given S is a sample space of equally likely sample outcomes and E is an event relative to S, the probability of E, written $P(E)$, is computed as $P(E) = \dfrac{n(E)}{n(S)}$, where $n(E)$ represents the number of elements in E, and $n(S)$ represents the number of elements in S.
• The complement of an event E is the set of sample outcomes in S, but not in E and is denoted $\sim E$.
• Given sample space S and any event E defined relative to S:
  (1) $P(\sim S) = 0$,   (2) $0 \leq P(E) \leq 1$,   (3) $P(S) = 1$,   (4) $P(E) = 1 - P(\sim E)$, and
  (5) $P(E) + P(\sim E) = 1$.
• Two events that have no outcomes in common are said to be mutually exclusive.
• If two events are not mutually exclusive, $P(E_1 \text{ or } E_2) \rightarrow P(E_1 \cup E_2) = P(E_1) + P(E_2) - P(E_1 \cap E_2)$.
• If two events are mutually exclusive, $P(E_1 \text{ or } E_2) \rightarrow P(E_1 \cup E_2) = P(E_1) + P(E_2)$.

EXERCISES

49. One card is drawn from a standard deck. What is the probability the card is a ten or a face card?

50. One card is drawn from a standard deck. What is the probability the card is a Queen or a face card?

51. One die is rolled. What is the probability the result is not a three?

52. Given $P(E_1) = \frac{3}{8}$, $P(E_2) = \frac{3}{4}$, and $P(E_1 \cup E_2) = \frac{5}{6}$, compute $P(E_1 \cap E_2)$.

53. Find $P(E)$ given that $n(E) = {_7C_4} \cdot {_5C_3}$ and $n(S) = {_{12}C_7}$.

54. To determine if more physicians should be hired, a medical clinic tracks the number of days between a patient's request for an appointment and the actual appointment date. The table given shows the probability that a patient must wait "d" days. Based on the table, what is the probability a patient must wait

Wait (days d)	Probability
0	0.002
$0 < d < 10$	0.07
$10 \leq d < 20$	0.32
$20 \leq d < 30$	0.43
$30 \leq d < 40$	0.178

    a. at least 20 days   **b.** less than 20 days

    c. 40 days or less   **d.** over 40 days

    e. less than 40 and more than 10 days   **f.** 30 or more days

SECTION 8.7   The Binomial Theorem

KEY CONCEPTS

- To expand $(a + b)^n$ for n of "moderate size," we can use Pascal's triangle and observed patterns.

- For any natural numbers n and r, where $n \geq r$, the expression $\binom{n}{r}$ (read "n choose r") is called the *binomial coefficient* and evaluated as $\binom{n}{r} = \dfrac{n!}{r!(n-r)!}$.

- If n is large, it is more efficient to expand using the binomial coefficients and binomial theorem.

- The following binomial coefficients are useful/common and should be committed to memory:

$$\binom{n}{0} = 1 \qquad \binom{n}{1} = n \qquad \binom{n}{n-1} = n \qquad \binom{n}{n} = 1$$

- We define $0! = 1$; for example $\binom{n}{n} = \dfrac{n!}{n!(n-n)!} = \dfrac{1}{0!} = \dfrac{1}{1} = 1$.

- The binomial theorem: $(a+b)^n = \binom{n}{0}a^n b^0 + \binom{n}{1}a^{n-1}b^1 + \binom{n}{2}a^{n-2}b^2 + \cdots + \binom{n}{n-1}a^1 b^{n-1} + \binom{n}{n}a^0 b^n$.

- The kth term of $(a+b)^n$ can be found using the formula $\binom{n}{r}a^{n-r}b^r$, where $r = k - 1$.

EXERCISES

55. Evaluate each of the following:

    a. $\binom{7}{5}$ **b.** $\binom{8}{3}$

56. Use Pascal's triangle to expand the binomials:

    a. $(x - y)^4$ **b.** $(1 + 2i)^5$

Use the binomial theorem to:

57. Write the first four terms of

    a. $(a + \sqrt{3})^8$ **b.** $(5a + 2b)^7$

58. Find the indicated term of each expansion.

    a. $(x + 2y)^7$; fourth   **b.** $(2a - b)^{14}$; 10th

MIXED REVIEW

1. Identify each sequence as arithmetic, geometric, or neither. If neither, try to identify the pattern that forms the sequence.

    a. $120, 163, 206, 249, \ldots$

    b. $4, 4, 4, 4, 4, 4, \ldots$

    c. $1, 2, 6, 24, 120, 720, 5040, \ldots$

    d. $2.00, 1.95, 1.90, 1.85, \ldots$

    e. $\frac{5}{8}, \frac{5}{64}, \frac{5}{512}, \frac{5}{4096}, \ldots$

    f. $-5.5, 6.05, -6.655, 7.3205, \ldots$

    g. $0.\overline{1}, 0.\overline{2}, 0.\overline{3}, 0.\overline{4}, \ldots$

    h. $525, 551.25, 578.8125, \ldots$

    i. $\frac{1}{2}, \frac{1}{4}, \frac{1}{6}, \frac{1}{8}, \ldots$

2. Compute by hand (show your work).

a. $10!$

b. $\dfrac{10!}{6!}$

c. $_{10}P_4$

d. $_{10}P_9$

e. $_{10}C_6$

f. $_{10}C_4$

3. The call letters for a television station must consist of four letters and begin with either a K or a W. How many distinct call letters are possible if repeating any letter is not allowed?

4. Given $a_1 = 9$ and $r = \frac{1}{3}$, write out the first five terms and the 15th term.

5. Given $a_1 = 0.1$ and $r = 5$, write out the first five terms and the 15th term.

6. One card is drawn from a well-shuffled deck of standard cards. What is the probability the card is a Queen or an Ace?

7. Two fair dice are rolled. What is the probability the result is not doubles (doubles = same number on both die)?

8. A house in a Boston suburb cost $185,000 in 1985. Each year its value increased by 8%. If this appreciation were to continue, find the value of the house in 2005 and 2015 using a sequence.

9. Evaluate each sum using summation formulas.

a. $\displaystyle\sum_{n=1}^{\infty} \left(\frac{2}{3}\right)^n$

b. $\displaystyle\sum_{n=1}^{10} (9 + 2n)$

c. $\displaystyle\sum_{n=1}^{5} 12n + \sum_{n=1}^{5}(-5) + \sum_{n=1}^{5} n^2$

10. Expand each binomial using the binomial theorem. Simplify each term.

a. $(2x + 5)^5$

b. $(1 - 2i)^4$

11. For $(a + b)^n$, determine:

a. the first three terms for $n = 20$

b. the last three terms for $n = 20$

c. the fifth term for $n = 35$

d. the fifth term for $n = 35$, $a = 0.2$, and $b = 0.8$

12. On average, bears older than 3 yr old increase their weight by 0.87% per day from July to November. If a bear weighed 110 kg on June 30th: (a) identify the type of sequence that gives the bear's weight each day; (b) find the general term for the sequence; and (c) find the bear's weight on July 1, July 2, July 3, July 31, August 31, and September 30.

13. Use a proof by induction to show that

$$3 + 6 + 9 + \cdots + 3n = \frac{3n(n+1)}{2}.$$

14. The owner of an arts and crafts store makes specialty key rings by placing five colored beads on a nylon cord and tying it to the ring that will hold the keys. If there are eight different colors to choose from, (a) how many distinguishable key rings are possible if no colors are repeated? (b) How many distinguishable key rings are possible if a repetition of colors is allowed?

15. Donell bought 15 raffle tickets from the Inner City Children's Music School, and five tickets from the Arbor Day Everyday raffle. The Music School sold a total of 2000 tickets and the Arbor Day foundation sold 550 tickets. For E_1: Donell wins the Music School raffle and E_2: Donell wins the Arbor Day raffle, find $P(E_1 \text{ or } E_2)$.

Find the sum if it exists.

16. $\frac{1}{3} + \frac{2}{3} + 1 + \frac{4}{3} + \cdots + \frac{20}{3}$

17. $0.36 + 0.0036 + 0.000036 + 0.00000036 + \cdots$

Find the first five terms of the sequences in Exercises 18 and 19.

18. $a_n = \dfrac{12!}{(12 - n)!}$

19. $\begin{cases} a_1 = 10 \\ a_{n+1} = a_n\left(\frac{1}{5}\right) \end{cases}$

20. A random survey of 200 college students produces the data shown. One student from this group is randomly chosen for an interview. Use the data to find

a. P(student works more than 10 hr)

b. P(student takes less than 13 credit-hours)

c. P(student works more than 20 hr and takes more than 12 credit-hours)

d. P(student works between 11 and 20 hr or takes 6 to 12 credit-hours)

	0-10 hr	11-20 hr	Over 20 hr	Total
1-5 credits	3	7	10	20
6-12 credits	21	55	48	124
over 13 credits	8	28	20	56
Total:	32	90	78	200

PRACTICE TEST

1. The general term of a sequence is given. Find the first four terms, the 8th term, and the 12th term.

   a. $a_n = \dfrac{2n}{n+3}$

   b. $a_n = \dfrac{(n+2)!}{n!}$

   c. $a_n = \begin{cases} a_1 = 3 \\ a_{n+1} = \sqrt{(a_n)^2 - 1} \end{cases}$

2. Expand each series and evaluate.

   a. $\displaystyle\sum_{k=2}^{6}(2k^2 - 3)$

   b. $\displaystyle\sum_{j=2}^{6}(-1)^j\left(\dfrac{j}{j+1}\right)$

   c. $\displaystyle\sum_{j=1}^{5}(-2)\left(\dfrac{3}{4}\right)^j$

   d. $\displaystyle\sum_{k=1}^{\infty}7\left(\dfrac{1}{2}\right)^k$

3. Identify the first term and the common difference or common ratio. Then find the general term a_n.

   a. $7, 4, 1, -2, \ldots$

   b. $-8, -6, -4, -2, \ldots$

   c. $4, -8, 16, -32, \ldots$

   d. $10, 4, \frac{8}{5}, \frac{16}{25}, \ldots$

4. Find the indicated value for each sequence.

   a. $a_1 = 4$, $d = 5$; find a_{40}

   b. $a_1 = 2$, $a_n = -22$, $d = -3$; find n

   c. $a_1 = 24$, $r = \frac{1}{2}$; find a_6

   d. $a_1 = -2$, $a_n = 486$, $r = -3$; find n

5. Find the sum of each series.

   a. $7 + 10 + 13 + \cdots + 100$

   b. $\displaystyle\sum_{k=1}^{37}(3k + 2)$

   c. For $4 - 12 + 36 - 108 + \cdots$, find S_7

   d. $6 + 3 + \frac{3}{2} + \frac{3}{4} + \cdots$

6. Each swing of a pendulum (in one direction) is 95% of the previous one. If the first swing is 12 ft, (a) find the length of the seventh swing and (b) determine the distance traveled by the pendulum for the first seven swings.

7. A rare coin that cost $3000 appreciates in value 7% per year. Find the value of after 12 yr.

8. A car that costs $50,000 decreases in value by 15% per year. Find the value of the car after 5 yr.

9. Use mathematical induction to prove that for $a_n = 5n - 3$, the sum formula $S_n = \dfrac{5n^2 - n}{2}$ is true for all natural numbers n.

10. Use the principle of mathematical induction to prove that S_n: $2 \cdot 3^{n-1} \le 3^n - 1$ is true for all natural numbers n.

11. Three colored balls (Aqua, Brown, and Creme) are to be drawn without replacement from a bag. List all possible ways they can be drawn using (a) a tree diagram and (b) an organized list.

12. Suppose that license plates for motorcycles must consist of three numbers followed by two letters. How many license plates are possible if zero and "Z" cannot be used and no repetition is allowed?

13. How many subsets can be formed from the elements in this set: $\{\text{🖼},\text{🖊},\text{✏},\text{🖌},\text{🗒},\text{✉}\}$.

14. Compute the following values by hand, showing all work: (a) $6!$ (b) $_6P_3$ (c) $_6C_3$

15. An English major has built a collection of rare books that includes two identical copies of *The Canterbury Tales* (Chaucer), three identical copies of *Romeo and Juliet* (Shakespeare), four identical copies of *Faustus* (Marlowe), and four identical copies of *The Faerie Queen* (Spenser). If these books are to be arranged on a shelf, how many distinguishable permutations are possible?

16. A company specializes in marketing various *cornucopia* (traditionally a curved horn overflowing with fruit, vegetables, gourds, and ears of grain) for Thanksgiving table settings. The company has seven fruit, six vegetable, five gourd, and four grain varieties available. If two from each group (without repetition) are used to fill the horn, how many different cornucopia are possible?

17. Use Pascal's triangle to expand/simplify:

    a. $(x - 2y)^4$

    b. $(1 + i)^4$

18. Use the binomial theorem to write the first three terms of (a) $(x + \sqrt{2})^{10}$ and (b) $(a - 2b^3)^8$.

19. Michael and Mitchell are attempting to make a nonstop, 100-mi trip on a tandem bicycle. The probability that Michael cannot continue pedaling for the entire trip is 0.02. The probability that Mitchell cannot continue pedaling for the entire trip is 0.018. The probability that neither one can pedal the entire trip is 0.011. What is the probability that they complete the trip?

20. The spinner shown is spun once. What is the probability of spinning

    a. a striped wedge

    b. a shaded wedge

    c. a clear wedge

    d. an even number

    e. a two or an odd number

    f. a number greater than nine

    g. a shaded wedge *or* a number greater than 12

    h. a shaded wedge *and* a number greater than 12

21. To improve customer service, a cable company tracks the number of days a customer must wait

until their cable service is installed. The table shows the probability that a customer must wait d days. Based on the table, what is the probability a customer waits

Wait (days d)	Probability
0	0.02
$0 < d < 1$	0.30
$1 \leq d < 2$	0.60
$2 \leq d < 3$	0.05
$3 \leq d < 4$	0.03

  a. at least 2 days   **b.** less than 2 days

  c. 4 days or less   **d.** over 4 days

  e. less than 2 or at least 3 days

  f. three or more days

22. An experienced archer can hit the rectangular target shown 100% of the time at a range of 75 m. Assuming the probability the target is hit is related to its area, what is the probability the archer hits within the

48 cm

64 cm

  a. triangle   **b.** circle

  c. circle but outside the triangle

  d. lower half-circle

  e. rectangle but outside the circle

  f. lower half-rectangle, outside the circle

23. A survey of 100 union workers was taken to register concerns to be raised at the next bargaining session. A breakdown of those surveyed is shown in the table in the right column. One out of the hundred will be

Expertise Level	Women	Men	Total
Apprentice	16	18	34
Technician	15	13	28
Craftsman	9	9	18
Journeyman	7	6	13
Contractor	3	4	7
Totals	50	50	100

selected at random for a personal interview. What is the probability the person chosen is a

  a. woman or a craftsman

  b. man or a contractor

  c. man and a technician

  d. journeyman or an apprentice

24. Cheddar is a 12-year-old male box turtle. Provolone is an 8-year-old female box turtle. The probability that Cheddar will live another 8 yr is 0.85. The probability that Provolone will live another 8 yr is 0.95. Find the probability that

  a. both turtles live for another 8 yr

  b. neither turtle lives for another 8 yr

  c. at least one of them will live another 8 yr

25. Use a proof by induction to show that the sum of the first n natural numbers is $\dfrac{n(n+1)}{2}$. That is, prove

$$1 + 2 + 3 + \cdots + n = \frac{n(n+1)}{2}.$$

CALCULATOR EXPLORATION AND DISCOVERY

Infinite Series, Finite Results

Although there were many earlier flirtations with infinite processes, it may have been the paradoxes of Zeno of Elea (~450 B.C.) that crystallized certain questions that simultaneously frustrated and fascinated early mathematicians. The first paradox, called the dichotomy paradox, can be summarized by the following question: How can one ever finish a race, seeing that one-half the distance must first be traversed, then one-half the remaining distance, then one-half the distance that then remains, and so on an infinite number of times? Although we easily accept that races can be finished, the subtleties involved in this question stymied mathematicians for centuries and were not satisfactorily resolved until the eighteenth century. In modern notation, Zeno's first paradox says $\frac{1}{2} + \frac{1}{4} + \frac{1}{8} + \frac{1}{16} + \cdots < 1$. This is a geometric series with $a_1 = \frac{1}{2}$ and $r = \frac{1}{2}$.

Illustration 1 ▶ For the geometric sequence with $a_1 = \frac{1}{2}$ and $r = \frac{1}{2}$, the nth term is $a_n = \frac{1}{2^n}$. Use the "**sum(**" and "**seq(**" features of your calculator to compute S_5, S_{10}, and S_{15} (see *Technology Highlight* from Section 8.1). Does the sum appear to be approaching some "limiting value"? If so, what is this value? Now compute S_{20}, S_{25}, and S_{30}. Does there still appear to be a limit to the sum? What happens when you have the calculator compute S_{35}?

Solution ▶ `CLEAR` the calculator and enter sum(seq (0.5^X, X, 1, 5)) on the home screen. Pressing `ENTER` gives

$S_5 = 0.96875$ (Figure 8.21). Press 2nd ENTER to recall the expression and overwrite the 5, changing it to a 10. Pressing ENTER shows $S_{10} = 0.9990234375$. Repeating these steps gives $S_{15} = 0.9999694824$, and it seems that "1" may be a

Figure 8.21

```
sum(seq(0.5^X,X,
1,5))
              .96875
sum(seq(0.5^X,X,
1,10))
         .9990234375
sum(seq(0.5^X,X,
1,15))
```

limiting value. Our conjecture receives further support as S_{20}, S_{25}, and S_{30} are closer and closer to 1, but do not exceed it.

Note that the sum of additional terms will create a longer string of 9's. That the sum of an infinite number of these terms *is* **1** can be understood by converting the repeating decimal $0.\overline{9}$ to its fractional form (as shown). For $x = 0.\overline{9}$, $10x = 9.\overline{9}$ and it follows that

$$
\begin{aligned}
10x &= & 9.\overline{9} \\
-x &= & -0.\overline{9} \\
\hline
9x &= & 9 \\
x &= & 1
\end{aligned}
$$

For a geometric sequence, the result of an infinite sum can be verified using $S_\infty = \dfrac{a_1}{1 - r}$. However, there are many nongeometric, infinite series that also have a limiting value. In some cases these require many, many more terms before the limiting value can be observed.

Use a calculator to write the first five terms and to find S_5, S_{10}, and S_{15}. Decide if the sum appears to be approaching some limiting value, then compute S_{20} and S_{25}. Do these sums support your conjecture?

Exercise 1: $a_1 = \frac{1}{3}$ and $r = \frac{1}{3}$

Exercise 2: $a_1 = 0.2$ and $r = 0.2$

Exercise 3: $a_n = \dfrac{1}{(n - 1)!}$

Additional Insight: Zeno's first paradox can also be "resolved" by observing that the "half-steps" needed to complete the race require increasingly shorter (infinitesimally short) amounts of time. Eventually the race is complete.

STRENGTHENING CORE SKILLS

Probability, Quick-Counting, and Card Games

The card game known as *Five Card Stud* is often played for fun and relaxation, using toothpicks, beans, or pocket change as players attempt to develop a winning "hand" from the five cards dealt. The various "hands" are given here with the higher value hands listed first (e.g., a full house is a better/higher hand than a flush).

Five Card Hand	Description	Probability of Being Dealt
royal flush	five cards of the same suit in sequence from Ace to 10	0.000 001 540
straight flush	any five cards of the same suit in sequence (exclude royal)	0.000 013 900
four of a kind	four cards of the same rank, any fifth card	
full house	three cards of the same rank, with one pair	
flush	five cards of the same suit, no sequence required	0.001 970
straight	five cards in sequence, regardless of suit	
three of a kind	three cards of the same rank, any two other cards	
two pairs	two cards of the one rank, two of another rank, one other card	0.047 500
one pair	two cards of the same rank, any three others	0.422 600

For this study, we will consider the hands that are based on suit (the flushes) and the sample space to be five cards dealt from a deck of 52, or $_{52}C_5$.

A flush consists of five cards in the same suit, a straight consists of five cards in sequence. Let's consider that an Ace can be used as either a high card (as in 10, J, Q, K, A) or a low card (as in A, 2, 3, 4, 5). Since the dominant characteristic of a flush is its *suit*, we first consider choosing one suit of the four, then the number of ways that the straight can be formed (if needed).

Illustration 1 ▶ What is the probability of being dealt a royal flush?

Solution ▶ For a royal flush, all cards must be of one suit. Since there are four suits, it can be chosen in $_4C_1$ ways. A royal flush must have the cards A, K, Q, J, and 10 and once the suit has been decided, it can happen in only this (one) way or $_1C_1$. This means

$$P \text{ (royal flush)} = \frac{_4C_1 \cdot {_1}C_1}{_{52}C_5} \approx 0.000\ 001\ 540.$$

Illustration 2 ▶ What is the probability of being dealt a straight flush?

Solution ▶ Once again all cards must be of one suit, which can be chosen in $_4C_1$ ways. A straight flush is any five cards in sequence and once the suit has been decided, this can happen in 10 ways (Ace on down, King on down, Queen on down, and so on). By the FCP, there are $_4C_1 \cdot {_{10}}C_1 = 40$ ways this can happen, but *four of these will be royal flushes that are of a higher value* and must be subtracted from this total. So in the intended context we have

$$P \text{(straight flush)} = \frac{_4C_1 \cdot {_{10}}C_1 - 4}{_{52}C_5} \approx 0.000\ 013\ 900$$

Using these examples, determine the probability of being dealt

Exercise 1: a simple flush (no royal or straight flushes)

Exercise 2: three cards of the same suit and any two other (nonsuit) cards

Exercise 3: four cards of the same suit and any one other (nonsuit) card

Exercise 4: a flush having no face cards

CUMULATIVE REVIEW CHAPTERS 1–8

1. Robot Moe is assembling memory cards for computers. At 9:00 A.M., 52 cards had been assembled. At 11:00 A.M., a total of 98 had been made. Assuming the production rate is linear
   a. Find the slope of this line and explain what it means in this context.
   b. Find a linear equation model for this data.
   c. Determine how many boards Moe can assemble in an eight-hour day.
   d. Determine the approximate time that Moe began work this morning.

2. Verify by direct substitution that $x = 2 + i$ is a solution to $x^2 - 4x + 5 = 0$.

3. Solve using the quadratic formula: $3x^2 + 5x - 7 = 0$. State your answer in exact and approximate form.

4. Sketch the graph of $y = \sqrt{x + 4} - 3$ using transformations of a parent function. Plot the x- and y-intercepts.

5. Write a variation equation and find the value of k: Y varies inversely with X and jointly with V and W. Y is equal to 10 when $X = 9$, $V = 5$, and $W = 12$.

6. Graph the piecewise function and state the domain and range.
$$y = \begin{cases} -2 & -3 \le x \le -1 \\ x & -1 < x < 2 \\ x^2 & 2 \le x \le 3 \end{cases}$$

7. Verify that $f(x) = x^3 - 5$ and $g(x) = \sqrt[3]{x + 5}$ are inverse functions. How are the graphs of f and g related?

8. For the graph of $g(x)$ shown, state where
   a. $g(x) = 0$ b. $g(x) < 0$
   c. $g(x) > 0$ d. $g(x)\uparrow$
   e. $g(x)\downarrow$ f. local max
   g. local min   h. $g(x) = 2$
   i. $g(4)$ j. $g(-1)$
   k. as $x \to -1^+$, $g(x) \to$ _____
   l. as $x \to \infty$, $g(x) \to$ _____
   m. the domain of $g(x)$

9. Compute the difference quotient for each function given.
   a. $f(x) = 2x^2 - 3x$ b. $h(x) = \dfrac{1}{x - 2}$

10. Graph the polynomial function given. Clearly indicate all intercepts. $f(x) = x^3 + x^2 - 4x - 4$

11. Graph the rational function $h(x) = \dfrac{2x^2 - 8}{x^2 - 1}$. Clearly indicate all asymptotes and intercepts.

12. Write each expression in logarithmic form:

    a. $x = 10^y$ **b.** $\dfrac{1}{81} = 3^{-4}$

13. Write each expression in exponential form:

    a. $3 = \log_x(125)$ **b.** $\ln(2x - 1) = 5$

14. What interest rate is required to ensure that $2000 will double in 10 yr if interest is compounded continuously?

15. Solve for x.

    a. $e^{2x-1} = 217$ **b.** $\log(3x - 2) + 1 = 4$

16. Solve using matrices and row reduction:
$$\begin{cases} 2a + 3b - 6c = 15 \\ 4a - 6b + 5c = 35 \\ 3a + 2b - 5c = 24 \end{cases}$$

17. Solve using a calculator and inverse matrices.
$$\begin{cases} 0.7x + 1.2y - 3.2z = -32.5 \\ 1.5x - 2.7y + 0.8z = -7.5 \\ 2.8x + 1.9y - 2.1z = 1.5 \end{cases}$$

18. Find the equation of the hyperbola with foci at $(-6, 0)$ and $(6, 0)$ and vertices at $(-4, 0)$ and $(4, 0)$.

19. Write $x^2 + 4y^2 - 24y + 6x + 29 = 0$ by completing the square, then identify the center, vertices, and foci.

20. Use properties of sequences to determine a_{20} and S_{20}.

    a. 262144, 65536, 16384, 4096, . . .

    b. $\dfrac{7}{8}, \dfrac{27}{40}, \dfrac{19}{40}, \dfrac{11}{40}, \cdots$

21. Empty 55-gal drums are stacked at a storage facility in the form of a pyramid with 52 barrels in the first row, 51 barrels in the second row, and so on, until there are 10 barrels in the top row. Use properties of sequences to determine how many barrels are in this stack.

22. Three $20 bills, six $10 bills, and four $5 bills are placed in a box, then two bills are drawn out and placed in a savings account. What is the probability the bills drawn are

    a. first $20, second $10

    b. first $10, second $20

    c. both $5

    d. first $5, second not $20

    e. first $5, second not $10

    f. first not $20, second $20

23. The manager of Tom's Tool and Equipment Rentals knows that 4% of all tools rented are returned late. Of the 12 tools rented in the last hour, what is the probability that

    a. exactly ten will be returned on time

    b. at least eleven will be returned on time

    c. at least ten will be returned on time

    d. none of them will be returned on time

24. Use a proof by induction to show
$$3 + 7 + 11 + 15 + \cdots + (4n - 1) = n(2n + 1)$$
for all natural numbers n.

25. A park ranger tracks the number of campers at a popular park from March ($m = 3$) to September ($m = 9$) and collects the following data (month, number of campers): (3, 56), (5, 126), and (9, 98). Assuming the data is quadratic, draw a scatter-plot and construct a 3×3 system of equations and solve to obtain a parabolic equation model. (a) What month had the maximum number of campers? (b) What was this maximum number? (c) How many campers might be expected in April? (d) Based on your model, what month(s) is the park apparently closed to campers (number of campers is zero or negative)?

More on Synthetic Division

As the name implies, synthetic division simulates the long division process, but in a condensed and more efficient form. It's based on a few simple observations of long division, as noted in the division $(x^3 - 2x^2 - 13x - 17) \div (x - 5)$ shown in Figure AI.1.

Figure AI.1

$$
\begin{array}{r}
x^2 + 3x + 2 \\
x - 5{\overline{\smash{\big)}\,x^3 - 2x^2 - 13x - 17}} \\
-\underline{(x^3 - 5x^2)} \\
3x^2 - 13x \\
-\underline{(3x^2 - 15x)} \\
2x - 17 \\
-\underline{(2x - 10)} \\
-7 \quad \text{remainder}
\end{array}
$$

Figure AI.2

$$
\begin{array}{r}
1 \quad\; 3 \quad\; 2 \\
x - 5{\overline{\smash{\big)}\,1 \quad -2 \quad -13 \quad -17}} \\
\underline{5} \\
3 \\
\underline{15} \\
2 \\
\underline{10} \\
-7 \quad \text{remainder}
\end{array}
$$

A careful observation reveals a great deal of repetition, as any term in red is a duplicate of the term above it. In addition, since the dividend and divisor must be written in decreasing order of degree, the variable part of each term is unnecessary as we can let the *position of each coefficient* indicate the degree of the term. In other words, we'll agree that

$$1 \quad -2 \quad -13 \quad -17 \quad \text{represents the polynomial} \quad 1x^3 - 2x^2 - 13x - 17.$$

Finally, we know in advance that we'll be subtracting each partial product, so we can "distribute the negative," shown at each stage. Removing the repeated terms and variable factors, then distributing the negative to the remaining terms produces Figure AI.2. The entire process can now be condensed by vertically compressing the rows of the division so that a minimum of space is used (Figure AI.3).

Figure AI.3

	1	3	2	quotient
$x - 5\overline{)}$ 1	−2	−13	−17	dividend
	5	15	10	products
	3	2	7	sums

Figure AI.4

	1	3	2	dividend
$x - 5\overline{)}$ 1	−2	−13	−17	
	5	15	10	products
1	3	2	−7	**remainder**
		quotient		

Further, if we include the lead coefficient in the bottom row (Figure AI.4), the coefficients in the top row (in **blue**) are duplicated and no longer necessary, since the quotient and remainder now appear in the last row. Finally, note all entries in the product row (in **red**) are five times the sum from the prior column. There is a direct connection between this multiplication by 5 and the divisor $x - 5$, and in fact, it is the *zero of the divisor* that is used in synthetic division ($x = 5$ from $x - 5 = 0$). A simple change in format makes this method of division easier to use, and highlights the location of the divisor and remainder (the **blue** brackets in Figure AI.5). Note the process begins by "dropping the lead coefficient into place" (shown in **bold**). The full process of synthetic division is shown in Figure AI.6 for the same exercise.

Figure AI. 5

divisor (use 5, not −5) → $\underline{5}|$ **1**   −2   −13   −17   ← coefficients of dividend

drop lead coefficient into place → **1**   ← quotient and remainder
                                                                                appear in this row

We then multiply this coefficient by the "divisor," place the result in the next column and add. In a sense, we "multiply in the diagonal direction," and "add in the vertical direction." Continue the process until the division is complete.

Figure AI. 6

$\underline{5}|$ 1   −2   −13   −17   ← coefficients of dividend

multiply by 5   5   15   10

       1   3   2 $\underline{-7}$ ← quotient and remainder
                                                    appear in this row

The result is $x^2 + 3x + 2 + \dfrac{-7}{x-5}$, read from the last row.

More on Matrices

Reduced Row-Echelon Form

A matrix is in reduced row-echelon form if it satisfies the following conditions:

1. All null rows (zeroes for all entries) occur at the bottom of the matrix.
2. The first non-zero entry of any row must be a 1.
3. For any two consecutive, nonzero rows, the leading 1 in the higher row is to the left of the 1 in the lower row.
4. Every column with a leading 1 has zeroes for all other entries in the column.

Matrices A through D are in reduced row-echelon form.

$$A = \begin{bmatrix} 0 & 1 & 0 & 0 & 0 & 5 \\ 0 & 0 & 0 & 1 & 0 & 3 \\ 0 & 0 & 0 & 0 & 1 & 2 \end{bmatrix} \quad B = \begin{bmatrix} 1 & 0 & 0 & 5 \\ 0 & 1 & 0 & 3 \\ 0 & 0 & 0 & 0 \end{bmatrix} \quad C = \begin{bmatrix} 1 & 0 & 0 & 5 \\ 0 & 1 & 3 & -2 \\ 0 & 0 & 0 & 0 \end{bmatrix} \quad D = \begin{bmatrix} 1 & 0 & 5 & 0 \\ 0 & 1 & 2 & 0 \\ 0 & 0 & 0 & 1 \end{bmatrix}$$

Where *Gaussian elimination* places a matrix in *row-echelon form* (satisfying the first three conditions), *Gauss-Jordan elimination* places a matrix in *reduced row-echelon form*. To obtain this form, continue applying row operations to the matrix until the fourth condition above is also satisfied. For a 3×3 system having a unique solution, the diagonal entries of the coefficient matrix will be 1's, with 0's for all other entries. To illustrate, we'll extend Example 4 from Section 6.1 until reduced row-echelon form is obtained.

EXAMPLE 4B ▶ Solving Systems Using the Augmented Matrix

Solve using Gauss-Jordan elimination: $\begin{cases} 2x + y - 2z = -7 \\ x + y + z = -1 \\ -2y - z = -3 \end{cases}$

$\begin{cases} 2x + y - 2z = -7 \\ x + y + z = -1 \\ -2y - z = -3 \end{cases}$ $\quad$ R1 ↔ R2 $\quad$ $\begin{cases} x + y + z = -1 \\ 2x + y - 2z = -7 \\ -2y - z = -3 \end{cases}$ $\quad$ matrix form → $\quad$ $\begin{bmatrix} 1 & 1 & 1 & | & -1 \\ 2 & 1 & -2 & | & -7 \\ 0 & -2 & -1 & | & -3 \end{bmatrix}$

$\begin{bmatrix} 1 & 1 & 1 & | & -1 \\ 2 & 1 & -2 & | & -7 \\ 0 & -2 & -1 & | & -3 \end{bmatrix}$ $\quad$ -2R1 + R2 → R2 $\quad$ $\begin{bmatrix} 1 & 1 & 1 & | & -1 \\ 0 & -1 & -4 & | & -5 \\ 0 & -2 & -1 & | & -3 \end{bmatrix}$ $\quad$ -1R2 $\quad$ $\begin{bmatrix} 1 & 1 & 1 & | & -1 \\ 0 & 1 & 4 & | & 5 \\ 0 & -2 & -1 & | & -3 \end{bmatrix}$

$\begin{bmatrix} 1 & 1 & 1 & | & -1 \\ 0 & 1 & 4 & | & 5 \\ 0 & -2 & -1 & | & -3 \end{bmatrix}$ $\quad$ 2R2 + R3 → R3 $\quad$ $\begin{bmatrix} 1 & 1 & 1 & | & -1 \\ 0 & 1 & 4 & | & 5 \\ 0 & 0 & 7 & | & 7 \end{bmatrix}$ $\quad$ $\dfrac{R3}{7} \to$ R3 $\quad$ $\begin{bmatrix} 1 & 1 & 1 & | & -1 \\ 0 & 1 & 4 & | & 5 \\ 0 & 0 & 1 & | & 1 \end{bmatrix}$

$\begin{bmatrix} 1 & 1 & 1 & | & -1 \\ 0 & 1 & 4 & | & 5 \\ 0 & 0 & 1 & | & 1 \end{bmatrix}$ $\quad$ $-$R2 + R1 → R1 $\quad$ $\begin{bmatrix} 1 & 0 & -3 & | & -6 \\ 0 & 1 & 4 & | & 5 \\ 0 & 0 & 1 & | & 1 \end{bmatrix}$ $\quad$ 3R3 + R1 → R1 $\quad$ -4R3 + R2 → R2 $\quad$ $\begin{bmatrix} 1 & 0 & 0 & | & -3 \\ 0 & 1 & 0 & | & 1 \\ 0 & 0 & 1 & | & 1 \end{bmatrix}$

The final matrix is in reduced row-echelon form with solution $(-3, 1, 1)$ just as in Section 6.1.

The Determinant of a General Matrix

To compute the determinant of a general square matrix, we introduce the idea of a **cofactor.** For an $n \times n$ matrix A, $A_{ij} = (-1)^{i+j}|M_{ij}|$ is the cofactor of matrix element a_{ij}, where $|M_{ij}|$ represents the determinant of the corresponding minor matrix. Note that $i + j$ is the sum of the row and column of the entry, and if this sum is even, $(-1)^{i+j} = 1$, while if the sum is odd, $(-1)^{i+j} = -1$ (this is how the sign table for a 3×3 determinant was generated). To compute the determinant of an $n \times n$ matrix, multiply each element in any row or column by its cofactor and add. The result is a tier-like process in which the determinant of a larger matrix requires computing the determinant of smaller matrices. In the case of a 4×4 matrix, each of the minor matrices will be size 3×3, whose determinant then requires the computation of other 2×2 determinants. In the following illustration, two of the entries in the first row are zero for convenience. For

$$A = \begin{bmatrix} -2 & 0 & 3 & 0 \\ 1 & 2 & 0 & -2 \\ 3 & -1 & 4 & 1 \\ 0 & -3 & 2 & 1 \end{bmatrix},$$

we have: $\det(A) = -2 \cdot (-1)^{1+1} \begin{vmatrix} 2 & 0 & -2 \\ -1 & 4 & 1 \\ -3 & 2 & 1 \end{vmatrix} + (3) \cdot (-1)^{1+3} \begin{vmatrix} 1 & 2 & -2 \\ 3 & -1 & 1 \\ 0 & -3 & 1 \end{vmatrix}$

Computing the first 3×3 determinant gives -16, the second 3×3 determinant is 14. This gives:

$$\det(A) = -2(-16) + 3(14)$$
$$= 74$$

Deriving the Equation of a Conic

The Equation of an Ellipse

In Section 7.2, the equation $\sqrt{(x + c)^2 + y^2} + \sqrt{(x - c)^2 + y^2} = 2a$ was developed using the distance formula and the definition of an ellipse. To find the standard form of the equation, we treat this result as a radical equation, isolating one of the radicals and squaring both sides.

$$\sqrt{(x + c)^2 + y^2} = 2a - \sqrt{(x - c)^2 + y^2} \qquad \text{isolate one radical}$$

$$(x + c)^2 + y^2 = 4a^2 - 4a\sqrt{(x - c)^2 + y^2} + (x - c)^2 + y^2 \qquad \text{square both sides}$$

We continue by simplifying the equation, isolating the remaining radical, and squaring again.

$$x^2 + 2cx + c^2 + y^2 = 4a^2 - 4a\sqrt{(x - c)^2 + y^2} + x^2 - 2cx + c^2 + y^2 \qquad \text{expand binomials}$$

$$4cx = 4a^2 - 4a\sqrt{(x - c)^2 + y^2} \qquad \text{simplify}$$

$$a\sqrt{(x - c)^2 + y^2} = a^2 - cx \qquad \text{isolate radical; divide by 4}$$

$$a^2[(x - c)^2 + y^2] = a^4 - 2a^2cx + c^2x^2 \qquad \text{square both sides}$$

$$a^2x^2 - 2a^2cx + a^2c^2 + a^2y^2 = a^4 - 2a^2cx + c^2x^2 \qquad \text{expand and distribute } a^2 \text{ on left}$$

$$a^2x^2 - c^2x^2 + a^2y^2 = a^4 - a^2c^2 \qquad \text{add } 2a^2cx \text{ and rewrite equation}$$

$$x^2(a^2 - c^2) + a^2y^2 = a^2(a^2 - c^2) \qquad \text{factor}$$

$$\frac{x^2}{a^2} + \frac{y^2}{a^2 - c^2} = 1 \qquad \text{divide by } a^2(a^2 - c^2)$$

Since $a > c$, we know $a^2 > c^2$ and $a^2 - c^2 > 0$. For convenience, we let $b^2 = a^2 - c^2$ and it also follows that $a^2 > b^2$ and $a > b$ (since $c > 0$). Substituting b^2 for $a^2 - c^2$ we obtain the standard form of the equation of an ellipse (major axis horizontal, since we stipulated $a > b$): $\frac{x^2}{a^2} + \frac{y^2}{b^2} = 1$. Note once again the x-intercepts are $(\pm a, 0)$, while the y-intercepts are $(0, \pm b)$.

The Equation of a Hyperbola

In Section 7.3, the equation $\sqrt{(x + c)^2 + y^2} - \sqrt{(x - c)^2 + y^2} = 2a$ was developed using the distance formula and the definition of a hyperbola. To find the standard form of this equation, we apply the same procedures as before.

$$\sqrt{(x + c)^2 + y^2} = 2a + \sqrt{(x - c)^2 + y^2} \qquad \text{isolate one radical}$$

$$(x + c)^2 + y^2 = 4a^2 + 4a\sqrt{(x - c)^2 + y^2} + (x - c)^2 + y^2 \qquad \text{square both sides}$$

$$x^2 + 2cx + c^2 + y^2 = 4a^2 + 4a\sqrt{(x - c)^2 + y^2} + x^2 - 2cx + c^2 + y^2 \qquad \text{expand binomials}$$

$$4cx = 4a^2 + 4a\sqrt{(x - c)^2 + y^2} \qquad \text{simplify}$$

$$cx - a^2 = a\sqrt{(x - c)^2 + y^2} \qquad \text{isolate radical; divide by 4}$$

$$c^2x^2 - 2a^2cx + a^4 = a^2[(x - c)^2 + y^2] \qquad \text{square both sides}$$

$$c^2x^2 - 2a^2cx + a^4 = a^2x^2 - 2a^2cx + a^2c^2 + a^2y^2 \qquad \text{expand and distribute } a^2 \text{ on the right}$$

$$c^2x^2 - a^2x^2 - a^2y^2 = a^2c^2 - a^4 \qquad \text{add } 2a^2cx \text{ and rewrite equation}$$

$$x^2(c^2 - a^2) - a^2y^2 = a^2(c^2 - a^2) \qquad \text{factor}$$

$$\frac{x^2}{a^2} - \frac{y^2}{c^2 - a^2} = 1 \qquad \text{divide by } a^2(c^2 - a^2)$$

From the definition of a hyperbola we have $0 < a < c$, showing $c^2 > a^2$ and $c^2 - a^2 > 0$. For convenience, we let $b^2 = c^2 - a^2$ and substitute to obtain the standard form of the equation of a hyperbola (transverse axis horizontal): $\dfrac{x^2}{a^2} - \dfrac{y^2}{b^2} = 1$. Note the x-intercepts are $(0, \pm a)$ and there are no y-intercepts.

The Asymptotes of a Central Hyperbola

From our work in Section 7.3, a central hyperbola with a horizontal axis will have asymptotes at $y = \pm\dfrac{b}{a}x$. To understand why, recall that for asymptotic behavior we investigate what happens to the relation for large values of x, meaning as $|x| \to \infty$. Starting with $\dfrac{x^2}{a^2} - \dfrac{y^2}{b^2} = 1$, we have

$$b^2x^2 - a^2y^2 = a^2b^2 \qquad \text{clear denominators}$$

$$a^2y^2 = b^2x^2 - a^2b^2 \qquad \text{isolate term with } y$$

$$a^2y^2 = b^2x^2\left(1 - \frac{a^2}{x^2}\right) \qquad \text{factor out } b^2x^2 \text{ from right side}$$

$$y^2 = \frac{b^2}{a^2}x^2\left(1 - \frac{a^2}{x^2}\right) \qquad \text{divide by } a^2$$

$$y = \pm\frac{b}{a}x\sqrt{1 - \frac{a^2}{x^2}} \qquad \text{square root both sides}$$

As $|x| \to \infty$, $\dfrac{a^2}{x^2} \to 0$, and we find that for large values of x, $y \approx \pm\dfrac{b}{a}x$.

Proof Positive—A Selection of Proofs from College Algebra

Proofs from Chapter 3

The Remainder Theorem

If a polynomial $p(x)$ is divided by $(x - c)$ using synthetic division, the remainder is equal to $p(c)$.

Proof of the Remainder Theorem

From our previous work, any number c used in synthetic division will occur as the factor $(x - c)$ when written as (quotient)(divisor) + remainder: $p(x) = (x - c)q(x) + r$. Here, $q(x)$ represents the quotient polynomial and r is a constant. Evaluating $p(c)$ gives

$$p(x) = (x - c)q(x) + r$$
$$p(c) = (c - c)q(c) + r$$
$$= 0 \cdot q(c) + r$$
$$= r \checkmark$$

The Factor Theorem

Given a polynomial $p(x)$,
(1) if $p(c) = 0$, then $x - c$ is a factor of $p(x)$, and
(2) if $x - c$ is a factor of $p(x)$, then $p(c) = 0$.

Proof of the Factor Theorem

1. Consider a polynomial p written in the form $p(x) = (x - c)q(x) + r$. From the remainder theorem we know $p(c) = r$, and substituting $p(c)$ for r in the equation shown gives:

$$p(x) = (x - c)q(x) + p(c)$$

and $x - c$ is a factor of $p(x)$, if $p(c) = 0$

$$p(x) = (x - c)q(x) \checkmark$$

2. The steps from part 1 can be reversed, since any factor $(x - c)$ of $p(x)$, can be written in the form $p(x) = (x - c)q(x)$. Evaluating at $x = c$ produces a result of zero:

$$p(c) = (c - c)q(x)$$
$$= 0 \checkmark$$

Complex Conjugates Theorem

Given $p(x)$ is a polynomial with real number coefficients, complex solutions must occur in conjugate pairs. If $a + bi$, $b \neq 0$ is a solution, then $a - bi$ must also be a solution.
    To prove this for polynomials of degree $n > 2$, we let $z_1 = a + bi$ and $z_2 = c + di$ be complex numbers, and let $\bar{z}_1 = a - bi$, and $\bar{z}_2 = c - di$ represent their conjugates, and observe the following properties:

1. The conjugate of a sum is equal to the sum of the conjugates.

sum: $z_1 + z_2$	sum of conjugates: $\bar{z}_1 + \bar{z}_2$
$(a + bi) + (c + di)$	$(a - bi) + (c - di)$
$(a + c) + (b + d)i$ $\rightarrow$ conjugate of sum $\rightarrow$	$(a + c) - (b + d)i \checkmark$

2. The conjugate of a product is equal to the product of the conjugates.

product: $z_1 \cdot z_2$ product of conjugates: $\bar{z}_1 \cdot \bar{z}_2$

$(a + bi) \cdot (c + di)$ $(a - bi) \cdot (c - di)$

$ac + adi + bci + bdi^2$ $ac - adi - cbi + bdi^2$

$(ac - bd) + (ad + bc)i$ $\quad \rightarrow$ conjugate of product $\rightarrow \quad$ $(ac - bd) - (ad + bc)i$ ✓

Since polynomials involve only sums and products, and the complex conjugate of any real number is the number itself, we have the following:

Proof of the Complex Conjugates Theorem

Given polynomial $p(x) = a_n x^n + a_{n-1} x^{n-1} + \cdots + a_1 x^1 + a_0$, where $a_n, a_{n-1}, \cdots, a_1, a_0$ are real numbers and $z = a + bi$ is a zero of p, we must show that $\bar{z} = a - bi$ is also a zero.

$$a_n z^n + a_{n-1} z^{n-1} + \cdots + a_1 z^1 + a_0 = p(z) \qquad \text{evaluate } p(x) \text{ at } z$$

$$a_n z^n + a_{n-1} z^{n-1} + \cdots + a_1 z^1 + a_0 = 0 \qquad p(z) = 0 \text{ given}$$

$$\overline{a_n z^n + a_{n-1} z^{n-1} + \cdots + a_1 z^1 + a_0} = \overline{0} \qquad \text{conjugate both sides}$$

$$\overline{a_n z^n} + \overline{a_{n-1} z^{n-1}} + \cdots + \overline{a_1 z^1} + \overline{a_0} = \overline{0} \qquad \text{property 1}$$

$$\bar{a}_n (\bar{z}^n) + \bar{a}_{n-1}(\bar{z}^{n-1}) + \cdots + \bar{a}_1(\bar{z}^1) + \bar{a}_0 = \overline{0} \qquad \text{property 2}$$

$$a_n (\bar{z}^n) + a_{n-1}(\bar{z}^{n-1}) + \cdots + a_1(\bar{z}^1) + a_0 = 0 \qquad \text{conjugate of a real number is the number}$$

$$p(\bar{z}) = 0 \quad \checkmark \quad \text{result}$$

An immediate and useful result of this theorem is that any polynomial of odd degree must have at least one real root.

Linear Factorization Theorem

If $p(x)$ is a complex polynomial of degree $n \geq 1$, then p has exactly n linear factors and can be written in the form $p(x) = a_n(x - c_1)(x - c_2) \cdot \ldots \cdot (x - c_n)$, where $a_n \neq 0$ and $c_1, c_2, \ldots, c_n$ are complex numbers. Some factors may have multiplicities greater than 1 ($c_1, c_2, \ldots, c_n$ are not necessarily distinct).

Proof of the Linear Factorization Theorem

Given $p(x) = a_n x^n + a_{n-1} x^{n-1} + \cdots + a_1 x + a_0$ is a complex polynomial, the Fundamental Theorem of Algebra establishes that $p(x)$ has a least one complex zero, call it c_1. The factor theorem stipulates $(x - c_1)$ must be a factor of P, giving

$$p(x) = (x - c_1)q_1(x)$$

where $q_1(x)$ is a complex polynomial of degree $n - 1$.

Since $q_1(x)$ is a complex polynomial in its own right, it too must also have a complex zero, call it c_2. Then $(x - c_2)$ must be a factor of $q_1(x)$, giving

$$p(x) = (x - c_1)(x - c_2)q_2(x)$$

where $q_2(x)$ is a complex polynomial of degree $n - 2$.

Repeating this rationale n times will cause $p(x)$ to be rewritten in the form

$$p(x) = (x - c_1)(x - c_2) \cdot \ldots \cdot (x - c_n)q_n(x)$$

where $q_n(x)$ has a degree of $n - n = 0$, a nonzero constant typically called a_n.

The result is $p(x) = a_n(x - c_1)(x - c_2) \cdot \ldots \cdot (x - c_n)$, and the proof is complete.

Proofs from Chapter 4

The Product Property of Logarithms

Given M, N, and b ≠ 1 are positive real numbers,
$\log_b(MN) = \log_b M + \log_b N.$

Proof of the Product Property

For $P = \log_b M$ and $Q = \log_b N$, we have $b^P = M$ and $b^Q = N$ in exponential form. It follows that

$$\log_b(MN) = \log_b(b^P b^Q) \qquad \text{substitute } b^P \text{ for } M \text{ and } b^Q \text{ for } N$$
$$= \log_b(b^{P+Q}) \qquad \text{properties of exponents}$$
$$= P + Q \qquad \text{log property 3}$$
$$= \log_b M + \log_b N \qquad \text{substitute } \log_b M \text{ for } P \text{ and } \log_b N \text{ for } Q$$

The Quotient Property of Logarithms

Given M, N, and b ≠ 1 are positive real numbers,

$\log_b\left(\dfrac{M}{N}\right) = \log_b M - \log_b N.$

Proof of the Quotient Property

For $P = \log_b M$ and $Q = \log_b N$, we have $b^P = M$ and $b^Q = N$ in exponential form. It follows that

$$\log_b\left(\frac{M}{N}\right) = \log_b\left(\frac{b^P}{b^Q}\right) \qquad \text{substitute } b^P \text{ for } M \text{ and } b^Q \text{ for } N$$
$$= \log_b(b^{P-Q}) \qquad \text{properties of exponents}$$
$$= P - Q \qquad \text{log property 3}$$
$$= \log_b M - \log_b N \qquad \text{substitute } \log_b M \text{ for } P \text{ and } \log_b N \text{ for } Q$$

The Power Property of Logarithms

Given M, N, and b ≠ 1 are positive real numbers and any real number x,
$\log_b M^x = x \log_b M.$

Proof of the Power Property

For $P = \log_b M$, we have $b^P = M$ in exponential form. It follows that

$$\log_b(M)^x = \log_b(b^P)^x \qquad \text{substitute } b^P \text{ for } M$$
$$= \log_b(b^{Px}) \qquad \text{properties of exponents}$$
$$= Px \qquad \text{log property 3}$$
$$= (\log_b M)x \qquad \text{substitute } \log_b M \text{ for } P$$
$$= x \log_b M \qquad \text{rewrite factors}$$

Student Answer Appendix

CHAPTER R

Exercises R.1, pp. 10–12

1. proper subset; element   **3.** positive; negative, 7, -7; principal
5. Order of operations requires multiplication before addition.
7. a. $\{1, 2, 3, 4, 5\}$ **b.** $\{\ \}$ **9.** True   **11.** True   **13.** True
15. $1.\overline{3}$

17. $2.\overline{5}$

19. ≈ 2.65

21. ≈ 1.73

23. a. i. $\{8, 7, 6\}$ **ii.** $\{8, 7, 6\}$ **iii.** $(-1, 8, 7, 6)$
iv. $\{-1, 8, 0, 75, \frac{9}{2}, 5.\overline{6}, 7, \frac{3}{5}, 6\}$ **v.** $\{\ \}$
vi. $\{-1, 8, 0.75, \frac{9}{2}, 5.\overline{6}, 7, \frac{3}{5}, 6\}$ **b.** $\{-1, \frac{3}{5}, 0.75, \frac{9}{2}, 5.\overline{6}, 6, 7, 8\}$
c.

25. a. i. $\{\sqrt{49}, 2, 6, 4\}$ **ii.** $\{\sqrt{49}, 2, 6, 0, 4\}$
iii. $\{-5, \sqrt{49}, 2, -3, 6, -1, 0, 4\}$ **iv.** $\{-5, \sqrt{49}, 2, -3, 6, -1, 0, 4\}$
v. $\{\sqrt{3}, \pi\}$ **vi.** $\{-5, \sqrt{49}, 2, -3, 6, -1, \sqrt{3}, 0, 4, \pi\}$
b. $\{-5, -3, -1, 0, \sqrt{3}, 2, \pi, 4, 6, \sqrt{49}\}$
c.

27. False; not all real numbers are irrational.   **29.** False; not all rational numbers are integers.   **31.** False; $\sqrt{25} = 5$ is not irrational.   **33.** c IV
35. a VI   **37.** d III   **39.** Let a represent Kylie's age: $a \geq 6$ years.
41. Let n represent the number of incorrect words: $n \leq 2$ incorrect.

43. 2.75 **45.** -4 **47.** $\frac{1}{2}$ **49.** $\frac{3}{4}$ **51.** 10 **53.** $-8, 2$ **55.** negative
57. $-n$ **59.** undefined, since $12 \div 0 = k$ implies $k \cdot 0 = 12$
61. undefined, since $7 \div 0 = k$ implies $k \cdot 0 = 7$ **63.**   **a.** positive
b. negative   **c.** negative   **d.** negative   **65.** $-\frac{11}{6}$ **67.** -2
69. $9^2 = 81$ is closest   **71.** 7 **73.** -2.185 **75.** $4\frac{1}{3}$ **77.** $-\frac{29}{12}$ or $-2\frac{5}{12}$
79. 0 **81.** -5 **83.** $-\frac{1}{10}$ **85.** $-\frac{7}{8}$ **87.** -4 **89.** $\frac{-11}{12}$ **91.** 64
93. 4489.70 **95.** $D \approx 4.3$ cm   **97.** $32°\text{F}$ **99.** $179°\text{F}$
101. Tsu Ch'ung-chih: $\frac{355}{113}$ **103.** negative

Exercises R.2, pp. 18–21

1. constant   **3.** coefficient   **5.** $-5 + 5 = 0, -5 \cdot (-\frac{1}{5}) = 1$
7. two; 3 and -5 **9.** two; 2 and $\frac{1}{4}$ **11.** three; $-2, 1,$ and -5
13. one; -1 **15.** $n - 7$ **17.** $n + 4$ **19.** $(n - 5)^2$ **21.** $2n - 13$
23. $n^2 + 2n$ **25.** $\frac{2}{3}n - 5$ **27.** $3(n + 5) - 7$ **29.** Let w represent the width. Then $2w$ represents twice the width and $2w - 3$ represents three meters less than twice the width.   **31.** Let b represent the speed of the bus. Then $b + 15$ represents 15 mph more than the speed of the bus.
33. $h = b + 150$ **35.** $L = 2W + 20$ **37.** $M = 2.5N$

39. $T = 12.50g + 50$ **41.** 14 **43.** 19 **45.** 0 **47.** 16 **49.** -36
51. 51 **53.** 2 **55.** 144 **57.** $\frac{-41}{5}$ **59.** 24
61.

x	Output
-3	14
-2	6
-1	0
0	-4
1	-6
2	-6
3	-4

-1 has an output of 0.

63.

x	Output
-3	-18
-2	-15
-1	-12
0	-9
1	-6
2	-3
3	0

3 has an output of 0.

65.

x	Output
-3	-5
-2	8
-1	9
0	4
1	-1
2	0
3	13

2 has an output of 0.

67. a. $7 + (-5) = 2$ **b.** $n + (-2)$ **c.** $a + (-4.2) + 13.6 = a + 9.4$
d. $x + 7 - 7 = x$ **69. a.** 3.2 **b.** $\frac{5}{6}$ **71.** $-5x + 13$
73. $-\frac{2}{15}p + 6$ **75.** $-2a$ **77.** $\frac{17}{12}x$ **79.** $-2a^2 + 2a$ **81.** $6x^2 - 3x$
83. $2a + 3b + 2c$ **85.** $\frac{29}{8}n + \frac{38}{5}$ **87.** $7a^2 - 13a - 5$ **89.** 10 ohms
91. a. $t = \frac{1}{2}j$ **b.** $t = 225$ mph   **93. a.** $L = 2W + 3$ **b.** 107 ft
95. $t = c + 22; 37¢$ **97.** $C = 25t + 43.50; \$81$
99. a. positive odd integer

Exercises R.3, pp. 31–34

1. power   **3.** $20x; 0$ **5. a.** cannot be simplified, unlike terms
b. can be simplified, like bases   **7.** $14n^7$ **9.** $-12p^5q^4$ **11.** $a^{14}b^7$
13. $216p^3q^6$ **15.** $32.768h^3k^6$ **17.** $\frac{p^2}{4q^2}$ **19.** $49c^{14}d^4$ **21.** $\frac{9}{16}x^6y^2$
23. $\frac{9}{4}x^3y^2$ **25. a.** $V = 27x^6$ **b.** 1728 units^3 **27.** $3w^3$ **29.** $-3ab$
31. $\frac{27}{8}$ **33.** $2h^3$ **35.** $\frac{-1}{8}$ **37.** -8 **39.** $\frac{4p^8}{q^6}$ **41.** $\frac{8x^6}{27y^9}$ **43.** $\frac{25m^4n^6}{4r^8}$
45. $\frac{25p^2q^2}{4}$ **47.** $\frac{3p^2}{-4q^2}$ **49.** $\frac{5}{3h^7}$ **51.** $\frac{1}{a^3}$ **53.** $\frac{a^{12}}{b^4c^8}$ **55.** $\frac{-12}{5x^4}$
57. $\frac{-2b^7}{27a^9c^3}$ **59.** 2 **61.** $\frac{7}{10}$ **63.** $\frac{13}{9}$ **65.** -4 **67.** 6.6×10^9
69. $0.000\,000\,006\,5$ **71.** 26,571 hrs; 1,107 days   **73.** polynomial, none of these, degree 3   **75.** nonpolynomial because exponents are not whole numbers, NA, NA   **77.** polynomial, binomial, degree 3
79. $-w^3 - 3w^2 + 7w + 8.2; -1$ **81.** $c^3 + 2c^2 - 3c + 6; 1$
83. $\frac{-2}{3}x^2 + 12; \frac{-2}{3}$ **85.** $3p^3 - 3p^2 - 12$ **87.** $7.85b^2 - 0.6b - 1.9$
89. $\frac{1}{4}x^2 - 8x + 6$ **91.** $q^6 + q^5 - q^4 + 2q^3 - q^2 - 2q$
93. $-3x^3 + 3x^2 + 18x$ **95.** $3r^2 - 11r + 10$ **97.** $x^3 - 27$
99. $b^3 - b^2 - 34b - 56$ **101.** $21v^2 - 47v + 20$ **103.** $9 - m^2$
105. $p^2 + 1.1p - 9$ **107.** $x^2 + \frac{3}{4}x + \frac{1}{8}$ **109.** $m^2 - \frac{9}{16}$
111. $6x^2 + 11xy - 10y^2$ **113.** $12c^2 + 23cd + 5d^2$
115. $2x^4 - x^2 - 15$ **117.** $4m + 3; 16m^2 - 9$
119. $7x + 10; 49x^2 - 100$ **121.** $6 - 5k; 36 - 25k^2$

123. $x - \sqrt{6}; x^2 - 6$ **125.** $x^2 + 8x + 16$ **127.** $16g^2 + 24g + 9$
129. $16p^2 - 24pq + 9q^2$ **131.** $16 - 8\sqrt{x} + x$
133. $xy + 2x - 3y - 6$ **135.** $k^3 + 3k^2 - 28k - 60$
137. a. 340 mg, 292.5 mg   **b.** Less, amount is decreasing.   **c.** after 5 hr
139 $F = kPQd^{-2}$ **141.** $5x^{-3} + 3x^{-2} + 2x^{-1} + 4$ **143.** $15   **145.** 6

Exercises R.4, pp. 42–45

1. product   **3.** binomial; conjugate   **5.** Answers will vary.
7. a. $-17(x^2 - 3)$ **b.** $7b(3b^2 - 2b + 8)$ **c.** $-3a^2(a^2 + 2a - 3)$
9. a. $(a + 2)(2a + 3)$ **b.** $(b^2 + 3)(3b + 2)$ **c.** $(n + 7)(4m - 11)$
11. a. $(3q + 2)(3q^2 + 5)$ **b.** $(h - 12)(h^4 - 3)$ **c.** $(k^2 - 7)(k^3 - 5)$
13. a. $-1(p - 7)(p + 2)$ **b.** $(q - 9)(q + 5)$ **c.** $(n - 4)(n - 5)$
15. a. $(3p + 2)(p - 5)$ **b.** $(4q - 5)(q + 3)$ **c.** $(5u + 3)(2u - 5)$
17. a. $(2s + 5)(2s - 5)$ **b.** $(3x + 7)(3x - 7)$ **c.** $2(5x + 6)(5x - 6)$
d. $(11h + 12)(11h - 12)$ **e.** $(b + \sqrt{5})(b - \sqrt{5})$
19. a. $(a - 3)^2$ **b.** $(b + 5)^2$ **c.** $(2m - 5)^2$ **d.** $(3n - 7)^2$
21. a. $(2p - 3)(4p^2 + 6p + 9)$ **b.** $(m + \frac{1}{2})(m^2 - \frac{1}{2}m + \frac{1}{4})$
c. $(g - 0.3)(g^2 + 0.3g + 0.09)$ **d.** $-2t(t - 3)(t^2 + 3t + 9)$
23. a. $(x + 3)(x - 3)(x + 1)(x - 1)$ **b.** $(x^2 + 9)(x^2 + 4)$
c. $(x - 2)(x^2 + 2x + 4)(x + 1)(x^2 - x + 1)$
25. a. $(n + 1)(n - 1)$ **b.** $(n - 1)(n^2 + n + 1)$
c. $(n + 1)(n^2 - n + 1)$ **d.** $7x(2x + 1)(2x - 1)$ **27.** $(a + 5)(a + 2)$
29. $2(x - 2)(x - 10)$ **31.** $-1(3m + 8)(3m - 8)$ **33.** $(r - 3)(r - 6)$
35. $(2h + 3)(h + 2)$ **37.** $(3k - 4)^2$ **39.** $-3x(2x - 7)(x - 3)$
41. $4m(m + 5)(m - 2)$ **43.** $(a + 5)(a - 12)$ **45.** $(2x - 5)(4x^2 + 10x + 25)$
47. prime   **49.** $(x - 5)(x + 3)(x - 3)$ **51. a.** H   **b.** E   **c.** C   **d.** F
e. B   **f.** A   **g.** I   **h.** D   **i.** G   **53.** $2\pi r(r + h)$, 7000π cm^2; 21,991 cm^2
55. $V = \frac{1}{3}\pi h(R + r)(R - r)$; 6π cm^3; 18.8 cm^3
57. $V = x(x + 5)(x + 3)$ **a.** 3 in.   **b.** 5 in.
c. $V = 24(29)(27) = 18{,}792$ in^3
59. $L = L_0\sqrt{\left(1 + \dfrac{v}{c}\right)\left(1 - \dfrac{v}{c}\right)}$ $L = 12\sqrt{(1 + 0.75)(1 - 0.75)}$
     $= 3\sqrt{7}$ in. ≈ 7.94 in.   **61. a.** $\frac{1}{8}(4x^4 + x^3 - 6x^2 + 32)$
b. $\frac{1}{18}(12b^5 - 3b^3 + 8b^2 - 18)$ **63.** $2x(16x - 27)(6x + 5)$
65. $(x + 3)(x - 3)(x^2 + 9)$
67. $(p + 1)(p^2 - p + 1)(p - 1)(p^2 + p + 1)$
69. $(q + 5)(q - 5)(q + \sqrt{3})(q - \sqrt{3})$

Exercises R.5, pp. 51–54

1. $1; -1$ **3.** common denominator   **5.** F; numerator should be -1
7. a. $-\dfrac{1}{3}$ **b.** $\dfrac{x + 3}{2x(x - 2)}$ **9. a.** simplified   **b.** $\dfrac{a - 4}{a - 7}$
11. a. -1 **b.** -1 **13. a.** $-3ab^9$ **b.** $\dfrac{x + 3}{9}$ **c.** $-1(y + 3)$ **d.** $\dfrac{-1}{m}$
15. a. $\dfrac{2n + 3}{n}$ **b.** $\dfrac{3x + 5}{2x + 3}$ **c.** $x + 2$ **d.** $n - 2$
17. $\dfrac{(a - 2)(a + 1)}{(a + 3)(a + 2)}$ **19.** 1   **21.** $\dfrac{(p - 4)^2}{p^2}$ **23.** $\dfrac{-15}{4}$ **25.** $\dfrac{3}{2}$
27. $\dfrac{8(a - 7)}{a - 5}$ **29.** $\dfrac{y}{x}$ **31.** $\dfrac{m}{m - 4}$ **33.** $\dfrac{y + 3}{3y(y + 4)}$ **35.** $\dfrac{x + 0.3}{x - 0.2}$
37. $\dfrac{n + \frac{1}{5}}{n + \frac{2}{3}}$ **39.** $\dfrac{3(a^2 + 3a + 9)}{2}$ **41.** $\dfrac{2n + 1}{n}$ **43.** $\dfrac{3 + 20x}{8x^2}$
45. $\dfrac{14y - x}{8x^2y^4}$ **47.** $\dfrac{2}{p + 6}$ **49.** $\dfrac{-3m - 16}{(m + 4)(m - 4)}$ **51.** $\dfrac{-5m + 37}{m - 7}$
53. $\dfrac{-y + 11}{(y + 6)(y - 5)}$ **55.** $\dfrac{2a - 5}{(a + 4)(a - 5)}$ **57.** $\dfrac{1}{y + 1}$
59. $\dfrac{m^2 - 6m + 21}{(m + 3)^2(m - 3)}$ **61.** $\dfrac{y^2 + 26y - 1}{(5y + 1)(y + 3)(y - 2)}$
63. a. $\dfrac{1}{p^2} - \dfrac{5}{p}; \dfrac{1 - 5p}{p^2}$ **b.** $\dfrac{1}{x^2} + \dfrac{2}{x^3}; \dfrac{x + 2}{x^3}$ **65.** $\dfrac{4a}{a + 20}$ **67.** $p - 1$

69. $\dfrac{x}{9x - 12}$ **71.** $\dfrac{-2}{y + 31}$ **73. a.** $\dfrac{1 + \frac{3}{m}}{1 - \frac{3}{m}}; \dfrac{m + 3}{m - 3}$ **b.** $\dfrac{1 + \frac{2}{x^2}}{1 - \frac{2}{x^2}}; \dfrac{x^2 + 2}{x^2 - 2}$
75. $\dfrac{f_2 + f_1}{f_1 f_2}$ **77.** $\dfrac{-a}{x(x + h)}$ **79.** $\dfrac{-(2x + h)}{2x^2(x + h)^2}$
81. a. $300 million; $2550 million   **b.** It would require many resources.   **c.** No
83. Price rises rapidly for first four days, then begins a gradual decrease. Yes, on the 35th day of trading.

P	$\dfrac{450P}{100 - P}$
40	300
60	675
80	1800
90	4050
93	5979
95	8550
98	22050
100	ERROR

Day	Price
0	10
1	16.67
2	32.76
3	47.40
4	53.51
5	52.86
6	49.25
7	44.91
8	40.75
9	37.03
10	33.81

85. $t = 8$ weeks   **87. b.** $20 \cdot n \div 10 \cdot n = 2n^2$, all others equal 2
89. $\dfrac{6}{23}; \dfrac{ac}{ad + bc}$

Exercises R.6, pp. 64–68

1. even   **3.** $(16^{\frac{1}{4}})^3$ **5.** Answers will vary.   **7.** 9   **9. a.** $7|p|$ **b.** $|x - 3|$
c. $9m^2$ **d.** $|x - 3|$ **11. a.** 4   **b.** $-5x$ **c.** $6z^4$ **d.** $\dfrac{v}{-2}$
13. a. 2   **b.** not a real number   **c.** $3x^2$ **d.** $-3x$ **e.** $k - 3$ **f.** $|h + 2|$
15. a. -5 **b.** $-3|n^3|$ **c.** not a real number   **d.** $\dfrac{7|v^5|}{6}$ **17. a.** 4
b. $\dfrac{64}{125}$ **c.** $\dfrac{125}{8}$ **d.** $\dfrac{9p^4}{4q^2}$ **19. a.** -1728 **b.** not a real number
c. $\dfrac{1}{9}$ **d.** $\dfrac{-256}{81x^4}$ **21. a.** $\dfrac{32n^{10}}{p^2}$ **b.** $\dfrac{1}{2y^{\frac{1}{4}}}$ **23. a.** $3m\sqrt{2}$ **b.** $10pq^2\sqrt[3]{q}$
c. $\dfrac{3}{2}mn\sqrt[3]{n^2}$ **d.** $4pq^3\sqrt{2p}$ **e.** $-3 + \sqrt{7}$ **f.** $\dfrac{9}{2} - \sqrt{2}$
25. a. $15a^2$ **b.** $-4b\sqrt{b}$ **c.** $\dfrac{x^4\sqrt{y}}{3}$ **d.** $3u^2v\sqrt[3]{v}$ **27. a.** $2m^2$
b. $3n$ **c.** $\dfrac{3\sqrt{5}}{4x}$ **d.** $\dfrac{18\sqrt[3]{3}}{z^3}$ **29. a.** $2x^2y^3$ **b.** $x^2\sqrt[4]{x}$ **c.** $\sqrt[12]{b}$
d. $\dfrac{1}{\sqrt[6]{6}} = \dfrac{\sqrt[6]{6^5}}{6}$ **e.** $b^{\frac{3}{4}}$ **31. a.** $9\sqrt{2}$ **b.** $14\sqrt{3}$ **c.** $16\sqrt{2m}$
d. $-5\sqrt{7p}$ **33. a.** $-x\sqrt[3]{2x}$ **b.** $2 - \sqrt{3x} + 3\sqrt{5}$
c. $6x\sqrt{2x} + 5\sqrt{2} - \sqrt{7x} + 3\sqrt{3}$ **35. a.** 98   **b.** $\sqrt{15} + \sqrt{21}$
c. $n^2 - 5$ **d.** $39 - 12\sqrt{3}$ **37. a.** -19
b. $\sqrt{10} + \sqrt{65} - 2\sqrt{7} - \sqrt{182}$
c. $12\sqrt{5} + 2\sqrt{14} + 36\sqrt{15} + 6\sqrt{42}$ **39.** Verified
41. Verified   **43. a.** $\dfrac{\sqrt{3}}{2}$ **b.** $\dfrac{2\sqrt{15x}}{9x^2}$ **c.** $\dfrac{3\sqrt[3]{6b}}{10b}$ **d.** $\dfrac{\sqrt[3]{2p^2}}{2p}$
e. $\dfrac{5\sqrt[3]{a^2}}{a}$ **45. a.** $-12 + 4\sqrt{11}$; 1.27   **b.** $\dfrac{6\sqrt{x} + 6\sqrt{2}}{x - 2}$
47. a. $\sqrt{30} - 2\sqrt{5} - 3\sqrt{3} + 3\sqrt{2}$; 0.05
b. $\dfrac{7 + 7\sqrt{2} + \sqrt{6} + 2\sqrt{3}}{-3}$; -7.60 **49.** 8.33 ft   **51. a.** $8\sqrt{10}$ m;
b. about 25.3 m   **53. a.** 365.02 days   **b.** 688.69 days   **c.** 87.91 days
55. a. 36 mph   **b.** 46.5 mph   **57.** $12\pi\sqrt{34} \approx 219.82$ m^2
59. a. $(x + \sqrt{5})(x - \sqrt{5})$ **b.** $(n + \sqrt{19})(n - \sqrt{19})$
61. a. $13\sqrt{3x} + 39\sqrt{x}$ **b.** Answers will vary.   **63.** $\dfrac{3\sqrt{2}}{2}$

Practice Test, pp. 70–71

1. a. True **b.** True **c.** False; $\sqrt{2}$ cannot be expressed as a ratio of two integers. **d.** True **2. a.** 11 **b.** -5 **c.** not a real number **d.** 20
3. a. $\frac{9}{8}$ **b.** $\frac{-7}{6}$ **c.** 0.5 **d.** -4.6 **4. a.** $\frac{28}{3}$ **b.** 0.9 **c.** 4 **d.** -7
5. a. ≈ 4439.28 **6. a.** 0 **b.** undefined **7. a.** 3; $-2, 6, 5$
b. $2; \frac{1}{3}, 1$ **8. a.** -13 **b.** ≈ 7.29 **9. a.** $x^3 - (2x - 9)$

b. $2n - 3\left(\dfrac{n}{2}\right)^2$ **10. a.** Let r represent Earth's radius. Then $11r - 119$

represents Jupiter's radius. **b.** Let e represent this year's earnings. Then $4e + 1.2$ million represents last year's earnings. **11. a.** $9v^2 + 3v - 7$
b. $-7b + 8$ **c.** $x^2 + 6x$ **12. a.** $(3x + 4)(3x - 4)$ **b.** $v(2v - 3)^2$

c. $(x + 5)(x + 3)(x - 3)$ **13. a.** $5b^3$ **b.** $4a^{12}b^{12}$ **c.** $\dfrac{m^6}{8n^3}$ **d.** $\dfrac{25}{4}p^2q^2$

14. a. $-4ab$ **b.** $6.4 \times 10^{-2} = 0.064$ **c.** $\dfrac{a^{12}}{b^4c^8}$ **d.** -6

15. a. $9x^4 - 25y^2$ **b.** $4a^2 + 12ab + 9b^2$
16. a. $7a^4 - 5a^3 + 8a^2 - 3a - 18$ **b.** $-7x^4 + 4x^2 + 5x$ **17. a.** -1

b. $\dfrac{2 + n}{2 - n}$ **c.** $x - 3$ **d.** $\dfrac{x - 5}{3x - 2}$ **e.** $\dfrac{x - 5}{3x + 1}$ **f.** $\dfrac{3(m + 7)}{5(m + 4)(m - 3)}$

18. a. $|x + 11|$ **b.** $\dfrac{-2}{3v}$ **c.** $\dfrac{64}{125}$ **d.** $-\dfrac{1}{2} + \dfrac{\sqrt{2}}{2}$ **e.** $11\sqrt{10}$

f. $x^2 - 5$ **g.** $\dfrac{\sqrt{10x}}{5x}$ **h.** $2(\sqrt{6} + \sqrt{2})$ **19.** $-0.5x^2 + 10x + 1200$;

a. 10 decreases of 0.50 or \$5.00 **b.** Maximum revenue is \$1250.
20. 58 cm

CHAPTER 1
Exercises 1.1, pp. 82–85

1. identity; unknown **3.** literal; two **5.** Answers will vary. **7.** $x = 3$

9. $v = -11$ **11.** $b = \dfrac{6}{5}$ **13.** $b = -15$ **15.** $m = -\dfrac{27}{4}$ **17.** $x = 12$

19. $x = 12$ **21.** $p = -56$ **23.** $a = -3.6$ **25.** $v = -0.5$

27. $n = \dfrac{20}{21}$ **29.** $p = \dfrac{12}{5}$ **31.** contradiction; { }

33. conditional; $n = -\dfrac{11}{10}$ **35.** identity; $\{x | x \in \mathbb{R}\}$ **37.** $C = \dfrac{P}{1 + M}$

39. $r = \dfrac{C}{2\pi}$ **41.** $T_2 = \dfrac{T_1 P_2 V_2}{P_1 V_1}$ **43.** $h = \dfrac{3V}{4\pi r^2}$ **45.** $n = \dfrac{2S_n}{a_1 + a_n}$

47. $P = \dfrac{2(S - B)}{S}$ **49.** $y = -\dfrac{A}{B}x + \dfrac{C}{B}$ **51.** $y = \dfrac{-20}{9}x + \dfrac{16}{3}$

53. $y = \dfrac{-4}{5}x - 5$ **55.** $a = 3; b = 2; c = -19; x = -7$

57. $a = -6; b = 1; c = 33; x = \dfrac{-16}{3}$
59. $a = 7; b = -13; c = -27; x = -2$ **61.** $h = 17$ cm **63.** 510 ft
65. 56 in. **67.** 3084 ft **69.** 48; 50 **71.** 5; 7 **73.** 11: 30 A.M.
75. 36 min **77.** 4 quarts; 50% O.J. **79.** 16 lb; \$1.80 lb **81.** 12 lb
83. 16 lb **85.** Answers will vary **87.** 69 **89.** -3
91. a. $(2x + 3)(2x - 3)$ **b.** $(x - 3)(x^2 + 3x + 9)$

Exercises 1.2, pp. 92–95

1. set; interval **3.** intersection; union **5.** Answers will vary.
7. $w \geq 45$ **9.** $250 < T < 450$

11.

13.

15.

17.

19. $\{x | x \geq -2\}; [-2, \infty)$ **21.** $\{x | -2 \leq x \leq 1\}; [-2, 1]$
23. $\{a | a \geq 2\};$

; $a \in [2, \infty)$

25. $\{n | n \geq 1\};$

; $n \in [1, \infty)$

27. $\{x | x < \frac{-32}{5}\};$

; $x \in (-\infty, \frac{-32}{5})$

29. { } **31.** $\{x | x \in \mathbb{R}\}$ **33.** $\{x | x \in \mathbb{R}\}$
35. $\{2\}; \{-3, -2, -1, 0, 1, 2, 3, 4, 6, 8\}$
37. $\{\}; \{-3, -2, -1, 0, 1, 2, 3, 4, 5, 6, 7\}$
39. $\{4, 6\}; \{2, 4, 5, 6, 7, 8\}$
41. $x \in (-\infty, -2) \cup (1, \infty);$

43. $x \in [-2, 5);$

45. no solution
47. $x \in (-\infty, \infty);$

49. $x \in [-5, 0];$

51. $x \in (\frac{-1}{3}, \frac{-1}{4});$

53. $x \in (-\infty, \infty);$

55. $x \in [-4, 1);$

57. $x \in [-1.4, 0.8];$

59. $x \in (-16, 8);$

61. $m \in (-\infty, 0) \cup (0, \infty)$ **63.** $y \in (-\infty, -7) \cup (-7, \infty)$
65. $a \in (-\infty, \frac{1}{2}) \cup (\frac{1}{2}, \infty)$ **67.** $x \in (-\infty, 4) \cup (4, \infty)$
69. $x \in [2, \infty)$ **71.** $n \in [4, \infty)$ **73.** $b \in [\frac{4}{3}, \infty)$ **75.** $y \in (-\infty, 2]$

77. a. $W = \dfrac{BH^2}{704}$ **b.** $W < 177.34$ lb **79.** $x \geq 81\%$ **81.** $b \geq \$2000$

83. $0 < w < 7.5$ m **85.** $7.2° < C < 29.4°$ **87.** $h > 6$
89. Answers may vary. **91.** $<$ **93.** $<$ **95.** $<$ **97.** $>$ **99.** $2n - 8$
101. $\frac{17}{18}x - 5$

Exercises 1.3, pp. 101–103

1. reverse **3.** $-7; 7$ **5.** no solution; answers will vary. **7.** $\{-4, 6\}$

9. $\{2, -12\}$ **11.** $\{-3.35, 0.85\}$ **13.** $\left\{-\dfrac{8}{7}, 2\right\}$ **15.** $\left\{-\dfrac{1}{2}, \dfrac{1}{2}\right\}$

17. { } **19.** $\{-10, -6\}$ **21.** $\{3.5, 11.5\}$ **23.** $\{-1.6, 1.6\}$

25. $[-5, 9]$ **27.** $\varnothing$ **29.** $\left(-1, \dfrac{3}{5}\right)$ **31.** $(-5, -3)$ **33.** $\left[\dfrac{8}{3}, \dfrac{14}{3}\right]$

35. $\varnothing$ 37. $\left[-\dfrac{7}{4}, 0\right]$ 39. $(-\infty, -10) \cup (4, \infty)$

41. $(-\infty, -3] \cup [3, \infty)$ 43. $\left(-\infty, -\dfrac{7}{3}\right] \cup \left[\dfrac{7}{3}, \infty\right)$

45. $\left(-\infty, \dfrac{3}{7}\right] \cup [1, \infty)$ 47. $(-\infty, \infty)$ 49. $(-\infty, 0) \cup (5, \infty)$.

51. $(-\infty, -0.75] \cup [3.25, \infty)$ 53. $\left(-\infty, -\dfrac{7}{15}\right) \cup (1, \infty)$

55. $45 \le d \le 51$ in. 57. in feet: [32,500, 37,600]; yes
59. in feet: $d < 210$ or $d > 578$ 61. a. $|s - 37.58| \le 3.35$
b. [34.23, 40.93] 63. a. $|s - 125| \le 23$ b. [102, 148]
65. a. $|d - 42.7| < 0.03$ b. $|d - 73.78| < 1.01$
c. $|d - 57.150| < 0.127$ d. $|d - 2171.05| < 12.05$

e. golf: $t \approx 0.0014$ 67. a. $x = 4$ b. $\left[\dfrac{4}{3}, 4\right]$ c. $x = 0$ d. $\left(-\infty, \dfrac{3}{5}\right]$

e. { } 69. $3x(2x + 5)(3x - 4)$ 71. $\dfrac{-3 + \sqrt{3}}{6} \approx -0.21$

Mid-Chapter Check, pp. 103–104

1. a. $r = -9$ b. $x = -6$ c. identity; $m \in \mathbb{R}$ d. $y = \dfrac{50}{13}$

e. contradiction: { } f. $x = 5.5$ 2. $v_0 = \dfrac{H + 16t^2}{t}$

3. $x = \sqrt{\dfrac{S}{\pi(2 + y)}}$

4. a. $x \ge 1$ or $x \le -2$

b. $16 < x \le 19$

5. a. $x \in \left(-\infty, \dfrac{5}{2}\right) \cup \left(\dfrac{5}{2}, \infty\right)$ b. $x \in \left(-\infty, \dfrac{17}{6}\right]$
6. a. $\{-4, 14\}$ b. { } 7. a. $q \in (-8, 0)$ b. $\{-6\}$
8. a. $d \in (-\infty, 0] \cup [4, \infty)$ b. $y \in \left(-\infty, -\dfrac{19}{2}\right) \cup \left(\dfrac{23}{2}, \infty\right)$
c. $k \in (-\infty, \infty)$ 9. 1 hr, 20 min 10. $w \in [8, 26]$; yes

Reinforcing Basic Concepts pp. 104

Exercise 1: $x = -3$ or $x = 7$
Exercise 2: $x \in [-5, 3]$
Exercise 3: $x \in (-\infty, -1] \cup [4, \infty)$

Exercises 1.4, pp. 111–114

1. $3 - 2i$ 3. $2; 3\sqrt{2}$ 5. (b) is correct. 7. a. $4i$ b. $7i$ c. $3\sqrt{3}$
d. $6\sqrt{2}$ 9. a. $-3i\sqrt{2}$ b. $-5i\sqrt{2}$ c. $15i$ d. $6i$ 11. a. $i\sqrt{19}$
b. $i\sqrt{31}$ c. $\dfrac{2\sqrt{3}}{5}i$ d. $\dfrac{3\sqrt{2}}{8}i$ 13. a. $1 + i; a = 1, b = 1$
b. $2 + \sqrt{3}i; a = 2, b = \sqrt{3}$ 15. a. $4 + 2i; a = 4, b = 2$
b. $2 - \sqrt{2}i; a = 2, b = -\sqrt{2}$ 17. a. $5 + 0i; a = 5, b = 0$
b. $0 + 3i; a = 0, b = 3$ 19. a. $18i; a = 0, b = 18$
b. $\dfrac{\sqrt{2}}{2}i; a = 0, b = \dfrac{\sqrt{2}}{2}$ 21. a. $4 + 5\sqrt{2}i; a = 4, b = 5\sqrt{2}$
b. $-5 + 3\sqrt{3}i; a = -5, b = 3\sqrt{3}$
23. a. $\dfrac{7}{4} + \dfrac{7\sqrt{2}}{8}i; a = \dfrac{7}{4}, b = \dfrac{7\sqrt{2}}{8}$ b. $\dfrac{1}{2} + \dfrac{\sqrt{10}}{2}i; a = \dfrac{1}{2}, b = \dfrac{\sqrt{10}}{2}$
25. a. $19 + i$ b. $2 - 4i$ c. $9 + 10\sqrt{3}i$ 27. a. $-3 + 2i$ b. 8
c. $2 - 8i$ 29. a. $2.7 + 0.2i$ b. $15 + \dfrac{1}{12}i$ c. $-2 - \dfrac{1}{8}i$
31. a. 15 b. 16 33. a. $-21 - 35i$ b. $-42 - 18i$

35. a. $-12 - 5i$ b. $1 + 5i$ 37. a. $4 - 5i; 41$ b. $3 + i\sqrt{2}; 11$
39. a. $-7i; 49$ b. $\dfrac{1}{2} + \dfrac{2}{3}i; \dfrac{25}{36}$ 41. a. 41 b 74 43. a. 11 b $\dfrac{17}{36}$
45. a. $-5 + 12i$ b $-7 - 24i$ 47. a. $-21 - 20i$ b $7 + 6\sqrt{2}i$
49. no 51. yes 53. yes 55. yes 57. yes 59. Answers will vary.

61. a. 1 b. -1 c. $-i$ d. i 63. a. $\dfrac{2}{7}i$ b. $\dfrac{-4}{5}i$

65. a. $\dfrac{21}{13} - \dfrac{14}{13}i$ b. $\dfrac{-10}{13} - \dfrac{15}{13}i$ 67. a. $1 - \dfrac{3}{4}i$ b. $-1 - \dfrac{2}{3}i$
69. a. $\sqrt{13}$ b. 5 c. $\sqrt{11}$ 71. $A + B = 10$ $AB = 40$
73. $7 - 5i\ \Omega$ 75. $25 + 5i$ V 77. $\dfrac{7}{4} + i\ \Omega$ 79. a. $(x + 6i)(x - 6i)$
b. $(m + i\sqrt{3})(m - i\sqrt{3})$ c. $(n + 2i\sqrt{3})(n - 2i\sqrt{3})$
d. $(2x + 7i)(2x - 7i)$ 81. $-8 - 6i$ 83. a. $P = 4s; A = s^2$
b. $P = 2L + 2W; A = LW$ 85. John

Exercises 1.5, pp. 124–128

1. descending; 0 3. quadratic; 1 5. GCF factoring: $x = 0, x = \dfrac{5}{4}$
7. $a = -1; b = 2; c = -15$ 9. not quadratic
11. $a = \dfrac{1}{4}; b = -6; c = 0$ 13. $a = 2; b = 0; c = 7$ 15. not quadratic
17. $a = 1; b = -1; c = -5$ 19. $x = 5$ or $x = -3$ 21. $m = 4$
23. $p = 0$ or $p = 2$ 25. $h = 0$ or $h = \dfrac{-1}{2}$ 27. $a = 3$ or $a = -3$
29. $g = -9$ 31. $m = -5$ or $m = -3$ or $m = 3$ 33. $c = -3$ or $c = 15$
35. $r = 8$ or $r = -3$ 37. $t = -13$ or $t = 2$ 39. $x = 5$ or $x = -3$
41. $w = -\dfrac{1}{2}$ or $w = 3$ 43. $m = \pm4$ 45. $y = \pm2\sqrt{7}; y \approx \pm5.29$
47. no real solutions 49. $x = \pm\dfrac{\sqrt{21}}{4}; x \approx \pm1.15$ 51. $n = 9; n = -3$
53. $w = -5 \pm \sqrt{3}; w \approx -3.27$ or $w \approx -6.73$ 55. no real solutions
57. $m = 2 \pm \dfrac{3\sqrt{2}}{7}; m \approx 2.61$ or $m \approx 1.39$ 59. $9; (x + 3)^2$
61. $\dfrac{9}{4}; \left(n + \dfrac{3}{2}\right)^2$ 63. $\dfrac{1}{9}; \left(p + \dfrac{1}{3}\right)^2$ 65. $x = -1; x = -5$
67. $p = 3 \pm \sqrt{6}; p \approx 5.45$ or $p \approx 0.55$
69. $p = -3 \pm \sqrt{5}; p \approx -0.76$ or $p \approx -5.24$
71. $m = \dfrac{-3}{2} \pm \dfrac{\sqrt{13}}{2}; m \approx 0.30$ or $m \approx -3.30$
73. $n = \dfrac{5}{2} \pm \dfrac{3\sqrt{5}}{2}; n \approx 5.85$ or $n \approx -0.85$
75. $x = \dfrac{1}{2}$ or $x = -4$ 77. $n = 3$ or $n = \dfrac{-3}{2}$
79. $p = \dfrac{3}{8} \pm \dfrac{\sqrt{41}}{8}; p \approx 1.18$ or $p \approx -0.43$
81. $m = \dfrac{7}{2} \pm \dfrac{\sqrt{33}}{2}; m \approx 6.37$ or $m \approx 0.63$
83. $x = 6$ or $x = -3$ 85. $m = \pm\dfrac{5}{2}$
87. $n = \dfrac{2 \pm \sqrt{5}}{2}; n \approx 2.12$ or $n \approx -0.12$ 89. $w = \dfrac{2}{3}$ or $w = \dfrac{-1}{2}$
91. $m = \dfrac{3}{2} \pm \dfrac{\sqrt{6}}{2}i; m \approx 1.5 \pm 1.12i$ 93. $n = \pm\dfrac{3}{2}$
95. $w = \dfrac{-4}{5}$ or $w = 2$ 97. $a = \dfrac{1}{6} \pm \dfrac{\sqrt{23}}{6}i; a \approx 0.16 \pm 0.80i$
99. $p = \dfrac{3 \pm 2\sqrt{6}}{5}; p \approx 1.58$ or $p \approx -0.38$
101. $w = \dfrac{1 \pm \sqrt{21}}{10}; w \approx 0.56$ or $w \approx -0.36$
103. $a = \dfrac{3}{4} \pm \dfrac{\sqrt{31}}{4}i; a \approx 0.75 \pm 1.39i$
105. $p = 1 \pm \dfrac{3\sqrt{2}}{2}i; p \approx 1 \pm 2.12i$
107. $w = \dfrac{-1}{3} \pm \dfrac{\sqrt{2}}{3}; w \approx 0.14$ or $w \approx -0.80$
109. $a = \dfrac{-6 \pm 3\sqrt{2}}{2}; a \approx -0.88$ or $a \approx -5.12$
111. $p = \dfrac{4 \pm \sqrt{394}}{6}; p \approx 3.97$ or $p \approx -2.64$
113. two rational; factorable 115. two complex 117. two rational;
factorable 119. two complex 121. two irrational 123. one repeated;
factorable 125. $x = \dfrac{3}{2} \pm \dfrac{1}{2}i$ 127. $x = -\dfrac{1}{2} \pm \dfrac{i\sqrt{3}}{2}$
129. $x = \dfrac{5}{4} \pm \dfrac{3i\sqrt{7}}{4}$ 131. $t = \dfrac{v \pm \sqrt{v^2 - 64h}}{32}$

133. $t = \dfrac{6 + \sqrt{138}}{2}$ sec, $t \approx 8.87$ sec 135. 30,000 ovens
137. a. $P = -x^2 + 120x - 2000$ b. 10,000 139. $t = 2.5$ sec, 6.5 sec
141. $x \approx 13.5$, or the year 2008 143. 36 ft, 78 ft
145. a. $7x^2 + 6x - 16 = 0$ b. $6x^2 + 5x - 14 = 0$
c. $5x^2 - x - 6 = 0$ 147. $x = -2i; x = 5i$ 149. $x = \dfrac{-3}{4}i; x = 2i$
151. $x = -1 - i; x = -13 - i$ 153. a. $P = 2L + 2W, A = LW$
b. $P = 2\pi r, A = \pi r^2$ c. $A = \dfrac{1}{2}h(b_1 + b_2), P = c + h + b_1 + b_2$
d. $A = \dfrac{1}{2}bh, P = a + b + c$ 155. 700 \$30 tickets; 200 \$20 tickets

Exercises 1.6, pp. 137–142

1. excluded 3. extraneous 5. Answers will vary.
7. $x = -2, x = 0, x = 11$ 9. $x = -3, x = 0, x = \dfrac{2}{3}$
11. $x = -\dfrac{3}{2}, x = 0, x = 3$ 13. $x = 0, x = 2, x = -1 \pm i\sqrt{3}$

15. $x = \pm 2, x = 5$ **17.** $x = 3, x = \pm 2i$ **19.** $x = \pm\sqrt{5}, x = 6$
21. $x = 0, x = 7, x = \pm 2i$ **23.** $x = \pm 3, x = \pm 3i$
25. $x = \pm 4, x = \pm 4i$ **27.** $x = \pm\sqrt{2}, x = \pm 1, \pm i$
29. $x = \pm 1, x = 2, x = -1 \pm i\sqrt{3}$
31. $x = -\frac{1}{2} \pm \frac{i\sqrt{3}}{2}, x = \frac{1}{2} \pm \frac{i\sqrt{3}}{2}, x = \pm 1$ **33.** $x = 1$ **35.** $a = \frac{3}{2}$
37. $y = 12$ **39.** $x = 3; x = 7$ is extraneous **41.** $n = 7$

43. $a = -1, a = -8$ **45.** $f = \dfrac{f_1 f_2}{f_1 + f_2}$ **47.** $r = \dfrac{E - IR}{I}$ or $\dfrac{E}{I} - R$

49. $h = \dfrac{3V}{\pi r^2}$ **51.** $r^3 = \dfrac{3V}{4\pi}$ **53. a.** $x = \frac{14}{3}$

b. $x = 8, x = 1$ is extraneous **55. a.** $m = 3$ **b.** $x = 5$ **c.** $m = -64$
d. $x = -16$ **57. a.** $x = 25$ **b.** $x = 7; x = -2$ is extraneous
c. $x = 2, x = 18$ **d.** $x = 6; x = 0$ is extraneous **59.** $x = -32$
61. $x = 9$ **63.** $x = -32, x = 22$ **65.** $x = -27, 125$
67. $x = \pm 5, x = \pm i$ **69.** $x = \pm 1, \pm 2$ **71.** $x = -1, \frac{1}{4}$
73. $x = \pm\frac{1}{3}, \pm\frac{1}{2}$ **75.** $x = -4, 45$

77. $x = -6; x = \frac{-74}{9}$ is extraneous **79. a.** $h = \sqrt{\left(\dfrac{S}{\pi r}\right)^2 - r^2}$

b. $S = 12\pi\sqrt{34}\,\text{m}^2$ **81.** $x = \pm 3, x = -2$ **83.** $x = 2, 4, 6$ or
$x = -2, 0, 2$ **85.** 11 in. by 13 in. **87.** $r = 3$ m; $r = 0$ and $r = 12$ do
not fit the context **89.** either \$50 or \$30 **91. a.** 32 ft, $(h = -32)$
b. 11 sec **c.** pebble is at canyon's rim **93.** 12 min **95.** $v = 6$ mph
97. $P \approx 52.1\%$ **99. a.** 36 million mi **b.** 67 million mi
c. 93 million mi **d.** 142 million mi **e.** 484 million mi
f. 887 million mi **101.** The constant "3" was not multiplied by the LCD,
$3x(x + 3) - 8x = x + 3; x = -1, 1$ **103.** $x \in [1, 2) \cup (2, \infty)$
105. a. $x = -5, -3, 5, 7$ **b.** $x = -2, -1, 6, 7$ **c.** $x = -2, 1, 3$
d. $x = -4, -2, 3$ **e.** $x = -1, 1, 7$ **f.** $x = -1, 1, 2, 7$
107. $2\sqrt{11}$ cm **109.** $-1 < x < 5$;

Summary and Concept Review, pp. 142–146

1. a. yes **b.** yes **c.** yes **2.** $b = 6$ **3.** $n = 4$ **4.** $m = -1$

5. $x = \frac{1}{6}$ **6.** no solution **7.** $g = 10$ **8.** $h = \dfrac{V}{\pi r^2}$ **9.** $L = \dfrac{P - 2W}{2}$

10. $x = \frac{c - b}{a}$ **11.** $y = \frac{2}{3}x - 2$ **12.** 8 gal **13.** $12 + \frac{9}{8}\pi$ ft$^2 \approx 15.5$ ft^2
14. $\frac{2}{3}$ hr $= 40$ min **15.** $a \geq 35$ **16.** $a < 2$ **17.** $s \leq 65$
18. $c \geq 1200$ **19.** $(5, \infty)$ **20.** $(-10, \infty)$ **21.** $(-\infty, 2]$
22. $(-9, 9]$ **23.** $(-6, \infty)$ **24.** $\left(-\infty, \frac{-8}{5}\right) \cup \left(\frac{23}{10}, \infty\right)$
25. a. $(-\infty, 3) \cup (3, \infty)$ **b.** $\left(-\infty, \frac{3}{2}\right) \cup \left(\frac{3}{2}, \infty\right)$ **c.** $[-5, \infty)$
d. $(-\infty, 6]$ **26.** $x \geq 96\%$ **27.** $\{-4, 10\}$ **28.** $\{-7, 3\}$ **29.** $\{-5, 8\}$
30. $\{-4, -1\}$ **31.** $(-\infty, -6) \cup (2, \infty)$ **32.** $[4, 32]$ **33.** $\{\,\}$ **34.** $\{\,\}$
35. $(-\infty, \infty)$ **36.** $[-2, 6]$ **37.** $(-\infty, -2] \cup [\frac{10}{3}, \infty)$
38. a. $|r - 2.5| \leq 1.7$ **b.** highest: 4.2 in., lowest: 0.8 in. **39.** $6\sqrt{2}i$
40. $24\sqrt{3}i$ **41.** $-2 + \sqrt{2}i$ **42.** $3\sqrt{2}i$ **43.** i **44.** $21 + 20i$
45. $-2 + i$ **46.** $-5 + 7i$ **47.** 13 **48.** $-20 - 12i$
49. $(5i)^2 - 9 = -34$ $\qquad (-5i)^2 - 9 = -34$
$\ 25i^2 - 9 = -34$ $\qquad 25i^2 - 9 = -34$
$\ -25 - 9 = -34\checkmark$ $\qquad -25 - 9 = -34\checkmark$
50. $(2 + i\sqrt{5})^2 - 4(2 + i\sqrt{5}) + 9 = 0$
$(2 - i\sqrt{5})^2 - 4(2 - i\sqrt{5}) + 9 = 0$
$4 + 4i\sqrt{5} + 5i^2 - 8 - 4i\sqrt{5} + 9 = 0$
$4 - 4i\sqrt{5} + 5i^2 - 8 + 4i\sqrt{5} + 9 = 0$
$5 + (-5) = 0\checkmark$ $\quad 5 + (-5) = 0\checkmark$
51. a. $2x^2 + 3 = 0; a = 2, b = 0, c = 3$ **b.** not quadratic
c. $x^2 - 8x - 99 = 0; a = 1, b = -8, c = -99$
d. $x^2 + 16 = 0; a = 1, b = 0, c = 16$ **52. a.** $x = 5$ or $x = -2$
b. $x = -5$ or $x = 5$ **c.** $x = -\frac{5}{3}$ or $x = 3$ **d.** $x = -2$ or $x = 2$ or $x = 3$
53. a. $x = \pm 3$ **b.** $x = 2 \pm \sqrt{5}$ **c.** $x = \pm\sqrt{5}i$ **d.** $x = \pm 5$
54. a. $x = 3$ or $x = -5$ **b.** $x = -8$ or $x = 2$
c. $x = 1 \pm \frac{\sqrt{10}}{2}; x \approx 2.58$ or $x \approx -0.58$ **d.** $x = 2$ or $x = \frac{1}{3}$
55. a. $x = 2 \pm \sqrt{5}\,i; x \approx 2 \pm 2.24i$ **b.** $x = \frac{3 \pm \sqrt{2}}{2}; x \approx 2.21$ or
$x \approx 0.79$ **c.** $x = \frac{3}{2} \pm \frac{1}{2}i$ **56. a.** 1.34 sec **b.** 4.66 sec **c.** 6 sec

57. a. 0.8 sec **b.** 3.2 sec **c.** 5 sec **58.** \$3.75; 3000 **59.** 6 hr
60. $x = \pm\sqrt{3}, x = 7$ **61.** $x = -2, x = 0, x = \frac{1}{3}$
62. $x = 0, x = 2, x = -1 \pm \sqrt{3}i$ **63.** $x = \pm\frac{1}{2}, x = \pm\frac{1}{2}i$ **64.** $x = \frac{-1}{2}$
65. $h = -\frac{5}{3}, h = 2$ **66.** $n = 13; n = -2$ is extraneous
67. $x = -3; x = 3$ **68.** $x = -4; x = 5$
69. $x = -1; x = 7$ is extraneous **70.** $x = \frac{5}{2}$ **71.** $x = -5.8; x = 5$
72. $x = -2, x = -1, x = 4, x = 5$
73. $x = -3, x = 3, x = -i\sqrt{2}, x = i\sqrt{2}$ **74. a.** 12,000 kilocalories
b. 810 kg **75.** width, 6 in.; length, 9 in. **76.** 1 sec; 244 ft; 8 sec
77. \$24 per load; \$42 per load

Mixed Review, pp. 147–147

1. a. $x \in (8, \infty)$ **b.** $x \in (-\infty, \frac{-4}{3}) \cup (\frac{-4}{3}, \infty)$ **3. a.** $x = 2, x = \pm 5i$
b. $x = 0, x = -5, x = \frac{5}{2} \pm \frac{5i\sqrt{3}}{2}$ **c.** $x = -\frac{7}{3}, x = \frac{5}{3}$
d. $(-\infty, 3] \cup [27, \infty)$ **e.** $v = \pm 27$ **f.** $x = 80$ **5.** $y = \frac{-3}{4}x - 3$
7. a. $x = -2$ **b.** $n = 5$ **9.** $x = 7, 11$ **11.** $x = -\sqrt{6}, \sqrt{6}$
13. $x = \frac{4}{5}$ **15.** $x = \pm\sqrt{5}, \pm i\sqrt{5}$ **17. a.** $v = 6, 2$ is extraneous
b. $x = -5; x = 4$ **c.** $x = 2; x = 18$ is extraneous **19.** $6'10''$

Practice Test, pp. 147

1. a. $x = 27$ **b.** $x = 2$ **c.** $C = \frac{P}{1 + k}$ **d.** $x = -4, x = -1$
2. 30 gal **3. a.** $x > -30$ **b.** $-5 \leq x < 4$ **c.** $x \in \mathbb{R}$
d. $x = 2, x = 4$ **e.** $x < -4$ or $x > 2$ **4.** $S \geq 177$
5. $z = -3, z = 10$ **6.** $x = \pm 5i$ **7.** $x = 1 \pm i\sqrt{3}$
8. $x = \pm 1, x = \pm 4$ **9.** $x = \frac{2}{3}, x = 6$ **10.** $x = -2, x = \pm\frac{3}{2}$
11. $x = 6, x = -2$ is extraneous **12.** $x = -\frac{3}{2}, x = 2$
13. $x = 16, x = 4$ is extraneous **14.** $x = -11, x = 5$
15. a. \$4.50 per tin **b.** 90 tins **16. a.** $t = 5$ (May) **b.** $t = 9$ (Sept.)
c. July; \$3000 more **17.** $-\frac{4}{3} \pm \frac{i\sqrt{5}}{3}$ **18.** $-i$ **19. a.** 1 **b.** $i\sqrt{3}$
c. 1 **20.** $-\frac{3}{2} + \frac{3}{2}i$ **21.** 34 **22.** $(2 - 3i)^2 - 4(2 - 3i) + 13 = 0$
$-5 - 12i - 8 + 12i + 13 = 0$ $\quad 0 = 0\checkmark$ **23. a.** $x = 5 \pm \frac{\sqrt{2}}{2}$
b. $x = \frac{5}{4} \pm \frac{i\sqrt{7}}{4}$ **24. a.** $x = \frac{3 \pm \sqrt{3}}{3}$ **b.** $x = 1 \pm 3i$
25. a. $F \approx 64.8$ g **b.** $W \approx 256$ g

Strengthening Core Skills pp. 149–150

Exercise 1: $\dfrac{7}{2} + (-1) = \dfrac{5}{2} = -\dfrac{b}{a}\checkmark$ $\quad \dfrac{7}{2} \cdot (-1) = \dfrac{-7}{2} = \dfrac{c}{a}\checkmark$

Exercise 2: $\dfrac{2 + 3\sqrt{2}}{2} + \dfrac{2 - 3\sqrt{2}}{2} = \dfrac{4}{2} = \dfrac{-b}{a}\checkmark$ $\quad \dfrac{2 + 3\sqrt{2}}{2} \cdot \dfrac{2 - 3\sqrt{2}}{2}$

$= \dfrac{-14}{4} = \dfrac{-7}{2} = \dfrac{c}{a}\checkmark$

Exercise 3: $(5 + 2\sqrt{3}i) + (5 - 2\sqrt{3}i) = 10 = \dfrac{-b}{a}\checkmark$

$(5 + 2\sqrt{3}i)(5 - 2\sqrt{3}i) = 25 + 12 = 37 = \dfrac{c}{a}\checkmark$

CHAPTER 2
Exercises 2.1, pp. 161–164

1. first, second **3.** radius, center **5.** Answers will vary.
7.

$D = \{1, 2, 3, 4, 5\}$
$R = \{2.75, 3.00, 3.25, 3.50, 3.75\}$

9. $D = \{1, 3, 5, 7, 9\}; R = \{2, 4, 6, 8, 10\}$
11. $D = \{4, -1, 2, -3\}; R = \{0, 5, 4, 2, 3\}$
13.

x	y
-6	5
-3	3
0	1
3	-1
6	-3
8	$\frac{-13}{3}$

15.

x	y
−2	0
0	2, −2
1	3, −3
3	5, −5
6	8, −8
7	9, −9

17.

x	y
−3	8
−2	3
0	−1
2	3
3	8
4	15

19.

x	y
−4	3
−3	4
0	5
2	$\sqrt{21}$
3	4
4	3

61. $(x-5)^2 + (y-4)^2 = 9$

63. $(2,3), r = 2, x \in [0,4], y \in [1,5]$

65. $(-1,2), r = 2\sqrt{3}, x \in [-1-2\sqrt{3}, -1+2\sqrt{3}],$ $y \in [2-2\sqrt{3}, 2+2\sqrt{3}]$

21.

x	y
10	3, −3
5	2, −2
4	$\sqrt{3}, -\sqrt{3}$
2	1, −1
1.25	0.5, −0.5
1	0

23.

x	y
−9	−2
−2	−1
−1	0
0	1
4	$\sqrt[3]{5}$
7	2

67. $(-4,0), r = 9, x \in [-13,5], y \in [-9,9]$

69. $(x-5)^2 + (y-6)^2 = 57, (5,6), r = \sqrt{57}$

71. $(x-5)^2 + (y+2)^2 = 25, (5,-2), r = 5$

25. $(3,1)$ **27.** $(-0.7, -0.3)$ **29.** $\left(\frac{1}{20}, \frac{1}{24}\right)$ **31.** $(0,-1)$

33. $(-1,0)$ **35.** $2\sqrt{34}$ **37.** 10   **39.** not a right triangle

41. not a right triangle   **43.** right triangle

45. $x^2 + y^2 = 9$ **47.** $(x-5)^2 + y^2 = 3$

73. $x^2 + (y+3)^2 = 14, (0,-3), r = \sqrt{14}$

49. $(x-4)^2 + (y+3)^2 = 4$ **51.** $(x+7)^2 + (y+4)^2 = 7$

75. $(x+2)^2 + (y+5)^2 = 11, (-2,-5), r = \sqrt{11}$

53. $(x-1)^2 + (y+2)^2 = 9$ **55.** $(x-4)^2 + (y-5)^2 = 12$

77. $(x+7)^2 + y^2 = 37, (-7,0), r = \sqrt{37}$

57. $(x-7)^2 + (y-1)^2 = 100$ **59.** $(x-3)^2 + (y-4)^2 = 41$

79. $(x-3)^2 + (y+5)^2 = 32, (3,-5), r = 4\sqrt{2}$

81. a. $(1, 71.5), (2, 84), (3, 96.5), (5, 121.5), (7, 146.5)$; yes **b.** $159
c. 2011 **d.**

83. a. $(x - 5)^2 + (y - 12)^2 = 625$ **b.** no
85. Red: $(x - 2)^2 + (y - 2)^2 = 4$;
Blue: $(x - 2)^2 + y^2 = 16$;
Area blue $= 12\pi$ units^2
87.

No, distance between centers is less than sum of radii.   **89.** Answers will vary.

91. a. center: $(6, -2)$; $r = 0$ (degenerate case) **b.** center: $(1, 4)$; $r = 5$
c. $r^2 = -1$; degenerate case
93. a. 0   **b.** not possible   **c.** 0.3; many answers possible
d. not possible   **e.** not possible   **f.** $\sqrt{3}$; many answers possible
95. $n = 1$ is a solution, $n = -2$ is extraneous

Exercises 2.2, pp. 174–177

1. 0, 0   **3.** negative, downward   **5.**

7.

x	y
-6	6
-3	4
0	2
3	0

9.

x	y
-2	1
0	4
2	7
4	10

11. $-0.5 = \frac{3}{2}(-3) + 4$ **13.**
$-0.5 = -\frac{9}{2} + 4$
$-0.5 = -0.5\checkmark$
$\frac{19}{4} = \frac{3}{2}(\frac{1}{2}) + 4$
$\frac{19}{4} = \frac{3}{4} + 4$
$\frac{19}{4} = \frac{19}{4}\checkmark$

15.

17.

19.

21.

23.

25.

27.

29.

31.

33. $m = 1$;
$(2, 4)$ and $(1, 3)$

35. $m = \frac{4}{3}$;
$(7, -1)$ and $(1, -9)$

37. $m = \frac{-15}{4}$;
$(1, -8)$ and $(-1, -\frac{1}{2})$

39. $m = \frac{-4}{7}$;
$(-10, 10)$ and $(11, -2)$

41. a. $m = 125$, cost increased $125,000 per 1000 sq ft **b.** $375,000
43. a. $m = 22.5$, distance increases 22.5 mph   **b.** about 186 mi
45. a. $m = \frac{23}{6}$, a person weighs 23 lb more for each additional 6 in. in height   **b.** 3.8
47. In inches: $(0, -6)$ and $(576, -18)$: $m = \frac{-1}{48}$. The sewer line is 1 in. deeper for each 48 in. in length.
49.

51.

53. $L_1: x = 2$; $L_2: y = 4$; point of intersection $(2, 4)$
55. a. For any two points chosen $m = 0$, indicating there has been no increase or decrease in the number of supreme court justices.
b. For any two points chosen $m = \frac{1}{10}$, which indicates that over the last 5 decades, one nonwhite or nonfemale justice has been added to the court every 10 yr.
57. parallel   **59.** neither   **61.** parallel   **63.** not a right triangle
65. not a right triangle   **67.** right triangle
69. a. 76.4 yr   **b.** 2010
71. $v = -1250t + 8500$
a. $3500
b. 5 yr
73. $h = -3t + 300$ **a.** 273 in.   **b.** 20 months
75. Yes they will meet, the two roads are not parallel: $\frac{38}{12} \neq \frac{30}{9.5}$.
77. a. $3789 **b.** 2012   **79. a.** 23%   **b.** 2005
81. $a = -6$ **83. a.** 142 **b.** -83 **c.** 9 **d.** $\frac{27}{2}$
85. perimeter of a rectangle, volume of a rectangular prism, volume of a right circular cylinder, circumference of a circle
87. 2 hr

Exercises 2.3, pp. 186–190

1. $\frac{-7}{4}$; $(0, 3)$ **3.** 2.5   **5.** Answers will vary

7. $y = \frac{-4}{5}x + 2$ **9.** $y = 2x + 7$ **11.** $y = \frac{-5}{3}x - 5$

x	y
-5	6
-2	$\frac{18}{5}$
0	2
1	$\frac{6}{5}$
3	$\frac{-2}{5}$

x	y
-5	-3
-2	3
0	7
1	9
3	13

x	y
-5	$\frac{10}{3}$
-2	$\frac{-5}{3}$
0	-5
1	$\frac{-20}{3}$
3	-10

13. $y = 2x - 3$: $2, -3$ **15.** $y = \frac{-5}{3}x - 7$: $\frac{-5}{3}, -7$
17. $y = \frac{-35}{6}x - 4$: $\frac{-35}{6}, -4$

19.   **21.**   **23.**

115. (1) d   (2) a   (3) c   (4) b   (5) f   (6) h
117. $x = \frac{5 \pm 2\sqrt{13}}{3}; x \approx -0.74$ or $x \approx 4.07$ **119.** 113.10 yd^2

Exercises 2.4, pp. 200–205

1. first   **3.** range   **5.** Answers will vary.   **7.** function   **9.** Not a function. The Shaq is paired with two heights.   **11.** Not a function; 4 is paired with 2 and -5.   **13.** function   **15.** function   **17.** Not a function; -2 is paired with 3 and -4.   **19.** function   **21.** function   **23.** Not a function; 0 is paired with 4 and -4.   **25.** function   **27.** Not a function; 5 is paired with -1 and 1.   **29.** function

25. a. $\frac{-3}{4}$ **b.** $y = \frac{-3}{4}x + 3$ **c.** The coeff. of x is the slope and the constant is the y-intercept.
27. a. $\frac{2}{5}$ **b.** $y = \frac{2}{5}x - 2$ **c.** The coeff. of x is the slope and the constant is the y-intercept.
29. a. $\frac{4}{5}$ **b.** $y = \frac{4}{5}x + 3$ **c.** The coeff. of x is the slope and the constant is the y-intercept.   **31.** $y = \frac{-2}{3}x + 2, m = \frac{-2}{3}$, y-intercept $(0, 2)$
33. $y = \frac{-5}{4}x + 5, m = \frac{-5}{4}$, y-intercept $(0, 5)$ **35.** $y = \frac{1}{3}x, m = \frac{1}{3}$, y-intercept $(0, 0)$ **37.** $y = \frac{-3}{4}x + 3, m = \frac{-3}{4}$, y-intercept $(0, 3)$
39. $y = \frac{2}{3}x + 1$ **41.** $y = 3x + 3$ **43.** $y = 3x + 2$
45. $y = 250x + 500$ **47.** $y = \frac{75}{2}x + 150$ **49.** $y = 2x - 13$
51. $y = -\frac{3}{5}x + 4$ **53.** $y = \frac{2}{3}x - 5$ **55.**

31. function   **33.** function

35. function, $x \in [-4, 5], y \in [-2, 3]$ **37.** function, $x \in [-4, \infty)$, $y \in [-4, \infty)$ **39.** function, $x \in [-4, 4], y \in [-5, -1]$ **41.** function, $x \in (-\infty, \infty), y \in (-\infty, \infty)$ **43.** Not a function, $x \in [-3, 5]$, $y \in [-3, 3]$ **45.** Not a function, $x \in (-\infty, 3], y \in (-\infty, \infty)$
47. $x \in (-\infty, 5) \cup (5, \infty)$ **49.** $x \in [\frac{-5}{3}, \infty)$
51. $x \in (-\infty, -5) \cup (-5, 5) \cup (5, \infty)$
53. $v \in (-\infty, -3\sqrt{2}) \cup (-3\sqrt{2}, 3\sqrt{2}) \cup (3\sqrt{2}, \infty)$
55. $x \in (-\infty, \infty)$ **57.** $x \in (-\infty, \infty)$ **59.** $x \in (-\infty, \infty)$
61. $x \in (-\infty, -2) \cup (-2, 5) \cup (5, \infty)$ **63.** $x \in [2, \frac{5}{2}) \cup (\frac{5}{2}, \infty)$
65. $x \in (2, \infty)$ **67.** $x \in (-4, \infty)$ **69.** $f(-6) = 0, f(\frac{3}{2}) = \frac{15}{4}, f(2c) = c + 3,$

57.   **59.**   **61.**

63. $y = \frac{2}{5}x + 4$ **65.** $y = \frac{-5}{3}x + 7$ **67.** $y = \frac{-12}{5}x - \frac{29}{5}$

69. $y = 5$ **71.** perpendicular   **73.** neither   **75.** neither
77. a. $y = \frac{-3}{4}x - \frac{5}{2}$ **b.** $y = \frac{4}{3}x - \frac{20}{3}$
79. a. $y = \frac{4}{9}x + \frac{31}{9}$ **b.** $y = \frac{-9}{4}x + \frac{3}{4}$
81. a. $y = \frac{-1}{2}x - 2$ **b.** $y = 2x - 2$
83. $y + 5 = 2(x - 2)$ **85.** $y + 4 = \frac{3}{8}(x - 3)$

$f(c + 1) = \frac{1}{2}c + \frac{7}{2}$ **71.** $f(-6) = 132, f(\frac{3}{2}) = \frac{3}{4}, f(2c) = 12c^2 - 8c,$

$f(c + 1) = 3c^2 + 2c - 1$ **73.** $h(3) = 1, h(\frac{-2}{3}) = \frac{-9}{2}, h(3a) = \frac{1}{a},$

$h(a - 2) = \frac{3}{a - 2}$ **75.** $h(3) = 5, h(\frac{-2}{3}) = -5, h(3a) = -5$ if $a < 0$ or

5 if $a > 0, h(a - 2) = 5\left(\frac{|a - 2|}{a - 2}\right)$

77. $g(4) = 8\pi, g\left(\frac{3}{2}\right) = 3\pi, g(2c) = 4\pi c, g(c + 3) = 2\pi(c + 3)$

79. $g(4) = 16\pi, g\left(\frac{3}{2}\right) = \frac{9}{4}\pi, g(2c) = 4\pi c^2, g(c + 3) = (c^2 + 6c + 9)\pi$

81. $p(5) = \sqrt{13}, p\left(\frac{3}{2}\right) = \sqrt{6}, p(3a) = \sqrt{6a + 3}, p(a - 1) = \sqrt{2a + 1}$

83. $p(5) = \frac{14}{5}, p\left(\frac{3}{2}\right) = \frac{7}{9}, p(3a) = \frac{27a^2 - 5}{9a^2},$

$p(a - 1) = \frac{3a^2 - 6a - 2}{a^2 - 2a + 1}$

85. a. $D = \{-1, 0, 1, 2, 3, 4, 5\}$ **b.** $R = \{-2, -1, 0, 1, 2, 3, 4\}$ **c.** 1 **d.** -1 **87. a.** $D = [-5, 5]$ **b.** $y \in [-3, 4]$ **c.** -2 **d.** -4 and 0
89. a. $D = [-3, \infty)$ **b.** $y \in (-\infty, 4]$ **c.** 2   **d.** -2 and 2

87. $y + 3.1 = 0.5(x - 1.8)$

91. a. 186.5 lb   **b.** 37 lb   **93.** $A = \frac{1}{2}(8) + 22 - 1 = 25$ units^2

95. a. $N(g) = 2.5g$ **b.** $g \in [0, 5]; N \in [0, 12.5]$ **97. a.** $[0, \infty)$
b. 750π **c.** 800   **99. a.** $c(t) = 42.50t + 50$ **b.** $156.25   **c.** 5 hr
d. $t \in [0, 10.6]; c \in [0, 500]$ **101. a.** Yes. Each x is paired with exactly one y.   **b.** 10 P.M.   **c.** 0.9 m   **d.** 7 P.M. and 1 A.M.
103. a. $\frac{\Delta \text{fertility}}{\Delta \text{time}} = \frac{-1}{20}$, negative, fertility is decreasing by one child every 20 yr   **b.** 1940 to 1950: $\frac{\Delta f}{\Delta t} = \frac{0.8}{10}$; positive, fertility is increasing by less than one child every 10 yr   **c.** 1940 to 1950: $\frac{\Delta f}{\Delta t} = \frac{0.8}{10}$; 1980 to 1990: $\frac{\Delta f}{\Delta t} = \frac{0.2}{10}$, the fertility rate was increasing four times as fast from 1940 to 1950.
105. negative outputs become positive

89. $y - 2 = \frac{6}{5}(x - 4)$; For each 5000 additional sales, income rises $6000.
91. $y - 100 = \frac{-20}{1}(x - 0.5)$; For every hour of television, a student's final grade falls 20%.   **93.** $y - 10 = \frac{35}{2}(x - \frac{1}{2})$; Every 2 in. of rainfall increases the number of cattle raised per acre by 35.   **95.** C   **97.** A
99. B   **101.** D   **103.** $m = \frac{-a}{b}$, y-intercept $= \frac{c}{b}$ **a.** $m = \frac{-3}{4}$, y-intercept $(0, 2)$ **b.** $m = \frac{-2}{5}$, y-intercept $(0, -3)$ **c.** $m = \frac{5}{6}$, y-intercept $(0, 2)$
d. $m = \frac{2}{3}$, y-intercept $(0, 3)$ **105. a.** As the temperature increases 5°C, the velocity of sound waves increases 3 m/s. At a temperature of 0°C, the velocity is 331 m/s.   **b.** 343 m/s   **c.** 50°C   **107. a.** $V = \frac{20}{3}t + 150$ **b.** Every 3 yr the value of the coin increases by $20; the initial value was $150.
c. $223.33   **d.** 15 years, in 2013   **e.** 3 yr   **109. a.** $N = 7t + 9$
b. Every 1 yr the number of homes with Internet access increases by 7 million.   **c.** 1993   **d.** 86 million   **e.** 13 yr   **f.** 2010
111. a. $P = 58{,}000t + 740{,}000$ **b.** Each year, the prison population increases by 58,000.   **c.** 1,726,000   **113.** Answers will vary.

107. a. $x \in (-\infty, -2) \cup (2, \infty)$; $x = \frac{2y + 3}{1 - y}$; $y \in (-\infty, 1) \cup (1, \infty)$
b. $x \in \mathbb{R}$ $x = \pm \sqrt{y + 3}$; $y \in [-3, \infty)$ **109. a.** $19\sqrt{6}$ **b.** 1
111. a. $(x - 3)(x - 5)(x + 5)$ **b.** $(2x + 3)(x - 8)$
c. $(2x - 5)(4x^2 + 10x + 25)$

Mid-Chapter Check, p. 205

1.

2. $\frac{-18}{7}$ **3.** positive, loss is decreasing (profit is increasing); $m = \frac{3}{2}$, yes; $\frac{1.5}{1}$, each year Data.com's loss decreases by 1.5 million.
4.

$y = \frac{3}{2}x + \frac{5}{2}$

5. $x = -3$; no; input -3 is paired with more than one output.
6. $y = \dfrac{-4}{3}x + 4$; yes **7. a.** 0 **b.** $x \in [-3, 5]$ **c.** -1
d. $y \in [-4, 5]$ **8.** from $x = 1$ to $x = 2$; steeper line $\rightarrow$ greater slope
9. $F(p) = \frac{3}{4}p + \frac{5}{4}$, For every 4000 pheasants, the fox population increases by 300: 1625. **10. a.** $x \in \{-3, -2, -1, 0, 1, 2, 3, 4\}$
$y \in \{-3, -2, -1, 0, 1, 2, 3, 4\}$ **b.** $x \in [-3, 4]$ $y \in [-3, 4]$
c. $x \in (-\infty, \infty)$ $y \in (-\infty, \infty)$

Reinforcing Basic Concepts, p. 206

1. a. $\frac{1}{3}$, increasing **b.** $y - 5 = \frac{1}{3}(x - 0)$ **c.** $y = \frac{1}{3}x + 5$
d. $x - 3y = -15$ **e.** $(0, 5)$, $(-15, 0)$

2. a. $\frac{-7}{3}$, decreasing **b.** $y - 9 = \frac{-7}{3}(x - 0)$ **c.** $y = \frac{-7}{3}x + 9$
d. $7x + 3y = 27$ **e.** $(0, 9)$, $(\frac{22}{7}, 0)$

3. a. $\frac{1}{2}$, increasing **b.** $y - 2 = \frac{1}{2}(x - 3)$ **c.** $y = \frac{1}{2}x + \frac{1}{2}$
d. $x - 2y = -1$ **e.** $(0, \frac{1}{2})$, $(-1, 0)$

4. a. $\frac{3}{4}$, increasing **b.** $y + 4 = \frac{3}{4}(x + 5)$ **c.** $y = \frac{3}{4}x - \frac{1}{4}$
d. $3x - 4y = 1$ **e.** $(0, \frac{-1}{4})$, $(\frac{1}{3}, 0)$

5. a. $\frac{-3}{4}$, decreasing **b.** $y - 5 = \frac{-3}{4}(x + 2)$ **c.** $y = \frac{-3}{4}x + \frac{7}{2}$
d. $3x + 4y = 14$ **e.** $(0, \frac{7}{2})$, $(\frac{14}{3}, 0)$

6. a. $\frac{-1}{2}$, decreasing **b.** $y + 7 = \frac{-1}{2}(x - 2)$ **c.** $y = \frac{-1}{2}x - 6$
d. $x + 2y = -12$ **e.** $(0, -6)$, $(-12, 0)$

Exercises 2.5, pp. 218–224

1. linear; bounce **3.** increasing **5.** Answers will vary.
7. **9.** even **11.** even

13.

15. odd **17.** not odd **19.** neither **21.** odd **23.** neither
25. $x \in [-1, 1] \cup [3, \infty)$ **27.** $x \in (-\infty, -1) \cup (-1, 1) \cup (1, \infty)$
29. $p(x) \geq 0$ for $x \in [2, \infty)$ **31.** $f(x) \leq 0$ for $x \in (-\infty, 2]$
33. $V(x)\uparrow: x \in (-3, 1) \cup (4, 6)$ $V(x)\downarrow: x \in (-\infty, -3) \cup (1, 4)$
constant: none **35.** $f(x)\uparrow: x \in (1, 4)$ $f(x)\downarrow: x \in (-2, 1) \cup (4, \infty)$
constant: $x \in (-\infty, -2)$ **37. a.** $p(x)\uparrow: x \in (-\infty, \infty)$ $p(x)\downarrow$: none
b. down, up **39. a.** $f(x)\uparrow: x \in (-3, 0) \cup (3, \infty)$
$f(x)\downarrow: x \in (-\infty, -3) \cup (0, 3)$ **b.** up, up
41. a. $x \in (-\infty, \infty), y \in (-\infty, 5)$ **b.** $x = 1, 3$
c. $H(x) \geq 0: x \in [1, 3]$ $H(x) \leq 0: x \in (-\infty, 1] \cup [3, \infty)$
d. $H(x)\uparrow: x \in (-\infty, 2)$ $H(x)\downarrow: x \in (2, \infty)$ **e.** local max: $y = 5$ at $(2, 5)$
43. a. $x \in (-\infty, \infty), y \in (-\infty, \infty)$ **b.** $x = -1, 5$
c. $g(x) \geq 0: x \in [-1, \infty)$ $g(x) \leq 0: x \in (-\infty, -1] \cup [0, 3.5]$
d. $g(x)\uparrow: x \in (-\infty, 1) \cup (5, \infty)$ $g(x)\downarrow: x \in (1, 5)$ **e.** local max: $y = 6$ at
$(1, 6)$; local min: $y = 0$ at $(5, 0)$ **45. a.** $x \in [-4, \infty), y \in (-\infty, 3]$
b. $x = -4, 2$ **c.** $Y_1 \geq 0: x \in [-4, 2]$ $Y_1 \leq 0: x \in [2, \infty]$
d. $Y_1\uparrow: x \in (-4, -2)$ $Y_1\downarrow: x \in (-2, \infty)$ **e.** local max: $y = 3$ at $(-2, 3)$
47. a. $x \in \mathbb{R}, y \in \mathbb{R}$ **b.** $x = -4$ **c.** $p(x) \geq 0: x \in [-4, \infty); p(x) \leq 0$:
$x \in (-\infty, -4]$ **d.** $p(x)\uparrow: x \in (-\infty, -3) \cup (-3, \infty); p(x)\downarrow$: never
decreasing **e.** local max: none; local min: none
49. a. $x \in (-\infty, -3] \cup [3, \infty), y \in [0, \infty)$ **b.** $(-3, 0), (3, 0)$
c. $f(x)\uparrow: x \in (3, \infty)$ $f(x)\downarrow: x \in (-\infty, -3)$ **d.** even
51. a. $x \in [0, 260], y \in [0, 80]$ **b.** 80 ft **c.** 120 ft **d.** yes **e.** $(0, 120)$
f. $(120, 260)$ **53. a.** $x \in (-\infty, \infty); y \in [-1, \infty)$ **b.** $(-1, 0), (1, 0)$
c. $f(x) \geq 0: x \in (-\infty, -1] \cup [1, \infty); f(x) < 0: x \in (-1, 1)$;
d. $f(x)\uparrow: x \in (0, \infty), f(x)\downarrow: x \in (-\infty, 0)$ **e.** min: $(0, -1)$

55. a. $t \in [72, 96], I \in [7.25, 16]$
b. $I(t)\uparrow$: $t \in (72, 74) \cup (77, 81) \cup (83, 84) \cup (93, 94)$
$I(t)\downarrow$: $t \in (74, 75) \cup (81, 83) \cup (84, 86) \cup (90, 93) \cup (94, 95)$ $I(t)$
constant: $t \in (75, 77) \cup (86, 90) \cup (95, 96)$　**c.** max: $(74, 9.25)$, $(81, 16)$
(global max), $(84, 13)$, $(94, 8.5)$, min: $(72, 7.5)$, $(83, 12.75)$, $(93, 7.25)$
d. Increase: 80 to 81; Decrease: 82 to 83 or 85 to 86
57. zeroes: $(-8, 0)$, $(-4, 0)$, $(0, 0)$, $(4, 0)$;
min: $(-2, -1)$, $(4, 0)$; max: $(-6, 2)$, $(2, 2)$

59. a. 7　**b.** 7　**c.** They are the same.
d. Slopes are equal.

61. a. 176 ft　**b.** 320 ft　**c.** 144 ft/sec　**d.** -144 ft/sec; The arrow is
going down.　**63. a.** 17.89 ft/sec; 25.30 ft/sec　**b.** 30.98 ft/sec; 35.78
ft/sec　**c.** Between 5 and 10.　**d.** 1.482 ft/sec, 0.96 ft/sec　**65.** 2

67. $2x + h$　**69.** $2x + 2 + h$　**71.** $\dfrac{-2}{x(x + h)}$

73. a. $\dfrac{\Delta g}{\Delta x} = 2x + 2 + h$　**b.** $\dfrac{\Delta g}{\Delta x} = -3.9$　**c.** $\dfrac{\Delta g}{\Delta x} = 3.01$

d. The rates of change have opposite sign, with the secant line to the left being slightly more steep.

75. a. $\dfrac{\Delta g}{\Delta x} = 3x^2 + 3xh + h^2$　**b.** $\dfrac{\Delta g}{\Delta x} \approx 12.61$　**c.** $\dfrac{\Delta g}{\Delta x} \approx 0.49$

d. Both lines have a positive slope, but the line at $x = -2$ is much steeper.

77. a. $\dfrac{\Delta d}{\Delta h} \approx 0.25$　**b.** $\dfrac{\Delta d}{\Delta h} \approx 0.05$

c. As height increases you can see farther, the sight distance is increasing much slower.

79. no; no; Answers will vary.　**81.** Answers will vary.

83. $x = -2, x = 10$　**85.** $y = \dfrac{2}{3}x - 1$

Exercises 2.6, pp. 234–239

1. stretch; compression　**3.** $(-5, -9)$; upward　**5.** Answers will vary.
7. a. quadratic;　**b.** up/up, $(-2, -4)$, $x = -2$, $(-4, 0)$, $(0, 0)$, $(0, 0)$;
c. $D: x \in \mathbb{R}, R: y \in [-4, \infty)$　**9. a.** quadratic;　**b.** up/up, $(1, -4)$,
$x = 1$, $(-1, 0)$, $(3, 0)$, $(0, -3)$; **c.** $D: x \in \mathbb{R}, R: y \in [-4, \infty)$
11. a. quadratic;　**b.** up/up, $(2, -9)$, $x = 2$, $(-1, 0)$, $(5, 0)$, $(0, -5)$;
c. $D: x \in \mathbb{R}, R: y \in [-9, \infty)$　**13. a.** square root;　**b.** up to the right,
$(-4, -2)$, $(-3, 0)$, $(0, 2)$; **c.** $D: x \in [-4, \infty), R: y \in [-2, \infty)$
15. a. square root;　**b.** down to the left, $(4, 3)$, $(3, 0)$, $(0, -3)$;
c. $D: x \in (-\infty, 4], R: y \in (-\infty, 3]$　**17. a.** square root;　**b.** up to the
left, $(4, 0)$, $(4, 0)$, $(0, 4)$; **c.** $D: x \in (-\infty, 4], R: y \in [0, \infty)$

19. a. absolute value;　**b.** up/up, $(-1, -4)$, $x = -1$, $(-3, 0)$, $(1, 0)$, $(0, -2)$;
c. $D: x \in \mathbb{R}, R: y \in [-4, \infty)$　**21. a.** absolute value;　**b.** down/down,
$(-1, 6)$, $x = -1$, $(-4, 0)$, $(2, 0)$, $(0, 4)$;　**c.** $D: x \in \mathbb{R}, R: y \in (-\infty, 6]$
23. a. absolute value;　**b.** down/down, $(0, 6)$, $x = 0$, $(-2, 0)$, $(2, 0)$, $(0, 6)$;
c. $D: x \in \mathbb{R}, R: y \in (-\infty, 6]$　**25. a.** cubic;　**b.** up/down, $(1, 0)$, $(1, 0)$,
$(0, 1)$;　**c.** $D: x \in \mathbb{R}, R: y \in \mathbb{R}$　**27. a.** cubic;　**b.** down/up, $(0, 1)$,
$(-1, 0)$, $(0, 1)$;　**c.** $D: x \in \mathbb{R}, R: y \in \mathbb{R}$　**29. a.** cube root;　**b.** down/up,
$(1, -1)$, $(2, 0)$, $(0, -2)$;　**c.** $D: x \in \mathbb{R}, R: y \in \mathbb{R}$　**31.** square root function;
y-int $(0, 2)$; x-int $(-3, 0)$; initial point $(-4, -2)$; up on right;
$D: x \in [-4, \infty), R: y \in [-2, \infty)$　**33.** cubic function; y-int $(0, -2)$;
x-int $(-2, 0)$; inflection point $(-1, -1)$; up, down; $D: x \in \mathbb{R}, R: y \in \mathbb{R}$

35. 　**37.** 　**39.**

41. 　**43.** 　**45.**

47. 　**49.** 　**51.**

53. 　**55.** 　**57.**

59. 　**61.**

63. g　**65.** i　**67.** e　**69.** j　**71.** l　**73.** c
75. 　left 2, down 1　**77.** 　left 3, reflected across x-axis, down 2

79. 　left 3, down 1　**81.** 　left 1, down 2

83. left 3, reflected across x-axis, down 2

85. left 1, reflected across x-axis, stretched vertically, down 3

87. left 2, reflected across x-axis, compressed vertically down 1,

89. left 1, reflected across y-axis, reflected across x-axis, stretched vertically up 3,

91. right 3, compressed vertically up 1

93. a.

b.

c.

d.

95. a.

b.

c.

d.

97. $f(x) = -(x - 2)^2$ **99.** $p(x) = 1.5\sqrt{x + 3}$

101. $f(x) = \frac{4}{5}|x + 4|$ **103.**

≈ -4.2, about 65 units^3, 65.4 units^3, yes

105.

compressed vertically, 2.25 sec

107.

a. compressed vertically,   **b.** 216 W,
c. $\approx$15.6, 161.5, power increases dramatically at higher windspeeds

109.

a. vertical stretch by a factor of 2,   **b.** 12.5 ft,
c. 5, 13, distance fallen by unit time increases very fast

111.

$x \in (0, 4)$; yes, $x \in (4, \infty)$; yes

113. Any points in Quadrants III and IV will reflect across the x-axis and move to Quadrants I and II.

115. $p = 140$ in. $A = 1168$ in^2
117. $f(x)\downarrow : x \in (-\infty, 4)$ $f(x)\uparrow : x \in (4, \infty)$

Exercises 2.7, pp. 248–253

1. continuous   **3.** smooth   **5.** Each piece must be continuous on the corresponding interval, and the function values at the endpoints of each interval must be equal. Answers will vary.

7. a. $f(x) = \begin{cases} x^2 - 6x + 10 & 0 \le x \le 5 \\ \frac{3}{2}x - \frac{5}{2} & 5 < x \le 9 \end{cases}$ **b.** $y \in [1, 11]$

9. $-2, 2, \frac{1}{2}, 0, 2.999, 5$ **11.** $5, 5, 0, -4, 5, 11$
13. $D: x \in [-6, \infty); R: y \in [-4, \infty)$

15. $D: x \in (-2, \infty); R: y \in (-4, \infty)$

17. $D: x \in (-\infty, \infty); R: y \in (-\infty, 3), \cup (3, \infty)$

19. $x \in (-\infty, 9); y \in [2, \infty)$

21. $x \in (-\infty, \infty); y \in [0, \infty)$

23. $x \in (-\infty, \infty); y \in (-\infty, -6) \cup (-6, \infty)$; discontinuity at $x = -3$, redefine $f(x) = -6$ at $x = -3; c = -6$

25. $x \in (-\infty, \infty); y \in [0.75, \infty);$
discontinuity at $x = 1$,
redefine $f(x) = 3$ at $x = 1; c = 3$

27. $f(x) = \begin{cases} \frac{1}{2}x - 1 & -4 \leq x < 2 \\ 3x - 6 & x \geq 2 \end{cases}$

29. $p(x) = \begin{cases} x^2 + 2x - 3 & x \leq 1 \\ x + 1 & x > 1 \end{cases}$

31. Graph is discontinuous at
$x = 0; f(x) = 1$ for $x > 0; f(x) = -1$
for $x < 0$.

33. a. $S(t) = \begin{cases} -t^2 + 6t & 0 \leq t \leq 5 \\ 5 & t > 5 \end{cases}$ **b.** $S(t) \in [0, 9]$

35. a.

Year (0 → 1950)	Percent
5	7.33
15	14.13
25	14.93
35	22.65
45	41.55
55	60.45

b. Each piece gives a slightly different value due to rounding of coefficients in each model. At $t = 30$, we use the "first" piece: $P(30) = 13.08$.

37. $C(h) = \begin{cases} 0.09h & 0 \leq h \leq 1000 \\ 0.18h - 90 & h > 1000 \end{cases}$
$C(1200) = \$126$

39. $C(t) = \begin{cases} 0.75t & 0 \leq t \leq 25 \\ 1.5t - 18.75 & t > 25 \end{cases}$
$C(45) = \$48.75$

41. $S(t) = \begin{cases} -1.35t^2 + 31.9t + 152 & 0 \leq t \leq 12 \\ 2.5t^2 - 80.6t + 950 & 12 < t \leq 22 \end{cases}$

$\$498$ billion, $\$653$ billion, $\$782$ billion

43. $c(m) = \begin{cases} 3.3m & 0 \leq m \leq 30 \\ 7m - 111 & m > 30 \end{cases}$;
$\$2.11$

45. $C(a) = \begin{cases} 0 & a < 2 \\ 2 & 2 \leq a < 13 \\ 5 & 13 \leq a < 20 \\ 7 & 20 \leq a < 65 \\ 5 & a \geq 65 \end{cases}$

$\$38$

47. a. $C(w - 1) = 17[w - 1] + 80$,   **b.** $0 < w \leq 13$;   **c.** $80¢$,
d. $165¢$,   **e.** $165¢$,   **f.** $165¢$,   **g.** $182¢$

49. yes; $h(x) = \begin{cases} 5 & x \leq -3 \\ -2x - 1 & -3 < x < 2 \\ -5 & x \geq 2 \end{cases}$

51. Y_1 has a removable discontinuity at $x = -2$; Y_2 has a discontinuity at $x = -2$ **53.** $x = -7, x = 4$ **55. a.** $4\sqrt{5}$ cm   **b.** $16\sqrt{5}$ cm^2
c. $V = 320\sqrt{5}$ cm^3

Exercises 2.8, pp. 264–270

1. $(f + g)(x); A \cap B$ **3.** intersection; $g(x)$ **5.** Answers will vary.
7. a. $x \in \mathbb{R}$ **b.** $f(-2) - g(-2) = 13$ **9. a.** $h(x) = x^2 - 6x - 3$
b. $h(-2) = 13$ **c.** they are identical   **11. a.** $x \in [3, \infty)$
b. $h(x) = \sqrt{x - 3} + 2x^3 - 54$ **c.** $h(4) = 75, 2$ is not in the domain
of h.   **13. a.** $x \in [-5, 3]$ **b.** $r(x) = \sqrt{x + 5} + \sqrt{3 - x}$
c. $2(7) = \sqrt{7} + 1, 4$ is not in the domain of r.   **15. a.** $x \in [-4, \infty)$
b. $h(x) = \sqrt{x + 4}(2x + 3)$ **c.** $h(-4) = 0, h(21) = 225$
17. a. $x \in [-1, 7]$ **b.** $r(x) = \sqrt{-x^2 + 6x + 7}$ **c.** 15 is not in the
domain of $r, r(3) = 4$ **19. a.** $x \in (-\infty, -4) \cup (-4, \infty)$
b. $h(x) = x - 4, x \neq -4$ **21. a.** $x \in (-\infty, -4) \cup (-4, \infty)$
b. $h(x) = x^2 - 2, x \neq -4$ **23. a.** $x \in (-\infty, 1) \cup (1, \infty)$
b. $h(x) = x^2 - 6x, x \neq 1$ **25. a.** $x \in (-\infty, 5) \cup (5, \infty)$
b. $h(x) = \dfrac{x + 1}{x - 5}, x \neq 5$ **27. a.** $x \in (-\infty, -2)$ **b.** $r(x) = \dfrac{2x - 3}{\sqrt{-2 - x}}$
c. 6 is not in the domain of r. $r(-6) = -\dfrac{15}{2}$ **29. a.** $x \in (5, \infty)$
b. $r(x) = \dfrac{x - 5}{\sqrt{x - 5}}$ **c.** $r(6) = 1; -6$ is not in the domain of r.
31. a. $x \in \left(-\dfrac{13}{2}, \infty\right)$ **b.** $r(x) = \dfrac{x^2 - 36}{\sqrt{2x + 13}}$ **c.** $r(6) = 0, r(-6) = 0$
33. a. $h(x) = \dfrac{2x + 4}{x - 3}$ **b.** $x \in (-\infty, 3) \cup (3, \infty)$ **c.** $x \neq -2, x \neq 0$
35. sum: $3x + 1, x \in (-\infty, \infty)$; difference: $x + 5, x \in (-\infty, \infty)$; product:
$2x^2 - x - 6, x \in (-\infty, \infty)$; quotient: $\dfrac{2x + 3}{x - 2}, x \in (-\infty, 2) \cup (2, \infty)$
37. sum: $x^2 + 3x + 5, x \in (-\infty, \infty)$; difference: $x^2 - 3x + 9$,
$x \in (-\infty, \infty)$; product: $3x^3 - 2x^2 + 21x - 14, x \in (-\infty, \infty)$;
quotient: $\dfrac{x^2 + 7}{3x - 2}, x \in \left(-\infty, \dfrac{2}{3}\right) \cup \left(\dfrac{2}{3}, \infty\right)$
39. sum: $x^2 + 3x - 4, x \in (-\infty, \infty)$; difference: $x^2 + x - 2$,
$x \in (-\infty, \infty)$; product: $x^3 + x^2 - 5x + 3, x \in (-\infty, \infty)$;
quotient: $x + 3, x \in (-\infty, 1) \cup (1, \infty)$
41. sum: $3x + 1 + \sqrt{x - 3}, x \in [3, \infty)$; difference: $3x + 1 - \sqrt{x - 3}$,
$x \in [3, \infty)$; product: $(3x + 1)\sqrt{x - 3}, x \in [3, \infty)$;
quotient: $\dfrac{3x + 1}{\sqrt{x - 3}}, x \in (3, \infty)$ **43.** sum: $2x^2 + \sqrt{x + 1}, x \in [-1, \infty)$;
difference: $2x^2 - \sqrt{x + 1}, x \in [-1, \infty)$; product:
$2x^2\sqrt{x + 1}, x \in [-1, \infty)$;   quotient: $\dfrac{2x^2}{\sqrt{x + 1}}, x \in (-1, \infty)$
45. sum: $\dfrac{7x - 11}{(x - 3)(x + 2)}, x \in (-\infty, -2) \cup (-2, 3) \cup (3, \infty)$;
difference: $\dfrac{-3x + 19}{(x - 3)(x + 2)}, x \in (-\infty, -2) \cup (-2, 3) \cup (3, \infty)$;
product: $\dfrac{10}{(x^2 - x - 6)}, x \in (-\infty, -2) \cup (-2, 3) \cup (3, \infty)$;
quotient: $\dfrac{2x + 4}{(5x - 15)}, x \in (-\infty, -2) \cup (-2, 3) \cup (3, \infty)$
47. $0; 0; 4a^2 - 10a - 14$ $a^2 - 9a$ **49. a.** $h(x) = \sqrt{2x - 2}$
b. $H(x) = 2\sqrt{x + 3} - 5$ **c.** D of $h(x)$: $x \in [1, \infty)$; D of $H(x)$:
$x \in [-3, \infty)$ **51. a.** $h(x) = \sqrt{3x + 1}$ **b.** $H(x) = 3\sqrt{x - 3} + 4$
c. D of $h(x)$: $x \in [-\frac{1}{3}, \infty)$ D of $H(x)$: $x \in [3, \infty)$

53. a. $h(x) = x^2 + x - 2$ **b.** $H(x) = x^2 - 3x + 2$ **c.** D of $h(x)$:
$x \in (-\infty, \infty)$ D of $H(x)$: $x \in (-\infty, \infty)$ **55. a.** $h(x) = x^2 + 7x + 8$
b. $H(x) = x^2 + x - 1$ **c.** D of $h(x)$: $x \in (-\infty, \infty)$ D of $H(x)$:
$x \in (-\infty, \infty)$ **57. a.** $h(x) = |-3x + 1| - 5$ **b.** $H(x) = -3|x| + 16$
c. D of $h(x)$: $x \in (-\infty, \infty)$ D of $H(x)$: $x \in (-\infty, \infty)$
59. a. $(f \circ g)(x)$: For $g(x)$ to be defined, $x \neq 0$.

$$\text{For } f[g(x)] = \frac{2g(x)}{g(x) + 3}, g(x) \neq -3 \text{ so } x \neq -\frac{5}{3}.$$

$$\text{domain: } \left\{ x | x \neq 0, x \neq -\frac{5}{3} \right\}$$

b. $(g \circ f)(x)$: For $f(x)$ to be defined, $x \neq -3$.

$$\text{For } g[f(x)] = \frac{5}{f(x)}, f(x) \neq 0 \text{ so } x \neq 0.$$

$$\text{domain: } \{x | x \neq 0, x \neq -3\}$$

c. $(f \circ g)(x) = \frac{10}{5 + 3x}$; $(g \circ f)(x) = \frac{5x + 15}{2x}$;
the domain of a composition cannot always be determined from the composed form
61. a. $(f \circ g)(x)$: For $g(x)$ to be defined, $x \neq 5$.

$$\text{For } f[g(x)] = \frac{4}{g(x)}, g(x) \neq 0 \text{ and } g(x) \text{ is never zero}$$

$$\text{domain: } \{x | x \neq 5\}$$

b. $(g \circ f)(x)$: For $f(x)$ to be defined, $x \neq 0$.

$$\text{For } g[f(x)] = \frac{1}{f(x) - 5}, f(x) \neq 5 \text{ so } x \neq \frac{4}{5}.$$

$$\text{domain: } \left\{ x | x \neq 0, x \neq \frac{4}{5} \right\}$$

c. $(f \circ g)(x) = 4x - 20$; $(g \circ f)(x) = \frac{x}{4 - 5x}$; the domain of a composition
cannot always be determined from the composed form
63. a. 41 **b.** 41 **65.** $g(x) = \sqrt{x - 2} + 1, f(x) = x^3 - 5$
67. $p(x) = 2(x + 4)^2 - 3, q(x) = (2x + 7)^2 - 1$ **69. a.** 6000
b. 3000 **c.** 8000 **d.** $C(9) - T(9)$; 4000 **71. a.** \$1 billion
b. \$5 billion **c.** 2003, 2007, 2010 **d.** $t \in (2000, 2003) \cup (2007, 2010)$
e. $t \in [2003, 2007]$ **f.** $R(5) - C(5)$; \$4 billion **73. a.** 4 **b.** 0 **c.** 2
d. 3 **e.** $-\frac{1}{3}$ **f.** 6 **g.** -3 **h.** 1 **i.** 1 **j.** undefined **k.** 0.5 **l.** 2

75. $h(x) = -\frac{2}{3}x + 4$ **77.** $h(x) = 4x - x^2$

79. $A = 2\pi r (20 + r); f(r) = 2\pi r, g(r) = 20 + r; A(5) = 250\pi$ units^2
81. a. $P(x) = 12{,}000x - 108{,}000$; **b.** nine boats must be sold
83. a. $p(n) = 11.45n - 0.1n^2$ **b.** \$123 **c.** \$327
d. $C(115) > R(115)$ **85.** $h(x) = x - 2.5$; 10.5 **87. a.** 4160
b. 45,344 **c.** $M(x) = 453.44x$; yes **89. a.** 6 ft **b.** 36π ft^2
c. $A(t) = 9\pi t^2$; yes **91. a.** 1995 to 1996; 1999 to 2004 **b.** 30; 1995
c. 20 seats; 1997 **d.** The total number in the senate (50); the number of
additional seats held by the majority **93.** Answers will vary.
95.

no, yes

x	$f(x)$	$g(x)$	$(f - g)(x)$
-2	27	15	12
-1	18	11	7
0	11	7	4
1	6	3	3
2	3	-1	4
3	2	-5	7
4	3	-1	4
5	6	3	3
6	11	7	4
7	18	11	7
8	27	15	12

97. a.   **b.**

c.   **99.** $y = -\frac{3}{2}x$

Summary and Concept Review, pp. 270–277

1. $x \in \{-7, -4, 0, 3, 5\}$ $y \in \{-2, 0, 1, 3, 8\}$

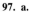

2. $x \in [-5, 5]$ $y \in [0, 5]$

x	y
-5	0
-4	3
-2	$\sqrt{21} \approx 4.58$
0	5
2	$\sqrt{21} \approx 4.58$
4	3
5	0

3. 65 mi **4.** $\left(\frac{5}{2}, -3\right)$ **5.**

6.

7. $(x + 1.5)^2 + (y - 2)^2 = 6.25$
8. a.

b.

$\frac{-5}{9}, (14, -7)$ $\frac{1}{3}, (0, 3)$

9. a. parallel **b.** perpendicular
10. a.

b.

11. a.

b.

12. a. vertical
b. horizontal
c. neither

13. yes   **14.** $m = \frac{2}{3}$, y-intercept (0, 2) when the rodent population increases by 3000, the hawk population increases by 200.

15. a. $y = \frac{-4}{3}x + 4$, $m = \frac{-4}{3}$, y-intercept (0, 4)   **b.** $y = \frac{5}{3}x - 5$, $m = \frac{5}{3}$, y-intercept (0, -5)

16. a.   **b.**

17. a.   **b.**

18. $y = 5, x = -2$; $y = 5$ **19.** $y = \frac{-3}{4}x + \frac{11}{4}$ **20.** $f(x) = \frac{4}{3}x$

21. $m = \frac{2}{5}$, y-intercept (0, 2), $y = \frac{2}{5}x + 2$. When the rabbit population increases by 500, the wolf population increases by 200.

22. a. $y - 90 = \frac{-15}{2}(x - 2)$ **b.** (14, 0), (0, 105)   **c.** $f(x) = \frac{-15}{2}x + 105$

d. $f(20) = -45, x = 12$ **23. a.** $x \in [-\frac{5}{4}, \infty)$
b. $x \in (-\infty, -2) \cup (-2, 3) \cup (3, \infty)$ **24.** 14; $\frac{26}{9}$; $18a^2 - 9a$ **25.** It is a function.   **26. I. a.** $D = \{-1, 0, 1, 2, 3, 4, 5\}$,
$R = \{-2, 1, 0, 1, 2, 3, 4\}$ **b.** 1   **c.** 2   **II. a.** $x \in (-\infty, \infty)$,
$y \in (-\infty, \infty)$ **b.** -1   **c.** 3   **III. a.** $x \in [-3, \infty), y \in [-4, \infty)$
b. -1   **c.** -3 or 3   **27.** $D: x \in (-\infty, \infty), R: y \in [-5, \infty)$,
$f(x)\uparrow: x \in (2, \infty), f(x)\downarrow: x \in (-\infty, 2), f(x) > 0: x \in (-\infty, -1) \cup (5, \infty)$,
$f(x) < 0: x \in (-1, 5)$ **28.** $D: x \in [-3, \infty), R: y \in (-\infty, 0], f(x)\uparrow$:
none, $f(x)\downarrow: x \in (-3, \infty), f(x) > 0$: none, $f(x) < 0: x \in (-3, \infty)$
29. $D: x \in (-\infty, \infty), R: y \in (-\infty, \infty), f(x)\uparrow: x \in (-\infty, -3) \cup (1,\infty)$,
$f(x)\downarrow: x \in (-3, 1), f(x) > 0: x \in (-5, -1) \cup (4, \infty)$,
$f(x) < 0: x \in (-\infty, -5) \cup (-1, 4)$
30. a. odd   **b.** even   **c.** neither   **d.** odd   **31. a.** $\frac{1}{4}$; the graph is rising to the right.   **b.** $2x - 1 + h$; 3.01

32.

zeroes: -6, 0), (0, 0),
(6, 0) (9, 0)
min: (-3, -8),
(7.5 -2)
max: (-6, 0), (3, 4)

33. squaring function   **a.** up on left/up on the right;   **b.** x-intercepts:
(-4, 0), (0, 0); y-intercept: (0, 0)   **c.** vertex (-2, -4)
d. $x \in (-\infty, \infty), y \in [-4, \infty)$ **34.** square root function   **a.** down on the right;   **b.** x-intercept: (0,0); y-intercept: (0, 0)   **c.** initial point (-1, 2);
d. $x \in [-1, \infty), y \in (-\infty, 2]$ **35.** cubing function   **a.** down on left/up on the right   **b.** x-intercepts: (-2, 0), (1, 0), (4, 0); y-intercept: (0, 2)
c. inflection point: (1, 0)   **d.** $x \in (-\infty, \infty), y \in (-\infty, \infty)$
36. absolute value function   **a.** down on left/down on the right
b. x-intercepts: (-1, 0), (3, 0); y-intercept: (0, 1)   **c.** vertex: (1, 2);
d. $x \in [-\infty, \infty), y \in (-\infty, 2]$ **37.** cube root   **a.** up on left, down on right   **b.** x-intercept: (1, 0); y-intercept: (0, 1)   **c.** inflection point: (1,0)
d. $x \in (-\infty, \infty), y \in (-\infty, \infty)$
38. quadratic   **39.** absolute value

40. cubic   **41.** square root   **42.** cube root

43. a.   **b.**   **c.**

44. a. $f(x) = \begin{cases} 5 & x \le -3 \\ -x + 1 & -3 < x \le 3 \\ 3\sqrt{x - 3} - 1 & x > 3 \end{cases}$ **b.** $R: y \in [-2, \infty)$

45.

$D: x \in (-\infty, \infty)$,
$R: y \in (-\infty, -8) \cup (-8, \infty)$,
discontinuity at $x = -3$;
define $h(x) = -8$ at $x = -3$

46. $-4, -4, -4.5, -4.99, 3\sqrt{3} - 9, 3\sqrt{3.5} - 9$
47. $D: x \in (-\infty, \infty)$ $R: y \in [-4, \infty)$

48. $\begin{cases} 20x & x \le 2 \\ 30x - 20 & 2 < x \le 4 \\ 40x - 60 & x > 4 \end{cases}$
For 5 hrs the total cost is $140.

49. $a^2 + 7a - 2$ **50.** 147   **51.** $x \in (-\infty, \frac{2}{3}) \cup (\frac{2}{3}, \infty)$
52. $4x^2 + 8x - 3$ **53.** 99   **54.** $x; x$
55. $f(x) = \sqrt{x} + 1; g(x) = 3x - 2$
56. $f(x) = x^2 - 3x - 10; g(x) = x^{\frac{1}{3}}$ **57.** $A(t) = \pi(2t + 3)^2$ **58. a.** 4
b. 7   **c.** 6   **d.** $\frac{-1}{5}$ **e.** 14

Mixed Review, pp. 277–278

1. $y = -\frac{4}{3}x + 4$ **3. a.** $(-\infty, 1) \cup (1, 4) \cup (4, \infty)$ **b.** $\left(\frac{3}{2}, \infty\right)$

5. $y = -\frac{3}{2}x - 2$ **7.** $(2, 2); (x - 2)^2 + (y - 2)^2 = 50$

9.

11. a.

rate of change is positive in $[-2, -1]$ since p is increasing in $(-\infty, 2)$; less; $\frac{\Delta y}{\Delta x} = \frac{14}{1}$ in $[-2, -1]$; $\frac{\Delta y}{\Delta x} = \frac{2}{1}$ in [1, 2]

b. In the interval [15, 15.01], $\dfrac{\Delta A}{\Delta t} \approx 200.1$

13. $\dfrac{1}{3x^2 - 4x + 1}$; $\left(-\infty, \dfrac{1}{3}\right) \cup \left(\dfrac{1}{3}, 1\right) \cup (1, \infty)$

15. $\dfrac{\Delta f}{\Delta x} = 2x + h, \dfrac{\Delta g}{\Delta x} = 3$; For small h, $2x + h = 3$ when $x \approx \dfrac{3}{2}$.

17. $D: x \in (-\infty, 6]$; $R: y \in (-\infty, 3]$ $g(x)\uparrow: x \in (-\infty, -6) \cup (3, 6)$ $g(x)\downarrow: x \in (-3, 3)$ $g(x)$ constant: $x \in (-6, -3)$ $g(x) > 0: x \in (-7, -1)$ $g(x) < 0: x \in (-\infty, -7) \cup (-1, 6)$ max: $y = 3$ for $x \in (-6, -3)$; $y = 0$ at $(6, 0)$ min: $y = -3$ at $(3, -3)$ **19.** $f(x) = -2x^2 + x + 3$

Practice Test, pp. 279–280

1. a. a and c are nonfunctions, they do not pass the vertical line test
2. neither   **3.**

4.

$(2, -3)$; $r = 4$

5. $y = -\dfrac{6}{5}x + \dfrac{2}{5}$ **6. a.** (7.5, 1.5),   **b.** ≈ 61.27 mi

7. $L_1: x = -3$ $L_2: y = 4$ **8. a.** $x \in \{-4, -2, 0, 2, 4, 6\}$
$y \in \{-2, -1, 0, 1, 2, 3\}$ **b.** $x \in [-2, 6]$ $y \in [1, 4]$ **9. a.** 300   **b.** 30
c. $W(h) = \dfrac{25}{2}h$ **d.** Wages are \$12.50 per hr.
e. $h \in [0, 40]$; $w \in [0, 500]$ **10. I. a.** square root
b. $x \in [-4, \infty), y \in [-3, \infty)$ **c.** $(-2, 0), (0, 1)$
d. up on right   **e.** $x \in (-2, \infty)$ **f.** $x \in [-4, -2)$
II. a. cubic   **b.** $x \in (-\infty, \infty)$ $y \in (-\infty, \infty)$ **c.** $(2, 0), (0, -1)$
d. down on left, up on right   **e.** $x \in (2, \infty)$ **f.** $x \in (-\infty, 2)$
III. a. absolute value   **b.** $x \in (-\infty, \infty)$ $y \in (-\infty, 4]$
c. $(-1, 0), (3, 0), (0, 2)$ **d.** down/down   **e.** $x \in (-1, 3)$
f. $x \in (-\infty, -1) \cup (3, \infty)$ **IV. a.** quadratic   **b.** $x \in (-\infty, \infty)$;
$y \in [-5.5, \infty)$ **c.** $(0, 0), (5, 0), (0, 0)$ **d.** up/up
e. $x \in (-\infty, 0) \cup (5, \infty)$ **f.** $x \in (0, 5)$

11. a. $\dfrac{7}{2}$ **b.** $\dfrac{-a^2 - 6a - 7}{a^2 + 6a + 9}$ **c.** $-\dfrac{31}{25} - \dfrac{8}{25}i$

12. $3x + 1$; $x \in [\frac{1}{3}, \infty)$ **13. a.** No, new company and sales should be growing   **b.** 19 for [5, 6]; 23 for [6, 7]

c. $\dfrac{\Delta s}{\Delta t} = 4t - 3 + 2h$. For small h, sales volume is approximately

$\dfrac{37{,}000 \text{ units}}{1 \text{ mo}}$ in month 10, $\dfrac{69{,}000 \text{ units}}{1 \text{mo}}$ in month 18, and $\dfrac{93{,}000 \text{ units}}{1 \text{ mo}}$ in month 24

14.

15.

16. a. $V(t) = \frac{4}{3}\pi(\sqrt{t})^3$ **b.** 36π in^3 **17. a.** $D: x \in [-4, \infty)$;
$R: y \in (-3, \infty)$ **b.** $f(-1) \approx 2.2$ **c.** $f(x) < 0: x \in (-4, -3)$
$f(x) > 0: x \in (-3, \infty)$ **d.** $f(x)\uparrow: x \in (-4, \infty)$ $f(x)\downarrow$: none
e. $f(x) = 3\sqrt{x + 4} - 3$ **18. a.** 4, $-4, 6.25$
b.

19.

20.

Strengthening Core Skills, p. 281

Exercise 1: $h(x) = x^2 - 28$; $x = 4 \pm 2\sqrt{7}$
Exercise 2: $h(x) = x^2 + 1$; $x = -2 \pm i$
Exercise 3: $h(x) = 2x^2 - \frac{3}{2}$; $x = \dfrac{5}{2} \pm \dfrac{\sqrt{3}}{2}$

Cumulative Review, p. 282

1. $x^2 + 2$ **3.** 29.45 cm   **5.** $x = 1$ **7. a.** $\frac{-1}{3}$ **b.** $\frac{3}{5}$
9.

$y = \frac{1}{2}x + \frac{7}{2}$ **11.** $(f \cdot g)(x) = 3x^3 - 12x^2 + 12x$;
$\left(\dfrac{f}{g}\right)(x) = 3x, x \neq 2$; $(g \circ f) = 22$

13. a. $D: x \in (-\infty, 8]$, $R: y \in [-4, \infty)$ **b.** 5, $-3, -3, 1, 2$
c. $(-2, 0)$ **d.** $f(x) < 0: x \in (-2, 2)$ $f(x) > 0: x \in (-\infty, -2) \cup [2, 8]$
e. min: $(0, -4)$, max: $(8, 7)$ **f.** $f(x)\uparrow: x \in (0, 8)$ $f(x)\downarrow: x \in (-\infty, 0)$

15. a. $\dfrac{x - 7}{(x - 5)(x + 2)}$ **b.** $\dfrac{b^2 - 4ac}{4a^2}$
17. a. False; $\mathbb{Z} \not\subset \mathbb{W}$ **b.** False; $\mathbb{W} \not\subset \mathbb{N}$ **c.** True   **d.** False; $\mathbb{R} \not\subset \mathbb{Z}$
19. $x = -5 \pm \dfrac{\sqrt{2}}{2}$; $x \approx -5.707$; $x \approx -4.293$

21. $W = 31$ cm, $L = 47$ cm   **23. a.** $x = \frac{-4}{3}, \frac{5}{2}$ **b.** $x = -5, -\sqrt{3}, \sqrt{3}$
25. $p = 15 + \sqrt{97}$ units ≈ 24.8 units. No, it is not a right triangle.
$5^2 + (\sqrt{97})^2 \neq 10^2$

MODELING WITH TECHNOLOGY I

Exercises, pp. 288–292

1.

positive

3.

a. linear   **b.** positive

5. a.

b. positive   **c.** $m \approx 1$

7. a.

    b. positive   **c.** $y = 2.4x + 69.4$, 74,200, 112,600

9. a.

    b. linear   **c.** positive
    d. $y = 0.96x + 1.55$, 63.95 in.

11. a.

    b. linear   **c.** positive
    d. $y = 9.55x + 70.42$; about 271,000
    The number of applications, since the line has a greater slope.

13. a.

    b. women: linear
    c. positive
    b. men: linear
    c. negative
    d. yes, |slope| is greater

15.

    a. linear   **b.** $y = 108.2x + 330.2$
    c. \$1736.8 billion; about \$2602.4 billion

17. a. $h(t) = -14.5t^2 + 90t$ **b.** $v = 90$ ft/sec   **c.** Venus

CHAPTER 3
Exercises 3.1, pp. 300–304

1. $\frac{25}{2}$ **3.** $0, f(x)$ **5.** Answers will vary.
7. left 2, down 9   **9.** right 1, reflected across x-axis, up 4

11. left 1, stretched vertically, down 8   **13.** right 2, stretched vertically, reflected across x-axis, up 15

---

15. right $\frac{7}{4}$, stretched vertically, down $\frac{25}{8}$ **17.** left $\frac{7}{6}$, stretched vertically, reflected across x-axis, up $\frac{121}{12}$

19. right $\frac{5}{2}$, down $\frac{17}{4}$ **21.** left 1, down 7

23. right 2, reflected across x-axis, up 6   **25.** left 3, compressed vertically, up $\frac{5}{2}$

27. right $\frac{5}{2}$, reflected across x-axis, stretched vertically, up $\frac{11}{2}$ **29.** right $\frac{3}{2}$, stretched vertically, down 6

31. left 3, compressed vertically, down $\frac{19}{2}$

33. $y = 1(x - 2)^2 - 1$ **35.** $y = -1(x + 2)^2 + 4$
37. $y = -\frac{3}{2}(x + 2)^2 + 3$ **39. i.** $x = -3 \pm \sqrt{5}$ **ii.** $x = 4 \pm \sqrt{3}$
iii. $x = -4 \pm \frac{\sqrt{14}}{2}$ **iv.** $x = 2 \pm \sqrt{2}$ **v.** $t = -2.7, t = 1.3$
vi. $t = -1.4, t = 2.6$ **41. a.** $(0, -66{,}000)$; when no cars are produced, there is a loss of \$66,000.   **b.** $(20, 0), (330, 0)$; no profit will be made if less than 20 or more than 330 cars are produced.   **c.** 175   **d.** \$240,250
43. a. 6 mi   **b.** 3600 ft   **c.** 3200 ft   **d.** 12 mi   **45. a.** $(0, -3300)$; if no appliances are sold, the loss will be \$3300.   **b.** $(20, 0), (330, 0)$; if less than 20 or more than 330 appliances are made and sold, there will be no profit.   **c.** $0 \le x \le 200$; maximum capacity is 200   **d.** 175, \$12,012.50
47. a. 288 ft   **b.**   **c.** 484 ft; 5.5 sec   **d.** 11 sec
49. a. $h(t) = -16t^2 + 32t + 5$ **b.** (i) 17 ft (ii) 17 ft
c. it must occur between $t = 0.5$ and $t = 1.5$ **d.** $t = 1$ sec
e. $h(1) = 21$ ft   **f.** 2 sec   **51.** 155,000; \$16,625   **53. a.** 96 ft $\times$ 48 ft
b. 32 ft $\times$ 48 ft   **55.** $f(x) = x^2 - 4x + 13$

57. a. radicand will be negative—two complex zeroes. **b.** radicand will be positive—two real zeroes. **c.** radicand is zero—one real zero. **d.** two real, rational zeroes.

e. two real, irrational zeroes. **59.** $\dfrac{x-2}{x-5}$ **61.** $x \in \left[-3, \frac{2}{3}\right]$

Exercises 3.2, pp. 312–315

1. synthetic; zero **3.** $P(c)$; remainder **5.** Answers will vary.
7. $x^3 - 5x^2 - 4x + 21 = (x-2)(x^2 - 3x - 10) + 3$
9. $2x^3 + 5x^2 + 4x + 17 = (x+3)(2x^2 - x + 7) - 4$
11. $x^3 - 8x^2 + 11x + 20 = (x-5)(x^2 - 3x - 4) + 0$

13. a. $\dfrac{2x^2 - 5x - 3}{x-3} = (2x+1) + \dfrac{0}{x-3}$

b. $2x^2 - 5x - 3 = (x-3)(2x+1) + 0$

15. a. $\dfrac{x^3 - 3x^2 - 14x - 8}{x-2} = (x^2 - 5x - 4) + \dfrac{0}{x+2}$

b. $x^3 - 3x^2 - 14x - 8 = (x+2)(x^2 - 5x - 4) + 0$

17. a. $\dfrac{x^3 - 5x^2 - 4x + 23}{x-2} = (x^2 - 3x - 10) + \dfrac{3}{x-2}$

b. $x^3 - 5x^2 - 4x + 23 = (x-2)(x^2 - 3x - 10) + 3$

19. a. $\dfrac{2x^3 - 5x^2 - 11x - 17}{x-4} = (2x^2 + 3x + 1) + \dfrac{-13}{x-4}$

b. $2x^3 - 5x^2 - 11x - 17 = (x-4)(2x^2 + 3x + 1) - 13$
21. $x^3 + 5x^2 + 7 = (x+1)(x^2 + 4x - 4) + 11$
23. $x^3 - 13x - 12 = (x-4)(x^2 + 4x + 3) + 0$
25. $3x^3 - 8x + 12 = (x-1)(3x^2 + 3x - 5) + 7$
27. $n^3 + 27 = (n+3)(n^2 - 3n + 9) + 0$
29. $x^4 + 3x^3 - 16x - 8 = (x-2)(x^3 + 5x^2 + 10x + 4) + 0$

31. $(2x+7) + \dfrac{-7x+5}{x^2+3}$ **33.** $-(x^2 - 4) + \dfrac{-4x+3}{x^2-1}$

35. a. -30 **b.** 12 **37. a.** -2 **b.** -22 **39. a.** -1 **b.** 3
41. a. 31 **b.** 0 **43. a.** -10 **b.** 0 **45. a.** yes **b.** yes **47. a.** no
b. yes **49. a.** yes **b.** yes

51.
$$\begin{array}{r|rrrr} -3 & 1 & 2 & -5 & -6 \\ & & -3 & 3 & 6 \\ \hline & 1 & -1 & -2 & 0 \end{array}$$

53.
$$\begin{array}{r|rrrr} 2 & 1 & 0 & -7 & 6 \\ & & 2 & 4 & -6 \\ \hline & 1 & 2 & -3 & 0 \end{array}$$

55.
$$\begin{array}{r|rrrr} \frac{2}{3} & 9 & 18 & -4 & -8 \\ & & 6 & 16 & 8 \\ \hline & 9 & 24 & 12 & 0 \end{array}$$

57. $P(x) = (x+2)(x-3)(x+5)$, $P(x) = x^3 + 4x^2 - 11x - 30$
59. $P(x) = (x+2)(x-\sqrt{3})(x+\sqrt{3})$, $P(x) = x^3 + 2x^2 - 3x - 6$
61. $P(x) = (x+5)(x - 2\sqrt{3})(x + 2\sqrt{3})$, $P(x) = x^3 + 5x^2 - 12x - 60$
63. $P(x) = (x-1)(x+2)(x - \sqrt{10})(x + \sqrt{10})$,
$P(x) = x^4 + x^3 - 12x^2 - 10x + 20$ **65.** $P(x) = (x+2)(x-3)(x-4)$
67. $p(x) = (x+3)^2(x-3)(x-1)$ **69.** $f(x) = 2(x - \frac{3}{2})(x+2)(x+5)$
71. $p(x) = (x+3)(x-3)^2$ **73.** $p(x) = (x-2)^3$
75. $p(x) = (x+3)(x-3)^3$ **77.** $p(x) = (x+3)(x-3)^2(x+4)^2$
79. 4-in. squares; 16 in. × 10 in. × 4 in. **81. a.** week 10, 22.5 thousand
b. one week before closing, 36 thousand **c.** week 9
83. a. 198 ft^3 **b.** 2 ft **c.** about 7 ft **85.** $k = 10$ **87.** $k = -3$
89. The theorems also apply to complex zeroes of polynomials.
91. $S_3 = 36$; $S_5 = 225$ **93.** yes, John wins.
95. $G(t) = 1400t + 5000$

Exercises 3.3, pp. 325–330

1. coefficients **3.** $a - bi$ **5.** b; 4 is not a factor of 6
7. $P(x) = (x+2)(x-2)(x+3i)(x-3i)$
$x = -2, x = 2, x = 3i, x = -3i$
9. $Q(x) = (x+2)(x-2)(x+2i)(x-2i)$
$x = -2, x = 2, x = 2i, x = -2i$ **11.** $P(x) = (x+1)(x+1)(x-1)$
$x = -1, x = -1, x = 1$ **13.** $P(x) = (x-5)(x+5)(x-5)$
$x = 5, x = -5, x = 5$
15. $(x-5)^3(x+9)^2$; $x = 5$, multiplicity 3; $x = -9$, multiplicity 2

17. $(x-7)^2(x+2)^2(x+7)$; $x = 7$, multiplicity 2; $x = -2$,
multiplicity 2; $x = -7$, multiplicity 1
19. $P(x) = x^3 - 3x^2 + 4x - 12$ **21.** $P(x) = x^4 - x^3 - x^2 - x - 2$
23. $P(x) = x^4 - 6x^3 + 13x^2 - 24x + 36$
25. $P(x) = x^4 + 2x^2 + 8x + 5$ **27.** $P(x) = x^4 + 4x^3 + 27$
29. a. yes **b.** yes **31. a.** yes **b.** yes
33. $\{\pm 1, \pm 15, \pm 3, \pm 5, \pm \frac{1}{4}, \pm \frac{15}{4}, \pm \frac{3}{4}, \pm \frac{5}{4}, \pm \frac{1}{2}, \pm \frac{15}{2}, \pm \frac{3}{2}, \pm \frac{5}{2}\}$
35. $\{\pm 1, \pm 15, \pm 3, \pm 5, \pm \frac{1}{2}, \pm \frac{15}{2}, \pm \frac{3}{2}, \pm \frac{5}{2}\}$
37. $\{\pm 1, \pm 28, \pm 2, \pm 14, \pm 4, \pm 7, \pm \frac{1}{6}, \pm \frac{14}{3}, \pm \frac{1}{3}, \pm \frac{2}{3}, \pm \frac{7}{6}, \pm \frac{1}{2}, \pm \frac{7}{2}, \pm \frac{28}{3}, \pm \frac{4}{3}\}$
39. $\{\pm 1, \pm 3, \pm \frac{1}{32}, \pm \frac{1}{2}, \pm \frac{1}{16}, \pm \frac{1}{4}, \pm \frac{1}{8}, \pm \frac{3}{32}, \pm \frac{3}{2}, \pm \frac{3}{16}, \pm \frac{3}{4}, \pm \frac{3}{8}\}$
41. $(x+4)(x-1)(x-3)$, $x = -4, 1, 3$
43. $(x+3)(x+2)(x-5)$, $x = -3, -2, 5$
45. $(x+3)(x-1)(x-4)$, $x = -3, 1, 4$
47. $(x+2)(x-3)(x-5)$, $x = -2, 3, 5$
49. $(x+4)(x+1)(x-2)(x-3)$, $x = -4, -1, 2, 3$
51. $(x+7)(x+2)(x+1)(x-3)$, $x = -7, -2, -1, 3$
53. $(2x+3)(2x-1)(x-1)$; $x = -\frac{3}{2}, \frac{1}{2}, 1$
55. $(2x+3)^2(x-1)$; $x = -\frac{3}{2}, 1$
57. $(x+2)(x-1)(2x-5)$; $x = -2, 1, \frac{5}{2}$
59. $(x+1)(2x+1)(x - \sqrt{5})(x + \sqrt{5})$; $x = -1, -\frac{1}{2}, \sqrt{5}, -\sqrt{5}$
61. $(x-1)(3x-2)(x-2i)(x+2i)$; $x = 1, \frac{2}{3}, 2i, -2i$
63. $x = 1, 2, 3, \frac{-3}{2}$ **65.** $x = -2, 1, \frac{-2}{3}$ **67.** $x = -2, -\frac{3}{2}, 4$
69. $x = 3, -1, \frac{2}{3}$ **71.** $x = 1, 2, -3, \pm \sqrt{7}\,i$ **73.** $x = -2, \frac{2}{3}, 1, \pm \sqrt{3}\,i$
75. $x = 1, 2, 4, -2$ **77.** $x = -3, 1, \pm \sqrt{2}$ **79.** $x = -1, \frac{3}{2}, \pm \sqrt{3}\,i$
81. $x = \frac{1}{2}, 1, 2, \pm \sqrt{3}\,i$ **83. a.** possible roots: $\{\pm 1, \pm 8, \pm 2, \pm 4\}$;
b. neither -1 nor 1 is a root; **c.** 3 or 1 positive roots, 1 negative root;
d. roots must lie between -2 and 2 **85. a.** possible roots: $\{\pm 1, \pm 2\}$;
b. -1 is a root; **c.** 2 or 0 positive roots, 3 or 1 negative roots; **d.** roots
must lie between -3 and 2 **87. a.** possible roots: $\{\pm 1, \pm 12, \pm 2, \pm 6,$
$\pm 3, \pm 4\}$; **b.** $x = 1$ and $x = -1$ are roots; **c.** 4, 2, or 0 positive roots,
1 negative root; **d.** roots must lie between -1 and 4 **89. a.** possible
roots: $\pm 1, \pm 20, \pm 2, \pm 10, \pm 4, \pm 5, \pm \frac{1}{2}, \pm \frac{5}{2}$; **b.** $x = 1$ is a root;
c. 1 positive root, 1 negative root; **d.** roots must lie between -2 and 1
91. $(x-4)(2x-3)(2x+3)$; $x = 4, \frac{3}{2}, -\frac{3}{2}$
93. $(2x+1)(3x-2)(x-12)$; $x = -\frac{1}{2}, \frac{2}{3}, 12$
95. $(x-2)(2x-1)(2x+1)(x+12)$; $x = 2, \frac{1}{2}, -\frac{1}{2}, -12$
97. a. 5 **b.** 13 **c.** 2 **99.** yes **101.** yes
103. a. 4 cm × 4 cm × 4 cm **b.** 5 cm × 5 cm × 5 cm
105. length 10 in., width 5 in., height 3 in.
107. 1994, 1998, 2002, about 5 yr **109. a.** 8.97 m, 11.29 m, 12.05 m,
12.94 m; **b.** 9.7 m, $+3.7$ **111. a.** yes, **b.** no, **c.** about 14.88
113A. a. $(x + 5i)(x - 5i)$ **b.** $(x + 3i)(x - 3i)$
c. $(x + i\sqrt{7})(x - i\sqrt{7})$ **113B. a.** $x = -\sqrt{7}, \sqrt{7}$
b. $x = -2\sqrt{3}, 2\sqrt{3}$ **c.** $x = -3\sqrt{2}, 3\sqrt{2}$
115. a. $C(z) = (z - 4i)(z + 3)(z - 2)$
b. $C(z) = (z - 9i)(z + 4)(z + 1)$
c. $C(z) = (z - 3i)(z - 1 - 2i)(z - 1 + 2i)$
d. $C(z) = (z - i)(z - 2 - 5i)(z - 2 + 5i)$
e. $C(z) = (z - 6i)(z - 1 - \sqrt{3}\,i)(z - 1 + \sqrt{3}\,i)$
f. $C(z) = (z + 4i)(z - 3 - \sqrt{2}\,i)(z - 3 + \sqrt{2}\,i)$
g. $C(z) = (z - 2 + i)(z - 3i)(z + i)$
h. $C(z) = (z - 2 + 3i)(z - 5i)(z + 2i)$ **117. a.** $w = 150$ ft, $l = 300$;
b. $A = 15{,}000$ ft^2 **119.** $r(x) = 2\sqrt{x + 4} - 2$

Exercises 3.4, pp. 340–343

1. zero; m **3.** bounce; flatter **5.** Answers will vary.
7. polynomial, degree 3 **9.** not a polynomial, sharp turns
11. polynomial, degree 2 **13.** up/down **15.** down/down
17. down/up; $(0, -2)$ **19.** down/down; $(0, -6)$ **21.** up/down; $(0, -6)$
23. a. even **b.** -3 odd, -1 even, 3 odd **c.** $f(x) = (x + 3)$
$(x + 1)^2(x - 3)$, deg 4 **d.** $x \in \mathbb{R}, y \in [-9, \infty)$ **25. a.** even
b. -3 odd, -1 odd, 2 odd, 4 odd
c. $f(x) = -(x + 3)(x + 1)(x - 2)(x - 4)$, deg 4

d. $x \in \mathbb{R}, y \in (-\infty, 25]$ **27. a.** odd   **b.** -1 even, 3 odd
c. $f(x) = -(x + 1)^2(x - 3)$, deg 3   **d.** $x \in \mathbb{R}, y \in \mathbb{R}$
29. degree 6; up/up; $(0, -12)$ **31.** degree 5; up/down; $(0, -24)$
33. degree 6; up/up; $(0, -192)$ **35.** degree 5; up/down; $(0, 2)$
37. b   **39.** e   **41.** c

43.   **45.**   **47.**

49.   **51.**   **53.**

55.   **57.**   **59.**

61.   **63.**   **65.**

67.   **69.**   **71.**

73.   **75.**

77. $h(x) = (x + 4)(x - \sqrt{3})(x + \sqrt{3})(x - \sqrt{3}i)(x + \sqrt{3}i)$
79. $f(x) = 2(x + \frac{5}{2})(x - \sqrt{2})(x + \sqrt{2})(x - \sqrt{3})(x + \sqrt{3})$
81. $P(x) = \frac{1}{6}(x + 4)(x - 1)(x - 3)$, $P(x) = \frac{1}{6}(x^3 - 13x + 12)$
83. $P(x) = x^4 - 2x^3 - 13x^2 + 14x + 24$
85. a. 280 vehicles above average, 216 vehicles below average, 154 vehicles below average   **b.** 6:00 A.M. $(t = 0)$, 10:00 A.M. $(T = 4)$, 3:00 P.M. $(t = 9)$, 6:00 P.M. $(t = 12)$
c. max: about 300 vehicles above average at 7:30 A.M.; min: about 220 vehicles below average at 12 noon

87. c. $B(x) = \frac{1}{4}x(x - 4)(x - 9)$, $-\$80,000$
89. a. $f(x) \to \infty, f(x) - \infty$ **b.** $g(x) \to \infty, g(x) \to \infty$; $x^4 \geq 0$ for all x
91. verified   **93.** $h(x) = \dfrac{1 - 2x}{x^2}$; $D : x \in \{x | x \neq 0\}$; $H(x) = \dfrac{1}{x^2 - 2x}$;
$D : x \in \{x | x \neq 0, x \neq 2\}$ **95. a.** $x = 2$ **b.** $x = 8$ **c.** $x = 4, x = -6$

Mid Chapter, p. 344

1. a. $x^3 + 8x^2 + 7x - 14 = (x^2 + 6x - 5)(x + 2) - 4$
b. $\dfrac{x^3 + 8x^2 + 7x - 14}{x + 2} = x^2 + 6x - 5 - \dfrac{4}{x + 2}$
2. $f(x) = (2x + 3)(x + 1)(x - 1)(x - 2)$ **3.** $f(-2) = 7$
4. $f(x) = x^3 - 2x + 4$ **5.** $g(2) = -8$ and $g(3) = 5$ have opposite signs
6. $f(x) = (x - 2)(x + 1)(x + 2)(x + 4)$
7. $x = -2, x = 1, x = -1 \pm 3i$
8. **9.**

10. a. degree 4; three turning points   **b.** 2 sec
c. $A(t) = (t - 1)^2(t - 3)(t - 5)$ $A(t) = t^4 - 10t^3 + 32t^2 - 38t + 15$
$A(2) = 3$; altitude is 300 ft above hard-deck $A(4) = -9$; altitude is 900 ft below hard-deck

Reinforcing Basic Concepts, pp. 344–345

Exercise 1: 1.532
Exercise 2: -2.152, 1.765

Exercises 3.5, pp. 356–362

1. as $x \to -\infty, y \to 2$ **3.** denominator; numerator   **5.** about $x = 98$
7. a. as $x \to -\infty, y \to 2$ **9. a.** as $x \to -\infty, y \to 1$
   as $x \to \infty, y \to 2$ as $x \to \infty, y \to 1$
b. as $x \to 1^-, y \to -\infty$ **b.** as $x \to -2^-, y \to \infty$
   as $x \to 1^+, y \to \infty$ as $x \to -2^+, y \to \infty$

11. reciprocal quadratic, $S(x) = \dfrac{1}{(x + 1)^2} - 2$

13. reciprocal function, $Q(x) = \dfrac{1}{x + 1} - 2$

15. reciprocal quadratic, $v(x) = \dfrac{1}{(x + 2)^2} - 5$

17. $\to -2$ **19.** $\to -\infty$ **21.** $-1; \pm\infty$
23. $x = 3, x \in (-\infty, 3) \cup (3, \infty)$
25. $x = 3, x = -3, x \in (-\infty, -3) \cup (-3, 3) \cup (3, \infty)$
27. $x = \frac{-5}{2}, x = 1, x \in (-\infty, -\frac{5}{2}) \cup (-\frac{5}{2}, 1) \cup (1, \infty)$
29. No V.A., $x \in (-\infty, \infty)$ **31.** $x = 3$, yes; $x = -2$, yes
33. $x = 3$, no   **35.** $x = 2$, yes; $x = -2$, no   **37.** $y = 0$, crosses at $(\frac{3}{2}, 0)$
39. $y = 4$, crosses at $(-\frac{21}{4}, 4)$ **41.** $y = 3$, does not cross
43. $(0, 0)$ cross, $(3, 0)$ cross   **45.** $(-4, 0)$ cross, $(0, 4)$
47. $(0, 0)$ cross, $(3, 0)$ bounce

49.   **51.**   **53.**

55.   **57.**   **59.**

61.   **63.**   **65.**

13. $P(x) = \begin{cases} \dfrac{x^3 - 8}{x - 2} & x \neq 2 \\ 12 & x = 2 \end{cases}$ **15.** $q(x) = \begin{cases} \dfrac{x^3 - 7x - 6}{x + 1} & x \neq -1 \\ -4 & x = -1 \end{cases}$

67. $f(x) = \dfrac{(x - 4)(x + 1)}{(x + 2)(x - 3)}$ **69.** $f(x) = \dfrac{x^2 - 4}{9 - x^2}$

71. a. Population density approaches zero far from town.   **c.** 4.5 mi, 704 people per square mi

73. a. $20,000, $80,000, $320,000; cost increases dramatically
b.   **c.** as $p \to 100^-$, $C \to \infty$

17. $R(x) = \begin{cases} \dfrac{x^3 + 3x^2 - x - 3}{x^2 + 2x - 3} & x \neq -3, x \neq 1 \\ -2 & x = -3 \\ 2 & x = 1 \end{cases}$

75. a. 5 hr; about 0.28   **b.** -0.019, -0.005; As the number of hours increases, the rate of change decreases.   **c.** $h \to \infty$, $C \to 0^+$; horizontal asymptote

77.   **79. a.**

19.   **21.**   **23.**

b. 35%; 62.5%; 160 gal;   **c.** 160 gal; 200 gal;   **d.** 70%; 75%
81. a. $225; $175   **b.** 2000 heaters   **c.** 4000   **d.** The horizontal asymptote at $y = 125$ means the average cost approaches $125 as monthly production gets very large. Due to limitations on production (maximum of 5000 heaters) the average cost will never fall below $A(5000) = 135$.
83. a. 5   **b.** 18   **c.** The horizontal asymptote at $y = 95$ means her average grade will approach 95 as the number of tests taken increases; no
d. 6   **85. a.** 16.0 28.7 65.8 277.8   **b.** 12.7, 37.1, 212.0   **c. a.** 22.4, 40.2, 92.1, 388.9   **b.** 17.8, 51.9, 296.8; answers will vary.
87. a. $q(x) = 3$, horizontal asymptote at $y = 3$; $r(x) = -7x + 10$, graph crosses HA at $x = \dfrac{10}{7}$ **b.** $q(x) = -2$, horizontal asymptote at $y = -2$; $r(x) = 7$, no zeroes—graph will not cross
89. $y = \dfrac{-4}{3}x - \dfrac{1}{3}$ **91.** $39, \frac{3}{2}, 1$

25.   **27.**   **29.**

31.   **33.**   **35.**

Exercises 3.6, pp. 371–375

1. nonremovable   **3.** two   **5.** Answers will vary.

7. $F(x) = \begin{cases} \dfrac{x^2 - 4}{x + 2} & x \neq -2 \\ -4 & x = -2 \end{cases}$

37.   **39.**   **41.**

9. $G(x) = \begin{cases} \dfrac{x^2 - 2x - 3}{x + 1} & x \neq -1 \\ -4 & x = -1 \end{cases}$

43.   **45.**   **47.**

49.   **51.** 119.1

53. a. $a = 5$, $y = 3a + 15$ **b.** 60.5   **c.** 10
55. a. $A(x) = \dfrac{4x^2 + 53x + 250}{x}$; $x = 0$, $g(x) = 4x + 53$
b. cost: $307, $372, $445, Avg. cost: $307, $186, $148.33   **c.** 8, $116.25

11. $H(x) = \begin{cases} \dfrac{3x - 2x^2}{2x - 3} & x \neq \dfrac{3}{2} \\ \dfrac{-3}{2} & x = \dfrac{3}{2} \end{cases}$

d.

57. a. $S(x, y) = 2x^2 + 4xy$; $V(x, y) = x^2y$ **b.** $S(x) = \dfrac{2x^3 + 48}{x}$

c. $S(x)$ is asymptotic to $y = 2x^2$.   **d.** $x = 2$ ft 3.5 in.; $y = 2$ ft 3.5 in.

59. a. $A(x, y) = xy$; $R(x, y) = (x - 2.5)(y - 2)$ **b.** $y = \dfrac{2x + 55}{x - 2.5}$

$A(x) = \dfrac{2x^2 + 55x}{x - 2.5}$ **c.** $A(x)$ is asymptotic to $y = 2x + 60$

d. $x \approx 11.16$ in.; $y = 8.93$ in.   **61. a.** $h = \dfrac{V}{\pi r^2}$ **b.** $S = 2\pi r^2 + \dfrac{2V}{r}$

c. $S = \dfrac{2\pi r^3 + 2V}{r}$ **d.** $r \approx 5.76$ cm, $h \approx 11.51$ cm; $S \approx 625.13$ cm^2

63. Answers will vary.   **65.** $S = \dfrac{\pi r^3 + 2V}{r}$; $r = 3.1$ in., $h = 3$ in.

67. $y = \frac{3}{4}x - 4$, $m = \frac{3}{4}$, $(0, -4)$ **69. a.** $P = 30$ cm,   **b.** $\overline{CD} = \frac{60}{13}$ cm,

c. 30 cm^2,   **d.** $A = \frac{750}{169}$ cm^2, and $A = \frac{4320}{169}$ cm^2

Exercises 3.7, pp. 384–388

1. vertical; multiplicity   **3.** empty   **5.** Answers will vary.
7. $x \in (0, 4)$ **9.** $x \in (-\infty, -5] \cup [1, \infty)$ **11.** $x \in (-1, \frac{7}{2})$
13. $x \in [-\sqrt{7}, \sqrt{7}]$ **15.** $x \in [-\frac{3}{2} - \frac{\sqrt{33}}{2}, -\frac{3}{2} + \frac{\sqrt{33}}{2}]$
17. $x \in (-\infty, -\frac{5}{3}] \cup [1, \infty)$ **19.** $x \in (-\infty, \infty)$ **21.** { }
23. $x \in (-\infty, 5) \cup (5, \infty)$ **25.** { }   **27.** $x \in (-\infty, \infty)$
29. $x \in (-\infty, \infty)$ **31.** $x \in (-\infty, -5] \cup [5, \infty)$
33. $x \in (-\infty, 0] \cup [5, \infty)$ **35.** { }   **37.** $x \in (-3, 5)$
39. $x \in [4, \infty) \cup \{-1\}$ **41.** $x \in (-\infty, -2] \cup \{2\} \cup [4, \infty)$
43. $x \in (-2 - \sqrt{3}, -2 + \sqrt{3})$ **45.** $x \in [-\infty, -3] \cup -\{1\}$
47. $x \in (-3, 1) \cup (2, \infty)$ **49.** $x \in (-\infty, -3) \cup (-1, 1) \cup (3, \infty)$
51. $x \in (-\infty, -2) \cup (-2, 1) \cup (3, \infty)$ **53.** $x \in [-1, 1] \cup \{3\}$
55. $x \in [-3, 2)$ **57.** $x \in (-\infty, -2) \cup (-2, -1)$
59. $x \in (-\infty, -2) \cup [2, 3)$ **61.** $x \in (-\infty, -5) \cup (0, 1) \cup (2, \infty)$
63. $x \in (-4, -2] \cup (1, 2] \cup (3, \infty)$ **65.** $x \in (-7, -3) \cup (2, \infty)$
67. $x \in (-\infty, -2] \cup (0, 2)$ **69.** $x \in (-\infty, -17) \cup (-2, 1) \cup (7, \infty)$
71. $x \in (-3, \frac{-7}{4}] \cup (2, \infty)$ **73.** $x \in (-2, \infty)$ **75.** $x \in (-1, \infty)$
77. $(-\infty, -3) \cup (3, \infty)$ **79.** $x \in (-\infty, -3] \cup [5, \infty)$
81. $x \in [-3, 0] \cup [3, \infty)$ **83.** $x \in (-\infty, -2) \cup (2, 3)$
85. $x \in (-\infty, -2] \cup (-1, 1) \cup [3, \infty)$ **87.** b   **89.** b   **91. a.** verified
b. $D = -4(p + \frac{3}{4})(p + 3)^2$, $p = -3$, $q = -2$; $p = \frac{-3}{4}$, $q = \frac{1}{4}$
c. $(-\infty, -3) \cup (-3, \frac{-3}{4})$ **d.** verified
93. $d(x) = k(x^3 - 192x + 1024)$ **a.** $x \in (5, 8]$ **b.** 320 units
c. $x \in [0, 3)$ **d.** 2 ft   **95. a.** verified   **b.** horizontal: $r_2 = 20$, as r_1
increases, r_2 decreases to maintain $R = 40$ vertical: $r_1 = 20$, as r_1 decreases,
r_2 increases to maintain $R = 40$ **97.** $R(t) = 0.01t^2 + 0.1t + 30$
a. $[0°, 30°)$ **b.** $(20°, \infty)$ **c.** $(50°, \infty)$ **99. a.** $n \geq 4$ **b.** $n \leq 9$
c. 13   **101. a.** yes, $x^2 \geq 0$ **b.** yes, $\dfrac{x^2}{x^2 + 1} \geq 0$
103. $x(x + 2)(x - 1)^2 > 0$; $\dfrac{x(x + 2)}{(x - 1)} > 0$
105. $R(x) < 0$ for $x \in (2, 8) \cup (8, 14)$
107. $F(x) = \begin{cases} f(x) & x \neq -4 \\ -6 & x = -4 \end{cases}$

109.

Exercises 3.8, pp. 394–399

1. constant   **3.** $y = \frac{k}{x^2}$ **5.** Answers will vary.   **7.** $d = kr$ **9.** $F = ka$

11. $y = 0.025\,x$

x	y
500	12.5
650	16.25
750	18.75

13. $w = 9.18$; $h\ \$321.30$; the hourly wage; $k = \$9.18/\text{hr}$

15. a. $k = \frac{192}{47}$ $S = \frac{192}{47}h$

b.   **c.** 330 stairs   **d.** $S = 331$; yes

17. $A = kS^2$ **19.** $P = kc^2$

21. $k = 0.112$; $p = 0.112\,q^2$

q	p
45	226.8
55	338.8
70	548.8

23. $k = 6$, $A = 6s^2$; 55,303,776 m^2

25. a. $k = 16$ $d = 16t^2$ **b.**

c. about 3.5 sec   **d.** 3.5 sec; yes   **e.** 2.75 sec   **27.** $F = \frac{k}{d^2}$ **29.** $S = \frac{k}{L}$

31. $Y = \dfrac{12,321}{Z^2}$

Z	Y
37	9
74	2.25
111	1

33. $w = \dfrac{3,072,000,000}{r^2}$; 48 kg   **35.** $l = krt$

37. $A = kh(B + b)$ **39.** $V = ktr^2$

41. $C = \dfrac{6.75R}{S^2}$

R	S	C
120	6	22.5
200	12.5	8.64
350	15	10.5

43. $E = 0.5mv^2$; 612.50 J
45. cube root family; answers will vary; 0.054 or 5.4%

Amount A	Rate R
1.0	0.000
1.05	0.016
1.10	0.032
1.15	0.048
1.20	0.063
1.25	0.077

47. $T = \dfrac{48}{V}$; 32 volunteers   **49.** $M = \dfrac{1}{6}E$; ≈ 41.7 kg

51. $D = 21.6\sqrt{S}$; ≈ 144.9 ft   **53.** $C = 8.5LD$; $76.50

55. $C \approx (4.4 \times 10^{-4})\dfrac{p_1 p_2}{d^2}$; about 223 calls   **57. a.** about 23.39 cm³,

b. about 191%   **59. a.** $M = kwh^2(\frac{1}{L})$ **b.** 180 lb

61. For f: $\dfrac{\Delta y}{\Delta x} = \dfrac{-10}{3}$ For g: $\dfrac{\Delta y}{\Delta x} = \dfrac{-110}{9}$; less; for both f and g, as

$x \to \infty, y \to 0$ **63. a.** about 3.5 ft   **b.** about 6.9 ft
65. $x = 0, x = -2 \pm 2i$
67.

Summary and Concept Review, pp. 399–404

1.

2.

3.

4. a. 0 ft   **b.** 108 ft   **c.** 2.25 sec   **d.** 144 ft, $t = 3$ sec
5. $q(x) = x^2 + 6x + 7; R = 8$ **6.** $q(x) = x + 1; R = 3x - 4$
7. $\underline{-7|}$ 2   13 -6 9   14
              -14 7 -7 -14
          2   -1 1   2 $\underline{|0}$
Since $R = 0, -7$ is a root and $x + 7$ is a factor.
8. $x^3 - 4x + 5 = (x - 2)(x^2 + 2x) + 5$ **9.** $(x + 4)(x + 1)(x - 3)$
10. $h(x) = (x - 1)(x - 4)(x^2 + 2x + 2)$
11. $\frac{1}{2}|$ 4   8 -3 -1
                2   5   1
          4   10   2 $\underline{|0}$
Since $R = 0, \frac{1}{2}$ is a root and $(x - \frac{1}{2})$ is a factor.
12. $3i|$ 1 -2 9 -18
            $3i$ $-9 - 6i$ 18
          1    $-2 + 3i$ $-6i$ $\underline{|0}$
Since $R = 0, 3i$ is a zero
13. $-7|$ 1   9   13 -10
              -7 -14 7
          1   2 -1 $\underline{|-3}$
$h(-7) = -3$
14. $P(x) = x^3 - x^2 - 5x + 5$ **15.** $C(x) = x^4 - 2x^3 + 5x^2 - 8x + 4$
16. a. $C(0) = 350$ customers   **b.** more at 2 P.M., 170
c. busier at 1 P.M., $760 > 710$
17. $\{\pm 1, \pm\frac{1}{2}, \pm\frac{1}{4}, \pm 5, \pm 10, \pm\frac{5}{2}, \pm\frac{5}{4}, \pm 2\}$ **18.** $x = -\frac{1}{2}, 2, \frac{5}{2}$
19. $p(x) = (2x + 3)(x - 4)(x + 1)$ **20.** only possibilities are $\pm 1, \pm 3$,
none give a remainder of zero   **21.** [1, 2], [4, 5]; verified   **22.** one sign
change for $g(x) \to 1$ positive zero; three sign changes for $g(-x) \to 3$ or 1
negative zeroes; 1 positive, 3 negative, 0 complex, or 1 positive, 1 negative, 2 complex; verified   **23.** degree 5; up/down; $(0, -4)$
24. degree 4; up/up; $(0, 8)$
25.

26.

27.

28. a. even   **b.** $x = -2$, odd; $x = -1$, even; $x = 1$, odd
c. deg 6: $P(x) = (x + 2)(x + 1)^2(x - 1)^3$

29. a. $\{x | x \in \mathbb{R}; x \neq -1, 4\}$ **b.** HA: $y = 1$; VA: $x = -1, x = 4$
c. $V(0) = \frac{9}{4}$ (y-intercept); $x = -3, 3$ (x-intercepts)   **d.** $V(1) = \frac{4}{3}$
31.

32.

33. $V(x) = \dfrac{x^2 - x - 12}{x^2 - x - 6}$; $V(0) = 2$

34. a. $y = 15$; as $|x| \to \infty$ $A(x) \to 15^+$. As production increases, average
cost decreases and approaches 15.   **b.** $x > 2000$
35. removable discontinuity at $(2, -5)$;

36. $H(x) = \begin{cases} \dfrac{x^2 - 3x - 4}{x + 1} & x \neq -1 \\ -5 & x = -1 \end{cases}$

37.

38.

39. a.

b. about 2450 favors **c.** about $2.90 ea.
40. factored form $(x + 4)(x - 1)(x - 2) > 0$

When $x = 0$

Neg   Pos   Neg   Pos
   -4 0   1   2

outputs are positive for
$x \in (-4, 1) \cup (2, \infty)$

41. $\dfrac{x^2 - 3x - 10}{x - 2} = \dfrac{(x - 5)(x + 2)}{x - 2} \geq 0$

When $x = 0$

Neg   Pos   Neg   Pos
   -2 0   2   5

outputs are positive or zero for
$x \in [-2, 2) \cup [5, \infty)$

42. $\dfrac{(x + 2)(x - 1)}{x(x - 2)} \leq 0$

When $x = -1$

Pos   Neg   Pos   Neg   Pos
   -2 -1 0   1   2

outputs are negative or zero for
$x \in [-2, 0) \cup [1, 2)$

43. $k = 17.5$; $y = 17.5\sqrt[3]{x}$

x	y
216	105
0.343	12.25
729	157.5

44. $k = 0.72$; $z = \dfrac{0.72v}{w^2}$

v	w	z
196	7	2.88
38.75	1.25	17.856
24	0.6	48

45. $t = 160$ **46.** 4.5 sec

Mixed Review, pp. 404–405

1. $y = -2(x - \frac{1}{2})^2 + \frac{9}{2}$ **3.** 80 GB, $40.00
5. $q(x) = x^3 - 2x^2 + x + 3; R = -7$ **7. a.** $P(-1) = 42$
b. $P(1) = -26$ **c.** $P(5) = 6$ **9. a.** $x = 9; x = \frac{8}{3}$
b. $P(x) = (x - 2)(x + 1)(x^2 + 9); x = 2, x = -1, x = -3i, x = 3i$

11.   **13.**

15. $x \in (-\infty, 3) \cup (-2, 2)$
17. a. $V(x) = (24 - 2x)(16 - 2x)(x) = 4x^3 - 80x^2 + 384x$
b. $512 = 4x^3 - 80x^2 + 384x$ $0 = x^3 - 20x^2 + 96x - 128$
c. for $0 < x < 8$, possible rational zeroes are 1, 2, and 4
d. $x = 4$ **e.** $x = 8 - 4\sqrt{2} \approx 2.34$ in. **19.** $R = kL(\frac{1}{A})$

Practice Test, pp. 405–406

1. a. $f(x) = -(x - 5)^2 + 9$ **b.** $g(x) = \frac{1}{2}(x + 4)^2 + 8$

2. $(-2, 0), y = 2x^2 + 4x$ **3. a.** 40 ft, 48 ft **b.** 49 ft **c.** 14 sec
4. $x - 5 + \dfrac{14x + 3}{x^2 + 2x + 1}$ **5.** $x^2 + 2x - 9 + \dfrac{-2}{x + 2}$

6.
$$\underline{-3|}\ \ 1\quad\ 0\quad -15\quad -10\quad\ \ 24$$
$$\qquad\qquad -3\quad\ \ \ 9\quad\ \ 18\ \ -24$$
$$\overline{\ \ \ \ \ 1\ \ -3\quad -6\quad\ \ \ 8\qquad 0\ \ R = 0}✓$$

7. -1 **8.** $P(x) = x^3 - 2x^2 + 9x - 18$
9. $Q(x) = (x - 2)^2(x - 1)^2(x + 1)$, 2 mult 2, 1 mult 2, -1 mult 1
10. a. $\pm 1, \pm 18, \pm 2, \pm 9, \pm 3, \pm 6$ **b.** 1 positive zero, 3 or 1 negative
zeroes; 2 or 0 complex zeroes **c.** $C(x) = (x + 2)(x - 1)(x - 3i)(x + 3i)$
11. a. 1992, 1994, 1998 **b.** 4 yr **c.** surplus of \$2.5 million

12.   **13.**   **14.**

15. a. removal of 100% of the contaminants **b.** \$500,000; \$3,000,000;
dramatic increase **c.** 88%
16. a.   **b.**

17. 800 **18. a.** $x \in (-\infty, -3] \cup [-1, 4]$ **b.** $x \in (-\infty, -4) \cup (0, 2)$
19. a.   **b.** $h = -\sqrt[3]{55}$; no **c.** 28.6% 29.6%
d. ≈ 11.7 hr **e.** 4 hr 43.7%
f. The amount of the chemical in the bloodstream becomes neglible.

20. 520 lb

Strengthening Core Skills, pp. 407–408

Exercise 1: $x \in (-\infty, 3]$
Exercise 2: $x \in (-2, -1) \cup (2, \infty)$
Exercise 3: $x \in (-\infty, -4) \cup (1, 3)$
Exercise 4: $x \in [-2, \infty)$
Exercise 5: $x \in (-\infty, -2) \cup (2, \infty)$
Exercise 6: $x \in [-3, 1] \cup [3, \infty)$

Cumulative Review chapter R–3, pp. 408–409

1. $R = \dfrac{R_1 R_2}{R_1 + R_2}$ **3. a.** $(x - 1)(x^2 + x + 1)$
b. $(x - 3)(x + 2)(x - 2)$ **5.** all reals **7.** verified
9. $y = \dfrac{11}{60}x + \dfrac{1009}{60}$; 39 min, driving time increases 11 min every 60 days

11. Month 9 **13.** $f^{-1}(x) = \dfrac{x^3 + 3}{2}$

15.   **17.** $X = 63$ **19.**

CHAPTER 4
Exercises 4.1, pp. 420–424

1. second; one **3.** $(-11, -2), (-5, 0), (1, 2), (19, 4)$ **5.** False, answers
will vary. **7.** one-to-one **9.** one-to-one **11.** not a function **13.** one-
to-one **15.** not one-to-one, fails horizontal line test: $x = -3$ and $x = 3$
are paired with $y = 1$ **17.** not one-to-one, $y = 7$ is paired with $x = -2$
and $x = 2$ **19.** one-to-one **21.** one-to-one **23.** not one-to-one;
$p(t) > 5$, corresponds to two x-values **25.** one-to-one **27.** one-to-one
29. $f^{-1}(x) = \{(1, -2), (4, -1), (5, 0), (9, 2), (15, 5)\}$
31. $v^{-1}(x) = \{(3, -4), (2, -3), (1, 0), (0, 5), (-1, 12), (-2, 21), (-3, 32)\}$
33. $f^{-1}(x) = x - 5$ **35.** $p^{-1}(x) = \dfrac{-5}{4}x$ **37.** $f^{-1}(x) = \dfrac{x - 3}{4}$
39. $Y_1^{-1} = x^3 + 4$ **41.** $f^{-1}(x) = x^3 + 2$ **43.** $f^{-1}(x) = \sqrt[3]{x - 1}$
45. $f^{-1}(x) = \dfrac{8}{x} - 2$ **47.** $f^{-1}(x) = \dfrac{x}{1 - x}$ **49. a.** $x \geq -5, y \geq 0$
b. $f^{-1}(x) = \sqrt{x} - 5, x \geq 0, y \geq -5$ **51. a.** $x > 3, y > 0$
b. $v^{-1}(x) = \sqrt{\dfrac{8}{x}} + 3, x > 0, y > 3$ **53. a.** $x \geq -4, y \geq -2$
b. $p^{-1}(x) = \sqrt{x + 2} - 4, x \geq -2, y \geq -4$
55. $(f \circ g)(x) = x, (g \circ f)(x) = x$ **57.** $(f \circ g)(x) = x, (g \circ f)(x) = x$
59. $(f \circ g)(x) = x, (g \circ f)(x) = x$ **61.** $(f \circ g)(x) = x, (g \circ f)(x) = x$
63. $f^{-1}(x) = \dfrac{x + 5}{3}$ **65.** $f^{-1}(x) = 2x + 5$ **67.** $f^{-1}(x) = 2x + 6$
69. $f^{-1}(x) = \sqrt[3]{x - 3}$ **71.** $f^{-1}(x) = \dfrac{x^3 - 1}{2}$ **73.** $f^{-1}(x) = 2\sqrt[3]{x} + 1$
75. $f^{-1}(x) = \dfrac{x^2 - 2}{3}, x \geq 0; y \in \left[-\dfrac{2}{3}, \infty\right)$
77. $p^{-1}(x) = \dfrac{x^2}{4} + 3, x \geq 0; y \in [3, \infty)$
79. $v^{-1}(x) = \sqrt{x - 3}, x \geq 3; y \in [0, \infty)$
81.   **83.**   **85.**

87.

89.

$D: x \in [0, \infty), R: y \in [-2, \infty);$
$D: x \in [-2, \infty), R: y \in [0, \infty)$

91.

$D: x \in (0, \infty), R: y \in (-\infty, \infty);$
$D: x \in (-\infty, \infty), R: y \in (0, \infty)$

93.

$D: x \in (-\infty, 4], R: y \in (-\infty, 4];$
$D: x \in (-\infty, 4], R: y \in (-\infty, 4]$

95. a. 31.5 cm   **b.** The result is 80 cm. It gives the distance of the projector from the screen.   **97. a.** $-63.5°F$ **b.** $f^{-1}(x) = \frac{-2}{7}(x - 59)$; independent: temperature, dependent: altitude   **c.** 22,000 ft   **99. a.** 144 ft

b. $f^{-1}(x) = \frac{\sqrt{x}}{4}$, independent: distance fallen, dependent: time fallen

c. 7 sec   **101. a.** 28,260 ft³   **b.** $f^{-1}(x) = \sqrt[3]{\frac{3x}{\pi}}$, independent: volume,

dependent: height   **c.** 9 ft   **103.** Answers will vary.   **105.** d
107. $x \in [-1, 2]$ **109. a.** $P = 2l + 2w$ **b.** $A = \pi r^2$ **c.** $V = \pi r^2 h$
d. $V = \frac{1}{3}\pi r^2 h$ **e.** $C = 2\pi r$ **f.** $A = \frac{1}{2}bh$ **g.** $A = \frac{1}{2}(b_1 + b_2)h$
h. $V = \frac{4}{3}\pi r^3$ **i.** $a^2 + b^2 = c^2$

Exercises 4.2, pp. 432–436

1. b^x; b; b; x **3.** a; 1   **5.** False; for $|b| < 1$ and $x_2 > x_1$, $b^{x_2} < b^{x_1}$ so
function is decreasing   **7.** 40,000; 5000; 20,000; 27,589.162   **9.** 500;
1.581; 2.321; 221.168   **11.** 10,000; 1975.309; 1487.206; 1316.872
13. increasing   **15.** decreasing

increasing

decreasing

17. up 2   **19.** left 3

21. reflect across y-axis   **23.** reflect across y-axis, up 3

25. left 1, down 3   **27.** up 1

29. right 2   **31.** down 2

33. e   **35.** a   **37.** b   **39.** 2.718282   **41.** 7.389056   **43.** 4.481689
45. 4.113250

47.   **49.**   **51.**

53. 3   **55.** $\frac{3}{2}$ **57.** $-\frac{1}{3}$ **59.** 4   **61.** -3 **63.** 3   **65.** 2   **67.** -2
69. 2   **71.** 3
73. a. 1732, 3000, 5196, 9000   **b.** yes   **c.** as $t \to \infty, P \to \infty$

75. no, they will have to wait about 10 min   **77. a.** \$100,000   **b.** 3 yr
79. a. $\approx$ \$86,806   **b.** 3 yr   **81. a.** \$40 million   **b.** 7 yr   **83.** 32%
transparent   **85.** 17% transparent   **87.** $\approx$ \$32,578   **89. a.** 8 g

b. 48 min   **91.** 9.5×10^{-7}; answers will vary   **93.** 9   **95.** $\frac{3}{2}$

97. a. $\frac{\Delta y}{\Delta x}$ = 0.3842, 0.056, 0.011, 0.003; the rate of growth seems to

be approaching zero   **b.** 16,608   **c.** yes, the secant lines are becoming
virtually horizontal
99.

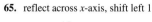

101. a. volume of a sphere   **b.** area of a
triangle   **c.** volume of a rectangular prism
d. Pythagorean theorem

Exercises 4.3, pp. 445–449

1. $\log_b x$; b; b; greater   **3.** (1, 0); 0   **5.** 5; answers will vary   **7.** $2^3 = 8$
9. $7^{-1} = \frac{1}{7}$ **11.** $9^0 = 1$ **13.** $8^{\frac{1}{3}} = 2$ **15.** $2^1 = 2$ **17.** $7^2 = 49$
19. $10^2 = 100$ **21.** $e^4 \approx 54.598$ **23.** $\log_4 64 = 3$ **25.** $\log_3 \frac{1}{9} = -2$
27. $0 = \log_e 1$ **29.** $\log_3 27 = -3$ **31.** $\log 1000 = 3$ **33.** $\log \frac{1}{100} = -2$
35. $\log_4 8 = \frac{3}{2}$ **37.** $\log_4 \frac{1}{8} = \frac{-3}{2}$ **39.** 1   **41.** 2   **43.** 1   **45.** $\frac{1}{2}$
47. -2 **49.** -2 **51.** 1.6990   **53.** 0.4700   **55.** 5.4161   **57.** 0.7841
59. shift up 3   **61.** shift right 2, up 3

63. shift left 1   **65.** reflect across x-axis, shift left 1

67. II   **69.** VI   **71.** V   **73.** $x \in (-\infty, -1) \cup (3, \infty)$ **75.** $x \in (\frac{3}{2}, \infty)$
77. $x \in (-3, 3)$ **79.** pH ≈ 4.1; acid   **81. a.** ≈ 4.7 **b.** ≈ 4.9
83. about 3.2 times   **85. a.** ≈ 2.4 **b.** ≈ 1.2 **87. a.** 20 dB   **b.** 120 dB
89. about 3162 times   **91.** 6,194 m   **93. a.** about 5434 m
b. 4000 m   **95. a.** 2225 items   **b.** 2732 items   **c.** \$117,000
d. verified   **97. a.** about 58.6 cfm   **b.** about 1605 ft^2 **99. a.** 95%
b. 67%   **c.** 39%   **101.** ≈ 4.3; acid   **103.** Answers will vary.   **a.** 0 dB
b. 90 dB   **c.** 15 dB   **d.** 120 dB   **e.** 100 dB   **f.** 140 dB   **105. a.** $\frac{-2}{3}$
b. $\frac{-3}{2}$ **c.** $\frac{-5}{2}$ **107.** D: $x \in \mathbb{R}$ R: $y \in \mathbb{R}$

109. $x \in (-\infty, -5); f(x) = (x + 5)(x - 4)^2 = x^3 - 3x^2 - 24x + 80$

Mid-Chapter Check, pp. 449–450

1. a. $\frac{2}{3} = \log_{27}9$ **b.** $\frac{5}{4} = \log_{81}243$ **2. a.** $8^{\frac{5}{3}} = 32$ **b.** $1296^{0.25} = 6$
3. a. $x = 5$ **b.** $b = \frac{5}{4}$ **4. a.** $x = 3$ **b.** $b = 5$ **5. a.** \$71,191.41
b. 6 yr   **6.** $F(x) = 4 \cdot 5^{x-3} + 2$ **7.** $f^{-1}(x) = (x - 1)^2 + 3$, D:
$x \in [1, \infty)$; R: $y \in [3, \infty)$; verified   **8. a.** $4 = \log_3 81$, verified
b. $4 \approx \ln 54.598$, verified   **9. a.** $27^{\frac{2}{3}} = 9$, verified   **b.** $e^{1.4} \approx 4.0552$,
verified   **10.** ≈ 7.9 times more intense

Reinforcing Basic Concepts p. 450

Exercise 1: about 158 times
Exercise 2: about 501 times
Exercise 3: about 12,589 times
Exercise 4: about 398 times
Exercise 5: about 39,811 times

Exercises 4.4, pp. 462–466

1. e **3.** extraneous   **5.** 2.316566275   **7.** $x \approx 29.964$ **9.** $x \approx 1.778$
11. $x \approx 2.200$ **13.** $x \approx 1.260$ **15.** $x \approx 4.7881$ **17.** $x \approx -3.1079$
19. $x = -\frac{\ln 2.32}{0.75}, x \approx -1.1221$ **21.** $x = e^{\frac{8}{3}} - 4, x \approx 10.3919$
23. $x = 5 - 10^{1.25}, x \approx -12.7828$ **25.** $x = \frac{e^{0.4} - 5}{2},$
$x \approx -1.7541$ **27.** $\ln(2x^2 - 14x)$ **29.** $\log(x^2 - 1)$ **31.** $\log_3 4$
33. $\log\left(\frac{x}{x + 1}\right)$ **35.** $\ln\left(\frac{x - 5}{x}\right)$ **37.** $\ln(x - 2)$ **39.** $\log_2 42$
41. $\log_5(x - 2)$ **43.** $(x + 2)\log 8$ **45.** $(2x - 1)\ln 5$ **47.** $\frac{1}{2}\log 22$
49. $4\log_5 3$ **51.** $3\log a + \log b$ **53.** $\ln x + \frac{1}{4}\ln y$ **55.** $2\ln x - \ln y$
57. $\frac{1}{2}[\log(x - 2) - \log x]$
59. $\ln 7 + \ln x + \frac{1}{2}\ln(3 - 4x) - \ln 2 - 3\ln(x - 1)$
61. $\frac{\ln 60}{\ln 7}$; 2.104076884   **63.** $\frac{\ln 152}{\ln 5}$; 3.121512475
65. $\frac{\log 1.73205}{\log 3}$; 0.499999576   **67.** $\frac{\log 0.125}{\log 0.5}$; 3
69. $f(x) = \frac{\log(x)}{\log(3)}; f(5) \approx 1.4650; f(15) \approx 2.4650; f(45) \approx 3.4650$;
outputs increase by 1; $f(3^3 \cdot 5) = 4.465$
71. $h(x) = \frac{\log(x)}{\log(9)}; h(2) \approx 0.3155; h(4) \approx 0.6309; h(8) \approx 0.9464$;
outputs are multiples of 0.3155; $h(2^4) = 4(0.3155) \approx 1.2619$
73. $x = 32$ **75.** $x = 6.4$ **77.** $x = 20, -5$ is extraneous
79. $x = 2, -\frac{5}{2}$ is extraneous   **81.** $x = 0$ **83.** $x = \frac{5}{2}$ **85.** $x = \frac{2}{3}$
87. $x = \frac{3}{2}$ **89.** $x = \frac{-19}{9}$ **91.** $x = \frac{e^2 - 63}{9}$
93. $x = 2; -9$ is extraneous   **95.** $x = 3e^3 - \frac{1}{2}; x \approx 59.75661077$
97. no solution   **99.** $t = -\frac{1}{2}; -4$ is extraneous

101. $x = 2 + \sqrt{3}, x = 2 - \sqrt{3}$ is extraneous
103. $x = \frac{\ln 231}{\ln 7} - 2; x \approx 0.7968$ **105.** $x = \frac{\ln 128,965}{3 \ln 5} + \frac{2}{3}; x \approx 3.1038$
107. $x = \frac{\ln 2}{\ln 3 - \ln 2}; x \approx 1.7095$ **109.** $x = \frac{\ln 9 - \ln 5}{2 \ln 5 - \ln 9}; x \approx 0.5753$
111. $x \approx 46.2$ **113.** $t = \frac{\ln\left(\frac{\frac{C}{P} - 1}{a}\right)}{-k}, t \approx 55.45$

115. a. 30 fish   **b.** about 37 months   **117.** about 3.2 cmHg
119. about 50.2 min   **121.** \$15,641   **123.** 6 hr, 18.0%
125. $M_f = 52.76$ tons   **127. a.** 26 planes   **b.** 9 days
129. a. $\log_3 4 + \log_3 5 = 2.7268$ **b.** $\log_3 4 - \log_3 5 = -0.203$
c. $2\log_3 5 = 2.9298$ **131. a.** d   **b.** e   **c.** b   **d.** f   **e.** a   **f.** c
133. $x = 0.69314718$ **135. a.** $(f \circ g)(x) = 3^{(\log_3 x + 2) - 2} = 3^{\log_3 x} = x$;
$(g \circ f)(x) = \log_3(3^{x-2}) + 2 = x - 2 + 2 = x$
b. $(f \circ g)(x) = e^{(\ln x + 1) - 1} = e^{\ln x} = x$;
$(g \circ f)(x) = \ln e^{x-1} + 1 = x - 1 + 1 = x$
137. a. $y = e^{x \ln 2} = e^{\ln 2^x} = 2^x$;
$y = 2^x \Rightarrow \ln y = x \ln 2, e^{\ln y} = e^{x \ln 2} \Rightarrow y = e^{x \ln 2}$
b. $y = b^x, \ln y = x \ln b, e^{\ln y} = e^{x \ln b}, y = e^{xr}$ for $r = \ln b$
139. Answers will vary.   **141.** b   **143.**

Exercises 4.5, pp. 475–480

1. Compound   **3.** $Q_0 e^{-rt}$ **5.** Answers will vary.   **7.** \$4896   **9.** 250%
11. \$2152.47   **13.** 5.25 yr   **15.** 80%   **17.** 4 yr   **19.** 16 yr
21. \$7561.33   **23.** about 5 yr   **25.** 7.5 yr   **27.** no   **29. a.** no
b. 9.12%   **31.** 7.9 yr   **33.** 7.5 yr   **35. a.** no   **b.** 9.4%   **37. a.** no
b. approx 13,609 euros   **39.** No; \$234,612.01   **41.** about 7 yr
43. 23 yr   **45. a.** no   **b.** \$302.25   **47. a.** $t = \frac{A - P}{pr}$ **b.** $p = \frac{A}{1 + rt}$
49. a. $r = n\left(\sqrt[nt]{\frac{A}{p}} - 1\right)$ **b.** $t = \frac{\ln\left(\frac{A}{p}\right)}{n \ln\left(1 + \frac{r}{n}\right)}$ **51. a.** $Q_0 = \frac{Q(t)}{e^{rt}}$
b. $t = \frac{\ln\left(\frac{Q(t)}{Q_0}\right)}{r}$ **53.** \$709.74   **55. a.** 5.78%   **b.** 91.67 hr   **57.** 0.65 g
59. 816 yr   **61.** about 12.4%   **63.** \$17,027,502.21   **65.** 7.93%
67. 2548.8 m   **69.** $P(x) = x^4 - 4x^3 + 6x^2 - 4x - 15$

Summary and Concept Review, pp. 480–484

1. no   **2.** no   **3.** yes   **4.** $f^{-1}(x) = \frac{x-2}{-3}$ **5.** $f^{-1}(x) = \sqrt{x + 2}$
6. $f^{-1}(x) = x^2 + 1; x \geq 0$
7. $f(x)$: D: $x \in [-4, \infty)$, R: $y \in [0, \infty); f^{-1}(x)$: D: $x \in [0, \infty)$,
R: $y \in [-4, \infty)$ **8.** $f(x)$: D: $x \in (-\infty, \infty)$, R: $y \in (-\infty, \infty)$;
$f^{-1}(x)$: D: $(-\infty, \infty)$, R: $y \in (-\infty, \infty)$ **9.** $f(x)$: D: $x \in (-\infty, \infty)$,
R: $y \in (0, \infty); f^{-1}(x)$: D: $x \in (0, \infty)$, R: $y \in (-\infty, \infty)$ **10. a.** \$3.05
b. $f^{-1}(t) = \frac{t - 2}{0.15}$ $f^{-1}(3.05) = 7$ **c.** 12 days
11.   **12.**   **13.**

14. 2   **15.** -2 **16.** $\frac{5}{2}$ **17.** 12.1 yr   **18.** $3^2 = 9$ **19.** $5^{-3} = \frac{1}{125}$
20. $e^{3.7612} \approx 43$ **21.** $\log_5 25 = 2$ **22.** $\ln 0.7788 \approx -0.25$
23. $\log_3 81 = 4$ **24.** 5   **25.** -1 **26.** $\frac{1}{2}$

27.

28.

29.

30. $x \in (-\infty, 0) \cup (6, \infty)$ **31.** $x \in (-\frac{3}{2}, \infty)$ **32. a.** 4.79   **b.** $10^{7.3} I_0$

33. a. $x = e^{32}$ **b.** $x = 10^{2.38}$ **c.** $x = \ln 9.8$ **d.** $x = \frac{1}{2}\log 7$

34. a. $x = \dfrac{\ln 4}{0.5}$, $x \approx 2.7726$ **b.** $x = \dfrac{\ln 19}{0.2}$, $x \approx 6.3938$

c. $x = \dfrac{10^3}{3}$, $x \approx 33.3333$ **d.** $x = e^{-2.75}$, $x \approx 0.0639$ **35. a.** $\ln 42$

b. $\log_9 30$ **c.** $\ln(\frac{x+3}{x-1})$ **d.** $\log(x^2 + x)$ **36. a.** $2\log_5 9$ **b.** $2\log_7 4$
c. $(2x-1)\ln 5$ **d.** $(3x+2)\ln 10$ **37. a.** $\ln x + \frac{1}{4}\ln y$
b. $\frac{1}{3}\ln p + \ln q$ **c.** $\frac{5}{3}\log x + \frac{4}{3}\log y - \frac{5}{2}\log x - \frac{3}{2}\log y$

d. $\log 4 + \frac{5}{3}\log p + \frac{4}{3}\log q - \frac{3}{2}\log p - \log q$ **38. a.** $\dfrac{\log 45}{\log 6} \approx 2.215$

b. $\dfrac{\log 128}{\log 3} \approx 4.417$ **c.** $\dfrac{\ln 124}{\ln 2} \approx 6.954$ **d.** $\dfrac{\ln 0.42}{\ln 5} \approx -0.539$
39. $x = \frac{\ln 7}{\ln 2}$ **40.** $x = \frac{\ln 5}{\ln 3} - 1$ **41.** $x = \frac{2}{1-\ln 3}$ **42.** $x \approx 6.389$
43. $x = 5$; -2 is extraneous   **44.** $x = 4.25$ **45. a.** 17.77%
b. 23.98 days   **46.** 38.6 cmHg   **47.** 18.5%   **48.** Almost, she needs
$42.15 more.   **49. a.** no   **b.** $268.93   **50.** 55.0%

Mixed Review, pp. 484–485

1. a. $\dfrac{\log 30}{\log 2} \approx 4.9069$ **b.** -1.5 **c.** $\frac{1}{3}$ **3. a.** $2\log_{10} 20$ **b.** $0.05x$
c. $(x-3)\ln 2$

5.

7.

9. a. $5^4 = 625$

b. $e^{0.45} = 0.15x$ **c.** $10^7 = 0.1 \times 10^8$ **11. a.** $x \in [1, \infty), y \in [2, \infty)$
b. $g^{-1}(x) = (x-2)^2 + 1, x \in [2, \infty), y \in [1, \infty)$ **c.** Answers will
vary.   **13.** $6 + \log 2$ **15.** $\frac{9}{4} + \frac{\sqrt{129}}{4}$ **17.** $I \approx 6.3 \times 10^{17}$
19. 1.6 m, 1.28 m, 1.02 m, 0.82 m, 0.66 m, 0.52 m

Practice Test, pp. 485–486

1. $3^4 = 81$ **2.** $\log_{25} 5 = \frac{1}{2}$ **3.** $\frac{5}{2}\log_b x + 3\log_b y - \log_b z$
4. $\log_b \dfrac{m\sqrt{n^3}}{\sqrt{p}}$ **5.** $x = 10$ **6.** $x = \dfrac{-5}{3}$ **7.** 2.68   **8.** -1.24

9.

10.

11. a. 4.19   **b.** -0.81

12. f is a parabola (hence not one-to-one), $x \in \mathbb{R}, y \in [-3, \infty)$; vertex is
at $(2, -3)$, so restricted domain could be $x \in [2, \infty)$ to create a one-to-one
function; $f^{-1}(x) = \sqrt{x+3} + 2, x \in [-3, \infty), y \in [2, \infty)$.

13. $x = 1 + \dfrac{\ln 89}{\ln 3}$ **14.** $x = 1, x = -5$ is extraneous   **15.** ≈ 5 yr
16. ≈ 8.7 yr   **17.** 19.1 months   **18.** 7% compounded semi-annually
19. a. no   **b.** $54.09   **20. a.** 10.2 lb   **b.** 19 weeks

Strengthening Core Skills, p. 488

Exercise 1: Answers will vary.
Exercise 2: **a.** $\log(x^2 + 3x)$ **b.** $\ln(x^2 - 4)$ **c.** $\log\frac{x}{x+3}$
Exercise 3: Answers will vary.
Exercise 4: **a.** $x\log 3$ **b.** $5\ln x$ **c.** $(3x-1)\ln 2$

Cumulative Review chapters 1–4, pp. 488–489

1. $x = 2 \pm 7i$ **3.** $(4 + 5i)^2 - 8(4 + 5i) + 41 = 0$ **5.** $f(g(x)) = x$
$g(f(x)) = x$ Since $(f \circ g)(x) = (g \circ f)(x)$, they are inverse functions.

7. a. $T(t) = 455t + 2645$ (1991 → year 1)   **b.** $\dfrac{\Delta T}{\Delta t} = \dfrac{455}{1}$, triple births

increase by 455 each year   **c.** $T(6) = 5375$ sets of triplets,
$T(17) = 10,380$ sets of triplets
9.

$D: x \in [-10, \infty), R: y \in [-9, \infty)$
$h(x)\uparrow: x \in (-2, 0) \cup (3, \infty)$ $h(x)\downarrow: x \in (0, 3)$

11. $x = 3, x = 2$ (multiplicity 2); $x = -4$ **13.** $\sqrt{\frac{2V}{\pi a}} = b$

15. a. $f^{-1}(x) = \dfrac{5x-3}{2}$ **b.**

c. $f^{-1}(f(x)) = x$

17. $x = 5, x = -6$ is an extraneous root   **19. a.** ≈ 88 hp for sport
wagon, ~81 hp for minivan   **b.** ≈ 3294 rpm
c. minivan, 208 hp at 5800 rpm

MODELING WITH TECHNOLOGY II
Modeling with Technology Exercises, pp. 495–502

1. e   **3.** a   **5.** d   **7.** linear   **9.** exponential   **11.** logistic
13. exponential
15. As time increases, the amount of radioactive
material decreases but will never truly reach 0 or
become negative. Exponential with
$b < 1$ and $k > 0$ is the best choice.
$y \approx (1.042)0.5626^x$

17. Sales will increase rapidly, then level off as the
market is saturated with ads and advertising becomes
less effective, possibly modeled by a logarithmic function.
$y \approx 120.4938 + 217.2705\ln(x)$
19. a.   **b.** about 1750   **c.** $y \approx \dfrac{1719}{1 + 10.2e^{-0.11x}}$

logistic

21. 4.95   **23.** 6.25   **25.** 5.75   **27.** 6.84
29. logarithmic, $y \approx -27.4 + 13.5\ln x$

a. 9.2 lb   **b.** 29 days   **c.** 34.8 lb

31. logarithmic, $y = 78.8 - 10.3 \ln x$
  a. 51,000 **b.** 1977 **c.** 30,400

33. exponential, $y = 50.21(1.07)^x$
  a. 86,270 **b.** 272,511 **c.** 1990

35. exponential, $y = 346.79(0.94)^x$
  a. 155,142 **b.** 78,548 **c.** 1993

37. quadratic $y \approx 0.576x^2 - 8.879x + 394$
  a. 360 million **b.** about 513 million
  c. from 1984 to 1990

39. linear, $y \approx 6.555x + 165.308$
  a. 224 million **b.** 264 million **c.** 2010

41. linear, $P(t) = 0.51t + 22.51$, 2005: 40.4%, 2010: 43%

43. linear, $y \approx 509.18x - 7.96$; about $6100; the next July ($x \approx 19.7$)

45. exponential, $y \approx 103.83(1.0595)^x$
  a. 220 **b.** The 22nd note, or F sharp
  c. frequency doubles, yes

47. exponential, $y \approx 8.02(1.0564)^x$
  $41.59/mo, $54.72/mo

49. quadratic, $y \approx 1.18x^2 - 10.99x + 4.60$; month 8

51. logistic, $y \approx \dfrac{222.133}{1 + 32.280e^{-0.336x}}$;

about 55 million, about 184 million, about 214 million; 2014

53.

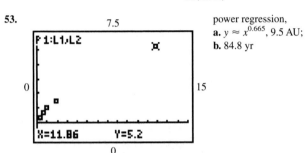

power regression,
  a. $y \approx x^{0.665}$, 9.5 AU;
  b. 84.8 yr

55.

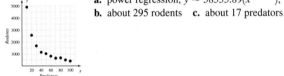

  a. power regression, $y \approx 58555.89(x^{-1.056})$;
  b. about 295 rodents **c.** about 17 predators

57. a.

linear, $W = 1.24L - 15.83$, 32.5 lb, 35.3 in.

57. b.

logarithmic, $C(a) \approx 37.9694 + 3.4229 \ln(a)$, about 49.3 cm, about 34 mo

CHAPTER 5

Exercises 5.1, pp. 511–515

1. inconsistent **3.** consistent; independent **5.** Multiply the first equation by 6 and the second equation by 10. **7.** $y = \frac{7}{4}x - 6$, $y = \frac{-4}{3}x + 5$
9. $y = x + 2$ **11.** $x + 3y = -3$ **13.** $y = x + 2$, $x + 3y = -3$
15. yes **17.** yes
19.   **21.**

23. $(-4, 1)$ **25.** $(3, -5)$ **27.** second equation, y, $(4, -3)$
29. second equation, x, $(10, -1)$ **31.** second equation, x, $\left(\frac{5}{2}, \frac{7}{4}\right)$
33. $(3, -1)$ **35.** $(-2, -3)$ **37.** $\left(\frac{11}{2}, 2\right)$ **39.** $(-2, 3)$ **41.** $(-3, 4)$
43. $(-6, 12)$ **45.** $(2, 8)$; consistent/independent
47. $\varnothing$; inconsistent **49.** $\{(x, y)|6x + y = 22\}$; consistent/dependent
51. $(4, 1)$; consistent/independent **53.** $(-3, -4)$; consistent/independent

55. $\left(\frac{-1}{2}, \frac{4}{3}\right)$; consistent/independent　**57.** $\left(-2, \frac{5}{2}\right)$
59. $(2, -1)$　**61.** 1 mph　4 mph　**63.** 2318 adult tickets; 1482 child tickets　**65.** premium: \$3.97, regular: \$3.87　**67.** nursing student \$6500; science major \$3500　**69.** 150 quarters, 75 dimes　**71. a.** 100 lawns/mo, **b.** \$11,500/mo　**73. a.** 1.6 billion bu, 3 billion bu, yes; **b.** 2.7 billion bu, 2.25 billion bu, yes; **c.** \$6.65, 2.43 billion bu　**75. a.** 3 mph, **b.** 5 mph
77. a. 3.6 ft/sec, **b.** 4.4 ft/sec　**79.** 1776; 1865　**81.** Tahiti: 402 mi^2, Tonga: 290 mi^2　**83.** $m_1 \neq m_2$; consistent/independent　**85.** \$6552 at 8.5%; \$11,551 at 6%　**87.** $(x - 5)(x - 2)(x + 1)(3x - 1) = 0$
89. $x \in [-2, 8]$

Exercises 5.2, pp. 524–527

1. triple　**3.** equivalent; systems　**5.** $z = 5$　**7.** Answers will vary.
9. Answers will vary.　**11.** yes; no　**13.** $(5, 7, 4)$
15. $(-2, 4, 3)$　**17.** $(1, 1, -2)$　**19.** $(4, 0, -3)$　**21.** $(3, 4, 5)$
23. $(1, 6, 9)$　**25.** no solution, inconsistent　**27.** $(p, 2 - p, 2 - p)$
29. $\left(-\frac{5}{3}p - \frac{2}{3}, -p - 2, p\right)$, other solutions possible
31. $(p, 2p, p + 1)$
33. $(p + 9, p - 4, p)$　**35.** $\{(x, y, z) | x - 6y + 12z = 5\}$　**37.** $(1, 1, 2)$
39. $\left\{(x, y, z) | x - \frac{5}{2}y - 2z = 3\right\}$　**41.** $\left(2, 1, \frac{-1}{3}\right)$
43. $(p + 5, p - 2, p)$　**45.** $(18, -6, 10)$
47. $\left(\frac{11}{3}, \frac{10}{3}, \frac{7}{3}\right)$　**49.** $(1, -2, 3)$　**51.** $\left(\frac{1}{2}, \frac{1}{3}, 3\right)$　**53.** ≈ 3.464 units
55. Monet \$1,900,000; Picasso \$1,100,000; van Gogh \$4,000,000
57. elephant, 650 days; rhino, 464 days; camel, 406 days
59. Albatross: 3.6 m, Condor: 3.0 m, Quetzalcoatlus: 12.0 m　**61.** 175 \$5 gold pieces; 50 \$10 gold pieces; 25 \$20 gold pieces
63. $A = -1, B = 1, C = -2$; verified
65. $x^2 + y^2 - 4x + 6y + 9 = 0$　**67.** $\left[-1, \frac{3}{2}\right]$　**69.** $x = 1$

Mid-Chapter Check, pp. 527–528

1. $(1, 1)$ consistent　**2.** $(5, 3)$ consistent　**3.** 20 oz　**4.** No
5. 2R1 = R2　**6.** $(1, 2, 3)$　**7.** $(1, 2, 3)$　**8.** $(p, p - 5, -p - 4)$
9. Morphy: 13, Mozart: 8, Pascal: 16　**10.** prelude: 2.75 min, storm: 2.5 min, sunrise: 2.5 min, finale: 3.25 min

Reinforcing Basic Concepts, pp. 528

Exercise 1: Premium: \$4.17/gal,　$\begin{cases} 15.3R + 35.7P = 211.14 \\ P = R + 0.10 \end{cases}$　Regular: \$4.07/gal
Exercise 2: Verified

Exercises 5.3, pp. 533–536

1. a. 3 or 4 not possible

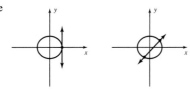

b. 3 or 4 not possible

c.

d.

e. 3 or 4 solutions not possible

f.

3. region; solutions　**5.** Answers will vary.

7.

line, parabola; $(-1, 5), (2, 2)$

9.

circle, absolute value; $(-6, 8), (8, 6)$

11.

parabola, parabola;
$(-1, -2), (2, 1)$

13. $(-4, -3), (3, 4)$ **15.** $(2, 5), (-4, -7)$ **17.** $(-3, 4), (-4, -3)$,
$(3, 4), (4, -3)$ **19.** $(4, -3), (-4, -3)$ **21.** no solution
23. $(-8, 1), (-7, 4)$ **25.** $(5, \log 5 + 5)$
27. $(-3, \ln 9 + 1), (4, \ln 16 + 1)$ **29.** $(0, 10), (\ln 6, 45)$
31. $(-3, 1), (2, 1024)$ **33.** $(-3, -21), (1, -1), (2, 4)$ **35.** $(2, -4)$,
$(6, 4)$ **37.** $(3, 5), (3, -5)$ **39.** $(-2.43, -2.81), (2, 1)$ **41.** $(0.72, 2.19)$,
$(2, 3), (4, 3), (5.28, 2.19)$

43.

45.

47. no solution

49.

51. $h \approx 45.8$ ft; $h = 40$ ft; $h = 30$ ft   **53.** The company breaks even if
either 18,400 or 48,200 cars are sold.
55. $1.83; $3
90,000 gal
$\begin{cases} 10P^2 + 6D = 144 \\ 8P^2 - 8P - 4D = 12 \end{cases}$
57. 8.5 m × 10 m   **59.** 5 km, 9 km   **61.** 8 × 8 × 25 ft
63. Answers will vary.   **65.** 18 in. by 18 in. by 77 in.
67. a. $x = \frac{-2 \pm 2\sqrt{10}}{3}$ **b.** $x = 0, x = 8$ **c.** $x = 2$ **69. a.** $m = \frac{-400}{1}$,
the copier depreciates by $400 a year   **b.** $y = -400x + 4500$ **c.** $1700
d. 9.5 yr

Exercises 5.4, pp. 545–548

1. half; planes   **3.** solution   **5.** The feasible region may be bordered by
three or more oblique lines, with two of them intersecting outside and
away from the feasible region.   **7.** No, No, No, No   **9.** No, Yes, Yes, No

11.

13.

15. No, No, No, Yes

17.

19.

21.

23.

25.

27.

29.

31.

33.

35.

37.

39. $\begin{cases} y - x \le 1 \\ x + y > 3 \end{cases}$

41. $\begin{cases} y - x \le 1 \\ x + y < 3 \\ y \ge 0 \end{cases}$ **43.** $(5, 3)$ **45.** $(12, 11)$ **47.** $(2, 2)$ **49.** $(4, 3)$

51. $5 < H < 10$

53.

$J + A \le 50,000$
$J \ge 20,000$
$A \le 25,000$

55. 300 acres of corn; 200 acres of soybeans
57. 240 sheet metal screws; 480 wood screws
59. 65 traditionals, 30 Double-T's
61. 220,000 gallons from Tulsa to Colorado; 100,000 gal from Tulsa to
Mississippi; 0 thousand gal from Houston to Colorado; 150,000 gal
from Houston to Mississippi

63.

$(3, 3)$; optimal solutions occur at vertices

65. $x = 5, \pm \sqrt{3}\,i$ **67.** 324 Ω

Summary and Concept Review, pp. 549–551

1.

$(4, 4)$

2.

$\left(\frac{1}{2}, -3\right)$

3.

$\left(\frac{8}{5}, \frac{-6}{5}\right)$

4. no solution; inconsistent   **5.** $(5, -1)$; consistent   **6.** $(7, 2)$; consistent
7. $(3, -1)$; consistent   **8.** $(2, 2)$; consistent   **9.** $\left(\frac{11}{4}, \frac{-1}{6}\right)$; consistent
10. Sears Tower is 1450 ft; Hancock Building is 1127 ft.
11. $(0, 3, 2)$ **12.** $(1, 1, 1)$ **13.** no solution, inconsistent
14. 3 aces, 4 face cards, 5 numbered cards   **15.** 1530 quarters, 1180
dimes, 710 nickels   **16.** circle, line, $(4, 3), (-3, -4)$ **17.** parabola, line,
$(3, -2)$ **18.** parabola, circle, $(\sqrt{3}, 2), (-\sqrt{3}, 2)$ **19.** circle, parabola,
$(1, 3), (-1, 3)$

20.

21. note the open circle showing noninclusion at $(0, -3)$; circle, parabola

parabola, circle

22. **23.** **24.**

25. Maximum of 270 occurs at both $(0, 6)$ and $(3, 4)$.

26. 50 cows, 425 chickens

Mixed Review, pp. 551–552

1. a. $\begin{cases} y = \frac{3}{5}x + 2 \\ y = \frac{3}{5}x + 2 \end{cases}$; consistent/dependent **b.** $\begin{cases} y = \frac{4}{3}x - 3 \\ y = \frac{2}{3}x - 2 \end{cases}$; consistent/independent **c.** $\begin{cases} y = \frac{1}{3}x - 3 \\ y = \frac{1}{3}x - \frac{5}{3} \end{cases}$; inconsistent

3. $(-2\ 3)$ **5.** 21 veggie, 33 beef **7.** $(9, 1, 1)$
9. $\{(x, y, z) | x \in \mathbb{R}, y = -7x + 7, z = -5x + 6\}$
11. **13.** no solution

15.

(x, y)	$P(x, y) = 2.5x + 3.75y$
$(0, 0)$	0
$(0, 7)$	26.25
$(7.5, 0)$	18.75
$(2, 6)$	27.5
$(6, 2)$	22.5

max value 27.5 at $(2, 6)$
17. $(2, 5), (2, -5); (-2, 5), (-2, -5)$
19.

Practice Test, pp. 552–553

1. $(2, 3)$ **2.** $\left(\frac{2}{5}, \frac{-4}{5}\right)$ **3.** $(-3, 2)$ **4.** $(2, -1, 4)$

5. $\{(x, y, z) | x = 2z - 1, y = 5z - 6, z \in \mathbb{R}\}$ **6.** $a = 5, b = 2$
7. 21.59 cm by 35.56 cm **8.** Tahiti 402 mi²; Tonga 290 mi²
9. Corn 25¢ Beans 20¢ Peas 29¢ **10.** $15,000 at 7%, $8000 at 5%, $7000 at 9%
11. **12.** $(5, 0)$

13. 30 plain; 20 deluxe

14. $(-1 - \sqrt{7}, 1 - \sqrt{7}), (-1 + \sqrt{7}, 1 + \sqrt{7})$
15. $(\sqrt{3}, 1), (-\sqrt{3}, 1)$ **16.** 15 ft, 20 ft
17. **18.** $\begin{cases} y > 0 \\ x^2 + y^2 < 9 \end{cases}$

19. the solution is $(0, 1)$

20. Answers may vary. Possible solution: $\begin{cases} x^2 + y^2 > 1 \\ x^2 + y^2 < 4 \\ x > 0, y < 0 \end{cases}$

Strengthening Core Skills, pp. 554–555

Exercise 1: $(-1, 4)$, elimination

Cumulative Review Chapters 1–5, pp. 555–556
1. **3.** **5.**

7. a. $D: x \in (-\infty, \infty)$ **b.** $R\ y \in (-\infty, 4]$ **c.** $f(x)\uparrow: x \in (-\infty, -1)$
$f(x)\downarrow: x \in (-1, \infty)$ **d.** n/a **e.** max: $(-1, 4)$ **f.** $f(x) > 0: x \in (-4, 2)$
$f(x) < 0: x \in (-\infty, -4) \cup (2, \infty)$ **g.** $\frac{\Delta y}{\Delta x} = \frac{7}{4}$
9. $x = 9, \pm\sqrt{2}\,i$ **11.** $3x^2 + 5$ **13.** 1
15. $(x - 2)(x + 2)(x - 3 - 3\sqrt{2})(x - 3 + 3\sqrt{2})$
17. $x \in (-\infty, \frac{-11}{2}] \cup (-3, \infty)$ **19.** ≈ 9.7 yr
21. **23.** 80 arrows, 25 balls, 15 bats

25. $x = 5, y \approx 4.7$

CHAPTER 6
Exercises 6.1, pp. 565–568

1. square **3.** 2; 3; 1 **5.** Multiply R1 by -2 and add that result to R2. This sum will be the new R2. **7.** $3 \times 2, 5.8$ **9.** $4 \times 3, -1$

11. $\begin{bmatrix} 1 & 2 & -1 & | & 1 \\ 1 & 0 & 1 & | & 3 \\ 2 & -1 & 1 & | & 3 \end{bmatrix}$; diagonal entries 1, 0, 1

13. $\begin{cases} x + 4y = 5 \\ y = \frac{1}{2} \end{cases} \rightarrow (3, \frac{1}{2})$ **15.** $\begin{cases} x + 2y - z = 0 \\ y + 2z = 2 \\ z = 3 \end{cases} \rightarrow (11, -4, 3)$

17. $\begin{cases} x + 3y - 4z = 29 \\ y - \frac{3}{2}z = \frac{21}{2} \\ z = 3 \end{cases} \rightarrow (-4, 15, 3)$ **19.** $\begin{bmatrix} 1 & -6 & | & -2 \\ 0 & -28 & | & -6 \end{bmatrix}$

21. $\begin{bmatrix} 1 & -3 & 3 & | & 2 \\ 0 & 23 & -12 & | & -15 \\ -2 & 1 & 0 & | & 4 \end{bmatrix}$ **23.** $\begin{bmatrix} 3 & 1 & 1 & | & 8 \\ 0 & -3 & -3 & | & -6 \\ 0 & -10 & -13 & | & 34 \end{bmatrix}$

25. $2R_1 + R_2 \rightarrow R_2$ **27.** $-5R_1 + R_2 \rightarrow R_2$ **29.** $(20, 10)$
$-3R_1 + R_3 \rightarrow R_3$ $\quad 4R_1 + R_3 \rightarrow R_3$

31. $(1, 6, 9)$ **33.** $(1, 1, 2)$ **35.** $(1, 1, 1)$ **37.** $(-1, \frac{-3}{2}, 2)$
39. linear dependence $(p - 4, -2p + 8, p)$ **41.** coincident dependence $\{(x, y, z) | 3x - 4y + 2z = -2\}$ **43.** no solution **45.** linear dependence, $(-\frac{5}{4}p - 3, \frac{1}{8}p - \frac{1}{2}, p)$ **47.** 28.5 units^2 **49.** Heat: 95, Mavericks: 92
51. Poe, \$12,500; Baum, \$62,500; Wouk, \$25,000 **53.** $A = 35°$, $B = 45°, C = 100°$ **55.** \$.4 million at 4%; \$.6 million at 7%; \$1.5 million at 8% **57.** $x = 84°; y = 25°$ **59.** $x^3 + x - 10; x^3 - x - 6$; $x^4 - 2x^3 - 8x + 16; x^2 + 2x + 4$ **61.** $C > 30,000$ in the year 2011 $(t \approx 6.39)$

Exercises 6.2, pp. 576–580

1. $a_{ij}; b_{ij}$ **3.** scalar **5.** Answers will vary. **7.** $2 \times 2, a_{12} = -3$, $a_{21} = 5$ **9.** $2 \times 3, a_{12} = -3, a_{23} = 6, a_{22} = 5$ **11.** $3 \times 3, a_{12} = 1$, $a_{23} = 1, a_{31} = 5$ **13.** true **15.** conditional, $c = -2, a = -4, b = 3$

17. $\begin{bmatrix} 10 & 0 \\ 0 & 10 \end{bmatrix}$ **19.** different orders, sum not possible

21. $\begin{bmatrix} 20 & -15 \\ -25 & -10 \end{bmatrix}$ **23.** $\begin{bmatrix} \frac{5}{2} & -1 & 0 \\ 0 & \frac{-7}{2} & 1 \\ 2 & \frac{3}{2} & -6 \end{bmatrix}$ **25.** $\begin{bmatrix} 1 & 2 & 0 \\ 0 & -1 & 2 \\ 4 & 3 & -6 \end{bmatrix}$

27. $\begin{bmatrix} 1 & 0 \\ 0 & 1 \end{bmatrix}$ **29.** $\begin{bmatrix} 6 & -3 & 9 \\ 12 & 0 & -6 \end{bmatrix}$ **31.** $\begin{bmatrix} 12 & -24 & 90 \\ -6 & 15 & -57 \end{bmatrix}$

33. $\begin{bmatrix} 79 & -30 \\ -50 & 19 \end{bmatrix}$ **35.** $\begin{bmatrix} 42 & 18 & -60 \\ -12 & -42 & 36 \end{bmatrix}$ **37.** $\begin{bmatrix} 0.71 & 0.65 \\ 1.78 & 3.55 \end{bmatrix}$

39. $\begin{bmatrix} 1 & -1.25 & 0.25 \\ -0.5 & -0.63 & 2.13 \\ 3.75 & 3.69 & -5.94 \end{bmatrix}$ **41.** $\begin{bmatrix} 1 & 0 \\ 0 & 1 \end{bmatrix}$ **43.** $\begin{bmatrix} 1 & 0 & 0 \\ 0 & 1 & 0 \\ 0 & 0 & 1 \end{bmatrix}$

45. $\begin{bmatrix} \frac{-3}{19} & \frac{4}{57} \\ \frac{1}{19} & \frac{5}{57} \end{bmatrix}$ **47.** $\begin{bmatrix} 0 & \frac{3}{4} & \frac{1}{4} \\ \frac{-1}{2} & \frac{3}{8} & \frac{1}{8} \\ \frac{-1}{4} & \frac{11}{16} & \frac{1}{16} \end{bmatrix}$ **49.** $\begin{bmatrix} 1.75 & 2.5 \\ 7.5 & 13 \end{bmatrix}$

51. $\begin{bmatrix} -0.26 & 0.32 & -0.07 \\ 0.63 & 0.30 & 0.10 \end{bmatrix}$ **53.** verified **55.** verified
57. $P = 21.448$ cm; $A = 27.7269$ cm^2
59. a.

$$V = \begin{matrix} & T & S \\ S & \begin{bmatrix} 3820 & 1960 \\ D & 2460 & 1240 \\ P & 1540 & 920 \end{bmatrix} \end{matrix} \quad M = \begin{matrix} & T & S \\ S & \begin{bmatrix} 4220 & 2960 \\ D & 2960 & 3240 \\ P & 1640 & 820 \end{bmatrix} \end{matrix}$$

b. 3900 more by Minsk
c. $V = \begin{bmatrix} 3972.8 & 2038.4 \\ 2558.4 & 1289.6 \\ 1601.6 & 956.8 \end{bmatrix}$ **d.** $\begin{bmatrix} 8361.6 & 5116.8 \\ 5636.8 & 4659.2 \\ 3307.2 & 1809.6 \end{bmatrix}$
$M = \begin{bmatrix} 4388.8 & 3078.4 \\ 3078.4 & 3369.6 \\ 1705.6 & 852.8 \end{bmatrix}$

61. $[22,000 \quad 19,000 \quad 23,500 \quad 14,000]$;
total profit
North: \$22,000
South: \$19,000
East: \$23,500
West: \$14,000
63. a. \$108.20 **b.** \$101
c. $\begin{matrix} \text{Science} \\ \text{Math} \end{matrix} \begin{bmatrix} 100 & 101 & 119 \\ 108.2 & 107 & 129.5 \end{bmatrix}$
First row, total cost for science from each restaurant; Second row, total cost for math from each restaurant.
65. a. 10 **b.** 20
c.
$$\begin{matrix} & \text{Spanish} & \text{Chess} & \text{Writing} \\ \text{Female} & \begin{bmatrix} 32.4 & 10.3 & 21.3 \\ \text{Male} & 29.9 & 9.6 & 19.5 \end{bmatrix} \end{matrix},$$
the approximate number of females expected to join the writing club
67. $\begin{bmatrix} 2^{n-1} & 0 & 2^{n-1} \\ 2^n - 1 & 1 & 2^n - 1 \\ 2^{n-1} & 0 & 2^{n-1} \end{bmatrix}$ **69.** $a = 2, b = 1, c = -3, d = -2$
71. $x \in [-3, -1] \cup [1, 3]$ **73.** $x^2 + 2x - 5$

Mid-Chapter Check, pp. 580–581

1. $3 \times 3, -0.9$ **2.** $2 \times 4, 0$ **3.** $(2, -3)$ **4.** $(2, 0, -5)$
5. $(p - 3, 2p - 8, p)$
6. a. $\begin{bmatrix} -13 & -17 \\ 35 & 9 \end{bmatrix}$ **b.** $\begin{bmatrix} 4 & 6 \\ -12 & -2 \end{bmatrix}$ **c.** $\begin{bmatrix} -5 & 5 \\ -5 & 15 \end{bmatrix}$
7. a. $\begin{bmatrix} 0.8 & 0.5 & 2.2 \\ -0.1 & 0.8 & -1 \\ 2.1 & 0.3 & 1.9 \end{bmatrix}$ **b.** $\begin{bmatrix} -3 & -1.5 & -6 \\ 1.5 & 0 & 3 \\ -6 & -1.5 & -6 \end{bmatrix}$ **c.** $\begin{bmatrix} 1 & 0 & 0 \\ 0 & 1 & 0 \\ 0 & 0 & 1 \end{bmatrix}$
8. a. $\begin{bmatrix} 17 & -32 & -13 \\ 5 & 0 & -5 \end{bmatrix}$ **b.** $\begin{bmatrix} -26 & -18 & 24 \\ 2 & -10 & -4 \end{bmatrix}$ **c.** $\begin{bmatrix} 24 & 4 \\ 0 & -5 \\ 16 & -39 \end{bmatrix}$
d. $\begin{bmatrix} -4 & -33 \\ -2 & 29 \end{bmatrix}$ **9.** used: \$80, new: \$125 **10.** $\begin{bmatrix} 4375 & 110 \\ 2400 & 59 \end{bmatrix}$,
P_{11}: total rebates paid by individuals, P_{21}: total rebates paid by business, P_{12}: free AAA years given to individuals, P_{22}: free AAA years given to business

Reinforcing Basic Concepts, p. 581

Exercise 1: P_{32}
Exercise 2:
1st row of A with 3rd column of B
2nd row of A with 2nd column of B
Exercise 3:
$[A] \rightarrow 3 \times 1; [B] \rightarrow 1 \times 3$
$[A] \rightarrow 3 \times 2; [B] \rightarrow 2 \times 3$
$[A] \rightarrow 3 \times 3; [B] \rightarrow 3 \times 3$
$[A] \rightarrow 3 \times n; [B] \rightarrow n \times 3; n \in \mathbb{N}$
Exercise 4: only $FE = P$

Exercises 6.3, pp. 591–595

1. diagonal; zeroes **3.** identity **5.** Answers will vary. **7.** verified
9. verified **11.** verified **13.** verified **15.** $\begin{bmatrix} \frac{1}{9} & \frac{2}{9} \\ \frac{-1}{9} & \frac{5}{18} \end{bmatrix}$

17. $\begin{bmatrix} -5 & 1.5 \\ -2 & 0.5 \end{bmatrix}$ **19.** verified **21.** verified **23.** $\begin{bmatrix} \frac{-2}{39} & \frac{1}{13} & \frac{10}{39} \\ \frac{1}{3} & 0 & \frac{1}{3} \\ \frac{-4}{39} & \frac{2}{13} & \frac{-19}{39} \end{bmatrix}$

25. $\begin{bmatrix} \frac{-9}{80} & \frac{31}{400} & \frac{27}{400} \\ \frac{1}{80} & \frac{41}{400} & \frac{-3}{400} \\ \frac{-1}{20} & \frac{-1}{100} & \frac{-17}{100} \end{bmatrix}$ **27.** $\begin{bmatrix} 2 & -3 \\ -5 & 7 \end{bmatrix} \begin{bmatrix} x \\ y \end{bmatrix} = \begin{bmatrix} 9 \\ 8 \end{bmatrix}$

29. $\begin{bmatrix} 1 & 2 & -1 \\ 1 & 0 & 1 \\ 2 & -1 & 1 \end{bmatrix}\begin{bmatrix} x \\ y \\ z \end{bmatrix} = \begin{bmatrix} 1 \\ 3 \\ 3 \end{bmatrix}$

31. $\begin{bmatrix} -2 & 1 & -4 & 5 \\ 2 & -5 & 1 & -3 \\ -3 & 1 & 6 & 1 \\ 1 & 4 & -5 & 1 \end{bmatrix}\begin{bmatrix} w \\ x \\ y \\ z \end{bmatrix} = \begin{bmatrix} -3 \\ 4 \\ 1 \\ -9 \end{bmatrix}$ **33.** $(4, 5)$ **35.** $(12, 12)$

37. no solution **39.** $(1.5, -0.5, -1.5)$ **41.** no solution
43. $(-1, -0.5, 1.5, 0.5)$ **45.** 1, yes **47.** 0, no **49.** 1
51. singular matrix **53.** singular matrix **55.** -34 **57.** 7

59. $\det(A) = -5; (1, 6, 9)$ **61.** $\det(A) = 0$ **63.** $A^{-1} = \begin{bmatrix} \frac{1}{13} & \frac{5}{13} \\ \frac{-2}{13} & \frac{3}{13} \end{bmatrix}$

65. singular **67.** 31 behemoth, 52 gargantuan, 78 mammoth, 30 jumbo
69. Jumpin' Jack Flash: 3.75 min
Tumbling Dice: 3.75 min
You Can't Always Get: 7.5 min
Wild Horses: 5.75 min
71. 30 of clock A; 20 of clock B; 40 of clock C; 12 of clock D
73. $p_1 = 72.25°, p_2 = 74.75°, p_3 = 80.25°, p_4 = 82.75°$
75. $y = x^3 + 2x^2 - 9x - 10$ **77.** 2 oz Food I, 1 oz Food II, 4 oz Food III
79. Answers will vary. **81. a.** -45 **b.** 52 **c.** -19 **d.** -4

83. $x = 6, x = 3, x = -2$ **85.** $x \in \left(-\infty, -\frac{9}{2}\right] \cup \left[-\frac{1}{2}, \infty\right)$

Exercises 6.4, pp. 606–609

1. $a_{11}a_{22} - a_{21}a_{12}$ **3.** constant **5.** Answers will vary.

7. $D = \begin{vmatrix} 2 & 5 \\ -3 & 4 \end{vmatrix}; D_x = \begin{vmatrix} 7 & 5 \\ 1 & 4 \end{vmatrix}; D_y = \begin{vmatrix} 2 & 7 \\ -3 & 1 \end{vmatrix}$ **9.** $(-5, 9)$

11. $\left(\frac{-26}{3}, \frac{25}{3}\right)$ **13.** no solution

15. a. $D = \begin{vmatrix} 4 & -1 & 2 \\ -3 & 2 & -1 \\ 1 & -5 & 3 \end{vmatrix}$ $D_x = \begin{vmatrix} -5 & -1 & 2 \\ 8 & 2 & -1 \\ -3 & -5 & 3 \end{vmatrix}$

$D_y = \begin{vmatrix} 4 & -5 & 2 \\ -3 & 8 & -1 \\ 1 & -3 & 3 \end{vmatrix}$ $D_z = \begin{vmatrix} 4 & -1 & -5 \\ -3 & 2 & 8 \\ 1 & -5 & -3 \end{vmatrix}$

$D = 22$, solutions possible
b. $D = 0$, Cramer's rule cannot be used.

17. $(1, 2, 1)$ **19.** $\left(\frac{3}{4}, \frac{5}{3}, \frac{-1}{3}\right)$ **21.** $(0, -1, 2, -3)$

23. $\frac{A}{x + 3} + \frac{B}{x - 2}$ **25.** $\frac{A}{x - 1} + \frac{B}{x + 2} + \frac{C}{x - 3}$

27. $\frac{A}{x} + \frac{B}{x - 3} + \frac{C}{x + 1}$ **29.** $\frac{A}{x} + \frac{B}{x^2} + \frac{C}{x + 2}$

31. $\frac{A}{x + 1} + \frac{Bx + C}{x^2 + 2} + \frac{Dx + E}{(x^2 + 2)^2}$ **33.** $\frac{4}{x} - \frac{5}{x + 1}$

35. $\frac{-4}{2x - 5} + \frac{3}{x + 3}$ **37.** $\frac{7}{x} + \frac{2}{x + 1} - \frac{1}{x - 1}$

39. $\frac{-1}{x} + \frac{4}{x + 1} + \frac{5}{(x + 1)^2}$ **41.** $\frac{3}{2 - x} - \frac{4}{x^2 + 2x + 4}$

43. $\frac{5}{x + 2} + \frac{x - 1}{x^2 + 3}$ **45.** $\frac{1}{x} + \frac{3x - 2}{(x^2 + 1)^2}$

47. $\frac{3}{x + 1} - \frac{2}{x - 3} + \frac{1}{(x - 3)^3}$ **49.** $320 + 32\pi \approx 420.5 \text{ in}^2$

51. 8 cm^2 **53.** 27 ft^2 **55.** 19 m^3 **57.** yes **59.** no **61.** yes, yes, yes

63. $\begin{cases} 15,000x + 25,000y = 2900 \\ 25,000x + 15,000y = 2700 \end{cases}$; 6%, 8%

65. $(-1, -1, 2)$; answers will vary. **67.** $x^2 + y^2 - 4x - 6y - 12 = 0$
69. left graph; graph shifts left 1 **71.** left graph; right shows four roots

Summary and Concept Review, pp. 609–611

1. Answers will vary. **2.** $(-2, -4)$ **3.** $(1, 6, 9)$ **4.** $(-2, 7, 1, 8)$

5. $\left\{(x, y, z) | x = 3y + 2, y \in \mathbb{R}, z = -\frac{5}{2}y - \frac{5}{2}\right\}$

6. $\begin{bmatrix} -7.25 & 5.25 \\ 0.875 & -2.875 \end{bmatrix}$ **7.** $\begin{bmatrix} -6.75 & 6.75 \\ 1.125 & -1.125 \end{bmatrix}$ **8.** not possible

9. $\begin{bmatrix} -2 & -6 \\ -1 & -7 \end{bmatrix}$ **10.** $\begin{bmatrix} 1 & 0 \\ 0 & 1 \end{bmatrix}$ **11.** $\begin{bmatrix} 1 & 0 & 4 \\ 5.5 & -1 & -1 \\ 10 & -2.9 & 7 \end{bmatrix}$

12. $\begin{bmatrix} 3 & -6 & -4 \\ -4.5 & 3 & -1 \\ -2 & 3.1 & 3 \end{bmatrix}$ **13.** not possible **14.** $\begin{bmatrix} -8 & 12 & 0 \\ -2 & -4 & 4 \\ -16 & -0.4 & -20 \end{bmatrix}$

15. $\begin{bmatrix} 15.5 & 6.4 & 17 \\ 9 & -17 & 2 \\ 18.5 & -20.8 & 13 \end{bmatrix}$ **16.** D **17.** It's an identity.

18. It's the inverse of B. **19.** E **20.** It is the inverse of F.
21. Matrix multiplication is not generally commutative. **22.** $(-8, -6)$

23. $(2, 0, -3)$ **24.** $\left(\frac{-19}{35}, \frac{25}{14}\right)$ **25.** $\left(\frac{-37}{19}, \frac{36}{19}, \frac{31}{19}\right)$ **26.** $(1, -1, 2)$

27. $\frac{91}{2}$ units^2 **28.** $\frac{5}{x - 2} + \frac{2x - 1}{x^2 + 3}$

Mixed Review, pp. 611–612

1. $(-10, 12)$ **3.** $\{(x, y, z) | x = z - 1, y = 2z - 2, z \in \mathbb{R}\}$

5. a. $\begin{bmatrix} -8 & 16 & -10 \\ 12 & 0 & 6 \end{bmatrix}$ **b.** $\begin{bmatrix} 9 & -6 & -7 \\ -7 & -1 & 2 \end{bmatrix}$

7. a. 3 **b.** 3 **c.** -4 **d.** 1 **9.** $(-9, -3, 2)$ **11.** $A^{-1} = \begin{bmatrix} 8 & -3 \\ -5 & 2 \end{bmatrix}$

13. $\begin{cases} x + 2y - z = 1 \\ x + z = 3 \\ 2x - y + z = 3 \end{cases}$ **15.** $\left(\frac{33}{31}, \frac{-10}{31}, \frac{-57}{31}\right)$ **17.** $A = 4.5$ units^2

19. 137 m by 82 m

Practice Test, pp. 612–613

1. $\left(-3, \frac{1}{2}\right)$ **2.** $\left\{(x, y, z) | x = -3y + \frac{16}{3}, y \in \mathbb{R}, z = 2y - 3\right\}$

3. $\left(2, 1, \frac{-1}{3}\right)$ **4. a.** $\begin{bmatrix} -6 & -5 \\ 8 & 9 \end{bmatrix}$ **b.** $\begin{bmatrix} 1.2 & 1.2 \\ -1.2 & -2 \end{bmatrix}$ **c.** $\begin{bmatrix} -3 & 1 \\ 3 & -5 \end{bmatrix}$

d. $\begin{bmatrix} -2 & -1 \\ 2.5 & 1.5 \end{bmatrix}$ **e.** -2 **5. a.** $\begin{bmatrix} 0 & -0.1 & 0 \\ 0.5 & -0.6 & 0 \\ -0.2 & -0.8 & -0.9 \end{bmatrix}$

b. $\begin{bmatrix} -0.3 & -0.06 & -0.12 \\ 0.06 & -0.06 & 0 \\ -0.18 & -0.24 & -0.48 \end{bmatrix}$ **c.** $\begin{bmatrix} 0.31 & -0.13 & 0.08 \\ -0.01 & -0.05 & -0.02 \\ 0.39 & -0.52 & -0.02 \end{bmatrix}$

d. $\begin{bmatrix} \frac{40}{17} & 0 & \frac{-10}{17} \\ \frac{40}{17} & 10 & \frac{-10}{17} \\ \frac{-35}{17} & -5 & \frac{30}{17} \end{bmatrix}$ **e.** $\frac{17}{500}$

6. $(-1, -6, 0), (1, -1, 1), (3, 4, 2)$, answers vary as $(2p - 1, 5p - 6, p)$

7. $\left(2, \frac{2}{3}\right)$ **8.** $(3, -2, 3)$ **9.** $\left(\frac{97}{34}, \frac{-18}{17}\right)$ **10.** $(1, 6, 9)$ **11.** $(1, -1, 2)$

12. $B = \begin{bmatrix} 6 \\ 13 \\ -11 \end{bmatrix}$ **13.** $(-1, 4), (2, 1), (4, -1)$ **14.** 5 mi^2

15. $r = -2, s = 1$ **16.** Dr. Brown owes $31,000; Dr. Stamper owes $124,000 **17.** 7.5 hr, 15.5 hr **18.** 11 one day, 6 two day, 3 five day
19. federal program: $200,000; municipal bonds: $1,300,000;
bank loan: $300,000 **20.** $\frac{1}{x - 3} + \frac{3x + 2}{x^2 + 3x + 9}$

Strengthening Core Skills, pp. 614–615

Exercise 1: $(1, -4, 1)$

Cumulative Review Chapters 1–6, pp. 616–617

1. a. $8 - i$ **b.** $16 - 30i$ **c.** $3 - 2i$ **d.** i **3.** $x = \dfrac{-3}{2}$ **5.** $x = \dfrac{1}{3}$

7. $y = \dfrac{-7}{5}x + \dfrac{4}{5}$ **9. a.** -18 **b.** 1 **c.** 7 **d.** -3 is not in the domain

11.   **13.**   **15.** $(-\infty, 1) \cup (5, \infty)$

17. $f(x)\uparrow: x \in (0, 2) \cup (4, \infty)$ $f(x)\downarrow: x \in (-\infty, 0) \cup (2, 4)$

constant: none   **19.** $2y - \dfrac{5}{2} + \dfrac{25}{2(2y + 1)}$ **21.** $x^3 - 4x^2 - 2x + 20$

23. $\left(-3, \dfrac{-5}{2}\right)$ **25.** condominium: \$400,000; delivery truck: \$60,000;

business: \$140,000

MODELING WITH TECHNOLOGY III

Modeling with Technology Exercises, pp. 624–627

1. 225 boards at \$400 a piece   **3.** 90,000,000 gal at \$3.07 per gallon
5. $(\sim410.07, \sim226.58)$ or about 227 boards at approximately \$410 a piece
7. $(\sim3.0442, \sim8.9964)$, or about 90,000,000 gal at approximately \$3.04
per gallon   **9.** 214.5 ft^2 of skin, 231.0 ft^2 of wood veneer, 516 tension rods,
and 498 ft of hoop   **11.** 955 ft^2 of skin, 1021.5 ft^2 of wood veneer, 2180
tension rods, and 2129.5 ft of hoop   **13.** 92,250 gal gasoline, 595,000 lb
corn, 227,500 oz yeast, and 134,750 gal water   **15.** 5 Silver, 9 Gold, and
2 Platinum   **17.** one bundle of first class = 9.25 measures of grain;
one bundle of second class = 4.25 measures of grain;
one bundle of third class = 2.75 measures of grain   **19.** Answers will
vary.   **21.** Answers will vary.   **23.** Answers will vary.   **25.** Answers
will vary.   **27.** Answers will vary.   **29.** Answers will vary.
31. Answers will vary.   **33.** Answers will vary.   **35.** Answers will vary.

CHAPTER 7

Exercises 7.1, pp. 635–637

1. geometry; algebra   **3.** perpendicular   **5.** point; intersecting
7. $(-2, -2)$; verified   **9.** $(2, -2)$; verified   **11.** $(\tfrac{13}{2}, -9)$; verified
13. $(x + 2)^2 + (y + 2)^2 = 5^2$ **15.** $(x - 2)^2 + (y + 2)^2 = 5^2$

17. $\left(x - \dfrac{13}{2}\right)^2 + (y + 9)^2 = \left(\dfrac{25}{2}\right)^2$ **19. a.** $d = 13$; B, C, E, G

b. $(13, 3 + 4\sqrt{3})$, $(14, 8)$; Many others   **21.** Verified, $d = \dfrac{8\sqrt{5}}{5}$

23. a. B, C, E **b.** Answers will vary.   **25.** Verified   **27.** $y = -\dfrac{1}{16}x^2$

29. $4x^2 + 3y^2 = 48$ **31.** Verified, verified   **33.** $3x^2 - y^2 = 3$

35. a. $\left(-\dfrac{12}{7}, -\dfrac{30}{7}\right)$,   **b.** $\left(-2, -\dfrac{4}{3}\right)$ **37.** Verified (both add to 8)

39. a. $x = 2\ln 9, x \approx 4.39$ **b.** $x = \dfrac{5}{2}\ln 54, x \approx 9.97$

41. $h(x) = \dfrac{(x + 3)(x - 3)}{(x + 2)(x - 2)}$

Exercises 7.2, pp. 645–650

1. $c^2 = |a^2 - b^2|$ **3.** $2a; 2b$ **5.** answers will vary.   **7.** $x^2 + y^2 = 49$
9. $(x - 5)^2 + y^2 = 3$ **11.** $(x - 1)^2 + (y - 5)^2 = 25$
13. $(x - 6)^2 + (y - 5)^2 = 9$ **15.** $(x - 2)^2 + (y + 5)^2 = 25$
center: $(6, 5), r = 3$ center: $(2, -5), r = 5$

17. $(x + 3)^2 + y^2 = 14$
center: $(-3, 0), r = \sqrt{14}$

19.   **21.**

23.

25. a. $\dfrac{x^2}{16} + \dfrac{y^2}{4} = 1, (0, 0), a = 4, b = 2$
b. $(-4, 0), (4, 0), (0, -2), (0, 2)$ **c.**

27. a. $\dfrac{x^2}{9} + \dfrac{y^2}{16} = 1, (0, 0), a = 3, b = 4$
b. $(0, -4), (0, 4), (-3, 0), (3, 0)$ **c.**

29. a. $\dfrac{x^2}{5} + \dfrac{y^2}{2} = 1, (0, 0), a = \sqrt{5}, b = \sqrt{2}$
b. $(-\sqrt{5}, 0), (\sqrt{5}, 0), (0, -\sqrt{2}), (0, \sqrt{2})$ **c.**

31. ellipse   **33.** circle

35. ellipse

37. $x^2 + \dfrac{(y+3)^2}{4} = 1$

39. $\dfrac{(x+2)^2}{16} + \dfrac{(y-1)^2}{4} = 1$

41. $\dfrac{(x-3)^2}{4} + \dfrac{(y+5)^2}{10} = 1$

43. $\dfrac{(x-3)^2}{25} + \dfrac{(y+2)^2}{10} = 1$

45. 20   **47.** 20   **49. a.** $(2, 1)$ **b.** $(-3, 1)$ and $(7, 1)$

c. $(2 - \sqrt{21}, 1)$ and $(2 + \sqrt{21}, 1)$ **d.** $(2, 3)$ and $(2, -1)$

e.

51. a. $(4, -3)$ **b.** $(4, 2)$ and $(4, -8)$ **c.** $(4, 0)$ and $(4, -6)$

     d. $(0, -3)$ and $(8, -3)$ **e.**

53. a. $(-2, -2)$ **b.** $(-5, -2)$ and $(1, -2)$ **c.** $(-2 + \sqrt{3}, -2)$ and

   $(-2 - \sqrt{3}, -2)$ **d.** $(-2, -2 + \sqrt{6})$ and $(-2, -2 - \sqrt{6})$

e.

55. $\dfrac{x^2}{36} + \dfrac{y^2}{20} = 1$ **57.** $\dfrac{(x-3)^2}{9} + \dfrac{(y+2)^2}{25} = 1$

59. $\dfrac{x^2}{16} + \dfrac{y^2}{9} = 1, (\pm\sqrt{7}, 0)$ **61.** $\dfrac{(x+3)^2}{4} + \dfrac{(y+1)^2}{16} = 1,$

$(-3, -1 \pm 2\sqrt{3})$ **63.** $A = 12\pi$ units^2 **65.** $\sqrt{7} \approx 2.65$ ft; 2.25 ft

67. 8.9 ft; 17.9 ft   **69.** $\dfrac{x^2}{15^2} + \dfrac{y^2}{8^2} = 1$; 6.4 ft

71. $\dfrac{x^2}{36^2} + \dfrac{y^2}{(35.25)^2} = 1$ **73.** $a \approx 142$ million miles,

$b \approx 141$ million miles, orbit time ≈ 686 days   **75.** $90,000\pi$ yd^2

77. $L = 8$ units; $(3\sqrt{5}, 4), (3\sqrt{5}, -4), (-3\sqrt{5}, 4), (-3\sqrt{5}, -4)$; verified

79. Verified

81. $(30 + 10\sqrt{3}) + (30 - 10\sqrt{3})i, \dfrac{(3 - \sqrt{3}) + (3 + \sqrt{3})i}{10}$

83. $R = \dfrac{kL}{d^2}$ $k = 0.003$ 250 Ω

Mid-Chapter Check, p. 650

1.

2.

3.

4.

5.

6.

7. $y = \dfrac{1}{12}x^2$

8. a. $\dfrac{(x+3)^2}{4} + \dfrac{(y-1)^2}{16} = 1$; $D: x \in [-5, -1]$; $R: y \in [-3, 5]$

b. $(x-3)^2 + (y-2)^2 = 16$; $D: x \in [-1, 7]$; $R: y \in [-2, 6]$

9. $\dfrac{y^2}{194} + \dfrac{x^2}{25} = 1$ **10.** $\dfrac{x^2}{16} + \dfrac{y^2}{4} = 1$

Reinforcing Basic Concepts, pp. 650–651

Exercise 1: $\dfrac{25(x-2)^2}{2} - \dfrac{9(y+3)^2}{4} = 1$

Exercise 2: $\dfrac{28(x-1)^2}{25} + \dfrac{48(y+2)^2}{25} = 1$

Exercises 7.3, pp. 661–664

1. transverse   **3.** midway   **5.** Answers will vary.

7.

9.

11.

13.

15.

17.

19.

21.

23. $(-4, -2), (2, -2), y = -2, (-1, -2), x = -1$

25. $(4, 1), (4, -3), x = 4, (4, -1), y = -1$

27.

29.

31.

33.

35.

37.

39.

41.

43.

45.

47.

49. circle   **51.** circle   **53.** hyperbola

55. hyperbola   **57.** circle   **59.** ellipse   **61.** $8, 2a = 8, 2b = 6$
63. $12, 2a = 16, 2b = 12$
65. a. $(3, 4)$ **b.** $(0, 4)$ and $(6, 4)$ **c.** $(3 - \sqrt{13}, 4)$ and $(3 + \sqrt{13}, 4)$
d. $2a = 6, 2b = 4$ **e.**

67. a. $(0, 3)$ **b.** $(-2, 3)$ and $(2, 3)$ **c.** $(-2\sqrt{5}, 3$ and $(2\sqrt{5}, 3)$
d. $2a = 4, 2b = 8$ **e.**

69. a. $(3, -2)$ **b.** $(1, -2)$ and $(5, -2)$ **c.** $(-1, -2)$ and $(7, -2)$
d. $2a = 4, 2b = 4\sqrt{3}$ **e.**

71. $\dfrac{x^2}{36} - \dfrac{y^2}{28} = 1$ **73.** $\dfrac{y^2}{9} - \dfrac{(x + 2)^2}{9} = 1$

75. $\dfrac{x^2}{4} - \dfrac{y^2}{9} = 1, (\pm \sqrt{13}, 0)$ **77.** $\dfrac{(y - 1)^2}{4} - \dfrac{(x - 2)^2}{5} = 1, 4 \text{ by } 2\sqrt{5}$

79. a. $y = \frac{2}{3}\sqrt{x^2 - 9}$ **b.** $x \in (-\infty, -3] \cup [3, \infty)$
c. $y = \frac{-2}{3}\sqrt{x^2 - 9}$ **81.** 40 yd   **83.** 40 ft

85. $\dfrac{x^2}{225} - \dfrac{y^2}{2275} = 1$, about $(24.1, 60)$ or $(-24.1, 60)$

87. a. $\dfrac{(x - 4)^2}{\frac{1}{4}} - (y - 2)^2 = 0$ **b.** $(x - 2)^2 + \dfrac{(x - 4)^2}{\frac{1}{3}} = 0$

89. a   **91.** $\dfrac{(x - 2)^2}{43} + \dfrac{(y - 3)^2}{9} = 1$

93.

     95. b and c

Exercises 7.4, pp. 671–674

1. horizontal; right; $a < 0$ **3.** $(p, 0); x = -p$ **5.** Answers will vary.
7. $x \in (-\infty, \infty), y \in [-4, \infty)$ **9.** $x \in (-\infty, \infty), y \in [-18, \infty)$

11. $x \in (-\infty, \infty),$ **13.** $x \in [-4, \infty), y \in (-\infty, \infty)$
$y \in [-10.125, \infty)$

15. $x \in (-\infty, 16], y \in (-\infty, \infty)$ **17.** $x \in (-\infty, 0], y \in (-\infty, \infty)$

19. $x \in [-9, \infty), y \in (-\infty, \infty)$ **21.** $x \in [-4, \infty), y \in (-\infty, \infty)$

23. $x \in (-\infty, 0], y \in (-\infty, \infty)$ **25.** $x \in [-6.25, \infty), y \in (-\infty, \infty)$

27. $x \in [-21, \infty), y \in (-\infty, \infty)$ **29.** $x \in (-\infty, 11], y \in (-\infty, \infty)$

31. $x \in (-\infty, \infty), y \in [3, \infty)$ **33.** $x \in [2, \infty), y \in (-\infty, \infty)$

35. $x \in [1, \infty), y \in (-\infty, \infty)$

37.   **39.**   **41.**

43.   **45.**   **47.**

49.   **51.**

53.   **55.**

57.   **59.**

61. $x^2 = 8y$ **63.** $y^2 = 16x$ **65.** $x^2 = -20y$
67. $(y + 2)^2 = -12(x - 2)$ **69.** $(x - 4)^2 = 12(y + 7)$
71. $(x - 3)^2 = 8(y - 2)$ **73.** $y^2 = 8(x + 1)$ vertex $(-1, 0)$ focus $(1, 0)$
75. $(y - 2)^2 = -8(x + 2)$ directrix: $x = 0$ endpoints $(-4, 6)$ and $(-4, -2)$

77.   **79.**

81.   **83.** 16 units^2

85.   **87.** 6 in.; $(13.5, 0)$ **89.** 14.97 ft, $(0, 41.75)$

91. $y^2 = 5x$ or $x^2 = 5y$, 1.25 cm
93. $(x - 2)^2 = \frac{1}{2}(y + 8)$; $p = \frac{1}{8}$; $(2, -8)$ **95.** 18 units^2 **97.** Answers will vary.   **99.** symmetric with respect to the y-axis, symmetric with respect to the origin

Summary and Concept Review, pp. 675–677

1. verified (segments are perpendicular and equal length)
2. $x^2 + (y - 1)^2 = 34$ **3.** verified   **4.** verified
5.   **6.**   **7.**

8.   **9.**

10. $\dfrac{x^2}{25} + \dfrac{y^2}{9} = 1$ **11. a.** $\dfrac{x^2}{169} + \dfrac{y^2}{25} = 1$ **b.** $\dfrac{x^2}{400} + \dfrac{y^2}{144} = 1$

12. $\dfrac{(x - 2)^2}{25} + \dfrac{(y - 1)^2}{4} = 1$ **13.**

14.   **15.**   **16.**

17.   **18.** $\dfrac{x^2}{9} - \dfrac{y^2}{16} = 1$

19. a. $\dfrac{x^2}{225} - \dfrac{y^2}{64} = 1$ **b.** $\dfrac{y^2}{16} - \dfrac{x^2}{9} = 1$

20. $\dfrac{(x-5)^2}{9} - \dfrac{(y-2)^2}{4} = 1$ **21.**

22.

23.

24.

Mixed Review, p. 677

1.

3.

5.

7.

9.

11.

13.

15.

17.

19.

21.

23. $d(P_1P_3) = d(P_2P_3)$

25. a. $(12, 0)$ **b.** $2a = 116; 2b \approx 113.5$ **c.** $\dfrac{x^2}{3364} + \dfrac{y^2}{3220} = 1$

Practice Test, p. 678

1. c   **2.** d   **3.** a   **4.** b

5.

6.

7.

8.

9.

10.

11.

12.

13. a. $\left(\dfrac{10}{3}, \dfrac{16}{3}\right), (-2, 0)$

b. $\left(\dfrac{2\sqrt{15}}{5}, \dfrac{2\sqrt{10}}{5}\right), \left(\dfrac{2\sqrt{15}}{5}, \dfrac{-2\sqrt{10}}{5}\right),$

$\left(\dfrac{-2\sqrt{15}}{5}, \dfrac{2\sqrt{10}}{5}\right), \left(\dfrac{-2\sqrt{15}}{5}, \dfrac{-2\sqrt{10}}{5}\right)$

14. 15 ft, 20 ft   **15.** $(x+2)^2 + (y-5)^2 = 8$ **16.** $\dfrac{x^2}{16} + \dfrac{y^2}{12} = 1$

17. 154.89 million miles; 128.41 million miles

18. $y = (x-1)^2 - 4$; D: $x \in (-\infty, \infty)$, R: $y \in [-4, \infty)$; focus: $\left(1, \dfrac{-15}{4}\right)$

19. $(x-1)^2 + (y-1)^2 = 25$; D: $x \in [-4, 6]$, R: $y \in [-4, 6]$

20. $\dfrac{(x+3)^2}{9} + \dfrac{y^2}{36} = 1$; D: $x \in [-6, 0]$, R: $y \in [-6, 6]$;

foci: $(-3, -3\sqrt{3}), (-3, 3\sqrt{3})$

Strengthening Core Skills, pp. 679–680

Exercise 1: $\dfrac{(x-2)^2}{\left(\frac{\sqrt{2}}{5}\right)^2} - \dfrac{(y+3)^2}{\left(\frac{2}{3}\right)^2} = 1$ $a = \dfrac{\sqrt{2}}{5}, b = \dfrac{2}{3}$

Exercise 2: $\dfrac{(x-1)^2}{\left(\frac{5\sqrt{7}}{14}\right)^2} + \dfrac{(y+2)^2}{\left(\frac{5\sqrt{3}}{12}\right)^2} = 1$ $a = \dfrac{5\sqrt{7}}{14}, b = \dfrac{5\sqrt{3}}{12}$

Exercise 3: $\dfrac{(x+3)^2}{\left(\frac{7}{2}\right)^2} + \dfrac{(y-1)^2}{\left(\frac{6}{5}\right)^2} = 1; a = \frac{7}{2}, b = \frac{6}{5}$

Exercise 4: $\dfrac{(x+3)^2}{\left(\frac{4\sqrt{5}}{3}\right)^2} - \dfrac{(y-1)^2}{\left(\frac{9}{2}\right)^2} = 1; a = \frac{4\sqrt{5}}{3} \approx 3, b = \frac{9}{2}$

Cumulative Review Chapters 1–7, p. 680

1. $x = 2, x = \pm 2i$ **3.** no solution **5.** $x = 3 \pm 2i$
7. $x = 2 + \dfrac{\ln 7}{\ln 3}$ **9.** $x = 3$
11.

13.

15.

17.

19.

21. a. $x \in (-\infty, \infty)$ **b.** $y \in (-\infty, 4]$
c. $f(x)\!\uparrow: x \in (-\infty, -1), f(x)\!\downarrow: x \in (-1, \infty)$ **d.** none **e.** max: $(-1, 4)$
f. $f(x) > 0: x \in (-4, 2)$ **g.** $f(x) < 0: x \in (-\infty, -4) \cup (2, \infty)$
23. $(3, 4), (3, -4), (-3, 4), (-3, -4)$ **25.** $3\frac{1}{3}$ L

CHAPTER 8
Exercises 8.1, pp. 688–690

1. pattern; order **3.** increasing **5.** formula defining the sequence uses the preceding term(s); answers will vary. **7.** 1, 3, 5, 7; $a_8 = 15$; $a_{12} = 23$ **9.** 0, 9, 24, 45; $a_8 = 189$; $a_{12} = 429$ **11.** $-1, 2, -3, 4$; $a_8 = 8$; $a_{12} = 12$ **13.** $\frac{1}{2}, \frac{2}{3}, \frac{3}{4}, \frac{4}{5}$; $a_8 = \frac{8}{9}$; $a_{12} = \frac{12}{13}$
15. $\frac{1}{2}, \frac{1}{4}, \frac{1}{8}, \frac{1}{16}$; $a_8 = \frac{1}{256}$; $a_{12} = \frac{1}{4096}$ **17.** $1, \frac{1}{2}, \frac{1}{3}, \frac{1}{4}$; $a_8 = \frac{1}{8}$; $a_{12} = \frac{1}{12}$
19. $\frac{-1}{2}, \frac{1}{6}, \frac{-1}{12}, \frac{1}{20}$; $a_8 = \frac{1}{72}$; $a_{12} = \frac{1}{156}$
21. $-2, 4, -8, 16$; $a_8 = 256$; $a_{12} = 4096$ **23.** 79 **25.** $\frac{1}{5}$ **27.** $\frac{1}{32}$
29. $\left(\frac{11}{10}\right)^{10}$ **31.** $\frac{1}{36}$ **33.** 2, 7, 32, 157, 782 **35.** $-1, 4, 19, 364, 132, 499$
37. 64, 32, 16, 8, 4 **39.** 336 **41.** 36 **43.** 28 **45.** $\frac{1}{2}, \frac{1}{3}, \frac{1}{4}, \frac{1}{5}$

47. $\frac{1}{3}, \frac{1}{120}, \frac{1}{15,120}, \frac{1}{3,991,680}$ **49.** $1, 2, \frac{9}{2}, \frac{32}{3}$ **51.** 15 **53.** 64 **55.** $\frac{137}{60}$
57. 10 **59.** 95 **61.** -4 **63.** 15 **65.** 50 **67.** $\frac{-27}{112}$ **69.** $\displaystyle\sum_{n=1}^{5}(4n)$
71. $\displaystyle\sum_{n=1}^{6}(-1)^n n^2$ **73.** $\displaystyle\sum_{n=1}^{5}(n+3)$ **75.** $\displaystyle\sum_{n=1}^{3}\frac{n^2}{3}$ **77.** $\displaystyle\sum_{n=3}^{7}\frac{n}{2^n}$ **79.** 35
81. 100 **83.** 35 **85.** $a_n = 6000(0.8)^{n-1}$; 6000, 4800, 3840, 3072, 2457.60, 1966.08 **87.** 5.20, 5.70, 6.20, 6.70, 7.20, \$13,824 **89.** ≈ 2690
91. verified **93.** approaches $\frac{1}{2}$ **95.** $3^{-x} = \frac{1}{81}; x = 4$
97. $(\sqrt{5}, -2), (\sqrt{5}, 2), (-\sqrt{5}, -2), (-\sqrt{5}, 2)$

Exercises 8.2, pp. 695–697

1. common; difference **3.** $\dfrac{n(a_1 + a_n)}{2}$; nth **5.** Answers will vary.
7. arithmetic; $d = 3$ **9.** arithmetic; $d = 2.5$
11. not arithmetic; all prime **13.** arithmetic; $d = \frac{1}{24}$
15. not arithmetic; $a_n = n^2$ **17.** arithmetic; $d = \frac{-\pi}{6}$ **19.** 2, 5, 8, 11
21. 7, 5, 3, 1 **23.** 0.3, 0.33, 0.36, 0.39 **25.** $\frac{3}{2}, 2, \frac{5}{2}, 3$ **27.** $\frac{3}{4}, \frac{5}{8}, \frac{1}{2}, \frac{3}{8}$
29. $-2, -5, -8, -11$ **31.** $a_1 = 2, d = 5, a_n = 5n - 3$, $a_6 = 27, a_{10} = 47, a_{12} = 57$ **33.** $a_1 = 5.10, d = 0.15$, $a_n = 0.15n + 4.95, a_6 = 5.85, a_{10} = 6.45, a_{12} = 6.75$
35. $a_1 = \frac{3}{2}, d = \frac{3}{4}, a_n = \frac{3}{4}n + \frac{3}{4}, a_6 = \frac{21}{4}, a_{10} = \frac{33}{4}, a_{12} = \frac{39}{4}$
37. 61 **39.** 1 **41.** 2.425 **43.** 9 **45.** 43 **47.** 21 **49.** 26
51. $d = 3, a_1 = 1$ **53.** $d = 0.375, a_1 = 0.65$ **55.** $d = \frac{115}{126}, a_1 = \frac{-472}{63}$
57. 1275 **59.** 601.25 **61.** -534 **63.** 82.5 **65.** 74.04 **67.** $210\sqrt{2}$
69. $S_6 = 21; S_{75} = 2850$ **71.** at 11 P.M. **73.** 5.5 in.; 54.25 in.
75. 220; 2520; yes **77. a.** linear function **b.** quadratic
79. $t \approx 3.6$ **81.** $f(x) = 49x + 972$; 1364

Exercises 8.3, pp. 704–707

1. multiplying **3.** $a_1 r^{n-1}$ **5.** Answers will vary. **7.** $r = 2$
9. $r = -2$ **11.** $a_n = n^2 + 1$ **13.** $r = 0.1$ **15.** not geometric; ratio of terms decreases by 1 **17.** $r = \frac{2}{5}$ **19.** $r = \frac{1}{2}$ **21.** $r = \frac{4}{x}$
23. not geometric; $a_n = \dfrac{240}{n!}$ **25.** 5, 10, 20, 40 **27.** $-6, 3, \frac{-3}{2}, \frac{3}{4}$
29. $4, 4\sqrt{3}, 12, 12\sqrt{3}$ **31.** 0.1, 0.01, 0.001, 0.0001 **33.** $-\frac{3}{8}$ **35.** $\frac{25}{4}$
37. 16 **39.** $a_1 = \frac{1}{27}, r = -3, a_n = \frac{1}{27}(-3)^{n-1}, a_6 = -9, a_{10} = -729$, $a_{12} = -6561$
41. $a_1 = 729, r = \frac{1}{3}, a_n = 729(\frac{1}{3})^{n-1}, a_6 = 3, a_{10} = \frac{1}{27}, a_{12} = \frac{1}{243}$
43. $a_1 = \frac{1}{2}, r = \sqrt{2}, a_n = \frac{1}{2}(\sqrt{2})^{n-1}, a_6 = 2\sqrt{2}, a_{10} = 8\sqrt{2}$, $a_{12} = 16\sqrt{2}$ **45.** $a_1 = 0.2, r = 0.4, a_n = 0.2(0.4)^{n-1}$, $a_6 = 0.002048, a_{10} = 0.0000524288, a_{12} = 0.000008388608$ **47.** 5
49. 11 **51.** 9 **53.** 8 **55.** 13 **57.** 9 **59.** $r = \frac{2}{3}, a_1 = 729$
61. $r = \frac{3}{2}, a_1 = \frac{32}{243}$ **63.** $r = \frac{3}{2}, a_1 = \frac{256}{81}$ **65.** $-10,920$
67. $\frac{3872}{27} \approx 143.41$ **69.** $\frac{2059}{8} = 257.375$ **71.** 728 **73.** $\frac{85}{8} = 10.625$
75. ≈ 1.60 **77.** 1364 **79.** $\frac{31,525}{2187} \approx 14.41$ **81.** $\frac{-387}{512} \approx -0.76$
83. $\frac{521}{25}$ **85.** $\frac{3367}{1296}$ **87.** $14 + 15\sqrt{2}$ **89.** no **91.** $\frac{27}{7}$ **93.** $\frac{125}{3}$
95. 12 **97.** 4 **99.** $\frac{1}{3}$ **101.** $\frac{1}{2}$ **103.** $-\frac{18}{5}$ **105.** 1296
107. about 6.3 ft; 120 ft **109.** \$18,841.60; 10 yr
111. 125.4 gpm; 10 months **113.** about 347.7 million
115. 51,200 bacteria; 12 half-hours later (6 hr) **117.** ≈ 0.42 m; 8 m
119. 35.9 in³; 7 strokes **121.** 6 yr **123.** $S_n = \log n!$
125. $x = \dfrac{-5}{2} \pm \dfrac{\sqrt{11}}{2}i$ **127.**

Exercises 8.4, pp. 713–715

1. finite; universally   **3.** induction; hypothesis   **5.** Answers will vary.

7. $a_n = 10n - 6$
$a_4 = 10(4) - 6 = 40 - 6 = 34;$
$a_5 = 10(5) - 6 = 50 - 6 = 44;$
$a_k = 10k - 6;$
$a_{k+1} = 10(k + 1) - 6 = 10k + 10 - 6 = 10k + 4$

9. $a_n = n$
$a_4 = 4;$
$a_5 = 5;$
$a_k = k;$
$a_{k+1} = k + 1$

11. $a_n = 2^{n-1}$
$a_4 = 2^{4-1} = 2^3 = 8;$
$a_5 = 2^{5-1} = 2^4 = 16;$
$a_k = 2^{k-1};$
$a_{k+1} = 2^{k+1-1} = 2^k$

13. $S_n = n(5n - 1)$
$S_4 = 4(5(4) - 1) = 4(20 - 1) = 4(19) = 76;$
$S_5 = 5(5(5) - 1) = 5(25 - 1) = 5(24) = 120;$
$S_k = k(5k - 1);$
$S_{k+1} = (k + 1)(5(k + 1) - 1) = (k + 1)(5k + 5 - 1) = (k + 1)(5k + 4)$

15. $S_n = \dfrac{n(n + 1)}{2}$
$S_4 = \dfrac{4(4 + 1)}{2} = \dfrac{4(5)}{2} = 10;$
$S_5 = \dfrac{5(5 + 1)}{2} = \dfrac{5(6)}{2} = 15;$
$S_k = \dfrac{k(k + 1)}{2};$
$S_{k+1} = \dfrac{(k + 1)(k + 1 + 1)}{2} = \dfrac{(k + 1)(k + 2)}{2}$

17. $S_n = 2^n - 1$
$S_4 = 2^4 - 1 = 16 - 1 = 15;$
$S_5 = 2^5 - 1 = 32 - 1 = 31;$
$S_k = 2^k - 1;$
$S_{k+1} = 2^{k+1} - 1$

19. $a_n = 10n - 6; S_n = n(5n - 1)$
$S_4 = 4(5(4) - 1) = 4(20 - 1) = 4(19) = 76;$
$a_5 = 10(5) - 6 = 50 - 6 = 44;$
$S_5 = 5(5(5) - 1) = 5(25 - 1) = 5(24) = 120;$
$S_4 + a_5 = S_5$
$76 + 44 = 120$
$120 = 120$
Verified

21. $a_n = n; S_n = \dfrac{n(n + 1)}{2}$
$S_4 = \dfrac{4(4 + 1)}{2} = \dfrac{4(5)}{2} = 10;$
$a_5 = 5;$
$S_5 = \dfrac{5(5 + 1)}{2} = \dfrac{5(6)}{2} = 15;$
$S_4 + a_5 = S_5$
$10 + 5 = 15$
$15 = 15$
Verified

23. $a_n = 2^{n-1}; S_n = 2^n - 1$
$S_4 = 2^4 - 1 = 16 - 1 = 15;$
$a_5 = 2^{5-1} = 2^4 = 16;$
$S_5 = 2^5 - 1 = 32 - 1 = 31;$
$S_4 + a_5 = S_5$
$15 + 16 = 31$
$31 = 31$
Verified

25. $a_n = n^3; S_n = (1 + 2 + 3 + 4 + \cdots + n)^2$
$S_1 = 1^2 = 1^3$
$S_5 = (1 + 2 + 3 + 4 + 5)^2$
$= 15^2$
$= 225$
$1 + 8 + 27 + 64 + 125 = 225$
$S_9 = (1 + 2 + \cdots + 9)^2$
$= 45^2$
$= 2025$
$1 + 8 + \cdots + 729 = 2025$
$\left[\dfrac{n(n + 1)}{2}\right]^2 = \dfrac{n^2(n + 1)^2}{4}$

27. 1. Show S_n is true for $n = 1$.
$S_1 = 1(1 + 1) = 1(2) = 2$
Verified
2. Assume S_k is true: $2 + 4 + 6 + 8 + 10 + \cdots + 2k = k(k + 1)$
and use it to show the truth of S_{k+1} follows. That is:
$2 + 4 + 6 + \cdots + 2k + 2(k + 1) = (k + 1)(k + 2)$
$S_k + a_{k+1} = S_{k+1}$
Working with the left hand side:
$2 + 4 + 6 + \cdots + 2k + 2(k + 1)$
$= k(k + 1) + 2(k + 1)$
$= k^2 + k + 2k + 2$
$= k^2 + 3k + 2$
$= (k + 1)(k + 2)$
$= S_{k+1}$
Since the truth of S_{k+1} follows from S_k, the formula is true for all n.

29. 1. Show S_n is true for $n = 1$.
$S_1 = \dfrac{5(1)(1 + 1)}{2} = \dfrac{5(2)}{2} = 5$
Verified
2. Assume S_k is true:
$5 + 10 + 15 + \cdots + 5k = \dfrac{5k(k + 1)}{2}$
and use it to show the truth of S_{k+1} follows. That is:
$5 + 10 + 15 + \cdots + 5k + 5(k + 1) = \dfrac{5(k + 1)(k + 1 + 1)}{2}$
$S_k + a_{k+1} = S_{k+1}$
Working with the left hand side:
$5 + 10 + 15 + \cdots + 5k + 5(k + 1)$
$= \dfrac{5k(k + 1)}{2} + 5(k + 1)$
$= \dfrac{5k(k + 1) + 10(k + 1)}{2}$
$= \dfrac{(k + 1)(5k + 10)}{2}$
$= \dfrac{5(k + 1)(k + 2)}{2}$
$= S_{k+1}$
Since the truth of S_{k+1} follows from S_k, the formula is true for all n.

31. 1. Show S_n is true for $n = 1$.
$S_1 = 1(2(1) + 3) = 5$
Verified
2. Assume S_k is true:
$5 + 9 + 13 + 17 + \cdots + 4k + 1 = k(2k + 3)$
and use it to show the truth of S_{k+1} follows. That is:
$5 + 9 + 13 + 17 + \cdots + 4k + 1 + 4(k + 1) + 1$
$= (k + 1)(2(k + 1) + 3)$
$S_k + a_{k+1} = S_{k+1}$
Working with the left hand side:
$5 + 9 + 13 + 17 + \cdots + 4k + 1 + 4k + 5$
$= k(2k + 3) + 4k + 5$
$= 2k^2 + 3k + 4k + 5 = 2k^2 + 7k + 5$
$= (k + 1)(2k + 5) = S_{k+1}$
Since the truth of S_{k+1} follows from S_k, the formula is true for all n.

33. 1. Show S_n is true for $n = 1$.

$$S_1 = \frac{3(3^1 - 1)}{2} = \frac{3(3 - 1)}{2} = \frac{3(2)}{2} = 3$$

Verified

2. Assume S_k is true:

$$3 + 9 + 27 + \cdots + 3^k = \frac{3(3^k - 1)}{2}$$

and use it to show the truth of S_{k+1} follows. That is:

$$3 + 9 + 27 + \cdots + 3^k + 3^{k+1}$$
$$= \frac{3(3^{k+1} - 1)}{2}$$

$$S_k + a_{k+1} = S_{k+1}$$

Working with the left hand side:

$$3 + 9 + 27 + \cdots + 3^k + 3^{k+1}$$
$$= \frac{3(3^k - 1)}{2} + 3^{k+1}$$
$$= \frac{3(3^k - 1) + 2(3^{k+1})}{2}$$
$$= \frac{3^{k+1} - 3 + 2(3^{k+1})}{2}$$
$$= \frac{3(3^{k+1}) - 3}{2}$$
$$= \frac{3(3^{k+1} - 1)}{2}$$
$$= S_{k+1}$$

Since the truth of S_{k+1} follows from S_k, the formula is true for all n.

35. 1. Show S_n is true for $n = 1$.

$$S_n = 2^{n+1} - 2$$
$$S_1 = 2^{1+1} - 2 = 2^2 - 2 = 4 - 2 = 2$$

Verified

2. Assume S_k is true:

$$2 + 4 + 8 + \cdots + 2^k = 2^{k+1} - 2$$

and use it to show the truth of S_{k+1} follows. That is:

$$2 + 4 + 8 + \cdots + 2^k + 2^{k+1} = 2^{k+1} - 2$$

$$S_k + a_{k+1} = S_{k+1}$$

Working with the left hand side:

$$2 + 4 + 8 + \cdots + 2^k + 2^{k+1}$$
$$= 2^{k+1} - 2 + 2^{k+1}$$
$$= 2(2^{k+1}) - 2$$
$$= 2^{k+2} - 2$$
$$= S_{k+1}$$

Since the truth of S_{k+1} follows from S_k, the formula is true for all n.

37. 1. Show S_n is true for $n = 1$.

$$S_n = \frac{n}{2n + 1}$$

$$S_1 = \frac{1}{2(1) + 1} = \frac{1}{2 + 1} = \frac{1}{3}$$

Verified

2. Assume S_k is true:

$$\frac{1}{3} + \frac{1}{15} + \frac{1}{35} + \cdots + \frac{1}{(2k - 1)(2k + 1)} = \frac{k}{2k + 1}$$

and use it to show the truth of S_{k+1} follows. That is:

$$\frac{1}{3} + \frac{1}{15} + \frac{1}{35} + \cdots + \frac{1}{(2k - 1)(2k + 1)}$$
$$+ \frac{1}{(2(k + 1) - 1)(2(k + 1) + 1)} = \frac{k + 1}{2(k + 1) + 1}$$

$$S_k + a_{k+1} = S_{k+1}$$

Working with the left hand side:

$$\frac{1}{3} + \frac{1}{15} + \frac{1}{35} + \cdots + \frac{1}{(2k - 1)(2k + 1)} + \frac{1}{(2k + 1)(2k + 3)}$$

$$= \frac{k}{2k + 1} + \frac{1}{(2k + 1)(2k + 3)}$$
$$= \frac{k(2k + 3) + 1}{(2k + 1)(2k + 3)}$$
$$= \frac{2k^2 + 3k + 1}{(2k + 1)(2k + 3)}$$
$$= \frac{(2k + 1)(k + 1)}{(2k + 1)(2k + 3)}$$
$$= \frac{k + 1}{2k + 3}$$
$$= S_{k+1}$$

Since the truth of S_{k+1} follows from S_k, the formula is true for all n.

39. 1. Show S_n is true for $n = 1$.

S_1:
$$3^1 \geq 2(1) + 1$$
$$3 \geq 2 + 1$$
$$3 \geq 3$$

Verified

2. Assume $S_k : 3^k \geq 2k + 1$ is true and use it to show the truth of S_{k+1} follows. That is: $3^{k+1} \geq 2k + 3$.

Working with the left hand side:
$$3^{k+1} = 3(3^k)$$
$$\geq 3(2k + 1)$$
$$\geq 6k + 3$$

Since k is a positive integer,
$$6k + 3 \geq 2k + 3$$

Showing $S_{k+1} : 3^{k+1} \geq 2k + 3$

Verified

41. 1. Show S_n is true for $n = 1$.

S_1:
$$3 \cdot 4^{1-1} \leq 4^1 - 1$$
$$3 \cdot 4^0 \leq 4 - 1$$
$$3 \cdot 1 \leq 3$$
$$3 \leq 3$$

Verified

2. Assume $S_k : 3 \cdot 4^{k-1} \leq 4^k - 1$ is true and use it to show the truth of S_{k+1} follows. That is: $3 \cdot 4^k \leq 4^{k+1} - 1$.

Working with the left hand side:
$$3 \cdot 4^k = 3 \cdot 4(4^{k-1})$$
$$= 4 \cdot 3(4^{k-1})$$
$$\leq 4(4^k - 1)$$
$$\leq 4^{k+1} - 4$$

Since k is a positive integer, $4^{k+1} - 4 \leq 4^{k+1} - 1$

Showing that $3 \cdot 4^k \leq 4^{k+1} - 1$

43. $n^2 - 7n$ is divisible by 2

1. Show S_n is true for $n = 1$.
$$S_n : n^2 - 7n = 2m$$
S_1:
$$(1)^2 - 7(1) = 2m$$
$$1 - 7 = 2m$$
$$-6 = 2m \quad \text{Verified}$$

2. Assume $S_k : k^2 - 7k = 2m$ for $m \in Z$ and use it to show the truth of S_{k+1} follows. That is: $(k + 1)^2 - 7(k + 1) = 2p$ for $p \in Z$.

Working with the left hand side:
$$= (k + 1)^2 - 7(k + 1)$$
$$= k^2 + 2k + 1 - 7k - 7$$
$$= k^2 - 7k + 2k - 6$$
$$= 2m + 2k - 6$$
$$= 2(m + k - 3)$$

is divisible by 2.

45. $n^3 + 3n^2 + 2n$ is divisible by 3

  1. Show S_n is true for $n = 1$. $S_n : n^3 + 3n^2 + 2n = 3m$

    $S_1:$

    $(1)^3 + 3(1)^2 + 2(1) = 3m$

          $1 + 3 + 2 = 3m$

               $6 = 3m$

               $2 = m$

    Verified

  2. Assume $S_k : k^3 + 3k^2 + 2k = 3m$ for $m \in Z$ and use it to show the truth of S_{k+1} follows.

    That is: $S_{k+1} : (k+1)^3 + 3(k+1)^2 + 2(k+1) = 3p$ for $p \in Z$.

    Working with the left hand side:

    $(k+1)^3 + 3(k+1)^2 + 2(k+1)$ is true.

    $= k^3 + 3k^2 + 3k + 1 + 3(k^2 + 2k + 1) + 2k + 2$

    $= k^3 + 3k^2 + 2k + 3(k^2 + 2k + 1) + 3k + 3$

    $= k^3 + 3k^2 + 2k + 3(k^2 + 2k + 1) + 3(k + 1)$

    $= 3m + 3(k^2 + 2k + 1) + 3(k + 1)$ is divisible by 3.

47. $6^n - 1$ is divisible by 5

  1. Show S_n is true for $n = 1$. $S_n : 6^n - 1 = 5m$

    $S_1:$

    $6^1 - 1 = 5m$

      $6 - 1 = 5m$

          $5 = 5m$

          $1 = m$ Verified

  2. Assume $S_k : 6^k - 1 = 5m$ for $m \in Z$ and use it to show the truth of S_{k+1} follows.

    That is: $S_{k+1} : 6^{k+1} - 1 = 5p$ for $p \in Z$.

    Working with the left hand side:

    $= 6^k - 1$

    $= 6(6^k) - 1$

    $= 6(5m + 1) - 1$

    $= 30m + 6 - 1$

    $= 30m + 5$

    $= 5(6m + 1)$

    is divisible by 5, Verified

49. Verified

51. $A = \begin{bmatrix} -1 & 2 \\ 3 & 1 \end{bmatrix}; B = \begin{bmatrix} 2 & -1 \\ 4 & 3 \end{bmatrix}$

$A + B = \begin{bmatrix} 1 & 1 \\ 7 & 4 \end{bmatrix}$

$A - B = \begin{bmatrix} -3 & 3 \\ -1 & -2 \end{bmatrix}$

$2A - 3B = \begin{bmatrix} -8 & 7 \\ -6 & -7 \end{bmatrix}$

$AB = \begin{bmatrix} 6 & 7 \\ 10 & 0 \end{bmatrix}$

$BA = \begin{bmatrix} -5 & 3 \\ 5 & 11 \end{bmatrix}$

$B^{-1} = \begin{bmatrix} 0.3 & 0.1 \\ -0.4 & 0.2 \end{bmatrix}$

53. $(x - 3)^2 + (x - 4)^2 = 25$

Mid-Chapter Check, pp. 715–716

1. $3, 10, 17, a_9 = 59$ **2.** $4, 7, 12, a_9 = 84$ **3.** $-1, 3, -5, a_9 = -17$

4. 360   **5.** $\displaystyle\sum_{k=1}^{6}(3k - 2)$ **6.** d   **7.** e   **8.** a   **9.** b   **10.** c

11. a. $a_1 = 2, d = 3, a_n = 3n - 1$ **b.** $a_1 = \frac{3}{2}, d = \frac{3}{4}, a_n = \frac{3}{4}n + \frac{3}{4}$

12. $n = 25, S_{25} = 950$ **13.** $n = 16, S_{16} = 128$ **14.** $S_{10} = -5$

15. $S_{10} = \dfrac{-29{,}524}{27}$ **16. a.** $a_1 = 2, r = 3, a_n = 2(3)^{n-1}$

b. $a_1 = \frac{1}{2}, r = \frac{1}{2}, a_n = \left(\frac{1}{2}\right)^n$ **17.** $n = 8, S_8 = \frac{1640}{27}$ **18.** $\frac{-343}{6}$ **19.** 1785

20. ≈ 4.5 ft; ≈ 127.9 ft

Reinforcing Basic Concepts pp. 716

Exercise 1: $71,500

Exercises 8.5, pp. 724–728

1. experiment; well-defined   **3.** distinguishable   **5.** Answers will vary.

7. a. 16 possible

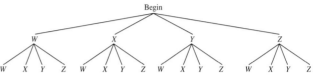

b. WW, WX, WY, WZ, XW, XX, XY, XZ, YW, YX, YY, YZ, ZW, ZX, ZY, ZZ

9. 32   **11.** 15,625   **13.** 2,704,000   **15. a.** 59,049   **b.** 15,120

17. 360 if double veggies are not allowed, 432 if double veggies are allowed.   **19. a.** 120   **b.** 625   **c.** 12   **21.** 24   **23.** 4   **25.** 120

27. 6   **29.** 720   **31.** 3024   **33.** 40,320   **35.** 6; 3   **37.** 90   **39.** 336

41. a. 720   **b.** 120   **c.** 24   **43.** 360   **45.** 60   **47.** 60   **49.** 120

51. 30   **53.** 60, BANANA   **55.** 126   **57.** 56   **59.** 1   **61.** verified

63. verified   **65.** 495   **67.** 364   **69.** 252   **71.** 40,320   **73.** 336

75. 15,504   **77.** 70   **79. a.** $\approx 1.2\%$ **b.** $\approx 0.83\%$ **81.** 7776   **83.** 324

85. 800   **87.** 6,272,000,000   **89.** 518,400   **91.** 357,696   **93.** 6720

95. 8   **97.** 10,080   **99.** 5040   **101.** 2880   **103.** 5005   **105.** 720

107. 52,650, no   **109. a.** $\dfrac{10!}{2!3!5!}$ **b.** $\dfrac{9!}{2!3!4!}$ **c.** $\dfrac{11!}{4!5!2!}$ **d.** $\dfrac{8!}{2!3!3!}$

111.

113. $A + B = \begin{bmatrix} 1.5 & 0.2 & -4 \\ -11 & 5.1 & 9 \\ 3.2 & 1 & 10 \end{bmatrix}$, $AB = \begin{bmatrix} 4.1 & 0.2 & 11 \\ -44.8 & 0.1 & 60 \\ -3.2 & 0.5 & 18 \end{bmatrix}$,

$A^{-1} = \begin{bmatrix} \frac{-19}{17} & \frac{-3}{17} & \frac{15}{17} \\ \frac{-10}{17} & \frac{2}{17} & \frac{7}{17} \\ \frac{12}{17} & \frac{1}{17} & \frac{-5}{17} \end{bmatrix}$

Exercises 8.6, pp. 735–741

1. $n(E)$ **3.** 0; 1; 1; 0   **5.** Answers will vary.

7. $S = \{\text{HH, HT, TH, TT}\}, \frac{1}{4}$

9. $S = \{\text{coach of Patriots, Cougars, Angels, Sharks, Eagles, Stars}\}, \frac{1}{6}$

11. $P(E) = \frac{4}{9}$ **13. a.** $\frac{1}{13}$ **b.** $\frac{1}{4}$ **c.** $\frac{1}{2}$ **d.** $\frac{2}{13}$

15. $P(E_1) = \frac{1}{8}, P(E_2) = \frac{5}{8}, P(E_3) = \frac{3}{4}$ **17. a.** $\frac{3}{4}$ **b.** 1   **c.** $\frac{1}{4}$ **d.** $\frac{1}{2}$

19. $\frac{3}{4}$ **21.** $\frac{6}{7}$ **23.** 0.991   **25. a.** $\frac{1}{12}$ **b.** $\frac{11}{12}$ **c.** $\frac{8}{9}$ **d.** $\frac{5}{6}$ **27.** $\frac{10}{21}$

29. $\frac{60}{143}$ **31.** b, about 12%   **33. a.** 0.3651   **b.** 0.3651   **c.** 0.3969

35. 0.9   **37.** $\frac{7}{24}$ **39.** 0.59   **41. a.** $\frac{1}{6}$ **b.** $\frac{7}{36}$ **c.** $\frac{1}{9}$ **d.** $\frac{4}{9}$

43. a. $\frac{2}{25}$ **b.** $\frac{9}{50}$ **c.** 0   **d.** $\frac{2}{25}$ **e.** 1   **45.** $\frac{3}{4}$ **47.** $\frac{11}{15}$

49. a. $\frac{1}{18}$ **b.** $\frac{2}{9}$ **c.** $\frac{8}{9}$ **d.** $\frac{3}{4}$ **e.** $\frac{1}{36}$ **f.** $\frac{5}{12}$ **51.** $\frac{1}{4}; \frac{1}{36}$; answers will vary.

53. a. 0.33   **b.** 0.67   **c.** 1   **d.** 0   **e.** 0.67   **f.** 0.08

55. a. $\frac{1}{2}$ **b.** $\frac{1}{2}$ **c.** 0.2165   **57. a.** $\frac{9}{16}$ **b.** $\frac{1}{4}$ **c.** $\frac{1}{16}$ **d.** $\frac{5}{16}$

59. a. $\frac{3}{26}$ **b.** $\frac{3}{26}$ **c.** $\frac{1}{13}$ **d.** $\frac{9}{26}$ **e.** $\frac{2}{13}$ **f.** $\frac{11}{26}$ **61. a.** $\frac{1}{8}$ **b.** $\frac{1}{16}$ **c.** $\frac{3}{16}$

63. a. $\frac{47}{100}$ **b.** $\frac{2}{25}$ **c.** $\frac{3}{100}$ **d.** $\frac{9}{50}$ **e.** $\frac{11}{100}$ **65. a.** $\frac{5}{429}$ **b.** $\frac{8}{2145}$ **67.** $\frac{1}{3360}$

69. $\frac{1}{1{,}048{,}576}$; answers will vary; 20 heads in a row.   **71.** $(9, 1, 1)$

73. $x \in [-1, 0) \cup [1, \infty)$

Exercises 8.7, pp. 747–748

1. one   **3.** $(a + (-2b))^5$ **5.** Answers will vary.
7. $x^5 + 5x^4y + 10x^3y^2 + 10x^2y^3 + 5xy^4 + y^5$
9. $16x^4 + 96x^3 + 216x^2 + 216x + 81$ **11.** $41 + 38i$ **13.** 35   **15.** 10
17. 1140   **19.** 9880   **21.** 1   **23.** 1
25. $c^5 + 5c^4d + 10c^3d^2 + 10c^2d^3 + 5cd^4 + d^5$
27. $a^6 - 6a^5b + 15a^4b^2 - 20a^3b^3 + 15a^2b^4 - 6ab^5 + b^6$
29. $16x^4 - 96x^3 + 216x^2 - 216x + 81$
31. $-11 + 2i$ **33.** $x^9 + 18x^8y + 144x^7y^2 + \cdots$
35. $v^{24} - 6v^{22}w + \frac{33}{2}v^{20}w^2$ **37.** $35x^4y^3$ **39.** $1792p^2$ **41.** $264x^2y^{10}$
43. ≈ 0.25 **45. a.** $\approx 17.8\%$ **b.** $\approx 23.0\%$ **47. a.** $\approx 0.88\%$
b. $\approx 6.9\%$ **c.** $\approx 99.0\%$ **d.** $\approx 61.0\%$ **49. a.** 99.33% **b.** 94.22%

51. $\dbinom{6}{6} = \dbinom{6}{0} = 1; \dbinom{6}{5} = \dbinom{6}{1} = 6; \dbinom{6}{4} = \dbinom{6}{2} = 15; \dbinom{6}{3} = 20$

53.

$f(3) = 1$
55. $g(x) > 0: x \in (-2, 0) \cup (3, \infty)$

Summary and Concept Review, pp. 749–753

1. $1, 6, 11, 16; a_{10} = 46$ **2.** $1, \frac{3}{5}, \frac{2}{5}, \frac{5}{17}; a_{10} = \frac{11}{101}$
3. $a_n = n^4; a_6 = 1296$ **4.** $a_n = -17 + (n-1)(3); a_6 = -2$ **5.** $\frac{255}{256}$
6. -112 **7.** 140   **8.** 35   **9.** not defined, 2, 6, 12, 20, 30
10. $\frac{1}{2}, \frac{3}{4}, \frac{5}{4}, \frac{9}{4}, \frac{17}{4}$ **11.** $\displaystyle\sum_{n=1}^{7} (n^2 + 3n - 2); 210$
12. $a_n = 2 + 3(n-1); 119$ **13.** $a_n = 3 + (-2)(n-1); -65$
14. 740   **15.** 1335   **16.** 630   **17.** -11.25 **18.** 875   **19.** 3240
20. 3645   **21.** 32   **22.** 2401   **23.** 10.75   **24.** 6560   **25.** $\frac{819}{512}$
26. does not exist   **27.** $\frac{50}{9}$ **28.** 4   **29.** $\frac{63,050}{6561}$ **30.** does not exist
31. 5   **32.** $a_9 = \$36,980; S_9 = \$314,900$ **33.** $\approx 7111.1 \text{ ft}^3$
34. $a_9 \approx 2105$ credit hrs; $S_9 \approx 14,673$ credit hours
35. verified

(1) Show S_n is true for $n = 1: S_1 = \dfrac{1(1+1)}{2} = 1$ ✓
(2) Assume S_k is true:

$$1 + 2 + 3 + \cdots + k = \frac{k(k+1)}{2}$$

Use it to show the truth of S_{k+1}:

$$1 + 2 + 3 + \cdots + k + (k+1) = \frac{(k+1)(k+2)}{2}$$

left hand side: $1 + 2 + 3 + \cdots + k + (k+1)$

$$= \frac{k(k+1)}{2} + \frac{2(k+1)}{2} = \frac{k(k+1) + 2(k+1)}{2}$$

$$= \frac{(k+1)(k+2)}{2}$$

36. verified

(1) Show S_n is true $n = 1: S_1 = \dfrac{1[2(1)+1](1+1)}{6} = 1$ ✓
(2) Assume S_k is true:

$$1 + 4 + 9 + \cdots + k^2 = \frac{k(2k+1)(k+1)}{6}$$

Use it to show the truth of S_{k+1}:

$$1 + 4 + 9 + \cdots + k^2 + (k+1)^2 = \frac{(k+1)(2k+3)(k+2)}{6}$$

left hand side: $1 + 4 + 9 + \cdots + k^2 + (k+1)^2$

$$= \frac{k(k+1)(2k+1)}{6} + \frac{6(k+1)^2}{6} = \frac{(k+1)[(2k^2 + k + 6k + 6]}{6}$$

$$= \frac{(k+1)(2k^2 + 7k + 6)}{6} = \frac{(k+1)(2k+3)(k+2)}{6}$$

37. verified
(1) Show S_n is true for $n = 1: S_1: 4^1 \geq 3(1) + 1$ ✓
(2) Assume S_k is true: $4^k \geq 3k + 1$
Use it to show the truth of S_{k+1}:
$$4^{k+1} \geq 3(k+1) + 1 = 3k + 4$$
left hand side: $4^{k+1} = 4(4^k)$
$$\geq 4(3k + 1) = 12k + 4$$
Since k is a positive integer, $12k + 4 \geq 3k + 4$ showing
$$4^{k+1} \geq 3k + 4$$

38. verified
(1) Show S_n is true for $n = 1: S_1: 6 \cdot 7^{1-1} \leq 7^1 - 1$ ✓
(2) Assume S_k is true: $6 \cdot 7^{k-1} \leq 7^k - 1$
Use it to show the truth of S_{k+1}:
$$6 \cdot 7^k \leq 7^{k+1} - 1$$
left hand side: $6 \cdot 7^k = 7 \cdot 6 \cdot 7^{k-1}$
$$\leq 7 \cdot 7^k - 1$$
$$\leq 7^{k+1} - 1$$

39. verified
(1) Show S_n is true for $n = 1: S_1: 3^1 - 1 = 2$ or $2(1)$ ✓
(2) Assume S_k is true: $3^k - 1 = 2p$ for $p \in Z$
Use it to show the truth of S_{k+1}:
$$3^{k+1} - 1 = 2q \text{ for } q \in Z$$
left hand side: $3^{k+1} - 1 = 3 \cdot 3^k - 1$
$$= 3 \cdot 2p$$
$$= 2(3p) = 2q \text{ is divisible by 2}$$

40. 6 ways

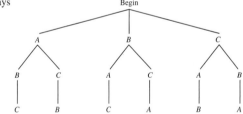

41. 720; 1000   **42.** 24   **43.** 220   **44.** 32   **45. a.** 5040   **b.** 840
c. 35   **46. a.** 720   **b.** 120   **c.** 24   **47.** 3360   **48. a.** 220   **b.** 1320
49. $\frac{4}{13}$ **50.** $\frac{3}{13}$ **51.** $\frac{5}{6}$ **52.** $\frac{7}{24}$ **53.** $\frac{175}{396}$ **54. a.** 0.608   **b.** 0.392
c. 1   **d.** 0   **e.** 0.928   **f.** 0.178   **55. a.** 21   **b.** 56
56. a. $x^4 - 4x^3y + 6x^2y^2 - 4xy^3 + y^4$ **b.** $41 - 38i$
57. a. $a^8 + 8\sqrt{3}a^7 + 84a^6 + 168\sqrt{3}a^5$
b. $78,125a^7 + 218,750a^6b + 262,500a^5b^2 + 175,000a^4b^3$
58. a. $280x^4y^3$ **b.** $-64,064a^5b^9$

Mixed Review, pp. 753–754

1. a. arithmetic   **b.** $a_n = 4$ **c.** $a_n = n!$ **d.** arithmetic   **e.** geometric
f. geometric   **g.** arithmetic   **h.** geometric   **i.** $a_n = \dfrac{1}{2n}$ **3.** 27,600

5. 0.1, 0.5, 2.5, 12.5, 62.5; $a_{15} = 610,351,562.5$ **7.** $\frac{5}{6}$ **9. a.** 2   **b.** 200
c. 210   **11. a.** $a^{20} + 20a^{19}b + 190a^{18}b^2$ **b.** $190a^2b^{18} + 20ab^{19} + b^{20}$
c. $52,360a^{31}b^4$ **d.** 4.6×10^{-18} **13.** verified   **15.** 0.01659   **17.** $\frac{4}{11}$
19. $10, 2, \frac{2}{5}, \frac{2}{25}, \frac{2}{125}$

Practice Test, pp. 755–756

1. a. $\frac{1}{2}, \frac{4}{5}, 1, \frac{8}{7}; a_8 = \frac{16}{11}, a_{12} = \frac{8}{5}$ **b.** 6, 12, 20, 30; $a_8 = 90, a_{12} = 182$
c. $3, 2\sqrt{2}, \sqrt{7}, \sqrt{6}; a_8 = \sqrt{2}, a_{12} = i\sqrt{2}$

2. a. 165   **b.** $\frac{311}{420}$ **c.** $\frac{-2343}{512}$ **d.** 7
3. a. $a_1 = 7, d = -3, a_n = 10 - 3n$
b. $a_1 = -8, d = 2, a_n = 2n - 10$ **c.** $a_1 = 4, r = -2, a_n = 4(-2)^{n-1}$
d. $a_1 = 10, r = \frac{2}{5}, a_n = 10\left(\frac{2}{5}\right)^{n-1}$
4. a. 199   **b.** 9   **c.** $\frac{3}{4}$ **d.** 6   **5. a.** 1712   **b.** 2183   **c.** 2188   **d.** 12
6. a. ≈ 8.82 ft   **b.** ≈ 72.4 ft   **7.** $6756.57   **8.** $22,185.27   **9.** verified
10. verified
11. a.

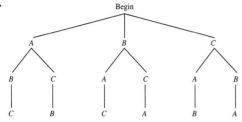

b. ABC, ACB, BAC, BCA, CAB, CBA
12. 302,400   **13.** 64   **14.** 720, 120, 20   **15.** 900,900   **16.** 302,400
17.   a. $x^4 - 8x^3y + 24x^2y^2 - 32xy^3 + 16y^4$ **b.** -4
18. a. $x^{10} + 10\sqrt{2}\,x^9 + 90x^8$ **b.** $a^8 - 16a^7b^3 + 112a^6b^6$
19. 0.989   **20. a.** $\frac{1}{4}$ **b.** $\frac{5}{12}$ **c.** $\frac{1}{3}$ **d.** $\frac{1}{2}$ **e.** $\frac{7}{12}$ **f.** $\frac{1}{4}$ **g.** $\frac{5}{12}$ **h.** 0
21. a. 0.08   **b.** 0.92   **c.** 1   **d.** 0   **e.** 0.95   **f.** 0.03   **22. a.** 0.1875
b. 0.589   **c.** 0.4015   **d.** 0.2945   **e.** 0.4110   **f.** 0.2055   **23. a.** $\frac{59}{100}$
b. $\frac{53}{100}$ **c.** $\frac{13}{100}$ **d.** $\frac{47}{100}$ **24. a.** 0.8075   **b.** 0.0075   **c.** 0.9925
25. verified

Strengthening Core Skills pp. 757–758

Exercise 1. $\dfrac{{}_4C_1 \cdot {}_{13}C_5 - 40}{{}_{52}C_5} \approx 0.001\,970$

Exercise 2. $\dfrac{4 \cdot {}_{13}C_3 \cdot {}_{39}C_2}{{}_{52}C_5} \approx 0.326\,170$

Exercise 3. $\dfrac{4 \cdot {}_{13}C_4 \cdot {}_{39}C_1}{{}_{52}C_5} \approx 0.042\,917$

Exercise 4. $\dfrac{4 \cdot {}_{10}C_5}{{}_{52}C_5} \approx 0.000\,388$

Cumulative Review Chapters 1–8, pp. 758–759

1. a. 23 cards are assembled each hour.   **b.** $y = 23x - 155$
c. 184 cards   **d.** $\approx 6:45$ A.M.
3. $x = \dfrac{-5 \pm \sqrt{109}}{6}; x \approx 0.91; x \approx -2.57$ **5.** $Y = \dfrac{kVW}{X}; Y = \dfrac{3VW}{2X}$
7. verified; reflections across $y = x$

9.   a. $4x + 2h - 3$ **b.** $\dfrac{-1}{(x + h - 2)(x - 2)}$

11.

13. a. $x^3 = 125$ **b.** $e^5 = 2x - 1$ **15. a.** $x \approx 3.19$ **b.** $x = 334$
17. $(5, 10, 15)$ **19.** $(-3, 3); (-7, 3), (1, 3); (-3 - 2\sqrt{3}, 3),$
$(-3 + 2\sqrt{3}, 3)$ **21.** 1333   **23. a.** $\approx 7.0\%$ **b.** $\approx 91.9\%$ **c.** $\approx 98.9\%$

d. $\binom{12}{0}(0.04)^{12}(0.96)^0$; virtually nil

25. $\begin{cases} 9a + 3b + c = 56 \\ 25a + 5b + c = 126; \ y = -7x^2 + 91x - 154 \quad \textbf{a.} \ \text{mid-June} \\ 81a + 9b + c = 98 \end{cases}$
b. ≈ 142 **c.** 98   **d.** November to February

Index

▼ The Toolbox and Other Functions

linear

$m < 0, b > 0$

linear
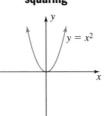
$y = mx + b$
$(0, b)$
$m > 0, b > 0$

identity

$y = x$
$m = 1, b > 0$

constant
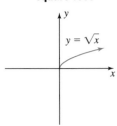
$y = b$
$m = 0, b > 0$

absolute value

$y = |x|$

squaring

$y = x^2$

cubing
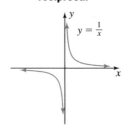
$y = x^3$

square root
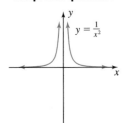
$y = \sqrt{x}$

cube root

$y = \sqrt[3]{x}$

greatest integer

$y = \lfloor x \rfloor$

reciprocal

$y = \frac{1}{x}$

reciprocal quadratic

$y = \frac{1}{x^2}$

exponential

$y = b^x$

exponential
$y = b^{-x}$

logarithmic
$y = \log_b x$

logistic
$y = \frac{c}{1 + ae^{\angle bx}}$
$y = c$
$\left(0, \frac{c}{1+a}\right)$

▼ Transformations of Basic Graphs

Given Function

$$y = f(x)$$

Transformation of Given Function

$$y = af(x \pm h) \pm k$$

vertical reflections
vertical stretches/compressions

horizontal shift h units,
opposite direction of sign

vertical shift k units,
same direction as sign

▼ Average Rate of Change of $f(x)$

For linear function models, the average rate of change on the interval $[x_1, x_2]$ is constant, and given by the slope formula: $\frac{\Delta y}{\Delta x} = \frac{y_2 - y_1}{x_2 - x_1}$. The average rate of change for other function models is non-constant. By writing the slope formula in function form using $y_1 = f(x_1)$ and $y_2 = f(x_2)$, we can compute the average rate of change of other functions on this interval:

$$\frac{\Delta y}{\Delta x} = \frac{f(x_2) - f(x_1)}{x_2 - x_1}$$